Physical constants and data

Speed of light	$c = 2.997925 \times 10^8$ m/s
Gravitational constant	$G = 6.67 \times 10^{-11}$ N $\cdot$ m²/kg²
Avogadro's number	$N_A = 6.022 \times 10^{26}$ particles/kmol
Boltzmann's constant	$k = 1.38066 \times 10^{-23}$ J/K
Gas constant	$R = 8314$ J/kmol $\cdot$ K $= 1.9872$ kcal/kmol $\cdot$ K
Planck's constant	$h = 6.6261 \times 10^{-34}$ J $\cdot$ s
Electron charge	$e = 1.60218 \times 10^{-19}$ C
Electron rest mass	$m_e = 9.1094 \times 10^{-31}$ kg $= 5.486 \times 10^{-4}$ u
Proton rest mass	$m_p = 1.6726 \times 10^{-27}$ kg $= 1.007276$ u
Neutron rest mass	$m_n = 1.6749 \times 10^{-27}$ kg $= 1.008665$ u
Permittivity constant	$\epsilon_0 = 8.85419 \times 10^{-12}$ C²/N $\cdot$ m²
Permeability constant	$\mu_0 = 4\pi \times 10^{-7}$ N/A²
Standard gravitational acceleration	$g = 9.80665$ m/s² $= 32.17$ ft/s²
Mass of earth	5.98×10^{24} kg
Average radius of earth	6.37×10^6 m
Average density of earth	$5{,}570$ kg/m³
Average earth-moon distance	3.84×10^8 m
Average earth-sun distance	1.496×10^{11} m
Mass of sun	1.99×10^{30} kg
Radius of sun	7×10^8 m
Sun's radiation intensity at the earth	0.032 cal/cm² $\cdot$ s $= 0.134$ J/cm² $\cdot$ s

PRINCIPLES OF PHYSICS

Photo Credits appear on pages I-9 to I-12, and on this page by reference.

This book is printed on acid-free paper.

3 4 5 6 7 8 9 0   VNH   VNH   9 0 9 8 7 6

ISBN 0-07-008986-8

This book was set in Century Old Style by York Graphics Services, Inc.
The editors were Anne C. Duffy and David A. Damstra;
the designer was Joan E. O'Connor;
the production supervisor was Richard A. Ausburn.
The cover photographer was Mark Gamba/The Stock Market.
The photo editor was Safra Nimrod;
the photo researcher was Mira Schachne.
New drawings were done by Fine Line Illustrations, Inc.
Von Hoffmann Press, Inc., was printer and binder.

Library of Congress Cataloging-in-Publication Data

Bueche, Frederick J.
   Principles of physics / Frederick J. Bueche, David A. Jerde.—
6th ed.
      p.   cm.
   ISBN 0-07-008986-8
   1. Physics.   I. Jerde, David A.   II. Title.
QC23.B8496   1995
530—dc20   94-19586

ABOUT THE
AUTHORS

Frederick J. Bueche, University of Dayton Distinguished Professor at Large, obtained his B.S. degree from the University of Michigan and his Ph.D. in physics from Cornell. After postdoctoral work in chemical physics at Cornell, he held professorships at the Universities of Wyoming, Akron, and Dayton. His extensive research in the physics of polymers, plastics and elastomers resulted in the publication of nearly 100 papers and a graduate-level text in the field. In recognition of his research, he has been elected a Fellow of the American Physical Society.

Primarily a teacher, Bueche has taught physics at all levels throughout his career, including two years with the Peace Corps in Turkey. He has authored several elementary physics texts used by many students throughout the world.

David A. Jerde is professor and chair of the Department of Physics, Astronomy, and Engineering Science at St. Cloud (Minnesota) State University. He received an M.S. degree in physics at the University of Minnesota and a Ph.C. at the University of Washington. From 1957 to 1969 he was a research physicist at the Boeing Company, Seattle, involved in basic research in plasma physics and applied research in infrared sensing and laser technology. In 1969 Professor Jerde joined the faculty of St. Cloud State, where he has taught physics for the past 25 years. He is the co-author of publications in plasma physics and has authored two editions of the Study Guide to accompany the text *Introductory College Physics* by J. Mulligan.

CONTENTS

PART FOUR   LIGHT AND OPTICS   671

PREFACE

Physics is the most fundamental of the sciences. Its principles underlie phenomena studied in all other technical fields, which is one reason why its study is required by so many academic programs. As teachers, we are fully aware of the difficulties today's students have in learning to use language in a precise and operational way. As the authors of this text, it is our intent to facilitate both student and instructor in the process of learning, emphasizing the conceptual simplicity and power of these ideas which make up our understanding of the physical world. Although the ability to solve problems is the ultimate test of learning physics, in this text we emphasize the development of conceptual reasoning more than mathematical manipulation. An overriding goal of this text is to impart to the student an understanding of the *process* of problem analysis and solution. It is that understanding that empowers the student and will lead her or him to a greater degree of confidence and a true mastery of the subject.

This text presents a one-year, introductory, laboratory-based course in physics at the college algebra and elementary trigonometry level. Traditionally, the areas of study requiring this course include biology, allied health, pre-medicine, pre-dentistry, pharmacy, environmental studies, engineering technology, and architecture. The point of this requirement is that a student in any of these disciplines needs to acquire:

1 A basic understanding of physics principles and an appreciation of scientific methodology and scientific literacy

2 The ability to apply critical and analytical reasoning, both qualitative and quantitative, to the solution of a wide variety of problem situations

The purpose of this text is to help both the student and the instructor toward the fulfillment of these two goals.

BASIC PHILOSOPHY OF THE TEXT

First of all, the text is *not* intended to be encyclopedic, nor does it contain lengthy mathematical derivations or historical biographies. Each basic principle is first highlighted conceptually, then written mathematically, and then immediately applied in illustrations and worked examples. The development of principles is accomplished by narratives that appeal, wherever possible, to common observations familiar to students.

The following assumptions form the basis for the specific features employed in the text:

1 The student should be able to demonstrate having mastered the above-stated goals in a number of ways. One way, which traditionally forms the basis of most course examinations, is to be able to solve quantitative problems. Another way is to be able to arrive at correct answers to *qualitative* questions involving the application of physics principles.

2 The ability to solve problems is based on the ability to formulate critical questions that, when answered, reveal the solution. Formulation of these questions involves the ability to identify (1) the essential known factors of a problem and (2) the principles that relate these known factors to the unknown ones. The student should learn that **a good question is the best initial response to a problem.**

3 Students almost universally have trouble with "word problems." Even if they develop the ability to formulate questions, they may be unable to translate those questions into mathematical form. Rather than just stating that mathematics is the language of physics, we place emphasis on developing the understanding that, because physics concepts are defined in quantifiable terms, **each definition and principle applied to a problem creates an equation.**

4 Working a large number of varied problems is one way of gaining experience in the application of principles.

5 Summarizing material is a way of unifying it and focusing on the interrelation of concepts.

6 Because the usual algebra-based physics course concentrates most of its time on classical physics, students finish the course learning very little about the developments made during the past 100 years. Understanding current applications of physics and the motivation for continued research necessitates some exposure to modern perspectives. These perspectives should accompany those classical principles that the modern developments have modified.

TOPICAL CHANGES IN THE SIXTH EDITION

This edition is still organized in a traditional manner into five parts:

Mechanics

Mechanical and thermal properties of matter; oscillations and waves

Electricity and magnetism

Light and optics

Modern physics

However, the following changes in topical coverage have been made:

1 The first four chapters have been re-ordered into a sequence more traditional than that contained in the fifth edition. The first chapter provides increased attention to limits of measurement and calculating with measured quantities. As part of this emphasis, increased attention is given to translating verbal statements into mathematical form.

2 Thermodynamics has been split into two chapters: one on the first law and one on the second law. Additional coverage on thermodynamic processes in gases and the specific heats of gases has been included.

3 A section on Gauss' law and the electric fields produced by symmetrical charge distributions has been added.

4 In the coverage of wave optics, diffraction and interference now precede optical devices.

NEW TO THIS EDITION

QUESTION/ANSWER FORMAT IN WORKED EXAMPLES

The most noticeable change in this edition is the addition of **dialogues** which accompany the worked examples. After each new physics principle is developed conceptually and written mathematically, it is followed by one or more worked examples. Rather than the usual approach of explaining the solution to the student using the author's experience and hindsight, a unique question-and-answer section articulates a series of questions the student should ask in order to translate the problem into solvable form. In the answers to these questions, the student is led through the structure of the solution—seeing how equations are created through the application of definitions and principles.

We do not pretend that the specific sequence of questions is unique to a particular problem—many variations could be employed—but the questions voiced are the ones the student is going to have to ask somewhere along the route to the solution. Seeing the explicit questioning process encourages the development of qualitative understanding and reduces the tendency to randomly try various "formulas" in the hope that one will magically provide the solution.

MODERN PHYSICS CONCEPTS

One third of the chapters on classical physics now end with a section called **A Modern Perspective,** which gives the student a glimpse of how twentieth-century physics has modified the classical principles of that particular chapter. Examples include "Mass at High Speeds" in Chapter 3 (Newton's Laws of Motion) and "The Minimum Quantity of Angular Momentum" in Chapter 8 (Rotational Work, Energy, and Momentum). The limits of validity on the assumptions of classical physics are explored, and some of the concepts of relativity and quantum theory are described. It is our contention that these glimpses into modern physics in the context of the corresponding classical principles convey the ongoing vitality of physics to the student. As useful as classical physics continues to be, covering only it in the course because of time constraints portrays physics as a moribund subject.

GUEST ESSAYS

Rather than include the traditional historical biographies, fascinating as they are, we have asked a number of current physicists to contribute brief autobiographies, emphasizing why they chose to become physicists and what continues to motivate them. These essays are called **Physicists at Work** and are intended to convey to the student the personal and human side of men and women who continue to explore the frontiers of knowledge and consequent applications.

GREAT CONTROVERSIES

The text contains three essays entitled **Great Controversies in Physics.** These are historical vignettes that demonstrate that our present understanding of physics is based on struggles between competing ideas and experimental observations, often over long spans of time. The topics covered are the controversies about falling objects, the nature of heat, and the nature of light. The role of critical questions in deciding the outcome of these controversies, asked in the form of definitive experiments, is emphasized.

OTHER FEATURES

Of course the strengths of previous editions have been retained in this revision. They include the following:

EMPHASIS ON CONCEPTUAL ANALYSIS

Throughout the narrative of each chapter, the student is repeatedly asked "Why?" or "Can you explain this?" as the authors make certain assertions based on previously developed ideas, and at the end of each chapter there are a number of conceptual questions called **Questions and Guesstimates.** Both of these features emphasize the importance of developing the ability to apply physics principles *qualitatively*. This is often a more difficult task for the student than finding the formal solution to a mathematical problem. Mastering this ability is both a necessary prerequisite to successful problem solving and a basic indicator of true understanding.

WORKED EXAMPLES AND EXERCISES

As described above, the format of the worked examples has been changed to include a dialog between instructor and student. In addition, at the conclusion of most examples, there is a related variation of the problem called an **Exercise,** where only the answer is given. The student thus has an immediate opportunity to test her or his understanding of the previous solution.

BUILT-IN STUDY GUIDE

Each chapter contains a comprehensive summary of the definitions, concepts, and mathematical expressions developed in the chapter. An important and unique feature of these summaries is the **Insights** section, in which potential problems in interpretation are anticipated and clarified. These extensive summary sections provide the student with a built-in study guide that clearly shows the relatedness of the principles covered in the chapter.

LEARNING GOALS

Detailed learning goals are articulated in each chapter. They are placed at the end of the chapter, where, along with the Questions and Guesstimates and Chapter Summaries, they provide a convenient, focused, comprehensive review for the student.

EXTENSIVE PROBLEM SETS

This new edition contains approximately 50 percent more end-of-chapter problems than the previous edition. Most of the problems are new, and many of those retained from the previous edition have been revised. Problems are grouped by chapter section and are graded on three levels of difficulty. In addition, each chapter contains a section of additional problems that involve a more integrated approach than do the problems grouped by section.

ILLUSTRATIONS AND PHOTOGRAPHS

More than 500 simple-to-understand, full-color drawings and graphs illustrate the development of new concepts throughout the text. Hundreds of photographs show the student examples of devices and their application, illustrating ways in which the principles of physics are an intimate part of everyday life.

SUPPLEMENTS SUPPORTING THE SIXTH EDITION

The following ancillary material has been prepared to accompany publication of this latest edition of *Principles of Physics.*

An **Instructor's Resource Guide,** prepared by Patrick Briggs of The Citadel and John Swez of Indiana State University, contains:

suggestions for lectures

conceptual problems and quantitative problems that can be duplicated and assigned for homework and/or classroom discussion

medical and allied health applications, including premed, biology, ecology and architecture examples

suggestions for organized activities for group study, including descriptions of "home experiments" that can be carried out with limited, commonplace equipment a listing of videocassettes, CD ROMs, and computer software germane to the college physics course

an instructor's guide to "Physics At Work," a videodisk available from McGraw-Hill (see below)

A **Solutions Manual** prepared by V. K. Saxena of Purdue University provides instructors with fully worked out solutions to all end-of-chapter problems in the text.

Fully colored Overhead Transparencies will be available for many figures in the textbook.

A videodisk, **"Physics At Work,"** from Videodiscovery, is a comprehensive program designed to help students understand and visualize physics principles. The double-sided CAV (standard-play) laser videodisk is supported by a quick reference card and a bar-coded image directory indexed by name, concept, and frame number. A correlation sheet will be included in the Instructor's Resource Guide which lists sections in the Physics at Work Videodisk which can be used in a college physics course.

A **Test Bank** by John Snyder (Southern Connecticut University) comprising more than 1000 problems in multistep and multiple-choice format is available both in a print version and as software for either IBM or Macintosh.

ACKNOWLEDGMENTS

A very large number of people are responsible for making the sixth edition of *Principles of Physics* possible. A list of professors who reviewed this edition is on the next page.

We would also like to thank Drs. John Harlander, Mark Nook, and Richard Schoenberger of St. Cloud State University for illuminating discussions on many pedagogical points. Dr. V. K. Saxena, Purdue University, has done an admirable job of creating most of the new problems as well as providing the solutions manual for the text.

We are grateful to the exceptionally capable staff at McGraw-Hill who have helped and supported us in countless ways, including tolerating the many delays imposed by other duties. Those who most certainly deserve special mention include Anne C. Duffy, David A. Damstra, Safra Nimrod, Sylvia Warren, and Joan O'Connor. Irene Nunes spent many hours and much red ink honing the rough manuscript into readable form.

Despite all efforts to eliminate them, errors are bound to remain, and we invite comments and corrections in order that future editions may be improved.

Frederick J. Bueche
David A. Jerde

REVIEWERS

George C. Alexandrakis	*University of Miami*
Richard Beedle	*North Hennepin Community College*
Walter Bevenson	*Michigan State University*
Kenneth Brown	*Kean College*
Darry S. Carlstone	*University of Central Oklahoma*
R. M. Catchings	*Howard University*
Larry Coleman	*University of Arkansas–Little Rock*
Brent Cornstubble	*United States Military Academy–West Point*
Melvin Davidson	*Western Washington University*
Peter G. Debrunner	*University of Illinois*
Mehri Fadavi	*Mississippi State University*

Kyle Forinash	*Indiana University–Southeast*
Charles Gale	*McGill University*
Michael Grady	*State University of New York–Fredonia*
Ira L. Hawk	*Cameron University*
E. Thomas Henkel	*Wagner College*
George Lehmberg	*Passaic County Community College*
Gerard P. Lietz	*DePaul University–Chicago*
William M. McCord	*Valencia Community College*
William Massano	*State University of New York–Maritime College*
Michael Matkovich	*Oakton Community College*
John W. Milton	*DePaul University*
Marvin Morris	*San Jose State University*
Mel Oakes	*University of Texas at Austin*
E. W. Prohofsky	*Purdue University*
Christopher Roddy	*Purdue University*
Frederick H. C. Schultz	*University of Wisconsin–Eau Claire*
Paul Sokol	*Pennsylvania State University*
Carey E. Stronach	*Virginia State University*
John A. Swez	*Indiana State University*
Franklin D. Trumpy	*Des Moines Area Community College*
Richard Vawter	*Western Washington University*

CHAPTER 1

INTRODUCTION

1.1 WHAT IS PHYSICS?

We humans all have varied reactions to the world. The artist in us will admire a sunset and wish we could communicate its beauty in a painting. The poet in us will try to find words with which to express the experience of the sunset. Another side of us, which might be the physicist trying to get out, is curious as to how far away the sun is, how big it is, and how it generates all that light and heat. Once we start asking such questions, it is hard to stop, and our philosophical or religious side might ask, "What is the meaning of the sunset?" In fact, we humans have the ability to experience all these reactions to some degree. When we say we are an artist, a poet, a philosopher, or a physicist, we are simply choosing to emphasize and develop our talent in one of these directions.

Physicists, to put it simply, are people who are motivated to ask questions and therefore to seek answers about how the physical world around us operates. Throughout this text you will find many personal accounts by people about what led them to become physicists and what continues to fascinate them in their chosen profession.

Physics, then, is a body of knowledge that provides organized answers to our questions about the physical world. It is also a process of obtaining these answers, commonly referred to as the scientific method. The two main tools of physics are logic and experimentation. From the laser to the integrated chip, from the electrical generator to the jet engine, from radio and television to lifesaving medicines and

Newton with a prism.

machines, developments made possible by scientific inquiry influence every moment of our lives.

Our efforts to understand natural processes by a combination of logical reasoning and controlled experimentation called the **scientific method** represent a recent chapter in human history. Prior to about 1600, questions of truth and falsehood were most often determined by political or religious dictates. The efforts of such great scientists as Galileo Galilei, Robert Boyle, Isaac Newton, and others introduced this scientific method to the world, often at great personal risk resulting from clashes with the religious or political authorities of the time.

Behind our belief in the scientific method as the way to understand nature are two basic assumptions. One is that experimental results are *reproducible.* Reproducibility means that the same set of circumstances will always produce the same observed results in the same experiment, no matter who the observer is. The second assumption is that nature exhibits *causality,* that is, cause-and-effect relationships determine *what* will happen as a result of certain starting conditions. Without these two principles, experimental observation would be fruitless because the results could not be generalized to show fundamental, *predictable* patterns of behavior. We would be living in a chaotic universe that would not, even in principle, be understandable.

Among all the sciences, physics is the most basic. It is a *quantitative* science. Its goal is to describe all phenomena in the physical world in terms of a few fundamental relationships between measurable properties of matter and energy. These fundamental relationships are called the *laws* of physics. They are statements of great generality, derived from and applying to a wide variety of phenomena. In order to develop quantitative laws, we must define the properties involved in the laws in terms that allow them to be measured. The goal of physics, then, is to express the fundamental relationships—these laws—in mathematical form. Doing so enables physicists to use the logical rules of mathematics to apply the laws to specific cases, thus obtaining quantitative results.

In the scientific method, laws begin as ideas, or *theories,* which are subject to experimental verification. Those theories whose quantitative predictions are borne out by experiment are strengthened; those whose predictions are contradicted are discarded. Eventually theories that are the most general and universal in their application gain the status of a physical law. Physics now contains many branches. Among them are mechanics, optics, atomic physics, nuclear physics, thermodynamics, electricity and magnetism, acoustics, quantum mechanics, and relativity. Some laws, such as the law of energy conservation, are used in all branches. Others, though equally valid, have a more restricted use.

Let's now begin our journey into the world of physics with a look at some of the tools we'll need along the way. Since physics is inherently mathematical, this course will require you to be comfortable using college algebra and some elementary trigonometry. This chapter and Appendix 3 provide a brief review of the mathematics you will encounter in your study of physics.

1.2 COUNTING AND MEASURING: ACCURACY AND PRECISION

One of the simplest methods of quantifying is to *count.* This method is applicable wherever we have individual units, such as apples, oranges, people, or atoms. In principle, counting is an *exact* process of quantifying because we are using whole

FIGURE 1.1

The length of the book is seen to be 26 cm to the nearest mark on the meterstick. The actual length could be between 25.5 and 26.5 cm. Thus the precision of the measurement is within a 1-cm range. We indicate the limits of precision in this case by writing 26 ± 0.5 cm.

numbers, or integers, to express a quantity. Of course, when we are confronted with a very large number of objects, such as the number of people in the United States or the number of atoms in a substance, there are practical limits to being exact. In such cases we must be satisfied to know the number to within some acceptable uncertainty. Nevertheless, we know that in principle the number could be known exactly.

Another method of quantifying is to *measure*. Unlike counting, though, the process of measurement is, in principle, inexact. When we measure, we are no longer using integers to determine quantity. Instead, we are using the markings on a meter stick or thermometer, say, or the ticks of a clock to measure quantities of length, temperature, and time. All such marks and ticks have an inherent limit of *precision,* even if the measurement is electronically converted to a digital form of readout. This limit is determined by the design and construction of the measuring device. No matter how carefully we measure, we can never obtain a result more precise than the limit of our measuring device. A general guideline is that a given measuring device has a limit of precision equal to one-half of the smallest division of measurement built into the device. When you make a measurement, you read the measured quantity to the nearest mark on the device. The "true" value of this measurement thus lies within a range of one-half of the smallest division of the device above or below the indicated mark.

The limit of precision of a measuring device is $\pm\frac{1}{2}$ the smallest division of measurement the device is able to display.

Thus a meterstick ruled in millimeters (mm) has a limiting precision of ± 0.5 mm, while a vernier caliper that can be read directly to the nearest 0.1 mm has a limiting precision of ± 0.05 mm (see Fig. 1.1). A stopwatch whose face is ruled in 0.5-second (0.5-s) intervals has a precision of ± 0.25 s, and a digital stopwatch that reads to the nearest 0.1 s has a limiting precision of ± 0.05 s.

A different kind of measurement uncertainty involves the possibility of incorrect design or calibration of the instrument, or incorrect reading or interpretation of the instrument. Such errors are called **systematic errors** and cause the measurement to be consistently higher or consistently lower than the true value. Such a measurement is said to be inaccurate.

We use many devices to measure physical quantities, such as length, time, and temperature. Some are analog devices, some digital. They all have some limit of precision.

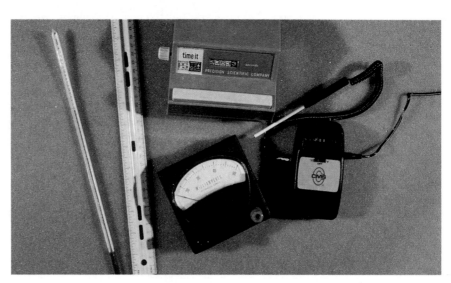

Accuracy is the extent to which systematic errors make a measured value differ from its true value.

Great care in instrument design, calibration, and reading can reduce systematic errors to a level of inaccuracy smaller than the limit of precision of the instrument.

Finally, multiple measurements of the same quantity using the same instrument often differ by more than the precision of the instrument. Such errors are called either **random errors** or **statistical errors**. They are caused by fluctuations in the physical property being measured, such as changes in temperature, electrical voltage, gas pressure, and the like. Statistical errors cannot be eliminated. They can be reduced by increasing the number of measurements, however, and their effect on the accuracy of the measured quantity can be calculated by statistical analysis. We do not use statistical analysis in this book.

1.3 DIMENSIONS AND UNITS IN MEASUREMENT

When measuring a physical quantity, we first have to identify what *kind* of physical property we are measuring. Do we want to determine the length of a swimming pool, for example, or the time it takes us to swim one lap? There are only seven basic kinds of physical properties necessary to describe all physical measurements. These properties, called **dimensions**, are length, mass, time, temperature, electric current, number of particles, and luminous intensity. All other quantities we deal with, such as force, energy, and momentum, can be derived from these seven basic dimensions.

A standard quantity must be defined for each basic physical dimension. These definitions are arbitrary, but each is based on a highly precise physical measurement. There is international agreement on both the definitions of each of the seven standard quantities and the design of the experiments used to measure them.

Once we've identified the kind of property we want to measure, our second task is to choose a system of measurement **units** for expressing the quantity we measure. Several systems of units have been used at various times and places to express measured quantities of the seven basic dimensions. Today, however, there are only two basic systems of measurement used in the world. The more universally used, and the one used almost exclusively in science, is the International System of Units (SI).* The second system, used commonly in the United States, is called the British System (although it no longer is the official system used in Great Britain). The SI will be the basis for this book, and we shall make only passing reference to comparisons with the British System.

The seven fundamental dimensions are expressed in the basic SI units shown in Table 1.1. The many physical quantities that are *combinations* of basic units are expressed in *derived* SI units, many of which are given their own special name. Examples of derived units are joules (of energy) and newtons (of force). A fairly complete list of basic and derived SI units is contained inside the front cover of the text. We shall describe the definition of specific units more fully as they are introduced.

TABLE 1.1

Fundamental dimensions and basic SI units

Dimension	Unit	Symbol
Length	Meter	m
Mass	Kilogram	kg
Time	Second	s
Temperature	Kelvin	K
Electric current	Ampere	A
Number of particles	Mole	mol
Luminous intensity	Candela	cd

*The abbreviation SI is from the French name, Le Système International d'Unités.

1.4 CALCULATING WITH UNITS AND CONVERTING BETWEEN SYSTEMS OF UNITS

Calculating with measured quantities always involves *two* processes: (1) doing the numerical calculation and (2) calculating the units of the resulting quantity. In calculating the latter, it is important to realize that *the units in a calculation are treated just like any other algebraic quantities*. Thus, for example, dividing 60 miles (mi) by 2 hours (h) gives

$$\frac{60\,\text{mi}}{2\,\text{h}} = 30\,\text{mi/h}$$

Similarly, multiplying 3 kilograms (kg) by 12 meters per second (m/s) gives

$$(3\,\text{kg})(12\,\text{m/s}) = 36\,\text{kg} \cdot \text{m/s}$$

The units used in various systems to measure a dimension usually have different names and represent different amounts of the dimension. For example, the meter (SI) and the yard (British) both measure length. The kilogram (SI) and the slug (British) both measure mass. We can convert any measurement from one system to another by using the appropriate equivalencies, called **conversion factors**. Commonly used conversion factors are given inside the front cover of the text.

Mistakes in calculation are often the result of the inconsistent use of units or incorrect use of conversion factors. One almost mistake-proof way to convert from one system of units to another is to notice that any ratio of unit equivalencies always has the value of 1. For example, by dividing both sides of the equation 1.00 inch (in) = 2.54 centimeters (cm) by 2.54 cm, we get

$$\frac{1.00\,\text{in}}{2.54\,\text{cm}} = \frac{2.54\,\cancel{\text{cm}}}{2.54\,\cancel{\text{cm}}} = 1$$

A variety of calculating devices is shown: abacus, pencil and paper, slide rule, and pocket calculator.

Because 1.00 in/2.54 cm = 1, we can use this conversion factor and the fact that multiplying any quantity by 1 does not change the quantity to convert from metric units (centimeters) to British (inches). Thus a length of 17.3 cm is equivalent to (17.3 cm) × (1.00 in/2.54 cm) = 6.81 in.

Notice that in using this conversion factor we are *not* saying that 1/2.54 = 1. Remember: we are calculating with *units*, not just numbers. The ratios 1.00 in/2.54 cm and 2.54 cm/2.54 cm are both *dimensionless* (length/length), and so the resulting answer is indeed a pure (and *exact*) number, 1. Thus multiplying any measured quantity by a conversion factor ratio has the effect of changing the units of the quantity and adjusting the numerical value to the new units. You just have to choose which units in the measured quantity you want to get rid of (cancel) and which units you want to replace them with. For example, to convert 20.0 feet (ft) to meters (m):

$$20.0 \, ft \times \frac{0.305 \, m}{1.00 \, ft} = 6.10 \, m$$

Notice how the units of feet cancel algebraically, leaving only the unit of meters. The numerical part of the calculation adjusts the original number of feet to the correct number of meters.

Similarly, to convert a speed of 60.0 mi/h to m/s:

$$60.0 \, mi/h \times \frac{1610 \, m}{1.00 \, mi} \times \frac{1.00 \, h}{3600 \, s} = 26.8 \, m/s$$

Keeping track of the units in an equation and doing correct conversions are two of the most important tasks in physics calculations. One important point to remember:

Every term in an equation must have the same units.

By *term* we mean each quantity being added or subtracted in an equation. A corollary to this is that both sides of an equation must have the same units.

1.5 SIGNIFICANT DIGITS IN CALCULATIONS

Since measuring instruments always have a limit of precision and since statistical errors are often present, every measurement in physics has a limit on how many digits in the result are known with certainty. The digits that are known with certainty are called **significant digits**. Whenever you work a problem in physics, you must use the right number of significant digits to express the results of both your measurement and your calculation.

Zeros may or may not be significant, depending on whether they are known values or are just being used to locate the decimal place. The ambiguity of zeros can be eliminated by using scientific notation, that is, by letting the exponent locate the decimal place and the number in front of the exponent contain the significant digits.

Examples

Measurement	Significant Digits	Remarks
3.1 cm	2	
4.36 m/s	3	
5.003 mm	4	Both zeros significant.
0.00875 kg	3	Zeros merely locate decimal.
8.75×10^{-3} kg	3	Same quantity as previous example.
4500 ft	2, 3, or 4	Ambiguous. Can't tell whether zeros are measured or only showing decimal place.
4.5×10^3 ft	2	Ambiguity of previous example removed.
4.500×10^3	4	Ambiguity of previous example removed.

In doing calculations, it is important to know how many significant digits to keep in the result. Pocket calculators give answers having ten or so digits, even when the input involves quantities with only two or three significant digits. A couple of simple rules can clarify this problem in most of the instances we will encounter in this course.

SIGNIFICANT DIGITS IN ADDITION OR SUBTRACTION

When adding or subtracting measured quantities, the precision of the answer can be only as great as the *least* precise term in the sum or difference. All digits up to this limit of precision are significant.

SIGNIFICANT DIGITS IN MULTIPLYING AND DIVIDING

When multiplying or dividing measured quantities, the number of significant digits in the result generally can be only as great as the *least* number of significant digits in any factor in the calculation.

Illustration 1.1

Suppose you make three measurements of length, using instruments of various precision, and get 3.76 cm, 46.855 cm, and 0.2 cm. What is the sum of these values?

Reasoning

Calculation:

$$
\begin{array}{r}
3.76 \ \text{cm} \\
+46.855 \ \text{cm} \\
+ \ 0.2 \quad \text{cm} \\
\hline
\end{array}
$$

The calculator gives: 50.815 cm

However, the rule for significant digits in addition and subtraction tells us that our answer is known only to the nearest 0.1 cm, since the least precise quantity (0.2) is known only to that precision. The *correct* answer therefore is 50.8 cm.

To see that this is so, let's look at the meaning of the precision of the individual numbers. Applying our $\pm\frac{1}{2}$ rule from page 3, we see that the first value is uncertain in a range from 3.755 to 3.765. The second value can be as large as 46.8555 and as small as 46.8545, while the third has a range from 0.15 to 0.25. To find the uncertainty in the sum, we can calculate first the largest sum, using the top of the range for all three numbers, and then the smallest sum, using the bottom of the range for each number:

$$
\begin{array}{ll}
\text{Largest sum:} & \text{Smallest sum:} \\
\quad\quad
\begin{array}{r}
3.765 \\
+46.8555 \\
+ \ 0.25 \\
\hline
50.8705
\end{array}
& \quad\quad
\begin{array}{r}
3.755 \\
+46.8545 \\
+ \ 0.15 \\
\hline
50.7595
\end{array}
\end{array}
$$

Thus the sum has an uncertainty range of slightly more than 0.1 cm. In this illustration, it is clear that even the third digit is in question. Certainly it is unjustified to claim any more precision than 50.8 cm.

Illustration 1.2

What is the volume of a box whose sides are measured to be 31.3 cm, 28 cm, and 51.85 cm?

Reasoning   First, remember that the volume of a box is found by multiplying length times width times height. The calculator gives

$$\text{Volume} = (31.3 \ \text{cm})(28 \ \text{cm})(51.85 \ \text{cm}) = 45{,}441.34 \ \text{cm}^3$$

But the significant digit rule allows us to keep only *two* digits (we are limited by the two significant digits in 28 cm):

$$\text{Volume} = 45{,}000 \ \text{cm}^3 = 4.5 \times 10^4 \ \text{cm}^3$$

Again it seems we are being too hard on ourselves when we throw away all those other digits. Looking again at the meaning of precision, however, we see that the three numbers could be as big as 31.35, 28.5, and 51.855. This would give a *maximum* volume

$$\text{Maximum volume} = (31.35 \ \text{cm})(28.5 \ \text{cm})(51.855 \ \text{cm}) = 46{,}300 \ \text{cm}^3$$

The *minimum* volume would be obtained by using the minimum values of the three numbers:

Minimum volume = (31.25 cm)(27.5 cm)(51.845 cm) = 44,600 cm³

The measurements can tell us only that the calculated volume must lie *within* this range. Thus even the second digit is uncertain, with 45,000 cm³ being roughly the two-digit average. ∎

In summary, it is important to remember the following:

Calculation cannot increase either the precision or the number of significant digits of measured quantities.

1.6 PHYSICS PRINCIPLES AS MATHEMATICAL EQUATIONS

Many students (you may be one of them) have little difficulty solving algebraic equations but stumble on so-called "word" problems, where the equations have to be derived from the statement of the problem. In other words, the process of constructing an equation from concepts expressed in language very often poses a great deal of difficulty for students. Yet this ability to construct a mathematical statement of a verbal problem is absolutely central to learning and understanding physics.

Constructing an equation from words boils down to the following:

1 Stripping away irrelevant parts of the word statement, or, put another way, abstracting the **essential quantities** from the sentence.
2 Letting simple **symbols** (for example, x, y) represent quantities whose values are not given.
3 Identifying the **mathematical form** of the **basic principles** that relate the essential quantities. Most often these principles are not given explicitly in the statement of the problem.

In short,

The definitions and laws of physics provide us with the relationships between physical properties that enable us to convert descriptive statements into mathematical equations.

Illustration 1.3

You plan to spend $10.00 on hamburger and steak. If you buy 3.00 pounds of hamburger at $1.29 per pound, how much steak can you buy if it costs $3.99 per pound?

Reasoning   Here the essential quantities are the total cost and the cost per pound and weight of the hamburger and steak. You are given both costs per pound, the weight of hamburger, and the total cost. What is unknown is the weight of the steak (denote it by the symbol x) you can afford after you buy the hamburger. The basic principle relating these quantities is one we all understand from daily life: price per pound multiplied by weight equals the cost of each item. We also know that the cost of the hamburger plus that of the steak must add up to $10.00. Writing all this out in mathematical form, we get an equation:

$$(3.00 \text{ lb})(\$1.29/\text{lb}) + (x \text{ lb})(\$3.99/\text{lb}) = \$10.00$$

It is a simple matter to solve this equation for the weight of the steak, x:

$$(x \text{ lb})(\$3.99/\text{lb}) = \$10.00 - \$3.87$$

Verify that $x = 1.54$ lb to three significant digits.

Illustration 1.4

A race car is traveling around a racetrack at a speed of 215 km/h. The distance around the track is 2.00 km. How much time will it take the car to complete 125 laps?

Reasoning   The essential quantities we are given are the number of laps to be traveled, the distance per lap, and the speed of the car. We are asked to find the total time, which we can represent by the symbol t. The basic principle relating time to speed and distance is again one that is probably familiar:

$$\text{Speed} = \frac{\text{distance traveled}}{\text{time taken}}$$

If we use v as the symbol for speed and d for distance traveled, we can translate this word equation into mathematical form:

$$v = \frac{d}{t}$$

It is important that you see this equation *not* as just a formula for v but rather as a *relationship* between the three quantities that can be manipulated according to the rules of algebra. For instance, multiplying both sides of the equation by t, we get

$$vt = \left(\frac{d}{\cancel{t}}\right)\cancel{t} = d$$

Dividing both sides of this latter equation by v, we get

$$\frac{\cancel{v}t}{\cancel{v}} = t = \frac{d}{v}$$

We are not given the total distance d traveled by the car, but the relation between

d and the given quantities is also something you have probably learned without studying physics:

Total distance = (distance per lap)(number of laps)
$$d = (l)(n)$$

where we let l symbolize distance per lap and n stand for the number of laps.

We have created two equations involving knowns and unknowns by applying two simple basic principles. Now the rest of the solution is purely mathematical. First we calculate d:

$$d = (l)(n) = \left(\frac{2.00 \text{ km}}{\text{lap}}\right)(125 \text{ laps}) = 250 \text{ km}$$

Next we calculate t:

$$t = \frac{d}{v} = \frac{250 \text{ km}}{215 \text{ km/h}} = 1.16 \text{ h}$$

In the last calculation, notice that km/(km/h) = h. ■

Even though many problems in this book are more involved than these two, the process outlined is the basis of "doing" physics. The more you learn of the basic principles of physics, the more you will appreciate how they are the key to being able to translate a word problem into a mathematical equation. Once again, you should never regard the principles as "formulas for" some quantity; they are *relationships* between physical properties as determined by observation. The ability to select and apply the appropriate principles to a specific problem is the essence of understanding physics. The rest of the process of solution is simply mathematics.

MATHEMATICS USED IN THIS COURSE

This course in physics you are now beginning requires a working knowledge of college algebra and some elementary trigonometry. In addition, knowledge of the formulas for the perimeter, area, and volume of common geometric shapes is assumed. Appendix 2 contains a detailed mathematical review, including some worked examples.

The types of algebraic equations you will encounter are:

1 Linear: $ax + b = 0$
2 Quadratic: $ax^2 + bx + c = 0$
3 Simultaneous linear equations in two or three unknowns; for example:

$$ax + by + c = 0 \qquad kx + ly + m = 0$$

Functional relationships you will encounter are:

1 Linear proportionality: $y = ax + b$
2 Quadratic proportionality: $y = ax^2 + bx + c$

3 Inverse proportionality: $y = \dfrac{k}{x}$

4 Inverse square proportionality: $y = \dfrac{k}{x^2}$

5 Logarithmic proportionality:

Base 10: $y = \log x$ $x = 10^y$
Natural (base e): $y = \ln x$ $x = e^y$

Each of these functional relationships can be displayed visually by means of a graph. It helps to be able to identify and interpret them readily.

The trigonometric functions and angular measurement we will use are:

1 $\sin x$, $\cos x$, and $\tan x$
2 Radian and degree angle measurement
3 Law of sines
4 Law of cosines

They are reviewed on the endpapers.

1.7 VECTOR AND SCALAR QUANTITIES

Whenever you measure a quantity, you express your result in terms of a number. For example, your height might be 165 cm, a quantity having both a numerical value, 165 (called the **magnitude** of the quantity), and a *unit of measure,* centimeters in this case. Equivalently, your height could be expressed as 65 in or 5.4 feet. In each case, the quantity has a magnitude and a unit of measure. Like such other quantities as the volume of a box or the number of candies in a jar, height has no

We use vectors every day to denote directions in which we travel.

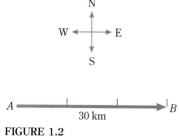

FIGURE 1.2

The vector arrow indicates a displacement of 30 km east.

direction associated with it. Quantities that have no direction associated with them are called **scalar quantities**.

Other quantities do have a direction associated with them. For example, a police officer is interested in more than just the magnitude of your car's motion on a one-way street; the officer becomes very upset if the direction of the motion is not correct. Motion is a quantity that involves direction as well as magnitude. We might give its magnitude as 40 km/h and its direction as eastward; the direction is needed if the motion is to be described fully. It is obvious, for example, that the physical result of going eastward at 40 km/h is very different from the physical result of going northward at the same speed. There are many other familiar quantities that involve direction. Among them are forces (pushes and pulls) and the movement you undergo in traveling from one city to another. Such quantities that have direction as well as magnitude are called **vector quantities**.

A convenient way to represent a vector quantity graphically is to draw a line whose length is proportional to the vector's magnitude and indicate the vector's direction by an arrow at one end of the line. For example, suppose a car travels 30 km east. We say that the car has undergone a *displacement* of 30 km east. Obviously the displacement is a vector quantity. Not only does it have magnitude, 30 km, but it also has direction, east. We can represent the displacement by a vector arrow, as shown in Fig. 1.2. The arrow is made three units long to represent its 30-km magnitude and is directed eastward to show the direction of the displacement.

1.8 ADDITION OF VECTORS

Everyone knows that when you add two apples and three apples, you have a total of five apples. This is an example of how scalar quantities are added. The sum of two scalars is simply the sum of their two magnitudes, assuming, of course, that they are expressed in the same units. Adding 40 cm³ of water to 20 cm³ of water gives 60 cm³. Again, scalar quantities add numerically.

Vector quantities do not add this way, however. Let us first illustrate this point using displacements.

The **displacement** from a point A to a point B is a vector quantity. Its magnitude is the straight-line distance from A to B; its direction is that of an arrow that points from A to B.

Let us consider what happens as you undergo a displacement of 30 km eastward and then 10 km northward, as shown in Fig. 1.3. We are interested in the total displacement resulting from these two displacements, namely, the displacement from A to C. This displacement, represented by the arrow labeled **R**, is called the **resultant displacement**. It is the sum of the two displacement vectors.

Obviously the resultant displacement from A to C is a vector and has a direction different from that of either of the original displacements. Moreover, its magnitude is certainly not 30 km + 10 km = 40 km. Instead, the pythagorean theorem gives the magnitude of the resultant displacement as

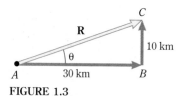

FIGURE 1.3

The vector diagram of a trip in which the traveler went 30 km east and then 10 km north.

$$\text{Magnitude of } \mathbf{R} = \sqrt{(10 \text{ km})^2 + (30 \text{ km})^2} = \sqrt{1000 \text{ km}^2} = 32 \text{ km}$$

As we see, vector addition is quite different from scalar addition.

Often the direction of the resultant vector is as important as its magnitude. One way to find its direction is to measure the angle θ in Fig. 1.3 with a protractor. If the drawing is exactly to scale, the measured angle $\theta = 18°$. We are able to state that the resultant displacement is 32 km at 18° north of east.

Before proceeding, we must mention how we indicate the vector nature of a quantity. Suppose we are concerned with a displacement of 40 m directed north. Let us represent the displacement by the letter D. If we want to consider only its magnitude, we designate the displacement as D. Thus we would write in this case $D = 40$ m. If we are concerned with the displacement's direction as well as its magnitude, however, we emphasize this point by representing the displacement in boldface type: **D** (or, when writing by hand, $\overline{D}$ or $\underset{\sim}{D}$). Be careful, then. When we write a symbol in boldface type, we are telling you that it represents a vector quantity and that you must pay attention to direction.

GRAPHICAL ADDITION OF VECTORS

We can always find the resultant displacement of several successive displacements by using a scale diagram. This was done in Fig. 1.3 for two such displacements. Notice that the method consists of laying out the vectors to scale at appropriate angles. The tail of the second vector must be placed at the tip of the first, and, of course, the resultant points from the tail of the first vector to the tip of the second.

This method for finding the resultant, called the **graphical method**, is easily extended to more than two displacements. For example, suppose we add the following successive displacements: 10 km east, 16 km south, 14 km east, 6 km north, and 4 km west. Drawing the vectors to scale, we add them tip to tail and obtain the diagram shown in Fig. 1.4. The resultant displacement **R** extends from the tail of the first vector to the tip of the last. Be sure you understand this diagram. Measurement on the scale diagram shows **R** to have a magnitude of 22 km and a direction of $\theta = 26°$ south of east.

This result does not depend on the order in which you add the vectors. For example, try changing the sequence in Fig. 1.4 to 16 km south, then 4 km west, followed by 10 km east, 6 km north, and finally 14 km east. You should be able to verify that the resultant displacement has the same value as obtained previously.

The result of adding vectors does not depend on the order in which you add them.

Figure 1.5 shows how we use the graphical method to add two displacements that are not at right angles to each other: 10 km directed 45° east of north and 5 km directed south. As before, we lay out the vectors to scale at their appropriate angles. The resultant points from the tail of the first vector to the tip of the second.

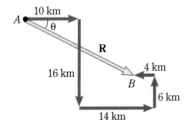

FIGURE 1.4

The graphical addition of five successive displacements.

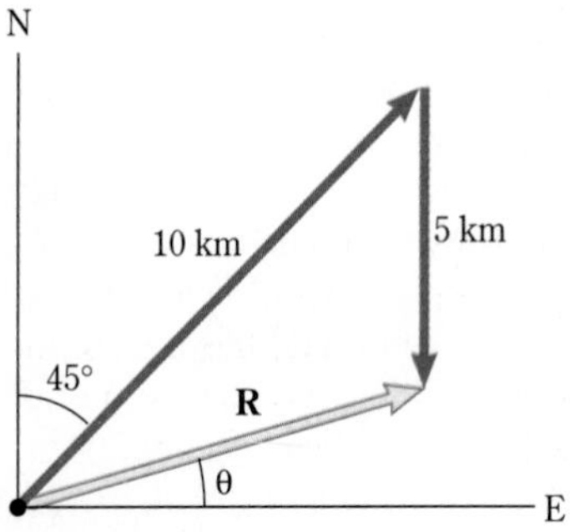

FIGURE 1.5

A vector diagram of a trip of 10 km northeast followed by one of 5 km south.

Illustration 1.5

Add the following displacements graphically:

Displacement (cm)	25	10	30
Angle (degrees)	30	90	120

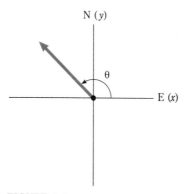

FIGURE 1.6

It is customary to measure angles relative to the east (or *x*) axis, as shown.

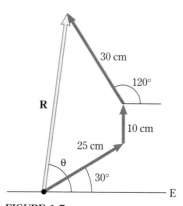

FIGURE 1.7

Addition of the displacements given in Illustration 1.5.

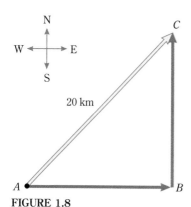

FIGURE 1.8

A displacement of 20 km northeast is resolved into component displacements **AB** east and **BC** north. Both **AB** and **BC** are rectangular components of vector **AC**.

The angles are measured relative to the east, as indicated in Fig. 1.6. It is customary to measure angles in this way.

Reasoning We lay out the vector diagram as in Fig. 1.7. (It would be a good idea to sketch the diagram yourself from the data given and compare your sketch with Fig. 1.7.) Measurements on the diagram show $R = 49$ cm and $\theta = 82°$. ∎

RECTANGULAR COMPONENTS OF VECTORS

Although the graphical method for adding vectors is simple and straightforward, it is cumbersome and only as accurate as our scale drawings. We therefore need another method that does not have these drawbacks. Such a method is called the *rectangular component method* for adding vectors. Before we describe the method, however, we must learn how to find rectangular components.

Suppose a person travels from point A to a point C that is 20 km northeast of A. The appropriate vector arrow representing this displacement is the arrow from A to C in Fig. 1.8. You can also go from A to C by the path ABC. In other words, you can undergo first the displacement from A to B and then the displacement from B to C. The net effect is the same either way: you have undergone a displacement from A to C. The displacement vector from A to C can be replaced by the two vectors **AB** and **BC** which are at right angles to each other. We call these two vectors the **rectangular components** of the original vector. In the next section, we shall see that vectors are easily added through their rectangular components, but we must first learn how to use trigonometry to find these rectangular components.

Let us briefly review the simple trigonometric functions of a right triangle. If you

People on both sides of the escalator have the same speed but different velocities.

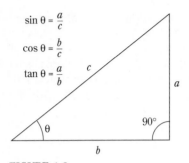

$$\sin \theta = \frac{a}{c}$$

$$\cos \theta = \frac{b}{c}$$

$$\tan \theta = \frac{a}{b}$$

FIGURE 1.9

The trigonometric functions of a right triangle.

$$\frac{c_y}{c} = \sin 37°$$

$$\frac{c_x}{c} = \cos 37°$$

$$c_x = (20 \text{ cm})(0.80) = 16 \text{ cm}$$

$$c_y = (20 \text{ cm})(0.60) = 12 \text{ cm}$$

$c = 20$ cm

c_y

90°

37°

c_x

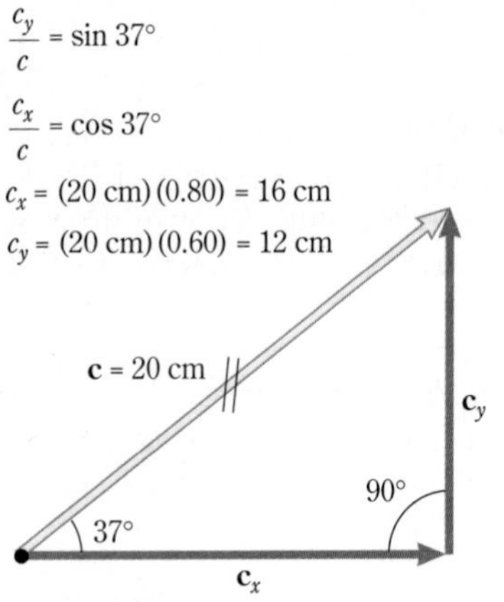

FIGURE 1.10

The two slash marks on the vector **c** indicate that it has been replaced by its components. Note that $\sin 37° = 0.60$ and $\cos 37° = 0.80$.

have not read the inside back cover yet, it is time to do so now. In terms of the sides of the right triangle in Fig. 1.9, these functions are defined as follows:

$$\sin \theta = \frac{\text{opposite side}}{\text{hypotenuse}} = \frac{a}{c} \qquad \cos \theta = \frac{\text{adjacent side}}{\text{hypotenuse}} = \frac{b}{c}$$

$$\tan \theta = \frac{\text{opposite side}}{\text{adjacent side}} = \frac{a}{b} \tag{1.1}$$

Most calculators give values of these functions for various angles. Notice that the functions are *dimensionless* ratios. It follows from Eqs. 1.1 that the two sides of the triangle can be found if we know the hypotenuse c and one angle:

$$a = c \sin \theta \qquad b = c \cos \theta$$

Let us now apply this knowledge to finding the components of a vector.

Figure 1.10 shows a 20-cm displacement vector that makes an angle of 37° with the x axis. (We shall now consider x and y directions instead of east and north. If you wish, you may associate x with east and y with north.) The original vector **c** is equivalent to the vector sum of its two rectangular components $\mathbf{c}_x$ and $\mathbf{c}_y$. We can find the magnitude of these components by using the sine and cosine relationships:

$$\mathbf{c}_x = \mathbf{c} \cos 37° = (20 \text{ cm})(0.80) = 16 \text{ cm}$$
$$\mathbf{c}_y = \mathbf{c} \sin 37° = (20 \text{ cm})(0.60) = 12 \text{ cm}$$

The 20-cm displacement at an angle of 37° to the x axis is equivalent to the sum of two rectangular component vectors: $\mathbf{c}_x = 16$ cm in the positive x direction and $\mathbf{c}_y = 12$ cm in the positive y direction.

It is possible in this way to replace any vector by its rectangular components. Once you have learned how to do this, it is a simple matter to add (or subtract) vectors of all types. Before proceeding, however, be sure you can find the x and y components of the vectors shown in Fig. 1.11. Notice that the direction of each

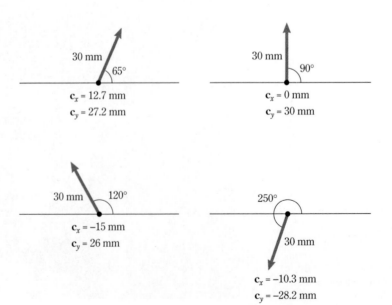

30 mm   65°
$\mathbf{c}_x = 12.7$ mm
$\mathbf{c}_y = 27.2$ mm

30 mm   90°
$\mathbf{c}_x = 0$ mm
$\mathbf{c}_y = 30$ mm

30 mm   120°
$\mathbf{c}_x = -15$ mm
$\mathbf{c}_y = 26$ mm

250°
30 mm
$\mathbf{c}_x = -10.3$ mm
$\mathbf{c}_y = -28.2$ mm

FIGURE 1.11

Verify that the components of these vectors are as stated.

component is indicated by an algebraic sign. When we write $c_x = -15$ mm, we mean that the component points in the negative x direction. Similarly, $c_y = 30$ mm means that the component points in the positive y direction. Thus the direction of a component vector is given by the algebraic sign appended to its numerical value.

TRIGONOMETRIC ADDITION OF VECTORS

Now that we know how to find vector components, it is easy to add displacements. Suppose, for example, that a bug crawling along a tabletop undergoes the displacements shown in Fig. 1.11:

30.0 mm at 65.0° to the positive x axis (east)
30.0 mm at 90.0°
30.0 mm at 120.0°
30.0 mm at 250.0°

where we measure angles as shown in Fig. 1.6.

We could find the bug's resultant displacement by using a scale diagram, as in Fig. 1.12, but the method is quite cumbersome in this case. A better approach is to use the components of the various vectors to find the components of the resultant. Let us call the x component of the resultant R_x. To obtain it, we simply add the x components of the individual vectors, components we already found in Fig. 1.11:

$$R_x = 12.7 + 0 + (-15.0) + (-10.3) \text{ mm}$$
$$= 12.7 + 0 - 15.0 - 10.3 = -12.6 \text{ mm}$$

Similarly, we can find the y component of the resultant, R_y, by adding the individual y components:

$$R_y = 27.2 + 30.0 + 26.0 - 28.2 = 55.0 \text{ mm}$$

These are the rectangular components of the resultant displacement. Notice that R_x is negative and therefore points in the negative x direction. It is very important that you take the signs of the components into account in determining the sum. Also notice that, as with graphical vector addition, you can add the components in any sequence without changing the result.

Figure 1.13 shows R plus its two rectangular components. The resultant is the hypotenuse of a right triangle whose other sides are $R_x = -12.6$ mm and $R_y = 55.0$ mm. To find the magnitude of R, we use the pythagorean theorem:

$$R = \sqrt{(55.0 \text{ mm})^2 + (12.6 \text{ mm})^2} = \sqrt{3184 \text{ mm}^2} = 56.4 \text{ mm}$$

To find the angle θ the resultant makes with the x axis, we first find angle ϕ in Fig. 1.13. Notice that

$$\tan \phi = \frac{\text{opposite side}}{\text{adjacent side}} = \frac{R_y}{R_x} = \frac{55.0}{12.6} = 4.37$$

We must now find the angle ϕ whose tangent is 4.37. This is called the **inverse** of

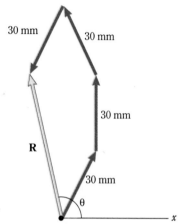

FIGURE 1.12

The resultant of the displacements in Fig. 1.11. By using a protractor and the same length scale as in Fig. 1.11, you should find that R represents a displacement of 56.4 mm at an angle of 103° relative to the x axis.

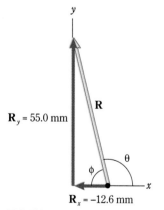

FIGURE 1.13

Compute the magnitude and direction of R from its components.

the tangent, written either as inv tan or $\tan^{-1}$. Trigonometric tables or a hand calculator give

$$\phi = \tan^{-1}(4.37) = 77.0°$$

Since $\theta + \phi = 180°$, we have

$$\theta = 180° - \phi = 103°$$

With a ruler and a protractor you can confirm these values from Figs. 1.12 and 1.13. When you use the trigonometric method, it is wise to at least sketch the situation graphically to see whether your result is realistic.

Illustration 1.6

Add the vector displacements in part *a* of Fig. 1.14.

Reasoning   We have labeled the vectors **a**, **b**, **c**, and **d**. The *x* and *y* components of **a** and **b** are obvious. We find the components of the other two vectors in parts *b* and *c* of the figure. Let us now tabulate the data in Fig. 1.14 in order to find $\mathbf{R}_x$ and $\mathbf{R}_y$:

	a	b	c	d
$\mathbf{R}_x$	+1.00	0	−4.00	−3.60
$\mathbf{R}_y$	0	+3.00	+3.00	−4.80

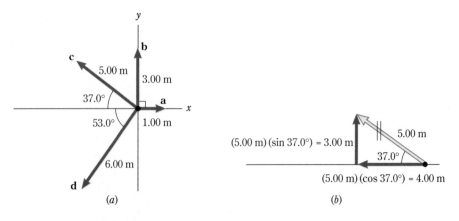

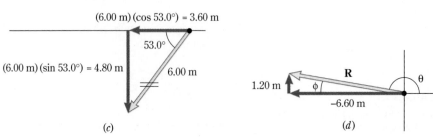

FIGURE 1.14

With the component method, the vectors in (*a*) can be added to give the resultant in (*d*).

We therefore have that

$\mathbf{R}_x = 1.00 + 0 - 4.00 - 3.60 = 1.00 - 7.60 = -6.60\,\text{m}$
$\mathbf{R}_y = 0 + 3.00 + 3.00 - 4.80 = +1.20\,\text{m}$

We now use these components to sketch $\mathbf{R}$ in Fig. 1.14d. From the sketch,

$$R = \sqrt{(6.60\,\text{m})^2 + (1.20\,\text{m})^2} = 6.71\,\text{m}$$

Also from Fig. 1.14d,

$$\tan \phi = \frac{1.20}{6.60} = 0.182$$

from which $\phi = 10°$. Then, from Fig. 1.14d,

$$\theta = 180° - 10° = 170°$$

Exercise What is the vector sum of the 5.00-m vector of Fig. 1.14b and the 6.00-m vector of Fig. 1.14c? *Answer: 7.81 m at 193°.*

1.9 SUBTRACTION OF VECTORS

Many physical situations lend themselves to analysis by vector subtraction. For example, if you walk 10 blocks east and then retrace your path by going 4 blocks west, you are subtracting a 4-block displacement from a 10-block displacement. If you wish, you could say you *added* a 10-block eastward displacement and a −4-block eastward displacement. The resultant displacement is 6 blocks eastward in either case (Fig. 1.15).

With this equivalency of the two descriptions in mind, we see that the subtraction of a vector is equivalent to the addition of the same vector with its direction reversed. The following rule applies to vector subtraction:

To subtract vector $\mathbf{B}$ from vector $\mathbf{A}$, reverse the direction of $\mathbf{B}$ and add it to $\mathbf{A}$.

In mathematical symbols,

$$\mathbf{A} - \mathbf{B} = \mathbf{A} + (-\mathbf{B})$$

where $-\mathbf{B}$ is simply vector $\mathbf{B}$ with its direction reversed.

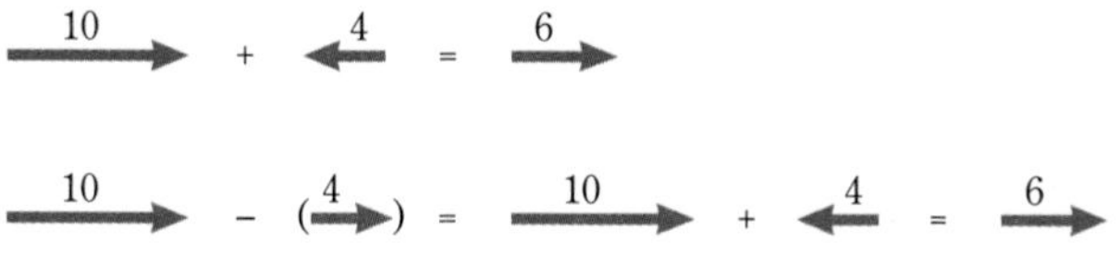

FIGURE 1.15

Two equivalent ways of describing a trip consisting of a 10-block eastward displacement and a 4-block westward displacement.

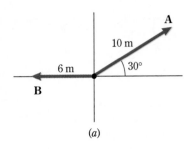

(a)

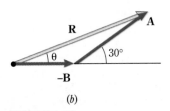

(b)

FIGURE 1.16
To find $\mathbf{A} - \mathbf{B}$, reverse the direction of $\mathbf{B}$ and add it to $\mathbf{A}$.

Illustration 1.7

Subtract vector $\mathbf{B}$ from vector $\mathbf{A}$ in Fig. 1.16*a*.

Reasoning   You should be able to show that the components of these vectors are

$$A_x = 8.70 \text{ m} \qquad A_y = 5.00 \text{ m}$$
$$B_x = -6.00 \text{ m} \qquad B_y = 0 \text{ m}$$

We wish to compute $\mathbf{R}$, where $\mathbf{R} = \mathbf{A} + (-\mathbf{B}) = \mathbf{A} - \mathbf{B}$.

$$R_x = A_x - B_x = 8.70 \text{ m} - (-6.00 \text{ m}) = 14.70 \text{ m}$$
$$R_y = A_y - B_y = 5.00 \text{ m} - 0 \text{ m} = 5.00 \text{ m}$$

we have

$$R = \sqrt{R_x^2 + R_y^2} = \sqrt{(14.70 \text{ m})^2 + (5.00 \text{ m})^2} = 15.5 \text{ m}$$

The angle $\mathbf{R}$ makes with the $+x$ axis is given by

$$\tan \theta = \frac{5.00}{14.70} = 0.340 \qquad \theta = \tan^{-1}(0.340)$$

from which $\theta = 18.8°$. We check our answer graphically using the diagram in Fig. 1.16*b*. Notice that we have reversed the direction of $\mathbf{B}$ and added it to $\mathbf{A}$. ∎

LEARNING GOALS

Now that you have finished this chapter, you should be able to

1 Define (*a*) limit of precision, (*b*) systematic errors, (*c*) accuracy, (*d*) statistical errors, (*e*) dimension, (*f*) measurement unit, (*g*) conversion factor, (*h*) significant digit, (*i*) scalar quantity, (*j*) vector quantity, (*k*) rectangular component, (*l*) resultant vector.

2 Identify the correct number of significant digits (*a*) in a given measured quantity, (*b*) in the result of adding or subtracting measured quantities, (*c*) in the result of multiplying or dividing measured quantities.

3 Give the correct derived unit that results from a mathematical calculation involving measured numbers having units.

4 Find the resultant of several displacement vectors by the graphical method.

5 When a displacement and its angle are given, find its x and y components.

6 Find the magnitude and angle of a vector from its x and y components.

7 Use the trigonometric method to add several vectors.

8 Subtract one vector from another.

SUMMARY

DEFINITIONS AND BASIC PRINCIPLES

Sources of Error in Measurement
SYSTEMATIC ERRORS: Errors due either to incorrect design or calibration of the measuring device or to incorrect reading or interpretation of the device.

STATISTICAL ERRORS: Differences in individual measurements of a quantity which are greater than the precision of the

measuring device. These differences arise from fluctuations in the quantity being measured.

Limit of Precision and Accuracy
The limit of precision of a measuring device is one-half the smallest division of measurement the device is able to display.

The accuracy of a measurement is the extent to which

systematic errors make a measured value differ from its true value.

Dimension and Unit of Measurement

DIMENSION: One of the seven basic measurable physical properties: length, mass, time, temperature, electrical current, number of particles, and luminous intensity. All other physical properties can be derived as combinations of the basic dimensions.

UNIT OF MEASUREMENT: A basic unit of measurement is the amount of a physical quantity defined by the standard of measurement for each basic dimension. Derived units are defined by the mathematical combination of basic units involved in the definition of a derived physical property. The two systems of units currently in use are the International System of Units (SI) and the British System.

Significant Digits

The significant digits in a measured or calculated quantity are those digits that are known with certainty.

RULES FOR CALCULATING WITH SIGNIFICANT DIGITS

1. When adding or subtracting measured quantities, the precision of the answer can be only as great as the least precise term in the sum or difference. All digits up to this limit of precision are significant.
2. When multiplying or dividing measured quantities, the number of significant digits in the result can generally be only as great as the least number of significant digits in any factor in the calculation.

INSIGHTS

1. Zeros can be ambiguous, often appearing simply to locate the decimal point. Scientific notation usually clarifies this ambiguity.
2. Your calculator cannot increase either the precision of or the number of significant digits in a measured quantity. Make sure you observe the above rules, rounding off your calculator answer to the correct number of digits.

Vectors and Scalars

A scalar is a quantity that has magnitude only.

A vector is a quantity that has both direction and magnitude.

Addition and Subtraction of Vectors

GRAPHICAL METHOD

1. Select a length scale to represent the magnitude of the vectors.
2. Select a reference axis from which to measure the directions of the vectors.
3. Select one vector. Draw it to scale in the correct direction. Draw a second vector to scale in the correct direction, starting with the tail of the second vector at the head of the first vector. Repeat this process with each vector in succession.
4. To find the sum, or resultant, vector, draw the straight line from the tail of the first vector to the tip of the last one. The scale length of this line is the magnitude of the resultant, and the direction of the resultant can be measured from the reference axis.

TRIGONOMETRIC METHOD

1. Select a convenient set of coordinate axes.
2. Resolve each vector into its rectangular coordinates, using the sine and cosine functions.
3. Add all the x coordinates together (taking account of signs), and all the y coordinates together. These sums are the x and y components, respectively, of the resultant.
4. Use the pythagorean theorem to find the magnitude of the resultant.
5. Find the direction θ of the resultant from the relation

$$\theta = \tan^{-1}\frac{R_y}{R_x}$$

INSIGHTS

1. You can add vectors in any order.
2. To subtract one vector from another, merely reverse the direction of the vector you want to subtract and follow the rules for addition.
3. In the trigonometric method of addition, observing the signs of $\mathbf{R}_x$ and $\mathbf{R}_y$ will enable you to make a sketch of the resultant triangle and identify what angle you are calculating in step 5 above.

QUESTIONS AND GUESSTIMATES

1 When you lie down on your bed tonight, what will be the resultant displacement your body has undergone since you rose this morning?

2 Two helicopter landing pads are a few kilometers apart. A woman takes off from one in a helicopter and eventually lands at the other. In the meantime, her husband walks from one to the other. Compare the wife's total displacement with that of the husband.

3 The sum of two vectors is zero. What can you conclude about their rectangular components?

4 Two displacements **A** and **B** are added. What must be the relation of **A** to **B** if the magnitude of their sum is (*a*) *A* + *B*, (*b*) zero?

5 Estimate the total resultant displacement you have undergone in the last (*a*) 1.5 h, (*b*) 24 h.

6 What are some physical situations in which vector quantities are subtracted? Can these quantities be thought of as being added rather than subtracted?

7 Represent each person in a city of 200,000 by a vector extending from toe to nose. Estimate the resultant of these vectors at (*a*) noon, (*b*) midnight.

8 Vector **A** lies in the *xy* plane. What range can its angle θ take on if (*a*) its *x* component is negative? (*b*) its *x* and *y* components are both negative? (*c*) its *x* and *y* components have opposite signs?

PROBLEMS

The problems at the end of every chapter are divided into three levels of difficulty: standard, somewhat difficult (indicated by one square ■), and most difficult (two squares). Problems labeled (G) should be solved graphically. Solve all others analytically. Angles are measured from the positive *x* axis unless otherwise stated. Please read the note on the next page.

Section 1.4

1 Carry out the following conversions of units, using the conversion factors found on the inside front cover of this book: (*a*) 60 mi/h to m/s, (*b*) 1.000 yr to s, (*c*) 440 yd to m, (*d*) 1500 m to ft, (*e*) 40 km/h to m/min.

2 Carry out the following conversions of units, using the conversion factors found on the inside front cover of this book: (*a*) 80 km/h to ft/s, (*b*) 220 days to s, (*c*) 2600 m to ft, (*d*) 8 mi/s to km/h, (*e*) 1300 km to inches.

Section 1.5

3 Using scientific notation, write the following lengths in meters with one digit to the left of the decimal place: (*a*) 62.8 km, (*b*) 0.00226 mm, (*c*) 33.3 nanometers (nm), (*d*) 135.8 micrometers (μm), (*e*) 3.0002×10^3 cm.

4 Using scientific notation, write the following masses in grams (g) with one digit to the left of the decimal: (*a*) 745 kg, (*b*) 0.0669 μg, (*c*) 32.55 ng, (*d*) 231 picograms (pg), (*e*) 74,800 mg, (*f*) 0.41 gigagrams (Gg).

5 Carry out the following computation and write the answer in the notation used in Probs. 1 and 2: $(732 \times 10^{-3}) \times (9.82 \times 10^5) \div (0.545 \times 10^7)$.

6 Carry out the following computation and write the answer in the notation used in Probs. 1 and 2: $(7.88 \times 10^5) \times (20.01) \div (341 \times 10^{-20})$.

7 Give the number of significant digits in each of the following quantities: (*a*) 3.649 cm, (*b*) 20.030 mi, (*c*) 0.000927 g, (*d*) 15 apples, (*e*) 3400 s.

8 Give the number of significant digits in each of the following quantities: (*a*) 14.67 mm, (*b*) 3.000×10^4 km, (*c*) 0.001 hours (h), (*d*) 1100 s, (*e*) $\pi/2$ radians (rad), (*f*) 3.77×10^{-6} kg.

9 Carry out the computation $(3.44 \times 10^8) \div (0.05899)$. Express your answer in scientific notation with the correct number of significant digits.

10 Carry out the computation $(0.44 \times 10^{-11}) \times (34.9 \times 10^3) \div (0.009)$. Express your answer in scientific notation with the correct number of significant digits.

11 Carry out the computation 120 in + 39.6 in + 13.55 in − 21 in. Express your answer in scientific notation with the correct precision.

12 Carry out the computation 13.37×10^3 m − 0.0933 m + 64 m. Express your answer in scientific notation with the correct precision.

■13 Evaluate the expressions (*a*) $(9.1 \times 10^{-31}) \times (14.7 \times 10^6) \div (331 \times 10^{-8})$, (*b*) $(13.6 \times 10^{-19})^{1/2}$, (*c*) $(3 \times 10^8)^2 \div (1.6 \times 10^{-13})$, (*d*) $(87.66 \times 10^{-5})^{1/2}$.

■14 Evaluate the expressions (*a*) $(0.088 \times 10^{-7})^{3/2}$, (*b*) $(20.3 \times 10^6) \times (3.15 \times 10^{-17})^3 \div (0.844 \times 10^{12})$, (*c*) $(27 \times 10^9)^{1/3}$, (*d*) $(81 \times 10^3)^{2/3}$.

Sections 1.7–1.9

15 To go from your house to a certain store, you must go six blocks east and three blocks south. What is your resultant displacement (magnitude and angle) as you make this trip? (G)

16 Find the resultant displacement of a car that goes 13.5 km north and then 30 km east. (G)

17 A treasure map says, "Start at the large tree. Go 125 paces straight south, then 40 paces at 45° north of west, then 60 paces straight west, and finally 30 paces at 30° south of east to where the treasure lies." How far from the tree and in what direction is the treasure? (G)

18 Hicksville is 220 km in a direction 40° north of west from Klutztown. A road goes 30 km straight north from Hicksville and then ends. When you reach the end of this road, how far and at what angle would you have to travel in a straight line in order to reach Klutztown? (G)

19 To go from St. Louis to Miami, a plane must fly 1780 km in a direction 47° south of east. To go from Ottawa to Miami, the plane must fly straight south for 2060 km. How far and in what direction must the plane fly to go from St. Louis to Ottawa? (G)

20 A displacement of 35 m is made in the *xy* plane at an angle of 57°. Find its *x* and *y* components. Repeat for angles of 122° and 240°.

21 In an *xy* coordinate system, point *P* is 85 cm from the origin and its *y* coordinate is −33 cm. Find *P*'s *x* coordinate and the direction of P's displacement from the origin. There are two possible answers. Find both.

22 Starting at the origin, you move an object in the xy plane as follows: 70 cm along $\theta = 15°$, then 25 cm along $\theta = 220°$. Find the distance and the displacement through which you moved the object.

23 Suppose you walk 610 m at 20° north of west from point A, followed by a walk 260 m at 45° north of east, ending at point B. What is the displacement from A to B? From B to A?

■24 From point A you ride your bicycle a distance of 4.55 km due east. You then follow a circular path centered on point A until you are directly south of A. You turn directly north and ride for 1.80 km, ending at point B. What is your displacement from A? What distance have you traveled?

25 Solve Prob. 17 using trigonometry.

26 Solve Prob. 18 using trigonometry.

27 Solve Prob. 19 using trigonometry.

■■28 A room has a ceiling that is 2.35 m high and a floor that measures 4.75 m by 5.50 m. Find the length of the diagonal line from a ceiling corner to the opposite floor corner. What angle does this line make with the floor?

■29 Vector A has a magnitude of 40 m at $\theta = 225°$. If we want to add to A a vector B so as to produce a resultant along the positive x axis that has a magnitude of 20 m, what must be the components of B?

30 Two displacements A and B lie in the xy plane. A is 49 cm at $\theta = 42°$, and B is 32 cm at $\theta = 115°$. What are the displacements $A + B$ and $B - A$?

■■31 When displacement B is added to displacement A the result is a displacement C that has components $C_x = -3.70$ cm, $C_y = +2.25$ cm, and $C_z = +4.60$ cm. Displacements A and B are in the same direction, but the magnitude of A is only one-third that of B. Find the components of A.

General Problems

■■32 A bug climbs up the north wall of a house a distance of 6.5 ft in a straight line at an angle of 65° relative to the floor. This brings the bug to the intersection of the north wall and the wall facing east. The bug then proceeds to travel along the east wall a distance of 2.5 ft in a direction 25° below horizontal. At this point, what is the bug's displacement from where it started? What angle does its displacement make with respect to the floor? With respect to the north wall?

■■33 A mine shaft goes straight down for 110 m. From its lower end, a horizontal tunnel goes 35 m east and then 70 m south to its end. What is the displacement from the top of the shaft to the end of the tunnel? What angle does this displacement make with respect to a vertical line?

■34 A boat travels a straight-line distance of 4.3 mi. At the end of this displacement, the boat is 1.6 mi west of where it started. Find the direction of the boat's travel and how far north or south of its starting point the boat finished. Two possible answers exist. Find both. (G)

■■35 Minneapolis is 400 mi northwest (45° west of north) of Chicago. An airplane flies from Minneapolis at a heading of 10° west of south while a second airplane flies from Chicago at an angle 45° west of south. What is the displacement of the point where their paths cross relative to Chicago? Relative to Minneapolis?

Note concerning the data given in the problems: To facilitate the statement of problems, we sometimes abbreviate data. When we state that "a book is 4 cm thick", you should consider the data to be reasonably accurate, 4.00 cm thick. Similarly, "a 5 kg object" should be interpreted to have a mass of 5.00 kg. Unless otherwise indicated, all data should be assumed accurate within the limits of normal measurements.

PART ONE

MECHANICS

"Science is a little bit like the air you breathe — it is everywhere"

DWIGHT EISENHOWER

Our study of physics begins with the subject of mechanics. In mechanics we seek to understand and explain the motion of material objects and also the conditions under which objects remain at rest. At first sight, you may wonder why such a study is important. However, with the few basic principles of mechanics we can describe and explain the motions of stars and planets, build bridges and skyscrapers, fly airplanes and put satellites into orbit. Moreover, many of the concepts of mechanics, such as force, energy, and momentum, play an important role in the study of other branches of physics as well.

Although many ancient philosophers attempted to explain why and how things move, a systematic theory of motion was not developed until the 17th century. The most notable contributions to this development were those of Galileo and Newton. Newton set out the first mathematical laws of motion in his *Principia* in 1687 in which he introduced the concept of mass as a quantity of matter and the concept of forces between objects which cause changes in their motion. Newton also put forth a mathematical description of gravity as a fundamental force by which all objects attract each other. This concept of universal gravity showed that the motions of the planets and the motion of objects falling to earth were governed by the same single principle.

Newton's laws of mechanics provided a reliable description of all known mechanical phenomena for over 200 years. Toward the close of the 19th century, physics began to delve into the realm of very small and very fast phenomena, such as the structure of atoms and the behavior of objects moving nearly as fast as the speed of light. It became apparent as the 20th century began that Newton's theory had to be modified in order to explain these new phenomena, which are far removed from our everyday experience. The results of these modifications, Relativity and Quantum Mechanics, have been very successful in explaining motion and mechanical structure in these cases.

CHAPTER 2

UNIFORMLY ACCELERATED MOTION

One of the most obvious of physical phenomena is the movement of objects, and so motion is an excellent place to begin our study of physics. Before we can understand the causes of motion, however, we must be able to describe motion quantitatively. And before we can describe motion quantitatively, we must define certain properties of motion, such as displacement, velocity, and acceleration, in terms of the dimensions of length and time. The quantitative description of motion without reference to physical causes is called **kinematics** and is the subject of this chapter. In later chapters, when we examine force and energy, we will study the causes of motion. The study of the relationship between motion and its causes is called **dynamics**.

2.1 UNITS OF LENGTH AND TIME

In order to define quantities which describe motion, we must first define the basic units of measurement of length and time.

The basic SI unit of length is the **meter**. Originally the meter was defined as the length of a standard metal bar kept at the International Bureau of Weights and Measures at Sevres, France. This bar represented one ten-millionth of the distance between the north pole and the equator, measured along the longitude line passing through Paris. You can imagine the difficulties in actually measuring that basic distance.

With the development of the laser and modern optical devices, light now provides us with the most precise method of measurement. Since 1983, therefore, the meter has been defined in terms of the speed of light through a vacuum:

1 meter = the distance light travels in vacuum in 1/299,792,458 second

The SI unit of time is the **second**, defined in terms of the frequency of light emitted in a specific atomic process:

1 second = the time taken for exactly 9,192,631,770 cycles of a certain wavelength of light to be emitted by the cesium atom.

If these definitions seem arbitrary, it's because they are! However, they are defined by precise (note the large number of significant digits) and accessible experiments. Scientists anywhere in the world (or universe) can duplicate the measurement of these units without transporting some standard object for comparison.

2.2 SPEED

Time and length standards. This cesium clock (*left*) is the primary standard for measuring time at the National Institute of Standards and Technology (NIST). This device keeps time with an accuracy of about 0.000 003 s per year. NIST uses this iodine-stabilized helium-neon laser (*right*) as its standard for length. The laser approximates the ideal meter to an accuracy of about 0.000 000 000 1 m.

When you say that a car is moving at a speed of 80 km/h, everyone knows what you mean; the car will go a distance of 80 km in 1 h provided it maintains this speed. In 0.5 h the car will go $0.5 \times 80 = 40$ km, and in 2 h it will go $2 \times 80 = 160$ km. In general, the distance a car travels if its speed is constant is

Distance traveled = speed × time taken

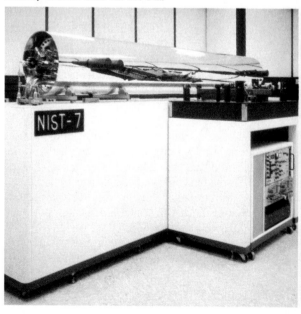

Solving for speed, we find that the defining equation for this quantity is

$$\text{Speed} = \frac{\text{distance traveled}}{\text{time taken}} \qquad (2.1)$$

We use this same equation to define the average speed of a car even when its speed is not constant. If the car goes 200 km in 4.0 h, its average speed is

$$\text{Average speed} = \frac{200 \text{ km}}{4.0 \text{ h}} = 50 \text{ km/h}$$

As you see, the units of speed are a distance unit divided by a time unit. For example, the average speed of a snail might be 1.5 mi/yr. Average speed is always the distance traveled divided by the time taken.

Notice that speed has no direction; it is a scalar quantity. A car's speedometer measures how fast the car is going, but tells us nothing about the direction of travel. The car may be traveling on a straight road across the prairie, or it may be circling a racetrack; its average speed is still 100 km/h if it travels 200 km in 2 h.

2.3   DISPLACEMENT AND AVERAGE VELOCITY

In everyday conversation, we use the terms *speed* and *velocity* interchangeably. In science, however, these two quantities have different meanings. As we shall see, velocity is a vector quantity (unlike speed, which is a scalar). Let us now develop the definition of velocity.

Suppose A and B represent two cities, with B being 800 km directly east of A, as shown in Fig. 2.1. There are many roads by which we might travel from A to B, each involving a different *distance* of travel. One such road is the green one shown in Fig. 2.1, which is 1200 km long. The *shortest* distance, however, which is a *straight line* from A to B, is 800 km. This shortest distance is represented by the blue vector **s** in Fig. 2.1. As we discussed in Chap. 1, we call **s** the displacement from A to B.* For the sake of emphasis, let us repeat the definition of displacement which we used in Chap. 1:

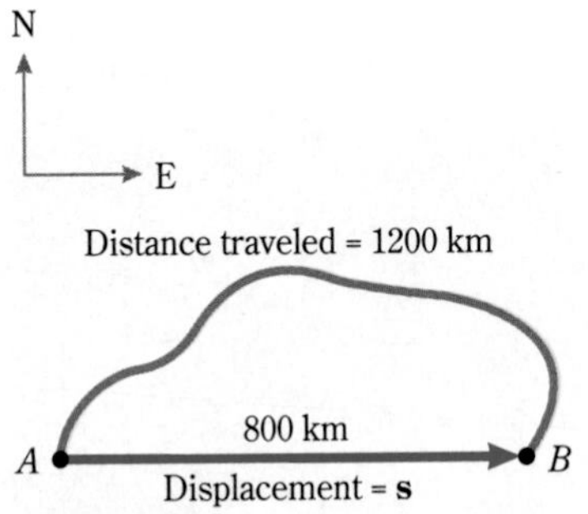

The displacement between any two points is a vector directed from one point to the other; the magnitude of this vector is the straight-line distance between the two points.

From Fig. 2.1, you can see the difference between distance traveled and displacement. To specify distance traveled, you must specify the *path* taken between the two points. Displacement is independent of the path taken. For example, whether you take the green path from A to B or the blue one, your displacement is 800 km. If you take the blue path, the distance you travel is equal to your displacement; if you take the green path, however, the distance you travel is 1200 km, but your displacement is still 800 km from where you started.

The difference between average speed and average velocity is similarly defined. In Sec. 2.2 we saw that average speed is defined in terms of distance traveled and so

N

↑

→ E

Distance traveled = 1200 km

FIGURE 2.1

The displacement **s** from A to B is 800 km eastward.

---

*At other times we shall use other symbols, such as **x** and **r**, to represent displacement. With displacement, as with all quantities, we are free to choose any algebraic symbol we wish to represent it.

Both the *Concord* and the *Queen Elizabeth 2* cross the Atlantic, but at very different speeds. The *Concord* takes less than 3 hours, while the *QE2* requires more than 3 days.

depends on the path taken. Average velocity, on the other hand, is a *vector* defined as the *displacement* from start to finish divided by the time taken:

$$\text{Average velocity} = \frac{\text{displacement vector}}{\text{time taken}}$$

In symbols,

$$\bar{\mathbf{v}} = \frac{\mathbf{s}}{t} \qquad \mathbf{s} = \bar{\mathbf{v}}t \tag{2.2}$$

The bar above the **v** is used to indicate average velocity. Notice that since $\bar{\mathbf{v}}$ is proportional to **s**, velocity is a vector quantity; it has direction, the same direction as the displacement vector. In Fig. 2.1, the velocity is eastward because the displacement **s** is eastward.

To illustrate the difference between average speed and average velocity, let us use a numerical example. Suppose a car takes 20 h to go from city A to city B along the green route in Fig. 2.1. Since **s** = 800 km eastward and t = 20 h, the average velocity of the car is

$$\bar{\mathbf{v}} = \frac{800 \text{ km east}}{20 \text{ h}} = 40 \text{ km/h, east}$$

(Notice that the average velocity, being a vector, has both magnitude, 40 km/h, and direction, east.) The average *speed* of the car, however, is

$$\text{Average speed} = \frac{\text{distance traveled}}{\text{time taken}} = \frac{1200 \text{ km}}{20 \text{ h}} = 60 \text{ km/h}$$

An important point to realize is that an object's speed need not equal the magnitude of its average velocity.

One last observation before we go on: *if you come back to your starting point,* your displacement and hence your average velocity are both *zero*, regardless of the

An object can change its direction of motion, following a curved path.

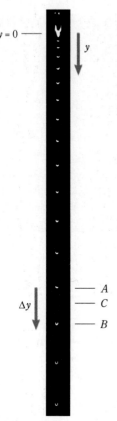

$y = 0$ ——

y

Δy —— A
—— C
—— B

FIGURE 2.2

A strobe light shows the ball's location at successive times. The ball falls from A to B in a time Δt. (*Education Development Center*)

A train on the tracks of the Nullarbor plain in South Australia is an example of one-dimensional motion. The tracks do not change direction for over 200 miles.

distance you traveled. You may have covered considerable distance at a certain average speed, but if you start and finish at the same point, your displacement is zero.

2.4   INSTANTANEOUS VELOCITY

Let us now consider the motion of a falling ball; a photograph of this type of motion is shown in Fig. 2.2. The position of such a ball at uniformly spaced times is captured by illumination from a flashing strobe light with equal time intervals between flashes. We shall call the time interval between successive flashes Δt (read "delta tee"). Notice that the ball speeds up as it falls. This is obvious from the fact that it falls a larger distance during each successive time interval. Let us now discuss how we can determine the velocity of the ball as it passes the arbitrary point C. Such a velocity at a single point is called the **instantaneous velocity**.

The direction of the velocity is clear. It is the same as the direction of motion, downward. To find a rough magnitude of the ball's velocity at C, we can compute its average velocity from A to B. We call the coordinate that measures the ball's position y. As the ball goes from A to B, its displacement is Δy. Since we are calling the time between strobe flashes Δt, the time it takes the ball to move from A to B is Δt. Therefore, for the region from A to B, the ball's average velocity is

$$\bar{\mathbf{v}} = \frac{\text{displacement}}{\text{time taken}} = \frac{\Delta y}{\Delta t}$$

This is not the exact velocity of the ball at C, however, because the velocity is continuously increasing. If we speed up the strobe light (in other words, make Δt smaller), the images of the ball will be closer to each other and points A and B will be much closer to C. If we perform our calculation for these new points A and B, the average velocity we obtain should be closer to the ball's velocity at C than the first value we calculated.

We can imagine a case in which the strobe light is flashing so rapidly that the time interval between flashes approaches zero, which we represent by $\Delta t \rightarrow 0$. Then A and B will be so close to C that the average velocity we compute is almost exactly equal to the velocity at C. We call the velocity at C the instantaneous velocity at that point and represent it by $\mathbf{v}$ (without the overbar). It is defined mathematically in terms of the experimental procedure we have just outlined and is therefore

$$\text{Instantaneous velocity} = \mathbf{v} = \lim_{\Delta t \to 0} \frac{\Delta y}{\Delta t} \qquad (2.3)$$

The symbol $\lim_{\Delta t \to 0}$ is read "in the limiting case where Δt approaches zero." It represents mathematically the experimental procedure in which Δt is made so small that the average velocity between A and B becomes essentially the instantaneous velocity at C, to whatever precision we require.

There is an interesting relation between the magnitudes of the instantaneous velocity at a point such as C and the speed at C. When we make Δt very small, the object cannot change its direction of motion appreciably during the time it takes to go from A to B. As a result, the straight-line distance from A to B equals the distance traveled by the object as it goes from A to B. Therefore, because the

A sprinter accelerates out of the starting blocks.

distance traveled and the displacement have the same magnitude, the instantaneous velocity and the speed at *C* also have the same magnitude. Thus,

The magnitude of the instantaneous velocity at a point is equal to the instantaneous speed at that point.

2.5 ONE-DIMENSIONAL MOTION

For most of the remainder of this chapter, we restrict our discussion to motion along a straight line, called one-dimensional motion. We shall learn in later chapters how to generalize our results to two-dimensional motion.

As an example of one-dimensional motion, consider the car in Fig. 2.3*a*. At the instant shown, its motion is in the positive *x* direction. Therefore the vector representing its velocity is also in this direction. If the car were going in the reverse direction, its velocity vector would point in the negative *x* direction. Hence, in one-dimensional motion, it is possible to show the direction of vectors by means of plus and minus signs.

Let us discuss the motion of the car in Fig. 2.3*a*. At a time *t*, the magnitude of its displacement from the coordinate origin is **x**. Suppose that the car was at $x = 0$ when $t = 0$ and that it is moving at 20 m/s. If we observe its position once every second, its position as a function of time yields the data

t (s):	0	1	2	3	4	5	6
x (m):	0	20	40	60	80	100	120

In other words, the magnitude of the car's displacement increases by 20 m each second. We can plot these data to give the graph of *x* as a function of *t* shown in Fig. 2.3*b*.

We can attach important meaning to the two small triangles in *b*. Notice that the vertical length of each is 20 m and the horizontal length is 1 s. The triangles therefore tell us that the car goes 20 m in the positive *x* direction every second. The vertical side, of length **Δx**, is the displacement the car undergoes during a time interval Δ*t*. Thus the average velocity of the car is

FIGURE 2.3

Motion along a straight line can be shown by a graph. In this case, the car's speed is constant at 20 m/s.

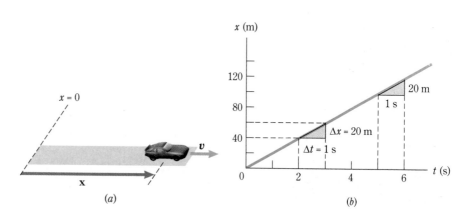

$$\overline{\mathbf{v}} = \frac{\text{displacement}}{\text{time taken}} = \frac{\Delta \mathbf{x}}{\Delta t}$$

where $\Delta \mathbf{x}$ is the vector displacement in the x direction. If $\Delta \mathbf{x}$ is positive, the velocity is in the positive x direction; if it is negative, the velocity is in the negative x direction. Thus either triangle in Fig. 2.3b can be used to find the velocity of the car.

As another example of straight-line motion, let us return to the falling ball of Fig. 2.2. In this case, the velocity is continually increasing instead of being constant. We can use the photograph to measure the position y of the falling ball as a function of time. Typical data obtained from such a photograph are

t (s):	0	0.02	0.04	0.06	0.08	0.10	0.12	0.14	0.16
y (cm):	0	0.20	0.78	1.76	3.14	4.90	7.06	9.60	12.5

Notice that the displacement $\Delta \mathbf{y}$ is a vector whose positive direction is taken to be downward. These data are plotted in Fig. 2.4. To find the ball's average velocity between points A and B on the graph, we must compute $\Delta \mathbf{y}/\Delta t$. From the graph, $t_B - t_A = \Delta t = 0.100 - 0.080 = 0.020$ s. From the table or graph, $y_B - y_A = \Delta \mathbf{y} = 4.90 - 3.14 = +1.76$ cm, so that

$$\overline{\mathbf{v}}_{AB} = \frac{\Delta \mathbf{y}}{\Delta t} = \frac{\mathbf{y}_B - \mathbf{y}_A}{t_B - t_A} = \frac{+1.76 \text{ cm}}{0.020 \text{ s}} = +88 \text{ cm/s}$$

This, within experimental error, is the average velocity between A and B. Since it is positive, $\mathbf{v}_{AB}$ is in the positive direction, downward.

The vertical side of the triangles in Figs. 2.3b and 2.4, called the **rise** of the graph, divided by the horizontal side, called the **run** of the graph, gives the average velocity. You may remember from your math classes that this ratio is the *slope* of the line that forms the third side of the triangle. In Fig. 2.4, therefore, $\Delta \mathbf{y}/\Delta t$ is the slope of the line joining A and B. We thus arrive at the following conclusion: the average velocity between any two points A and B on a graph of displacement versus time is the slope of the line joining these two points.

If we pass to the limiting case of A and B being very close together, the line connecting these two points becomes tangent* to the graph. Hence

The slope of the graph of displacement versus time at any point is equal to the instantaneous velocity at that point.

We therefore see that the slope of the graph of displacement versus time has great importance. It tells us the instantaneous velocity of a moving object.

Illustration 2.1

Figure 2.5a shows a ball after it has been tossed straight up, and Fig. 2.5b shows the ball's y coordinate as a function of time. Let us find the ball's instantaneous velocity

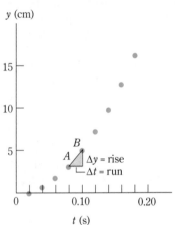

y (cm)

FIGURE 2.4

A plot of the data for an experiment such as the one in Fig. 2.2.

---

*A tangent line to a point on a curve is that line (there is only one for each point) which passes through the point but does not touch or intersect any adjacent points of the curve.

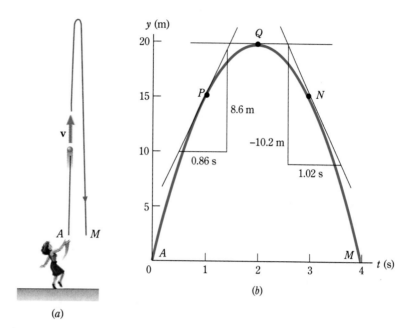

FIGURE 2.5

The essentially straight-line motion in (a) is shown in the graph in (b).

at (**a**) P, (**b**) Q, and (**c**) N. Let us also find the average velocity (**d**) between A and Q and (**e**) between A and M.

Reasoning   The graph shows that the ball rises to a height of 20 m and then falls back down. We use the fact that $\mathbf{v}$ at any point is given by the slope of the tangent line at that point.

(**a**) Draw the tangent to the curve at P:

$$\mathbf{v}_P = \text{slope at } P = \frac{8.6 \text{ m}}{0.86 \text{ s}} = 10 \text{ m/s}$$

(**b**) $\mathbf{v}_Q = \text{slope at } Q = 0$

The ball stops at Q and then begins to fall.

(**c**) $\mathbf{v}_N = \text{slope at } N = \dfrac{-10.2 \text{ m}}{1.02 \text{ s}} = -10.0 \text{ m/s}$

The sign here is negative because the slope is negative at N. The ball is now moving in the negative y direction; in other words, it is falling. The slope of the curve gives both the magnitude and the direction of the velocity. A negative slope means the velocity is in the negative y direction.

(**d**) Draw a straight line (a chord) from A to Q (not shown). It rises 20 m in 2.0 s. Then, because $\bar{\mathbf{v}} = \text{rise/run}$,

$$\bar{\mathbf{v}}_{AQ} = \text{slope of chord from } A \text{ to } Q = \frac{20 \text{ m}}{2.0 \text{ s}} = 10 \text{ m/s}$$

(e) $\bar{\mathbf{v}}_{AM}$ = slope of chord from A to M = $\dfrac{0 \text{ m}}{4.0 \text{ s}}$ = 0 m/s

We can see that this result is correct because the ball is at the same position for A and M. Thus its total displacement is zero. Then

$$\bar{\mathbf{v}}_{AM} = \frac{\text{displacement}}{\text{time taken}} = \frac{0 \text{ m}}{4.0 \text{ s}} = 0 \text{ m/s}$$

As we have said, the scientific definition of average velocity differs from the common meaning of average speed. ∎

2.6 ACCELERATION

Suppose that at a certain instant an object has a velocity (not speed) $\mathbf{v}_o$, and that its velocity at a later time t is $\mathbf{v}_f$. (The subscripts o and f stand for "original" and "final.")

The average acceleration $\bar{\mathbf{a}}$ of the object during this time interval is defined to be

$$\bar{\mathbf{a}} = \frac{\text{change in velocity}}{\text{time taken}} = \frac{\mathbf{v}_f - \mathbf{v}_o}{t} \qquad (2.4)$$

In other words, acceleration is the change in velocity (not speed) per unit time. The units of acceleration are a velocity unit divided by a time unit. Therefore acceleration has units of length divided by time squared, which in the SI is m/s^2.

To see what this definition means in a practical situation, consider a car that starts from rest and attains a speed of 20 m/s in 12 s as it travels in the positive x direction. We have $\mathbf{v}_o = 0$ and $\mathbf{v}_f = 20$ m/s, both in the positive x direction, and the time taken $t = 12$ s. Therefore

$$\bar{\mathbf{a}} = \frac{\mathbf{v}_f - \mathbf{v}_o}{t} = \frac{20 \text{ m/s} - 0 \text{ m/s}}{12 \text{ s}} = 1.7 \text{ m/s}^2$$

An example of free-fall motion.

The positive sign shows that this is a vector in the positive x direction.

Suppose that the car continues to move in the positive x direction, but now slows from 20 m/s to 0 m/s in 12 s. The average acceleration would then be

$$\bar{\mathbf{a}} = \frac{\mathbf{v}_f - \mathbf{v}_o}{t} = \frac{0 \text{ m/s} - 20 \text{ m/s}}{12 \text{ s}} = -1.7 \text{ m/s}^2$$

Notice that the sign is now negative. Remember, the sign of a vector indicates its direction. In this case, we have already decided to designate positive vectors as those pointing in the positive x direction. The negative sign of $\mathbf{a}$ indicates it is directed toward *negative x, opposite* to the direction of motion. This describes motion in which the object is slowing down, which we commonly call a *deceleration*, but we prefer to use the term **negative acceleration**. Let us emphasize the basic idea here:

For one-dimensional vectors, you are free to choose which of the two possible directions will represent positive vectors. Once you have made this choice in a specific problem, you must use the correct sign for all vectors that enter into a vector calculation. The sign of the vector resulting from the calculation indicates that vector's direction.

Example 2.1

Figure 2.5*b* shows the time variation of the vertical (*y*) position of a ball thrown straight upward. Graph the ball's velocity and find its acceleration.

R *Reasoning*

Question How do I get velocity data from Fig. 2.5*b*?
Answer As explained in Illustration 2.1, the velocity at any instant is the *slope* of the *y* versus *t* graph at that instant. The slope data are given at *P*, *Q*, and *N*, which correspond to times of 1.0, 2.0, and 3.0 s, respectively. Pick a few more points (say every 0.5 s), draw the tangent lines as carefully as you can at each point, and calculate each slope. You should be able to verify the following data:

Time (s)	→	0	0.5	1.0	1.5	2.0	2.5	3.0	3.5	4.0
Velocity (m/s)	→	20	15	10	5	0	−5	−10	−15	−20

As we have discussed, the negative signs for some of the velocities indicate that then the object is traveling toward the negative *y* direction.

Question How do I lay out a graph for this data?
Answer By plotting the values of **v** on the *y* axis and those of *t* on the horizontal axis. (This is what we mean when we say plot **v** versus *t*, or **v** as a function of *t*.) Choose the scales of values such that the axes can accommodate the ranges of your data. The result should be as shown in Fig. 2.6.

Question How is acceleration related to this graph?
Answer Acceleration is represented by the slope of this graph, just as velocity is represented by the slope of a graph which plots position as a function of time (Fig. 2.5*b*).

Question My graph appears to be a straight line with a negative slope. What does this mean?
Answer A straight line has the same slope everywhere. In this particular case, the straight line means that the motion has constant acceleration. The negative slope indicates that the direction of **a** is negative. In our table of data, we chose the upward direction to be positive. This applies to *all* vector quantities— displacement, velocity, and acceleration. Thus a negative acceleration points *downward*.

Question What is the value of this acceleration?
Answer One set of slope data is shown in Fig. 2.6, from which we calculate

$$\mathbf{a} = \frac{\text{rise}}{\text{run}} = \frac{-14 \text{ m/s}}{1.4 \text{ s}} = -10 \text{ m/s}^2$$

Try this calculation from your graph, using another pair of data points.

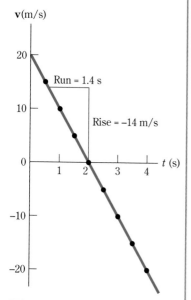

FIGURE 2.6

The time dependence of the velocity of the ball in Fig. 2.5*a*. What is the ball's acceleration?

Solution and Discussion During its entire trip (both up and down), the ball's acceleration is about 10 m/s² directed downward. When it is moving upward, the ball slows down 10 m/s each second, and when it is moving downward, its speed increases 10 m/s each second. As we shall see in Sec. 2.9, more exact measurements give the ball's acceleration as 9.8 m/s². ∎

2.7 UNIFORMLY ACCELERATED LINEAR MOTION

Situations in which acceleration varies are often difficult to analyze mathematically. For that reason, we shall restrict ourselves to situations in which acceleration is constant as in Example 2.1. (We say that an object in such a situation is uniformly accelerated.) Although this may seem like a drastic simplification, many physical systems approximate this condition. For example, objects falling freely under the action of gravity near the earth's surface have constant acceleration. Let us now see how to describe the linear motion of objects that undergo uniform (constant) acceleration.

Because the motion is along a straight line, we can simplify our discussion by using plus and minus signs to show direction. Moreover, we shall represent the vector displacement by **x**, the x-directed velocity by **v**, and the x-directed acceleration by **a**. For example, the object in Fig. 2.7 is moving with a constant acceleration a in the x direction. It passes point A with velocity $\mathbf{v}_o$ and passes point B with velocity $\mathbf{v}_f$ at a time t later. The displacement from A to B is **x**.

For the trip from A to B, we can state the following results:

1 The average velocity $\bar{\mathbf{v}}$ for the trip is

$$\bar{\mathbf{v}} = \frac{\text{displacement}}{\text{time}} = \frac{\mathbf{x}}{t}$$

from which

$$\mathbf{x} = \bar{\mathbf{v}}t \qquad (2.5)$$

Equation 2.5 involves only a single vector on each side of the equals sign. This equation can thus be written without the vector notation, since the directions of **x** and $\bar{\mathbf{v}}$ (and hence their signs) must always be the same:

$$\bar{v} = x/t \qquad (2.5)$$

2 Because the acceleration is constant, the average and instantaneous accelerations are the same. The definition of acceleration becomes

$$\mathbf{a} = \frac{\mathbf{v}_f - \mathbf{v}_o}{t} \qquad \mathbf{v}_f = \mathbf{v}_o + \mathbf{a}t \qquad (2.6)$$

3 Since the object is undergoing uniform acceleration, its velocity changes linearly with time from $\mathbf{v}_o$ to $\mathbf{v}_f$. As a result, the average velocity between A and B is

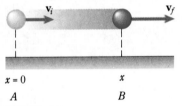

$x = 0$ 　　　　　 x

A 　　　　　　　 B

FIGURE 2.7

It takes a time interval t for the ball to move from A to B.

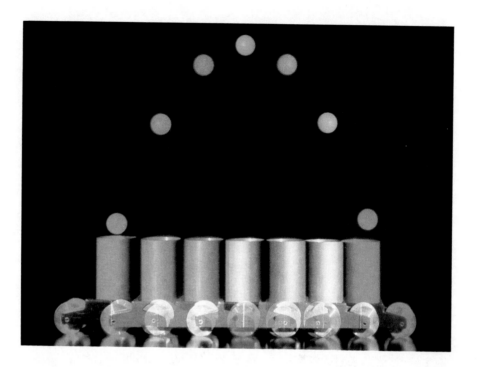

The path of a projectile is a parabola when gravity is constant and air resistance is insignificant.

simply the average of these two values:

$$\overline{\mathbf{v}} = \frac{\mathbf{v}_o + \mathbf{v}_f}{2}$$

(2.7)

We now have three equations that apply to uniformly accelerated motion, Eqs. 2.5, 2.6, and 2.7. They are sufficient to describe motion in any ordinary situation in which acceleration is uniform.

We are now beginning to expand the number of concepts and definitions at our disposal. One of the most frequently occurring complaints about word problems is: "I always have trouble in changing a word problem into equation form. How do I know which equation to use?" Much of the difficulty comes in translating the words of the problem first into the precise concepts physics uses and then into the corresponding symbols used in the equations. Here is a brief guide to help you translate problems involving motion:

Question or statement	Translation
When?	What value of t?
Where?	What value of position? (x, y, or s, for example)
Starts from rest	$\mathbf{v}_o = 0$
How fast?	What value of v?
How long does it take?	What value of Δt?
How far?	What value of $x_f - x_o$? (or $y_f - y_o$, $s_f - s_o$, etc.)
Comes to rest	$\mathbf{v}_f = 0$

Example 2.2

Suppose a car starts from rest and accelerates uniformly to 5.0 m/s in 10 s as it travels along the x axis. Find its acceleration and the distance it travels in this time.

R

Reasoning

Question   In terms of the symbols used in the previous list of equations, what information am I given?
Answer

1 "Starts from rest" means $\mathbf{v}_o = 0$.
2 "Accelerates uniformly" means that Eqs. 2.5, 2.6, and 2.7 apply to this situation.
3 "To 5.0 m/s in 10 s" means that $\mathbf{v}_f = 5.0$ m/s at $t = 10$ s.
4 "Travels along the x axis" means that this is motion in one dimension and so values of x describe the car's position.

Question   What quantities am I supposed to determine?
Answer   The value of the acceleration, $\mathbf{a}$, and the distance the car covers, x.

Question   Which of the equations should I use?
Answer   The ones that involve both the known quantities ($\mathbf{v}_o$, $\mathbf{v}_f$, and t) and the unknown quantities (x and $\mathbf{a}$). One is Eq. 2.6:

$$\mathbf{a} = (\mathbf{v}_f - \mathbf{v}_o)/t$$

Since the equation for x (Eq. 2.5) involves average velocity, you must determine this quantity before you can use the equation. Average velocity is given by Eq. 2.7:

$$\bar{\mathbf{v}} = \tfrac{1}{2}(\mathbf{v}_f + \mathbf{v}_o)$$

Solution and Discussion   The acceleration is

$$\mathbf{a} = \frac{5.0 \text{ m/s} - 0 \text{ m/s}}{10 \text{ s}} = 0.50 \text{ m/s}^2$$

The average velocity is

$$\bar{\mathbf{v}} = \tfrac{1}{2}(0 \text{ m/s} + 5.0 \text{ m/s}) = 2.5 \text{ m/s}$$

Thus the distance the car travels in 10 s is

$$x = (2.5 \text{ m/s})(10 \text{ s}) = 25 \text{ m}$$

Its acceleration during that time interval is 0.50 m/s². Notice again how the units are treated consistently as algebraic symbols in the calculations.

Example 2.3

Suppose a car traveling at 5.00 m/s is brought to rest in a distance of 20.0 m. Find its acceleration and the time it takes to stop. Assume that motion is along the x axis and that acceleration is constant.

R

Reasoning

Question $\quad$ What information am I given? How do I interpret the words?
Answer

1 "Traveling at 5.00 m/s" means $v_o = 5.00$ m/s.
2 "Brought to rest" means $v_f = 0$.
3 "In a distance of 20.0 m" means that the change of velocity (at constant acceleration) takes place over a distance of 20.0 m.

Question $\quad$ What am I required to find?
Answer $\quad$ The acceleration **a** and the time t it takes the car to stop.

Question $\quad$ How can I find t? I don't have a formula for it.
Answer $\quad$ You don't have a *formula* for anything. You have *relationships* between various quantities used in describing motion. Some of these relationships involve t.

Question $\quad$ If I use Eq. 2.6 to calculate **a**, I will need to know the value of t. What other equation involves t?
Answer $\quad$ Equation 2.5, $x = \bar{v}t$, which can be rearranged to $t = x/\bar{v}$.

Question $\quad$ How can I determine $\bar{v}$ from what is given?
Answer $\quad$ From the relationship described by Eq. 2.7: $\bar{v} = \frac{1}{2}(v_f + v_o)$.

Solution and Discussion $\quad$ From Eq. 2.7, you find that $\bar{v} = 2.50$ m/s. Then the time the car takes to stop is

$$t = \frac{x}{\bar{v}} = \frac{20.0 \text{ m}}{2.50 \text{ m/s}} = 8.00 \text{ s}$$

Once you know t, you can calculate the acceleration from Eq. 2.6:

$$\mathbf{a} = \frac{\mathbf{v}_f - \mathbf{v}_o}{t} = \frac{0 \text{ m/s} - 5.00 \text{ m/s}}{8.00 \text{ s}}$$

$$= -0.625 \text{ m/s}^2$$

The negative sign tells you that the direction of **a** is opposite the direction of **v** and hence describes the slowing down of the car.

Now let us look at an example that requires a little more manipulation of the equations. Working through this example shows how important it is to be able to use the rules of algebra confidently.

Example 2.4

A car starts from rest and accelerates at 4.00 m/s^2 through a distance of 20.0 m. (*a*) How fast is it then going? (*b*) How long did it take to cover the 20.0 m?

R

Reasoning

Question   What do I know and what do I need to solve for?
Answer   You know that $\mathbf{v}_o = 0$, $\mathbf{a} = 4.00$ m/s^2, and $\mathbf{x} = 20.0$ m. You need to find the velocity $\mathbf{v}_f$ at the instant the car has traveled 20.0 m and the time t at that instant.

Question   What relationships do I have to work with?
Answer   Again, Eqs. 2.5, 2.6, and 2.7 apply. In the context of the present problem, each of these equations contains two unknowns. Therefore none of them is useful by itself. You must solve them *simultaneously* by combining them, which results in two more very useful equations. Let us hold off solving this example in order to develop these new equations in a general way.

2.8   TWO DERIVED EQUATIONS FOR UNIFORMLY ACCELERATED MOTION

Example 2.4 can be solved easily if we obtain two more equations to use along with Eqs. 2.5, 2.6, and 2.7. To derive the new equations, we must solve the three known equations simultaneously. Once we have done this, we shall not have to repeat the process, but shall simply add the new equations to the list which we are able to call upon in solving future problems.

If we substitute the value for $\bar{\mathbf{v}}$ given by Eq. 2.7 for $\bar{\mathbf{v}}$ in Eq. 2.5, we obtain

$$\mathbf{x} = \tfrac{1}{2}(\mathbf{v}_f + \mathbf{v}_o)t \tag{2.8}$$

Now we replace t by its value from Eq. 2.6 to get

$$\mathbf{x} = \left(\frac{\mathbf{v}_f + \mathbf{v}_o}{2}\right)\left(\frac{\mathbf{v}_f - \mathbf{v}_o}{\mathbf{a}}\right) \qquad \text{or} \qquad (\mathbf{v}_f)^2 - (\mathbf{v}_o)^2 = 2\mathbf{a}\mathbf{x}$$

Here we have a situation where two vectors are multiplied, which has not been discussed. In one dimension we can readily see how to interpret this. Each vector can only have a positive or a negative value. A vector times itself is just the square of its magnitude: $(\mathbf{v}_f)^2 = v_f^2$, and $(\mathbf{v}_o)^2 = v_o^2$. The one-dimensional product of $\mathbf{a}$ times $\mathbf{x}$ can have only the values $+ax$ or $-ax$, depending on whether $\mathbf{a}$ and $\mathbf{x}$ have the same or opposite signs. Thus we can write the above equation in terms of the magnitudes of the vectors:

$$v_f^2 = v_o^2 + 2ax \qquad \text{(\textbf{a} and \textbf{x} same sign)} \tag{2.9a}$$
$$v_f^2 = v_o^2 - 2ax \qquad \text{(\textbf{a} and \textbf{x} opposite signs)} \tag{2.9b}$$

We derive our second equation by making a different substitution in Eq. 2.8.

Replacing $\mathbf{v}_f$ in that equation by its value obtained from Eq. 2.6, we get

$$\mathbf{x} = \tfrac{1}{2}\mathbf{v}_o t + \tfrac{1}{2}(\mathbf{v}_o + \mathbf{a}t)t$$

which simplifies to

$$\mathbf{x} = \mathbf{v}_o t + \tfrac{1}{2}\mathbf{a}t^2 \tag{2.10}$$

We now have five equations to use in solving problems involving uniformly accelerated motion along a straight line:

$$\mathbf{x} = \bar{\mathbf{v}}t \tag{2.11a}$$

$$\bar{\mathbf{v}} = \frac{\mathbf{v}_o + \mathbf{v}_f}{2} \tag{2.11b}$$

$$\mathbf{v}_f = \mathbf{v}_o + \mathbf{a}t \tag{2.11c}$$

$$v_f^2 = v_o^2 + 2ax \tag{2.11d}$$

$$\mathbf{x} = \mathbf{v}_o t + \tfrac{1}{2}\mathbf{a}t^2 \tag{2.11e}$$

R

Example 2.4 (Continued)

Question Which equations best fit the specific problem?
Answer Since a, v_o, and x are known, Eq. 2.11d gives v_f immediately: $v_f^2 = v_o^2 + 2ax$. Knowing v_f, you can find t from 2.11a and 2.11b as you did in Example 2.3.

Question Is there a more direct, easier way to find t?
Answer Yes. The advantage in having developed the additional equations in general form is that you can now use them directly. For instance, when applied to this problem, both Eqs. 2.11c and 2.11e involve t as the only unknown. Since 2.11c is a linear equation, whereas Eq. 2.11e is quadratic, Eq. 2.11c is easier to solve for t: $t = (v_f - v_o)/a$.

Solution and Discussion

(a) $v_f^2 = v_o^2 + 2ax = 0 + 2\ (4.00\ \text{m/s}^2)(20.0\ \text{m}) = 160\ \text{m}^2/\text{s}^2$

Then $v_f = \pm\sqrt{160\ \text{m}^2/\text{s}^2} = \pm 12.6$ m/s. As always, there are two solutions to a quadratic equation. You have assumed that the motion is in the positive x direction, and so you take $+12.6$ m/s to be the correct answer. (The -12.6 m/s solution refers to the fact that if a had been negative, the car would have been going 12.6 m/s in the $-x$ direction.)

(b) $t = \dfrac{v_f - v_o}{a} = \dfrac{12.6\ \text{m/s} - 0}{4.00\ \text{m/s}^2}$

$\qquad = 3.15$ s

Example 2.5

A car is moving at 60.0 km/h when it begins to slow down with a deceleration of 1.50 m/s². How long does it take to travel 70.0 m as it slows down?

R

Reasoning

Question   The only quantity I need to determine is the time t. What am I given?
Answer   Initial velocity v_o = 60.0 km/h, deceleration = 1.50 m/s², and distance x = 70.0 m.

Question   What does the term *deceleration* mean?
Answer   It means a negative acceleration, in other words, an acceleration whose direction is opposite the direction of the velocity. Thus if velocity is to be taken as +60.0 km/h, we must set a = −1.50 m/s².

Question   The units of velocity are different from those of a and x. What must I do to reconcile this?
Answer   You need to convert the 60.0 km/h to m/s:

$$60 \text{ km/h} = (60.0 \text{ km/h})(1000 \text{ m/1 km})(1 \text{ h/3600 s})$$
$$= 16.7 \text{ m/s}$$

You must always check that every quantity is expressed in the same units before you do any numerical computation. (Chapter 1 reviews conversion of units.)

Question   Which equation or equations fit this problem?
Answer   It has to be one of the three that involve t. Equations 2.11a and c both need v_f to be of use. Equation 2.11e is the one equation that involves t as the only unknown. It is a quadratic equation, however, and solving it is more tedious than solving a linear equation.

Question   Is there a way I can find v_f?
Answer   Yes, from Eq. 2.11d, $v_f^2 = v_o^2 + 2ax$.

Solution and Discussion

1 Equation 2.11d gives

$$v_f^2 = v_o^2 + 2ax = (16.7 \text{ m/s})^2 + 2(-1.50 \text{ m/s}^2)(70.0 \text{ m})$$
$$= 279 \text{ m}^2/\text{s}^2 - 210 \text{ m}^2/\text{s}^2 = 69.0 \text{ m}^2/\text{s}^2$$
$$v_f = \pm 8.30 \text{ m/s}$$

We are interested in motion to the right, and so we choose v_f = +8.30 m/s.
2 Equation 2.11c gives

$$t = \frac{v_f - v_o}{a} = \frac{+8.30 \text{ m/s} - +16.7 \text{ m/s}}{-1.50 \text{ m/s}^2}$$

$$= \frac{-8.4 \text{ m/s}}{-1.50 \text{ m/s}^2} = +5.60 \text{ s}$$

Notice the consistent use of the algebraic sign of the acceleration a. Because of this consistent usage both t and v_f come out with the correct sign.

Controversies in Physics: Theories of Free-Fall

The study of the behavior of falling objects has a long and interesting history. It is a prime example of the difference between good science and poor science. We begin the story at the time of the famous philosopher Aristotle (384–322 B.C.).

In Aristotle's time, it was known that a light object falls through air more slowly than a heavy object. Noting this fact, Aristotle based his theory of falling objects on the idea that all objects are composed of four elements: earth, air, fire, and water. Objects composed primarily of earth and water try to reach their natural resting place, the earth; hence, when they are allowed to do so, they fall to the earth. Objects composed of air try to rise to *their* natural resting place, the sky. According to Aristotle, a stone falls fast because it is composed primarily of earth and thus strongly seeks its natural resting place. A feather, however, is mostly air and hence seeks the earth less strongly; it thus falls more slowly than a stone. Moreover, Aristotle concluded that the speed of a falling object is constant. If you let a feather (or a sheet of facial tissue) fall through the air, you will see how he came to this conclusion. The

fact that a stone falls with ever-increasing speed eluded him, however, because he had no way to measure the flight of such swiftly falling objects. Aristotle was a highly respected philosopher, and so few people were inclined to question his theories and conclusions. For this reason, little progress was made in understanding how falling objects behave until the time of Galileo, nearly 2000 years later.

By the year 1250, science as we know it, began to appear. Roger Bacon (1214–1294) was among the first to espouse the idea that experience (that is, experiment) is necessary for the development of reliable theories about natural behavior. However, it appears that even he did not recognize the importance of controlling the variables that influence the results of an experiment. As late as 1605, in his noted treatise *The Advancement of Learning*, Francis Bacon (1561–1626) insisted—contrary to then existing practice—that theories be based upon experimentally determined facts.

It was Galileo Galilei (1564–1642) who finally led the way to the development of true science, performing important experiments in astronomy, optics,

and mechanics. The most important aspect of his work was his recognition that meaningful experiments must be controlled. By this we mean that, insofar as possible, only one variable at a time should be changed during an experiment. Thus Galileo recognized that comparing the way a feather and a stone fall is a nearly uninterpretable experiment because there are so many differences between the two objects. He designed ingenious experiments to time accurately the way similar objects of different weights fall and was able to establish that an object's weight does not influence its acceleration, provided the effects of air friction are negligible. Furthermore, he found that freely falling objects do not fall with constant speed, as Aristotle believed, but instead undergo constant acceleration.

Over the years, the methods of science have been further refined, but experiment remains at the heart of all good science. Without carefully controlled experiments to provide us with unambiguous results, we can only guess about the behavior of the world about us. To be of value, scientific theories must be based upon experimental fact.

Before going on, you should solve this problem by means of Eq. 2.11e, to convince yourself that you can work with the quadratic solution. Sometimes quadratics are unavoidable. Review the method in Appendix 2, and then use these hints:

1 Using the values from this problem, Eq. 2.11e is

$$70.0 = 16.7t + \tfrac{1}{2}(-1.50)t^2 = 16.7t - 0.750t^2$$

We have suppressed the units for the moment so that you can more easily see the form of the equation.

2 The standard form of the quadratic is $at^2 + bt + c = 0$. In this form, we have $-0.750t^2 + 16.7t - 70.0 = 0$, from which we can identify the standard coefficients:

$$a = -0.750 \qquad b = +16.7 \qquad c = -70.0$$

Show that the quadratic formula gives $t = 5.6$ s and 16.7 s. Why must we discard the latter solution?

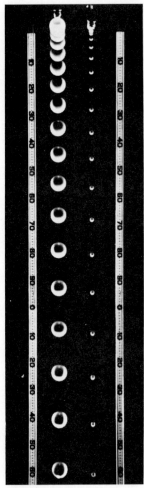

FIGURE 2.8

The falling objects are made visible at equal time intervals by means of a strobe light. Although the objects are of different size and weight, they fall in unison. (*Education Development Center*)

TABLE 2.1

Acceleration due to gravity g

Place	$g \; (m/s^2)$
Beauford, N.C.	9.7973
New Orleans	9.7932
Galveston	9.7927
Seattle	9.8073
San Francisco	9.7997
St. Louis	9.8000
Cleveland	9.8024
Denver	9.7961
Pikes Peak	9.7895

2.9 FREELY FALLING BODIES

Consider the experiment shown in Fig. 2.8: two objects falling freely under the pull of gravity. A strobe light captures the objects' positions at equally spaced time intervals. Notice that, despite differences in size and weight, the objects fall with the same acceleration. This fact was first pointed out by Galileo (1564–1642). Measurements show that, near earth's surface, a **freely falling** object accelerates downward with an acceleration of 9.8 m/s². In other words, after it is dropped, a freely falling object has the following speeds after successive 1-s intervals: 9.8 m/s, 19.6 m/s, 29.4 m/s, and so on. The downward-directed velocity increases by 9.8 m/s each second; to put it another way, acceleration is 9.8 m/s² downward.

Despite this assertion, we know that a marble, a feather, and a piece of facial tissue all fall in different ways. These are not *freely* falling bodies, however. As the feather falls, the friction of the air against it retards its fall. This friction force nearly balances the pull of gravity on the feather and hence the feather is decidedly not free to fall. Similarly, a facial tissue falls slowly because of air effects. A marble, on the other hand, weighs much more than a feather or a facial tissue. The pull of gravity on it is far larger than the air friction retarding its fall. We can therefore consider the marble to be falling freely under the action of gravity unless its speed becomes so high that the force of air friction becomes very large.

The fall of objects subject to no appreciable force except the pull of gravity is exceptionally simple to analyze. Experiment shows that they fall with a downward acceleration (on earth) of 9.80 m/s². We call this the **acceleration due to gravity** and represent it by the symbol *g*. The value of *g* varies slightly from place to place on the earth. Typical values are given in Table 2.1.

Look back to the situation in Fig. 2.5, which shows the motion of a ball subject only to the force of gravity. We analyzed this motion in Example 2.1 and in Figs. 2.5*b* and 2.6. The ball's acceleration is about 10 m/s² downward, whether the ball is rising or falling. This is another example of the fact that a freely falling body always has an acceleration of 9.8 m/s² downward. It does not matter whether the body is rising or falling; its acceleration is still *g* downward. While it is rising, the ball of Example 2.1 has a velocity that is *decreasing* at the rate of 9.8 m/s every second, until the ball reaches the top of its flight. At this *instant*, its velocity is zero. Then, while it is falling, the ball's velocity is *increasing* by 9.8 m/s every second.

We shall soon analyze the free-fall motion of objects in several examples, but before we do so, notice the following facts. First, if we choose *up* as positive, then the acceleration due to gravity, being directed downward, is −9.8 m/s². It is very important that we keep track of the signs of displacement, velocity, and acceleration because the signs tell us the direction of these quantities. Second, because the acceleration is constant (9.8 m/s² downward), motion under the action of gravity is uniformly accelerated motion and our five motion equations apply. However, we shall replace *x* by *y* in the equations to emphasize the vertical nature of the motion.

You should be very careful in applications that involve up-and-down motion. It is absolutely necessary to decide at the beginning which direction is positive. The choice is arbitrary, but once you have made it in a particular problem, you must retain that choice throughout the problem.

Example 2.6

You drop a stone from a bridge. If it takes 3.0 s for the stone to hit the water below, how high above the water was your hand when you let go of the stone?

Ignore air friction. (Notice here that the problem ends the instant just before the stone hits the water. It is only during this interval that the stone is a freely falling object.)

Reasoning

Question   What quantities do I know?
Answer   How long it takes the stone to fall, the zero initial velocity, and the fact that the motion is free-fall, meaning that acceleration is 9.8 m/s^2 downward.

Question   What am I to find?
Answer   How far the stone has traveled vertically in the 3.0 s. You can call this distance y.

Question   Here the motion is entirely downward. Should I take that direction to be positive or negative?
Answer   You can make *either* choice, but once your choice of sign convention is made, it has to apply to all vectors throughout the problem. For example:

If you choose *up positive*, you must use $\mathbf{a} = -9.8$ m/s^2. You should anticipate that this choice will give you a negative value for y since the stone now has a negative (downward) displacement.
If you choose *down positive*, $\mathbf{a} = +9.8$ m/s^2. You should expect that y will now come out to be positive.

Question   Which of the equations of motion best fits this problem?
Answer   Equation 2.11e is the equation that relates position directly to time. Although the position is written as x, we certainly can use a different symbol such as y for vertical position if we choose to do so. Then Eq. 2.11e can be written as

$$y = v_o t + \tfrac{1}{2}at^2$$

Solution and Discussion   Putting in the numbers, with down chosen as positive, you get

$$y = v_o t + \tfrac{1}{2}at^2 = (0)(3.0\text{ s}) + \tfrac{1}{2}(+9.8\text{ m/s}^2)(3.0\text{ s})^2$$
$$= 44\text{ m}$$

Exercise   What is the velocity of the stone just before it strikes the water?
Answer: 29 m/s

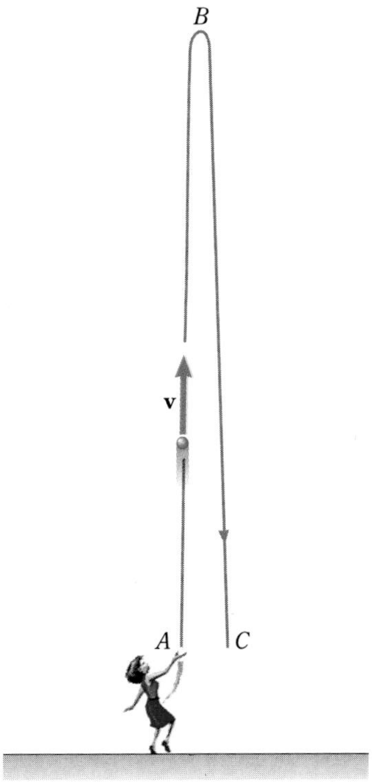

Example 2.7

A ball is thrown upward with an initial speed of 15.0 m/s and then falls and is caught by the person who threw it. The path of the ball is shown in Fig. 2.9. (*a*) How high does the ball go? (*b*) How fast is it going the instant before it is caught? (*c*) How long is it in the air?

FIGURE 2.9

The ball is thrown upward from point *A* with a speed of 15 m/s. Since the ball stops for an instant at point *B*, its velocity there is zero.

R   *Reasoning   Part (a)*

Question   What kind of motion is this?
Answer   Free-fall, as in Example 2.6, but here you have different initial conditions.

Question   What quantities do I know?
Answer $v_o = +15.0$ m/s if you choose up as positive. For free-fall, you then have $a = -9.80$ m/s^2.

Question   How do I interpret question *a*? What physical condition defines the highest point of the flight?
Answer   At point *B* in Fig. 2.9, the ball has stopped for an instant. So the condition for the highest point is that $v = 0$. If you consider for now only the part of the flight from *A* to *B*, you can treat the velocity at *B* as the final velocity, i.e., $v_f = 0$.

Question   What do I want to find at this instant when $v_f = 0$?
Answer   The value of vertical position *y*. A convenient point to choose for $y = 0$ is point *A*, the start of the flight.

Question   What equation connects the distance *y* with the quantities I am given?
Answer   Since you know the magnitudes v_o, v_f, and a, you could use Eq. 2.11d: $v_f^2 = v_o^2 + 2ay$, where *y* replaces *x* as the distance variable.

Solution and Discussion   Solving Eq. 2.11*d* for *y* and putting in the given numbers,

$$y = \frac{v_f^2 - v_o^2}{2a} = \frac{0 \text{ m}^2/\text{s}^2 - (15.0 \text{ m/s})^2}{2(-9.8 \text{ m/s}^2)} = +11.5 \text{ m}$$

You should be able to verify that all signs are consistent with the choice of up as the positive direction.

R   *Reasoning   Part (b)*

Question   What does the phrase "the instant before it is caught" mean?
Answer   The ball is caught at the same height it was thrown from. Thus the phrase means the instant *t* just before the ball has returned to its starting height ($y = 0$).

Question   Can I use the same initial conditions for this part of the problem as I used for part (*a*)?
Answer   Yes, since part (*b*) is simply a continuation of the same motion. Thus $v_o = +15.0$ m/s, $a = -9.8$ m/s^2, and $y_o = 0$.

Question   What is the relation between *y* and v_f?
Answer   Once again, $v_f^2 = v_o^2 + 2ay$.

Question   Under what condition do I want to solve this equation?
Answer   This time you want to solve for v_f when $y = 0$.

Solution and Discussion When you put in $y = 0$, you get $v_f^2 = v_o^2 = (15.0 \text{ m/s})^2$. The equation looks trivial, but remember, a quadratic always has two solutions, and it is up to you to interpret them. The solutions are

$$v_f = +15 \text{ m/s} \quad \text{and} \quad -15 \text{ m/s}$$

The -15 m/s represents a velocity *downward* and thus is the correct solution to (b).

There is an alternate way to arrive at the solution: treat point B as the starting point for a ball dropped from rest through a distance of 11.5 m. The problem is similar to the one in Example 2.6.

R

Reasoning Part (c)

Question What equation connects t to what I now know?
Answer $\mathbf{y} = \mathbf{v}_o t + \frac{1}{2}\mathbf{a}t^2$.

Question Under what condition do I want to solve this equation?
Answer You want to find t when $y = 0$.

Solution and Discussion The equation then becomes

$$0 = (15.0 \text{ m/s})t - \tfrac{1}{2}(-9.8 \text{ m/s}^2)t^2$$

You should be able to show that the two solutions are

$$t = 0 \quad \text{and} \quad t = \frac{15.0}{4.90} = 3.06 \text{ s}$$

There are two times when $y = 0$: when the ball is thrown ($t = 0$) and when it is caught ($t = 3.06$ s).

Example 2.8

If a ball is thrown straight upward as in Fig. 2.9 and is caught 5 s after being thrown, how fast must it have been going when it left the person's hand?

R

Reasoning

Question I can see that this is another situation of free-fall motion, with $\mathbf{a} = -9.8 \text{ m/s}^2$. What can I say about the specific conditions in this example?
Answer The time of flight $t = 5.0$ s, and the initial and final positions are the same ($y_f = 0$ and $y_o = 0$).

Question I'm asked to find the initial velocity $\mathbf{v}_o$. What equation connects $\mathbf{v}_o$ to $\mathbf{a}$, t, and y?
Answer Equation 2.11e: $y = v_o t + \frac{1}{2}at^2$

Solution and Discussion   Solving 2.11*e* for $\mathbf{v}_o$:

$$\mathbf{v}_o t = \mathbf{y} - \tfrac{1}{2}\mathbf{a}t^2 \qquad \mathbf{v}_o(5.0\ \text{s}) = 0 - \tfrac{1}{2}(-9.8\ \text{m/s}^2)(5.0\ \text{s})^2$$

This gives $\mathbf{v}_o = +24$ m/s. Be sure you can identify the choice of positive direction chosen here and can interpret what the plus sign in the answer means.

2.10   PROJECTILE MOTION

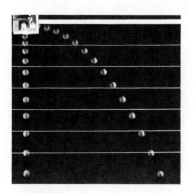

FIGURE 2.10

A flash photograph of two golf balls, one falling from rest and the other projected horizontally. The interval between light flashes is $\tfrac{1}{30}$ s, and the horizontal lines are 15 cm apart. (*Educational Development Center*)

Very seldom does a baseball, bullet, or golf ball follow a straight-line path. These objects move in two dimensions, and their motion is called **projectile motion**. To illustrate this type of motion, let us examine Fig. 2.10. There we see ball 1 falling straight downward with downward acceleration of 9.8 m/s², as we have already discussed. Ball 2, however, is projected horizontally at the same time that ball 1 is dropped. The strobe light records the projectile motion of ball 2 and the straight-line motion of ball 1. *Notice that the vertical positions of the two balls are the same at each flash of the strobe.* Apparently, ball 2 falls vertically with the same acceleration of 9.8 m/s² even though it is moving horizontally at the same time. This observation provides us with a description of projectile motion:

If air resistance is neglected, a projectile moves horizontally at constant speed as it falls vertically with acceleration *g*.

In Chap. 3 we will justify this behavior in terms of Newton's laws. For now, we simply accept as experimental fact that, when air resistance is negligible, a projectile's velocity vector can be separated into two components:

1 The projectile moves vertically with constant acceleration *g*.
2 Simultaneously, the projectile moves horizontally with constant horizontal *velocity*.

PROJECTILE FIRED HORIZONTALLY

In Fig. 2.11, a baseball is projected horizontally from point *A* with velocity $\mathbf{v}_o$. If air resistance is negligible, the baseball maintains this same horizontal velocity until it strikes something. In other words, there is no horizontal component of acceleration. Simultaneously, the downward vertical velocity of the ball increases by 9.8 m/s each second as the ball falls freely. Let us now analyze this type of motion.

Because the two perpendicular motions are independent of each other, we can analyze them separately. Consider first the horizontal motion, which is extremely simple because it is linear motion with constant velocity v_x. Thus with zero horizontal acceleration, the equations that describe the horizontal component of the ball's motion are

$$v_o = v_f = \overline{v} = v_x \qquad x = \overline{v}t = v_x t \tag{2.12}$$

Vertically, the ball moves in the *y* direction under the free-fall acceleration due to gravity. Our previous equations of uniformly accelerated motion thus apply to this

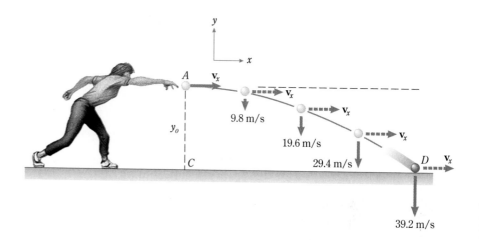

FIGURE 2.11

The projected ball undergoes two independent motions at right angles to each other.

component of the ball's motion. From Fig. 2.11, we can see that the initial vertical component of velocity is zero, $v_{oy} = 0$. If we choose ground level as $y = 0$, then we can say that the initial vertical position of the ball is designated as y_o. The equations which describe the vertical motion of the ball then become

$$v_y = 0 + (-9.8 \text{ m/s}^2)t$$
$$y - y_o = 0 + \tfrac{1}{2}(-9.8 \text{ m/s}^2)t^2 \tag{2.13}$$

Thus

$$y = y_o - (4.9 \text{ m/s}^2)t^2 \tag{2.14}$$

This is the first time we have used a starting position (x_o or y_o) other than zero. Doing so presents no problem, however, because choice of starting position is always arbitrary.

Our procedure, then, is to recognize that the motion of any projectile near the earth's surface contains two independent motions. If air resistance is negligible, the horizontal motion is motion with constant velocity, and the vertical motion is that of a freely falling body along a vertical line. We calculate each motion separately as components of the total motion, and then combine the solutions to obtain the complete answer.

Example 2.9

Consider the situation in Fig. 2.11. Suppose the ball leaves the thrower's hand at point A with a velocity of 15 m/s directed horizontally. Take point A to be 2.0 m above the ground. Where will the ball hit the ground?

R

Reasoning

Question What does the question mean in the terms used in the equations I have?

Answer It means how far is it from point C (directly below point A) to the impact point D in Fig. 2.11? More precisely, if you make the convenient choice of $x = 0$ at point C, the question becomes, "What is the value of x where the ball hits the ground?" This distance is called the **range** of the projectile.

Question What does the phrase "hits the ground" mean in terms of my equations?

Answer The ground is 2.0 m below the starting point A of the motion. If you choose point A to be at $x = 0$ *and* $y = 0$, the ball hits the ground when its vertical position is $y = -2.0$ m.

Question Is there an equation that connects my unknown, x, to my known, y?

Answer None that we have developed yet.

Question If I don't have any equations that apply, how can I solve the problem?

Answer By realizing that x and y are indirectly connected through the time variable, which appears in the equations of motion for both components of velocity (Eqs. 2-12 and 2.14). You therefore need to find the "time of flight" of the ball.

Question How do I interpret the time of flight in the terms used in my equations?

Answer This is the time it takes for the ball to go from $y = 0$ to $y = -2.0$ m, starting with zero initial vertical velocity. This part of the motion is the same as dropping the ball through 2 m from rest.

Question What equation determines this time interval?

Answer From Eq. 2.14 we know that, in general, $y = y_o - (4.9 \text{ m/s}^2)t^2$. In this case, the starting point is $y_o = 0$. You want the value of t that results from putting $y = -2.0$ m.

Question Once I know t, what equation determines the position x where the ball hits the ground?

Answer As long as the ball doesn't hit the ground, it keeps going horizontally at 15 m/s. The equation describing this is Eq. 2.12: $x = v_x t$. The value of t for time of flight will give the value of x at impact, which is the range.

Solution and Discussion

1 Time of flight is determined from the relationship $(2.0 \text{ m}) = (4.9 \text{ m/s}^2)t^2$, giving $t = 0.64$ s.
2 The range is then $x = (15 \text{ m/s})(0.64 \text{ s}) = 9.6$ m.

PROJECTILE FIRED AT AN ANGLE

Another general type of projectile motion is that of an object thrown or shot from ground level with an initial velocity v_o and aimed at some angle θ_o above the horizontal. Suppose, for instance, the cannon in Fig. 2.12a fires a shell. As it moves to the right, the shell rises to some maximum height H above the ground and then descends. It finally strikes the ground at some horizontal distance (again, the projectile's *range*) from its firing point. The shell's motion is governed by the principles we just discussed for horizontal projectiles, but here the initial conditions are

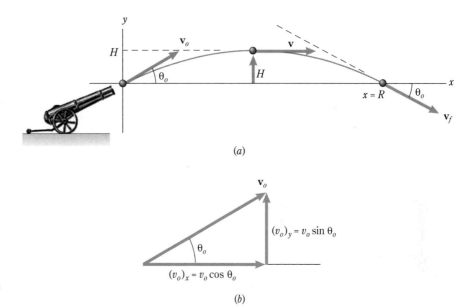

(a)

(b)

FIGURE 2.12

(a) The trajectory of a projectile fired at an angle. (b) The components of the initial velocity.

different. Let us examine this situation in more detail.

The horizontal component of $\mathbf{v}_o$ is $v_o \cos \theta_o$ (Fig. 2.12b). As in the previous example, this part of the motion remains constant since there is no horizontal component of acceleration. Therefore the equation governing horizontal motion is

$$x = (v_o \cos \theta_o)t$$

where we have assumed $x = 0$ at the firing point.

The vertical component of motion is precisely that discussed in Example 2.7, with the initial upward velocity equal to $v_o \sin \theta_o$. We can immediately write the equations describing the vertical motion:

$$y = (v_o \sin \theta_o)t + \tfrac{1}{2}(-9.8 \text{ m/s}^2)t^2$$
$$v_y = v_o \sin \theta_o + (-9.8 \text{ m/s}^2)t$$

Notice that the path taken by the shell, called its trajectory, is symmetric about the midpoint of its flight. One consequence of this symmetry is that the time the shell takes to reach maximum height is half the total time of flight. Also, the symmetry means that the angle and speed with which the shell hits the ground are the same as the initial values, except that the direction of the velocity is into the ground instead of away from it.

In Example 2.9, we noted that we do not yet know any equation which *directly* relates y to x. Using the above equations, however, we can eliminate the time variable and derive such a relationship, called the **trajectory equation**. From the equation for x, we have $t = x/(v_o \cos \theta_o)$. Putting this expression in place of t in the equation for y, we get

$$y = (\tan \theta_o)x - \left(\frac{g}{2v_o^2 \cos^2 \theta_o} \right)x^2 \tag{2.15}$$

(We have made use of the fact that $\sin\theta/\cos\theta = \tan\theta$.) This is a quadratic relationship of the form $y = ax^2 + bx$, with $a = \tan\theta_o$ and $b = g/2v_o^2\cos^2\theta_o$.

Example 2.10

Suppose you have a gun that fires a shell at an initial velocity ("muzzle velocity") of 0.800 km/s. If you aim the gun 30.0° above the horizontal, how far away will the shell hit the ground, assuming it hits at the same level you fired from? How long is the shell in the air, and how high does it rise? Neglect air resistance.

R

Reasoning

Question What quantities are given?
Answer $v_o = 0.800$ km/s $g = -9.8$ m/s^2 $\theta_o = 30.0°$

Question Are these units consistent with one another?
Answer No. Before using numbers, you must change 0.800 km/s to 800 m/s.

Question Can I find the range of the shell directly from the information I have?
Answer Yes, now that you have the trajectory equation.

Question How do the words "hit the ground" relate to the terms in the trajectory equation?
Answer You want to find the value of x where the shell strikes the ground, i.e., where $y = 0$.

Solution and Discussion When you put the value $y = 0$ into the trajectory equation, you get

$$0 = (\tan 30.0°)x - \left[\frac{4.9 \text{ m/s}^2}{(800 \text{ m/s})^2(\cos^2 30.0°)}\right]x^2$$

Notice that there are two solutions (two values of x where $y = 0$). One is at $x = 0$, where the shell started. By dividing the above equation by x, you can show that the other is at

$$x = \frac{(\tan 30.0°)(\cos^2 30.0°)(800 \text{ m/s})^2}{4.90 \text{ m/s}^2} = 56,600 \text{ m} = 56.6 \text{ km}$$

This is approximately 34 mi!

Question What equation determines the time of flight?
Answer Either the equation of x versus t (solve for t when $x = 56.6$ km), or the equation of y versus t (solve for t when $y = 0$).

Solution and Discussion From the equation of y versus t,

$$0 = v_o(\sin 30.0°)t - \tfrac{1}{2}gt^2$$

we get $t = 0$ and

$$t = \frac{2v_o(\sin 30.0°)}{g} = \frac{(1600 \text{ m/s})(0.500)}{9.80 \text{ m/s}^2}$$

$$= 81.5 \text{ s} = 1.36 \text{ min}$$

From the x versus t equation, we have $56.6 \times 10^3 \text{ m} = v_o \, (\cos 30.0°)t$, which gives $t = (56.6 \times 10^3 \text{ m})/(800 \text{ m/s})(0.866) = 81.7 \text{ s}$. The difference between answers is due to round-off error.

Question What condition gives us the maximum height?
Answer Maximum height is achieved at the instant that $v_y = 0$. Before that instant, v_y is positive, and after that instant, it is negative.

Question Is there an equation that directly relates y to v_y?
Answer Yes. It is essentially Eq. 2.11d, applied to variables in the y direction: $(v_y)_f^2 = (v_y)_o^2 - 2gy$.

Solution and Discussion Maximum height ($y = H$) is obtained from

$$0 = (800 \text{ m/s})^2(\sin^2 30.0°) - 2(9.80 \text{ m/s}^2)H$$
$$H = 8160 \text{ m} = 8.16 \text{ km}$$

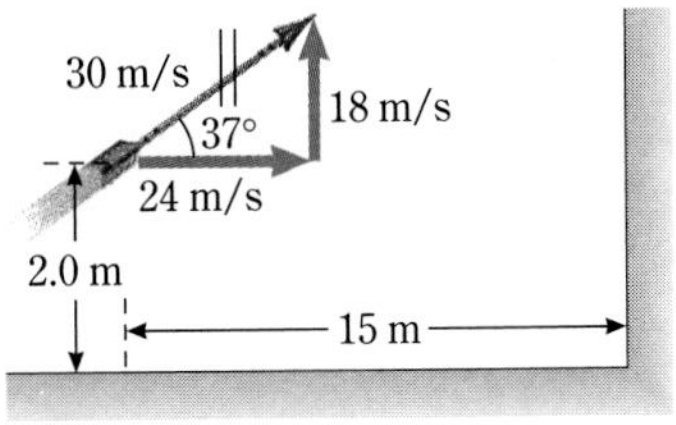

FIGURE 2.13

Where will the arrow hit the wall? Is it going up when it hits, or is it already on the way down?

Example 2.11

An arrow is shot with a velocity of 30.0 m/s at an angle of 37.0° above the horizontal. The arrow is initially 2.00 m above the ground and 15.0 m from a wall, as shown in Fig. 2.13. (*a*) At what height above the ground does it hit the wall? (*b*) Is it still going up just before it hits, or is it already on its way down? Neglect air friction.

Reasoning

Question What is the translation of question (a) into the terms used in the equations of motion?
Answer It asks, "What is the value of y ("at what height?") when $x = 15.0 \text{ m}$ (where the wall is)?"

Question Does the trajectory equation apply?
Answer Yes. Even though the trajectory equation was derived while referring to a situation where the launch and impact heights were the same, *any* pair of values x and y which lie on the trajectory satisfy Eq. 2.15. Thus you can put $x = 15.0 \text{ m}$ in the equation and solve for the height y that corresponds to that point on the trajectory.

Question What quantities in the trajectory equation am I given?
Answer $v_o = 30.0 \text{ m/s}$, $g = 9.80 \text{ m/s}^2$, $\theta_o = 37.0°$, and $y_o = 0$, assuming you choose the launch height to be the reference height.

Question What height should I choose for y_o?
Answer That's an arbitrary choice. Here either the ground level or the height of

the launch point would be good choices. Whichever choice you make, be sure to stick with it throughout.

Question   What tells me whether the arrow is going up or coming down at the moment of impact?
Answer   The sign of v_y at that instant. If v_y is positive, the arrow is going up. If it is negative, the arrow is going down.

Question   The trajectory equation doesn't involve v_y. What other relation can I apply that does?
Answer   One of the equations of motion along the y-axis is $v_y = v_{oy} - gt$. If the time of impact can be found, this will give you the value of v_y, including its sign.

Question   What condition determines how long it takes for the arrow to hit the wall?
Answer   It is the time t when $x = 15.0$ m. These are related by the equation of horizontal motion: $x = v_{ox}t$.

Solution and Discussion

1 The value of y when $x = 15.0$ m is

$$y = (\tan 37.0°)(15.0 \text{ m}) - \frac{9.8 \text{ m/s}^2}{2(30.0 \text{ m/s})^2 \cos^2 37.0°}$$
$$= 11.3 \text{ m} - 1.9 \text{ m} = 9.4 \text{ m}$$

2 The time of impact with the wall is

$$t = \frac{x}{v_o \cos 37.0°} = \frac{15.0 \text{ m}}{(30.0 \text{ m/s})(0.800)} = 0.625 \text{ s}$$

3 The vertical component of velocity at this time is

$$v_y = v_o \sin 37.0° - gt$$
$$= (30.0 \text{ m/s})(0.600) - (9.80 \text{ m/s}^2)(0.625 \text{ s}) = +11.9 \text{ m/s}^2$$

Thus the arrow hits the wall while still going upward, before it has reached the peak of its flight.

Exercise   Knowing the velocity components of the arrow in Example 2.11, find the magnitude and direction of the velocity vector at the instant the arrow hits the wall.   *Answer: v = 26.8 m/s, at an angle of 26.4° above horizontal.*

Exercise   How far away must the wall be if the arrow is to hit at this same height above launch (9.3 m), but on the way down?   *Answer: 73.2 m.*

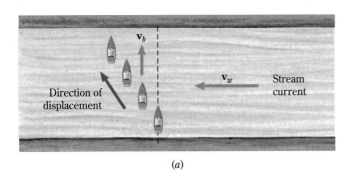

(a)

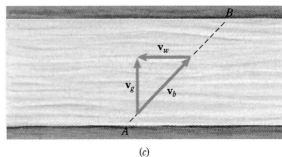

(c)

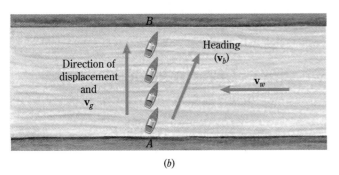

(b)

FIGURE 2.14

(a) The velocity of the water causes the boat to travel at an angle relative to its heading. (b) By heading slightly into the current, the boat can travel straight across the stream. (c) The addition of velocities for a boat traveling directly across the stream. The velocity of the water adds to $\mathbf{v}_b$ to cause the displacement of the boat to be along AB, at a velocity $\mathbf{v}_g$.

2.11 ADDITION OF VELOCITIES IN TWO DIMENSIONS: RELATIVE VELOCITY

A common situation involving addition of velocity vectors arises when a boat moves across a flowing river or when an airplane flies in moving air. As shown in Fig. 2.14a, a boat whose bow is pointed directly across a stream will be carried downstream as it crosses the stream. If the person in the boat wishes to reach the shore directly opposite her starting point, she must compensate for the stream velocity by pointing the boat at a certain angle upstream (Fig. 2.14b). Similarly, wind velocity must be taken into account in choosing the heading of an airplane as it flies from one city to another. Let us see how we describe this type of motion by addition of velocities.

Let us take the example of an airplane which must travel in a straight line between cities A and B in the presence of a constant prevailing wind. We can identify three velocities. There is the velocity of the wind relative to the ground, $\mathbf{v}_w$. The plane has a certain "airspeed" along its heading (the direction it is pointed). This can be designated $\mathbf{v}_p$, which represents the velocity the plane would have in still air. Finally, there is the velocity the plane has relative to the ground, $\mathbf{v}_g$, which is in the direction of the plane's displacement. As shown in Fig. 2.15, $\mathbf{v}_g$ is simply the resultant of the other two velocities:

$$\mathbf{v}_g = \mathbf{v}_w + \mathbf{v}_p \tag{2.16a}$$

The same vector addition applies to the boat crossing the stream, shown in Fig. 2.14c. We use the symbol $\mathbf{v}_b$ for the velocity of the boat relative to the water and note that $\mathbf{v}_w$ would now refer to the stream velocity of the water:

$$\mathbf{v}_g = \mathbf{v}_w + \mathbf{v}_b \tag{2.16b}$$

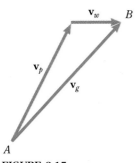

FIGURE 2.15

The addition of velocities for an airplane flying from A to B. $\mathbf{v}_p$ is the velocity in the direction the plane is pointed. $\mathbf{v}_g$ is the velocity of the plane relative to the ground, along its displacement.

Let us summarize the analysis of this type of problem:

1 $\mathbf{v}_g$ and the displacement of the boat or plane are in *the same direction* relative to the ground. Thus you can identify the direction of $\mathbf{v}_g$ by knowing in what direction you want the boat or plane to actually travel relative to points on the ground. Once you determine this direction, remember that the vector diagrams for Eqs. 2.15 involve *velocity* vectors, not displacements.

2 $\mathbf{v}_b$ or $\mathbf{v}_p$ is along the direction the boat or plane is *pointed* (the *heading*). In general, this is *not* the same direction $\mathbf{v}_b$ or $\mathbf{v}_p$ is traveling relative to the ground. The speed along the heading is the speed the boat or plane would have in still water or air.

Illustration 2.2

In Fig. 2.16*a* a stream is shown flowing with a velocity of 6.00 mi/h. You want to steer your boat across the stream so as to reach point *B* on the opposite shore, directly across from your starting point, *A*. Your boat is capable of traveling 15.0 mi/h in still water. At what angle upstream should you point your boat?

Reasoning

Graphical method   The path you want to travel, from *A* to *B*, determines the direction of the velocity $\mathbf{v}_g$. You know that this vector is the sum of $\mathbf{v}_b$ and $\mathbf{v}_w$. $\mathbf{v}_w$

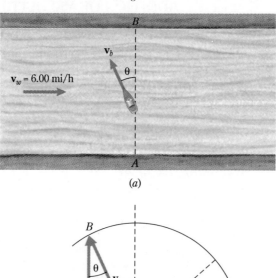

(a)

(b)

FIGURE 2.16

(*a*) What heading θ is required for the boat to travel from *A* to *B*? The speed of the boat is $\mathbf{v}_b = 15.0$ mi/h. (*b*) The boat's heading θ can be determined graphically. Notice also that $\mathbf{v}_b \sin \theta = \mathbf{v}_w$ and $\mathbf{v}_b \cos \theta = \mathbf{v}_g$.

Radius of circle represents $\mathbf{v}_b = 15.0$ mi/h

is given, both in direction and speed. You also know the magnitude, but not the direction, of $\mathbf{v}_b$. You can put this together graphically in a velocity diagram as follows:

1 Draw a line in the direction of AB. This will be the direction of $\mathbf{v}_g$.
2 Choose a convenient scale to represent speeds, such as 10.0 cm = 10.0 mi/h. Starting at the beginning of the line you just drew, draw the vector $\mathbf{v}_w$. Using the suggested scale, this would be a line perpendicular to AB in the direction of the current having a length 6.00 cm.
3 From the tip of the vector $\mathbf{v}_w$, draw a circle whose radius represents the magnitude of $\mathbf{v}_b$, 15.0 mi/h. This would be a 15.0-cm radius on your scale drawing. The intersection of this circle with the line AB represents the vector $\mathbf{v}_b$ which adds to $\mathbf{v}_w$ to form $\mathbf{v}_g$, as shown in Fig. 2.16*b*.

Both the heading of the boat (the direction of $\mathbf{v}_b$) and the magnitude of $\mathbf{v}_g$ can be read off the vector diagram.

Analytical method   In order to have a resultant velocity vector in the direction of AB, the component of $\mathbf{v}_b$ parallel to the stream must be equal and opposite to the stream velocity $\mathbf{v}_w$. Thus if θ is the angle between $\mathbf{v}_b$ and $\mathbf{v}_g$,

$$v_b \sin \theta = v_w$$

Thus

$$\theta = \sin^{-1} \frac{v_w}{v_b} = \sin^{-1} \frac{6.00}{15.0} = \sin^{-1} 0.400$$
$$= 23.6°$$

We can then find the magnitude of $\mathbf{v}_g$:

$$v_g = v_b \cos \theta = (15.0 \text{ mi/h}) \cos 23.6° = 13.7 \text{ mi/h}$$

Exercise   If the stream is 1/8 mi wide, how long does it take your boat to reach the opposite side?   *Answer: 32.8 s.*

Example 2.12

Your airplane is capable of flying 220 mi/h in still air. You want to fly from your hometown to a city that lies 350 mi directly northeast. The wind is blowing steadily toward the west at 25 mi/h. What heading should you choose and how long will it take you to make the trip?

R

Reasoning

Question   How can I construct the vector diagram?
Answer   Very much the same as in the previous illustration. In this case, you will not get a right triangle, however. Fig. 2.17 shows a sketch of the situation.

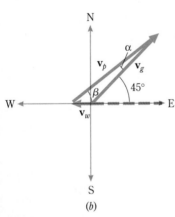

FIGURE 2.17

(a) The displacement vector of the airplane in Example 2.12. The direction *AB* is the same as the direction of the velocity $\mathbf{v}_g$. (b) The addition of velocity vectors, from which we find the angle θ and the groundspeed $\mathbf{v}_g$.

Question   What angle is the heading?
Answer   The angle θ tells you how much north of east you must steer. If you want to express this as a compass heading, where north is taken to be 0°, you would subtract θ from 90°.

Question   What determines the time of the trip?
Answer   You want to travel 350 mi in the direction of $\mathbf{v}_g$. The time will be $t = 350 \text{ mi}/v_g$.

Question   Without a right triangle, how can I find an analytical solution?
Answer   The **law of sines** (see inside back cover) is a very useful and simple relationship between the angles and the sides of any triangle. If you know any two angles and one side of a triangle, it allows you to find the other sides.

Question   How much am I given in the vector triangle?
Answer   You know the sides v_w and v_p, and the angle opposite v_p. The law of sines can be used twice. First you can find the angle opposite v_w. Then the angle θ opposite v_g can be found by remembering that the sum of the three angles of any triangle must be 180°. A second application of the law of sines then yields v_g.

Solution and Discussion   The angle opposite v_p is 135°. The law of sines gives

$$\frac{v_w}{\sin \alpha} = \frac{v_p}{\sin 135°}$$

$$\sin \alpha = \left(\frac{25}{220}\right) \sin 135° = 0.080 \qquad \alpha = 4.61°$$

Angle β is thus

$$\beta = 180.0° - 135.0° - 4.6° = 40.4°$$

Applying the law of sines a second time,

$$\frac{v_g}{\sin 40.4°} = \frac{v_p}{\sin 135°}$$

This yields $v_g = 0.917 v_p = 202 \text{ mi/h}$. The 350-mile trip will then take

$$t = \frac{350 \text{ mi}}{202 \text{ mi/h}} = 1.73 \text{ h} = 1 \text{ h}, 44 \text{ min}$$

Notice in still air the trip would take

$$\frac{350 \text{ mi}}{220 \text{ mi/h}} = 1.59 \text{ h} = 1 \text{ h}, 35 \text{ min}.$$

LEARNING GOALS

Now that you have finished this chapter, you should be able to

1 Define (*a*) speed, (*b*) velocity, (*c*) acceleration, (*d*) gravitational acceleration, (*e*) free fall.

2 Describe how you measure (*a*) the average velocity of an object as it moves from *A* to *B* and (*b*) the instantaneous velocity at any point along the way.

3 Calculate an object's velocity at any time when you are given the graph of its motion showing position as a function of time.

4 Calculate the acceleration of an object when you are given the graph of its velocity as a function of time.

5 State the five uniform-motion equations, explain the symbols in them, and state the condition under which the equations can be applied.

6 Solve simple problems involving uniformly accelerated motion, including free-fall.

7 Find the distance traveled by and the time of flight of (*a*) a projectile shot horizontally from a given height over level ground, and (*b*) a projectile shot over level ground at a given initial angle above horizontal.

8 Find the heading and ground velocity of a boat or airplane moving in the presence of a known stream current or wind, when the desired displacement is given.

SUMMARY

DERIVED UNITS AND PHYSICAL CONSTANTS

Gravitational Acceleration (g)
The free-fall acceleration of objects near the earth's surface is

$$g = 9.80 \text{ m/s}^2$$

DEFINITIONS AND BASIC PRINCIPLES

Average Speed ($\bar{v}$)

$$\text{Average speed} = \bar{v} = \frac{\text{distance traveled}}{\text{time taken}} = \frac{x}{t} \tag{2.1}$$

Average Velocity ($\bar{v}$)

$$\text{Average velocity} = \bar{v} = \frac{\text{displacement vector}}{\text{time taken}} = \frac{\mathbf{s}}{t} \tag{2.2}$$

Instantaneous Velocity
When the time interval over which an average velocity is measured approaches zero, the average velocity becomes equal to the instantaneous velocity at that instant.

$$\text{Instantaneous velocity} = \mathbf{v} = \lim_{\Delta t \to 0} \frac{\Delta x}{\Delta t} \tag{2.3}$$

INSIGHTS
1. *Magnitude.* The magnitude of the instantaneous velocity is the speed at that instant.
2. *Direction.* The direction of the velocity is the direction of the displacement.
3. *Graphical interpretation (one-dimensional motion)*: On a graph of *x* versus *t*, the slope of the graph at any instant of time is equal to the velocity at that instant.

Average Acceleration ($\bar{\mathbf{a}}$)
The average acceleration is the change in instantaneous velocity divided by the time taken for that change to occur.

$$\bar{\mathbf{a}} = \frac{\mathbf{v}_f - \mathbf{v}_o}{t} \tag{2.4}$$

INSIGHTS
1. The direction of the acceleration is in the direction of the change of velocity.
2. Since velocity is a vector, it can change either in magnitude *or* in direction. Thus an object is accelerating when *either* its speed or its direction is changing.
3. *Graphical interpretation (one-dimensional motion)*: On a graph of velocity versus time, the slope of the graph at any instant represents the instantaneous acceleration at that time.

Equations Describing Uniformly Accelerated Motion in One Dimension

$$\mathbf{x} = \bar{\mathbf{v}}t \tag{2.11a}$$

$$\bar{\mathbf{v}} = \tfrac{1}{2}(\mathbf{v}_f + \mathbf{v}_o) \tag{2.11b}$$

$$\mathbf{v}_f = \mathbf{v}_o + \mathbf{a}t \tag{2.11c}$$

$$v_f^2 = v_o^2 + 2ax \tag{2.11d}$$

$$\mathbf{x} = \mathbf{v}_o t + \tfrac{1}{2}\mathbf{a}t^2 \tag{2.11e}$$

INSIGHTS
1. Gravitational free-fall is one example of uniformly accelerated motion, with $\mathbf{a} = g = 9.80 \text{ m/s}^2$ at earth's surface.
2. Acceleration in a direction opposite that of velocity represents slowing down; acceleration in the same direction as that of velocity represents speeding up.

Equations of Projectile Motion
PROJECTILE LAUNCHED HORIZONTALLY

x component: $\quad v_x = v_o = v_f \quad$ (no horizontal acceleration)

$$x = v_x t$$

y component: $\quad v_y = gt \qquad (v_{oy} = 0)$

$$y - y_o = \tfrac{1}{2}gt^2 \qquad (v_{oy} = 0)$$

PROJECTILE LAUNCHED AT AN ANGLE θ_0 WITH VELOCITY $\mathbf{v}_o$.

x component: $\quad v_x = v_o \cos\theta_o = \text{constant}$

$$x = (v_o \cos\theta_o)t \qquad (x_o = 0)$$

y component: $\quad v_y = v_o \sin\theta_o - gt$

$$y = (v_o \sin\theta_o)t - \tfrac{1}{2}gt^2 \qquad (y_o = 0)$$

Trajectory equation: $\quad y = (\tan\theta_o)x - \left[\dfrac{g}{2v_o^2(\cos^2\theta)}\right]x^2 \qquad (2.15)$

INSIGHTS

1. The range of a projectile is the value of x at which impact with the ground is made (usually $y = 0$).

2. The time of flight is the time from launch until impact. It is thus the value of t which corresponds to the value of x at impact (the range).

3. For a projectile launched with an upward component of velocity, the maximum height is reached when $v_y = 0$.

Addition of Velocities in Two Dimensions
A boat or an airplane moving with a heading velocity $\mathbf{v}_b$ (or $\mathbf{v}_p$) in the presence of a stream or wind velocity $\mathbf{v}_w$ will have a velocity $\mathbf{v}_g$ relative to the ground given by

$$\mathbf{v}_g = \mathbf{v}_w + \mathbf{v}_b \qquad \text{or} \qquad \mathbf{v}_g = \mathbf{v}_w + \mathbf{v}_p$$

INSIGHTS

1. The displacement of the boat or plane relative to the ground will be in the direction of v_g.
2. If the velocity $\mathbf{v}_w$, the speed v_b (or v_p), and the direction of $\mathbf{v}_g$ are given, the direction of $\mathbf{v}_p$ (the heading) and the ground speed (magnitude of $\mathbf{v}_g$) can be found.

QUESTIONS AND GUESSTIMATES

1 Give an example of a case in which an object's velocity is zero but its acceleration is not.

2 Can an object's velocity ever be in a direction other than the direction of its acceleration? Explain.

3 Sketch graphs of velocity and acceleration as a function of time for a car as it strikes a telephone pole. Repeat for a billiard ball in a head-on collision with the edge of the billiard table.

4 Are any of the following true? (a) An object can have a constant velocity even though its speed is changing. (b) An object can have a constant speed even though its velocity is changing. (c) An object can have zero velocity even though its acceleration is not zero. (d) An object subjected to a constant acceleration can reverse its velocity.

5 A rabbit enters the end of a drainpipe of length L. Its motion from that instant is shown in Fig. P2.1. Describe the motion in words.

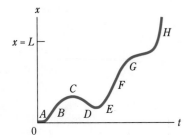

0   FIGURE P2.1

6 A high school girl who is an average runner completes a 100-m dash by running twice around an indoor circular track that is 50 m in circumference. Estimate her average speed and average velocity.

7 A stone is thrown straight up in the air. It rises to a height h and then returns to the thrower. For the time the stone is in the air, sketch the following graphs: y versus t; v versus t; a versus t.

8 Under what condition is it wrong to say that an object's acceleration is negative when the object is thrown upward? Does the sign of the acceleration depend at all on the direction of motion? Can an object's acceleration be positive when the object is slowing down?

9 The acceleration due to gravity on the moon is only about one-sixth that on earth. Estimate the ratio of the height to which you could throw a baseball on the moon to the corresponding height on earth.

10 How could you best analyze Fig. 2.8 to obtain the value of g? Assume the time interval between strobe flashes is known.

11 Airplane enthusiasts hold meets at which they show off their skills. One event is to drop a sack of sand exactly in the center of a circle on the ground while flying at a predetermined height and speed. What is so difficult about that? Don't they just drop the sack when they are directly above the circle?

PROBLEMS (See note on page 23.)

Sections 2.2–2.5

1 If it takes an airplane 2 h and 30 min to fly the 1200 air miles from Minneapolis–St. Paul to New York City, what is the average speed of the plane in mi/h? In m/s?

2 A particle accelerator shoots out electrons moving at 2.99×10^8 m/s. How long does it take such particles to travel 5.0 mm?

3 In a television tube, electrons are emitted from an electrode at one end of the tube and strike a light-emitting coating on the picture screen at the other end of the tube. If the electrons are emitted with a velocity of 1.25×10^8 m/s, how long does it take the electrons to hit the screen 16.7 cm away?

4 Sound travels approximately 340 m/s in still air. If you shout across a canyon and hear the echo reflected from the opposite wall 3.5 s after you shout, how far away is the opposite wall?

5 During a video game, a point on the screen moves 9.6 cm in the positive y direction and then 3.6 cm in the negative x direction, all in a time of 3.9 s. (*a*) What was the point's average velocity during this time? (*b*) What was its average speed?

6 To get to work, you drive 2.2 mi east, then 1.5 mi south, followed by 3.7 mi at 45° south of east. You make the trip in 21 min. What is (*a*) your average velocity, and (*b*) your average speed?

7 Figure P2.2 shows the motion of an ant along a straight line. Find the ant's average velocity from (*a*) *A* to *E*, (*b*) *B* to *E*, (*c*) *C* to *E*, (*d*) *D* to *E*, and (*e*) *C* to *D*.

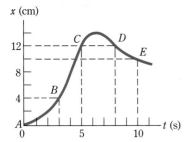

FIGURE P2.2

8 The motion of a bug along a wire strung along the x axis is shown in Fig. P2.2. Find the bug's average velocity from (*a*) *B* to *D*, (*b*) *D* to *E*, (*c*) *A* to *D*, and (*d*) *A* to *B*.

▪ 9 Mary can run at a top speed of 4.2 m/s, whereas Kim can run only at 3.4 m/s. They are to race a distance of 200 m, starting at the same point. If the race is to end in a tie, how much sooner should Kim start running before Mary does?

▪10 As an alternative plan to the situation in Prob. 9, Kim is to start at the same time as Mary, but from a point a distance s ahead of Mary. (Mary is to run the full 200 m.) For the race to end in a tie, what should the distance s be?

▪11 A girl walks eastward along a street; Fig. P2.3 is a graph of her displacement from home. Find her average velocity for the whole time interval shown and her instantaneous velocity at *A*, *B*, and *C*.

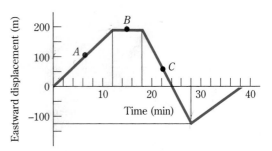

FIGURE P2.3

▪12 For the girl in Prob. 11, (*a*) find her average velocity for the time interval $t = 7$ min to $t = 14$ min. Also find her instantaneous velocity at (*b*) $t = 13.5$ min and (*c*) $t = 23$ min.

13 The graph of a particle's motion along the x axis is shown in Fig. P2.4. Find the average velocity for the interval *A* to *C*. Also find the instantaneous velocity at *D* and at *A*.

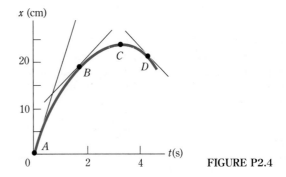

FIGURE P2.4

14 For the particle motion graphed in Fig. P2.4, find the average velocity from *C* to *D*. Find the instantaneous velocity at *B* and at *C*.

15 Two dogs start running toward each other from points 135 m apart. One dog runs at 6.75 m/s, and the other runs at 5.25 m/s. When they meet, how far are they from where the slower dog started?

16 A truck traveling east at 18.8 m/s is 1.56 km ahead of a car also traveling east, at 25.5 m/s. Assuming both speeds remain constant, how long will it take for the car to pull even with the truck?

Sections 2.6–2.8

17 A car on a straight road accelerates from 2.18 m/s to

7.75 m/s in a time of 5.77 s. What is the car's average acceleration?

18 An airplane in straight-line flight changes speed from 460 km/h to 325 km/h in 52.5 s. Express the plane's average acceleration in m/s².

19 A car moving at 23.7 m/s skids to a stop in 10.8 s. Find the car's average acceleration and the distance it travels while stopping.

20 A drag racer claims to be able to go from rest to 200 mi/h in 5.0 s. What is the average acceleration of this car in m/s²? How far does the racer go in this time?

21 A certain drag racer covers a distance of one-quarter mile in 4.87 s, starting from rest. What is the average acceleration of this racer, and what is its speed at the quarter-mile mark?

■**22** A bullet traveling 220 m/s strikes a tree and penetrates 4.33 cm before stopping. Find the average acceleration of the bullet and the time it takes to stop.

■**23** In a television tube such as that mentioned in Prob. 3, electrons are accelerated from rest to 1.25×10^8 m/s in a distance of 1.12 cm. How long does this take, and what is the average acceleration of the electrons?

■**24** A truck traveling at 22.5 m/s decelerates at 2.27 m/s². (*a*) How much times does it take for the truck to stop? (*b*) How far does it travel while stopping? (*c*) How far does it travel during the third second after the brakes are applied?

■**25** A bullet traveling 190 m/s strikes a piece of wood 2.54 cm thick and passes through, emerging with a speed of 80 m/s. Find the average acceleration of the bullet and the time it takes to pass through the wood.

■**26** A rubber ball traveling at a speed of 31.5 m/s strikes a concrete wall and bounces straight backward with a speed of 28.5 m/s. The collision with the wall took 0.15 s. Find the average acceleration the ball experienced during the collision.

■■**27** A locomotive is pulling a train 580 m long, including the locomotive. The locomotive accelerates uniformly from rest and reaches a road crossing 1.35 km from its starting point in 9.66 min. (*a*) How long after the locomotive reaches the crossing does the caboose reach the crossing, assuming the train maintains its uniform acceleration? (*b*) How fast is the train going when the caboose reaches the crossing?

■**28** The first car of a stationary train is blocking a crossing. Just as the train begins to move, a waiting motorist notices that it takes 18.8 s for one railcar to move through a distance equal to its length *L*. Find the train's acceleration in terms of *L*. Assuming constant acceleration, how long after the train starts will the first 50 railcars have passed the waiting motorist?

■■**29** A car is traveling at 27 m/s along a road parallel to a railroad track. How long does it take the car to pass a 920-m-long train traveling at 18.3 m/s (*a*) in the same direction as the car and (*b*) in the opposite direction?

■**30** Just as a car starts to accelerate from rest at a constant 2.44 m/s², a bus moving at a constant speed of 19.6 m/s passes the car in a parallel lane. How long before the car overtakes the bus? How fast is the car going then? How far has the car gone at that point?

■■**31** Two cars, both traveling at 30.5 m/s, are headed toward each other in the same traffic lane. When they are 250 m apart they see each other and begin to decelerate at the same rate. What must be the magnitude of this deceleration if the cars are to barely avoid collision?

Section 2.9

32 A loose brick falls from a window ledge to the street 21.3 m below. How fast is the brick going just as it hits the street? How much time does it take the brick to fall?

33 A person falls off a plank which hangs over a stream. She hits the stream 1.32 s later. How far above the stream is the plank? How fast is the person going when she hits the water?

■**34** You throw a baseball straight upward with an initial speed of 23.9 m/s. How high above the point of release from your hand does the ball go before starting to fall? How much time does it take the ball to reach its maximum height?

■**35** A stone is thrown straight downward from the top of a building 26.0 m high with an initial speed of 18.6 m/s. How long does it take the stone to reach the ground? How fast is the stone moving just as it hits the ground?

■**36** A batter hits a popup where the baseball goes straight upward. The fielder catches the ball 9.3 s after it was hit, at the same level the ball left the bat. How high did the baseball go? How fast was the ball going just as it was caught?

■**37** A girl standing on top of a 22-m-high building throws a coin upward with a speed of 8.8 m/s. How long does it take the coin to hit the ground? How fast is the coin going just as it hits the ground?

■**38** A 1.9-m-tall burglar is running along a sidewalk at a constant speed of 3.77 m/s. Your apartment window is 17.8 m directly above this sidewalk. You drop a flowerpot from rest that hits the burglar on the head directly below you. How far away from the point of impact was the burglar at the time you released the flowerpot?

■**39** Two balls are dropped from different heights. One is dropped 0.85 s before the other, but both strike the ground at the same time, 6.25 s after the first ball was dropped. From what heights were the two balls released?

■■**40** An elevator in which a woman is standing is moving upward at a constant speed of 3.35 m/s. The woman drops a coin from a height of 1.25 m above the elevator floor. How long does it take the coin to strike the elevator floor?

■■**41** Repeat Prob. 40 if the elevator is at rest at the instant the coin is dropped, but is accelerating upward at 3.5 m/s².

Section 2.10

42 A marble rolls horizontally off a table at a speed of 3.7 m/s. A second marble is dropped vertically from the table at the same instant. If the table is 1.20 m high, how far apart do the marbles land? What difference is there in the times of impact of the two marbles?

43 A fire hose shoots water horizontally from the top of a building toward the wall of a building 31 m away. The water leaves the nozzle of the hose at a speed of 6.4 m/s. How far below the level of the nozzle does the water hit the wall? (*Hint:* Consider the water to be a stream of particles leaving the hose.)

■44 At a circus, the "human cannonball" is shot out of a cannon at a speed of 24.4 m/s. The cannon barrel is pointed 50° above the horizontal. (*a*) How far (horizontally) from the end of the cannon should the net used to catch the person be placed? (*b*) How long is the person's time of flight? Assume the net and the end of the cannon are at the same level.

■■45 Suppose you fire a projectile 35° above horizontal with an initial velocity of 200 m/s. It lands in a valley 300 m below the launch point. What is the range of the projectile, and what is the projectile's time of flight?

■■46 The stunt driver in Fig. P2.5 wishes to shoot off the ramp and land on the platform. How fast must the motorcycle be going as it leaves the ramp if the stunt is to succeed?

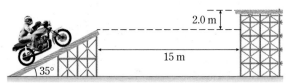

FIGURE P2.5

Section 2.11

■47 A helicopter is headed (pointed in the direction of) due north. The helicopter can fly 75 mi/h in still air, and the wind is coming from the northeast at 20 mi/h. What is the helicopter's velocity relative to the ground? How far does the helicopter travel in 20 minutes?

■48 You want to row your boat directly across a river, whose current flows at 0.85 m/s. You are capable of rowing the boat at a speed of 2.1 m/s. (*a*) In which direction must you head your boat in order to move directly across the river? (*b*) If the river is 45 m wide, how long does it take you to cross?

■49 An airplane has an airspeed (in still air) of 650 km/h. The plane is heading 25° west of north, but the pilot notices that the plane is actually traveling 18° west of north. What eastward wind velocity is the plane encountering?

General Problems

■50 You are traveling along a highway at 95 ft/s, following a car going the same speed. Suppose the maximum deceleration for both cars is 22.7 ft/s². Suddenly the car ahead of you slams on the brakes, stopping as quickly as possible. It takes you 0.40 s to respond by slamming on your brakes, also stopping as quickly as possible. What minimum distance should you have maintained to avoid a collision?

■51 A police car is sitting one-quarter of a mile off a main highway. The policeman receives a report of a car speeding along the highway at 75.0 mi/h. The situation is shown in Fig. P2.6. The police car has a maximum acceleration of 28.0 ft/s². What is the minimum distance from the intersection that the car can be when the police car begins to accelerate if the policeman wants to arrive at the intersection 30 s ahead of the car?

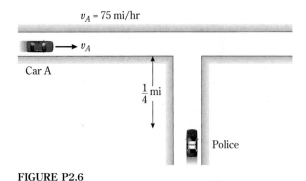

FIGURE P2.6

■■52 A physics student comes up with the following scheme to measure a building's height. A timing mechanism is set up to measure the time it takes for a lead weight dropped from the top of the building to fall the last 1.5 m before hitting the ground. Observations for a certain building show that the weight takes 0.109 s to fall this last 1.5 m. How high is the building?

■■53 A ball is thrown straight upward with a speed v_o from a point h m above the ground. Show that the time the ball takes to reach the ground is given by the expression

$$\frac{v_o}{g}\left(1 + \sqrt{1 + \frac{2hg}{v_o^2}}\right)$$

■■54 A railcar is moving horizontally with a speed of 24 m/s and decelerating at 3.65 m/s² when a light bulb 2.55 m above the floor comes loose and drops. Where, relative to the point directly below its original position, will the bulb strike the floor?

■55 A lead weight is dropped from rest into a lake from a platform 10 m above the water. When it hits the water, its velocity is reduced to one-tenth of the velocity it had acquired just before impact. It sinks at this constant reduced velocity until it reaches the lake bottom 6.5 s after hitting the water. How deep is the lake?

■■56 You are standing on an observation deck 100 m above a city street and drop a rock from rest. A friend stands on the street directly below you and throws a rock vertically upward at the same instant that you drop the rock. The initial upward velocity of his rock is 50 m/s. Assuming that they are moving along the same vertical line, and that air resistance can be neglected, calculate: (*a*) at what height they will collide, (*b*) when they will collide, and (*c*) whether your friend's rock will be rising or descending when they collide.

""57 Suppose you have a sports car which has a maximum acceleration of $a = 24$ ft/s^2 and a maximum deceleration from braking of $a = -32$ ft/s^2. You want to find the minimum time it would take you to start from rest, cover $\frac{1}{4}$ mi, and come to a stop at the $\frac{1}{4}$-mi mark. You do this by accelerating as much as possible for part of the $\frac{1}{4}$ mi, followed by a period of maximum deceleration to the final stop. What is this minimum time?

CHAPTER 3

NEWTON'S LAWS OF MOTION

In Chap. 2, we defined and discussed velocity and acceleration without considering what causes an object's motion. We now investigate how accelerations are caused by forces, and in so doing we state and discuss Newton's three laws of motion, laws of primary importance in physics.

3.1 THE DISCOVERY OF PHYSICAL LAWS

Two persons closely associated with the origin of the scientific method are Galileo Galilei and Isaac Newton. Even though he had to work with instruments of very limited precision, Galileo was one of the earliest persons to insist that nature could be understood through controlled experiments. In the early 1600s, Galileo developed the concept of the *inertia* of objects and gave the first correct description of the acceleration of objects falling near the earth's surface. His results in both of these discoveries contradicted ideas developed by the Greek philosopher Aristotle (about 350 B.C.), ideas held as correct by many of Galileo's contemporaries. It is instructive to examine the two competing ideas in each case, in order to show by example the nature of scientific reasoning and physical law.

An apple and a feather fall together in an evacuated chamber. When the effects of air resistance are negligible, all objects fall with the same acceleration.

INERTIA

According to Aristotle, the "natural" state of any object was to be at rest: any object put into motion "naturally" came to rest. Up until Galileo's time, this phenomenon was considered a fundamental rule of nature. Galileo, however, observed that whenever a moving object came to rest, it was always because of some "force," like friction, which retarded the motion and eventually stopped the object. He observed that the weaker the retarding force, the longer the object took to stop. The nature of retarding forces could be different for each separate situation, and Galileo could make no useful generalization about them. His genius, however, was to reason that, if *no* retarding force were to act, objects would *keep going indefinitely*! This tendency of moving objects to keep moving Galileo called the *principle of inertia*. As we shall see in Sec. 3.2, Newton later developed a more formal description of inertia, which included objects at rest as well as those in motion.

FALLING OBJECTS

Another famous claim of Aristotle's was that heavy objects fall to earth faster than light objects. As we learned in Sec. 2.9, however, Galileo believed that falling bodies *all* accelerate at the same rate and, when dropped from the same height, fall to the ground in the same amount of time.

It is not a trivial matter to sort out who is right here, since we often see heavier objects fall faster than light ones. A cannonball and a feather provide a good example. Moreover, the same irregularly shaped object—a sky diver for instance— can fall at different speeds depending on whether it falls flat or edge-on. Galileo surmised that the controlling factor in the way things fall is the degree to which they are affected by air drag. The effect of drag camouflages the underlying reality. Take away the air, he reasoned, and you will reveal the *fundamental* principle governing the behavior of falling objects, which is that of constant and equal acceleration. This is the situation which we studied in Sec. 2.9.

Thus in both of these controversies, Galileo, by stripping away the irrelevancies and secondary influences that concealed a simpler behavior of nature, was able to extract the most fundamental and general laws. This kind of unifying insight is a major characteristic of the scientific method.

Isaac Newton (1642–1727) is credited for establishing a truly mathematical basis for physical law. His laws of motion, which we study in this chapter, are mathematical statements of extreme simplicity. Yet they represent great generality, applying to all situations that involve moving objects (with the exception of very high speeds, where Einstein's work modifies Newton's equations). Newton also developed the universal theory of gravitation, which we will take up in Chap. 7. With this theory, such diverse phenomena as projectiles moving near the earth's surface and the orbits of the planets around the sun can be understood as examples of a single principle.

3.2 THE CONCEPT OF FORCE AND NEWTON'S FIRST LAW OF MOTION

We begin our study of the work of Isaac Newton by examining his three **laws of motion,** first published in 1687 in a classic compendium entitled Principia Mathematica Philosophiae Naturalis (*Mathematical Principles of Natural Philosophy*). In this work, Newton introduces the concepts of **mass** and **force,** connecting these concepts to the acceleration of objects. Let us proceed to discuss forces, and return to the concept of mass later in connection with Newton's second law.

We are familiar in a casual way with the notion of forces, experiencing various pushes and pulls in everyday life. We recognize that the earth exerts a force on objects that we call *gravity,* and that we have to exert a force on an object to lift it against gravity. Our experience also tells us that forces have direction, and therefore are *vector* quantities. Many forces may act on an object at a given time, in various directions.

One way of exerting a force on an object is by attaching a spring to the object and pulling on the other end of the spring. Let us use this simple example to show hypothetically how a standard amount of force can be defined. A pointer attached to the spring (Fig. 3.1*a*) indicates a certain amount of stretch, and hence a certain amount of force exerted by the stretched spring. Every time the spring is stretched by this amount, we can assume the same amount of force is being exerted. This arbitrarily chosen amount of stretch can be used to indicate a reproducible standard amount of force exerted by the spring.

In order to double or triple the standard force, we could attach two or three identical springs to the object, each stretched by the standard amount, as in Fig. 3.1*b*. We could also observe that when we attach two of these springs to a single one and stretch them both to the standard length, the single spring is stretched to twice the standard length (Fig. 3.1*c*). Repeating this experiment with three springs attached to a single one will show that the single spring is stretched three times the standard amount. We can thus conclude that the amount of force exerted by a single spring is linearly proportional to the amount of stretch, and proceed to **calibrate** a scale on the spring to indicate multiples of the standard force. Thus even without specifying a particular unit of force, we see that conceptually we have identified a way to exert measurable amounts of force on an object by applying such springs to the object.

Table 3.1 lists some categories of forces that we encounter in our daily lives. We shall discuss many applications of these forces in more detail as we proceed.

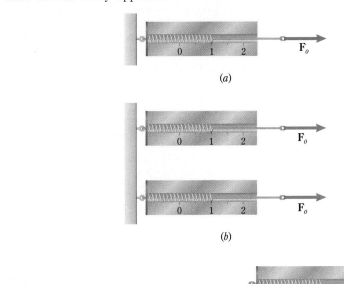

(*a*)

(*b*)

(*c*)

FIGURE 3.1

(*a*) $\mathbf{F}_o$ is the amount of force required to stretch the spring by a certain amount, say 1 cm. (*b*) Two springs identical to the one above, each stretched by 1 cm, produce a force on the wall of $2\mathbf{F}_o$. (*c*) Applying a force $2\mathbf{F}_o$ stretches the spring twice as far. Thus the single spring exerts a force of $\mathbf{F}_o$, $2\mathbf{F}_o$, $3\mathbf{F}_o$ when stretched by 1 cm, 2 cm, and 3 cm, respectively.

Cables are used to lift objects by exerting tension forces.

TABLE 3.1 *Some common categories of forces*

Category	Examples
Forces involving *tension*	Forces that *pull* on objects via strings, cables, ropes, or other material attached to the objects
Forces of *compression*	Forces involving rigid* objects which support and bear weight (shelves, floors, benches, etc.)
	Forces arising from fluid pressure
	Forces that occur when solid* objects collide
	Forces that occur *perpendicular* to the areas of contact when solid objects are pushed together
Forces of *friction* or *viscosity*	Forces which resist the slipping or sliding motion between surfaces in contact and are *parallel* to the surfaces
Fundamental forces that act between objects separated in space	The attractive force of gravity between all material objects
	The electrical force between objects possessing electrical charge
	Magnetic forces between electrical currents

*Objects are rigid or solid because of the forces between the atoms or molecules of which the objects are made. These forces are fundamentally electrical in nature. When we speak of tension or compression forces we are speaking of situations in which the atomic or molecular forces of objects, such as a rope or a table top, are strong enough to exert these forces without breaking.

Newton's first law of motion pertains to situations in which there is zero resultant force acting on an object. This means that, although there may be a number of individual forces acting, the *vector sum* of those forces is zero. We often say that the **net force** is zero in this case. If the object is at rest, Newton's first law states:

An object at rest remains at rest if there is zero resultant force acting on it.

We are familiar with many such situations. A book lying on a table experiences a downward pull of gravity, balanced by the upward force of support supplied by the rigid table. During a tug-of-war, the flag in the middle of the rope remains stationary if both sides are pulling in opposite directions with equal force. You may

Bodies at rest.

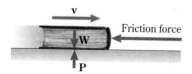

FIGURE 3.2

Friction causes the object to slow to a stop.

wonder why we hold Newton in such esteem for such a seemingly obvious conclusion. We do so in part because the first law also applies to moving objects in a far less obvious way.

In formalizing Galileo's observations about moving objects (Sec. 3.1), Newton reasoned as follows. The book lying on the table has no net force acting on it. As mentioned above, the forces on it in the vertical direction add up to zero. If we give the book a horizontal shove to set it in motion, nothing has changed vertically: the vertical forces still balance. We observe, however, that the book comes to rest after traveling some distance across the table. Corroborating what Galileo had surmised earlier, Newton stated that an unbalanced horizontal force, called **friction,** acts on the book to retard its motion and bring it to a stop (see Fig. 3.2). If we make the surface smoother, thereby reducing the force of friction, the book slides further before stopping. Newton concluded that, in the *absence* of this net force, the book wouldn't slow at all.

Although it is impossible to eliminate friction completely in everyday situations, both Newton and Galileo were able to *idealize* the actual situation. By asking, *"What if* friction were absent?" they could perceive the basic principle of motion hidden behind the complication introduced by friction. Newton further concluded that, in order for a moving object to be deflected from its direction of motion, an unbalanced force in the direction of the deflection must act on the object. These conclusions are summarized in the more general form of Newton's first law:

A moving object continues to move with constant velocity if the vector sum of the external forces acting on the object is zero.

Notice that the word used is *velocity,* not *speed.* This law states that neither the magnitude *nor the direction* of the object's velocity will change. In other words, the object will continue to move along a straight line. Of course, this statement holds for $v = 0$ as well as for any other value of v.

3.3 INERTIA AND MASS

Closely linked with the first law is the concept of inertia which we met in Sec. 3.1. A common definition of this term is

Inertia is the tendency of an object at rest to remain at rest and of an object in motion to remain in motion with its original velocity.

We have much common experience concerning inertia. We know, for example, that a cement truck has much more inertia than a child's wagon, since the wagon is much easier to set in motion than the truck. Moreover, the wagon is much easier to stop than the truck when they are both moving at the same velocity. It is difficult to alter the state of motion of an object that has large inertia.

To make the concept of inertia quantitative, we define a quantity called mass. In the SI system of units, mass is defined in the following way. A certain metal cylinder carefully preserved near Paris, France, is defined to have a mass of exactly one kilogram (1 kg). (Figure 3.3 shows a duplicate cylinder kept at the U.S. National Bureau of Standards in Washington, D.C.) By definition, any object that has the same inertia as this standard kilogram object has a mass of 1 kg. An object having three times as much inertia is *defined* as having a mass of 3 kg, and so on. Newton's second law of motion describes how the mass of an object is involved in determining how the object reacts when the net force acting on it is *not* zero.

FIGURE 3.3

The platinum-iridium cylinder shown here is a copy of the standard kilogram mass. The cylinder is kept at the U.S. National Bureau of Standards, whose responsibility it is to preserve this secondary standard of mass. *(National Bureau of Standards)*

Physicists at Work ALAN LIGHTMAN *Massachusetts Institute of Technology*

One of my ecstatic experiences with scientific research happened about 1980, in a small room in my home in Massachusetts. For about six months, I had been struggling with an unsolved problem in theoretical physics. The problem goes like this: Put some protons and electrons in a spherical container of a certain temperature and a certain size. The particles will whiz around and, if the temperature is high enough, new particles can be created from the energy of motion. The question is: How many new particles will be created? The answer could have implications for the behavior of black holes. My only equipment was a large stock of white unlined paper and a wastebasket, which I had made much use of.

I soon realized that my problem was not fundamental. I was not going to discover a new law of nature. But I was confronting a problem that had never been solved, and the reliance on myself to discover a truth, no matter how small, I found exhilarating. There was so much of the world that I had just taken on faith. I had been told that I was once the size of a hollyhock seed. I had been told that the earth is not flat as it seems but bends on itself in a huge, mottled ball. I understood that I had to trust others for much that I knew—no one can come close to verifying all the facts he or she believes. But every unverified fact extracts a small price. Little by little, I had paid away my confidence. By contrast, nothing would build confidence so much as discovering something on my own, from the beginning, taking no one's word for it. I relished my problem with the particles and the sphere, and I carried my calculations around with me as if they were heirloom letters.

One morning at dawn, I awoke with an odd feeling and went to my study. Suddenly, I could see through my problem. I don't know how I found my way, but it wasn't by going from one equation to the next. Somehow, my subconscious mind had been studying the problem in a different way, and it all fit together, clean as a new dollar bill.

It is difficult to describe that lifting feeling in a creative act when everything suddenly falls into place. It has similarities to sailing a round-bottomed boat in strong wind. Normally, the hull of the boat stays down in the water, with the resulting frictional drag greatly limiting the speed of the boat. But in high wind, every once in a while the hull lifts out of the water, and the drag instantly drops to near zero. It feels like a gigantic hand has suddenly grabbed hold of your mast and flung you across the surface like a skimming stone. It's called planing.

I planed on that early morning and on a few other occasions in my scientific career. That swift soaring moment of discovery is worth all the months of frustration. And for a little while you, the discoverer, are the only one in the world who knows this new thing. You will soon go to the office and tell your colleagues, you will publish your results, but for those first few moments you know something true that no one else knows, you have vast power, and that feeling of specialness you felt as a child and then lost becomes briefly as real as the coffee cup you are holding in your hand.

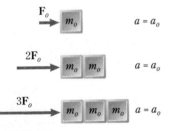

$F_o \rightarrow \boxed{m_o}$ $a = a_o$

$2F_o \rightarrow \boxed{m_o}\,\boxed{m_o}$ $a = a_o$

$3F_o \rightarrow \boxed{m_o}\,\boxed{m_o}\,\boxed{m_o}$ $a = a_o$

FIGURE 3.4

For constant acceleration, **F** is proportional to m.

3.4 NEWTON'S SECOND LAW

We know from experience that it is more difficult to change either the speed or the direction of motion of a massive object than a less massive object. To quantify this experience, we can carry out the experiment shown schematically in Fig. 3.4. In Sec. 3.2, we saw how we could define a standard quantity of force by producing a calibrated spring. Let us now call this standard force $\mathbf{F}_o$. The identical objects, each of mass m_o, might be 1-kg masses floating with negligible friction on an air table, for example. As indicated in Fig. 3.4, we find that to produce the same acceleration $\mathbf{a}_o$, the net force must increase in proportion to the increased mass. We can therefore conclude that

A bobsled is accelerated by forces
exerted on it by the team.

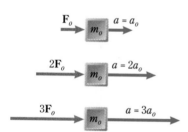

FIGURE 3.5

For constant mass, **F** is proportional
to a.

$\mathbf{F}_{\text{net}} \sim$ mass

for constant acceleration. (The symbol $\sim$ is read "is proportional to.")

A variation on this experiment is shown in Fig. 3.5, where increases in the net force on the same mass m_o cause increases in the acceleration proportional to the net force. Thus we conclude

$\mathbf{F}_{\text{net}} \sim \mathbf{a}$

for constant mass. We also observe that the acceleration is in the direction of the net force.

We can combine these two observations into an equation:

$$\mathbf{F}_{\text{net}} = km\mathbf{a} \tag{3.1a}$$

where k is some constant of proportionality.

This simple result is known as **Newton's second law of motion.** Despite its simplicity, it is an enormously general statement. It covers all kinds of forces and all kinds of objects. It reduces the complexities of different forces and various objects to the fundamental properties that determine motion in all possible specific instances: the measurable amounts of force and mass. In this way, the second law unifies an extremely wide range of situations into a common framework and thus qualifies as a fundamental physical law.

We must now evaluate the proportionality constant. We do so by deciding to define the basic SI unit of force as that amount of net force which, if applied to a 1-kg mass, will result in the mass having an acceleration of 1 m/s^2 (Fig. 3.6). If this definition seems arbitrary, it is. We are free to define the *unit* of force any way we want; we are *not* free to invent the way force is related to mass and acceleration, however. With this choice for our definition of a force unit, the proportionality

FIGURE 3.6

A net force of 1 N will give a 1-kg mass an acceleration of 1 m/s^2.

constant in Eq. 3.1a is simply unity (a value of 1). The name given to this amount of force is 1 **newton** (N). We now can redefine force more quantitatively:

A net force of one newton is that force which gives a one-kilogram mass an acceleration of one meter per second per second.

The newton is an example of a *derived* unit of measurement. From $\mathbf{F} = m\mathbf{a}$, we have

$$1\,\mathrm{N} = (1\,\mathrm{kg})(1\,\mathrm{m/s^2}) = 1\,\mathrm{kg \cdot m/s^{2}}*$$

Although the newton is the SI force unit, two other force units are sometimes used. They are the **dyne** and the **pound** (lb), where

$$1\,\mathrm{dyne} = 10^{-5}\,\mathrm{N}\ \text{exactly}$$

and

$$1\,\mathrm{pound\ (lb)} = 4.4482\,\mathrm{N}$$

We can separate the vectors in Eq. 3.1a into their rectangular components, giving one equation along each coordinate axis:

$$\begin{aligned}
(\mathbf{F}_{net})_x &= \Sigma \mathbf{F}_x = ma_x \\
(\mathbf{F}_{net})_y &= \Sigma \mathbf{F}_y = ma_y \\
(\mathbf{F}_{net})_z &= \Sigma \mathbf{F}_z = ma_z
\end{aligned}$$

(3.1b)

The symbol Σ is the summation sign, telling us in the first equation to sum up the x components of the individual forces acting, and similarly for the y and z components in the other equations. In doing the sum, it is necessary to take account of the signs of the components of each force, of course.

Example 3.1

A 900-kg car is to accelerate from rest to 12.0 m/s in 8.00 s along a straight road. How large a force is required?

R

Reasoning

Question What principle determines the force required?
Answer Newton's second law: $\mathbf{F}_{net} = m\mathbf{a}$.

Question Mass is given. How can I find acceleration?
Answer We assume the acceleration is constant. Then we can use the equations of motion developed in Chap. 2. We know $\mathbf{v}_o = 0$, $\mathbf{v}_f = 12.0$ m/s, and that the time it takes for this change is $t = 8.00$ s. Therefore we can use Eq. 2.11 in the form $\mathbf{a} = (\mathbf{v}_f - \mathbf{v}_o)/t$.

---

*This is our first encounter with a derived unit that is given a special name. It is important to remember the combination of basic units (dimensions) which defines a derived unit, since only then can you see which units cancel when the unit is used in a particular calculation.

Solution and Discussion

1 The acceleration is

$$\mathbf{a} = \frac{12.0\,\text{m/s} - 0}{8.00\,\text{s}} = +1.50\,\text{m/s}^2$$

2 The force is

$$\mathbf{F} = (900\,\text{kg})(1.50\,\text{m/s}^2) = +1350\,\text{N}$$

Notice that the signs are positive. The car is "speeding up"; that is, **a** is in the direction of **v**, and **F** must be in the direction of **a**.

Make sure you understand that the units of $\text{kg} \cdot \text{m/s}^2$ are newtons.

Exercise   How far does the car go in the 8.00 s?   *Answer: 48 m*

FREE-BODY DIAGRAMS

When we apply Newton's laws to specific situations, many forces can be acting at the same time. Some act on the object whose acceleration we are seeking to determine. Others act on the object's surroundings. For example, Fig. 3.7a shows a child pulling on a wagon. There are many forces acting on the wagon: the rope, gravity, and the upward compression force the rigid ground exerts on the wagon's wheels. There are also forces acting on the ground and on the child, but if we are interested only in the motion of the wagon, these are irrelevant. *In general, no forces acting on an object's surroundings will determine what will happen to the object, except as they help provide the forces that act directly on it.*

In order to clarify the situation, it is useful to draw a picture that isolates and identifies only those forces acting on the object of interest. Such a picture is called a **free-body diagram.** Even when the magnitude of some of the forces acting on the object are unknown, we can usually indicate them on the free-body diagram by a symbol and identify their directions. Figure 3.7b shows the free-body diagram for the wagon. Such a diagram simplifies writing the force sums in Eqs. 3.1b that apply to the wagon.

A common source of error in vector calculations involves denoting direction incorrectly. Usually we can see from the free-body diagram the directions unknown forces must have. We can then supply their correct signs in the component equations. Once we have introduced these signs into the equations, we have taken direction into account. The solution of the equations then should give us positive values which represent the *magnitudes* of the vectors.

(a)

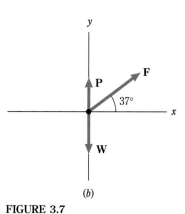

(b)

FIGURE 3.7

Forces exerted on a wagon in (*a*) are shown by the wagon's free-body diagram (*b*).

Example 3.2

Suppose that the girl pulls on the wagon as shown in Fig. 3.7a with a force of 25.0 N. As a result, the wagon accelerates horizontally. The wagon has a mass of 10.4 kg, and the downward force of gravity on the wagon, its *weight*, is 102 N. Assume that there is no friction to impede the wagon. Find the wagon's acceleration and the upward force of compression **P** that the ground exerts on the wagon under these conditions.

R

Reasoning

Question   The forces on the wagon are identified in its free-body diagram, Fig. 3.7*b*. How do I know that there is a compression force **P**?
Answer   Examine the vertical component of Newton's second law (Eq. 3.1*b*). If the motion of the wagon is entirely horizontal, $\mathbf{a}_y$ must be zero, and the vertical forces must add up to zero. You can easily see that the upward component of the girl's force isn't enough to oppose the 102-N weight. The ground must supply whatever additional force is needed, otherwise the wagon would accelerate vertically.

Question   What is the equation relating vertical force components?
Answer $P + (25.0\,\text{N})(\sin 37.0°) - 102\,\text{N} = 0$

Question   What determines the horizontal acceleration?
Answer   The net horizontal force, which is $(25.0\,\text{N})(\cos 37.0°) = 20.0\,\text{N}$.

Solution and Discussion   The acceleration is

$$\mathbf{a}_x = \frac{(\mathbf{F}_{\text{net}})_x}{m} = \frac{20.0\,\text{N}}{10.4\,\text{kg}} = 1.92\,\text{m/s}^2$$

The upward force of compression is

$$P = 102\,\text{N} - (25.0\,\text{N})(\sin 37.0°) = 87.0\,\text{N} \quad \blacksquare$$

Before proceeding to more applications of Newton's second law, let us discuss the third law and examine weight and friction more fully.

3.5   ACTION AND REACTION: THE THIRD LAW

The wrestlers exert equal and opposite forces on one another.

You may know that the earth orbits the sun because of the gravitational force exerted on the earth by the sun. Newton was able to treat this type of motion successfully after he discovered the law of gravitation, a topic we discuss in Chap. 7. Did you ever wonder, however, about the gravitational force the earth exerts on the sun? To measure this force directly, we would have to carry out measurements on the sun—a seemingly impossible task. Fortunately, it is often possible to state the value of such inaccessible forces by using another law discovered by Newton, the action-reaction law.

Push your finger against a wall and the wall pushes back on your finger. As another example, consider what happens when you kick a football. Your foot exerts a force on the ball, but you can also feel an oppositely directed force exerted by the ball on your foot. An object sitting on a table pushes down on the table, but the table pushes up on the object.

Newton examined a multitude of such situations and arrived at a quantitative conclusion, **Newton's third law:**

If object *A* exerts a force **F** on object *B*, then object *B* exerts a force −**F** on object *A*, equal in magnitude to **F** but opposite in direction.

One of these forces (either one) is called the *action force;* the other is called the *reaction force.* The third law states that the reaction force is exactly equal in magnitude to the action force and opposite in direction. It states more than that, however, because it tells us that the forces act on two different objects. The action force is exerted by one body on a second body, and that second body exerts the reaction force back on the first body.

Because of the third law, we can say that the action and reaction forces are equal in magnitude and opposite in direction in each of the examples in Table 3.2. Remember, *the action and reaction forces act on different objects.* From time to time we shall use this law to tell us the force on one body when the force on another is known.

To illustrate the third law, suppose a compact car and a semi-trailer truck collide head-on. Which is hit "harder," i.e., with greater force? Most people will look at the results of the collision and conclude that the car was certainly hit harder. Yet Newton's third law states that the force exerted by the car on the truck was exactly equal in magnitude (and in the opposite direction) to the force exerted by the truck on the car. How can we reconcile these seemingly conflicting conclusions?

First, our everyday language tends to obscure precise meanings. The phrase "hit harder," although we think we know exactly what it means, confuses the *force* of the collision with the *result* of the collision. In other words, we assume that greater damage has to be caused by greater force. To see what really determines the damage, let's look at Newton's second law in a different form: $\mathbf{F} = m\mathbf{a}$ can be written as

$$\mathbf{a} = \mathbf{F}/m$$

TABLE 3.2 *Situations involving Newton's third law*

Action	Reaction	Comments
Your weight pressing down on a chair	The rigid chair pushing up on you, which supports your weight	If the seat of the chair rips or breaks, down you go!
The rearward frictional force of a car's tires on the road when the car accelerates	The forward frictional force of the road on the tires (and hence on the car), which causes the car to accelerate	If there is ice (i.e., no friction), the wheels spin, but no acceleration of the car results.
The forward force of the car seat on you, which causes you to accelerate with the car	The backward force you exert on the car seat, which causes you to "sink" into it	If it is a reclining seat that is not locked into place, you'll end up in a horizontal position as the car accelerates.
The force of a bat on a ball, which causes the ball to fly over the fence for a home run	The ball hitting the bat backward with an equal force	Sometimes this reaction force is large enough to break the bat.
The rearward force on the anchor when you throw it horizontally over the stern of a boat	The forward force the anchor exerts on you (and hence on the boat), which causes you and the boat to lunge forward	This is the principle by which jet engines and rockets operate. They are called "reaction engines."

The advantage of this form is that it shows how the *result* (acceleration) is determined by the *cause* (force). *When equal forces are applied to two objects, the objects' masses determine the result.* The larger mass will undergo a lesser acceleration than the smaller mass. Thus the truck's velocity undergoes a relatively small change during the collision, slowing down but continuing in the same direction. The much less massive car, *struck with the same force,* undergoes a much larger change of velocity. The car is not only stopped, but is shoved backward in a reversal of its direction of motion. This larger acceleration puts a much greater stress on the structural frame of the car, resulting in more damage to the car than to the truck and the appearance of having been hit harder.

3.6 MASS AND ITS RELATION TO WEIGHT

We have defined mass in terms of the 1-kg standard mass, and other masses are defined by comparison with this standard. Suppose a given force is applied first to the standard 1-kg mass and then to an unknown object. If the force gives the same acceleration to the two objects and if we assume the objects to be free of all other unbalanced forces, then the masses of the two objects are the same. This follows directly from $\mathbf{F}_{net} = m\mathbf{a}$, for if the two $\mathbf{F}$'s and the two $\mathbf{a}$'s are the same, then the masses must also be the same. Similarly, an object of mass n kilograms will acquire an acceleration of only $1/n$th that imparted to the standard kilogram by the same force. We see, then, that the unknown mass of an object can be determined by comparing its acceleration with that of the standard kilogram when both are subject to the same force.

In practice, however, we usually determine masses by "weighing" them on some sort of scales. If we use a balance beam scale, we are comparing the force of gravity on an unknown mass on one end of the beam to that on a known standard on the other end of the beam. If we use a spring scale, we are measuring the amount the

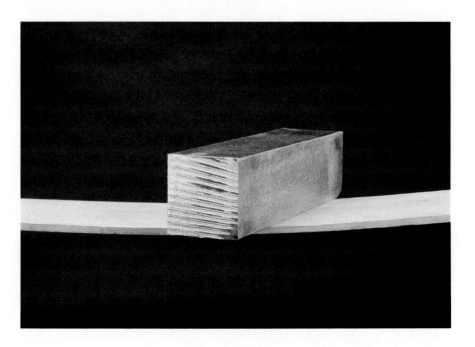

Heavy object on plank or beam, showing bending of the plank under the weight of the object.

spring must be stretched in order to exert an upward force on the mass equal to the force of gravity downward on the mass. Weight can thus be defined as follows:

The weight of an object is the force that gravity exerts on the object.

It is very important to realize that, although related, the mass and weight of an object are distinctly different physical properties. Weight is a *force,* whereas mass is one of the fundamental dimensions.

There is a simple experiment that tells the relation between mass and weight. When the only force acting on an object is its weight (the force of gravity on it), the object undergoes free-fall acceleration, g (Fig. 3.8). If we symbolize weight by W, Newton's second law for a freely falling body is

$$\mathbf{F}_{net} = W = mg \qquad (3.2)$$

Even when the object is sitting on a table or the floor, the force of gravity is unchanged. Thus Eq. 3.2 states that weight is *proportional* to mass.

The gravitational force on an object depends on where the object is located. Even at earth's surface, g varies slightly from equator to pole and from sea level to mountain altitudes. As we shall see in Chap. 7, gravity varies greatly from planet to planet. For instance, the moon's surface gravity is about one-sixth that of earth's. Thus *the weight of an object can vary, depending on the strength of the gravitational force at the location of the object.* The object's mass, on the other hand, is the same, regardless of gravitational conditions.

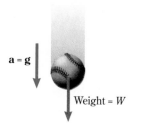

FIGURE 3.8

The unbalanced force on the object is W, giving it a free-fall acceleration g.

a = g

Weight = W

Illustration 3.1

What is the weight of a 5.25-kg mass? What is the mass of an object weighing 14.6 N? Assume the value of g is 9.80 m/s² in both cases.

Reasoning Since $W = mg$, the weight of a 5.25-kg mass is

$$W = (5.25 \, \text{kg})(9.80 \, \text{m/s}^2) = 51.5 \, \text{N}$$

Equation 3.2 can be rearranged to give $m = W/g$. Thus the mass corresponding to a 14.6-N weight is

$$m = \frac{14.6 \, \text{N}}{9.80 \, \text{m/s}^2} = 1.49 \, \text{kg} \quad \blacksquare$$

3.7 FRICTION FORCES

Before we illustrate the use of Newton's second law, let us discuss friction because friction forces play an important part in many applications of Newton's laws.

Try the experiment shown in Fig. 3.9. Push lightly against your textbook with a horizontal force. The book does not move. Because it remains stationary, we know that $\mathbf{F}_{net} = 0$. Thus at least one other force must be acting opposite the force you are applying. This opposing force must be supplied by the table where it contacts

FIGURE 3.9

The friction force f opposes the sliding of the book.

F

f

We know from everyday experience that friction forces between surfaces cause the surfaces to heat up—a fact often applied to start fires.

Pushing force

Book starts to slide

f_c

Book continues to slide

f_k

Time

FIGURE 3.10

The book in Fig. 3.9 begins to slide when the pushing force equals or exceeds f_c.

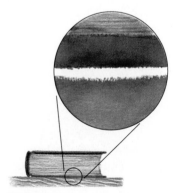

FIGURE 3.11

When they are magnified, the surfaces are seen to be rough.

the book, shown as the force **f** in the figure. We call this the force of static friction, which you can see has the following properties: *it opposes the attempt to make the object slide and is directed parallel to the surfaces in contact.*

Slowly increase the force with which you push on the book, as shown in the graph of Fig. 3.10. When the magnitude of the pushing force reaches a certain critical value f_c, the book suddenly begins to move. Then, to keep it moving at constant speed, a smaller force of magnitude f_k is sufficient. (The subscript k stands for *kinetic,* which means "moving.") We see from this simple experiment that two friction forces are important. The maximum (critical) static friction force f_c must be overcome to start the object moving, and then a smaller friction force f_k opposes the motion of the sliding object. Remember, f_c symbolizes the maximum value that static friction, f_s, can have. For any value of f_s up to this critical value, the static friction force prevents the start of sliding motion.

The major reasons for this behavior can be seen in Fig. 3.11: the surfaces in contact are far from smooth. Even highly polished surfaces look like this when observed at high magnification. The jagged points from one surface penetrate the valleys of the other surface, and this causes the surfaces to resist sliding. Once sliding has begun, however, the surfaces do not have time to "settle down" onto each other completely. As a result, less force is required to keep them moving than to start the motion.

As you might expect from this model of its origin, the friction force depends on how forcefully two surfaces are pushed together. We describe this feature of the situation by what is called the *normal force* $\mathbf{F}_N$ (*normal* meaning *perpendicular* in this context). An example of a normal force is the perpendicular force a supporting surface exerts on any surface resting on it. In Fig. 3.12a, the block pushes down on the supporting surface with a force equal to the block's weight. The supporting surface pushes back with an equal, opposite force, and so $F_N = W_1$ in this case. In Fig. 3.12b, the force pushing down on the supporting surface is the sum of the weights of the two blocks, and so the supporting force is $F_N = W_1 + W_2$ in this case.

Experiment shows that the magnitudes f_c and f_k are often directly proportional to F_N. In equation form,

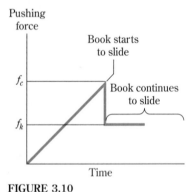

$$f_c = \mu_s F_N \qquad f_k = \mu_k F_N \tag{3.3}$$

where μ is the Greek letter mu. The factors μ_s and μ_k are called the *static* and *kinetic coefficients of friction,* respectively. They vary widely depending on what the surfaces are made of and how clean and dry they are. Typical values are given in Table 3.3.

Although friction forces depend significantly on the smoothness and cleanliness of the surfaces, the following approximate statements can be made: (1) at low speeds, f_k does not change much with speed as one surface slides over the other, and (2) for a given value of F_N, f_c and f_k are fairly insensitive to the size of the area of contact between the surfaces.

The direction of the friction force is always parallel to the surfaces, but the magnitude of the force is proportional to the magnitude of the force of compression perpendicular (normal) to the surfaces.

Another example of a normal force is pictured in Fig. 3.13. A block of wood is being pressed against a wall by a horizontal force **H**. The wall pushes back on the block with a normal force $\mathbf{F}_N$. You can easily verify that you can hold the block in place if you push hard enough horizontally. Newton's second law tells us the following in this case:

1 Since there is no horizontal acceleration, $F_N = H$.
2 If the block remains stationary, there must be enough static friction upwards to balance gravity downward. Thus $f_s = mg$.

In this case $f_s = \mu_s F_N = \mu_s H$. This is an example showing that a normal force need not always be vertical. It depends on the orientation of the surfaces.

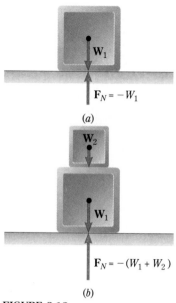

FIGURE 3.12

The normal force F_N is the perpendicular force exerted by the supporting surface on the object it supports.

An example of the low coefficient of friction between ice and plastic.

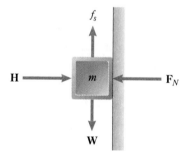

FIGURE 3.13

A horizontal force can provide enough friction to keep the block from falling.

TABLE 3.3 *Coefficients of friction*

Materials in contact	μ_s	μ_k
Rubber on dry concrete	~0.9	~0.7
Rubber on wet concrete	0.7	0.5
Wood on snow	0.08	0.06
Steel on Teflon	0.04	0.04
Steel on steel	0.75	0.57
Steel on ice	0.02	0.01
Wood on wood	0.7	0.4
Metal on metal (lubricated)	0.10	0.07
Glass on glass	0.9	0.4

Illustration 3.2

Referring to Fig. 3.13, what is the least amount of force H you must exert on the block to keep it in place? The block has a mass of 2.2 kg and the coefficient of static friction between the wall and the block is 0.65.

Reasoning The weight of the block is $W = mg = (2.2 \text{ kg})(9.8 \text{ m/s}^2) = 22 \text{ N}$. The static friction force must equal this in magnitude: $f_s = 22 \text{ N}$, where f_s can be any amount up to

$$f_s \leqslant f_c = \mu_s F_N = \mu_s H$$

Thus the applied force **H** must be

$$H \geqslant \frac{f_s}{\mu_s} = \frac{22 \text{ N}}{0.65} = 34 \text{ N}$$

The *least* force that will create enough friction to keep the block in place is 34 N.

3.8   APPLICATION OF NEWTON'S SECOND LAW

We now have the necessary background for applying Newton's second law to a variety of situations. Before we exhibit its use through examples, let us point out the general procedure you should follow.

1 Sketch a picture of the problem.
2 Isolate the object for which you wish to write $\mathbf{F} = m\mathbf{a}$.
3 Draw a free-body diagram for the isolated object, showing all forces acting on it. Do not include forces that do not act directly on the object.
4 Choose a convenient coordinate system for the free-body diagram and find the components of the forces.
5 Write the $\mathbf{F} = m\mathbf{a}$ equations in component form for the forces in the free-body diagram. When you do so, $\mathbf{F}$ should be in newtons, m in kilograms, and $\mathbf{a}$ in m/s^2. Remember that $m = W/g$.
6 Solve the component equations for the unknowns.
7 Check the reasonableness of the results.

Sometimes, when more than one object is moving, you may have to repeat steps 2 to 5 for objects other than the isolated one. Although, in the interest of brevity, we do not show each step in the following examples, this omission does not diminish their importance.

Example 3.3

A woman pushes on a box weighing 500 N with a force **F** directed at 30° below horizontal, as shown in Fig. 3.14a. **(a)** What must **F** be in order that she can make the box start to slide? **(b)** If she maintains this same force once the box starts to slide, what will its acceleration be? Assume the box and the floor are wood, and use the values for coefficients of friction given in Table 3.3.

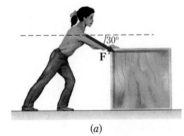

(a)

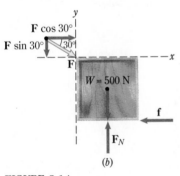

(b)

FIGURE 3.14

Notice that the normal force on the box equals 500 N + **F** sin 30°.

R

Reasoning   Part (a)

Question   Under what condition will the box start sliding?
Answer   When a horizontal force is applied that is equal to the critical force of static friction, f_c.

Question   What do I have to know to find f_c?
Answer $f_c = \mu_s F_N$. $\mu_s = 0.7$, as seen in Table 3.3.

Question   What principle determines F_N?
Answer   The vertical component of acceleration is zero, so Newton's second law requires that $\Sigma F_y = 0$. ($F_N - W - F \sin 30° = 0$) Notice that there are two forces downward, and F_N upward. This equation allows you to find F_N.

Question   What does the free-body diagram of the box look like?
Answer   It is shown in Fig. 3.14*b*. When the box just begins to slide, $f = f_c$.

Question   What is the condition on the horizontal forces in order to start the box sliding?
Answer $F \cos 30° \geqslant f_c = (0.7)F_N$

Solution and Discussion   We have two simultaneous equations in the unknowns F and F_N. We first solve for F_N in terms of F:

$$F_N = W + F \sin 30°$$

Notice that the floor has to support *more* than just the weight. According to Newton's third law, this same amount of force is bearing down on the floor.
   Substituting the expression for F_N into the horizontal force equation gives us

$$F \cos 30° = (0.7)(F \sin 30° + 500 \text{ N})$$

Collecting terms,

$$F[\cos 30° - 0.7(\sin 30°)] = 0.7(500 \text{ N})$$

$$F(0.866 - 0.35) = 350 \text{ N}$$

$$F = \frac{350 \text{ N}}{0.516} = 678 \text{ N}$$

We can now find F_N if we want to.

$$F_N = F \sin 30° + W = (678 \text{ N})(0.500) + 500 \text{ N} = 839 \text{ N}$$

Check the equality of the horizontal forces:

$$F \cos 30° = (678 \text{ N})(0.866) = 587 \text{ N}$$
$$f_c = \mu_s F_N = 0.7(839 \text{ N}) = 587 \text{ N}$$

R

Reasoning   Part (b)

Question   Why will the box accelerate?
Answer   Because once the box starts moving, friction decreases to $f_k = \mu_k F_N$. If the woman continues to exert the force found above, there will be a net force horizontally.

Question   Will F_N change?
Answer $F_N = F \sin 30° + W$. Nothing in this expression changes.

Question   What will the net horizontal force be?
Answer $587\,\text{N} - (0.4)(839\,\text{N}) = 587\,\text{N} - 336\,\text{N} = 251\,\text{N}$.

Question   What principle will determine the acceleration?
Answer   Newton's second law. $a = F_{\text{net}}/m$, where m is the mass of the box.

Question   What is the value of m?
Answer   Mass is related to weight by $W = mg$, or $m = W/g$. Here $m = (500\,\text{N})/(9.8\,\text{m/s}^2) = 51\,\text{kg}$.

Solution   and   Discussion   Putting   the   numbers   in   gives $a = (251\,\text{N})/(51\,\text{kg}) = 4.9\,\text{m/s}^2$.

Example 3.4

Two blocks having masses $m_1 = 1.0$ and $m_2 = 2.0\,\text{kg}$ are in contact with each other on a horizontal table, as shown in Fig. 3.15a. Friction between each block and the table is negligible. A horizontal force **F** is applied to m_1, causing the blocks to accelerate to the right with $\mathbf{a} = 3.0\,\text{m/s}^2$. (**a**) What is the magnitude of F? (**b**) What are the forces of compression between the blocks?

(a)

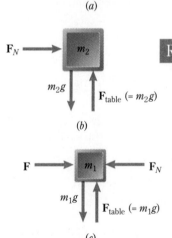

(b)

(c)

FIGURE 3.15

The individual free-body diagrams reveal the normal forces of compression between the two blocks.

R

Reasoning

Question   The blocks move together. Can I treat them as one object, with $M = 3.0\,\text{kg}$?
Answer   Yes, for part (*a*).

Question   What principle determines F?
Answer   Newton's second law: $\mathbf{F} = m\mathbf{a} = (3.0\,\text{kg})(3.0\,\text{m/s}^2) = 9.0\,\text{N}$

Question   The compressional forces aren't shown in Fig. 3.15a. How can I determine them?
Answer   These forces will appear if you isolate each block in its own free-body diagram. There must be some kind of horizontal normal force between the two blocks since they are being pushed together.

Question   What does the free-body diagram of m_2 look like?
Answer   It is shown in Fig. 3.15b. **F** is applied to m_1 only, so it does not appear in the diagram for m_2.

Question   What principle determines $\mathbf{F}_N$?
Answer $\mathbf{F}_N$ is a force of compression between the rigid blocks. It is the only horizontal force acting on m_2 and therefore is solely responsible for m_2's acceleration, according to Newton's second law.

Question   What is the equation giving $\mathbf{F}_N$?
Answer $\mathbf{F}_N = m_2\mathbf{a} = (2.0\,\text{kg})(3.0\,\text{m}/s^2) = 6.0\,\text{N}$.

Question   What determines the compressional force on m_1?
Answer   Newton's third law says that it must be equal to and opposite that acting on m_2.

Question   What is the free-body diagram for m_1?
Answer   It is shown in Fig. 3.15c.

Solution and Discussion   Notice that the net force acting on m_1 alone is $\mathbf{F} - \mathbf{F}_N$. (Why the minus sign)? Thus for m_1,

$$\mathbf{F} - \mathbf{F}_N = m_1\mathbf{a} = (1.0\,\text{kg})(3.0\,\text{m}/s^2) = 3.0\,\text{N}$$

This gives $\mathbf{F}_N = \mathbf{F} - 3.0\,\text{N} = 6.0\,\text{N}$, in agreement with the result for m_2.

Example 3.5

A 3300-lb car is traveling at 38 mi/h when its brakes are applied and it skids to rest. The skidding tires experience a friction force about 0.70 times the weight of the car. How far does the car go before stopping? Take the motion to be along the x axis.

R

Reasoning

Question   What quantity am I to determine?
Answer   The distance x the car travels while its speed is going from $+38$ mi/h to zero.

Question   What is causing the car to stop?
Answer   A constant friction force equal to 0.70 times the car's weight.

Question   What principle connects this force to the change in velocity?
Answer   Newton's second law: $\mathbf{a} = \dfrac{\mathbf{F}_{\text{net}}}{m}$. Friction is the only horizontal force, and so $\mathbf{F}_{\text{net}} = 0.70\,W_{\text{car}}$.

Question   To use Newton's second law, we need the car's *mass*. Do I know this value?
Answer   We can get it from $m = W/g$.

Question   I then have everything necessary to determine a from the second law. The time of stopping is not given. Is there a principle that directly connects the change in speed to the distance required?

Answer The equation of uniformly accelerated motion which directly relates x to change of speed is

$$v^2 = v_o^2 + 2ax$$

Once you have a from the second law, you can solve this for x.

Question There seems to be a confusing variety of units. Do I need to convert them?

Answer Yes. For consistency, we always use SI units in this book. You have to convert weight to newtons, after which mass will come out in kilograms. You also have to convert mi/h to m/s.

Solution and Discussion

1 The conversion of units gives us

$$W_{car} = (3300 \, \text{lb})(4.45 \, \text{N/1 lb}) = 1.5 \times 10^4 \, \text{N}$$
$$v_o = (38 \, \text{mi/h})(1.61 \, \text{km/mi})(1.00 \, \text{h/3600 s}) = 1.7 \times 10^{-2} \, \text{km/s}$$
$$= 17 \, \text{m/s}$$

2 The car's mass is

$$M_{car} = \frac{W_{car}}{g} = \frac{1.5 \times 10^4 \, \text{N}}{9.8 \, \text{m/s}^2} = 1.5 \times 10^3 \, \text{kg}$$

3 The friction force is

$$\mathbf{F}_{net} = -0.70(1.5 \times 10^4 \, \text{N}) = -1.0 \times 10^4 \, \text{N}$$

The negative sign is consistent with the direction of the friction force, in the $-x$ direction.

4 Acceleration is

$$\mathbf{a} = \frac{F_{net}}{m} = \frac{-1.0 \times 10^4 \, \text{N}}{1.5 \times 10^3 \, \text{kg}} = -6.9 \, \text{m/s}^2$$

5 Then the distance the car takes to stop is

$$x = \frac{v_f^2 - v_o^2}{2a} = \frac{0 - (17 \, \text{m/s})^2}{2(-6.9 \, \text{m/s}^2)} = 21 \, \text{m}$$

This example shows how two basic principles combine to determine the solution. Pay particular attention to how the words of the problem generate the equations through the use of these principles.

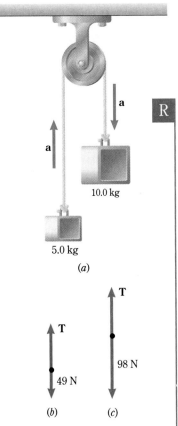

FIGURE 3.16

The blocks have the same magnitude of acceleration, in the directions shown.

Example 3.6

The two masses in Fig. 3.16 are tied to opposite ends of a massless rope, and the rope is hung over a massless and frictionless pulley.* Find the acceleration of the masses. (This device is called an *Atwood's machine*.)

Reasoning

Question Do the two masses have different accelerations?
Answer No. We assume that the rope does not stretch, and so the masses are constrained to move with the same acceleration.

Question What principle determines the acceleration?
Answer Newton's second law, applied to each mass individually.

Question What forces are acting on the masses?
Answer Their weights, mg, downward, and the tension force T in the rope, always directed away from an object in the direction of the attached rope.

Question What do the free-body diagrams look like?
Answer As shown in Figs. 3.16*b* and *c*. Notice there is no reference to the pulley, which simply serves to support the rope.

Question As the system moves, one mass is going to go up while the other comes down. How will I choose a positive direction for vectors?
Answer The second law applies to each mass individually. You can choose the positive direction of motion for each mass to be the direction in which it moves. Because it is larger, the 10-kg mass will move downward.

Question What equations does the second law give in these cases?
Answer $98\,\text{N} - T = (10\,\text{kg})a$
$T - 49\,\text{N} = (5\,\text{kg})a$

Notice there are two unknowns, a and T, and two equations to be solved simultaneously.

Solution and Discussion If we add these two equations together, T will be eliminated, leaving one equation to be solved for a:

$$98\,\text{N} - \cancel{T} + \cancel{T} - 49\,\text{N} = (10\,\text{kg})a + (5\,\text{kg})a = (15\,\text{kg})a$$

Then

$$a = \frac{49\,\text{N}}{15\,\text{kg}} = 3.3\,\text{m/s}^2$$

If we want to, we can put this value of a into either of the above equations and find the tension in the rope:

$$T = (5\,\text{kg})(3.3\,\text{m/s}^2) + 49\,\text{N} = 65\,\text{N}$$

*We specify that the rope and pulley be massless so that we can neglect their inertias. Because the pulley is both massless and frictionless, the tension in the rope is the same on both sides of the pulley.

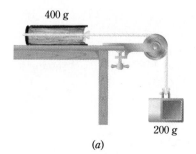

400 g

200 g

(a)

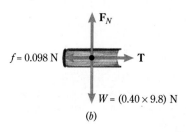

F_N

$f = 0.098$ N

T

$W = (0.40 \times 9.8)$ N

(b)

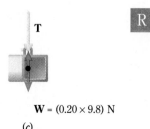

T

R

$W = (0.20 \times 9.8)$ N

(c)

FIGURE 3.17

Although the friction force retards the motion of the book, the weight of the 200-g mass is large enough to cause both objects to move. The weight of the book is balanced by the push of the table.

Exercise   What is the general expression for the acceleration of this system if you call the larger mass m_1 and the smaller mass m_2?

$$\textit{Answer:} \quad a = \left(\frac{m_1 - m_2}{m_1 + m_2}\right)g.$$

Example 3.7

In Fig. 3.17a, a string is attached to a 400-g book on a table, then passes over a massless, frictionless pulley and is attached to a hanging 200-g mass. The coefficients of friction are $\mu_s = 0.4$ and $\mu_k = 0.2$. The free-body diagrams for the book and the mass on the string are shown in Figs. 3.17b and 3.17c. **(a)** If the system is let go from rest, will it begin to move? **(b)** If it moves, what is the acceleration of the book?

Reasoning   Part (a)

Question   What is implied by the pulley being massless and frictionless?
Answer   No force is required to set the pulley into rotation. Its sole purpose is to change the direction of whatever tension force is in the string.

Question   What is necessary to start the book moving?
Answer   The tension force exerted by the string on the book must be at least equal to the critical force of static friction, f_c.

Question   What determines the amount of tension in the string?
Answer   It will be determined by applying the second law to both the book and the 200-g mass. Notice that the string constrains them to move together, and hence whenever in motion, they must have the same magnitude of acceleration.

Question   Is the tension force the same at the two ends of the string?
Answer   As long as the pulley is massless and frictionless, yes. Then the two tension forces are seen to be equal by applying Newton's third law. In later chapters we shall deal with "real" pulleys.

Question   What equations do I get from the second law applied to the book?
Answer   Vertically, $F_N = W = (0.400\,\text{kg})(9.80\,\text{m/s}^2) = 3.92\,\text{N}$.   Horizontally, $T - f = (0.400\,\text{kg})a$.

Question   What equation does the second law give me for the hanging mass?
Answer $(0.200\,\text{kg})(9.80\,\text{m/s}^2) - T = (0.200\,\text{kg})a$. Notice that in both this equation and the one in the previous answer, we assumed positive vectors in the direction that each object would move.

Question   In the static case, what would the tension be?
Answer   In that case, $a = 0$, so from the previous answer $T = 1.96\,\text{N}$.

Question   What is the critical friction force?
Answer $f_c = \mu_s F_N = (0.40)(3.92\,\text{N}) = 1.6\,\text{N}$.

Solution and Discussion Notice that this force alone cannot hold the book against the 1.92-N tension force. The crucial question you asked was the "what if this were a static case?" question. The answer turns out to be physically impossible, which tells you the book will slide unless another force is supplied to help friction.

R

Reasoning Part (b)

Question What changes as a result of the book moving?
Answer The friction force will be *kinetic* friction, $f = \mu_k F_N = (0.20)(3.92\,\text{N}) = 0.78\,\text{N}$, which is less than f_c. Also, since a is no longer zero, T will *not* be equal to the hanging weight. We have already seen the two second-law equations that involve a and T.

Solution and Discussion The two equations involving a and T are

$$T - 0.78\,\text{N} = (0.400\,\text{kg})a \qquad \text{and} \qquad 1.96\,\text{N} - T = (0.200\,\text{kg})a$$

By adding these equations, T is eliminated:

$$1.96\,\text{N} - 0.78\,\text{N} = 1.18\,\text{N} = (0.600\,\text{kg})a$$

This gives $a = 1.96\,\text{m/s}^2$. Put this value into either equation to find T:

$$T - 0.78\,\text{N} = (0.400\,\text{kg})(1.96\,\text{m/s}^2) = 0.784\,\text{N}$$
$$T = 0.78\,\text{N} + 0.784\,\text{N} = 1.56\,\text{N}$$

Verify that the correct number of significant digits has been kept in these calculations.

Example 3.8

A police officer investigating a highway accident notices that one car made skid marks 20.0 m long on the dry, level concrete pavement. She assumed the driver had locked the brakes at the start of the skid, which caused the car to be uniformly decelerated to a stop. The accident happened in a zone where the speed limit was 50 km/h. Could the officer file speeding charges against the driver?

R

Reasoning

Question What principle connects the data given to the driver's speed before applying the brakes?
Answer The equation of motion involving v_o, v_f, a, and x: $v_f^2 = v_o^2 + 2ax$, with $v_f = 0$ and $x = $ distance of skid.

Question How many unknowns are there?
Answer Two: a and v_o.

Question What other principle can I apply that involves at least one of these?

Answer   Newton's second law of motion. The acceleration is caused by a force of sliding friction between the tires and the road.

Question   What equation does this information give me?
Answer $m\mathbf{a} = \mathbf{F}_{net} = \mathbf{f}$, where the magnitude $f = \mu_k F_N = \mu_k W_{car}$, and $m =$ mass of the car.

Question   Do I need to find the weight and mass of the car?
Answer   Since $W_{car} = mg$, m appears on both sides of the second-law equation, and thus cancels out. You don't need to know what it is.

Question   What is the coefficient of friction?
Answer   Table 3.3 shows that for rubber on dry concrete, $\mu_k = 0.7$.

Solution and Discussion   The two equations we've developed for this case are:

$$v_o^2 + 2ax = 0 \quad \text{and} \quad \cancel{m}a = -\mu_k \cancel{m}g \text{ or } a = -\mu_k g$$

The direction of the acceleration is along $-x$, and so the correct sign is introduced when we change from vector notation to magnitude notation. The second equation gives

$$a = -(0.7)(9.8 \text{ m/s}^2) = -7 \text{ m/s}^2$$

Then we find that

$$v_o = [2(7 \text{ m/s}^2)(20.0 \text{ m})]^{1/2} = 17 \text{ m/s}$$

Converting this to km/h:

$$v_o = (17 \cancel{m}/\cancel{s})(3600 \cancel{s}/\text{h})(1 \text{ km}/1000 \cancel{m}) = 61 \text{ km/h}$$

The driver was exceeding the speed limit at the time the brakes were applied.

3.9   WEIGHT AND WEIGHTLESSNESS

A fascinating physical phenomenon called weightlessness is sometimes observed when objects are accelerating. Although we postpone discussing weightlessness in orbiting spaceships until we have studied motion in a circle, we can discuss other examples of weightlessness here. A great deal of insight into this phenomenon can be obtained by considering an object suspended from the ceiling of an elevator, as in Fig. 3.18. The reading on the spring scale is the quantity commonly called the object's weight. However, since we have defined weight as the force of gravity on an object, we call the spring-scale reading the *apparent weight* of the object.

   The free-body diagram in Fig. 3.18b shows the forces acting on a pail. There are only two: the pull of gravity (the weight of the pail) W and the upward tension force, call it **T**, with which the hook pulls on the pail.* The upward pull of the hook is equal to the reading on the scale and therefore to the pail's apparent weight.

*We neglect the small weight of the hook.

FIGURE 3.18

The spring scale reads the force T with which the hook pulls on the pail, and this is equal to the apparent weight of the object.

Case 1: Elevator at rest
Since $\mathbf{a}_y = 0$ in this case, $\Sigma \mathbf{F}_y = m\mathbf{a}_y$ becomes

$$\Sigma \mathbf{F}_y = 0 \qquad \text{or} \qquad T - W = 0$$

So $T = W$ in this case, the scale reads W, and the apparent weight of the pail equals the gravitational force on it.

Case 2: Elevator with constant velocity
Because the velocity is constant, the acceleration is zero. Thus the analysis used in case 1 applies here also, and the scale reading is W. The apparent weight equals the actual weight.

Case 3: Elevator accelerating upward
Let us call the acceleration $\mathbf{a}_y$. If up is taken as the positive direction, $\Sigma \mathbf{F}_y = m\mathbf{a}_y$ becomes

$$T - W = ma_y$$

from which

$$\text{Apparent weight} = T = W + ma_y$$

Here the apparent weight of the pail is more than the rest value. The hook must not only balance the gravitational force but also provide an unbalanced force $T - W$ upward in order to cause the upward acceleration. (Notice how important it is to define a positive direction for the forces and acceleration.)

Case 4: Elevator accelerating downward
If up is still taken as the positive direction, the acceleration now becomes negative. We have from $\Sigma \mathbf{F}_y = m\mathbf{a}_y$,

$$T - W = -ma_y$$

from which

$$\text{Apparent weight} = T = W - ma_y$$

Clearly, the apparent weight of the pail is less than the pull of gravity on it. ■

An interesting case occurs when an object is falling freely, so that $a_y = g$, the acceleration due to gravity. Since $W = mg$, we find

$$T = mg - mg = 0$$

and the pail appears to be *weightless*. In our elevator example, when the pail is in free-fall, so is the scale, and the hook attached to the pail cannot exert an upward force to hold the pail in place. The reading on the scale goes to zero, and the pail appears to be weightless. The same result would be observed if we were using a balance beam scale to measure the pail's weight. In a free-fall environment, both sides of the balance (and the pail on the balance) would have the same acceleration, g, and no weights would be required to balance the beam.

Although this situation is hypothetical, it does show that the apparent weight of an object depends critically on the object's acceleration. In fact, the generalization of the weightless condition during free-fall is the following:

An object is weightless (zero apparent weight) whenever the force of gravity is the only force acting on the object.

We shall see in Chap. 7 that this condition applies to satellites and all their contents in gravitational orbits around the earth (or other planets as well). Thus even though we define weight as the force of gravity on an object, we must remember that the weight we measure, which we call the apparent weight, *differs from this if the object being weighed is accelerating.* In the majority of cases we experience, this acceleration is zero.

3.10 MOTION ON AN INCLINE

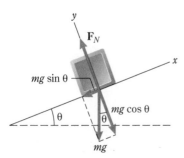

FIGURE 3.19

In dealing with an object on an incline, it is convenient to take the *x* and *y* axes parallel and perpendicular to the incline, as shown. The forces are then split into components along these axes.

An important type of one-dimensional motion is motion on an incline, or ramp. The ramp in Fig. 3.19 makes an angle θ relative to the horizontal. The weight mg of an object on the ramp is still vertically downward, and the normal force exerted by the ramp on the object is (by definition) perpendicular to the ramp. Motion is constrained to take place *along the ramp,* and so the convenient choice of coordinate axes is x along the ramp and y perpendicular to it.

To proceed as we have before, all forces in the object's free-body diagram must be resolved into components along these axes. Notice in the figure that the weight, mg, is resolved into an x component $mg \sin \theta$ down the ramp ($-x$ direction) and a y component $mg \cos \theta$ along the $-y$ direction. Can you see why the angle θ is where it is in the vector diagram? F_N is entirely in the $+y$ direction. If friction exists, it must be entirely in the x direction, positive or negative depending on which direction is opposite the direction of the object's motion (or, in the stationary case, *tendency* to move).

Let us summarize the conditions that govern the two axes:

1 Since no motion is allowed perpendicular to the ramp, Newton's first law requires that the forces in the y direction must add up to zero.
2 The motion is entirely along the x direction and is governed by Newton's second law.

Motion on an incline.

Illustration 3.3

Suppose friction is negligible and no other forces are applied to the object shown in Fig. 3.19. (*a*) Calculate how long it would take the object to slide down a 40° ramp 1 m in length. (*b*) Calculate how fast the object would be going at the bottom.

Reasoning To derive the pertinent equations, we apply the two conditions summarized above. Perpendicular to the ramp, we must have $F_N = mg \cos \theta$ (*not* $= mg$ anymore!). The object will have a net force $mg \sin \theta$ down the ramp, which will give an acceleration in that direction of

$$a = \frac{F_{\text{net}}}{m} = \frac{mg \sin \theta}{m} = g \sin \theta$$

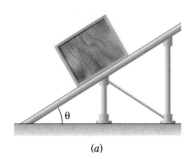

(a)

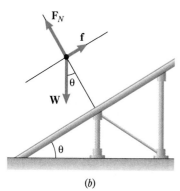

(b)

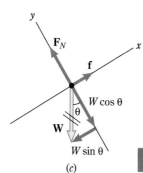

(c)

FIGURE 3.20

This acceleration can be used in the same equations of one-dimensional motion that we have used before:

$$v_f^2 = v_o^2 + 2ax = 0 + 2(g \sin \theta)x$$

where the + direction is conveniently chosen down the ramp in the direction of motion. Also, with $v_o = 0$,

$$x = 0 + \tfrac{1}{2}at^2 = \tfrac{1}{2}(g \sin \theta)t^2$$

and

$$v_f = 0 + at = (g \sin \theta)t$$

(a) $x = \tfrac{1}{2}(g \sin \theta)t^2$ gives $1 = \tfrac{1}{2}(6.3)t^2$ from which $t = 0.56$ s

(b) $v_f = [2(9.8 \text{ m/s}^2)(\sin 40°)(1 \text{ m})]^{1/2} = 3.55 \text{ m/s}$

Example 3.9

A box of mass m is placed on an incline as shown in Fig. 3.20a. **(a)** Find the general expression, in terms of m, θ, and μ_s, for the largest angle θ for which the box will remain stationary. **(b)** For angles larger than this, find the general expression for the acceleration of the box down the incline.

Reasoning Part (a)

Question What does the free-body diagram of the box look like?
Answer It is shown in Fig. 3.20b. Remember that friction always acts parallel to the surfaces in contact, and opposite the motion. Thus in this case friction is directed *up* the ramp.

Question For the box to stay in place, what condition must be satisfied?
Answer The net force on it must be zero. This means the x and y components of the net force must each be zero.

Question What equations does this condition give me?
Answer Along the ramp: $mg \sin \theta = f$
Normal to the ramp: $F_N = mg \cos \theta$

Question What determines the force of static friction, f?
Answer f is determined by the force of compression F_N between the surfaces in contact. f can have any value necessary to balance $mg \sin \theta$ up to a maximum of $\mu_k F_N$.

Solution and Discussion At the maximum angle, we have that $f_c = \mu_s F_N$ must just equal the component of the weight directed down the plane $mg \sin \theta_c$:

$$\mu_s F_N = \mu_s \, mg \, \cos \theta_c = mg \sin \theta_c$$

Dividing both sides of the equation by $mg \cos \theta_c$, we have

$$\frac{\sin \theta_c}{\cos \theta_c} = \tan \theta_c = \mu_s$$

The maximum angle, called the **angle of repose,** is then

$$\theta_c = \tan^{-1} \mu_s$$

R

Reasoning *Part (b)*

Question If θ_c is exceeded, which physical property changes?
Answer Static friction changes to kinetic friction. Thus

$$f = \mu_k mg \, \cos \theta \qquad \text{for } \theta > \theta_c$$

Question What is the net force along the ramp when the box is sliding?
Answer It depends on which direction you choose to have a positive sign. Choosing positive to be *down* the ramp, you have

$$F_{\text{net}} = W \sin \theta - f = mg \sin \theta - \mu_k mg \, \cos \theta$$

Question What principle allows me to calculate acceleration from this information?
Answer Newton's second law:

$$mg \sin \theta - \mu_k mg \, \cos \theta = ma$$

All these quantities are directed along the ramp.

Solution and Discussion Solving for a gives

$$a = g(\sin \theta - \mu_k \cos \theta)$$

Notice that the mass has canceled out. This means that all masses will have the same acceleration as long as the coefficients of friction are the same. Let us inspect this general result for some interesting special cases:

1 *No friction.* In this case, $\mu_k = 0$ and $a = mg \sin \theta$, which is the same result we obtained in Illustration 3.1.
2 *Vertical ramp.* If $\theta = 90°$, $\sin \theta = 1$ and $\cos \theta = 0$. Then $a = g$, giving us the case of free-fall, as we should expect.

It is always instructive to examine *limiting* cases of the general algebraic solution.

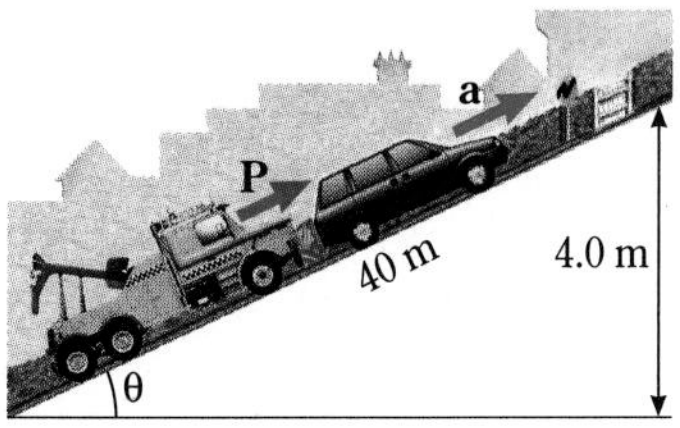

(a)

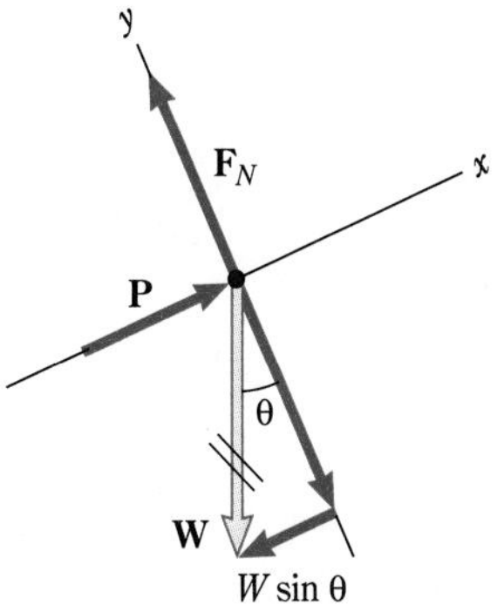

(b)

FIGURE 3.21

The force **P** is partly balanced by the component of the weight acting down the incline. Acceleration up the incline results from the unbalanced portion of **P**.

Example 3.10

The 1200-kg automobile in Fig. 3.21*a* is to accelerate at 0.50 m/s² up a hill that rises 4.0 m in each 40 m. How large a force must push on the car to cause this acceleration? Ignore friction.

R

Reasoning

Question   How is this situation different from the situation in previous examples?
Answer   This time there is an applied force, P (for *push*), *up* an incline.

Question   In order to find the components of the car's weight, I need to know the angle of the hill. What is the relation between the distances given in the figure and this angle?
Answer   From the definition of the sine,

$$\sin \theta = \frac{4.0 \text{ m}}{40 \text{ m}} = 0.10$$

Therefore $\theta = \sin^{-1} 0.10 = 5.7°$. The free-body diagram for the car is shown in Fig. 3.21*b*.

Question   What principle connects the push *P* with the acceleration?
Answer   Newton's second law applied to the axis along the hill.

Question   What equation does this principle give me?
Answer   Choosing the positive direction to be up the hill in the direction of *a*, you have

$$P - mg \sin \theta = ma$$

(You're told that you can neglect friction.)

Solution and Discussion   Solving for *P*, we have

$$P = ma + mg \sin \theta$$
$$= (1200 \text{ kg})(0.50 \text{ m/s}^2) + (1200 \text{ kg})(9.8 \text{ m/s}^2)(0.10)$$
$$= 600 \text{ N} + 1200 \text{ N} = 1800 \text{ N}$$

Notice that we did not need to find the value of θ. All we used was $\sin \theta$, given directly by the ratio of sides in Fig. 3.21. If we wanted to find the force F_N, we would need θ so that we could calculate $\cos \theta$. The two terms in the solution show what the push *P* must accomplish. It takes 1200 N just to counteract the component of the car's weight down the hill and another 600 N to produce the required acceleration.

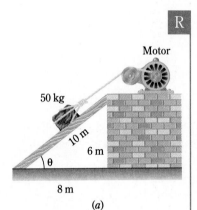

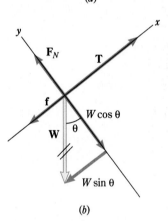

FIGURE 3.22
Since the block is moving up the incline with constant speed, the pull resulting from the motor must exactly balance the sum of the friction force and the component of the weight down the incline.

Example 3.11

A motor is to pull a 50-kg block up the incline in Fig. 3.22. The coefficient of friction between block and incline is 0.70. What is the tension in the rope if the block is moving at constant speed?

Reasoning

Question   What does the condition of constant speed imply?
Answer   It is one way of saying that acceleration is zero.

Question   What principle then applies to the problem?
Answer   Newton's first law: $F_{\text{net}} = 0$, both along the incline and perpendicular to it. The first law treats zero velocity as simply one example of the more general condition of constant velocity.

Question   What equations does the first law give in this case?
Answer   Figure 3.21b shows the free-body diagram for the block. From it you obtain

$$T - W\sin\theta - f = 0 \qquad \text{(along the incline)}$$
$$F_N - W\cos\theta = 0 \qquad \text{(perpendicular to the incline)}$$

Question   Can I determine the force f from the given data?
Answer   Since the block is sliding,

$$f = \mu_k F_N = \mu_k mg \cos\theta$$

Solution and Discussion   From the equation along the incline,

$$T = (0.70)(50\,\text{kg})(9.8\,\text{m/s}^2)(8/10) + (50\,\text{kg})(9.8\,\text{m/s}^2)(6/10)$$

Can you see why the fractions represent the cosine and sine of the angle? The final answer is

$$T = 270\,\text{N} + 290\,\text{N} = 560\,\text{N}$$

Example 3.12

A system of blocks is placed on a 37° incline as shown in Fig. 3.23a. Assume that the coefficients of friction between the incline and the 5-kg block are $\mu_s = 0.70$ and $\mu_k = 0.50$. (*a*) Show that the system will start to slide once it is let go. (*b*) What acceleration will the blocks have?

Reasoning   Part (a)

Question   How can I show that the system will start to slide?

Answer Assume that it will stick, and show that the numbers produced for that case are impossible or inconsistent.

Question What are the free-body diagrams for the two blocks?
Answer They are shown in Figs. 3.23*b* and *c*. When you assume a static case, *f* is the force of *static* friction.

Question What equations apply to the static case?
Answer For the 7-kg block:

$$T - (7.0\,\text{kg})(9.8\,\text{m/s}^2) = 0$$

immediately giving $T = 69\,\text{N}$. For the 5-kg block:

$$T - f - (5.0\,\text{kg})(9.8\,\text{m/s}^2)(\sin 37°) = 0$$

With $T = 69\,\text{N}$, this becomes

$$f = 69\,\text{N} - 29\,\text{N} = 40\,\text{N}$$

Question How can I tell whether or not a force of friction this large is possible?
Answer *f* has a maximum possible value, f_c, given by

$$f_c = \mu_s F_N = \mu_s mg\,\cos 37°$$

Solution and Discussion From the last equation, $f_c = (0.70)(49\,\text{N})(0.80) = 27\,\text{N}$, yet the condition for sticking would require that $f_c = 40\,\text{N}$. Therefore we must conclude that sticking is not possible in this situation, and the system will slide.

Reasoning Part (*b*)

Question Once the blocks start sliding, what changes from the previous assumption?
Answer *f* is now given by $f = \mu_k mg\,\cos 37°$, and *T* is no longer equal to the weight of the 7-kg block.

Question What principle now applies?
Answer Newton's second law, applied individually to each block.

Question What equations result?
Answer For the 7-kg block:

$$69\,\text{N} - T = (7.0\,\text{kg})a$$

For the 5-kg block:

$$T - (49\,\text{N})(0.60) - (0.50)(49\,\text{N})(0.80) = (5.0\,\text{kg})a$$

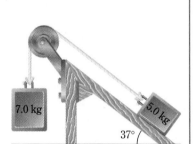

(*a*)

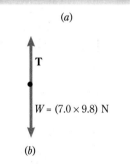

$$T$$
$$W = (7.0 \times 9.8)\,\text{N}$$

(*b*)

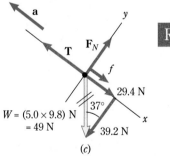

(*c*)

FIGURE 3.23

(*a*) The 7-kg block falls downward, pulling the 5-kg block up the incline. (*b*) The free-body diagram for the 7-kg block. (*c*) The free-body diagram for the 5-kg block.

Solution and Discussion   Notice again that the two blocks have the same acceleration. As in previous examples, adding the two equations will eliminate T and allow us to find a:

$$69\,\text{N} - (49\,\text{N})(0.60) - (0.50)(49\,\text{N})(0.80) = (7.0\,\text{kg} + 5.0\,\text{kg})a$$

This gives

$$a = \frac{20\,\text{N}}{12\,\text{kg}} = 1.7\,\text{m/s}^2$$

You should be able to show that putting this value into either of the second-law equations gives the tension: $T = 57\,\text{N}$.

3.11   A MODERN PERSPECTIVE: MASS AT HIGH SPEEDS

Physics as it was known at the close of the nineteenth century is referred to as *classical* physics. Many people felt at that time that all the basic principles necessary to describe physical phenomena had been discovered. As the twentieth century began, however, physicists began to experiment on the atom, discovering the extremely small particles that make up its structure. The principles of the nineteenth century were not able to explain how these particles behaved. Also, Einstein put forth his theory of relativity, which modifies Newton's laws when the velocities of objects approach the speed of light. As the horizons of experiment were extended to smaller and faster phenomena, revolutionary modifications of classical physics were required to explain the results. These new developments are called *modern* physics, even though they started almost a century ago.

The main subject of this course is classical physics, which is still a valid and

The two-mile long Stanford Linear Accelerator. Electrons in this accelerator achieve speeds that approach, but cannot be made to exceed, the speed of light. The behavior of high speed particles in particle accelerators such as this is in agreement with Einstein's relativity theory.

powerful description of the world in many practical ways. It is also necessary to understand classical principles before you can fully understand the modern modifications. However, throughout this text, we present some of the modern perspectives where they are relevant to classical topics, without pretending to be complete or rigorous. These modern topics are then treated more thoroughly in the final chapters.

As our first side trip into the world of modern physics, then, let's look at how the mass of an object behaves at extremely high speeds.

The definition of mass and its use in Newton's second law of motion seems to imply that mass is an inherent and constant property of an object. We have observed that the *weight* of an object may change, depending on the object's acceleration or on changes in the force of gravity acting on it, but we have assumed that mass always remains constant. Indeed, to Newton and other classical physicists, mass was a measure of the amount of matter an object contained and therefore constant by definition.

With this view of mass, combined with $v = at$, Newton's second law predicts that an object's velocity increases without limit as long as a net force continues to provide acceleration to the object:

$$v = at = \frac{F}{m} t \tag{3.4}$$

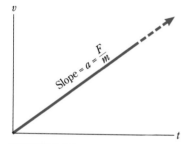

FIGURE 3.24

The graph of v versus t as given by Newton's second law with a constant force. The constant slope represents a constant increase in v without limit as long as the force F is applied.

A graph of v versus t shows a straight-line increase in v as long as F continues to be applied, as shown in Fig. 3.24. At the beginning of the twentieth century, Albert Einstein developed the theory of relativity, which seemed to contradict some fundamental ideas of classical physics. One of these contradictions was his prediction that no object could be accelerated to speeds greater than the speed of light (given the symbol c). Great as the speed of light is ($c = 300{,}000$ km/s, or a little more than 186,000 mi/s), there is nothing in Newton's laws that places any such limit on the speed of objects. Yet experiments have borne out the correctness of Einstein's prediction. Electrons, for example, have been accelerated in one experiment by strong forces applied for sufficient times to have given them speeds much greater than the speed of light if Newton's laws were right. Yet their measured speeds showed them to be traveling at "merely" $0.99999992c$.

To make sense out of this seeming confusion, we must look at Einstein's view of mass. The theory of relativity predicts that the mass of an object increases as its speed increases! The mathematical statement is

$$m = \frac{m_o}{\sqrt{1 - v^2/c^2}} \tag{3.5}$$

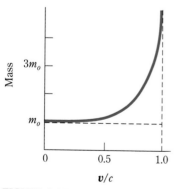

FIGURE 3.25

The mass of a moving object approaches infinity as the object's speed approaches the speed of light.

where m_o is called the **rest mass** and is equivalent to the "regular" mass we have been using throughout this chapter. We can plot a graph of m versus v/c to get a picture of what Eq. 3.5 means (Fig. 3.25). As you can see by examining the equation, as long as v is much smaller than c, so that $v^2/c^2 \ll 1$, then the square-root term is practically equal to 1, meaning that $m \approx m_o$. This condition corresponds to the essentially horizontal part of the graph in Fig. 3.25, running from $v/c = 0$ to about $v/c = 0.4$. Only when v approaches c does m begin to differ significantly from m_o. Notice that, as v approaches c (in other words, as v/c approaches 1), the denominator in Eq. 3.5 becomes zero, which means that the mass becomes *infinitely* large.

Does this mean that somehow the object is getting bigger or is accumulating more material? Not at all. To understand what is going on, we must go back to the fundamental concept of mass as a measure of the inertia of an object: the object's "resistance" to changes in velocity when forces are applied. Relativity tells us that when an object's speed approaches the speed of light, more and more force is required to change its velocity, i.e., it has more inertia.

Should we then conclude that Newton was wrong? Before you answer this question, remember that in almost all of our (and Newton's) practical experience, the speeds we observe are *very* slow compared with c. Newton's laws work extremely well in all these cases. Einstein's equation is entirely consistent with Newton's laws at "low" speeds. The beauty of Einstein's equation comes in that it shows explicitly how Newton's laws must be modified when the speeds are in a range beyond our everyday experience.

We can see precisely how Newton's second law is modified when the applied force is constant. It too predicts c as a limit on speeds, which we can see by substituting into Eq. 3.4 the expression for m given by Eq. 3.5:

$$v = \frac{F}{m}t = \frac{Ft}{m_o/\sqrt{1 - v^2/c^2}} = \frac{Ft}{m_o}\sqrt{1 - \frac{v^2}{c^2}}$$

Notice that the equation now involves v on both sides. We must therefore rearrange terms in order to solve for v. First square both sides to remove the square-root sign:

$$v^2 = \left(\frac{Ft}{m_o}\right)^2\left(1 - \frac{v^2}{c^2}\right) = \left(\frac{Ft}{m_o}\right)^2 - \left(\frac{Ft}{m_oc}\right)^2 v^2$$

Now collect terms in v^2 and factor v^2 out of them:

$$v^2\left[1 + \left(\frac{Ft}{m_oc}\right)^2\right] = \left(\frac{Ft}{m_o}\right)^2$$

Finally, take the square root of the result.

$$v = \frac{Ft/m_o}{\sqrt{1 + (Ft/m_oc)^2}} \qquad (3.6)$$

How speed depends on the duration of the force F is now more complicated than before, containing t in the denominator as well as in the numerator. Figure 3.26 is a graph of v versus t as given by Eq. 3.6. At low speeds, v has a *linear* behavior, with a slope = F/m_o, which is exactly the acceleration in Newton's second law. Notice that as time gets arbitrarily large, so that the 1 in the square root term can be neglected, we have the *limiting value* for v:

$$v \text{ (as } t \to \infty) = \frac{Ft/m_o}{\sqrt{(Ft/m_oc)^2}} = c$$

Einstein's theory of relativity also predicts that measurements of the other two basic quantities of mechanics, length and time, change at very high speeds. The surprising predictions of relativity are discussed more fully in Chap. 26.

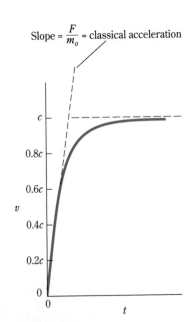

FIGURE 3.26

The relativistic behavior of v as a function of t under application of a constant force. Notice that the initial slope is the "classical" acceleration, F/m_o. As v approaches c, the slope decreases, indicating an increase in mass.

LEARNING GOALS

Now that you have finished this chapter, you should be able to

1 Define (*a*) inertia, (*b*) mass, (*c*) net force, (*d*) newton, (*e*) normal force, (*f*) friction force, (*g*) coefficient of friction.
2 State Newton's first law and give several examples to illustrate it.
3 State Newton's second law in words and in equation form. Identify what is meant by F_{net}, *m*, and *a*. Explain why it is so important to isolate an object when applying this law.
4 State Newton's third law and point out the action-reaction pair of forces in simple situations.
5 Identify the forces acting on an object in simple situations and draw the free-body diagram for the object. The situations should include those involving forces of friction, compression, and tension.
6 Find the normal force exerted on an object by a rigid surface in contact with the object.

7 Combine Newton's second law with the equations of uniformly accelerated motion to determine the motion of objects when constant forces act on them.
8 Identify the force of friction (including direction) acting on an object in various situations, given the coefficients of friction between the object and a surface.
9 Give the relation between the mass and weight of an object. State the condition under which the apparent weight is equal to the force of gravity on an object. Give the condition under which an object is weightless.
10 Resolve the forces acting on an object supported by an incline into components parallel and perpendicular to the incline. Apply Newton's second law to an object on incline in terms of these force components.

SUMMARY

DERIVED UNITS AND PHYSICAL CONSTANTS

Force
1 newton (N) = 1 kg · m/s^2

DEFINITIONS AND BASIC PRINCIPLES

Mass
The mass of an object is a measure of the object's inertia, or resistance to change in the state of motion of the object. It is one of the basic physical dimensions, defined by the international standard kilogram.

Force
Force is a physical interaction which, if applied alone to an object, will produce an acceleration of that object.

One newton of net force will give a 1-kilogram mass an acceleration of 1 m/s^2.

Newton's Laws of Motion
FIRST LAW: If the vector sum of the external forces acting on an object is zero, the velocity of the object will remain constant. This is also known as the principle of inertia.

SECOND LAW: A net force applied to an object will produce an acceleration proportional to and in the direction of the net

force. The proportionality constant is the reciprocal of the object's mass:

$$\mathbf{F}_{net} = m\mathbf{a} \quad \text{or} \quad \mathbf{a} = \frac{\mathbf{F}_{net}}{m}$$

THIRD LAW: If object *A* exerts a force **F** on object *B*, then *B* exerts an equal and opposite force, −**F**, on *A*.

INSIGHTS
1. The property of an object tending to maintain the object's state of motion is referred to as the object's inertia. The more inertia, the stronger is this tendency in the presence of a force. The quantitative measure of inertia is the object's mass.
2. Mass is a fundamentally defined dimension of physics, measured in SI units of kilograms. It is a scalar quantity.
3. The second law implies that if the net force on an object at rest is zero, the object will remain at rest, since this is the special case of $v = 0$.
4. Changing an object's state of motion requires a net external force. An object cannot change speed or direction by internal forces.
5. The second law is a vector equation and can be applied separately to each of the rectangular components of motion.
6. Examples of uniformly accelerated motion in one dimension from Chap. 2 can now be interpreted as resulting from the application of a constant net force along

the direction of motion, since **a** is constant and the objects don't change direction.

7. In the third law, the two equal and opposite forces are not applied to the same object. Each is applied to one of the two interacting objects.

The Relation of Weight to Mass

Weight (*W*) is proportional (not equal!) to mass. The proportionality constant depends on the magnitude of the gravitational force acting on the object. This magnitude can vary, depending on whether the object is on earth, on the moon, or in outer space. On earth, the proportionality is the free-fall acceleration, *g*. Mathematically,

$$W = mg \quad \text{or} \quad m = W/g$$

INSIGHTS

1. Weight is a force, dimensionally different from mass. In SI units, it is measured in newtons.
2. Weight is a vector quantity. The direction of weight is the direction of the gravitational force on the object.
3. Mass does not depend on the gravitational environment an object is placed in, but weight does.
4. An object's apparent weight is not *mg* if the object is accelerating. Its apparent weight is greater than *mg* if it is accelerating in a direction opposite to gravity, and is less than *mg* if accelerating in the direction of gravity. If gravity is the only force acting on an object, the object is in free-fall and is "weightless" (apparent weight = 0).

Normal Force

A normal force between two surfaces in contact with each other is the force of compression perpendicular to the surfaces.

Friction Forces

STATIC FRICTION: A force between stationary surfaces in contact, opposite in direction of forces which attempt to start the surfaces sliding along each other. The static friction force is thus a force parallel to the surfaces and can take on any value up to a certain critical maximum f_c, after which the surfaces start to slip. The magnitude of this maximum is given by

$$f_c = \mu_s F_N$$

where F_N is the normal force on the surfaces. μ_s is the coefficient of static friction, which depends on the nature and substances of the surfaces.

KINETIC (MOVING) FRICTION: A force between sliding surfaces that are in contact, opposite in direction to the sliding motion. It is also parallel to the surfaces, and has a magnitude given by

$$f_k = \mu_k F_N$$

μ_k is the coefficient of kinetic friction. It also depends on the nature and substances of the surfaces, and always has a smaller value than μ_s.

INSIGHTS

1. The magnitudes of f_c and f_k depend on the force perpendicular to the surfaces, while their direction is parallel to the surfaces.
2. The two coefficients are dimensionless quantities, having no units.
3. f_k usually does not depend significantly on the speed of sliding between the surfaces.
4. Usually neither f_c nor f_k depends significantly on the area of contact between the surfaces.

Motion on Inclines

Any motion on an incline is constrained to be along the ramp, and so the algebraic sum of force components perpendicular to the incline must be zero.

The algebraic sum of force components parallel to the incline is responsible for the acceleration along the incline:

$$\Sigma \mathbf{F}_x = m\mathbf{a}$$

If the incline has an angle of θ relative to the horizontal, the component of weight down the incline is $mg \sin \theta$, and the component of weight perpendicular to the incline is $mg \cos \theta$.

Friction forces must be along the ramp and directed opposite the direction of motion or the tendency to move.

Mass at High Speeds

Objects cannot be accelerated to speeds at or greater than the speed of light, *c*. As the speed of an object approaches *c*, its mass (inertia) increases, making further increases in speed more difficult. The dependence of mass on speed is given by

$$m = \frac{m_o}{\sqrt{1 - v^2/c^2}}$$

QUESTIONS AND GUESSTIMATES

1 Why does a passenger tend to slide across the seat as a car quickly turns a corner? Why does a carton of eggs fall off the seat if the car stops too quickly?

2 Distinguish clearly between mass, weight, and inertia.

3 Clearly identify the action-reaction forces in each of the following: a child kicks a can; the sun holds the earth in orbit; a ball breaks a window; a parent spanks a child; a ball bounces on a table; a boat tows a water skier.

4 For objects on the moon, the acceleration due to gravity is about 1.67 m/s^2. How much does an object that has a measured mass of 2 kg on earth weigh on the moon? On earth? What is its mass on the moon?

5 Because objects on the moon weigh only about one-sixth as much as they do on earth, you would almost certainly be able to lift a heavy football player if the two of you were on the moon. Could you easily stop him if he was running at a fair rate across the moon's surface?

6 Is it possible for an object on earth to accelerate downward at a rate greater than g?

7 Suppose a brick is dropped from a height of several centimeters onto your open hand when it is lying flat on a tabletop. Why will your hand probably be injured in this situation even though you could easily catch the brick in your free hand without injury?

8 It is generally believed that, on the average, an intoxicated person receives slighter injuries from a fall than a sober person. Why might this belief be valid?

9 Consider the large mops used to sweep the halls in a school. It is easy to slide the mop along the floor if the handle makes only a small angle with the floor. If the angle between handle and floor is too large, however, you cannot move the mop along no matter how hard you push. Explain. Can you find a relation between the critical angle for sliding and the coefficient of friction between floor and mop?

10 An object is being weighed in an elevator. If the elevator suddenly begins to accelerate upward, explain what happens if the weighing device is (a) a spring balance, (b) an analytical two-pan beam balance, (c) a triple-beam balance (an unequal-arm balance).

11 A car at rest is struck from the rear by a second car. The injuries (if any) incurred by the two drivers are of distinctly different character. Explain what happens to each driver.

12 Estimate the minimum distance in which a car can be accelerated from rest to 10 m/s if its motor is extremely powerful.

13 When a high jumper leaves the ground, where does the force that accelerates her or him upward come from? Estimate the force that must be applied to the jumper in a 2-m-high jump.

14 Estimate the force your ankles must exert as you strike the floor after jumping from a height of 2.0 m. Why should you flex your legs in such a situation?

PROBLEMS (See note on page 23.)

Section 3.4

1 How large a horizontal force must be exerted on an 8.5-g bullet to give it an acceleration of $18,500 \text{ m/s}^2$? With this acceleration, how fast will the bullet be going after it has moved 2.35 cm from rest?

2 A horizontal unbalanced force of 4600 N accelerates a 1650-kg car from rest along a horizontal straight roadway. (a) What is the acceleration of the car? (b) How long does it take the car to reach a speed of 21.2 m/s?

3 A certain 1350-kg car can accelerate from rest to 23.4 m/s in 7.7 s. (a) What is its acceleration? (b) How large a force is needed to produce this acceleration?

4 A horizontal force of 26.7 N is required to slide a box along a level floor at a constant speed of 0.485 m/s. How large is the friction force opposing the motion?

5 If a tow rope is pulled upward at an angle of 27° to the horizontal with a force of 365 N, it can slide a 55.2-kg box along the floor at a constant speed of 20.5 cm/s. How large a friction force opposes the box's motion?

6 A water skier is being pulled by a boat at a constant speed of 13.5 m/s. The tension in the cable pulling the skier is 165 N. How large is the retarding force exerted on the skier by the water and the air?

7 A 72-kg parachutist is gliding to earth with a constant speed of 9.1 m/s. The parachute has a mass of 6.6 kg. (a) How much does the parachutist weigh? (b) How large an upward force does the air rushing past the parachutist and chute exert on them?

8 A horizontal force of 4770 N is required to give a 1270-kg car an acceleration of 0.175 m/s^2 on a level road. How large a retarding force opposes the motion?

9 An advertisement claims that a certain 1060-kg car can be accelerated from rest to 80 km/h in 9.4 s. How large a net force must act on the car to give it this acceleration?

10 A 1730-kg car is to be towed by another car. If the towed car is to be accelerated uniformly from rest to 2.3 m/s in 10.3 s, how large a force must the tow rope be able to exert?

11 A 1570-kg car moving at 17.5 m/s is to be stopped in 94.5 m. How large must the stopping force be? Assume uniform deceleration.

Section 3.5

12 A ball weighing 5 N falls to earth. (a) What is the net force on the ball as it falls? (b) What is the force (include direction) exerted by the ball on the earth as it falls?

13 Assume the ball in Prob. 12 is sitting on a table. (a) What is

the net force on the ball? (b) What are the forces (including direction) exerted by the ball on the table and on the earth?

14 A truck crashes into a small car, exerting a force of 26,000 N on the car. How large a force does the car exert on the truck? Why does the car suffer greater damage?

■15 A rifle is firmly attached to a heavy bench, with its 75-cm-long barrel pointing horizontally. It fires a 9.0-g bullet, which leaves the muzzle at a speed of 970 m/s. Assuming the acceleration of the bullet down the length of the barrel to be constant, what horizontal force does the gun exert on the bench as it fires?

■16 Two blocks with masses $m_1 = 3.2$ kg and $m_2 = 4.1$ kg are touching each other on a frictionless table, as in Fig. P3.1. If the force shown acting on m_1 is 6.8 N, (a) what is the acceleration of the two blocks and (b) with what force does m_1 push against m_2? (c) Repeat a and b if **F** is in the reverse direction and pushes on m_2 rather than on m_1.

FIGURE P3.1

Section 3.6

17 What is the weight (in newtons and pounds) of the following objects: (a) a 1.0-kg ball, (b) a 60-kg person, (c) a 1350-kg car, (d) a 1-ton moose, (e) 454 g of butter?

18 What is the mass in kilograms of the following objects: (a) 1/2 lb of flour, (b) a 15-N lamp, (c) a 160-lb person, (d) a 1750-N log, (e) 1 metric ton of coal?

19 A rope pulls upward on a briefcase weighing 54 N. The briefcase is accelerating upward at $a = 0.77$ m/s². What is the tension in the rope?

20 A rope is lowering a 20.5-kg bag of potatoes. The bag is accelerating downward at $a = 0.155$ m/s². What is the tension in the rope?

■21 Freely falling objects near the surface of the moon are observed to accelerate downward at $a = 1.63$ m/s². A certain astronaut (including spacesuit) weighs 960 N on earth. (a) What does the astronaut weigh at the surface of the moon? (b) What is her mass on the moon? (c) What is her mass on the earth?

Section 3.7

22 Each block in Fig. P3.2 weighs 70 N, and $T = 35$ N. Find the normal force in each case.

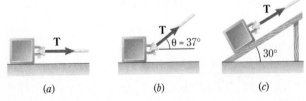

FIGURE P3.2

23 Each block in Fig. P3.3 weighs 47 N, and $P = 28$ N. Find the normal force in each case.

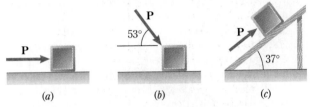

FIGURE P3.3

■24 In Fig. P3.3, suppose the block weighs 66 N, $P = 42$ N, and the appropriate coefficient of friction is 0.22. (a) What is the friction force in each case? (b) What is the block's acceleration?

■25 In Fig. P3.2, the weight of the block is 54 N, $T = 39$ N, and the appropriate coefficient of friction is 0.42. (a) What is the friction force in each case? (b) What is the block's acceleration?

26 A 5.5-kg box slides down a 27° incline under the action of gravity. If the box slides with constant speed, how large is the friction force that impedes its motion?

27 A 27-g block is sitting on an adjustable incline. The angle of the incline is slowly increased, and the block begins to slide when the angle is 38.5°. What is the coefficient of friction between block and incline? Is this the static or the kinetic coefficient?

■28 In Fig. P3.3b, the static coefficient of friction is 0.50. If the block weighs 165 N, at what value of P will it begin to move?

■29 If the coefficient of friction between a car's tires and a roadway is 0.62, what is the least distance in which the car can accelerate from rest to 20.7 m/s?

■30 A boy is running across a slippery floor at 3.55 m/s when he decides to slide. If the coefficient of friction between his shoes and the floor is 0.15, how far will he slide before stopping?

■31 What is the shortest distance in which a car going 34.2 m/s can stop on a level roadway if the maximum friction coefficient (the static coefficient) between its tires and the pavement is 0.83?

Section 3.8

32 An electron ($m = 9.1 \times 10^{-31}$ kg) in a television tube is accelerated from rest to 6.25×10^7 m/s in a distance of 0.88 cm. Find the average accelerating force of the electron. How many times larger than mg is it?

33 An 1130-kg car traveling at 16.7 m/s collides with a tree and stops in a distance of 0.77 m. How large is the average force exerted by the tree on the automobile?

■34 A 9.1-g bullet enters a 2.3-cm-thick piece of plastic with a speed of 165 m/s. It passes through the plastic and emerges with a speed of 92 m/s. How large is the average force exerted by the bullet on the plastic?

■35 If you pull straight up on a 3.2-kg mass with a cord that is just capable of holding a 15.0-kg mass at rest, what is the maximum upward acceleration you can impart to the 3.2-kg mass?

■36 A book sits on the roof of a car as the car accelerates horizontally from rest. If the static coefficient of friction between car and book is 0.36, what is the maximum acceleration the car can have if the book is not to slip?

■37 A carton of eggs rests on the seat of a car moving at 22.5 m/s. What is the least distance in which the car can be uniformly slowed to a stop if the eggs are not to slide? The value of μ between carton and seat is 0.24.

■■38 A cement block sits on the floor of a station wagon descending a 23.5° incline while decelerating at 1.15 m/s². How large must the static coefficient of friction between floor and block be if the block is not to slide?

■■39 In Fig. P3.4, the tension in the rope pulling the two blocks is 58 N. Find the acceleration of the blocks and the tension in the connecting cord if the friction force on the blocks is negligible. Repeat if the coefficient of friction between blocks and surface is 0.33.

FIGURE P3.4

■40 In Fig. P3.4, how large must T be to give the blocks an acceleration of 0.62 m/s² (*a*) if friction forces are negligible, (*b*) if the coefficient of friction between blocks and surface is 0.43? Also find the tension in the connecting cord in each case.

■41 In Fig. P3.5, block 1 has a mass of 3.25 kg and block 2 has a mass of 1.90 kg. (*a*) Ignoring friction, what are the acceleration of the blocks and the tension in the connecting cord? (*b*) Repeat for a friction force of 10.2 N retarding block 1.

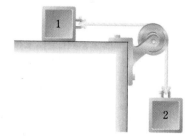

FIGURE P3.5

■42 In Fig. P3.5, object 1 is a 2650-g mass and object 2 is a 1650-g mass. When the system is released, object 2 falls 65 cm in 1.44 s. How large a friction force opposes the motion of object 1? Assume that no friction forces exist in the rest of the system.

■43 In Fig. P3.6, find the tension in the cord and the time needed for the masses to move 220 cm starting from rest. Assume the pulley to be frictionless and massless.

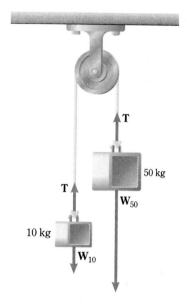

FIGURE P3.6

■■44 The pulley in Fig. P3.7 is massless and frictionless. Find the acceleration of the mass in terms of F if there is no friction between the surface and the mass. Repeat for friction force f.

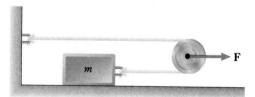

FIGURE P3.7

■■45 There is negligible friction between the blocks and table of Fig. P3.8. Compute the tension in the cord and the acceleration of m_2 if $m_1 = 375$ g, $m_2 = 275$ g, and $F = 0.72$ N. *Hint:* Note that $a_2 = 2a_1$.

FIGURE P3.8

46 In Fig. P3.9, the coefficient of friction is the same at the top and bottom of the 700-g block. If $a = 135$ cm/s^2 when $F = 1.90$ N, how large is the coefficient of friction?

FIGURE P3.9

47 Find the tensions in the two cords and the accelerations of the blocks in Fig. P3.10 if friction is negligible. The pulleys are massless and frictionless, $m_1 = 215$ g, $m_2 = 500$ g, and $m_3 = 365$ g.

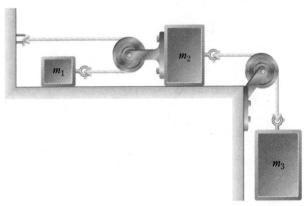

FIGURE P3.10

48 Find the acceleration of the blocks in Fig. P3.11 and the tension in the cord (*a*) if there is negligible friction, (*b*) if $\mu = 0.25$. Find the general expression for a in terms of m_1 on the ramp, m_2, g, and μ.

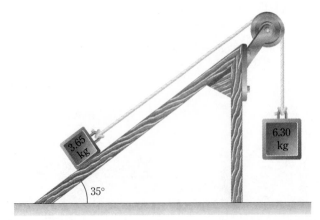

FIGURE P3.11

49 The force **F** in Fig. P3.12 pushes a block of mass M, which in turn pushes a block of mass m. There is no friction between M and the supporting surface. If the friction coefficient between the two blocks is μ, how large must F be if the block of mass m is not to slip?

FIGURE P3.12

Sections 3.9–3.10

50 The friction force retarding the motion of an 85-kg box across a level floor is 365 N. (*a*) What is the coefficient of friction between box and floor? (*b*) Assuming that the friction coefficient does not change as the speed increases, how large an acceleration can you give the box by pulling it with a force of 660 N that is inclined at an angle of 48° above the horizontal?

51 (*a*) Find the acceleration of the 2.85-kg block in Fig. P3.13 if the coefficient of friction between block and surface is 0.77. (*b*) Repeat if the 50-N force is pushing *down* on the block at an angle of 22.5° below the horizontal (that is, if the force shown in the figure is reversed in direction).

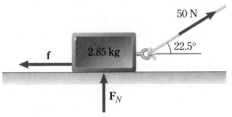

FIGURE P3.13

52 How large a force parallel to a 37° incline is needed to give a 3.35-kg box an acceleration of 1.85 m/s^2 up the incline (*a*) if friction is negligible, (*b*) if the coefficient of friction is 0.45?

53 A 10.6-kg box is released on a 22° incline and accelerates down the incline at 0.37 m/s^2. Find the friction force impeding its motion. How large is the coefficient of friction?

54 A 45-kg woman stands on a spring scale inside an elevator. (The scale reads the force with which it pushes upward on the woman.) What does the scale read when the elevator is accelerating (*a*) upward at 3.65 m/s^2, (*b*) downward at 2.70 m/s^2?

55 A 220-g mass is hung from a thread; from the bottom of the 220-g mass, a 275-g mass is hung by a second thread. Find the tensions in the two threads if the masses are (*a*) standing still, (b) accelerating upward at 16.5 m/s², (c) moving downward at a constant acceleration of 7.8 m/s², (d) falling freely under the action of gravity, (e) moving downward at a constant speed of 10 m/s.

■56 A 0.95-kg block starts to slide from rest down a 32° incline. How far does it slide in the first 2.7 s (a) if friction is negligible, (b) if μ between block and surface is 0.50?

■57 A 1250-kg car is at rest on a hill that is inclined 8.5° to the horizontal. How far does the car move in the first 8.0 s after the brakes are released (a) if the car rolls freely down the hill, (b) if a 1600-N friction force retards its motion?

General Problems

■■58 Two frictionless carts of masses M_1 and M_2 sit at rest on a straight horizontal track. They are a distance D apart, and a rope stretches between them. The occupants of cart 1 pull on the rope in such a way that the tension in it is constant, and the carts move toward each other. (a) Where, relative to the original position of cart 2, do the two carts collide? (b) What is the ratio of their speeds just before collision?

■■59 Prove that the acceleration of a car moving on a horizontal road cannot exceed μg, where μ is the coefficient of friction between tires and road. What is the similar expression for the acceleration of a car going up an incline of angle θ? Why is it counterproductive to cause the car to "burn rubber" when it is "scratching off"? Does it matter whether the car has two-wheel or four-wheel drive?

■■60 A passenger in a large ship sailing in a quiet sea hangs a ball from the ceiling of her cabin by means of a long thread. She notes that whenever the ship accelerates, the pendulum ball lags behind the point of suspension, and the pendulum no longer hangs vertically. How large is the ship's acceleration when the pendulum stands at an angle of 6.5° to the vertical?

■61 In Fig. P3.14, a 4-kg box is on a horizontal surface where the coefficients of static and kinetic friction are 0.8 and 0.6 respectively. You pull on the box with a force of 50 N at an angle of 30° above the horizontal. (a) What is the normal force on the box? (b) What is the acceleration of the box? (c) Answer questions a and b, assuming you reverse your force, so that you *push* on the box at an angle of 30° below horizontal. (*Hint:* Don't assume the box is already moving when you push on it.)

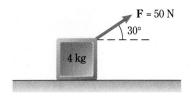

FIGURE P3.14

■62 As shown in Fig. P3.15, you are exerting a horizontal force on a block of wood, pushing it against a vertical wooden wall. You are pushing hard enough so that the block does not fall. If the coefficient of static friction between the wall and the block is 0.65, what is the least amount of force you must be exerting?

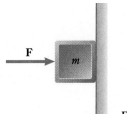

FIGURE P3.15

■■63 In Fig. P3.16, a 50-g mass sits on top of a 200-g mass. The coefficient of static friction between these two masses is 0.3. The 200-g mass is free to move along a frictionless horizontal table. A string connects the 200-g mass via a massless, frictionless pulley to a mass m_1. What is the largest value of m_1 for which the 50-g mass will remain on top of the 200-g mass as the system accelerates?

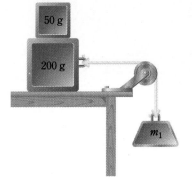

FIGURE P3.16

CHAPTER 4

STATIC EQUILIBRIUM

An important part of physics has to do with objects and systems at rest. This branch of physics, called **statics,** is of central importance to those who design and build bridges, buildings, and any other structures whose stability we depend on. It is also an important and instructive application of the laws of mechanics we studied in the previous chapter. In our study of statics in this chapter, we discover that two basic conditions must be satisfied if an object is to remain at rest. We learn how to apply these conditions and become acquainted with their consequences.

4.1 THE FIRST CONDITION FOR EQUILIBRIUM

When an object is at rest and remains at rest, we say it is **in static equilibrium.** There are *two* conditions for equilibrium, the first of which we can derive from Newton's second law. An object at rest is a specific case of having a constant velocity, namely *zero*. Thus an object remaining at rest experiences no acceleration, and both Newton's first law and his second law tell us that the net force on the object must therefore be zero. This is the first condition for equilibrium:

For an object to be in equilibrium, the vector sum of the forces acting on it must be zero.

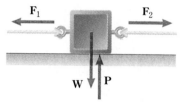

FIGURE 4.1

For the block to remain at rest, forces must cancel in both horizontal and vertical directions.

Stating that the vector sum of the forces acting on an object are zero is equivalent to saying that all rectangular components of acceleration in Newton's second law (Eqs. 3.1*b*) are equal to zero:

$$\Sigma F_x = 0 \qquad \Sigma F_y = 0 \qquad \Sigma F_z = 0 \tag{4.1}$$

A two-dimensional example is shown in Fig. 4.1. For the box to stay at rest with the four forces acting on it, the horizontal and vertical components of the forces must each add up to zero. Applying Eqs. 4.1 to this case gives us

$$F_1 - F_2 = 0 \qquad \text{and} \qquad P - W = 0$$

Here we have taken direction into account by supplying signs (right and up positive, left and down negative). The symbols F_1, F_2, W, and P then represent the magnitudes of the forces.

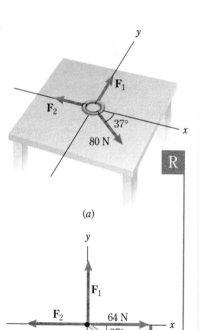

(a)

(b)

FIGURE 4.2

Find $\mathbf{F}_1$ and $\mathbf{F}_2$ if the ring is to be at equilibrium.

Example 4.1

The ring in Fig. 4.2 is lying on a table and being pulled by three strings that hold it at rest. One of the strings is exerting a tension force of 80 N. Find the tensions in the other two strings. (Remember from Chap. 3 that a tension is a force directed along a string or rope and away from the object to which it is attached.)

Reasoning

Question Since these three forces are all horizontal, how does gravity play a role in determining the tension?
Answer The downward pull of gravity and the upward push of the table must cancel each other in order for the ring to remain on the table. Also, these two vertical forces have no horizontal components, and so they cannot affect the tension in the horizontal strings.

Question What principle determines the tensions F_1 and F_2?
Answer The first condition for equilibrium: both *x* components and *y* components of all forces must add up to zero.

Question F_1 has only a *y* component and F_2 only an *x* component. What are the components of the 80-N vector?
Answer As shown in Fig. 4.2*b*, the components are +64 N along *x* and −48 N along *y*.

Question What equations does the equilibrium condition give me?
Answer $\Sigma F_x = 0$: $64\,\text{N} + F_2 = 0$
$\qquad\quad\ \ \Sigma F_y = 0$: $F_1 + (-48\,\text{N}) = 0$

Solution

$$F_1 = +48\,\text{N} \qquad F_2 = -64\,\text{N}$$

Exercise Replace the 80-N force in Fig. 4.2*a* by an unknown force F_3. If $F_1 = 42\,\text{N}$, find F_2 and F_3. *Answer:* $F_2 = 56\,\text{N}$, $F_3 = 70\,\text{N}$

A suspension bridge depends on all forces acting on it being in static equilibrium.

4.2   SOLVING PROBLEMS IN STATICS

With a little practice, you will be able to use Eq. 4.1 to solve many problems in statics. You should, however, follow a few simple rules so that you do not become confused:

1 Isolate the object you are going to talk about. The forces acting on this object are the *only* ones you need in writing Eq. 4.1.
2 Draw the forces acting on the object you have isolated and label them in a free-body diagram. (Use symbols such as $\mathbf{F}_1$, $\mathbf{P}$, and $\mathbf{Q}$ for any forces whose values are not yet known.)
3 Split each force into its x, y, and z components and label the components in terms of the symbols given in rule 2 and the proper sines and cosines.
4 Write down Eq. 4.1.
5 Solve the equations for the unknowns.

Example 4.2

The object in Fig. 4.3*a* weighs 400 N and hangs at rest. Find the tensions in the two cords.

R   *Reasoning*

Question   How do I identify the forces acting on the object?
Answer   The weight of the object is downward and equal to 400 N. From the definition of tension, the directions of the other forces must be along the cords and away from the object. Call one of these forces $\mathbf{F}_1$ and the other $\mathbf{F}_2$. The free-body diagram using these symbols is shown in Fig. 4.3*b*.

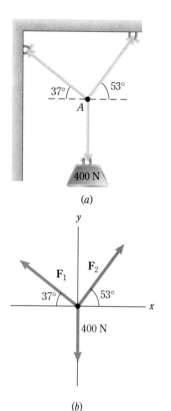

(a)

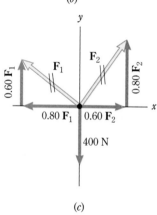

(b)

(c)

FIGURE 4.3

Since the junction point of the cords in (a) is at equilibrium, the *y* forces must cancel each other in (c). The same holds true for the *x* forces.

Question What principle determines the tensions $\mathbf{F}_1$ and $\mathbf{F}_2$?
Answer The first condition for equilibrium.

Question Since I don't know $\mathbf{F}_1$ or $\mathbf{F}_2$, how can I write down their components?
Answer Remember, the values of the sine and cosine represent the fractions of $\mathbf{F}_1$ and $\mathbf{F}_2$ that are exerted in the *y* and *x* directions, respectively:

$$(\mathbf{F}_1)_x = \mathbf{F}_1 \cos 37° = (0.80)\mathbf{F}_1 \qquad (\mathbf{F}_1)_y = \mathbf{F}_1 \sin 37° = (0.60)\mathbf{F}_1$$
$$(\mathbf{F}_2)_x = \mathbf{F}_2 \cos 53° = (0.60)\mathbf{F}_2 \qquad (\mathbf{F}_2)_y = \mathbf{F}_2 \sin 53° = (0.80)\mathbf{F}_2$$

These components are shown in Fig. 4.3c.

Question What equations does the first condition give me?
Answer In vector notation, the conditions $\Sigma \mathbf{F}_x = 0$ and $\Sigma \mathbf{F}_y = 0$ become

$$(0.8)\mathbf{F}_1 + (0.6)\mathbf{F}_2 = 0 \qquad (0.6)\mathbf{F}_1 + (0.8)\mathbf{F}_2 - 400\,\text{N} = 0$$

In magnitude notation, we note the directions of the components from Fig. 4.3c and apply signs to the magnitudes:

$$-(0.8)F_1 + (0.6)F_2 = 0 \tag{a}$$
$$(0.6)F_1 + (0.8)F_2 - 400\,\text{N} = 0 \tag{b}$$

Notice that you have two equations in the two unknowns.

Solution

Method 1 Elimination of one unknown by addition or subtraction Multiply Eq. *a* by 0.6 and Eq. *b* by 0.8. Thus

$$0.36F_2 - 0.48F_1 = 0 \tag{c}$$
$$0.64F_2 + 0.48F_1 - 320\,\text{N} = 0 \tag{d}$$

Addition of Eqs. *c* and *d* gives

$$1.00F_2 - 320\,\text{N} = 0 \qquad F_2 = 320\,\text{N}$$

Substituting this value for F_2 in Eq. *c* gives

$$0.48F_1 = (0.36)(320\,\text{N}) \qquad F_1 = 240\,\text{N}$$

Method 2 Substitution for one unknown in favor of the other Solve Eq. *a* for F_1 in terms of F_2. This gives $F_1 = 0.75F_2$. Substitute this value in Eq. *b* to obtain

$$0.80F_2 + (0.60)(0.75F_2) - 400\,\text{N} = 0$$

from which $F_2 = 320\,\text{N}$. Substituting this value for F_2 in Eq. *a* gives $F_1 = 240\,\text{N}$.

Example 4.3

Figure 4.4 shows a picture hanging on a wall by two cords, each making an angle of 20° relative to the horizontal. Each cord will break if the tension in it exceeds 60 N. What is the *maximum* picture weight the cords can support in this arrangement?

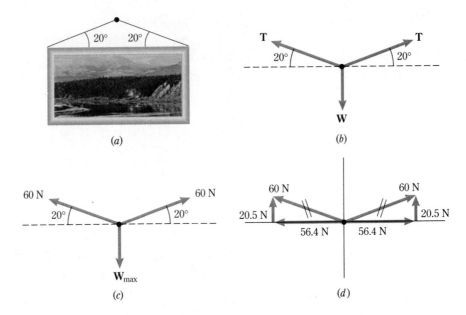

FIGURE 4.4

A hanging picture and its free-body diagrams. Notice how the free-body diagram simplifies the drawing: the picture is reduced to a point and its weight and the two tensions in the cords emanate from that point.

R

Reasoning

Question   How does the maximum weight of the picture relate to the tensions in the cords?

Answer   Notice in Fig. 4.4a that the cords play a symmetrical role. Thus you can assume that both will have equal tension, whatever the weight of the picture. Thus the maximum picture weight is that weight which causes a tension of 60 N in each cord. The free-body diagram for the picture is shown in Fig. 4.4b for the general case of tensions T in the cords. The tension in the cord attached to the left side of the picture is shown as directed upward and to the right. The tension in the other cord is shown upward and to the left.

Question   What are the rectangular components of all the forces?
Answer W is entirely in the y direction. Each cord has components of magnitude

$$T_x = T\cos 20° = T(0.94) \quad \text{and} \quad T_y = T\sin 20° = T(0.34)$$

Question   What is the principle relating W_{max} to the tensions?
Answer   The first condition for equilibrium applies, with $T = 60$ N. With this value of T, the component equations yield $T_x = 56.4$ N and $T_y = 20.5$ N.

Question   What equations does the first condition give me?
Answer: $\Sigma F_x = 0$ shows that the 56.4-N horizontal components cancel, as shown in Fig. 4.4d. $\Sigma F_y = 0$ becomes

$$20.5\,\text{N} + 20.5\,\text{N} - W_{max} = 0$$

Solution and Discussion $W_{max} = 41.0$ N. Notice that the cords cannot support a weight equal to their breaking strength when arranged like this. The closer the cords get to being horizontal, the less weight they can support without breaking.

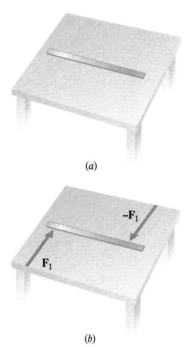

(a)

(b)

FIGURE 4.5

Even though $\Sigma\mathbf{F} = 0$ for it, the stick in (*b*) is not in equilibrium.

FIGURE 4.6

How must $\mathbf{F}_1$ and $\mathbf{F}_2$ be related if the wheel is not to turn?

4.3 TORQUE

It is possible for an object to move even when the first condition for equilibrium is satisfied. There is a second condition that must be satisfied if an object is to be in static equilibrium. It is easy to show this by referring to Fig. 4.5. We see there a meter stick supported by a tabletop. The stick is at equilibrium in part *a* because the pull of gravity on it is balanced by the upward push of the table and we have $\Sigma\mathbf{F} = 0$.

Now consider what happens when you push near its two ends with equal but oppositely directed forces $\mathbf{F}_1$ and $-\mathbf{F}_1$: the meterstick does not remain at rest. Even though $\mathbf{F}_1$ balances $-\mathbf{F}_1$ and therefore the condition $\Sigma\mathbf{F} = 0$ is satisfied, the stick begins to rotate. There must be another condition, one involving rotation, that must be satisfied if the object is to be in static equilibrium. We discuss that second (and final) condition for equilibrium in the next section. First, however, we must discuss how forces cause rotation.

To learn how forces and rotations are related, we can perform the experiment in Fig. 4.6. We see there a wheel that consists of two disks cemented together. The wheel is free to rotate on a stationary axle that we call the *axis* of rotation. By hanging objects from the two cords, we can determine the turning effect of a force. The force $\mathbf{F}_2$ tries to turn the wheel clockwise, while $\mathbf{F}_1$ tries to turn the wheel counterclockwise. By experimenting with different radii r_1 and r_2 for the two disks, we find that the two turning effects balance whenever

$$r_1F_1 = r_2F_2$$

Therefore it is clear that the turning effect depends both on the magnitude of the force and on its distance from the axis.

We can learn more about turning effects from Fig. 4.7. A meterstick free to rotate about an axis through its center is subjected to two forces $\mathbf{F}_1$ and $\mathbf{F}_2$, as shown. Experiment shows that the system is balanced whenever $\mathbf{F}_2$ has a magnitude that satisfies the condition

$$(0.5L)(F_1) = \text{lever arm} \times F_2$$

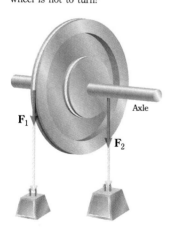

(a) Perspective view

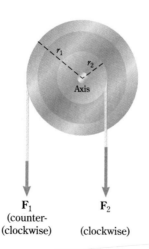

(b) End-on view

F₁
(counter-
(clockwise)

F₂
(clockwise)

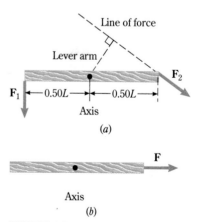

FIGURE 4.7

The turning effect of a force about the axis is dependent on the product of the force and the lever arm. What is the turning effect of the force in (*b*)?

The tangential force of the falling water creates a torque about the axle of the waterwheel.

where the meaning of the term "lever arm" is shown in Fig. 4.7. In terms of the line labeled "line of force" (which is the endless line along which the force vector lies), the lever arm is defined as follows:

The **lever arm** of a force is the perpendicular distance from a given axis to the line along which the force acts.

The turning effect of a force about an axis is called the torque* about the axis, defined by the following:

The **torque** produced by a force about an axis is equal to the product of the force and its lever arm about the axis: $\tau = $ lever arm $\times$ force

It is of particular interest to notice what the torque is when the line of force passes through the axis, as in Fig. 4.7*b*. Then the lever arm is zero, and so

$$\tau = 0 \times F = 0$$

When the line of force goes through the axis, the torque due to the force about that axis is zero.

The units of torque are those of distance times force. In SI units, this is the meter-newton (m · N).

In Figs. 4.6 and 4.7 $\mathbf{F}_1$ and $\mathbf{F}_2$ tend to turn the objects in opposite directions, so we must somehow treat the torques the forces produce as opposing each other. In other words, a torque has a *direction* associated with it. For a fixed axis, however, there are only two possible (and opposite) senses of rotation about the axis, which we can describe as clockwise and counterclockwise. We can take the direction of torques into account by assigning one sign to those tending to turn an object in one sense of rotation, and the opposite sign to those torques tending to produce an opposite rotation. Traditionally, we make the following choice:

Torques that tend to produce a counterclockwise (ccw) rotation about an axis will be assigned a positive value. Those that tend to produce a clockwise (cw) rotation will be assigned a negative value.

Illustration 4.1

Find the lever arms and torques for the forces in Fig. 4.8.

Reasoning   From its definition, the lever arm is zero for $\mathbf{F}_1$, *a* for $\mathbf{F}_2$ and $\mathbf{F}_3$, and *b* for $\mathbf{F}_4$. Using the sign convention discussed above, the torques are

$\mathbf{F}_1$ 0
$\mathbf{F}_2$ $+aF_2$
$\mathbf{F}_3$ $-aF_3$
$\mathbf{F}_4$ $+bF_4$ ■

---

*Torque is pronounced "tork," as in fork, and is represented by the greek letter tau (τ).

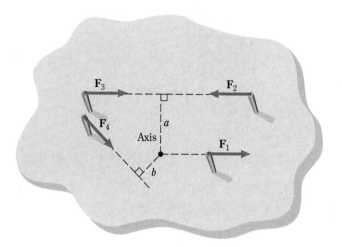

FIGURE 4.8

Find the lever arms and torques for the forces with respect to the axis.

4.4 THE SECOND CONDITION FOR EQUILIBRIUM

Now that we know how to express the turning effect of a force in terms of torque, it is a simple matter to state the second, and final, condition for static equilibrium. Careful experiments show that, for an object to remain motionless, the clockwise torques acting on it must be balanced by the counterclockwise torques. Using the sign convention for torques from the previous section, we can state that

For an object to be in static equilibrium, the algebraic sum of the clockwise and counterclockwise torques acting on it must be zero.

This statement is the **second condition for equilibrium.**

We can write this condition in mathematical form by using the symbolism $\Sigma\tau$ to represent "the sum of all the torques." Then the second condition for equilibrium becomes $\Sigma\tau = 0$.

All the requirements for a body to be in equilibrium are now known.* To summarize, in two dimensions,

$$\Sigma\mathbf{F}_x = 0 \qquad \Sigma\mathbf{F}_y = 0 \qquad \Sigma\tau = 0 \tag{4.2}$$

The terms **moment** and **moment of force** are sometimes used instead of torque. In that case, the lever arm is frequently referred to as the **moment arm.** The concepts are the same, of course.

In our previous applications of Newton's second law and in applying the first condition for equilibrium, it didn't matter *where* we showed the various forces attached to an object on its free-body diagram. That is no longer the case when we calculate torques and apply the second condition for equilibrium. It is very important to remember the following:

When using the second condition for equilibrium, it is necessary to show the correct placement of forces on an object's free-body diagram.

*Throughout this discussion, we have tacitly assumed that the motion of the object under consideration is restricted to a plane, that is, to two dimensions. A large share of the cases of interest are of this type.

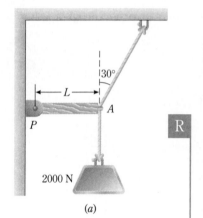

(a)

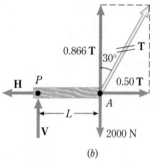

(b)

FIGURE 4.9
The beam is isolated as the object for discussion, and its free-body diagram is shown in (*b*). We assume the beam's weight to be negligible.

Example 4.4

In Fig. 4.9, we see a beam of length L able to rotate about one end (P) and attached to a 2000-N object at the other. Find the tension T in the red supporting cable. Assume that the weight of the beam is negligible.

Reasoning

Question   For which object should I draw a free-body diagram?
Answer   Since you want to find the tension in the red cable, you should choose some part of the system to which that cable is attached, either the beam or the ceiling. Since the forces on the beam are easier to identify, it is the best choice.

Question   What forces act on the beam?
Answer   The tensions in the two cables and whatever forces are exerted by the wall on the axis P. (The problem statement tells you to neglect the weight of the beam.)

Question   How can I know the forces exerted by the wall?
Answer   You can't at the outset, but in general you can specify some vertical force V and some horizontal force H.

Question   What if I specify their directions wrong on the free-body diagram?
Answer   If you do, the *magnitude* will come out negative, telling you you should have chosen the opposite direction. In other words, it is all right to choose the wrong direction; doing so won't affect anything but the sign of your answer. You can change signs as needed when you are done calculating.

Question   Can I find the tension in the lower cable?
Answer   Yes. It is the only force that supports the 2000-N weight. Thus this tension must be 2000 N. The free-body diagram you come up with after answering all these questions is shown in Fig. 4.9*b*. Notice that the entire beam is shown, so that the forces can be placed correctly in the diagram.

Question   What equations result from the first condition for equilibrium?
Answer $\Sigma F_x = 0$: $-H + (0.50)T = 0$
          $\Sigma F_y = 0$: $(0.866)T + V - 2000\,\text{N} = 0$

Because you have three unknowns, H, V, and T, you need a *third* equation involving them.

Question   What other principle can I apply?
Answer   The second condition for equilibrium, $\Sigma \tau = 0$.

Question   What axis should I choose for calculating torques?
Answer   Any axis will do,* but if you choose an axis perpendicular to the page through point P, the forces **H** and **V** and the horizontal component of **T** act through the axis and hence create *zero* torques.

Question   What equation results?
Answer   Using the torque sign convention,

$$-(2000\,\text{N})L + (0.866T)L = 0$$

*A detailed justification of this statement is the subject of Sec. 4.6.

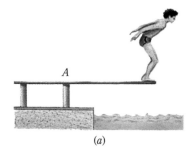

(a)

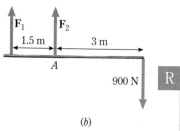

(b)

FIGURE 4.10

A 900-N man stands at the end of a diving board. We guess that the forces exerted by the two pedestals on the board are as shown. It turns out that the direction of one of the forces is guessed incorrectly.

Solution and Discussion   The last equation gives $T = 2310$ N. In this case you have the desired result from the torque equation alone! If you wish, you can put this value of T into the x- and y-component equations and solve them for H and V.

Exercise   Find H and V in Fig. 4.9.   *Answer:* $H = 1150$ N, $V = 0$

Example 4.5

The 900-N man in Fig. 4.10 is about to dive from a diving board. Find the forces exerted by the two pedestals on the board. Assume that the board has negligible weight.

Reasoning

Question   What forces are acting on the board?
Answer   The man's weight downward and the vertical forces exerted by the pedestals.

Question   The man's weight is given, but the pedestal forces aren't. In what direction do the pedestal forces act?
Answer   You don't know the directions, but you do know that at least *one* of them has to be upward, or the diving board would collapse. Again, if you choose the wrong direction for an unknown force in your free-body diagram, all that will happen is that you will get a negative value for its magnitude. The free-body diagram of the board, Fig. 4.10b, shows one possible choice of directions for F_1 and F_2.

Question   What results from the first condition for equilibrium?
Answer   There are no horizontal forces here, so $\Sigma \mathbf{F}_y = 0$ gives

$$F_1 - 900\,\text{N} + F_2 = 0$$

In order to remain on the balance beam, a gymnast must keep her center of gravity directly above the beam. Once her center of gravity is displaced to the side of the beam, a fall is inevitable.

Question   What axis should I use for calculating torques?
Answer   Again, any axis will do. Choose an axis through point *A*, for example, where one of the pedestals meets the board.

Question   What result does the second condition give about this axis?
Answer $-(900\,\text{N})(3\,\text{m}) - F_1(1.5\,\text{m}) = 0$.

Notice that F_2 does not appear in this equation, because it creates no torque about the chosen axis.

Solution and Discussion   The torque equation gives $F_1 = -1800\,\text{N}$. This negative value tells you that the direction of F_1 is *opposite the choice made in the free-body diagram*. Substituting this value into the force equation then gives

$$F_2 = 900\,\text{N} - (-1800\,\text{N}) = 2700\,\text{N}$$

Even with the wrong choice of direction for F_1, you get the right answers as long as you remain consistent with signs throughout the algebra.

4.5   THE CENTER OF GRAVITY

We avoided two complications in Examples 4.4 and 4.5. One was the choice of an axis about which to calculate torques. Without justification, we asserted that any choice will work; this topic is discussed in Sec. 4.6. The other complication we avoided by assuming the beam and diving board to have negligible weight. Because this is not a generally valid assumption, we must now consider how to take the weight into account when applying the second condition for equilibrium. In particular, where can we consider the pull of gravity to be applied to an object so that we can calculate its lever arm relative to a chosen axis?

Of course, gravity pulls on all parts of any object. For torque purposes, however, the pull of gravity (the object's weight) appears to act at one point. We call this point the **center of gravity (c.g.)** of the object. Let us now see how this point can be located experimentally.

Suppose we wish to locate the center of gravity of the object in Fig. 4.11. We first support the object from a string connected to some point *A* on the object. The string is free to rotate about the axis through point *P*. The balance of forces and torques causes the object to take the equilibrium position shown. Only two forces act on the body: gravity vertically downward and the tension in the string vertically upward. We know the vector sum of these two forces to be zero, since equilibrium exists. Since the string passes through *P*, its tension exerts no torque about this axis. Thus for the *sum* of the torques about *P* to be zero, the torque about *P* due to gravity must also be zero. This can be true only if gravity has the *net effect* of acting along the line *AB* in Fig. 4.11*a*, which passes through *P*.

Let us now suspend the object from a second point, *M*, as shown in Fig. 4.11*b*. Using the same reasoning, we can conclude that gravity effectively acts along the line *MN*. There is one point which is common to both lines *AB* and *MN*, namely their point of intersection, *C*. Gravity acts through point *C* in *both* cases. A check may be made by selecting a third suspension point on the object. We would find that a vertical line through this third point would also pass through point *C*. We conclude then that *C* is the center of gravity of the object:

(a)

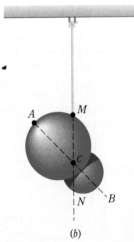

(b)

FIGURE 4.11

An experimental method for determining the center of gravity of an object.

This girder is being lifted by a single cable attached at the girder's center of gravity. Since there is no net torque on the girder, it remains level as it is lifted.

The center of gravity of an object is the point at which the force of gravity on an object can be considered to be applied for the purpose of calculating the torque it produces about any chosen axis.

For objects made of a uniform material and having simple symmetry, such as rods, spheres, and cubes, the center of gravity is at their *geometric* center. This point does not need to be a point physically within the material of the object. For example, a hoop made of uniform material has its c.g. at its geometric center, even though all its material is around the rim.

(a)

4.6 THE POSITION OF THE AXIS IS ARBITRARY

Often an object at equilibrium has an obvious axis of rotation, and it is common to refer to that axis when calculating torques. In other situations, no obvious axis exists. We see in this section that we are free to choose *any* axis we may find convenient when applying the second condition of equilibrium. One argument showing that any axis may be chosen is that, at equilibrium, an object is not rotating about *any* axis, either inside or outside of the object. Thus the torques produced by forces acting around *any* (and *every*) axis must add up to zero. Beyond this general argument, however, let's show the result mathematically.

Consider the situation in Fig. 4.12, where a sign painter of weight W_p stands in equilibrium on a uniform board of weight W_b and length L. The board's center of gravity is at its geometric center, so W_b is shown to act there in Fig. 4.12b. The tensions in the supporting cables are represented by T_1 and T_2. We now show that the final form of the torque equation for this equilibrium situation does not depend on what axis we choose.

Let us choose a line through A as the axis. You should verify that the torque equation $\Sigma \tau = 0$ then becomes

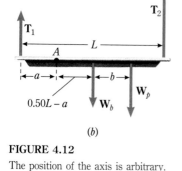

(b)

FIGURE 4.12

The position of the axis is arbitrary.

$$-T_1(a) - W_b(0.50L - a) - W_p(0.50L - a + b) + T_2(L - a) = 0$$

Let us group the terms involving the arbitrary length a:

$$-a(T_1 - W_b - W_p + T_2) - 0.50W_bL - W_p(0.50L + b) + T_2L = 0$$

We can easily show, however, that the factor multiplying a is zero *provided the system is at equilibrium.* At equilibrium, $\Sigma F_y = 0$, and so

$$T_1 + T_2 - W_b - W_p = 0$$

Because this is the factor that multiplies a in the equation, the term in question is zero. Therefore, the torque equation becomes

$$-0.50W_bL - W_p(0.50L + b) + T_2L = 0$$

which is independent of a and the position chosen for the axis. In this case at least, the position chosen for the axis is arbitrary.

Although we have obtained this result for a specialized situation, it is possible to give a more general proof. We therefore have the following general result:

In writing the torque equation for a body at equilibrium, the position chosen for the axis is arbitrary.

In practice, we usually choose the axis so that the line of an unknown force passes through it. Then the torque due to that force is zero and does not appear in the torque equation.

Example 4.6

The uniform beam in Fig. 4.13 weighs 50 N. The beam is secured to the wall by a hinge. If the beam is in static equilibrium, how large is the tension in the upper cable and what are the horizontal and vertical components of force exerted by the hinge on the beam?

FIGURE 4.13

The forces acting on the beam in (*a*) are shown in detail in (*b*). Notice that the force component of 0.6 T pulls upward on the beam at point S, and so the lever arm of this component about P is 100 cm.

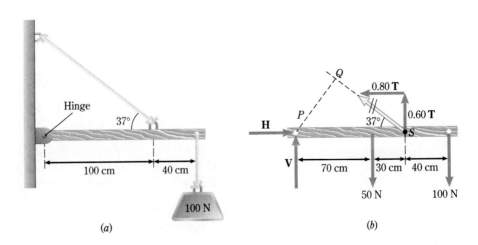

(*a*)

(*b*)

R

Reasoning

Question Can I identify all the forces acting on the beam and locate them on a free-body diagram?

Answer The tension in the upper cable acts at the point of attachment S in the direction of the cable. The 100-N weight acts downward at the end of the beam, and the beam's 50-N weight acts downward at the midpoint of the beam. The wall can exert a force on the beam by means of the hinge. We can represent this force generally by some vertical component V and some horizontal component H. The free-body diagram is shown in Fig. 4.13*b*. If the directions chosen for H and V are incorrect, the solutions of the equilibrium equations will yield negative values for them.

Question Is there an obvious choice for an axis?

Answer An axis passing through the hinge at P simplifies the torque calculations, since the forces H and V produce no torque around that axis.

Question How is the tension in the upper cable involved in the conditions for equilibrium?

Answer It contributes both a horizontal and vertical component to the first condition, and produces a ccw torque about an axis through the hinge.

Question What equations result from the first condition?

Answer Horizontally: $H - T_x = H - (0.80)T = 0$

Vertically: $V + T_y - 50\,\text{N} - 100\,\text{N} = 0$

or

$$V + (0.60)T - 150\,\text{N} \quad = 0$$

Question What equation do I get from the second condition?

Answer The weights both contribute cw torques about P, while T_y contributes a ccw torque:

$$T_y(1.0\,\text{m}) - (50\,\text{N})(0.70\,\text{m}) - (100\,\text{N})(1.4\,\text{m}) = 0$$

or

$$(0.60)T(1.0\,\text{m}) - 35\,\text{m} \cdot \text{N} - 140\,\text{m} \cdot \text{N} = 0$$

Notice that the horizontal component of T passes through the hinge and so contributes no torque. Also, the weights act at different points on the beam; their individual lever arms must be used.

Solution and Discussion   Notice that you have three equations in the three unknowns H, V, and T. The given data allow us to keep no more than two significant digits. The torque equation immediately gives the tension, $T = 290\,\text{N}$. Substitution of this value for T in the two force equations gives $H = 230\,\text{N}$ and $V = -24\,\text{N}$. Since we treated **V** as an *upward* vector, this result tells us that **V** actually points *downward*.

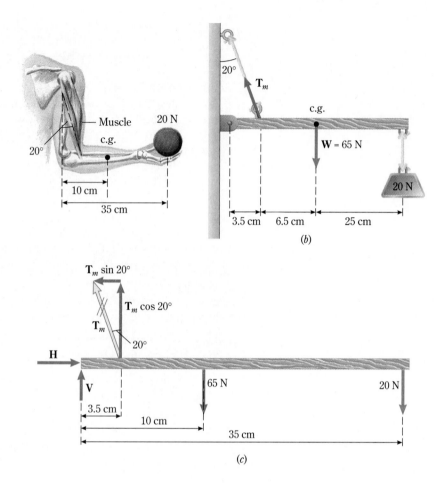

FIGURE 4.14

We can analyze the forces in the human arm by use of the models in (b) and (c).

Example 4.7

A person holds a 20-N weight as shown in Fig. 4.14a. Find the tension in the supporting muscle and the component forces exerted on the elbow. Typical characteristics of the forearm plus hand (from elbow to finger tips) are: weight, 65 N; length, 35 cm; c.g., located between the elbow and wrist 10 cm from the elbow; supporting muscle attached 3.5 cm from the elbow, making an angle of 20° relative to the vertical.

R   *Reasoning*

Question   What object do I want to consider as being in equilibrium?
Answer   The forearm plus hand, with an axis through the elbow.

Question   What forces act on the forearm, and where are they located in the free-body diagram?
Answer   See Figs. 4.14b and 4.14c. Here you are abstracting only the essential forces out of Fig. 4.14a. Notice the fundamental similarity with the beam in the previous example. Two seemingly different situations are reduced to the same

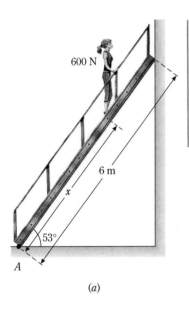

(a)

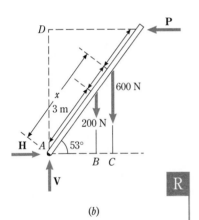

(b)

FIGURE 4.15

A 600-N woman stands on a 200-N ladder. On the assumption that the wall is smooth, the forces acting on the ladder are as shown in (b).

problem. The power of physics lies in its ability to simplify and unify through this sort of reduction to fundamentals.

Question What equations do the conditions for equilibrium give?
Answer Using the elbow as the axis for the calculation of torques,

$$\Sigma F_x = 0: \quad H - T_m \sin 20° = 0$$
$$\Sigma F_y = 0: \quad V + T_m \cos 20° - 65\,N - 20\,N = 0$$
$$\Sigma \tau = 0: \quad (T_m \cos 20°)(0.035\,m) - (65\,N)(0.10\,m) - (20\,N)(0.35\,m) = 0$$

Solution The torque equation gives $T_m = 410\,N$. Substituting this value into the two force equations gives

$$H = 140\,N \qquad V = -300\,N$$

The minus sign tell us that V is downward.

All these forces are much larger than the weight being held. Can you show that T_m becomes very large as the arm is outstretched? Why is it very tiring to hold a weight in your outstretched hand?

Example 4.8

In Fig. 4.15a a uniform ladder 6.0 m long and weighing 200 N stands against a wall at an angle of 53° above horizontal. There is no friction between the ladder and the wall, but the coefficient of static friction between the ladder and the floor is $\mu_s = 0.55$. A woman weighing 600 N (135 lb) slowly climbs the ladder. What is the maximum distance she can climb, measured up the ladder from its base, before the ladder will fall?

Reasoning

Question Why will the ladder fall if she climbs too high?
Answer As the woman climbs, she is changing the lever arm of the torque her weight creates about any chosen axis. This will affect the forces acting on the ladder at the wall and floor. One of these forces contributing to equilibrium, that of friction between the floor and the ladder, has a maximum possible value. If that maximum value is exceeded, the ladder will slip, causing it to rotate clockwise.

Question What forces are exerted on the ladder by the floor and the wall?
Answer At the floor, friction can exert a horizontal force H to the right and the floor supplies a vertical force V upward. The wall, being frictionless, can only exert a horizontal push P to the left.

Question I know where to place the weight of the ladder, but how do I decide where to place the woman's weight?
Answer Simply assign her a position on the ladder at a general distance x from the base. Then the free-body diagram of the ladder is as shown in Fig. 4.15b.

Question What am I ultimately looking for that will tell me the condition under which the ladder slips?

Answer   From the conditions for equilibrium, you want to find an expression showing how the friction force H depends on the position of the woman, x. Then you ought to be able to tell what value of x corresponds to the maximum possible value of H.

Question   What equations result from the first condition for equilibrium?
Answer $\Sigma \mathbf{F}_x = 0$: $H - P = 0$ so $H = P$
                  $\Sigma \mathbf{F}_y = 0$: $200\,\text{N} + 600\,\text{N} - V = 0$ so $V = 800\,\text{N}$

Question   What axis should I choose, and what equation results from the second condition?
Answer   As before, we can simplify the torque equation by choosing an axis through which the maximum number of forces act. This is an axis through point A in Fig. 4.15b. Verify that the lever arms of the forces about A are:

For P: $(6.0\,\text{m}) \sin 53° = 4.8\,\text{m}$

For the weight of the ladder: $(3.0\,\text{m}) \cos 53° = 1.8\,\text{m}$

For the weight of the woman: $x(\cos 53°) = 0.60x$

Then $\Sigma \tau = 0$ gives

$$(4.8\,\text{m})P - (1.8\,\text{m})(200\,\text{N}) - (0.60x)(600\,\text{N}) = 0$$

Question   How do I obtain a relation between H and x?
Answer   Notice that $H = P$ from one of the force equations. You can thus replace P with H in the torque equation and solve for x in terms of H:

$$(4.8\,\text{m})H - 360\,\text{m} \cdot \text{N} - 360x\,\text{N} = 0$$

$$x = \frac{(4.8\,\text{m})H - 360\,\text{m} \cdot \text{N}}{360\,\text{N}} = \left(\frac{H}{75}\right)\text{m/N} - 1\,\text{m}$$

Question   What is the condition for maximum x ($x_{\max}$)?
Answer   The last equation shows that x is directly proportional to H. Thus $x_{\max}$ corresponds to $H_{\max}$.

Question   What determines $H_{\max}$?
Answer $H_{\max} = \mu_s F_N$, where F_N is the normal force of the floor on the ladder. We have called this V in this problem, and have already found that $V = 800\,\text{N}$.

Solution and Discussion   We have $H_{\max} = (0.55)(800\,\text{N}) = 440\,\text{N}$. Then

$$x_{\max} = \frac{H_{\max}}{75} - 1 = \frac{440}{75} - 1 = 4.9\,\text{m}$$

The ladder will slip if the woman climbs to within 1.1 m of the top end of the ladder.

Exercise   What is the smallest value of μ_s for which the woman could climb all the way to the top of the ladder?   *Answer:* μ_s must be at least 0.66 in this case.

R

Example 4.9

To demonstrate that the choice of axis is arbitrary, return to Example 4.8 and choose an axis through point B in Fig. 4.15b. This axis is entirely outside the ladder. Verify that this choice gives the same result as the previous choice of an axis through point A.

Reasoning

Question What forces produce no torque about B?
Answer H and the ladder's weight, since they pass through point B.

Question What are the lever arms of the other forces about B?
Answer For P, the same: 4.8 m
For V: $(3\,\mathrm{m})\cos 53° = 1.8\,\mathrm{m}$
For the woman's weight: $(x - 3\,\mathrm{m})\cos 53° = (0.60)x - 1.8\,\mathrm{m}$

Our diagram shows $x > 3$ m. However, if we chose this incorrectly, and x comes out to be less than 3 m, the sign of the lever arm becomes negative, which automatically reverses the direction of the torque. As with incorrect guesses as to the direction of forces, incorrect guesses as to the sense of rotation will simply give you a reversed sign in the answer.

Question What is the torque equation about B?
Answer $(4.8\,\mathrm{m})P - (1.8\,\mathrm{m})V - (600\,\mathrm{N})(0.60x - 1.8\,\mathrm{m}) = 0$

Question Have the first-condition equations changed?
Answer Neither the choice of axis nor the placement of forces on the object affects the first condition.

Solution and Discussion Using the same results of the force equations as obtained in Example 4.8, we have $V = 800$ N and $H = P$. Notice that then the last equation above reduces to

$$(4.8\,\mathrm{m})H - (1.8\,\mathrm{m})(800\,\mathrm{N}) - (360\,\mathrm{N})x + 1080\,\mathrm{m \cdot N} = 0$$

which gives

$$(4.8\,\mathrm{m})H - (360\,\mathrm{N})x - 360\,\mathrm{m \cdot N} = 0$$

This is precisely the same relation between H and x as we obtained in Example 4.8.

4.7 BACK INJURY FROM LIFTING

You probably have been warned that there is a right and a wrong way to lift a heavy object. Let us apply what we have learned to see why this is true. As a typical situation, look at the man lifting a 60-N bowling ball in Fig. 4.16a. Back strain is likely to occur if the tension in the back muscle becomes too large or if the compression of the spine on the hip is excessive. We can easily calculate these forces by idealizing the situation, as shown in b. This model replaces the spine by

(a)

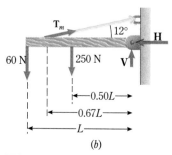

(b)

FIGURE 4.16

The forces in the man's back can be found using the model in (b).

In order for this "mobile" to remain stationary, not only must the first condition of equilibrium be satisfied, but also the second condition of equilibrium must be satisfied about any axis you may choose to consider. In particular, each point of support must be attached at the center of gravity of that part of the mobile that it supports.

a horizontal beam pivoted at the hip. Call the tension in the back muscle T_m and let the components of the force at the hip be H and V. Take the weight of the upper portion of the man's body to be 250 N with the dimensions shown.

As the man holds the ball at equilibrium, the appropriate equations become

$$\Sigma \mathbf{F}_x = 0: \qquad\qquad\qquad H - T_m \cos 12° = 0$$
$$\Sigma \mathbf{F}_y = 0: \qquad\qquad T_m \sin 12° + V - 60 - 250 = 0$$
$$\Sigma \tau = 0: \quad (250)(0.50\,L) + (60)(L) - (T_m \sin 12°)(0.67\,L) = 0$$

where all forces are in newtons. (Be sure you understand how the torque equation arises.) Divide through the last equation by L and solve to find $T_m = 1330$ N. Substitution in the other equations yields $V = 32$ N and $H = 1300$ N.

Notice how very large these forces are. Although the bowling ball weighs only 60 N, the tension in the back muscle is 1330 N and the force on the spine is about that large. Obviously, when you bend over to lift an object, you place tremendous strain on your back. If you squat down to lift an object and hold your back fairly erect, however, the forces are much smaller. You should be able to show this using the model in Fig. 4.16*b*.

LEARNING GOALS

Now that you have finished this chapter, you should be able to

1 Define (*a*) static equilibrium, (*b*) lever arm, (*c*) torque, (*d*) center of gravity.
2 Find the torque due to a given force with reference to a specified axis and apply the sign convention to determine the sign of the torque.
3 State the two conditions for static equilibrium in words and in equation form.

4 Give the position of the center of gravity of simple uniform objects and determine the position of the center of gravity of more complex objects.
5 Locate the force of gravity on an object in the object's free-body diagram.
6 Solve simple static problems by application of the two conditions for equilibrium.

SUMMARY

DEFINITIONS AND BASIC PRINCIPLES

Static Equilibrium
An object which is at rest and which remains at rest indefinitely is said to be in static equilibrium.

Lever Arm
The lever arm of a force about a selected axis is the perpendicular distance from the axis to the line along which the force acts.

Torque (τ)
The torque produced by a force about a selected axis is the product of the force times the lever arm of the force about the axis:

$$\tau = \text{lever arm} \times \text{force}$$

Torque has the SI units of m · N.

INSIGHT
The effect of a torque can be characterized as clockwise (cw) or counterclockwise (ccw), depending on whether that torque alone would tend to rotate the object in one or the other of these two opposite senses. To take account of these two opposite effects, we use the sign convention that ccw torques are positive and cw torques are negative. These effects can be identified by inspection of the free-body diagram of the object.

Center of Gravity (c.g.)
The point at which the force of gravity can be considered to

be applied for the purpose of calculating the torque it produces about any chosen axis.

INSIGHTS
1. This definition means that you can draw the weight of an object on a free-body diagram as acting through the object's center of gravity.
2. Objects made of uniform material and in symmetric shapes have their c.g.'s at their geometric centers.

First Condition for Equilibrium
The vector sum of all forces acting on an object in equilibrium must be zero: $\Sigma \mathbf{F} = 0$. This requires that $\Sigma \mathbf{F}_x = 0$, $\Sigma \mathbf{F}_y = 0$, and $\Sigma \mathbf{F}_z = 0$.

Second Condition for Equilibrium
The algebraic sum of cw and ccw torques acting about any axis on an object in equilibrium must be zero: $\Sigma \tau = 0$.

INSIGHTS
1. In applying the first condition for equilibrium, the place where the forces are attached to the object doesn't matter; only the direction of the forces is important.
2. In applying the second condition, it is necessary to know where the forces are attached to the object, so that torques about any chosen axis can be correctly calculated.
3. In applying the second condition for equilibrium, any axis may be chosen about which to calculate torques, even an axis outside the object.
4. Since torque is zero when the line of action of a force passes through an axis, it is advantageous to choose an axis through which as many forces pass as possible.

QUESTIONS AND GUESSTIMATES

1 A traffic light hung from a cable that stretches across a street invariably causes the cable to sag. Why don't the workers remove the sag when they hang the cable?

2 Draw the free-body diagrams for a 300-N girl in the following equilibrium situations: (*a*) she stands on one foot, (*b*) she hangs from a bar with one hand, (*c*) she stands on her head, (*d*) she does a handstand with a single hand resting on a stool.

3 Refer to Fig. 4.9. Will the tension in the upper cable increase or decrease as its angle with the vertical is decreased? What will be the tension in it when the cable is vertical?

4 The center of gravity of a hollow spherical shell is inside the shell. Name a few other objects for which the center of gravity is not on the object. Where, approximately, is the center of gravity of a mixing bowl? A clothes hanger?

5 We are told that slender people are less apt to have back trouble than obese people. Why should this be true?

6 A child watches a parade by sitting on its father's shoulders with its legs straddling his neck. Discuss the various ways the father could lower the child to the ground. Which ways could lead to serious injury to the father's back?

7 A strong horizontal wind topples a tree. Why is it not really correct to say that the wind pulled the tree out of the ground? Explain what actually happens.

8 A boy stands in a large garbage pail. Fastened to the handle of the pail is a rope that passes over a pulley suspended from the ceiling. The boy pulls on the free end of the rope and tries to lift the pail and himself. What happens to the tension in the rope and the force the boy exerts on the bottom of the pail as he slowly pulls harder and harder on the rope? Can the boy lift himself and the pail off the floor?

9 A woman is unable to loosen the nut that holds a lawn mower blade in place because she cannot apply enough force to the wrench she has available. She slips the handle of the wrench into a pipe 80 cm long and then removes the nut easily. Explain.

10 Each of the following tools makes use of torque. Describe the torques that exist in each case: wire cutters, wheelbarrow, crescent wrench, bottle opener, claw hammer, nutcracker.

PROBLEMS

Sections 4.1 and 4.2

1 A wood cube that weighs 25 N is fastened by a cord to the bottom of another 35-N wood cube. The 35-N cube is supported by a cord fastened to the ceiling. Find the tensions in the upper and the lower cords.

2 A physics book (weight = 12.0 N) sits at equilibrium on top of a 32-N dictionary which sits on a tabletop. Find the force with which (*a*) the table pushes up on the dictionary and (*b*) the dictionary pushes up on the physics book.

3 Three cords pull an object. Two cords have their forces in the xy plane and are 240 N at 30° and 320 N at 120°. (Angles are specified in usual way in the xy plane.) Find the force **F** of the third cord if the object is to remain in equilibrium.

4 An object is subjected to three forces in the xy plane: a force of 180 N at 105°, a force of 75 N at 240°, and a third force **F**. Find the force **F** if the object remains in equilibrium.

5 In Fig. P4.1, two objects weighing 90 N each hang from a cord that passes over a frictionless pulley. What are the tensions in the three cords if (*a*) the weight of the pulley is negligible and (*b*) the pulley weighs 25 N?

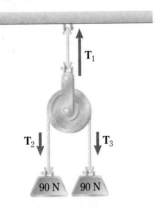

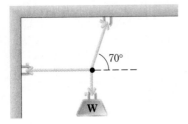

FIGURE P4.1

6 In Fig. P4.2 the weight $W = 1600$ N. What is the tension in (*a*) the horizontal cord and (*b*) the cord running to the ceiling?

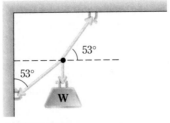

FIGURE P4.2

■ 7 If the tension in the horizontal cord of Fig. P4.2 is 390 N, what is the weight of the object?

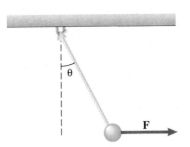

FIGURE P4.3

■ 8 In Fig. P4.3 the system is found to come to equilibrium for $\theta = 30°$ when horizontal force F is 240 N. What is the weight of the object which hangs at the end of the cord?

■ 9 If the object in Fig. P4.3 weighs 575 N, what is the angle θ if the system is at equilibrium with $F = 310$ N?

■10 In the previous problem, what is the tension in the cord?

11 A child holds a sled at rest on a frictionless snow-covered hill with an inclination of 30° by means of a rope held parallel to the hill. The sled weighs 100 N. Find the force the child must exert on the rope if the sled is to stay in equilibrium.

■12 In Fig. P4.4 the tension in the cord attached to the vertical wall is 72 N. Find (*a*) the tension in the cord attached to the ceiling and (*b*) weight W.

(figure, see FIGURE P4.4)

FIGURE P4.4

■13 In the previous problem if the weight $W = 300$ N, find the tensions in the two cords.

■14 Three weights W_1, W_2, and W_3 are in equilibrium as shown in Fig. P4.5. Pulleys are frictionless so that they do not affect the tensions in the cords. If $W_1 = 720$ N, find W_2 and W_3.

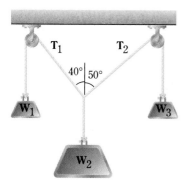

FIGURE P4.5

■**15** In the previous problem (Fig. P4.5) if $W_2 = 300$ N, what are the values of the weights W_1 and W_3 for the system to be in equilibrium?

■■**16** For the equilibrium situation shown in Fig. P4.6 $W_2 = 600$ N. The pulleys are frictionless so that they do not affect the tensions of the cords. Find the weights W_1 and W_3, and the tensions in the cords T_1 and T_2.

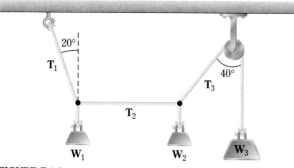

FIGURE P4.6

■■**17** For the equilibrium shown in Fig. P4.6 the tension in the cord $T_1 = 1200$ N. Find the weights W_1, W_2, W_3.

■**18** A jogger breaks her leg and is put in traction as shown in Fig. P4.7. Assume that the pulleys are frictionless; then the tension in the cord is everywhere the same, namely, 150 N. How large is the force stretching the leg? How large is the upward force exerted by the device on the foot and leg together?

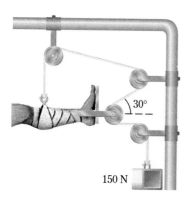

150 N

FIGURE P4.7

■**19** The pulleys in Fig. P4.8 are frictionless and have negligible weights. At equilibrium $W_1 = 600$ N. Find the weight W_2 and the tensions T_1, T_2, T_3, and T_4.

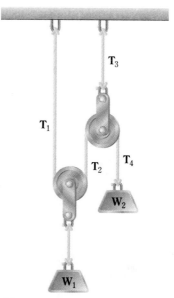

FIGURE P4.8

■**20** In Fig. P4.9 the pulleys are frictionless and have negligible weight. With what force must the 540-N man pull downward on the rope in order to support himself free from the floor?

FIGURE P4.9

Section 4.3

21 Find the torques about an axis through A for the forces shown in Fig. P4.10 if the length of the bar is $L = 5.0$ m.

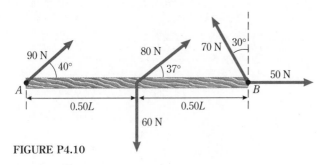

FIGURE P4.10

22 Find the torques about an axis through B for the forces shown in Fig. P4.10 if the length of the bar is $L = 8.0$ m.

23 For each force in Fig. P4.11, what is the (a) lever arm (b) torque about the axis through P? The square has sides of length 4 m.

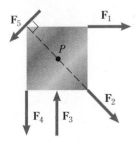

FIGURE P4.11

■24 The crank arm of a bicycle pedal is 16 cm long. If a 360-N girl puts all her weight on one pedal, how much torque is developed (a) when the crank is horizontal (b) when the pedal is 30° from the top?

25 The bolts on a motorcycle engine require 80 N · m of torque. How much force must a mechanic apply on a 20-cm wrench in order to open the bolt?

26 A 500-N diver stands at the end of a 4-m-long diving board. What torque does the weight of the diver produce about an axis at the midpoint of the diving board?

■27 A clock has a minute hand whose tip rubs against the inside of the glass cover. The frictional force between the tip and the glass cover is 0.04 N and the length of the hand is 5 cm. What is the minimum torque necessary to be applied to the minute hand if you do not want the clock to be stopped?

Sections 4.4 and 4.5

28 Two balls, weighing 200 N and 240 N respectively, are attached to the ends of a 1.2-m-long rigid bar of negligible weight. At what point must the bar be supported on a knife-edge so that it remains horizontal?

29 How large a force F applied vertically to the wheelbarrow handles in Fig. P4.12 will it take to lift a 600-N load at the center of gravity indicated? Take the lengths $a = 0.8$ m and $b = 0.2$ m.

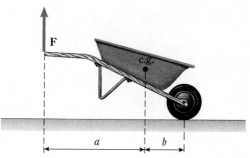

FIGURE P4.12

30 Two children are playing on a balanced seesaw. One 400-N child sits 1.2 m from the center. Where on the other side must the second child sit if her weight is 480 N?

■31 In Fig. P4.13 the vertical ropes with tensions T_1 and T_2 support two weights W_1 and W_2 at the ends of a weightless plank. If T_1 is 240 N and W_2 is 280 N, find the values of W_1 and T_2.

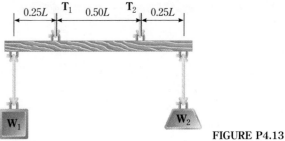

FIGURE P4.13

■32 The uniform 200-N plank in Fig. P4.13 is supported by two ropes. If each rope can withstand a tension of 900 N and if W_2 is to be twice as heavy as W_1, what is the greatest value W_1 can have? Assume the ropes holding the weights are strong enough not to break.

33 For the nail puller in Fig. P4.14, how large a force is applied to the nail when the force on the handle is 240 N? Assume the force applied to the nail to be vertical. Take the lengths $a = 0.3$ cm and $b = 5$ cm.

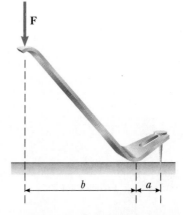

FIGURE P4.14

■**34** To locate a person's center of gravity, he is placed on two scales as shown in Fig. P4.15. The left and the right scales read 260 N and 200 N, respectively. Assume that the scale readings have been corrected by subtracting the readings when the person was not in place. Find the position x of the center of gravity if the length L is 2 m.

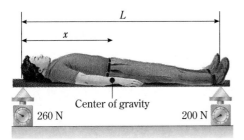

FIGURE P4.15

■**35** In Fig. P4.16, the uniform beam weighs 280 N. Find (*a*) the tension in the upper rope and (*b*) the horizontal and vertical components, H and V, of the force exerted by the pin if $W = 840$ N.

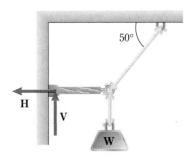

FIGURE P4.16

■**36** The uniform 450-N beam in Fig. P4.17 supports the load as shown. (*a*) How large can the load be if the horizontal rope is able to hold 2800 N? (*b*) What are the horizontal and vertical components of the force at the base of the beam?

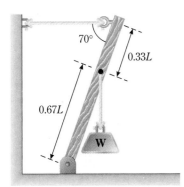

FIGURE P4.17

■■**37** An 8-m uniform ladder weighing 240 N rests against a smooth (frictionless) wall. The coefficient of static friction between the ladder and the ground is 0.7, and the ladder makes an angle of 45° with the ground. How far up the ladder can an 800-N fire fighter climb before the ladder begins to slip?

■**38** A window washer is standing on a scaffold supported by two vertical ropes at each end. The scaffold is uniform and weighs 300 N and is 4 m long. Find the tension in each rope when the 800-N washer stands 1.6 m from one end.

■■**39** When you stand on tiptoe, the situation is much like that shown in Fig. P4.18. The magnitude of F, the push of the floor, will be equal to the person's weight if the person is standing on one foot. If the weight of a person is 720 N, find (*a*) the tension in the Achilles tendon and (*b*) the components H and V at the ankle.

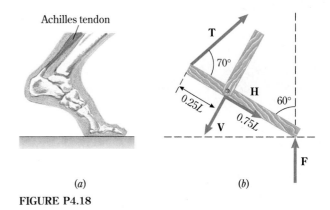

(*a*)   (*b*)

FIGURE P4.18

■■**40** In Fig. P4.19, the beam weighs 960 N and $T_3 = 840$ N. Find T_1, T_2, W, and the force with which the beam pushes down on the frictionless pin at its base.

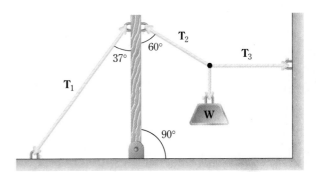

FIGURE P4.19

Sections 4.6 and 4.7

■**41** The uniform block in Fig. P4.20 is $2\frac{1}{2}$ times as tall as it is wide. Friction prevents the block from slipping. If the angle

θ is slowly increased, at what inclination will the block topple over?

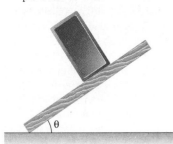

FIGURE P4.20

＂42 The bowling ball in Fig. P4.21 weighs 80 N and rests against the walls of a frictionless groove. How large are the forces that the walls of the groove exert on the ball? Consider the ball to be a uniform sphere.

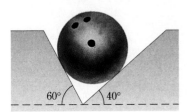

FIGURE P4.21

＂43 In Fig. P4.22, a uniform bar of length L and weight W protrudes from a wheel of radius b that can rotate freely on an axle. What weight w must an object hanging from the wheel's rim have if the system is to be at equilibrium in the position shown?

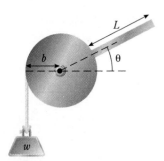

FIGURE P4.22

＂44 The rigid bar of negligible weight in Fig. P4.23 has a length L. It has very small frictionless wheels attached to its ends which can roll along the inclined sides of the equilateral triangle. Two weights, w and W, hang from the bar, each at a distance $0.25L$ from one of the ends. The bar comes to equilibrium at an angle $\theta = 12°$ to the horizontal. Find the ratio w/W.

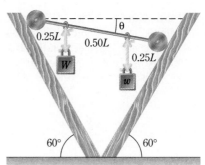

FIGURE P4.23

＂45 Three masses, as shown in Fig. P4.24, are arranged in an equilateral triangle. They are held in place by three thin rods of negligible weights. If we suspend this rigid system by a string attached to m_3, at what angle relative to the vertical will the side connecting m_3 to m_2 hang?

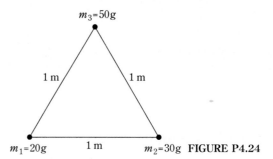

$m_1 = 20g$ 1 m $m_2 = 30g$ **FIGURE P4.24**

General Problems

＂46 The crane shown in Fig. P4.25 consists of a uniform 2-m-long beam which is allowed to pivot around a fixed axis through A. A cable is attached to the other end of the beam, B, and runs to a winch at C, 2 m directly above A. The crane is in equilibrium in the position shown, holding a 100-kg mass hanging from B. The beam itself has a mass of 20 kg. Find (*a*) the horizontal and vertical forces exerted on the beam at point A and (*b*) the tension in the cable BC.

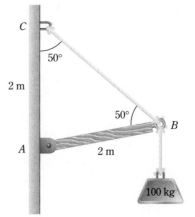

FIGURE P4.25

"47 You and a friend are sawing down a standing tree, and want to make sure the tree does not fall toward your house. You know you have the strength to exert a force of only 425 N, which might not be enough to keep the tree from falling on the house. Being a physics student who understands force components, you tie one end of a rope to the tree being sawed and the other end to a second tree located in the direction opposite your house. You then push sideways on the middle of the rope with a force of 425 N, as in Fig. P4.26. This causes the rope to make an angle of 8.0° relative to the straight line between the trees. As a result of your ingenuity, how much force are you able to exert on the tree in the direction opposite the house?

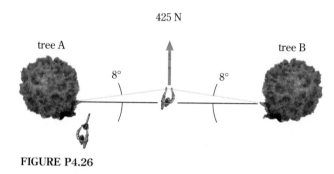

FIGURE P4.26

""48 You are rolling a barrel along level ground and come to a step 18.0 cm high, as shown in Fig. P4.27. In order to pull the barrel onto the step, you apply a horizontal force **F** at the top of the barrel as shown. If the barrel has a radius of 52.5 cm and weighs 1230 N, what is the minimum force **F** which will roll the barrel onto the step?

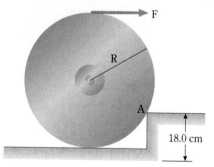

FIGURE P4.27

"49 A uniform board of mass 13.6 kg and length 4.4 m hangs over a platform by a distance of 1.4 m. A 9.6-kg puppy walks out on the board. How far can the puppy go beyond the edge of the platform before the board begins to tip?

""50 In moving a heavy chest up a flight of stairs, you and a friend grab opposite ends of the chest and apply vertical forces to the bottom. You tell your friend that you will lead going up the stairs, and you proceed, with the chest tipping so that its bottom surface makes a 37° angle above the horizontal. The forces the two of you apply are shown in Fig. P4.28. Assume the chest is uniform, has a mass M, a length L, and a height $h = 0.4 L$. Which of you exerts the greater force?

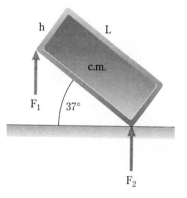

FIGURE P4.28

CHAPTER 5

WORK AND ENERGY

In principle, we can describe all motions in terms of the forces that cause them. In many instances, however, the concepts of work and energy, introduced in this chapter, greatly simplify the description of motion. For one thing, work and energy are *scalar* quantities and thus simpler to deal with mathematically than force *vectors*. Even more important, we shall see that energy has many forms and occurs in every branch of physics.

One of the most fundamental and unifying concepts in physics is the principle of **conservation of energy** in all physical processes. In order to understand this principle, we shall begin by studying the definitions of work and energy.

5.1 THE DEFINITION OF WORK

When you sit at your desk studying this book, you are not doing work. This does not mean that you are lazy or that learning physics is an effortless process. It is simply stating a fact that arises from the definition of work the scientist uses.

Scientists define the work done by a force in the following way. Suppose, as shown in Fig. 5.1, a force $\mathbf{F}$ pulls an object from A to B through a displacement $\mathbf{s}$. We represent the component of $\mathbf{F}$ in the direction of $\mathbf{s}$ by F_s. Then the work done by $\mathbf{F}$ during the displacement $\mathbf{s}$ is

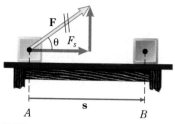

FIGURE 5.1

The work done by **F** in displacing the object from A to B is $F_s s = (F \cos \theta)s$.

Work done by $\mathbf{F} = F_s s$ $\qquad$ (5.1a)

To repeat a point made above, work is a scalar quantity; it has no direction associated with it.

In the SI, forces are measured in newtons and distances are measured in meters, and so the unit for work is the newton-meter (N · m). This unit is given a special name, the joule (J).

A **joule** is the work done by a force of one newton as it acts through a distance of one meter along the line of the force: $1\,J = 1\,N \cdot m$.

Other units sometimes used to measure work are the foot-pound (ft · lb), the erg, and the electronvolt (eV), where

$1\,ft \cdot lb = 1.356\,J$
$1\,erg = 1 \times 10^{-7}\,J \quad$ exactly
$1\,eV = 1.602 \times 10^{-19}\,J$

Quantities expressed in these other units must always be changed to joules before they can be used in our fundamental SI-based equations.

We can express the defining equation for work in a form different from Eq. 5.1a by noticing in Fig. 5.1 that

$$F_s = F \cos \theta$$

where θ is the angle between **F** and **s**. Substituting this value for F_s in Eq. 5.1a, we obtain

Work done by $\mathbf{F} = Fs \cos \theta$ $\qquad$ (5.1b)

In summary,

The work W done by a force **F** acting on an object during a displacement **s** is $F_s s$ or $Fs \cos \theta$.

In these equivalent expressions, F_s is the component of **F** in the direction of the displacement **s** and the angle θ is the angle between **F** and **s**.

Notice that the presence of $\cos \theta$ in Eq. 5.1b implies that work can be positive or negative. It is positive when $0° < \theta < 90°$ (**F** has a component in the direction of the displacement) and negative when $90° < \theta < 180°$ (**F** has a component opposite the direction of the displacement). This definition of work applies to each individual force that may be acting in a given situation. That is, the work done by each force may be calculated by applying Eq. 5.1.

Illustration 5.1

In Fig. 5.2, a person exerts a vertical force **F** on a pail as the pail is carried a horizontal distance of 8.0 m at constant speed. How much work does **F** do?

Reasoning The definition of work is $W = Fs \cos \theta$. In Fig. 5.2, **F** is vertical and **s** is horizontal. Hence $\theta = 90°$ and

FIGURE 5.2

No work is done on the pail by **F** since **F** has no component in the direction of the displacement.

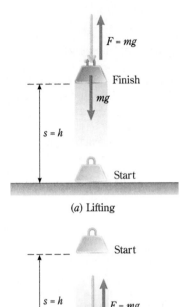

(a) Lifting

(b) Lowering

FIGURE 5.3

The work done by the lifting force **F** is *mgh* in (a) and −*mgh* in (b).

$$W = Fs \cos 90° = 0$$

No work is done by the vertical force because it has no component in the direction of motion. Notice also, that to begin the horizontal motion, there had to have been a momentary component of horizontal force, but no force is needed to maintain a constant horizontal velocity.

Illustration 5.2

How much work do you do on an object of weight *mg* as (**a**) you lift it a distance *h* straight up at constant speed and (**b**) you lower it through this same distance, again at constant speed?

Reasoning (**a**) The lifting situation is shown in Fig. 5.3*a*. To lift the object, you must pull up on it with a force equal to its weight *mg*.* The displacement is *h* upward, and the lifting force is in the same direction. Therefore, from the definition of work,

$$W = Fs \cos 0° = (mg)(h)(1) = mgh$$

This is the work you do as you lift the object a distance *h*.
(**b**) Figure 5.3*b* shows what happens when you lower the object. Now **F** and **s** are in opposite directions. Therefore *F* = *mg* and *θ* = 180°. We then find from *W* = *Fs* cos *θ* that

$$W = (mg)(h)(\cos 180°) = mgh(-1) = -mgh$$

The work you do is negative in this case because the force you exert, **F**, is in a direction opposite to that of **s**. Another way of looking at doing negative work is that the work is being done *on* you, not *by* you. Gravity is doing positive work on the object in this case. Similarly, in part (*a*) we could say that the force of gravity was doing negative work on the pail when you lifted it.

Exercise How much work does the pull of gravity do on the object of Illustration 5.2 as (*a*) it is lifted and (*b*) it is lowered? *Answer:* (*a*) −*mgh*; (*b*) *mgh*

Example 5.1

The box in Fig. 5.4 is being pulled along the floor at constant speed by a force **F**. Let us say that the friction force opposing the motion is 20 N and that the box has a mass of 30 kg. Find the magnitude of **F** and work done by **F** as the box is moved 5.0 m.

---

*A force slightly larger than *mg* is needed to give the object an initial acceleration upward, but once the object is moving, a force *mg* upward balances the pull of gravity and the object keeps moving with constant velocity.

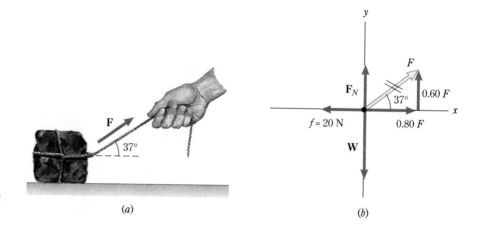

FIGURE 5.4

The horizontal component of **F** does work on the box, while the work done by the vertical component is zero.

(a)

(b)

R *Reasoning*

Question What must I know in order to calculate the work?
Answer The pulling force, or at least its component in the direction of displacement, the displacement, and the angle between **s** and **F**.

Question The displacement and angle are given, but the pulling force **F** is not. What clue does the problem statement give about **F**?
Answer **F** is such that the box moves at constant speed along the floor. This means that $\Sigma F_x = 0$, or $F_x = f = 20\,\text{N}$, in the direction opposite **f**.

Question What is the expression for the work done by **F** in this case?
Answer $W = F_x x$.

Question Does the box's mass play any role?
Answer No. Mass plays a role in determining weight and friction force, but weight is perpendicular to the displacement in this case, and so does no work on the box. *f* is given directly. Often you are given more information than you need to solve a problem. Part of the solution is to be able to recognize the relevant information.

Question What is the relation between F_x and F?
Answer $F_x = F \cos 37°$.

Solution and Discussion The magnitude of the applied force is

$$F = \frac{F_x}{\cos 37°} = \frac{20\,\text{N}}{0.80} = 25\,\text{N}$$

The work done by **F** is

$$W = F_x x = (20\,\text{N})(5.0\,\text{m}) = 100\,\text{J}$$

Remember that, by definition, the vertical component of **F** does no work on the box as long as the box's motion is entirely horizontal.

Exercise Calculate the work done by the force of friction. *Answer:* $-100\,\text{J}$.

Who is expending more power: the runner in (*a*) or the man walking up the stairs in (*b*)?

5.2 POWER

When you purchase an automobile, you may be interested in the horsepower of its engine. You know that usually an engine with large horsepower is most effective in accelerating the automobile. Let us now learn the precise meaning of power.

Power measures the *rate* at which work is being done. Its defining equation is

$$\text{Power} = \frac{\text{work done}}{\text{time taken to do the work}}$$

or, in symbols,

$$P = \frac{W}{t} \qquad\qquad (5.2)$$

When work is measured in joules and t in seconds, the unit for power is the joule per second, which is called the watt (W), after the Scottish inventor of the steam engine, James Watt.

$$1 \text{ watt} = \frac{1\,\text{J}}{\text{s}}$$

For motors and engines, power is often measured in horsepower (hp), where

$$1 \text{ hp} = 746\,\text{W}$$

Of course, since the watt is the SI unit for power, we must use it rather than horsepower in our equations. For example, an electric motor that has a power rating of $\frac{1}{4}$ hp is capable of producing

$$\left(\frac{1}{4}\,\text{hp}\right)\left(746\frac{\text{W}}{\text{hp}}\right) = 186\,\text{W}$$

of power. That means that the motor can perform 186 J of work each second.

We obtain another convenient relation for power when we notice that the work done on an object by a force F_x as the force displaces the object a distance x is $F_x x$. Using this expression in Eq. 5.2, we find

$$P = \frac{W}{t} = \frac{F_x x}{t} = F_x\left(\frac{x}{t}\right)$$

Now, since x/t is the speed v_x with which the object is moving in the x direction, we have that

$$P = F_x v_x \tag{5.3}$$

or

$$P = Fv \cos \theta$$

where θ is the angle between **F** and **v**. Equations 5.2 and 5.3 both assume that the power output is constant. If F_x and/or v_x varies with time, then Eq. 5.2 gives the average power during the time interval t and Eq. 5.3 gives the instantaneous power at the instant at which F_x and v_x have the values being used.

Equation 5.2 is used to define a frequently encountered unit used for work. Noting that

Work = power × time

we measure power in kilowatts and time in hours. Then the work done by the source of the power has the units of kilowatts × hours, and this unit of work is called the kilowatthour (kWh). It can be related to the joule through

$$1\,\text{kWh} = (1\,\text{kWh})\left(1000\frac{\text{W}}{\text{kW}}\right)\left(3600\frac{\text{s}}{\text{h}}\right) = 3.60 \times 10^6\,\text{W} \cdot \text{s} = 3.60 \times 10^6\,\text{J}$$

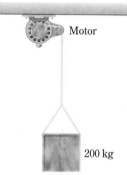

Motor

200 kg

FIGURE 5.5

We wish to find the power output of the motor as it lifts the object with a speed of 3.00 cm/s.

R

Example 5.2

The motor in Fig. 5.5 is lifting a 200-kg object at a constant speed of 3.00 cm/s. What power, in horsepower, is being developed by the motor?

Reasoning

Question What quantities must I know to calculate the power being developed by the motor?
Answer You can solve this problem with either Eq. 5.2 or Eq. 5.3. Since the velocity of the object is given, Eq. 5.3 is more convenient.

Question What condition determines the force being exerted by the motor?
Answer The load is being raised at constant speed. Thus there is no net force, and so the upward force exerted by the motor must equal the weight of the load: $F = mg$.

Question What is the expression for the power in this case?
Answer Since the velocity and force are in the same direction ($\theta = 0$), $P = Fv$.

Solution and Discussion   Putting in the given values,

$F = (200\,\text{kg})(9.8\,\text{m/s}^2) = 1960\,\text{N}$
$P = (1960\,\text{N})(0.0300\,\text{m/s}) = 58.8\,\text{N} \cdot \text{m/s} = 58.8\,\text{W}$

Converting to horsepower

$$58.8\,\text{W}\frac{1\,\text{hp}}{746\,\text{W}} = 0.0788\,\text{hp}$$

To see the connection between this method and Eq. 5.2, consider how far the load would travel in one second, namely $s = 3.00\,\text{cm}$. The work the motor would do is

$$W = Fs = (1960\,\text{N})(0.0300\,\text{m}) = 58.8\,\text{J}$$

Since this was done in 1 s, the power is $P = W/t = 58.8\,\text{J/s} = 58.8\,\text{W}$.

Exercise   What would the motor's power output be, in watts, if the load was being lowered at a speed of 3.0 cm/s?   *Answer:* $-58.8\,\text{W}$

5.3   KINETIC ENERGY

If an object can do work, we say that the object possesses energy. For that reason, *energy is often said to be the ability to do work.* Although, as we shall see, the concept of energy is too complex to be described fully in such a brief statement, associating energy and work still proves useful. There are many kinds of energy, and we begin our study of them by considering kinetic energy.

A moving baseball can break a window, a moving hammer can drive a nail, and a stone moving upward can rise against the force of gravity. Obviously, moving objects have the ability to do work; in other words, they possess energy. We call the energy possessed by an object because of its motion **kinetic* energy (KE).**

To take a concrete situation, suppose a loaded wagon of total mass m is coasting with velocity v_o, as in Fig. 5.6. As indicated, a person is pulling backward on the wagon with a constant force $-F_x$, trying to stop its motion. According to Newton's third law, the wagon exerts an equal force forward on the person. As the wagon

*From the Greek word *kinetikos,* to move. Recall that in Chap. 2 we used the term *kinematics* to describe our study of motion, and in Chap. 3, sliding (moving) friction was called *kinetic* friction.

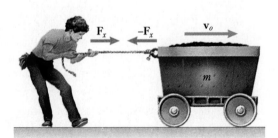

FIGURE 5.6

The wagon loses kinetic energy as the person pulling backward on it slows it.

An impressive example of kinetic energy.

and the person move through a distance x, the work done on the person by the wagon is

$$W(\text{on person}) = F_x x$$

Let us now relate this amount of work to the resulting change in the wagon's motion.

Because of the retarding force $-F_x$ being exerted on it, the wagon is decelerating. According to Newton's second law,

$$a_x = \frac{-F_x}{m}$$

Using the equation of motion $v_f^2 - v_o^2 = 2a_x x$ (Eq. 2.9) to replace a_x with $(v_f^2 - v_o^2)/2x$, we find that

$$F_x = -\left(\frac{m}{2x}\right)(v_f^2 - v_o^2)$$

If we place this expression for F_x into the equation for work done on the person, we get

$$W(\text{on person}) = \tfrac{1}{2}mv_o^2 - \tfrac{1}{2}mv_f^2 \qquad (5.4)$$

This expression tells us how much work a moving object can do as it slows from a speed of v_o to a speed v_f. If the wagon is brought to rest, so that $v_f = 0$, the work it will have done is $\tfrac{1}{2}mv_o^2$. We therefore conclude that an object of mass m traveling with speed v is capable of doing an amount of work equal to $\tfrac{1}{2}mv^2$ as it is brought to rest.

Using this observation as a rationale, we define an object's kinetic energy as follows:

The **kinetic energy (KE)** of an object of mass m moving with a speed v is

$$KE = \tfrac{1}{2}mv^2 \tag{5.5}$$

You can quickly verify from Eq. 5.5 that the SI unit of kinetic energy is the same as for work, namely joules. Notice that kinetic energy is a scalar, as are all other forms of energy. Also, since m and v^2 are always *positive* quantities, so is kinetic energy.

5.4   THE WORK-ENERGY THEOREM FOR NET FORCE

In this section, we obtain a relationship between the work done on an object and the change in the object's kinetic energy. We could do this by computing the work done on the wagon of Fig. 5.6, but instead, let us take the more general situation shown in Fig. 5.7: a wagon of mass m moving in the positive x direction under the influence of two forces. Call the resultant force acting on the wagon $\mathbf{F}_{net}$. Then, because the motion is in the x direction, $\mathbf{F}_{net} = m\mathbf{a}$ becomes

$$F_{net} = ma_x$$

As we did in the preceding section, we use Eq. 2.9 to replace a_x by the initial and final velocities of the object together with the distance moved (x) to obtain

$$F_{net}x = \tfrac{1}{2}mv_f^2 - \tfrac{1}{2}mv_o^2$$

However, $F_{net}x$ is simply the work done on the wagon by the resultant force acting on it. Therefore our result can be summarized as

$$\text{Work done on object by } F_{net} = \text{change in object's KE}$$
$$\text{Work done by } F_{net} = \tfrac{1}{2}mv_f^2 - \tfrac{1}{2}mv_o^2 = \Delta KE \tag{5.6}$$

This relationship is called the **work-energy theorem for net force.** In applying it, you should recognize that a net force in the direction of motion speeds up the motion; it increases the kinetic energy. Retarding forces such as friction, however, do negative work on the object. This follows from the fact that the direction of the retarding force is opposite the direction of the displacement, and so $F_x x \cos \theta$ becomes $F_x x \cos 180°$, which is $-F_x x$. Hence a net retarding force decreases the kinetic energy.

A net force in the direction of motion increases the kinetic energy of an object, whereas a net stopping force decreases it.

The work-energy theorem is an extremely powerful relationship, and we shall be using it throughout our study of physics.

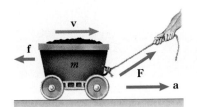

FIGURE 5.7

The resultant force acting on the wagon causes its kinetic energy to change.

FIGURE 5.8

The net force acting on the coasting car is $\mathbf{f}$.

Example 5.3

A 2000-kg car traveling at 20 m/s coasts to rest on level ground in a distance of 100 m. How large is the average friction force acting on the car? See Fig. 5.8.

R

Reasoning

Question   Are there any forces acting horizontally other than friction?
Answer   No.

Question   What principle relates the average friction force f to the stopping of the car?
Answer   You could use the equations of kinematics (2.11a to 2.11e) to find the acceleration of the car and then find f from Newton's second law as we did in Chap. 3, or you could use the work-energy theorem for net force, which equates the change in kinetic energy to the work done by the net force. One advantage of the work-energy approach is that it allows you to calculate with scalar quantities. Using scalars rather than vectors makes solving problems simpler in many cases.

Question   Do the data given allow me to calculate ΔKE?
Answer   Yes. $\Delta\text{KE} = \frac{1}{2}m(v_f^2 - v_o^2)$, where $v_f = 0$.

Question   What is the expression for work in this situation?
Answer $W = fs\cos 180°$, since f and s are in opposite directions.

Solution and Discussion   The work-energy theorem says

$$\tfrac{1}{2}[0 - (2000\ \text{kg})(20\ \text{m/s})^2] = f(100\ \text{m})(-1)$$

giving

$$f = \frac{\frac{1}{2}(2000\ \text{kg})(400\ \text{m}^2/\text{s}^2)}{100\ \text{m}} = 4000\ \text{kg} \cdot \text{m/s}^2 = 4000\ \text{N}$$

Exercise   If the friction force on the car of Example 5.3 is constant at 4000 N, use the work-energy theorem to find how fast the car is going after traveling 50 m.   *Answer:*   14.1 m/s.

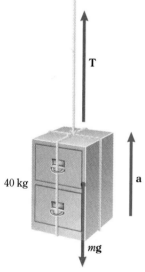

FIGURE 5.9

To accelerate the object upward, T must be larger than mg.

Example 5.4

A rope pulling upward on the 40-kg object in Fig. 5.9 is to accelerate the file cabinet from rest to a speed of 0.30 m/s in a distance of 50 cm. Use the work-energy theorem to find the required tension in the rope.

R

Reasoning

Question   How does the work-energy theorem involve the tension in the rope?
Answer   The tension is one of the forces that compose the net force. The net force does work equal to the change in kinetic energy.

Question   What is the net force on the cabinet?
Answer $T - mg$. This net force must be directed upward if the object is to be accelerated upward.

Question   How much work does the net force do?
Answer   Since F_{net} and the displacement s are parallel, $\cos\theta = 1$ and $W = (T - mg)s$.

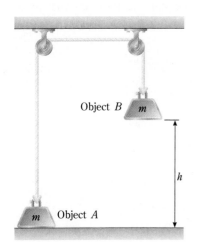

FIGURE 5.10

As object B falls, it does work by lifting object A.

Question   What equation does the work-energy theorem give here?
Answer $(T - mg)s = \frac{1}{2}mv_f^2 - 0$, with T the only unknown.

Solution and Discussion   Solving for T,

$$T = \frac{\frac{1}{2}mv_f^2}{s} + mg = \frac{\frac{1}{2}(40\,\text{kg})(0.30\,\text{m/s})^2}{0.50\,\text{m}} + (40\,\text{kg})(9.8\,\text{m/s}^2) = 396\,\text{N}$$

Notice that the work done by the tension is $Ts = (396\,\text{N})(0.50\,\text{m}) = 198\,\text{J}$. The work done by gravity is $-mgs = -(40\,\text{kg})(9.8\,\text{m/s}^2)(0.50\,\text{m}) = -196\,\text{J}$.

Exercise   If the rope will break when the tension in it exceeds 600 N, what maximum speed can be given to the cabinet in the 50 cm through which it is lifted?   *Answer:*   2.28 m/s.

5.5   GRAVITATIONAL POTENTIAL ENERGY

As we have seen, some objects are able to do work by virtue of their motion. They have kinetic energy. Other objects can do work because of either their position or their configuration. Such objects are said to have *potential energy*. Let us begin our study of potential energy by discussing the energy an object has because of gravitational forces.

Consider the system in Fig. 5.10, where the pulleys are assumed frictionless. Because both objects have the same weight mg, when object B is given a slight downward push to start it moving, it will fall slowly toward the floor with constant speed. At the same time, object A will rise slowly. By the time B has fallen the distance h to the floor, A has been lifted a distance h.

We now ask: How much work is done on object A by the rope as the object is raised from the floor with constant speed? Because the tension in the rope is equal to A's weight, mg, the lifting work done by the rope is, from the definition of work,

Lifting work done = (tension)(distance) = mgh

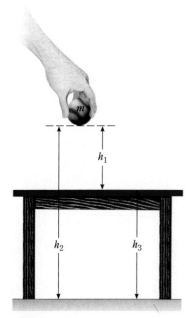

FIGURE 5.11

The floor and the table top provide two convenient choices for a reference height. Depending on which choice we make, the gravitational potential energy of the ball can be either mgh_1 or mgh_2. Notice that the two values differ by the additive constant mgh_3.

Who or what external agent does this work? The weight of the object B pulls object A up, and so B does the work. We must therefore conclude that object B possessed the ability to do work when it hung in its original position above the floor. The amount of work B can potentially do is mgh, where h is the distance through which B can fall. With this in mind, we define

Gravitational potential energy (GPE) = mgh (5.7)

Once again, as for all other forms of energy, the SI unit of GPE is the joule.

Gravitational potential energy is not defined in absolute terms. Instead, it depends on the vertical position used as a reference. In any given situation, two people may choose different reference levels, giving values of GPE that differ by an additive constant. For example, consider the ball in Fig. 5.11. For the person who considers the height of the ball above the table, the GPE of the ball is mgh_1. A person who chose the floor as a reference, however, would say its GPE is mgh_2. Both values are *correct,* as long as the reference level is specified.

What is meaningful from a physics standpoint are *changes* in potential energy when vertical position changes. If the ball in Fig. 5.11 falls 1 m, that change in position is the same in any and all choices of reference level.

Negative potential energies are possible. For example, suppose one chooses to measure distances from the top of the table in Fig. 5.11. When the ball is at a distance y above the table, its potential energy is mgy. As it is lowered to the tabletop, its gravitational potential energy decreases to zero. When it is lowered still farther, its y coordinate becomes negative, and so its gravitational potential energy becomes negative. This simply means that the ball has less potential energy below the table than it has at the table top, the position arbitrarily taken as the zero level. In order to restore it to the zero level of GPE, the ball must be lifted.

If the weightlifter is 1.6 m tall, can you estimate the gravitational potential energy the weights have relative to the floor?

Illustration 5.3

You are in a room that is 3.00 m high from floor to ceiling. A table top is 1.10 m above the floor. A 2.27-kg sack of flour is on the table top.

Part A   What is the gravitational potential energy of the sack relative to (*a*) the floor, (*b*) the table top, and (*c*) the ceiling?

Reasoning   In each case, the weight of the sack is $mg = 22.2$ N. The vertical positions relative to the reference levels are:

$$h_a = (1.10 \text{ m}) \qquad h_b = 0 \qquad h_c = -1.90 \text{ m}$$

The values of GPE are:

(*a*) GPE $= (22.2 \text{ N})(1.10 \text{ m}) = 24.4$ J
(*b*) GPE $= 0$
(*c*) GPE $= (22.2 \text{ N})(-1.90 \text{ m}) = -42.2$ J

Part B   If the sack of flour is moved from the table top to the floor, what is the change in its GPE relative to the three reference levels in part A?

Reasoning   In general, since mg is constant,

$$\Delta\text{GPE} = \Delta(mgh) = mg\,\Delta h$$

In each of the three cases above, $\Delta h = -1.10$ m. For each case,

$$\Delta\text{GPE} = (22.2 \text{ N})(-1.10 \text{ m}) = -24.4 \text{ J}$$

The specific changes in each case are:

(*a*) $\Delta\text{GPE} = 0 - (+24.4 \text{ J}) = -24.4 \text{ J}$
(*b*) $\Delta\text{GPE} = -24.4 \text{ J} - 0 = -24.4 \text{ J}$
(*c*) $\Delta\text{GPE} = -66.6 \text{ J} - (-42.2 \text{J}) = -24.4 \text{ J}$

Thus the *change in GPE does not depend on the reference level chosen.* It is only these changes that are physically significant. ∎

A high jumper curves her body in such a way that her center of mass actually stays below the bar.

5.6   THE CENTER OF MASS

So far, we have treated gravitational potential energy as if objects were point masses, with no size. When calculating GPE for real objects, a question arises as to what point to use to measure vertical position. If an object is lifted in such a way that it is not rotated, all points on it go through the same change of height, and so any point can be used to measure GPE. Suppose, however, we consider a uniform rectangular block lying on its largest face, as Fig. 5.12. How much work is necessary to set the block up on its smallest face?

From previous discussions, we can see that the work is going to equal the increase in GPE, since no other form of energy is changed:

$$W = \Delta\text{GPE} = mg\,\Delta h$$

Note, however, that not *all* points on the block are going to be lifted by the same amount. Because we have different parts of the block moving different vertical distances, it becomes crucial that we know what Δh refers to.

The key to knowing what value of Δh to use lies in what is called the **center of mass (c.m.)** of an object. In Chap. 4, we introduced the center of gravity as the point through which the force of gravity can be considered as acting on an object. When the force of gravity is constant across the dimensions of an object, the center of gravity and the center of mass coincide. This is certainly going to be the case in most problems we encounter in this book. We observed in Chap. 4 that for objects of uniform density and geometric symmetry, we can take the c.g. (and c.m.) to be at the object's *geometric center*. (For cases where this is not so, there is a way to calculate the c.m. from its definition, which we do not need to do here.)

We can now use the concept of center of mass to specify the meaning of Δh:

The change in gravitational potential energy of an object depends on the change in vertical position of the object's center of mass.

Near the surface of earth, this can be written as

$$\Delta\text{GPE} = mg\,\Delta h_{\text{c.m.}} \qquad\qquad (5.8)$$

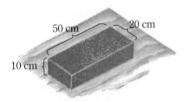

FIGURE 5.12

A uniform block lying on a table. How much work must we do to stand the block on its smallest face?

Illustration 5.4

Calculate the work necessary to raise the block in Fig. 5.12 so that it sits on its smallest face. The mass of the block is 10 kg.

Reasoning We need to identify the initial and final positions of the block's center of mass. Since the block is uniform, we take its c.m. to be at its geometric center. Inspection of Fig. 5.12 shows that this point is 5 cm above the table when the block is lying on its largest face. When the block is set on one of its smallest faces, the c.m. is 25 cm above the table. So we have that $\Delta h_{\text{c.m.}} = 20$ cm $= 0.20$ m, and the change in GPE is

$$\begin{aligned} \Delta \text{GPE} &= mg \, \Delta h_{\text{c.m.}} = (10 \text{ kg})(9.8 \text{ m/s}^2)(0.20 \text{ m}) \\ &= 19.6 \text{ J} \end{aligned}$$

This is how much work must be done to reposition the block. ∎

5.7 THE GRAVITATIONAL FORCE IS CONSERVATIVE

To lift an object straight up at constant speed, a force equal to the weight of the object mg is required. As a result, the work done in lifting the object straight up through a distance h is mgh. We now show that even if the object is not lifted straight up, this same result is true.

Suppose we wish to lift the pail in Fig. 5.13*a* from the floor to the tabletop. How much work must we do? Let us lift it along the path shown by the line from A to B, with the lifting force being directed vertically upward throughout the motion.

To compute the work done in lifting the pail from A to B, we approximate the actual path by the stepped path shown in *b*. If we make the step lengths very small, we can make the stepped path identical to the smooth path shown in Fig. 5.13*a*. We know that the lifting force is vertical. Therefore, it does no work in the horizontal movements of the stepped path. Work is done by the lifting force only in the vertical movements. When the pail is raised, positive work is done, but when it is lowered (as near point C), negative work is done. As a result, the work done in the downward movements cancels the work done in the equivalent upward movements. We therefore conclude that the work done depends only on the net effect of all the vertical movements. In going from A to B, the pail, of mass m, is lifted a net distance h. As a result, the work done is mgh, which is the same as the work done in lifting the pail straight up from A through a distance h and then moving it sideways to B. In fact, since the path shown from A to B is perfectly arbitrary, we conclude that

If point A is a distance h below point B, the work done against the force of gravity in lifting a mass m from A to B is mgh.

This result holds for any path taken between A and B, as long as g does not change in going from A to B. Of course, if the mass is lowered from B to A, the work done against the gravitational force is $-mgh$.

The gravitational force is an example of what we call a **conservative force.**

A force is said to be conservative if the work done in moving an object from point A to point B against the force is not dependent on the path taken for the movement.

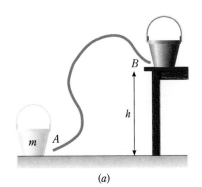

(a)

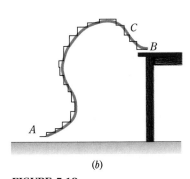

(b)

FIGURE 5.13

The path in (*a*) can be approximated by the series of horizontal and vertical steps in (*b*).

We shall see later that electrostatic and nuclear forces are also conservative. So are *elastic forces,* such as those exerted by a stretched or compressed spring. Friction forces, on the other hand, are not conservative. You can easily verify this by sliding your textbook from one point to another across a table. You obviously have to do more work when you slide it by way of a complicated long path than when you follow a straight-line path. If the work done by a force depends on the path taken between two points, as is the case for friction, that force is said to be *nonconservative.*

An equivalent way of distinguishing conservative from nonconservative forces is that it is possible to define a potential energy associated with conservative forces. This is not possible with nonconservative forces, since they are dependent on path, not solely on position.

To see why we call some forces conservative, let us define a system's **mechanical energy (ME)** as the sum of its kinetic and potential energies.

ME = KE + PE

The term *potential energy* can include more than one type of potential energy if more than one conservative force is acting. We find that if *only* conservative forces are acting on a system, the mechanical energy of the system is conserved, or remains constant, as the system moves. Thus we can summarize a very important property of conservative forces:

Conservative forces are those that conserve the mechanical energy of a system.

This is one form of a more general statement called the conservation of energy, which we shall discuss in succeeding sections. Conservation laws are among the most important statements in physics. They tell us which physical quantities remain constant when changes are taking place in a system.

5.8   THE INTERCONVERSION OF KINETIC AND POTENTIAL ENERGY

Each time you toss an object into the air or drop it, you see an example of the interchange of kinetic energy and gravitational potential energy. For example, when you toss a coin upward, its kinetic energy changes to gravitational potential energy, as we shall now prove.

In Fig. 5.14, a coin of mass m is thrown upward with an initial velocity v_o. At the instant it is at the top of its path, $y = h$ and $v_f = 0$. During the motion, the coin's acceleration remains constant, $a = -g$. Thus, from Eq. 2.9, $v_f^2 - v_o^2 = 2ay$, we have

$$0 - v_o^2 = -2gh$$

Solving this equation for h gives $h = v_o^2/2g$. We can use this value for h in the expression for the GPE of the coin at the top of its path:

$$GPE = mgh = mg\frac{v_o^2}{2g} = \tfrac{1}{2}mv_o^2$$

This shows that, when air drag can be neglected, *the gravitational potential energy of an object at the top of its path equals the kinetic energy it had at the bottom.*

The coin's original kinetic energy changes to GPE as the coin rises. A similar

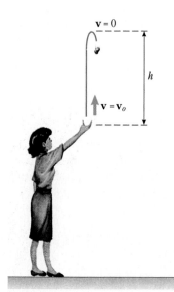

FIGURE 5.14

The kinetic energy of the coin is changed to gravitational potential energy as the coin rises. As it falls, the potential energy is changed back to kinetic energy.

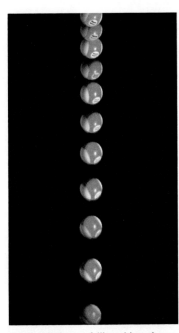

Another look at a falling object shows the conversion of gravitational potential energy into kinetic energy— as the ball loses height (GPE), it gains speed.

situation exists as the coin falls freely: it loses GPE but gains an equal amount of KE. This is an example of the conservation of mechanical energy. When gravity alone acts on an object, we can write this as follows:

$$\Delta ME = 0 = \Delta KE + \Delta GPE$$

Thus

$$\Delta KE = -\Delta GPE$$

If other conservative forces are present, changes in their corresponding potential energies would be included in the same way as ΔGPE.

5.9 THE LAW OF CONSERVATION OF ENERGY

When you keep in mind that energy is related to the ability to do work, it becomes clear that there are many other forms of energy. Coal, oil, gasoline, and other fuels possess energy, since they can undergo chemical combustion, converting some of the stored energy to perform work. This stored energy is referred to as *chemical energy*. Certain atomic nuclei can be split or *fissioned* in nuclear reactors, releasing energy to do the work of turning the turbines that generate electricity. Thus nuclei possess *nuclear energy*. Still another type of energy results from the fact that electric charges can do work; charges possess *electric energy*. Energy is often stored in elastic devices. A stretched spring or bowstring possesses *elastic potential energy* which can be transformed into the kinetic energy of a mass attached to the spring or an arrow on the bowstring.

A very important and common form of energy is found in the motions of the atoms or molecules of a substance. Although these motions involve the kinetic energy of individual atoms, the atoms move in random directions with various speeds. This behavior differs from the motion of an entire object, in which all of its atoms are moving together with the velocity of the object, and the object's kinetic energy can be characterized by its mass and speed ($\frac{1}{2}mv^2$). The random motions of atoms and molecules are a form of energy that is an internal property of a material,

The potential energy of the wrecking ball is about to be converted into kinetic energy.

which we call *thermal energy* (TE). The amount of thermal energy possessed by an object is related to the temperature of the object. The nature of this relationship will be explored in much greater detail in later chapters of this book. For now, let us simply realize that one effect work can have is to change the thermal energy of an object.

For example, if you slide your book across the floor, the kinetic energy you gave to the book disappears as the book comes to rest. Yet the book has not gained GPE, since the floor is level. What has happened to the book's original energy as it left your hand? The only force acting on the book in the direction of the displacement is that of kinetic friction. This force does work, as we have already seen. Our experience tells us that the book (and floor) "heat up" slightly when friction is present. This is a common way of referring to the fact that the thermal energy of these materials has been increased. Thus in this case we can answer the question of what has happened to the original KE. It has been entirely changed by the work of friction into the TE of the book and floor. In other words, the work done by friction appears as an increase in TE:

$$-W_{\text{fr}} = \Delta\text{TE}$$

The negative sign is necessary because W_{fr} is always negative, whereas the TE has increased.

In any physical process, there are always transformations of some forms of energy into other forms. All such energy transactions are subject to the following constraint:

Energy can be neither created nor destroyed. When a loss occurs in one form of energy, an equal increase occurs in other forms.

This statement is called the **law of conservation of energy.** Its validity rests on the fact that no experiment has ever violated it. It is one of the most general and powerful principles of physics. Also, since energy of one form or another is found in all branches of physics, this conservation law is one of the most *unifying* statements in all of physics.

In order for us to make practical use of this concept of energy conservation, we must (1) separate conservative from nonconservative forces and (2) carefully define the system whose energy we want to calculate. The only conservative force we have dealt with so far is gravity. Others we will encounter later include elastic forces and electrical forces between charges. All other forces, such as pushes and pulls, friction, and viscosity, are nonconservative. In these terms, we can write the law of conservation of energy as an extended form of the previous work-energy theorem:

Work done by nonconservative forces external to a system is equal to change in kinetic energy plus change in potential energy plus change in thermal energy:

$$W_{\text{ext}} = \Delta\text{KE} + \Delta\text{PE} + \Delta\text{TE} \tag{5.9}$$

ΔTE is the result of work done by frictional forces within the system, including fluid viscosity and air resistance.

This form of the work-energy theorem keeps track of all the energy transactions within and external to the system. If work is done on the system, some of that work can go into changing the motion of the system, some can go into changing the

Friction forces exerted by the material of the target stop the arrows, converting their kinetic energy into thermal energy.

position of parts of the system, and some can go into internal molecular motion (which is thermal).

Whenever no external nonconservative forces act on a system, Eq. 5.9 becomes

$$\Delta KE + \Delta PE + \Delta TE = 0 \qquad (5.9a)$$

This states that an increase in the thermal energy of a system comes at the expense of a decrease in the mechanical energy. If friction is negligible, $\Delta TE = 0$, and *mechanical* energy is conserved:

$$\Delta KE + \Delta PE = 0 \qquad (5.9b)$$

Equation 5.9 thus is a very general statement that includes all of the special cases. It is important to realize that the effect of all conservative forces acting on the system are taken into account by the potential energy term in Eq. 5.9.

Example 5.5

A 900-kg car is moving on a horizontal road at 20 m/s when its brakes are applied and the car skids to a stop in 30 m. Use the concepts of work and energy to find the friction force between the car's tires and the road.

R

Reasoning

Question   The extended work-energy theorem should apply to all cases. What is the system of interest here?
Answer   If you choose the car plus the road to be your system, then you can say that $W_{ext} = 0$.

Question   How is the force of friction involved in the work-energy theorem?
Answer   The negative of the work done by friction equals the increase in the thermal energy of the road plus the tires.

$$-W_{fr} = \Delta TE$$

Question   What changes are occurring in other forms of energy?
Answer   GPE is not changing, since the car is moving horizontally. KE is decreasing from its initial value to zero.

Question   What equation does the work-energy theorem give in this case?
Answer $\Delta KE + \Delta TE = 0$, which becomes

$$(0 - \tfrac{1}{2}mv_o^2) + fs = 0$$

where $s = 30$ m. Notice that $fs = -W_{fr}$.

Solution and Discussion   Solving for f:

$$f = \frac{mv_o^2}{2s} = \frac{(900 \text{ kg})(20 \text{ m/s})^2}{2(30 \text{ m})} = 6000 \text{ N}$$

Example 5.6

A 3.0-kg ball falls to the ground from a height of 4.0 m. Use energy concepts to determine how fast it is going just before it hits the ground. Neglect air resistance.

R

Reasoning

Question   What should I choose as the system of interest?
Answer   Just the ball, since there is no interaction between the ball and the air or the ground.

Question   Are there terms in the work-energy theorem that I can say are equal to zero?
Answer   Yes. $\Delta TE = 0$ when you can neglect air resistance. Also $W_{ext} = 0$ since no nonconservative forces are acting on the system (the ball).

Question   But isn't gravity external to the ball? How is that taken into account?
Answer   Gravity is a conservative force, and is taken into account by the PE term in the work-energy theorem.

Question   What specific equation does the work-energy theorem give in this case?
Answer   It is another example of the conservation of mechanical energy,

$$\Delta KE + \Delta GPE = 0$$

Solution and Discussion   If you take the ground to be the reference level for GPE, you have

$$\Delta KE = \tfrac{1}{2}mv^2 - 0 \quad \text{and} \quad \Delta GPE = 0 - mg(4.0\,\text{m})$$

This gives

$$\tfrac{1}{2}mv^2 - mg(4.0\,\text{m}) = 0$$

Notice that the mass of the ball cancels out of each term. Solving for v,

$$v_f = (2gh_o)^{1/2} = [2(9.8\,\text{m/s}^2)(4.0\,\text{m})]^{1/2} = 8.9\,\text{m/s}$$

Emergency truck runout on mountain grade, showing piles of energy-absorbing sand.

Example 5.7

A 50-kg crate falls off the roof of a building. By the time it hits the street 40 m below, it is moving at a speed of 20 m/s. Using energy concepts, find the average force of air drag during the crate's fall.

R

Reasoning

Question   Should I include the air as part of the system?
Answer   You can approach the problem either way. If the air is part of the system, then the work of air resistance appears as the positive ΔTE term in the work-energy theorem. If you choose only the crate to be the system, then the force of air resistance does work *external* to the system, W_{ext}. This is a negative amount of work, which appears on the left side of the work-energy equation. Both cases are mathematically the same. The important thing is to define your system carefully and then be consistent with it.

Question   I will choose to include the air as part of the system. What are the changes in each energy term in the work-energy equation?
Answer   The force of air resistance does work through the distance of the fall, *h*. Thus

$$\Delta \text{TE} = -W_{\text{fr}} = f_{\text{air}}(40\,\text{m})$$

KE is increasing from 0 to $\frac{1}{2}m(20\,\text{m/s})^2$. GPE is changing by $mg(-40\,\text{m})$.

Question   Are there any other nonconservative forces acting?
Answer   No. There is no other source of friction, and there are no external ropes or other forces attached to the crate.

Question   What equation results from the work-energy theorem?
Answer $0 = \frac{1}{2}m(20\,\text{m/s})^2 - mg(40\,\text{m}) + f_{\text{air}}(40\,\text{m})$

Be sure you understand the signs in these terms.

Solution and Discussion   Solving the equation for f_{air}, we get

$$f_{\text{air}} = mg - \frac{mv^2}{2h}$$

$$= (50\,\text{kg})(9.8\,\text{m/s}^2) - \frac{(50\,\text{kg})(20\,\text{m/s})^2}{2(40\,\text{m})}$$

$$= 240\,\text{N}$$

Exercise   Calculate the changes in each of the terms in the work-energy theorem in the above problem.   *Answer:* $\Delta \text{TE} = +9600\,\text{J}$; $\Delta \text{KE} = +10,000\,\text{J}$; $\Delta \text{GPE} = -19,600\,\text{J}$.

Example 5.8

A 300-kg roller coaster car starts from rest at point *A* in Fig. 5.15 and coasts down the track. If the retarding friction force is 20 N, how fast is the car going (*a*) at point *B*? (*b*) At point *C*?

R

Reasoning   Part (a)

Question   What changes occur in the KE and GPE of the car in going from *A* to *B*?

FIGURE 5.15

The gravitational potential energy the car has at A is changed to kinetic energy and thermal energy generated by friction as it moves to points B and C.

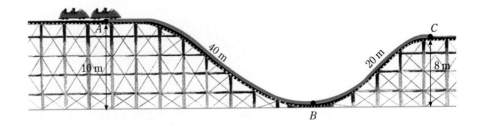

Answer   GPE is changing by $mg\,\Delta h$, where $\Delta h = -10$ m. KE is changing from 0 to $\frac{1}{2}mv_B^2$, where v_B is the unknown you're looking for.

Question   Should I include the track in my system?
Answer   As in the previous example, you are free to choose, as long as you take the force of friction into account correctly.

Question   This time I will take just the car as my system. Then what term in the work-energy theorem involves the friction?
Answer   If you treat friction as external, then $W_{ext} = -fs$, where $s = 40$ m along the track from A to B.

Question   What equation do I get from the work-energy theorem?
Answer $-fs = (\frac{1}{2}mv_B^2 - 0) + mg\,\Delta h$.

Solution and Discussion   Solving for v_B and putting in the values,

$$v_B = [2(9.8\,\text{m/s}^2)(10\,\text{m}) - 2(20\,\text{N})(40\,\text{m})/(300\,\text{kg})]^{1/2}$$
$$= 13.8\,\text{m/s}$$

R   ***Reasoning   Part (b)***

Question   To find v_C, should I start at A again?
Answer   You can start either at A or at B, using the conditions at either point as your initial conditions. If you choose A, you don't have to know what happens at B in order to solve for point C.

Question   How much does GPE change between A and B? Between B and C?
Answer $\Delta\text{GPE} = mg\,\Delta h$. From A to C, $\Delta h = -2$ m. From B to C, $\Delta h = +8$ m.

Question   How much work is done by friction between A and B? Between B and C?
Answer   Once again, W_{ext} depends on the path distance. From A to C, $W_{ext} = -(20\,\text{N})(60\,\text{m}) = -1200$ J; from B to C, $W_{ext} = -(20\,\text{N})(20\,\text{m}) = -400$ J.

Question   What is the change in KE from A to C and from B to C?
Answer   At B, we found that the car is moving at 13.8 m/s, which represents the initial speed for the B–C segment:

$$\Delta\text{KE}_{B-C} = \tfrac{1}{2}m[v_C^2 - (13.8\,\text{m/s})^2] \quad \text{and} \quad \Delta\text{KE}_{A-C} = \tfrac{1}{2}mv_C^2 - 0$$

Solution and Discussion   The work-energy theorem gives us the equations

A–C: $-1200\,\text{J} = \frac{1}{2}mv_C^2 + mg(-2\,\text{m})$
B–C: $-400\,\text{J} = \frac{1}{2}mv_C^2 - \frac{1}{2}m(13.8\,\text{m/s})^2 + mg(8\,\text{m})$

You should be able to show that $v_C = 5.6$ m/s in both cases.

Make sure you notice that ΔGPE depends only on the difference in the vertical positions of A and B, whereas W_{ext} (and ΔTE, if you chose the track as part of the system) depends on the actual *distance along the path* taken from A to B. In other words, energy changes due to conservative forces depend only on initial and final positions, but energy changes due to nonconservative forces depend on the path taken.

Exercise   How fast will the car be moving at C if it has a speed of 5.0 m/s at A and friction forces are negligible?   *Answer:*   14.9 m/s.

Example 5.9

Starting from rest, the two children in Fig. 5.16 push on a 50-kg sled. The total force they exert is 80 N as they push the sled for 10 m along level, frictionless ice to the top of the hill. They let go just as the sled plunges over the bank. On the way downhill, the sled encounters some gravel on the ice. It reaches the bottom of the slope, which is a 20-m vertical distance below the top, with a speed of 14 m/s. How much thermal energy was created by the friction with the gravel?

R   *Reasoning*

Question   I thought the work done by friction depends on the path of the motion. The path is not given here. How can I proceed without it?
Answer   Your statement is true if you are using the *definition* of work. However, you know that total energy is conserved. If you include the ground as part of your system, then the work done by friction shows up as TE, which is what you are asked to find.

Question   Do I have to find the speed of the sled at B, or can I use points A and C as the start and finish?
Answer   You can find the speed at B, but the law of energy conservation holds between any two points. Using A and C is the direct way to the answer.

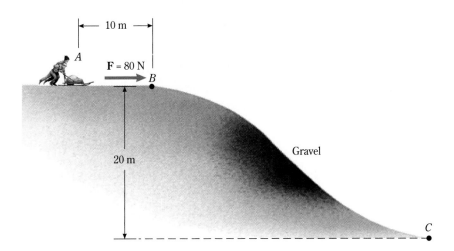

FIGURE 5.16

How much work is done on the sled by friction due to the gravel?

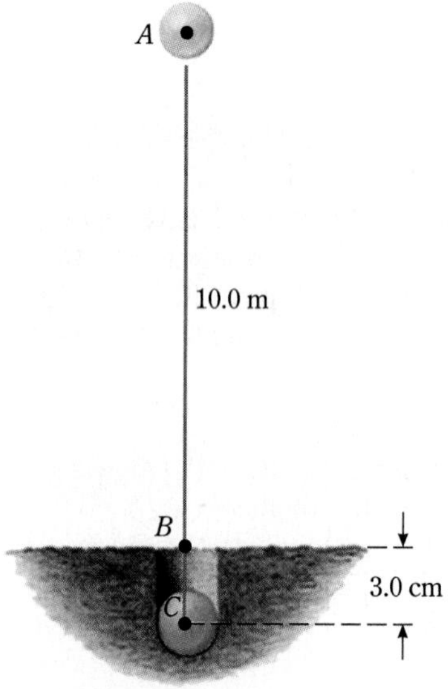

FIGURE 5.17

The gravitational potential energy of the ball at A has been lost doing friction work on the sand by the time the ball comes to rest at C.

R

Question   Between A and C, what is the change in KE?
Answer $KE = 0$ at A; $KE = \frac{1}{2}m(14 \text{ m/s})^2$ at C.

Question   What is the change in PE between A and C?
Answer $\Delta PE = mg(-20 \text{ m})$.

Question   What is ΔTE?
Answer ΔTE is the unknown you are asked to solve for.

Question   Is there nonconservative work done on the system?
Answer   Yes, the work supplied by the children between A and B. They are external to the system of the sled and hill and exert a nonconservative force that does an amount of work $W_{\text{ext}} = (80 \text{ N})(10 \text{ m}) = +800 \text{ J}$.

Question   What equation does the work-energy theorem give between A and C?
Answer $+800 \text{ J} = \frac{1}{2}m(14 \text{ m/s})^2 + mg(-20 \text{ m}) + \Delta TE$

Solution and Discussion   Solving for ΔTE, we get

$$\Delta TE = 800 \text{ J} - \frac{1}{2}(50 \text{ kg})(14 \text{ m/s})^2 + (50 \text{ kg})(9.8 \text{ m/s}^2)(20 \text{ m})$$
$$= 5700 \text{ J}$$

Looking at the individual terms, you see that the children gave the sled 800 J of kinetic energy and gravitation added 9800 J more during the descent. Friction took away 5700 J, leaving 4900 J of KE at the bottom, which means a speed of 14 m/s for a 50-kg object. Once again, energy is conserved.

Example 5.10

A 2.000-kg ball falls into a box of sand from a height of 10.00 m, as shown in Fig. 5.17. The ball comes to rest 3.00 cm beneath the surface of the sand. How large is the average force exerted on it by the sand?

Reasoning

Question   What principle involves the average force exerted by the sand?
Answer   If you choose the ball and the sand to comprise your system, the work-energy theorem will include the term

$$\Delta TE = f_{\text{sand}}(0.030 \text{ m})$$

Question   At what level—A, B, or C—should I take PE to be zero?
Answer   You can choose any one of them, but since you don't have to know the speed at B, level B would be an inconvenient choice.

Question   If I choose A as my reference, what are ΔKE and ΔGPE between A and C?
Answer   The ball is at rest at both points, and so $\Delta KE = 0$. As far as the work-energy theorem goes, it doesn't matter that it moves in between. $\Delta GPE = mg(h_C - h_A) = mg(-10.03 \text{ m})$

Question   What is W_{ext}?
Answer $W_{\text{ext}} = 0$, since you have included the sand in your system.

Question What equation is obtained from the work-energy theorem?
Answer $\Delta\text{GPE} + \Delta\text{TE} = 0$ with $\Delta\text{TE} = -W_{\text{fr}}$

Solution and Discussion In this case, with $\Delta\text{KE} = 0$, all the initial GPE is converted to thermal energy of the ball and the sand:

$$\Delta\text{TE} = -\Delta\text{GPE} = (2.000\,\text{kg})(9.800\,\text{m/s}^2)(10.03\,\text{m}) = 196.6\,\text{J}$$

Then

$$f_{\text{sand}} = \frac{196.6\,\text{J}}{0.030\,\text{m}} = 6550\,\text{N}$$

Example 5.11

Consider the pendulum (a ball at the end of a string) in Fig. 5.18*a*. It is released from rest at point *A*. How fast is the ball moving (*a*) at *B*, (*b*) at *C*? Ignore air drag and any friction at the point from which the pendulum swings.

Reasoning

Question Is any thermal energy being produced?
Answer No. Because you are told to ignore friction and air drag, you know you don't have to deal with thermal energy in this problem.

Question Is any external work being done on the ball?
Answer No. The only force on the ball other than gravity is the tension in the string. You can see that this tension is always perpendicular to the motion of the ball and so can do no work.

Question What is the form of the work-energy theorem here?
Answer $\Delta\text{KE} + \Delta\text{PE} = 0$

Question What is ΔPE between *A* and *B*? Between *A* and *C*?
Answer *A* and *C* are at the same level, and so $\Delta\text{PE}_{A-C} = 0$. As shown in Fig. 5.19*b*, *A* is $(1.50\,\text{m})\cos 40° = 1.15\,\text{m}$ below the attachment point. *B* is the length of the string, 1.50 m, below the attachment point. Therefore, *B* is $1.50\,\text{m} - 1.15\,\text{m} = 0.35\,\text{m}$ below *A*. Thus $\Delta\text{PE}_{A-B} = mg(-0.35\,\text{m})$.

Question What equations do we get from the work-energy theorem that determine v_B and v_C?
Answer Since $\Delta\text{PE}_{A-C} = 0$, there is no change in KE, and so $v_C = 0$. For *A–B*,

$$(\tfrac{1}{2}mv_B^2 - 0) + mg(-0.35\,\text{m}) = 0$$

Solution and Discussion The last equation gives

$$v_B = [2(9.8\,\text{m/s}^2)(0.35\,\text{m})]^{1/2} = 2.62\,\text{m/s}$$

This is an example of perpetual *oscillation,* or transfer back and forth, between potential and kinetic energies where no friction or external forces are present. It shows directly the meaning of a conservative force, as opposed to a heat-producing (nonconservative) force, which would cause the motion to diminish in time.

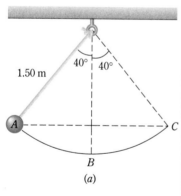

1.50 m   40° | 40°

A ⤏⤏⤏⤏⤏⤏⤏⤏⤏⤏⤏ *C*

B

(a)

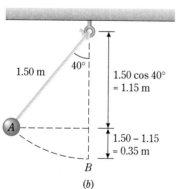

1.50 m   40°

1.50 cos 40°
= 1.15 m

A

1.50 – 1.15
= 0.35 m

B

(b)

FIGURE 5.18

As the pendulum swings back and forth, its kinetic energy and potential energy keep interchanging.

Example 5.12

Static friction between the tires of a car and the roadway enable the car to accelerate when the engine applies torque to the wheels. Suppose the 2000-kg car of Fig. 5.19 accelerates from rest to 15.0 m/s as it travels 80 m along a level road. During this time air and the car's bearings exert an average friction force of 500 N opposing the motion. (*a*) How large a force must the road exert on the car to cause this acceleration? (*b*) How much average power does this force deliver if the acceleration of the car is constant?

R

Reasoning

Question What shall I include in the system of interest?
Answer The car and the air. Then the heat-producing forces, which add up to 500 N, are internal, and are responsible for ΔTE.

Question What about the static friction between the road and the tires?
Answer Static friction produces no thermal energy. The patch of tire that meets the road does not slip along the road surface. The tire rolls, so that a new patch of tire continually comes in contact with the road as the car moves. If we treat the road as external, we can evaluate the work done by the static force of friction at this point of contact. This work appears as the term W_{ext} in the work-energy theorem.

Question What changes occur in the other forms of energy?
Answer GPE doesn't change, since the road is level.
ΔKE $= \frac{1}{2}(2000 \text{ kg})(15.0 \text{ m/s})^2 - 0$. ΔTE $= (500 \text{ N})(80 \text{ m})$.

Question What does the work-energy theorem give me?
Answer $W_{ext} = F(80 \text{ m}) = \Delta$KE $+ \Delta$TE

Question For part (*b*), how is the power delivered by the force related to the force?
Answer Power is energy per time, or the rate at which energy is produced. The power delivered in this case is the work done by F divided by the time taken to cover the 80 m.

Question What determines the time taken?
Answer The acceleration is assumed to be constant, so the equations for uniformly accelerated motion apply. In particular, $s = \bar{v}t$, where $s = 80$ m and $\bar{v} = v/2 = 7.5$ m/s.

Solution and Discussion   Part (a)

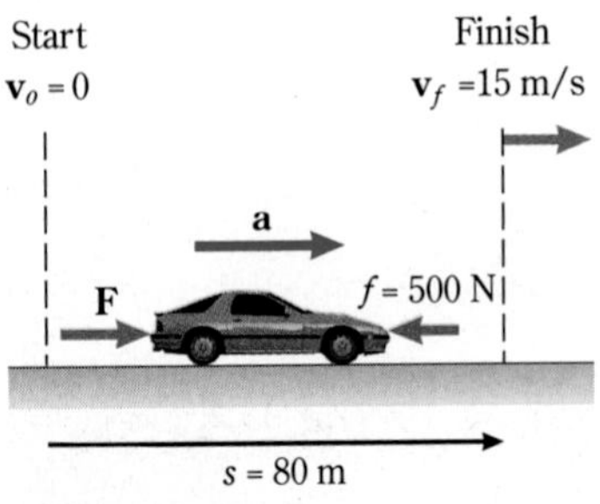

From the work-energy equation,

$$W_{ext} = 225{,}000 \text{ J} + 40{,}000 \text{ J} = 265{,}000 \text{ J}$$

The first term is the increase in the KE, and the second term is the thermal energy created by air resistance and friction within the car. The force F applied at the areas of contact between the road and tires is found from

$$W_{ext} = F(80 \text{ m}) = 265{,}000 \text{ J}$$

Thus $F = 3310$ N.

Start
$\mathbf{v}_o = 0$

Finish
$\mathbf{v}_f = 15$ m/s

a

F   f = 500 N

s = 80 m

FIGURE 5.19

How large is the force responsible for the acceleration?

Solution and Discussion **Part (b)**

The time taken to go the 80 m is

$$t = \frac{s}{v/2} = \frac{80\,\text{m}}{7.5\,\text{m/s}} = 10.7\,\text{s}$$

Then the average power delivered by F is

$$\bar{P} = \frac{W_{\text{ext}}}{t} = \frac{265{,}000\,\text{J}}{10.7\,\text{s}} = 24{,}800\,\text{W} = 33.2\,\text{hp}$$

Remember, this is *average* power. Since $P = Fv$, the power expended increases with speed.

Approximately 25 percent of an automobile engine's power actually gets converted into kinetic energy, so the engine would have to be capable of creating at least $4(28.7\,\text{hp}) = 115\,\text{hp}$ to accomplish the described motion.

5.10 SIMPLE MACHINES

Machines are devices we use to help us do work. A *simple machine* is a mechanical device that can exert a force on an object at one point when an external force is applied to the device at another point. Some examples are levers, pulleys, the wheel and axle, and jackscrews.

Simple machines cannot create energy. According to the principle of conservation of energy, a machine can provide no more work output than the amount of work supplied to it. In reality, machines are subject to some friction, and so work output is less than work input by an amount equal to the thermal energy produced. The *efficiency* of a machine measures the degree to which it converts work input to work output.

Window washers use pulley systems to be able to raise and lower their scaffolds.

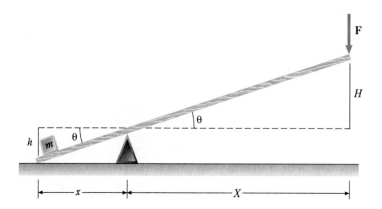

$$\% \text{ efficiency} = \frac{\text{work output}}{\text{work input}} \times 100 \qquad (5.10)$$

If a machine could operate at 100 percent efficiency, it would be called an *ideal* machine.

Although machines cannot create energy, they can magnify the input force, and this is their main usefulness. Consider the simple lever in Fig. 5.20. Let us assume that friction at the pivot, or *fulcrum,* is negligible, so that we are dealing with an ideal machine. As the applied force F is applied over the distance H, the work input is

Work input $= FH$

The weight mg, which we call the **load,** is lifted through a displacement h. Thus the work output of the lever is

Work output $= mgh$

Since we are assuming an *ideal* machine,

Work input $=$ work output   or $FH = mgh$

In Fig. 5.20 the two triangles to the right and left of the fulcrum are similar, and so $h/H = x/X$. Thus

$$F = mg\frac{h}{H} = mg\frac{x}{X}$$

This equation tells us that, to lift the load, we need to exert a force F that is less than mg by the ratio x/X. For example, if $x = \frac{1}{2}X$, F would have to be only $\frac{1}{2}mg$. The lever has thus multiplied the input force by 2.

Simple machines can multiply forces applied to them.

We call the ability of a simple machine to multiply forces the **mechanical advantage** of the machine. If F_o is the force output of the machine and F_i is the force applied to it (i.e., the input force), we define

$$\text{Actual Mechanical Advantage (AMA)} = \frac{F_o}{F_i} \qquad (5.11)$$

An automobile jack might typically require an input force of 100 N to lift a 5000-N load. Thus the AMA of the jack would be

$$\text{AMA} = \frac{F_o}{F_i} = \frac{5000\,\text{N}}{100\,\text{N}} = 50$$

The price you pay for magnifying a force with a simple machine is that the distance through which the load is moved is shorter than the distance through which you exert the applied force. In the case of our lever, to move a load a distance y, you have to exert the $\frac{1}{2}mg$-force over a distance $2y$. This difference in distance is simply a consequence of the conservation of energy. Thus for an *ideal* machine,

$$F_i s_i = F_o s_o \qquad \text{(ideal machine only)}$$

where s_i is the distance over which the applied force is exerted and s_o is the distance the load is moved.

The mechanical advantage of an ideal machine can be expressed as the ratio of the input to output displacements:

$$\text{Ideal Mechanical Advantage (IMA)} = \frac{s_i}{s_o} \qquad (5.12)$$

By combining the expressions for efficiency, AMA, and IMA, we can write the efficiency of a machine as

$$\% \text{ efficiency} = \frac{\text{AMA}}{\text{IMA}} \times 100 \qquad (5.13)$$

Let us illustrate the use of these equations by referring to the simple machine in Fig. 5.21. It is called a wheel and axle and is used to enable a small input force F to lift a heavy load W. We can compute the machine's IMA by noting that, when the wheel and axle turn through one revolution, the two cords wind or unwind a length equal to the circumferences of their respective circles. Thus $s_i = 2\pi b$ and $s_o = 2\pi a$. Then

$$\text{IMA} = \frac{s_i}{s_o} = \frac{2\pi b}{2\pi a} = \frac{b}{a}$$

If the machine were 100 percent efficient, a force F could lift a weight

$$W = \frac{b}{a}F$$

By making the radius of the wheel much larger than the radius of the axle, we obtain a very effective lifting device.

Pulley systems are also interesting simple machines. The one in Fig. 5.22 lifts an object of weight W when the force F pulls the cord out from the top pulley. This

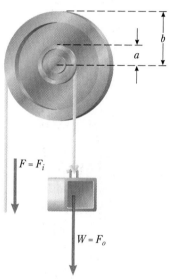

FIGURE 5.21

The wheel and axle has an IMA given by the ratio of the radius of the wheel to the radius of the axle.

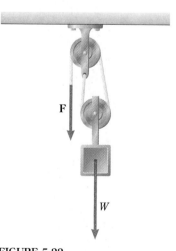

FIGURE 5.22

The IMA of this pulley system is 2.

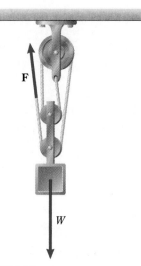

FIGURE 5.23

This system has an IMA of 4.

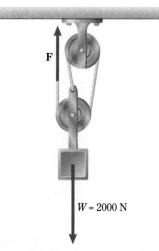

FIGURE 5.24

What is the IMA of this system?

pulley is fastened to the ceiling, while the lower pulley moves upward as the force F pulls the cord out. Notice that the lower pulley moves upward a distance $0.5s_i$ when the cord is pulled out a distance s_i from the top pulley. (Each of the two cords supporting the lower pulley shortens $0.5s_i$, giving a total shortening s_i for the cord between the pulleys.) Therefore we have

$$\text{IMA} = \frac{s_i}{s_o} = \frac{s_i}{0.5s_i} = 2.00$$

This pulley system has an IMA of 2. You should be able to show that the ideal mechanical advantage of the system in Fig. 5.23 is 4.

The actual mechanical advantage of these pulley systems is considerably less than their ideal mechanical advantage. Not only is there friction in the pulleys, but the system must also lift the useless weight of the movable pulley. Even so, pulley systems are widely used to lift heavy objects.

Example 5.13

To lift a 2000-N object with the pulley system in Fig. 5.24, an input force of 800 N is required. Find the IMA, AMA, and efficiency for the system.

Reasoning

Question   Which of the two kinds of mechanical advantage involves the input and output forces?

Answer $\text{AMA} = \dfrac{F_o}{F_i} = \dfrac{2000\,\text{N}}{800\,\text{N}} = 2.50$

Question   What must I know to calculate the IMA?
Answer   The ratio of the distance over which the applied force is exerted to the distance the load moves.

Question   How can I tell how much the load is raised when the free end of the cord is pulled?
Answer   In any diagram like this, count the number of cords that are involved in lifting the load, that is, the cords that exert an upward tension on the load. Any displacement of the free end of the cord is divided equally among this number of supporting cords in lifting the load. In Fig. 5.23, for example, the force is divided among four cords. Here, in Fig. 5.24, three cords are pulling up, and thus the distance the load moves is one-third the distance F moves. So

$$\text{IMA} = \frac{s_i}{s_o} = \frac{3s_o}{s_o} = 3.00$$

Question   How does the efficiency depend on the mechanical advantages?
Answer   % efficiency $= \dfrac{\text{AMA}}{\text{IMA}} \times 100$

$$= \frac{2.50}{3.00} \times 100 = 83\%$$

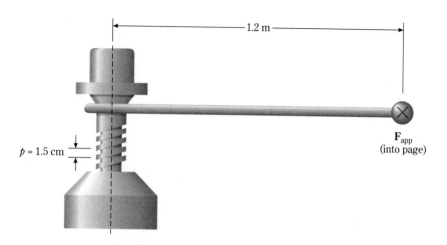

FIGURE 5.25

A jackscrew.

p = 1.5 cm

1.2 m

$\mathbf{F}_{app}$
(into page)

Example 5.14

The jackscrew in Fig. 5.25 is used to lift a load of 15,000 N. The handle turns in a horizontal circle which is perpendicular to the page. The × shown at the end of the handle indicates that a force is being applied into the page at that position. (*a*) If the AMA is 125 when a force is applied at the end of a handle 1.2 m long, how big a force is required to lift the load? (*b*) The pitch of the screw, which is the vertical separation between adjacent threads, is 1.5 cm. What is the IMA? (*c*) In raising the load a vertical distance of 30 cm, how much thermal energy is produced?

R *Reasoning   Part (a)*

Question   How is the applied force related to the AMA and the load?

Answer $\text{AMA} = 125 = \dfrac{\text{load}}{\text{applied force}}$

Solution and Discussion   The solution is easy:

$$\text{Applied force} = \frac{\text{load}}{125} = \frac{15{,}000\text{ N}}{125} = 120\text{ N}$$

Notice that the problem states that this force is applied at the end of the handle. If it were applied at another point on the handle, the force would have a different lever arm about the axle, so a different force would be required to move the load and the AMA would be different. In other words, the AMA of a machine depends on the specific details of how the machine is used.

R *Reasoning   Part (b)*

Question   How does the IMA involve the screw pitch?
Answer   IMA is the ratio of the distance over which the input force is applied to the distance the load travels. To have a pitch of 1.5 cm means that the screw lifts the load 1.5 cm for each complete turn of the handle. During each complete turn the input force is applied through the circumference of a circle 1.2 m in radius.

Solution and Discussion These answers give $s_i = 2\pi(1.2\,\text{m}) = 7.54\,\text{m}$ for a load displacement upward of $s_o = 1.5 \times 10^{-2}\,\text{m}$. Thus

$$\text{IMA} = \frac{7.54\,\text{m}}{1.5 \times 10^{-2}\,\text{m}} = 500$$

Reasoning   Part (c)

Question The work-energy theorem involves ΔTE. How does the theorem apply to this case?

Answer The work done by the input force is $W_{\text{ext}} = F_i s_i$. You can calculate this for each turn of the screw. The load is at rest at the beginning and end of the motion, so ΔKE = 0. Each turn increases the GPE of the load by ΔGPE = $(15,000\,\text{N})s_o$. ΔTE is the only unknown term in the work-energy theorem.

Solution and Discussion We have $W_{\text{ext}} = (120\,\text{N})(7.54\,\text{m}) = 905\,\text{J}$ per turn. ΔGPE = $(15,000\,\text{N})(1.5 \times 10^{-2}\,\text{m}) = 225\,\text{J}$ per turn. The work-energy theorem then looks like this:

$$905\,\text{J} = 225\,\text{J} + \Delta\text{TE} \qquad \text{(per turn)}$$

This gives ΔTE = 680 J per turn. For 20 turns, ΔTE = 13,600 J.

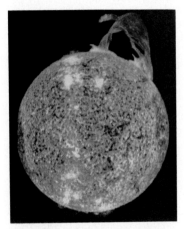

The power radiated by the sun is produced by the conversion of mass into energy through nuclear fusion deep in its core.

A MODERN PERSPECTIVE
THE EQUIVALENCE BETWEEN MASS AND ENERGY

Of all the equations in physics, the one most widely popularized in the media and therefore most widely recognized by the general public is probably $E = mc^2$. This equation was developed by Albert Einstein in his theory of relativity in the early part of this century. What does this simple yet profound statement mean?

First of all, as we learned in Sec. 3.12, c stands for the speed of light, $3.00 \times 10^8\,\text{m/s}$. This is a very large number and becomes much larger when squared. The symbol m represents the mass of an object or system of objects, while E stands for an amount of energy. The statement $E = mc^2$ says that there is some energy, called *mass energy,* associated with the presence of matter. The equation tells, for instance, that a 1-kg mass possesses

$$E = (1\,\text{kg})(3 \times 10^8\text{m/s})^2 = 9 \times 10^{16}\,\text{J}$$

of energy! Just doing this calculation, however, hardly gives insight into what form of energy this is or how the equation is to be interpreted.

It helps to look closely at how matter is structured. The objects we encounter in daily life are made up of atoms of various chemical elements, bound together by electromagnetic forces into molecules. Chemical reactions, such as combustion, can change molecular structures. As atoms are rearranged in molecules, work is done by the electromagnetic binding forces, resulting in a change in the potential energy of the system. Remember, we stated at the beginning of Sec. 5.5 that potential energy results from the position or *configuration* of interacting objects. A change in molecular structure is a change in configuration, and hence represents a change in the potential energy of the molecule, which we call its **binding energy.**

If the atoms in the new molecular structure are more tightly bound together than they were before the rearrangement, the potential energy of the system has decreased. Energy is given off by the system, generally in the form of heat and light. If the reaction results in new molecules in which the atoms are less tightly bound, the system must have gained some energy, perhaps as heat.

Einstein's equation relating mass to energy implies that changes in the energy of a system are accompanied by changes in the system's mass. Thus another version of $E = mc^2$ is

$$\Delta m = \frac{\Delta E}{c^2} \tag{5.14}$$

A typical value for the energy released by the complete combustion of common fuels is about 10^7 J per kg of matter (fuel plus oxygen) entering into the reaction. What does Einstein's equation tell us about how much each kilogram of matter changes when it undergoes combustion? Equation 5.14 predicts that each kilogram of mass will change by

$$\Delta m = \frac{1 \times 10^7 \, \text{J}}{(3 \times 10^8 \, \text{m/s})^2} = 1.1 \times 10^{-10} \, \text{kg}$$

Thus a typical chemical reaction would change the mass of the reactants by only one part in 10 billion! This is a mass change that cannot be measured even by the most precise methods known today. Thus in our everyday experience with chemical reactions, we are completely unaware of any mass changes.

However, if we investigate atomic nuclei, we find protons and neutrons, which we refer to as **elementary particles,** bound together in various structures by means of a nuclear binding force which is much stronger than the electromagnetic forces between atoms. Chemical reactions do not alter these nuclear structures, but *nuclear* reactions such as fission and fusion do. Fission is the process by which heavy nuclei such as uranium or plutonium are split into lighter fragments and is the source of energy in present commercial nuclear reactors. Fusion involves the fusing, or sticking together, of light nuclei into heavier, more complex nuclear structures. An important example is the fusion of four hydrogen nuclei into a nucleus of helium, which is the major source of energy production in the sun.

If we carefully measure the total mass of nuclei before and after a fission or fusion reaction, we find a measurable decrease. Furthermore, this mass decrease is related to the energy released by the reaction in precise agreement with Eq. 5.14. For fission, about 0.1 percent of the original mass of a heavy nucleus is converted to energy; for fusion the number is approximately 0.8 percent. These are both very measurable mass changes, in contrast to those resulting from typical chemical reactions. As a consequence, nuclear reactions release on the order of 10 to 100 million times more energy per kilogram of reacting mass than do chemical reactions!

The ultimate conversion of mass to energy would occur if by some process an initial amount of matter were to completely disappear, being entirely replaced by purely radiant energy (light), which has no mass. This 100 percent conversion of mass to energy has been observed in the laboratory in the process we call *matter-antimatter annihilation.* Elementary particles each have their antimatter counterparts, which are not found in stable forms of matter but which are formed briefly in nuclear reactions. An example is the antielectron, or positron, which has the same physical characteristics as the ordinary electron except for its electric charge, which is positive. When an electron and positron strike each other, they both cease to exist. They are observed to be replaced by the creation of two massless gamma rays, which are radiation energy, or "light," of extremely short wavelength. The total energy of the gamma rays is measured to be precisely equal to the total mass of the original electron and positron multiplied by c^2.

The reverse of this process has also been observed: namely the creation of particle-antiparticle pairs produced from pure gamma radiation. These observations provide a most impressive verification of Einstein's theory of relativity.

LEARNING GOALS

Now that you have finished this chapter, you should be able to

1 Define (*a*) work, (*b*) joule, (*c*) power, (*d*) watt, (*e*) kilowatt-hour, (*f*) kinetic energy, (*g*) work-energy theorem, (*h*) gravitational potential energy, (*i*) law of conservation of energy, (*j*) efficiency of a machine, (*k*) IMA and AMA of a machine.
2 Compute the work done on an object by a specified force when the object is moved through a given distance.
3 Compute power in simple situations. Change from watts to horsepower and vice versa.
4 Compute the change in kinetic energy of an object subjected to a known net force acting through a known distance.
5 Compute the change in gravitational potential energy of an object as it is moved from one place to another.
6 Distinguish between conservative and nonconservative forces.
7 Give several examples in which potential and kinetic energies are interchanged. Repeat for interchange of kinetic and thermal energies.
8 State what happens to the energy lost when work is done against friction forces.
9 Use the law of conservation of energy in the form of the extended work-energy theorem to solve simple problems which involve the interchange of kinetic, potential, and thermal energies in a system, including cases where external work is done on the system.
10 Compute the IMA, AMA, and efficiency of a simple machine when appropriate data are given.
11 Use the equation $E = mc^2$ to calculate the amount of energy released during a reaction in which the mass of a system is decreased by a given amount. Also find the amount of mass which is converted in a reaction to release a given amount of energy.

SUMMARY

DERIVED UNITS AND PHYSICAL CONSTANTS

Energy and Work
1 joule (J) = $1 \, \text{N} \cdot \text{m}$

Power
1 watt (W) = $1 \, \text{J/s}$

DEFINITIONS AND BASIC PRINCIPLES

Work
The work done by a force F acting on an object while the object undergoes a displacement s is

$$W = Fs \cos \theta$$

where θ is the angle between the force and the displacement vectors.

INSIGHTS
1. Even though force and displacement are vectors, work is a scalar.
2. Work can be zero in three ways: (a) force = 0, (b) displacement = 0, or (c) $\cos \theta = 0$, which means that the force is perpendicular to the direction of motion ($\theta = 90°$).
3. Work can be positive or negative, depending on the angle between F and s. If $\theta < 90°$, the work is positive; when $\theta = 90°$, work is zero, and if $\theta > 90°$, the work is negative. In the case of friction, $\theta = 180°$, which means that work done by frictional forces is $-fs$.
4. If many forces are acting on an object, you can calculate the work done by each force individually. The net work done on the object will be the algebraic sum of these individual contributions. This will give you the same result you would get if you found the net force first and calculated the work done by it.

Power

Power is the rate at which work is done:

$$P = \frac{W}{t}$$

INSIGHTS

1. Power is measured in watts (joules per second) in the SI and in horsepower (hp) in the British system: 1 hp = 746 W.
2. If the force F doing work is being applied to an object whose velocity is v, the power delivered by the force is

$$P = Fv \cos \theta$$

where θ is the angle between F and v.
3. A corollary to the definition of power is that

$$\text{Work} = \text{power} \times \text{time}$$

This gives rise to the common unit of energy used in the electrical industry, the kilowatt-hour (kwh):

$$1 \text{ kwh} = (1000 \text{ W})(3600 \text{ s}) = 3.6 \times 10^6 \text{ J}$$

Kinetic Energy

Kinetic energy (KE) is energy an object possesses due to its motion.

$$\text{KE} = \tfrac{1}{2}mv^2$$

INSIGHTS

1. As is the case for work and all forms of energy, KE is measured in joules in the SI.
2. Since m and v^2 can never have negative values, KE must always be positive.

Work-Energy Theorem for Net Force

Work done by a net force = $W_{\text{net}} = \Delta\text{KE}$

Gravitational Potential Energy

Gravitational Potential Energy (GPE) depends on the height or vertical position of an object relative to some chosen reference level. As long as the object experiences the constant gravitational force mg, this can be written

$$\text{GPE} = mgh$$

INSIGHTS

1. GPE can be positive, negative, or zero, depending on your choice of reference for measuring h.

2. Changes in GPE do not depend on the path by which an object changes position, but only on its initial and final vertical positions.
3. For dimensional objects, the GPE is defined in terms of the vertical position of their center of mass. For uniform, symmetrical objects, the center of mass is at their geometric center.

Conservative Force

If the work done by a force depends only on the endpoints of the object's change of position and not on the details of the path taken, the force is said to be conservative. Examples are gravity, elastic forces, and electrostatic forces. When a force is conservative, we can define a potential energy of position associated with it.

Thermal Energy

Thermal energy (TE) is the internal energy of a substance associated with the random motion of its atoms or molecules. If frictional forces, including air resistance or fluid viscosity, act on a system, the system's TE will be increased by an amount equal to the work done by these forces.

Law of Conservation of Energy

Energy can be neither created nor destroyed in any physical process. When a loss occurs in one form of energy, an equal increase occurs in other forms.

INSIGHTS

There is no law of conservation of any one form of energy. The law applies only to the sum of all forms of energy that may be present in a given case.

Extended Work-Energy Theorem

$$W_{\text{ext}} = \Delta\text{KE} + \Delta\text{PE} + \Delta\text{TE}$$

INSIGHTS

1. This theorem is simply a way of expressing the law of conservation of energy applied to a specified system.
2. When using the extended theorem, work done by conservative forces on the system being considered is taken into account in the ΔPE term, and work done by frictional forces within the system appears as a gain in the system's thermal energy, ΔTE. W_{ext} represents any non-conservative forces that might be doing work on the system from outside, such as someone pushing or pulling on the system. W_{ext} can be positive or negative.

Mechanical Advantage of Simple Machines

Actual Mechanical Advantage (AMA) = $\dfrac{F_o}{F_i}$

where F_o and F_i refer to the output and input forces.

Ideal Mechanical Advantage (IMA) = $\dfrac{s_i}{s_o}$

where s_i and s_o refer to the distances over which the input and output forces are exerted.

Efficiency of Simple Machines

% efficiency = $100 \times \dfrac{\text{work output}}{\text{work input}} = 100 \times \dfrac{\text{AMA}}{\text{IMA}}$

QUESTIONS AND GUESSTIMATES

1 A conscientious hobo in a boxcar traveling from Chicago to Peoria pushes on the front of the boxcar all the way. Having at one time been a physics student, he thinks his pushing does a great deal of work since both F_s and s are large. Where is the flaw in his reasoning?

2 A person holds a bag of groceries while standing still talking to a friend. A car sits stationary with its motor running. From the standpoint of work and energy, how are these two situations similar?

3 As a rocket reenters the atmosphere, its nose cone becomes very hot. Where does this heat energy come from?

4 Reasoning from the interchange of kinetic and potential energies, explain why the speed of a satellite in a noncircular orbit about the earth keeps changing. Is its speed largest at apogee (farthest point from earth) or at perigee (closest point)?

5 Describe a situation in which the gravitational potential energy of an object is negative. Will everyone agree that it is negative? Can an object have negative kinetic energy?

6 A car cannot accelerate along an extremely slippery roadway. Suppose a car of mass m accelerates from rest to a speed v along a horizontal roadway. Assume that its wheels do not slip. How much work does the friction force between wheels and pavement do in the process?

7 Is energy a vector or a scalar quantity?

8 The coefficient of sliding friction for a block on an incline is large enough so that the block does not move by itself. The block is pulled up the incline at constant speed by a force parallel to the incline. Compare the work done by (a) the pulling force, (b) the friction force, and (c) gravity. Repeat for the block moving down the incline.

9 Automobiles, bicycles, and many other devices have gear systems that can be changed by shifting. Considering these to be ideal machines, discuss why shifting is used.

10 About what horsepower is a human being capable of producing for a short period, as in climbing a flight of stairs?

11 Estimate the force a driver experiences when her car hits another car head on. Assume both cars to be similar and traveling at 25 m/s. Discuss the effect of seat belts and other safety factors.

12 A human heart consumes about 1 J of energy per heartbeat. About how many joules of energy must food furnish to a person each day to supply this energy? For comparison purposes, one nutritionist's calorie of food energy is equivalent to 4184 J.

PROBLEMS

Section 5.1

1 How much work is done in pulling a box 2 m across a table top with a horizontal force of 35 N?

2 A force of 240 N at an angle of 30° above the horizontal is required to pull a child's wagon. How much work is done if the wagon is pulled through 10 m?

3 A woman pushes a lawn mower with a force of 180 N at an angle of 24° downward from the horizontal. How much work does she do as she pushes the mower a horizontal distance of 50 m?

4 A 1250-kg car skids to a stop in 36 m. How large a friction force acts between its four skidding tires and the pavement if the coefficient of friction is 0.7? How much work does the friction force do on the car?

5 A weight lifter lifts a 400-N set of weights from the ground level to a height of 1.8 m. Assuming that he moves the weights at a constant speed, how much work does he do?

6 A man lifts a 200-N bucket at a constant speed in a vertical well shaft. If he does 8 kJ of work in bringing the bucket to the top of the well, how deep is the well?

7 A janitor does 360 J of work against frictional force of 20 N in pushing a power sweeper across the floor in 4.5 s. Assum-

ing that the sweeper moves with a constant speed, how fast is it moving?

8 A student applies a constant horizontal force F to a 30-kg book carton, pulling it across her dormitory floor at a constant speed. If the coefficient of friction between the carton and the floor is 0.5, how much work does she do as she moves the carton 8 m?

9 How much lifting work is done by a 60-kg athlete as she climbs a flight of stairs 6 m high?

10 An 80-kg crate is pushed 3.5 m up a frictionless inclined ramp that makes an angle of 24° with the horizontal. How much work would be required to push the crate? Assume that the crate is pushed with a constant speed.

11 In the previous problem how much work would be required if the coefficient of friction between the crate and the ramp were 0.3 and the pushing force is parallel to the ramp?

12 By changing the angle of a variable-angle inclined ramp, a dock worker finds that a 50-kg carton slides down the ramp with constant speed when the incline angle is 36°. How much work does the friction force do on the carton as it slides 2.5 m?

Section 5.2

13 What is the horsepower of a 100-W light bulb?

14 What power in watts is needed to push a loaded supermarket cart with a horizontal 50-N force through a horizontal distance of 20 m in 5 s?

15 A friction force of 20 N opposes the sliding of a 6-kg carton along the horizontal floor. What power is supplied to the carton as it is dragged along the floor at a constant speed of 0.6 m/s?

16 A machine lifts a 240-kg crate at a constant speed through a height of 5 m in 6 s. What is the power output of the machine?

17 A motor boat requires 100 hp to move at a constant speed of 16 m/s. What is the resistive force due to water at that speed?

18 A certain tractor can pull with a steady force of 12,000 N while moving at 2.5 m/s. How much power in watts and in horsepower is the tractor developing under these conditions?

19 At what average speed would a 64-kg student have to climb a 5-m rope to match the power output of a 150-W light bulb?

20 A pump is needed to lift water from a well through a height of 3.0 m at a rate of 0.6 kg/min. What must the minimum power of the pump be in watts and in horsepower?

21 An electric motor that can develop 1.6 hp is used to lift a 20-kg carton through a distance of 8 m. What is the minimum time needed to lift the carton?

22 An elevator has a 11-hp motor. What is the maximum weight it can raise with uniform speed through a height of 36 m in 10 s?

Sections 5.3 and 5.4

23 What is the kinetic energy of a 2000-kg car moving at 20 m/s?

24 What is the ratio of the kinetic energy of a car traveling at a speed of 100 km/h to the kinetic energy of another car with same mass traveling at 25 m/s?

25 How far does a 1.2-g bullet with kinetic energy of 1.2 J go in 2.0 s?

26 With what speed would a 72-kg jogger have to run to have the same kinetic energy as a 1200-kg car traveling at 2.0 km/h?

27 How much work is required to increase the speed of a 800-kg sedan from 15 to 20 m/s? Compare this with the work needed to further increase the speed by the same amount, that is, from 20 to 25 m/s. Neglect frictional forces.

28 How large a force is needed to accelerate a proton ($m = 1.67 \times 10^{-27}$ kg) from rest to 3×10^7 m/s in 2.0 cm? (A proton is a hydrogen atom that has lost its electron.)

29 A particle accelerator known as a van de Graaff generator can accelerate a beam of protons ($m = 1.67 \times 10^{-27}$ kg) from rest to 10^7 m/s. If one such machine accelerates 3.6×10^{16} protons per second, how many watts of power is it producing?

30 A baseball pitcher throws a baseball at a speed of 80 mi/h. If the mass of the baseball is 160 g, what is its kinetic energy?

31 A 1000-kg vehicle is traveling at a speed of 18 m/s. What amount of work must the brakes do in order to bring the vehicle to rest in 24 m?

32 A 1.5-g bullet with a speed of 400 m/s strikes a block of wood and comes to rest at a depth of 5 cm. (*a*) How large is the average decelerating force? (*b*) How long does it take to stop the bullet?

33 A 90-kg football player running at 6 m/s is stopped after traveling 1.8 m by a player on the opposite team. (*a*) How large is the average force the opposing player exerts? (*b*) How long does it take to stop the runner?

34 A child kicks his 8-kg sled on a frozen pond, giving an initial speed of 2 m/s. The coefficient of kinetic friction between the bottom of the sled and the ice is 0.12. Use the energy method to find the distance the sled moves before coming to rest.

Sections 5.5–5.7

35 Relative to the ground, what is the gravitational potential energy of a 12-kg bowling ball at the top of a 150-m-tall building?

36 A 2.0-kg vase rests on a shelf 0.5 m above a table, which in turn is 0.8 m from the ground. What is the gravitational potential energy of the vase (*a*) with respect to the tabletop? (*b*) with respect to the ground?

37 Two balls of 5.0 kg and 3.0 kg are hung over a pulley by means of a rope, with the 5-kg ball resting on a tabletop. What is the change in the potential energy of the system if the 5-kg ball is raised by 50 cm?

38 A 75-kg hiker climbs a 600-m-high hill. (*a*) How much work does the hiker do against gravity? (*b*) Does this amount of work depend on the path the hiker takes? (Neglect force of friction.) (*c*) If the hiker takes 96 min, what average horsepower is expended?

Sections 5.8–5.9

39 A 16,000-kg van takes 45 min to climb from an elevation of 1500 to one of 2700 m along a mountain road. (*a*) How much work does the van do against gravity? (*b*) What average horsepower does the van expend against gravity?

40 With what speed will a 0.5-kg ball hit the ground if it is dropped from a height of 40 m? (Neglect friction.)

41 A box of groceries slides from rest without friction down an inclined ramp that makes an angle of 30° with the horizontal. What is the speed of the box after it slides 2.0 m along the ramp?

42 When an object is thrown upward, it rises to a height *h*. How high is the object, in terms of *h*, when it has lost one-half of its original kinetic energy? What is its speed at that point?

43 A 3-kg box dropped from a height of 10 m is moving at 10 m/s just before it hits the ground. How large is the average friction force retarding the motion?

44 A motor is to lift a 960-kg elevator from rest at ground level such that it has a speed of 3.2 m/s at a height of 24 m. (*a*) How much work does the motor do? (*b*) What percentage of the total work goes into kinetic energy?

45 A 3.2-kg mass starts from rest at the top of a 30° incline that is 6.0 m long. Its speed as it reaches the bottom is 3.0 m/s. Use energy methods to find the average friction force that retards the sliding motion.

46 A 2.4-kg box has a speed of 5.0 m/s at the bottom of a 30° inclined ramp. (*a*) How far up the ramp does the box slide if the ramp is frictionless? (*b*) How far does it slide if the coefficient of kinetic friction is 0.2?

■47 Starting from rest, a locomotive pulls a series of boxcars up a 3° incline. After the train has moved 2.4 km, its speed is 45 km/h. Assume that the entire train has a mass of 6.4×10^5 kg. (*a*) How much work does the locomotive do? (*b*) What fraction of the work is done against gravity? (*c*) Assuming the acceleration is uniform, how long does it take for the train to achieve this speed? (*d*) What average horsepower does the locomotive expend?

■■48 An electric motor is to power a pump that lifts 1.0 kg of water originally in a tank through a height of 2.2 m in 200 s. Assume that the water is moving at 1.5 m/s when it exits at the top. What horsepower output should the motor have? The speed of the water in the tank is negligible.

■■49 A 240-g ball is thrown straight upward with a speed of 14 m/s. (*a*) How high does it rise if the friction forces are negligible? (*b*) If it rises only 6.5 m, how large is the average air resistance which impedes the motion? (*c*) Under the influence of the friction force in *b*, how fast is the ball when it returns to the thrower?

50 A 640-g block of ice is released from the top of a 30° incline and slides 160 cm down the plane to its bottom. What is the speed of the block when it reaches the bottom (*a*) if the incline is frictionless and (*b*) if the friction force is 1.0 N?

■51 A child starts from rest at the top of a slide of height 4 m. If she reaches the bottom with a speed of 6 m/s, what percentage of her total energy at the top of the slide has been lost as a result of friction?

52 A roller coaster car of mass *m* starts from rest at point *A* and travels along the tracks shown in Fig. P5.1. Assuming that the tracks are frictionless, find the speed of the car at points *B* and *C*.

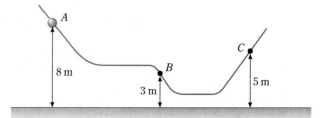

FIGURE P5.1

53 The car in the previous problem has a speed of 1.5 m/s to the left as it passes *A*. If the tracks are frictionless, find the speed of the car at points *B* and *C*.

■54 In Fig. P5.1, the 400-kg car starts from rest at *A* and passes point *B* with a speed of 3 m/s. If the distance from *A* to *B* along the tracks is 20 m, how large is the average friction force retarding the motion of the car?

55 A ball of mass *m* is suspended as a pendulum bob from a cord 3.6 m long. A force on the ball pulls the cord to one side until it makes an angle of 60° with the vertical, and then the system is released. With what speed is the mass moving as it passes directly underneath the point of suspension? (Ignore air friction.)

56 For the pendulum of the previous problem, what is the speed of the ball when the cord makes an angle of 30° with the vertical?

■57 At high speeds, the friction forces acting on a car increase in proportion to v^2, where *v* is the car's speed. If this is considered the major factor involved and if a car's gasoline mileage rating is 30 km/gal at 80 km/h, what is its rating at 100 km/h?

■58 A 625-g block starts to slide up a 30° incline with an initial speed of 2.2 m/s. It stops after sliding 40 cm and slides back down. Assuming the friction force impeding its motion to be constant, (*a*) how large is the friction force and (*b*) what is the block's speed as it reaches the bottom?

Section 5.10

59 A 640-kg object is to be lifted by a pulley system using a force of 440 N. The machine found suitable for this purpose lifts the load 0.45 m when the applied force moves 9.6 m. Find the (*a*) AMA, (*b*) IMA, and (*c*) efficiency of the machine.

60 A pulley system lifts a 240-kg load when a force of 180 N is applied to it. If the efficiency of the system is 87 percent, find (*a*) AMA, (*b*) IMA, and (*c*) s_i/s_o.

61 What must be the ratio of the radii of a wheel-and-axle device if the device is to lift 24 kg with an applied force of 28 N? Assume the efficiency of the device to be 89 percent.

62 For a particular type of car jack, the operator moves her hand (the input force) through 38 cm for every 1.0 cm of the load is lifted. (*a*) What is the IMA of the jack? (*b*) Assuming 22 percent efficiency, how large an applied force is needed to lift 3600 N?

63 An electric motor is labeled 0.5 kW. On the assumption that it is 88 percent efficient, how many horsepower can it deliver?

▪64 A ¼-hp motor has attached to its shaft a pulley of diameter 7.2 cm. If the shaft rotates at 1600 rev/min, how large a load is the belt running on the pulley capable of pulling? Assume the motor has an efficiency of 89 percent.

▪65 A certain 55-W motor runs with a shaft speed of 1800 rev/min. Because of reducing gears, the final (output) shaft rotates at 16 rev/min. (*a*) If the machine is 33 percent efficient, with what force can it pull the belt on a 3.2-cm-radius pulley at the output shaft? (*b*) If the gear system is reversed so that the output shaft rotates at 160,000 rev/min, what force is available to pull the belt on the same pulley? Assume the power output to the motor is 55 W.

General Problems

66 A 6.4-kg object is lifted vertically through a distance of 6 m by a light string under a tension of 84 N. The friction force is negligible. Find (*a*) the work done by the tension, (*b*) the work done by gravity, and the (*c*) final speed of the object if it starts from rest.

▪▪67 A child's 240-g toy car is driven by an electric motor that has a constant power output. The car can climb a 24° incline at 16 cm/s and can travel on a horizontal table at 39 cm/s. The friction force retarding the motion is *kv*, where *k* is a constant and *v* is its speed. How steep an incline can it climb with a speed of 28 cm/s?

▪▪68 The frictionless system in Fig. P5.2 is released from rest. After the right-hand mass has risen 72 cm, the object of mass 0.50*m* falls loose from the system. What is the speed of the right-hand mass when it returns to its original position?

▪▪69 A 2.4-kg block is moved up a 30° inclined ramp under the influence of a constant horizontal (not parallel to the ramp) force of magnitude 45 N. The coefficient of friction is 0.12, and the block is pushed 1.8 m up the ramp. Find (*a*) the work

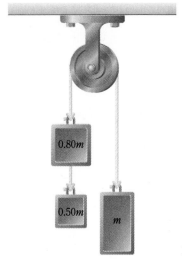

FIGURE P5.2

done by the applied force, (*b*) the work done by gravity, (*c*) the work done by friction, and (*d*) the change in kinetic energy of the block.

▪70 Two circus performers, Jack and Jill, whose total mass is 120 kg, start a swing which is 5 m long such that initially the rope attached to their swing makes an angle of 36° with the horizontal. At the bottom of the arc, Jill, whose mass is 52 kg, steps off. What is the maximum height of a landing ramp that Jack can reach as the swing continues?

▪71 A 60-kg sky diver falls from rest from a height of 2400 m above the ground. By the time she has fallen the first 1000 m, she has reached a constant speed of 60 m/s. (*a*) How much work has been done by the force of air resistance during first 1000 m of fall? (*b*) How much work does this force do during the next 800 m of fall?

72 A modern jet engine is capable of exerting a force (called the *thrust* of the engine) of 50,000 lb at full throttle. If the airplane is moving at 240 km/h at takeoff, what power is the engine developing, in watts and in horsepower?

▪73 A 72-kg person expends 420 W of power when walking on a level treadmill at a speed of 2.0 m/s. When the treadmill is inclined without changing the speed, the expended power increases to 640 W. Assuming that all the increased output power is used to overcome the force of gravity, find the angle of incline of the treadmill.

▪74 A 0.5-kg projectile is fired horizontally with an initial speed of 2.0 m/s from the top of a 100-m-tall building. At the instant before the projectile hits the ground, find (*a*) the work done on the projectile by the gravity, (*b*) the change in kinetic energy since the projectile was fired, and (*c*) the final kinetic energy of the projectile.

CHAPTER 6

LINEAR MOMENTUM

The law of conservation of energy, discussed in the previous chapter, is not the only conservation law obeyed by nature. A second example, the law of conservation of linear momentum, is the subject of this chapter. We will see that this law is a direct consequence of Newton's third law—action forces and reaction forces—and will examine its applications to collisions and rocket engines. In addition, we will define the center of mass of a system of objects and discuss the importance of this concept. Linear momentum and its conservation will prove to be a most useful tool as we continue our study of the laws of physics.

6.1 THE CONCEPT OF LINEAR MOMENTUM

An experience common to us all is that a moving object possesses a property that causes it to exert a force on anyone or anything trying to stop it. The faster the object is moving, the harder it is to stop. In addition, the more massive the object is, the more difficulty we have in stopping it. For example, an automobile moving at 2 m/s is stopped much less easily than is a tricycle traveling at the same speed. Newton called this property of a moving object its *quantity of motion*. Today we term it the **linear momentum** of the moving object.

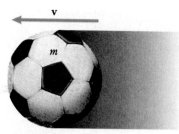

FIGURE 6.1

The linear momentum of this object is $m\mathbf{v}$, a vector quantity.

We define linear momentum in the following way. Consider the soccer ball of mass m and velocity $\mathbf{v}$ shown in Fig. 6.1. For the ball,

$$\text{Linear momentum} = \mathbf{p} = m\mathbf{v} \tag{6.1}$$

where $\mathbf{p}$ is the symbol used for linear momentum. Because linear momentum is a derived quantity, its units are those resulting from its definition. In the SI, these are kg · m/s. This is one case where we do *not* have a special name for the combination of basic units.

Notice that the momentum of an object is large if the object has large mass and large velocity. From its defining equation, momentum is a *vector* quantity, having the same direction as the velocity $\mathbf{v}$ of the object. Finally, notice that both linear momentum and kinetic energy depend on both speed and mass. The *magnitude* of an object's momentum is related to its kinetic energy in a simple way:

$$\text{KE} = \frac{1}{2}mv^2 = \frac{(mv)^2}{2m} = \frac{p^2}{2m}, \quad \text{and} \quad p^2 = 2m(\text{KE}) \tag{6.2}$$

6.2 NEWTON'S SECOND LAW RESTATED

There is an important relationship between the net force applied to an object and the change in linear momentum that the force causes. When a net force $\mathbf{F}$ accelerates an object, it obviously causes the velocity, and thus the momentum, of the object to change. Let us now investigate this relationship by seeing how Newton's second law looks when it is written in terms of linear momentum.

Consider the crate of mass m in Fig. 6.2. Because of the force $\mathbf{F}$ applied to it, the crate experiences an acceleration, and the mass and force are related through Newton's second law, $\mathbf{F} = m\mathbf{a}$. From the definition of acceleration as $\mathbf{a} = (\mathbf{v}_f - \mathbf{v}_o)/t$, however, $\mathbf{F} = m\mathbf{a}$ becomes

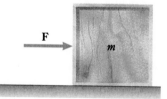

FIGURE 6.2

The net applied force $\mathbf{F}$ increases the momentum of the crate. Momentum has direction, and the increase will be in the direction of $\mathbf{F}$.

$$\mathbf{F} = \frac{m(\mathbf{v}_f - \mathbf{v}_o)}{t}$$

This can be rewritten as

$$\mathbf{F} = \frac{m\mathbf{v}_f - m\mathbf{v}_o}{t} \quad \text{or} \quad \mathbf{F} = \frac{\Delta\mathbf{p}}{t} \tag{6.3}$$

where $\Delta\mathbf{p}$ is the change in linear momentum that occurs in time t. Thus we have related the net force acting on the object to its change in linear momentum.

Equation 6.3 is the form in which Newton stated his second law, rather than as $\mathbf{F} = m\mathbf{a}$. In words, Eq. 6.3 says that the net force that acts on an object equals the time rate of change of the object's linear momentum. In some situations, Eq. 6.3 is preferable to $\mathbf{F} = m\mathbf{a}$ because this latter form applies only if the mass of the object is constant. Nowadays, for example, we often accelerate atomic particles to such high speeds that their mass increases. (This effect was first predicted by Einstein in his theory of relativity, discussed in Chaps. 4 and 25.) In such a situation, Eq. 6.3 is valid but $\mathbf{F} = m\mathbf{a}$ is not. Whenever the mass of the object being accelerated is changing, Eq. 6.3 is the form of Newton's second law that one should use. One

This high speed photograph reveals the momentary force given to the football by the kicker's foot. The product of this force times the duration of the force is the impulse given to the ball, equal to the resulting change in its momentum.

practical example of a changing-mass situation is rocket and jet propulsion, which is discussed later in this chapter.

Often we wish to apply the concept of momentum change to situations in which the applied force is not constant. For example, suppose a bat strikes a ball of mass m and changes the ball's velocity from $\mathbf{v}_o$ to $\mathbf{v}_f$ during the time t in which the ball is in contact with the bat. Then we use Eq. 6.3 to define the average force $\bar{\mathbf{F}}$ exerted by the bat on the ball. We have, after multiplying through by t,

$$\bar{\mathbf{F}}t = \Delta \mathbf{p} \qquad (6.4)$$

In the case of the bat hitting the ball, this becomes

$$\bar{\mathbf{F}}t = m\mathbf{v}_f - m\mathbf{v}_o$$

The product $\bar{\mathbf{F}}t$ is called the **impulse** of the force. Because change in momentum is fairly easy to measure, the impulse can be evaluated even though the average force and the time of contact may be extremely difficult to determine.

Illustration 6.1

In Fig. 6.3, a 1500-kg car traveling along a straight line reduces its speed from 20 m/s at A to 15 m/s at B in 3.0 s. How large is the average force retarding its motion?

Reasoning   From Eq. 6.3, the momentum form of Newton's second law, we can write

$$\bar{\mathbf{F}} = \frac{m\mathbf{v}_f - m\mathbf{v}_o}{t}$$

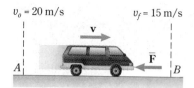

FIGURE 6.3

The car takes 3.0 s to travel from *A* to *B*. Determine $\bar{\mathbf{F}}$.

Take the direction of motion as the positive direction. Then $\mathbf{v}_o = +20$ m/s, $\mathbf{v}_f = +15$ m/s, and $t = 3.0$ s. After making these substitutions, we find that $\bar{\mathbf{F}} = -2500$ N. (Notice that we are using plus and minus signs to show direction.) The fact that $\bar{\mathbf{F}}$ is negative tells us that it is in the negative direction, a fact we show in Fig. 6.3.

Exercise   How far is it from *A* to *B*?   *Answer:*   52.5 m

Example 6.1

The 1200-kg car in Fig. 6.4, initially moving at 20 m/s, strikes a tree and comes to rest in a distance $s = 1.5$ m. Estimate the average stopping force the tree exerts on the car.

R

Reasoning

Question   How is the stopping force related to the change in the car's motion?
Answer   You have a choice in how you describe the change. You can calculate the car's deceleration as before, or, in the new terminology of this chapter, you can say the car's *momentum* has changed and relate the force directly to that change.

Question   What is the change in the car's momentum?
Answer $\Delta\mathbf{p} = m\mathbf{v}_f - m\mathbf{v}_o = 0 - (1200\text{ kg})(20\text{ m/s}) = -24\,000\text{ kg}\cdot\text{m/s}$

Notice the minus sign. It says the direction of the *change* in momentum is opposite the direction of the initial velocity.

Question   What relates the stopping force to $\Delta\mathbf{p}$?
Answer   The *impulse* produced by the force is equal to $\Delta\mathbf{p}$: (Eq. 6.4)

$$\mathbf{F}t = \Delta\mathbf{p}$$

Question   How can I determine the time duration of the force?
Answer   In the absence of other information, assume the deceleration to be constant during the collision. Then the average speed can be found and related to the stopping distance and the time:

$$\bar{v} = \frac{v_f - v_o}{2} = 10\text{ m/s} \quad\text{and}\quad s = \bar{v}t, \text{ giving } t = \frac{s}{\bar{v}} = \frac{1.5\text{ m}}{10\text{ m/s}} = 0.15\text{ s}$$

Solution and Discussion   The average stopping force can now be calculated:

$$\bar{\mathbf{F}} = \frac{-24000\text{ kg}\cdot\text{m/s}}{0.15\text{ s}} = -1.6\times10^5\text{ N}$$

Notice how large this force is (approximately 18 tons). Notice also that it depends strongly on the distance the car travels in coming to rest; the force decreases as the distance increases. It is for this reason that bumpers and exterior frame members on modern cars are designed to "give" during collisions, thus absorbing the "shock".

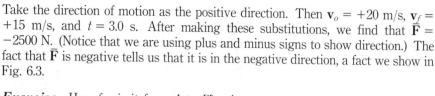

FIGURE 6.4

How large is the stopping force?

Illustration 6.2

To illustrate how air bags reduce injuries in automobile collisions, consider the following. Without an air bag or shoulder belt, the upper half of a driver is not stopped (or even slowed) until hitting the padded steering wheel. Thus the driver's head and upper torso will hit at about the same speed the car was going at the time of the collision. Assume the stopping distance, or "give," of the steering wheel is 1 cm. With an air bag, assume the give is 50 cm. The give of body tissue can be as much as 5 cm. Furthermore, assume that the upper half (30 kg) of a 60-kg driver strikes either the wheel or the air bag at the car's speed of 20 m/s. Calculate the forces exerted on the driver in each case.

Reasoning   In Example 6.1, we saw that the average retarding force during a collision depends inversely on the distance over which the stopping takes place. The car in Example 6.1 compresses over a relatively large distance (1.5 m). The driver, however, doesn't begin to stop until he or she encounters either the steering wheel or the air bag. Thus the body and head are required to stop over a shorter distance and hence a shorter time than the car. Putting the data given above into the equations of Example 6.1, we get, for the steering wheel,

$$t = \frac{0.06 \text{ m}}{10 \text{ m/s}} = 0.006 \text{ s}$$

Thus the body must stop in 6 ms! This requires an average force of

$$\bar{F} = \frac{0 - (30 \text{ kg})(20 \text{ m/s})}{0.006 \text{ s}} = -1.0 \times 10^5 \text{ N}$$

This is slightly more than 11 tons!
    For the air bag,

$$t = \frac{0.56 \text{ m}}{10 \text{ m/s}} = 0.056 \text{ s}$$

In this case the average stopping force is

$$\bar{F} = \frac{0 - (30 \text{ kg})(20 \text{ m/s})}{0.056 \text{ s}} = -1.1 \times 10^4 \text{ N}$$

This is still a large force, equal to about 1.25 tons. But when distributed over the area of the body contacting the air bag, it is similar to the effect of the force the body experiences when submerged in about 15 ft of water.

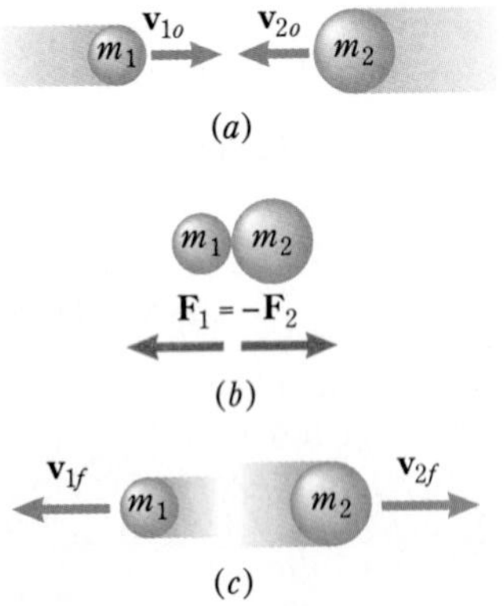

6.3 CONSERVATION OF LINEAR MOMENTUM

We saw in Chap. 5 that energy is conserved and that knowing this is important in understanding the world about us. Linear momentum also obeys a conservation law, as we now show.

    Consider the collision of two particles shown in Fig. 6.5a. The particles might be balls, molecules, or any other two objects. Newton's third law tells us that the particles exert forces on each other that are of equal magnitude but oppositely directed. Let us compute the change in momentum of the particle on the left in Fig. 6.5 as a result of the collision. From Eq. 6.3, Newton's second law in momentum form, we have for the *average* force

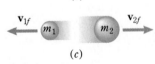

$$\mathbf{v}_{1o} \quad \mathbf{v}_{2o}$$
$$m_1 \rightarrow \quad \leftarrow m_2$$

(a)

$$m_1 \, m_2$$
$$\mathbf{F}_1 = -\mathbf{F}_2$$

(b)

$$\mathbf{v}_{1f} \quad \quad \mathbf{v}_{2f}$$
$$\leftarrow m_1 \quad \quad m_2 \rightarrow$$

(c)

FIGURE 6.5

When the two particles in (a) collide, they exert oppositely directed forces on each other, as in (b). Because the forces are of equal magnitude, what can we say about the momenta in (c) as compared to (a)?

Collisions in sports are partially inelastic. Notice the deformation of the colliding players, indicating some of the energy is absorbed internally.

$$\bar{\mathbf{F}}_1 t = m_1 \mathbf{v}_{1f} - m_1 \mathbf{v}_{1o} = \Delta \mathbf{p}_1$$

Similarly, for the particle on the right

$$\bar{\mathbf{F}}_2 t = m_2 \mathbf{v}_{2f} - m_2 \mathbf{v}_{2o} = \Delta \mathbf{p}_2$$

The same time interval t appears in both equations because that is the interval during which the balls are touching each other. If we add these two expressions, we obtain

$$(\bar{\mathbf{F}}_1 + \bar{\mathbf{F}}_2)(t) = (m_1 \mathbf{v}_{1f} - m_1 \mathbf{v}_{1o}) + (m_2 \mathbf{v}_{2f} - m_2 \mathbf{v}_{2o}) \qquad (6.5)$$
$$= \Delta \mathbf{p}_1 + \Delta \mathbf{p}_2 = \Delta \mathbf{P}_{\text{tot}}$$

where we identify the total momentum of the system as

$$\mathbf{P}_{\text{tot}} = \mathbf{p}_1 + \mathbf{p}_2$$

Since the vector force $\mathbf{F}_1$, the action force, is equal to the reaction force $\mathbf{F}_2$ but directed oppositely, we have $\mathbf{F}_1 = -\mathbf{F}_2$ and the left side of Eq. 6.5 is zero. Hence

$$\Delta \mathbf{P}_{\text{tot}} = 0$$

In words, this equation says that the *individual* momenta in the system can change, but only in a way that conserves the total momentum:

$$\Delta \mathbf{p}_1 = -\Delta \mathbf{p}_2$$

We can extend this line of reasoning to much more complicated systems. To do so, we define what is called an **isolated system:** An isolated system is a group of objects on which the net resultant force from outside is zero. For such a group (or system) of objects, if one object in the group experiences a force, there must exist an equal but opposite reaction force on some other object in the group. As a result, the change in linear momentum of the group of objects as a whole is always zero.

These considerations are summarized in the **law of conservation of linear momentum:**

The total linear momentum of an isolated system is a constant.

Even if the system we are considering is not isolated, the law is often useful. For example, in a collision of two cars, the skidding of the wheels along the pavement causes unbalanced external forces to act on the two-car system. Even so, the force of one car on the other at the instant of collision is often much larger than the skidding forces on the road. Therefore, the large changes in momentum that occur at the instant of collision are almost all the result of the force of one car on the other. As a result, the law of conservation of linear momentum can be applied to the two-car system at the instant of collision even though the system is not completely isolated.

In applying the conservation law, we must remember that momentum is a vector. To illustrate the importance of this, refer to Fig. 6.6. If we take the x direction to be positive, the total momentum of the system before collision (Fig. 6.6a) is

$$\text{Momentum before} = m_1\mathbf{v}_{1o} + m_2\mathbf{v}_{2o}$$
$$= (2\,\text{kg})(6\,\text{m/s}) + (3\,\text{kg})(-4\,\text{m/s})$$
$$= 12 - 12 = 0$$

where the value for $\mathbf{v}_{2o}$ is negative because $\mathbf{v}_{2o}$ is in the negative x direction. Even though each body has momentum, their combined momentum is zero. This is, of course, a very special case, chosen because it emphasizes so dramatically the fact that momentum is a vector. However, this particular case of zero total momentum is interesting in several other respects.

What must be true after the collision? The law of conservation of linear momentum tells us that the momentum of this isolated system is not changed by the collision. Hence, in this case, the momentum after collision must still be zero. One possible way in which this could be achieved is shown in Fig. 6.6b. Notice that the momentum of each body has a magnitude of 9 kg · m/s, one being positive, the other negative. This is definitely a possible solution for the problem, since momentum is conserved. We have the right, though, to ask whether this is the only solution to the problem.

It is a simple matter to show that the situation shown in Fig. 6.6b is not what happens in one certain case. Suppose that one of the bodies has a wad of chewing gum stuck on the side where the collision occurs. If the gum is sticky enough, the two bodies remain stuck together after the collision. What can these bodies do if they stick together?

The law of conservation of momentum allows only one answer in this case. Since the momentum of the system is zero before collision, it must still be zero after collision. Now, however, since the bodies are stuck together, they must move as a unit, and their velocities are in the same direction. Unless their final velocity is zero, the momentum after collision cannot have the required zero value. Therefore, in this case, the moving objects collide, stick together, and come to a complete stop. In this case, the kinetic energy of the colliding objects is lost during the collision; most of the lost kinetic energy appears as thermal energy in the chewing gum.

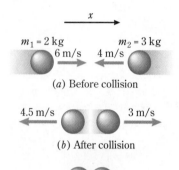

$m_1 = 2$ kg $m_2 = 3$ kg

6 m/s   4 m/s

(a) Before collision

4.5 m/s   3 m/s

(b) After collision

$v = 0$

(c) After collision

FIGURE 6.6

The situations in (b) and (c) are physically possible results of the collision of the bodies in (a). In both instances, the momentum is the same as before collision, namely zero. Hence momentum is conserved, although kinetic energy is not.

The situation in Fig. 6.6 points up an important difference between the conservation of linear momentum and that of energy. Kinetic energy itself doesn't have to be conserved, because there are many other forms of energy into which it can be changed and still conserve *total* energy. But there is only one kind of linear momentum; it cannot change into some other form. Thus conservation of linear momentum always applies in isolated systems, but you cannot say the same about kinetic energy.

Example 6.2

As shown in Fig. 6.7, a truck whose mass is 3.00×10^4 kg is traveling at 10.0 m/s and collides with a 1200-kg car traveling at 25.0 m/s in the opposite direction. If they stick together, how fast and in what direction are they moving after the collision?

R

Reasoning

Question What is the isolated system?
Answer As discussed above, the forces between road and car and those between road and truck are negligible compared with the forces of collision. Thus the car and truck can be treated as if they were an isolated system during the collision.

Question What principle applies to the collision?
Answer The conservation of linear momentum. You *cannot* assume that kinetic energy is conserved because there is no such principle.

Question What is the system's momentum before the collision?
Answer Arbitrarily choosing the direction of the truck's velocity to be positive, we have

$$(\mathbf{P}_i)_{\text{truck}} = (3.00 \times 10^4 \text{ kg})(+10.0 \text{ m/s}) = +3.00 \times 10^5 \text{ kg m/s}$$

$$(\mathbf{P}_i)_{\text{car}} = (1.20 \times 10^3 \text{ kg})(-25.0 \text{ m/s}) = -3.00 \times 10^4 \text{ kg m/s}$$
$$= -0.300 \times 10^5 \text{ kg m/s}$$

So

$$(\mathbf{P}_i)_{\text{tot}} = +2.70 \times 10^5 \text{ kg m/s}$$

(a) Before

FIGURE 6.7

Momentum is conserved in this collision, even though kinetic energy is not. Where do you think most of the lost kinetic energy goes?

(b) After

Question   What is the expression for the total momentum after collision?
Answer   The car and truck are stuck together, and so they have the same velocity, $\mathbf{v}_f$. The mass is the sum of their masses, and so

$$(\mathbf{P}_f)_{tot} = (3.00 \times 10^4\,\text{kg} + 1.20 \times 10^3\,\text{kg})\mathbf{v}_f = (3.12 \times 10^4\,\text{kg})\mathbf{v}_f$$

Question   What equation does the conservation of linear momentum give?
Answer $(3.12 \times 10^4\,\text{kg})\mathbf{v}_f = +2.70 \times 10^5\,\text{kg m/s}$

Solution and Discussion   Solving for $\mathbf{v}_f$, we obtain

$$\mathbf{v}_f = \frac{2.70 \times 10^5\,\text{kg m/s}}{3.12 \times 10^4\,\text{kg}} = +8.65\,\text{m/s}$$

The $+$ sign signifies that the wreckage is moving in the direction of the truck. Of course this is the speed immediately after collision. Friction will then bring the wreck to a stop. Also remember, the car and the truck hit each other with the *same* amount of force. The car's smaller mass requires that it undergo the greater change of velocity.

Exercise   Find the change in momentum of the car and the truck. *Answer:* $\Delta P_{car} = +4.04 \times 10^4\,kg\,m/s$, $\Delta P_{truck} = -4.04 \times 10^4\,kg\,m/s$.

Example 6.3

Figure 6.8*a* shows an x-ray photograph of a pistol just after a bullet has been fired. (You can see the bullet in the gun barrel if you look carefully.) The hot combustion gases from the exploded gunpowder are accelerating the projectile part of the bullet down the barrel. If the masses of the gun and bullet are M and m, respectively, and the exit velocity of the bullet is $\mathbf{v}_{bf}$, find the recoil velocity of the gun $\mathbf{v}_{gf}$.

FIGURE 6.8

Before the gun was fired, its momentum was zero. Hence the sum of the momenta must still be zero after it is fired. *(Hewlett-Packard)*

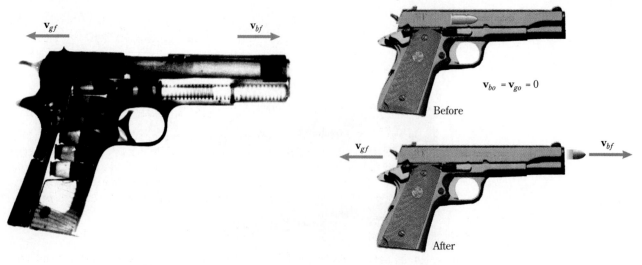

(a)

(b)

R

Reasoning

Question   What can I choose to be the isolated system?
Answer   The gun and the bullets it contains are essentially isolated, even though held in the hand. At the moment of firing, the forces caused by the explosion of the gunpowder are much larger than the force exerted by the hand. We are interested in finding the recoil velocity right at this moment.

Question   What is conserved during the explosion?
Answer   The conservation of linear momentum applies here, even though kinetic energy is obviously created by the explosion. Linear momentum must always be conserved when the system has no external forces acting on it.

Question   What is the momentum of the system before the shot is fired?
Answer   Zero. Both the gun and its bullets are at rest.

Question   What is the expression for the momentum just after firing?
Answer   Using vector notation,

$$\mathbf{P}_{tot} = M\mathbf{v}_{gf} + m\mathbf{v}_{bf}$$

Question   What equation do I get by applying the conservation of linear momentum?
Answer   You can equate the total momentum before to that after firing.

$$M\mathbf{v}_{gf} + m\mathbf{v}_{bf} = 0$$

Solution and Discussion   Solving algebraically, the recoil velocity of the gun is

$$\mathbf{v}_{gf} = -\frac{m}{M}\mathbf{v}_{bf}$$

The minus sign shows that the direction of the recoil is opposite to that of the bullet. The more massive the gun, the less is the magnitude of the gun's recoil velocity.

Exercise   What is the recoil velocity of a 2-kg gun if it shoots a 7-g bullet out of the muzzle at a speed of 500 m/s?   *Answer: 1.75 m/s.*

6.4   ELASTIC AND INELASTIC COLLISIONS

Kinetic energy is lost in many collisions. For example, when the two objects in Fig. 6.6c collide, they remain motionless after collision. All their original kinetic energy is changed to other forms of energy upon collision. Similarly, when two cars collide, a portion of their original kinetic energy is lost as work is done to mangle the cars. We call any collision during which kinetic energy is lost an *inelastic collision.*

An **inelastic collision** is one during which kinetic energy is lost.

Under certain special conditions, scarcely any kinetic energy is lost in a collision. In the ideal case, when no kinetic energy is lost, the collision is said to be a *perfectly*

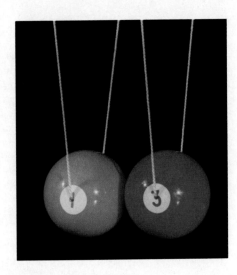

(a) An example of an inelastic collision. Notice the distortion of the tennis ball. (b) An elastic collision: the collision does not distort the surface of the billiard balls significantly.

elastic collision. Collisions between hard balls, such as billiard balls, are nearly perfectly elastic. Collisions between individual molecules, atoms, and subatomic particles often result in no loss of kinetic energy; these are perfectly elastic collisions.

A **perfectly elastic collision** is one during which kinetic energy is conserved.

Example 6.4

In Fig. 6.9, a 40-g ball traveling to the right at 30 cm/s collides head on with an 80-g ball that is at rest. If the collision is perfectly elastic, what is the velocity of each ball after collision? (By "head on," we mean that all motion takes place on a straight line.)

Reasoning

Question What does the term *perfectly elastic* imply?
Answer That both the momentum and the KE of the system of the two balls are conserved during the collision.

Question What is the momentum of this system before the collision?
Answer Ball 2 is at rest, and so its initial momentum is zero. The total momentum of the system is then just the initial momentum of ball 1:

$$(\mathbf{P}_{tot})_i = m_1\mathbf{v}_{1i} = (0.040\,\text{kg})(0.30\,\text{m/s}) = 0.012\,\text{kg}\,\text{m/s}$$

where the subscript i refers to initial values. Referring to Fig. 6.9, you can see that the direction of this vector is to the right (positive sign = to the right).

Question What is the expression for the momentum after collision?
Answer Designating the final velocities by subscript f, we have

$$(\mathbf{P}_{tot})_f = (0.040\,\text{kg})\mathbf{v}_{1f} + (0.080\,\text{kg})\mathbf{v}_{2f}$$

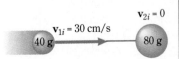

FIGURE 6.9

If the head-on collision is perfectly elastic, what are the velocities after collision?

Question   How do I know these are the right signs to use?
Answer   You don't know at this point. Because we gave both terms on the right side of the equation a plus sign, the above expression *assumes* both balls will be moving to the right. Ball 1 will either slow down and continue to the right or it will rebound to the left.

Question   How can I tell which is the case?
Answer   If you obtain a positive answer for $\mathbf{v}_{1f}$, your choice of signs was correct. A negative answer for $\mathbf{v}_{1f}$ means that ball 1 is going in the opposite direction, i.e., to the left. Thus the worst that can happen, regardless of your choice, is that you will get a negative number.

Question   What equation do I get from the conservation of linear momentum?
Answer $(\mathbf{P}_{\text{tot}})_i = (\mathbf{P}_{\text{tot}})_f$, which gives you

$$0.012 \,\text{kg m/s} = (0.040 \,\text{kg})(\mathbf{v}_{1f} + 2\mathbf{v}_{2f})$$

Question   Since there are two unknowns, I need a second equation. What other principle can I apply?
Answer   Because we are told that the collision is perfectly elastic, we know that kinetic energy is conserved. Thus we can say

$$\tfrac{1}{2}(0.040 \,\text{kg})(0.30 \,\text{m/s})^2 + 0 = \tfrac{1}{2}(0.040 \,\text{kg})(v_{1f})^2 + \tfrac{1}{2}(0.080 \,\text{kg})(v_{2f})^2$$

or

$$0.090 \,\text{m}^2/\text{s}^2 = 2v_{2f}^2 + v_{1f}^2$$

Solution and Discussion   We can solve these two equations by first finding $\mathbf{v}_{1f}$ in terms of $\mathbf{v}_{2f}$ from the momentum equation. For now, let us suppress the units notation in the interest of clarity.

$$\mathbf{v}_{1f} = 0.30 - 2\mathbf{v}_{2f}$$

Squaring,

$$v_{1f}^2 = 0.090 - 1.2v_{2f} + 4v_{2f}^2$$

Now substitute this into the kinetic energy equation:

$$2v_{2f}^2 + (0.090 - 1.2v_{2f} + 4v_{2f}^2) = 0.090$$

Collecting terms, we get

$$6v_{2f}^2 - 1.2v_{2f} = 0$$

The solutions to this quadratic equation are $\mathbf{v}_{2f} = 0$ and $0.20 \,\text{m/s}$. Putting these values into the momentum equation gives

$$\mathbf{v}_{1f} = 0.30 \,\text{m/s and} -0.10 \,\text{m/s}$$

The first pair of answers ($\mathbf{v}_{1f} = 0.30 \,\text{m/s}$, $\mathbf{v}_{2f} = 0$) implies that ball 1 just keeps going right through ball 2, which stays at rest. This is a *mathematically* possible

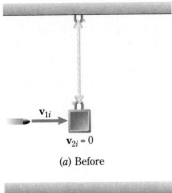

(a) Before

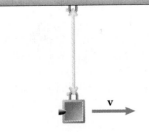

(b) Just after

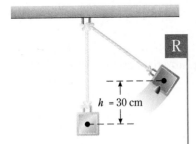

(c) Highest position

FIGURE 6.10

The momentum is the same in (a) and (b) but not in (c). Kinetic energy is changed to gravitational potential energy in going from (b) to (c).

solution, but clearly is not a *physically* possible one. The other, correct, solution shows that after the collision, ball 1 rebounds back to the left at a speed of 0.10 m/s. Ball 2 moves to the right with a speed of 0.20 m/s.

We encounter many examples in which the mathematical equations are satisfied by solutions which don't have a physical meaning. In those circumstances it is your job to examine the physical situation to see which of the solutions makes sense. For example, one of the solutions of a quadratic equation for the time of flight of a projectile may be negative. If we assumed that the projectile was launched at $t = 0$, a negative time clearly would not have physical meaning. The positive solution for t would be the physically correct one.

Exercise   What happens if the balls have equal masses m?   *Answer: They interchange velocities.*

Example 6.5

In Fig. 6.10, a 10-g pellet of unknown speed is shot into a 2.000-kg block of wood suspended from the ceiling by a cord. The pellet hits the block and becomes lodged in it. After the collision, the block and pellet swing to a height 30 cm above the original position. What is the speed of the pellet before collision? (This device is called a **ballistic pendulum.**)

Reasoning

Question   Is kinetic energy conserved in this situation?
Answer   You can tell that it is *not*, since there is inelastic collision when the pellet sticks in the block.

Question   Is momentum conserved?
Answer   If you choose an *isolated* system, always. At the moment of impact, the pellet plus the block of wood is the obvious system here. (Although this system is not truly isolated because of the forces exerted on it by gravity and the string, these forces cancel vertically at the moment of impact. Later, as the pendulum swings, this is no longer true, and momentum is no longer conserved.)

Question   What equation does conservation of linear momentum give me?
Answer   To solve this problem algebraically, let us call the pellet's mass m and the block's mass M. Conservation of linear momentum says

$$mv_{1i} + 0 = (m + M)V$$

where v_{1i} is the speed of the pellet before impact and V is the speed of the combined pellet and block after impact. Notice that both speeds are unknown.

Question   How is the height related to the speeds?
Answer   After the collision, gravity is the only force acting on the system. The work-energy theorem, with $\Delta TE = 0$ and $W_{\text{net}} = 0$, says that the KE the block acquires just *after* collision is transformed into GPE at the top of the travel.

Question   What equation do I get from the work-energy theorem?
Answer $\frac{1}{2}(m + M)V^2 = (m + M)gh$

Notice that this equation involves only *one* unknown, V.

Solution and Discussion The last equation gives V:

$$V = (2gh)^{1/2} = [2(9.8\,\text{m/s}^2)(0.30\,\text{m})]^{1/2} = 2.4\,\text{m/s}$$

Putting this value for V into the momentum equation then gives v_{1i}:

$$v_{1i} = \frac{(2.000 + 0.010\,\text{kg})(2.4\,\text{m/s})}{0.010\,\text{kg}} = 490\,\text{m/s}$$

Example 6.6

Suppose there is a beam of particles, each of mass m and traveling with a speed v, which hit a rigid wall perpendicularly, as shown in Fig. 6.11a. Each collision is perfectly elastic. Assume there are n particles per cubic meter in the beam and that the beam has a cross-sectional area A. Using the momentum form of Newton's second law, find the expression for the average force exerted on the wall by this beam.

Reasoning

Question What is the cause of the force on the wall?
Answer Each time a particle hits the wall, the particle bounces back in a perfectly elastic collision. In order for this rebounding to occur, the wall must exert a force on the particle, and by Newton's third law, the particle must exert an equal and opposite force on the wall. Over a time t, the average force on the wall is the number of collisions during that time multiplied by the momentum change in a single collision.

Question How do I interpret "purely elastic" here?
Answer Purely elastic means that KE doesn't change. The rigid wall doesn't move or give (essentially infinite mass compared to that of the particles), and so it doesn't have any KE. All the KE is therefore in the motion of the particles. So when a particle hits the wall with speed v, it must bounce back with the *same* speed. Remember, KE is a scalar, which means that the KE of a particle is the same after the collision.

Question Then what is the momentum change during any one collision?

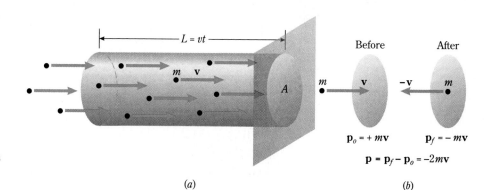

FIGURE 6.11

(*a*) A beam of particles striking an area A of a wall. (*b*) The momentum change of a purely elastic collision with a wall.

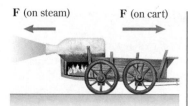

F (on steam)   **F** (on cart)

FIGURE 6.12
A jet-propelled cart.

Answer   As shown in Fig. 6.11*b*, the momentum of any particle before collision is $+m\mathbf{v}$, and afterward it is $-m\mathbf{v}$. The *change* in momentum (remember, it is a *vector*) is

$$\Delta\mathbf{p} = \mathbf{p}_f - \mathbf{p}_o = (-m\mathbf{v}) - (+m\mathbf{v}) = -2m\mathbf{v}$$

Remember, the direction of the force which causes a momentum change is in the direction of that change. In this case, $\Delta\mathbf{p}$ is negative, so the direction of $\Delta\mathbf{p}$ and hence of the force on the particle is to the *left*, and the force the particle exerts on the wall is to the *right*.

Question   How many collisions are there per second?
Answer   In Fig. 6.11*a*, all the particles in the cylinder of length $L = vt$ will impact in the time t. The volume of this cylinder is $AL = Avt$. Since n is the number of particles per cubic meter, the number of collisions in time t is

$$N = nAL = nAvt$$

The number of collisions per second is $N/t = nAv$.

Solution and Discussion   The magnitude of the average force the beam exerts on the wall is then

$$\bar{F} = +(2mv)(nAv) = 2mv^2nA$$

The force *per area* is called the **pressure** (*P*).

$$P = \frac{\bar{F}}{A} = 2mv^2n = 4(\text{KE})n$$

where KE is the kinetic energy of a single particle. Later on, in Chap. 10, we will use a slight variation of this to derive the pressure exerted by the molecules of an ideal gas on the walls of a container.

A rocket derives its thrust by exhausting gases at extremely high speed from the rocket nozzle. The rearward momentum given to these gases is equal to the forward momentum imparted to the space shuttle.

6.5 ROCKETS AND JET PROPULSION

Although we think of rockets and jet engines as being relatively new devices, Newton understood the principle of their operation. He devised a jet propulsion system like the one shown in Fig. 6.12 and explained how the law of conservation of momentum applies to it. The steam generated by the boiling water shoots out of the rear of the engine and has momentum in the rearward direction. Because the water and engine initially have zero momentum, however, the cart and engine must now move (that is, recoil) in the forward direction with a momentum equal in magnitude to that of the escaping steam. Thus the cart acquires forward momentum.

In modern rockets and jet engines of all types, fuel is burned and very hot gases are formed. The gas molecules shoot out of the rear of the engine much as a stream of bullets would shoot from a fantastically fast-repeating gun. Like the gun recoiling, the rocket or the jet plane recoils in the direction opposite the motion of the ejected gas. Since the momentum acquired by the gas molecules is directed toward the rear, the rocket must acquire an equal momentum in the opposite direction (forward) because momentum is conserved.

A close examination of this kind of propulsion system shows that the interior of the engine restricts the hot gas molecules in such a way that they are shot preferentially rearward. In the process, however, according to Newton's law of action and reaction, the molecules exert a forward force on the engine, thereby thrusting the rocket forward. Both these forces occur within the engine. No force is exerted on the craft from outside. The craft is not propelled by interaction of the expelled hot gases with the atmosphere, and, in fact, a rocket operates best in outer space, where there is no air. Air exerts a friction force that retards the motion of the rocket and is therefore undesirable.

Example 6.7

Consider the gun mentioned in the exercise after Example 6.3. If this gun is a machine pistol that can fire 10 rounds (shots) per second, determine the recoil force exerted on the gun, averaged over a 1-s burst of shots.

R

Reasoning

Question What causes this recoil force?
Answer The bullets are accelerated out of the barrel by forces exerted on them by the exploding gunpowder contained by the gun. By Newton's third law, the bullets must exert an equal and opposite force on the gun.

Question How is this force related to the velocity of the bullets?
Answer Equation 6.4 tells you that the average force exerted on the bullets multiplied by the time over which the averaging is done is equal to the change of momentum of the bullets.

$$\bar{\mathbf{F}}t = \Delta \mathbf{p}_{\text{bullets}}$$

Question Over what time interval should I average?
Answer Given the statement of the problem, 1 s seems convenient. In 1 s, 10 bullets are each given a momentum equal to $(0.007 \text{ kg})(500 \text{ m/s}) = 3.5$ kg m/s. This is a total momentum change of the bullets of 35 kg m/s in each second.

Solution and Discussion The average force on the bullets is then

$$\bar{\mathbf{F}} = \frac{\Delta \mathbf{p}_{\text{bullets}}}{t} = \frac{35 \text{ kg m/s}}{1 \text{ s}} = 35 \text{ N or } 7.9 \text{ lb}$$

The average force on the gun is equal to this, in the recoil direction.

As mentioned previously, jet or rocket engines operate on this same principle, but instead of shooting discrete bullets at a relatively slow rate, these engines exhaust molecules of gas at extremely high velocities. The exhaust gases can be treated as a continuous fluid, being expelled in an amount ΔM of mass per time interval Δt. This fluid is expelled at the exhaust velocity V_{ex}. We can rewrite Newton's second law in a form that is especially convenient for this situation, where the *rate of mass exhausted is a constant:*

$$\mathbf{F}_{\text{thrust}} = \frac{\Delta \mathbf{p}_{\text{gas}}}{\Delta t} = \frac{\Delta (M_{\text{gas}} V_{\text{ex}})}{\Delta t} = \frac{\Delta M_{\text{gas}}}{\Delta t} V_{\text{ex}}$$

The term after the second equal sign comes from the definition of momentum: $\mathbf{p} = m\mathbf{v}$.

Illustration 6.3

A Centaur rocket shoots hot gas from its engine at a rate of 1300 kg/s. The gas molecules leave the rocket at a speed of 50,000 m/s. How much thrust does the Centaur rocket develop?

Reasoning   According to the form of Newton's second law developed above, the thrust is

$$\mathbf{F}_{\text{thrust}} = \frac{\Delta M_{\text{gas}}}{\Delta t} V_{\text{ex}} = (1300 \text{ kg/s})(50,000 \text{ m/s})$$

$$= 65 \times 10^6 \text{ N or about 7000 tons of force!} \quad \blacksquare$$

Most rocket engines are designed to operate at a fixed rate of combustion, and so the thrust remains the same as long as the engine is operating. However, as the engine continues to burn, fuel mass is consumed and "thrown overboard" as exhaust. Thus the total mass of the rocket is decreasing. As a result, the acceleration of the rocket is not constant, but *increases,* even though the thrust is constant. This is an example of a force acting on a nonconstant mass.

FIGURE 6.13

The momentum vector can be replaced by its components.

6.6   MOMENTUM CONSERVATION IN TWO AND THREE DIMENSIONS

As with the other vector quantities we have dealt with, momentum can be resolved into its rectangular components once we choose a coordinate system. Figure 6.13 shows a vector $\mathbf{p}$ resolved into x, y, and z components, for example. For isolated systems, we can apply the conservation of linear momentum to *each component*. In effect, this means that the conservation of linear momentum creates two equations in a two-dimensional problem and three equations in a three-dimensional one. Let us see how these equations can be used.

FIGURE 6.14

A bomb at rest before exploding and its fragments afterward.

Example 6.8

Suppose a bomb of mass M is suspended at rest from a cable and detonated, exploding into three pieces. As shown in Fig. 6.14, half of its mass ($M/2$) is observed to have a speed v_o in the $+x$ horizontal direction right after the explosion, and another piece of mass $M/3$ is seen to have a speed of $2v_o$ at an angle of $37°$ above the horizontal. Determine the velocity of the third piece, whose mass is $M/6$.

R

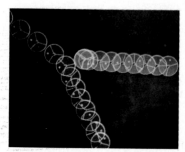

Conservation of momentum in a two-dimensional collision. Assuming the collision to be elastic, is there any way to tell which direction the disks are moving?

Reasoning

Question   What principle applies during the explosion?
Answer   Since the bomb is isolated, the momentum of the total mass of the bomb is conserved. In this two dimensional problem, this means that each component of the momentum is conserved.

Question   What is the original momentum?
Answer   Zero, in both x and y directions.

Question   What are the components of momentum after explosion?
Answer   Let $\mathbf{v}_x$ and $\mathbf{v}_y$ stand for the velocity components of the third piece. Then

$$\mathbf{p}_x = \frac{M}{6}\mathbf{v}_x + \frac{M}{2}\mathbf{v}_o + \frac{M}{3}2\mathbf{v}_o\cos 37°$$

and

$$\mathbf{p}_y = \frac{M}{6}\mathbf{v}_y + \frac{M}{3}2\mathbf{v}_o\sin 37°$$

Question   What equations does the conservation of momentum give here?
Answer   Because the initial momentum was zero, each of these components can be set equal to zero.

Solution and Discussion   For the x component we have

$$\frac{M}{6}\mathbf{v}_x + \frac{M}{2}\mathbf{v}_o + \frac{M}{3}2\mathbf{v}_o\cos 37° = 0$$

Notice that M cancels out. This gives

$$\frac{\mathbf{v}_x}{6} = -\left[\frac{\mathbf{v}_o}{2} + \frac{2(0.8)\mathbf{v}_o}{3}\right]$$

$$\mathbf{v}_x = -6.2\mathbf{v}_o$$

The y component is

$$\frac{M}{6}\mathbf{v}_y + \frac{M}{3}2\mathbf{v}_o\sin 37° = 0$$

giving

$$\mathbf{v}_y = -2.4\mathbf{v}_o$$

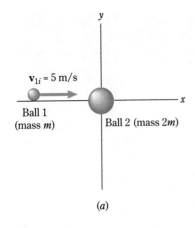

$\mathbf{v}_{1i} = 5$ m/s

Ball 1
(mass m)

Ball 2 (mass $2m$)

(a)

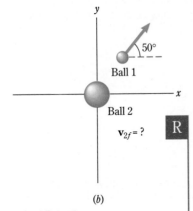

50°

Ball 1

Ball 2

$\mathbf{v}_{2f} = ?$ R

(b)

FIGURE 6.15

Balls before (a) and after (b) collision.
What does ball 2 do after collision?

The negative signs indicate that the components are in the $-x$ and $-y$ directions. The *magnitude* of the unknown $\mathbf{v}$ is

$$v = [(6.2)^2 + (2.4)^2]^{1/2}v_o = 6.65v_o$$

The direction of the third piece's velocity is

$$\theta = \tan^{-1}\left(\frac{2.4}{6.2}\right) = 21.2° \qquad \text{measured below the } -x\text{-axis}$$

Example 6.9

In Fig. 6.15a, ball 1 has mass m and is moving at 5 m/s. It collides with ball 2, of mass $2m$, which is at rest. After collision, ball 1 has a velocity of 2 m/s at an angle of 50° relative to its original direction as shown in Fig. 6.15b. (**a**) What is the velocity of ball 2 after the collision? (**b**) Determine whether this collision is elastic or inelastic. If KE is lost, what percentage is lost?

Reasoning Part (a)

Question If I don't know what kind of collision, how can I proceed?
Answer You can't tell at the outset, but it is best to assume any collision to be inelastic unless told otherwise. In other words, you cannot assume in general that kinetic energy is conserved.

Question What should I choose to be the system?
Answer The two balls form an isolated system, since the only forces acting are between them.

Question What principle can I apply?
Answer Momentum is conserved in any case. It can be applied to each component of momentum separately.

Question What is the initial momentum?
Answer $\mathbf{P}_{oy} = 0$ and $\mathbf{P}_{ox} = m(5$ m/s$)$
Vectors to the right and up will be chosen to be positive.

Question What is the final momentum?
Answer The final momentum of ball 1 is

$$\mathbf{P}_{1x} = m(2 \text{ m/s}) \cos 50° \qquad \text{and} \qquad \mathbf{P}_{1y} = m(2 \text{ m/s}) \sin 50°$$

The final momentum of ball 2 is

$$\mathbf{P}_{2x} = (2m)\mathbf{v}_{2x} \qquad \text{and} \qquad \mathbf{P}_{2y} = (2m)\mathbf{v}_{2y}$$

Question What equations does conservation of momentum give?
Answer In the x direction,

$$m(5 \text{ m/s}) = m(2 \text{ m/s}) \cos 50° + (2m)\mathbf{v}_{2x}$$

In the y direction,

$$0 = m(2 \text{ m/s}) \sin 50° + (2m)\mathbf{v}_{2y}$$

Solution and Discussion   Notice that the mass m cancels out of every term. The y equation gives

$$\mathbf{v}_{2y} = \frac{-(2 \text{ m/s})(0.766)}{2} = -0.766 \text{ m/s}$$

The x equation gives

$$\mathbf{v}_{2x} = \frac{5 \text{ m/s} - (2 \text{ m/s})(0.6431)}{2} = +1.86 \text{ m/s}$$

Thus the speed of ball 2 is

$$v_2 + [(-0.766)^2 + (1.86)^2]^{1/2} \text{ m/s} = 2.01 \text{ m/s}$$

The direction of $\mathbf{v}_2$ relative to the $+x$-axis is

$$\theta = \tan^{-1}\left(\frac{-0.766}{1.86}\right) = -22.4°$$

R

***Reasoning*   Part (b)**

Question   What is the initial KE?
Answer $(\text{KE})_i = \frac{1}{2}m(5 \text{ m/s})^2 = \frac{1}{2}m(25 \text{ m}^2/\text{s}^2)$

Question   What is the final KE?
Answer $(\text{KE})_f = \frac{1}{2}(2m)(2.01 \text{ m/s})^2 + \frac{1}{2}m(2 \text{ m/s})^2 = \frac{1}{2}m(12.1 \text{ m}^2/\text{s}^2)$

Question   Don't I need to know the mass now?
Answer   If you wanted to calculate $\Delta(\text{KE})$, yes. But not if you want only the fractional loss.

Question   What is the expression for the fraction of KE lost?
Answer $\dfrac{(\text{KE})_f - (\text{KE})_i}{(\text{KE})_i}$

Solution and Discussion   Putting in numbers, the fractional loss of KE is

$$\frac{\frac{1}{2}m(12.1 - 25)}{\frac{1}{2}m(25)} = -\frac{12.9}{25} = -0.516$$

Thus the collision is inelastic, with 51.6 percent of the original KE being converted into thermal energy of the balls.

6.7   CENTER OF MASS MOMENTUM

In momentum, as in many other situations, the concept of the center of mass of a system plays a special role. Previously we have utilized only the center of mass of

At the moment of explosion, the trajectories of the fireworks fragments are such that the velocity of their center of mass is the same as that of the fireworks just before explosion.

symmetric objects. To illustrate a broader perspective of this concept, we now define the center of mass of a system of N point masses in two dimensions. We let the masses be designated as m_1, m_2, m_3, ..., m_N, and their coordinates as x_1, x_2, x_3, ..., x_N, and y_1, y_2, y_3, ..., y_N.

The x and y coordinates of this system's center of mass are defined by the following:

$$X_{\text{c.m.}} = \frac{m_1 x_1 + m_2 x_2 + \cdots + m_N x_N}{m_1 + m_2 + \cdots + m_N} \qquad (6.6)$$

$$= \frac{m_1 x_1 + m_2 x_2 + \cdots + m_N x_N}{M_{\text{tot}}}$$

and

$$Y_{\text{c.m.}} = \frac{m_1 y_1 + m_2 y_2 + \cdots + m_N y_N}{m_1 + m_2 + \cdots + m_N} \qquad (6.7)$$

$$= \frac{m_1 y_1 + m_2 y_2 + \cdots + m_N y_N}{M_{\text{tot}}}$$

Illustration 6.4

Find the location of the center of mass of the system made up of the earth and the moon. Take the distance between them to be 240,000 mi and the mass of the moon m_M to be .0123 times that of the earth, m_E.

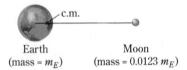

Earth
(mass = m_E)

Moon
(mass = 0.0123 m_E)

FIGURE 6.16

The center of mass of the Earth-moon system.

Reasoning We can take the line between earth and moon to be our x axis, so this is a one-dimensional problem. Furthermore, if we assume that both earth and moon are spheres, we can take their individual centers of mass to be at their geometric centers. Choose earth to be at $x = 0$; then the moon is at $x = 240{,}000$ mi. This is shown in Fig. 6.16. The position of the center of mass on the x axis is given by

$$X_{\text{c.m.}} = \frac{m_M x_M + m_E x_E}{m_M + m_E}$$

$$= \frac{(0.0123)m_E(240{,}000 \text{ mi}) + m_E(0)}{1.0123 m_E}$$

$$= \frac{(0.0123)(240{,}000 \text{ mi})}{1.0123} = 2930 \text{ mi}$$

measured from the center of the earth. Since the earth's radius is approximately 4000 mi, this point lies well within the earth's surface! ∎

If the masses in a system change position, the center of mass coordinates will generally change as a result. We can write down these changes from Eqs. 6.6 and 6.7:

$$\Delta X_{\text{c.m.}} = \frac{m_1\,\Delta x_1 + m_2\,\Delta x_2 + \cdots + m_N\,\Delta x_N}{M_{\text{tot}}}$$

$$\Delta Y_{\text{c.m.}} = \frac{m_1\,\Delta y_1 + m_2\,\Delta y_2 + \cdots + m_N\,\Delta y_N}{M_{\text{tot}}}$$

If we now divide both sides of these equations by a time interval Δt, we get expressions for the *velocity* components of the center of mass:

$$(\mathbf{V}_x)_{\text{c.m.}} = \frac{m_1\mathbf{v}_{1x} + m_2\mathbf{v}_{2x} + \cdots + m_N\mathbf{v}_{Nx}}{M_{\text{tot}}}$$

$$(\mathbf{V}_y)_{\text{c.m.}} = \frac{m_1\mathbf{v}_{1y} + m_2\mathbf{v}_{2y} + \cdots + m_N\mathbf{v}_{Ny}}{M_{\text{tot}}}$$

The numerators are simply the x and y components of the total momentum of the system, $(\mathbf{P}_{\text{tot}})_x$ and $(\mathbf{P}_{\text{tot}})_y$. Multiplying both sides by M_{tot}, we have an alternative way of writing the total momentum of a system: it is just the momentum of the system's center of mass.

$$\mathbf{P}_{\text{tot}} = M_{\text{tot}}\mathbf{V}_{\text{c.m.}}$$

Thus the conservation of linear momentum can be restated as follows:

In the absence of net external force, the velocity of the center of mass of an isolated system remains constant.

Illustration 6.5

Calculate the velocity of the center of mass of the system of balls in Fig. 6.15 before and after collision. Show that the momentum of the center of mass is conserved.

Reasoning Before collision, neither ball has a y component of velocity, so

$$(\mathbf{V}_{\text{c.m.}})_{xo} = \frac{m(5\text{ m/s}) + (2m)(0)}{m + 2m} = 1.67\text{ m/s}$$

$$(\mathbf{V}_{\text{c.m.}})_{yo} = 0$$

After collision,

$$(\mathbf{V}_{\text{c.m.}})_{xf} = \frac{m(2\text{ m/s})(\cos 50°) + 2m(1.86\text{ m/s})}{3m}$$

$$= 1.67\text{ m/s}$$

$$(\mathbf{V}_{\text{c.m.}})_{yf} = \frac{m(2\text{ m/s})(\sin 50°) + 2m(-0.766\text{ m/s})}{3m}$$

$$= \frac{+1.53\text{ m/s} - 1.53\text{ m/s}}{3m} = 0$$

Thus the collision did not alter the velocity of the center of mass.

A MODERN PERSPECTIVE
CONSERVATION OF MOMENTUM IN ATOMIC AND NUCLEAR COLLISIONS

The principles of conservation of momentum and conservation of kinetic energy in elastic collisions have been instrumental in expanding our understanding of physical interactions which occur at the extremely small scale of the atom and its nucleus. Many other concepts of classical physics have been modified by results of experiments in this realm, but these two principles have remained intact. We now discuss two examples of the application of these principles in modern physics. These are the discovery in 1932 of a new elementary particle called the neutron, and the observation in 1923 of the particle-like collisions between light and electrons.

THE DISCOVERY OF THE NEUTRON

In 1930, Walter Bothe discovered that beryllium atoms emit a very penetrating radiation when bombarded by high-energy particles. The nature of this radiation was first determined 2 years later by James Chadwick. Chadwick was not able to observe directly the particles that we now know to make up this radiation because they are uncharged and thus difficult to capture or even detect. Instead, he let the particles collide with hydrogen and nitrogen atoms. The motions of these atoms can be measured, as we shall see in later chapters. Whenever a particle collides with an atom, the atom is given energy and momentum. Such collisions are perfectly elastic, and so the kinetic energy before collision can be equated to the kinetic energy after collision. A second equation describing the collision can be obtained by equating the momenta before and after collision. Because he could measure the energy and momentum of the atoms, Chadwick had enough data to solve the energy and momentum equations for mass of the incoming particle, the neutron. It was in this way that he found the neutron mass to be 1.67×10^{-27} kg.

SCATTERING OF X-RAYS BY ELECTRONS

During the nineteenth century, both experimental and theoretical studies established that light is an electromagnetic wave phenomenon. Toward the end of that century, the discovery of radio waves and x-rays extended our knowledge of light to extremely long and extremely short wavelengths, respectively. By 1903 it had been established experimentally as well as theoretically that light waves carry energy and momentum.

The results of some experiments at the start of the twentieth century, however, in which light exchanged energy with atomic particles, could not be explained by wave-particle interactions. One set of experiments involved the way in which electrons are ejected from metal surfaces when the surfaces are illuminated by light, a phenomenon called the *photoelectric effect*. (The photoelectric effect is the principle by which solar cells operate, such as those used in photographic exposure meters and solar-powered calculators.) Another set of experiments involved the way in which x-rays are produced by bombarding a metal surface with energetic electrons. Both of these phenomena could be explained only in terms of light being a stream of particles. These particles had to have some very strange properties, however. They

Compton and Simon with equipment used to demonstrate particle aspect of x-rays.

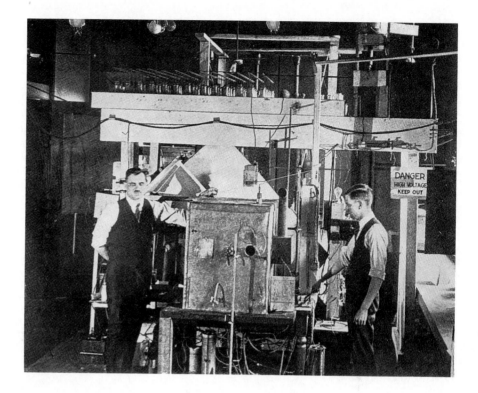

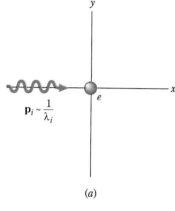

$$\mathbf{p}_i \sim \frac{1}{\lambda_i}$$

(a)

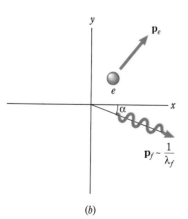

$$\mathbf{p}_e$$

$$\alpha$$

$$\mathbf{p}_f \sim \frac{1}{\lambda_f}$$

(b)

FIGURE 6.17

The Compton effect. Scattering of an x-ray by an electron, producing a scattered x-ray of longer wavelength.

had to travel at the speed of light, have no mass, and their energy and momentum both had to be inversely proportional to the wavelength of the light they represented. This latter suggestion was particularly strange, since it embodied the concept of a *particle* whose dynamic properties involved a *wave* property.

In 1923 the American physicist Arthur H. Compton carried out an experiment which showed that light, in the form of x-rays, scatters off electrons in elastic collisions like billiard balls. As the x-rays hit the electrons at rest, they transfer some of their energy and momentum to the electrons. A diagram of such a collision is shown in Fig. 6.17.

Since the energy and momentum of the x-rays are inversely proportional to the wavelength, this decrease in energy and momentum is detectable as an increase in the wavelength of the scattered x-ray compared to the wavelength of the incident x-ray. By applying the conservation principles for energy and momentum to the situation in Fig. 6.17, it is easy to derive a formula for this wavelength change. It turns out to be dependent on the angle through which the x-ray is scattered by the collision.* Compton's experimental results were precisely in agreement with this formula, which not only once again verified the conservation laws, but also gave direct evidence that in these collisions the x-rays display particle-like properties. Compton was later awarded the Nobel Prize in Physics for this work.

*What is observed in the Compton scattering experiment is the wavelength λ_f of the scattered x-ray and its direction relative to the incident x-ray, α, as shown in Fig. 6.17. Application of energy and momentum conservation to the scattering process gives the prediction that the change in the wavelength, $\lambda_f - \lambda_i$, should be proportional to $(1 - \cos \alpha)$.

LEARNING GOALS

Now that you have finished this chapter, you should be able to

1 Define (a) linear momentum, (b) impulse, (c) isolated system, (d) elastic versus inelastic collision, (e) recoil, (f) ballistic pendulum, (g) pressure, (h) center of mass of a system of masses.
2 State Newton's second law in terms of momentum.
3 Find the change in momentum of an object due to a given impulse, or vice versa.
4 State the law of conservation of linear momentum and use it in simple situations.
5 Analyze the collision of two objects that stick together on impact.
6 Analyze situations in which an object originally at rest explodes into multiple pieces.

7 Analyze situations in which two objects move along a straight line, undergo a perfectly elastic collision, and then continue to move along the same straight line.
8 Give plausible reasons for the fact that kinetic energy is not conserved in most collisions.
9 Explain the operating principle of rockets, jet engines, and similar devices which operate on the basis of recoil.
10 Calculate the position of the center of mass of a system of masses and the velocity of the center of mass.
11 Apply the conservation of momentum to the momentum of a system's center of mass.
12 Apply the conservation of momentum in two- and three-dimensional problems.

SUMMARY

DERIVED UNITS AND PHYSICAL CONSTANTS

Momentum
Basic SI Unit: $1 \, \text{kg} \cdot \text{m/s}$

DEFINITIONS AND BASIC PRINCIPLES

Linear Momentum
The linear momentum **p** of an object of mass m moving with a velocity **v** is

$$\mathbf{p} = m\mathbf{v} \qquad (6.1)$$

This is a vector quantity in the direction of the velocity.

Impulse
If an average net force $\bar{\mathbf{F}}$ acts on an object for a time t, the impulse of the force is defined as

$$\text{Impulse} = \bar{\mathbf{F}}t$$

The impulse a force imparts to an object is equal to the change in the objects' linear momentum.

$$\bar{\mathbf{F}}t = \Delta\mathbf{p} = m(\mathbf{v}_f - \mathbf{v}_o)$$

This is a direct consequence of the second law of motion.

The Principle of Conservation of Linear Momentum
The total linear momentum of an isolated system is a constant. This is a direct result of the third law of motion. It says that no matter what happens internally, internal forces cannot change a system's total momentum.

INSIGHTS
1. An isolated system is an assemblage of masses on which there are no external forces acting. In practical terms, this means that any external forces which may be acting have a negligible effect compared with the effect of internal forces.
2. The total momentum of a system is the *vector* sum of all the individual momenta of the masses in the system.
3. Individual momenta in an isolated system can change, but only in such a way that those changes balance out, leaving the total momentum constant.
4. The momentum of a system can be resolved into its rectangular components. The conservation principle can be applied to each component individually.

Types of Collisions
INELASTIC COLLISIONS
An *inelastic collision* is one in which some kinetic energy of the system is lost.

ELASTIC COLLISIONS
A *perfectly elastic* collision is one in which KE is conserved.

INSIGHTS
1. Usually the KE lost in an inelastic collision is converted mostly to thermal energy in the system.

2. Momentum must *always* be conserved in all collisions within isolated systems.
3. If a system has some momentum initially, all of its KE cannot be lost in an inelastic collision, since enough KE must remain to maintain the original momentum.

Center of Mass
The center of mass of a system of N masses is defined by

$$X_{c.m.} = \frac{m_1 x_1 + m_2 x_2 + \cdots + m_N x_N}{M_{tot}} \tag{6.6}$$

and

$$Y_{c.m.} = \frac{m_1 y_1 + m_2 y_2 + \cdots + m_N y_N}{M_{tot}} \tag{6.7}$$

where x_n and y_n are the coordinates of the nth mass.

Momentum of the Center of Mass
The momentum of the center of mass of a system is equal to the total momentum of the system.

$$\mathbf{P}_{c.m.} = M_{tot} \mathbf{V}_{c.m.} = \mathbf{P}_{tot}$$

Thus the velocity of the center of mass of an isolated system remains constant.

QUESTIONS AND GUESSTIMATES

1 When a cannon is fired, it recoils for some distance against a cushioning device. Why is it necessary to make the support so that it "gives" in this way?

2 A wad of gum is shot at a block of wood. In which case does the gum exert the larger impulse on the block, when it sticks or when it rebounds?

3 When a balloon filled with air is released so that the air escapes, the balloon shoots off into the air. Explain. Would the same thing happen if the balloon were released in a vacuum?

4 Explain why a rocket can accelerate even in outer space, where there is no air against which it can push.

5 An inventor constructs a sailboat with a large electric fan mounted on it. He directs the fan at the sail and blows air at it, expecting thereby to move in the direction of this artificial wind. To his surprise, the boat moves slowly in the opposite direction. Can you tell him why it does so?

6 A ball dropped onto a hard floor has a downward momentum, and after it rebounds, its momentum is upward. The ball's momentum is not conserved in the collision, even though it may rebound to the height from which it was dropped. Does this contradict the law of momentum conservation?

7 Reasoning from the impulse equation, explain why it is unwise to hold your legs rigidly straight when you jump to the ground from a wall or table. How is this related to the commonly held belief that a drunken person has less chance of being injured in a fall than one who is sober?

8 Explain, in terms of the impulse equation, the operating principle of impact-absorbing car bumpers and similar impact-absorbing devices.

9 A baseball player has the following nightmare. He is accidentally locked in a railroad boxcar. Fortunately, he has his ball and bat along. To start the car moving, he stands at one end and bats the ball toward the other. The impulse exerted by the ball as it hits the end wall gives the car a forward motion. Since the ball always rebounds and rolls along the floor back to him, the player repeats this process over and over. Eventually the car attains a very high speed, and the player is killed as the boxcar collides with another car sitting at rest on the track. Analyze this dream from a physics standpoint.

10 Explain how a Mexican jumping bean jumps.

11 Two blocks of unequal mass are connected by a spring, with the whole system lying on an essentially frictionless table. The blocks are pushed together and tied with a string so that the spring is compressed. Describe the motion of the blocks when the string is cut.

12 A 70-kg woman jumps from a roof 10 m above the ground. (*a*) What is her approximate speed just before she strikes the ground? (*b*) She lands on her feet but allows her legs to "give." About how long does it take her to come to rest? (*c*) About how large an average force does the ground exert on her?

13 Suppose you lay your hand flat on a tabletop and then drop a 1.0-kg mass squarely on it from a height of 0.50 m. Estimate the average force exerted on your hand by this mass. Why is injury very likely in this case even though you can *catch* the mass easily when it is dropped from this height?

PROBLEMS

Section 6.1

1 What is the linear momentum of (*a*) a 1350-kg car moving northward at 95 km/h, (*b*) a 12.5-g bullet moving up at 2450 ft/s, and (*c*) a 7.3×10^7-kg ocean liner moving west at 20 mi/h? Express your answers in SI units.

2 What is the linear momentum of a 7.50-kg stone after it has fallen from rest through a distance of 15.5 m?

3 Derive the general expression for the momentum of an object of mass *m* falling from rest through a distance *h*.

4 What is the linear momentum of a 1600-kg car whose kinetic energy is 8.50×10^5 J? What is the car's speed?

5 Derive the general expression relating the kinetic energy and linear momentum of a mass *m*.

Section 6.2 (use momentum and impulse methods)

6 How large a force is needed to stop a coasting 115-kg bike and rider in 2.1 s if the bike's original speed is 17.1 m/s?

7 Determine the average force needed to change the velocity of a 22,000-kg bus from rest to 13.6 m/s in 10.5 s.

■ 8 A jetliner having three engines and weighing 440,000 lb at takeoff requires 1750 m to reach its takeoff speed of 240 km/h. What average force must each of the engines exert during takeoff? Assume that friction can be ignored.

9 A 12.5-g bullet moving at 235 m/s passes through a sheet of plastic 3.4 cm thick and emerges with a speed of 125 m/s. The time it takes the bullet to pass through the sheet is 1.90×10^{-4} s. Find the average stopping force exerted on the bullet.

■■10 A 345-g ball moving at 15.5 m/s strikes a wall perpendicularly and rebounds straight back at a speed of 10.7 m/s. After the initial contact, the center of the ball moves 0.225 cm closer to the wall before rebounding. Assuming uniform deceleration, calculate the time the ball is in contact with the wall. How large an average force does the wall exert on the ball during this time?

■11 A proton ($m = 1.67 \times 10^{-27}$ kg) moving at 5.8×10^7 m/s shoots through a sheet of plastic foam 0.33 cm thick and emerges with a speed of 1.5×10^7 m/s. How long does it take to pass through the plastic, assuming uniform deceleration? What average force retarded the proton's motion?

■12 A 62-g arrow moving at 23.2 m/s strikes a watermelon and drills a 75-cm-long hole straight through it. It takes the arrow 0.0375 s to emerge from the melon. What average force opposes the arrow's motion?

■13 A stream of water from a hose is hitting a window. The window is vertical, the stream is horizontal, and the water stops when it hits the window. About 26 cm³ (that is, 26 g) of water with speed 2.10 m/s strikes the window each second. Find (*a*) the impulse on the window exerted in time *t* and (*b*) the average force exerted on the window.

■14 Pieces of coal drop vertically from the bottom of a chute at a rate of 7.5 kg/s onto a conveyor belt moving horizontally at a speed of 2.0 m/s. What force is necessary to drive the conveyor belt? Assume negligible friction in the drive mechanism.

Sections 6.3–6.4

15 During a switching operation, a train car of mass M_1 coasting along a straight track with velocity **v** strikes and couples to a car of mass M_2 that is sitting at rest. Find the speed of the cars after coupling.

16 While engaging in target practice, a woman shoots a 5.25-g bullet with horizontal velocity 185 m/s into a 5.5-kg log sitting on top of a post. The bullet lodges in the log. With what speed does the log fly off the post?

17 Two identical balls collide. Ball 1 is traveling to the right at 36 m/s, and ball 2 is moving to the left with a speed of 12 m/s. Find the direction and magnitude of their velocity after collision if they stick together.

18 (*a*) Repeat Prob. 17 if ball 2 has twice the mass of ball 1. (*b*) If the balls are observed to be at rest after collision, what must the mass of ball 2 be in terms of ball 1?

■19 The 17.5-g bullet in Fig. P6.1 moves with speed 5560 cm/s, strikes an 8.45-kg block resting on the table, and bounces straight back with a speed of 1260 cm/s. Find (*a*) the speed of the block immediately after collision and (*b*) the friction force between the block and table if the block moves 132 cm before stopping.

FIGURE P6.1

■20 A 2.6-kg block rests over a small hole in a table. A woman beneath the table shoots a 12.7-g bullet through the hole into the block, where it lodges. How fast was the bullet going if the block rises 55 cm above the table?

■21 A ball is falling freely. When its downward speed reaches 9.2 m/s, it explodes into two equal parts. One part goes straight up to a height of 13.7 m above the point of explosion. What is the velocity of the other part just after the explosion? Repeat for the case where the part moving upward has twice the mass of the other part.

■22 In Fig. P6.2, both balls have the same mass. The ball on the left is displaced to the outlined position and released; it collides with the stationary ball and sticks to it. (*a*) How fast are the balls moving just after collision? (*b*) What fraction of its kinetic energy did the first ball lose in the collision?

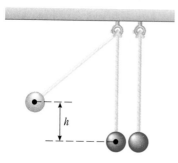

FIGURE P6.2

■**23** Suppose the two balls in Fig. P6.2 have different masses; the ball on the left has m_1. When it is let go from the height shown, it hits the second ball and sticks to it. The combination then swings to a height $h/6$. Find the mass m_2 of the second ball in terms of m_1.

24 In Fig. P6.2, two identical masses are displaced to a height h, one to the left and the other to the right. They are released simultaneously and undergo a perfectly elastic collision at the bottom. How high does each swing after the collision?

■■**25** The mass on the left in Fig. P6.2 is pulled aside and released. Its velocity at the bottom is $\mathbf{v}_o$ just as it collides with the ball on the right in a perfectly elastic collision. Find the velocities of the two balls just after collision if the ball on the left has a mass 3.5 times that of the ball on the right.

■**26** A neutron ($m = 1.67 \times 10^{-27}$ kg) moving with velocity $\mathbf{v}_o$ strikes a stationary particle of unknown mass in a perfectly elastic collision and rebounds straight back with a velocity of 0.7 $\mathbf{v}_o$. What is the mass of the struck particle?

■**27** A neutron (mass m_o) moving with velocity $\mathbf{v}_o$ strikes the stationary nucleus of an iron atom (mass $= 56m_o$) and rebounds straight back in a perfectly elastic collision. Find the velocity of the iron nucleus after the collision, assuming it is able to move freely.

■**28** What fraction of the neutron's original kinetic energy was lost to the iron nucleus in Prob. 27?

■■**29** When an object of mass m_1 moving at speed v_o collides perfectly elastically with a mass m_2 at rest, show that the largest fraction of m_1's original kinetic energy is transferred to m_2 when $m_1 = m_2$. (Hint: Assume $m_2 = km_1$, where k represents any number, then derive an expression for the amount of KE m_2 has after collision in terms of k. Show that this has the greatest value when $k = 1$.)

Section 6.5

30 The earliest true rocket used as a long-range weapon of war was the German V-2, developed toward the end of World War II. Its engine burned fuel and exhausted the combustion products at the rate of approximately 600 kg/s at an exhaust velocity of 2000 m/s. The totally fueled rocket had a mass at launch of 9×10^4 kg. (*a*) What is the thrust developed by the V-2? (*b*) With what initial acceleration did the V-2 rise from the launching pad? Express this in *g*'s.

■**31** You find yourself on a sheet of absolutely frictionless ice, carrying a 7.2-kg bowling ball. You are 21.5 m horizontally from the nearest bare ground. Your mass is 72 kg. In order to get off the ice, you throw the ball in a direction directly away from the nearest ground at a speed of 3.3 m/s. How long after throwing the ball do you reach the bare ground?

32 While coasting along a street at 0.65 m/s, a 13.9-kg girl in a 6.4-kg wagon sees a vicious dog in front of her. She has with her a 2.27-kg bag of sugar she is bringing home from the store, and she throws it at the dog with a forward velocity of 4.67 m/s relative to her original motion. What is the velocity of the girl and wagon after she throws the sugar?

■**33** A 1.25-kg pistol lies at rest on an essentially frictionless bench. It accidentally fires and shoots a 15-g bullet parallel to the bench. How far has the bullet moved by the time the gun has recoiled 3.50 mm?

34 A machine gun fires 100 13.5-g bullets per minute at a speed of 650 m/s. What is the average recoil force on the machine gun during a 1 min burst?

■**35** An 18,500-kg spaceship is traveling in outer space. The crew finds it necessary to reduce the ship's speed by 20 m/s. The rocket engine in the rear of the ship can burn fuel and oxidant at the rate of 85 kg/s and exhaust the gases at a speed of 2300 m/s. In what direction should the ship be positioned and for how long should the rocket engine burn in order to make the desired speed correction?

Section 6.6

■**36** A bomb of mass m_o is at rest when it suddenly explodes into three pieces of equal mass $m_o/3$. One piece flies out along the positive x axis at a speed of 42 m/s, while another flies in the negative y direction with a speed of 25 m/s. Find the velocity of the third piece. Repeat if the third piece has a mass $m_o/2$ and the other two each have mass $m_0/4$.

■**37** Car A (mass M_A) is traveling north at speed v_o, and car B (mass $2M_A/3$) is traveling at the same speed west. The two cars collide at an intersection, sticking together. What is their common velocity just after collision?

■**38** Two protons are moving along the x axis, one with velocity $\mathbf{v}_o$, the other with velocity $-\mathbf{v}_o$. They undergo a perfectly elastic collision. After colliding, one goes off at an angle of 50° to the positive x axis. What happens to the other? What are the velocities of the two protons after collision?

■**39** A particle has x, y velocity components ($-v_o$, 0) and a second particle of the same mass has velocity components ($v_o/2$, $v_o/2$). After colliding, one particle has velocity components $(0, v_o)$. Find the velocity components for the other. Is the collision perfectly elastic?

▪40 A hockey puck slides along the ice in the positive x direction with speed v_o. It strikes an identical puck at rest. After collision the pucks move away in directions of 30° and 60° from the positive x axis. What are the speeds of the pucks?

▪▪41 A ball of mass m is moving with speed v to the left along the x axis toward the origin. It strikes a glancing blow on a ball of mass $m/5$ at rest at the origin. After collision, the incoming ball is moving to the left with speed $v/2$ at an angle of 40° above the negative x axis. Find the speed and direction of the other ball.

▪▪42 Repeat Prob. 41 if the incoming ball is reflected back along a direction of 40° relative to the positive x axis with a speed of $v/4$.

▪43 A 1500-kg car is traveling north at a speed of 22 m/s. A second car whose mass is 1800 kg is traveling east at 32 m/s. They reach the intersection of their roads at the same instant, collide, and stick together after the collision. Find their combined velocity just after the collision.

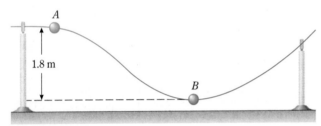

FIGURE P6.3

General Problems

▪44 How much work must be done to double the momentum of a 1250-kg car going 15.2 m/s?

▪▪45 A 65-kg astronaut becomes separated in space from her spaceship. She is 30.5 m away from it and drifting further away at a speed of 5.5 cm/s relative to it. In order to get back, she throws an 850-g wrench in the direction away from the ship with a speed of 6.9 m/s. Will this work? If so, how long will it take her to reach the spacecraft?

▪46 According to a police report, a car was sitting at rest at a stoplight when it was hit from the rear by a truck weighing 1.5 times as much as the car. At the time of collision both vehicles had their four-wheel brakes on, and skid marks showed that they skidded together 7.8 m in the original direction the truck was traveling before coming to rest. Assuming a coefficient of friction of 0.8, what must have been the approximate speed of the truck just before the collision?

▪47 Ball A in Fig. P6.3 is released from point A. It slides along the frictionless wire and collides with ball B. If the collision is perfectly elastic, find how high ball B rises after the collision. Assume the mass of B is 1/3 that of A.

▪▪48 The Atwood machine in Fig. P6.4 has a third mass attached to it by a limp string. After being released, the $2m$ mass falls a distance D before the limp string becomes taut. Thereafter, both masses on the left rise at the same speed. What is this speed? Assume the pulley to be massless and frictionless.

FIGURE P6.4

▪▪49 Suppose the $2m$ mass in Fig. P6.4 is supported so that it cannot fall. When the support for the lower mass on the left is removed, this mass falls freely a distance L before the limp string connecting it to the other mass becomes taut. Thereafter the three masses move in unison. Find the speed with which they move.

▪▪50 A vertical, uniform chain of total mass M and length L is being lowered onto a table at a constant speed v. At time $t = 0$, the lower end of the chain just touches the table. Derive an expression for the force exerted by the chain on the table as a function of t. Graph the relation between F and t for values of t while the chain is being lowered and after the chain has been completely deposited on the table.

▪51 A 50-g racketball hits the front wall of the court perpendicularly at a spot 0.5 m above the floor. The ball is traveling 50 m/s just before hitting and rebounds with an initial horizontal velocity. It hits the floor 12.4 m from the front wall. (a) With what speed did the ball rebound off the wall? (b) If the duration of the impact with the wall was 0.025 s, what average force did the ball exert on the wall?

CHAPTER 7

MOTION IN A CIRCLE

A spaceship orbiting the earth and the earth circling the sun are familiar examples of motion in a near-circular path. Objects that spin on an axis and rotating wheels are also well known to us. In this chapter, we learn how to describe motions such as these.

7.1 ANGULAR DISPLACEMENT θ

To describe the motion of an object along a line, we need a coordinate along the line, which we often take to be the x coordinate. To describe the motion of an object on a circular path or the rotation of a wheel on an axle (or axis), we need a coordinate to measure angles, the rotational counterpart of linear displacement. You are probably familiar with the usual ways for doing this, but let us summarize them in review.

Consider the two positions of the wheel in Fig. 7.1. In going from position a to position b, the wheel has turned through the angle θ. There are three common ways in which θ is measured. We can measure it in *degrees* (deg), and we know that one full circle is equivalent to 360°. Alternatively, we can measure it in *revolutions* (rev). One full circle is one revolution, and so we know that

$$1 \text{ rev} = 360°$$

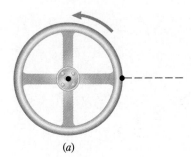

(a)

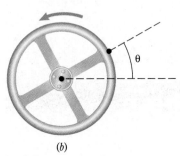

(b)

FIGURE 7.1

The angle θ describes the angular distance through which the wheel has turned.

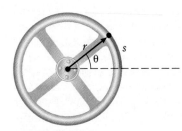

FIGURE 7.2

In radian measure, $\theta = s/r$.

TABLE 7.1
Some commonly used angles in degrees and radians

Degrees	Radians
20°	$\pi/9$
30°	$\pi/6$
36°	$\pi/5$
45°	$\pi/4$
60°	$\pi/3$
90°	$\pi/2$

The third method for measuring angles is that of *radian measure,* which was discussed in Chap. 1. Figure 7.2 summarizes the definition of radian measure. There the wheel has turned through an angle θ as a point on the rim has moved a distance s about the center. The value of θ in radians (rad) is just the ratio of s to the wheel's radius, r:

$$\theta \text{ (rad)} = \frac{s}{r} \tag{7.1}$$

Notice that 1 revolution corresponds to $s = 2\pi r$, giving $\theta = 2\pi r/r = 2\pi$ rad. The following are useful relations for you to remember:

$$1 \text{ rev} = 360° = 2\pi \text{ rad}$$

$$1 \text{ rad} = \frac{180}{\pi} \text{ degrees} \cong 57.3°$$

Notice that degrees, revs, and rads are all *dimensionless* quantities. That is, they do not involve any of the basic dimensions of physical measurement. Thus when they are involved in a calculation, they do not change the units of the terms in the equation. Nevertheless, it is important to stipulate in which way an angle is measured in order that the result of a calculation can be interpreted correctly. As we will see in Sec. 7.5, there are some cases where angles *must* be given in radians in order for a calculation to be correct.

Illustration 7.1

Convert a 70.0° angle to radians and revolutions.

Reasoning   Using the conversion factors 2π rad/360° and 1 rev/360°, we have

$$70.0° = (70.0 \text{ deg})\left(\frac{2\pi \text{ rad}}{360 \text{ deg}}\right) = 1.22 \text{ rad}$$

$$70.0° = (70.0 \text{ deg})\left(\frac{1 \text{ rev}}{360 \text{ deg}}\right) = 0.194 \text{ rev}$$

Exercise   Change 0.210 rad to degrees and revolutions.   *Answer: 12.0°, 0.0334 rev.*

7.2   ANGULAR VELOCITY ω

When we state that a phonograph record is turning at 33 rev/min, we are giving its angular velocity. We are describing how fast it rotates. Analogously to linear motion, where average velocity is defined to be displacement divided by time, we define

$$\text{Average angular velocity} = \frac{\text{angular displacement}}{\text{time taken}}$$

Modern windmills use their angular velocity to drive an electrical generator.

$$\bar{\omega} = \frac{\theta}{t} \qquad\qquad (7.2)$$

where ω (Greek omega) is angular velocity. Typical units for ω are radians per second, degrees per second, and revolutions per minute.

The wheels in Figs. 7.1 and 7.2 can rotate in two "directions": clockwise and counterclockwise. We discussed these two senses of rotation about an axis in Chap. 4 when torques and the second condition for equilibrium were introduced. Angular displacement θ and angular velocity ω about a fixed axis are vectors in the same sense as torques, in that they, too, can have two possible opposite senses of rotation. It is customary to designate counterclockwise rotations as having positive values and clockwise rotations as having negative values. This is the same sign choice we made for torques in Chap. 4. The algebraic equations involving angular quantities will then give answers that can be interpreted consistently with this choice.

Just as we did in linear motion, we make a distinction between average and instantaneous angular velocity. You will recall that instantaneous linear velocity is obtained by measuring the linear distance moved in a time so small that the velocity does not change appreciably. Doing the same in rotation, we define **instantaneous angular velocity** as

$$\omega = \lim_{\Delta t \to 0} \frac{\Delta \theta}{\Delta t} \qquad\qquad (7.3)$$

In this expression, $\Delta\theta$ is the small angular distance moved by a rotating wheel in the small time Δt, and the limit notation tells us to take the value of the ratio as the time interval Δt approaches zero, as discussed in Chap. 2.

Illustration 7.2

The wheel in Fig. 7.2 turns through 1800 rev in 1.0 min. Find its average angular velocity in rad/s.

Reasoning From the defining equation,

$$\bar{\omega} = \frac{\theta}{t} = \frac{1800\,\text{rev}}{60\,\text{s}} = 30\,\text{rev/s}$$

Then

$$30\,\text{rev/s} = \left(30\frac{\text{rev}}{\text{s}}\right)\left(\frac{2\pi\,\text{rad}}{\text{rev}}\right) = 60\pi\,\text{rad/s} = 190\,\text{rad/s}$$

Exercise Through how many radians does the wheel turn in 15 s?
Answer: $2.85 \times 10^3\ rad$

7.3 ANGULAR ACCELERATION α

Average linear acceleration, defined in Chap. 2 by the equation

$$\mathbf{a} = \frac{\mathbf{v}_f - \mathbf{v}_i}{t}$$

measures the rate at which the velocity of an object is changing. The quantity $\mathbf{v}_f - \mathbf{v}_i$ is the change in velocity during time t. You will recall that typical units for acceleration are meters per second squared and feet per second squared.

In the case of rotating objects, we are often interested in how they speed up or slow down. Hence we are concerned with *angular* acceleration, that is, the rate of change of *angular* velocity. We define the average angular acceleration α (alpha) of a rotating wheel or any other object by the relationship

$$\text{Average angular acceleration} = \frac{\text{change in angular velocity}}{\text{time taken}}$$

$$\bar{\alpha} = \frac{\omega_f - \omega_i}{t} \tag{7.4}$$

The units of angular acceleration are those of angular velocity divided by time. For example, if t is measured in seconds and ω in radians per second, the angular acceleration is expressed in radians per second per second. Although it is not wrong to measure ω in radians per second when t is in minutes, giving units of radians per second per minute, it is generally preferable to use the same unit for t in both places.

If the angular acceleration is uniform (constant), we know that, as with linear motion, the average angular velocity is

$$\bar{\omega} = \tfrac{1}{2}(\omega_f + \omega_i)$$

Illustration 7.3

A wheel starts from rest and attains a rotational velocity of 240 rev/s in 2.0 min. What is its average angular acceleration?

Reasoning We know that

$$\omega_i = 0 \qquad \omega_f = 240\ \text{rev/s} \qquad t = 2.00\ \text{min} = 120\ \text{s}$$

From the definition of angular acceleration,

$$\overline{\alpha} = \frac{\omega_f - \omega_i}{t} = \frac{(240 - 0)\ \text{rev/s}}{120\ \text{s}} = 2.00\ \text{rev/s}^2$$

Exercise What is the wheel's angular speed (in radians per second) 130 s after starting from rest? *Answer: 1630 rad/s.*

7.4 ANGULAR MOTION EQUATIONS

As you have probably recognized by now, there is a great deal of similarity between the linear and angular motion equations. The θ in angular motion corresponds to x in linear motion, ω corresponds to v, and α corresponds to a. Moreover, we have defined ω and α by equations that are identical to those for v and a except for the interchange of symbols. This leads us to the conclusion that all the motion equations we learned for linear uniformly accelerated motion carry over to uniformly accelerated angular motion as follows:

Linear	Angular	
$s = \overline{v}t$	$\theta = \overline{\omega}t$	(7.5a)
$v_f = v_i + at$	$\omega_f = \omega_i + \alpha t$	(7.5b)
$\overline{v} = \frac{1}{2}(v_f + v_i)$	$\overline{\omega} = \frac{1}{2}(\omega_f + \omega_i)$	(7.5c)
$2as = v_f^2 - v_i^2$	$2\alpha\theta = \omega_f^2 - \omega_i^2$	(7.5d)
$s = v_i t + \frac{1}{2}at^2$	$\theta = \omega_i t + \frac{1}{2}\alpha t^2$	(7.5e)

There is no need to learn new equations for angular motion. Simply replace the linear-motion variables with their angular counterparts. We shall see in the next chapter that even the equations for kinetic energy and momentum have easily guessed angular analogs. Let us now see how the problem solution methods we used for linear motion carry over to angular motion.

By pushing tangentially at the perimeter of a turntable, kids give the table an angular acceleration.

Example 7.1

A roulette wheel turning at 3.00 rev/s coasts to rest uniformly in 18.0 s. What is its deceleration? How many revolutions does it turn through while coming to rest?

R

Reasoning

Question   What quantities are known and what is to be found?
Answer   Given are $\omega_i = 3.00$ rev/s, $\omega_f = 0$, and $t = 18.0$ s. You need to find α and θ.

Question   What equations of motion relate the unknowns to the data given?
Answer   The *definition* of α (Eq. 7.4) involves ω and t.

$$\alpha = \frac{\omega_f - \omega_i}{t}$$

To find θ, you can use Eq. 7.5a without knowing α first. Once you know α, you could then choose to use either Eq. 7.5d or 7.5e to find θ.

Solution and Discussion   The angular acceleration is

$$\alpha = \frac{0 - 3.00 \text{ rev/s}}{18.0 \text{ s}} = -0.167 \text{ rev/s}^2$$

The minus sign is important, as it indicates a deceleration of the wheel. Using Eq. 7.5c, we have that

$$\bar{\omega} = \tfrac{1}{2}(0 + 3.00 \text{ rev/s}) = 1.50 \text{ rev/s}$$

Then Eq. 7.5a tells us that

$$\theta = \bar{\omega}t = (1.50 \text{ rev/s})(18.0 \text{ s}) = 27.0 \text{ rev}$$

Another approach to finding θ would be to use Eq. 7.5e:

$$\theta = (3.00 \text{ rev/s})(18.0 \text{ s}) + \tfrac{1}{2}(-0.167 \text{ rev/s}^2)(18.0 \text{ s})^2$$
$$= 27.0 \text{ rev}$$

Notice how important it is to observe the correct sign of α.

Exercise   Use Eq. 7.5d to determine θ.

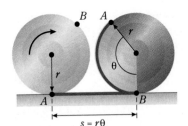

FIGURE 7.3

As the wheel turns through an angle θ, it lays out a tangential distance $s = r\theta$.

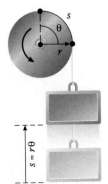

FIGURE 7.4

How much string does the spool wind as it turns through one revolution?

7.5   TANGENTIAL QUANTITIES

When a spool unwinds a string or a wheel rolls along the ground without slipping, both rotational and linear motions occur. We wish now to find out how these two types of motion are related. The relation between linear and angular distances, s and θ, is inherent in Eq. 7.1, the definition of angular measure. To see this, let us look at Fig. 7.3.

In the figure, we see that the linear distance the wheel rolls, s, is equal to the tangential distance traveled by a point on the rim. This allows us to relate linear motion to angular motion for the rolling wheel. As long as the wheel does not slip, we have $s = r\theta$, where θ is measured in radians. Furthermore, if we look at the spool in Fig. 7.4, we see that there is a similar relationship for the way in which the

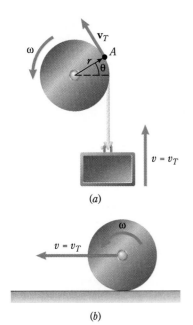

FIGURE 7.5

The angular velocity ω is related to the tangential velocity v_T by $v_T = \omega r$. In this relation, ω must be in radian measure.

string is wound on its rim. As the spool turns through an angular displacement of θ, a length s of string is wound on the spool's rim. In all such cases, we have

$$s = r\theta \qquad (\theta \text{ in radians}) \tag{7.6}$$

Note once again that in these cases θ must be measured in radians, because Eq. 7.6 is based on the definition of radian measure.

As the spool in Fig. 7.4 turns at a certain rate, the mass at the end of the string rises with a certain velocity. Similarly, as the wheel in Fig. 7.3 rolls without slipping along the ground, it rotates about its axis at a certain rate and its center moves with a certain linear velocity. The linear *speed* in each of these cases is the same as the speed of a point on the rim of the spool or the wheel. The point on the rim is traveling with this speed in a direction always *tangent* to the spool or wheel. We call this motion of a point on the rim the **tangential velocity** of the point, $\mathbf{v}_T$. Let us now relate $\mathbf{v}_T$ to the angular velocity ω of the wheel.

As the spool in Fig. 7.5a turns with constant speed through an angle θ in a time t, its rotational velocity is $\omega = \theta/t$. Because $\theta = s/r$, where r is the radius of the spool, we can substitute this value in the equation for ω and obtain

$$\omega = \frac{s/r}{t} = \frac{s}{t}\frac{1}{r}$$

However, s/t is simply the speed with which the mass in Fig. 7.5a is lifted, and that speed is equal to the tangential speed v_T of point A. Therefore this equation for ω yields $\omega = v_T/r$, or

$$\text{Tangential speed} = v_T = \omega r \tag{7.7}$$

Here, too, radian measure must be used. In a similar way, we can show that the center of the wheel in Fig. 7.5b moves with a speed $v_T = \omega r$, provided the wheel does not slip. Thus we see that Eq. 7.7 is an important relationship between the rotational motion of an object and the linear motions that result from the rotation.

Another quantity of interest is what we call the tangential acceleration. If ω is increasing for a rotating wheel, then v_T must also be increasing. The angular acceleration α is (Eq. 7.4)

$$\alpha = \frac{\omega_f - \omega_i}{t}$$

where $\omega_f - \omega_i$ is the change in the angular velocity during the time interval t. Because $\omega = v_T/r$, we can write this as

$$\alpha = \frac{v_{Tf} - v_{Ti}}{rt} \qquad \text{or} \qquad \frac{v_{Tf} - v_{Ti}}{t} = \alpha r$$

This, however, is simply the rate of change of tangential speed, or the magnitude of the *tangential acceleration* $\mathbf{a}_T$. Therefore the magnitude of $\mathbf{a}_T$ is related to the angular acceleration α by

$$a_T = \alpha r \tag{7.8}$$

This is also the linear acceleration of the center of a rolling wheel or of a given point on an unwinding string. Can you show this from a consideration of the fact that acceleration is the rate of change of velocity—tangential velocity in this case?

Equations 7.6, 7.7, and 7.8 show that although various points on a rotating object have different *linear* displacements, velocities, and accelerations (depending on how far each point is from the axis of rotation), *all points on a rigid object share the same angular motion.*

Example 7.2

A car with 80-cm-diameter wheels starts from rest and accelerates uniformly to 20 m/s in 9.0 s. Find the angular acceleration and final angular velocity of one wheel.

R

Reasoning

Question   What do the given data describe?
Answer   The linear acceleration of the car. You are also given the diameter of the wheels, which are assumed to be rolling without slipping on the road.

Question   If I find the linear acceleration of the car, how is it related to the angular acceleration of a wheel?
Answer   The linear acceleration of the car is the same as the linear acceleration of the wheel axle. Equations 7.7 and 7.8 and Fig. 7.5*b* show us that the angular motion is related to the linear motion by

$$\alpha = \frac{a_T}{r} \quad \text{and} \quad \omega = \frac{v_T}{r}$$

Solution and Discussion   Solving first for the linear acceleration of the car,

$$a_T = \frac{v_f - v_i}{t} = \frac{20\,\text{m/s} - 0}{9.0\,\text{s}} = 2.2\,\text{m/s}^2$$

The angular acceleration is then

$$\alpha = \frac{2.2\,\text{m/s}^2}{0.40\,\text{m}} = 5.6\,\text{s}^{-2} = 5.6\,\text{rad/s}^2$$

Notice that there is nothing that explicitly shows the radian measure for the angular quantity. It is implicit in the use of Eqs. 7.6, 7.7, and 7.8.
The final angular velocity is then

$$\omega = \alpha t = (5.6\,\text{rad/s}^2)(9.0\,\text{s}) = 50\,\text{rad/s}$$

Exercise   Through how many revolutions does each wheel turn during the 9.0 s?   *Answer: 36 rev.*

Example 7.3

In an experiment such as that shown in Fig. 7.5a, suppose that the mass starts from rest and accelerates downward at 8.6 m/s². If the radius of the spool is 20 cm, what is its rotation rate after 3.0 s?

R

Reasoning

Question How is the motion of the mass related to the rotation of the spool?
Answer By the radius of the spool, since the string holding the mass is wound around the spool's perimeter and unwinds without slipping.

Question In order for the mass to accelerate downward at a given acceleration a, what must be the corresponding angular acceleration of the spool?

Answer $\alpha = \dfrac{a_T}{r}$

Question How is the rotation rate connected to α?
Answer Rotation rate is angular velocity, given by

$$\omega = \alpha t$$

Solution and Discussion The numerical values are

$$\alpha = \frac{8.6 \, \text{m/s}^2}{0.20 \, \text{m}} = 43 \, \text{rad/s}^2$$

and

$$\omega = \alpha t = (43 \, \text{rad/s}^2)(3.0 \, \text{s}) = 130 \, \text{rad/s}$$

Notice once again that you have to understand that the radian is the angular measure involved.

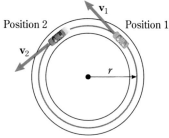

FIGURE 7.6

Even though the *speed* of the car is constant around the track, its *velocity* is continuously changing because the direction of the velocity vector is not constant.

7.6 CENTRIPETAL ACCELERATION

A very interesting situation arises when an object travels along a circular path with constant speed. For example, in Fig. 7.6 we see a car traveling with constant speed v around a circular track. Let us say that its speed is 20 m/s. Although the speed is 20 m/s at positions 1 and 2 and at all other points along the track, *the car is undergoing acceleration*. To understand this statement, we must remember two facts: (1) speed and velocity are not the same thing and (2) acceleration is defined to be the time rate of change of velocity (a vector), not of speed (a scalar). Because the *direction* of the velocity at position 1 is not the same as that at position 2, the velocity changes as the car moves along the track. From the definition of average acceleration, we have for the average acceleration of the car between 1 and 2

$$\overline{a} = \frac{\text{change in velocity}}{\text{time taken}}$$

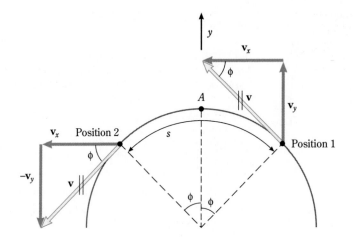

FIGURE 7.7

Notice that, in going from position 1 to position 2, the car's velocity changes by $-2v_y$. The minus sign indicates a direction along $-y$.

Let us now compute the acceleration of the car.

The situation is redrawn in Fig. 7.7. The y component of the car's velocity is v_y at position 1 and $-v_y$ at position 2. The x component of the velocity at 1 is the same as at 2. Hence we find that in going from 1 to 2,

$$(\text{Change in velocity})_y = v_{yf} - v_{yo} = -v_y - v_y = -2v_y$$

The time the car takes to go from 1 to 2 is $t = s/v$, where v is the constant tangential velocity of the car along the track and s is the arc length from 1 to 2. Moreover, from the definition of radian measure, $\theta = s/r$, we have that

$$2\phi = \frac{s}{r} \qquad \text{or} \qquad s = 2r\phi$$

since s subtends an angle 2ϕ in this case. Thus we find that

$$t = \frac{s}{v} = \frac{2r\phi}{v}$$

We now know the change in velocity, $-2v_y$, and the time taken, $2r\phi/v$. Hence we can compute

$$\overline{a} = \frac{\text{change in velocity}}{\text{time taken}} = \frac{-2v_y}{2r\phi/v} = -\frac{vv_y}{r\phi}$$

From Fig. 7.7, though, we see that $v_y = v \sin \phi$. Therefore

$$\overline{a} = -\frac{v^2 \sin \phi}{r\phi}$$

This is the *average* acceleration of the car as it moves from position 1 to position 2. However, we are interested in the *instantaneous* acceleration at a point such as A. To obtain it, we simply let ϕ shrink to a very small value. Then, because $\sin \phi \cong \phi$ in radians when ϕ is small (use your calculator to check that this is true), the instantaneous acceleration is

Locate the forces on the bike and rider in rounding the curve. Why must the bike and the rider bank into the curve?

$$a = -\frac{v^2 \sin \phi}{r\phi} \cong -\frac{v^2\phi}{r\phi} = -\frac{v^2}{r}$$

This is the acceleration of the car as it goes by point A. Since the speed is constant, all points on the circle are equivalent, and so the magnitude of $a = v^2/r$ no matter where we choose point A.

Let us now find the direction of this acceleration. Remember, the direction of **a** is by definition in the direction of $\Delta \mathbf{v}$. Figure 7.7 shows that at point A, $\Delta \mathbf{v} = -2v_y$. The negative sign indicates that this is a vector pointing from point A toward the negative y direction, which is toward the center of the circle. At A, therefore, $\Delta \mathbf{v}$ (and **a**) *points toward the center of the circle.* But the point A in Fig. 7.7 was chosen arbitrarily. We could always choose our y axis to pass through any other selected point. Thus the conclusion we have drawn from using this point is a perfectly general one, and would apply to all other points on the circle. To summarize:

An object moving with constant speed v along a circular path of radius r is undergoing an acceleration directed toward the center of the circle. This is called *centripetal* (literally "center-seeking") *acceleration,* $\mathbf{a}_c$. The magnitude of this acceleration is

$$a_c = \frac{v^2}{r} = \omega^2 r \tag{7.9}$$

where we have used the relation $v = \omega r$.

The acceleration $\mathbf{a}_c$ describes the rate of turning, in the sense that it represents the rate of change of the *direction* of the motion.

7.7 CENTRIPETAL FORCE

Newton's first law states that a net force must act on an object if the object is to be deflected from straight-line motion. Therefore an object traveling on a circular path must have a net force deflecting it from its straight-line path. For example, if the track in Fig. 7.6 is too slippery to provide the required friction force at the wheels, the car will slip off the track in a straight line tangent to the circle. Similarly, the ball in Fig. 7.8 being twirled in a circular path is compelled to follow this path by the centerward pull of the string. If the string breaks when the ball is at point B, the ball will follow the straight-line path indicated by the broken line tangent to the circle.

Now that we know about centripetal acceleration, computing the force needed to hold an object of mass m in a circular path is a simple task. When traveling on a circle, the object experiences an acceleration toward the center of the circle, the centripetal acceleration $a_c = v^2/r$, where r is the circle's radius and v is the tangential speed of the object on the circular path. A force in the same direction, toward the center of the circle, must pull on the object to furnish this acceleration. It is the force $\mathbf{F}_c$ in Fig. 7.8, for example. From $\mathbf{F}_{\text{net}} = m\mathbf{a}$, we find this required force, called *centripetal force,* to have a magnitude

$$F_c = ma_c = \frac{mv^2}{r} \tag{7.10}$$

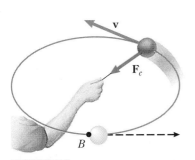

FIGURE 7.8

If the string breaks when the ball is at B, the ball will follow the tangential broken line.

The force required to hold an object of mass m and moving at speed v in a circular path of radius r is called the **centripetal force** and its magnitude is given by mv^2/r. It is directed toward the center of the circle.

We will encounter many additional examples of centripetal forces later, such as those provided by gravity producing a circular satellite orbit and magnetic forces causing electrically charged particles to move in circles.

It is of interest to notice that no work is done by the centripetal force. To do work, a force must have a component in the direction of motion. The centripetal force is directed along the radius of a circle, however, whereas the motion occurs tangent to the circle. Because the tangent is perpendicular to the radius, the centripetal force has no component in the direction of motion. It therefore does no work. It simply changes the direction of the motion.

To summarize the two effects forces can have in changing the velocity of an object:

Forces tangent, or parallel, to the direction of motion change only the *speed* of an object and can do work on the object. Forces perpendicular to the direction of motion change only the *direction* of motion and can perform no work.

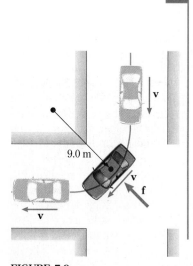

FIGURE 7.9

For the car to turn the corner as shown, the friction force **f** between the tires and the pavement must furnish the centripetal force needed to hold the car in a circular path.

Example 7.4

A 1200-kg car is turning a corner at 8.00 m/s, and it travels along an arc of a circle in the process (Fig. 7.9). (**a**) If the radius of the circle is 9.00 m, how large a horizontal force must the pavement exert on the tires to hold the car in the circular path? (**b**) What minimum coefficient of friction must exist in order for this car not to slip?

Reasoning

Question   What is the car's acceleration as it turns the corner?
Answer   The centripetal acceleration is

$$a_c = \frac{v^2}{r} = \frac{(8.00\,\text{m/s})^2}{9.00\,\text{m}} = 7.11\,\text{m/s}^2$$

Question   What force is necessary to accomplish this?
Answer $F = ma = (1200\,\text{kg})(7.11\,\text{m/s}^2) = 8530\,\text{N}$

Question   For part (*b*), how is this force related to the force of friction?
Answer   The driver must rely on *static* friction between the tires and the road to make the turn safely. If the tires skid, the force between the tires and the road becomes that of *kinetic* friction, which is usually less than static friction. The maximum force of static friction in this case is

$$f_s(\text{max}) = \mu_s F_N = \mu_s mg$$

Solution and Discussion μ_s must then be at least equal to $(8530\,\text{N})/mg$. Thus

$$\text{min } \mu_s = \frac{8530\,\text{N}}{(1200\,\text{kg})(9.8\,\text{m/s}^2)} = 0.725$$

Notice that the units in this equation cancel completely, giving a dimensionless number.

Example 7.5

A ball tied to the end of a string is swung in a vertical circle of radius r, as in Fig. 7.10. What is the tension in the string when the ball is at point A if the ball's speed is v at that point? Do not neglect the force of gravity.

R

Reasoning

Question   At point A what are all the forces acting on the ball?
Answer   At this point, there are only two forces on the ball. Gravity, mg, is acting downward, as is the tension in the string, T.

Question   What is the acceleration of the ball at point A?
Answer   At the instant it reaches point A the ball is moving in a circle of radius r at a speed v. The acceleration that describes this motion is $a_c = v^2/r$. At point A the center of the circle is downward, the same as the direction of both forces.

Question   What is the equation that results from the second law?
Answer $F_{net} = mg + T = mv^2/r$. Thus

$$T = \frac{mv^2}{r} - mg = m\left(\frac{v^2}{r} - g\right)$$

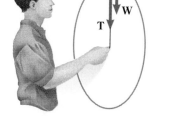

FIGURE 7.10

When the ball is in the position shown, its weight provides part of the necessary centripetal force.

Solution and Discussion
In the expression for tension above, notice that if $v^2/r < g$, T is negative, which is not possible physically, since a string can exert only a *pull* on whatever it is attached to. It cannot exert a *push*, since it would just collapse. Thus the ball must reach the top of the circle with at least a speed of $(gr)^{1/2}$ to remain in the circular path. If v is less than this, the ball will fall down out of the circle.

Exercise   What would the tension in the string be at the bottom of the circle if the ball was going at speed v?   *Answer:* $T = mv^2/r + W = m(v^2/r + g)$.

In order to swing the hammer in a circle, the thrower must be able to exert a sufficient centripetal force on the chain. Notice how the angle of his leg and foot enable him to do this.

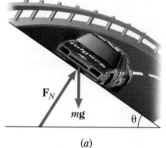

F_N

mg

θ

(a)

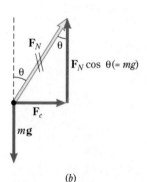

F_N θ

θ

$F_N \cos\ \theta(= mg)$

F_c

mg

(b)

FIGURE 7.11

For proper banking, the vertical component of normal force balances mg and the horizontal component supplies the centripetal force.

Example 7.6 *Banking curves*

A curve in a road has a 60-m radius. It is to be banked so that no friction force is required for a car going at 25 m/s to safely make the curve. At what angle should it be banked?

R

Reasoning

Question Without friction, what force can produce the centripetal acceleration?
Answer From Fig. 7.11a you can see that F_N is not completely vertical, but has a horizontal component directed toward the center of the car's circular path. It is this horizontal component of F_N that produces the centripetal acceleration.

Question What coordinates should I choose into which to resolve the forces?
Answer Figure 7.11b, the free-body diagram for the car, shows F_N resolved into horizontal and vertical components where θ is the angle of bank of the road. The reason for this choice is that the car is moving in a horizontal circle, and so its centripetal acceleration is directed horizontally toward the center of this circle.

Question What are the second-law statements in this situation?
Answer Vertically $a_y = 0$, and so

$$mg = F_N \cos\theta$$

This determines F_N. Horizontally, $a_x = a_c = v^2/r$, and so

$$F_N \sin\theta = F_c = \frac{mv^2}{r}$$

Question What condition determines the required angle?
Answer The angle can be found by eliminating F_N from the two component equations.

Solution and Discussion From the first equation, $F_N = mg/(\cos\theta)$. Putting this value for F_N into the second equation gives us

$$\frac{mg\sin\theta}{\cos\theta} = mg\tan\theta = \frac{mv^2}{r}$$

or

$$\theta = \tan^{-1}\left(\frac{v^2}{gr}\right)$$

Putting in the given values for v and r,

$$\theta = \tan^{-1}\left[\frac{(25\text{ m/s})^2}{(9.8\text{ m/s}^2)(60\text{ m})}\right] = 47°$$

Without friction, the car will slide *down* this bank if its speed is less than 25 m/s and *up* the bank if its speed is greater.

Example 7.7 *Banked curves with friction*

Let us now extend the previous example to the case where there is friction. For the same bank angle as before, find the *maximum* speed a car can have without skidding if the coefficient of static friction between tires and road is 0.8. The situation is shown in Fig. 7.12a, where friction is shown directed along the two surfaces in contact. The direction of the friction force is such that the force resists the tendency of the car to skid to the outside of the curve.

R ### Reasoning

Question How is the free-body diagram of the car changed from the last example?

Answer As shown in Fig. 7.12b, the force of friction is added parallel to the bank of the curve.

Question How do the components of **f** change the situation?

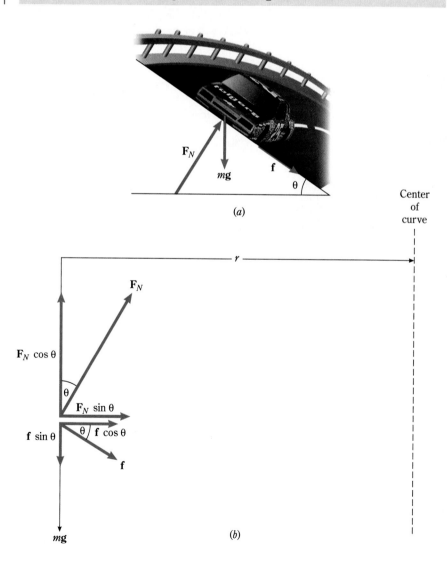

(a)

(b)

FIGURE 7.12

When there is friction on a banked curve, it adds a contribution to $\mathbf{F}_c$.

The extreme banking of curves on a racetrack allows the cars to maintain high speeds through the turn.

Answer   **f** has a horizontal component ($f\cos\theta$) which adds to the horizontal component of $\mathbf{F}_N$ to provide a larger centripetal force than before. This added force will allow the car to go around the curve at higher speeds. The vertical component of **f** ($f\sin\theta$) is downward, adding to $m\mathbf{g}$.

Question   What equations result from applying the second law?
Answer   The vertical forces must again balance:

$$F_N\cos\theta = mg + f\sin\theta$$

The net horizontal force causes the centripetal acceleration:

$$F_N\sin\theta + f\cos\theta = \frac{mv^2}{R}$$

Question   What condition determines the *maximum* speed allowed?
Answer   The maximum speed requires the maximum centripetal force. F_N cannot vary, but f can supply a force up to a maximum of $\mu_s F_N$.

Question   What equation determines the maximum speed?
Answer   *Two* equations:

$$F_N\sin\theta + \mu_s F_N\cos\theta = \frac{mv_{\text{max}}^2}{R}$$

and
$$F_N\cos\theta = mg + \mu_s F_N\sin\theta$$

Solution and Discussion   The last equation determines F_N:

$$F_N = \frac{mg}{\cos\theta - \mu_s\sin\theta}$$

We then substitute this value of F_N into the first equation and solve for v_{max}:

$$\frac{mg(\sin\theta + \mu_s\cos\theta)}{\cos\theta - \mu_s\sin\theta} = \frac{mv_{\text{max}}^2}{R}$$

Notice that the mass now cancels out.

$$v_{max}^2 = \frac{gR(\sin\theta + \mu_s \cos\theta)}{\cos\theta - \mu_s \sin\theta}$$

Putting in the numbers from the last example,

$$v_{max}^2 = \frac{(9.8)(60)(0.728 + 0.549)}{0.686 - 0.582}$$

$$= 7240 \ (m/s)^2$$

Then

$$v_{max} = 85 \ m/s = 310 \ km/h = 190 \ mi/h$$

7.8 A COMMON MISCONCEPTION

People sometimes jump to completely erroneous conclusions when interpreting their experiences. For example, a man seated in the center of a car seat sometimes thinks that he has been pushed to the side of the car as it rounds a corner. He might even assert that the force pushing him was so great that it threw him against the side hard enough to injure him. This is nonsense, of course. There was no mysterious ghost pushing him toward the side of the car. Certainly no material object was pushing him in that direction. He must therefore be mistaken.

The same man would not claim that a mysterious force suddenly threw him violently against the dashboard as the car stopped suddenly. He knows that his forward momentum can be lost only if some force retards his motion. Hence, when the car stopped suddenly, he continued going forward until the dashboard began to exert a force on him to stop him from moving forward. This is merely an example of Newton's idea that things continue in uniform motion until a force acts on them to stop them.

Similarly with the car turning the corner: the friction between pavement and tires pushed horizontally on the car and altered its straight-line motion. It is too bad about the man sitting in the middle of the nearly frictionless seat. The friction force between the seat of his pants and the car seat was too small to alter his straight-line motion. Hence he slid along in a straight line until he hit the side of the car, which then exerted a force on him so that he could travel in the same curved path as the car.

7.9 NEWTON'S LAW OF GRAVITATION

One of the most interesting examples of nearly circular motion, that of the planets around the sun, was the subject of intensive study by scientists about four centuries ago. From 1576 to 1597 the Danish astronomer Tycho Brahe compiled the most thorough and precise observations of planetary motion ever made up to that time. From those data, the German mathematician Johannes Kepler formulated his laws of planetary motion in the years 1609–1618. Those laws indicated that planetary

Once the forces supporting a building are removed, the force of gravity on the building becomes obvious.

orbits were almost circular and that the square of the time taken by a planet to orbit the sun (T) was proportional to the cube of the planet's distance from the sun (R):

$$T^2 \propto R^3$$

This last relationship is known as Kepler's third law.

Later in the seventeenth century, when Newton began his study of forces, this information was available to him, but a unifying physical law explaining planetary behavior was not yet known. Once Newton's laws of motion, including the concept of centripetal force and acceleration, had been developed, the way was clear to discover the nature of the force of gravity.

Newton reasoned that an attractive force between the sun and a planet provided the centripetal acceleration required by the planet's orbit. Thus, since $F_g = ma_c$, we can use Eq. 7.9 to write

$$F_g = \frac{m_p v^2}{R}$$

where m_p is the planet's mass. Newton also reasoned that the orbital time, or period T, was

$$T = \frac{2\pi R}{v} \qquad \text{giving} \qquad v \propto \frac{R}{T}$$

Squaring this relation and using Kepler's third law then gives us

$$v^2 \propto \frac{R^2}{R^3} \propto \frac{1}{R}$$

By combining all these relationships, Newton concluded that the force the sun exerts on the planet must be of the form

$$F_g \propto \frac{m_p}{R^2}$$

Using his own third law, Newton realized that the planet exerted an equal force on the sun (Fig. 7.13). This *symmetry* meant that the force must depend on both masses in the same way, that is, that the force must be of the form

$$F_g \propto \frac{m_s m_p}{R^2}$$

where m_s is the mass of the sun.

Newton also proposed that the same gravitational force that causes the moon to accelerate toward the earth (centripetal acceleration) causes objects (like the fabled apple in his orchard) to fall toward the earth with the acceleration g. Sensing that gravity was a fundamental and universal force, he generalized the above examples into his universal law of gravitation:

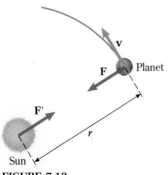

FIGURE 7.13

The sun and the planet attract each other with equal-magnitude forces.

Two uniform spheres with masses m_1 and m_2 that have a distance r between centers attract each other with a radial force of magnitude

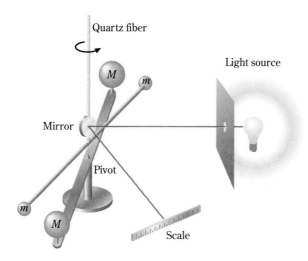

FIGURE 7.14

A schematic diagram of the Cavendish balance. Observe how the light beam is used to detect the twist of the fiber.

$$F = G\frac{m_1 m_2}{r^2}$$ (7.11)

The value of the gravitational constant G in Eq. 7.11 is not predicted by theory and can be determined only by experiment. Its value was first found by Henry Cavendish in 1798 with a device called the Cavendish balance (Fig. 7.14). Two identical masses m are suspended from an extremely thin and delicate quartz fiber. The two large masses M can be moved close to the small masses m, and the attraction between M and m causes the fiber to twist. After the system is calibrated so that the amount of force needed to produce a given twist is known, the attraction force between m and M can be computed directly from the observed twist of the fiber. Then, since m, M, r, and F are all known, these values can be used in Eq. 7.11 to solve for G. The value currently accepted for G is

$$G = 6.672 \times 10^{-11} \, \text{N} \cdot \text{m}^2/\text{kg}^2$$

Illustration 7.4

Two uniform spheres, both of 70.0-kg mass, hang as pendulums so that their centers of mass are 2.00 m apart. Find the gravitational force of attraction between them and compare it with their weight.

Reasoning   The gravitational force is given by Eq. 7.11:

$$F_g = \frac{Gm_1 m_2}{r^2}$$

$$= \frac{(6.67 \times 10^{-11} \, \text{N} \cdot \text{m}^2/\text{kg}^2)(70.0 \, \text{kg})(70.0 \, \text{kg})}{(2.00 \, \text{m})^2}$$

$$= 8.17 \times 10^{-8} \, \text{N}$$

The weight of each sphere is $mg = (70\,\text{kg})(9.8\,\text{m/s}^2) = 686\,\text{N}$. Thus the ratio of the gravitational force each sphere exerts on the other to their weight is

$$\frac{F_g}{W} = \frac{8.17 \times 10^{-8}}{686} = 1.19 \times 10^{-10}$$

On our scale of everyday phenomena, gravitational forces are significant only when at least one of the interacting masses is "astronomical." ∎

Every object on the earth is attracted by the mass of the earth. We have repeatedly calculated the force of this attraction as mg, which we have called the weight of an object. This calculation is based on our observation of the free-fall acceleration produced by gravity at the earth's surface. Now let us see how to interpret g using the law of gravity.

Figure 7.15 depicts a small mass m at or near the surface of the earth. The center of mass of the earth can be taken to be at its geometric center if we assume the earth to be a uniform sphere. The distance between m and m_E (the mass of the earth) required in Eq. 7.11 is then the radius of the earth, shown in Fig. 7.15 as R_E. The force the earth exerts on m according to the law of gravity is then

$$F_g = \frac{G m m_E}{R_E^2}$$

In comparing this expression with the object's weight mg, we see what physical quantities fundamentally determine the value of g:

$$F_g = \frac{G m m_E}{R_E^2} = \text{weight} = mg$$

Thus

$$g = \frac{G m_E}{R_E^2} \tag{7.12a}$$

Notice that the mass m cancels out, so that g is the same for all masses at the earth's surface.

As pointed out in Sec. 3.6, the weight of an object of mass m depends upon its location on the earth's surface. We notice from Eq. 7.12a that g, and therefore the weight, depend on the distance from the center of the earth. Since the earth bulges somewhat at the equator, there are slight variations in g and in weights from place to place on the earth. (In addition, the rotation of the earth causes an object's apparent weight to be less at the equator than at the poles.)

We can easily extend Eq. 7.12a to find the gravitational acceleration at the surface of any planet whose mass and radius are m_p and R_p:

$$g_p = \frac{G m_p}{R_p^2} \tag{7.12b}$$

Exercise   Using the value of G given above, $m_E = 6.0 \times 10^{24}\,\text{kg}$ and $R_E = 6400\,\text{km}$, show that Eq. 7.12a gives $9.8\,\text{m/s}^2$ for g.

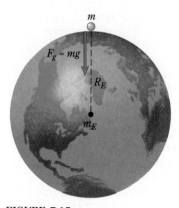

FIGURE 7.15

The force of gravity on a mass m at the surface of the earth.

Physicists at Work ROBERT H. MARCH *University of Wisconsin, Madison*

I first got interested in physics in my teens, babysitting for the kids of a neighbor who was a physicist. He seemed to enjoy his work a lot more than most adults I knew, and I found his library fascinating. I'm motivated mainly by curiosity and prefer to work on the frontiers of knowledge. For nearly 25 years I studied subatomic particles, but by 1980 it looked like we were beginning to understand that area a bit too well for my taste, and so I switched to astrophysics.

My current interest is the search for the origins of cosmic rays, which are particles and atomic nuclei that continuously bombard the Earth from outer space. They generate nearly half the background radiation in our environment. Though cosmic rays were discovered almost a century ago, we still don't know where they're coming from. That's because our Milky Way galaxy is full of weak magnetic fields that deflect electrically charged particles from a straight-line path so that they can't be traced back to their origin.

Cosmic rays have so much energy that they can't possibly come from normal stars like our Sun. We believe they originate in a few places in the universe where there are stupendous forces available to accelerate them, such as the gravity of a black hole or the electromagnetic forces near a rapidly-spinning pulsar. (A pulsar is a "neutron star," essentially a gigantic atomic nucleus with 50 percent more mass than the Sun squeezed into a sphere a few miles in diameter. Some pulsars have very strong magnetic fields.)

Though charged particles can't be traced back to their origin, neutral ones can. At present, I'm helping to build a detector for *neutrinos,* which are electrically neutral relatives of the electron. These particles interact so feebly with matter that they can pass right through the Earth without leaving a trace. To have any hope of detecting them you must monitor an immense quantity of matter, and even then you can detect only a small fraction of the ones that pass through. The detector can't be on the Earth's surface—it would be swamped by radiation from the charged cosmic rays. So we're building a device called DUMAND, for "Deep Underwater Muon and Neutrino Detector," three miles deep on the ocean floor off Hawaii. Muons are charged relatives of the neutrinos, and are similar to electrons but about 200 times heavier.

DUMAND consists of 216 very sensitive light detectors watching about two million tons of seawater, a volume much larger than Sears Tower. When neutrinos interact with nuclei, some of them convert into muons that radiate a faint blue glow as they pass through the water. The light detectors pick up this signal and feed data to computers on shore that reconstruct the path of the muon, which is close to that of its parent neutrino.

One thing I like about this project is that it is both international and interdisciplinary. The DUMAND team includes scientists from Japan, Germany, and Switzerland as well as Americans, and oceanographers as well as physicists. We have even made an interesting discovery in marine biology: that luminescent microorganisms in the deep ocean only emit light when stimulated by the motion of a nearby object.

DUMAND will open a new window on the cosmos. Nearly every time this has been done before—in radio, infrared, ultra-violet, x-ray, and gamma-ray astronomy—the most important discoveries have been complete surprises. I'm hoping that we will be as lucky, for it is the unexpected that really moves science forward.

By the way—that guy I used to babysit for worked in cosmic rays, so I guess I've come full circle!

7.10 ORBITAL MOTION

Perhaps the most majestic examples of circular motion are found in the heavens. The earth and other planets travel around the sun in nearly circular paths; the earth's moon follows a nearly circular path around the earth, and the moons of other planets do much the same around their planets. In addition, planets created by

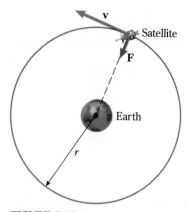

FIGURE 7.16

The centripetal force on the satellite is supplied by the gravitational attraction of the earth.

people—in other words, artificial satellites—trace nearly circular paths around the earth. Let us now examine this type of motion, *orbital motion*.

In Fig. 7.16, we see the moon or some other satellite following a circular path around the earth. The satellite's mass is m_s, and its speed is v. We represent the orbit radius by r and the mass of the earth by m_E. A centripetal force of magnitude $m_s v^2 / r$ is required to hold the satellite in orbit. This force is furnished by the gravitational attraction of the earth on the satellite:

$$\text{Gravitational force} = G \frac{m_s m_E}{r^2}$$

Applying Newton's second law to this circular motion, we get

$$G \frac{m_s m_E}{r^2} = \frac{m_s v^2}{r} \tag{7.13}$$

It is of interest to notice that m_s, the satellite mass, cancels from this equation. We therefore conclude that a satellite's orbit does not depend on its mass. At a given orbital speed v and radius r, the moon and a baseball will orbit in the same way. Thus any satellite orbiting at radius r must have the following speed, given by Eq. 7.13:

$$v = \sqrt{\frac{G m_E}{r}} \tag{7.14}$$

The period of a satellite in circular orbit is $T = 2\pi r / v$. If we substitute Eq. 7.14 for

The only force acting on an astronaut and spacecraft in orbit is the force of gravity exerted by the earth. All objects in orbit are thus in a state of free fall and hence their apparent weight is zero.

v into this expression for T and square the result, we find

$$T^2 = \left(\frac{2\pi r}{v}\right)^2 = \left(\frac{4\pi^2}{Gm_E}\right)r^3 = \text{constant} \times r^3 \qquad (7.15)$$

which is in agreement with Kepler's third law.

Example 7.8

Assuming the orbit of the earth about the sun to be circular (it is actually slightly elliptical) with radius 1.5×10^{11} m, find the mass of the sun.

R

Reasoning

Question What principle relates the earth-sun distance to the sun's mass?
Answer A combination of the law of gravity, which gives the amount of force on the earth, and Newton's second law applied to circular motion, which relates this force to the centripetal acceleration of the earth in its orbit.

Question What equation results?
Answer The force of gravity exerted by the sun (mass m_S) on the earth (mass m_E) can be written as $F_g = Gm_E m_S/r^2$, where r is the earth-sun distance. This force is directed toward the center of the circle in which we are assuming the earth travels. Thus we can identify this force as the centripetal force causing the earth's centripetal acceleration:

$$F_c = F_g = \frac{Gm_E m_S}{r^2} = \frac{m_E v^2}{r}$$

Question How do I find v?
Answer From the length of a year, which is the period of earth's orbit.

$$v = \frac{2\pi r}{T} \qquad \text{where} \qquad T = 365.25 \text{ days}$$

Once you find v, the only unknown is m_S.

Solution and Discussion Convert T to seconds:

$$T = (365.25 \text{ days})\left(\frac{24.0 \text{ h}}{1 \text{ day}}\right)\left(\frac{3600 \text{ s}}{1.00 \text{ h}}\right)$$

$$= 3.16 \times 10^7 \text{ s}$$

Then

$$v = \frac{2\pi(1.50 \times 10^{11} \text{ m})}{3.16 \times 10^7 \text{ s}} = 2.98 \times 10^4 \text{ m/s}$$

This is about 67,000 mi/h!

The mass of the sun is then found to be

$$m_S = v^2 r / G$$

$$= \frac{(2.98 \times 10^4 \text{ m/s})^2 (1.5 \times 10^{11} \text{ m})}{6.67 \times 10^{-11} \text{ N} \cdot \text{m}^2/\text{kg}^2}$$

$$= 2.00 \times 10^{30} \text{ kg}$$

Example 7.9

Radio and television signals are sent from continent to continent by "bouncing" them from *geosynchronous satellites*. These satellites circle the earth once each 24 h, and so if the satellite circles eastward above the equator, it always stays over the same spot on the earth because the earth is rotating at this same rate. Weather satellites are also designed to hover in this way. (*a*) What is the orbital radius for a geosynchronous satellite? (*b*) What is its speed?

Reasoning

Question   What is either given or known in this problem?
Answer   The period T is required to be 24 h = 86,400 s. You may also assume that G and the mass of the earth are known.

Question   Is there a direct relationship between T and the orbital radius?
Answer   Yes. Kepler's third law (Eq. 7.15), which was derived in the previous section.

Solution and Discussion   Equation 7.15 can be rearranged to give

$$r^3 = \frac{Gm_E}{4\pi^2} T^2$$

$$= \frac{(6.67 \times 10^{-11} \text{ N} \cdot \text{m}^2/\text{kg}^2)(5.98 \times 10^{24} \text{ kg})}{4\pi^2} \times (8.64 \times 10^4 \text{ s})^2$$

$$= 7.52 \times 10^{22} \text{ m}^3$$

Thus, the orbital radius (Part *a*) is

$$r = 4.22 \times 10^7 \text{ m} = 26{,}200 \text{ mi}$$

measured from the center of the earth. The orbital speed (Part *b*) is

$$v = \frac{2\pi r}{T} = \frac{2\pi(4.22 \times 10^7 \text{ m})}{8.64 \times 10^4 \text{ s}} = 3070 \text{ m/s}$$

Exercise   Determine the period and orbital speed of a "low orbit" satellite, that is, one where the orbital radius is essentially the same as the radius of the earth.   *Answer: $T = 5060\,s = 84.3\,min$; $v = 7910\,m/s = 17{,}700\,mi/h$.*

7.11 APPARENT WEIGHT AND WEIGHTLESSNESS

We often hear that objects appear to be weightless in a spaceship circling the earth or on its way to a distant point in space. Let us examine this effect in detail. First, we should restate our definition of weight: the pull of gravity on an object. On the earth, the weight of an object is the earth's gravitational pull on the object. Similarly, an object's weight on the moon is taken to be the moon's gravitational pull on the object.

Ordinarily we measure the weight of an object by placing it on a scale, usually at rest. The scale exerts a supporting force on the object equal to the force of gravity. What we measure as the object's weight is this supporting force. For example, if we heft the object in our hand to estimate its weight, we are trying to judge the amount of force we must exert to hold it.

As we shall see, the force we must exert to hold the object is equal to the force of gravity only when the object is *unaccelerated*. Therefore we reserve the term **apparent weight** for the scale reading and other similar methods of measuring the force supporting the object.

To illustrate this point, let us consider the apparent weight of an object of mass m in an elevator. In Fig. 7.17a, if the elevator is at rest, Newton's second law tells us that, since the acceleration is zero, the resultant force on the object is zero. Calling the gravitational force on the object (its weight) W and the tension in the string holding it T, we have

$$T - W = 0 \qquad \text{or} \qquad T = W$$

when $a = 0$. In this instance, the tension in the string is W and the apparent weight of the object (the scale reading) is equal to its actual weight W.

This situation prevails as long as $a = 0$. Under that condition, $T = W$ and the apparent weight is equal to the actual weight. Even if the elevator is moving up or down at constant speed, the acceleration is still zero and the apparent weight still equals the actual weight.

Let us now examine the situation in Fig. 7.17b, where the elevator is accelerating downward. If we apply Newton's second law as before, we find

$$W - T = ma$$

which gives

$$T = W - ma$$

Notice that the tension in the string, and therefore the scale reading, are less than W by the amount ma. To the person in the accelerating elevator, the object appears to weigh less than W. Its apparent weight is $W - ma$.

The most spectacular observation occurs when the elevator is falling freely so that $a = g$, the acceleration in free-fall. Then, since $W = mg$ and since $a = g$ for a freely falling body, the tension in the string

$$T = W - ma$$

becomes

$$T = mg - mg = 0$$

$a = 0$
$T = W$
(a)

a downward
$W - T = ma$
$T = W - ma$
(b)

FIGURE 7.17

The weight of an object in an elevator seems to a person in the elevator to vary, depending on the acceleration of the elevator.

The object appears weightless in a freely falling elevator! If we think about it a little, this is not strange at all. Since the elevator and everything in it are supposedly accelerating with the acceleration of free-fall, by the very definition of free-fall, there can be no force supporting the objects (elevator and everything in it) or in any way retarding their free-fall. Hence all support forces on the elevator and everything in it must be zero. The tension in the cord supporting the object must be zero. All objects within the elevator appear to be weightless.

Exercise Show that, in an elevator with an *upward* acceleration a, apparent weight would be larger than true weight: $T = W + ma$.

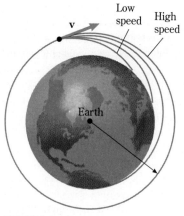

FIGURE 7.18

If an object is shot fast enough tangent to the earth, it will circle the earth. (Newton was probably the first to recognize this fact.)

We see from these considerations that *in accelerating systems, the apparent weight of an object is not necessarily equal to its true weight*. In particular, if the system is falling freely,* all support forces must be zero and all objects appear to be weightless. This means that whenever a spaceship is falling freely in space, that is, when its rocket engines are not being operated, everything within this freely falling system appears weightless. It does not matter where the system is, or whether it is falling under the force of attraction of the earth, the sun, or some distant star; as long as it is falling freely, everything in it appears weightless.

A satellite circling the earth is simply an example of a freely falling object. At first this statement may surprise you, but it is easily seen to be correct. Consider the behavior of a projectile shot parallel to the horizontal surface of the earth in the absence of air friction. (At satellite altitudes, the air is so thin as to be almost negligible.) The situation is shown in Fig. 7.18. The various paths are for a projectile shot tangent to the earth at successively larger speeds. We see that, during the projectile's free fall to the earth, the curvature of the path decreases with increasing horizontal speed. If the projectile is shot fast enough tangent to the earth, the curvature of its path will match the curvature of the earth, as shown. In this case, the projectile (a satellite perhaps) will simply circle the earth. Since it circles the earth, the satellite is accelerating toward the earth's center. Its radial acceleration is simply g, the free-fall acceleration. In effect, the satellite is falling toward the center of the earth at all times, but the curvature of the earth prevents it from hitting the surface. Since the satellite is in free fall, all objects within it also are, and hence appear weightless.

A MODERN PERSPECTIVE
THE INTERACTION BETWEEN GRAVITY AND LIGHT

Our study of mechanics so far has centered on understanding how forces acting on masses determine the motion or the equilibrium of those masses. The universal law of gravity discussed in this chapter describes a fundamental attractive force between any two masses. We have described the effect of gravity in determining circular orbits of planets and satellites, and in causing the acceleration of objects falling near the earth. One everyday phenomenon that we have not mentioned yet concerns the motion of light. There is no matter, and hence no mass, associated with light, although it contains energy and momentum, as we shall see in later chapters. Can the force of gravity affect the motion of something that has no material substance? Nothing in Newton's theory predicts any such effect.

*Recall that a freely falling object is one that is subject to only one unbalanced external force: the pull of gravity.

One common observation about light is that it travels in straight lines. In fact, we use this property to *define* straight lines for surveying and measuring distances. We often refer to light "rays" to describe the direction in which light is traveling. We recognize the fact that a light ray can be "bent," or refracted, when it passes from one transparent material to another, for example, when it enters glass or water from the air. But as light travels through empty space, or even through air of uniform temperature and pressure, it appears never to deviate from a straight-line path. For example, a beam of light directed parallel to the ground is not observed to follow the curved trajectory of a projectile. The light seems to be unaffected by the earth's gravity.

A second property of light is that it is always observed to travel through space at the same speed, 3.00×10^8 m/s. Methods of measuring this very high speed are described in later chapters. Thus our experience seems to tell us that light never undergoes an acceleration, but instead remains constant in speed and direction. Both of the above observations suggest that gravity does not exert a force on light.

However, in the years just before and during the first world war, Albert Einstein developed a complex theory of gravity which was more general than Newton's. Einstein's theory is known as the *general theory of relativity,* and in it, gravity is viewed as a consequence of the geometrical properties of space. To see what this puzzling assertion means, let us consider what we usually think of when we talk about straight lines.

As we mentioned in Chap. 2, a straight line can be defined as the shortest distance between two points. We often refer to such lines as **geodesics.** If we want to draw a straight line, we usually do so on a flat surface such as a sheet of paper. Suppose, however, you were presented a blank *sphere* with two points marked on it and asked to draw a straight line between them on the surface of the sphere. Your reaction might be to say that is impossible, because *every* line on the sphere's surface is curved. However, if you followed the shortest-distance definition, you would draw a line that is part of what is called a **great circle,** that is, a circle whose center coincides with the center of the sphere. The result, shown in Fig. 7.19, may still look to you very much like a curved line, but in the two-dimensional space defined by the curved surface of the sphere, the line fits the definition of "straight." The difference between the two-dimensional surfaces of the sphere and the flat sheet of paper lies in a property of space called **curvature.** Although we can view curvature in two dimensions, curvature of three dimensions is impossible to represent in a drawing. We can only try to use the two-dimensional description as a conceptual comparison.

Einstein's theory holds that empty space, that is, space without matter, is "flat" in three dimensions. Furthermore, Einstein proposed that the presence of mass introduces a curvature to space. The greater the mass, the greater the curvature of the space in the vicinity of the mass. The theory shows that it takes an astronomically large mass, such as a star, to produce a significant amount of curvature. This curvature causes the path of a moving object to depart from a straight line in flat space as it passes close to a large mass. Newton, *who assumed space had no curvature,* would view this "curved" path as an acceleration caused by the gravitational force exerted on the object. In contrast, Einstein's view of gravity is that the path's curve is related to the amount of curvature of space experienced by the object.

Now let us see how Einstein's ideas involve the trajectories followed by light. We have already discussed that light travels along straight lines (geodesics). In a curved space, a geodesic is a different path than it would be if the space were flat. Recall the comparison of the straight lines on a sphere compared to those on a flat

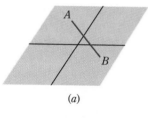

(*a*)

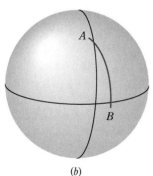

(*b*)

FIGURE 7.19

Geodesics on (*a*) a flat plane and (*b*) the surface of a sphere. Lines *AB* in both cases are straight in their respective two-dimensional spaces.

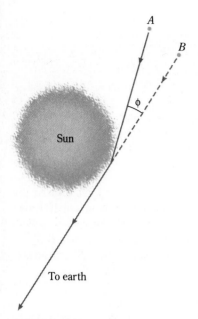

FIGURE 7.20

Bending of starlight by the sun. Light from the star A is deflected as it passes close to the sun on its way to earth. We see the star in the apparent direction B, shifted by the angle ϕ. Einstein predicted that $\phi = 1.745$ seconds of angle.

sheet. Thus Einstein proposed that if we could observe light moving along a geodesic near a large mass, we should see some evidence that the light was deflected from a flat space geodesic by the curvature produced by the large mass. One such possible observation would be to observe light from a distant star passing very near to the sun on its way to our telescope. If Einstein is correct, the curvature of space near the sun's mass should alter the path of the light and hence shift the apparent position of the star from the position it has when not in line with the sun. This effect is shown in Fig. 7.20. In the language of Newton's classical physics, we would say that the sun exerts a *force* on the light, causing its path to curve. However, nothing in Newton's law of gravity would predict such an interaction between mass and light.

In May of 1919 a total eclipse of the sun was to occur, when the sun was in line with a group of bright stars known as the Hyades. During an eclipse one can observe stars which appear very close to the sun's edge. Einstein had calculated that light grazing the sun should be shifted in direction by an angle of 1.745 seconds. The star's apparent position should thus be shifted by this angle. (One second of angle is 1/3600 of a degree. Modern telescopes can easily measure much smaller angles than this.) The British Royal Astronomical Society sent two groups, one to West Africa and the other to Northern Brazil, to test Einstein's theory. Both groups observed the effect, and subsequent measurements made at eleven different eclipses give an average value of the shift which is within 0.2 percent of Einstein's prediction.

In 1916 the German physicist Karl Schwarzschild derived an even more astonishing result of the curvature of space. He predicted that an extremely compact, massive object could produce such an extreme curvature of space close to the star that light passing within a certain critical distance from the mass would be captured, falling into the mass. This critical distance, known as the **event horizon,** is given by

$$R = \frac{2GM}{c^2}$$

where c stands for the speed of light, 3×10^8 m/s. R turns out to be about 3 km for a mass M equal to our sun's mass. In other words, if our sun was able to collapse into a sphere of this size or smaller, light passing closer to the collapsed sun than this distance could not escape from the enormous gravitational pull. Such objects, from which not even light could escape, would not be able to emit energy of any kind. Hence they are called **black holes.** To produce the necessary gravitational collapse to form a black hole, a star would have to have a mass greater than approximately 3 times the sun's mass. Many stars are observed to have greater mass than this, so astronomers believe that they, as similar stars must have done in the past, will eventually collapse into a black hole. Although such objects cannot be observed directly, the very great acceleration they give to material just outside their event horizons should produce intense x-rays, detectable by earth satellite telescopes. A growing number of such x-ray observations seem to provide evidence that black holes do exist.

We conclude that light is affected by the presence of mass, but in a way that is not included in Newton's law of gravity. Once again, it is necessary to use developments in the twentieth century which have modified and extended classical physics in order to explain new observations.

LEARNING GOALS

Now that you have finished this chapter, you should be able to

1 Define (*a*) radian, (*b*) angular velocity, (*c*) angular acceleration, (*d*) tangential distance, (*e*) tangential velocity, (*f*) tangential acceleration, (*g*) centripetal (radial) acceleration, (*h*) centripetal force, (*i*) apparent weight.
2 Convert an angle in degrees, radians, or revolutions to each of the other units.
3 State the five angular-motion equations and use them to solve problems.
4 Convert between tangential, angular, and linear quantities.
5 Relate angular quantities to linear quantities involved in rolling wheels and a cord wound on a spool.
6 Explain why an object moving around a circle at constant speed is accelerating. Give the direction and magnitude of the acceleration.

7 Analyze the free-body diagram of an object moving in a circle and apply Newton's second law relating centripetal force to centripetal acceleration.
8 Compute the gravitational force one object exerts on another.
9 Calculate the supporting force needed on an object of known mass if it is (*a*) moving with constant speed, (*b*) accelerating upward, (*c*) accelerating downward. Explain what is meant by apparent weight in such circumstances, and show why it differs from the object's weight.
10 Explain why an object orbiting the earth (or in some similar situation) is said to be falling freely. Use your explanation to point out why objects appear weightless under certain circumstances.

SUMMARY

DERIVED UNITS AND PHYSICAL CONSTANTS

Universal Gravitational Constant
$G = 6.67 \times 10^{-11} \text{ N} \cdot \text{m}^2/\text{kg}^2$

Radian Measure
$1 \text{ rad} = \dfrac{1}{2\pi} \text{ rev} \approx 57.3°$

DEFINITIONS AND BASIC PRINCIPLES

Angular Measurement
ANGULAR DISPLACEMENT (θ)

$$\theta \text{ (rad)} = \frac{\text{arc length}}{\text{radius}} = \frac{s}{r} \quad (7.1)$$

ANGULAR VELOCITY (ω)

$$\omega = \frac{\Delta\theta}{\Delta t} \quad (7.2)$$

ANGULAR ACCELERATION (α)

$$\alpha = \frac{\Delta\omega}{\Delta t} \quad (7.4)$$

Angular Equations of Motion (Constant α)
$$\theta = \overline{\omega}t \quad (7.5a)$$

$$\omega_f = \omega_i + \alpha t \quad (7.5b)$$
$$\overline{\omega} = \tfrac{1}{2}(\omega_f + \omega_i) \quad (7.5c)$$
$$2\alpha\theta = \omega_f^2 - \omega_i^2 \quad (7.5d)$$
$$\theta = \omega_i t + \tfrac{1}{2}\alpha t^2 \quad (7.5e)$$

INSIGHTS
1. Angular measures are dimensionless, but it is useful to keep the type of measurement (radians, revs, degrees) visible in your calculations as a reminder.
2. There are two opposite "directions" of rotation, and we must specify them in our equations. We use + to specify counterclockwise and − to specify clockwise rotations.

Centripetal Acceleration (a_c)
An object moving in a circle of radius r with a speed v has an acceleration directed toward the center of the circle:

$$a_c = \frac{v^2}{r} = \omega^2 r \quad (7.9)$$

Centripetal Force (F_c)
In order for circular motion to be possible, there must be a net force on an object directed toward the center of the circle.

$$F_c = ma_c = \frac{mv^2}{r} = m\omega^2 r \quad (7.10)$$

INSIGHT
F_c cannot do work on the object or change its speed, since it is always perpendicular to the velocity.

Universal Law of Gravity

The force of gravity between two objects of mass m_1 and m_2 separated by a distance r is

$$F_g = \frac{Gm_1m_2}{r^2} \qquad (7.11)$$

1. For spherically symmetric objects, the distance r is between the two centers of mass.
2. The gravitational force is always *attractive,* tending to pull each object toward the other.

QUESTIONS AND GUESSTIMATES

1 As a wheel spins on its axle at a constant speed ω, describe the following for a point P at a radius r from the center and state how each varies with r: (a) tangential velocity, (b) angular velocity, (c) angular acceleration, (d) tangential acceleration, (e) centripetal acceleration.

2 After the standard tires on a car are replaced by tires with a 15 percent larger diameter, the car's speedometer no longer reads correctly. Explain how to obtain the correct reading from the actual reading.

3 When mud flies off the tire of a moving bicycle, in what direction does it fly? Explain.

4 Figure P7.1 shows a simplified version of a cyclone-type dust remover used for purifying industrial waste gases before they are vented to the atmosphere. The gas is whirled at high speed around a curved path, and the dust particles collect at the outer edge and are removed by a water spray or by some other means. Explain the principle behind this method.

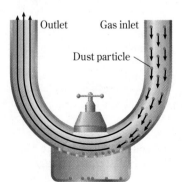

FIGURE P7.1

5 Discuss the principle of the spin cycle in an automatic washing machine.

6 An insect lands on a phonograph record that is on a turntable. Describe qualitatively the motion of the insect as the record begins to rotate. Assume that the insect is quite close to the axis and that there is some, but not much, friction between insect and record surface.

7 On the moon, the acceleration due to gravity is 1.67 m/s^2. How would this make a person's life there different from what we are accustomed to on earth?

8 To give a person a horizontal acceleration of $5g$, where $g = 9.8 \text{ m/s}^2$, a force of "$5\,g$'s" is required. What do we mean by this terminology? What do we mean when we say the pilot of an airplane experiences a force of several g's as the plane pulls out of a steep dive? Why might the pilot "black out" if the pullout is too fast?

9 From the fact that the moon circles the earth at a radius of 3.8×10^8 m, estimate the mass of the earth.

10 Can we obtain the masses of the other planets of the Solar System if we know the radii of their orbits and the mass of the earth?

11 About how fast can a car be going and still negotiate a turn from one street into a perpendicular street? Assume that each street is concrete with one lane in each direction.

12 During the 1970 flight of Apollo 13 to the moon, serious trouble developed when the ship was about halfway there, and it returned to earth without executing its moon mission. After the trouble developed, however, the ship continued toward the moon, passed behind it, and only then returned to earth. Why didn't the astronauts simply turn around when the trouble developed?

13 Suppose a huge mass, much larger than the mass of the Solar System or our galaxy, is created at this instant far away. The large gravitational force it would exert on the Solar System would cause us to accelerate toward this distant mass. After the first few seconds, what long-term effects would we notice on earth because of this acceleration? Assume that the earth's acceleration due to this cause is of the order of 10 m/s^2.

PROBLEMS

Sections 7.1–7.4

1 Express each of the following angles in degrees, revolution, and radians: (a) 32°, (b) 2.65 rad, (c) 0.67 rev.

2 Express each of the following angles in degrees, revolutions, and radians: (a) 0.29 rev, (b) 195°, (c) 1.35 rad.

3 A roulette wheel of radius 85 cm has two numbers on its rim. The distance along the rim between the numbers is 2.8 cm. Find the angle subtended at the center of the wheel by the numbers. Give your answer in radians, degrees, and revolutions.

4 A sphere of radius 33 cm has two dots on its surface. The distance along the surface between the dots is 4.1 cm. Find the angle subtended at the center of the sphere by the two dots. Express your answer in radians, degrees, and revolutions.

5 Calculate the angular velocity of the second hand of a watch in radians per second and revolutions per minute.

6 Calculate the angular velocity of the minute hand of a watch in degrees per second and radians per hour.

7 A phonograph record rotates at 33.3 rev/min. (a) What is its angular speed in radians per second? (b) Through how many degrees does it rotate in 0.225 s?

8 (a) What is the angular velocity of the hour hand of a clock in radians per second? (b) Through how many degrees does the hand turn in 18 s?

9 A phonograph turntable accelerates from rest to an angular speed of 33.3 rev/min in 0.77 s. What is its average angular acceleration in revolutions per minute squared and radians per second squared?

10 A phonograph turntable turning at 33.3 rev/min coasts to rest in 10.5 s. What is its average angular acceleration in revolutions per second squared and in radians per second squared?

11 A merry-go-round takes 22 s to accelerate from rest to its operating speed of 3.75 rev/min. Find (a) its acceleration in revolutions per second squared and (b) the number of revolutions turned in this time.

12 How large an angular acceleration (in radians per second squared) must be given to a wheel if it is to be accelerated from rest to a rotational speed of 540 rad/s after 7.0 rev?

13 A certain roulette wheel coasts to rest in 18.5 s. If the wheel turns through 9.5 rev in that time, how fast was it originally turning?

14 A wheel that is turning at 32 rev/min speeds up until its speed is 48 rev/min. The change takes 17.5 s. Find (a) its angular acceleration in radians per second squared and (b) the number of degrees through which it turned in this time.

Section 7.5

15 A ceiling fan is turning at 0.67 rev/s. The tip of its blade is 95 cm from the center. How fast, in centimeters per second, is the tip of the blade moving?

16 A merry-go-round is rotating with a speed of 3.65 rev/min. How fast (in meters per second) is a child at a radius of 2.75 m moving?

17 A bowling ball of diameter 23.5 cm rolls 16.5 m along the floor without slipping. Through how many revolutions does it roll?

18 If its diameter is 72 cm, through how many revolutions must the wheel of a car turn as the car travels 550 m?

19 What is the angular acceleration of a 65-cm-diameter wheel on a vehicle as the vehicle undergoes an acceleration of 0.375 m/s²?

20 An object is being lifted by a cord wound on the rim of a wheel whose diameter is 43 cm. If the wheel is accelerating at 0.36 rad/s², what is the acceleration of the object in meters per second squared?

21 The radius of the earth is 6.37×10^6 m. (a) How fast, in meters per second, is a tree at the equator moving because of the earth's rotation? (b) A polar bear at the North Pole?

22 The earth orbits the sun in 365.25 days. What is the speed, in meters per second, of the earth in the orbit? The earth-sun distance is 1.50×10^{11} m.

23 A wheel of diameter 35.5 cm is turning at a rate of 0.71 rev/s and winds a string on its rim. How long a piece of string is wound in 20 s?

24 A wheel turning at 2450 rev/min has a diameter of 7.8 cm. If a string is to be wound on the wheel, how much string is wound in 5.0 s?

25 A vehicle is traveling along the road at 25.5 m/s. If the diameter of its wheels is 106 cm, how fast are the wheels rotating in revolutions per second, radians per second, and degrees per second?

26 A 55-cm-diameter wheel comes loose from a car that is going 27 m/s, and the wheel rolls alongside the car. Find the angular speed of the wheel in revolutions per second, radians per second, and degrees per second.

27 A bicycle with 62.5-cm-diameter wheels is coasting at 6.6 m/s. It decelerates uniformly and stops in 38 s. (a) How far does it go in this time? (b) Through how many revolutions does each wheel turn as the bicycle comes to a stop?

28 A car with 72.5-cm-diameter wheels starts from rest and accelerates uniformly to 21.5 m/s in 36 s. Through how many revolutions does each wheel turn?

▪29 A motor turning at 1660 rev/min coasts uniformly to rest in 16 s. (a) Find its angular deceleration and the number of revolutions it turns before stopping. (b) If the motor has a wheel of radius 6.25 cm attached to its shaft, what length of belt does the wheel wind in the 16 s?

▪30 Two gear wheels that are meshed together have radii of 0.65 and 0.15 m. Through how many revolutions does the smaller wheel turn when the larger turns through 4.5 rev?

▪31 A car accelerates uniformly from rest to 17.5 m/s in 23.6 s. Find the angular acceleration of one of its wheels and the number of revolutions turned by a wheel in the process. The radius of the car wheel is 0.40 m.

▪32 A belt runs on a wheel of radius 44 cm. During the time the wheel takes to coast uniformly to rest from an initial speed of 1.8 rev/s, 29.5 m of belt length passes over the wheel. Find the deceleration of the wheel and the number of revolutions it turns while stopping.

Sections 7.6–7.7

33 A 1420-kg car moving at 21.2 m/s is rounding a curve of radius 37.5 m. How large a horizontal force is needed to hold the car in this path?

34 A 380-g mass at the end of a string is whirled in a horizontal circle of radius 75 cm. If its speed in the circle is 7.7 m/s, what must the tension in the string be? Neglect the force of gravity.

35 A carton of eggs sits on the horizontal seat of a car as the car rounds a 26-m-radius bend at 16.5 m/s. What minimum coefficient of friction must exist between carton and seat if the eggs are not to slip?

36 A 22.7-mg bug sits on the smooth edge of a 30-cm-radius phonograph record as the record is slowly brought up to its normal rotational speed of 33.3 rev/min. How large must the coefficient of friction between bug and record be if the bug is not to slip off? (It is a very compact bug, and so air friction can be ignored.)

37 In a certain research device, a man is subjected to an acceleration of 5.3g. This is done by rotating him in a horizontal circle at very high speed. The seat in which he is strapped is 11.3 m from the rotational axis. How fast must he be rotating, in revolutions per second?

▪38 An old trick is to swing a pail of water in a vertical circle. If the rotation rate is large enough, water will not fall out of the pail when the pail is upside down at the top of its path. What is the minimum speed your hand must have at the top of the circle if the trick is to succeed? Assume your arm is 0.72 m long.

▪39 The designer of a roller coaster wishes the riders to experience weightlessness as they round the top of one hill. How fast must the car be going if the radius of curvature at the hilltop is 30 m?

40 In an ultracentrifuge, a solution is rotated with an angular speed of 5000 rev/s at a radius of 15 cm. How large is the centripetal acceleration of each particle in the solution? Compare the centripetal force needed to hold a particle of mass m in the circular path with the weight of the particle mg.

41 The red blood cells and other particles suspended in blood are too light in weight to settle out easily when the blood is left standing. How fast (in revolutions per second) must a sample of blood be centrifuged at a radius of 8.5 cm if the centripetal force needed to hold one of the particles in a circular path is 12,000 times the weight of the particle mg? Why do the particles separate from the solution in a centrifuge?

42 A certain car of mass m has a maximum friction force of 0.85 mg between it and the pavement as it rounds a curve on a flat road. How fast can the car be moving if it is to successfully negotiate a curve of radius 31.5 m?

Section 7.9

43 A neutron is an uncharged particle with a mass of 1.67×10^{-27} kg and a radius of the order of 10^{-15} m. Find the gravitational attraction between two neutrons whose centers are 1.00×10^{-12} m apart. Compare this with the weight of the neutron on earth.

44 Find the force of gravity that the moon exerts on a 70-kg student sitting next to you on earth. The moon's mass is 7.3×10^{22} kg, and its distance from the earth is $3.8 \times$ 10^5 km. Compare this with the weight of the student on earth.

45 Compare the gravitational pull on a spaceship at the surface of the earth with the gravitational pull when the ship is orbiting 5000 km above the surface of the earth. (The radius of the earth is 6380 km.)

46 The planet Jupiter has a mass 314 times that of the earth. Its radius is 11.3 times larger than the earth's radius. Find the acceleration due to gravity on Jupiter.

47 The acceleration due to gravity on the moon is only one-sixth that on earth. Assuming that the earth and the moon have the same average composition, what would you predict the moon's radius to be in terms of the earth's radius R_E? (In fact, the moon's radius is 0.27R_E.)

▪48 A satellite orbits the earth in about 80 min if its orbital radius is 6500 km. Use these data to find the mass of the earth.

▪49 One satellite planet of Jupiter, called Callisto, circles Jupiter once each 16.8 days. Its orbital radius is 1.88×10^9 m. Use these data to find the mass of Jupiter.

General Problems

▪▪50 A 450-g ball at the end of a cord is whirled in an almost horizontal circle of radius 1.25 m. Its tangential speed in the circle is 8.5 m/s. Do *not* neglect the weight of the ball; the string cannot be perfectly horizontal. (*a*) What must the tension in the cord be? (*b*) What angle does the cord make with the horizontal?

▪▪51 As shown in Fig. P7.2, a man on a rotating platform holds a pendulum in his hand. The pendulum ball is 6.8 m from the center of the platform. The rotational speed of the platform is 0.045 rev/s, and the pendulum hangs at an angle θ to the vertical. Find θ.

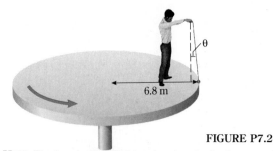

6.8 m

FIGURE P7.2

▪▪52 The bug in Fig. P7.3 has just lost its footing near the top of the bowling ball. It slides down the ball without significant friction. Show that it will lose contact with the surface at the angle θ shown, where cos θ = 2/3.

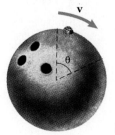

FIGURE P7.3

■■53 Figure P7.4 shows a possible design for a space colony. It consists of a cylinder of diameter 7 km and length 30 km floating in space. Its interior is provided with an earthlike environment, and to simulate gravity, the cylinder spins on its axis. What should the rotation rate of the cylinder be, in revolutions per hour, so that a person standing on the landmass will press down on the ground with a force equal to his or her weight on earth? (For details, see G. K. O'Neill, *Physics Today,* September 1974, p. 32, and February 1977, p. 30.)

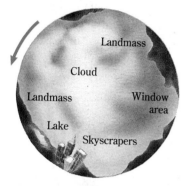

FIGURE P7.4

■■54 A particle is to slide along a horizontal path on the inside of the funnel shown in Fig. P7.5. The surface of the funnel is frictionless. What must the speed v of the particle be, in terms of r and θ, to execute this motion?

FIGURE P7.5

■■55 A pendulum consisting of a 140-g ball at the end of a string 225 cm long is released from rest at an angle of 65° to the vertical. Find the tension in the string at the instants where the angle is 25°.

■■56 The beads of masses M and m in Fig. P7.6 can slide freely on the wire circle. They are at rest in the positions shown when M is released and collides elastically with m. What is the largest value possible for m/M if bead m is to travel over the top without exerting a downward force on the wire there?

FIGURE P7.6

■■57 Suppose a rocket ship achieves a maximum upward acceleration of 40 m/s² during liftoff, and that this acceleration occurs within 10 mi of the earth's surface. What would the apparent weight of a 180-lb astronaut be?

■■58 Answer Prob. 57 in the case where the ship achieves the 40 m/s² acceleration at an altitude of 1500 m above the earth's surface. The direction of this acceleration is radially outward from earth.

■■59 Balls A and B, each of mass m, are tied together by a cord of length L. A second identical cord of length L is tied to mass A, and a woman swings the two balls in a horizontal circle, holding on to the other end of the second cord. The situation is pictured in Fig. P7.7. As she whirls the balls faster and faster, which of the two strings will break first, the one she is holding on to or the one joining A and B? At what angular velocity will this happen? Assume $m = 500$ g, $L = 0.6$ m, and the breaking strength of the cords is 235 N. Ignore the weight of the balls; that is, assume the circle is truly horizontal.

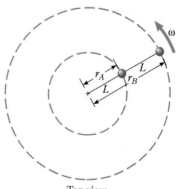

Top view   FIGURE P7.7

■■60 An 800-kg sports car with a 700-N driver hits a dip in the road. The dip has a vertical radius of curvature of 60 m. This causes the car's springs to "bottom out," or compress completely for an instant at the bottom of the dip. When standing still, it takes 5000 N of weight in addition to the car's weight to fully compress the springs. How fast was the car going when it encountered the dip? What was the apparent weight of the driver at that instant?

CHAPTER 8

ROTATIONAL WORK, ENERGY, AND MOMENTUM

Newton's second law relates the force acting on an object to the object's mass and linear acceleration: $\mathbf{F} = m\mathbf{a}$. When an object, such as a wheel, can rotate about an axis, *torques* can give it an *angular* acceleration. We see in this chapter that an analog to $\mathbf{F} = m\mathbf{a}$ exists for rotational motion. It relates the torque acting on an object to the product of the object's angular acceleration and a quantity that measures its rotational inertia. Moreover, we see that a rotating object has both kinetic energy and rotational momentum.

8.1 ROTATIONAL WORK AND KINETIC ENERGY

It is easy to see that a rotating object has kinetic energy. For example, the rotating wheel in Fig. 8.1 is composed of many tiny bits of mass, each moving as the wheel turns. A typical bit of mass, such as the one labeled m_1, has a velocity $\mathbf{v}_1$ and thus a kinetic energy $\frac{1}{2}m_1 v_1^2$. Let us begin our study of the properties of rotating objects by seeing how a wheel might acquire kinetic energy.

The wheel in Fig. 8.2 is originally at rest, but it can rotate freely on an axle through its center. When a force $\mathbf{F}$ pulls on the cord wound on the wheel's rim, the wheel begins to rotate. The work done by the force as it pulls the end of the string through a distance s is

FIGURE 8.1

The wheel is shown rotating in a clockwise sense (gold arrow). As the wheel rotates, each tiny bit of mass within it possesses some KE. For m_1, the KE is $\frac{1}{2}m_1v_1^2$.

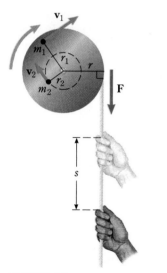

FIGURE 8.2

As the force **F** does work by pulling the string a distance s, the wheel gains an amount Fs of kinetic energy.

Work done by $\mathbf{F} = Fs$

As the length s of string is unwound, the wheel turns through an angle θ, which is related to s through $s = r\theta$ (Eq. 7.1). Substitution of this value for s yields the following expression for the work done:

Work done by $\mathbf{F} = Fr\theta$

We can better understand this expression by noticing that Fr is "force times lever arm" in Fig. 8.2, and this is simply the torque τ applied to the wheel.* Thus we find the relation between the work done on a wheel when it turns through an angle θ and the torque applied to it:

$$W = \tau\theta \tag{8.1}$$

It is interesting to notice that this is the result we would guess by analogy to linear motion. For linear motion, we had $W = F_x x$. In rotation, force is replaced by torque and linear distance by angular distance. Hence $F_x x$ becomes $\tau\theta$, as proved in Eq. 8.1.

According to the net work-energy theorem, the work done on the wheel by the net force must appear as kinetic energy. We call the kinetic energy of a rotating object its **rotational kinetic energy** KE_{rot}. You will recall that the kinetic energy of an object that is due to its linear motion is $\frac{1}{2}mv^2$, and we refer to this as translational kinetic energy KE_{trans}. Let us now compute the kinetic energy of a rotating object from a consideration of the kinetic energy of each of the little pieces of mass that compose the object.

Refer back to Fig. 8.1. As the wheel rotates, a typical tiny mass (such as m_1) that is part of the wheel possesses translational kinetic energy. For m_1, it is $\frac{1}{2}m_1v_1^2$. If there are N tiny masses ($m_1, m_2, m_3, \ldots, m_N$) that compose the wheel, their combined kinetic energy is

$$KE \text{ of wheel} = \tfrac{1}{2}m_1v_1^2 + \tfrac{1}{2}m_2v_2^2 + \tfrac{1}{2}m_3v_3^2 + \cdots + \tfrac{1}{2}m_Nv_N^2$$

But m_1, for example, travels around a circle of radius r_1. The tangential velocity m_1 on this circle is v_1. From Eq. 7.7, we know that the angular velocity of the wheel is related to this tangential velocity through $v_1 = \omega r_1$, and so we have

$$\tfrac{1}{2}m_1v_1^2 = \tfrac{1}{2}m_1\omega^2 r_1^2$$

Similar expressions exist for all the other tiny masses. We substitute these values in the kinetic energy equation and obtain

$$KE \text{ of wheel} = \tfrac{1}{2}m_1r_1^2\omega^2 + \tfrac{1}{2}m_2r_2^2\omega^2 + \cdots + \tfrac{1}{2}m_Nr_N^2\omega^2$$

Because all parts of the wheel have the same angular velocity ω, it can be factored out of every term, giving

$$KE \text{ of wheel} = \tfrac{1}{2}\omega^2(m_1r_1^2 + m_2r_2^2 + \cdots + m_Nr_N^2)$$

---

*You may wish to review the concept of torque, presented in Sec. 2.4.

The factor in parentheses is usually represented by the symbol I and is called the **moment of inertia** of the rotating object:

$$I = \text{moment of inertia} = m_1 r_1^2 + m_2 r_2^2 + \cdots + m_N r_N^2 \tag{8.2}$$

Notice that the SI units of I are $\text{kg} \cdot \text{m}^2$.

We shall discuss the moment of inertia soon; then you will see why it is indeed a measure of inertia of the wheel. Even now, however, you can see that it depends not only on the quantity of matter m in the body, but *also on how that matter is distributed.*

Our expression for the kinetic energy of the rotating wheel can now be rewritten in terms of I:

$$\text{KE}_{\text{rot}} = \text{rotational KE} = \tfrac{1}{2} I \omega^2 \tag{8.3}$$

This is the kinetic energy an object has because of its rotation. Notice that, once again, we could have guessed its general form. In analogy to $\tfrac{1}{2} m v^2$, we see that v is replaced by ω and I apparently is the rotational counterpart of the mass m.

As we mentioned earlier, the rotational energy is related to the work done on the wheel by the applied torque. To be specific, suppose the wheel is rotating with speed ω_o when a torque τ is suddenly applied to it. The torque acts while the wheel turns through an angle θ (so that the work the torque does is $\tau\theta$) and is then removed. At that time, the angular velocity of the wheel is ω_f. The work-energy theorem describes this process by telling us that

Work done on wheel = change in wheel's KE
$$\tau\theta = \tfrac{1}{2} I \omega_f^2 - \tfrac{1}{2} I \omega_o^2$$
$$\tau\theta = \tfrac{1}{2} I (\omega_f^2 - \omega_o^2)$$

where we have used Eq. 8.3 for the wheel's rotational kinetic energy.

This relation between work and rotational kinetic energy can be simplified by using the angular motion equation (Eq. 7.5d), $\omega_f^2 - \omega_o^2 = 2\alpha\theta$. After substituting this quantity, we can cancel θ from the equation and obtain

$$\tau = I\alpha \tag{8.4}$$

where α must be measured in radians per second squared. (Why?) We have thus arrived at a relation between the angular acceleration of the wheel and the torque that caused that acceleration. It is the rotational-motion analog to $F = ma$.

8.2 ROTATIONAL INERTIA

We know that rotating objects have inertia. After you turn off an electric fan, the blade coasts for some time as the friction forces of the air and the axle bearings slowly cause it to stop. The moment of inertia I of the fan blade measures its rotational inertia, as we can understand in the following way.

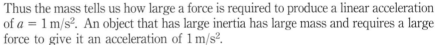

The driving rods of a steam locomotive are connected to the drive wheels at off-center points. The force exerted by the piston thus creates a torque about the axis of the wheels.

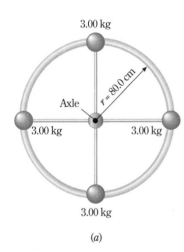

(a)

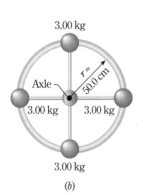

(b)

FIGURE 8.3

Which wheel is more difficult to set into rotation?

In linear motion, the inertia of an object is represented by the object's mass. From $F = ma$, we have

$$m = \frac{F}{a}$$

Thus the mass tells us how large a force is required to produce a linear acceleration of $a = 1 \, \text{m/s}^2$. An object that has large inertia has large mass and requires a large force to give it an acceleration of $1 \, \text{m/s}^2$.

The rotational analog of $F = ma$, which is $\tau = I\alpha$, gives similar information about I, the moment of inertia:

$$I = \frac{\tau}{\alpha}$$

which tells us how large a torque is needed to produce an angular acceleration of $\alpha = 1 \, \text{rad/s}^2$. Objects that have large values for I require large torques to change their rotation rate. Clearly, I measures the rotational inertia of any object.

Now let us examine the mathematical representation for moment of inertia. From Eq. 8.2,

$$I = m_1 r_1^2 + m_2 r_2^2 + \cdots + m_N r_N^2 = \sum_{i=1}^{N} m_i r_i^2$$

Let us apply this to the two wheels in Fig. 8.3. Each consists of four masses mounted on a frame of negligible mass. In part a, we have

$$\begin{aligned} I_a &= m_1 r_1^2 + m_2 r_2^2 + m_3 r_3^2 + m_4 r_4^2 \\ &= (3.00 \, \text{kg})(0.800 \, \text{m})^2 + (3.00 \, \text{kg})(0.800 \, \text{m})^2 + (3.00 \, \text{kg})(0.800 \, \text{m})^2 + (3 \, \text{kg})(0.800 \, \text{m})^2 \\ &= 7.68 \, \text{kg} \cdot \text{m}^2 \end{aligned}$$

For part b,

$$I_b = (3.00 \text{ kg})(0.500 \text{ m})^2 + (3.00 \text{ kg})(0.500 \text{ m})^2 + (3.00 \text{ kg})(0.500 \text{ m})^2$$
$$+ (3.00 \text{ kg})(0.500 \text{ m})^2$$
$$= 3.00 \text{ kg} \cdot \text{m}^2$$

As we see, the wheel in b has a much smaller moment of inertia than the one in a. Although the masses of the wheels are the same, their moments of inertia differ because the masses are farther from the axis of rotation in a than in b. Because I varies as r^2 (Eq. 8.2), the moment of inertia gets larger the farther the mass is from the axis. Thus, a greater torque is needed in a than in b.

As a more practical example, let us compute the moment of inertia of the hoop of mass M shown in Fig. 8.4. We assume it to rotate about an axis through its center and perpendicular to the plane of the hoop. In our mind's eye, we split the hoop into a large number of tiny masses as shown. Each mass is at a distance b from the axis. The moment of inertia becomes

$$I_{\text{hoop}} = m_1 r_1^2 + m_2 r_2^2 + \cdots + m_N r_N^2$$
$$= m_1 b^2 + m_2 b^2 + \cdots + m_N b^2 = b^2(m_1 + m_2 + \cdots + m_N)$$

The sum of the tiny masses composing the hoop is simply its total mass M, however. Therefore

$$I_{\text{hoop}} = b^2 M$$

In principle the moment of inertia for any object can be calculated in this same way. However, calculus is usually needed to carry out the summation. We list the results of such calculations for several simple objects in Table 8.1. In some cases,

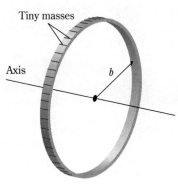

Tiny masses

Axis

b

FIGURE 8.4

What is I for the hoop about the axis shown?

TABLE 8.1   *Moments of inertia*

Object	Axis	I	Radius of gyration k
Point mass (moving in a circle of radius r)		mr^2	r
Hoop		mb^2	b
Solid disk (radius b)		$\frac{1}{2}mb^2$	$b/\sqrt{2}$
Solid sphere (radius b)		$\frac{2}{5}mb^2$	$b\sqrt{\frac{2}{5}}$
Solid cylinder (radius b)		$\frac{1}{2}mb^2$	$b/\sqrt{2}$
Solid thin cylinder (length L)		$\frac{1}{12}mL^2$	$L/\sqrt{12}$

different axes of rotation are possible. For example, a cylinder might rotate about either of the two axes shown in the table. Therefore *one must state the axis being used in order to know the appropriate moment of inertia.*

Another important feature concerning I can be seen from Table 8.1. In all cases, I is the product of the object's mass and the square of some dimension of the object. For example, I for a sphere is the mass of the sphere multiplied by $(\sqrt{\frac{2}{5}}\,b)^2$. Similarly, I for a disk is $m(b/\sqrt{2})^2$. For a hoop, $I = mb^2$. In general, we can write

$$I = mk^2 \tag{8.5}$$

where k is a length characteristic of the object, called the object's **radius of gyration.** The **radius of gyration** is the "effective" radius of any object which gives the same moment of inertia as a hoop of the same mass. For example, Table 8.1 shows that $k = b$ for a hoop, which is a reasonable value since each bit of mass composing the hoop is at a distance b from the axis. For a sphere, however, $k = \sqrt{\frac{2}{5}}b = 0.63b$ because only the farthest points on the sphere are a distance b from the axis. As another example, in Fig. 8.3*a* we have $k = 0.800$ m. For that object,

$$I = mk^2 = (12.0\,\text{kg})(0.800\,\text{m})^2 = 7.68\,\text{kg} \cdot \text{m}^2$$

as we found previously. Typical values for k are given in Table 8.1.

Let us summarize what we have found.

1 An object of mass m possesses rotational inertia. We represent this quantity by I, the moment of inertia. In equation form, $I = mk^2$, where k is the radius of gyration of the object, which depends on the object's shape and on the axis about which I is calculated.
2 A rotating object has rotational kinetic energy, $\text{KE}_{\text{rot}} = \frac{1}{2}I\omega^2$.
3 A torque τ applied to an object that is free to rotate gives the object an angular acceleration: $\tau = I\alpha$.
4 The work done by a torque τ when it acts through an angle θ is $\tau\theta$.

PARALLEL-AXIS THEOREM

The moments of inertia of the objects in Table 8.1 are calculated about the centers of mass of the objects. There is a very simple and useful theorem by which we can calculate the moments of inertia of these same objects about *any other axis* which is parallel to the center of mass axis. It is known as the **parallel-axis theorem,** and we state it here without proof:

The moment of inertia of an object about an axis O which is parallel to the center of mass axis of the object is

$$I_o = I_c + Md^2 \tag{8.5}$$

where I_c = moment of inertia about the center of mass

M = mass of object

d = distance between the parallel axes

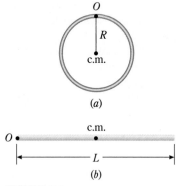

(a)

(b)

FIGURE 8.5

(a) A hoop of mass M and radius R. (b) A thin rod of mass M and length L. What are their moments of inertia about the axes perpendicular to the page and passing through the points labeled O?

Illustration 8.1

Determine the moment of inertia of **(a)** a hoop of radius R about an axis perpendicular to the plane of the hoop and through a point on its rim (Fig. 8.5a) and **(b)** a solid thin rod of length L about an axis through its end and perpendicular to length (Fig. 8.5b). Assume both objects have a mass M.

Reasoning   **(a)** The axis about O in Fig. 8.5a is a distance $d = R$ from an axis through the hoop's center of mass. From Table 8.1, we see the hoop has $I_c = MR^2$. The parallel-axis theorem gives

$$I_o = I_c + Md^2 = MR^2 + MR^2 = 2MR^2$$

(b) In Fig. 8.5b, the axis is a distance $L/2$ from the center of mass axis. Table 8.1 gives the value $\frac{1}{12}ML^2$ for I_c. By the parallel-axis theorem,

$$I_o = \tfrac{1}{12}ML^2 + M(L/2)^2 = ML^2(\tfrac{1}{12} + \tfrac{1}{4}) = \tfrac{1}{3}ML^2$$

Exercise   Determine the moment of inertia of a solid disk of radius R and mass M about an axis running through a point on its rim and perpendicular to the plane of the disk.   *Answer:* $\frac{3}{2}MR^2$.

Example 8.1

Find the rotational kinetic energy of the earth due to its daily rotation on its axis. Assume it to be a uniform sphere, $m = 5.98 \times 10^{24}$ kg, $r = 6.37 \times 10^6$ m.

R

Reasoning

Question   What information do I need to calculate KE_{rot}?
Answer   The moment of inertia of the object and its angular velocity. $\text{KE}_{\text{rot}} = \frac{1}{2}I\omega^2$.

Question   What is the moment of inertia of a sphere?
Answer   From Table 8.1, we have $I = \frac{2}{5}MR^2$.

Question   How do I find the angular velocity of earth?
Answer   You know the earth rotates 1 revolution every 24 h.

Question   Do I have to convert this to other units?
Answer   Yes. ω must be expressed in radians per second.

Solution and Discussion   Converting units of angular velocity,

$$\omega = (1.00 \text{ rev/day})(1.00 \text{ day}/24.0 \text{ h})(1.00 \text{ h}/3600 \text{ s})(2\pi \text{ rad/rev})$$
$$= 7.27 \times 10^{-5} \text{ rad/s}$$

(Data on the physical characteristics of the earth are found inside the front cover of the text.) The moment of inertia is

$$I = \tfrac{2}{5}(5.98 \times 10^{24} \text{ kg})(6.37 \times 10^6 \text{ m})^2 = 9.71 \times 10^{37} \text{ kg} \cdot \text{m}^2$$

Finally, the rotational energy is

$$KE_{rot} = \tfrac{1}{2}(9.71 \times 10^{37}\,kg \cdot m^2)(7.27 \times 10^{-5}\,rad/s)^2$$
$$= 2.56 \times 10^{29}\,J$$

Be sure you understand how the units work out to be joules.

Example 8.2

A certain wheel with a radius of 40 cm has a mass of 30 kg and a radius of gyration of 25 cm. A cord wound around its rim supplies a tangential force of 1.8 N to the wheel, which turns freely on an axle through its center. (See, for example, Fig. 8.2.) Find the angular acceleration of the wheel.

R

Reasoning

Question What determines angular acceleration?
Answer From Eq. 8.4, net applied torque and the object's moment of inertia.

Question Is enough information given to calculate net torque?
Answer Yes. There is only one force, applied tangentially 40 cm from the axis. Thus

$$\tau = (1.8\,N)(0.40\,m) = 0.72\,N \cdot m$$

Question In order to find moment of inertia, don't I have to know the shape of the wheel?
Answer Whenever you are supplied with an object's radius of gyration k, you can immediately use $I = mk^2$.

When the brake caliper is compressed, a tangential torque is applied to the rim of the wheel, causing a negative angular acceleration.

Question What is the equation that determines α?

Answer $\alpha = \dfrac{\tau}{I}$ (Eq. 8.4)

Solution and Discussion   Calculate I:

$$I = (30\,\text{kg})(0.25\,\text{m})^2 = 1.9\,\text{kg} \cdot \text{m}^2$$

Then

$$\alpha = 0.72\,\text{N} \cdot \text{m}/(1.9\,\text{kg} \cdot \text{m}^2) = 0.38/\text{s}^2$$

The units are shown in this way to demonstrate that radians do not automatically show up in the derived units. It is very important that you understand that rad/s^2 is implied in the answer.

Example 8.3

The larger wheel in Fig. 8.6 has a mass of 80 kg and a radius r of 25 cm. It is driven by a belt as shown. The tension in the upper part of the belt is 8.0 N, and that in the lower part is essentially zero. (***a***) How long does it take for the belt to accelerate the larger wheel from rest to a speed of 2.0 rev/s? (***b***) How far does the wheel turn in this time? (***c***) What is the work done by the belt on the wheel? Assume the wheel to be a uniform disk.

R

Reasoning   Part (a)

Question What equation relates time to a change of angular velocity?
Answer   If angular acceleration is constant,

$$\omega = \omega_o + \alpha t \quad (\text{Eq. 7.5}b)$$

In this case, $\omega_o = 0$.

Question Is enough information given to calculate α?
Answer   If you have moment of inertia and net torque, you can use Eq. 8.4. The mass and radius of the wheel are given, as well as the fact that it is a uniform disk. The tension exerted tangentially by the belt at the wheel's perimeter is also given.

Question What is the moment of inertia of a disk?
Answer   From Table 8.1, it is given by

$$I = \tfrac{1}{2}MR^2$$

Solution and Discussion   The moment of inertia is

$$I = \tfrac{1}{2}(80\,\text{kg})(0.25\,\text{m})^2 = 2.5\,\text{kg} \cdot \text{m}^2$$

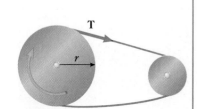

FIGURE 8.6

Angular acceleration is imparted to the large-radius wheel by the torque resulting from the tension T in the upper part of the belt. Notice that the lower part of the belt is slack.

The net torque about the axis is

$$\tau = \text{force} \times \text{lever arm} = (8.0\,\text{N})(0.25\,\text{m}) = 2.0\,\text{N} \cdot \text{m}$$

The angular acceleration is

$$\alpha = \frac{\tau}{I} = \frac{2.0\,\text{N} \cdot \text{m}}{2.5\,\text{kg} \cdot \text{m}^2} = 0.80\,\text{rad/s}^2$$

Then the time to get up to speed is

$$t = \frac{\omega}{\alpha} = \frac{2(2\pi\,\text{rad/s})}{0.80\,\text{rad/s}^2} = 16\,\text{s}$$

R

Reasoning   Part (b)

Question   What is meant by "how far does the wheel turn"?
Answer   It means "what is the angular displacement, θ?"

Question   What is the relation between θ and t?
Answer   Starting from rest ($\omega_o = 0$), we have

$$\theta = \tfrac{1}{2}\alpha t^2 \qquad \text{(Eq. 7.5e)}$$

Solution and Discussion   Putting in the above numbers,

$$\theta = \tfrac{1}{2}(0.80\,\text{rad/s}^2)(16\,\text{s})^2 = 99\,\text{rad}$$

R

Reasoning   Part (c)

Question   What is the definition of work applied to rotation?
Answer   Work done by a torque = torque $\times$ angular displacement:

$$W = \tau\theta \qquad \text{(Eq. 8.1)}$$

Solution and Discussion   Using the above numbers,

$$W = (2.0\,\text{N} \cdot \text{m})(99\,\text{rad}) = 200\,\text{N} \cdot \text{m} = 200\,\text{J}$$

This is an especially tricky situation involving units, so pay attention! *Even though torque has the unit of newton · meters, it does not itself represent an amount of work.* The force and lever arm distance used in its calculation are *perpendicular* to each other. Yet when multiplied by angular displacement (*dimensionless*), the resulting quantity *does* represent work, even though the units seem not to have changed!

Exercise   From I and ω, find the rotational kinetic energy of the wheel.
Answer: 200 J.

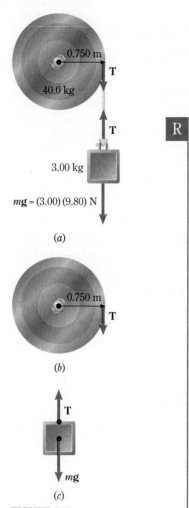

0.750 m

T

40.0 kg

T

3.00 kg

$mg = (3.00)(9.80)$ N

(a)

0.750 m

T

(b)

T

mg

(c)

FIGURE 8.7

As the 3-kg block accelerates under the pull of gravity, the tension in the rope imparts an angular acceleration to the wheel.

Example 8.4

The 3.00-kg block in Fig. 8.7a hangs from a cord wound on a 40.0-kg wheel. The wheel has a radius of 0.750 m and a radius of gyration of 0.600 m. Find **(a)** the angular acceleration of the wheel and **(b)** the distance the block falls in the first 10.0 s after it is released.

Reasoning   Part (a)

Question   What determines α?
Answer   From Eq. 8.4, the net torque on the wheel and its moment of inertia. The free-body diagram (Fig. 8.7b) shows that the net torque is provided by the tension in the string wound around the wheel.

Question   The wheel is not a simple disk. What is its moment of inertia?
Answer   When you know the radius of gyration k of any object, you can write the moment of inertia as $I = Mk^2$.

Question   Which equation determines α?
Answer   Newton's second law in rotational form gives you

$$\alpha = \frac{\tau}{I} = \frac{rT}{Mk^2}$$

Reasoning   Part (b)

Question   What determines the tension?
Answer   The end of the string attached to the block must exert a tension T on the block, and so you must look at the motion of the block also. The free-body diagram of the block (Fig. 8.7c) shows that the net force on the block is $mg - T$.

Question   What equation determines the motion of the block?
Answer $F_{\text{net}} = mg - T = ma$

Question   Is there a connection between the angular acceleration of the wheel and the downward acceleration of the block?
Answer   Yes. As the wheel turns through an angular displacement θ, the block descends a linear distance $r\theta$. Thus $a = \alpha r$.

Question   How do I combine the two second-law equations?
Answer   With the substitution of $r\alpha$ for a, the two equations become

$$Tr = (Mk^2)\alpha \qquad \text{and} \qquad mg - T = m(r\alpha)$$

If you multiply the second equation by r and add the two together, the tension cancels, giving

$$mgr = (Mk^2 + mr^2)\alpha \qquad \text{or} \qquad \alpha = \frac{mgr}{Mk^2 + mr^2}$$

Solution and Discussion   We can find α directly:

$$\alpha = \frac{(3.00\text{ kg})(9.8\text{ m/s}^2)(0.750\text{ m})}{(40.0\text{ kg})(0.600\text{ m})^2 + (3.00\text{ kg})(0.750\text{ m})^2} = 1.37\text{ rad/s}^2$$

We find the acceleration a from $a = r\alpha$:

$$a = (0.750\,\text{m})(1.37\,\text{rad/s}^2) = 1.03\,\text{m/s}^2$$

The equation relating distance fallen to time (with $v_o = 0$) is

$$y = \tfrac{1}{2}at^2 = \tfrac{1}{2}(1.03\,\text{m/s}^2)(10.0\,\text{s})^2 = 51.5\,\text{m}$$

This distance is of course measured downward relative to the block's original position. If the distance and time of fall had been measured, this analysis could be reversed to find the moment of inertia and hence the radius of gyration of the wheel. This latter method is commonly used to make such determinations.

Example 8.5

Find the angular velocity of the wheel of Example 8.4 after the block has dropped 80.0 cm. Use energy considerations, assuming no friction.

R

Reasoning

Question What basic principle applies to this situation?
Answer With no friction and no external force other than gravity, the sum of KE and PE is conserved.

Question How is the angular velocity of the wheel involved in the energy of the system?
Answer The wheel's rotational KE, $\tfrac{1}{2}I\omega^2$, is part of the total KE of the system.

Question What equation does conservation of energy give me?
Answer Since the system starts from rest, $KE_o = 0$, and so

$$\tfrac{1}{2}mv^2 + \tfrac{1}{2}I\omega^2 + mg\,\Delta h = 0$$

where $\Delta h = -80.0$ cm.

Question Is there a relation between v and ω?
Answer Yes. As a string unwinds without slipping from an axle of radius r, the connection between the linear distance Δh and the corresponding angular displacement θ is $\Delta h = r\,\Delta\theta$. It follows that $v = r\omega$.

Question What, then, is the final equation to solve for ω?
Answer $\tfrac{1}{2}(mr^2 + I)\omega^2 = mg\,\Delta h$, with $I = Mk^2$

Solution and Discussion We can solve this equation algebraically for ω^2, and then put in the data:

$$\omega^2 = 2mg\,\frac{\Delta h}{mr^2 + Mk^2}$$

$$= \frac{2(3.00\,\text{kg})(9.80\,\text{m/s}^2)(0.800\,\text{m})}{(3.00\,\text{kg})(0.750\,\text{m})^2 + (40.0\,\text{kg})(0.600\,\text{m})^2}$$

$$\omega = 1.71\,\text{rad/s}$$

The rollers on this paving machine have very large moments of inertia. When the machine is moving, a large fraction of its kinetic energy is in the rotation of the rollers.

Since $\frac{1}{2}\omega^2$ is a common factor in the two parts of the KE, notice that the ratio of translational to rotational KE in the system at any time is just $(mr^2)/(Mk^2)$.

Exercise   Calculate the total KE and the fraction of the total involved in the wheel's rotation in the above example.   *Answer: $KE_{tot} = 23.5\,J$; $KE_{rot} = 0.894(KE_{tot}) = 21.0\,J$.*

8.3   COMBINED ROTATION AND TRANSLATION

In Fig. 8.8 we see a wheel rolling without slipping. Each little piece of material within the wheel is undergoing two types of motion at the same time. The center of the wheel, which is the wheel's center of mass, is moving horizontally with speed $v_{c.m.}$. At the same time, however, the wheel is rotating about its center of mass with angular speed ω. Therefore, the wheel possesses both translational and rotational kinetic energy.

It is possible to show that the total kinetic energy of the wheel can be expressed quite simply if we restrict ourselves to rotation about one particular axis, that through the wheel's center of mass. This is the usual axis for a rolling object. Then we can state:

The total kinetic energy of an object undergoing both translational and rotational motion about its center of mass axis can be written as the sum of translational KE of the center of mass plus the rotational KE about the center of mass:

$$KE_{tot} = \tfrac{1}{2}Mv_{c.m.}^2 + \tfrac{1}{2}I_c\omega^2$$

where M = mass of object
$v_{c.m.}$ = velocity of object's center of mass
I_c = moment of inertia about center of mass axis

Let us see how this fact about KE helps to solve problems involving both KE_{tot} and KE_{rot}.

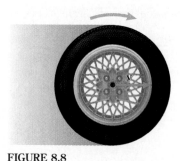

FIGURE 8.8

As the wheel rolls, it has both rotational and translational kinetic energy.

Example 8.6

A uniform sphere of radius r and mass m starts from rest at the top of an incline of height h and rolls down (Fig. 8.9). How fast is the sphere moving when it reaches the bottom? (Assume that it rolls smoothly and that frictional energy losses are negligible.)

R

Reasoning

Question   What principle applies most directly to this situation?
Answer   The conservation of mechanical energy.

Question   What are the initial and final values of PE?
Answer   If you choose the bottom of the incline as your reference, $PE_o = mgh$ and $PE_f = 0$.

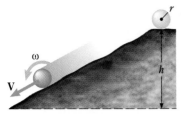

FIGURE 8.9

As the sphere rolls to the bottom of the incline, its gravitational potential energy is changed to kinetic energy of rotation and translation.

Question What are the initial and final values of KE?
Answer $KE_o = 0$ $KE_f = \frac{1}{2}mv_{c.m.}^2 + \frac{1}{2}I_c\omega^2$

Question What is the value of I_c?
Answer From Table 8.1, $I_c = \frac{2}{5}mr^2$ for a sphere.

Question Is there a connection between $v_{c.m.}$ and ω?
Answer As long as the sphere is *rolling without slipping*, $v_{c.m.} = r\omega$ (Eq. 7.7). This is not true unless this condition is satisfied.

Question What equation does this conservation of energy give me?
Answer Because all the initial energy of the sphere is PE and all its energy as it reaches the bottom is KE, you can write $PE_i = KE_f$. Using $v_{c.m.}/r = \omega$, you then get

$$mgh = \frac{1}{2}mv_{c.m.}^2 + \frac{1}{2}(\tfrac{2}{5})(mr^2)\left(\frac{v_{c.m.}}{r}\right)^2$$

Solution and Discussion Notice that the radius r cancels out in the last term. If I_c had been given simply as mk^2, this would not be true, of course. Also notice that m cancels out of every term. Now solve for $v_{c.m.}$ algebraically:

$$v_{c.m.}^2 = \frac{2gh}{1 + \frac{2}{5}} = \tfrac{10}{7}gh$$

If we examine the denominator of the middle expression, we see that the second term, $\frac{2}{5}$, shows the effect of *rotational* inertia. It adds to the first term, 1, which is the translational term. Rolling means that the original PE gets *divided up* between rotation and translation, so that the final speed of the center of mass is less than if frictionless sliding had occurred. If the sphere didn't roll at all, but just slid down the incline, its speed would be $v_{c.m.}^2 = 2gh$. This is the same speed that we obtained in previous problems involving free-fall and falling through a height h.

Example 8.7

Suppose three uniform bodies having the same mass m and the same radius r are resting at the top of a hill at a height h above the bottom. One body is a sphere, another a hoop, and the third a solid disk. Starting at the same time, all three roll down the hill without slipping. Which reaches the bottom first?

R *Reasoning*

Question What does "reaching the bottom first" imply?
Answer Each body has the same distance to cover in going down the hill. Thus the one that gains the greatest translational speed will reach bottom first.

Question Why will they reach bottom with different speeds?
Answer When the bodies are rolling without slipping, they have to rotate one revolution for every $2\pi r$ distance their centers of mass can move. Their moments of inertia will determine what fraction of the PE can go into translation of their centers of mass.

Question   What is the general equation which shows the effect of the moments of inertia?

Answer   Refer to the previous example and generalize it.

Solution and Discussion   Instead of putting in specific moments of inertia, the equation of energy conservation generally states that

$$v_{\text{c.m.}}^2 = \frac{2gh}{1 + I_c/mr^2} = \frac{2gh}{1 + N}$$

where N is the numerical factor in the general expression for I_c. For example, $N = 1$ for a hoop, $\frac{1}{2}$ for a disk, and $\frac{2}{5}$ for a sphere.

Thus we have

$$v_{\text{c.m.}}^2(\text{hoop}) = 2gh/(1 + 1) = gh$$
$$v_{\text{c.m.}}^2(\text{disk}) = 2gh/(1 + \tfrac{1}{2}) = \tfrac{4}{3}gh = 1.33gh$$
$$v_{\text{c.m.}}^2(\text{sphere}) = 2gh/(1 + \tfrac{2}{5}) = \tfrac{10}{7}gh = 1.43gh$$

The object with the *smallest* I_c (sphere) will gain the *least* rotational KE, and so have the greatest translational KE. It will reach the bottom of the hill first. ∎

8.4   ANGULAR MOMENTUM

In view of the many analogies found thus far between linear and rotational phenomena, it should come as no surprise that linear momentum has a rotational counterpart. Rotational, or angular, momentum is associated with the fact that a rotating object persists in rotating. Linear momentum is defined by the product of the amount of translational inertia m times translational velocity **v**. The analogous quantities for rotation are rotational inertia, I, and angular velocity, **ω**. Thus you might predict that angular momentum **L** will be given by

$$\mathbf{L} = \text{angular momentum} = I\boldsymbol{\omega} \tag{8.6}$$

Previously, we have characterized the direction of quantities associated with rotation, such as torque, angular displacement, and angular velocity, as simply either clockwise or counterclockwise about a chosen fixed axis. An alternative and often more convenient way of denoting rotational direction is *to represent the direction by a vector along the axis around which the rotation is taking place* (Fig. 8.10a). The connection between these two descriptions is shown in Fig. 8.10b. If we curl the fingers of our right hand around the axis in the sense that the object is rotating, our right thumb points in one of the two directions along the axis of rotation. We take this to be the direction of the angular velocity and hence the direction of the angular momentum. Changing from clockwise to counterclockwise rotation reverses the direction of the thumb, as you can verify. Assigning the direction of rotational vectors in this way along the axis of rotation is called the **right-hand rule.**

Angular momentum obeys a conservation law much like the one obeyed by linear momentum. The **law of conservation of angular momentum** can be stated as follows:

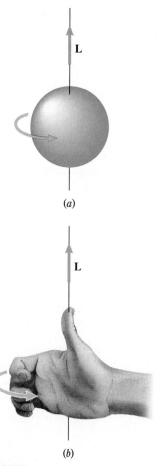

(a)

(b)

FIGURE 8.10

The sphere in *(a)* is rotating in the sense given by the gold arrow. Its angular velocity and angular momentum are taken to be upward along the rotational axis, as shown by the right-hand rule in *(b)*.

(*a*) A skater begins a spinning jump with arms outstretched for balance. Once in the spin, the skater's angular momentum $I\omega$ remains constant. (*b*) In order to spin as fast as possible (large ω), the skater reduces his moment of inertia I about the vertical axis to a minimum by drawing his arms and legs as close to the axis as he can.

If no net torque acts on a body or system, its angular momentum remains constant in both magnitude and direction:

$$I\omega = \text{constant} \qquad \text{if} \qquad \Sigma\tau = 0$$

Notice that, if no unbalanced torque acts on the object, the *direction* of the angular-momentum vector does not change. This is equivalent to saying that *the axis of rotation of a spinning object does not alter its orientation unless a net torque acts on it.* You can demonstrate this orientation effect using a simple gyroscope or a swiftly spinning wheel. For example, a large wheel set rotating about a north-south axis does not change its orientation readily unless very large forces are applied to it. When a torque is applied to a rotating system such as this, the resulting motion is interesting, since it appears to contradict what we expect to happen. Although the analysis of these effects is too complicated for us to pursue in this course, the effects are easily demonstrated, and your instructor may show you some.

Conservation of angular momentum is very important in any system whose moment of inertia changes through the action of some internal forces, such as a collapsing star or an ice skater who starts a spin with outstretched arms and then brings the arms in closer to the body. In both cases, since mass is being redistributed closer to the axis of rotation, *the moments of inertia are decreasing,* even though the mass remains the same. Since this change occurs without any external torques, $I\omega$ must remain constant, requiring the rate of spin ω to increase. Similarly, if the moment of inertia were to *increase,* the angular velocity would have to decrease proportionately.

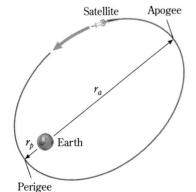

FIGURE 8.11

Find the ratio of the satellite's speed at perigee to that at apogee.

Example 8.8

Consider a satellite orbiting the earth as shown in Fig. 8.11. Find the ratio of its speed at the point of closest approach to the earth (*perigee*) to that at the point of furthest distance (*apogee*).

As the tether ball winds up on its cord around the post, its angular velocity increases. Can you explain why?

FIGURE 8.12

Why does the coasting rotating system slow down as water drips into the beaker?

R

Reasoning

Question   What principle would relate the velocities at these two points?
Answer   If angular momentum is conserved, you can equate the satellite's angular momenta and relate the angular velocities at these points. These angular velocities are related to the corresponding linear velocities.

Question   How can I tell if angular momentum is conserved in this case?
Answer   The test is whether there is a net torque applied to the satellite. This requires you to identify an axis about which to calculate torque.

Question   How do I make a choice of an axis?
Answer   The gravitational force exerted by the earth on the satellite is along a line that passes through the earth. Thus if you choose an axis through the earth perpendicular to the satellite's orbit, you can say that the torque produced by gravity about this axis is zero, and so the angular momentum of the satellite relative to this axis is constant.

Question   What equation does the conservation theorem give me?
Answer   Using the subscripts p and a to refer to perigee and apogee respectively, you have $L_p = L_a$, or $I_p\omega_p = I_a\omega_a$.

Question   What are the moments of inertia of the satellite at apogee and perigee?
Answer   Using r_p and r_a as the distances from earth to the satellite,

$$I_p = mr_p^2 \quad \text{and} \quad I_a = mr_a^2$$

where m is the satellite's mass.

Question   What is the relation between angular and linear speeds at these two points?
Answer   Since linear velocity is perpendicular to the radial distance at both points, you can write

$$v_a = r_a\omega_a \quad \text{and} \quad v_p = r_p\omega_p$$

Solution and Discussion   The conservation theorem yields

$$\frac{\omega_p}{\omega_a} = (r_a/r_p)^2 = \frac{v_p/r_p}{v_a/r_a}$$

So

$$\frac{v_p}{v_a} = \frac{r_a}{r_p}$$

Thus the satellite moves proportionately more slowly as its distance from the earth increases.

Example 8.9

The beaker in Fig. 8.12 sits over the axis of a rotating turntable. The table is coasting on a frictionless bearing, and $I = 8.0 \times 10^{-4}\,\text{kg} \cdot \text{m}^2$ for the table-beaker combination. Water drips slowly into the beaker along the axis. If the beaker was rotating at 2.0 rpm when empty, what is its rotational speed when it contains 300 g of water? The inner radius of the beaker is 3.5 cm.

R

Reasoning

Question   Is angular momentum conserved?
Answer   Yes. Since the water is entering along the axis of rotation, it cannot exert a torque on the rotating system.

Question   What property is changing?
Answer   The mass on the turntable is increasing, and because of that the rotational inertia of the system is increasing. Thus, in order for L to be conserved, the rate of rotation must decrease proportionately.

Question   What is the moment of inertia of the water?
Answer   You can assume at this slow rate of rotation that the water is essentially a disk having the same radius as the inside of the beaker. Thus

$$I_w = \tfrac{1}{2}(0.30\,\text{kg})(0.035\,\text{m})^2 = 1.8 \times 10^{-4}\,\text{kg} \cdot \text{m}^2$$

for a final value.

Question   What equation results from the application of conservation of angular momentum?
Answer $I_o\omega_o = (I_o + I_w)\omega_f$

Solution and Discussion   The last equation gives

$$
\begin{aligned}
\omega_f &= \frac{\omega_o I_o}{I_o + I_w} \\[2mm]
&= \frac{(2.0\,\text{rpm})(8.0 \times 10^{-4}\,\text{kg} \cdot \text{m}^2)}{(8.0 \times 10^{-4} + 1.8 \times 10^{-4})\text{kg} \cdot \text{m}^2} \\[2mm]
&= 2.0\,\text{rpm}\,\frac{8.0}{9.8} = 1.6\,\text{rpm}
\end{aligned}
$$

Exercise   Suppose the kinetic energy of the falling water can be neglected. Show that the final system has 19 percent less kinetic energy than the original system. What happened to this lost energy?

8.5 A MODERN PERSPECTIVE: THE MINIMUM QUANTITY OF ANGULAR MOMENTUM

How small is small? The notion of a smallest or minimum unit in which something can exist is a common concept. Take, for example, a bathtub full of water. We can divide the water in a bathtub into gallon containers or milliliters, and further divide it into drops. But when we divide the water up into individual molecules, we have reached the most basic amount in which water can exist. If we split the water molecule into its component oxygen and hydrogen atoms, we no longer have water.

Similarly, an atom of oxygen is the smallest quantity in which oxygen can exist. As we will see later in the course, it seems that electric charge cannot exist in a smaller amount than that carried by a single electron or proton.*

However, there is no obvious limit on smallness of length or time, apart from the difficulty we may have of measuring quantities precisely enough. Classical physics throughout the nineteenth century treated distance and time as though they were infinitely divisible, or *continuous,* properties of nature. Thus we speak of a *point* mass and the concept of *instantaneous* position, velocity, and acceleration, implying that space and time can be shrunk indefinitely without reaching a finite minimum value.

The same reasoning follows when we consider dynamic properties such as energy and angular momentum. Even though matter may have a basic quantum of mass, such as the mass of the elementary particles composing the atom, if position and time can be shrunk to zero, then a finite mass can have values of speed and kinetic energy in a continuous range all the way down to zero. However, at the beginning of the twentieth century the idea was proposed that certain dynamic properties of matter *did* occur in discrete amounts. This was one of the conceptual revolutions that marked the end of the classical period and the beginning of what is called modern physics.

It was in the years 1900 and 1905 that the German physicists Max Planck and Albert Einstein separately proposed that the emission (Planck) and absorption (Einstein) of radiation energy (i.e., light) by matter occurred in "packets," or **quanta,** of energy which were proportional to the frequency of the light. With this idea, Planck was able to explain experimental observations of the way in which heated objects emit light, and Einstein was able to explain the results of experiments involving the absorption of light by metal surfaces. The principles of classical physics were completely unable to explain either of these phenomena, which we discuss in more detail in Chap. 26.

The constant of proportionality which defined the quantum of radiation energy in Planck's theory is called **Planck's constant,** h. It has the extremely small value of

$$h = 6.63 \times 10^{-34} \text{ J} \cdot \text{s}$$

Notice that the units of this constant are the same as those for angular momentum L:

$$1 \text{ J} \cdot \text{s} = 1 \text{ (N} \cdot \text{m) s} = 1 \text{ (kg} \cdot \text{m/s}^2) \cdot \text{m} \cdot \text{s} = 1 \text{ (kg} \cdot \text{m}^2)/\text{s}$$

It is tempting to see if the value of h represents a fundamental quantum of the amount of angular momentum L and hence of the rotational kinetic energy $L^2/2I$ that an object can have. In other words, is it true that the angular momentum of a rotating object is some multiple of this basic amount? That is, is $L = I\omega = nh$, where $n = 1, 2, 3, \ldots$, etc.? Furthermore, is the rotational kinetic energy of the object governed by the following relation?

$$\text{KE}_{\text{rot}} = \frac{L^2}{2I} = \frac{(nh)^2}{2I} = n^2 \frac{h^2}{2I}$$

*Evidence has recently been found for the existence of fundamental particles called **quarks,** which theory has predicted bear charges of one-third and two-thirds of the electronic charge. This does not alter the fact that electric charge cannot be subdivided beyond a certain finite minimum quantum, however. It merely changes the size of the quantum.

If these relations are true, they predict nonzero values of h/I for the minimum possible angular velocity and $h^2/2I$ for minimum rotational KE. Thus in order to determine whether the angular velocity and rotational KE of an object are indeed quantized or can become zero as Newton's classical laws predict, we have to be able to detect experimentally the difference between a value of zero and a value of h/I for minimum angular velocity and between zero and $h^2/2I$ for KE_{rot}.

For objects common to our experience, a typical moment of inertia is large enough so that $h^2/2I$ becomes an exceedingly small number, so small that we cannot possibly distinguish its value from zero. For example, according to the above expression for KE_{rot}, a 50-g meterstick rotating about its center of mass would have a quantum of rotational energy of approximately 5×10^{-65} J and a minimum angular velocity of about 1.6×10^{-31} rad/s! From the standpoint of measurement, these values are essentially zero, bearing out our experience that the meterstick can be at rest. Thus if angular momentum is going to exhibit quantized behavior, the extremely small value of h requires that we look at objects having extremely small moments of inertia. Examples of such objects are an electron's angular momentum as it orbits the nucleus of the hydrogen atom and the angular momenta of individual diatomic molecules such as N_2 and H_2.

It was the Danish physicist Niels Bohr who first applied the quantization of angular momentum to the hydrogen atom in 1911, thereby explaining the observed pattern of emission and absorption of light by hydrogen. Bohr assumed that the electron's angular momentum could only have values given by multiples of $h/2\pi$:

$$L \text{ (electron)} = mr^2\omega = n\frac{h}{2\pi}$$

This strange and controversial assumption proved to be the key to the subsequent development of modern atomic theory.

The quantized nature of angular momentum in diatomic molecules was used by Einstein to explain the absorption of heat by molecular gases, a topic discussed in Chap. 12. Yet another successful application of the idea of quantized angular momentum came in 1925, when the Dutch physicists Uhlenbeck and Goudsmit predicted theoretically that the electron itself had rotation, or spin, in the amount of $\frac{1}{2}(h/2\pi)$. Their prediction explained the behavior of hydrogen atoms when in the presence of a magnetic field.

Thus during the first three decades of the twentieth century a new era of physics saw the rapid development of the revolutionary idea that the dynamic behavior of very small masses is governed by quanta of rotational energy and angular momentum. This area of study is known as **quantum mechanics,** and has proved to be extremely successful in explaining the behavior of matter and energy on the atomic and subatomic scale.

LEARNING GOALS

Now that you have finished this chapter, you should be able to

1 Define (*a*) kinetic energy of rotation, (*b*) moment of inertia, (*c*) radius of gyration, (*d*) parallel axis theorem, (*e*) angular momentum.

2 Write the rotational analogs of $F = ma$, $KE_{trans} = \frac{1}{2}mv^2$, $p = mv$, and $W = F_x x$.

3 Identify the moment of inertia of simple objects, such as those in Table 8.1, about their center of mass and calculate their moment of inertia about any axis parallel to the center of mass axis.

4 Use the relation $\tau = I\alpha$ in simple situations involving accelerated rotational motion.

5 Relate the work done on an object by a torque to the object's change in rotational kinetic energy in simple situations.

6 Find the total kinetic energy of an object that is rotating and translating simultaneously.

7 Solve simple situations involving the conservation of energy for rolling objects.

8 State the law of conservation of angular momentum and use it in simple problems.

SUMMARY

DERIVED UNITS AND PHYSICAL CONSTANTS

Angular Momentum (L)
$L = I\omega$ kg · m²/s (or N · s)   (Eq. 8.6)

DEFINITIONS AND BASIC PRINCIPLES

Moment of Inertia (I)
$I = m_1 r_1^2 + m_2 r_2^2 + \cdots + m_n r_n^2$ kg · m²   (Eq. 8.2)

The results of such calculations for some simple objects are given in Table 8.1.

Radius of Gyration (k)
The radius of gyration of an object is that length for which the object's moment of inertia can be written

$I = Mk^2$ (Eq. 8.5)

Newton's Second Law Applied to Rotation
$\tau = I\alpha$ (Eq. 8.4)

INSIGHT

1. In using this rotational form of Newton's second law, α must be expressed in rad/s².

Parallel-Axis Theorem
The moment of inertia of a rigid body about an axis O a distance d from the center of mass axis is

$I_o = I_c + Md^2$

Kinetic Energy of Rotation
An object with angular velocity ω and moment of inertia I about an axis has rotational kinetic energy given by

$KE_{rot} = \frac{1}{2}I\omega^2$ (Eq. 8.3)

An object which is both rotating and translating has a total KE equal to

$KE_{tot} = \frac{1}{2}Mv_{c.m.}^2 + \frac{1}{2}I_c\omega^2$

A uniform object of radius r rolling without slipping has the following relation between ω and $v_{c.m.}$:

$v_{c.m.} = r\omega$

Conservation of Angular Momentum
If there is no net external torque acting on a system, the system's angular momentum L remains constant. This means that the product of $I\omega$ remains constant even if I and/or ω change.

INSIGHT

1. I can change in two ways: total rotating mass may change or the distribution of existing mass may change, because of a change of radius.

QUESTIONS AND GUESSTIMATES

1 Devise a demonstration that shows that a rotating wheel can do work because of its rotational kinetic energy.

2 Two bicycle wheels are identical except that one has its rubber tire replaced by a solid ring of metal of the same size. They are mounted on identical stationary axles so that they can coast relatively freely. If they start with the same speed, which wheel will coast to rest first?

3 Three wheels have the same mass and rim radius, but wheel *a* is a solid uniform disk, wheel *b* has a heavy rim with light spokes, and wheel *c* is a typical automobile wheel with tire. Compare their moments of inertia about their rotation axes.

4 Estimate your moment of inertia when you are standing erect, using as axis (*a*) a vertical line that passes through the center of your body and (*b*) a horizontal line perpendicular to your stomach.

5 One proposal for storing energy is by means of a massive, rapidly rotating flywheel. Discuss the pros and cons of this device as applied to (*a*) an automobile and (*b*) an electric power station.

6 Refer to Fig. 8.7. Assume friction is negligible. (*a*) The tension in the connecting cord is less than *mg*. Why? (*b*) What effect does changing the moment of inertia of the wheel have on the tension in the cord?

7 A tiny bug is at rest on the rim of a turntable that rotates

without friction. What happens to the turntable as the bug (*a*) runs radially in toward the center and (*b*) runs clockwise around the rim? Take into account when the bug starts, when it runs at constant speed, and when it stops.

8 Which will roll down an incline faster, a hollow sphere or a solid sphere? Will the radius of the sphere affect its speed? Repeat for a hoop and a uniform solid disk.

9 In order to keep a football or any other projectile from wobbling, the thrower causes it to spin about an axis in line with the direction of motion. Explain.

10 A "do-it-yourselfer" builds a helicopter with a single propeller on a vertical axis. In the helicopter's maiden flight, the operator becomes sick because the whole helicopter tends to spin about a vertical axis. What is wrong? How is this difficulty overcome in more sophisticated machines?

11 Suppose that the sun's attraction for the earth suddenly doubled. What effect would this have on the earth's rate of rotation and its orbit about the sun?

PROBLEMS

Section 8.1

1 A force of 6 N is applied to a string wound around the rim of a 9-cm-radius wheel. How much work is done by this force as it turns the wheel through 36°?

2 The friction torque on a certain wheel-axle system is 0.060 N · m. How much work does this torque do on the system as the wheel makes four complete turns?

3 How much work must be done on a wheel, with a moment of inertia $I = 0.4$ kg · m², to accelerate it from rest to an angular speed of 150 rev/min?

4 A potter's wheel having a moment of inertia of 1.5 kg · m² is spinning at 36 rev/min when it begins to coast to rest. How much work do the friction forces do in stopping the wheel?

5 A phonograph wheel of moment of inertia 0.0015 kg · m² is spinning with 45 rev/min. (*a*) How much work will the friction forces do to stop the wheel once the power is turned off? (*b*) How large is the average torque exerted by friction forces if the wheel coasts uniformly to rest in 25 s?

6 How large a torque is required to give an angular acceleration of 2.4 rad/s² to a wheel that has a moment of inertia of 0.25 kg · m²?

7 A torque of 15 N · m is applied to a heavy wheel whose moment of inertia is 30 kg · m². What is the angular acceleration of the wheel?

8 A wheel whose moment of inertia is 24 kg · m² is subjected to a clockwise torque of 18 N · m. If the wheel is spinning counterclockwise at 6 rev/min when the torque is applied, how long will it take to stop the wheel completely?

9 A tangential force of 40 N applied to the rim of a 16-cm-radius wheel gives the wheel an angular acceleration of 0.5 rad/s². What is the moment of inertia of the wheel?

10 In an experiment a torque of 0.200 N · m is applied to a thin, rigid rod, causing it to rotate about an axis through its center and perpendicular to its length with an angular acceleration of 0.45 rad/s². What is the moment of inertia of the rod?

11 A 120-kg merry-go-round in the shape of a horizontal disk of moment of inertia 175 kg · m² is set into motion by pulling a rope wrapped about its rim. What constant force must be exerted on the rope in order to bring it from rest to an angular speed of 30 rev/min in 3 s?

12 The shaft of a 0.3-hp (output) motor rotates at 5 rev/s. (*a*) How much work can the motor do each second? (*b*) How large an output torque can the motor produce when it is operating at its rated speed?

■13 The output shaft of a gear system is attached to a motor that has a rated power of 0.2 W. Mounted on the shaft is a wheel that has a moment of inertia of 0.8 kg · m². Estimate the minimum time it takes the motor to accelerate the wheel from rest to 24 rev/min.

■14 A wheel mounted on an axle has a cord wound around its rim. The moment of inertia of the wheel is 0.1 kg · m². The wheel is accelerated from rest by a force of 25 N which pulls the end of the cord for a distance of 0.8 m. What is the final angular speed of the wheel?

15 A wheel of radius 10 cm with a moment of inertia $I = 0.08$ kg · m² is rotating at 180 rev/min when a tangential friction force of 1.0 N is applied to the rim. Through how many revolutions does the wheel turn before coming to a stop?

16 What is the kinetic energy of a phonograph disk with a moment of inertia of 0.012 kg · m² rotating at 45 rev/min?

Section 8.2

17 What is the length of the rod described in Prob. 10 if it has a mass of 0.50 kg?

18 The 0.5-m-long spokes in Fig. P8.1 have negligible mass compared with the eight 3-kg masses. Find the moment of inertia of the system (*a*) about an axis through its center and perpendicular to the plane of the figure and (*b*) about the line *AA'* as axis.

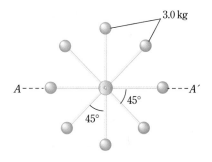

FIGURE P8.1

19 Each of the four masses in Fig. P8.2 has mass m. The connecting rods have masses negligible as compared to m. Find the moment of inertia of the system of four masses (a) about AA' as axis, (b) about BB' as axis, and (c) about an axis through the center O and perpendicular to the plane of the figure. Consider the balls to be point masses.

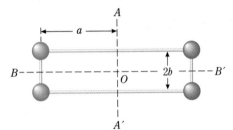

FIGURE P8.2

20 The two hoops in Fig. P8.3 are held together by spokes of negligible mass. The inner hoop has a mass of m_i and that of the outer hoop is m_o. The radii of the inner and outer hoops are a and b, respectively. Find the moment of inertia of the system about an axis through the center and perpendicular to the plane of the hoops.

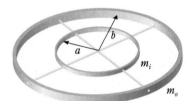

FIGURE P8.3

21 The hoop of mass M in Fig. P8.4 has a very light rod fastened through it, along its diameter, with identical masses m attached at its end. Find the moment of inertia of the system about an axis through the center of the hoop C and perpendicular to its plane.

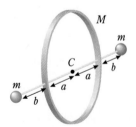

FIGURE P8.4

22 A wheel in the form of a uniform circular disk has a moment of inertia I_d about an axis through its center and perpendicular to its plane. A rim of radius 40 cm and mass 1.8 kg is slipped onto the wheel. Find the moment of inertia of the combinations about the same axis.

23 For (a) a hoop and (b) a uniform disk, of mass M and radius R as shown in Fig. P8.5, determine the moment of inertia about an axis through a point O on the rim and perpendicular to the plane of the object.

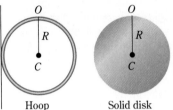

Hoop   Solid disk   FIGURE P8.5

24 A sphere of mass M and radius R is suspended by a massless cord of length L as in Fig. P8.6. Determine the moment of inertia of the sphere about an axis perpendicular to the plane of the page and through the point of suspension O.

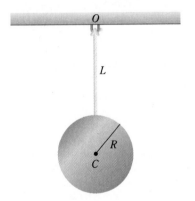

FIGURE P8.6

25 Determine the moment of inertia of a long, thin uniform cylindrical rod through one end of the rod (at 0) and perpendicular to its length (See Fig. P8.7). What is the moment of inertia of this rod about an axis at a distance of $L/3$ from O and perpendicular to its length?

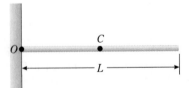

FIGURE P8.7

26 A cable passes over a pulley, which can be considered as a uniform disk of mass 2.4 kg and radius 0.6 m. Because of friction between the pulley and the cable, tension in the cord is not the same on opposite sides of the pulley. The force on one side is 150 N and the force on the other side is 120 N. Determine the angular acceleration of the pulley.

27 The wheel of an automobile can be considered as a solid disk of radius 35 cm and a mass of 6.5 kg. What is its rotational kinetic energy when rotating at 3 rev/s?

28 What angular speed (in revolutions per second) must a solid uniform cylindrical wheel of radius 0.5 m and mass 4 kg have if it has to have the same rotational kinetic energy as a solid uniform sphere of the same radius and mass spinning at 6 rev/s?

29 What is the rotational kinetic energy of a 60-cm-diameter bicycle wheel of mass 4.0 kg when the bicycle is traveling at 4 m/s? Assume the wheel has a radius of gyration $k = 50$ cm.

30 (a) A certain wheel has a mass of 54 kg and a radius of gyration of 30 cm. How large a torque is required to accelerate this wheel from rest to 0.5 rev/s in 25 s? (b) How far does the wheel turn in this time?

31 A solid cylinder of mass 1.8 kg and radius 20 cm is rotating about an axis along its length and through its center at an angular velocity of 2 rev/s. What torque is required to stop it in 15 s?

32 A force of 2.2 N acts tangentially to the rim of a solid 52-kg disk of radius 32 cm. (a) How long does it take to accelerate the disk, rotating about an axis through its center and perpendicular to its plane, from rest to 210 rev/min? (b) How many revolutions does the disk turn during this time?

33 A horizontal 100-kg merry-go-round of radius 1.6 m is started from rest by a constant horizontal force of 60 N applied tangentially to its rim. Find the kinetic energy of the system after 3 s.

34 A 5.0-cm-radius, 6.0-kg solid cylinder is mounted on an axle along its own axis. A cord wound around it supplies a tangential force of 3.6 N for 3 s. If the cylinder starts from rest, (a) how fast (in revolutions per second) is it rotating at the end of the 3 s? (b) How much kinetic energy does it have at that moment?

■35 An 8.0-cm-radius wheel is mounted on a horizontal axis. A cord of negligible mass is wound around the rim, and a 60-g mass is suspended from the cord. After the mass is released (from rest), the system accelerates in such a way that the mass drops 3 m in 5 s. What is the moment of inertia of the wheel? What is the tension in the cord as the mass is descending?

■36 A cylinder of radius 24 cm is mounted on a horizontal axis coincident with its own axis. A cord is wound on the cylinder's rim, and a 100-g mass is hung from it. After being released from rest, the mass drops 180 cm in 1.5 s. Find the moment of inertia of the cylinder and the tension in the cord while the mass is falling.

■37 An 80-g mass is suspended from the rim of a 100-cm-diameter wheel by a string wound around the wheel. The wheel has a moment of inertia $I = 0.1$ kg · m² and is mounted on a frictionless axle. It is accelerated from rest by letting the mass fall. (a) How fast (in revolutions per second) is the wheel turning when the mass has fallen 1.0 m? (b) What is the rotational kinetic energy of the wheel at that moment?

■38 A cylindrical wheel with moment of inertia $I = 900$ kg · m² is spinning at 21.0 rev/min when a mechanism is engaged that causes a cord on the rim to lift a 6.0-kg mass as the wheel coasts to rest. How high can the wheel lift the mass? Neglect any change in rotational kinetic energy during engagement.

■39 The system in Fig. P8.8 is released from rest. (a) How fast is the frictionless wheel (with moment of inertia $I = 0.008$ kg · m² and radius $r = 8.0$ cm) turning after the 520-g mass has fallen 2.4 m? (b) How long does it take the mass to drop that far?

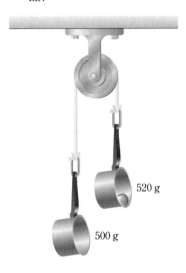

FIGURE P8.8

■40 The system in Fig. P8.9 is released from rest. There is no friction between the block and the table, and the frictionless pulley has a moment of inertia $I = 0.008$ kg · m² and a radius of 8.0 cm. (a) How fast is the mass on the right moving after it has fallen 100 cm? (b) How long does it take the mass to fall this distance? (c) What is the rotational kinetic energy of the pulley at that moment?

FIGURE P8.9

Section 8.3

■41 A hoop of radius 6 cm starts from rest and rolls without slipping down a slope. (a) What is its linear speed when it reaches a point 50 cm vertically lower than its starting point? (b) How fast is it rotating (in revolutions per second) at that time?

■42 Repeat the last problem for (a) a wheel (disk) that has a radius of 6 cm and a radius of gyration 5 cm and (b) a uniform disk of radius 6 cm.

■43 A steel ball bearing of radius 0.6 cm is rolling without slipping along a table at 45 cm/s when it starts to roll up an incline. How high above the table level does it rise before stopping? Ignore friction losses.

■44 A solid sphere has a radius of 30 cm and a mass of 80 kg. How much work is required to get the sphere rolling with an angular speed of 40 rad/s on a horizontal surface? (Assume the sphere starts from rest and rolls without slipping.)

■45 A solid bowling ball of radius 12 cm and a mass of 8 kg rolls without slipping along a bowling alley with a linear speed of 1.6 m/s. What is the total kinetic energy of the ball?

■46 A uniform disk released from the top of an inclined plane reaches the bottom with a speed of 12 m/s. How high is the upper end of the inclined plane from its bottom? Assume that the disk rolls without slipping and ignore friction.

■47 A 2.2-kg solid ball of radius 0.6 m starts at the height of 3.2 m above the ground level and rolls down an incline which makes an angle of 24° with the horizontal. A solid disk and a ring having exactly the same mass and radius as the ball also start at the same time from the same height and roll down the same incline. If the three objects roll without slipping, which of the three reaches the bottom first? Which reaches the bottom last?

■■48 A solid ball, a solid disk, and a hoop all with same moment of inertia $I = 0.05$ kg · m² start simultaneously from rest from the top of an incline 3 m above the ground level. If they all roll without slipping, which of the three will win the race to the bottom of the incline?

Section 8.4

49 Determine the magnitude of angular momentum of a solid uniform disk of radius 50 cm and mass 2.4 kg spinning at 6 rev/s about an axis through its center and perpendicular to its plane.

50 Repeat the previous problem for a solid sphere of the same mass and radius rotating with the same speed as in Prob. 49.

51 In Fig. P8.10, two identical small balls, each of mass 1.2 kg, are fastened to the ends of a light metal rod 1.0 m long. The rod is pivoted at its center and is rotating at 10 rev/s. An internal mechanism can move the balls in toward the pivot. (a) Find the moment of inertia of the original device. (b) If the balls are suddenly moved in until they are 30 cm from the pivot, what is the new speed of rotation?

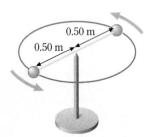

FIGURE P8.10

■52 A woman stands over the center of a horizontal disklike platform that is rotating freely at 2 rev/s about a vertical

axis through its center and straight up through the woman. She holds two 2-kg masses in her hands close to her body. The combined moment of inertia of platform, woman, and masses is 1.8 kg · m². The woman now extends her arms so as to hold the masses far from her body. In so doing, she increases the moment of inertia of the system to 2.4 kg · m². (a) What is the final rotational speed of the platform? (b) Is the kinetic energy of the system changed during the process? Explain.

■53 A phonograph record in the shape of a disk of 12-cm radius and mass of 0.1 kg rotates freely about a vertical axis through its center at 45 rev/min. An 18-g bug drops onto the disk at a point 4 cm from the center of the disk. What is the new angular speed of the disk?

■54 An ice skater spins with an angular speed of 3 rev/s with her arms outstretched. She lowers her arms so that her moment of inertia decreases by 15 percent. Determine (a) her new speed of spin, and (b) the percentage change in her kinetic energy.

■55 An ice skater moving with speed v_o past a post grabs the end of a rope tied to the post. The original length of the rope is L_o. As the skater circles the post, the rope winds around the post, becoming progressively shorter. Assuming that the skater coasts and does not try to stop, how fast will she be moving when the length of the rope (the circle radius) is (a) $3L_o/4$, (b) $L_o/2$, and (c) $L_o/3$? Assume that the post radius is much smaller than L_o.

■56 A merry-go-round consists of an essentially uniform 150-kg disk of radius 6.0 m rotating at 15 rev/min about a vertical axis through its center. An 80-kg person is standing on the outer edge of the merry-go-round. (a) How fast will the disk be rotating if the person walks 3 m toward the center of the disk? (b) What is the change in the kinetic energy of the system?

■57 Suppose no one is standing on the merry-go-round in Prob. 56, and it is rotating at 12 rev/min. If an 80-kg person quickly sits down on its outer edge, what is its new angular speed?

■■58 Figure P8.11 shows a disk and a shaft (moment of inertia I_1) coasting with angular speed ω_1. A nonrotating disk of moment of inertia I_2 is dropped onto the first disk and couples to it. (a) Find the angular speed of the disks after cou-

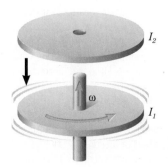

FIGURE P8.11

pling. (b) Repeat if the dropped disk has an initial angular speed ω_2 in the same direction as ω_1. (c) Repeat if $\omega_2 = \omega_1$, but ω_2 and ω_1 are in opposite directions. (d) What happens to the kinetic energy of the system? Determine the ratio of the final to the initial kinetic energy of the system.

Additional Problems

""59 Stars with mass greater than about 1.5 times that of our sun are unstable. Driven by gravitational forces, they sometimes collapse to form neutron stars. These are incredibly dense stars in which all atoms are collapsed; in effect, the atomic electrons and protons have combined to form neutrons. The final star has a radius of only about 10^{-5} that of the original star. Our sun rotates on its axis about once each 25 days. (a) How long would the sun take to complete one revolution if it were to undergo such a collapse? Find the ratio of the star's final rotational kinetic energy to its original kinetic energy. Explain this change of kinetic energy. Assume the sun to be a uniform sphere.

""60 A 25-g block in Fig. P8.12 revolves in a circle on a frictionless table. It is held by a cord that passes through a tiny hole at the center of the circle. The angular speed of the block is 30 rev/min and the radius $r = 72$ cm. (a) What is the magnitude of the force **F**? (b) If the cord is pulled down 12 cm, what is the new angular speed of the block? (c) How much work does the force **F** do to shorten the circle radius to 60 cm? Assume that the block is quite small relative to the radius of the circle and can be considered as a point mass.

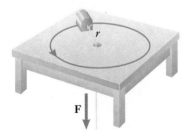

FIGURE P8.12

""61 A cylindrical winch of mass M and radius R is spinning with an angular speed ω_o while winding up a slack string on its rim. The slack string is attached to a mass m sitting on the floor beneath the winch. Eventually the slack is all taken up, and the mass is suddenly lifted from the floor. Show that the fraction of the total kinetic energy lost in the process of bringing the mass up to speed is $2m/(M + 2m)$. Neglect changes in gravitational potential energy.

""62 A solid uniform cylinder has a wide ribbon wrapped around its circumference. One end of the ribbon is attached to the ceiling (Fig. P8.13). The cylinder is released from rest, and as it falls the ribbon unwinds without slipping. The mass of the cylinder is 0.60 kg and its radius is 20 cm. Determine (a) the angular acceleration of the cylinder, (b) the tension in the ribbon, and (c) the angular velocity at the instant the cylinder has fallen 2.5 m from its initial starting position.

FIGURE P8.13

""63 Use energy methods to determine the speed of the center of mass of the cylinder in Prob. 62 at the instant it has fallen 2.5 m. Show that this result is in accord with the answer to part (c) of Prob. 62.

""64 Two identical disks of mass M and radius R are on a frictionless table, each spinning about its central axis with angular velocity ω_o (Fig. P8.14a). They gradually move toward each other, and when they touch, they stick together at the point of contact O. As a result, they rotate about the point O with angular velocity ω_f (Fig. P8.14b). Determine ω_f in terms of ω_o.

""65 Two cylinders start from rest at the 3-m (from the ground) high top of a plane inclined at an angle of 30° with the horizontal. Both cylinders have a mass of 1 kg and a radius of 10 cm. One of the cylinders is solid and the other is a thin shell. When the first one (solid) reaches the bottom of the incline, how far down the incline has the other one moved?

""66 As the earth moves around the sun in an elliptical orbit its angular momentum remains constant. Use this information to show that the angular speed of the earth in its orbit is greatest when it is closest to the sun.

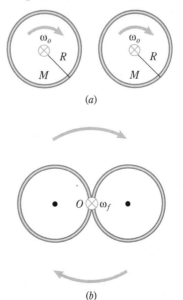

(a)

(b)

FIGURE P8.14

PART TWO

MECHANICAL AND THERMAL PROPERTIES OF MATTER; OSCILLATIONS AND WAVES

To see heat pass from a cold body to a warm one, it will not be necessary to have the acute vision, the intelligence, and the dexterity of Maxwell's demon; it will suffice to have a little patience.

HENRI POINCARÉ

Now that we have developed the concepts of mass and force and have learned some principles which describe the motion of matter, we will proceed to inquire about the internal properties of matter. Long before anything was known about atoms and molecules, the large-scale, or macroscopic, properties of matter were investigated. In the third century B.C., the Greek engineer Archimedes explained the buoyant force a fluid exerts on an object immersed in it. In the seventeenth and eighteenth centuries, the laws which describe the effects of pressure and temperature on gases were developed. The existence of different physical states of matter (solid, liquid, and gas) and the degree to which matter can be stretched and compressed when forces are applied were also investigated during this period. Other phenomena involving macroscopic properties include the manner in which fluids flow, and the connection between heat added to a substance and the resulting temperature change or change in state.

The study of heat and thermal properties of matter proceeded separately from the study of mechanics until the middle of the nineteenth century. One of the major accomplishments of that century came in the understanding that heat is a form of energy and that units measuring quantities of heat have an equivalence to mechanical energy units. This development is described in the Great Controversies essay in Chap. 11. The laws of thermodynamics, which describe the ability to convert heat into work and work into heat, are the basic principles upon which heat engines and refrigerators operate.

A wide variety of phenomena involve oscillations, or vibrations, in which motion repeats in regular cycles, or periods. Such regular motion, as for example a pendulum, provides us with a way to measure time. The bulk properties of matter determine how such vibrations travel through materials as waves, providing us with an understanding of sound and the operation of musical instruments.

CHAPTER 9

MECHANICAL PROPERTIES OF MATTER

All materials are composed of atoms. The force between atoms is basically electrical in nature, stemming from the fact that atoms are composed of charged particles (electrons and protons). The way these atoms are arranged and in what combination ultimately determines a material's bulk behavior. For most practical purposes, it is these bulk properties of matter, often referred to as **mechanical** properties, instead of the detailed atomic descriptions, that are of interest. In this chapter, we study mechanical properties such as density, elasticity, and fluid pressure and flow.

9.1 STATES OF MATTER

The world around us is composed of three recognizable types of materials: solids, liquids, and gases. We call these types the three *states of matter*. The fundamental difference between them lies in the way that forces act between the atoms or molecules composing the substance. In gases, the interatomic forces are practically nonexistent, allowing individual gas atoms (or molecules) to move independently except during collisions with one another. This freedom of motion also allows the gas to fill any available volume. In liquids and solids these interatomic forces are

Solid

+ heat

Liquid

+ heat

Gas

FIGURE 9.1

Water can exist in three states.

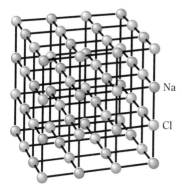

FIGURE 9.2

A small portion of a salt (NaCl) crystal.

Na

Cl

strong enough so that external forces cannot significantly change the volume occupied by a given solid or liquid sample. Thus we say that solids and liquids are nearly *incompressible*. In solids, the interatomic forces arrange the atoms in a rigid three-dimensional array, or lattice structure. As a result, solids not only are incompressible, but also *rigid*, resisting attempts to change their shape. This three-dimensional structure is lacking in liquids, however. As a result, liquids are readily *deformable* in shape, conforming to the shape of any container, and are able to flow in response to applied forces.

The state in which a particular substance exists depends on the temperature of the substance and on the external pressure surrounding it. You are perhaps most familiar with water and its changes of state from solid to liquid to gas (steam) as heat is added (Fig. 9.1).

As simple as this classification seems, there are many cases where the distinction between states becomes blurred. For example, most solids possess an orderly three-dimensional lattice structure. These solids are referred to as *crystalline solids*. Figure 9.2 shows the cubic symmetry of table salt, a crystalline solid. Solids of another type have their atoms arranged randomly, without such long-range order. These solids are called *amorphous* solids, and often they flow very, very slowly, changing their shape over periods of many years. Glass and many plastics are examples. Unlike crystalline solids, amorphous solids do not possess a definite and abrupt melting point; instead they gradually become more and more like a liquid as they are heated, flowing more readily. A similar blurring of the transition between states occurs at high pressures, where the transition between gases and liquids becomes indistinct for many materials.

At high enough temperatures, glass flows as a viscous fluid.

TABLE 9.1
Densities

Material	Density (kg/m^3)
Gases (at 1 atm and 0°C unless otherwise specified)	
Air	1.29
Air (20°C)	1.20
Helium	0.179
Carbon dioxide	1.98
Liquids (at 20°C unless otherwise specified)	
Water (4°C)	1.00×10^3
Water	0.998×10^3
Seawater	1.025×10^3
Alcohol, ethyl	0.79×10^3
Mercury (0°C)	13.6×10^3
Gasoline	0.680×10^3
Solids (at 20°C)	
Aluminum	2.70×10^3
Bone (approximate)	1.8×10^3
Brass	8.7×10^3
Copper	8.89×10^3
Glass (approximate)	2.6×10^3
Gold	19.3×10^3
Granite	2.7×10^3
Ice (0°C)	0.92×10^3
Iron	7.86×10^3
Lead	11.3×10^3
Osmium	22×10^3

TABLE 9.2
The density of water

Temperature (°C)	State	Density (g/cm^3)
0	Solid	0.917
0	Liquid	0.9998
3.98	Liquid	1.0000
10	Liquid	0.9997
25	Liquid	0.9971
100	Liquid	0.9584

9.2 DENSITY AND SPECIFIC GRAVITY

We frequently make use of a property called the **density** of a material, defined as

$$\text{Density} = \frac{\text{mass of substance}}{\text{volume of substance}}$$

Density is represented by the Greek letter rho (ρ). Thus, if an object of volume V has a mass m, its density is

$$\rho = \frac{m}{V} \tag{9.1}$$

The SI unit of density is kilograms per cubic meter, but densities are sometimes given in grams per cubic centimeter. Typical densities are given in Table 9.1.

Because most materials expand with increasing temperature, densities usually decrease as materials are heated. A most notable exception is that of water between 0°C and 4°C. In ice, the H_2O molecules are arranged in a lattice in which the oxygen atoms form tetrahedrons. This arrangement in three dimensions gives rise to a honeycomb of hexagonal empty spaces between the tetrahedrons, and a relatively low density for ice. As ice melts, individual tetrahedrons still persist at 0°C, but can move with respect to their neighbors, filling in some of the empty hexagonal spaces. This results in approximately a 10 percent increase in density upon melting. Above 4°C the increased thermal energy of the molecules causes the average distance of separation to increase as is the case with other substances. Table 9.2 summarizes the peculiar behavior of the density of water around its freezing point.

This property of water has profound consequences for the world around us. It means that in winter ice forms on top of lakes and rivers rather than on the bottom. This in turn allows the ice to melt in the spring, when it is exposed to the warmth of

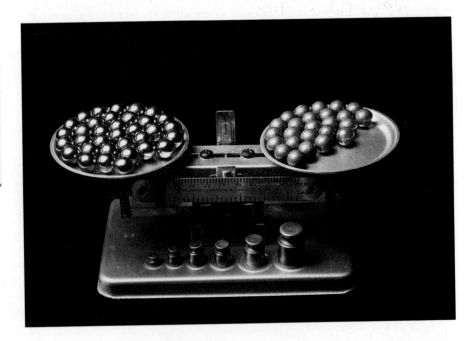

The lead pellets (right) and steel pellets (left) are the same size. Since lead has greater density than steel, it takes fewer of them to weigh the same as the larger number of steel pellets.

the sun and wind. In the process of freezing, the coldest water at the surface of a lake sinks, allowing the warmer bottom water to rise. This "turnover" provides for oxygenation of all levels of the lake twice a year.

A property closely related to density is **specific gravity** (SG), defined as the ratio of a material's density to that of water at 4°C:

$$SG = \frac{\rho}{\rho_{H_2O}} \tag{9.2}$$

Notice that specific gravity is a *dimensionless number*. For example, from Table 9.1, the specific gravities of lead and aluminum are 11.3 and 2.70, respectively.

Illustration 9.1

A cube of uranium ($\rho = 18{,}680 \text{ kg/m}^3$) is 2.00 cm on each side. (**a**) Find its mass. (**b**) How large a cube of ice ($\rho_i = 920 \text{ kg/m}^3$) has the same mass?

Reasoning (**a**) From the definition of density, $\rho = m/V$, we have that

$$m_u = \rho_u V_u = (18{,}680 \text{ kg/m}^3)(8.00 \times 10^{-6} \text{ m}^3) = 0.149 \text{ kg}$$

(**b**) Again from the definition of density,

$$V_i = \frac{m_i}{\rho_i} = \frac{0.149 \text{ kg}}{920 \text{ kg/m}^3} = 162 \times 10^{-6} \text{ m}^3$$

Taking the cube root of this gives the side of the ice cube to be 5.45 cm. ∎

9.3 HOOKE'S LAW; ELASTIC MODULI

Many objects, such as a coiled spring or a metal rod, exhibit a property called **elasticity.** When stretched or compressed by an applied force, they tend to return to their original length when the force is removed. For example, the spring of original length L_o in Fig. 9.3 is stretched an amount ΔL by the applied force **F**. Robert Hooke (1635–1703) investigated such behavior and found that, provided the spring is not stretched too far, it stretches twice as far under twice the force. In general, $\Delta L \propto F$. Hooke described his findings by a rule now known as **Hooke's law:**

When an elastic object is stretched or otherwise distorted, the amount of distortion is linearly proportional to the distorting force.

If a spring is stretched too much, however, beyond what is called its **elastic limit,** it deviates from this direct proportionality between ΔL and F. Furthermore, the spring then does not retract to its original length when the force is removed.

When the spring in Fig. 9.3 is replaced by a solid rod, the rod also obeys Hooke's law. Although the rod's fractional elongation is far smaller than that of the spring, the rod nonetheless stretches in conformity with Hooke's law at small elongations. The behavior observed in a typical experiment is shown in Fig. 9.4. Hooke's law is

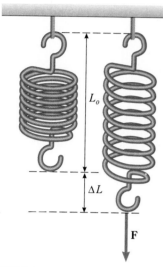

FIGURE 9.3

For a spring that obeys Hooke's law, the deformation ΔL varies in proportion to the applied force **F**.

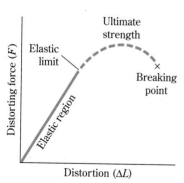

FIGURE 9.4

A typical stress-strain curve. Hooke's law applies in the elastic region only. The maximum force that the object being distorted can sustain is its ultimate strength. Often the material yields (softens) somewhat before breaking.

The Hooke's law behavior of springs makes them excellent exercise devices. The more you stretch them, the harder you have to pull.

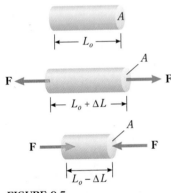

FIGURE 9.5

Tensional and compressional stress applied to a rod. The stress is F/A and the strain is $\Delta L/L_o$.

obeyed in the elastic region only. In the following discussion, we assume that the forces and elongations are small enough that the material is not distorted beyond its elastic limit.

To use Hooke's law in describing the elastic properties of solids, we use the terms *stress* and *strain*. Let us define these two quantities in terms of the stretching (or *tensile*) experiment of Fig. 9.5. A tensile (stretching) force **F** pulls perpendicularly to the end area A of a rod of original length L_o and causes the rod to stretch an amount ΔL. We define the stress due to **F** as

$$\text{Stress} = \frac{\text{force}}{\text{area}} = \frac{F}{A} \tag{9.3}$$

The SI unit of stress is newtons per square meter.

We define the strain in the rod of Fig. 9.5 to be

$$\text{Strain} = \frac{\text{elongation}}{\text{original length}} = \frac{\Delta L}{L_o} = \text{fractional change in length} \tag{9.4}$$

Strain is defined as $\Delta L/L_o$, rather than just ΔL, because any elastic object stretches in proportion to its original length. By dividing ΔL by L_o, we eliminate the effect of the length of the object, an effect that tells us nothing about the inherent properties of the material the object is made of.

Because strain is the ratio of two lengths, it has no units. We see later in this section that there are other types of strain, depending on the geometry of the situation. In the present case, we are speaking about tensile strain. If the rod were being compressed lengthwise, the strain would still be the ratio of length change to original length.

We are now in a position to restate Hooke's law. Stress is a measure of distorting force, and strain is a measure of distortion. Thus Hooke's law becomes

$$\text{Stress} = (\text{constant})(\text{strain}) \tag{9.5}$$

In this form, the law may be applied to many situations other than the stretching of a rod. Hooke's experiments proved its applicability to the stretching, bending, and twisting of numerous springs and other objects. Of course, as was pointed out earlier, Hooke's law applies only in the elastic range of deformations.

The proportionality constant in Eq. 9.5 depends on the material and on the type of deformation involved. It is called the **elastic modulus** of the material. Thus, by definition,

$$\text{Elastic modulus} = \frac{\text{stress}}{\text{strain}} \tag{9.6}$$

Because strain has no units, modulus has the units of stress. Notice that the modulus is large if a large stress produces only a small strain. *Hence modulus is a measure of a material's rigidity.* There are several types of moduli, depending on the exact way in which the material is being stretched, bent, or otherwise distorted. We now discuss the most common of these moduli.

YOUNG'S MODULUS

When stress is applied perpendicularly to an area along one dimension, as in Fig. 9.5, it is called **longitudinal stress.** Longitudinal stress can be either tensile (causing elongation) or compressional (causing shortening) along the one dimension. The elastic modulus which describes the fractional change in length in these situations is called **Young's modulus,** Y:

$$Y = \frac{F/A}{\Delta L/L_o} \tag{9.7}$$

Typical values of Y are given in Table 9.3. Notice that values for the elastic limit and tensile strength are also given. If a stress which exceeds the elastic limit for a material is applied, the material will not return to its original length, but will remain

TABLE 9.3 *Approximate elastic properties*

Material	Young's modulus ($10^9\,N/m^2$)	Shear modulus ($10^9\,N/m^2$)	Bulk modulus ($10^9\,N/m^2$)	Elastic limit ($10^9\,N/m^2$)	Tensile strength ($10^9\,N/m^2$)
Aluminum	70	23	70	0.13	0.14
Brass	90	36	60	0.35	0.45
Copper	110	42	140	0.16	
Glass	55	23	37		
Iron (wrought)	90	70	100	0.17	0.32
Lead (rolled)	16	6	8		0.02
Polystyrene	1.4	0.5	5		0.05
Rubber	0.004	0.001	3		0.03
Steel	200	80	160	0.24	0.48
Tungsten	350	120	20		0.41
Benzene			1.0		
Mercury			28		
Water			2.2		
Air			1×10^{-4}		

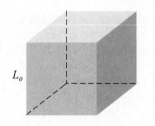

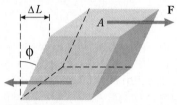

FIGURE 9.6

Here ΔL is exaggerated so that it can be seen. The shear modulus is given by $(F/A)/(\Delta L/L_o) = (F/A)/\tan \phi \cong (F/A)/\phi$.

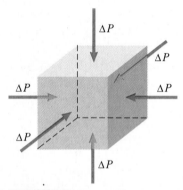

FIGURE 9.7

The cube, of original volume V_o, will contract by an amount ΔV under the action of the increased external pressure ΔP. The arrows show the direction of the force components producing the pressure.

permanently stretched when the stress is removed. The tensile strength is the amount of applied tensile stress that causes the material to break.

SHEAR MODULUS

Suppose we try to distort a cube of material in the manner shown in Fig. 9.6. A force is applied *parallel* to the cube's top face, which has an area A. The effect of this force is to move the top face and the bottom face in opposite directions parallel to each other, which we call *shear*. In this case we define the **shear stress** to be F/A and the **shear strain** to be $\Delta L/L_o$, but pay special attention to how these symbols are defined in the figure. L_o is the thickness of the material, measured along a vertical line in Fig. 9.6. As the shear force is applied, this vertical line becomes distorted through an angle ϕ, called the **angle of shear.** ΔL measures the amount of movement of one end of the line relative to its original position. From Fig. 9.6 we can see that the shear strain $\Delta L/L_o = \tan \phi$. The **shear modulus,** S, is

$$S = \frac{F/A}{\Delta L/L_o} = \frac{F/A}{\tan \phi} \qquad (9.8)$$

When the shearing angle is small (a few degrees or less), we can use the small angle approximation $\tan \phi \approx \phi$, and write

$$S = \frac{F/A}{\phi}$$

Typical values of S are given in Table 9.3. Liquids have values of $S = 0$, since they flow ($\Delta L/L_o = \infty$) under the action of shearing forces.

BULK MODULUS

Suppose a block of volume V_o is subjected to a pressure increase ΔP on all sides (Fig. 9.7). The cube's volume change ΔV is a negative number since the volume shrinks. In this case, the strain is defined to be $-\Delta V/V_o$, and the stress, F/A, is the applied pressure increase ΔP. As with the other types of moduli, bulk modulus B is defined to be the ratio of stress to strain:*

$$\text{Bulk modulus} = B = -\frac{\Delta P}{\Delta V/V_o} \qquad (9.9)$$

BULK COMPRESSIBILITY

The compressibility k of a material is a measure of how easily the material is compressed. In other words, compressibility is just the reciprocal of bulk modulus. Usually, the equation by which compressibility is defined is written

*The negative sign accounts for the fact that ΔV is negative when ΔP is positive.

$$-\frac{\Delta V}{V_o} = k \, \Delta P$$

Compressibility has the same units as reciprocal pressure. Liquids generally have a much higher compressibility than do solids.

Example 9.1

In a large lecture hall, a pendulum is to be made by suspending a 40-kg ball from the end of a steel wire 15 m long. (*a*) What cross-sectional area should the wire have if the applied stress in it is to be only 10 percent of its breaking stress? (*b*) How far will the ball stretch the wire?

R

Reasoning

Question   How do I know the breaking stress of steel?
Answer   The breaking stress of a material is its tensile strength. From Table 9.3, the tensile strength of steel is $0.48 \times 10^9 \, \text{N/m}^2$.

Question   What determines the applied stress on the wire?
Answer   The weight of the 40-kg ball, 390 N. The stress is this force divided by the cross-sectional area of the wire.

Question   What equation will determine the wire's area A?
Answer $\dfrac{F}{A} = (0.10)(0.48 \times 10^9 \, \text{N/m}^2)$

where $F = 390 \, \text{N}$ and the factor 0.10 represents the 10 percent limit stated in the problem.

Question   How is the stretch of the wire related to this applied stress?
Answer   The fractional stretch is related to the stress by Young's modulus ($Y = 200 \times 10^9 \, \text{N/m}^2$ for steel):

$$\frac{\Delta L}{L_o} = \frac{F/A}{Y}$$

Solution and Discussion   (*a*) The cross-sectional area is

$$A = \frac{390 \, \text{N}}{0.48 \times 10^8 \, \text{N/m}^2} = 8.1 \times 10^{-6} \, \text{m}^2$$

Using $A = \pi R^2$, we find that this is a radius of about 1.6 mm.

(*b*) The fractional change in length is

$$\frac{\Delta L}{L_o} = \frac{0.48 \times 10^8 \, \text{N/m}^2}{200 \times 10^9 \, \text{N/m}^2}$$

$$= 2.4 \times 10^{-4}$$

Then

$$\Delta L = (2.4 \times 10^{-4})(15 \text{ m}) = 3.6 \text{ mm}$$

Exercise How large a stress is needed to produce a 0.020 percent elongation in an aluminum wire? *Answer: 1.4 × 10⁷ N/m²*

9.4 PRESSURE IN FLUIDS

Consider the fluid in the container of Fig. 9.8. It is at rest and exerts an outward force on the container walls. The outward force on the area A is $\mathbf{F}_\perp$, where the subscript alerts you that the force is perpendicular to the wall. The average pressure on area A is defined to be

$$\text{Average pressure} = \bar{P} = \frac{F_\perp}{A} \tag{9.10}$$

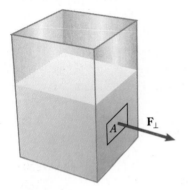

FIGURE 9.8
The average pressure on the area A is $F_\perp/A$.

Even though pressure is a scalar quantity, we must remember that the force producing it does have direction, even though we often omit the subscript on $\mathbf{F}_\perp$. The SI unit for pressure is the same as that previously discussed for stress, N/m². In fact, pressure is an example of compressional stress, as we saw in the previous section. When we describe pressure, however, we usually give a special name to N/m², the **pascal (Pa).** That is,

$$1 \text{ N/m}^2 = 1 \text{ Pa}$$

We shall encounter many other units for measuring pressure, perhaps more than for any other physical quantity. To help avoid confusion, these units are summarized on the inside cover of the text.

To measure the pressure within a fluid, the device shown in Fig. 9.9 could be used. The fluid exerts a force $\mathbf{F}$ on the piston, and the piston moves until the force exerted by the spring balances the force due to the fluid. When the device is suitably calibrated, the displacement of the piston can be used to measure F. If the area of the piston is A, then the pressure is simply F/A. By making the area of the piston very small, we can obtain the pressure very close to any point within the fluid. It is this quantity we refer to when we speak of the pressure at a point in the fluid.

Let us now discuss a number of important facts about pressure and forces within fluids. Strictly speaking, these facts apply to fluids that are *incompressible*. In practical terms, this means that the bulk compressibilities of the fluids are so small that the pressures we consider cause insignificant changes in volumes. Virtually all liquids qualify, whereas gases do not.

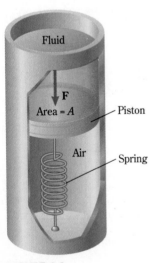

FIGURE 9.9
A simple device for measuring pressure.

1. In a fluid at rest, the forces exerted by the fluid are always perpendicular to surfaces in contact with the fluid, regardless of the orientation of the surfaces.

From Newton's third law, forces exerted by the surface on the fluid must be equal to and opposite those exerted by the fluid on the surface. There can be no component of force parallel to the surface, since fluids cannot remain at rest when they are subjected to shearing forces.

2. In a fluid at rest, the net force on any volume element of the fluid must be zero.

This follows from Newton's second law. If there were a net force on any part of the fluid, it would cause that part to flow.

3. At any point a depth h below the surface of a fluid of density ρ, the pressure due to the weight of the fluid is $\rho g h$.

To see that $P = \rho g h$, consider the fluid of density ρ in the cylinder in Fig. 9.10. The weight of the fluid at the bottom, a depth h below the surface, is

$$\text{Weight} = Mg = \rho V g$$

where M is the mass of fluid in the column and we have used $M = \rho V$. This weight is distributed over the area A of the bottom of the column, producing a pressure P:

$$P = \frac{F}{A} = \frac{\text{weight}}{A} = \frac{\rho V g}{A}$$

The volume of the cylinder is $V = Ah$. Substituting this for V in the above equation, we get the pressure at a depth h below the surface of a fluid due to the weight of the fluid:

$$P = \frac{\rho A h g}{A} = \rho g h \tag{9.11}$$

4. If an external force causes an increase in pressure at any point in a confined incompressible fluid, *every* point in the fluid experiences the same pressure change. This is known as **Pascal's principle.**

For example, a container of fluid open at the top experiences the pressure of the atmosphere, P_a, pressing down at its surface. Pascal's principle states that *every* point in the container experiences this added pressure. At a depth h, the total pressure in the fluid is then

$$P = P_a + \rho g h$$

Usually when we use a gauge to measure pressure in a container, we are doing so with P_a surrounding us and the gauge. What the gauge measures is the *difference* between the pressure in the container and P_a. Thus the difference between total pressure in a container and the surrounding pressure, P_a, is called **gauge pressure, P_G:**

$$P_G = P - P_a \tag{9.12}$$

The gauge pressure at a depth h in a fluid open to the atmosphere is thus

$$P_G = P - P_a = \rho g h$$

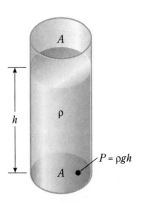

FIGURE 9.10

The pressure due to a column of fluid. $P = \rho g h$ at a depth h below the surface.

FIGURE 9.11

A liquid at equilibrium in open connected containers has its open surfaces at the same level.

Pascal's principle forms the basis for the operation of hydraulic lifts and presses, as well as hydraulic braking systems. We will look at some examples presently.

5. The pressure in a volume of fluid at rest is the same at all points at the same depth.

This is a consequence of statement 3, since we did not have to specify any particular horizontal position in the fluid to derive $P_G = \rho gh$. A corollary to this statement is that a liquid at rest in a series of open connected containers has all its open surfaces at the same level (Fig. 9.11).

With these five statements, we can proceed to examine a number of applications.

Illustration 9.2

The apparatus shown in Fig. 9.12 is one version of a hydraulic press. When a force F_1 is applied to the piston of area A_1, how large a force F_2 on the other piston (of area A_2) is needed to balance F_1?

Reasoning   The pressure F_1 causes on A_1 is $P = F_1/A_1$. By Pascal's principle this pressure is exerted at all points throughout the fluid, including the surface A_2. So at the large piston, we can write $P = F_2/A_2$, and thus derive the equation

$$\frac{F_2}{A_2} = \frac{F_1}{A_1}$$

Solving for F_2, we get

$$F_2 = F_1 \frac{A_2}{A_1} \tag{9.13}$$

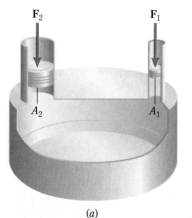

(a)

FIGURE 9.12

The principle of a hydraulic lift. (*a*) A small force on the small piston can raise a large weight on the large piston. (*b*) Pressure is created in the hydraulic fluid by a pump (not visible). This pressure is transmitted through the hydraulic lines to the working pistons. The large hydraulic pistons multiply the force created by the pump, enabling the backhoe's claw to exert very large forces.

(b)

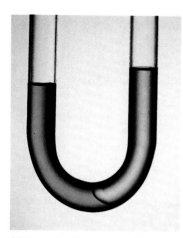

Oil (orange) and water (purple) establish an equilibrium in a U tube. Since the oil is less dense, it takes a higher column to exert as much pressure at the interface as a shorter column of water does.

The applied force is thus multiplied by the ratio of the two areas. A hydraulic press is an example of a lever, a device that allows us to lift very large weights by the application of modest forces.

You should not interpret the multiplying of force in a hydraulic device as somehow implying a multiplication of *work* done by the device. This would violate the principle of conservation of energy. To see that $W_{in} = W_{out}$ (neglecting frictional forces), we begin by using the definition of work:

$$W_{in} = F_1 h_1 \qquad \text{and} \qquad W_{out} = F_2 h_2$$

where h_1 and h_2 are the distances the two pistons travel. Let us form the ratio of the two amounts of work:

$$\frac{W_{in}}{W_{out}} = \frac{F_1/F_2}{h_2/h_1} = \frac{A_1}{A_2}\frac{h_1}{h_2} \qquad (9.14)$$

We have used Eq. 9.13 for the ratio of forces. Now consider what *incompressibility* means. A volume element of the fluid cannot be changed in size. Whatever volume is displaced in one piston must be moved to the other piston. In piston 1, the displaced volume is $V_1 = A_1 h_1$; in piston 2 it is $V_2 = A_2 h_2$. Incompressibility requires that

$$A_2 h_2 = A_1 h_1 \qquad \text{or} \qquad \frac{A_2}{A_1} = \frac{h_1}{h_2}$$

Putting this condition into Eq. 9.14, we see that

$$\frac{W_{in}}{W_{out}} = 1$$

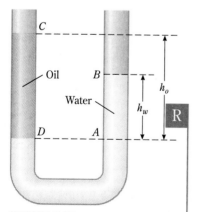

FIGURE 9.13

Because the water column *BA* balances the oil column *CD*, the density of the oil can be determined.

Example 9.2

Water and oil are placed in the two arms of the glass U tube of Fig. 9.13. If they come to rest as shown, what is the density of the oil?

Reasoning

Question What condition determines the equilibrium of the two liquids?
Answer The crucial point is the oil-water interface, marked *D* in Fig. 9.13. The oil is exerting a downward force on that interface that must be balanced by the upward force due to the water.

Question Does this mean that the *pressures* exerted by each liquid on the other are equal at the interface?
Answer Fundamentally, equilibrium means a balance of forces. Since the two fluids share a common area, and since $P = F/A$, the pressures are also equal.

Question What effect does the atmosphere have?
Answer Both ends of the tube are open, and so P_a is exerted in both arms and hence cancels out. The condition for equilibrium in this case is equal *gauge* pressures at *D*.

Question   What is the pressure at D due to the oil?
Answer   The gauge pressure is $P_{\text{oil}} = \rho_{\text{oil}} g h_o$.

Question   What is the pressure at D due to the water?
Answer   Since D is at the same level as A, the pressure of the water must be the same at D as at A. This is (gauge pressure) $P_w = \rho_w g h_w$.

Solution and Discussion   Equating the two gauge pressures, we get

$$\rho_{\text{oil}} = \rho_w \frac{h_w}{h_o}$$

Because we know that $\rho_w = 1.00 \times 10^3 \, \text{kg/m}^3$, the measurement of the two levels allows us to find ρ_{oil}.

Exercise   If the oil has a density of 800 kg/m³, how long a column of water is needed to balance an oil column 30 cm long?   *Answer: 24 cm*

9.5   PRESSURE IN GASES; THE ATMOSPHERE

The earth's atmosphere, which we have previously referred to without elaboration, is the most important example of gas pressure for humans. We live at the bottom of a sea of air which exerts a considerable pressure on us and everything in our environment.

There is a difference between the pressure of the atmosphere and that exerted by the columns of fluids we have been discussing. The air, being a gas, is compressible and therefore does not have a constant density throughout all levels of the atmosphere. Also, the atmosphere does not have a well-defined top surface. It simply thins out with increasing height. At 20 km above the earth's surface the density of the atmosphere has decreased to about 8 percent of its sea-level value. Thus we cannot apply a simple expression such as $\rho g h$ to calculate P_a accurately. We can, however, make an *approximation* of P_a by this means and compare it with the measured value.

Let us take the atmosphere to be a fluid 20 km deep with a constant density equal to half the sea-level density of air, $\frac{1}{2}(1.29 \, \text{kg/m}^3)$. Then our estimate of the pressure of the atmosphere is

$$P_{\text{est}} = \tfrac{1}{2}(1.29 \, \text{kg/m}^3)(9.8 \, \text{m/s}^2)(20 \times 10^3 \, \text{m}) = 1.3 \times 10^5 \, \text{Pa}$$

The measured value of P_a at 0°C and at sea level, called 1 standard atmosphere, or 1 atm, is

$$P_a \text{ (standard)} = 1.01325 \times 10^5 \, \text{Pa} = 1 \, \text{atm}$$

Remember that the actual P_a varies with temperature and altitude as well as with variations in weather. Thus the pressure of the atmosphere where you are at any given time will usually not be precisely this standard value.

To measure atmospheric pressure, we use a *barometer*. There are many different devices used for this purpose, but the mercury barometer is one of the most funda-

Physicists at Work PATRICK HAMILL *San Jose State University*

I started college as an English major; somewhere in the back of my mind was a vague idea of writing the Great American Novel or at least living in a garret in Paris and leading the bohemian life. Well, my advisor told me that even English majors had to take a science course. He suggested Physics 12, a course commonly known as "Tinker-Toy Twelve." Since I was reasonably good at math, I suggested that I might take a more challenging course. With an evil grin, he said, "Sure," and signed me into the physics course for physics and engineering majors.

I can't say why, but I really enjoyed the course. One of the things about physics that especially intrigued me was the idea that a physical system, like a ball rolling down an incline, could be described by mathematical equations. That is called "making a model" of a physical system. Nowadays, making a model usually involves writing a complicated program and running it on a supercomputer, rather than using simple math to solve a few equations, but the idea is the same. I am still in the business of making models. The models I make are for NASA and involve analyzing the ozone hole. I also teach physics at San Jose State University, where I frequently

teach the freshman physics class that got me started. It's one of my favorite classes.

As you probably know, in the upper atmosphere at altitudes from 20 to 50 km there is a layer of ozone-rich air that blankets the earth like an invisible layer of clouds. When you look up at the sky on a cloudy day, you sometimes see a hole in the cloud layer where there is clear sky. When scientists in Antarctica looked up at the sky and saw there was no ozone overhead, they called it an "ozone hole," in analogy with a hole in a cloud layer.

The discovery of the ozone hole is an interesting story. For years the British government supported a small group of scientists who camped out on the bleak Palmer Peninsula of Antarctica and measured the amount of ozone in the atmosphere. Starting about 1975 they noticed that the ozone over Antarctica was behaving in a very strange way. Each October the amount of ozone was less than in the previous October! This trend has essentially continued to the present. Now during October at certain altitudes there is no ozone at all over Antarctica.

The ozone hole was a mystery that required physicists, meteorologists, chemists, as well as a few engineers, to figure out. It was not the fictional mystery of a cheap whodunit, but rather a real-life mystery that *had* to be solved. In fact many people believe that understanding the ozone hole is the most important social/scientific problem facing our industrial society.

Ozone is a molecule made up of three oxygen atoms. It is denoted by O_3. In the atmosphere, oxygen is usually found as molecular oxygen, O_2. At very high altitudes, ultraviolet light from the sun is absorbed by O_2 and breaks the bond between the two oxygen atoms. Some of the single oxygen atoms combine with oxygen molecules to form ozone. Ozone is very important

because it absorbs ultraviolet light. It is actually doubly important because ultraviolet light is absorbed in both the production and the destruction of ozone. There is a delicate balance between ozone production and destruction and, as a result, ultraviolet light levels at the surface of the earth are quite tolerable. Ultraviolet light causes suntans, and sometimes sunburns, and occasionally skin cancer. If it were not for ozone, the surface of the earth would be bathed in ultraviolet light and all living organisms would perish. Obviously, any drastic change in the ozone layer has to be considered a serious threat to humanity.

The basic questions were: why is the ozone disappearing, why in Antarctica, and why only in October? The first question was soon answered. The ozone was disappearing because people were releasing chlorofluorocarbons (CFCs for short) into the atmosphere. CFCs are very useful compounds. Some are used as refrigerants, others are used in making foam plates and cups, and yet others are used in industrial processes such as manufacturing computer chips. CFCs are chemically inert until they drift up to very high altitudes, where ultraviolet light causes them to break apart and release chlorine. It turns out that chlorine is deadly to ozone. One chlorine atom can destroy nearly a million ozone molecules!

So CFCs are the villain in our mystery story. But why Antarctica? And why only in October? Well, that's where my research entered the picture. For some years I had been working with NASA scientists, studying satellite data. We noticed an interesting phenomenon. Every winter, high over Antarctica a tenuous mist or cloud appeared. (Remember that winter in Antarctica is June, July and August.) As you might suspect, at twenty kilometers above Antarctica in the winter it gets pretty cold, as low as minus ninety degrees Celsius. It is the coldest region in the atmosphere. As my friend Brian Toon of NASA pointed out, it is cold enough for nitric acid to freeze out of the atmosphere and form clouds. Clouds made of frozen nitric acid? The idea was interesting enough for NASA to send an ER-2 research airplane loaded with instruments to the tip of South America and fly it over Antarctica. The nitric acid clouds were there!

But what do nitric acid clouds have to do with ozone disappearing in October? The answer is that the clouds, which form only in the Antarctic win-

ter absorb nitric acid, thus changing the composition of the air around them. They then act like miniature chemical factories and convert inert chlorine substances to active species that destroy ozone. The nitric acid droplets eventually fall to lower altitudes, removing nitrogen compounds and leaving the atmosphere in a state where ozone depletion can occur. When the long Antarctic night is over, the sun shines on this "processed" air and ozone depletion begins. By October, there is practically no ozone left in the region where the polar stratospheric clouds were formed.

Solving the mystery of how the

ozone hole forms does not mean the problem is solved. Governments, industry, and private citizens have to cooperate to make sure we keep the protective ozone shield in place. Nevertheless, solving the mystery was the crucial first step that had to be taken.

My research is very exciting. I think it is great to do something that is not only fun but also meaningful. I guess the advisor who signed me up for the "hard" physics class really did me a terrific favor.

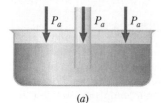

(a)

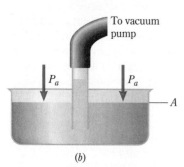

To vacuum pump

A

(b)

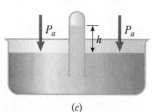

(c)

FIGURE 9.14

When the tube is evacuated, the mercury rises until $\rho gh = P_a$. Hence the device, a barometer, is capable of measuring atmospheric pressure.

mental. We can understand its operation by referring to Fig. 9.14. In part *a*, an open tube is partly immersed in a beaker of mercury. Because the air pressure inside the tube is equal to the air pressure outside, the mercury stands at the same level inside and outside.

Now suppose that a vacuum pump evacuates the air from the tube, as in Fig. 9.14*b*, and the tube is sealed off, as in *c*. Once all the air is removed, the pressure on the mercury surface inside the tube is zero. (Recall that the pressure exerted by a gas on a surface is the result of collisions of the gas molecules with the surface. If no air molecules are present, we have a perfect vacuum and the pressure is zero.) Now the pressure at level *A* inside the tube is due only to the height *h* of mercury in the tube and is ρgh, where ρ is the density of mercury. Note that the pressure on the mercury surface at level *A* outside the tube is still atmospheric pressure P_a. Moreover, statement 5 in Sec. 9.4 tells us that the pressure inside the tube at level *A* is the same as it is outside the tube. Therefore

Pressure at *A* outside = pressure at *A* inside

$$P_a = \rho gh \tag{9.15}$$

Atmospheric pressure can support a column of mercury whose height is given by Eq. 9.15. To find the column height of any other fluid that can be supported by the atmosphere, we need only use the density of that fluid in Eq. 9.15.

The column height of mercury corresponding to a standard atmosphere is (to three significant digits)

$$h = \frac{1.01325 \times 10^5 \,\text{Pa}}{(13.6 \times 10^3 \,\text{kg/m}^3)(9.80 \,\text{m/s}^2)}$$

$$= 0.760 \,\text{m} = 760 \,\text{mm}$$

This is equal to 29.9 in. You probably have heard weather reports refer to barometric pressures of about 30 in or 760 mm.

There are two common units of pressure worth mentioning at this point. One, called the **torr,** is named after the inventor of the barometer, the Italian physicist

When air is pumped out of a sealed metal can, the pressure from the outside atmosphere causes the can to collapse.

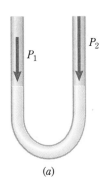

(a)

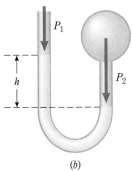

(b)

FIGURE 9.15

In a manometer, the pressure difference $P_2 - P_1$ is measured by the height difference h.

Evangelista Torricelli (1608–1647). The other, the **bar,** is commonly used in the science of meteorology. These units have the values:

1 torr = 1 mmHg = (1/760) atm
1 bar = 10^5 Pa (exactly)

One bar is approximately equal to typical atmospheric pressure. Changes in atmospheric pressure due to weather systems are typically measured in millibars.

Commercial mercury barometers are more refined than the simple device shown in Fig. 9.15. They have an accurate scale beside the mercury column and special devices to adjust the level of the mercury in the beaker. There are other types of barometers based on different principles, but for accurate work the mercury barometer is preferred. However, it must be at least 76 cm long (why?), and so there is often good reason to replace it by a smaller, but less accurate, device.

Another device often used to measure gas pressures is a **manometer** (Fig. 9.15). Although it has many variations, a manometer is basically a U-shaped tube partly filled with a liquid, often mercury. If the mercury in a manometer stands at equal levels in the two arms of the tube, as shown in *a*, we know that the two gas pressures P_1 and P_2 above the columns must be equal. However, if P_2 is larger than P_1, the columns will adjust as in *b*.

The difference in the heights h, measured in millimeters, gives us the pressure difference $P_2 - P_1$ directly in torr as long as the fluid is mercury. If column 1 is open to the atmosphere, this measurement is the gauge pressure in column 2. If P_2 is less than P_1, h as defined in Fig. 9.15*b* will be negative. Thus a negative gauge pressure indicates a pressure in a container which is less than the ambient atmospheric pressure.

For small pressure differences, precision can be gained by using a fluid less dense than mercury. The difference in heights is increased by a factor of 13,600/ρ, where ρ is the SI density of the fluid used instead of mercury. Notice that if water were used in a barometer to measure P_a, the water column would be about (76 cm)(13,600/1000) = 1034 cm high, nearly the height of a three-story building.

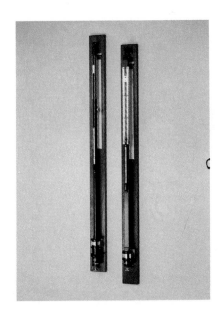

A mercury barometer.

Example 9.3

A simple test to determine the capabilities of a person's lungs is to have the person blow with full force into one end of a manometer, as shown in Fig. 9.16. Suppose that in this case a water manometer is used and the difference in the water levels is 80.0 cm, as shown. What is the pressure in the lungs?

R

Reasoning

Question   What does the 80.0-cm level difference tell us?
Answer   It allows us to calculate the gauge pressure, $P_G = \rho g h$, where ρ is the density of water and $h = 80.0$ cm.

Question   What is the total pressure in the lungs?
Answer $P_{tot} = P_G + P_{atm}$. We need to know the surrounding atmospheric pressure to calculate P_{tot}. If we assume 1 atm, we have

$$P_{tot} = P_G + 1.01 \times 10^5 \, \text{Pa}$$

Solution and Discussion   You should understand that P_{atm} is not always going to be 1 atm. A barometric reading in the room where the experiment is being done is needed to determine the actual value of P_{atm} in any given case.
The gauge pressure is

$$P_G = (1.00 \times 10^3 \, \text{kg/m}^3)(9.80 \, \text{m/s}^2)(0.800 \, \text{m}) = 7.84 \times 10^3 \, \text{Pa}$$

Total pressure is

$$P_{tot} = (101 + 7.84) \times 10^3 \, \text{Pa} = 109 \times 10^3 \, \text{Pa}$$

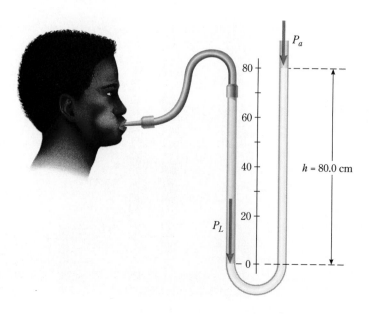

FIGURE 9.16

By blowing into the manometer, the person is able to support a column of fluid 80.0 cm high. How large is P_L?

Example 9.4

Suppose a solid steel anchor sank to the bottom of one of the deepest ocean trenches, 6.90 mi below the surface. Calculate the change in the density of the steel anchor produced by the pressure of the water.

R

Reasoning

Question Why will the density be affected by sinking?
Answer Density = mass/volume. The mass of the anchor remains the same, but the pressure of the water will decrease its volume.

Question What relates the change of volume to the pressure applied?
Answer The bulk modulus of the steel: $\Delta V/V_o = -\Delta P/B$

Question What is ΔP in this case?
Answer It represents the difference in pressure between the atmospheric pressure on the anchor at sea level and the total pressure on the anchor at the ocean bottom. In other words, ΔP is the gauge pressure $\rho g h$ produced at a depth h of 6.9 mi of seawater.

Question Once $\Delta V/V_o$ is found, how can I relate it to $\Delta \rho$?
Answer If you assume the mass of the anchor to be m, the original density can be written $\rho_o = m/V_o$. The underwater density is $\rho = m/V$, where $\Delta V = V - V_o$.

Solution and Discussion
The gauge pressure of 6.90 mi of seawater is

$$P_G = (1.025 \times 10^3 \text{ kg/m}^3)(9.80 \text{ m/s}^2)(6.90 \text{ mi})(1610 \text{ m/mi})$$
$$= 1.12 \times 10^8 \text{ Pa} = 1100 \text{ atm}$$

The bulk modulus of steel is $16 \times 10^{10} \text{ N/m}^2$. The gauge pressure produces a volume change given by

$$\frac{\Delta V}{V_o} = \frac{-\Delta P}{B} = \frac{-(1.12 \times 10^8 \text{ Pa})}{16 \times 10^{10} \text{ N/m}^2}$$
$$= -7.00 \times 10^{-4}$$

Notice that Pa and N/m² cancel. The new volume is then

$$V = (1.0000 - 0.0007)V_o = 0.9993 V_o$$

The submerged density is

$$\rho = \frac{m}{V} = \frac{m}{0.9993 V_o} = \frac{\rho_o}{0.9993} = 1.0007 \rho_o$$

This enormous pressure has increased the density of steel by only seven-hundredths of one percent!

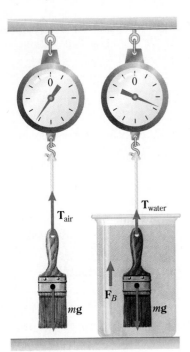

FIGURE 9.17

The water exerts an upward buoyant force F_B on the brush. The scale reads T_{air} when the brush is in air and T_{water} when it is in water.

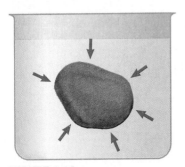

FIGURE 9.18

What does Archimedes' principle tell us about the buoyant force on the object?

The salt water in the Dead Sea has a greater density than fresh water. As a result, a swimmer floats with a smaller fraction of her body submerged.

9.6   ARCHIMEDES' PRINCIPLE; BUOYANCY

The experiment shown in Fig. 9.17 is probably not new to you. It demonstrates the well-known fact that objects appear to weigh less when they are submerged in a fluid. If you have ever tried to support a person while you were swimming, you know full well that the supporting force required is far less than the weight of the person. Similarly, in Fig. 9.17, the supporting force T is decreased when the brush is placed in water. Apparently, the water exerts an upward force F_B on the brush. We call this the **buoyant force.**

The law of fluids that describes the buoyant force is called **Archimedes' principle.** To arrive at it, consider the object in Fig. 9.18. It is buoyed up by the fluid around it. Apparently, the net effect of the fluid forces acting on the object is to give it an upward force F_B. Basically F_B is a result of the fact that pressure increases with depth, so that the upward force on the bottom of the object is larger than the downward force on its top.

To see how large the buoyant force is, notice what would happen if the object were made of the same material as the fluid. We would not be able to distinguish the object from the fluid, and so it would simply remain at rest with no outside supporting force required. This means that F_B is just large enough to support the object in this case. That is, $F_B = mg$, where mg is the weight of the fluid object.

Of course, the buoyant force due to the fluid cannot depend on what the object is made of. Hence F_B is always the same, equal to the weight of the volume of fluid displaced by the object. We are therefore led to the statement of Archimedes' principle:

A body partially or wholly immersed in a fluid is buoyed up by a force equal to the weight of the fluid that it displaces.

You should go back through this reasoning to assure yourself that we made no use of the fact that the object in Fig. 9.18 is *wholly* immersed.

Example 9.5

Suppose the brush in Fig. 9.17 has a mass M and a density ρ. Find its apparent weight (the reading W_{app} of the scale on the right) when it is submerged in a fluid of density ρ_f.

R *Reasoning*

Question   What does the scale reading measure?
Answer   It measures the *net* downward force on the brush. This is the difference between the force of gravity downward and the buoyant force F_B upward:

$$W_{app} = M_b g - F_B$$

The subscript b refers to properties of the brush.

Question   What does F_B depend on?
Answer   The brush is totally submerged, and so F_B is equal to the weight of fluid displaced by its *entire* volume.

Question   What is the brush's volume?
Answer   From the definition of density, $V_b = M_b/\rho_b$. This is also the volume of the displaced fluid.

Question   What is the weight of this volume of fluid?
Answer $W_f = M_f g = \rho_f V_f g = \rho_f V_b g = F_B$

Solution and Discussion   Putting all these parts together along with $M_b g = \rho_b V_b g$, we get

$$W_{app} = \rho_b V_b g - \rho_f V_f g = (\rho_b - \rho_f) V_b g$$

Notice the following:

1 If $\rho_b > \rho_f$, the net force is downward and if the brush is released, it will sink.
2 If $\rho_b < \rho_f$, the net force is upward and the brush will rise through the fluid if released.
3 If $\rho_b = \rho_f$, the submerged brush has *neutral* buoyancy, and will neither sink nor rise.

Example 9.6

A queen's gold crown has a mass of 1.30 kg. However, when it is weighed while it is completely immersed in water, its apparent mass is 1.14 kg. Is the crown solid gold?

R *Reasoning*

Question   What is the key to knowing whether the crown is solid gold?
Answer   Being made of solid gold would mean the crown would have the density of gold. If it was made of a mixture of materials or of some other uniform material, or was hollow, its density would be different from that of pure gold.

Question   How can we calculate density without measuring the crown's volume?

Answer   By applying Archimedes' principle to the data supplied. This was done in Example 9.5. You can algebraically rearrange the result of that example into the form

$$W_{app} = W_c\left(1 - \frac{\rho_f}{\rho_c}\right)$$

where ρ_c is the density of the crown and W_c is the weight of the crown in air.

Question   What is the weight of the crown in air?

Answer $W_c = Mg = (1.30\,\text{kg})(9.80\,\text{m/s}^2) = 12.7\,\text{N}$

Solution and Discussion   The above equation can be solved for ρ_c:

$$\rho_c = \frac{\rho_f W_c}{W_c - W_{app}}$$

Putting in the weight values and the density of water,

$$\rho_c = \frac{(1.00 \times 10^3\,\text{kg/m}^3)(12.7\,\text{N})}{12.7\,\text{N} - (1.14\,\text{kg})(9.80\,\text{m/s}^2)} = 8.31 \times 10^3\,\text{kg/m}^3$$

The density of gold is much larger, $19.3 \times 10^3\,\text{kg/m}^3$. The crown certainly is not solid gold.

Example 9.7

Ice has a density of $0.92 \times 10^3\,\text{kg/m}^3$ and therefore floats in water. What fraction of a floating piece of ice is submerged?

R   *Reasoning*

Question   What physical condition describes floating?

Answer   A floating object must experience a buoyant force equal to its weight in order to remain in equilibrium at the fluid's surface.

Question   What is the equation for this condition?

Answer $F_B = Mg$, where F_B is the weight of water displaced and M is the mass of the floating object.

Question   How much water is displaced?

Answer   A volume equal to the submerged (not total) volume of the ice. Call this V_s.

Solution and Discussion   When we substitute $\rho_w V_s g$ for F_B and $\rho_{ice} V_{ice}$ for M_{ice}, the equation for floating becomes

$$\rho_w V_s g = \rho_{ice} V_{ice} g$$

Concrete is denser than water, yet these concrete bridges float and support the weight of many cars. Can you explain how this can be?

Thus the fraction of volume that is submerged is

$$\frac{V_s}{V_{ice}} = \frac{\rho_{ice}}{\rho_w} = \frac{0.92}{1.00} = 92\%$$

Truly one sees only the tip of the iceberg!

9.7 VISCOSITY AND FLUID FLOW

Molasses and honey are examples of what we call very *viscous* fluids. They flow very slowly when poured. Water and alcohol, which are much less viscous, flow quite freely. The degree to which a fluid resists flowing is a property called **viscosity.** To give quantitative meaning to viscosity, we refer to the shearing experiment of Fig. 9.19. Two parallel plates, each of area A, are separated by a distance L, and the region between the plates is filled with a fluid whose viscosity we shall denote by η (Greek eta). The force F causes the top plate to move past the lower one with a speed v. We characterize the rapidity of the shearing motion by the **shear rate** of the two plates and the fluid between them:

$$\text{Shear rate} = \frac{\text{speed of top plate past bottom plate}}{\text{distance between plates}} = \frac{v}{L}$$

Thus a shear stress F/A applied to the upper plate causes a shear rate v/L in the fluid.

We define the viscosity η of the fluid as the ratio of shear stress to shear rate:

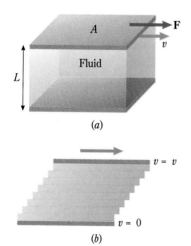

FIGURE 9.19

As the upper plate moves, layers of the fluid slide over one another. Viscous energy losses arise because of the friction forces that retard the motion of these layers.

TABLE 9.4

Viscosities of liquids and gases at 30°C

Material	Viscosity (mPl)*
Air	0.019
Acetone	0.295
Methanol	0.510
Benzene	0.564
Water	0.801
Ethanol	1.00
Blood plasma	~1.6
SAE No. 10 oil	200
Glycerin	629
Glucose	6.6×10^{13}

*$1 \text{ mPl} = 10^{-3} \text{ Pa} \cdot \text{s} = 1 \text{ cP}$.

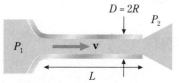

FIGURE 9.20

The flow rate through the tube is given by Poiseuille's law. Here **v** is shown for $P_1 > P_2$.

$$\eta = \text{viscosity} = \frac{\text{shear stress}}{\text{shear rate}} \qquad (9.16a)$$

As you see, a highly viscous fluid is one that requires a large shear stress to cause it to flow with a given shear rate.

In terms of the experiment of Fig. 9.19, we have stress $= F/A$ and rate $= v/L$. Using these measured quantities, we can compute the viscosity of the fluid:

$$\eta = \frac{\text{shear stress}}{\text{shear rate}} = \frac{F/A}{v/L} \qquad (9.16b)$$

We see from its defining equation that the SI unit for viscosity is the pascal · second (Pa · s). This unit is given the special name **poiseuille** (Pl). Other common units for viscosity are the poise (P), where 1 P = 0.1 Pl, and the centipoise (cP). This latter unit is easily remembered because it is equal to the millipoiseuille: 1 cP = 1 mPl. Typical viscosities are given in Table 9.4.

We can gain further insight into the meaning of viscosity by examining Fig. 9.19b. Notice that the fluid layers next to the two plates remain attached to the plates. We can think of the fluid between the plates as consisting of many thin layers, many more than are shown. As the upper plate moves, these layers must slide over one another. In a high-viscosity fluid, the layers do not slide easily, and so a great amount of work must be done to shear the fluid.

The flow of water and similar fluids through a pipe or tube is of particular interest, as we shall see. The volume Q of fluid flowing through a pipe each second is called the *flow rate* in the pipe. For example, if 50 cm³ of water flows out of the pipe in Fig. 9.20 each second, $Q = 50 \text{ cm}^3/\text{s}$.

In Fig. 9.20, the pressures in the fluid at the two ends of the pipe are P_1 and P_2. We call $P_1 - P_2$ the *pressure differential,* and, as we might guess, the flow rate through the pipe is proportional to it for simple fluids. You would also expect the flow rate to be large for a pipe with a large radius R and small length L. The equation for flow rate in a situation such as this was found by the French physician Jean Louis Marie Poiseuille (1799–1869). For flow rates that are not too large,

$$Q = \left(\frac{\pi R^4}{8\eta L}\right)(P_1 - P_2) \qquad (9.17)$$

This is called **Poiseuille's law.** Notice that Q increases as the fourth power of R, the pipe's radius.

Illustration 9.3

Older people often develop blood-circulation difficulties because of deposits building up in their arteries. By what factor is the blood flow in an artery reduced if the artery's radius is cut in half? Assume the same pressure differential in the two cases.

Reasoning Poiseuille's law tells us that the volume of blood Q flowing through the artery each second is related to R by

$$Q \propto R^4$$

In the original artery, $Q_o = $ (constant)(R_o^4), but in the constricted artery, $Q = $ (constant)$(R_o/2)^4$. Taking the ratio Q/Q_o, we find $Q/Q_o = 1/16$. The flow rate is reduced by a factor of 16. It is clear from this strong dependence of Q on R why blood-circulation difficulties result from arterial deposits.

Exercise   Find the flow rate of water through a 20-cm-long capillary tube that has a diameter of 0.15 cm if the pressure difference along this length of the tube is 4.0×10^3 Pa. Use 0.80 mPl for the viscosity of water.   *Answer: 3.1 cm³/s*

9.8 BERNOULLI'S EQUATION

As we have seen, all liquids have a characteristic viscosity. If the viscosity is large a great deal of work is needed to push the liquid through a pipe. As the layers of the liquid flow past each other, viscous friction forces between them cause the liquid to heat up. Many liquids, however, have such small viscosity that friction energy losses are negligible, at least for certain purposes. When this is the case, an important relationship called *Bernoulli's equation* can be found for the pressure in a moving fluid. It was first published by Daniel Bernoulli in 1738.

Consider the pipe system shown in Fig. 9.21. It is completely filled with an incompressible fluid between the two frictionless pistons. We shall say that piston 1 is being pushed to the right with constant speed v_1 and that piston 2 is moving to the right with speed v_2. The force F_1 on piston 1 is balanced by the force P_1A_1 that results from the pressure of the fluid, where A_1 is the area of piston 1. (The forces on the piston must balance, or else it would be accelerating, and we have already specified that it is moving with constant speed.) Similarly, at piston 2, $F_2 = P_2A_2$. In a time t, piston 1 moves a distance v_1t, thereby displacing a volume of fluid $(v_1t)(A_1)$. Since the fluid is incompressible, however, piston 2 must make way for an equal volume of fluid. Hence $(v_1t)(A_1) = (v_2t)(A_2)$, or

$$v_1A_1 = v_2A_2 \qquad (9.18)$$

Bernoulli asked what happens as a result of the work done by piston 1. This work is just $F_1(v_1t)$, and since $F_1 = P_1A_1$, we have

Input work $= P_1A_1v_1t$

Because piston 2 does an amount of work $F_2(v_2t)$, some of the input work is used there.

In addition, the fluid pressed to the right by piston 1 is essentially transferred to

FIGURE 9.21

The work done by $F_1(=P_1A_1)$ must equal the work done against $F_2(=P_2A_2)$ plus the sum of the changes in the gravitational potential energy and kinetic energy of the fluid.

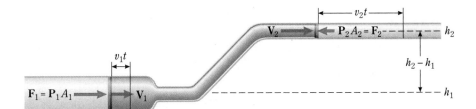

the upper tube. As a consequence, that fluid (with mass M and volume A_1v_1t) is given some gravitational potential energy. Moreover, since it is now traveling with a different speed v_2, its kinetic energy is changed. Of course, some energy is transformed to thermal energy by viscosity-related friction heating, but we are assuming this to be negligible. We therefore have the following equation that tells us what happened to the input work:

Input work = output work + change in GPE + change in KE

or, using the symbols of Fig. 9.21,

$$P_1A_1v_1t = P_2A_2v_2t + Mg(h_2 - h_1) + \tfrac{1}{2}Mv_2^2 - \tfrac{1}{2}Mv_1^2$$

where the volume of the fluid involved, A_1v_1t, has a mass M. From the definition of density,

$$M = \rho A_1v_1t = \rho A_2v_2t$$

Substituting this in the above equation gives, after rearrangement,

$$P_1 + \tfrac{1}{2}\rho v_1^2 + \rho gh_1 = P_2 + \tfrac{1}{2}\rho v_2^2 + \rho gh_2 \qquad (9.19)$$

This is **Bernoulli's equation.** Because the pistons need not really be present, points 1 and 2 can be any two points in the fluid. Notice, however, that the equation is applicable only if friction forces can be neglected.

Illustration 9.4   Torricelli's theorem

A simple application of Bernoulli's principle is shown in Fig. 9.22. A large tank has a small spigot on it at a height h_2 from the bottom of the tank. The tank is filled with fluid up to a height h_1. The top of the fluid is open to the atmosphere. Find the speed with which the fluid flows from the spigot.

Reasoning   The two places in the fluid where we want to apply Bernoulli's principle are (1) at the top of the fluid and (2) at the spigot. Since the spigot is so small, the efflux speed v_2 is much larger than the speed v_1 with which the top surface of the water drops. We can therefore approximate v_1 as zero. Bernoulli's equation can then be written

$$P_1 + \rho gh_1 = P_2 + \tfrac{1}{2}\rho v_2^2 + \rho gh_2$$

Because P_1 and P_2 are both nearly equal to atmospheric pressure, we can consider them equal.
    Then

$$\rho gh_1 = \tfrac{1}{2}\rho v_2^2 + \rho gh_2$$

from which we get

$$v_2 = \sqrt{2g(h_1 - h_2)} \qquad (9.20)$$

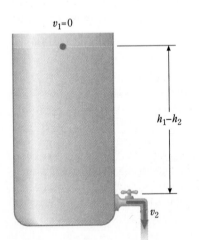

FIGURE 9.22

Torricelli's theorem tells us how fast the liquid is moving as it flows out of the spigot.

This is **Torricelli's theorem.** Notice that the efflux speed is the same that an object freely falling through a height $h_1 - h_2$ would have. This points up the fact that when water flows from the spigot, it is as though the same amount of water had been taken from the top of the tank and dropped to the spigot level. The top level of the tank is a little lower, and the gravitational potential energy lost has gone to kinetic energy in the efflux water. If the spigot had been pointed upward, this kinetic energy would allow the ejected water to rise to the height of the top of the water in the tank before stopping. In practice, viscous energy losses would alter the result somewhat.

Exercise What is v_2 if the tank is closed at the top and the air pressure in it is kP_t, where k is a constant? *Answer:* $\sqrt{2g(h_1 - h_2) + 2(k - 1)(P_t)/\rho}$

Illustration 9.5 Pressure in a horizontal pipe

FIGURE 9.23

Since the fluid velocity is greatest at B, the pressure is lowest at that point.

Suppose water flows through the pipe system of Fig. 9.23. Because the same amount of water must flow past points A, B, and C each second, the speed of the water in the narrow pipe at B must be larger than the speed at A and C. Assuming the flow speed at A and C to be 0.200 m/s and that at B to be 2.00 m/s, compare the pressure at B with that at A.

Reasoning Applying Bernoulli's equation and noting that the average gravitational potential energy is the same at all three points, we have

$$P_A + \tfrac{1}{2}\rho v_A^2 = P_B + \tfrac{1}{2}\rho v_B^2$$

Substituting $v_A = 0.200$ m/s, $v_B = 2.00$ m/s, and $\rho = 1000$ kg/m^3 gives $P_A - P_B = 1980$ Pa. Hence *the fluid pressure within the constriction is much less than that in the large pipes on either side.* This is probably opposite what you would guess at first. However, it is true and has wide application. Aspirators, for example, obtain a partial vacuum by forcing water through a constriction where the pressure is greatly reduced due to the increase in flow velocity.

We can see qualitatively that the pressure at A must be larger than that at B. Because each little volume of fluid is accelerated as it moves from A to B, an unbalanced force to the right must act on it. To supply this unbalanced force, the pressure must decrease as one goes from A to B. You should be able to reverse this line of reasoning to show that the pressure at C is larger than that at B.

This result—that *pressure is low where speed is high*—provides an explanation of such diverse facts as the lift on an airplane wing and the curve ball pitched by a ballplayer. The flow around an airplane wing is shown in Fig. 9.24. Because the air

FIGURE 9.24

The airplane wing experiences a force that is directed from the region beneath the wing, where the air has low velocity and high pressure, to the high-velocity (low-pressure) region above the wing.

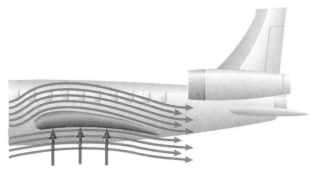

has to travel farther over the top of the wing than under it, the air is moving faster on the upper side than on the lower. The pressure is therefore lower at the top, and the wing is forced upward. This effect is also used on race cars; winglike fins are used to produce a downward force, thereby increasing the normal force and hence the friction force between tires and track. This allows the car to travel around curves faster than would otherwise be possible. ∎

9.9   LAMINAR VERSUS TURBULENT FLOW

Let us examine how a fluid flows through a pipe. Friction forces exerted on the fluid by the pipe walls tend to restrain the flow, as do the viscous forces within the fluid. As a result, the fluid close to the walls flows more slowly than that near the center of the pipe. We show this effect in Fig. 9.25*a*, where the lengths of the arrows indicate the magnitude of the velocity at various positions in the pipe. (In Illustrations 9.4 and 9.5, *v* was the average velocity on the pipe cross section.)

Another feature of flow through a pipe is shown in Fig. 9.25*b*. Suppose a tiny speck of dust, like the one at point *A*, is flowing with the fluid. If the flow rate is low, the speck follows the line shown as it moves through the pipe. Other specks, and the fluid as well, follow similar smooth lines. We call these flow lines **stream-lines**, and this type of flow is called **laminar flow.** In laminar flow, each element of the fluid follows a repeatable streamline.

If the speed of the fluid becomes high enough, a rather abrupt change in the pattern of flow occurs. Rather than flowing in smooth streamlines, the fluid develops contorted and fluctuating flow patterns such as shown in Fig. 9.25*c*. This type of flow is called **turbulent flow.** With turbulent flow, friction (or viscous) energy losses become greater than in the case of laminar flow. This in turn causes an increased frictional drag on surfaces in contact with the flowing fluid. Poiseuille's law is no longer valid when flow becomes turbulent.

Fluid does not have to be confined to a pipe for these types of flow behavior to occur. The same behavior is observed when fluid flows past any surface, such as

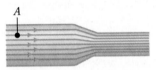

(*a*) Fluid velocity

(*b*) Streamlines (laminar flow)

(*c*) Turbulent flow

FIGURE 9.25

Examples of various features of flow in a pipe: (*a*) velocity profile; (*b*) laminar flow; (*c*) turbulent flow.

Pieces of tape show the pattern of wind flow over the surface of a car during a wind tunnel test.

an airplane wing or the contours of an automobile. Because of the increased friction associated with the onset of turbulence, the makers of autos and airplanes wish to design the surfaces of cars and planes so that turbulent effects are minimized. Thus a means of predicting when the onset of turbulence occurs has obvious practical importance.

When fluid flow around an object is laminar, the retarding force, or *drag force*, F_D, is linearly proportional to the fluid speed v. The exact value of the drag force is difficult to calculate mathematically, and is usually measured experimentally. Wind tunnels, for example, are used to measure drag forces due to air flowing past cars and airplanes. In 1843 the English physicist G. Stokes was able to calculate the relationship between F_D and v in the case of a sphere of radius r moving slowly through a fluid whose viscosity is η. The result is known as **Stokes' law:**

$$F_D = 6\pi\eta r v \tag{9.21}$$

At speeds high enough to induce turbulence, the drag force is no longer simply proportional to speed. Instead, it varies as a complicated series of terms in higher powers of v. In most cases involving cars and airplanes, F_D is found to be proportional to v^2:

$$F_D = \tfrac{1}{2}\rho C_D A v^2 \tag{9.22}$$

where A is the frontal area of the car or airplane. The dimensionless constant C_D is called the *drag coefficient*. Some values of the drag coefficient are given in Table 9.5.

Although turbulence is very difficult to treat mathematically, there is a unifying concept that simplifies the situation. Experiment shows that flow changes from laminar to turbulent when a critical value is reached for a dimensionless constant called the **Reynolds number,** N_R. The Reynolds number is given by the expression

$$N_R = \rho v d / \eta \tag{9.23}$$

where ρ, v, and η are the fluid's density, speed of flow, and viscosity, respectively, and d is a characteristic dimension of the flow system, which depends on the specific application. For example, for flow through a pipe, d is the pipe diameter. For a sphere moving through a fluid, d is the sphere's diameter. In the case of an irregular object such as an airplane, d would be some average of the dimensions of the plane. Table 9.6 lists a few examples of critical Reynolds numbers.

TABLE 9.5 *Typical wind-tunnel values of drag coefficients*

Object	Drag coefficient (C_D)
Flat plate	1.2
Skydiver (Stretched horizontally)	1.0
Motorcycle and rider	0.9
Automobile (sedan)	0.5
Automobile (streamlined sports car)	0.25
Streamlined train	0.15

TABLE 9.6 *Approximate critical values of the Reynolds number*

N_R	Transition phenomenon
10	Maximum N_R for laminar flow around a sphere (Stokes' law)
1000–1200	Onset of turbulence in a cylindrical pipe with irregular inlet
2000–3000	Onset of turbulence in a long cylindrical pipe (limit of validity of Poiseuille's law)
20,000–40,000	Onset of turbulence for pipes with optimized entrance nozzles
3×10^5	Upper limit for the $F_D \propto v^2$ behavior

An example of the transition from laminar flow to turbulent flow.

Despite the lack of precision in critical values of the Reynolds number, these values are very useful in determining what we call *scaling laws*. For example, two systems, one of which is a scale model of the other, will experience the same flow patterns if N_R is the same for both. Such systems are said to be *dynamically similar*. This concept forms the basis for wind-tunnel tests on small-scale cars and airplanes. The flows are similar if the product vd (and hence N_R) is the same. Thus a slow fluid flow (small v) around a large object (large d) will be the same as for the same fluid flowing at twice the speed around an object half as large.

Example 9.8

How fast can a raindrop having a diameter of 3.0 mm fall before the flow of air around it becomes turbulent?

Reasoning

Question What is the falling raindrop an example of?
Answer We can approximate the raindrop as a sphere. As it falls at speed v, the air flows past it with that same speed.

Question What principle will determine if this flow becomes turbulent?
Answer In the case of a sphere, the critical Reynolds number is given in Table 9.6 as $N_R = 10$.

Question Do I have enough data to evaluate v_{max}?
Answer You can look up the viscosity and density of air:

$$\eta = 0.019 \times 10^{-3}\,Pl \quad \text{and} \quad \rho = 1.29\,kg/m^3$$

The factor d in the case of a falling sphere is the sphere's diameter, 3.0 mm.

Solution and Discussion Solve Eq. 9.24 for v:

$$v = N_R\eta/\rho d$$

The critical velocity above which turbulence will occur is (using $N_R = 10$)

$$v_{max} = \frac{(10)(1.9 \times 10^{-5}\,Pl)}{(1.29\,kg/m^3)(3.0 \times 10^{-3}\,m)}$$

$$= 4.9 \times 10^{-2}\,m/s = 4.9\,cm/s$$

Notice how small this speed is. Also notice that the speed is inversely proportional to the raindrop's diameter.

Example 9.9

Approximately what volume of water per second can flow through a pipe 2.0 cm in diameter before turbulent flow will occur?

R *Reasoning*

Question What condition will indicate when turbulence will occur?
Answer Turbulence will occur when the Reynolds number exceeds the critical value of 2000 to 3000 given in Table 9.6. You can choose $N_R = 2000$ for this example.

Question What connection is there between N_R and the volume of flow per second?
Answer The critical value of N_R will give you the maximum flow speed, v. The volume rate of flow is $\Delta V/\Delta t = vA$.

Solution and Discussion Putting $N_R = 2000$ into Eq. 9.23 along with the value of the viscosity of water from Table 9.4 gives the maximum speed for laminar flow:

$$v_{max} = \frac{(2000)(0.801 \times 10^{-3}\,\text{Pl})}{(1.00 \times 10^3\,\text{kg/m}^3)(2.00 \times 10^{-2}\,\text{m})} = 0.0801\,\text{m/s} = 8.01\,\text{cm/s}$$

The area of the pipe is $A = \pi d^2/4 = 3.14 \times 10^{-4}\,\text{m}^2 = 3.14\,\text{cm}^2$. Thus the maximum volume rate of laminar flow is

$$\frac{\Delta V}{\Delta t} = (8.01\,\text{cm/s})(3.14\,\text{cm}^2) = 25.2\,\text{cm}^3/\text{s}$$

Exercise What would the maximum speed and volume rate of laminar flow of water be through a pipe 10 cm in diameter? *Answer:* $v_{max} = 1.60\,cm/s;$ $\Delta V/\Delta t = 126\,cm^3/s$

Example 9.10

What horsepower is required to move a car through air ($\rho = 1.29\,\text{kg/m}^3$) at a constant speed of 60.0 mi/h on a level road? Assume that the frontal area A of the car is 2.30 m² and that the mass of the car is 1250 kg. Also assume that at this speed the critical Reynolds number for the car is exceeded.

R *Reasoning*

Question What is the power requirement related to?
Answer To move at constant speed, the engine must provide enough force via the tires on the drive wheels to equal the drag force on the car due to the air

flowing around it. You should recall that the *power* produced by a force is the product of the force times the speed of the object to which the force is applied.

Question   What allows me to calculate the drag force?
Answer   If the critical value of N_R is exceeded, the flow is turbulent, and the drag force is given by Eq. 9.22. The value 0.50 for the drag coefficient C_D can be found from Table 9.5.

Question   What is the equation for power that results?
Answer   From Eq. 9.22, $F_{app} = F_D = \frac{1}{2}\rho A C_D v^2$. Thus

$$\text{Power} = F_{app}v = (\tfrac{1}{2}\rho A C_D v^2)v = \tfrac{1}{2}\rho A C_D v^3$$

Solution and Discussion   First, convert 60 mi/h to 26.8 m/s. Then, using the data given,

$$\text{Power} = \tfrac{1}{2}(1.29\,\text{kg/m}^3)(2.30\,\text{m}^2)(0.50)(26.8\,\text{m/s})^3 = 1.4 \times 10^4\,\text{W}$$

Since 1 hp = 746 W, this is power = $(1.4 \times 10^4\,\text{W})(1\,\text{hp}/746\,\text{W})$ = 19 hp. Notice the strong dependence of power on the car's speed ($P \propto v^3$). The same car going at 30 mi/h would require only 1/8 as much power to balance the force of air drag! This is a major reason that gas consumption depends greatly on speed.

9.10   TERMINAL VELOCITY

Until now we have usually treated falling objects as freely accelerating with constant acceleration g. Yet there are many examples where objects fall at constant velocity over much of their fall rather than at constant acceleration. In these cases we call this constant velocity the object's *terminal velocity*. Of course constant velocity implies zero net force acting on the object. In this case the drag force

Sky divers accelerate to a steady terminal velocity, at which the upward air resistance equals the force of gravity on them.

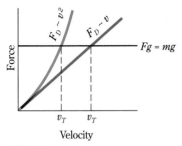

FIGURE 9.26

Forces on a falling object. Terminal velocity occurs at the speed where air drag is equal to the object's weight, *mg*.

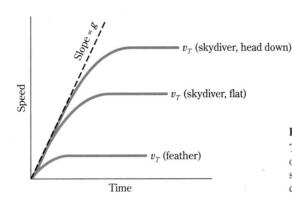

FIGURE 9.27

Terminal velocities vary between objects and depend on many factors, such as frontal area and drag coefficient.

upward created by the object's speed through the air is equal to the force of gravity downward. We can visualize this situation in a graph of force versus velocity (Fig. 9.26). In the last section we saw that drag force F_D is often proportional to v, or in some cases v^2. These dependencies are shown on the graph. At the same time, the force of gravity, mg, is independent of v. Thus mg appears as a horizontal line on the graph. As the speed of the object increases, F_D approaches the value mg. The velocity v_T at which $F_D = mg$ is the condition for net force = 0. We can make graphical sketches of velocity versus time for these situations. Figure 9.27 shows cases of small, medium, and large terminal velocities. These could represent a feather, a skydiver falling with body flat to direction of fall, and a skydiver falling head first with arms at her side, for example. In each case, notice that the object starts to fall with the same acceleration, g. Terminal velocity is determined by many properties, such as density, frontal area, shape, etc. Let us now study a sphere falling in air when the flow is laminar.

Example 9.11

Tiny particles in suspension in a liquid slowly fall through the liquid with a terminal velocity. This velocity is referred to as the *sedimentation rate*. Find the sedimentation rate for spherical particles of radius $r = 2.00 \times 10^{-3}$ cm as they fall through water at 20.0°C. The density of the material composing the particles is 1050 kg/m³, and the viscosity of water is 1.00 mPl.

R | *Reasoning*

Question What is the basic principle that determines sedimentation rate?
Answer Sedimentation rate is a terminal velocity, so the condition is that the net force on the particles is zero.

Question What are the various forces acting on the particles?
Answer Gravity acts downward. Two forces act upward: a buoyant force and the viscous drag force.

Question What are the expressions for each of these forces?
Answer $F_g = mg$; $F_B = \rho_f V g$ (Archimedes' principle); $F_D = 6\pi \eta r v_T$ (Stokes' law).

Question What equation does net force = 0 give me?

Answer $mg = \rho_f V g + 6\pi\eta r v_T$

Here $V = \frac{4}{3}\pi r^3$ and $m = \rho_p(\frac{4}{3})\pi r^3$

Solution and Discussion When we make the above substitutions and collect terms, the equation for zero net force becomes

$$(\rho_p - \rho_f)\tfrac{4}{3}\pi r^3 g - 6\pi\eta r v_T = 0$$

When we solve for v_T, we get

$$v_T = \frac{2r^2 g}{9\eta}(\rho - \rho_f) = 4.36 \times 10^{-3}\,\text{cm/s}$$

This is quite a low velocity. Even so, a beaker containing this solution would clarify by sedimentation in a few hours.

You can see that the sedimentation rate depends on the difference between particle and fluid densities. It also depends on the cross-sectional area (r^2) of the particles. Notice also that v_T is proportional to g.

LEARNING GOALS

Now that you have finished this chapter, you should be able to

1 Define (*a*) fluid, (*b*) crystalline and amorphous solids, (*c*) density, (*d*) Hooke's law, (*e*) stress and strain, (*f*) modulus, (*g*) Young's modulus, (*h*) shear modulus, (*i*) bulk modulus, (*j*) pascal, (*k*) Pascal's principle, (*l*) buoyant force, (*m*) Archimedes' principle, (*n*) laminar and turbulent flow, (*o*) Bernoulli's equation, (*p*) drag force, (*q*) terminal velocity, (*r*) viscosity, (*s*) Reynolds number.

2 Use the definition of density in simple computations.

3 Use Hooke's law in its stress-strain form to compute the deformation of an elastic material in tension, shear, and bulk compression when the appropriate modulus is given.

4 Find force from pressure, and vice versa.

5 Given appropriate data, compute the absolute and gauge pressures at a given depth in a fluid.

6 Explain the operation of a barometer and a manometer and use them to calculate gas pressures.

7 Express pressures in pascals, torr, atmospheres, and bars.

8 Explain the theory of the hydraulic press.

9 Use Archimedes' principle to find the buoyant force on an object of known mass and density (or volume).

10 Identify each quantity in Poiseuille's equation and use it in simple calculations.

11 Use Bernoulli's equation to derive Torricelli's theorem and to show that pressure is least where flow rate is greatest.

12 Relate the drag force of an object to its terminal velocity in free fall.

13 Use the Reynolds number and other appropriate data to calculate approximately the critical velocity for the onset of turbulent flow in a fluid.

14 Calculate the drag force due to viscous flow around the object for various flow speeds and fluids, given the critical Reynolds number and the dimensions and drag coefficient of the object.

SUMMARY

DERIVED UNITS AND PHYSICAL CONSTANTS

Pressure Units
$1\,\text{N/m}^2 = 1\,\text{Pa}$
$1\,\text{atm} = 1.01325 \times 10^5\,\text{Pa}$
$1\,\text{torr} = 1\,\text{mmHg} = 133.3\,\text{Pa} = (1/760)\,\text{atm}$
$1\,\text{bar} = 10^5\,\text{Pa}$

Viscosity Units
$1\,\text{Pa} \cdot \text{s} = 1\,\text{poiseuille (Pl)}$
$1\,\text{poise (P)} = 0.1\,\text{Pl}$
$1\,\text{centipoise (cP)} = 10^{-3}\,\text{Pl} = 1\,\text{mPl}$

DEFINITIONS AND BASIC PRINCIPLES

Mass Density (ρ)

$$\rho = \frac{\text{mass}}{\text{volume}} = m/V \,(\text{kg/m}^3) \tag{9.1}$$

Specific Gravity (SG)

$$\text{SG} = \frac{\rho}{\rho_{\text{H}_2\text{O}}} \tag{9.2}$$

Stress

1. Longitudinal stress $= \dfrac{F}{A}$

   where **F** is perpendicular to the plane of A.

2. Shear stress $= \dfrac{F}{A}$

   where **F** is parallel to the plane of A.

3. Volume stress $= -\Delta P$

$$\tag{9.3}$$

Strain

1. Longitudinal strain $= \dfrac{\Delta L}{L_o}$

   where ΔL is parallel to L_o.

2. Shear strain $= \dfrac{\Delta L}{L_o} = \phi$ (shear angle)

   where ΔL is perpendicular to L_o.

3. Volume strain $= \dfrac{\Delta V}{V_o}$

Elastic Modulus (N/m² or Pa)

$$\text{Modulus} = \frac{\text{stress}}{\text{strain}}$$

1. Young's modulus $Y = \dfrac{F/A}{\Delta L/L_o}$ $\tag{9.7}$

2. Shear modulus $S = \dfrac{F/A}{\Delta L/L_o} = \dfrac{F/A}{\phi}$ $\tag{9.8}$

3. Bulk modulus $B = \dfrac{-\Delta P}{\Delta V/V_o}$ $\tag{9.9}$

Pressure (P)

$$P = \frac{F_{\perp}}{A} \,(\text{N/m}^2 = \text{Pa}) \tag{9.10}$$

Gauge Pressure (P_G)

$$P_G = P_{\text{tot}} - P_a \tag{9.12}$$

Gauge pressure exerted by a column of fluid:
$$P_G = \rho_f g h$$

where ρ_f is the fluid density and h is the depth.

Archimedes' Principle

Buoyant force F_B = weight of the fluid displaced by a body partially or wholly immersed in the fluid.

INSIGHTS

1. For a body of volume V wholly submerged,
   $F_B = \rho_f V g$
2. The condition of a body of mass M *floating* requires that
   $F_B = Mg$

Fluid Flow

EQUATION OF INCOMPRESSIBILITY

For an incompressible fluid

$$vA = \text{constant at all points in the fluid} \tag{9.18}$$

where v = flow velocity
A = cross-sectional area of flow

VISCOSITY (η)

$$\eta = \frac{\text{shear stress}}{\text{shear rate}} = \frac{F/A}{v/L} \quad (\text{Pa} \cdot \text{s} = \text{Pl}) \tag{9.16}$$

where v is the relative velocity of two layers of the fluid which are separated by a distance L.

POISEUILLE'S LAW

Flow rate (Q) in a viscous fluid:

$$Q = \left(\frac{\pi R^4}{8\eta L}\right)(P_1 - P_2) \quad (\text{m}^3/\text{s}) \tag{9.17}$$

where R = pipe radius
L = pipe length
$P_1 - P_2$ = pressure difference over length L

BERNOULLI'S PRINCIPLE

For nonviscous flow of a fluid of constant density

$$P + \tfrac{1}{2}\rho v^2 + \rho g h = \text{constant} \tag{9.19}$$

at all points of the fluid.

INSIGHT

1. Bernoulli's principle has the consequence that, for a horizontal pipe, fluid pressure is smallest where flow velocity is greatest.

REYNOLDS NUMBER (N_R)

$$N_R = \frac{\rho v d}{\eta} \tag{9.23}$$

where v = flow velocity
d = pipe diameter or diameter of spherical object placed in flow

ρ = fluid density
η = fluid viscosity

INSIGHT

1. As a general rule, flow velocity through a pipe becomes turbulent when N_R exceeds approximately 2000. For a sphere moving through a fluid, this transition in flow occurs when N_R is approximately 10.

DRAG FORCE

If flow is laminar, the drag force on a sphere of radius r moving through a fluid with velocity v is

$$F_D = 6\pi\eta r v \quad \text{(Stokes' law)} \qquad (9.21)$$

If flow is turbulent, the drag force is proportional to v^2:

$$F_D = \tfrac{1}{2}\rho A C_D v^2 \qquad (9.22)$$

where ρ = fluid density
    A = frontal area of object
    C_D = drag coefficient

QUESTIONS AND GUESSTIMATES

1 How could you determine the density of (a) a cubical block of metal, (b) a liquid, (c) an oddly shaped piece of rock?

2 How could you measure (a) the tensile modulus of the rubber in a rubber band? (b) The shear modulus of gelatin? (c) The bulk modulus of foam rubber?

3 Does the water pressure at the base of a dam depend on the size of the lake behind the dam?

4 A partly filled bottle of mercury has a screw cap. It is taken aloft on a spaceship. When the bottle is orbiting the earth in the ship, what is the pressure at a depth of 2.0 cm in the mercury? What is the pressure at this depth when the ship lands on the moon?

5 How can one determine the density of an irregular object that (a) sinks in water and (b) floats in water?

6 Estimate the average density of the human body. How could you measure your density to within 1 percent using simple equipment at a swimming pool? Some people are able to float more easily than others. Explain what factors are involved.

7 How is it possible for a ship made from steel to float? Won't steel always sink in water? How does a submarine move to various depths?

8 A glass filled to the brim has an ice cube floating partly above the water. Does the water overflow as the ice cube melts?

9 A glass filled to the brim with water sits on a scale. A block of wood is gently placed in the water so that it floats. Some water overflows, but it is wiped away. How does the final reading of the scale compare with the initial reading?

10 Blood contains many tiny particles too small to be seen with a microscope. Sedimentation rates can be used to determine whether or not these particles have clumped into groups. Explain how this can be done and examine the assumptions you make.

11 Why don't people use water barometers? After all, mercury is poisonous and expensive.

12 We can imagine the molecules of an ideal gas as acting like tiny balls in continual motion. An ideal gas of colloidal-size particles can also exist. However, glass beads and pool balls do not act like ideal gases. At what size does the dividing line come, and to what is it due?

13 The composition of the atmosphere changes with altitude. As one goes farther above the earth, the percentage of hydrogen molecules in the air increases and the percentage of nitrogen molecules decreases. Why?

PROBLEMS

Unless otherwise stated, assume that atmospheric pressure is 101 kPa.

Sections 9.1 and 9.2

1 A solid sphere made of a certain material has a radius of 3.0 cm and a mass of 98.0 g. What is the density of the material?

2 A solid cube is 2.0 cm on each side and has a mass of 24 g. What is the density of the cube?

3 At 20°C, what is the approximate mass of air in a box-shaped room that is $6.0 \times 5.0 \times 2.5 \, \text{m}^3$?

4 A 200-g bottle has a mass of 340 g when filled with water and 344 g when filled with blood plasma. What is the density of the plasma?

5 What is the mass of an ice cube 4.0 cm on each side?

6 A king orders a gold crown of mass 2.00 kg to be made. When the crown arrives the king has doubts about its purity and orders that its volume be measured. The volume is found to be 190 cm^3. Is the crown pure gold?

▪ **7** Suppose that the crown in Prob. 6 is a mixture of brass and gold. What percentage of the crown's mass is pure gold?

▪ **8** The density within a neutron star is about 1×10^{19} kg/m^3. What would the radius of the earth be if earth were compressed to this density? Mass of the earth $M_e = 5.98 \times 10^{24}$ kg.

9 To determine the density of an unknown fluid, an empty 100-cm^3 volumetric flask that has a mass of 56.5 g is filled with the fluid. The mass of the filled flask is 231.3 g. What is the density of the fluid?

▪ **10** To determine the approximate density of a 52.2-g stone, a student uses a 50.0-cm^3 graduated cylinder that has a mass of 36.7 g. She places the stone in the cylinder and fills the cylinder with water to the 50.0-cm^3 mark. The combined mass of the system is 130.0 g. What is the density of the stone?

11 If silver costs $150.00/kg, how long is each side of a $1 million cube of silver? (The density of silver is 10.5×10^3 kg/m^3.)

12 A water bed in the form of a rectangular box has dimensions 2.0 m × 1.8 m × 30 cm. Find the weight of the water inside the water bed. (Neglect the dimensions of the outer covering.)

Section 9.3

13 A 3.2-m-long wire has a radius of 0.36 mm. When a 7.2-kg load is hung from it, the wire stretches 1.58 mm. What is Young's modulus for the material of the wire?

14 A 24-kg load stretches a steel wire that is originally 160 cm long. The radius of the wire is 0.56 mm. How far does the wire stretch under this load?

15 A cylindrical aluminum pillar 6.0 m high has a radius of 30 cm. If a 2200-kg statue is placed on top of the pillar, by how much will the pillar be compressed?

16 A 6.0-m steel beam 2.0 cm in radius is used to support a part of a bridge. The beam is designed so as not to stretch more than 6×10^{-5} m. What is the maximum load that it can withstand?

17 What is the force necessary to compress a 3.0-cm brass cube to 99.8 percent of its original height? (Assume that the cube is compressed only in one direction.)

18 In Prob. 17 what is the force required to compress the cube in all three dimensions by 99.8 percent?

19 A cube of gelatin 4.0 cm on each edge is subjected to a shearing force of 0.50 N on its top surface. Because of this force, the top surface displaces 2.7 mm. What is the shear modulus for gelatin?

20 What increase in pressure will be needed to decrease the volume of a sample of water by 2 percent?

21 A block of foam rubber shrinks by 12 percent when subjected to a pressure of 1000 kPa. What is the bulk modulus for the rubber?

▪ **22** Shear stress in excess of about 4.0×10^5 kPa ruptures steel. Determine the minimum shear force required to punch a 1-cm-radius hole in a steel plate of 1.0 cm thickness.

Sections 9.4 and 9.5

23 Atmospheric pressure is about 100 kPa. By what fraction does the volume of a glass ball change as the air around it is removed in a vacuum chamber?

24 How large a pressure increase above atmospheric pressure is needed to decrease a volume of mercury by 0.1 percent?

25 Suppose there is a perfect vacuum inside a sealed coffee can. What force must the 8.0-cm-diameter cover support when it is exposed to the atmosphere? Use $P_a = 100$ kPa.

26 How much force does the atmosphere exert on a person's back? Assume that $P_a = 100$ kPa and that the back has an area of about 320 cm^2. Why doesn't this enormous force crush the person?

27 What is the pressure due to the water at the bottom of a 12-m-deep lake? Compare this with atmospheric pressure, about 100 kPa.

28 What is the absolute pressure at the bottom of the lake in Prob. 27?

29 What is the pressure due to mercury at the base of a column of mercury 765 mm high? Compare this with atmospheric pressure, about 100 kPa.

▪ **30** The pressure in a sealed water pipe on the ground floor of a high-rise building is 2.8×10^5 Pa above atmosphere. The pressure in the same pipe on the top floor is only 1.2×10^5 Pa. How tall is the building?

▪ **31** (a) What is the pressure due to water 1600 m beneath the ocean's surface? Take the density of seawater to be 1025 kg/m^3. (b) If the bulk modulus of the seawater is the same as that of pure water, by what percent does the density of water increase in going from the ocean surface to this depth?

▪ **32** A 1250-kg automobile is supported by four tires inflated to a gauge pressure of 180 kPa. How large an area of each tire is in contact with the pavement? Assume the wheels share the equal load.

▪ **33** The gauge pressure at the bottom of a reservoir is five times that at a depth of 1.2 m. How deep is the reservoir?

▪ **34** A container has a 12-cm-thick layer of oil floating on 25 cm of water. The density of the oil is 850 kg/m^3. What is the combined pressure due to the fluids at the bottom of the container?

35 A glass tube is bent into a U shape, as in Fig. 9.15. Water is poured into the tube until it stands 12 cm high on each side. Kerosene ($\rho = 870$ kg/m^3) is added slowly to one side until the water on the other side rises 5 cm. What is the length of the kerosene column?

36 In the previous problem, suppose a 3.0-cm column of benzene is poured into one side. How far will the water in the other side rise?

37 If the column of mercury in a barometer is 74.6 cm, what is the atmospheric pressure?

38 Hydraulic stamping machines exert tremendous forces on a sheet of metal to form it into the desired shape. Suppose the input force is 900 N on a piston that has a diameter of 1.80 cm. The output force is exerted on a piston that has a diameter of 36 cm. How large a force does the press exert on the sheet being formed?

▪39 The plunger of a certain hypodermic needle has a cross-sectional area of 0.76 cm². How large a force must be applied to the plunger if the liquid in the needle is to move into a vein where the pressure is 18.6 kPa above atmosphere?

40 Water is confined in a strong container by means of a piston with a cross-sectional area of 0.60 cm². How large a force on the piston is required to increase the density of the water by 0.01 percent?

41 Suppose a water barometer is used to measure atmospheric pressure. How tall is the water column on a day when a mercury barometer stands at 76 cm?

▪42 Denver, the "mile-high city," often has an atmospheric pressure of only 60 cmHg. How tall a column of oil (density = 879 kg/m³) will this pressure support?

43 When the atmospheric pressure is 100 kPa, with how large a force does the atmosphere press down on a 20 cm × 28 cm book resting on a table? If the book has a mass of 2.1 kg, what is the ratio of this force to the weight of the book?

▪44 How large a force does the atmosphere exert on the surface of a ball of diameter 24 cm? Assume the atmospheric pressure to be 98 kPa.

Section 9.6

45 A metal cube is 2.0 cm on each edge. What is the buoyant force on it when it is completely submerged in oil of density 864 kg/m³?

46 A 2.40-g object has an apparent mass of 1.62 g when it is completely submerged in water at 20°C. What is (*a*) the volume of the object and (*b*) its density?

47 A 6.24-g object has an apparent mass of 5.39 g when completely submerged in oil. If the density of the object is 6.4 g/cm³ find the density of the oil.

48 A 4.923-g object has an apparent mass of 2.241 g when completely submerged in water and an apparent mass of 2.612 g when completely submerged in oil. What is the density of the oil?

49 A downward force of 18 N must be applied to a woman weighing 480 N to keep her completely submerged in water. What is the density of her body?

50 A block of foam plastic has a volume of 25 cm³ and a density of 800 kg/m³. How large a force is required to hold it under water?

51 A solid cube of unknown material is floating upright in water; 25 percent of the cube is above the surface of water. What is the density of the material?

52 Icebergs are made of freshwater ice, which has a density of 920 kg/m³. The density of ocean water, in which they float, is approximately 1.03 × 10³ kg/m³. What fraction of an iceberg lies below the surface of water?

▪53 A raft 6 m × 4 m in area is floating on a river. When a loaded car pulls onto the raft, it sinks 3.0 cm lower in the water. What is the weight of the car?

▪54 What must the minimum volume of a block of material (density = 810 kg/m³) be if it is to hold a 64-kg man entirely above the surface of a lake when he stands on the block?

▪55 When a beaker partially filled with water is placed on an accurate scale, the scale reads 22 g. If a piece of wood with a density of 905 kg/m³ and a volume of 2.1 cm³ is floated on the water, what will the scale read?

▪56 When a beaker partially filled with water is placed on an accurate scale, the scale reads 22 g. If a piece of metal with density 3800 kg/m³ and volume 2.4 cm³ is suspended by a thin string so that the metal is submerged in the water but does not rest at the bottom of the beaker, what does the scale read?

▪57 A 240-cm³ block of foam plastic (density = 600 kg/m³) is to be weighted with aluminum so that it just sinks in water. What mass of aluminum is required to be hung from the block?

▪58 A cube of metal (density = 6 × 10³ kg/m³) has a cavity inside it. It weighs 2.4 times as much in air as it does when completely submerged in water. What fraction of the cube's volume is the cavity?

Section 9.7

59 By what factor does the rate of fluid flowing through a capillary tube change when the length of the tube is increased to five times its original value and its radius is increased to three times its original value? Assume that the pressure difference across the ends of the tube remains unchanged.

60 A hypodermic needle is replaced by a needle the length of which is two-thirds that of the original and the diameter of which is one-third that of the original. By what factor must the pressure differential across the needle change if the flow rate is to remain unchanged?

▪61 A hypodermic needle 3.6 cm long has an internal diameter of 0.24 mm. Its plunger has an area of 0.084 cm². When a force of 6.4 N is applied to the plunger, at what rate does water at 30°C flow through the needle?

▪62 A needle 4.0 cm long with a radius of 0.3 mm is used for blood transfusion. The pressure differential across the needle is achieved by elevating the blood bottle 1 m above the patient's arm. The blood pressure in the patient's vein is 10 cmHg. (*a*) What is the rate of flow of blood through the needle? (*b*) At this rate of flow, how long will it take to inject 1 liter of blood into the patient? The density of the blood is 1050 kg/m³ and its coefficient of viscosity is 4 × 10⁻³ Pa.

▪63 The blood pressure of a certain person is 125/85 mmHg. The average pressure is about 105 mmHg, which is about

1.40×10^4 Pa. Assume that a 4.0-cm-long needle with a radius of 0.3 mm is inserted into the person's bloodstream where the pressure has this average value. At what rate does blood flow from the needle? Use $\eta_{blood} = 4$ mPl.

■64 A cubical block 3.0 cm on each edge sits on a flat plate. There is a 0.04-mm-thick oil film between block and plate ($\eta_{oil} = 0.40$ mPl). Find the force required to pull the block across the plate at a speed of 0.3 m/s.

■65 At 20°C, the water pressure in a horizontal pipe decreases at the rate of 60 kPa per 100 m when the water flows at 3.0 liter/min. What is the radius of the pipe?

Sections 9.8 and 9.9

66 A pipe near the lower end of a large water-storage tank springs a small leak, and a stream of water shoots from it. The top of the water tank is 10 m above the leak. (*a*) With what speed does the water gush from the hole? (*b*) If the hole has an area of 0.08 cm², how much water flows out in 1 min?

67 Water is flowing through a closed pipe system. At one point, the speed of the water is 2.8 m/s; at a point 4.0 m higher than the first the speed is 4.2 m/s. (*a*) If the pressure is 84 kPa at the lower point, what is the pressure at the upper point? (*b*) What is the pressure at the upper point if the water stops flowing there when the pressure at the lower point is 62 kPa? Assume all pressures are absolute.

68 An airplane wing is designed so that the speed of the air below the wing is 300 m/s when the speed of the air across the top is 360 m/s. What is the pressure difference between the top and bottom of the wing?

69 In Prob. 68 if the area of the wing is 20 m², what is the net upward force on the wing?

70 A horizontal 4.0-cm-diameter pipe is joined to a second pipe of diameter 3.0 cm. There is a pressure difference of 7.2 kPa between the two pipes. (*a*) Which pipe has the higher water pressure? (*b*) What volume of water flows through the tubes per minute?

71 The nozzle of a lawn sprinkler sprays water vertically upward to a height of 5 m. What is the gauge pressure in the nozzle?

■72 In an aorta with a cross-sectional area of 1.6 cm² the blood (density = 1050 kg/m³) flows with a speed of 30 cm/s. What is the flow rate, in kilograms per second, of blood in the aorta? The aorta branches to form a large number of fine capillaries with a combined cross-sectional area of 2.0×10^3 cm². What is the speed of blood flow in the capillaries?

■73 A scale model of a 1.8-m-high automobile is 18.0 cm high. If the model is tested in a wind tunnel, how fast should the air be moving to simulate the actual car's motion at 80 km/h?

■74 Show that the Reynolds number can be written as $N_R = 2Q\rho/\pi\eta r$ for flow through a cylindrical pipe of radius r.

75 How fast can a 3.6-mm-diameter water drop fall through air before turbulent flow sets in? Take $N_R = 10$.

76 Determine the velocity at which the flow of water through a 1.0-cm-diameter tube becomes turbulent. Take $N_R = 3000$.

Section 9.10

77 If you assume that Stokes' law applies, what is the terminal velocity of a 4.0-mm-diameter water drop falling through air? Does Stokes' law actually apply in this situation?

78 Small, solid 1-mm-diameter spheres fall through water with a terminal velocity of 1.2 cm/s. What is the density of the spheres?

79 An oil drop (density = 850 kg/m³) is falling through air with a terminal velocity of 0.05 mm/s. Determine the radius of the drop. Take the density of air to be 1.29 kg/m³ and its viscosity $\eta_{air} = 0.019$ mPl.

80 A solid aluminum sphere of radius 0.4 mm falls through water at 30°C. Find (*a*) the buoyant force on it and (*b*) its terminal speed. Assume laminar flow.

81 Calculate the ratio of sedimentation rates for a mixture of small spheres that are all made up of the same material but the diameters of which are in the ratio 1:2:3.

■82 A piece of wood ($\rho = 840$ kg/m³) is fashioned into a sphere of diameter 0.6 cm and is released deep in a lake. Assuming laminar flow, what is its terminal velocity as it rises toward the surface? Is the assumption of laminar flow justified?

Additional Problems

■■83 A rubber sheet 3.2 mm thick is cemented between two parallel metal plates. The rubber sheet is 10.0 cm × 10.0 cm and is the same size as the plates. A shear stress is applied to the rubber by pulling the two plates in opposite directions, each with a force of 45 N. How far does one plate move relative to the other if the shear modulus of rubber is 1.20 MPa?

■■84 A 10-kg block is to be pulled along a horizontal surface by a horizontal steel wire whose radius is 2.0 mm². If friction is ignored, what maximum acceleration can be given to the block? The tensile strength of steel is 0.50 GPa.

■■85 A car with all windows closed runs off a bridge into a river. It comes to rest with the center of the driver's door 3.6 m below the surface of water. How large a force must the driver exert on the door to push it open? The area of the door is about 0.9 m².

■■86 In a hydraulic car lift in a garage, compressed air is used to exert a force on a small piston having a radius of 4 cm. This pressure is used to lift a 12,500-N car on another piston of radius 20 cm. What air pressure on the smaller piston must be present to lift the car?

■■87 A mercury manometer is used to monitor the pressure in a chemical reaction chamber. In the end that is open to the atmosphere, the mercury level is 2.83 cm higher than in the end leading to the chamber. The barometric reading is 74.82 cmHg. What is the pressure inside the reaction chamber?

■■88 A manometer that uses oil of density 864 kg/m³ is used to measure the pressure in an environmental testing chamber. In the end open to the atmosphere, the oil level is 11.6 cm higher than in the end leading to the chamber. The barometric reading is 74.23 cmHg. What is the pressure in the chamber?

"89 (*a*) Water is flowing upward in the pipe system of Fig. P9.1. At points 1 and 2, the pipe radii are r_1 and r_2, respectively. The water speed at point 1 is 30 cm/s. What is the pressure difference $P_2 - P_1$ between the two points? (*b*) Repeat if the flow is reversed.

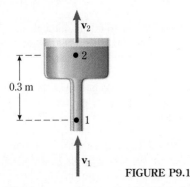

FIGURE P9.1

"90 An aluminum sphere of radius b is being lifted with constant speed v through a fluid of density ρ and viscosity η by means of a thin wire attached to it. Find the tension in the wire.

"91 A fire fighter's water hose spews water from the nozzle at a rate of 0.02 m³/s. When the nozzle is directed upward, the water shoots to a height of 32 m. Suppose the hose lies in a straight line along the ground when a fire fighter tries to hold the nozzle end vertically. Describe the horizontal force that the fire fighter must exert on the nozzle to hold it stationary.

"92 A manometer using alcohol in the U tube shows a difference $h_1 - h_2 = 80$ cm in the heights of the alcohol when end 1 is open to the room and end 2 is connected to a container of gas. In the same room, a mercury barometer shows a reading of 740 mm. What are the gauge and absolute pressures of the gas in the container in torr and SI units?

"93 The bag of a balloon when filled with helium is a sphere 40.0 m in diameter. What total weight, including bag, gondola, and contents, can the balloon lift in air at standard temperature and pressure? If you wanted to lift more than this, would you wait for a cooler or a warmer day?

CHAPTER 10

TEMPERATURE AND THE KINETIC THEORY OF GASES

In Chap. 9 we learned about measuring gas pressure and some of the properties of flowing gases. We now turn our attention to the concept of temperature and the dependence of gas pressure on temperature. We shall derive a fundamental physical explanation of temperature in terms of the kinetic energy of the atoms or molecules of a gas. The molecular model used to obtain this relationship is called the *kinetic theory of gases*. Let us begin with a discussion of temperature in familiar terms involving our experience with thermometers.

10.1 THERMOMETERS AND TEMPERATURE SCALES

As mentioned in Chap. 1, temperature is one of the seven fundamentally defined dimensions in physics. Although we will not give a formal definition of temperature until later, we can say here that, in a very simple way, we experience temperature as the "hotness" or "coldness" of an object. A common demonstration that gas pressures depend on temperature is that hot tires have higher pressure than cold tires. Temperature affects our lives in many other ways. We depend on accurate temperature measurements of the atmosphere to determine clothing and heating requirements, for example. Devices by which we measure temperature are

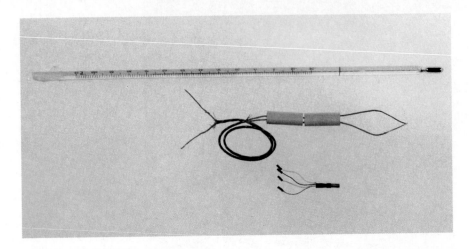

Thermometers can use any physical property that depends on temperature. Here are examples that use (1) the thermal expansion of a liquid, (2) the temperature-dependent voltages at the junction between different metals (called a thermocouple), and (3) the temperature dependence of electrical resistance (a resistance thermometer).

called **thermometers.** There are many such devices and they can be calibrated in various temperature scales.

The most common type of thermometer is shown in Fig. 10.1. A liquid—usually mercury or alcohol—is sealed in a glass capillary tube that has a bulb at one end which serves as a reservoir of the liquid. Because these liquids expand as the temperature increases, the liquid level in the capillary rises as the temperature increases. (The glass also expands, but much less than the liquid.) The thermometer historically was marked into divisions in the following way.

Two reference points are marked on the capillary. One is the position of the top of the liquid when the thermometer is at the temperature of ice and water at equilibrium under standard atmospheric pressure. This is the freezing level in Fig. 10.1. The second reference point is obtained when the thermometer is at the boiling point of water (under standard atmospheric pressure). This is the boiling level in the figure.

Two temperature scales are most often encountered in daily life in the United States: **Celsius** and **Fahrenheit.** The Celsius scale (proposed in 1742 by the Swedish scientist Anders Celsius) places the freezing point of pure water at 0°C (degrees Celsius) and the boiling point at 100°C. Both of these phenomena are

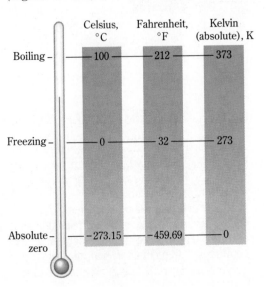

FIGURE 10.1

The boiling and freezing points of water can be used to illustrate how the three usual temperature scales are related.

measured at standard atmospheric pressure. Hence there are 100 degrees between the two reference points, which is why this scale is sometimes referred to as the centigrade scale. The German physicist Gabriel Fahrenheit originally proposed a sort of centigrade scale. He referenced 0°F (degrees Fahrenheit) as the freezing temperature of a saline solution and 100°F as the temperature of the human body (actually now known to be 98.6°F). On this scale, the freezing and boiling temperatures of pure water at 1 atm pressure are 32°F and 212°F, respectively. Thus 180 Fahrenheit degrees and 100 Celsius degrees span the same temperature range. The sizes of the degrees are hence related by $1\,F° = 100/180 = 5/9\,C°$. Notice that the latter notation refers to a temperature *range,* whereas °F and °C refer to specific temperature readings.

A third scale, the **Kelvin,** or **absolute,** scale, is of greatest importance in science. Its SI unit of temperature is the fundamentally defined unit called the *kelvin* (K). Changing temperature by one kelvin (we do not use the term *degree* kelvin) is the same as changing it by $1\,C°$. The freezing and boiling points of water are at 273.15 K and 373.15 K, respectively. We will see presently why the Kelvin scale is of basic scientific importance.

These historical definitions of the temperature scales have been supplanted, as we will see later. The new definitions, however, were chosen so as to preserve the scales essentially as they had been defined originally. As you can see from Fig. 10.1, there is a simple relationship between the Celsius temperature T_C and the absolute (Kelvin) temperature T:

$$T = T_C + 273.15$$

Although we do not use the Fahrenheit scale in this book, readings on it can be converted using the equations

$$T_C = (T_F - 32)(5/9)$$
$$T = 273.15 + (T_F - 32)(5/9)$$

We use thermometers routinely in our lives. Even so, there is a fundamental law of physics involving them that may have escaped your notice. When you place a thermometer in intimate contact with an object, the thermometer soon reaches a steady reading called the temperature of the object, and we say that the object and the thermometer are in *thermal equilibrium* with each other. If you now place this object in contact with an object that is at a higher temperature, the temperatures of the two objects will change and the objects will eventually reach thermal equilibrium at an intermediate temperature. We say that heat has flowed from the hotter to the colder object. These facts are well known. However, the following variation of this experiment is extremely important.

Suppose a thermometer reads the same temperature for two objects. What happens when these two objects are placed in intimate contact with each other? The answer is that nothing happens; the temperature of neither object changes. The two objects are in thermal equilibrium with each other. *Objects or systems that have the same temperature are in thermal equilibrium with each other.* This seemingly obvious statement is one form of the **zeroth law of thermodynamics,** which we state formally as follows:

Two bodies or systems that are in thermal equilibrium with a third body are in thermal equilibrium with each other.

As we see, temperatures are equal for objects that are in thermal equilibrium.

Galaxies such as this contain hundreds of billions of stars. It would take approximately a *trillion* such galaxies to contain one mole of stars.

10.2 THE MOLE AND AVOGADRO'S NUMBER

In the next section, we discuss how the pressure of a gas depends on its temperature and density. To facilitate the discussion, however, we need to use a few terms that are ordinarily learned in chemistry. Because you may be unfamiliar with these terms, let us now spend a short time discussing them.

The number of carbon atoms in 12 g of carbon* is called **Avogadro's number** N_A. Experiments show this number to be 6.02214×10^{23} atoms per 12 g of carbon and it is used to define a measure of the quantity of any substance, a quantity called the **mole** (abbreviation mol):

A mole of substance is the amount of substance that contains N_A particles.

For example, one mole of baseballs consists of 6.022×10^{23} baseballs. Similarly, one mole of water consists of N_A water molecules. As you see, the mole is a measure not of mass but of number of entities. To summarize,

Avogadro's number $= N_A = 6.02214 \times 10^{23}$ particles per mole

Because we most often use the SI units kilograms and kilomoles in discussions, we usually replace this value for N_A by its equivalent:

$N_A = 6.02214 \times 10^{26}$ particles/kmol

Two related terms that we should be familiar with are *atomic mass* and *molecular mass*. Both are represented by M.

The **molecular** (or **atomic**) **mass** M of a substance is the mass in kilograms of one kilomole of the substance.

*Precisely, in 12 g of the carbon-12 isotope.

For instance, since 12 kg of carbon 12 is defined to contain N_A atoms, 1 kmol of ^{12}C has an atomic mass $M = 12$ kg/kmol precisely. Some other examples of approximate values of M are: hydrogen, $M = 1$ kg/kmol; oxygen gas (O_2), $M = 32$ kg/kmol; water (H_2O), $M = 18$ kg/kmol; nitrogen gas (N_2), $M = 28$ kg/kmol. More precise values of M for all the elements can be found in Appendixes 1 and 2.

Illustration 10.1

The atomic mass of copper is 63.5 kg/kmol. Find the mass of a single copper atom.

Reasoning Because $M = 63.5$ kg/kmol, 63.5 kg of copper contains 6.022×10^{26} atoms. Therefore the mass of one atom is

$$\text{Mass per atom} = \frac{63.5 \text{ kg}}{6.022 \times 10^{26} \text{ atoms}} = 1.05 \times 10^{-25} \text{ kg/atom}$$

This same method can be used to find the mass of any atom or molecule for which M is known. Because M kilograms contains N_A entities, we have that

$$\text{Mass per entity} = \frac{M}{N_A}$$

Exercise Find the mass of an oxygen molecule, O_2. *Answer: 5.31×10^{-26} kg*

Example 10.1

Find the volume associated with a mercury atom in liquid mercury. For mercury, $\rho = 13{,}600$ kg/m^3 and $M = 201$ kg/kmol.

Reasoning

Question What assumption can I make about the way atoms are structured in mercury?
Answer Since we are talking about a liquid, you can assume that the atoms are "touching" one another. Thus the volume per atom can be calculated by finding the ratio of the total volume of a sample to the total number of atoms in that sample.

Question What sample should I take?
Answer The most convenient one would be a kilomole, since you know there are N_A atoms in it.

Question How can I find the volume of 1 kmol?
Answer You are given the density and M for mercury. You want to find cubic millimeters per kilomole from this. Notice that

$$\frac{\text{Volume}}{\text{Kmol}} = \frac{M \text{ (kg/kmol)}}{\rho \text{ (kg/m}^3)}$$

Question How do I find the volume of one atom?

Answer The volume of one atom is $1/N_A$ times the volume per kilomole.

Solution and Discussion Plugging in numbers,

$$\frac{\text{Volume}}{\text{Kmol}} = \frac{201 \text{ kg/kmol}}{1.36 \times 10^4 \text{ kg/m}^3}$$

$$= 1.48 \times 10^{-2} \text{ m}^3/\text{kmol}$$

Then

$$\frac{\text{Vol}}{\text{Atom}} = \frac{1.48 \times 10^{-2} \text{ m}^3/\text{kmol}}{6.022 \times 10^{26} \text{ atoms/kmol}}$$

$$= 2.45 \times 10^{-29} \text{ m}^3/\text{atom}$$

In order to get a better feeling for the size of something this small, let us use the formula for the volume of a sphere to find the radius of each atom. This formula is $V = \frac{4}{3}\pi r^3$. Thus

$$r = \left(\frac{3 \times 2.45 \times 10^{-29}}{4\pi}\right)^{1/3} = 1.8 \times 10^{-10} \text{ m}$$

The diameter of the atom is twice this, or 3.6×10^{-10} m. Thus even one of the most massive atoms has a diameter of only about 0.36 nm. One way to try to visualize this is to realize that 1 million of these atoms side by side would only span 0.36 mm!

10.3 THE IDEAL-GAS LAW

Some of the earliest investigators into the nature of temperature were concerned with how the pressure of a gas changes with temperature. Definitive experiments were carried out centuries ago, and today students still perform these basic experi-

Bubbles of gas in a liquid grow larger as they rise toward the surface. Why?

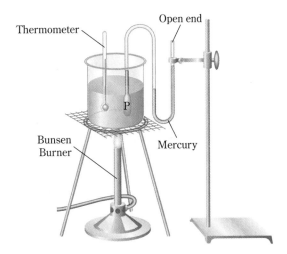

FIGURE 10.2

A simple device for measuring how temperature affects the pressure of a fixed volume of gas.

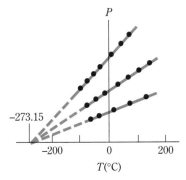

FIGURE 10.3

The pressure of a dilute gas at constant volume decreases as the temperature is lowered (Gay-Lussac's law). The three curves are for the same gas, but with different amounts of gas in the volume.

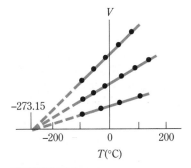

FIGURE 10.4

The volume of a dilute gas varies linearly with temperature, provided P is constant (Charles' law). The three curves are for the same gas, but at different pressures.

ments. A typical simple apparatus is shown in Fig. 10.2. The pressure of the gas is measured as a function of temperature while the volume of gas is held constant. When data from such an experiment are plotted, they lead to graphs such as the one shown in Fig. 10.3.

As you see from the graph, the data lead to a linear relation between absolute pressure (gauge pressure plus P_a) and temperature. Different straight lines are observed for different initial conditions within the container. In every case, however, *provided the gas is far from conditions under which it condenses, or liquefies,* a linear relation exists between temperature and pressure at constant volume.

Another informative experiment is to measure the volume of a gas as a function of temperature with the pressure held constant. Typical results are shown in Fig. 10.4. Here, too, a linear relation is found; the volume varies linearly with temperature. Again, this is true as long as the gas is far removed from condensation conditions.

Figures 10.3 and 10.4 show another interesting feature: *all the linear plots extrapolate to the same temperature intercept,* $-273.15°C$.

The equations that represent these linear experimental relationships are

P (at constant V) = (constant)$(T_c + 273.15\,\mathrm{C°})$
V (at constant P) = (constant)$(T_c + 273.15\,\mathrm{C°})$

It is important to realize that this behavior is observed for *any* ideal gas. According to these equations, both P and V go to zero when $T_c = -273.15°C$. This unique temperature is called **absolute zero,** and it forms the basis of the Kelvin temperature scale mentioned in Sec. 10.1. We cannot obtain data near this temperature on most gases because they condense to the liquid state at considerably higher temperatures. Nevertheless, the fact that such a unique temperature exists indicates that it may have some fundamental importance, which we will discuss in more detail later.

One further experimental result is that when T is held constant and P or V is changed, the product PV remains constant according to the following equation:

PV = (constant)(amount of gas)$(T_c + 273.15\,\mathrm{C°})$

You can check that this is consistent with the other two equations.

The amount of gas in a sample is usually measured by the number of moles, n, given by

$$n = \frac{m}{M}$$

where m is the mass of the gas sample and M is the atomic or molecular mass of the gas. The constant in the above equation for PV is another example of a universal physical constant whose value must be experimentally determined. It is called the **gas constant** and given the symbol R. Putting all these symbols together in the equation for PV, we obtain

$$PV = nRT \tag{10.1}$$

where we have used T (kelvins) $= T_c + 273.15\,C°$.

This is called the **ideal-gas law,** and gases that obey it are called *ideal gases*. All gases that are far removed from the conditions under which they condense show nearly ideal behavior. Repeated measurements have established the following SI value for the gas constant R:

$$R = 8314\,\text{J/kmol} \cdot \text{K} = 8.314\,\text{J/mol} \cdot \text{K}$$

You should check the units of R to see that they are consistent with the other quantities in Eq. 10.1.

10.4   USING THE GAS LAW

Now that we understand the meaning of the quantities in the gas law, we are in a position to apply it to various problems. In using this relation, it is very important that you pay attention to units. *Absolute temperatures must be used for T.* In the SI, the pressure P is in pascals (that is, N/m^2) and volumes are measured in cubic meters. R then has one of the values given in Sec. 10.3, depending on whether n is expressed in moles or kilomoles.

Example 10.2

Standard atmospheric pressure and temperature are $1.01325 \times 10^5\,\text{Pa}$ and $0.000\,C°$. Find the volume that 1.000 kilomole of an ideal gas occupies at these values of P and T.

R

Reasoning

Question   What basic principle determines the volume?
Answer   The gas law relates four quantities: P, V, T, and n. If we are given any three, it can be solved for the remaining one.

Question   How do the given data translate into the symbols in the gas law?
Answer   We are given $T_c = 0.000\,C°$, but we must use the Kelvin temperature, $T = 273.15 + T_c = 273.15\,\text{K}$. We also have $P = 1.000\,\text{atm} = 1.013 \times 10^5\,\text{Pa}$ and $n = 1.000\,\text{kmol}$.

Solution and Discussion   The gas law can be solved algebraically for V:

$$V = \frac{nRT}{P}$$

The data give (to four significant digits):

$$V = (1.000\,\text{kmol})(8314\,\text{J/kmol} \cdot \text{K})(273.15\,\text{K})/(1.013 \times 10^5\,\text{Pa})$$

$$= 22.42\,\text{m}^3/\text{kmol}$$

This is a convenient fact to remember:

One kilomole of any ideal gas occupies a volume of $22.4\,\text{m}^3$ under standard conditions.

Example 10.3

If 14.0 mg of nitrogen gas ($M = 28.0\,\text{kg/kmol}$) is held at 27.0°C in a container that has a volume of $5.00 \times 10^3\,\text{cm}^3$, what is the gas pressure in the container?

R

Reasoning

Question   Do I have enough data to solve for P from the gas law?
Answer   You have values for T_C, V, m, and M. With $m/M = n$, you have three of the four quantities given.

Question   Are the given units all in SI?
Answer   No. You must convert T_C to T, V to cubic meters, and m to kilograms.

Solution and Discussion   The gas law can be solved for P:

$$P = \frac{nRT}{V} = \frac{m}{M}\frac{RT}{V}$$

The SI quantities given are

$$T = 27.0 + 273 = 300\,\text{K} \qquad m = 14.0 \times 10^{-6}\,\text{kg} \qquad V = 5.00 \times 10^{-3}\,\text{m}^3$$

Then

$$P = \frac{(14.0 \times 10^{-6}\,\text{kg})(8314\,\text{J/kmol} \cdot \text{K})(300\,\text{K})}{(28.0\,\text{kg/kmol})(5.00 \times 10^{-3}\,\text{m}^3)}$$

$$= 249\,\text{N/m}^2 = 249\,\text{Pa}$$

Example 10.4

Use the gas law to determine the mass of air contained in a 50.0-cm³ flask at 700 torr pressure and 20°C. Air consists of approximately 80% N_2 and 20% O_2 by mass.

R

Reasoning

Question   With P, V, and T_C given, do I have enough information to obtain m?
Answer   You have enough to obtain n, but to obtain m you must also know the molecular mass, M.

Question   Since air is a mixture, not a pure substance, how can I find M?
Answer   You know from Sec. 10.2 that M (N_2) = 28 kg/kmol and M (O_2) = 32 kg/kmol. You also know the percentage of the total mixture contributed by each gas. Thus

$$M \text{ (air)} = (0.80)(28 \text{ kg/kmol}) + (0.20)(32 \text{ kg/kmol})$$
$$= 29 \text{ kg/kmol}$$

Question   What are the other quantities in SI units?
Answer $T = 273 + 20 = 293 \text{ K}$
$P = (1.103 \times 10^5 \text{ Pa/atm})(700 \text{ torr})/(760 \text{ torr/atm})$
$= 9.33 \times 10^4 \text{ Pa}$
$V = 50.0 \times 10^{-6} \text{ m}^3$

Solution and Discussion   In terms of m, the gas law is $PV = (m/m)RT$. Solving for m:

$$m = \frac{PVM}{RT}$$

$$= (9.33 \times 10^4 \text{ Pa})(50.0 \times 10^{-6} \text{ m}^3) \frac{(29 \text{ kg/kmol})}{(8314 \text{ J/kmol K})(293 \text{ K})}$$

$$= 5.5 \times 10^{-5} \text{ kg}$$

Example 10.5

An oil drum containing only air at 20°C is sealed off. It is then set out in the sun, where it heats up to 60°C. If the original pressure is 1.0 atm, what is the final pressure in the drum? Assume the volume of the drum remains constant as the temperature changes.

R

Reasoning

Question   Here I don't know either n or V. Is there a way to use the gas law without explicitly solving for them?
Answer   Yes. You know that n and V remain constant, since the drum remains the same size and is assumed to be airtight. This condition allows you to use the gas law to form ratios of the before and after conditions. The quantities n, V, and R will cancel in these ratios.

Question   How do I form the ratios?
Answer   Write the gas law twice, once for the original conditions and once for the final ones.

$$P_1 V = nRT_1 \quad \text{and} \quad P_2 V = nRT_2$$

Taking the ratio of the first equation to the second gives the simple result

$$\frac{P_1}{P_2} = \frac{T_1}{T_2}$$

Question Must these units all be in SI?
Answer You must *always* use the Kelvin scale T for temperatures. This is because T and T_C or T_f are related by *additive* numbers, which won't cancel in a ratio. All other quantities (P, V, and n) can be used in any units in ratios, since the multiplying conversion factors will cancel. Just be sure that before and after values are in the *same* units.

Solution and Discussion We are given $T_1 = 20 + 273 = 293\,\text{K}$ and $T_2 = 60 + 273 = 333\,\text{K}$. Also $P_1 = 1.0\,\text{atm}$. Then

$$P_2 = P_1 \frac{T_2}{T_1} = \frac{(1.0\,\text{atm})(333\,\text{K})}{293\,\text{K}} = 1.1\,\text{atm}$$

Notice what a different (and wrong) answer using T_C would give.

Example 10.6

A pressure gauge gives a reading of 190 kPa on your car tire on a day when the temperature is $-10°C$ and the barometric pressure is 800 torr. What will the pressure gauge read after the car has been driven and the tire temperature (including the air in the tire) is 35°C? Assume the volume of the tire does not change.

R *Reasoning*

Question This is the same type of problem as the previous example. Can I use gauge pressure directly in the gas law?
Answer No, for the same reason you cannot use T_C. The gas law requires total pressure, which differs from P_G by an additive amount. You can use any pressure units in a ratio, but both pressures must be total, not gauge, pressures.

Question What is the initial total pressure?
Answer $P_1 = P_a + P_G$

$$= (800/760)(1.01 \times 10^5\,\text{Pa}) + 1.90 \times 10^5\,\text{Pa}$$
$$= (1.06 + 1.90) \times 10^5\,\text{Pa} = 2.96 \times 10^5\,\text{Pa}$$

Question What is the equation which determines P_2?
Answer $\dfrac{P_2}{P_1} = \dfrac{T_2}{T_1}$ or $P_2 = P_1 \dfrac{T_2}{T_1}$

with $T_1 = 273 + (-10) = 263\,\text{K}$ and $T_2 = 35 + 273 = 308\,\text{K}$.

Solution and Discussion   Using the data, we obtain

$$P_2 = (2.96 \times 10^5 \, \text{Pa})(308/263) = 3.47 \times 10^5 \, \text{Pa}$$

Remember, this is *total* pressure. To find what the gauge reading is, you must subtract P_a:

$$(P_2)_G = 3.47 \times 10^5 \, \text{Pa} - (1.06 \times 10^5 \, \text{Pa})$$

$$= 241 \, \text{kPa}$$

Example 10.7

A diesel engine ignites its fuel-air mixture by compressional heating rather than by the use of spark plugs. To demonstrate this effect, consider a diesel engine whose compression ratio is 18:1. This means that when the engine is running, the piston changes the cylinder volume from an initial volume V_1 to a final volume $V_2 = \frac{1}{18}V_1$. Suppose the gaseous fuel-air mixture enters the cylinder at 300 K at a pressure of 740 torr when the cylinder volume is V_1. What is the temperature of the gas when the piston has changed the cylinder volume to V_2 and the pressure has risen to 37,000 torr?

R | *Reasoning*

Question   Can I use ratios of the gas law again?
Answer   Yes. Although here P, V, and T are all changing, they do so in a way such that PV/T remains constant ($PV/T = nR$).

Question   What is the equation for the ratio of temperatures?
Answer $P_1V_1/T_1 = P_2V_2/T_2$ gives

$$\frac{T_2}{T_1} = \frac{P_2V_2}{P_1V_1}$$

Solution and Discussion   The above equation yields

$$T_2 = T_1\frac{P_2}{P_1}\frac{V_2}{V_1}$$

$$= 300 \, \text{K}\left(\frac{37,000 \text{ torr}}{740 \text{ torr}}\right)\left(\frac{1}{18}\right) = 833 \, \text{K}$$

This is a temperature sufficiently high to ignite the fuel.

10.5 THE MOLECULAR BASIS FOR THE GAS LAW

The gas law, $PV = nRT$, expresses the pressure of an ideal gas in terms of its temperature. Let us now consider a little more thoroughly what we mean by an ideal gas. We know that a gas is composed of atoms or molecules of a substance (or a mixture of substances) which are free to fill any volume that contains them. More precisely, an ideal gas is one for which the following conditions hold:

1 The atoms or molecules composing the gas can be considered as *point* masses, taking up negligible volume compared with their container's volume V.
2 No significant forces act between the atoms or molecules except when they collide with each other and with the boundaries of the container. These collisions are all assumed to be *perfectly elastic*.

We shall now derive the connection between the temperature of a gas and the mechanical properties of its molecules. The model we use to obtain this relationship is a simplified version of the **kinetic theory of gases.**

To start, let us refer to Example 6.7. There we used the principles of energy and momentum conservation to show that a beam of particles exert a pressure on a wall when they collide with it. We assumed there that all the particles had the same mass m and same velocity **v.** We further assumed that all collisions were perfectly elastic. We obtained the result that the pressure on the wall could be written as

$$P = 4(KE)n_v$$

where KE = the kinetic energy of the particles (all the same) and $n_v = N/V$ = the number of particles per unit volume in the beam. (We have changed this last symbol from the n without subscript used in Chap. 6 so that we do not confuse it with the symbol for the number of moles.)

Only minor changes need to be made to this result to have it represent the pressure exerted by the molecules of a gas at temperature T rather than a directed beam of particles. These changes involve the following considerations:

1. Unlike the beam, the molecules in a gas do not all have the same speed. However, we can adequately describe the gas in terms of the *average* speed of the molecules. Thus the pressure the gas exerts will be expressed in terms of the *average* KE of its molecules.

2. In a gas, the particles are moving in all directions in three dimensions. On average, the x, y, and z components of their velocities must be equal, since there is nothing that defines a special direction in the container. This means that the contributions to $\langle KE \rangle$ from each of the three components of motion will be equal:

$$\overline{\tfrac{1}{2}mv_x^2} = \overline{\tfrac{1}{2}mv_y^2} = \overline{\tfrac{1}{2}mv_z^2}$$

and

$$\overline{\tfrac{1}{2}mv_x^2} + \overline{\tfrac{1}{2}mv_y^2} + \overline{\tfrac{1}{2}mv_z^2} = \overline{\tfrac{1}{2}mv^2} = \overline{KE}$$

Combining these expressions, we obtain

$$\overline{\tfrac{1}{2}mv_x^2} = \overline{\tfrac{1}{2}mv_y^2} = \overline{\tfrac{1}{2}mv_z^2} = \tfrac{1}{3}\overline{KE}$$

3. For the same reason given in item 2 above, as many particles on average will be traveling in the positive x, y, and z directions as in the negative directions. Consider the wall of the container that is perpendicular to the $+x$ axis (Fig. 10.5). Only particles having a component of velocity along positive x will strike this wall and contribute to the pressure. These particles possess an average kinetic energy of $\tfrac{1}{2}(\tfrac{1}{2}mv_x^2)$, or $(\tfrac{1}{6})\overline{KE}$.

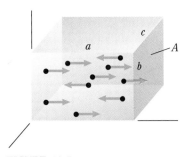

FIGURE 10.5

Half the molecules (those traveling in the $+x$ direction) will strike the area A.

Thus the needed modification to the result of Example 6.7 is to replace the KE of the beam by $\frac{1}{6}\overline{KE}$ of the molecules of the gas. Thus instead of $P = 4(KE)n_v$, we have, for an ideal gas

$$P = 4(\tfrac{1}{6})(\overline{KE})n_v = \tfrac{2}{3}(\overline{KE})n_v \qquad (10.2)$$

We are now ready to interpret temperature in terms of molecular energies in the gas. The expression for pressure in Eq. 10.2 can be equated to the pressure in the gas law:

$$\tfrac{2}{3}(\overline{KE})n_v = \frac{nRT}{V}$$

Now let us reconcile some of these symbols. The number of moles, n, is related to the total number of molecules, N, by $n = N/N_A$. Then $n/V = (N/V)/N_A = n_v/N_A$. With these substitutions and a little more algebraic manipulation, we can write

$$\overline{KE} = \frac{3}{2}\frac{R}{N_A}T = \tfrac{3}{2}kT \qquad (10.3)$$

where $k = R/N_A$ is called **Boltzmann's constant** and has the value

$$k = 1.38 \times 10^{-23}\,\text{J/K}$$

Equation 10.3 is one very important result of the kinetic theory of gases. It gives the meaning of gas temperature as a measure of the average kinetic energy of the gas molecules.

Absolute temperature is a measure of the average translational kinetic energy of the molecules in an ideal gas.

Notice that the classical meaning of absolute zero (0 K) is that temperature at which molecular motion ceases.

There is a second important observation we can now make about the meaning of thermal equilibrium. You may recall that substances in thermal equilibrium with each other have the same temperature.

If two ideal gases are in thermal equilibrium with each other, the average translational kinetic energy per molecule is the same in both.

This is true whether or not the gases have the same composition.

We can go one step further and calculate the average of v^2 for the molecules. Assuming all the gas molecules have the same mass m, we have

$$\overline{KE} = \overline{\tfrac{1}{2}mv^2} = \tfrac{1}{2}m\overline{v^2} = \tfrac{3}{2}kT \qquad (10.4)$$

Thus

$$\overline{v^2} = 3kT/m$$

If we take the square root of this quantity we obtain a kind of average speed, called the *root-mean-square (or rms) speed*.

$$v_{\text{rms}} = \sqrt{\frac{3kT}{m}} \qquad\qquad (10.5)$$

The rms speed is *not* the same as the usual average, or mean, speed. *Rather, it is the speed that a molecule with the average kinetic energy possesses.* It is important to understand that this value of speed is an indication of the average speed between collisions. Collisions are always interrupting and redirecting individual motions.

Although these statements apply only to an ideal gas, we shall see in later chapters that absolute temperature is a measure of kinetic energy per molecule even in liquids and solids. However, it is not a simple measure.

Before leaving this section, we should point out that our results apply to real gases only at moderate and high temperatures. Near absolute zero, very strange things happen; some metals become resistanceless conductors of electricity, and certain fluids flow without friction (that is, their viscosity becomes zero). The behavior of molecules at low temperatures must be dealt with in terms of quantum mechanics, a topic we consider in the last few chapters of this book as well as in some of the Modern Perspectives featured throughout the book.

Illustration 10.2

What is the root-mean-square speed of a nitrogen molecule in air at 27.0°C?

Reasoning   Equation 10.5 requires that we know the mass per molecule, m, as well as the temperature. The mass per molecule is the molecular mass of the gas, M, divided by the number of molecules in a mole, N_A. Nitrogen, N_2, has a molecular mass of 28.0 kg/kmol. Thus

$$m = \frac{M}{N_A} = \frac{28.0\ \text{kg/kmol}}{6.02 \times 10^{26}/\text{kmol}} = 4.65 \times 10^{-26}\ \text{kg}$$

Equation 10.5 then yields

$$v_{\text{rms}} = \sqrt{\frac{3kT}{m}} = \sqrt{\frac{3(1.38 \times 10^{-23}\ \text{J/K})(300\ \text{K})}{4.65 \times 10^{-26}\ \text{kg}}} = 517\ \text{m/s}$$

Notice that this is a very high speed, about one-third of a mile per second! In view of this, can you explain why it takes so long for the odor of a gas—perfume molecules, for example—to cross a room?

10.6   DISTRIBUTION OF MOLECULAR SPEEDS

We implied in the previous section that not all molecules in a gas have the same speed, but we did not specify what the distribution of speeds is. That is, we did not state what fraction of molecules have a given speed or what range of speeds occur in a gas. In 1860 the Scottish physicist James Clerk Maxwell used the kinetic theory of gases to derive a theoretical expression that describes the relative number of molecules in a gas at a given temperature T that have a given speed. The result,

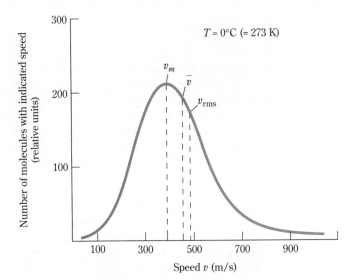

FIGURE 10.6

Maxwell distribution of speeds in a sample of O_2 gas at 273 K. The values of the most probable speed (v_m), the average speed ($\bar{v}$), and the root-mean-square speed (v_{rms}) are shown.

which we call a **Maxwellian distribution,** is shown graphically in Fig. 10.6 for the case of a gas of O_2 molecules at 273 K. Notice that, in addition to v_{rms}, two other statistically important speeds are shown. One, v_m, is the *most probable speed,* that is, the speed that more molecules have than any other speed. The other, $\bar{v}$, is the *average speed* of the distribution. Let us summarize the values of these three speeds:

$$v_m = \sqrt{2}\sqrt{\frac{kT}{m}} = 1.414\sqrt{\frac{kT}{m}}$$

$$\bar{v} = \sqrt{\frac{8}{\pi}}\sqrt{\frac{kT}{m}} = 1.596\sqrt{\frac{kT}{m}}$$

$$v_{rms} = \sqrt{3}\sqrt{\frac{kT}{m}} = 1.732\sqrt{\frac{kT}{m}}$$

The field of marathon runners shows a distinct distribution in speeds.

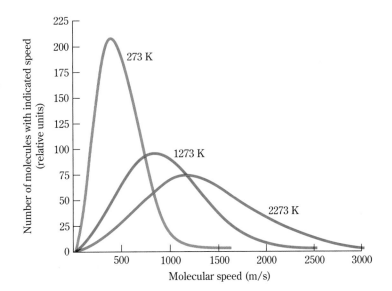

FIGURE 10.7

The distribution of speeds in a gas of N_2 molecules. The peak of the distribution moves to higher speeds and the spread in speeds widens as the temperature of the gas increases.

Thus once you find one of these values, the others follow easily.

Figure 10.7 shows how the distribution of speeds in a given sample of N_2 gas changes as the temperature is varied. As the temperature is increased, the distribution of speeds spreads and the peak, v_m, moves toward higher values. Due to the width of the distribution curve, there are always some (relatively few) molecules moving very slowly and others with speeds many times larger than v_{rms}.

Maxwell's theory was the subject of a great deal of controversy, since an experimental test of his theory required a vacuum chamber in which to observe molecular speeds without collisions continually changing them. Vacuum techniques had not been developed at the time Maxwell announced his theory, and it was not until 1926 that the German physicist Otto Stern was able to perform the experiment that confirmed Maxwell's predicted speed distribution. Maxwell's theory and its confirmation mark an important step in achieving a fundamental understanding of the thermal properties of matter, a topic we investigate in the next few chapters.

LEARNING GOALS

Now that you have finished this chapter, you should be able to

1 Define (a) thermal equilibrium, (b) thermometer, (c) Celsius scale, (d) Fahrenheit scale, (e) absolute zero, (f) Kelvin scale, (g) zeroth law of thermodynamics, (h) Avogadro's number, (i) mole and kilomole, (j) atomic mass and molecular mass, (k) gas constant R, (l) Boltzmann's constant k, (m) ideal gas, (n) ideal-gas law, (o) kinetic theory of gases, (p) root-mean-square speed

2 Sketch a diagram showing the three common temperature scales. On each, locate absolute zero and the freezing and boiling points of water. Convert temperatures between these three scales.

3 Compute the mass of an individual atom or molecule of a substance, given the substance's atomic or molecular mass, M.

4 Compute the number of moles or kilomoles in a sample of known mass, when the atomic or molecular mass of the substance is known.

5 Given values for two of the three quantities P, V, and T, use the ideal-gas law to find the third one.

6 State under what conditions a gas is an ideal gas.

7 Compute the average translational kinetic energy of the atoms or molecules of an ideal gas when the gas temperature is given.

8 Compute the root-mean-square speed of the atoms or molecules in a given mass of an ideal gas when the temperature and the atomic or molecular mass of the gas are given.

SUMMARY

DERIVED UNITS AND PHYSICAL CONSTANTS

Avogadro's Number: The Mole (N_A)
$N_A = 6.02214 \times 10^{23}$ particles/mol

Gas Constant (R)
$R = 8314\,\text{J/kmol} \cdot \text{K}$

Boltzmann's Constant (k)
$k = R/N_A = 1.38 \times 10^{-23}\,\text{J/K}$

DEFINITIONS AND BASIC PRINCIPLES

Temperature Scales
The following reference points are for pure water at an ambient pressure of 1 atm:

	°C	°F	K
Boiling point	100	212	373.15
Freezing point	0	32	273.15

INSIGHTS
1. The kelvin is the basic SI unit of temperature.
2. The Celsius degree is equal in size to the kelvin.
3. The Fahrenheit degree is 5/9 as large as the Celsius degree.
4. 0 K is absolute zero.
5. The relation between T_C and T_F is $T_C = (T_F - 32)(5/9)$

The Mole and Avogadro's Number
Avogadro's number (N_A) is defined as the number of atoms in precisely 12 g of the isotope ^{12}C.

One mole is any collection of N_A entities.

The molecular (or atomic) mass of a substance is the mass of one mole of the molecules (or atoms) of the substance.

Ideal-Gas Law
$PV = nRT = NkT$

INSIGHTS
1. Temperature must always be expressed in kelvins, even when using ratios of this equation to compare different conditions.
2. In the ideal-gas law, n represents the number of moles or kilomoles, whereas N represents the number of atoms or molecules.
3. The gas constant, R, can be expressed in many different units. You must always be sure that the units of P and V are consistent with those in the value of R.
4. The pressure P is the total gas pressure, not gauge pressure.

Kinetic Theory of Gases
The average translational kinetic energy per atom or molecule in an ideal gas is related to the gas temperature by

$$\overline{\text{KE}} = \tfrac{1}{2}m\overline{v^2} = \tfrac{3}{2}kT$$

The root-mean-square (rms) speed of the atoms or molecules in the gas is

$$v_{\text{rms}} = \sqrt{\overline{v^2}} = \sqrt{\frac{3kT}{m}}$$

QUESTIONS AND GUESSTIMATES

1 Compare the gravitational potential energy of a nitrogen molecule that is 1 m above the ground with its translational kinetic energy when the temperature is (*a*) 0°C and (*b*) −270°C.

2 Although the air is composed mostly of N_2 molecules, there is some O_2 present, of course. Do both kinds of molecules have the same average speed? What is the exact relation between their average speeds?

3 To escape from the earth, an object must be shot out from it with a speed of at least 11,200 m/s. Use this fact and an approximate atmospheric temperature to explain why only a tiny amount of hydrogen exists in the atmosphere, even though billions of years ago there may have been more hydrogen than nitrogen in it.

4 Boyle's law for gases states that the volume of a gas varies inversely with pressure, provided the mass and temperature of the gas are maintained constant. Show that Boyle's law is a special case of the ideal-gas law.

5 Charles' law for gases states that the volume of a gas increases in direct proportion to the absolute temperature, provided the pressure and mass of the gas are maintained constant. Show that this is a special case of the ideal-gas law.

6 Dalton's law of partial pressures states that the total pressure of a mixture of gases is equal to the sum of the partial pressures of the gases in the mixture. Using the ideal-gas law and kinetic theory, justify Dalton's law.

7 Hydrogen and oxygen gas are sealed off at atmospheric pressure in a strong glass jar containing two electrodes. A spark from the electrodes ignites the gases so that the reaction

$2H_2 + O_2 \rightarrow 2H_2O$ results. Will the pressure in the tube have changed after the temperature has come back to its original value (200°C)? Explain. What if the original temperature is 200°C and the final temperature is 20°C?

8 The earth's atmosphere contains about 10^{19} kg of gas. When Julius Caesar gasped, "Et tu, Bruté!" as he lay mortally

wounded, he exhaled a certain volume of nitrogen molecules. Estimate how many of these historic molecules you breathe in with each breath.

9 What would a mercury barometer read in a spaceship orbiting the earth if the air pressure in the spaceship is 75 cmHg?

PROBLEMS

Section 10.1

1 Convert the following to the other two temperature scales: (*a*) 74°F, (*b*) −28°C, (*c*) 780 K.

2 Convert the following to the other two temperature scales: (*a*) 72°C, (*b*) −22°F, (*c*) 230 K.

3 The boiling point of liquid hydrogen is −252.87°C. Express this temperature in degrees Fahrenheit and in kelvins.

4 On a certain day the range between the lowest and the highest temperatures is 62 F°. Calculate this change in temperature in degrees Celcius and in kelvins.

5 The boiling point of a material is 486.60°C and its melting point is 528.4°F below its boiling point. (*a*) What is the melting point in degrees Celsius? (*b*) Determine the boiling and melting points on the Fahrenheit scale.

6 If the temperature of a substance on the Celsius scale changes by ΔT_C, show that the corresponding change on the Fahrenheit scale is $\Delta T_F = \frac{9}{5}\Delta T_C$.

7 It is believed that the highest temperature ever recorded on earth, which occurred in Libya in 1922, was 136°F. The lowest recorded temperature was −128.56°F, at the Vostok Station, Antarctica in 1983. Convert these temperature extremes to degrees Celsius and to kelvins.

8 At what temperature do the Fahrenheit and Celsius scales have the same numerical value?

9 What temperature of an object would be the same on the Fahrenheit and Kelvin scales?

10 The body temperature of a healthy person is about 98.6°F. Express this in degrees Celsius and kelvins.

Section 10.2

11 What are the masses of individual (*a*) gold, (*b*) silver, and (*c*) iron atoms?

12 The chemical formula for ammonia gas is NH_3. What is the mass of a molecule of ammonia?

13 Benzene has the chemical formula C_6H_6. How many molecules of benzene are there in a sample of 50 g?

14 How many atoms are there in a 20-g block of pure copper?

15 A beaker contains 80 g of pure water. How many molecules of water are there in the beaker? The chemical formula for water is H_2O.

16 The molecular mass of nylon is 10,000 kg/kmol. The density of nylon is 1100 kg/m³. (*a*) Find the mass of a nylon molecule. How many of these molecules are there in (*b*) 1 kg and (*c*) 1 m³ of nylon?

17 Ethyl alcohol (C_2H_5OH) has a density of about 790 kg/m³.

Find (*a*) the mass of an ethyl alcohol molecule and (*b*) the number of such molecules in 1 liter of ethyl alcohol.

18 Consider a 60-kg man to be one huge molecule. What is his molecular mass?

Sections 10.3 and 10.4

19 A 1-liter tank contains oxygen O_2 at 22°C and a gauge pressure of 2.2×10^6 Pa. What mass of oxygen is in the tank?

20 A 2-liter tank contains helium gas (He) at 33°C and 1200 kPa. What mass of helium is in the tank?

■21 Estimate the total mass of air in an unheated room of size 6 m × 8 m × 10 m on a winter day when the room temperature is 20°F. Take the average molecular mass of air to be 28.8 kg/kmol. How much air will enter or leave the room if the temperature is raised to 75°F? Assume the room remains at atmospheric pressure.

22 A test tube filled with an ideal gas at a gauge pressure of 180 kPa at 27°C is completely sealed. What will be the gauge pressure in the tube if it is heated to 384°C?

■23 A half-liter flask weighs 568 mg more when it is filled with an unknown gas than when it is evacuated. When the flask is filled, the gas temperature is 23°C and the pressure is 80 kPa. What is the molecular mass of the gas molecules?

24 A sealed test tube contains nitrogen gas N_2 at 27°C; the gauge pressure is 240 kPa. What is the gauge pressure of the gas when it is cooled to a temperature of −88°C?

25 What volume of air originally at 100 kPa is needed to fill an automobile tire of volume V_o to a gauge pressure of 160 kPa? Assume the temperature of the air does not change.

■26 A spherical air bubble is released by a submarine at the bottom of a lake. The volume of the bubble triples by the time it rises to the surface of the lake. Assuming that the temperature of the lake and the air does not change during the ascent of the bubble, estimate the depth of the lake.

27 A 1-liter tank contains oxygen gas at a gauge pressure of 840 kPa. What volume would the gas occupy if it is expanded until it reaches atmospheric pressure, 100 kPa? Assume the gas temperature remains constant.

28 A gas at room temperature (27°C) and atmospheric pressure 100 kPa is compressed to a volume one-tenth as large as its original volume and an absolute pressure of 2500 kPa. What is the new temperature of the gas?

29 The pressure of a gas in a tank is tripled while its volume is

halved. Find the ratio of the final temperature of the gas to its original temperature.

■30 A tank contains 1 mol of oxygen gas at an absolute pressure of 500 kPa and a temperature of 27°C. (*a*) If the gas is heated at a constant volume until the pressure becomes four times that of the original, what is the new temperature of the gas? (*b*) If the gas is heated such that both the volume and pressure are doubled, what is the new temperature of the gas?

■31 In a diesel engine, the piston compresses air at 30° from approximately atmospheric pressure to a pressure of about 5400 kPa and a volume of about one-fifteenth the original volume. What is the final temperature of the compressed air?

■32 One way to cool a gas is to let it expand. In a typical cooling process, a gas at 27°C and 4000 kPa is expanded to atmospheric pressure and a volume 36 times larger. What is the final temperature of the cooled gas?

■33 A gas at 27°C and an absolute pressure of 1000 kPa is suddenly expanded into a chamber having 12 times the volume. Its new temperature is −10°C. What is the final pressure of the gas?

■34 A fish at a depth of 10 m in a freshwater lake exhales an air bubble whose volume is V_o. Find the volume of the bubble just before it reaches the surface. Assume that the temperature of the bubble remains constant.

■35 A 16-cm-long cylindrical test tube open at one end is inverted and pushed vertically down into water. When the closed end is at the water surface, how high has the water risen inside the tube? Assume the air pressure at the water's surface (and in the tube before submerging) is 1.00 atm. Also assume that the air in the tube remains at constant temperature as it is immersed.

■36 A balloon used for weather forecasting is designed to expand to a maximum radius (assuming that it can be considered as a hollow sphere) of 24 m when in flight at the altitude, where the air pressure is only 3 kPa and the temperature is −73°C. If the balloon is filled with helium at atmospheric pressure and is at 27°C, what is the volume of the balloon at the lift-off?

37 One liter of water is converted to steam (water vapor) at atmospheric pressure and 100°C. What is the volume of the resulting water vapor?

38 Use the ideal-gas law and the definition of mole in terms of mass of a gas to find the density of a gas.

39 Using the ideal-gas law determine the density of oxygen gas O_2 at standard temperature and pressure.

Sections 10.5 and 10.6

40 The temperature of the sun's interior is estimated to be about 14×10^6 K. Protons ($m = 1.67 \times 10^{-27}$ kg) comprise most of its mass. Estimate the rms speed of a proton by assuming that the protons act as particles in an ideal gas.

41 What is the temperature at which the rms speed of nitrogen molecules equals the rms speed of helium at 27°C?

42 At what temperature do the molecules of an ideal gas have an rms speed that is eight times their rms speed at 0°C?

43 What is the average translational kinetic energy of oxygen molecules at room temperature (27°C)?

44 The escape velocity for a projectile on earth is about 11.2 km/s. (*a*) At what temperature do hydrogen molecules have an rms speed equal to this speed? (*b*) Repeat for nitrogen N_2 and oxygen O_2 molecules.

45 The escape velocity on the surface of moon is about 2.37 km/s. At what temperature do helium atoms have an rms speed equal to this speed?

46 The temperature of outer space is about 3 K and consists mainly of a gas of single hydrogen atoms. On the average, there is about one such atom per cubic centimeter of volume. (*a*) Find the pressure of this gas in outer space; express your answer in atmospheres. (*b*) Find the average kinetic energy of one of these hydrogen atoms. (*c*) What speed does an atom of this energy have?

47 Show that the pressure of an ideal gas can be written $P = \rho v^2/3$.

■48 Find the density of water vapor (steam) at 1 atm and 100°C if it is considered as an ideal gas. For comparison the actual density of steam is 0.598 kg/m³. Justify any difference you may note.

49 If the rms speed of a molecule in a gas at 27°C is 80 m/s, what is its mass? Is this a realistic molecular example?

■50 A beam of particles, each of mass m_o and speed v, is directed along the x axis. The beam strikes an area 1 mm square, with 1×10^{16} particles striking per second. Find the pressure on the area due to the beam if the particles stick to the area when they hit. Evaluate for an electron beam in a television tube, where $m_o = 9.1 \times 10^{-31}$ kg and $v = 8 \times 10^7$ m/s.

■51 A 2.5-L cubic container contains a mixture of helium (He) and hydrogen gases in equilibrium at 120°C. (*a*) What is the average kinetic energy of each type of molecule? (*b*) What are the rms speeds of the molecules? (*c*) If the container has 1 mol of helium and 2 mol of hydrogen, what is the total pressure inside the container?

Additional Problems

■52 A sealed cubical container 24 cm on each edge contains twice Avogadro's number of molecules at a temperature of 27°C. What is the force exerted by the gas on one of the walls of the container?

■53 A vertical right circular cylinder of height 36.00 cm and base area 10.0 cm² is sitting open under standard temperature and pressure. A gas-tight 4.8-kg piston is placed in the cylinder and allowed to fall to an equilibrium height. At equilibrium, what are the height of the piston and the pressure in the cylinder? Assume the final temperature to be 0°C.

■54 A 1-meter-long narrow glass tube is sealed at one end and placed in a horizontal position. A drop of mercury large enough to close off the tube is placed at the midpoint of the tube when the temperature is 27°C. The sealed end is then immersed in boiling water (temperature 100°C). What will be the new position of the mercury drop in the tube?

55 A vertical capillary tube has a 6-cm column of mercury filling its lower sealed end. The tube is sealed off (at atmospheric pressure) at a point 20 cm above the top of the mercury. If the tube is inverted, how long is the air column at its bottom?

56 A piece of dry ice (CO_2) is placed in a test tube, which is then sealed off. If the mass of dry ice is 0.4 g and the sealed test tube has a volume of 22 cm³, what is the final pressure of the CO_2 in the tube if all the CO_2 vaporizes and reaches thermal equilibrium with the surrounding at 27°C?

57 When a 24-cm³ test tube is sealed off at a very low temperature, a few drops of liquid nitrogen condense in the tube from the air (the boiling point of nitrogen is −210°C). What is the nitrogen pressure in the tube when the tube is warmed to 27°C if the mass of the drops is 0.08 g?

58 In Fig. P10.1, a frictionless piston of area A and mass M separates two equal volumes V_o of ideal gas in which the pressure is P_o. The cylinder is now set on end. Find the equilibrium upper volume in terms of P_o and V_o.

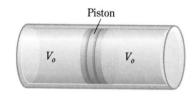

FIGURE P10.1

59 A spherical balloon ($V = 5\,m^3$) is filled with helium gas ($M = 4.0\,kg/kmol$) on a day on which atmospheric conditions have their standard values. (*a*) How many kilograms of helium are in the balloon if it floats in air? Neglect the mass of the balloon. (*b*) What is the pressure of the helium?

60 In Fig. P10.2, a uniform tube with an open stopcock is lowered into mercury so that 12 cm of the tube remains unfilled. After the stopcock is closed, the tube is lifted 8 cm. What is the height y in the tube? Assume standard atmospheric conditions.

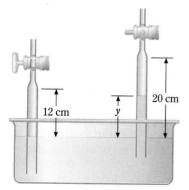

FIGURE P10.2

61 At a constant pressure of 1 atm, a container of air is observed to expand from 22 liters to 53.6 liters as the temperature is increased from 27°C to 750 K. Is there a leak in the container? If so, how much air has leaked in or out during this process?

62 Suppose you have an insulated rectangular box 1 m in length, having cross-sectional area A. The box is divided by a movable, insulated gas-tight partition as shown in Fig. P10.3. In the left compartment you place 105 g of argon gas at 300 K. In the right compartment you place 15 g of helium gas at 260 K. What position will the movable partition assume if the temperatures of the gases remain the same?

FIGURE P10.3

63 The atmosphere of Venus is practically all (96%) CO_2. At the surface the temperature is approximately 750 K and the pressure is approximately 90 earth atmospheres. Find the density of CO_2 and the rms speed of CO_2 molecules on Venus.

64 A 2.0-liter container of an ideal gas at 240 kPa and 20°C is connected through a small tube to an 8-liter container of the same gas at 27°C and 100 kPa. After the gas has come to equilibrium, its temperature is 23°C. What is the final pressure of the gas?

CHAPTER 11

THERMAL PROPERTIES OF MATTER

The discussion in the preceding chapter, which described the effect of heat on gases, considered gaseous atoms and molecules as acting like balls darting here and there. We ignored the fact that the atoms and molecules have internal structure and that they may possess forms of energy other than translational kinetic energy. Using these same simplifications, early investigators were able to achieve good agreement between theory and experiment for many gases. In liquids and solids, however, many other complications influence the thermal behavior of the atoms and molecules. The assumptions made in describing ideal gases no longer provide adequate descriptions of experimental results. Let us now see how we describe the thermal properties of these more complex systems.

11.1 THE CONCEPT OF HEAT

It has long been known that hot objects can be used to heat cooler objects. Only since the middle of the ninteenth century, however, has there been any real understanding of the processes involved. Not surprisingly, our understanding of the nature of heat developed rapidly as the kinetic theory of gases evolved. As we saw in the preceding chapter, the kinetic theory leads at once to a physical meaning for temperature. The absolute temperature T of a gas is proportional to the average

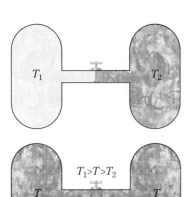

FIGURE 11.1

When two gases are brought into contact with each other, collisions between molecules originally at T_1 and molecules at T_2 cause the average molecular kinetic energy in both cylinders to change, thereby equalizing the temperatures.

translational kinetic energy of a molecule in the gas. We concluded that in a gas, the average translational kinetic energy of a molecule of mass m_o can be found from the relation

$$(\tfrac{1}{2}m_o v^2)_{\text{av}} = \tfrac{3}{2}kT \tag{10.4}$$

where $k = 1.38 \times 10^{-23}$ J/K is Boltzmann's constant.

Suppose now that two containers of gas, originally at temperatures T_1 and T_2 ($T_1 > T_2$), are brought together and the gases allowed to mix, as in Fig. 11.1. We know from experiment that the temperature of the mixed gas comes to some final value T intermediate between T_1 and T_2. The interpretation of this behavior on the basis of the kinetic theory is that the hot, higher-average-energy molecules from container 1 collide with the cold, lower-average-energy molecules from container 2. In these collisions the higher-energy molecules lose energy (decreasing in temperature) and the lower-energy molecules gain energy (increasing in temperature). This exchange of energy proceeds until the mixture of gases has a single average energy, after which collisions do not result in a net gain or loss of energy on the average. This equilibrium condition means that the gases have achieved a single, steady temperature T, where $T_1 > T > T_2$. The same process occurs when liquids or solids at different temperatures are brought into contact.

From this and many similar considerations, we conclude that, when two bodies at different temperatures are brought into contact with each other, energy is transferred, or *flows*, from the hotter to the cooler body. *The energy that is transferred in a situation such as this is what we refer to as heat energy.*

Heat energy is the energy transferred from a warm body to a cooler one as a result of the temperature difference between the two.

There is a corollary to this statement:

If two objects are at the same temperature, no heat energy is exchanged between them when they are brought into contact.

This condition when objects at the same temperature in contact with each other do not exchange energy, is a condition known as **thermal equilibrium.** This concept forms the basis for a statement which we call the **zeroth law of thermodynamics:**

If two objects are each in thermal equilibrium with a third object, they are in thermal equilibrium with each other.

This statement may seem obvious,* but it is the basis for being able to measure temperature. If a thermometer (the third object) comes to thermal equilibrium with two different objects and reads the same temperature in both cases, we can conclude that the other objects are at the same temperature without having to bring them into contact with each other.

---

*In fact, this statement was considered so obvious that it wasn't until after the *first* law of thermodynamics had been stated that the need for an explicit definition of temperature based on thermal equilibrium became apparent. Thus the name "zeroth" law was given to this definition, emphasizing that it underlies the first law.

11.2 THERMAL ENERGY

Let us now investigate what happens when heat energy flows into a substance. First let us consider a monatomic gas, such as helium. To a first approximation, each atom acts like a compact ball that darts here and there. Although it has some rotational kinetic energy because of its spinning motion on its axis, this energy ($\frac{1}{2}I\omega^2$) is negligibly small relative to the translational kinetic energy because the atom's moment of inertia is very, very small. Therefore we can neglect the rotational KE and treat the atom as if it has only translational KE.

Diatomic molecules, such as N_2 and O_2 (Fig. 11.2), have much larger moments of inertia due to the distance between the two atoms making up each molecule. As a result, their rotational energy can be comparable to their translational energy.

In addition to translational and rotational energy, diatomic molecules have a third kind: vibrational. The chemical bond between atoms in Fig. 11.2 is shown as a spring because the atoms can vibrate along the line joining them much like masses on the end of an elastic spring. This vibration involves both kinetic energy of motion and the potential energy associated with stretching and compressing the bond. Unlike the situation in a monatomic gas, then, heat energy added to a diatomic gas does not all appear as translational energy of the molecules. Some of the heat can go into these other modes (rotational and vibrational) of internal energy.

In gases composed of more complex molecules than diatomic, there are additional ways the molecules can rotate and vibrate. Then an even smaller share of added heat energy becomes translational kinetic energy. This leads us to conclude that the more complex the molecules making up a gas, the more heat energy must be added to increase the temperature of the gas by a given amount. The precise relation between heat added and the resulting temperature rise is the subject of Sec. 11.4.

In liquids and solids, the situation becomes very complex. In addition to bonds within molecules, there are bonds between adjacent molecules, and so added heat can cause a multitude of internal motions within the bulk of the material. In all cases, these motions are continually altered by random collisions of the moving atoms and as a result have no set, repeatable direction; they are random and are called *thermal motions*. We term the energy resident in these random motions *thermal energy*, which we encountered in Chap. 5 when we discussed the effect of friction forces.

There is an important difference between *heat* and *thermal energy*. Heat is the energy that flows from one substance to another because of a difference in temperature. Thermal energy is the energy contained within a substance by virtue of the random motions of its atoms and molecules. When heat is added to a substance, some of it can do mechanical work, as in the case of a piston moved by heated gas. Thus not *all* of the heat added necessarily becomes thermal energy.

Thermal energy is the energy associated with the random motion of atoms and molecules.

When heat flows into a substance, it most often becomes thermal energy; other possibilities are discussed later. Also, the thermal energy of a substance can be increased by mechanical means as well as by adding heat.

Prior to the end of the eighteenth century, the study of heat and the study of mechanics were completely separate. In the 1780s the American physicist Benjamin Thompson first realized there was a connection between mechanical work and the production of heat. He was involved in the boring of cannon in Bavaria and noticed that as long as the boring machines were operating, the cannon barrels were

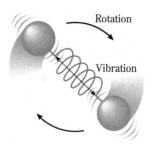

FIGURE 11.2

A diatomic gas molecule has both translational and rotational energy. It also has vibrational energy associated with the springlike bond between its atoms.

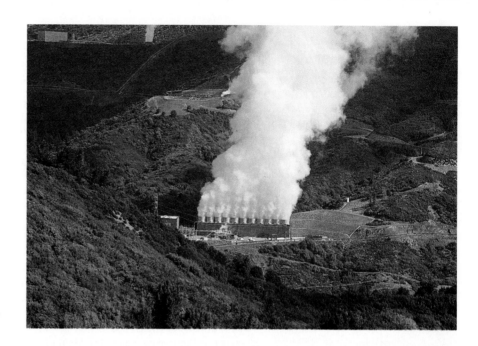

At some locations such as this in California, thermal energy from the earth's interior (geothermal energy) is close enough to the surface to be used to produce electricity.

being heated. Prior to this, the prevailing view of heat was that it was a fluid, called *caloric;* hot objects possessed caloric, and cold objects did not. As hot and cold objects came in contact, the caloric flowed from the hot to the cold, resulting in a leveling of temperature. Thompson's observations showed that heat could be *produced* by the friction of mechanical forces. In the middle of the nineteenth century, the work of the English physicist James Prescott Joule accurately determined the equivalence between mechanical units of energy and thermal units of heat.

Every Scout knows that you can start a fire by rubbing pieces of wood together vigorously. Friction work sets molecules at the two wood surfaces into violent

Meteors, sometimes called shooting stars, are dramatic examples of the conversion of kinetic energy to thermal energy. As these small bits of matter traveling at high speed enter the earth's atmosphere, friction with the air causes them to heat and vaporize.

Controversies in Physics: The Nature of Heat

We have few sensory perceptions that are as basic and common as that of feeling hot and cold. References to the nature of heat go back at least as far as Lucretius (first century B.C.), a Roman poet who wrote of heat as a material substance. Yet the awareness that heat is a form of energy did not come about until the middle of the nineteenth century. The story of competing ideas about heat and the arguments put forth to support them not only illustrates the true nature of scientific progress, but is one of great importance. The historian Cajori calls the first law of thermodynamics "the greatest generalization in physics of the nineteenth century." We live in an era that is so dependent on the conversion of heat into mechanical work (the internal combustion engine and steam turbine, for example) that our economy can be characterized as a "thermodynamic economy."

Basically, the two competing theories of heat were those of a material substance, called *caloric,* and the opposing concept of heat as the motion of the particles of which objects are made. Among the most prominent names in science during the seventeenth century who looked upon heat as vibratory motion of the particles of a substance were Descartes, Boyle, and Newton. However, this theory did not have a sound experimental basis, and was overthrown in the eighteenth century by proponents of the caloric theory. It was during this same time that the steam engine was developed by Thomas Newcomen in England and James Watt in Scotland.

The fundamental assumptions of the caloric theory were: (1) caloric was a fluid capable of penetrating all spaces and flowing into and out of all bodies, and (2) caloric was strongly attracted to matter, but exerted a repulsive force on itself. According to this theory the structure of matter was determined by a balance between the gravitational attraction of atoms toward each other and the self-repulsion of the caloric contained in a body. (Remember, at this time the electromagnetic structure of atoms was not known, and the strength of the gravitational force, G, wasn't measured until the close of the century.) The idea of an "imponderable" (literally, "unable to be weighed") fluid permeating matter has been applied many times throughout history as an attempt to explain physical phenomena.

The caloric theory successfully explained many observed facts. Hot objects contained more caloric, while cold objects contained less. The heating and cooling of bodies was explained by caloric flowing into and out of them. As the temperature of a body increased, the increased amount of caloric caused the body to expand due to the self-repulsive nature of caloric. The melting of a solid was explained by the repulsive force due to the caloric becoming great enough at the melting temperature to overcome the gravitational force holding the atoms in place. In a gaseous substance gravitational effects were considered negligible.

In Scotland, Joseph Black introduced the concept of two forms of caloric, termed *latent* and *sensible.* Sensible caloric related to changes in temperature. The heat involved in a phase change such as freezing was explained by caloric actually combining with the atoms, losing its sensible form and becoming latent. The latent caloric was returned to sensible form when a body underwent the reverse phase change. Production of heat by hammering or rubbing was explaining as "squeezing" sensible caloric out of the solid. The well-known rise in the temperature of the boiling point of a substance as pressure is increased was similarly explained: applying pressure to a substance near its boiling point squeezes out some of the sensible caloric and a higher temperature has to be reached before the substance has recovered enough caloric to vaporize.

It was an American, Benjamin Thompson, better known as Count Rumford, who made the first concerted experimental attack on the caloric theory in the late eighteenth century. In 1775, Thompson left America for Europe, where he was named a count by the Elector of Bavaria in 1790. While employed to oversee the boring of cannon barrels, he conducted experiments that showed the connection between the mechanical work of the boring machine and the unlimited production of heat. As long as the machine worked, heat was produced, which led Rumford to reject the notion that this heat came from a finite supply of caloric contained in the metal of the barrel.

Rumford also carried out experiments designed to measure the weight of the caloric fluid. He did this by attempting to measure the difference in weight of hot and cold bodies, especially the difference in weight of water as it changed phase. His experiments were extremely precise, and Rumford showed that there was no perceptible weight change as caloric supposedly flowed into and out of his samples. These and other experiments on the conduction of heat convinced Rumford that heat was due to molecular *motion* rather than an inexhaustible and weightless material substance. Ironically, proponents of the caloric theory gained in number through the first half of the nineteenth century, despite the support of Rumford by such eminent scientists as Sir Humphrey Davy and Thomas Young.

The man who first established the quantitative equivalence between mechanical work and the production of heat was the Englishman James Prescott Joule (1818–1889). Joule performed experiments on the production of heat by electrical current, the friction of flowing water, the compression of air, and the action of paddle wheels stirring water. He presented his measurements of the mechanical equivalent of heat in 1847 at Oxford. His talk would have passed without comment except for the interest shown by the young William Thomson, later Lord Kelvin, one of England's most renowned men of science. Others, notably the American physicist Henry Rowland, refined the results of Joule's original experiments. The year 1847 remains the date at which the first law of thermodynamics, incorporating heat as an internal mechanical energy, was definitively established. The seemingly ordinary statement of the mechanical equivalence of heat, 1 kilocalorie = 4184 N · m, is in fact one of the most profound statements in classical physics. It is little wonder that today the unit of newton-meter is given the name *joule*.

random motion, which constitutes additional thermal energy. In general, friction-related losses of mechanical energy appear as thermal energy. The law of conservation of energy assures us that the mechanical energy lost gives rise to an equal amount of thermal energy.

11.3 HEAT UNITS

Because heat and thermal energy are forms of energy, their basic SI unit is the joule. However, there are other units commonly used in heat measurements, called *thermal units,* that were in widespread use long before it became known that heat is a form of energy. Because they are still used extensively, we introduce them here.

A unit called the *calorie* (cal) was originally defined to be the amount of heat required to raise the temperature of one gram of water one Celsius degree (1 C°). The food calorie is actually a *kilocalorie* (kcal), equal to 1000 cal, and is written with a capital letter: Calorie (Cal). Another thermal unit, the *British thermal unit* (Btu), was originally defined as the amount of heat required to raise the temperature of one pound of water one Fahrenheit degree (1 F°).

Once it was realized that heat was a form of energy, Thompson and Joule performed many measurements to determine the **mechanical equivalent of heat,** by which the traditional thermal units could be converted to joules. Today the calorie and the Btu are defined in terms of the joule:

1 cal = 4.184 J
1 Btu = 1054 J

11.4 SPECIFIC HEAT CAPACITY

To increase the temperature of an object, we must increase the thermal energy of its molecules. We can do this by letting heat flow into the object from a hotter substance. Similarly, if we wish to cool an object, we can allow heat energy to flow from it to a cooler substance. To describe cooling and heating quantitatively, we must know how much energy is required to change the temperature of an object.

The quantity of heat that must flow into or out of a unit mass of a substance to change its temperature by one degree is called the **specific heat capacity** c of the substance.

When a quantity Q of heat flows into a mass m of substance, the temperature of the mass will increase by an amount ΔT. Then, by definition,*

$$\text{Specific heat capacity } c = \frac{Q}{m \, \Delta T}$$

After clearing fractions, we have

$$Q = cm \, \Delta T \tag{11.1}$$

From the definition, the units of specific heat capacity are $J/kg \cdot C°$, although the unit $cal/g \cdot C°$ is in more common use. You should be able to show that

$$1 \, cal/g \cdot C° = 4184 \, J/kg \cdot C°$$

Typical values for c are listed in Table 11.1. Notice that for water, $c = 1.000 \, cal/g \cdot C°$. As we shall see, specific heat capacity varies somewhat with temperature. The values given in the table apply at temperatures near room temperature. Notice that large values of c indicate that it takes relatively large amounts of heat per gram to change the temperature of a substance by a given amount. Small values of c denote large changes of T for relatively small amounts of heat transfer. From our discussion in Sec. 11.2, we might thus expect gases composed of complex molecules to have larger specific heats than simple monatomic gases, since added heat is shared by many forms of internal energy. We investigate the specific heats of gases in Chap. 12.

TABLE 11.1 *Specific heat capacities*

Substance	c (cal/g $\cdot$ C°)	c (J/kg $\cdot$ C°)
Water	1.000	4184
Human body	0.83	3470
Ethanol	0.55	2300
Paraffin	0.51	2100
Ice (0°C)	0.50	2100
Steam (100°C)*	0.46	1920
Aluminum	0.21	880
Glass	0.15	600
Iron	0.11	460
Copper	0.093	390
Mercury	0.033	140
Lead	0.031	130

*At constant volume.

*The symbol Q represents an amount of heat transfer. Positive values of Q denote heat added to a substance; negative values denote heat given up by the substance. The notation ΔT represents the resulting change in temperature.

Example 11.1

How much heat is required to change the temperature of (*a*) 400 g of water from 18.0 to 23.0°C? (*b*) 400 g of copper from 23.0 to 18.0°C?

Reasoning

Question   What is the relationship between the amount of heat added and the change in temperature?
Answer   It involves the mass of material and its specific heat:

$$Q = cm\,\Delta T$$

Question   What units can I use?
Answer   The units you choose for specific heat must agree with the units you use for *m* and *Q*. Table 11.1 gives you two choices.

Solution and Discussion

(*a*) $Q = (1.00\ \text{cal/g} \cdot \text{C}°)(400\ \text{g})(+5.00\ \text{C}°) = 2000\ \text{cal}$

In SI units this is

$$Q = (4184\ \text{J/kg} \cdot \text{C}°)(0.400\ \text{kg})(+5.00\ \text{C}°) = 8370\ \text{J}$$

(*b*) Notice here that $\Delta T = -5.00\ \text{C}°$ and that we use the value of *c* for copper:

$$Q = (0.093\ \text{cal/g} \cdot \text{C}°)(400\ \text{g})(-5.00\ \text{C}°) = -190\ \text{cal} = -780\ \text{J}$$

In (*a*) heat is added (+ sign on *Q*) to the water and in (*b*) heat is given up (− sign on *Q*) by the copper.

Exercise   Determine the final temperature of 700 g of copper, originally at 16.0°C, to which 400 J of heat energy is added. *Answer: 17.5°C.*

11.5   BOILING AND HEAT OF VAPORIZATION

Let us now discuss what happens when a liquid evaporates. Molecules of a liquid are held rather tightly to each other by their mutual attractions. (These attractions are mostly electrical in nature.) For the most part, the molecules at the surface of the liquid are unable to escape into the region above the surface. As in a gas, however, thermal motion causes a few of the molecules to have extraordinarily high energy, as we discussed in Sec. 10.6. A molecule with enough energy can escape from the surface and move from the liquid state to the gaseous state. In this way, the liquid *evaporates,* or *vaporizes.*

Because only the most energetic molecules can escape from the liquid, the average energy of the molecules left behind decreases as vaporization proceeds. Therefore, because temperature is a measure of energy, the temperature of an isolated liquid must decrease as vaporization proceeds. We thus arrive at an explanation for the well-known fact that evaporation can cool a liquid.

FIGURE 11.3

When the vapor is saturated in a closed container, the number of molecules evaporating from the liquid in a given time is exactly equal to the number condensing from the vapor into the liquid.

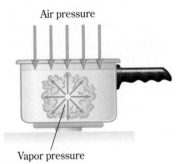

FIGURE 11.4

The boiling temperature is the temperature at which the vapor pressure in the bubble equals the external pressure on the liquid. (The size of the bubble is exaggerated.)

Energy must be furnished to the molecules of a liquid if they are to escape from the liquid surface. The amount of energy required to do this, which varies from one material to another, is called the heat of vaporization and is defined as follows:

The energy required to change a unit mass of molecules of a substance from the liquid phase to the vapor (gas) phase is called the **heat of vaporization** (H_v) of the substance.

$$Q = mH_v \tag{11.2}$$

When a unit mass of a substance condenses from the vapor to the liquid phase, the same amount of energy is liberated from the substance. Values of H_v are given in Table 11.2.

A liquid boils when vapor bubbles form and grow within it. To see how this happens, we must first understand the term **vapor pressure.** Suppose you have a liquid and its vapor in a closed container, as in Fig. 11.3. At any given temperature, equilibrium occurs when the number of molecules evaporating from the liquid is balanced by the number condensing from the vapor to the liquid. We call the pressure of the liquid's vapor under this equilibrium condition the vapor pressure of the liquid. It, of course, increases with increasing temperature. Why?

Now suppose we have a container of liquid which is open to the atmosphere, as in Fig. 11.4. Because of the random motions of the molecules in a liquid, several molecules may pull away from each other and form an empty space, a hole, within the liquid. Molecules evaporate into the hole, and so the vapor pressure within it rises. If time allows, the vapor pressure within the hole becomes equal to the vapor pressure of the liquid at this temperature. At low temperatures, the vapor pressure within the hole is small and the much larger atmospheric pressure applied to the liquid causes the hole to collapse. When the temperature of the liquid is high, however, the vapor pressure within the hole is large, larger perhaps than the pressure within the liquid due to the atmosphere. Then the excess pressure within the hole, now a vapor-filled bubble, causes the bubble to expand. The buoyancy of the bubble causes it, and many others like it, to rise to the surface of the liquid and explode, a phenomenon we recognize as boiling. Thus we see that a critical situation is reached when the temperature becomes high enough for the vapor pressure

TABLE 11.2   *Heats of vaporization and fusion*

Substance	Melting point (°C)	Boiling point (°C)	H_v		H_f	
			kJ/kg	cal/g	kJ/kg	cal/g
Helium	−270	−269	21	5.0	5.2	1.25
Oxygen	−219	−183	210	51	13.8	3.3
Nitrogen	−210	−196	200	48	25.5	6.1
Ethanol	−114	78	854	204	105	25
Mercury	−39	357	270	65	11.7	2.8
Water	0	100	2260	539	335	80
Lead	327	1750	858	205	23	5.9
Aluminum	660	2450	10500	2520	397	95
Gold	1063	2660	1580	377	64	15.4
Copper	1083	2595	4810	1150	205	49

*At 1 atm pressure.

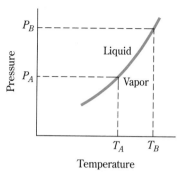

FIGURE 11.5

A typical vaporization curve. At pressure P_A, boiling occurs at temperature T_A. At a higher pressure P_B, boiling occurs at a higher temperature T_B.

Two different changes of state: (a) water changing from solid to liquid (melting); (b) carbon dioxide changing from solid to gas (subliming).

of the liquid to equal the pressure of the atmosphere above it. Then vapor-filled bubbles which form in the liquid grow, thereby producing boiling.

A liquid boils at a temperature where its vapor pressure just equals the external pressure on it.

The vapor pressure of water is 101 kPa at 100°C. Since 1 atm = 101 kPa, water normally boils at 100°C. High in the mountains, however, where atmospheric pressure may be only 80 kPa, boiling occurs at about 94°C. Boiling points for typical liquids (at P_a = 101 kPa) are given in Table 11.2. We can measure the boiling temperature of a substance at various ambient pressures and plot a graph of the results (Fig. 11.5 shows the example of water). The line separating liquid from vapor is known as the **vaporization curve.** We can choose a certain pressure on the diagram and follow that pressure to the right until it intersects the vaporization curve. At this pressure and the temperature corresponding to this intersection, the substance will boil. Boiling is an example of what we call a **change of phase,** and Fig. 11.5 is known as a **phase diagram.** Notice from Fig. 11.5 that increasing the pressure on the water will increase the temperature at which it boils.

It is important to understand that *when a sample of material undergoes a change of phase, the heat added to or given up by the substance does not change the temperature of the substance until the entire sample has changed phase.* Turning up the burner under a kettle of boiling water causes the water to boil more vigorously, *but not at a higher temperature.* The heat accompanying the change of phase from liquid to gas is determined by the mass of the sample and the heat of vaporization of the substance, as given by Eq. 11.2.

11.6 MELTING AND HEAT OF FUSION

Ice crystals melt at 0°C under standard pressure. Before melting occurs, the water molecules of the ice are ordered in a crystalline lattice. They are held in place by rather strong intermolecular forces. To melt the crystal, one must tear the molecules out of this tight arrangement and cause them to become disordered. This process requires energy, which is usually supplied by heat.

We therefore find that when a crystalline material is heated, it begins to melt at a certain temperature. As heat is slowly added to the crystal-liquid mixture, the temperature remains constant until all the crystals have melted. The substance has a definite melting temperature, and a definite amount of heat—called the heat of fusion—must be furnished to melt the crystals at this temperature.

The energy required to change a unit mass of a substance from the solid phase to the liquid phase is called the **heat of fusion** (H_f) of the substance.

$$Q = mH_f \tag{11.3}$$

When a unit mass of a substance changes from the liquid to the solid phase, the same amount of energy is liberated from the substance.

As in the case of vaporization, heat added to or lost from a substance during the change of phase from solid to liquid or liquid to solid does not change the temperature of the substance until the entire sample has changed phase.

The heat of fusion for water is 335 kJ/kg (80 cal/g). Values for other materials are

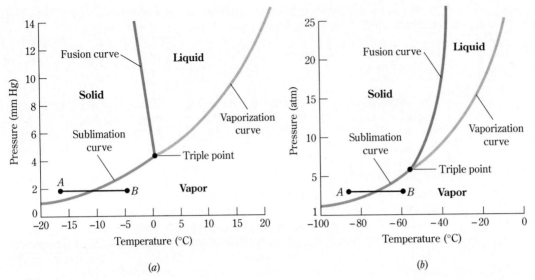

(a) (b)

FIGURE 11.6

Phase diagrams for (a) water and (b) carbon dioxide, showing their triple points.

listed in Table 11.2. Notice that the hydrogen-bonded materials water and ethanol have much higher heats of fusion and vaporization than the others. Why?

The freezing point of a liquid can be altered somewhat by applying large pressures to the system. Materials that contract upon freezing have their melting points raised by increased pressure. Most materials behave in this way. A few materials, such as water, expand when they freeze. Increased pressure decreases the freezing point of such substances. The pressure of an ice skater's blade causes the ice below it to melt. In this case, the skater is skating on ice lubricated with a thin film of water. This behavior can be seen in a plot of a substance's **fusion curve,** which is a graph showing how melting temperature depends on pressure. Examples are shown in Fig. 11.6 for water and carbon dioxide. Since melting temperature depends only slightly on pressure, these curves are almost vertical. Most materials, like carbon dioxide, have a fusion curve with a positive slope. The fusion curve of water, however, has a slightly negative slope, which indicates that added pressure lowers the melting temperature, reflecting the fact that water expands as it freezes.

A complete phase diagram also shows that below a certain pressure, a substance can change directly from a solid to a gas without going through a liquid phase at all. We call this process **sublimation.** The sublimation curves of water and carbon dioxide are included in Fig. 11.6. Notice the large difference in the values of pressure along the vertical axes of the two graphs.

Figure 11.6 shows that there is a single point at which the three curves separating the phases intersect. This point, found only at a single combination of temperature and pressure unique to each substance, is known as the **triple point** of that substance. For water, the triple point occurs at 0.01°C and 4.58 torr (0.006 atm). For carbon dioxide the values are −56.6°C and 5.11 atm.

We can see from Fig. 11.6 that in order for sublimation to occur, the pressure on the substance must be below the triple-point pressure of the substance. Lines AB are examples of sublimation processes. We observe the sublimation of carbon dioxide at ordinary pressures, since 1 atm is well below the triple-point pressure for CO_2. Thus we refer to solid carbon dioxide as "dry" ice. It would require pressures higher than 5.11 atm to produce a liquid phase of CO_2. There is a heat of sublimation associated with this phase change, just as the heats of fusion and vaporization are associated with the other phase changes we discussed.

Illustration 11.1

How much heat is released from 50 g of water as it (*a*) changes from liquid to the crystalline phase at 0°C and (*b*) changes from steam to liquid water at 100°C?

Reasoning

(*a*) When a mass m crystallizes, it liberates an energy mH_f. Thus

$$Q = mH_f = (50 \text{ g})(80 \text{ cal/g}) = 4000 \text{ cal} = 16{,}700 \text{ J}$$

(*b*) The heat liberated by a mass m of gas as it condenses is mH_v. Therefore

$$Q = mH_v = (50 \text{ g})(539 \text{ cal/g}) = 27{,}000 \text{ cal} = 113{,}000 \text{ J}$$

Notice that the steam-to-water transition liberates much more heat than does the water-to-ice transition.

Exercise How much heat is needed to melt 500 g of lead at 327°C? *Answer: 4.29×10^5 J.*

11.7 CALORIMETRY

Many experiments involving heat are carried out in a container called a *calorimeter*. This is a device that thermally isolates materials so that heat cannot flow out of or into them from the surroundings. The common vacuum thermos flask is a fairly good calorimeter. Heat is prevented from flowing through its double glass walls by the shiny metal coating on them and by the vacuum between them. We will see in Secs. 11.9 to 11.11 why such a design is effective.

Suppose two or more materials at various temperatures are placed together inside a calorimeter. The materials will share thermal energies until they all reach the same temperature, that is, until thermal equilibrium is established. Because no energy flows into or out of the container, the law of conservation of energy leads us to a very important conclusion: if heat gains are taken as positive changes and heat losses are taken as negative changes, then

The sum of the heat exchanges within the calorimeter is zero.

In other words, the total energy of the isolated system inside the calorimeter is unchanged.

Before we apply this idea to various examples, let us review what types of heat changes we may encounter.

1 When a mass m undergoes a temperature change from T_o to T_f, Eq. 11.1 tells us that the amount of heat gained or lost is

$$Q = mc(T_f - T_o)$$

where c is its specific heat capacity. Remember, this applies only to temperature ranges which do not involve a change of phase of the material.

FIGURE 11.7

As heat is added to a solid substance, its temperature rises until the fusion (melting) temperature T_f is reached. Further addition of heat changes the phase of the substance without changing the temperature. Once all the substance has become liquid, addition of heat increases its temperature until the vaporization temperature T_v is reached. Another plateau of temperature occurs until all the substance has vaporized, after which additional heat causes the temperature of the gas to rise.

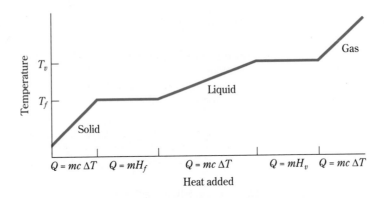

2 When a mass m melts, Eq. 11.2 shows the heat exchange is $Q_f = +mH_f$; when it crystallizes, $Q_f = -mH_f$.

3 When a mass m vaporizes, Eq. 11.3 shows the heat exchange is $Q_v = +mH_v$; when it condenses, $Q_v = -mH_v$.

Figure 11.7 summarizes the amounts of heat associated with temperature rises and changes of phase of a substance. Each phase has a different specific heat; ice and steam, for example, have a different value of c than does liquid water. Notice also in Fig. 11.7 that, as we have discussed, heat added or subtracted during a change of phase does not change the temperature of a substance.

Example 11.2

A cup contains 200 g of coffee at 98°C. What mass M of ice at 0°C must be added to change the coffee temperature to 60°C? Neglect heat flow from the coffee to the cup; that is, the cup is assumed to be a perfect calorimeter.

R

Reasoning

Question   What heat exchanges are going to take place?
Answer   The coffee is going to lose heat since its temperature is decreasing by 38 C°. Assume the coffee is essentially water, so that you can use $c = 1.0\,\text{cal/g}\cdot\text{C}°$. Then all necessary data are given to calculate the amount of heat lost. The ice is going to gain this same amount of heat. You are to find how much ice this will take.

Question   As the ice gains heat, what happens?
Answer   First it will absorb heat as it melts. Then, as liquid water, its temperature increases as it absorbs more heat (Fig. 11.7).

Question   How high will its temperature go?
Answer   The water and coffee must reach the *same* temperature to be in equilibrium. This is given as 60°C.

Question   What is the mathematical expression for the heat absorbed by the ice and water?
Answer   Heat absorbed = $Q_{gain} = MH_f + cM(60°C - 0°C)$ where c = specific heat of water.

Solution and Discussion   The amount of heat lost by the coffee is

$$Q_{lost} = (1.0 \text{ cal/g} \cdot \text{C}°)(200 \text{ g})(-38 \text{ C}°) = -7600 \text{ cal}$$

Equating this to the expression for heat gained by the ice:

$$Q_{gain} = M(80 \text{ cal/g}) + M(1.0 \text{ cal/g} \cdot \text{C}°)(+60 \text{ C}°) = 7600 \text{ cal}$$

Solving for M gives 54.3 g. Notice that $(54.3 \text{ g})(80 \text{ cal/g}) = 4344$ cal is involved in the melting. Then 3256 cal goes to raise the temperature of the melted ice to 60°C.

Exercise   Find the final temperature if only 40 g of ice is added. *Answer: 68°C.*

Example 11.3

An 80.0-g piece of metal at 100°C is dropped into a perfect calorimeter that contains 400 g of oil at 18.0°C. If the final temperature of the system is 23.1°C, what is the specific heat capacity c_m of the metal? For the oil, $c = 0.650 \text{ cal/g} \cdot \text{C}°$.

R

Reasoning

Question   What are the heat exchanges in this problem?
Answer   The metal is going to lose heat in cooling from 100°C down to 23.1°C. The oil is going to gain the same amount of heat in going from 18.0°C to the same final temperature, 23.1°C. Enough data are given to calculate this amount of heat.

Question   What is the specific equation for this situation?
Answer $(80.0 \text{ g})(c_m)(-76.9 \text{ C}°) + (400 \text{ g})(0.650 \text{ cal/g} \cdot \text{C}°)(+5.10 \text{ C}°) = 0.$

Solution and Discussion   This equation becomes

$$-(6150 \text{ g} \cdot \text{C}°)c_m + 1330 \text{ cal} = 0$$

Solving for c_m gives $c_m = 0.216 \text{ cal/g} \cdot \text{C}°$.

Example 11.4

A large glass container holds 500 g of mercury at 20°C. An electric heater immersed in the mercury delivers 70 W of power. How long does it take the heater to boil away 30 g of mercury? Ignore the mass of the heater, and assume that all the electric power is transferred into heating the mercury alone.

R

Reasoning

Question   What do I have to know to calculate how much energy it takes to boil the 30 g of mercury away?

Answer   You need the specific heat of liquid mercury, its boiling temperature, and its latent heat of vaporization.

Question   What is the expression for the amount of heat required?
Answer   All of the 500 g must be heated to boiling before *any* can be vaporized. Then an additional amount of latent heat must be supplied to boil 30 g. Thus

$$Q = (500 \text{ g})(c)(T_{\text{boil}} - 20°\text{C}) + (30 \text{ g})H_v$$

Question   How does the 70 W of power relate to the time?
Answer   Remember, power = energy/time. Since $1 \text{ W} = 1 \text{ J/s}$, we have that $70 \text{ W} = 70 \text{ J/s}$. Because this information is in SI units, you'll want to use c in SI units too. The equation that gives you the required time is $Q \text{ (J)} = (70 \text{ W})t$.

Solution and Discussion   Calculating Q, using data from Tables 11.1 and 11.2,

$$Q = (0.500 \text{ kg})(140 \text{ J/kg} \cdot \text{C}°)(357 - 20)\text{C}° + (0.30 \text{ kg})(2.7 \times 10^5 \text{ J/kg}) = 31,000 \text{ J}$$

The time it takes is

$$t = Q/70 \text{ W} = (31,000 \text{ J})/(70 \text{ J/s}) = 450 \text{ s} = 7.5 \text{ min}$$

Exercise   How long would this same heater require to vaporize 50 g of water in a vat that is already at 100°C?   *Answer:*   *27 min.*

Example 11.5

A 10-g lead bullet is traveling at 100 m/s when it strikes and embeds itself in a wooden block. By about how much does its temperature rise on impact if all the kinetic energy of the bullet changes to thermal energy in the bullet?

R   *Reasoning*

Question   How much heat is created as the bullet comes to rest?
Answer   An amount equal to all of its initial KE: ΔKE lost = Q gained.

Question   What equation relates the temperature rise to the bullet's KE?
Answer $\frac{1}{2}mv^2 = mc\,\Delta T$

Solution and Discussion   Using the value of c for lead from Table 11.1, you get

$$\Delta T = \frac{(1/2)v^2}{c} = \frac{(0.5)(100 \text{ m/s})^2}{1.3 \times 10^2 \text{ J/kg} \cdot \text{C}°}$$

$$= 39 \text{ C}°$$

You should verify that the units cancel as indicated. Notice that the ΔT depends on the *square* of the speed.

Hence, if the original temperature of the bullet is 20°C, its final temperature is about 59°C. If the bullet were traveling at 600 m/s, ΔT would be 36 times as large, and the final temperature would be about 1430°C. Of course, the bullet would melt before this temperature was reached, and so the above computation would no longer be correct. How could the computation be carried out in this latter case?

Illustration 11.2

When nutritionists state that 1 kg of bread has a food value of 2600 Cal, they mean that if the dried bread is burned in pure oxygen, it will give off 2600 kcal of heat energy. (Basically, the body generates heat from food in a somewhat similar chemical reaction.) Estimate how much heat energy a human body gives off each day.

Reasoning Depending on the person, the nutritional calorie intake each day is 2000 to 3000 Cal. Since these are actually kilocalories, the body's metabolism generates on the order of 2×10^6 to 3×10^6 cal of heat. Because body temperature must remain nearly constant, the body must lose this energy as it is generated. The air we exhale and the evaporation of perspiration from the skin are well-known mechanisms for cooling the body, but others are important as well.

Exercise If a 60-kg person retained all the heat energy from the 1800 Cal she consumed in a day, how much would her body temperature rise? Use 0.83 cal/g · C° for the specific heat of her body. *Answer: 36 C°.*

11.8 THERMAL EXPANSION

As we have seen, the temperature of a substance is a measure of the energy resident in its molecules. As the temperature of a liquid or solid is raised, the molecules, having greater energy, generally vibrate through larger distances. This increased amplitude of vibration of a given molecule forces its neighboring molecules to

The edges of concrete slabs in a roadway must be separated by expansion joints to allow the slabs to expand toward each other without buckling when temperature rises.

Extremely hot temperatures expanded these rails so much that they exceeded the expansion gap between rail sections. As a result, the rails buckled sideways, causing a train to derail.

remain at a greater average distance from it. Hence, the solid or liquid expands. Although there are some notable exceptions to this rule over small temperature ranges (for example, water contracts in going from 0 to 4°C),* it is generally true that substances expand with increasing temperature, provided a phase change does not occur.

Clearly, the thermal expansion of the metal in a building or bridge can be a matter of considerable practical importance. If provision were not made for thermal expansion, railway tracks and concrete highways would buckle under the action of the hot summer sun. Therefore, it is necessary to know exactly how a material expands with temperature.

Suppose a bar of initial length L_o undergoes a temperature change ΔT that causes its length to change by an amount ΔL. The fractional change in the length of the bar is $\Delta L/L_o$. For many solids we find that over a certain range of temperature, this fractional change in length is linearly proportional to the temperature change. We define the **coefficient of linear thermal expansion** α for the material by

$$\alpha = \frac{\text{fractional change in length}}{\text{temperature change}} = \frac{\Delta L/L_o}{\Delta T}$$

which gives

$$\Delta L = \alpha L_o \Delta T \tag{11.4}$$

The units of α are reciprocal degrees, either $1/C°$ or $1/K$. A few typical values for α are listed in Table 11.3.

As an example of the use of the linear expansion coefficient, suppose a brass rod 75 cm long is subjected to a temperature change of $+50\,C°$. Its length increase is (see Table 11.3 for α)

$$\Delta L = \alpha L_o \Delta T = (19 \times 10^{-6}/C°)(0.75\,\text{m})(50\,C°) = 7.1 \times 10^{-4}\,\text{m}$$

TABLE 11.3 *Coefficients of thermal expansion (per Celsius degree at 20°C)*

Substance	$\alpha \times 10^6$	$\gamma \times 10^6$
Diamond	1.2	3.5
Glass (heat-resistant)	~3	~9
Glass (soft)	~9	~27
Iron and steel	12	36
Brick and concrete	~10	~30
Brass	19	57
Aluminum	25	75
Mercury		182
Rubber	~80	~240
Glycerin		500
Gasoline		~950
Methanol		1200
Benzene		1240
Acetone		1490

*In water, hydrogen bonding binds the molecules into groups of several molecules each in a definite structure even above the melting point of ice. As the temperature increases, these groups break up, causing a new, more compact arrangement of the molecules.

Because this change in length is very small, the value of L_o used to determine ΔL is not sufficiently temperature-dependent to cause worry about the temperature at which it is measured. Actually, α does vary somewhat with temperature, and for very precise work one should use the value appropriate to a given temperature range. In practice, however, this complication is seldom of any consequence.

A useful analogy to thermal expansion is that of photographic enlargement. Every linear dimension of the object undergoes the same fractional change, including holes in the material. The perimeter of a hole will change in length by the same amount whether or not the substance fills the hole. Thus increases in temperature will *expand* holes, not reduce them.

Thermal expansion of volumes is also important, especially in the case of liquids. In a manner analogous to the way we defined the linear expansion coefficient, we define a **coefficient of volume thermal expansion** γ as the relative change in volume per unit change in temperature:

$$\gamma = \frac{\Delta V / V_o}{\Delta T}$$

which yields

$$\Delta V = \gamma V_o \Delta T \tag{11.5}$$

The units of γ are also reciprocal degrees. As an example of its use, suppose that $100\ cm^3$ of benzene at 20°C is heated to 25°C. According to Eq. 11.5, its volume will change by an amount (see Table 11.3 for γ)

$$\Delta V = (1.24 \times 10^{-3}/C°)(100\ cm^3)(5\ C°) = 0.62\ cm^3$$

This is a 0.6 percent change in volume, which is an appreciable change in V for many purposes. It is therefore necessary to stipulate the temperature at which V should be measured if the γ coefficients in Table 11.3 are to apply. The values given there are for $T = 20°C$. Of course, for small temperature changes not too far removed from 20°C, we can compute ΔV to fairly good precision using V measured anywhere in this small temperature range.

Table 11.3 shows that the linear expansion coefficient for a solid is essentially one-third its volume expansion coefficient. This is a general rule for solids that expand to the same extent in all directions. Problem 52 asks you to show why this rule follows from the definitions of α and γ.

Example 11.6

Suppose slabs of concrete 20 m long are laid end to end to form a roadway. How large a gap should be allowed between adjacent slabs at −20°C so that they don't buckle when temperatures reach +50°C?

R

Reasoning

Question What does the condition of "not buckling" require?
Answer The slabs cannot buckle unless they touch each other. Therefore a safe answer would be that they just barely touch at the higher temperature.

Question What is the equation determining the amount a slab will expand in this temperature range?
Answer $\Delta L = L_o \alpha \Delta T$ where $\Delta T = +70\,C°$

Question Is this ΔL the same as the gap length we must allow?
Answer In order to touch, each adjacent slab must expand to fill *half* the gap separating them. Thus each slab has to expand half a gap *on each end*. The total expansion of each slab will therefore be equal to the gap.

Solution and Discussion Table 11.3 gives the expansion coefficient for concrete. Then

$$\Delta L = (20\,\text{m})(10 \times 10^{-6}/\text{C}°)(+70\,C°) = 0.014\,\text{m} = 1.4\,\text{cm}$$

Example 11.7

Suppose at 20°C you have a 1.000-m-long piece of brass wire. You bend the wire into a circle, leaving a gap of 1 mm between the ends. If you increase the temperature of the wire to 73°C, what will happen to the size of the gap?

R

Reasoning

Question What change of length results from this temperature increase?
Answer Using data from Table 11.3,

$$\Delta L = (1.000\,\text{m})(19 \times 10^{-6}/\text{C}°)(+53\,C°) = 1.0 \times 10^{-3}\,\text{m} = 1.0\,\text{mm}$$

Question Does this mean that the 1 mm gap *closes?*
Answer Remember the analogy to photographic enlargement, which tells you that the gap *increases* by the same *fractional* amount (10^{-3}) as any other linear dimension. So the gap would enlarge by 10^{-3} mm.

Question Beside the analogy argument, how can I show that the gap gets bigger?
Answer The original circumference C of the circle is *1.001 m*, not 1.000 m. This circumference increases by a fractional amount $\Delta C / C_o = 10^{-3}$, so the new circumference is

$$C = C_o + (0.001)C_o = 1.001\,\text{m} + .001001\,\text{m} = 1.002001\,\text{m}$$

The wire has increased in length by 1 mm, but the circumference of the circle it is a part of has increased by slightly more than 1 mm. We have exaggerated the number of significant digits in the last calculation in order to demonstrate this increase in C.

Example 11.8

Suppose you have a 50.0-ml container made of soft glass that is full to the brim with benzene at 0.0°C. Will some benzene spill if the temperature is raised to 30.0°C? If so, how much?

R

Reasoning

Question What will tell me whether or not some benzene will spill?
Answer Both the container and the benzene start with the same volume (that's what "filled to the brim" means). Also, they are undergoing the same temperature change. The interior volume (the capacity) of the container will expand according to the volume coefficient of soft glass. If the volume coefficient of benzene is greater than that of soft glass, the increased capacity won't accommodate all the excess benzene. Some will have to spill over.

Question Which coefficient of volume expansion is greater?
Answer Table 11.3 shows that the coefficient of benzene is much greater than that of soft glass. Thus some benzene will overflow the container.

Question What equation gives the excess volume of benzene?
Answer It is the difference in the expanded volumes:

Volume spilled = ΔV (benzene) $- \Delta V$ (glass)

Solution and Discussion Calculating the individual volume expansions:

$$\Delta V \text{ (glass)} = (50.0 \text{ ml})(27 \times 10^{-6}/\text{C}°)(+30.0 \text{ C}°)$$
$$= 0.040 \text{ ml}$$

$$\Delta V \text{ (benzene)} = (50.0 \text{ ml})(1240 \times 10^{-6}/\text{C}°)(+30.0 \text{ C}°)$$
$$= 1.86 \text{ ml}$$

Subtracting these, we find that 1.82 ml of the benzene will spill over.

11.9 TRANSFER OF HEAT: CONDUCTION

When you hold a metal spoon in hot water, heat flows along its handle to your hand. The interpretation of this is simple. Heat energy enters the spoon from the hot water and causes the atoms in that end of the spoon to acquire large thermal

Materials that have poor heat conduction find many practical applications.

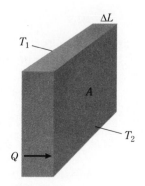

FIGURE 11.8

Because $T_1 > T_2$, heat flows through the slab in the direction shown.

energy. They then vibrate with very large energy and, when they collide with their cooler neighbors, pass energy along to them. Thus thermal energy is passed along the spoon from the hot end to the cool end. Eventually the whole spoon becomes hot. This method of heat transfer is called thermal conduction.

In **thermal conduction**, heat is transferred through a material by the collisions of adjacent atoms or molecules.

Conduction occurs at different rates in different materials. A wooden stick can burn at one end, and still the other end remains relatively cool, but a metal knife or spoon transmits heat rapidly from one end to the other. The ability of a material to conduct heat depends on its atomic structure. Metals have electrons within them that can move rather freely throughout the metal. As they move, they carry thermal energy from one part of the metal to another. Hence metals, because they contain many free electrons, are excellent heat conductors.

We use the experiment sketched in Fig. 11.8 to describe heat conduction quantitatively. A slab of material of thickness ΔL and face area A has a temperature differential $T_1 - T_2 = \Delta T$ impressed across it. The rate of heat flow $Q/\Delta t$ through the slab depends on ΔT, on A, and on ΔL. It is directly proportional to ΔT and to A (in other words, the rate of heat flow increases as ΔT and/or A increases) and inversely proportional to ΔL (the rate decreases as ΔL increases). Experiment shows that

$$\frac{Q}{\Delta t} = k\frac{A\,\Delta T}{\Delta L} \qquad (11.6)$$

The quantity $\Delta T/\Delta L$ is often called the *temperature gradient*. The constant k depends on the material of which the slab is made and is called the *thermal conductivity* of the material. Typical values for k are given in Table 11.4 for $Q/\Delta t$ in watts, A in square meters, ΔL in meters, and ΔT in kelvins. As you see, k is large for good thermal conductors, such as metals, and small for poor thermal conductors, which are referred to as *insulators*.

The thermal conductivity of an object influences how hot the object feels to the touch. A hot metal burns your hand easily because heat can flow readily from it to your hand. A piece of wood at the same temperature does not burn your hand nearly as badly. Because of the much lower thermal conductivity of wood, only the thermal energy at the point of contact can easily reach your hand. In effect, your hand quickly cools the wood at the point of contact. Using similar reasoning, can you explain why a cold tile floor seems warmer to your bare feet if you stand on a carpet that lies on it?

TABLE 11.4
Thermal conductivities*

Material	k $(W/K \cdot m)$†
Silver	430
Copper	400
Aluminum	240
Brass	105
Concrete	0.8
Glass	0.8
Brick	0.6
Asbestos paper	0.2
Rubber	0.2
Wood	0.08
Bone	0.042
Muscle	0.042
Glass wool (fiberglass)	0.04
Plastic foam	0.03
Fat	0.021

*These are approximate values; k varies somewhat with temperature.

†1 W/K · m = (1/418.4)(cal/s)/C° · cm = 6.94 Btu · in/h · ft² · F°

Example 11.9

The interior dimensions of a cubical soft drink cooler are $30 \times 30 \times 30$ cm. Each wall is 4.0 cm thick and is made of plastic for which $k = 0.032$ W/K · m. Ice inside the box maintains the interior temperature at 0°C. How much ice melts each hour when the outside temperature is 25°C?

R

Reasoning

Question What determines how much ice melts?
Answer The amount of heat that flows into the cooler each hour. Every 80 cal will melt 1.0 g of ice.

Question What determines the rate at which heat flows into the cooler?
Answer Three things:

1 The temperature gradient, $\Delta T/\Delta L$, between the inside and outside of the cooler.
2 The area of the walls of the cooler.
3 The thermal conductivity of the plastic.

Question What is the expression for the rate of heat flow?
Answer Equation 11.6 gives $\dfrac{Q}{\Delta t} = kA\dfrac{\Delta T}{\Delta L}$

Question What area should I use?
Answer The cooler has six equal sides, each having an area of $(0.30\,\text{m})(0.30\,\text{m}) = 0.090\,\text{m}^2$. The total area is then $0.54\,\text{m}^2$.

Solution and Discussion Calculating the rate of heat flow:

$$\frac{Q}{\Delta t} = (0.032\,\text{W/K}\cdot\text{m})(0.54\,\text{m}^2)(25\,\text{K}/0.040\,\text{m}) = 11\,\text{W}$$

$$= (11\,\text{J/s})(1.0\,\text{cal}/4.184\,\text{J}) = 2.6\,\text{cal/s}$$

Thus in 1 h $3600(2.6) = 9300$ cal flows into the cooler. This is enough to melt

$$\frac{9300\,\text{cal}}{80\,\text{cal/g}} = 120\,\text{g of ice.}$$

(a)

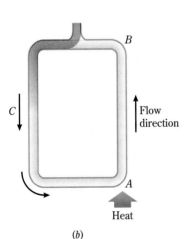

B

C

Flow direction

A

Heat

(b)

FIGURE 11.9

When heat is applied to the liquid in the tube, the dye shows that the liquid circulates counterclockwise. Heat is carried along by the liquid in a process known as convection.

11.10 TRANSFER OF HEAT: CONVECTION

A simple experiment that illustrates convection is shown in Fig. 11.9. When the glass tube is filled with water, a little colored dye placed near the neck remains nearly motionless (part *a*). However, when the tube is heated at one corner, as in part *b*, the liquid begins to flow counterclockwise around the tube, carrying the dye with it.

The reason for this motion is quite simple. A heated liquid or gas expands, and so the water in the lower right corner of the tube at *A* expands when heated. It is now less dense than the rest of the liquid. The more dense column of liquid on the left can no longer be supported by the less dense column on the right. The left column falls, pushing the water along in the tube. Hence the liquid on the right flows upward. This means of heat transfer is called convection.

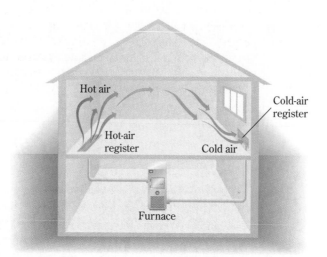

FIGURE 11.10

In convection, a moving fluid transports the heat energy. In the case of this home heating system, the fluid is air.

In **convection,** fluid flow carries heat from place to place.

Conduction does not involve the motion of the molecules over large distances. Heat is transferred from molecule to molecule by collision. In convection, however, the molecules of the transferring material move along with the heat. Only liquids and gases can transfer heat by convection because only in these materials can the molecules move over large distances.

Many homes are heated by air convection (Fig. 11.10). Even in heating systems without fans, the circulatory movement of the air is appreciable. For example, to a person standing over a hot-air register above an air furnace, the rush of hot air from the register is often quite noticeable. Proper design of such convection systems must allow the cool air to return to the furnace much as the cool liquid circulates back to point A in Fig. 11.9b. This is the purpose of the cold-air registers in such heating systems.

Weather phenomena are partly the result of convective air currents. Thermal

Convective heat transfer in the atmosphere can often become turbulent and violent.

currents that circulate air near the edges of mountain ranges are particularly interesting. Quite large effects are noticed at various fixed times of day as the cool air from the mountains flows down and causes the warm air on the nearby plains to rise. The Gulf Stream and the Japan current are other interesting examples of large-scale transfer of heat by convection.

11.11 TRANSFER OF HEAT: RADIATION

We know that the sun warms the earth. It is, in fact, our major source of heat. We can easily see that the heat from the sun is not transferred to us by either conduction or convection because there are very few molecules in the vast reaches of space between us and the sun. Hence vibrational transfer by conduction and circulatory transfer by convection are impossible. We have here a case of heat transfer through a vacuum, that is, through empty space. This method of heat transfer is called **radiation.**

As we will see when we study electricity and magnetism, radiation is energy in the form of electromagnetic waves traveling through space at the speed of light. All objects emit radiation. At ordinary temperatures, most of this radiation is *infrared* radiation. Infrared radiation is strongly absorbed by water molecules, including those in our body cells. When we feel warmed by infrared radiation from an electrical heater, for example, it is because the radiation is being converted to heat as it is absorbed by our bodies. Although infrared radiation is sometimes referred to as heat radiation, it is incorrect to think of infrared radiation as being heat until the energy is thermalized by an absorption process such as this.

The rate at which objects emit radiant energy depends very strongly on their temperature. It also depends on their surface area and the nature of the surface. This is summarized in a principle of physics known as **Stefan's law.** The radiant energy emitted per second by an object is

$$\frac{Q}{\Delta t} = e\sigma A T^4 \tag{11.7}$$

The Solar One facility at Barstow, California, is an example of converting solar energy to electrical energy. The individual mirrors all focus sunlight onto the collector at the top of the tower, producing high temperature steam. The steam then powers turbines connected to electrical generators.

where A is the surface area of the object and T is its kelvin temperature. The constant σ is called the **Stefan-Boltzmann constant** and has the value

$$\sigma = 5.67 \times 10^{-8}\,\text{W/m}^2 \cdot \text{K}^4$$

The factor e is called the *emissivity* of the object. It ranges in value from 0 to 1 and depends on the nature of the emitting surface. Dark, rough surfaces have emissivities close to 1, while bright, shiny surfaces have e closer to zero. An object with $e = 1.00$ is a "perfect" emitter and is referred to as a **blackbody.** For polished copper, e is about 0.3. Generally, good emitters are also good absorbers of radiation.

This last observation makes it possible for us to discuss *net* absorption or loss of radiant energy between an object and its surroundings. An object placed in an environment where the temperature of surrounding objects is T_s will absorb radiation at the rate

$$\left(\frac{Q}{\Delta t}\right)_{\text{abs}} = e\sigma A T_s^4$$

At the same time, if the object is at a temperature T, it will emit energy at the rate

$$\left(\frac{Q}{\Delta t}\right)_{\text{emit}} = e\sigma A T^4$$

Thus the rate at which energy is absorbed or lost by the object is the difference between $(Q/\Delta t)_{\text{abs}}$ and $(Q/\Delta t)_{\text{emit}}$:

$$\left(\frac{Q}{\Delta t}\right)_{\text{net}} = e\sigma A (T^4 - T_s^4)$$

When $T > T_s$, there is a net loss of energy and the object cools. When $T < T_s$, there is a net gain and the object heats up. Of course at the same time an object may gain or lose energy by convection and/or conduction as well.

Example 11.10

The approximate surface temperature of the sun is 6000 K. Taking the sun to be a sphere with radius 7×10^8 m and assuming an emissivity of 0.93, calculate the total power radiated from its surface.

R

Reasoning

Question   What physical properties determine the power radiated by an object?
Answer   The object's area, temperature, and emissivity.

Question   What basic principle gives me the equation relating these quantities?
Answer   Stefan's law, which is

$$\frac{Q}{\Delta t} = P = \sigma e A T^4$$

Question   How do I find the surface area of the sun?
Answer   For a sphere, $A = 4\pi R^2$.

Solution and Discussion   The area of the sun is

$$A = 4\pi(7 \times 10^8 \, \text{m})^2 = 6 \times 10^{18} \, \text{m}^2$$

Then the radiated power is

$$P = (5.67 \times 10^{-8} \, \text{W/m}^2 \cdot \text{K}^4)(0.93)(6 \times 10^{18} \, \text{m}^2)(6000 \, \text{K})^4$$
$$= 4 \times 10^{26} \, \text{W}$$

This is an enormous amount of power, as you might expect, due both to the sun's large size and to its high temperature.

Exercise   Find the power radiated *per square meter* of the sun's surface. This would be the same for any object having the same emissivity at 6000 K. *Answer:*   *70 MW/m²*.

11.12   BUILDING INSULATION

Anyone who has to pay to heat or cool a home is interested in thermal insulation. It is obvious from a glance at Table 11.4 that metals are the worst insulators and plastic foam is among the best. Most modern buildings are heavily insulated by layers of foamed plastic or glass wool (fiberglass). These materials are very good insulators because they trap air, one of the very best insulators. Air by itself, however, undergoes convection, and its true potential is realized only when it is held in place by a porous material such as glass wool.

Most walls are layered structures. Suppose a wall consists of several layers whose conductivities and thicknesses are k_1 and L_1, k_2 and L_2, and so on. It turns out that the heat flow through a three-layer wall is

$$\frac{Q}{\Delta t} = \frac{A \, \Delta T}{(L_1/k_1) + (L_2/k_2) + (L_3/k_3)}$$

where ΔT is the temperature difference across the wall. Additional layers simply add more terms to the denominator. The quantities L_1/k_1 and so on measure each layer's resistance to heat flow and are referred to as the ***R* values** for each layer. In terms of them,

$$\frac{Q}{\Delta t} = \frac{A \, \Delta T}{R_1 + R_2 + \ldots + R_N} = \frac{A \, \Delta T}{R_{\text{tot}}} \tag{11.8}$$

for N layers. Typical R values of interest are given in Table 11.5. We give R in both SI and British units ($\text{ft}^2 \cdot \text{F}° \cdot \text{h/Btu}$), since the latter units are used frequently. The conversion factor is $1 \, \text{ft}^2 \cdot \text{F}° \cdot \text{h/Btu} = 0.176 \, \text{m}^2 \cdot \text{K/W}$.

To see the utility of R values, consider a wall that consists of three layers: 2.00 cm of wood, 9.0 cm of glass wool, and 1.0 cm of gypsum board. Its total R value is the sum of the individual values. From Table 11.5,

$$R_{\text{tot}} = 0.185 + 1.95 + 0.06 = 2.20 \, \text{m}^2 \cdot \text{K/W}$$

Whether you live in a hot or cold climate, it is important to insulate buildings to minimize heat transfer between the interior and exterior. This helps to stabilize the desired inside temperature and reduces the cost and fuel consumption required by heating and cooling devices.

TABLE 11.5   *R factors (approximate)*

Material	Thickness (cm)	$R(m^2 \cdot K/W)$	$R\ (ft^2 \cdot F° \cdot h/Btu)$
Solid wood	2.00	0.185	1.05
Plywood	1.30	0.111	0.63
Fiberboard (insulating)	1.90	0.370	2.1
Gypsum board	1.00	0.060	0.34
Carpet plus pad	....	0.35	2.0
Asphalt shingles	....	0.070	0.4
Concrete (cast)	20	0.11	0.64
Concrete block:			
Normal	20	0.20	1.1
Lightweight	20	0.35	2.0
Glass wool (fiberglass)	2.5	0.65	3.7
	9.0	1.95	11
	15.0	3.3	19
Window (single-pane)	....	0.18	1
Window (double-pane)	....	0.35	2

We can now use Eq. 11.8 with this total R value to compute heat flow through the wall. Notice that most of the thermal resistance is due to the glass wool insulation. Even a layer of concrete blocks has much less effect than a 6-in-thick batt of glass wool.

Illustration 11.3

The wall for which we have just calculated an R value of $2.20\ m^2 \cdot K/W$ has an area of $5.0 \times 3.0\ m^2$. How much heat is lost through it each hour when the temperature is 20°C inside and -10°C outside?

Reasoning   The rate at which heat is lost by conduction is given by

$$\frac{Q}{\Delta t} = \frac{A\,\Delta T}{R_1 + R_2 + R_3} = \frac{A\,\Delta T}{R_{tot}}$$

$$= \frac{(15\ m^2)(30\ K)}{(220\ m^2 \cdot K/W)} = 200\ W = 200\ J/s$$

In 1 h, $\Delta t = 3600$ s, so

$$Q = (200\ J/s)(3600\ s) = 7.2 \times 10^5\ J$$

LEARNING GOALS

Now that you have finished this chapter, you should be able to

1 Define (*a*) heat and thermal energy, (*b*) thermal equilibrium and the zeroth law of thermodynamics, (*c*) calorie, Calorie, and Btu, (*d*) specific heat capacity, (*e*) heats of vaporization and fusion, (*f*) change of phase, (*g*) phase diagram, (*h*) calorimeter, (*i*) thermal expansion coefficients, (*j*) heat conduction, (*k*) heat convection, (*l*) heat radiation, (*m*) Stefan's law, (*n*) thermal conductivity and R factor, (*o*) triple point, (*p*) fusion curve, (*q*) vaporization curve, (*r*) sublimation curve

2 Explain how the zeroth law allows us to use a thermometer to measure temperature.

3 Use the equation $Q = mc\,\Delta T$ to solve simple problems in calorimetry.

4 Explain why evaporation leads to cooling.

5 Explain why the boiling point of a liquid changes as the external pressure on the liquid changes.

6 Use a phase diagram to interpret the phase changes of a substance and the dependence of those phase changes on pressure and temperature.

7 Describe qualitatively how the temperature of a crystalline substance changes as it is slowly heated, melted, heated further, and vaporized.

8 Solve simple problems in calorimetry involving heats of fusion and vaporization. Explain why the law of conservation of energy is basic to the solution.

9 Use thermal expansion coefficients in simple situations.

10 Determine how much heat flows through a slab of material when the temperatures of the two faces of the slab are given.

11 Determine the rate at which an object radiates energy.

12 Find the R factor to determine heat flow through a wall consisting of several layers.

SUMMARY

DERIVED UNITS AND PHYSICAL CONSTANTS

Mechanical Equivalent of Heat
1 calorie = 4.184 J

Stefan-Boltzmann Constant
$\sigma = 5.67 \times 10^{-8}\,\text{W/m}^2 \cdot \text{K}^4$

DEFINITIONS AND BASIC PRINCIPLES

Heat
Heat is the energy transferred from a warm object to a cooler one as a result of the temperature difference between the two.

Thermal Equilibrium
If two objects are at the same temperature, they are in thermal equilibrium and no heat energy is exchanged between them when they are brought into contact.

Zeroth Law of Thermodynamics
If two objects are each in thermal equilibrium with a third object, they are in thermal equilibrium with each other.

Thermal Energy
Thermal energy is the energy associated with the random motions of the atoms and molecules of a substance.

Specific Heat Capacity (c)
The specific heat capacity of a substance relates the heat added to or subtracted from a sample to the resulting temperature change.

$Q = mc\,\Delta T$

Heats of Vaporization and Fusion
The heat of vaporization (H_v) is the amount of heat per unit mass necessary to change the phase of a substance from liquid to gas.

$Q = mH_v$

The heat of fusion (H_f) is the heat per unit mass necessary to change the phase of a substance from solid to liquid.

$Q = mH_f$

Phase Diagram
The phase diagram of a substance is a graph of pressure versus temperature which shows the values of P and T for which phase changes of the substance occur. The solid and liquid phases are separated by the fusion curve, liquid and gas phases are separated by the vaporization curve, and solid and gas phases are separated by the sublimation curve.

Triple Point
The triple point of a substance is the value of pressure and temperature at which all three phases can coexist in equilibrium. On a phase diagram it is the intersection of the fusion, vaporization, and sublimation curves.

Coefficients of Thermal Expansion
The coefficient of linear thermal expansion (α) is the ratio of fractional change in length of an object to the temperature difference which caused the change.

$$\frac{\Delta L}{L_o} = \alpha\,\Delta T$$

The coefficient of volume thermal expansion (γ) is the ratio of fractional change in the volume of an object to the temperature difference which caused the change.

$$\frac{\Delta V}{V_o} = \gamma\,\Delta T$$

INSIGHT

1. For a given ΔT every linear dimension or element of

volume of an object undergoes the same fractional change, just like a photographic enlargement. This applies to the size of holes or cavities in objects as well.

Transfer of Heat by Conduction
The rate at which heat is conducted through a slab of thickness ΔL and surface area A is given by

$$\frac{Q}{\Delta t} = kA\frac{\Delta T}{\Delta L}$$

where ΔT = temperature difference between two sides of slab
k = thermal conductivity of slab material

INSIGHTS
1. The ratio $\Delta T/\Delta L$ is known as the temperature gradient across the slab.
2. An alternative description of heat conductivity involves the definition of the R factor of the material:

$$R = \frac{\Delta L}{k}$$

Then

$$\frac{Q}{\Delta t} = \frac{A\,\Delta T}{R}$$

The advantage of using the R factor occurs when a wall is composed of a number of thicknesses of different materials. The rate of heat conduction through the laminated wall is

$$\frac{Q}{\Delta t} = \frac{A\,\Delta T}{R_{tot}}$$

where R_{tot} is the sum of the R factors of the individual slabs making up the wall.

Transfer of Heat by Radiation
The rate at which an object loses thermal energy by radiation depends on the object's absolute temperature and the area and nature of its surface:

$$\left(\frac{Q}{\Delta t}\right)_{rad} = e\sigma AT^4$$

where e = emissivity of surface
σ = Stefan-Boltzmann constant

INSIGHTS

1. The emissivity is a dimensionless number from 0 to 1. It depends on the nature of the object's surface. Highly reflecting, polished surfaces have low values. Dark, rough surfaces have higher values.
2. An object that is a good emitter (e close to 1) is also a good absorber of radiation. Its absorptivity is the same as its emissivity. The rate of absorption of radiant energy by an object in an environment at temperature T_s is given by

$$\left(\frac{Q}{\Delta t}\right)_{abs} = e\sigma AT_s^4$$

QUESTIONS AND GUESSTIMATES

1 Which would you expect to have the larger specific heat capacity, 10 g of oxygen (O_2) gas or 10 g of argon (Ar) gas? Both contain about the same number of molecules. (Why?)

2 A student has a thermos jug containing an unknown substance at temperature T_1. After hot water at $T_2 > T_1$ is added, the temperature in the jug is still T_1. The student concludes that the material in the jug has an infinite specific heat capacity. Explain why the experiment implies that $c = \infty$. What is the probable explanation of these experimental results?

3 Can heat be added to something without its temperature changing? What if the "something" is a gas? A liquid? A solid?

4 A certain type of wax melts at 60°C. Describe an experiment by which you could determine its heat of fusion.

5 It is possible to make water boil furiously by cooling a flask of water that has been stoppered when the water was boiling at 100°C. Explain.

6 Why does cold metal feel cooler than wood at the same temperature?

7 When temperatures slightly below freezing are expected, farmers sometimes protect their fruits and vegetables by misting them with water. What is the principle behind this procedure?

8 Why is a burn caused by steam at 100°C apt to be much worse than one caused by hot water at 100°C?

9 Temperature fluctuations are much less pronounced on land close to large bodies of water than they are in the central regions of large land masses. Explain.

10 It is well known that a room filled with people becomes very warm unless it is properly ventilated. Assuming that a person gives off heat equivalent to the calories he or she burns throughout the day, estimate how much the temperature of your classroom would rise in 1 h if there were no heat loss out of the room.

11 About how much water would have to evaporate from the skin of an average-size person to cool his or her body by 1 C°? How does this fit in with what you have heard about the effect of perspiration on the body? ($c_{body} \cong$ 0.83 cal/g · C°)

12 If ice is subjected to high pressure, its melting point is decreased to below 0°C. To a rough approximation, the melting temperature decreases by about 5 C° for each 6.0 ×

10^7 Pa of applied pressure. Estimate the melting temperature of ice beneath an ice skater's skate.

13 Estimate the temperature of the sun's surface from the following facts: radiation reaching the earth from the sun is 1340 J/m² · s, the radius of the sun is 7×10^8 m, and the sun-to-earth distance is 1.5×10^{11} m.

PROBLEMS

Section 11.4

1 How much heat (in calories and joules) must be added to raise the temperature of 475 g of water from 5°C to 30°C?

2 How much heat (in calories and joules) must be removed from 1.65 kg of water to cool it from 73°C to 18°C?

3 How much heat (in calories and joules) is removed from 135 g of copper to change its temperature from 150°C to −25°C?

4 How much heat (in calories and joules) is required to raise the temperature of 2.80 kg of aluminum from 29°C to 122°C?

Sections 11.5–11.7

5 How much heat is given off as 25 g of ethanol vapor at 78°C is condensed and then cooled to 15°C?

6 How much heat is required to heat 1.35 kg of mercury from −12°C to 357°C and then to vaporize it?

■7 How much heat must be removed from 275 g of steam at 100°C to condense it and lower its temperature to create ice at a final temperature of −35°C? Assume the pressure of the steam to be 1 atm.

8 How much heat must be added to 240 g of aluminum to change it from a solid at 27°C to a liquid at 660°C?

■9 A 26-g ice cube at −10°C is dropped into a plastic cup containing 375 g of water at 37°C. What is the final temperature of the mixture? Neglect any heat transfer with the cup.

■10 45 g of molten lead at 327°C is poured into a hole in a block of ice which is at 0°C. How much ice is melted by the time the lead has come into thermal equilibrium with the ice block?

■11 Suppose that 36 g of solid mercury at −39°C is dropped into a large container of water-ice mixture at 0°C. The equilibrium temperature remains at 0°C. How much additional ice was formed by the addition of the mercury?

■12 How much perspiration must evaporate from a 4.5-kg baby to reduce its body temperature by 2.2 C°? The heat of vaporization of water at body temperature is about 580 cal/g.

■13 The average power from the sun which is incident upon the earth's atmosphere is approximately 0.138 W/cm². Most of this power is absorbed by the atmosphere before it reaches earth's surface. Assume that 0.09 percent of the original incident power is absorbed by the surface of a lake. How much mass of water evaporates from each square millimeter

of surface area per hour? Use the same heat of vaporization as in Prob. 12.

■14 Suppose 225 g of lead at 120°C is dropped into 75 g of water at 25°C contained in a 30-g aluminum cup at the same initial temperature as the water. What equilibrium temperature results?

■15 To 90 g of water contained in a 40-g copper cup, both initially at 40°C, enough ice at −15°C is added to bring the final equilibrium temperature to 20°C. How much ice was added?

■16 A 60-g iron can contains 45.0 g of water and 15.0 g of ice at 0°C. To this 275 g of hot lead is slowly added. The final temperature of the can and contents is observed to be 14°C. What was the original temperature of the lead?

17 Assuming that the total heat of vaporization of water is used to separate 1 g of water molecules from each other at the boiling point, how much energy per molecule does this amount to? Compare this energy to kT at the boiling temperature.

18 Through what vertical height would a 10-g lead bullet at 250°C have to fall in order to melt upon impact with the street? Assume all the mechanical energy of the bullet was absorbed as heat by the bullet.

19 An electric water heater supplies energy to the water in a tank at the rate of 2500 W. How long do you have to wait for 250 kg of water at 15°C to be raised in temperature to 70°C? Assume the tank to be perfectly insulated from its surroundings.

■20 Cool water is drawn into a hot-water tank at 18.0°C. Hot water at 75°C is being drawn from the tank at a rate of 400 cm³/min. What power output must the water heater have to provide this supply of hot water on a continual basis? Assume the tank to be perfectly insulated from its surroundings.

■21 A 60-kg woman consumes 2500 kcal of food energy per day. If the average rate of heat lost by the woman to her surroundings over a 24-h period is 110 W, how much exercise must she do in order to "work off" the rest of her food intake? If she did this exercise by climbing stairs, what height of stairs would she have to climb?

■22 A 50-kg girl running at 4.8 m/s while playing basketball falls and skids on her leg until she stops. How much heat is generated by this skid? Assume that all this heat is absorbed in a volume of flesh 20 cm^2 in area and 1.0 mm deep. What will be the temperature rise of the flesh? Assume $c = 0.83$ cal/g and $\rho = 950$ kg/m^3 for flesh.

··23 A 50-g copper cup that is insulated from its surroundings contains 125 g of water at 20.5°C. A 310-g mixture of brass and gold filings at 145°C is placed into the water. It is found that an equilibrium temperature of 42.3°C results. What fraction of the filings is gold? Use c (gold) = 0.031 cal/g · C° and c (brass) = 0.090 cal/g · C°.

·24 Space probes have gathered the following data about Mars and Venus: (a) the surface temperature of Venus is approximately 458°C, and (b) the atmospheric pressure at the surface of Mars is approximately 0.006 standard Earth atmospheres. From this information and the phase diagram of water (Fig. 11.6), what can you imply about the state of water on these planets?

Section 11.8

25 An aluminum meterstick is heated from 10°C to 45°C. What is its fractional change in length?

26 A brass sphere has a radius of 3.5500 cm at -12°C. What is its radius at 55°C?

27 Steel rails 12.5 m long when $T = -30$°C are to be laid end-to-end with enough gap between adjacent rails that they will just touch when the temperature is 45°C. Without the gap the rails would buckle as the temperature increased. How long should each gap be?

··28 You have two measuring tapes, one made of steel and the other of aluminum. Both are calibrated to measure correctly (to four significant digits) at 20°C. When measuring the length of a pipe at -15°C, the steel tape reads 2.630 m. What measurement does the aluminum tape give? What is the correct length of the pipe (at -15°C) to four significant digits?

29 A 1-m length of copper wire at 110°C is bent into a circle with a 1-mm gap separating the ends (Fig. P11.1). What happens to the gap as you heat the wire? Is there a temperature at which the gap will close?

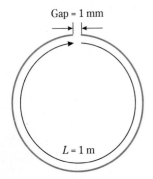

Gap = 1 mm

$L = 1$ m

30 It is common machine-shop practice to shrink-fit cylindrical rods into holes in wheels, block, and plates. Suppose a rod 2.0125 cm in diameter is to be fitted into a 1.9975-cm-diameter hole in a brass block. These dimensions are at 20°C. To what temperature must the block be heated for the unheated rod to just fit into the hole?

31 A heat-resistant glass flask is calibrated to hold 100.0 cm^3 of a liquid when at 20°C. How much more will it hold when at 50°C? *Hint:* Remember that every dimension of the flask expands as though it were a solid volume.

·32 Suppose you have a steel container with a 500 cm^3 capacity at -10.0°C. In it you have a brass sphere of radius 3.50 cm. The container is then filled to the top with methanol. By the time the container and its contents have warmed to room temperature of 27.0°C, how much methanol has spilled? (The volume of a sphere is $V = \frac{4}{3}\pi R^3$.)

·33 It is desired to have a steel wire and a brass wire lengthen by the same amount when subjected to the same temperature change. What should be the ratio of the lengths of the two wires?

Section 11.9

34 How much heat flows through a 1.3 m × 2.7 m sheet of 2.1-cm-thick plywood ($k = 0.083$ W/K · m) in 1 h if the temperatures on its two faces are -10°C and 27°C?

35 How much heat flows through a 30-cm-thick concrete wall (area = 15 m^2) in 1 h if the temperature is 0.0°C on one side and 22.3°C on the other?

·36 Deep boreholes into the earth show that the temperature increases about 1°C for each 30 m of depth. If you assume the earth's crust has $k = 1.5$ W/K · m, how much heat flows out each second through each square meter of the earth's crust?

·37 A brass pipe has an inside diameter of 7.5 cm and a wall thickness of 0.20 cm. It is used to carry steam at 120°C. If the surrounding air is at 27°C, what is the rate of heat loss per meter of the pipe's length?

·38 A plasticfoam ice chest in the form of a rectangular box has the outside dimensions of 45 cm × 35 cm × 30 cm. The chest's wall thickness is 3.75 cm. If the box is to maintain an inside temperature of 0°C when the outside temperature is 30°C, how much ice will melt inside the box each hour?

Section 11.11

39 A metal sphere with radius 1.8 cm and emissivity 0.55 is heated to 550°C and then suspended by a thin wire in a room whose temperature is 25°C. (a) At what rate does the sphere initially radiate energy, neglecting absorption of energy from the room? (b) What is the initial net rate of energy loss by the sphere?

■**40** A hot tungsten wire filament has a radius of 0.060 cm, a temperature of 3000 K, and an emissivity of 0.74. Calculate the rate of energy emission by a 1-m length of the wire. Ignore the radiation received from the surrounding environment.

■**41** Suppose a black plate ($e = 0.90$) is used as a solar collector. The plate is placed in direct sunlight, where it absorbs energy at the rate of 800 W for each square meter of its surface. What equilibrium temperature does the plate reach? Assume the back side of the plate is perfectly insulated, and that the front side loses energy only by radiation.

■**42** A solar collector for a hot water system absorbs solar radiation at the rate of 660 W/m². Its collecting area is 3.8 m². Cold water at 15°C enters the collector. What volume of water per minute at an output temperature of 60°C can this collector deliver?

Section 11.12

43 What is the R value for a 1.4-cm-thick layer of (a) glass and (b) plywood? ($k_g = 0.80$, $k_p = 0.085$ W/K · m)

■**44** If the pipe in Prob. 37 is wrapped in a 3.0-cm-thick blanket of fiberglass, by what factor is the pipe's heat loss reduced?

45 Compare the rate of heat loss through the following walls, assuming that the difference between inside and outside temperatures is the same for all cases: (a) a layer of 15.0 cm of fiberglass between two 1.75-cm-thick sheets of gypsum board, (b) a 30-cm-thick concrete wall sheathed between two sheets of 2.0-cm-thick plywood, and (c) a double-pane window.

General Problems

46 Water flows in a continual stream over a waterfall 70 m high. If all the water's gravitational potential energy is converted into heat, how much warmer is the water at the bottom of the falls than at the top?

■**47** A 2.5-g lead bullet is moving at 210 m/s when it strikes a bag of sand and is brought to rest. (a) Assuming that all the work of friction with the sand is converted to thermal en-

ergy of the bullet, what is the rise in the temperature of the bullet as it is brought to rest? (b) Answer the same question if the bullet lodges in a 90-g block of wood that is free to move when hit.

■■**48** An iron beam 8.5 m long with a cross-sectional area of 85 cm² has its ends embedded in concrete walls. The structure was constructed when the temperature was 10°C. What force does the beam exert on the walls when the temperature rises to 34°C? (Take Young's modulus for iron to be $Y = 19 \times 10^{10}$ N/m².)

■■**49** A brass wire is attached between two fixed points when the wire's temperature is 700°C. At what temperature does the wire break when it cools? The tensile breaking strength of brass is 0.45×10^9 N/m².

■■**50** A block of steel has a volume of 1.25 m³ at sea level at 20°C. This block sinks to the bottom of an ocean trench whose depth is 11,500 m, where the water temperature is 5.5°C. Calculate the resulting volume change of the block.

■■**51** A uniform steel sphere of radius R_o and mass M is set spinning about its center with angular velocity ω_o when its temperature is 27°C. If its temperature is raised to 350°C without disturbing the sphere, what are its new values of angular velocity and rotational kinetic energy?

■**52** A cube of metal with original edge length L_o has a volume $(L_o + \Delta L)^3$ after its temperature is increased by ΔT. The linear coefficient of thermal expansion for the metal is α. Use this information and the binomial theorem to show that the volume coefficient of thermal expansion for the metal is, to a first-order approximation, 3α.

■**53** A 55-kg aluminum disk has a radius of 17.5 cm. It is spinning at 9.9 rev/s when a brake exerts a friction force at the rim of the disk, bringing the disk to a stop. If 75 percent of the work done by friction is transferred as heat to the disk, how much is the temperature of the disk increased?

■■**54** A steel sphere of radius 0.22 cm is sinking through oil at its terminal velocity, given by Stokes' law. At what rate is heat being produced by the viscous force the water exerts on the sphere?

CHAPTER 12

THE FIRST LAW OF THERMODYNAMICS

Long before the nature of atoms and molecules was known, a powerful way of discussing heat, work, and internal energy had been found. It involves the description of matter in terms of macroscopic properties* such as pressure, temperature, volume, and heat flow. This way of discussing the behavior of objects and substances is called **thermodynamics.** Today, even though we understand quite well how atoms and molecules behave, thermodynamics is still widely used in all branches of science. This chapter introduces this important and useful area of study.

12.1 STATE VARIABLES

In thermodynamics, we most frequently discuss the behavior of a definite sample of matter we call a **system.** A typical system may be the molecules in a gas-filled container or those in a solution, or even such a complex system as the molecules in a rubber band. For any meaningful thermodynamic discussion, the system must be well specified. Only then can we give an unambiguous description of it. For example, to design a steam turbine that will be used to generate electricity, engineers

---

*Macroscopic properties are those that involve the average effects of a very great number of molecules.

need to know what pressures and temperatures the steam will have, as well as the volume the steam occupies as it passes through the turbine. Only then will the engineer know the amount of mechanical power the turbine can produce from a given amount of thermal energy.

To describe a system, we use quantities that apply either to the whole system or to some well-defined portion of it. Typical measurable quantities of any system are pressure, temperature, and volume. In thermodynamics, we also use such quantities as internal energy, heat, work, and a quantity we shall encounter later called entropy. As the condition of a system changes, these quantities may change. It is important that we know which quantities are suitable for representing the exact condition of the system. Let us now see what they are.

When a container having n moles of an ideal gas reaches equilibrium, the gas has a definite temperature, pressure, and volume. If any two of the quantities T, P, and V are given, the third can be calculated from the ideal-gas law (Eq. 10.1), and therefore is also known. This particular situation, where the gas (the system) has specified values of T, P, and V, is called a **thermodynamic state** of the system. Whenever the gas is returned to these same values of T, P, and V, its state will be the same. Even though each molecule within the system may not be doing the same thing whenever the system is brought to a particular state, the system as a whole still has the same properties when described macroscopically.

To put it another way, *certain measurable properties of a system are always the same whenever the system is in a given thermodynamic state. The variables that describe these properties are called* **state variables.** For example, P, V, and T are state variables for a system that consists of a gas. No matter how a gas reaches a particular equilibrium state, that state is characterized by the same values of these state variables.

Another important quantity used to characterize a system is its internal energy:

The **internal energy** *(U)* of a system is the sum of all kinetic and potential energies possessed by its atoms or molecules.

Internal energy is an example of a physical property that we call a **state function,** defined as a physical property which can be defined entirely in terms of state variables. It follows from this statement that the value of a state function, such as internal energy, does not depend on the *processes* by which a system reaches a particular state.

In contrast to internal energy, heat and work are not state functions because the amounts of heat added or work done on a system to change its state by a certain amount *do* depend on the **process** employed. Asking how much heat a system contains, for example, is a meaningless question. Systems do not "contain" either heat or work. These concepts both represent processes of energy *transfer* into or out of the system. Heat represents a transfer of thermal energy, which can change a system's internal energy. Such a transfer is, however, only one way to change internal energy. Mechanical work, such as friction or compression, done on a system is another.

12.2   THE FIRST LAW OF THERMODYNAMICS

Early workers in thermodynamics developed the idea that energy is conserved. They became convinced that heat is a form of energy and therefore must be considered when accounting for energy gains and losses. Thus they were led to a funda-

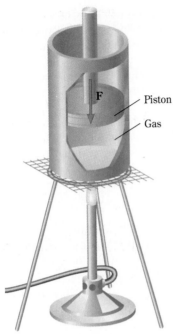

FIGURE 12.1

As heat is added to the gas in the container, the gas can gain internal energy and the gas can do work by expanding its volume against the external force exerted by the piston.

mental relation between heat, work, and internal energy. Let us now see what this relation is.

A system in a given state has a definite amount of internal energy. We now ask what happens to the system as an amount of heat flows into it. This added energy can be utilized in only two ways: (1) it can lead to an increase in the internal energy of the system or (2) it can provide the system with the energy it needs to do an amount of work W on its surroundings. For example, our system might be the gas in the cylinder of Fig. 12.1. Heat added to this system can cause two changes: (1) it can raise the temperature of the gas and thus increase its internal energy and (2) it can cause the gas to expand, thereby lifting the piston and allowing the gas to do work on the piston.

If you examine any other system, you will find that adding heat to it results in a similar situation. We conclude that, for a system,

$$\left(\begin{array}{c}\text{Heat added}\\\text{to system}\end{array}\right) = \left(\begin{array}{c}\text{increase in internal}\\\text{energy of system}\end{array}\right) + \left(\begin{array}{c}\text{external work}\\\text{done by system}\end{array}\right)$$

This statement is called the **first law of thermodynamics.** In equation form,

$$Q = \Delta U + W \tag{12.1}$$

Notice that the first law is a statement of the law of conservation of energy which includes thermal energy.

In using the first law, we must be very careful about signs. The quantity Q is always the heat that flows *into* the system. If heat flows out of the system, Q is negative. The quantity ΔU is the *increase* in the internal energy of the system, and W is the work done *by* the system. If the gas in Fig. 12.1 lifts the piston, then the gas does external work and W is positive. If the piston is pushed down by an outside force, however, then W is negative because the *gas* does negative work. To understand this latter statement, recall that work = force × displacement × cos θ, where θ is the angle between the force vector and the displacement vector. In Fig. 12.1, the *upward* force the gas exerts on the piston is equal to **F** (if we assume the piston is being pushed at constant speed). When the piston moves *downward* a distance Δs, the work done by the gas is

$$W = F\,\Delta s \cos 180° = -F\,\Delta s$$

Therefore, *when a gas is compressed, the work done by it is negative.*

We have already observed that heat and work depend on the way the state of a gas is changed. In order to use the first law, we must now investigate ways in which we can calculate Q and W for a variety of processes.

12.3   THE WORK DONE BY A CHANGE IN THERMODYNAMIC STATE

Consider a system consisting of a gas enclosed in a cylinder by a movable piston, as in Fig. 12.2. Suppose the gas just supports the weight of the piston, so that the pressure of the gas is maintained constant at a value given by

(a)   (b)

FIGURE 12.2

If the piston has a surface area A, then $\Delta V = A\,\Delta y$.

$$P = \frac{F}{A} = \frac{\text{piston weight}}{\text{piston surface area}}$$

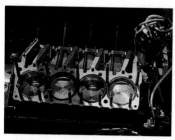

The modern internal combustion engine is a prime example of a heat engine. Thermal energy in the ignited air-fuel mixture does work on the pistons, causing the crankshaft to turn, moving the car. Much of the thermal energy is then lost as waste heat in the exhaust.

When the gas is heated, it expands an amount ΔV, as indicated in part *b*. During the expansion, the piston rises a distance Δy and the work $(F \Delta y \cos \theta)$ done on the piston by the gas is, because $\theta = 0°$ in this case,

$$W = F \Delta y = PA \Delta y$$

Since $A \Delta y$ is ΔV, the increase in the volume of the gas, we find that

$$W = P \Delta V \qquad\qquad (12.2)$$

If heat is lost by the system, the gas contracts; ΔV is negative, and thus the work done by the system is negative. In that case, the surroundings have done work on the system.

Of course, expansion at constant pressure is only one of many ways volume can change. In a constant-pressure case the calculation of work is simple: $W = P \Delta V$. However, work is done by or on a gas whenever its volume changes, regardless of by what process. We need a way to calculate work in all processes, not just constant-pressure ones. To do this, we make use of a graph of pressure versus volume, called a **PV diagram** (Fig. 12.3). *Any point on a PV diagram represents a thermodynamic state of the gas.* This is so because, if we specify the values of P and V, the temperature T of the gas is then determined by the ideal-gas law.

In Fig. 12.3, we show two states of a sample of gas by the points A and B. We have chosen them to be at the same pressure $P_A = P_B = P$. The line connecting A and B represents the *process* by which the state changes. The arrow on the line shows which way the change is proceeding. The straight horizontal line indicates that the change takes place at constant pressure. It is important that you realize that any number of other paths could connect A and B. Each one would represent a different thermodynamic process and would involve a different amount of work.

We already know how to calculate the work during the constant-pressure process shown in Fig. 12.3:

$$W = P \Delta V = P(V_B - V_A)$$

Notice that $P(V_B - V_A)$ is the *area under the line AB*, the blue rectangle in Fig. 12.3. If the process were a compression from state B to state A, ΔV and hence W would be negative, denoting work done *on* the system. The area under the line would be the same, and so you have to supply the correct algebraic sign by noticing whether V increases $(+)$ or decreases $(-)$.

Let us generalize this result. Consider the arbitrary process marked by the curve

FIGURE 12.3

The work done by a system as it expands at constant pressure equals the area under the *PV* curve.

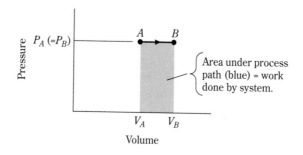

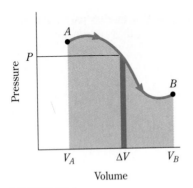

FIGURE 12.4

The work done by a system in going from state A to state B by any process is equal to the area under the PV curve.

AB in Fig. 12.4. Here P, V, and T are all changing. The dark blue region in the figure is a small enough portion of the process that the pressure can be considered constant. Its area is again $P\Delta V$, which represents the work done in this portion of the process. We can proceed from A to B by a series of similar small volume changes and sum up the work done in each. This sum will amount to the total work done from A to B and will also equal the total area under the process curve (the light blue area). We therefore conclude:

The work done during a change of thermodynamic state is equal to the area under the process curve in a PV diagram.

The work is positive if the volume increases and negative if it decreases.

Illustration 12.1

Weights are added to the piston of Fig. 12.5a as the temperature of the gas in the cylinder changes so that the gas contracts in the way shown by the PV diagram in Fig. 12.5b. Find the work done by the gas in going from the situation represented by point A, through point B, to point C.

Reasoning We must compute the area under the curve. Notice that this irregular shape consists of three simple shapes: the green and blue rectangles and the yellow triangle. We can calculate these three simple areas and add them to get the total area we need. The area under the green portion AB is

$$(5.0 \times 10^5 \, \text{Pa})[(800 - 500) \times 10^{-6} \, \text{m}^3] = 150 \, \text{J}$$

[Notice that $1 \, \text{Pa} = 1 \, \text{N/m}^2$, and so $(1 \, \text{Pa})(1 \, \text{m}^3) = 1 \, \text{N} \cdot \text{m} = 1 \, \text{J}$.] Similarly, the area under the curve from B to C is

$$(2.0 \times 10^5 \, \text{Pa})(200 \times 10^{-6} \, \text{m}^3) + \tfrac{1}{2}(3.0 \times 10^5 \, \text{Pa})(200 \times 10^{-6} \, \text{m}^3) = 70 \, \text{J}$$

FIGURE 12.5

How much work is done by a gas as it moves from the conditions at point A to those at point C along the path ABC?

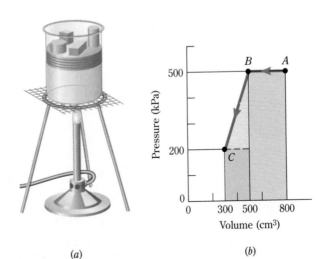

(a)

(b)

where we have made use of the fact that the area of a triangle is one-half the base times the height. Therefore

Area under PV curve $= 150\,\text{J} + 70\,\text{J} = 220\,\text{J}$

Because the process we are considering involves a decrease in volume, the work done by the gas is negative. We therefore conclude that the work done by the gas in going from A to C via B is $-220\,\text{J}$.

Exercise How much work would the system do if the PV diagram were a straight line from A to C? *Answer: -175 J.*

For paths on a PV diagram that create areas which have no simple geometric formula, we can closely approximate the area by plotting the process on graph paper and counting the squares under the curve.

12.4 INTERNAL ENERGY OF AN IDEAL GAS

In Chap. 10 we learned that the total translational kinetic energy of an ideal gas depends on the temperature of the gas:

$$\text{KE}_{\text{trans}} = N(\overline{\text{KE}}) = N(\tfrac{3}{2})kT = n(\tfrac{3}{2}RT) \tag{10.4}$$

where N = the number of molecules
$\quad n$ = number of moles
$\quad k$ = Boltzmann's constant

We would like to understand the connection between this relationship and the internal energy U of the gas.

Gases composed of single atoms, such as helium and monatomic oxygen, have no forms of internal energy other than translational kinetic energy.* Thus for these monatomic gases, we can equate internal energy to translational kinetic energy:

$$U = \text{KE}_{\text{trans}} = \tfrac{3}{2}nRT \qquad \text{(monatomic gas)}$$

From this it follows that changes in temperature are associated with changes in internal energy by

$$\Delta U = \tfrac{3}{2}nR\,\Delta T \tag{12.3}$$

For gases composed of molecules, internal energy can consist of rotational and vibrational energy in addition to translational energy. This is because the atoms making up the molecules can oscillate along the direction of the bonds holding them together. The molecules also have significant moments of inertia about axes perpendicular to these bonds. As a result, diatomic molecules (two atoms per molecule) and more complicated molecules have a larger value of U than monatomic gases at

---

*We neglect the fact that an atom has internal energy associated with its electrons, protons, and neutrons. Changes in these energies occur only at temperatures much higher than those we consider here.

the same temperature. Detailed discussion of complex molecules lies beyond the scope of this course. It turns out, however, that the expression for U can always be written in the form of some whole number K times $\frac{1}{2}nRT$:

$$U = K(\tfrac{1}{2}nRT)$$

For monatomic gases, $K = 3$, and we arrive at Eq. 12.3. For other gases, K is an integer equal to or greater than 3, depending on the type of gas and on the temperature.

In all ideal gases, U is dependent on only *one* state variable: T. Thus U is a state function, and we can conclude the following:

As the state of an ideal gas changes, the change in the internal energy depends only on the initial and final temperatures, and not on the process of change.

12.5   HEAT TRANSFER AND SPECIFIC HEATS OF IDEAL GASES

The amount of heat transferred into or out of a gas, like work, depends on the details of the process being used. (Hence Q is not a state function.) There are two types of processes for which we can easily calculate heat transfer directly: constant-*volume* processes and constant-*pressure* processes.

CONSTANT-VOLUME PROCESSES

If heat is added to a gas kept at constant volume, *no work is done* (since $\Delta V = 0$). The first law of thermodynamics then states that all the heat must go into increasing internal energy:

$$Q = \Delta U \qquad \text{(constant volume)} \tag{12.4}$$

From Eq. 12.3, we know that the relation between ΔU and ΔT for a monatomic gas is $\Delta U = \frac{3}{2}nR\,\Delta T$. Thus we can relate the heat transfer to the resulting ΔT:

$$Q = \Delta U = \tfrac{3}{2}nR\,\Delta T$$

Up to this point, the only quantity we've met which relates Q to ΔT for a given amount of matter is the specific heat c of the substance, where $c = Q/m\,\Delta T$, where m is the sample's mass (Eq. 11.1). In the case of gases we very often measure the amount of gas in moles. We define the **molar specific heat** C in a similar way:

$$C = \frac{Q}{n\,\Delta T} \tag{12.5}$$

where n = number of moles. This ratio depends on which heat-transfer process is used, and so we must add a symbol to denote which process we are talking about. For a constant-volume process, we use the symbol C_V.

Using Eqs. 12.3 and 12.4 to obtain $Q = \frac{3}{2}nR\,\Delta T$, we obtain a particularly simple form of Eq. 12.5:

$$C_V = \tfrac{3}{2}R \qquad \text{(monatomic gas)}$$

In the case of more complex molecules, the same method gives the more general result:

$$C_V = K\frac{R}{2}$$

where K is an integer, as mentioned in the previous section.

CONSTANT-PRESSURE PROCESS

We have seen that $W = P\Delta V$ in a constant-pressure process. The first law then gives us

$$Q = \Delta U + W = \Delta U + P\Delta V \tag{12.6a}$$

When P is constant, the ideal-gas law tells us that

$$P\Delta V = nR\,\Delta T$$

Thus

$$Q = \Delta U + nR\,\Delta T \qquad \text{(constant } P\text{)} \tag{12.6b}$$

In the same way we defined C_V, we now define the molar specific heat at constant pressure, C_P:

$$C_P = \frac{Q}{n\,\Delta T} = \frac{\Delta U + P\Delta V}{n\,\Delta T}$$

$$= \frac{\Delta U}{n\,\Delta T} + \frac{nR\,\Delta T}{n\,\Delta T} = C_V + R$$

This result does not depend on the type of gas, and so we have

$$C_P = \tfrac{3}{2}R + R = \tfrac{5}{2}R \qquad \text{(monatomic gas)}$$

$$C_P = K\frac{R}{2} + R = (K + 2)R \qquad \text{(molecular gas)}$$

It should not surprise you that C_P is always larger than C_V. At constant pressure some of the heat must go into doing work (raising the piston in Fig. 12.1, for instance), leaving less available to increase internal energy and hence temperature. A larger specific heat denotes a smaller temperature response for the same heat transfer.

We denote the *ratio* of specific heats for these two processes by the symbol γ:

$$\gamma = \frac{C_P}{C_V} \tag{12.7}$$

TABLE 12.1   *Molar and mass specific heat capacities of gases*

Gas	$\dfrac{C_V}{R}$	$\dfrac{C_P}{R}$	$\gamma = \dfrac{C_P}{C_V}$	$\dfrac{C_P - C_V}{R}$	$c_V(J/kg \cdot K)$
He	1.50	2.49	1.66	0.99	3,130
Ne	1.50	2.46	1.64	0.96	620
Ar	1.50	2.50	1.67	1.00	310
Kr	1.50	2.52	1.68	1.02	150
Xe	1.50	2.49	1.66	0.99	95
Hg (360°C)	1.50	2.50	1.67	1.00	62
O_2	2.48	3.48	1.40	1.00	650
N_2	2.48	3.48	1.40	1.00	740
H_2	2.40	3.39	1.41	0.99	10,000
CO	2.46	3.48	1.41	1.02	730
HCl	2.51	3.54	1.41	1.03	810
CO_2	3.37	4.37	1.30	1.00	640
H_2O (200°C)	3.23	4.23	1.31	1.00	1,500
CH_4	3.24	4.24	1.31	1.00	1,690

All gases at 15°C unless stated otherwise.

Experimentally determined values of C_V, C_P, and γ are given in Table 12.1 for a number of gases. Notice that for monatomic gases $\gamma \approx 1.67$. The value Eq. 12.7 would predict is $\gamma = \frac{5}{3}/\frac{3}{2} = 1.67$! The values of γ for the other gases give us information about the value of K applicable to them. Our equations predict the general result that $\gamma = (K + 2)/K$. For diatomic molecules $\gamma = 1.40 = \frac{7}{5}$, which indicates $K = 5$. More complex molecules have $\gamma = 1.3 = \frac{4}{3}$, which corresponds to $K = 6$. Thus experiment bears out the expectation that K is an integer. Further discussion on the meaning of the value of K is given in Sec. 12.8.

Another test of the equations we derived above lies in the following relation:

$$C_P - C_V = R \qquad \text{or} \qquad \frac{C_P}{R} - \frac{C_V}{R} = 1$$

The fifth column in Table 12.1 shows that this prediction holds true for all gases.

Illustration 12.2

Calculate the amount of heat necessary to raise the temperature of 2.00 moles of He from 20.0°C to 50.0°C using (*a*) a constant-volume process and (*b*) a constant-pressure process. Repeat these calculations for CO_2.

The relevant equations are:

For (*a*): $Q = nC_V \Delta T$
For (*b*): $Q = nC_P \Delta T$

From Table 12.1 we get the values $C_V = 1.50R$ and $C_P = 2.49R$ for He. Thus for He:

(*a*) $Q = (2.00 \text{ mol})(1.50)(8.315 \text{ J/mol} \cdot \text{C°})(30.0°\text{C}) = 748 \text{ J}$

(*b*) $Q = (2.00 \text{ mol})(2.49)(8.315 \text{ J/mol} \cdot \text{C°})(30.0°\text{C}) = 1240 \text{ J}$

Since the temperature rise is the same in both cases, the change in internal energy is the same: $\Delta U = 748 \text{ J}$. In (*b*) the excess heat of 492 J went into the work of expansion.

For CO_2 we have $C_V = 3.37R$ and $C_P = 4.37R$. Thus the amounts of heat for the same two processes are

(a) $Q = \left(\dfrac{3.37}{1.50}\right)(748\,\text{J}) = 1680\,\text{J}$

(b) $Q = \left(\dfrac{4.37}{2.49}\right)(1242\,\text{J}) = 2180\,\text{J}$

Notice that there is essentially the same difference, 500 J, available to do the work of expansion. More heat is required to raise the temperature of CO_2 because the molecules can absorb energy by rotating.

12.6   TYPICAL THERMODYNAMIC PROCESSES IN GASES

When we draw a PV diagram, which is simply a graph showing how P varies with V, we assume that the changes that are occurring in the system are slow enough for the pressure and temperature to be uniform throughout the whole system at any instant.

We have already discussed two processes in which a thermodynamic quantity is held constant. The constant-volume process is called **isovolumetric** (sometimes called **isochoric**). Isovolumetric processes are represented by *vertical* straight lines on a PV diagram. Constant-pressure processes are called **isobaric,** and are represented by *horizontal* straight lines. Let us now examine two more processes in which some quantity is held constant.

CONSTANT-TEMPERATURE PROCESS

If we change the state of a gas in such a way that its temperature remains con-

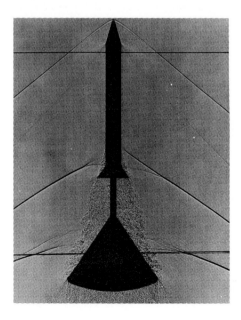

This photo of a shock wave in a supersonic wind tunnel shows the sharp density change across the wave. Sometimes the compression, which is adiabatic (why?), is severe enough to cause the gas behind the shock to become luminous. This can be seen for instance in the shock waves caused by detonating explosives.

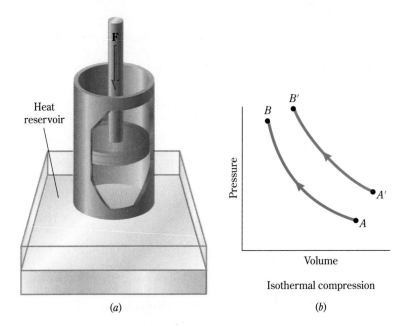

FIGURE 12.6

The PV diagram for an isothermal compression. The isotherm $A'B'$ is at a higher temperature than AB. Why?

(a)   (b)

stant,* we have an **isothermal** process. Since internal energy depends only on temperature, it follows that $\Delta U = 0$ during an isothermal process. The first law then becomes

$$Q = W \qquad \text{(isothermal process)} \tag{12.8}$$

Thus all heat added to the gas goes into the work of isothermal expansion. Conversely, all work of isothermal compression done on the gas must be lost as heat to the surroundings. Figure 12.6a shows a container of gas in good thermal contact with a heat reservoir (an oven, cooling bath, or any other device which can supply or take away heat while maintaining a constant temperature). As weights are slowly added to the piston, the pressure on the gas increases and its volume decreases.

The ideal-gas law states that if T is constant, so is the product PV. This in turn means that P and V are *inversely* related during an isothermal process:

$$PV = \text{constant} \qquad \text{or} \qquad P = \frac{\text{constant}}{V} \qquad \text{(isothermal process, ideal gas)} \tag{12.9}$$

This equation gives us the path of an isothermal process (called an **isotherm**) on the PV diagram, shown in Fig. 12.6b. The higher the temperature of an isotherm, the farther it lies from the coordinate axes. For example, the green isotherm $A'B'$ in Fig. 12.6b represents a higher temperature than the blue isotherm AB.

An expression for the work done during an isothermal process can be derived using the methods of calculus. Although the derivation is beyond the mathematical level of this course, the result is easy to use, and so we present it here without proof:

$$W = nRT \ln \frac{V_f}{V_i} \qquad \text{(isothermal process, ideal gas)}$$

*When we draw a PV diagram, we assume that changes which occur in the system are slow enough for the pressure and temperature to be uniform throughout the system at any instant.

Here T is the absolute temperature of the isotherm and V_f and V_i are the final and initial volumes. The function ln is the natural logarithm (to base e). This expression correctly supplies the sign of the work, since ln of a number less than 1 is negative. This is the case for compression of the gas, when $V_f < V_i$.

ZERO HEAT-TRANSFER PROCESS

Our fourth process is one in which the thermodynamic state changes with *no heat transfer* between the system and its surroundings. This is called an **adiabatic** process. For example, if a system is well insulated from its surroundings, heat transfer is often negligible, and all processes taking place within the system are adiabatic. Or if a process is carried out extremely rapidly (such as a very sudden compression of a gas), no appreciable heat can flow into or out of the system in that short time. Such a process is also adiabatic.

For an adiabatic process, $Q = 0$ and the first law ($Q = \Delta U + W$) becomes

$$\Delta U = -W \quad \text{(adiabatic)} \quad (12.10)$$

This relation tells us that *if a system does adiabatic work, its internal energy must decrease.* The work is done at the expense of internal energy. If adiabatic work is done *on* the system, however, the internal energy increases. Illustration 12.3 and Example 12.1 will show two practical uses of adiabatic processes. First, however, we examine the adiabatic behavior of an ideal gas in more detail.

An adiabatic process for an ideal gas is not described in terms of $PV = nRT$ alone because all three state variables (P, V, and T) change during the process. To find how these changes are related, we need a second equation. We find this by noticing that the work done on the gas goes completely into increased *internal* energy. This increase causes a temperature change in the system, but this same temperature change could have been carried out by adding heat to the system. Hence, a relation between heat, temperature change, and work can be found even for an adiabatic process. For an ideal gas, this line of thought leads to the following result:

if an ideal gas undergoes an adiabatic change from P_1, V_1, T_1 to P_2, V_2, T_2, then

$$P_1 V_1^\gamma = P_2 V_2^\gamma \quad (12.11)$$

where $\gamma = C_P/C_V$ for the gas.

We can write this adiabatic relation as $P = \text{constant}/V^\gamma$. Since $\gamma > 1$ in all cases, P diminishes faster as V increases in an adiabatic process than it does in the isothermal relation $P = \text{constant}/V$. This behavior is shown in Fig. 12.7.

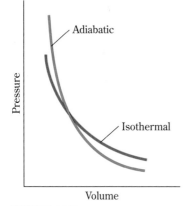

FIGURE 12.7

Comparison of adiabatic and isothermal changes for an ideal gas.

Example 12.1

In a cylinder of a diesel engine, the air is suddenly (therefore adiabatically) compressed by a piston, a process that causes a sudden rise in temperature. The new temperature is high enough that injected fuel ignites immediately without the necessity of a spark plug. Suppose the piston compresses the air to one-fifteenth of its original volume. Take the initial conditions to be 1.00 atm and $T_1 = 27.0°C$. Find the final temperature and pressure, T_2 and P_2.

R

Reasoning

Question   Can I use the ideal-gas law?
Answer   Yes, but you have *two* unknowns, P_2 and T_2. Thus you need a second relation, one that depends on the process by which the state of the gas is changing.

Question   What condition applies to an adiabatic process?
Answer PV^{γ} = constant. This means you can write

$$P_1 V_1^{\gamma} = P_2 V_2^{\gamma}$$

Note that P_2 is the only unknown in this equation, since you know the *ratio* of the volumes ($V_2 = V_1/15$). From Table 12.1, you can see that $\gamma = 1.40$ for N_2 and O_2, which are the gases that make up the air in the cylinder. Thus you can write

$$P_2 = P_1 \left(\frac{V_1}{V_2}\right)^{\gamma} = (1.00 \text{ atm})\left(\frac{15}{1}\right)^{1.40}$$

Question   Once I determine P_2, how can I use the ideal-gas law to find T_2? The number of moles, n, is not given.
Answer   Again, forming ratios is the simplest approach. Since n remains constant, it will cancel out of ratios, whatever its value:

$$\frac{T_2}{T_1} = \frac{P_2 V_2}{P_1 V_1} = \frac{P_2}{P_1}\frac{V_2}{V_1}$$

Question   Do I have to be concerned about the units of temperature in this ratio?
Answer   The ideal-gas law *always* requires absolute temperatures.

Solution and Discussion   Your calculator should have an x^y key. To calculate $(15)^{1.4}$, enter 15, press the x^y key, enter 1.4, press =. You should get 44.3. Thus

$$P_2 = (44.3)P_1 = 44.3 \text{ atm} = 4.48 \times 10^6 \text{ Pa}$$

Finally, the equation for T_2 gives

$$T_2 = T_1 \frac{44.3}{1}\frac{1}{15} = 2.95(T_1) = 2.95(300 \text{ K}) = 886 \text{ K} = 613°C$$

This temperature is high enough to ignite the fuel-air mixture.

Exercise   What would the final pressure be if the gas were compressed to one-fifteenth of its volume *isothermally*?   *Answer:*   *15 atm.*

TABLE 12.2 *Summary of thermodynamic processes (ideal monatomic gas)*

Process	Constant	Heat transfer (Q)	Work done (W)	Change in internal energy (ΔU)	Form of the first law
Isobaric	P(or V/T)	$nC_P \Delta T$	$P\Delta V$	$\frac{3}{2}nR\,\Delta T$	$Q = \Delta U + P\Delta V$
Isovolumetric (or isochoric)	V(or T/P)	$nC_V \Delta T$ $= \frac{3}{2}nR\,\Delta T$	0	$\frac{3}{2}nR\,\Delta T$	$Q = \Delta U$
Isothermal	T(or PV)	$nRT \ln \dfrac{V_2}{V_1}$	$nRT \ln \dfrac{V_2}{V_1}$	0	$Q = W$
Adiabatic	PV^γ	0	$-\frac{3}{2}nR\,\Delta T$	$\frac{3}{2}nR\,\Delta T$	$\Delta U = -W$

12.7 APPLICATIONS OF THE FIRST LAW

The first law of thermodynamics applies to all possible thermodynamic processes, relating the three quantities Q, W, and ΔU. We have discussed four special processes involving ideal gases for which direct calculations of these three quantities are simplified. The results are summarized in Table 12.2. One of our goals is to be able to calculate Q, W, and ΔU for any process we may encounter. If we are able to determine any *two* of them, the first law will yield the remaining one. If we are given the description of a process as a path AB on a PV diagram, we can proceed in the following way:

1 Work (W_{AB}) can always be found by determining the area under the path AB. If AB is composed of straight lines, this step amounts to calculating the areas of rectangles or triangles. If AB is a curved path, we can plot the curve on graph paper and count squares under the curve.

2 For ideal gases, the temperature of any state (a point on the PV diagram) can be found from the ideal-gas law. Thus T_A and T_B can be calculated. Since internal energy *does not depend on the process by which the state is changing, but only on the temperatures at the end points A and B,* we can calculate ΔU from Eq. (12.3):

$$\Delta U = \tfrac{3}{2}nR(T_B - T_A) \qquad \text{(monatomic gas)}$$

3 The first law (Eq. 12.1) can then be used to determine the heat transferred into or out of the gas during the process, Q_{AB}:

$$Q_{AB} = \Delta U + W_{AB}$$

We must always remember to apply the correct signs to Q and W. Positive signs denote heat *added to* the gas and work *done by* the gas (expansion). Negative signs denote heat *lost by* the gas and work *done on* the gas (compression).

Let us now see how these rules apply in some examples.

Example 12.2

Assume you have 1 mol of a monatomic ideal gas, initially at $P_1 = 1$ atm and $T_1 = 27°C$. Calculate Q, W, and ΔU for (**a**) adiabatic, (**b**) isothermal, and

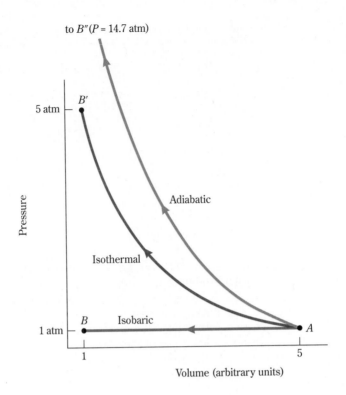

FIGURE 12.8

Three compressions from the same
initial state to the same final volume.

(c) isobaric compressions in which the final volume is one-fifth of the initial
volume in each case. The three processes are shown in Fig. 12.8.

R

Reasoning

Question   Which of the three quantities is easiest to start with?
Answer   You can calculate ΔU if you know the initial and final temperatures.
You can also calculate the work either by a formula or by finding the area under
the process path in the PV diagram. Determination of Q usually follows by
application of the first law.

Question   What is the expression for ΔU?
Answer   In *all* cases for a monatomic gas, $\Delta U = \frac{3}{2}nR(T_2 - T_1)$.

Question   What are the final temperatures in the three cases?
Answer

(a) Refer to the adiabatic calculation in Example 12.1. The difference in the
present case is that the gas is monatomic, with $\gamma = 1.67$ (from Table 12.1). You
have

$$P_2 = P_1(\tfrac{5}{1})^{1.67} = (1 \text{ atm})(14.7) = 14.7 \text{ atm}$$

Then

$$T_2(\text{adiabatic}) = T_1\frac{P_2}{P_1}\frac{V_2}{V_1} = (14.7)\left(\frac{1}{5}\right) = 2.94T_1 = 882 \text{ K}$$

(b) In the isothermal case,

$$T_2 \text{ (isothermal)} = T_1 = 27°C = 300 \text{ K}$$

(c) For the isobaric case, T is proportional to V, since P is constant. Thus if $V_2 = V_1/5$,

$$T_2 \text{ (isobaric)} = \frac{T_1}{5} = 60 \text{ K}$$

Question   What are the internal energy changes in each case?
Answer

(a) Adiabatic: $\Delta U = \frac{3}{2}nR\,\Delta T$

$$= \tfrac{3}{2}(1 \text{ mol})(8.314 \text{ J/mol} \cdot \text{K})(882 \text{ K} - 300 \text{ K}) = +7260 \text{ J}$$

(b) Isothermal: $\Delta U = 0$ since $\Delta T = 0$
(c) Isobaric: $\Delta U = \frac{3}{2}(1)(8.314 \text{ J/mol} \cdot \text{K})(60 \text{ K} - 300 \text{ K}) = -2990 \text{ J}$

Question   What are the expressions for the work done in each case?
Answer   From Table 12.2 you have

(a) W (adiabatic) $= -\Delta U = -7260 \text{ J}$
(b) W (isothermal) $= nRT \ln \dfrac{V_2}{V_1}$

$$= (1)(8.314 \text{ J/mol K})(300 \text{ K}) \ln (\tfrac{1}{5}) = -4010 \text{ J}$$

(c) W (isobaric) $= P(V_2 - V_1) = nR(T_2 - T_1)$

$$(1)(8.314 \text{ J/mol} \cdot \text{K})(60 \text{ K} - 300 \text{ K}) = -1995 \text{ J}$$

Question   What are the expressions for heat transferred?
Answer

(a) Q (adiabatic) $= 0$
(b) Q (isothermal) $= W = -4010 \text{ J}$
(c) Q (isobaric) $= \Delta U + W$ (isobaric) $= \Delta U + P\Delta V = \Delta U + nR\,\Delta T$

$$= -2990 \text{ J} + (-1995 \text{ J}) = -4980 \text{ J}$$

Solution and Discussion   Notice the following important points:

1 The minus sign indicates that work is done *on* the gas in all three cases, as you would expect in any type of compression.
2 In the adiabatic case, all the work done on the gas goes into the increase in internal energy.
3 To keep the temperature from rising in the isothermal case, as much heat must be allowed to escape as there is work done on the gas.
4 You should verify that the amount of heat lost in the isobaric case is the same as you would get by using $Q = nC_P\,\Delta T$.

Example 12.3

Two moles of argon gas are taken through the thermodynamic process ABC shown in Fig. 12.9. Determine the change in internal energy, the work, and the heat transfer during this process.

R Reasoning

Question   Is the process ABC any one of the four previously discussed?
Answer   No. None of the previous expressions for Q or W apply to this process.

Question   How can I determine the work done?
Answer   Work is *always* the area under the process path in the PV diagram.

Question   What is the area under path ABC?
Answer   It is the area of the green triangle ABC plus the red rectangle below the line AC:

$$\text{Area} = W = \tfrac{1}{2}(0.250 \text{ atm})(50.0 \text{ liters}) + (0.500 \text{ atm})(50.0 \text{ liters})$$

Question   How can I tell whether this is work done by or on the gas?
Answer   By noting whether the process is a compression ($\Delta V < 0$) or an expansion ($\Delta V > 0$). In this case the gas is expanding and so is doing work. Thus the area calculation must be interpreted as a *positive* amount of work.

Question   Since this is not a simple process, how can I calculate ΔU?
Answer   The internal energy ΔU does not depend on the process. It is always equal to $\tfrac{3}{2}nR\,\Delta T$ for a monatomic gas.

Question   How do I calculate the temperatures at A and C?
Answer   By using the ideal-gas law: $T = PV/nR$.

Question   What relation can I use to determine Q_{ABC}?
Answer   Once you have found W and ΔU, the first law of thermodynamics gives you Q: $Q = \Delta U + W$.

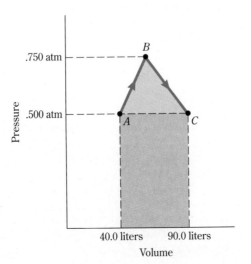

FIGURE 12.9

The thermodynamic process ABC for Example 12.3.

Solution and Discussion   The area under path *ABC* gives

$$W = +31.2 \, \text{atm} \cdot \text{liter} = +3160 \, \text{J}$$

The temperature at *A* is

$$T_A = \frac{P_A V_A}{nR} = \frac{(0.500 \, \text{atm})(40.0 \, \text{liter})}{(2.00 \, \text{mol})(0.0820 \, \text{atm liter/mol} \cdot \text{K})}$$

$$= 122 \, \text{K}$$

(Notice the choice of *R* to achieve consistency of units.) Since $P_A = P_C$,

$$T_C = T_A \frac{V_C}{V_A} = (122 \, \text{K})\frac{90.0}{40.0} = 274 \, \text{K}$$

Then

$$\Delta U = \tfrac{3}{2}(2.00 \, \text{mol})(8.314 \, \text{J/mol} \cdot \text{K})(274 \, \text{K} - 122 \, \text{K})$$
$$= +3790 \, \text{J}$$

The heat transfer is the algebraic sum of ΔU and *W*:

$$Q = +3790 \, \text{J} + 3160 \, \text{J} = +6950 \, \text{J}$$

This is the amount of heat *added* to the gas during the process.

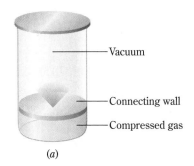

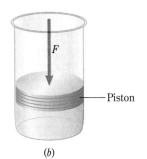

FIGURE 12.10

When a hole is made in the partition in (*a*), the gas expands into a vacuum. In (*b*), the partition has been replaced by a piston. The gas expands slowly as the piston moves upward. Under which condition will an ideal gas be cooled more?

Illustration 12.3   ***The throttling process***

The insulated container in Fig. 12.10*a* is sectioned into two parts, with gas at high pressure in the bottom and vacuum in the much larger top part. A small hole is opened in the connecting wall so that the gas expands adiabatically into the vacuum chamber. (*a*) Describe the resulting temperature change of the gas. (*b*) Suppose the high-pressure section instead contained a liquid which vaporizes as it expands into the vacuum. Describe the temperature change of the material.

Reasoning   This type of process, in which a gas freely expands through a small opening or a porous disk, is called a *throttling process*. Since the process is adiabatic, the first law tells us that $\Delta U = -W$, where *W* is the work done by the gas.

(*a*) We are asked to deal with the specific case of a gas, which we shall assume to be an ideal gas. The gas does no work as it expands into the vacuum because there is zero external pressure resisting the expansion, and so $P \Delta V = 0$. It follows from the first law that the internal energy of the gas doesn't change. Since $T \sim U$ for an ideal gas, *the gas temperature remains constant.*

(*b*) The result is quite different when the compressed material is a liquid, for example, butane or freon, which vaporizes as it expands into the vacuum. Since the material changes phase, the heat of vaporization must be supplied from some source of energy. Since the process is adiabatic, there is no external heat supplied. The energy required to vaporize a mass m of the fluid must then come from the thermal energy already possessed by the fluid: $\Delta U = -mL_V$. As a consequence, the average kinetic energy of the molecules decreases during the expansion. The temperature of the gas is therefore lower than that of the original liquid. This is similar to the cooling that takes place during evaporation.

One of the most common instances of this phenomenon involves aerosol spray cans, which contain a liquid propellant under pressure. You have probably noticed that when you depress the valve on such a can, allowing the contents to vaporize, the valve and the can become quite cool. Even though in this case the expansion takes place into the atmosphere rather than a vacuum, the cooling effect of the change of phase is still dramatic. This throttling process utilizing the very large heat of vaporization is the basis of all refrigeration devices, including air conditioners and refrigerator/freezers. We will examine these in more detail in the next chapter.

Finally, in certain cases even an ideal gas cools upon adiabatic expansion. For example, suppose the partition in Fig. 12.10a is replaced by a movable piston, as in Fig. 12.10b. If the top section of the container now contains air at some pressure lower than that in the bottom section, the expanding gas must do work against this pressure. As long as the expansion is adiabatic, this work must be supplied from the internal energy of the gas, resulting in a decrease in temperature. ■

A MODERN PERSPECTIVE
TEMPERATURE DEPENDENCE OF MOLAR SPECIFIC HEATS OF GASES

In Sec. 12.5 we observed that monatomic ideal gases have measured values of C_V and C_P that agree with classical theory. Furthermore, classical theory does not predict any temperature dependence for specific heats of ideal gases, monatomic or not. Experimental results show, however, that specific heats of diatomic and polyatomic gases do depend on temperature and have values at low to moderate temperatures that disagree with classical predictions. To understand what causes this discrepancy, we have to turn once again to the concept of quantized energy, a topic we discussed in Sec. 8.5.

A fundamental principle of thermal equilibrium is that each component of motion, in the x, y and z directions, shares equally in the internal energy of a gas. This principle is called the **equipartition theorem.** Thus each component of an atom's translational motion has an average of one-third of the translational kinetic energy ($3kT/2$) of the atom, or $\frac{1}{2}kT$. We call these independent components of motion the **degrees of freedom** of the gas. A monatomic gas has 3 degrees of freedom, one for each of the three components of its velocity vector.

To treat molecular gases, we extend the equipartition theorem to all the independent motions (degrees of freedom) by which the molecules can possess energy. Linear diatomic molecules such as H_2 can rotate about two independent axes perpendicular to the line joining the atoms. The equipartition theorem gives an average energy share of $\frac{1}{2}kT$ to each rotational degree of freedom. Also, both kinetic and potential energies are involved in the vibration of the bond joining the two atoms. Each of these forms of energy are again predicted to possess an average value equal to $\frac{1}{2}kT$. We are thus led to predict that linear diatomic molecules should have a total

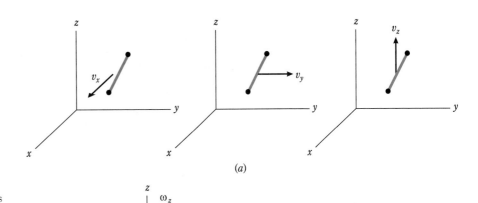

FIGURE 12.11

A diatomic molecule has seven degrees of freedom. According to the classical equipartition theorem, the average energy per molecule should be $7(\frac{1}{2}kT)$. (a) Three translational degrees of freedom:

$E = \frac{1}{2}mv_x^2 + \frac{1}{2}mv_y^2 + \frac{1}{2}mv_z^2$

$\overline{E} = \frac{1}{2}kT + \frac{1}{2}kT + \frac{1}{2}kT = \frac{3}{2}kT$

(b) Two rotational degrees of freedom:

$E = \frac{1}{2}I\omega_y^2 + \frac{1}{2}I\omega_z^2$

$\overline{E} = \frac{1}{2}kT + \frac{1}{2}kT = kT$

(c) Two vibrational degrees of freedom:

$E = \frac{1}{2}mv_r^2 + \frac{1}{2}k(r - r_0)^2$

$\overline{E} = \frac{1}{2}kT + \frac{1}{2}kT = kT$

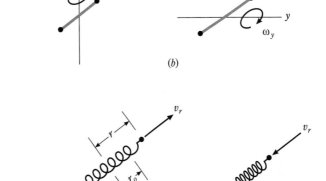

internal energy of $7(\frac{1}{2}kT)$ per molecule (see Fig. 12.11). Thus for n moles,

$$U = \frac{7}{2}nRT = nC_VT$$

which gives

$$C_V = \frac{7}{2}R \qquad \text{and} \qquad C_P = C_V + R = \frac{9}{2}R$$

Now we can interpret the notation used in Sec. 12.5, where we wrote the general expression $C_V = K(R/2)$ and $\gamma = C_P/C_V = (K + 2)/K$. We now see that the integer K represents the number of degrees of freedom the gas has available for the sharing of thermal energy. For monatomic gases, $K = 3$ and $\gamma = \frac{5}{3} = 1.67$, in agreement with experiment. For diatomic gases, classical theory gives $K = 7$ and $\gamma = 9/7 = 1.28$. However, most diatomic gases (Table 12.1) show $\gamma = 1.4$ instead, which indicates only five degrees of freedom ($K = 5$). Despite the fact that there is no temperature dependence in these results, the measured values of C_V and C_P in fact do vary with temperature for molecular gases.

Let us look at the experimental results for a gas of H_2 molecules. As Fig. 12.12 shows, for temperatures below about 50 K, C_V for H_2 has the same value, $\frac{3}{2}R$, as a

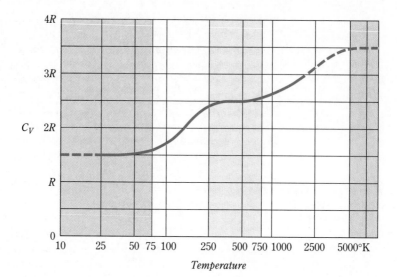

FIGURE 12.12

Experimental values of C_V for diatomic hydrogen gas as a function of temperature, plotted on a logarithmic scale.

monatomic gas. Above 250 K but below 750 K, $C_V = \frac{5}{2}R$. Only above 5000 K does C_V finally approach its classically predicted value of $\frac{7}{2}R$. This behavior seems to indicate that none of the rotational or vibrational energy modes are present at extremely low (50 K) temperatures, and that only two of these modes become activated in the middle temperature range. Classically, the equipartition principle holds that molecular collisions distribute internal energy equally among *all* possible degrees of freedom, *independent of temperature.*

This behavior of C_V, clearly at odds with classical theory, remained a puzzle until explained by Einstein in 1907. Once again the explanation required a revision of some underlying assumption of classical physics. We have seen (Sec. 8.5) that classical theory assumes there is no limit on the "smallness" of the angular momentum a rotating object can have. We also saw that this assumption must be rejected in the case of atomically small objects. There is nothing in our experience with macroscopic rotators that would have made this assumption questionable. The wheel on a car, for instance, seems to be able to rotate more and more slowly as the car stops, coming to a perfectly smooth stop. Similarly, nothing in our experience with oscillating systems, such as springs and pendulums, would lead us to believe that there was a nonzero limit on the minimum oscillation possible. It is often the case, however, that the most "obvious" assumptions we make are the most difficult to question.

In Sec. 8.5, we mentioned that the angular momenta of very small rotating bodies have a quantum of $L_1 = h/2\pi$ and hence a minimum rotational energy given by $E_1 (\text{rot}) = L_1^2/2I = h^2/8\pi^2 I$. Here h is Planck's constant ($6.6.2 \times 10^{-34}$ J · s) and I is the moment of inertia about the axis of rotation. Notice that Planck's constant h is *squared* in the energy expression, which makes the numerator extremely small. However, the moments of inertia of individual molecules are also very small, since they involve both small masses and small distances between the atoms. For example, the moment of inertia of the H_2 molecule about an axis perpendicular to the bond between H atoms is on the order of $I \approx 10^{-47}$ kg m^2. This is extremely small in macroscopic terms, and when we put this value of I into the expression for $E_1(\text{rot})$, we get $E_1(\text{rot}) \approx 10^{-21}$ J. Again, this seems to be so small on the macroscopic scale that we might feel it is indistinguishable from zero energy. However, we should notice that Boltzmann's constant, which dictates the amount of thermal energy available to each degree of freedom, is even smaller, on the order of 10^{-23} J/K. Viewed from this perspective, the quantum of rotational energy for H_2 seems large, approximately equal to kT for $T = 100$ K.

In 1907 Einstein postulated that the possible energies a vibrating molecule could have are also quantized. That is, the vibrational energies cannot be arbitrarily small, but are multiples of a fundamental, indivisible quantity of energy. Einstein postulated that the quantum of energy for an oscillator is proportional to the frequency f of the oscillation, using Planck's constant as the proportionality constant. Thus $E_{osc} = n(hf)$, where n is an integer and again $h = 6.62 \times 10^{-34}$ J · s (or J/Hz). According to this idea, *an oscillator cannot have less energy than* $E_1(vib) = hf$. The frequencies of large macroscopic oscillators are so low as to make hf a very small quantum, once again an impossibly small threshold energy to detect. Molecular vibrations, however, occur at very high frequencies, and the quantity hf represents a sizable "chunk" of energy on that scale.

Now let's see how the existence of rotational and vibrational energy quanta allows us to explain the behavior of the specific heats. It is collisions between molecules that cause thermal energy to be distributed statistically among the various modes of molecular energy. The energy transfer in an average collision is approximately kT. At very low temperatures this energy becomes very small. If the temperature of the gas is low enough, the energy involved in an average collision ($\approx kT$) is less than the quantum of rotational energy ($h^2/8\pi^2 I$), and hence is not sufficient to start the molecule rotating at all. Thus for temperatures below approximately

$$kT_{rot} = \frac{h^2}{8\pi^2 I} \qquad \text{or} \qquad T_{rot} \approx \frac{h^2}{8\pi^2 I k}$$

the two rotational degrees of freedom are "frozen out," and do not contribute to the specific heat of the gas. Some values of T_{rot} are shown in Table 12.3. Notice that T_{rot} for H_2 is entirely consistent with the previous discussion in which we estimated $E_1(rot)$.

Similarly, if the temperature of the gas is low enough so that the kT transferred in an average collision is below the quantum hf of the vibration of the bond between atoms, the average collisions can't "activate" the vibrations of the molecules, and the two vibrational modes of energy do not contribute to the specific heat. Statistically, this means that until the gas reaches a temperature where

$$kT_{vib} = hf \qquad \text{or} \qquad T_{vib} = \frac{hf}{k}$$

TABLE 12.3

Temperatures for activation of rotational and vibrational energies of diatomic molecules

Substance	T_{vib} (K)	T_{rot} (K)
H_2	6100	85
OH	5400	27
HCl	4300	15
CO	3100	2.8
NO	2750	2.5
O_2	2300	2.1
Cl_2	800	0.35

the vibrational degrees of freedom are frozen out. Examples of T_{vib} are also shown in Table 12.3.

To summarize, classical theory assumes that all possible internal energy degrees of freedom always get an equal share ($\frac{1}{2}kT$) of the thermal energy. For diatomic ideal gases, there are seven degrees of freedom, and so the specific heat predicted by classical theory is $C_V = 7(kT/2)$. Modern quantum theory predicts that there are "threshold" temperatures required to set quantized energy modes into operation. There is no quantum for translational motion, so at any temperatures above $T = 0$ K all gases have three active degrees of freedom, and hence have $C_V = \frac{3}{2}R$. This temperature range is shown as the blue section of Fig. 12.12. As T approaches T_{rot}, an increasing fraction of collisions are able to activate the two degrees of rotation in diatomic molecules. The specific heat C_V thus undergoes a transition from $\frac{3}{2}R$ to $\frac{5}{2}R$, shown as the yellow section of Fig. 12.12. When T approaches T_{vib}, C_V undergoes another transition as collisions are able to set the molecule into vibration. Above T_{vib} (the red section of Fig. 12.12) the specific heat of the diatomic gas is $\frac{7}{2}R$, demonstrating that all seven degrees of freedom share in the distribution of thermal energy. Notice that at room temperatures most of the gases shown in Table 12.3 have rotational energies active but not vibrational energies. Thus they have 5 degrees of freedom, and exhibit a value of 1.4 for γ.

The puzzling behavior of specific heats discussed at the beginning of this section is successfully explained by the existence of extremely small quanta of rotational and vibrational energy. Small as the individual quantum is, its effects manifest themselves in macroscopic behavior. This was a great triumph of the new "quantum" mechanics in the early part of the twentieth century.

LEARNING GOALS

Now that you have finished this chapter, you should be able to

1 Define (*a*) state of a system, (*b*) state variable, (*c*) state function, (*d*) internal energy, (*e*) *PV* diagram (*f*) molar specific heat, (*g*) isobaric, isovolumetric, adiabatic, and isothermal processes, (*h*) throttling process.
2 State the first law of thermodynamics in equation form and explain the meaning of each term, including the interpretation of signs.
3 State what quantity is held constant during each of the following processes: (*a*) isobaric, (*b*) isovolumetric, (*c*) isothermal, and (*d*) adiabatic.
4 Calculate the work done by a system during an arbitrary process involving volume change if a *PV* diagram of the process is given.

5 Calculate the change in internal energy of an ideal gas when the initial and final states are given.
6 Explain why C_P is larger than C_V for a gas. Calculate C_P and γ when C_V is given. Calculate C_V and C_P when γ is given.
7 Apply the first law of thermodynamics to calculate the heat transfer during changes of state when W and ΔU are known.
8 Use the first law of thermodynamics to explain (*a*) why a gas warms when compressed adiabatically, (*b*) why a gas does not change temperature during a free expansion, (*c*) why a fluid is often cooled when it undergoes a throttling process.
9 Given a value for C_V, C_P, or γ, identify the number of active degrees of freedom in an ideal gas.

SUMMARY

DEFINITIONS AND BASIC PRINCIPLES

Thermodynamic State Variables
Thermodynamic state variables are those quantities which specify the macroscopic thermodynamic state of a system. Each set of their values corresponds to one specific state. For an ideal gas, the state variables are P, V, and T.

Thermodynamic State Functions
A thermodynamic state function is a property which depends only on state variables. It has a unique value for each state, which does not depend on the process by which the system reaches that state.

Internal Energy (U)
The internal energy of a system is the sum of all kinetic and potential energies possessed by its atoms and molecules. It is a thermodynamic state function.

INSIGHTS
1. For an ideal monatomic gas, $U = \frac{3}{2}nRT = \frac{3}{2}NkT$.
2. Since internal energy is a state function, changes in U depend only on the initial and final states of a system, and not on the process of change.

The First Law of Thermodynamics
The first law of thermodynamics is a statement of conservation of energy which includes heat transfer into or out of a system:

$$Q = \Delta U + W$$

INSIGHTS
1. Positive Q means heat is added to the system. Positive W means work is done by the system.
2. Positive work always indicates an expansion in volume. Negative work means a compression, in which work is done on the system by some external force.

Special Cases of Change of Thermodynamic State
Some changes of thermodynamic state occur in which a quantity is held constant. These changes simplify the first law in various ways. Four such changes are

1. Isobaric (constant pressure)
2. Isovolumetric (constant volume)
3. Isothermal (constant temperature)
4. Adiabatic (no heat transfer between system and surroundings)

The characteristics of these changes are summarized in Table 12.2.

PV Diagrams
A *PV* diagram is a graph of the pressure versus volume for a system, used to display changes of state for gases, where volume changes are significant. Each point on such a graph represents a thermodynamic state. A line or curve on the graph traces a specific process of change of state.

INSIGHTS
1. For ideal gases, the ideal-gas law can be used to calculate the temperature for any point on the *PV* diagram.
2. A horizontal line on a *PV* diagram represents an isobaric process.
3. A vertical line represents an isovolumetric process.

Calculation of ΔU in Changes of State
ΔU can be calculated for any change of state when the initial and final temperatures are known:

$$\Delta U = nC_V \, \Delta T$$

Molar Specific Heats of Ideal Gases
CONSTANT-VOLUME PROCESS (C_V)

$$C_V = \frac{3}{2}R \quad \text{(monatomic gas)}$$

$$C_V = \frac{K}{2}R \quad \text{(molecular gas)}$$

K is approximately an integer whose value depends on the specific gas and its temperature. (See Table 12.1 for actual values of C_V.)

CONSTANT-PRESSURE PROCESS (C_P)

$$C_P = C_V + R \quad \text{(all gases)}$$

INSIGHTS
1. A useful parameter is the ratio of these specific heats, symbolized by γ.

$$\gamma = \frac{C_P}{C_V}$$

2. For adiabatic processes, P and V change such that PV^γ is constant.

Calculation of Work in Thermodynamic Processes
Work depends on the type of process. For the four special processes we can calculate W algebraically:

1. Isovolumetric: $W = 0$
2. Isobaric: $W = P\Delta V$
3. Isothermal: $W = nRT \ln \dfrac{V_2}{V_1}$
4. Adiabatic: $W = -(\Delta U)$

For all other processes work can be determined graphically as the area under the process curve in a *PV* diagram. The sign of W is found by noting whether the process involves an increase or decrease of volume.

Calculation of Heat Transfer in Thermodynamic Processes
Heat can be calculated directly for the four special processes:

1. Isovolumetric: $Q = nC_V \, \Delta T$
2. Isobaric: $Q = nC_P \, \Delta T$
3. Isothermal: $Q = W = nRT \ln \dfrac{V_2}{V_1}$
4. Adiabatic: $Q = 0$

For other processes, Q can be calculated from the first law once W and U have been found by methods outlined above:

$$Q = \Delta U + W$$

Definition of Throttling Process
A throttling process (also known as a free expansion) is a process in which a gas under high pressure expands adiabatically through a small opening into a region of vacuum or much lower pressure.

INSIGHTS
1. Since the gas does no work and $Q = 0$ during a free expansion, the temperature of an ideal gas remains constant.
2. If a fluid expands into the vacuum by changing state to a gas, the heat of vaporization is provided by the internal energy of the fluid, and so the temperature of the substance is lowered.

QUESTIONS AND GUESSTIMATES

1 An inventor claims that he has an engine that is started by a battery but then runs without an outside power source as it recharges the battery and does external work. What law of nature does this invention disprove? What does the first law tell you about perpetual motion machines?

2 The relation $\Delta U = Q - W$ is not always equivalent to the relation $\Delta U = Q - P\,\Delta V$. Give an example in which the latter relation does not apply even though the former does.

3 In each of the following processes, point out what is meant by each quantity in the equation $\Delta U = Q - W$: an ice cube slowly melts in water at 0°C; ice heats from $-30°C$ to $-10°C$; steam in a closed boiler cools from 120°C to 110°C; solid CO_2 (dry ice) sublimates (changes from a solid to a gas directly, without ever passing through the liquid phase) in dry air in a large container; a bottle of pop freezes and the bottle cracks.

4 Although $C_P - C_V = R$ for any ideal gas, the specific heat per unit mass, $c_P - c_V$ varies from gas to gas. Why the difference?

5 How can one determine the molecular mass of an ideal gas by measuring c_P and c_V for it?

6 A container of ideal gas is to be compressed to half its original volume. Under which condition would the work done on the gas be larger: isothermal or adiabatic?

7 Two cylinders sit side by side and are enclosed by a movable piston. They are identical in all respects except that one contains oxygen (O_2) and the other contains helium (He). Both are compressed adiabatically to one-fifth their original volume. Which gas shows the larger temperature rise?

PROBLEMS

Section 12.2

1 A 2.2-kg block of ice melts to water at 0°C. By how much has the internal energy of the ice changed? Neglect the small change in volume.

2 If a 75-g piece of copper is heated from 27°C to 115°C, how much has its internal energy changed? Neglect the small change in volume.

3 By how much must one lower the temperature of a 65-g piece of aluminum to change its internal energy by 350 J? Neglect any volume change.

4 What is the internal energy change of 265 g of molten lead as it solidifies at its melting temperature? Neglect any volume change.

Section 12.3

5 Figure P12.1 shows the PV diagram for a gas confined to a cylinder by a piston. How much work does the gas do as it expands from state A to state C along the path shown?

6 In the PV diagram shown in Fig. P12.1, how much work does the gas do as it is compressed from state D to state A along the path shown?

7 An ideal gas is compressed isothermally to one-fifth of its original volume. The work done to compress the gas is 167 J. (a) By how much does the internal energy of the gas change? (b) How much heat has been added to or lost by the gas?

8 An ideal gas expands to three times its original volume. The work done by the gas is 350 J, and 570 J of heat is added during expansion. (a) Does the temperature of the gas increase, decrease, or remain constant? (b) By how much does the internal energy change?

9 A certain quantity of helium gas in a closed, rigid container is heated from $-95°C$ to 70°C. This requires the addition of 130 J of heat. How much helium (in grams and moles) is in the container?

■10 Fig. P12.2 shows a thermodynamic cycle in which an ideal gas is changed from state A to state B, then changed from B to C, finally returning to its original state A. Calculate the work done by the gas during the complete cycle. *Hint*: Be sure you observe the correct signs of W for each step in the cycle.

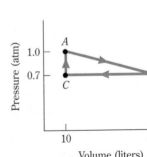

FIGURE P12.1

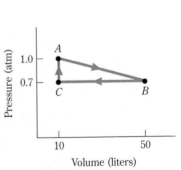

FIGURE P12.2

Section 12.5

11 Suppose that 2.3 mol of argon gas is heated from $-45°C$ to 90°C. Find the change in internal energy of the gas and the work it does if the heating occurs at (a) constant volume and (b) constant pressure.

12 Nitrogen gas (N_2) at standard temperature and pressure fills a 700-liter rigid container. How many joules of heat are required to raise its temperature to 27°C? What is the pressure of the gas at 27°C?

13 Using the fact that the atmosphere consists of approximately 79% N_2 and 21% O_2 by mass, calculate c_V for air.

14 Must heat be added to or extracted from an ideal monatomic gas if it is to be compressed from 795 to 260 cm³ under a constant pressure of 155 kPa? How much heat? Take the initial temperature of the gas to be 230°C.

15 A balloon is filled with 4.5 m³ of helium at standard temperature and pressure. How much heat is required to raise the temperature of the gas to 37°C if the balloon expands at atmospheric pressure?

Sections 12.6 and 12.7

16 How much work is necessary to compress an ideal gas isothermally from 125 to 60 liters if, during the process, 35 cal of heat flows from the gas?

17 How much work would be required to compress 3.3 mol of O_2 gas at 25°C isothermally from 90 to 40 cm³? How much heat would be gained or lost by the gas?

18 Repeat Prob. 17 if the compression is adiabatic instead of isothermal.

19 If an ideal gas has a value of 1.28 for γ, find the values of C_P and C_V for the gas.

20 If an ideal gas has a value of $C_V = 5R/2$, find the value of γ for the gas.

21 The temperature of 90 g of N_2 gas is raised from 10°C to 100°C at a constant pressure of 1 atm. Find ΔU, W, and Q for this process.

22 Return to Prob. 10 and calculate the change in internal energy of the gas and the heat gained or lost in each of the processes AB, BC, and CA. Assume you have 5 g of helium.

23 Two mol of an ideal gas with $\gamma = 1.40$ are taken through the thermodynamic process ABC shown in Fig. P12.3. Find the work done, the change of internal energy, and the heat gained or lost for this process.

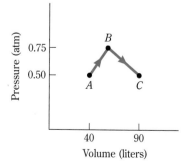

FIGURE P12.3

24 Two-thirds of a mole of ideal gas is compressed adiabatically in such a way that its temperature rises by 45 C° when 370 J of work is done on the gas by the compressor. (a) What is the change in internal energy of the gas during the compression? (b) If the gas subsequently cools to its original temperature while remaining at its new volume, how much heat flows from the gas? (c) What are the values of C_V and γ for this gas?

25 Suppose that 30 g of hydrogen (H_2) is confined to a cylinder fitted with a movable piston. The gas pressure is maintained at 4.4 atm as the gas temperature is changed from 20°C to 270°C. How much heat is required for this process?

26 A 16,000-cm³ cylinder closed at one end by a piston contains 1.1 mol of CO_2 at 30°C. The piston is suddenly pushed in so as to compress the gas adiabatically to a new volume of 1600 cm³. Find the final temperature of the gas and the work done on it.

27 Nitrogen gas (N_2) is expanded adiabatically from an original pressure of 25 atm and a temperature of 27°C to such a volume that its final temperature is $-25°C$. By what factor must its volume increase?

28 A quantity of oxygen (O_2) is confined to a volume of 3 liters at 10 atm pressure at 27°C. Find the final pressure if the gas is allowed to expand (a) isothermally and (b) adiabatically to a new volume of 10 liters.

29 Helium gas at 27°C and a pressure of 1.6 atm is compressed adiabatically to one-fourth of its original volume. Find the final temperature and pressure of the gas.

General Problems

30 A sample of air ($\gamma = 1.40$) is slowly compressed from 1.0 atm pressure to 2.0 atm. The original volume and temperature are $V_1 = 20$ L and $T_1 = 290$ K. The temperature remains constant during this compression. Next the air is suddenly (adiabatically) expanded back to its original pressure of 1 atm. (a) Sketch a PV diagram of these processes, (b) find the final volume and temperature, and (c) find ΔU, Q, and W for each process.

31 At 1 atmosphere pressure and a temperature of 100°C, 1 kg of liquid water becomes 1.67 m³ of steam when it is boiled. (a) How much of the heat supplied to vaporize the water goes into the work of expansion? (b) How much of the heat goes into the internal energy of the steam?

32 Suppose you have 36 g of water initially at 20°C under 1 atm of pressure. How much heat is required to raise the water temperature to the boiling point, boil the water, and raise the temperature of the steam to 150°C, while maintaining the pressure at 1 atm? *Hint:* Water vapor is a triatomic molecular gas possessing three active rotational degrees of freedom at these temperatures, in addition to the usual translational degrees of freedom.

33 For a gas having $\gamma = 1.28$, what fraction of heat added during a constant-pressure expansion goes into internal energy and what fraction does the work of expansion?

34 30.0 kg of CO_2 undergoes an isothermal compression at $T = 500$ K. The density of the gas goes from 3.75 to 30.0 kg/m^3. (a) What is the original pressure of the gas? (b) Calculate the work done on the gas, its change in internal energy, and the heat gained or lost.

▪35 One kg of aluminum is heated from 25°C to 600°C. How much work does the aluminum do in expanding against the surrounding 1 atm pressure? Calculate the ratio of work to heat added, W/Q, and determine to what precision the approximation $W \approx 0$ is valid.

▪▪36 Argon gas is confined inside a vertical cylinder of diameter 8.0 cm by a 15.0-kg piston that is free to move vertically. The cylinder is inside a large vacuum chamber and is thermally insulated from its surroundings. When the argon in the cylinder is at 35°C and the air pressure in the outside chamber is 760 torr, the equilibrium position of the piston is 22.5 cm from the bottom of the cylinder. (a) How many moles of argon are in the cylinder? (b) The air is now evacuated from the vacuum chamber (the chamber pressure is reduced to less than 0.001 torr). What is the new equilibrium position of the piston and the new temperature of the argon?

▪▪37 For a certain diatomic ideal gas, $c_V = 920$ J/kg · K. Find the approximate values of (a) molecular mass M, (b) molar specific heats C_V and C_P, and (c) γ.

▪▪38 At the base of a mountain, the air pressure is 1.0 atm and the temperature is 300 K. (a) If the air rises adiabatically to the top of the mountain, where $P = 0.94$ atm, what is its temperature? (Assume the air is almost entirely N_2 and O_2.) (b) If some water vapor condenses from the air, will the air temperature rise or fall?

CHAPTER 13

THE SECOND LAW OF THERMODYNAMICS

As has been mentioned, the first law of thermodynamics is a statement of the principle of conservation of energy. Any process which would violate it cannot occur. There are, however, many processes that would not violate the first law but which nevertheless do not occur. For instance, a stone resting on the ground never converts the thermal energy in it and its surroundings into kinetic energy, spontaneously shooting up into the air. The first law does not rule out such a possibility, but still it never happens. If you put some ice into a container of hot water, the mixture comes to some intermediate equilibrium temperature. The ice never gets colder and the hot water hotter, even though energy could be conserved that way also. Nature has a preferred direction for the course of spontaneous events. It is as though nature has decreed that time is not reversible. Time is like an arrow that points in only one direction, and all spontaneous natural processes must follow the path nature has chosen.

We will see that a second law of thermodynamics is necessary to explain the direction of time's arrow. This law tells us that order in the universe progresses relentlessly to disorder. Why this is true will become apparent as we examine the subjects of order and disorder.

Each of the ten numbered balls of this lottery machine has an equal probability of being chosen. Can you calculate the total number of microstates there are?

13.1   ORDER AND DISORDER

As any gambler knows, the odds on an event happening are increased if the event can occur in many different ways. To illustrate this fact, let us consider a game in which five identical coins are tossed onto a table after being shaken well. Only six possible events can result from such a toss (Table 13.1).

At first guess, you might think that each event listed in Table 13.1 is equally likely to occur, but that is not correct. There is only one way in which event 1 or event 6 can occur. However, there are five ways in which event 2 could occur. If we call the five coins A, B, C, D, and E, these ways are as listed in Table 13.2. Because there are 5 times as many ways for event 2 to occur, it is 5 times as likely to occur as event 1. Event 5 can also happen in five ways. As a result, events 2 and 5 are equally likely to occur. And both these events are 5 times as likely as events 1 and 6.

Let us take a moment to summarize these observations in general terms. In defining each event in Table 13.1, we consider the individual coins to be equivalent in the sense that it makes no difference *which ones* come up heads or tails. We call each event a **macrostate** of the possible coin arrangements. In Table 13.2 we identify the various ways individual coins can form a single macrostate, specifically event 2 in Table 13.1. Each individual arrangement in Table 13.2 is called a **microstate.** Table 13.3 summarizes the number of microstates for each event in Table 13.1.

The probability of a given macrostate occurring is based on a simple assumption:

Each **microstate** has an equal probability of occurring.

Thus the probability of a specific event (macrostate) occurring is simply the ratio of the number of microstates that can form that event to the *total* number of microstates that can occur. For example, the total number of microstates available for five coins is $2^5 = 32$. Thus the probability of event 2 occurring is $5/32 = 15.6$ percent. Table 13.3 shows the probability for each of the six events in Table 13.1.

We can extend this reasoning to cases involving more coins. Suppose 100 coins

TABLE 13.1
Six possible results (events) when five coins are tossed

Event	Heads	Tails
1	0	5
2	1	4
3	2	3
4	3	2
5	4	1
6	5	0

TABLE 13.2
Possible ways for event 2 to occur

Way	A	B	C	D	E
1	H	T	T	T	T
2	T	H	T	T	T
3	T	T	H	T	T
4	T	T	T	H	T
5	T	T	T	T	H

TABLE 13.3
Probability table for five coins

Event (macrostate)	Number of ways (microstates)	Probability of macrostate
1	1	$\frac{1}{32} = 0.03$
2	5	$\frac{5}{32} = 0.16$
3	10	$\frac{10}{32} = 0.31$
4	10	$\frac{10}{32} = 0.31$
5	5	$\frac{5}{32} = 0.16$
6	1	$\frac{1}{32} = 0.03$

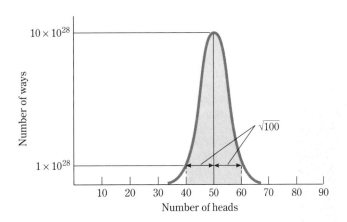

FIGURE 13.1

The number of ways in which the indicated number of heads come up when 100 coins are tossed. For fewer than 30 (or more than 70) heads, the number of ways is too small to be shown on this graph and can be approximated by zero. Approximately 96 percent of the possible ways lie in the blue area, between 40 and 60 heads.

are tossed. Now the total number of microstates available is $2^{100} = 1.3 \times 10^{30}$! Only *one* of these microstates corresponds to the macrostate of all heads, and only one corresponds to all tails. On the other hand, there are approximately 10×10^{28} microstates that can form the macrostate of 50 heads and 50 tails (Fig. 13.1). *The microstate of 100 heads is just as possible as any of the other individual microstates,* but the probability of the 100-head *macrostate* is only 10^{-29} as great as the 50 head–50 tail macrostate. Fig. 13.1 summarizes the probabilities for 100 coins.

We can summarize all such results in a very simple way. Notice in Fig. 13.1 that the graph line decreases to about one-tenth of its maximum value at 40 and 60 heads. To give an estimate of the width of the graph's peak, we could say that it extends from $50 - 10$ to $50 + 10$. In other words, if you throw 100 coins, the number of heads that should come up is 50 ± 10. The general result is

If N coins are tossed, the expected number of heads is $\frac{1}{2}N \pm \sqrt{N}$.

The number following the $\pm$ sign is called the expected deviation, and tells us the range within which the number of heads will occur. A more detailed statistical analysis shows that only about 4 percent of the 100 coin tosses will result in a number of heads outside of this range.

In the case of 1 million coins (10^6), we should expect $500,000 \pm 1000$ heads to come up. Notice how precise this result is. It says the number of heads will lie between 501,000 and 499,000. As the number of coins becomes very large, the percent deviation from the average become very small.

This example with the coins is typical of the universe in general. When things are left to happen spontaneously, the results are governed by the laws of probability. For example, suppose you have a box containing 10^{20} gas molecules, as in Fig. 13.2. What are the chances of finding all the molecules bunched up in one half of the box? We can answer this question by using the results we derived from the coin tosses. The two halves of the box represent two equal possibilities for any single gas molecule, similar to the heads-tails possibilities for a coin. Our previous result tells us that the number of molecules in one side of the box will be

$$\tfrac{1}{2}(10^{20}) \pm \sqrt{10^{20}} = 5 \times 10^{19} \pm 10^{10} = (5,000,000,000 \pm 1) \times 10^{10}$$

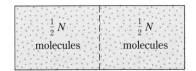

FIGURE 13.2

What is the likelihood that all the molecules will appear in one end of the box?

For all practical purposes, the number of molecules in one half of the box is equal to the number in the other half. The expected deviation is only one part in 5 billion! And, of course, there is essentially no chance at all that all the molecules will spontaneously be found in one half of the box, since such a macrostate is represented by only one microstate (out of $2^{10^{20}}$).

These considerations have fundamental importance for all spontaneous processes. Reasoning from them, we can predict that thermal motion (as well as other random-type disturbances) causes systems to change from order to disorder. As a crude example, consider 100 coins again. Suppose we carefully arrange all with heads up. They then have a high degree of order. Now let us give them a type of motion similar to random thermal motion by shaking them. *They quickly disorder and never return to their original improbable state of order.*

Similarly, we can give the gas molecules in Fig. 13.2 order by placing them all in one end of the box. If, however, we allow them to adjust with spontaneous thermal motion, they become disordered and fill the whole box. They will never spontaneously return to their original ordered state.

Basic to this discussion are the concepts of order and disorder. A state of highest order can occur in only one microstate, with each coin or gas molecule placed in a single exact way. There are more ways of achieving disordered states, and they are thus more *probable*. Spontaneous changes in a system move the system toward states of higher disorder because such states are more probable. To summarize:

If an isolated system composed of many parts is allowed to undergo spontaneous change, it changes in such a way that its disorder increases or, at best, does not decrease.

This law of nature, applicable to large numbers of molecules, is one form of the **second law of thermodynamics.** It explains the tendency of systems to reach thermodynamic equilibrium, even though the first law doesn't require such changes. The equilibrium state, from which there is no tendency to change, is the state of maximum probability and hence of maximum disorder.

13.2 ENTROPY

The implications of order and disorder in a system can be approached in two quite different ways. Both approaches use a quantity called **entropy,** which was introduced in the mid-1800s by R. Clausius to describe the consequences of the fact that heat is always observed to flow from hot to cold. Because the existence of atoms was still quite speculative at that time, Clausius described the behavior of systems in terms of their macroscopic state variables, P, V, T, and U.

Suppose an amount of heat Q is added in a reversible way to a system whose temperature is maintained at a value T. The resulting change in the entropy of the system, denoted by ΔS, is defined to be

$$\Delta S = \frac{Q}{T} \tag{13.1}$$

From this we can see that the system *gains* entropy (ΔS is positive) when heat flows *into* the system. Observe from Eq. 13.1 that the SI units of entropy are J/K. Occasionally the practical thermal units of kcal/K or cal/K are used.

Notice that ΔS is defined only for an isothermal process. However, Clausius was able to show that entropy, like internal energy U, is a state function. Thus if two systems are in the same macroscopic state (same P, V, and T), they have the same entropy. Furthermore, being a state function means that ΔS does not depend on the

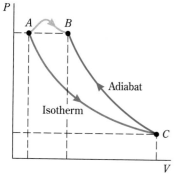

FIGURE 13.3

Because entropy is a state function, the change in the entropy of a system as it changes state along path AB is the same as the sum of the entropy changes along AC and CB.

process by which the state of the system changes. This may seem to contradict Eq. 13.1, since Q *does* depend on process. One way to resolve this seeming contradiction is as follows:

1 *Any* change from a state A to a state B can be made by a combination of an isothermal process to an intermediate state C followed by an adiabatic process from C to B.
2 By definition, Q is zero during an adiabatic change, and so $\Delta S_{CB} = 0$.
3 For the isothermal process AC, $\Delta S_{AC} = Q/T$.
4 Thus $\Delta S_{AB} = \Delta S_{AC} + \Delta S_{CB} = \Delta S_{AC}$, *whatever* the path from A to B.

Point 4 is the characteristic which defines a state function. Of course, to calculate ΔS_{AC} we would have to find the intermediate state C, but this can always be done.

Example 13.1

By how much does the entropy of the system change as a 20.0-g cube of ice melts at 0.00°C? Assume that heat can flow into the ice cube but otherwise it is an isolated system.

Reasoning

Question What sort of process is this?
Answer Ice melts at constant temperature (Sec. 11.6), and so it is an isothermal process.

Question What determines the isothermal change of entropy?
Answer The amount of heat transferred and the temperature at which the process takes place: $\Delta S = Q/T$.

Question How can I find the amount of heat transferred?
Answer It depends on the mass of the ice and the heat of fusion for water (Table 11.2): $Q = mH_f$.

Solution and Discussion Using $T = 273\,\text{K}$, we have

$$\Delta S = \frac{mH_f}{T} = \frac{(20.0\ \text{g})(80.0\ \text{cal/g})}{273\ \text{K}} = 5.86\ \text{cal/K} = 24.5\ \text{J/K}$$

This increase in entropy is a measure of the increased disorder in the arrangement of the water molecules as they lose their orderly solid structure.

Exercise If the temperature change is small, you can use the *average* temperature in the isothermal expression to calculate the entropy change. What is the entropy change if the ice is originally at $-10°C$? *Answer: 24.7 J/K.*

The connection between entropy and the degree of disorder of a system was provided by the Austrian physicist Ludwig Boltzmann. As we discussed previously, each macrostate of a system can occur in a definite number of microstate arrangements of its molecules. Let us symbolize the number of microstates contributing to a given macroscopic state by the Greek omega, Ω. The larger the value of

Ω, the greater the probability of that macrostate occurring. The equilibrium state (the state with the greatest probability) is that state for which Ω is a maximum.

From these concepts, Boltzmann showed that the relation between entropy S and Ω is

$$S = k \ln \Omega \tag{13.2}$$

where k is the same Boltzmann's constant that occurs in the kinetic theory of gases. If a macrostate can occur as a result of only one microstate, $\Omega = 1$. Since $\ln 1 = 0$, Eq. 13.2 tells us that the entropy of such an unlikely (highly ordered) state is zero. As we consider states of greater probability (and hence greater disorder), $\ln \Omega$ and S become larger. Thus we see from Boltzmann's approach that *entropy measures the disorder* of the macrostate of a system. The second law of thermodynamics can then be restated as follows:

If an isolated system undergoes change, it changes in such a way that its entropy increases or at best remains constant.

Example 13.2

Suppose you have a box containing 100 molecules. Consider two macrostates of the molecular distribution in the box. State A has 60 molecules in one half of the box and 40 in the other. State B has the molecules divided equally in the two halves. Use Fig. 13.1 to calculate the entropy change as the box goes from state A to state B.

R

Reasoning

Question   What does the entropy of the states depend on?
Answer   Their probability, measured by the number of microstates that can form the macrostate.

Question   How do I get this information from Fig. 13.1?
Answer   As previously mentioned, molecular distribution in two halves of a box is the same as the problem of coins landing heads or tails. Figure 13.1 shows that state B has 10^{29} microstates, while state A has approximately one-tenth as many.

Question   What expression gives the entropy of a state?
Answer   The Boltzmann definition, $S = k \ln \Omega$. The difference in the entropy of the two states, ΔS, is then

$$\Delta S = S_B - S_A = k(\ln \Omega_B - \ln \Omega_A)$$

Solution and Discussion   The entropies of the two states are

$$S_A = (1.38 \times 10^{-23} \text{ J/K}) \ln(1 \times 10^{28}) = 8.90 \times 10^{-22} \text{ J/K}$$
$$S_B = (1.38 \times 10^{-23} \text{ J/K}) \ln(1 \times 10^{29}) = 9.21 \times 10^{-22} \text{ J/K}$$

Then

$$\Delta S = (9.21 - 8.90) \times 10^{-22}\,\text{J/K} = 0.31 \times 10^{-22}\,\text{J/K}$$

Notice that this is an *increase* in entropy, denoting the greater degree of disorder and hence higher probability of the state of equal distribution (state *B*). A shorter way to solve this problem would have been to remember that the difference in the logarithms of two numbers is the same as the logarithm of their ratio:

$$\Delta S = k(\ln \Omega_B - \ln \Omega_A) = k \ln \frac{\Omega_B}{\Omega_A}$$

$$= (1.38 \times 10^{-23}\,\text{J/K}) \ln 10 = 0.32 \times 10^{-22}\,\text{J/K}$$

13.3 HEAT ENGINES; CONVERSION OF THERMAL ENERGY TO WORK

The development of thermodynamics began in the late 1700s, the time of the Industrial Revolution. It was then that the invention of the steam engine led to a momentous change in our civilization. Because early steam engines were primitive devices that operated at low efficiency, scientists of the time were called upon to examine the physical laws governing these machines. It was this call that provided the major impetus for the early work in thermodynamics. The results of that work were far-reaching and today permeate both the physical and the biological sciences.

A steam engine is only one example of a **heat engine,** defined as *any device that converts some thermal energy to mechanical work*. The steam engine fits this description. So does a gasoline engine, which uses the thermal energy given off by the combustion of fuel. More exotic engines that use heat from the sun or from nuclear reactors are also heat engines. Let us now look at the laws all such engines obey.

The jet engines on this aircraft convert thermal energy to work, but the visible exhaust clearly shows that a considerable amount of thermal energy is lost as waste heat.

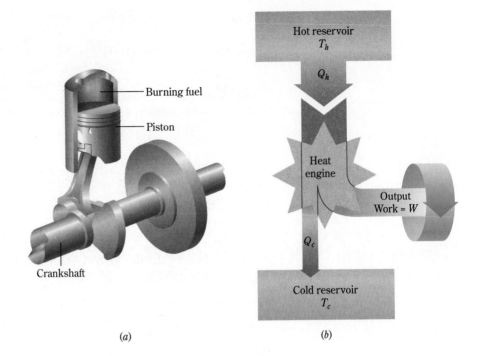

Burning fuel

Piston

Crankshaft

Hot reservoir
T_h

Q_h

Heat engine

Output
Work = W

Q_c

Cold reservoir
T_c

FIGURE 13.4

In a heat engine, the input energy Q_h must equal the sum of the exhaust energy Q_c and the output work W.

(a)

(b)

A diagram of a simple heat engine is shown in Fig. 13.4a. The burning fuel creates high-pressure gas that causes the piston to move downward. This linear motion is changed to rotational motion by the crankshaft, and the engine goes through the same cycle of motion over and over. Of course, many details, such as the required valves and sparkplugs, are not shown. The essential feature, however, is that thermal energy is transformed to mechanical energy.

Figure 13.4b gives a more general representation of a heat engine. Heat Q_h flows from a high-temperature (hot) reservoir to the engine. This is the input energy for the engine. Part of the input energy is used in doing work, and the remainder, Q_c (the exhaust heat), flows to a low-temperature (cold) reservoir. The low-temperature reservoir is often simply the atmosphere, as when a car exhausts spent fuel through its tail pipe.

Because the engine must obey the law of conservation of energy, the first law of thermodynamics applied to it becomes, for one cycle of its motion,

$$Q_{net} = Q_h - Q_c = W + \Delta U$$

where W is the output work of the engine per cycle. However, each time the engine goes through a thermodynamic cycle, there is no net change in its internal energy. Hence, for a complete cycle, $\Delta U = 0$ and the above equation becomes

$$W = Q_h - Q_c$$

We shall use this relation to compute the efficiency of the engine. As with any machine, we define efficiency as the ratio of output work to input energy. In the present case,

$$\text{Efficiency} = \frac{W}{Q_h}$$

Substituting for W the value given above, we have

$$\text{Efficiency} = \frac{Q_h - Q_c}{Q_h} = 1 - \frac{Q_c}{Q_h} \tag{13.3}$$

We see that the exhaust heat Q_c, which represents thermal energy that has not been converted to work, is responsible for the inefficiency of a heat engine.

If we could make Q_c zero, the engine would be 100 percent efficient. However, we now use the concept of entropy to show that this is impossible. A very definite limit exists on the efficiency of any engine.

Let us compute the entropy change of the system of Fig. 13.4b as heat flows. The engine is unchanged by the heat flow, and so its entropy does not change. However, the hot reservoir loses heat Q_h while the cold reservoir gains heat Q_c.

$$\Delta S_h = \frac{-Q_h}{T_h} \quad \text{and} \quad \Delta S_c = \frac{Q_c}{T_c}$$

According to the second law, the total change in entropy must be greater than or equal to zero:

$$\Delta S_c + \Delta S_h \geqslant 0$$

$$\frac{Q_c}{T_c} - \frac{Q_h}{T_h} \geqslant 0$$

If we transpose the negative term, divide through by Q_h, and multiply through by T_c, we have

$$\frac{Q_c}{Q_h} \geqslant \frac{T_c}{T_h} \tag{13.4}$$

We can now substitute this value in Eq. 13.3 to obtain

$$\text{Efficiency} \leqslant 1 - \frac{T_c}{T_h} \tag{13.5}$$

According to Eq. 13.5, the maximum efficiency of any heat engine is

$$\text{Maximum efficiency} = 1 - \frac{T_c}{T_h} \tag{13.6}$$

Thus we are led to a startling result: even the best-designed heat engine has a limiting efficiency, constrained by the temperature extremes between which it operates. From Eq. 13.6 we see that the limiting efficiency can be increased either by getting Q_h from a very high temperature or by exhausting Q_c at a very low temperature. Notice that only if Q_c could be exhausted at 0 K would an engine be able to operate at 100 percent efficiency, converting all the input of heat into work. Since even the empty space of the universe has a temperature of approximately 3 K, such an engine is impossible. This result is a direct consequence of the second law of thermodynamics, and is often quoted as an alternate statement of the second law:

A device which converts 100 percent of input heat into mechanical work is a physical impossibility.

In Chap. 12 we saw how work and heat transfer can be calculated during a thermodynamic cycle by using a *PV* diagram of the processes involved in the cycle. A pioneer in thermodynamics, the French physicist Sadi Carnot, showed that an idealized engine using only adiabatic and isothermal expansions and compressions yields the maximum efficiency given by Eq. 13.6. Such an engine is known as **Carnot engine** and exists only as an ideal. Real engines fall considerably short of this theoretical maximum of efficiency for many reasons, including friction and other heat losses. For example, automobile engines have an efficiency of approximately 25 percent, though the operating temperatures would allow a theoretical maximum of 80 percent. Steam turbines used to generate electricity have an efficiency limit of about 60 to 65 percent, but actually convert only about 45 percent of the thermal energy of the hot steam from the boiler into the work of turning the generator.

High-temperature thermal energy can be converted to work with greater efficiency than can low-temperature thermal energy. This fact is why temperature is often referred to as a measure of the *quality* of thermal energy. Two substances at different temperatures represent a thermodynamic system that is more ordered than the system which exists after the substances exchange heat and come to an equilibrium temperature. Work represents a highly ordered state of molecular behavior (all molecules moving in the direction of motion, for example), and hence a state of low entropy. The Carnot engine can thus be interpreted as the heat engine that increases entropy by the smallest possible amount. When high-temperature thermal energy simply mixes with lower-temperature energy without work being produced, entropy increases by a maximum amount. Once that happens, the opportunity to extract work from the original ordered situation is irretrievably lost.

Example 13.3

A modern coal-fired electrical power plant utilizes a steam turbine to turn an electrical generator. The turbine receives steam at 800 K and exhausts it at 300 K. Consider a power plant designed to produce electrical power at a rate of 1000 megawatts (MW). If the turbine operates at its theoretical maximum efficiency, at what rate does it produce exhaust heat?

R

Reasoning

Question   What determines the turbine's maximum theoretical efficiency?
Answer   Its temperature extremes, according to Carnot's analysis (Eq. 13.6):

$$\text{Maximum efficiency} = 1 - \frac{T_c}{T_h}$$

Question   How is this maximum efficiency related to the rate of heat production?
Answer   Efficiency is the ratio of work produced (output) to heat input, Q_h. The first law of thermodynamics also tells us that $Q_h = W + Q_c$, where Q_c is the waste heat. These same relations can be expressed in terms of power.

Question What does the 1000 MW represent?

Answer The electrical power output available to do work.

Question What equation relates the operating temperatures to W and Q?

Answer $1 - \dfrac{T_c}{T_h} = \dfrac{W}{Q_h} = \dfrac{W}{W + Q_c} = \dfrac{P_{out}}{P_{out} + P_{waste}}$

Solution and Discussion Let us first find the maximum efficiency:

$$\text{Maximum efficiency} = 1 - \frac{300\,\text{K}}{800\,\text{K}} = 0.625 = 62.5\%$$

(Always remember to use Kelvin scale temperatures.) Now we have

$$0.625 = \frac{P_{out}}{P_{out} + P_{waste}}$$

from which we get

$$P_{waste} = \frac{P_{out}}{0.625} - P_{out} = \frac{1000\,\text{MW}}{0.625} - 1000\,\text{MW} = 600\,\text{MW}$$

This means that energy in the form of high-temperature steam must be supplied to the turbine at the rate of 1000 MW + 600 MW = 1600 MW.

Exercise The efficiency of modern steam engines is actually about 45 percent. What would be realistic values of the rates of heat exhaust and heat input for such a turbine? *Answer: P_{waste} = 1200 MW and P_{in} = 2200 MW. Notice that a drop of 17.5 percent in efficiency increases the waste heat by 39 percent for the same level of power output.*

This air conditioner is having its Freon supply checked. The compressor is the black object in the background. One of the heat exchangers, including a fan, is housed inside the blue structure in the foreground.

13.4 REFRIGERATION SYSTEMS

There are many instances when, instead of extracting work from thermal energy, we want to cool a substance without having a cooler material to mix with it. The second law will not permit this to happen spontaneously, since this process would require the cool object to become cooler than its surroundings. While the second law prohibits heat from flowing from cold to hot, we can do work on a system to force heat to flow "uphill" in temperature, much in the same way that we can pump water uphill against gravity. A process that utilizes work to lower the temperature of a substance is called a **refrigeration cycle.** Examples in common use include refrigerators, air conditioners, and heat pumps.

The energy flow in a refrigeration cycle is fundamentally the reverse of that in a heat engine, as shown in Fig. 13.5a. The device operates between two temperature extremes, T_h and T_c. An input of work, W, allows the device to take in an amount of heat Q_c at the cooler temperature and to exhaust an amount of heat Q_h at the higher temperature. Once again, the first law of thermodynamics requires that energy in = energy out, or

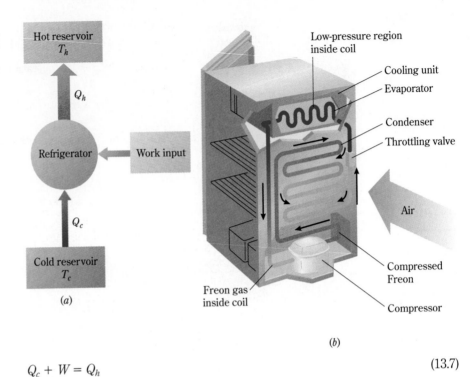

FIGURE 13.5

(a) The energy flow in a refrigeration system. (b) A schematic diagram of a refrigerator.

$$Q_c + W = Q_h \tag{13.7}$$

A schematic diagram of a refrigerator is shown in Fig. 13.5b. Such a system makes use of a low-boiling-point fluid, such as Freon, which boils at $-30°C$ at atmospheric pressure. The compressor at the bottom of the unit compresses the Freon gas to a pressure high enough that it will liquefy when cooled somewhat. During this nearly adiabatic compression, the work (W) done on the gas causes it to become quite hot. The hot Freon passes through the coils of the condenser, where it loses some of its high temperature heat to the surrounding air. (When you reach behind a refrigerator you can feel the heated air in the vicinity of the coils.) This cooling causes the Freon to become liquid, giving up its heat of vaporization to the surrounding air. The heat given up by cooling plus the heat of vaporization add up to Q_h. The near-room-temperature liquid Freon under high pressure is then sent to a throttling valve, where it evaporates and expands into the low-pressure region called the *evaporator*. (See Illustration 12.3.) The coiled tubing of the evaporator is inside the refrigerator, and as the now very cold gaseous Freon flows through these coils, heat (Q_c) flows from the warmer contents of the refrigerator into the Freon, cooling the interior. Eventually the (now warmer) Freon leaves the evaporator coils and returns to the compressor. There it is compressed again, and both the heat it absorbed in the cooling coils and the heat generated during its compression are removed in the condenser coils as the cycle repeats.

Air conditioners operate in the same way. The cooling coil is inside the house, the condensing coil is outside, and heat is transferred from the house to the outdoors. This cools the interior of the house and warms the outdoors. (Put your hand near the outside of an operating air conditioner and you will feel the high-temperature heat, Q_h, being exhausted.)

Equation 13.7 shows that $Q_h > Q_c$, by an amount equal to the work done by the compressor:

$$W = Q_h - Q_c$$

As a measure of the effectiveness of the refrigerator, we define a **coefficient of performance (COP)** as the ratio of the heat extracted at low temperature to the work input required:

$$COP = \frac{Q_c}{W} \tag{13.8}$$

We can use Eq. 13.7 to eliminate W and obtain

$$COP = \frac{Q_c}{Q_h - Q_c} \tag{13.9}$$

Notice from Eq. 13.9 that all values of COP—the ratio of heat extracted to work input—are greater than 1. This indicates that a small amount of work can extract a larger amount of heat.

As we did for a heat engine, we can use the entropy considerations of the second law to relate the amounts of heat transferred to the temperatures of their respective heat reservoirs. We then find that the coefficient of performance of a refrigerator has a maximum possible value given by

$$\text{Maximum COP} = \frac{T_c}{T_h - T_c} \tag{13.10}$$

Notice that the performance is best (COP is greatest) when the temperature difference is small. This makes sense because it takes less work to force heat to flow to a reservoir that is at a slightly higher temperature than to one at a much higher temperature.

Heat pumps are another example of the use of a refrigeration cycle. They are made so that the two sets of cooling coils can be reversed. This enables them to be used as an air conditioner, with the evaporator coils inside a building, or to be used as a heating unit, with the condenser coils inside, providing the exhaust heat to the room. In this latter mode, the building is being heated by colder thermal energy from outdoors being raised in temperature through the work done by the compressor.

The purpose of a heat pump used for heating is slightly different than that of an air conditioner. Its function is to deliver high-temperature heat Q_h to an environment, rather than to remove a low-temperature heat Q_c. Since the COP is an indicator of how well a device performs its intended function, the COP for a heat pump must be defined as

$$\text{COP (heat pump)} = \frac{Q_h}{W} = \frac{Q_h}{Q_h - Q_c} \tag{13.11}$$

The maximum COP for a heat pump is then given by

$$\text{Maximum COP (heat pump)} = \frac{T_h}{T_h - T_c} \tag{13.12}$$

Notice the slight difference between Eqs. 13.11 and 13.12 and the COP for the refrigerator (Eqs. 13.9 and 13.10).

Physicists at Work   KAREN ST. GERMAIN, *University of Nebraska, Lincoln*

For the past six years I have been working in a field called "remote sensing," which is part physics and part engineering. Generally speaking, *remote sensing* means collecting physical information about an object or a location without actually having to go to the object or location.

One way to accomplish this is by transmitting electromagnetic energy and waiting for it to be reflected from the object(s). A familiar example of this is the weather radar picture we see on the evening news, where the reflecting objects are rain droplets and the information gained is the amount of rain in the atmosphere. A second approach is simply to measure the natural radiation coming from the object or scene of interest with instruments called radiometers. Perhaps the most common example of this type of remote sensing is an infrared receiver which measures the physical temperature of a scene, and is used in night-vision instruments to see warm things, such as people (remember the infrared glasses used in *The Silence of the Lambs*?), animals, and engines.

My interests lie in contributing to the study of the global environment by using remote sensing techniques to answer geophysical questions. This includes both short-term interests, like early warning of natural disasters, and longer-term climate and habitat studies.

My first research project as a graduate student was in the short-term goal category. The problem at hand was the difficulty of predicting how hurricanes, moving quickly over the ocean, will intensify and where they will make landfall. Currently, for each hurricane that makes landfall, a hurricane warning is issued over an average of 300 miles of coastline, at a cost of $30,000 per mile. Improving landfall prediction by only 10 percent for one storm would save enough money to fund hurricane research for a whole year.

According to the Hurricane Research Center, the key piece of missing information is the wind speed at the ocean surface. Wind speed could be measured by sending a ship into the storm, but for obvious reasons this is a very dangerous option. Reconnaissance aircraft have been flying into storms for years, but it isn't safe for them to fly at the low altitudes necessary to estimate the sea surface wind speed. The remote sensing solution was to design a radiometer that can be mounted looking downward from the belly of the aircraft. This instrument measures the natural radiation coming up from the ocean, which is closely related to the ocean roughness, which is, in turn, a function of the wind blowing across the surface. After several seasons of testing, we can now successfully measure hurricane surface winds from a safe aircraft altitude. It was for this project that I flew through my first hurricane, Gilbert, in the fall of 1988. I can assure you that no rollercoaster ride has seemed exciting since!

Obviously, the goal of my hurricane research was an immediate improvement in the prediction of hurricane landfall and intensity, but for much of remote sensing the goals are part of longer-term environmental studies. For instance, as discussions of global climate change and greenhouse warming intensify, it is generally accepted that the extent and thickness of ice in the polar regions should be very sensitive to even small changes in the average global temperature. From satellites we can take daily measurements of the *extent* of polar ice, but the thickness of the ice remains elusive. We have laboratory evidence that the natural microwave radiation coming from ice over water is related to the ice's thickness. This suggests that measuring the natural radiation via satellite will allow us to map out ice thickness.

Before such a large satellite project is mounted, however, "proof of concept" fieldwork is necessary. In July of 1989 we mounted a highly sensitive radiometer on the side of a German icebreaker scheduled to travel to Antarctica the following August. As the ship moved through the sea ice, the receiver measured the radiation, which was later successfully compared to actual thickness measurements. During this project, in addition to accomplishing my research goals, I had the opportunity to interact with scientists from Germany, Russia, Colombia, the United Kingdom, and Canada. Because it was springtime in the southern hemisphere, we were also treated to many spectacular displays by emperor penguins, seals, killer whales, and the beautiful petrels. To top off a successful research trip, after reaching port in Africa, several of my new friends and I went on safari through the African wilderness.

I was originally attracted to physics and engineering by the love of understanding how things work. I could never have anticipated how exciting the pursuit of such understanding would be. By this I mean the travel associated with fieldwork, access to organizations like NASA, working with scientists in other disciplines, and learning a little bit about climate cycles, hurricanes, and penguins.

Illustration 13.1

How much work must be put into a heat pump in order to supply 1000 J of heat to a room if the condenser temperature is 40°C and the outside temperature is 0°C? Assume the heat pump operates at the maximum COP (not really possible).

Reasoning   Here $T_h = 313\,\mathrm{K}$ and $T_c = 273\,\mathrm{K}$. The maximum COP for these temperatures is

$$\frac{313\,\mathrm{K}}{313\,\mathrm{K} - 273\,\mathrm{K}} = 7.8$$

This is the ratio of heat delivered, Q_h, to work input, W. So with $Q_h = 1000\,\mathrm{J}$, we have

$$W = \frac{Q_h}{\mathrm{COP}} = \frac{1000\,\mathrm{J}}{7.8} = 130\,\mathrm{J}$$

The heat pump has delivered 7.8 times as much heat to the room as the work paid for in the form of the electricity running the compressor. Actual heat pumps operating at the same temperatures might achieve a COP of about 3–4.

LEARNING GOALS

Now that you have finished this chapter, you should be able to

1 Define (*a*) entropy, (*b*) microstate and macrostate, (*c*) Carnot engine, (*d*) heat engine, (*e*) refrigeration system, (*f*) efficiency of a heat engine, and (*g*) coefficient of performance of a refrigeration system.

2 Give several examples of physical systems which, left to themselves, become more disordered. Explain in each case why the reverse process is not observed.

3 Calculate the entropy change of a simple system during an isothermal process.

4 Explain the relation between entropy and probability and use the Boltzmann relation to calculate the entropy and entropy change of simple systems.

5 State the second law of thermodynamics in terms of (*a*) the direction of heat transfer between systems at two temperatures, (*b*) the phenomenon of thermodynamic equilibrium, (*c*) the degree of disorder of a system, and (*d*) the conversion of heat into work by a heat engine.

6 Define heat engine and refrigeration system in terms of their function and energy flow.

7 Perform simple calculations with the concepts of efficiency and coefficient of performance.

8 Identify the components of a refrigeration cycle. Explain the differences in the application of the refrigeration cycle to refrigerators, air conditioners, and heat pumps.

SUMMARY

DEFINITIONS AND BASIC PRINCIPLES

The Second Law of Thermodynamics
There are many ways of stating the second law:

1. Heat transfer always occurs from high temperature to low temperature.
2. An isolated system will tend to the state of maximum disorder. This is also the state of maximum probability.
3. When an isolated system undergoes changes, its entropy change is greater than or equal to zero.
4. It is impossible for a heat engine to convert thermal energy to work with 100% efficiency.

Entropy (S)
Entropy is a thermodynamic state function, defined in terms of the probability Ω of a particular state occurring:

$$S = k \ln \Omega$$

Entropy increases when heat is added to a system and decreases when heat is removed. The change of entropy for an isothermal process is:

$$\Delta S = \frac{Q}{T} \quad \text{(isothermal process)}$$

The SI units of entropy are J/K.

Efficiency of a Heat Engine

$$\text{Efficiency} = \frac{\text{work}}{\text{heat input}} = \frac{W}{Q_h}$$

The maximum efficiency of a heat engine operating between temperatures T_c and T_h is

$$\text{Maximum efficiency} = 1 - \frac{T_c}{T_h}$$

Coefficient of Performance (COP) of Refrigerators and Heat Pumps

$$\text{COP (refrigerator)} = \frac{Q_c}{W_{\text{in}}}$$

$$\text{COP (heat pump)} = \frac{Q_h}{W_{\text{in}}}$$

The maximum coefficients of performance of refrigerators and heat pumps operating between temperatures T_c and T_h are:

$$\text{Maximum COP (refrigerator)} = \frac{T_c}{T_h - T_c}$$

$$\text{Maximum COP (heat pump)} = \frac{T_h}{T_h - T_c}$$

QUESTIONS AND GUESSTIMATES

1 Suppose a box has such a good vacuum within it that it contains only five gas molecules. Sometimes all five will be in one half of the box. How can you reconcile this with the second law and our discussion of disorder?

2 Some people claim you can cool a watermelon by placing it in a wet blanket and letting it sit in a breeze even though the air temperature is very high. Doesn't this contradict the second law?

3 Estimate the rate at which a person's entropy changes while she or he is lounging around. The average person's metabolic rate under these circumstances is about 100 W (that is, the person uses stored energy at this rate).

4 Consider the simple heat engine in Fig. P13.1. The heated

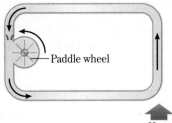

Paddle wheel

Heat   **FIGURE P13.1**

liquid on the right expands and is lifted by the cooler liquid on the left. As a result, the liquid circulates counterclockwise in the tube. As it does so, it rotates the paddle wheel, which is then coupled to external devices to do output work. Explain what factors affect the efficiency of this engine. What should be done to make it most efficient?

5 Each of a pair of dice has six sides, labeled 1 to 6. When a pair is tossed, what is the ratio of the chance that the up sides will total x to the chance that they will total y when (*a*) $x = 2$, $y = 3$ and (*b*) $x = 2$, $y = 4$?

6 A child who wishes to cool off the kitchen of a home opens the refrigerator door and leaves it open. Will this work? Answer the question from both short- and long-term considerations. Would the situation be any different if an old-fashioned ice box were used rather than a refrigerator?

7 At the present time, the sun is the major source of the energy we use here on earth. Trace this solar energy from its source through our uses of it and show that no conflict exists with the second law. Pay particular attention to the ordering process that occurs in photosynthesis.

PROBLEMS

Section 13.1

1 Three pennies painted different colors are tossed at random. (*a*) In how many different head-tail combinations can the pennies be arranged? (*b*) What is the probability that all

three will come up tails? (*c*) What is the probability for two heads and one tail?

2 (*a*) When a pair of dice are thrown, in how many ways can the sum of the two up sides be 5? What is the probability of

rolling a 5? (*b*) In how many ways can the sum be 11? What is the probability of rolling an 11? (*c*) What is the most probable sum, and what is its probability?

■ 3 You are invited to a game of dice on an alien planet, where the "dice" are in the form of four-sided tetrahedrons. The sides are numbered 1 through 4. The game is played with *three* of these dice, and the sum is calculated by adding up the scores of the sides facing *down* after each throw. (*a*) Construct a probability table listing every possible combination for the three dice. How many different arrangements are possible? (*b*) In how many ways can the sum of the three down sides be 5? In how many ways can it be 11? What are the probabilities for these two results? (*c*) What is the most probable sum and what is its probability?

4 Construct a graph of the probability distribution for the dice in Probs. 2 and 3 by plotting the probability for each sum versus the sums.

■ 5 When N labeled coins are tossed, 2^N combinations of heads and tails are possible. How many combinations are possible for (*a*) 3 coins, (*b*) 5 coins, (*c*) 50 coins?

6 Nine ants have fallen into a box and move randomly. (*a*) Using the explanation of Prob. 5, determine the probability that all nine are in the left half of the box. (*b*) What is the probability of eight ants being in the left half and one ant in the right half?

Section 13.2

7 What is the entropy change in 315 g of mercury as it changes from liquid to solid at its normal melting point of $-39°C$?

8 When 2.3 g of water freezes at 0°C, what is the entropy change of the water?

9 An average adult person generally emits energy at the rate of approximately 105 W while sedentary. At what rate is the person's entropy changing?

10 Five kilograms of water is slowly heated from 27°C to 37°C. What is the approximate value of the change in the entropy of the water?

■ 11 A sample of helium gas expands isothermally at $-90°C$ to 3.75 times its original volume. There are 9 g of helium in the sample. What is the change in entropy of the helium?

■ 12 Consider a system of two containers, one at 350 K and the other at 290 K, each holding 6.5 moles of hydrogen gas (H_2). The two containers are insulated from their surroundings, but make contact with each other so that heat is free to flow from the hotter gas to the cooler one. (*a*) Find the entropy change of each of the two samples as the temperature of the hotter gas falls to 340 K. (*b*) Repeat part (*a*) when the two gases have come to an equilibrium temperature. (*c*) Find the total entropy change of the system for parts (*a*) and (*b*).

13 Five coins are tossed. What are the entropies of the following results: (*a*) 1 tail, 4 heads, (*b*) 3 heads, 2 tails, (*c*) 5 heads?

Section 13.3

14 A heat engine uses the interior of a hot furnace at 850°C as its hot thermal energy reservoir and air at 65°C as its cold reservoir. What is the maximum possible efficiency of the engine under these conditions?

15 In modern steam-turbine engines, heat input is in the form of steam at about 600°C. Heat is exhausted to a condenser at about 70°C. What is the highest efficiency possible for such a steam turbine?

16 Actual modern steam turbines operate at efficiencies of about 46 percent. If one such engine produced 500 MW of power, (*a*) how much heat would it give to its low-temperature environment in a 24-h period, and (*b*) how much energy from high-temperature steam would it require in the same period?

17 Suppose you leave a 100-W light bulb on continuously for 1 month (30 days). If the electricity is generated and delivered by the power company to the bulb at 30 percent efficiency, how much thermal energy does the power plant exhaust to the environment due to your oversight?

■ 18 When gasoline is burned, it gives off 50,000 J/g (called its *heat of combustion*). If a car's engine is 25 percent efficient, how much gasoline per hour must it burn in order to develop an output of 50 hp? Express this result in kilograms per hour and gallons per hour.

■ 19 Modern nuclear power plants have an overall efficiency of about 30 percent. Fossil-fuel plants, due to slightly higher steam turbine temperatures, achieve overall efficiencies of 40 percent. Compare the rate of heat emission from a nuclear plant with that from a fossil-fuel plant, each of which produces an output power of 1000 MW.

■ 20 A heat engine contains helium originally in a state where $V = 0.100\,m^3$, $P = 1\,atm$, and $T = 300\,K$. The thermodynamic state of the gas is changed along the cycle *ABCDA* as shown in Fig. P13.2. (*a*) Calculate the heat inputs and outputs for the four parts of the cycle. Carefully observe the sign of Q in each case. (*b*) Calculate the work outputs and inputs for the four parts of the cycle. Carefully observe the sign of W in each case. (*c*) Calculate the efficiency, W_{out}/Q_{in}, for this engine.

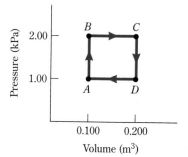

FIGURE P13.2

"21 A heat engine containing 2 mol of an ideal monatomic gas operates in the thermodynamic cycle shown in Fig. P13.3. Process AB is isochoric, BC is adiabatic, and CA is isobaric. (*a*) Calculate Q and W for each of the processes. (*b*) Calculate the efficiency of this engine. (*c*) Calculate the maximum efficiency for any engine working at the temperatures in this cycle.

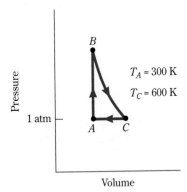

$T_A = 300$ K

$T_C = 600$ K

Pressure

1 atm

Volume　　　　**FIGURE P13.3**

Section 13.4

"22 A certain refrigerator requires 0.90 kW for its operation. It is capable of removing 560 cal/s from the interior of the refrigerator. What is its COP? At what rate does it give off heat to the room around it?

"23 An air conditioner requires 0.90 kW for its operation. It exhausts 560 calories of heat to the outdoors each second. How many calories does it take each second from the room it is cooling? Express this in Btu per hour. What is the COP of the air conditioner?

"24 Suppose a refrigerator has a COP of 5.5. (*a*) How much energy does it consume in removing 1850 cal from its interior? (*b*) What power rating will this refrigerator have if it can remove 1850 cal from its interior each minute?

25 You have installed a heat pump in your house. It delivers heat to the interior of the house at a temperature of 40°C. Compare the maximum possible COP of the heat pump when the outside temperature (the cold thermal reservoir) is (*a*) 0°C and (*b*) −30°C.

General Problems

"26 A dish of hot food is placed in a refrigerator maintained at 5°C. To cool to this temperature, the food must lose 220,000 J. (*a*) How much electric energy is needed to operate the compressor if we assume that room temperature is 23°C and the refrigerator runs at half its theoretical maximum COP? (*b*) If electricity sells for $0.075/kWh, how much money does this cooling cost?

"27 A scientist on an earthlike planet, who has no knowledge of the existence of earth or its science, decides to establish a temperature scale based on the principle of the maximum conversion of thermal energy to work permitted by the second law of thermodynamics. We earthlings know this as the Carnot statement of the second law, $T_h/T_c = Q_h/Q_c$. The scientist decides to have 100 temperature degrees between the boiling and freezing points of water. From measurements on a Carnot cycle at the boiling and freezing points of water at the standard pressure of the planet's atmosphere, the scientist finds $Q_c/Q_h = 0.732$. What temperatures must the scientist assign to the freezing and boiling points of water? Can you infer anything about the planet's atmosphere compared to that of earth?

"28 A 1500-kg car is to accelerate from rest to 8.3 m/s in 6.6 s. (*a*) What is the minimum horsepower the engine must deliver if all friction losses are ignored? (*b*) Assuming the car uses its fuel with an efficiency of 22 percent, determine the amount of gasoline consumed in the 6.6-s time period. Gasoline delivers about 50,000 J per gram burned.

"29 Suppose a gas turbine has a combustion temperature (T_h) of 2400°C and an exhaust temperature (T_c) of 400°C. It operates at one-third of its maximum possible efficiency. The exhaust of this turbine is used to produce 400°C steam as input heat to turn a lower temperature steam turbine. This steam turbine exhausts heat at 70°C and achieves 70 percent of its maximum possible efficiency. This is an example of what is called a *combined cycle* engine. (*a*) What are the individual efficiencies of each engine? (*b*) What is the total efficiency of the combined cycle operation? (*c*) If both engines were ideal Carnot engines, what would be the maximum possible efficiency of this combination?

"30 Generalize Prob. 29 by considering two ideal heat engines connected in series, the first operating between temperatures T_1 and T_2, and the second between T_2 and T_3. (Assume $T_1 > T_2 > T_3$.) Show that the efficiency of this combination can be written as $1 − (T_3/T_1)$.

"31 Suppose you can buy electricity for $0.075/kWh and fuel oil for $1.25/gal. Each gallon of oil will produce 36,000 kcal of heat when burned. Suppose you have the following choices for heating your home: (*a*) install an oil burner delivering heat at 75 percent efficiency, (*b*) install electric baseboard heaters, which convert 100 percent of the delivered electric energy to heat, (*c*) use electricity to run a heat pump which will operate at a COP of 4. Determine what it would cost to deliver 100,000 kcal of heat into your home using each of these methods.

■■32 The thermodynamic cycle of the modern internal combustion engine can be reasonably approximated as consisting of two adiabatic and two isochoric processes as shown in Fig. P13.4. Processes ab and cd are adiabatic. The volume compression ratio of the engine is V_1/V_2. Assume the air-fuel mixture to be an ideal gas with $\gamma = 1.4$. (a) Using the basic definition of efficiency, $W_{\text{out}}/Q_{\text{in}}$, analyze this cycle to show that its efficiency can be expressed as

$$\text{Efficiency} = 1 - \left(\frac{V_1}{V_2}\right)^{\gamma-1}$$

(b) Calculate this efficiency for the cases where $V_2 = 6V_1$ and $V_2 = 20V_1$.

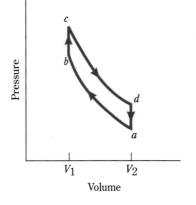

FIGURE P13.4

CHAPTER 14

VIBRATION AND WAVES

In the preceding chapters, we discussed mechanics and the properties of matter. In the next two chapters, we apply many of these concepts to the study of vibration and wave motion. *Wave* is a term that applies to a wide range of phenomena produced by objects vibrating in periodic motion. Vibrating guitar strings or vocal cords produce sound waves, for example, and electric charges vibrating on an antenna produce radio waves.

In this chapter we describe wave motion in general and give some simple examples of periodic motions that produce waves in a taut string or in a spring. In Chap. 15, we study sound waves, in which the vibration medium is not a string or a spring, but air molecules. In later chapters we encounter electromagnetic waves, such as radio and light waves. As should be evident, the topic of waves is of great importance in our everyday world.

14.1 PERIODIC MOTION

All vibrating systems undergo the same motion over and over again. For example, the pendulum in Fig. 14.1 vibrates (or oscillates) back and forth time after time after time. We say that such motion is *periodic* and define the period of the motion as follows:

FIGURE 14.1

A pendulum undergoes periodic motion. It executes one half cycle of vibration as it moves from the leftmost position to the rightmost position.

The **period** of vibration τ (Greek tau) is the time taken to make one complete vibration.

In the case of the pendulum of Fig. 14.1, the period is the time it takes the pendulum to swing from A to C and back to A. Notice that the period is the *total* time the pendulum ball is away from A during one complete vibration. We call the motion undergone in one period a *vibration cycle*.

Often we speak of the frequency of vibration, defined in the following way:

The **frequency** of vibration f is the number of vibration cycles completed by the system per unit of time.

We often express frequencies in cycles per second (s^{-1}). For example, a guitar string might undergo 330 vibration cycles in 1 second, and its frequency is thus $330\,s^{-1}$. The SI unit for frequency is the hertz (Hz), which is just another name for cycles per second: $1\,Hz = 1\,s^{-1}$. Notice that "cycles" is a term that has no physical units. Remember, however, that Hz always implies that you are counting in cycles per second.

There is an important relation between frequency f and period τ. Because frequency is the number of vibrations per unit time and because one complete vibration takes a time τ,

$$f = \frac{\text{number of vibrations}}{\text{time taken}} = \frac{1\ \text{vibration}}{\text{period}}$$

Thus we have the general relation

$$f = \frac{1}{\tau} \tag{14.1}$$

This relation applies to all periodic motions. If the period of a certain motion is 0.020 s, say, then its frequency is 50 Hz.

Another property of periodic motion is the amplitude of the motion:

The **amplitude** is the maximum displacement from the equilibrium position the object maintains when it is not vibrating.

For the pendulum in Fig. 14.1, the amplitude is the distance AB or BC. Notice that the amplitude is only half of the full distance through which the system swings.

Another important feature of vibrating systems is the way they interchange energy between potential and kinetic. For example, when the pendulum ball in Fig. 14.1 is at A or at C, it is momentarily at rest and thus has no kinetic energy. At these points, it possesses only gravitational potential energy. As it swings toward B, however, it loses potential energy and gains an equal amount of kinetic energy. Thus, as it swings back and forth, it retains a constant energy, but the energy keeps changing between kinetic and potential.

Another typical vibrating system is shown in Fig. 14.2. It consists of a mass at the end of a spring, and we assume that the mass can slide back and forth without friction. The spring-mass system is shown in its equilibrium position in part (*a*). There is no horizontal force acting on the mass in this position. (The pull of gravity down is balanced by the push of the table up, and so the net vertical force on the mass is always zero.)

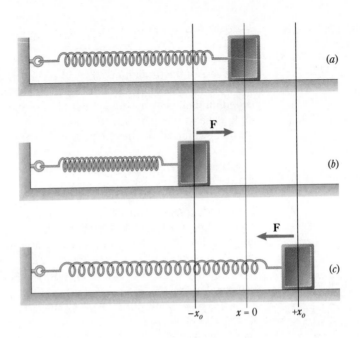

FIGURE 14.2

(*a*) The mass is at equilibrium at $x = 0$ before the system is set into motion. At that position, the spring exerts no force on the mass. (*b*) The compressed spring has potential energy stored in it and exerts a restoring force on the momentarily stationary mass. (*c*) The stretched spring has the same amount of potential energy stored in it as in (*b*). It exerts a restoring force on the momentarily stationary mass.

Suppose we compress the spring by moving the mass to the position $-x_o$ shown in Fig. 14.2*b*. During this process, we do work on the spring, and so we store potential energy in it. The compressed spring exerts a force on the mass, a force that tends to push the mass back to the $x = 0$ position. If the mass is now released so that it can move freely under the force applied to it by the spring, the spring will accelerate it to the right until the position $x = 0$ is reached. The mass is now moving to the right quite swiftly, and at the $x = 0$ point the spring has lost all the potential energy stored in it when it was compressed. It is clear that the potential energy stored in the spring now appears as the kinetic energy of the moving mass.

The mass does not stop at $x = 0$, however, because it must lose its kinetic energy by doing work before it can come to rest. As the mass proceeds to the right of $x = 0$, it begins to stretch the spring and to store energy in it. By the time the mass reaches the position $+x_o$ shown in Fig. 14.2*c*, it has lost all its kinetic energy by doing work against the spring. The kinetic energy of the mass has been changed completely to potential energy in the stretched spring. Therefore, when the mass reaches $x = x_o$, its velocity momentarily becomes zero.

The spring, now stretched, accelerates the mass to the left. When the mass reaches $x = 0$, all the energy is once again in the form of kinetic energy. The mass again compresses the spring to the position $x = -x_o$, at which point all the kinetic energy is changed to potential energy stored in the compressed spring. Now the process repeats itself, and the mass will vibrate back and forth between $x = +x_o$ and $x = -x_o$ forever if there are no friction losses. Notice that as the mass oscillates, the energy oscillates back and forth between kinetic and potential but the total energy remains constant; energy is conserved.

A similar situation occurs if we hang a mass on a spring and allow the system to stretch downward to a position where the upward spring force balances the downward weight. Although we do not pause to prove it here, the motion of the mass oscillating vertically on the spring about this position is identical to the horizontal motion described above.

The pendulum and spring-mass combination are only two examples of vibrating systems. All such systems involve a similar continuous interchange of energy. Because many of the systems of interest involve springs of one sort or another, let us take a moment to find the energy stored in a spring.

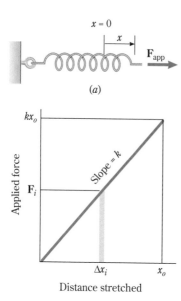

$x = 0$

x

$\mathbf{F}_{app}$

(a)

kx_o

F_i

Slope = k

Applied force

Δx_i x_o

Distance stretched

(b)

FIGURE 14.3

In order to stretch a spring, a force must be applied which is opposite to and at least as great as the restoring force exerted by the spring. The spring force is proportional to the amount of stretch x, so $\mathbf{F}_{app}$ must also be, as indicated in (b). The work done by $\mathbf{F}_{app}$ is equal to the area under the graph of $\mathbf{F}_{app}$ versus x.

14.2 HOOKE'S LAW AND ELASTIC POTENTIAL ENERGY

We saw in Chap. 9 that many elastic (springlike) systems obey Hooke's law, which states that a distorting force is proportional to the distortion it causes. For a spring being stretched by an applied force $\mathbf{F}_{app}$, as in Fig. 14.3a, the displacement $\mathbf{x}$ the spring is stretched is related to $\mathbf{F}_{app}$ through

$$\mathbf{F}_{app} = k\mathbf{x} \tag{14.2}$$

where k is called the **spring constant** and has SI units of newtons per meter. The spring constant measures the "stiffness" of a spring. Larger values imply that larger forces are required to produce a given amount of stretch.

Figure 14.3b shows how force varies with distortion for the spring of Fig. 14.3a. The graph is a straight line with slope k whose equation (Hooke's law) is given by Eq. 14.2. Let us now compute the energy stored in a stretched or compressed spring that obeys Hooke's law.

We can show that the work done in stretching the spring from $x = 0$ to $x = x_o$ is the area under the graph line in Fig. 14.3b. To do that, we examine the shaded rectangle shown. Its area is $F_i\,\Delta x_i$, where F_i is the stretching force that prevails during the small increase in distortion Δx_i. Because $W = F_s\,\Delta s$, this area is also the work done by the stretching force during this small increase in displacement. Imagine the region under the line from $x = 0$ to $x = x_o$ to be filled by many such rectangles. The sum of their areas gives the total work done in stretching the spring from $x = 0$ to $x = x_o$. Hence

The work done in stretching or compressing an elastic element is equal to the area under its F-versus-x graph line.

The coil springs of a car exert a force proportional to the amount they are stretched or compressed. The shock absorber running down the middle damps out the vibration that would otherwise continue after hitting a bump.

This is similar to our previous calculations (Sec. 12.3) when we used a *PV* diagram to determine the work done on a gas when its volume changed. You should be able to extend this discussion to confirm the compression portion of this statement.

Since the area of a triangle is one-half its base times its height, we see from Fig. 14.3 that the area under the graph line is $(\frac{1}{2}x_o)(kx_o)$. However, this equals the work done in stretching the spring, and so it is equal to the potential energy stored in the spring. We therefore conclude that the potential energy stored in a spring with constant k that has been stretched or compressed a distance x is

$$\text{Elastic PE} = \text{EPE} = \tfrac{1}{2}kx^2 \tag{14.3}$$

Now that we know how much elastic energy is stored in a spring (or any other system that obeys Hooke's law), we can use the law of energy conservation to learn much about the vibration of the system shown in Fig. 14.2. Because we are assuming friction losses to be negligible, the sum of the potential energy stored in the spring and the kinetic energy of the mass must remain constant. To place this statement in equation form, consider once again the system in Fig. 14.2 when the spring is stretched to $x = x_o$ and released. Because its initial total energy at the moment of release is $\tfrac{1}{2}kx_o^2$, its total energy at any later time is

$$\text{EPE} + \text{KE} = \tfrac{1}{2}kx_o^2$$

Substituting gives

$$\tfrac{1}{2}kx^2 + \tfrac{1}{2}mv^2 = \tfrac{1}{2}kx_o^2 \tag{14.4}$$

where m and v pertain only to the mass attached to the spring, because the spring is assumed to have negligible mass. You should recognize that x_o is the amplitude of the motion.

Equation 14.4, despite its simplicity, is a very powerful tool. We can use it to find the velocity of the mass at any point x in the motion:

$$v = \pm\sqrt{\frac{k}{m}(x_o^2 - x^2)}$$

Do not memorize this equation because it is simply Eq. 14.4 rearranged. Notice that $v = 0$ at $x = x_o$, when the mass is at the end of its vibration; when $x = 0$, the velocity has its largest value, $x_o\sqrt{k/m}$. We already know these facts from our qualitative discussion of energy interchange. Now, however, we are able to find the velocity of the vibrating mass at any position x.

It remains yet to find the acceleration of the vibrating mass. When the system is vibrating freely, the situation is as shown in Fig. 14.4. The only unbalanced force acting on the mass is **F**, the pull of the spring on it. We call this a restoring force because it always acts in such a direction as to pull or push the system back to its

FIGURE 14.4

The force **F** of the spring on the mass is a restoring force and is given by **F** = −k**x**.

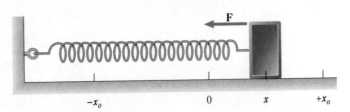

equilibrium position. The magnitude of **F** is kx, the same as the force required to stretch the spring an amount x. However, its direction is *opposite* the direction of stretch, and so its value is $\mathbf{F} = -k\mathbf{x}$; the negative sign signifies that it is a restoring force, that is, one directed opposite the displacement **x**. Since **F** is the unbalanced force on the mass, the acceleration of the mass is, from $\mathbf{F} = m\mathbf{a}$,

$$\mathbf{a} = -\frac{k}{m}\mathbf{x} \qquad (14.5)$$

Notice that the magnitude of the acceleration is maximum when $x = \pm x_o$ because there the restoring force is maximum; when $x = 0$, the restoring force is zero and so is the acceleration. As you see, we can use Eqs. 14.4 and 14.5 to find the velocity and acceleration of the mass at any displacement x.

Example 14.1

A particular spring stretches 20.0 cm when a 500-g mass is hung from it. Suppose this mass is replaced by a 2.00-kg mass and the system is then vibrated horizontally as in Fig. 14.4 by displacing the mass 40.0 cm from its equilibrium position and releasing it. Find (**a**) the maximum velocity of the mass, (**b**) its maximum acceleration, and (**c**) its velocity and acceleration when $x = 10.0$ cm.

R *Reasoning*

Question What is the condition for maximum velocity?
Answer It occurs when all the energy of the system is kinetic. This occurs at the position where the spring is not stretched or compressed, i.e., at $x = 0$.

Question What physical law relates velocity to position?
Answer The conservation of energy:

$$\tfrac{1}{2}kx^2 + \tfrac{1}{2}mv^2 = \tfrac{1}{2}kx_o^2$$

At $x = 0$,

$$v = v_{\max} = \sqrt{\frac{k}{m}}\,x_o$$

Question The value of k is not given. What do I have to know in order to calculate it?
Answer The spring constant k is the ratio of applied force to the resulting stretch of the spring. All the data you need are given in the statement of the problem.

Question For part (*b*), what is the condition for maximum acceleration?
Answer Maximum acceleration occurs when the net force is a maximum, which occurs at the points of maximum stretch and compression, where $x = \pm x_o$. This is also the condition for zero kinetic energy, or $v = 0$.

Question For part (*c*), what relates velocity to any intermediate position x?
Answer Once again, the conservation of energy (Eq. 14.4). For any intermediate position *both* KE and PE are present. The relation can be written

$$\tfrac{1}{2}mv^2 = \tfrac{1}{2}k(x_o^2 - x^2)$$

Question   How is acceleration related to a particular position?
Answer   The force exerted by the spring depends on the position according to
$\mathbf{F} = -k\mathbf{x}$. Since it is the net force on m, this force is solely responsible for the
acceleration according to $\mathbf{F} = m\mathbf{a}$.

Solution and Discussion   We can solve Eq. 14.2 for k first, using the weight
of the 500-g mass for $\mathbf{F}_{app}$:

$$k = \frac{(0.500 \text{ kg})(9.80 \text{ m/s}^2)}{0.200 \text{ m}} = 24.5 \text{ N/m}$$

(*a*) The magnitude of the maximum velocity can then be calculated directly:

$$v_{max} = x_o \sqrt{\frac{k}{m}} = 0.400 \text{ m} \sqrt{\frac{24.5 \text{ N/m}}{2.00 \text{ kg}}} = 1.40 \text{ m/s}$$

You should verify the units.

(*b*) The maximum acceleration is

$$a_{max} = \frac{kx_o}{m} = \frac{(24.5 \text{ N/m})(0.400 \text{ m})}{2.00 \text{ kg}} = 4.90 \text{ m/s}^2$$

(*c*) The acceleration at $\mathbf{x} = +10.0$ cm is

$$\mathbf{a} = -\frac{k\mathbf{x}}{m} = -\frac{(24.5 \text{ N/m})(0.100 \text{ m})}{2.00 \text{ kg}} = -1.22 \text{ m/s}^2$$

Notice that the direction of $\mathbf{a}$ is opposite the direction of the displacement $\mathbf{x}$.
Also remember that k is a property of the spring that remains constant for any
given spring. It can be found by measuring F/x for any displacement as long as
Hooke's law applies.
  Finally, the velocity at $x = 10.0$ cm is found from

$$\mathbf{v} = \pm \sqrt{\frac{k(x_o^2 - x^2)}{m}}$$

$$= \pm \sqrt{\frac{(24.5 \text{ N/m})[(0.400 \text{ m})^2 - (0.100 \text{ m})^2]}{2.00 \text{ kg}}}$$

$$= \pm 1.36 \text{ m/s}$$

Both signs are necessary since the mass can be going toward x_o or returning
from it at $x = 10.0$ cm.

Exercise   Find $\mathbf{v}$ and $\mathbf{a}$ when $x = -5.00$ cm. *Answer:* $\pm 1.39 \text{ m/s};$
$0.613 \text{ m/s}^2.$

A clock pendulum performs simple harmonic motion. Since the pendulum's period is constant, the clock is able to keep correct time.

14.3 SIMPLE HARMONIC MOTION

There are many types of periodic motion, of which the mass on a spring is only one. However, the description of the motion of the mass on a spring is particularly simple. Furthermore, there are many other examples, including pendulums, whose periodic motion has the same description. The essential feature of these simple periodic systems is that, whenever the system is displaced from an equilibrium position, *there arises a restoring force* that is *linearly proportional* to the magnitude of the displacement.

In the case of the spring-mass system, we have seen that Hooke's law (Eq. 14.2) governs the motion:

$$\mathbf{F} = -k\mathbf{x}$$

where k is the spring constant.

Generalizing this expression, we have the basic form of the force law:

Restoring force = −(constant)(displacement from equilibrium)   (14.6)

When the restoring force is the only force acting, the acceleration of the vibrating mass is

$$\mathbf{a} = \frac{-(\text{constant})(\text{displacement})}{\text{mass}}$$ (14.7)

The motion of any system governed by a force of the form of Eq. 14.6 is called simple harmonic motion (SHM).

Simple harmonic motion results when a system moves in response to a restoring force that is linearly proportional to the magnitude of the system's displacement from equilibrium.

By analyzing the restoring force in any given situation, we can identify the proportionality constant of Eqs. 14.6 and 14.7, called the *force constant* for the system. Once we recognize that a system is subject to a restoring force with this property, we don't have to derive the equations of the system's motion all over again. This force constant plays exactly the same role in the motion as the spring constant k does in the spring-mass system already analyzed. We will encounter other examples of SHM as we proceed. Let us first discuss the time dependence of SHM and derive an expression for the frequency of the motion.

14.4 FREQUENCY OF SIMPLE HARMONIC MOTION

Finding an expression for the frequency of simple harmonic motion (SHM) is straightforward when calculus is used. We shall use a graphic method, however, since calculus is not a requirement for this course.

We start by imagining a particle Q moving in a circle of radius x_o at constant speed v_o. We call this circle a **reference circle,** and the diagram of this motion is

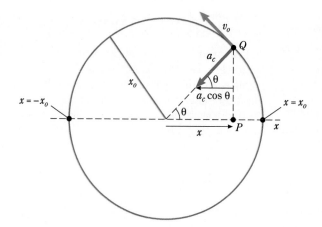

FIGURE 14.5

As Q moves around the circle of radius x_o with constant speed v_o, point P undergoes simple harmonic motion from $-x_o \leq x \leq x_o$. Because the radius of the circle is x_o, $x = x_o \cos \theta$.

shown in Fig. 14.5. The motion of Q can also be described as having constant angular velocity $\Delta\theta/\Delta t = \omega$, given by $\omega = v_o/x_o$ (Eq. 7.7). Remember from Sec. 7.2 that ω is measured in radians per second. The period (τ) in which Q completes a cycle of motion is the time it takes Q to move once around the circle, or

$$\tau = \frac{2\pi x_o}{v_o} = 2\pi \left(\frac{x_o}{v_o}\right)$$

The frequency f of the motion, the number of cycles per second, is just the reciprocal of the period:

$$f = \frac{1}{\tau}$$

Notice in Fig. 14.5 that the point P is the x-coordinate position of the point Q. For any value of θ, P is at $x = x_o \cos \theta$. *As Q goes around the reference circle, P travels along the x axis from $+x_o$ to $-x_o$ and back to $+x_o$ with the same period and frequency as Q.* We now show that the motion of P is SHM.

The circular motion of Q has, from Eq. 7.9, the centripetal acceleration

$$a_c = \frac{v_o^2}{x_o} = \omega^2 x_o$$

The inward radial direction of a_c is shown in Fig. 14.5. The corresponding acceleration of point P is the x component of a_c:

$$\mathbf{a}(P) = -a_c \cos \theta$$

The negative sign shows the direction of $\mathbf{a}(P)$.

Using the expression for a_c and the relation $x/x_o = \cos \theta$, we get

$$\mathbf{a}(P) = -\omega^2 x \tag{14.8}$$

where ω has a constant value. This proves that P performs SHM, since $\mathbf{a} = -k\mathbf{x}$ is the general form of acceleration that defines SHM.

Now it is an easy step to identify the frequency of SHM in general. Using both Eqs. 14.7 and 14.8, we have

$$\mathbf{a} = -\omega^2 \mathbf{x} = -\left(\frac{k}{m}\right)\mathbf{x}$$

where k is the force constant in Eq. 14.7. From this we can identify ω:

$$\omega = \sqrt{\frac{k}{m}} \tag{14.9}$$

The frequency of the SHM of point P is then

$$f = \frac{\omega}{2\pi} = \frac{1}{2\pi}\sqrt{\frac{k}{m}} \tag{14.10}$$

and the period of SHM is

$$\tau = \frac{1}{f} = 2\pi\sqrt{\frac{m}{k}} \tag{14.11}$$

Our derivation did not need to refer to any particular example of SHM. Therefore we conclude that Eqs. 14.10 and 14.11 are general expressions for the frequency and period of any system undergoing SHM. *Once the force constant k has been identified for a specific system, f and τ can immediately be found.*

Illustration 14.1

Find the frequency of vibration for the system of Example 14.1.

Reasoning In that example, the spring constant was 24.5 N/m and the mass at the end of the spring was 2.00 kg. From Eq. 14.10,

$$f = \frac{1}{2\pi}\sqrt{\frac{24.5\,\text{N/m}}{2.00\,\text{kg}}} = 0.557\,\text{s}^{-1} = 0.557\,\text{Hz}$$

14.5 SINUSOIDAL MOTION

A simple mathematical equation can be written for an object vibrating with simple harmonic motion. Point P in Fig. 14.5 has an x-coordinate position given by

$$x = x_o \cos\theta$$

Let us look at a graph of the functions $\sin\theta$ and $\cos\theta$, as shown in Fig. 14.6. This shows that both functions vary from $+1$ to -1 cyclically with a period of 360°, or 2π radians. As $\cos\theta$ varies between these limits, x goes from $+x_o$ to $-x_o$, which are

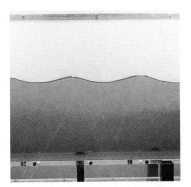

The photograph of the cross-section of a water wave shows the sinusoidal form of the wave.

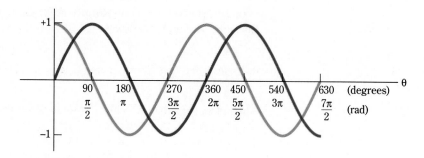

FIGURE 14.6

Graphs of sin θ versus θ (blue) and cos θ versus θ (green).

the amplitude positions of our SHM. The angle θ is called the **phase** of sin θ and cos θ. Notice that the two curves are the same in every respect except that sin θ lags behind cos θ by one-fourth of a cycle. We say that the sine is a quarter cycle, or 90°, *out of phase* with the cosine.

In the description of SHM in the previous section, the reference angle θ was changing in time at a constant rate ω: $\theta = \omega t$. This leads us to the description of the position of P at any instant of time:

$$x = x_o \cos(\omega t) = x_o \cos(2\pi f t) = x_o \cos\left(\frac{2\pi t}{\tau}\right)$$ (14.12)

Notice the equivalency of these three expressions. *It is vital to remember that the quantity in parentheses in all three expressions is in radians.*

Motion that is described by a cosine (or sine) function of time such as Eq. 14.12 is known as **sinusoidal motion.** Sinusoidal motion and simple harmonic motion are the same. A demonstration that will help you visualize the sinusoidal nature of SHM is shown in Fig. 14.7. An object with an attached pen is suspended from a spring. When the object is raised to y_o and released, it undergoes simple harmonic motion with amplitude y_o. Behind the vibrating object is a sheet of paper moving to the left at constant speed. The pen marks on this paper the position of the mass as it vibrates up and down.

Let us count time from the instant the object is released. This is the left end of the trace in (*b*). We take this point as $t = 0$. The position of the object at the instant shown in the figure occurs at some later time. Hence, this trace can be considered a plot of the displacement y of the object as a function of time. According to Eq. 14.12, the equation of this trace is

$$y = y_o \cos(2\pi f t) = y_o \cos(\omega t) = y_o \cos\left(\frac{2\pi t}{\tau}\right)$$

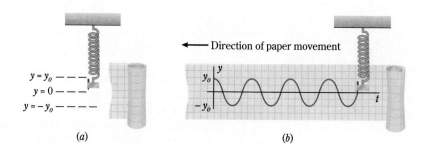

FIGURE 14.7

The vibrating mass traces a cosine curve as a function of time.

Since, for SHM, $a(t) = -(k/m)x(t)$*, the time dependence of a is also described by the same sinusoidal function, except with the minus sign:

$$a = -a_{max} \cos(2\pi ft) \qquad \text{where} \qquad a_{max} = \left(\frac{k}{m}\right)x_o = 4\pi^2 f^2 x_o \qquad (14.13)$$

The function $-\cos(2\pi ft)$ is just one-half of a cycle behind $+\cos(2\pi ft)$. We say that the acceleration is a half cycle, or 180°, *out of phase* with x.

Finally, let us see how the velocity of the mass varies with time. We saw in Sec. 14.2 from energy considerations that the speed of the mass is a maximum when $x = 0$ and zero when x is at a maximum (i.e., when $x = x_o$). We might expect v to oscillate between v_{max} and $-v_{max}$ in a way that is similar to the way in which x and a oscillate. This turns out to be true, but v is one-fourth cycle (90°) out of phase with x and a, and thus is described by $\sin(2\pi ft)$:

$$v = -v_{max} \sin(2\pi ft) \qquad (14.14)$$

where we previously found that $v_{max} = x_o \sqrt{k/m} = 2\pi f x_o$.

Example 14.2

Refer again to the system in Example 14.1. Write the expressions for position and velocity as a function of time. Calculate the position, velocity, and acceleration of the mass at $t = 1.00$ s.

R

Reasoning

Question What do I have to know in order to write the expressions for x and v?
Answer The general forms of SHM when m is released from $x = +x_o$ at $t = 0$ are

$$x = x_o \cos(2\pi ft)$$
$$v = -2\pi f x_o \sin(2\pi ft)$$
$$a = -4\pi^2 f^2 x_o \cos(2\pi ft)$$

Examining these three equations shows that all you need are the amplitude x_o and the frequency f. These are given in Example 14.1 and Illustration 14.1.

Question How do I evaluate the sin and cos functions at $t = 1.00$ s?
Answer The important thing to remember is that $2\pi ft$ is in *radians*, not degrees.

Solution and Discussion

Illustration 14.1 gave us $f = 0.557$ Hz. At $t = 1.00$ s,

$$2\pi ft = 2\pi(0.557 \text{ s}^{-1})(1.00 \text{ s}) = 3.50 \text{ rad}$$

---

*Remember that the notation $x(t)$ and $a(t)$ refers to the fact that x and a are dependent on the values of the time t. The symbol $x(t)$ is read: "x as a function of t."

A calculator gives

$$\sin(3.50\,\text{rad}) = -0.351 \qquad \cos(3.50\,\text{rad}) = -0.936$$

With the amplitude $x_o = 0.40\,\text{m}$, at $t = 1.00\,\text{s}$ we have

$$x = (0.40\,\text{m})(-0.936) = -0.37\,\text{m}$$
$$v = -2\pi(0.557\,\text{Hz})(0.40\,\text{m})(-0.351) = +0.49\,\text{m/s}$$
$$a = -4\pi^2(0.557\,\text{Hz})^2(0.40\,\text{m})(-0.936) = +4.6\,\text{m/s}^2$$

The signs show that, at this instant, x is to the left of equilibrium in Fig. 14.4, it is moving to the right (returning from $-x_o$), and its acceleration is to the right.

Example 14.3

A 250-g mass is moving with SHM according to the expression $x = (1.30\,\text{m})\cos(2.09t)$. (*a*) What are the amplitude and frequency of this motion? (*b*) What is the force constant of the system? (*c*) Find the time at which the mass first reaches the position $+\frac{1}{2}x_o$ after the system is released.

R

Reasoning

Question Where do amplitude and frequency appear in the given expression?
Answer From the general expression for SHM, $x = x_o\cos(2\pi ft)$, you can identify the amplitude (x_o) as that number which multiplies the cosine function. You can also see that the number which multiplies t inside the cosine function is $2\pi f$. From the data given, this tells you that $2\pi f = 2.09$.

Question How is the force constant involved?
Answer Along with the mass m, it determines f:

$$f = \frac{1}{2\pi}\sqrt{\frac{k}{m}}$$

Thus once you have identified f, you can calculate k.

Question What is the significance of the phrase "the first time x reaches $\frac{1}{2}x_o$"?
Answer It should remind you that the mass is going to cross this position many times as the $\cos(2\pi ft)$ goes through its cyclical values. You need to find the lowest value of t for which $x = \frac{1}{2}x_o$.

Question What is the expression that describes when this will occur?
Answer Solve Eq. 14.12 for the smallest value of time at which $x(t) = \frac{1}{2}x_o$:

$$\tfrac{1}{2}x_o = x_o\cos(2.09t) \qquad \text{or} \qquad 0.500 = \cos(2.09t)$$

To solve for t, you must calculate the inverse cosine:

$$\cos^{-1}(0.500) = 2.09t \qquad \text{(radians)}$$

Solution and Discussion
(*a*) From the equation of motion we can identify

$$x_o = 1.3 \, \text{m} \qquad \text{and} \qquad 2\pi f = 2.09/\text{s}$$

The latter gives $f = 0.333 \, \text{Hz}$, and hence $\tau = 1/f = 3.00 \, \text{s}$.

(b) From $k/m = (2\pi f)^2 = (2.09)^2$, you get

$$k = (0.250 \, \text{kg})(4.37 \, \text{Hz}^2) = 1.09 \, \text{N/m}$$

(c) The lowest value of t for which $\cos^{-1}(0.500) = 2.09t$ rad is

$$t = \frac{\cos^{-1}(0.500)}{2.09 \, \text{Hz}} = \frac{1.05}{2.09 \, \text{Hz}} = 0.500 \, \text{s}$$

From the expression for $x(t)$ you can see that at $t = 0$, $\cos(2\pi ft) = \cos 0 = 1$, which tells you that the initial position of the mass is $+x_o$. You know that the mass will reach $x = 0$ one quarter cycle, or $\tau/4 = 3.00 \, \text{s}/4 = 0.750 \, \text{s}$, later. On its way to $x = 0$, the mass will have to cross the position $x = \frac{1}{2}x_o$ for the first time. This is consistent with the answer of $t = 0.500 \, \text{s}$ obtained above.

14.6 THE SIMPLE PENDULUM

We know that a simple pendulum like that shown in Fig. 14.8 oscillates with periodic motion. If we can show that the restoring force is proportional to the displacement from equilibrium, we can conclude that the pendulum moves with SHM.

The equilibrium position is where the pendulum hangs vertically. We now displace the pendulum from equilibrium so that its cord makes an angle θ with the vertical. As shown in Fig. 14.8, there are two forces acting on the mass m: the tension $\mathbf{T}$ and the weight $m\mathbf{g}$. The tension $\mathbf{T}$ is always directed along the cord toward the point of suspension, and $m\mathbf{g}$ is always vertically downward. The net

FIGURE 14.8

The simple pendulum. The restoring force is $mg \sin \theta \approx mg \, \theta$. Note that $\theta = s/L$, where s is the arc length between points A and B.

Examples of pendulums: playground swings vibrate at a frequency determined by their length.

radial force along the cord $(T - mg \cos \theta)$ constrains m to move in a circular arc whose radius is the length of the pendulum, L. The tangential component of $m\mathbf{g}$, $mg \sin \theta$, is directed along the circle back toward equilibrium. Thus we can write

$$F_{\text{restoring}} = -mg \sin \theta$$

where the minus sign indicates a direction opposite increasing θ. Notice that this force is not proportional to the angular displacement θ. For small angles, however, we can use the fact that $\sin \theta \approx \theta$, where θ is expressed in radians. (This approximation is precise to three significant digits if $\theta \leqslant 10°$, which is 0.174 rad.) From the definition of radian measure, we can also write $\theta = s/L \approx x/L$. Then the restoring force is

$$F = -mg\theta = -\left(\frac{mg}{L}\right)x \tag{14.15}$$

which is the form of the force-displacement equation which produces SHM.

We can immediately identify the force constant k from the general form $F = -kx$:

$$k = \frac{mg}{L}$$

Then the frequency of vibration of the pendulum is directly obtained:

$$f = \frac{1}{2\pi}\sqrt{\frac{k}{m}} = \frac{1}{2\pi}\sqrt{\frac{g}{L}} \tag{14.16}$$

Notice that *the frequency of a simple pendulum does not depend on the pendulum's mass*. It depends only on the length L and on g. This result offers a simple yet precise means for measuring g. A pendulum of known length can be timed over many periods of vibration to give a very precise value of f. Then g can easily be calculated from Eq. (14.16).

We can write the equation of motion for the pendulum as

$$\theta = \theta_o \cos (2\pi f t) = \theta_o \cos \left(\sqrt{\frac{g}{L}}\, t \right)$$

Remember that the above results are valid only for small-amplitude swings of the pendulum, where $\sin \theta \approx \theta$.

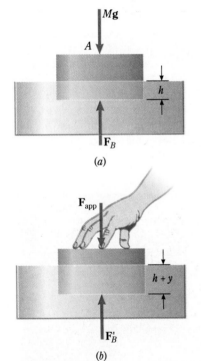

FIGURE 14.9

(a) A block floating in water. A depth h is submerged; $F_B = Mg$. (b) The block is pushed down an additional depth y by F_{app}. This increases the buoyant force by an amount F'_B, which is proportional to y.

Example 14.4

Consider a block of wood of mass M and cross-sectional area A floating in water as shown in Fig. 14.9a. At equilibrium the block is floating with a depth h submerged. Recalling our work on buoyant forces (Chap. 9), show that if the block is pushed downward a small distance y (Fig. 14.9b) and then released, it will bob up and down with SHM. Assume viscosity is negligible. Find an expression for the bobbing frequency.

R

Reasoning

Question   What must I show in order to conclude that the motion is SHM?
Answer   You must show that the net restoring force on the block is proportional to the displacement y.

Question   What forces act on the block?
Answer   At equilibrium, the weight of the block is balanced by the buoyant force of the water.

Question   What determines the buoyant force?
Answer   The weight of the water displaced by the block is the buoyant force. Thus

$$Mg = \rho_{H_2O} Ahg$$

at equilibrium.

Question   If I push down an additional distance y, how much additional buoyant force results?
Answer   A force upward equal to the weight of the additional water you caused to be displaced, which is $\rho_{H_2O} Ayg$.

Question   When I let go of the block, what is the net force on it?
Answer   Just the opposite of the additional force you exerted:

$$F = -(\rho_{H_2O} Ag)y$$

Since the term in parentheses is a constant, this is the general expression that will produce SHM. You should be able to show that, if you had *lifted* the block by a small amount y, you would get the same result.

Question   What determines the frequency of the motion?
Answer   The force constant k and the mass M:

$$f = \frac{1}{2\pi}\sqrt{\frac{k}{M}}$$

Question   What is the force constant in this case?
Answer   It is always the constant of proportionality between force and displacement. In this case you can make the identification

$$k = \rho_{H_2O} Ag$$

Solution and Discussion   Now you have all the pieces necessary to write an expression for f. Remember the above expression for the mass of the block:

$$M = \rho_{H_2O} Ah$$

Putting this and k into the expression for f, you get

$$f = \frac{1}{2\pi}\sqrt{\frac{\rho_{H_2O} Ag}{\rho_{H_2O} Ah}} = \frac{1}{2\pi}\sqrt{\frac{g}{h}}$$

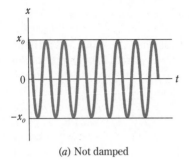

(a) Not damped

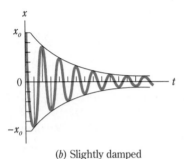

(b) Slightly damped

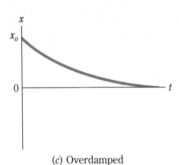

(c) Overdamped

FIGURE 14.10
The way a system vibrates depends on the extent of the energy losses within it.

An example of resonance: when the person standing behind the swing pushes on the swing in phase with its vibration frequency, the amplitude of the vibration quickly increases.

This very interesting result shows that the frequency has the same form as that of the simple pendulum. The submerged depth at equilibrium takes the place of the length of the pendulum.

14.7   FORCED AND DAMPED VIBRATIONS

In any vibrating system, there is always some loss of energy to friction forces. As a result, a vibrating pendulum or mass at the end of a spring vibrates with constantly decreasing amplitude as time goes by. This fact is illustrated in Fig. 14.10. Part (*a*) shows the vibration in the ideal case with no friction. It is the situation we discussed in preceding sections. A more realistic situation is shown in *b*, where the vibration is influenced by friction forces. We say that such a system is *damped* and that, in this case, the amplitude of the vibration *damps down* fairly quickly.

When the friction forces are very large, the system does not vibrate at all; instead, it simply returns slowly to its equilibrium position, as shown in Fig. 14.10*c*. Such a system is said to be *overdamped*. This situation exists, for example, when the mass at the end of a spring is immersed in a very viscous liquid. The mass does not move beyond the equilibrium position in such a case. When the friction forces are just large enough that the system returns to the equilibrium position without overshooting it, we say that the system is *critically damped*.

If any system is to vibrate for an extended time, energy must be added continually to replace the energy lost doing work against friction forces. For example, to keep a child swinging at constant amplitude on a swing, you must push the swing from time to time to add energy to the system.

Everyone knows that there is a right and a wrong way to push a swing if it is to swing high. You must push with the motion of the swing and not against it. Only in this way can energy be added to the system effectively. If you push against the motion, you can stop the vibration, since the vibrating object must then do work on you, the pushing agent. These simple facts have importance in all forced, or *driven,* vibrating systems.

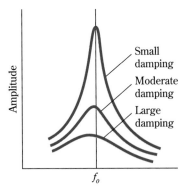

FIGURE 14.11

The amplitude of a driven vibration as a function of equal driving forces at various frequencies f. The resonance frequency of the undamped vibration is f_o. The three curves are for the same oscillator at different extents of damping.

In a driven system, such as a grandfather clock, the vibration is usually sustained by a repetitive force acting on the system. This force has a frequency f, which may or may not be the same as the natural frequency of vibration of the system f_o. When $f = f_o$, the driving agent is most effective in adding energy to the system. At all other frequencies, the driving force is not quite in step with the motion of the system, and so the action of the force is less effective in adding energy. How the amplitude of a vibrating system varies with the frequency of the applied force is shown in Fig. 14.11. Notice, as we said above, that *the driving force is most effective when its frequency f equals the natural frequency f_o of the system.* In that case, we say that the force is in *resonance* with the system. More will be said about f_o, the **resonance frequency** of the system, in Sec. 14.10.

14.8 WAVE TERMINOLOGY

Many vibrating objects act as sources for waves. For example, sound waves can originate from a vibrating tuning fork or from a vibrating guitar string. We begin our study of waves by referring to one easily visualized type: a wave on a string.

A disturbance can be sent down a string as shown in Fig. 14.12. The disturbance, or **pulse,** is initiated by a sudden up-and-down motion of the hand holding the string and travels with speed v along the string. Note two very important features of such a pulse. First, it carries energy down the string. When the pulse strikes a given point on the string, it causes that portion of the string to momentarily acquire both kinetic and potential energy. This energy was given to the pulse by the source that initiated it. The energy moves with speed v down the string along with the pulse.

Second, the pulse is a record of what the source has done. We can see in Fig. 14.12 that the hand moved to initiate the pulse at a definite time in the past. Indeed,

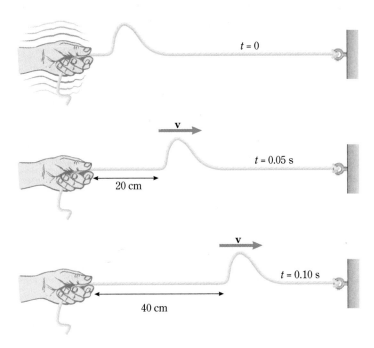

FIGURE 14.12

A pulse carries energy down the string. What is the speed of the pulse?

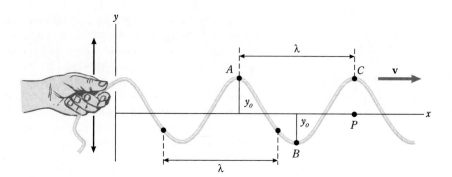

FIGURE 14.13

A source that vibrates with simple harmonic motion sends a sinusoidal wave down the string.

what the source was doing at any past time t is shown by the string at a distance $x = vt$ from the source. In other words, the string at a distance x from the source is performing the same motion that the source initiated at a time $t = x/v$ previously.

Let us now see what happens when the source vibrates with simple harmonic motion. The situation is shown in Fig. 14.13. As we expect, the string shows the past history of the way its end was vibrated. The up-and-down motion of the string's end is transmitted down the string with speed v, which we call the **wave speed.** As a result, the string has a sinusoidal form at any given instant, and this sinusoidal pattern on the string travels to the right with speed v. As it moves, the pattern carries energy down the string, energy furnished by the source of vibration.

There are certain words we use to describe such a wave. The points A and C, the tops of the wave, are called **wave crests.** Points such as B are called **wave troughs.** We call the maximum displacement of the string from its equilibrium position the **amplitude** of the wave. The amplitude of the wave in Fig. 14.13 is y_o. Notice that it is only half the total vertical displacement of the string.

The distance between two adjacent crests on the wave—between A and C, for example—is equal to a distance called the **wavelength** of the wave, indicated by λ (Greek lower-case lambda) in Fig. 14.13. The wavelength of a wave is the distance between any two adjacent points along the wave which have the same phase. One wavelength of a wave is sent out by the wave source as it executes one complete vibration.

At a fixed point such as P, the string repeatedly moves up and down as the wave moves to the right. During the time it takes the source to send out one wavelength, one wavelength must pass through P to make way for a new one. As a result, P undergoes one complete cycle of motion in the same time the source takes to undergo one complete vibration. Thus the period of the vibrating source is the same as the period of vibration of a point in the path of the wave. This time taken for a complete vibration of a point in the path of the wave is called the period of the wave and is represented by τ. As for an oscillator, the frequency of the wave is related to its period through $f = 1/\tau$. Furthermore, the frequency is equal to the number of wave crests passing through point P each second.

A very important relation exists between wavelength and frequency. Referring again to Fig. 14.13, note that a length λ of the wave is sent out during the time τ that the wave source takes to undergo one complete vibration. Therefore, the wave moves a distance λ in a time τ, and so we find, from $v = x/t$, that $v = \lambda/\tau$, where v is the speed of the wave. Thus

$$\lambda = v\tau \quad \text{and} \quad \lambda = \frac{v}{f} \tag{14.17}$$

This relation is true for all waves, not just for waves on a string. Frequency is determined physically by the frequency of the source of the wave. The wave speed is determined by the properties of the medium through which the wave passes. The wavelength is then v/f by definition.

The speed of a wave on a string is given by a relation that we state without derivation. If the tension in the string is T and if the mass of a length L is m, then the speed of the wave along the string is

$$v = \sqrt{\frac{T}{m/L}} \tag{14.18}$$

We can justify that speed should depend on tension and mass per unit length. The tension in the string is responsible for the force that accelerates a piece of the string as the pulse passes through the region. The greater the tension, the greater the acceleration, and so the motion of the pulse is swift if the tension is high. On the other hand, the more massive the string, the more inertia it has. The mass per unit length therefore affects the speed with which the pulse moves. A massive string has large inertia, and the speed of a pulse on it will be relatively low.

Illustration 14.2

A certain guitar string has a mass of 2.0 g and a length of 60 cm. What must the tension in the string be if the speed of a wave on it is to be 300 m/s?

Reasoning From Eq. 14.18, $T = (m/L)(v^2)$, and since $v = 300$ m/s, $m = 0.0020$ kg, and $L = 0.60$ m, $T = 300$ N. Notice that this tension is quite large: it is equivalent to the weight of a mass of about 30 kg pulling on the string. ∎

14.9   REFLECTION OF A WAVE

In order for a wave to travel along it, the string in Fig. 14.13 must be held taut at its right-hand end. A wave traveling to the right cannot continue beyond the support. We must be able to account for what happens to the energy carried by the wave, since energy cannot just disappear. Two things may happen: (1) some of the energy may be absorbed by the material of the support, and (2) some of the energy may be reflected back to the left along the string. For the purpose of this discussion, let us assume that all of the energy is reflected, which may be approximately correct in many cases.

To study this effect, let us consider only a single wave pulse propagating along the string, as shown in Fig. 14.14a. When this pulse reaches the support, it exerts an upward force on the support. Since the support is fixed in place, it cannot move. However, it exerts an equal and opposite force *downward* on the string. This force accelerates the string downward to the extent that its momentum carries it below the equilibrium line. The result is that the pulse is turned upside down as it hits the support, and the reflected pulse appears as shown in Fig. 14.14b. If the string had been completely free to move up and down at its right end, the pulse would not have been inverted, although it still would have been reflected, since the energy

v (pulse)

(a)

v (pulse)

(b)

FIGURE 14.14

A pulse on a string is inverted when it is reflected from a fixed end. The vertical arrows show motion of individual portions of the string.

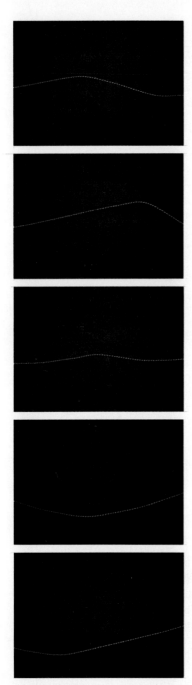

In this sequence, a wave pulse travels to the right down a rope and reflects from a fixed end. Notice that the reflected pulse, moving to the left, is inverted relative to the incident pulse.

could not just disappear at the end of the string. In summary, *a pulse is inverted by reflection at a fixed end, and is reflected but not inverted at a free end.*

Next, let us consider what happens when a reflected pulse traveling to the left along the string meets a pulse moving to the right. Suppose two rectangular-shaped pulses* are going in opposite directions on the same string, as in Fig. 14.15*a*. After they begin to overlap, the situation is that shown in Fig. 14.15*b*. The position each would have if the other pulse were not there is shown by dashed lines, and the actual displacement of the string is shown by the green line. The net displacement is the vector sum of the individual wave displacements. This is an example of the **principle of superposition:**

A point subjected to two or more wave pulses simultaneously is displaced an amount equal to the vector sum of the individual displacements.

All waves we deal with in this text conform to this principle.

We can now apply this principle to see what happens when a sinusoidal wave traveling down a string is reflected by a rigid support at its end. The situation is shown in Fig. 14.16. When an incident pulse in (*a*) reaches the support, the pulse is inverted and reflected. Part (*b*) shows the hypothetical incident and reflected waves. We term them "hypothetical" because the string itself obeys neither wave. Instead, it sums the two waves and takes the form shown in (*c*) at the instant the incident and reflected waves are positioned as in (*b*). Notice that the displacement of the string at the support is zero, as it must always be. Moreover, the displacement at this instant is zero at several other points as well.

Now comes the really interesting part. Suppose we redraw Fig. 14.16*b* for any other instant. Even though the incident and reflected waves are at different positions in the new sketch, their sum is still zero at the same points as in (*c*). In other words, the string never moves at all at the positions labeled *N*. If you were to watch the string, it would appear blurred as it oscillates back and forth between the limits shown in (*d*). The points *N* along the string, which never move, are called **nodes.** The points *A*, which are midway between the nodes and move most, are called **antinodes.** This type of vibration, in which the string vibrates back and forth within a well-defined envelope (or limiting curve), is called a **standing wave.** We shall have more to say about standing waves in a moment.

If you look at the instantaneous position of the string in (*d*), you can see that the nodes are half a wavelength apart. Similarly, the distance between adjacent antinodes is $\frac{1}{2}\lambda$. This is a convenient fact to remember:

The distance between adjacent nodes or adjacent antinodes in a standing wave is $\frac{1}{2}\lambda$.

*You well might ask how one could possibly make such a pulse shape on a string. Arbitrarily shaped pulses, including rectangular ones, can be made by using a combination of many different wave frequencies simultaneously. This is most often done with electronic circuits to shape pulses of voltage. Here, we hypothesize pulses of rectangular shape because they have constant amplitude, and, as the pulses overlap, the addition of these amplitudes is simpler to illustrate than for other shapes.

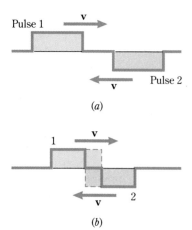

FIGURE 14.15

The principle of superposition. The green line shows the actual shape of the string as the blue and red pulses travel along it in opposite directions. In (*a*), before the pulses overlap, the string has the shape of the separate pulses. (*b*) In the region of overlap, the amplitudes of the two pulses add algebraically, giving zero displacement of the string in this case.

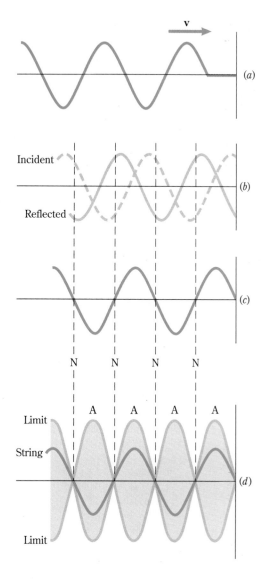

FIGURE 14.16

Standing waves, or resonances, occur on a vibrating string when incident and reflected waves precisely reinforce one another. The combined incident and reflected waves give rise to nodes and antinodes on the string. (The amplitudes of the waves in (*a*) through (*d*) are much exaggerated.) (*e*) Photo of (*d*) on string.

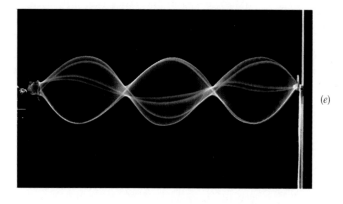

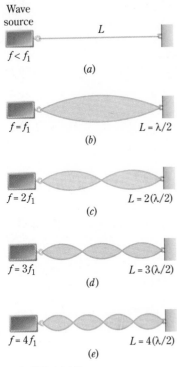

Wave source

$f < f_1$

(a)

$f = f_1$ $L = \lambda/2$

(b)

$f = 2f_1$ $L = 2(\lambda/2)$

(c)

$f = 3f_1$ $L = 3(\lambda/2)$

(d)

$f = 4f_1$ $L = 4(\lambda/2)$

(e)

FIGURE 14.17

Resonances of a taut string.

14.10 WAVE RESONANCE: STANDING WAVES ON A STRING

When a pendulum, a child on a swing, or a mass at the end of a spring is set into motion by a periodic force, the system moves with largest amplitude when the frequency of the force equals the natural vibration frequency of the free system. In Sec. 14.7, we used the example of pushing a child on a swing to show that a system undergoes resonance, that is, it vibrates most strongly when the frequency of the driving force is equal to the frequency of the system's free vibration. A similar situation exists for vibration of a string, as shown in Fig. 14.17. If you vibrate it with too low a frequency, the string vibrates so little that it appears motionless, as in Fig. 14.17a. If you slowly increase the frequency of vibration, the string begins to vibrate widely at a certain frequency, as in Fig. 14.17b. At this fundamental resonance frequency, f_1, the string vibrates widely and appears as a blur between the limits shown. This is an impressive example of the standing-wave phenomenon mentioned earlier. Experiment shows that the string also resonates to other higher frequencies, as shown in (c), (d), and (e) of the figure. In summary, the string resonates to the fundamental frequency f_1 and to higher frequencies $2f_1, 3f_1, 4f_1$, and so on.

There is an easy way to specify the conditions under which resonance occurs. Looking at Fig. 14.17, we see that the string always resonates in whole segments, where a *segment* is the distance between adjacent nodes or antinodes. The fixed ends are always at nodes. Therefore the string will resonate only if it is one segment long, two segments long, and so on. Since the length of a segment is $\frac{1}{2}\lambda$, however, the string can resonate only if it is $\lambda/2$ long, or $2(\lambda/2)$ long, or $3(\lambda/2)$ long, and so on. Indeed, we can state in general that a string fastened firmly at its two ends resonates only if it is a whole number of half wavelengths long. For example, in Fig. 14.17b–e, the length of the string is equal to $\lambda/2, 2(\lambda/2), 3(\lambda/2)$, and $4(\lambda/2)$. In general, then, for resonance of a string fastened at both ends,

$$L = n\frac{\lambda_n}{2} \qquad \text{where } n = 1, 2, 3, \ldots \tag{14.19}$$

and where λ_n is the wavelength when the string resonates in n segments. Since wavelength is related to frequency by Eq. 14.17, we see at once that a string of fixed length resonates only to certain very special frequencies. We say that the resonant frequencies f_n of the string are *quantized*, meaning that they occur at discrete values, separated by frequency gaps. These discrete values of resonance frequency are integer multiples of the fundamental resonance frequency f_1:

$$f_n = \frac{v}{\lambda_n} = \frac{v}{2L/n} = n\frac{v}{2L} = nf_1$$

Often the resonance frequencies of a taut string are connected with the production of music by string instruments. Although we will defer a detailed discussion of sound until the next chapter, this connection gives rise to additional vocabulary used in describing standing waves. The fundamental frequency f_1 is sometimes referred to as the **first harmonic,** with f_2, f_3, f_4, and f_n as the second, third, fourth, and nth harmonics. Thus the term *harmonic* in general refers to a vibration of a *single sinusoidal wave frequency,* and the term *simple harmonic motion* refers to periodic motion that can be described by a sine or cosine function with a single frequency.

Example 14.5

The speed of a wave on a particular string is 24 m/s. If the string is 6.0 m long, to what driving frequencies will it resonate? Draw a picture of the string for the first three harmonics.

Reasoning

Question What is the condition for resonance waves on the string?
Answer The length of the string must be an integral number of half wavelengths. In equation form,

$$L = n\left(\frac{\lambda_n}{2}\right) \quad \text{or} \quad \lambda_n = \frac{2L}{n}$$

Question How are these resonance wavelengths related to the resonance frequencies?
Answer The frequencies and wavelengths of all waves are related by $v = f\lambda$. In this case,

$$f_n = \frac{v}{\lambda_n}$$

Question What will the first three resonance waves look like?
Answer Resonance wave is another term for *standing* wave. For the first three standing waves, the string will have a wave pattern of one, two and three loops between its fixed ends. This is shown in Fig. 14.17*b*, *c*, and *d*.

Solution and Discussion Using the data given, we find for the first three wavelengths:

$$\lambda_1 = \frac{12\,m}{1} = 12\,m$$

$$\lambda_2 = \frac{12\,m}{2} = 6.0\,m$$

$$\lambda_3 = \frac{12\,m}{3} = 4.0\,m$$

The corresponding frequencies are

$$f_1 = \frac{24\,\text{m/s}}{12\,\text{m}} = 2.0\,\text{Hz}$$

$$f_2 = \frac{24\,\text{m/s}}{6.0\,\text{m}} = 4.0\,\text{Hz} \quad \text{and} \quad f_3 = \frac{24\,\text{m/s}}{4.0\,\text{m}} = 6.0\,\text{Hz}$$

Exercise If the string resonates in three segments to $f = 11\,\text{Hz}$, what is the speed of the waves? *Answer: 44 m/s.*

14.11   TRANSVERSE AND LONGITUDINAL WAVES

We have spent a great deal of time discussing waves on a string because we can easily see the wave shape of the string, and the principles that apply to them apply to many other vibrating systems. Waves on a string are an example of **transverse waves**. They are given this name because the particles of the string—or, in general, of the *medium* through which the wave passes—move perpendicular (or *transverse*) to the direction of wave propagation. For example, the wave on a string propagates from left to right while the string moves up and down.

Another type of wave can be produced by the following experiment. Place a long spring on a smooth tabletop for support and clamp it at one end. In Fig. 14.18a, the spring is shown at equilibrium. If it is suddenly compressed as in part (b), the loops near the end at which the compression force is applied are compressed before the rest of the spring experiences the disturbance. The compressed loops then exert a force on the loops to the right of them, and the compression travels down the spring. At the fixed end, the compressional energy is reflected; thus the compression is reversed and ends up traveling to the left, as shown in (d).

This type of wave is not a transverse wave since the particles of the spring vibrate back and forth in the same direction as that in which the wave is propagated, along the spring. A compressional wave such as this, where the motion of the particles is along the direction of wave propagation, is called a **longitudinal wave.**

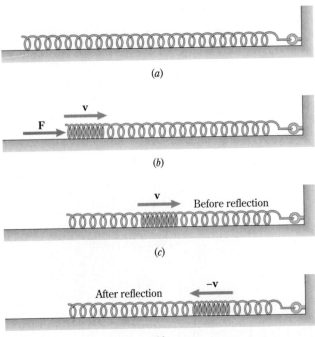

FIGURE 14.18

A longitudinal pulse travels down the spring and is reflected at the fixed end.

Physicists at Work VICTOR A. STANIONIS *Iona College*

COMPUTER MUSIC: The Science and Technology of a New Art

What is MIDI? Wake up! Switch on the music! Listen to it! Is it real or MIDI music? In all probability the sounds of the acoustical string, wind, and percussion instruments you heard were electronically synthesized, orchestrated, and controlled by a computer, and played through an audio reproduction system. It is computer music.

Electronically produced whining and howling sounds have been experimented with since the turn of the century as a possible source of generating music. They were rarely used outside of university and industrial laboratories. It wasn't until the early 1960s, with the success of the recording "Switched-on-Bach," that the general public became familiar with synthesized sound. Since then, the musical world has been transformed and catapulted into the electronic age and the world of computers.

The sound in "Switched-on-Bach" was simple and "thin," and relied on sound generated by electronic oscillators in an analog synthesizer called the MOOG. The complex sounds produced by traditional instruments with all their accompanying harmonics were lacking. Musicians tried to solve the problem by hooking up several synthesizers to be played at once but were faced with incompatibility problems that made such arrangements impossible.

In the early 1980s, physicists and engineers conjured up a means of synthesizing the acoustical instruments of the past and generating new instrumental sounds by releasing the genie of digital computer technology. The howling of the analog synthesizer was replaced by musicians using the MIDI music system and the computer. It is as if you had a band or orchestra locked in a computer that would play music at your bidding, exactly the way you wanted it played. In digital synthesis, complex voltages that are fed to the speakers are generated from mathematical equations built into the synthesizer. Every digital synthesizer has at least one microprocessor.

MIDI is an acronym for musical instrument digital interface. It is a hardware and software standard established by the makers of electronic instruments to provide for compatibility among instruments of different manufacturers. Each synthesizer uses a different method to generate sound and to imitate different instrumental sounds. Some are better replicating a piano, others at imitating a guitar. The need to obtain the best sounds from each led to a means of connecting them in a compatible fashion and to the establishment of the MIDI standard.

The MIDI standard specifies such things as the format of information transmitted through it and also the type of physical connection—the MIDI port—that is built into the instrument and the accompanying connections. It specifies voltage and transmission rates of signals. Instruments that can receive and send MIDI codes are called MIDI devices. The MIDI can transmit messages indicating when a key on a keyboard is first pressed, when it is released, and the nuances of the note played. Most computers do not come with a built-in MIDI port, but the serial port is easily converted by connecting a MIDI interface with a single plug.

The MIDI computer music system—made up of a computer, MIDI interface, synthesizer, and audio system—as shown in the figure, puts sophisticated instruments and the potential of generating a rock concert within the price range and hands of the casual consumer. With the appropriate software and hardware, the personal computer is transformed into a state-of-the-art music center.

As a college professor, I am always seeking new ways to excite and motivate my students and enhance my teaching. The advent of the computer age and the magical attraction of music has given me a vehicle to teach physics, especially to nonscience majors, using a stimulating and fascinating medium, computer music.

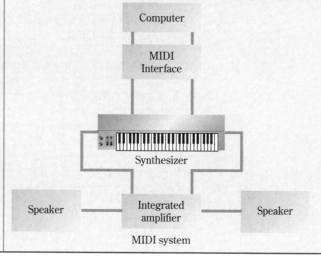

MIDI system

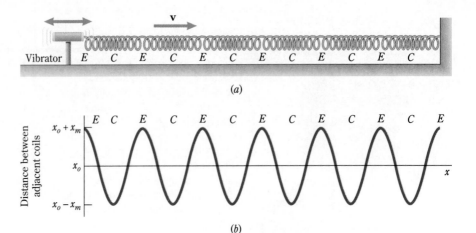

FIGURE 14.19

(*a*) A longitudinal wave produced by alternating compressions and extensions of a spring. (*b*) A graph of the compressions (*C*) and extensions (*E*) of the spring in 14.19*a*. The value x_o represents the separation between coils when the spring is undisturbed.

We can create a continuous longitudinal wave if we connect the free end of the spring to a vibrating source that alternately pushes and pulls the end with a frequency f. Regions of closely spaced coils are sent down the spring alternating with regions of stretched-out coils, as shown in Fig. 14.19*a*. Suppose the vibrator causes the end of the spring to move in simple harmonic motion; in this case the distance between adjacent coils along the spring is represented by the graph shown in Fig. 14.19*b*. *Notice that the variation in extension and compression of the coils follows a sinusoidal curve.*

Furthermore, this wave pattern of compression and extension travels along the spring at a certain speed determined by the properties of the spring. With the help of Fig. 14.19*b*, we can describe the longitudinal wave in the same terms we used for transverse waves. The wavelength is the distance between any two successive compressions or any two successive extensions. The amplitude is the difference between the adjacent-coil distance at maximum compression (or maximum extension) and the adjacent-coil distance of the undisturbed spring. The same relation holds between wave speed v, frequency f, and wavelength λ: $v = f\lambda$.

One of the most important examples of longitudinal waves is sound waves, which we will study in the next chapter.

14.12   STANDING COMPRESSIONAL WAVES ON A SPRING

A longitudinal wave on a spring has many features in common with a transverse wave on a string. If a longitudinal wave is sent down a spring, the wave and its energy are usually reflected at the end of the spring. This reflected wave can interfere with later waves being sent down the spring from the source. If the proper relation is maintained between the frequency of the source and the various parameters of the spring, resonance occurs. It is this feature of the spring system that we now investigate.

As with resonance on a string, the driving source in a spring system is usually close to a node (a point of zero motion) since at resonance the spring moves much more than the source. Also, if the other end of the spring is held motionless, it too

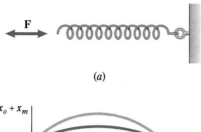

(a)

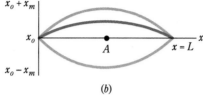

(b)

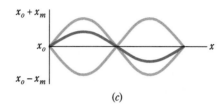

(c)

(d)

FIGURE 14.20

Standing waves result from longitudinal vibration of a spring. These graphs depict the distances between adjacent coils as in Fig. 14.19. This is not the shape of the spring.

must be at a node. We represent the resonance motion of a spring by the diagrams in Fig. 14.20. Remember that these graphs do not show the actual form of the longitudinal wave on the spring. (In contrast, such graphs *do* show the actual waveform for transverse waves.) Instead, they show the displacement of each point on the spring from its equilibrium position. These displacements along the x axis vary sinusoidally with x. Nodes occur at points where waves traveling to the right and to the left always cancel, leaving the spring neither compressed nor extended. Standing wave conditions occur when the length of the spring is an integral multiple of the distance between successive nodes. Thus the condition for resonance of longitudinal waves on a spring fixed at both ends is the same as for transverse waves:

$$n\frac{\lambda_n}{2} = L \qquad \text{where } n = 1, 2, \ldots$$

This relation, when combined with the one for wavelength and frequency, $\lambda = v/f$, tells us at once that the spring resonance frequencies (the harmonic frequencies) are

$$f_n = n\frac{v}{2L} \qquad \text{where } n = 1, 2, \ldots$$

R

Example 14.6

A spring 300 cm long resonates in three segments (nodes at both ends) when the driving frequency is 20.0 Hz. What is the speed of the wave in the spring?

Reasoning

Question How can I derive the wave speed from the description of the wave?
Answer As for all other waves, $v = f\lambda$. You also know that the frequency of the wave is the same as the driving frequency.

Question What is the wavelength of the described wave?
Answer Three segments means that three half wavelengths equal the length of the spring.

Solution and Discussion The wavelength is obtained from $L = 3(\lambda/2)$. Thus

$$\lambda = \frac{2L}{3} = \frac{600 \text{ cm}}{3} = 200 \text{ cm} = 2.00 \text{ m}$$

Then, with $f = 20.0 \text{ Hz} = 20.0 \text{ s}^{-1}$,

$$v = f\lambda = (20.0 \text{ s}^{-1})(2.00 \text{ m}) = 40.0 \text{ m/s}$$

We could, of course, have obtained this result by simply substituting in the relation using $n = 3$. However, most physicists prefer not to memorize a different relation for each case. They ordinarily use the number of half wavelengths on the total spring to find λ and then the relation $f = v/\lambda$ to find the unknown. As a matter of fact, almost all the resonance situations we encounter can be described by the use of this relation and an examination of the resonant system. It is not necessary to memorize an equation for each case.

Exercise What is the wave speed when the wave vibrates at the same frequency in five segments? *Answer: 24.0 m/s.*

LEARNING GOALS

Now that you have finished this chapter, you should be able to

1 Define or explain (a) amplitude, period, and frequency of vibration, (b) hertz, (c) spring constant, (d) simple harmonic motion, (e) sinusoidal motion, (f) damping, (g) resonance, (h) sinusoidal wave, (i) wavelength, (j) wave crest and wave trough, (k) amplitude, period, and frequency of a wave, (l) node and antinode, (m) standing wave, (n) wave resonance, (o) relationship between segment length and λ, (p) transverse wave, (q) longitudinal wave, (r) harmonic.

2 Use energy considerations to find the speed of a simple harmonic motion oscillator at any position in its path. State where the speed is greatest and where it is least.

3 Use Newton's second law to find the acceleration of a simple

harmonic motion oscillator at any point in its path. State where the acceleration is largest and where it is smallest.

4 Explain how one can ascertain whether or not a motion is simple harmonic and how the test method is related to Hooke's law.

5 Explain how motion in a reference circle provides a description of simple harmonic motion.

6 Given sufficient data, find the natural frequency of vibration of (a) a spring-mass system and (b) a pendulum.

7 Explain why simple harmonic motion is called sinusoidal motion. Give the equation for a sinusoidal motion and explain the quantities in it.

8 Point out what causes the restoring force in the case of a simple pendulum and explain why the motion is only ap-

proximately simple harmonic. Give the equation for the period of the motion.

9 Sketch several standing-wave forms for a string fixed at both ends. Given the length of the string and either f or v, use the standing-wave pattern to compute v or f.

10 Draw the longitudinal standing-wave resonance form for a spring fixed at both ends.

SUMMARY

DERIVED UNITS AND PHYSICAL CONSTANTS

Unit of Frequency
1 hertz (Hz) = 1 cycle/second = $1\,\text{s}^{-1}$

DEFINITIONS AND BASIC PRINCIPLES

Frequency (f)
Frequency (f) is the number of vibration cycles completed per unit of time. If time is in seconds, the unit of f is Hz.

Period (τ)
Period (τ) is the time a vibrating system takes to complete one cycle. It is the reciprocal of the frequency: $\tau = 1/f$.

Amplitude of Periodic Motion
Amplitude is the maximum displacement of a system from its equilibrium position.

Simple Harmonic Motion (SHM)
Simple harmonic motion results when a system moves in response to a restoring force that is linearly proportional to the amount of displacement of the system from equilibrium: $\mathbf{F} = -k\mathbf{x}$.

Frequency of Vibration in SHM
The frequency of vibration of SHM is

$$f = \frac{1}{2\pi}\sqrt{\frac{k}{m}}$$

where k is the force constant tending to restore the system to equilibrium and m is the mass of the vibrating object.

The Mathematical Form of SHM: Sinusoidal Motion
The expression for the time dependence of the position of an object in SHM is

$$x = x_o \cos(\omega t) = x_o \cos(2\pi ft) = x_o \cos\left(\frac{2\pi t}{\tau}\right)$$

where x_o = amplitude
f = frequency (Hz)
ω = angular frequency (rad/s)
τ = period (s)

The expressions for velocity and acceleration are

$$v = -(2\pi f x_o)\sin(2\pi ft)$$
$$a = -(2\pi f)^2 x_o \cos(2\pi ft)$$

INSIGHTS
1. Notice that $2\pi f = \sqrt{\dfrac{k}{m}}$, where k is the force constant of the system and m is the vibrating mass.
2. At any particular time t, the quantity $\omega t = 2\pi ft = 2\pi t/\tau$ is called the phase of the motion. It tells which part of the cycle the system is in at that instant. The phase is measured in radians. Each cycle is composed of 2π rad.
3. The above expressions are for a system released from its amplitude position at $t = 0$.

The Simple Pendulum
For small angles of swing, a simple pendulum executes SHM with a frequency given by

$$f = \frac{1}{2\pi}\sqrt{\frac{g}{L}}$$

INSIGHT
1. For angles under $\approx 10°$, this expression is obeyed to at least three significant digits of precision.

Wave Terminology
Wave speed v is the speed with which a wave pulse is transmitted through the medium which carries the wave. Wavelength (λ) is the distance between any two adjacent points along the wave which have the same phase.

The following relation holds for all waves:

$$v = f\lambda \qquad \text{where } f = \text{frequency of vibration}$$

INSIGHTS
1. The properties of the medium determine wave speed, and the frequency is determined by the source frequency. The above relationship then determines what the wavelength must be.
2. For waves on a string, the wave speed is

$$v = \sqrt{\frac{\text{tension}}{\text{mass per length}}} \qquad (14.18)$$

Reflections of Waves
A wave reflected from a fixed end of the medium carrying the wave is inverted relative to the original wave. A wave reflected from a free end of the medium is reflected upright.

INSIGHT
1. An inverted wave is the same as one that has had its phase changed by one half cycle (π rad) relative to the original.

The Principle of Superposition
A point subjected to two or more waves simultaneously is displaced an amount equal to the vector sum of the individual disturbances.

Standing Waves on a String
For a string fixed at both ends, standing waves (resonances) occur when the length of the string is an integral number of half wavelengths:

$$L = n\frac{\lambda_n}{2}$$

INSIGHTS
1. Since wave speed is the same for all frequencies for a given set of conditions of the medium, the resonance wavelengths have frequencies given by

$$f_n = v/\lambda_n = n\frac{v}{2L}$$

2. Resonance frequencies are a macroscopic example of the quantization of a physical quantity. That is, f cannot take on all values; instead, it can take on only a restricted set, given by whole-number multiples of a fundamental amount. In this case, the fundamental frequency is $f_1 = v/2L$.

Transverse and Longitudinal Waves
Transverse waves are those in which the movement of the medium is perpendicular to the direction of the propagation of the wave.

Longitudinal waves are those in which the movement of the medium is the same as the direction of the propagation of the wave.

QUESTIONS AND GUESSTIMATES

1 On the same horizontal axis, sketch graph lines that relate the horizontal position of a pendulum bob to (*a*) its kinetic energy, (*b*) its potential energy, and (*c*) its total energy.

2 On the same horizontal axis, sketch graph lines that relate the position of the mass in a spring-mass system to (*a*) its velocity and (*b*) its acceleration.

3 Two equal-weight masses hang together at the end of the same spring, and the system is set vibrating. What happens to the amplitude, frequency, and maximum speed of the end of the spring if one of the masses falls off when (*a*) the spring is at its largest extension and (*b*) the mass is passing through the equilibrium position?

4 A precocious student says that she can predict the frequency of a spring-mass system even though she knows neither the spring constant nor the mass. All she needs to know is how far the spring stretches when the mass is hung from it. Should you bet money that she cannot do this?

5 How does the period of a pendulum change if the pendulum is in an accelerating elevator? Consider both upward and downward acceleration.

6 How could you compute the up-and-down resonance frequency of a car by using data on how much the car lowers as the load on it is increased? Estimate this frequency for an automobile. When might it be important?

7 Sometimes an automatic washer vibrates strongly during the spin-dry cycle. Why? Is an unbalanced load the whole story? What should a designer do to minimize this problem?

8 The value of g on the moon is one-sixth that on earth. How would the frequency of vibration of each of the following change as each is moved from earth to the moon: (*a*) a horizontal spring-mass system; (*b*) a vertical spring-mass system; (*c*) a simple pendulum? How would each behave in a spaceship orbiting the earth?

9 The two idealized pulses in Fig. P14.1 are moving down the string at 20 m/s. Sketch how the string will look 0.40 s later. Repeat for 0.20 s later.

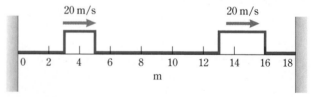

FIGURE P14.1

10 Is it possible for two identical waves traveling in the same direction down a string to give rise to a standing wave?

11 If you watch people trying to carry a large pan full of water, you will see that some are quite successful at it, but for others, who are equally careful, the water sloshes badly. What makes the difference?

PROBLEMS

Sections 14.1 and 14.2

1 A mass hanging at the end of a spring is 45 cm above the floor when it is at equilibrium. When the mass is pulled down 9.6 cm and released, it reaches its lowest point 19 more times in the first 97.3 s after release. What are the (a) frequency, (b) period, and (c) amplitude of the motion?

2 A pendulum is drawn aside by a small angle from the vertical position and released. The end of the pendulum swings between points 8.75 cm apart. It takes 268 s for the pendulum to reach its starting point for the sixtieth time after being released. What are the (a) frequency, (b) period, and (c) amplitude of the motion?

3 A certain spring that obeys Hooke's law stretches 42 cm when a load of 0.28 N is added to it. How much potential energy is stored in the spring when it is compressed by 3.35 cm?

4 The spring on a popgun obeys Hooke's law and requires a force of 300 N to compress it 12.5 cm to its cocked position. How much potential energy does the spring possess in its cocked position?

5 A 250-g mass is attached to the end of a spring whose spring constant is $k = 120$ N/m. The spring is stretched 5.0 cm from its equilibrium position and released. Find (a) the speed of the mass as it passes through the equilibrium position and (b) its acceleration just as it is released.

6 A spring-mass system slides on a horizontal frictionless surface. The mass is 75 g. A horizontal force of 0.66 N applied to the spring stretches it 7.8 cm. Find (a) the speed of the mass as it passes through the equilibrium position and (b) its acceleration just as it is released.

7 If the force constant for a spring in a popgun is 1650 N/m and the spring is compressed by 9.0 cm when cocked, what maximum speed can a 22-g pellet have as it leaves the gun?

Sections 14.3 and 14.4

8 A 3.5-kg mass oscillates in simple harmonic motion at the end of a spring. The amplitude of the motion is 40 cm, and the spring constant is 150 N/m. Find the speed and acceleration of the mass when its displacement is (a) 40 cm, (b) 0 cm, and (c) 20 cm.

9 The 450-g mass in a spring-mass system has a maximum speed of 21 cm/s as it vibrates with an amplitude of 4.2 cm. Find (a) the spring constant, (b) the maximum acceleration of the mass, and (c) the speed and acceleration of the mass when it is 3.0 cm from its equilibrium position.

■10 A circle of radius 26 m is marked out on the center of a football field, and a girl runs around the perimeter at a constant speed of 3.75 m/s. At the same time a boy runs up and down the sideline so as to keep pace with her motion along that direction. Find (a) the frequency of the boy's motion, (b) the acceleration at the end points of his motion, and (c) his maximum speed.

■11 A satellite circles the earth at a speed of 3100 m/s in an orbit that passes over both poles and has a radius of 4.2×10^7 m. Consider a spot, moving through the center of the earth along the north-south axis, that keeps pace with the satellite's north-south component of motion. Find (a) the frequency of the spot's motion, (b) the acceleration of the spot at the end points of its motion, and (c) its maximum speed.

12 When a 160-g mass hangs at the end of a spring, the system vibrates in such a way that it completes 33 cycles in 80.5 s. What is the spring constant of the spring?

13 Two children notice that they can make a car vibrate up and down through 12 cycles in 19.5 s. (a) Assuming the car has a mass of 1450 kg, find the spring constant of its suspension system. (b) If the children have a total mass of 45 kg, how much should the level of the car rise when they get out of the car?

■14 A 0.85-kg block rests on a horizontal frictionless surface and is connected to two walls by two springs having spring constants k_1 and k_2. The situation is shown in Fig. P14.2. If $k_1 = 44$ N/m and $k_2 = 34$ N/m, with what frequency will the mass vibrate if pulled aside and released?

FIGURE P14.2

■15 A mass m is hung at the end of a length L of wire that has a cross-sectional area A and a Young's modulus Y. Show that the mass will vibrate up and down with a frequency $f = (1/2\pi)\sqrt{AY/Lm}$.

Section 14.5

■16 A mass vibrates back and forth on a spring, obeying the equation $x = 18 \sin(3.7t)$ cm. Find its (a) amplitude, (b) frequency, and (c) period of motion. (d) If the mass is 520 g, what is the spring constant?

■17 As it vibrates up and down at the end of a spring, a 165-g mass obeys the relation $y = 9.4 \sin(6.8t)$ cm. Find (a) the spring constant and the (b) amplitude, (c) frequency, and (d) period of the motion.

18 Write the mathematical description of the position of the mass in Prob. 5 as a function of time; i.e., write down the form of $x(t)$. Use SI units.

■19 A 0.88-kg mass attached to a spring is pulled 2.95 cm from equilibrium along the positive x-axis and released from rest. The spring has a spring constant $k = 40$ N/m. (a) Write down the time dependence of the position of the mass, $x(t)$, and the velocity, $v(t)$. (b) Find the values of x and v at the instants $t = 0.5$ s, 1.0 s, and 2.0 s. (c) Find the value of t when the mass reaches its starting point for the third time after being released. (*d*) Find the time when the mass first reaches the position $x = -1.50$ cm.

Section 14.6

20 How long must a pendulum be if its period is to be 2.0 s (*a*) on earth and (*b*) on the moon? Objects weigh one-sixth as much on the moon as on earth.

21 Two pendulums are to have frequencies such that $f_1 = 3f_2$. What should be the ratio L_1/L_2 of their lengths?

∎22 A pendulum is drawn aside to a certain angle and released. When the ball passes the low point of the arc, the tension in the string is twice the weight of the ball. Show that the original displacement angle is 60°.

23 At sea level at the North Pole, a 99.2-cm-long pendulum is timed for 1000 s and completes 499.0 oscillations. The same pendulum makes 500.5 complete swings in 1000 s when it is at sea level at the equator. Calculate the values of *g* at the pole and at the equator.

24 You find yourself on an alien planet and, among other things, you want to find out the strength of gravity on the planet. Fashioning a pendulum 1.0 m in length, you find that 100 swings take 178 s. On earth you weigh 635 N. What do you weigh on this planet?

∎25 A light spring has an unstretched length of 30.5 cm. You hang a 300-g mass on the spring and set the stretched spring into small-amplitude motion as a simple pendulum. The pendulum has a period of 1.45 s. Assuming a value of $g = 9.80$ m/s², find the spring constant of the spring.

Sections 14.8–14.10

26 The wave in Fig. P14.3 is moving to the right at 25 cm/s. Find its (*a*) wavelength, (*b*) amplitude, (*c*) frequency, and (*d*) period.

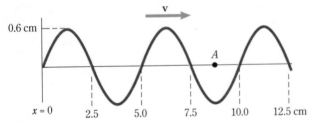

FIGURE P14.3

27 When the wave on a string passes through *A* in Fig. P14.3, the string vibrates as $y = y_o \sin(2\pi ft)$. What are y_o and *f* if the speed of the wave is 38 cm/s?

28 All radio waves travel through air with a speed of 3.0×10^8 m/s. What is the wavelength of a typical wave from a station that emits waves of frequency 1450 kHz?

29 Light waves travel through air with a speed of 3.0×10^8 m/s. Green light has a wavelength of about 520 nm. What is the frequency of these waves?

30 Refer to Fig. P14.1. Sketch the situation 2.2 s later.

31 In Fig. P14.1, how long does it take for the pulses to return to the same positions?

32 How large a mass must be hung on the end of a thread 175 cm long in order to make the speed of transverse waves along the thread 46.5 m/s? A 5-m length of the thread has a mass of 0.855 g.

33 A rope is stretched between two poles a distance of 34 m apart. The rope has a mass of 55 g per meter of length. A transverse pulse given to the middle of the rope takes 0.37 s to reach either pole. How much tension is there in the rope?

34 A standing wave pattern consisting of three loops is set up on a taut string by a 180-Hz vibrator. The length of the string is 2.20 m. What are (*a*) the wavelength and (*b*) the speed of the waves?

35 The string in Prob. 34 has a mass of 1.70 g/m. What tension in the string is required to produce the described wave pattern?

∎36 What tension is required to produce a four-loop pattern in the string of Probs. 34 and 35?

∎37 A wire stretched between poles 12.5 m apart is seen to vibrate in the wind with one node in the middle. (The ends are nodes as well, of course.) The sound produced by the vibrating wire has a frequency of 43 Hz. If the wire is known to have a mass density of 4.5 g/m, what tension must the wire have?

38 A certain string clamped at both ends resonates at a fundamental frequency of 256 Hz. What are the next three higher resonance frequencies?

39 A certain string resonates in three segments at a frequency of 145 Hz. Give four other resonance frequencies for this string.

∎40 A string has one of its resonances at 760 Hz and the next highest is at 950 Hz. What is the string's fundamental resonance frequency?

∎41 A violinist changes the pitch produced from a string by moving her finger along the string, thus changing the position of one of the end nodes of the string. (*a*) If the free string has a fundamental frequency of 440 Hz, what fundamental frequency is produced when she places her finger one-fifth of the way from the upper end of the string? (*b*) Where would she place her finger to produce a fundamental frequency of 1100 Hz?

42 A spring is stretched to a length of 3.60 m and set into longitudinal vibration by an oscillator at one end. When the driving frequency is 4.5 Hz, the spring vibrates in resonance with five nodes (including its ends) along its length. What is the speed of the longitudinal waves?

∎43 A small transverse vibrator is connected to one end of a horizontal string. The vibrator operates at a frequency of 120 Hz and moves with an amplitude small enough that the end of the string can be considered to be a node of standing wave patterns. The string, with a linear density of 0.65 g/m, passes over a pulley at a distance of 1.80 m from the vibrator. Various masses are hung on the end of the string. What masses are required to produce resonances of (*a*) four, (*b*) five, and (*c*) six loops?

General Problems

■**44** A piston undergoes vertical simple harmonic motion with amplitude 21.5 cm and frequency *f*. A washer sits freely on top of the piston. At low piston frequencies, the washer moves up and down with the piston. However, at very high frequencies, the washer momentarily floats above the piston as the piston starts its downward motion. (*a*) What is the maximum acceleration of the piston when the washer begins to separate from it? (*b*) What is the lowest frequency at which separation occurs?

■**45** A compressed spring with a mass attached to its end is immersed in a container of water at 19.500°C. The container, spring, mass, and water together are thermally equivalent to 95 g of water. After the spring is released, it vibrates back and forth with decreasing amplitude as a result of friction (viscous) forces imposed by the liquid. When the system stops vibrating, the temperature is 19.625°C. (*a*) How much energy was stored in the spring? (*b*) If the spring was compressed 5.8 cm, what is its spring constant?

■■**46** A nonviscous liquid is placed in an open-ended U tube of uniform bore, as shown in Fig. P14.4. The total distance from *A* to *B* through the tube is *L*. The liquid is set in oscillation by a person blowing momentarily on the end above *A*. Show that the motion is simple harmonic and has a frequency $(1/\pi)\sqrt{g/2L}$.

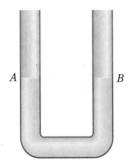

FIGURE P14.4

■■**47** Find the frequency with which the pendulum in Fig. P14.5 oscillates for small oscillations.

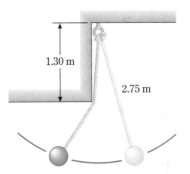

FIGURE P14.5

■■**48** Two wires of the same cross section are stretched between the same two supports. One wire is made of steel, the other of aluminum. The tension in the steel wire, T_1, is such that the wire resonates in its fundamental frequency of transverse vibrations. How much tension, in terms of T_1, would the aluminum wire require in order to resonate in (*a*) its fundamental and (*b*) its third harmonic?

■**49** A clock has a pendulum made of a small, heavy weight at the end of a steel rod whose mass can be ignored. At a temperature of 27°C the clock keeps accurate time and the pendulum has a period of 0.3333 s. During a heat wave the air conditioning fails and the room temperature rises to 38°C. Does the clock then run too fast or too slow? How much error will accumulate in 12 h at the higher temperature?

■■**50** A uniform, flat wooden raft whose specific gravity is 0.85 floats in fresh water. It sinks until its upper surface is just at water level when a 90-kg man stands on it. (*a*) Show that the additional buoyant force on the block obeys Hooke's law. (*b*) Find the spring constant of this system and the frequency of vertical vibration of the raft when the man jumps off. Assume that damping effects of viscosity can be ignored.

CHAPTER 15

SOUND

The concepts involving wave motion discussed in the preceding chapter are now applied to one particular form of wave motion, sound. A study of sound is important not only in its own right but also because it affords us a valuable means of consolidating our knowledge of wave motion. Many of the ideas discussed in connection with sound will also be important in our study of light and other types of wave motion.

15.1 THE ORIGIN OF SOUND

Sound waves are longitudinal waves that are transmitted through almost any substance—solid, liquid, or gas. The waves are created by any mechanism which produces compressional vibrations of the surrounding medium. Some examples are the vibrating string of a guitar, vibrating vocal chords, and exploding gas in a firecracker. Sound cannot travel through a vacuum because in a vacuum there is no material to transmit the compressions. A common demonstration of this is to show that a ringing bell cannot be heard if the bell is in a vacuum chamber. The bell is vibrating, but there is no surrounding material to carry the vibration to our ears.

We are most often concerned with sound waves in air since that is the basis of our sense of hearing. However, sound travels faster and with less energy loss in

432

liquids and solids than in air. This is why putting your ear to a steel railroad track allows you to hear the approach of a train long before you can hear it in the air. Although we usually refer to sound as those waves we can hear, sound can have frequencies far above and far below those to which our ears are sensitive. We shall discuss the human ear as a sound detector in later sections.

15.2   SOUND WAVES IN AIR

Let us now consider the action of a loudspeaker generating a simple sound. A simple loudspeaker consists of a cone-shaped sheet of flexible material, called a *diaphragm,* that can be oscillated back and forth by means of an applied force **F**, as in Fig. 15.1. (We will see how the force is obtained when we study magnetic forces in Chap. 19.)

When the diaphragm in Fig. 15.1 moves to the right, it compresses the air in front of it and a **compression** travels out through the air. An instant later, the diaphragm is moving to the left, leaving a region of decreased air pressure in its wake, called a **rarefaction.** This disturbance also travels out from the loudspeaker. Hence a series of pressure disturbances, the compressions and rarefactions, travel out from the loudspeaker. This is very similar to the compressional waves on a spring discussed in the preceding chapter.

The wave produced by a loudspeaker is sketched in Fig. 15.2, where compressions are shown as *A*, *B*, and *C* and rarefactions as *P*, *Q*, and *R*. A plot of the air pressure along this sound wave at a given instant is also given in Fig. 15.2. The pressure at the horizontal line in this graph is average atmospheric pressure. The compressions and rarefactions cause only slight variations in air pressure. Even for very loud sounds, the pressure variations are only about 0.01 percent of atmospheric pressure.

The sound waves sent out by a loudspeaker or any other sound source are not usually confined to a straight-line path. Instead, they spread out in all directions from the source. To better understand this feature of wave motion, refer to Fig. 15.3*a*, which shows a water wave traveling out from a source. A diagrammatic way of representing this situation is shown in Fig. 15.3*b*. As we see, the wave crests (called **wavefronts** in this context) become larger and larger circles as they move away from the source. At great distances from the source, the circles are so large that they have very little curvature. A wave crest viewed from a very distant

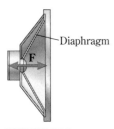

FIGURE 15.1

The flexible loudspeaker diaphragm vibrates back and forth to send compressions and rarefactions out into the air.

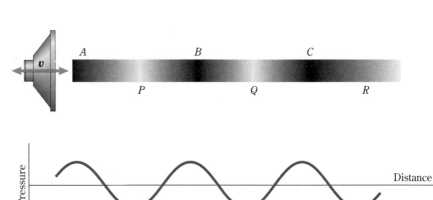

FIGURE 15.2

The sound wave sent out by a loudspeaker consists of alternate high- and low-pressure regions in the air. In practice, the pressure changes by only about 0.01 percent or less.

FIGURE 15.3

(*a*) A wave source sends circular waves across the surface of the water. (*b*) A diagram used to represent a situation such as (*a*). *(Education Development Center)*

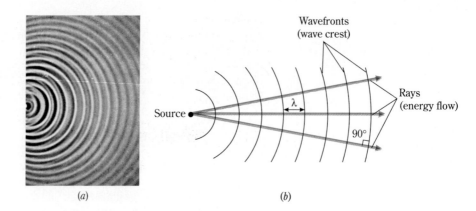

(*a*)

(*b*)

source appears to be a nearly straight line as it sweeps past a point on the water. Waves far from their source are thus called **plane waves,** a terminology carried over from three-dimensional waves, as we shall soon see.

The water waves in Fig. 15.3 carry energy away from the source. Because energy moves in the direction of wave propagation, the energy carried by the wave travels out along radial lines, such as those labeled **ray** in the figure. Notice that the *rays are by definition perpendicular to the wavefronts.* Because wavefronts far from a source form nearly straight lines and because rays are perpendicular to wavefronts, *rays are parallel to each other when they are far from a wave source, that is, in a plane wave.*

An analogous situation exists for sound waves in air. Because this is a three-dimensional situation, however, the wavefronts are **spherical** surfaces centered on the sound source rather than the circles in the two-dimensional case. These **spherical waves** have decreasing curvature as they move farther away from the source and appear essentially flat at a distant point; the waves are then essentially flat planes, and they, too, are called plane waves. As before, the rays are perpendicular to the wavefronts, and so the rays are parallel to each other in plane waves.

Notice one other feature of the waves in Fig. 15.3*a*: their amplitude, as indicated by the amount of contrast between crests and troughs, decreases with increasing distance from the source. This reflects the fact that the energy carried by the wave is spread out over an increasingly larger wavefront as the wave proceeds outward from the source. Hence a unit length of a wavefront contains less energy as the front moves away from the source. This feature does not occur with waves on strings, springs, or rods because there all the energy propagates along the same line, in only one dimension. With two-dimensional and three-dimensional waves, however, the rays *diverge* from the source. As they spread out, the original amount of energy is likewise spread over a greater line or area. Only in plane waves is this decrease in energy absent. The rays of plane waves are parallel, and so the energy is carried in only one direction and therefore does not diminish.

15.3　THE SPEED OF SOUND

In Sec. 14.8 we learned that the speed of transverse waves on a taut string is given by

$$v = \sqrt{\frac{T}{m/L}}$$

(14.18)

Airplanes allow us to travel through the air at high speed. The plane in (*a*) flies nearly as fast as sound waves; the one in (*b*) is designed to fly faster than sound.

(a)   *(b)*

This has the general form

$$v = \sqrt{\frac{\text{restoring force}}{\text{inertial factor}}}$$

We might expect that the speed of longitudinal waves through a medium follows a similar relation. This is indeed the case, where the restoring force for compressions is related to the elastic modulus of the medium and the inertial factor is the medium's density. For a one-dimensional medium, such as a wire or a track, the appropriate modulus is Young's modulus, Y. In the case of two- and three-dimensional media, we must use the bulk modulus, B. Thus we have the following expressions for the speed of sound waves:

$$v = \sqrt{\frac{Y}{\rho}} \qquad \text{(one-dimensional medium)} \tag{15.1}$$

$$v = \sqrt{\frac{B}{\rho}} \qquad \text{(two- or three-dimensional medium)} \tag{15.2}$$

Let us now apply Eq. 15.2 to the case of the speed of sound through a gas.

For an ideal gas, the value of B depends on the process by which the gas is compressed. For instance, B is equal to the gas pressure P if the compression is isothermal. However, compressions occur very suddenly when a sound wave passes through a small volume in the gas, so suddenly that no heat transfer has a chance to take place. Thus sound wave compressions are *adiabatic*. A little calculus applied to the ideal-gas law (Eq. 10.1) shows that $B = \gamma P$ for adiabatic compressions, where $\gamma = C_p/C_v$.

We have the result that the speed of sound in an ideal gas is

$$v = \sqrt{\frac{\gamma P}{\rho}} \tag{15.3}$$

The ideal-gas law, however, gives the gas pressure in terms of the gas temperature:

$$P = \frac{nRT}{V} = \frac{m}{V}\frac{RT}{M} = \rho\frac{RT}{M}$$

where m is the mass of the n moles in the gas sample and M is the atomic or molecular mass of the gas. Substitution of this expression for P into Eq. 15.3 gives

TABLE 15.1
Speed of sound*

Material	v (m/s)
Air†	331.45
Oxygen	316
Helium	965
Hydrogen	1284
Water	1402
Water (20°C)	1482
Water (50°C)	1543
Aluminum	5100
Copper	3560
Iron	5130

*Values are at 0°C unless noted otherwise.
†For air near room temperature,
$v = 331.45 + 0.61T$ m/s, where T is
the Celsius temperature.

$$v = \sqrt{\frac{\gamma RT}{M}} \tag{15.4}$$

It is interesting to note that the wave-speed dependence on P and ρ indicated in Eq. 15.3 has disappeared. Equation 15.4 shows that the only thermodynamic state variable that determines the speed of sound is the temperature of the gas.

Typical values for the speed of sound at 0° are given in Table 15.1. Notice the footnote in that table which states how v in air changes with T.

Illustration 15.1

Find the speed of sound in neon gas at 0°C.

Reasoning We can use Eq. 15.4 with $M = 20.18$ kg/kmol. For a monatomic gas, $\gamma = 1.66$ (Table 12.1). Then

$$v = \sqrt{\frac{\gamma RT}{M}} = \sqrt{\frac{(1.66)(8314 \text{ J/kmol} \cdot \text{K})(273 \text{ K})}{20.18 \text{ kg/mol}}} = 432 \text{ m/s}$$

Exercise Two ideal gases have the same molecular mass M, but gas A is monatomic and gas B is diatomic. Find the ratio v_A/v_B. *Answer: 1.09*

15.4 INTENSITY AND INTENSITY LEVEL

We saw in Chap. 14 that a source sending a wave down a string also sends energy with the wave. Indeed, all waves carry energy along with them. Sound waves are no exception. For example, the loudspeaker in Figs. 15.1 and 15.2 sends out sound-wave energy, which flows in the direction of wave propagation.

Suppose a sound wave is traveling in the propagation direction shown in Fig. 15.4. We define the intensity of the wave in terms of the energy it carries. To be precise, we consider a unit area perpendicular to the propagation direction, as shown. We then define the intensity I of the wave to be the energy the wave carries per second through this unit area. Since power is energy produced per unit of time,

Sound intensity I is the power passing through a unit area perpendicular to the direction of wave propagation.

$$I = \frac{\text{power}}{\text{area}}$$

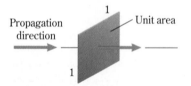

FIGURE 15.4

Sound intensity is the amount of energy flowing through a unit area per second. The area must be perpendicular to the direction of propagation, as shown.

The SI units of sound intensity are watts per square meter. Representative sound intensities are listed in Table 15.2. Notice what a wide range of intensities can be heard by the ear, a truly remarkable measuring device.

One of the many interesting properties of the ear is that its response to various levels of intensity is roughly proportional to the *logarithm* of I. That is, our perception of the relative *loudness* of two sounds is not I_2/I_1 but rather $\log(I_2/I_1)$. Thus a

TABLE 15.2 *Approximate sound intensities and intensity levels*

Type of sound	Intensity (W/m²)	Intensity level (dB)
Pain-producing	1	120
Jackhammer or riveter*	10^{-2}	100
Busy street traffic*	10^{-5}	70
Ordinary conversation*	10^{-6}	60
Average whisper*	10^{-10}	20
Rustle of leaves*	10^{-11}	10
Barely audible sound	10^{-12}	0

*For a person near the sound source.

convenient scale for expressing loudness (called **intensity level** or **sound level**) is the **decibel** scale, defined by

$$\text{Intensity level in decibels (dB)} = 10 \log \frac{I}{I_o} \tag{15.5}$$

where I is the intensity (in watts per square meter) of the sound under consideration and I_o is often, but not always, the intensity of the least audible sound, 10^{-12} W/m². Notice that the intensity level of the least audible sound is

$$10 \log \frac{I}{I_o} = 10 \log \frac{10^{-12}}{10^{-12}} = 10 \log 1 = 0 \, \text{dB}$$

while a pain-producing sound whose intensity is 1 W/m² has an intensity level of

$$10 \log \frac{I}{I_o} = 10 \log \frac{1}{10^{-12}} = 10 \log 10^{12} = 120 \, \text{dB}$$

This scale compresses the 12 orders of magnitude of audible intensity into a scale of 0 to 120 dB. Table 15.3 shows the dB values corresponding to various intensities. Table 15.2 also shows the dB value for various sources of sound we encounter.

TABLE 15.3
The decibel scale*

Intensity (W/m²)	Intensity level (dB)
10^{-12}	0
10^{-11}	10
10^{-10}	20
10^{-9}	30
.	.
.	.
.	.
10^{-1}	110
1	120
10	130

*1 B (bel) = 10 dB and is named after Alexander Graham Bell, the inventor of the telephone.

Illustration 15.2

Find the sound level in decibels of a sound wave that has an intensity of 10^{-5} W/m².

Reasoning From Eq. 15.5,

$$\text{Sound level (in dB)} = 10 \log \frac{I}{I_o} = 10 \log \frac{10^{-5}}{10^{-12}} = 10 \log 10^7 = (10)(7) = 70 \, \text{dB}$$

Exercise Find the sound level equivalent to an intensity of 4.0×10^{-8} W/m². *Answer: 46 dB.*

Example 15.1

Find the intensity of a sound that has an intensity level of 35.0 dB.

R | *Reasoning*

Question   What is this intensity level referred to?
Answer   Unless specified otherwise, assume the reference level is the least audible sound.

Qustion   What is the expression involving the unknown intensity?
Answer $35.0\,\text{dB} = 10 \log \left(\dfrac{I}{I_o}\right)$ where $I_o = 10^{-12}\,\text{W/m}^2$

Question   How do I extricate I from the log?
Answer   By taking the inverse log (antilog) of both sides of the equation. Divide by 10 first and then remember that antilog $(\log x) = x$.

Solution and Discussion   Dividing by 10, we have $3.50 = \log(I/I_o)$. Taking antilogs,

$$\text{antilog} (3.50) = 10^{3.50} = 3160$$

$$\text{antilog} \left[\log \left(\frac{I}{I_o}\right)\right] = \frac{I}{I_o}$$

Thus $\dfrac{I}{I_o} = 3160$, which gives

$$I = 3160 I_o = 3160\,(10^{-12}\,\text{W/m}^2) = 3.16 \times 10^{-9}\,\text{W/m}^2$$

15.5   INTENSITY FROM A POINT SOURCE: THE INVERSE-SQUARE LAW

In Sec. 15.2 we mentioned that the amplitude and hence the energy content of a wave in three dimensions generally decrease with distance from the wave source. Let us now derive an expression for this decrease when the waves are radiating equally in all directions from a point source. Realistically, a **point source** is one whose dimensions are very small compared with the distance at which we are measuring intensity.

Consider two imaginary spheres, of radii R_1 and R_2, centered on a source which is emitting P watts of power in sound waves. This is shown in Fig. 15.5. We assume that the source is radiating *isotropically,* that is, with equal intensity in all directions. On the surface of sphere 1, the power P is spread out uniformly over the surface, which has an area $A_1 = 4\pi R_1^2$. Thus at any place a distance R_1 from the source the intensity of the sound is

$$I_1 = \frac{P}{4\pi R_1^2}$$

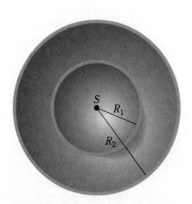

FIGURE 15.5

The power P of the source is spread over an area of $4\pi R_1^2$ at a distance R_1 and over an area $4\pi R_2^2$ at R_2. Intensity $= P/4\pi R^2$.

Similarly, at a distance R_2, the intensity is

$$I_2 = \frac{P}{4\pi R_2^2}$$

The ratio of intensities is

$$\frac{I_2}{I_1} = \left(\frac{R_1}{R_2}\right)^2$$

The general statement of how intensity depends on distance is called an **inverse-square law:**

The intensity of waves radiating isotropically from a point source is inversely proportional to the square of the distance from the source.

Suppose there are a number of independent sources sending out waves at the same time from various locations. The total wave intensity measured by an observer is just the sum of the individual intensities (I_1, I_2, etc.) at the observer's location:

$$I_{\text{tot}} = I_1 + I_2 + I_3 + \cdots \tag{15.6}$$

Example 15.2

Suppose the sound from a buzzer reaches you with an intensity I_1. A second buzzer producing the same amount of sound energy is only half as far from you as the first one. Assume that you are so far from both buzzers that they are effectively point sources of sound. (*a*) What is the total intensity reaching you, in terms of I_1, if both buzzers are sounding? (*b*) What intensity level (in decibels) would you measure with both buzzers sounding compared to just the first buzzer sounding by itself?

Reasoning

Question What intensity ratio exists between two sources of equal power, one twice as far away as the other?
Answer The intensities are inversely proportional to the squares of the distances. In this case $I_2/I_1 = (2/1)^2 = 4$. Thus I_2 (the intensity of the closer buzzer) will be 4 times the intensity of I_1 (the intensity of the farther buzzer).

Question How do intensities add?
Answer Arithmetically, as in Eq. 15.6: $I_{\text{tot}} = I_1 + I_2$.

Question How does the expression for intensity level apply in comparing two sound levels when one is not the threshold of hearing, I_o?
Answer You can use Eq. 15.5 to compare *any* two values of intensity.

Solution and Discussion

(*a*) The total intensity is

$$I_{\text{tot}} = I_1 + 4I_1 = 5I_1$$

(b) The dB difference between this total intensity and just that of the first buzzer is

$$dB = 10 \log \left(\frac{5I_1}{I_1} \right) = 10 \log 5 = +7\,dB$$

Exercise   Show that for each doubling of sound intensity, the intensity level increases by approximately 3 dB. *Hint:* Notice that $\log 2 = 0.30103$.

15.6   THE FREQUENCY RESPONSE OF THE EAR

People vary in their ability to hear sounds. We all know persons whose hearing has been in some way impaired. The sensitivity of their ears has decreased to considerably below that of a person with normal hearing. However, most people agree fairly well on the intensity of a sound that is just audible and also on how loud a sound must be before it causes pain. We can therefore set average limits of audibility for the human ear.

The response of the ear to sound depends on the frequency of the sound as well as its intensity. The ear is more sensitive to some frequencies than to others. Most people cannot hear sound waves that have a frequency higher than about 20,000 Hz. Waves of higher frequency are called **ultrasonic waves**—meaning "beyond" or "above" sound. Similarly, most people are not able to hear sounds below a frequency of about 20 Hz.

The ear is most sensitive near 3000 Hz. At frequencies other than this, the sound must be made more intense before it is audible. This variation of ear sensitivity with frequency is shown in Fig. 15.6. The lower curve shows the minimum audible intensity level as a function of frequency. For example, a sound wave with a frequency of 1000 Hz can be heard when it has an intensity level of about 5 dB. At a frequency of 100 Hz, however, the sound level must be about 30 dB if the sound is to be audible. Of course, near the frequency limits of audible sound (20 and 20,000 Hz), the intensity must be very large for a sound to be heard.

The upper curve in Fig. 15.6 shows how intense a sound must be to produce

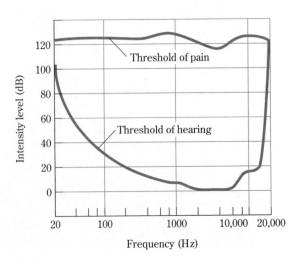

FIGURE 15.6

The normal ear can hear sounds that have intensities above the lower curve.

Physicists at Work THOMAS D. ROSSING *Northern Illinois University*

APPLIED PHYSICS: Using Physics to Solve Problems

Physicists study an exceedingly wide range of objects, all the way from quarks to galaxies. Those who explore these extremes—particle physicists and astrophysicists—derive great satisfaction from extending the frontiers of human knowledge. Other physicists derive equal satisfaction from applying the physical principles to solving practical problems. Most of my own research has been in applied physics, an area that combines elements of physics and engineering.

My first job after graduate school was with a large computer company, where I researched the properties of thin magnetic films, destined to replace ferrite cores in high-speed computer memories. Much of our research em-phasized the practical (e.g., how to get the films to switch states more rapidly), but I was also able to do some basic research (e.g., spin-wave resonance).

Some years later, after I moved to college teaching, I became interested in the physics of musical instruments. This led me to shift my main area of research from magnetism to acoustics. Over the years, my students and I have investigated a number of muscial instruments, from guitars to handbells, from snare drums to gamelans. By applying basic physics principles, we have been able to determine how these instruments create musical sound and, in some cases, have been able to suggest ways in which the instruments could be improved.

In studying the acoustics of musical instruments, we have used techniques that depend upon a variety of physical principles. Holographic interferometry, for example, reveals the vibration modes of a sound-radiating surface, such as the surface of a Chinese bell. Piezoelectric transducers are used to measure force and acceleration in a computer-aided technique called *modal analysis*. Describing the sound radiation field of a musical instrument is not too different from describing the electromagnetic field radiated by a complex antenna.

The whole area of physics and the arts is an interesting and rewarding one for study. There is a small but close-knit international community of musical acousticians within which I have met some of my closest friends.

The same is true, I am told, for physicists who apply physics to the visual arts, dance, theater arts, etc. Unfortunately, support for research in this area is difficult to obtain. (Perhaps this is one reason we're such a close-knit group; we're all motivated by the love of our research alone.)

Recently, I have again focused some of my attention on the field of magnetism. I have collaborated in research at the Argonne National Laboratory on magnetic levitation using superconductors. We have looked at two future applications of magnetic levitation: high-speed maglev vehicles and magnetically levitated flywheels for energy storage. Not only is the physics exciting, but both of these applications offer great potential for improving our environment, a long-time concern of mine.

Often it is difficult to distinguish between basic and applied physics. What begins as research to solve a practical problem may lead to new discoveries, even to Nobel prizes (e.g., high-temperature superconductors, the laser, the transistor, the tunnel diode, etc.). By the same token, basic research in physics often leads to practical applications that were not anticipated when the research was undertaken.

Whether your own interests, combined with future employment opportunities (jobs are generally more plentiful in applied physics), lead you into basic research or into problem solving, rest assured that there is no lack of challenge and excitement in either the study or the practice of physics!

A vocal quartet is an example of four sounds of varying frequencies and different quality blending together to make music.

pain. It is not very frequency-dependent and shows that an intensity level of about 120 dB is painful. Such high sound levels can cause permanent damage to the ear. Indeed, long-term exposure to sounds in the range down to about 90 dB can cause hearing impairment. Of course, loss of hearing can be caused by other factors as well.

15.7 SOUND PITCH AND QUALITY

Pitch is our qualitative perception of whether a musical sound (a tone) is high, like that of a soprano, or low, like that of a bass singer. To investigate pitch and its relation to other features of sound, it is convenient to use a simple experiment. If a high-quality loudspeaker is driven by an electrical system that generates a sinusoidal force, the sound wave given off is an almost pure sine wave that has the frequency being generated by the electrical system. A common example of such a single-frequency sound is the one we hear on the radio as the emergency broadcast network test signal. A listener can compare the pitch of this sound with that of another sound. If the frequency of the driving force is increased so as to increase the frequency of the sound given off by the loudspeaker, the listener will notice at once that the new sound has a higher pitch than the first sound. In such cases, the terms *pitch* and *frequency* are nearly synonymous. Conversely, if the frequency is decreased, the pitch is lowered.

Single-frequency sound waves are not very common, however. When a violin string is bowed, for example, the sound wave given off is not a pure sine wave. This is readily apparent to anyone who has ever compared the tone from a violin being played by an expert with that from one being played by a beginner. In the former case, the tone is full and melodious, whereas the beginner may obtain rasp-

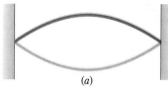

(a)

Fundamental, 1st harmonic: f_1

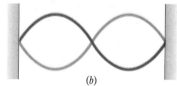

(b)

2nd harmonic, 1st overtone: $2f_1$

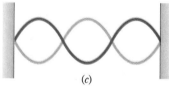

(c)

3rd harmonic, 2nd overtone: $3f_1$

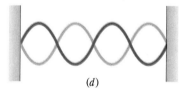

(d)

4th harmonic, 3rd overtone: $4f_1$

FIGURE 15.7

The four simplest vibration patterns for standing waves on a string.

ing sounds on the same string. We say that the **quality** of the tone is different in the two cases.

As we saw in Sec. 14.10, a string may resonate in more than one way. Typical simple vibration patterns are shown and named in Fig. 15.7. Since the wavelengths in the cases shown are in the ratio $1:\frac{1}{2}:\frac{1}{3}:\frac{1}{4}$ and since $f = v/\lambda$, the vibration frequencies are in the ratio $1:2:3:4$.

It is very difficult, however, to cause a string to vibrate exactly as shown in any single pattern of Fig. 15.7. Instead, if the string is bowed near one end, as usually occurs, it vibrates in several ways at once, causing several harmonics to occur at the same time. To find the resulting vibration, it is necessary to add the waves for the various harmonics excited. Of course, because each harmonic is excited to a different amplitude, the proper amplitude must be used for each harmonic when they are added.

A typical example for a vibrating violin string is shown in Fig. 15.8. The amplitudes of vibration of the various harmonics are indicated by the lengths of the vertical bars. In this case, all but the first two harmonics are relatively weak. Even so, the tone heard by the ear is different from what it would be if only the first or the second harmonic alone were present.

Figure 15.8 also shows similar diagrams for the sounds of various other instruments. The piano string shows many more harmonics than the violin string. This is probably the result of the way in which the strings are set into vibration. The violinist pulls a bow slowly across the string, whereas the piano string is excited by a hammer blow.

The quality of a sound depends on which harmonics occur in it and their relative amplitudes. If all sounds were pure sine waves, much of the variety of sound would be lost. The tone of all human voices would be the same, and a voice could be recognized only by a characteristic frequency or inflection. Much of the beauty of music would be lost if the qualities of all sounds were the same.

In a complex sound, such as that of a piano or clarinet, pitch is not always easily defined. It can no longer be taken as identical to frequency since the sound contains several waves that are nearly equal in amplitude but vary in frequency. Some people have an unusual hearing impairment, often unknown to themselves: they are unable to hear sound frequencies above perhaps 6000 Hz. Since most of the sounds we hear consist, partly at least, of frequencies below this, these people are still able to hear sounds that are audible to other people. However, the *quality* of the sounds they hear is quite unlike that of sounds heard by a person with normal hearing. Quality and pitch of sound are thus complex, subjective properties.

FIGURE 15.8

Each musical instrument has its own characteristic sound. The quality of the sound is influenced by the harmonics that compose it and their relative amplitudes. The heights of the vertical bars show the relative strength (amplitude) of each harmonic wave.

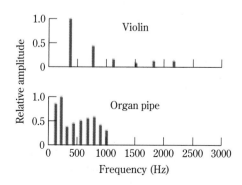

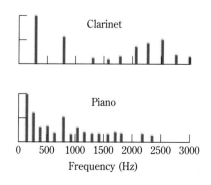

15.8 INTERFERENCE OF SOUND WAVES

Suppose we have a pipe system such as that shown in Fig. 15.9. A single-frequency sine wave is sent into the pipe at the left by a loudspeaker. The sound splits, with half the sound intensity going up through the upper pipe and the remaining half through the lower pipe. Each pipe carries half the sound, and this sound is a wave motion in the air, that is, a series of compressions and rarefactions.

Eventually the two waves are reunited at the outlet on the right (D), where a sound detector, such as an ear or microphone, is placed. The sound emitted at D can be made loud or very faint, depending on the position of the sliding upper pipe EAF. Moreover, as this pipe is slowly pulled upward, the sound intensity at D becomes alternately large and small. We now investigate the reasons for this phenomenon, called *interference*.

When the air is compressed by a rightward movement of the loudspeaker diaphragm, a region of high pressure (a compression) starts into the pipe at C. This compression causes compressions to move in both pipes, toward A and toward B. In other words, the original compression at C splits into two equal parts; one part goes up toward A, and the other goes down toward B. Since the compressions, represented by the red dots, propagate through the pipes with the speed of sound, they reach D at the same time, provided that L_A, the pipe length from C to D through A, is the same as L_B, the pipe length from C to D through B. The compressions are reunited at D, giving the original compression, and this is what exists at D. This situation is shown in Fig. 15.9a. The colored dots represent the rarefactions.

We can represent the situation in Fig. 15.9a by the graph in Fig. 15.10a. The waves in each half of the pipe are plotted. At the moment they divided, they were *in phase* with each other. When they recombine after traveling equal distances, they are still in phase. This means that crests meet each other and troughs meet each other at D. The principle of superposition mentioned in Chap. 14 says that the amplitude of the recombined wave is the algebraic sum of the two amplitudes. The resulting large amplitude is also shown in Fig. 15.10a.

The situation we have just described is an example of **constructive interference,** where the amplitudes of the two waves *reinforce* each other. The result is a relatively high sound intensity at D.

Let us now look at Fig. 15.9b, where the path CAD has been lengthened by

FIGURE 15.9

A sound wave from the loudspeaker is split into two parts. Compressions of the sound wave are denoted by red dots, while rarefactions are shown as green dots. When the two parts of the wave reunite at the detector (D), either a loud sound or a weak sound results, depending on the path lengths traveled by each part. In (a), the wave compressions reinforce, creating a loud sound. In (b), the upper path length is $\lambda/2$ longer than the lower path, causing a compression and rarefaction to combine, resulting in cancellation and a weak or zero sound level.

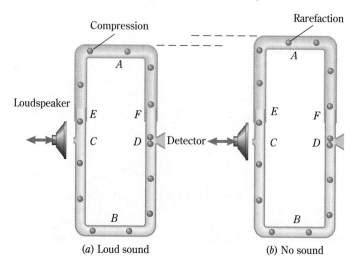

(a) Loud sound   (b) No sound

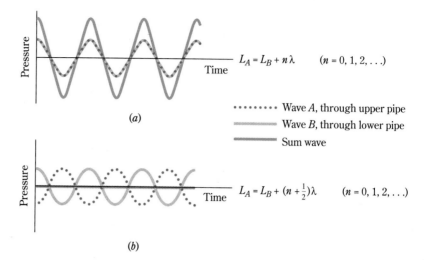

$$L_A = L_B + n\lambda \qquad (n = 0, 1, 2, \ldots)$$

·········· Wave A, through upper pipe
——— Wave B, through lower pipe
——— Sum wave

(a)

(b)

$$L_A = L_B + (n + \tfrac{1}{2})\lambda \qquad (n = 0, 1, 2, \ldots)$$

FIGURE 15.10

Waves A and B can reinforce or cancel depending on their positions relative to each other. The waves in (a) are in phase, while in (b) waves A and B are 180° (or one half wavelength) out of phase.

sliding the upper pipe away from the source and detector. The upper path is now one half wavelength longer than the lower path. Consequently, the half crest which started at C and traveled through the upper pipe has one half wavelength left to travel when its twin arrives at D after traveling the lower path. This means that the wave in the upper path arrives one half cycle *out of phase* with the wave in the lower path. In other words, crests of one wave always meet troughs of the other. The amplitudes then *cancel*, and no sound is detected at D. This situation, shown graphically in Fig. 15.10*b*, is an example of **destructive interference.**

We can generalize these results by noting that, when the upper pipe is extended to where it is one wavelength longer than the lower pipe, constructive interference occurs again. Now any two crests which divide at C will no longer meet each other at D. Instead, the crest going through B will meet a crest which left C one cycle earlier. It makes no difference *which* crests or troughs meet. The interference that results is the same. Constructive interference will occur whenever path L_A is an integral number of wavelengths longer or shorter than path L_B:

Constructive interference (loud sound): $\quad L_A = L_B \pm n\lambda \qquad$ where $n = 0, 1, 2, 3, \ldots$

Similar reasoning leads to the condition for destructive interference. This occurs whenever the path difference is an odd integer multiple of half wavelengths.

Destructive interference (no sound): $\quad L_A = L_B \pm n\lambda/2 \qquad$ where $n = 1, 3, 5, \ldots$

It is not necessary to have a pipe system such as this to obtain interference. All we need are two waves that are exactly the same in frequency and shape. If these waves are combined after traveling different distances, they will interfere with each other. The following example explains another situation involving interference.

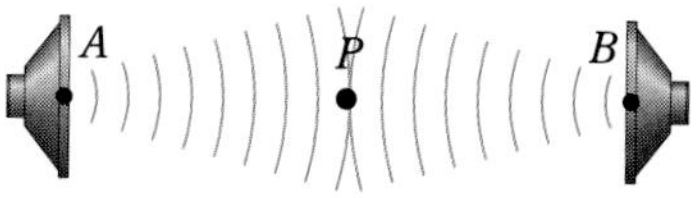

FIGURE 15.11

When $\overline{AP} = \overline{PB}$, the waves reinforce at P.

Example 15.3

The two identical sound sources in Fig. 15.11 vibrate in phase to send identical waves ($\lambda = 70$ cm) toward each other. A loud sound is heard by an observer at P midway between the sources. The observer slowly moves toward source B. How far does she have to move before the sound she hears becomes very weak?

R

Reasoning

Question   What condition will make the sound very weak?
Answer   When the waves from the two sources arrive at the observer one half cycle out of phase, destructive interference will occur.

Question   Why doesn't the sound intensity become *zero* if there is destructive interference?
Answer   Remember, the intensity of three-dimensional waves decreases with distance from the source. At *P*, midway between the sources, the intensities from the two sources are equal. Since the observer then moves toward *B*, the intensity from *B* will be slightly greater than that from *A* at the point of destructive interference.

Question   At what position will this happen?
Answer   Where the distance from source *A* to the observer is $\lambda/2$ longer than the distance from source *B* to her.

Question   How much difference in these distances results from the observer moving a distance *x* toward *B*?
Answer   The distance *AP* increases by *x* and *PB* decreases by *x*. So the *difference*, *AP − PB*, is equal to 2*x*.

Solution and Discussion   One half wavelength is 35 cm. We have

$$AP - PB = 2x = 35 \text{ cm}$$

Thus the observer has to move $35 \text{ cm}/2 = 17.5 \text{ cm}$ toward *B*.

15.9   BEATS

When people tune a string on a piano, they do not merely listen to see whether the tone of the string is the same as that of the standard tuning fork used for comparison. Instead, they use a much more precise way of judging the accuracy to which the string is adjusted. They listen for *beats* between the sounds of the two vibrating objects. This is a very sensitive method for determining equal frequencies and is widely used for that purpose.

Let us begin by examining what happens when two vibrating sources emit waves of exactly the same frequency and in phase (synchronized) with each other. As we see in Fig. 15.12, two 1000-Hz sources produce a combined 1000-Hz wave whose amplitude is constant. Suppose now that the frequency of source *B* is decreased to 999 Hz, as shown in Fig. 15.13.

At time $t = 0$, the loudspeakers are in phase, that is, both are sending out a compression, as indicated by the rightward-pointing arrow. If a person's ear is equidistant from the speakers, the compressions arrive at the ear together, and a large compression results; a loud sound is heard. As time goes on, *B*, vibrating at a slightly lower frequency than *A*, begins to fall behind. After 0.5 s, *A* has vibrated 500.00 times and is just sending out a compression, as shown for $t = 0.5$ s, in Fig. 15.13. Loudspeaker *B*, however, has vibrated only 499.50 times and is exactly one half cycle behind *A*. It is sending out a rarefaction (the leftward-pointing arrow). The compression from *A* now reaches the ear at the same time as the rarefaction

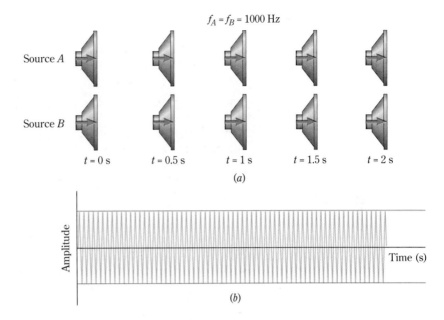

FIGURE 15.12

Two sources vibrating in phase with each other at the same frequency create a sound with constant amplitude.

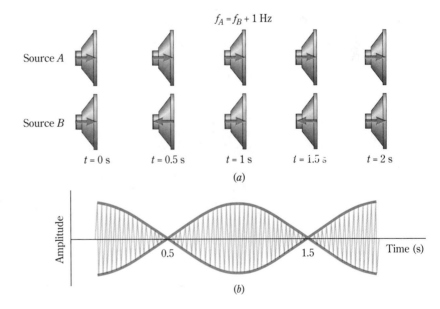

FIGURE 15.13

Beats result when two similar sources vibrate with slightly different frequencies. The rapidly varying green wave represents (not to scale) 1000-Hz and 999-Hz source frequencies. The red curves show the amplitude variation which results from the interference of the two frequencies. It is this amplitude variation which is heard as beats.

from B, and they exactly cancel each other. Hence, at this instant, no sound is heard.

As time continues, loudspeaker B falls still farther behind A. After 1 s, B has vibrated 999 times while A has vibrated 1000 times. Source B has now fallen one cycle behind A. Hence once again they are both sending out compressions together, and a loud sound is heard.

This process continues, as shown in the succeeding portions of Fig. 15.13a. At times of $0, 1, 2, 3, \ldots$ s, the sources are in phase and the sound heard is loudest. At times of $0.5, 1.5, 2.5, \ldots$ s, nothing is heard because the sources are 180° out of phase.

A piano tuner adjusts the tension in a piano string to change its frequency. One technique is to listen for beats between the string and a calibrated frequency standard.

The combined sound wave is shown in Fig. 15.12*b* as a function of time. Notice that the amplitude of the combined wave varies in time, going from one maximum to the next in 1 s. A listener hears these amplitude pulses at a frequency of 1/s. We call these pulses **beats** and can draw the following conclusion:

The number of beats per second (the beat frequency) equals the difference in frequencies of the two sources.

For example, two sources emitting sound at 100 Hz and 97 Hz produce a beat frequency of 3/s. Sources emitting 5000 Hz and 5010 Hz produce 10 beats per second.

Listening for beats provides an extremely sensitive way of tuning musical instruments. A piano tuner will use a frequency source emitting a desired frequency and adjust the tension in the piano string until the sounds from the string and the standard produce no perceptible beats. Using this method, it is as easy to tune a 5000-Hz string to within 1 Hz as it is to tune a 50-Hz string.

Sometimes the beat frequency between two sound waves gives rise to a third distinct sound. For example, suppose the frequencies of two sounds are 1000 Hz and 1200 Hz. Their beat frequency is 200 Hz, well within the frequency range of the ear, and the listener will hear a sound of this frequency in addition to the two original sounds.

15.10   RESONANCE IN AIR COLUMNS

Reservoir

FIGURE 15.14

When the water in the tube is just the right height, resonance will occur.

If you hold a vibrating tuning fork over the open end of a glass tube that is partly filled with water, the sound of the tuning fork can be greatly amplified under certain conditions. While the fork is held as shown in Fig. 15.14, the reservoir is lowered so that the water level in the tube falls. When the tube water level is at a certain height, the air column above the water in the tube resonates loudly to the

Organ pipes of different lengths resonate at different frequencies. Can you state other physical conditions that the resonant frequency depends on?

sound being sent into it by the tuning fork. In fact, there are usually several heights at which the air column in the tube resonates.

The situation here is much like the case of standing waves on a vibrating string. In place of the string excited by an oscillator at one end, we have an air column with a sound source at its open end. The oscillator sends a wave down the string, and the wave is reflected at the other end. Similarly, the sound source sends a sound wave down the air column, and the wave is reflected by the surface of the water. As we learned in Chap. 14, a string resonates only when the wavelength of the wave can produce a standing wave pattern on the length of the string. In particular, nodes must occur at the two ends of the string, and so the string resonates only if its length is $n(\frac{1}{2}\lambda)$, where n is an integer and $\frac{1}{2}\lambda$ is the distance between nodes.

Resonance of the air column in Fig. 15.14 differs in a major respect from that of a string. The water surface in the air-filled tube acts as a fixed, closed end for the tube. Longitudinal motion of the air in the tube is prevented at this closed end by the water. Hence the closed end of the air column (the water surface) must be the location of a node in the sound-wave resonance pattern. At the open end of the tube, however, the air is free to move out into the open space above, and so at that point there is maximum motion, that is, an antinode.* Therefore, if the air column in Fig. 15.14 is to resonate, it must have a node at its closed end and an antinode at its open end. The air column will resonate only if the sound wavelength is such that the wave fits into the column with a node at one end and an antinode at the other. Several of these resonating waves are shown schematically in Fig. 15.15.

Notice that the graphs in Fig. 15.15 are *not* a picture of the waveform as they were for a vibrating string. *Instead, they depict the amplitude of displacement of air molecules along the length of the tube.* Where there is a node, no longitudinal displacement occurs. Maximum displacement occurs at the antinodes. Remember that the distance between two adjacent nodes or antinodes is $\lambda/2$. Hence, the dis-

---

*The antinode is not precisely at the end of the tube. However, for tubes with radii much smaller than λ, this complication can usually be ignored.

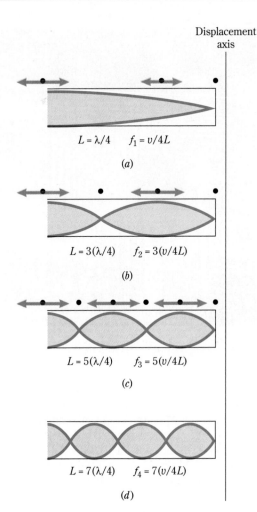

Displacement axis

$L = \lambda/4 \qquad f_1 = v/4L$

(a)

$L = 3(\lambda/4) \qquad f_2 = 3(v/4L)$

(b)

$L = 5(\lambda/4) \qquad f_3 = 5(v/4L)$

(c)

$L = 7(\lambda/4) \qquad f_4 = 7(v/4L)$

(d)

FIGURE 15.15

Simple modes of vibration for a resonating pipe closed at one end. These are graphs of longitudinal displacement versus position along the pipe. Relative displacements are shown above graphs (a), (b), and (c).

tance from a node to an antinode is $\lambda/4$. If we call the length of the air-filled tube L, the length in Fig. 15.15a is from a node to an antinode, or $L = \lambda/4$. In Fig. 15.15b, the tube is equal to three node-antinode lengths; hence $L = 3(\lambda/4)$, and so on.

The resonance frequencies (harmonics) shown in Fig. 15.15 can be found from $f = v/\lambda$. These frequencies are easily computed by using the values of the wavelengths required for standing wave patterns as discussed above. Notice that the first resonant frequency above the fundamental f_1 is $3f_1$. This is often called the first *overtone*. The second overtone is $5f_1$, the third overtone $7f_1$, and so on. In this case of a tube closed at one end, *only the odd-numbered harmonics* resonate.

A pipe or tube need not be closed at one end to resonate. For example, you can use a piece of glass tube as a whistle by blowing across either of its ends. The simplest possible resonances of a tube that is open at both ends are shown in Fig. 15.16. In each case, the ends of the tube are antinodes because the air is free to move there. The resonance frequencies are computed in the usual way by using the fact that $f = v/\lambda$, with λ being determined as indicated in the figure. Notice that these resonance conditions for the tube open at both ends are the same as for a string fixed at both ends. Since the frequency of a tuning fork or other source of vibration is usually known, resonances in a tube such as the one in Fig. 15.14 can be used to measure the speed of sound.

Displacement axis

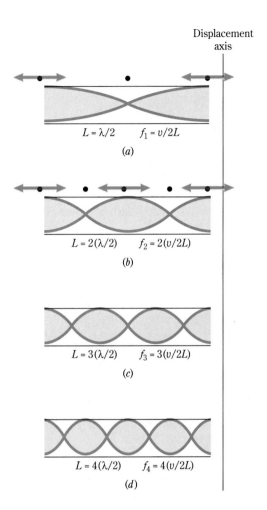

$L = \lambda/2$ $f_1 = v/2L$

(a)

$L = 2(\lambda/2)$ $f_2 = 2(v/2L)$

(b)

$L = 3(\lambda/2)$ $f_3 = 3(v/2L)$

(c)

FIGURE 15.16

Displacement graphs of simple modes of vibration for a resonating pipe open at both ends.

$L = 4(\lambda/2)$ $f_4 = 4(v/2L)$

(d)

To summarize, the resonance wavelengths and frequencies in air columns are:

One end open, one closed: $\lambda_n = \dfrac{4L}{n}$ $f_n = \dfrac{nv}{4L}$ (n = odd integers)

Both ends open: $\lambda_n = \dfrac{2L}{n}$ $f_n = \dfrac{nv}{2L}$ (n = all integers)

When you blow across the end of a tube, this very complex process sends waves of many frequencies down the tube. Of this multitude of frequencies, the tube resonates to only one or two. For this reason, the resonating tube usually gives off a loud sound of a single frequency. However, if you try hard enough, you can often cause the tube to resonate to two frequencies at the same time and thereby give off two tones simultaneously.

Many musical instruments make use of resonating air columns. The flute and piccolo are basically tubes whose length can be changed by means of holes along the tube. A clarinet is similar, but in it the sound waves are generated by the vibration of a reed in the mouthpiece. More complex tube resonance systems are seen in the trumpet, trombone, and tuba. In these instruments, the player elicits various resonance tones by changing the length of the resonating pipe. In addition, the sound waves in these instruments are generated by the vibration of the player's lips in the mouthpiece.

Example 15.4

An organ pipe 60.0 cm long is open at both ends. Assume the air in the pipe is at 20°C. (**a**) Find the frequency of the fundamental resonance and that of the first overtone. (**b**) Repeat part (*a*) for the same pipe closed at one end. (**c**) If the original pipe is filled with helium at 20°C, what is the frequency of its fundamental resonance?

Reasoning

Question What relation determines the fundamental wavelength?
Answer For a pipe open at both ends, $L = \lambda_1/2$. So

$$\lambda_1 = 2L$$

Question What else must I know in order to calculate f_1?
Answer Since $f_1 = v/\lambda_1$, you must also know the wave speed v.

Question What does the wave speed depend on?
Answer On the kelvin temperature, the molecular mass of the gas, and γ, the ratio of the specific heats of the gas:

$$v = \sqrt{\frac{\gamma RT}{M}}$$

Question Since air is a mixture, how do I find its molecular mass?
Answer Air is approximately 80% N_2 ($M = 28$ kg/kmol) and 20% O_2 ($M = 32$ kg/kmol). Thus the value of M for air is

$$M_{\text{air}} = (0.80)28 + (0.20)32 = 28.8 \text{ kg/kmol}$$

Question What harmonic is the first overtone?
Answer A pipe that is open at both ends has resonances at *all* harmonics, both odd and even. So the first overtone is f_2.

Question Is there an easy way to find this, knowing f_1?
Answer Yes. Since v and L are constant and each resonant frequency f_n is proportional to n, you can easily use *ratios*. If n and m denote two different harmonics, the ratio is simply

$$\frac{f_n}{f_m} = \frac{n}{m}$$

Question When the pipe is closed at one end, what changes?
Answer The wavelength of the fundamental and the harmonics that can resonate.

Question What are the new fundamental wavelength and frequency?

Answer $\lambda_1 = 4L$ and $f_1 = \dfrac{v}{4L}$

Question Which harmonic forms the first overtone?
Answer In the pipe with one end closed and one end open, only *odd* harmonics resonate. Thus the first overtone is the third harmonic, $f_3 = 3v/4L = 3f_1$.

Question What changes when helium fills the tube instead of air?
Answer The molecular mass. Also, helium is a monatomic gas and has a different value of γ.

Solution and Discussion

(a) In part (*a*) the fundamental wavelength is simply

$$\lambda_1 = 2L = 1.2 \, \text{m}$$

The molecular mass of air is 28.8 kg/mol. Air at 20°C has a value of 1.4 for γ. Then

$$v = \left(\frac{\gamma R T}{M} \right)^{1/2}$$

$$= [(1.4)(8314 \, \text{J/kmol} \cdot \text{K})(293 \, \text{K})/(28.8 \, \text{kg/mol})]^{1/2} = 344 \, \text{m/s}$$

You could also have used the approximate formula (Table 15.1)

$$v = 331 \, \text{m/s} + 0.61T = 331 + 12.2 = 343 \, \text{m/s}$$

(Remember that T in this formula is the Celsius temperature.) Then

$$f_1 = v/\lambda_1 = \frac{343 \, \text{m/s}}{1.20 \, \text{m}} = 286 \, \text{Hz}$$

The first overtone is $f_2 = 2f_1 = 572 \, \text{Hz}$.

(b) The speed of sound is the same, so

$$\lambda_1 = 4L = 2.40 \, \text{m} \quad \text{and} \quad f_1 = \frac{343 \, \text{m/s}}{2.40 \, \text{m}} = 143 \, \text{Hz}$$

This is *half* the fundamental frequency of the pipe that is open at both ends. The first overtone is

$$f_3 = 3f_1 = 3(143 \, \text{Hz}) = 429 \, \text{Hz}$$

You see that the same pipe has a very different harmonic content depending on whether one end is closed or open.

(c) Finally, with helium, the atomic mass is 4 and $\gamma = 1.67$. This makes the speed of sound considerably faster. We can use ratios of γ and M inside the square root:

$$v_{\text{He}} = v_{\text{air}} \sqrt{\left(\frac{1.67}{1.40}\right)\left(\frac{29}{4.0}\right)} = 2.94 v_{\text{air}}$$

$$= (343\,\text{m/s})(2.94) = 1010\,\text{m/s}$$

Since f is proportional to v and L remains the same, we have

$$f_1(\text{He}) = \left(\frac{1010\,\text{m/s}}{343\,\text{m/s}}\right) f_1(\text{air}) = 2.94\, f_1(\text{air})$$

$$= 841\,\text{Hz}$$

This effect of helium on the speed of sound is the reason that a person who speaks immediately after breathing helium makes a high-pitched sound.

15.11   THE DOPPLER EFFECT

We now turn to a different phenomenon common to waves of all kinds and to sound waves in particular: the **Doppler effect.*** You have certainly experienced this effect, although you may have been unaware of what caused it. As an example, whenever a fast-moving ambulance approaches and passes you, the sound of its siren seems to behave in a peculiar way. The siren's frequency seems to change from a higher to a lower tone as the ambulance rushes by. In other words, the frequency of the sound is raised when the source of the sound approaches you and lowered when the source is moving away from you. A similar effect occurs with light and electromagnetic waves. When radar is reflected from a moving car, its frequency is shifted relative to the transmitted frequency. The amount of frequency

*The phenomenon is named after the Austrian physicist Christian Johann Doppler, who showed in 1842 that the effect should be observed for sound and light waves.

A common example of the Doppler effect in sound waves occurs when a fire truck rushes past us. The sound from its siren or horn seems to lower in frequency as the truck changes from approaching us to receding from us.

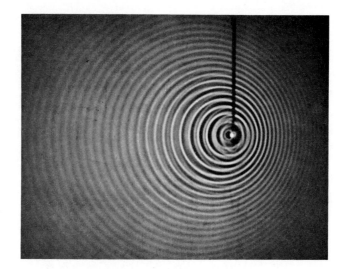

FIGURE 15.17

The vibrating vertical rod is a source of water waves. Because the source is moving to the right, the wavelengths in that direction are shortened. *(Education Development Center)*

shift depends on the car's speed, enabling a law officer to determine whether or not a car is traveling under the legal speed limit. In general, *motion* between the source of any type of wave and an observer has an effect on the wave frequency measured by the observer.

The Doppler effect can be understood by referring to Fig. 15.17. Shown there is a water wave source that is moving to the right through the water. Although the source sends out circular waves, the centers of successive circles move to the right with the source. This motion causes the wave crests to be closer together at the right than at the left. The moving source, in effect, causes the wavelength of the waves to be different in various directions.

A similar phenomenon occurs with sound waves, as we can see from Fig. 15.18. The wave source in Fig. 15.18*a* is stationary. If a listener at *P* is also stationary, the frequency of the disturbance striking the listener's ear is identical to the source frequency *f*.

Figure 15.18*b* shows what happens when the source is moving and the listener is stationary. The motion of the source causes the wavelengths of the waves it sends out to differ in different directions. The motion of the source does *not* affect the wave speed. Thus the changed wavelength produces the effect of a changed frequency heard by the listener. If the source moves toward the listener, she hears a higher frequency than *f*. If the source moves away, she hears a lower frequency.

FIGURE 15.18

The frequency of sound that a listener hears depends on the velocity of both the source and the listener. In (*b*), the source is moving to the right. When it is at *A*, it sends out the crest labeled *A*; when at *B*, it sends out crest *B*; and so on.

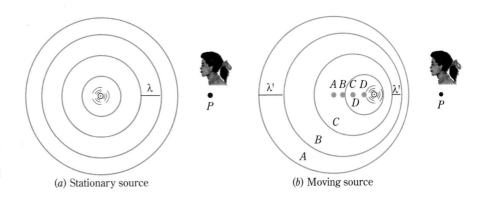

(*a*) Stationary source   (*b*) Moving source

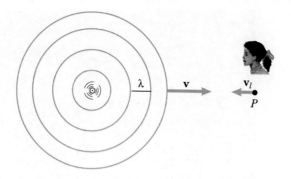

FIGURE 15.19

The waves strike the approaching listener with relative velocity $v + v_l$. When the listener moves away from the source, the relative velocity is $v - v_l$.

A different situation occurs when the listener is moving relative to a stationary source, as shown in Fig. 15.19. When moving toward the source, the listener encounters more wavefronts per second than the source is producing. Hence she hears a frequency higher than f. When she is moving away from the source, fewer wavefronts reach her per second, and she measures a frequency less than f.

This effect can be summarized qualitatively:

The observed frequency of sound is increased when the source and listener are approaching each other and is decreased when they are receding from each other.

As mentioned before, this effect applies to all forms of waves, not just to sound waves.

Let us now examine the Doppler effect quantitatively. In Fig. 15.18b, we can see that the distance between crests in the direction of the listener is *shortened* by the distance the source travels in the period between the emission of wavefronts. This is just the period of the waves, τ. Thus the effective measured wavelength, λ', is

$$\lambda' = \lambda - v_s\tau$$

where v_s is the speed of the source. Similarly, in the direction *away* from the listener, the distance between crests is lengthened by $v_s\tau$, giving

$$\lambda' = \lambda + v_s\tau$$

Using $\lambda = v/f$ and $\tau = 1/f$, we get

$$\frac{v}{f'} = \frac{v}{f} \pm \frac{v_s}{f}$$

or

$$f' = f\frac{v}{v \pm v_s} \qquad \text{(moving source)} \qquad (15.7)$$

where v is the wave speed in the medium. The plus sign refers to the case of the receding source, and the minus sign is for an approaching source.

Now suppose we have the situation where the listener is moving with speed v_l. In Fig. 15.19, we see that the relative speed between listener and wave is $v + v_l$ when the listener is moving *toward* the source. If she is moving *away*, the relative speed is $v - v_l$. We assume that the listener does not travel faster than the wave speed. The observed period between waves is no longer λ/v, but rather

$$\tau' = \frac{1}{f'} = \frac{\lambda}{v \pm v_l} = \frac{v/f}{v \pm v_l}$$

Thus

$$f' = f\frac{v \pm v_l}{v} \qquad \text{(moving listener)} \tag{15.8}$$

Here the plus sign indicates movement *toward* the source, and the minus sign means movement *away from* the source. If you forget which signs to use, it always helps to remember the general qualitative rule expressed previously. You should also realize that Eqs. 15.7 and 15.8 give *different* frequency shifts for the same velocity, depending on which is moving, the source or the listener.

Einstein showed, however, that Eqs. 15.7 and 15.8 are not correct for light waves when the source or the observer is moving with a speed close to the speed of light. The difficulty arises because of the relativity principle, which states that the speed of light in vacuum is independent of the motion of the observer or the light source. In cases involving such extremely high speeds, the amount of frequency shift is the same whether it is the source or the observer that is moving.

Example 15.5

On a cold winter day ($T = 0°C$), a car is moving at 20.0 m/s along a straight road, sounding its 500-Hz horn. You are standing at the side of the road. What frequency do you hear as the car is (*a*) approaching you and (*b*) receding from you?

R

Reasoning

Question Which Doppler equation pertains to this situation?
Answer You, the listener, are stationary. Thus Eq. 15.7 with the minus sign pertains to part (*a*) and with the plus sign pertains to (*b*).

Question What is the speed of sound, *v*?
Answer At 0°C Table 15.1 gives us 331 m/s.

Solution and Discussion For part (*a*) we have

$$f' = 500\,\text{Hz}\left(\frac{331\,\text{m/s}}{331\,\text{m/s} - 20.0\,\text{m/s}}\right)$$

$$= 500\,\text{Hz}\frac{331}{311} = 532\,\text{Hz}$$

In part (*b*),

$$f' = 500\,\text{Hz}\left(\frac{331\,\text{m/s}}{331\,\text{m/s} + 20\,\text{m/s}}\right)$$

$$= 500\,\text{Hz}\frac{331}{351} = 472\,\text{Hz}$$

This difference of $532 - 472 = 60\,\text{Hz}$ as the car passes you is certainly a very obvious effect.

Exercise Find the frequencies you would hear when you are (*a*) approaching and (*b*) moving away from a stationary 500-Hz horn at 20.0 m/s. *Answer: Approaching: $f' = 530\,\text{Hz}$; moving away: $f' = 470\,\text{Hz}$.*

Example 15.6

A car is approaching you, traveling at a speed v_s. In the car is a loudspeaker emitting a 440-Hz tone. At the side of the road, you have an identical 440-Hz source. You hear 20 beats per second between your source and the one in the car. How fast is the car traveling?

R

Reasoning

Question How is the beat frequency related to the speed of the car?
Answer If the car were stationary, you would hear the two tones at the same frequency. Since the car is moving, the frequency you hear from its loudspeaker is shifted by the Doppler effect. This shifted frequency beats against the source you have.

Question What does the 20 beats per second represent?
Answer The difference between your frequency f and the Doppler-shifted frequency f':

$$\text{Beat frequency} = f' - f \tag{1}$$

Question What expression tells me what f' is?
Answer The car is approaching you, so you should use Eq. 15.7 with the minus sign:

$$f' = f\frac{v}{v - v_s} \tag{2}$$

Question What equation do I get when I combine expressions (1) and (2)?
Answer $20\,\text{Hz} = f' - f = f\dfrac{v}{v - v_s} - f$

Since f is known and $v = 343\,\text{m/s}$ at 20°C, the single unknown is the car's speed, v_s.

Solution and Discussion The solution requires a bit of careful algebraic manipulation:

$$20\,\text{Hz} = 440\,\text{Hz}\frac{343\,\text{m/s}}{343\,\text{m/s} - v_s} - 440\,\text{Hz}$$

Factoring the 440-Hz term:

$$20\,\text{Hz} = 440\,\text{Hz}\left(\frac{343\,\text{m/s}}{343\,\text{m/s} - v_s} - 1\right)$$

Dividing both sides by 20 Hz and doing the subtraction inside the parentheses yields

$$1 = 22\frac{v_s}{343\,\text{m/s} - v_s}$$

Solving for v_s:

$$22v_s = 343\,\text{m/s} - v_s \qquad v_s = \frac{343\,\text{m/s}}{23} = 15\,\text{m/s}$$

15.12 SUPERSONIC SPEED

An interesting situation arises when the speed of a sound source approaches or equals the speed of sound. Then, from Eq. 15.7, we see that the sound frequency f' approaches infinity. This simply means that a nearly infinite number of wave crests reach the listener in a very short time. We can easily understand this by referring once again to Fig. 15.18*b*.

Suppose the moving source has a speed equal to the speed of sound. Then all the wave crests in front of the source lie one upon the other, concentrating the wave energy into a very small region in front of the source. A compressional wave can move through the air only at the speed of sound, v. When an aircraft is moving through air **faster** than v, the abrupt compression of air it causes moves outward from the plane at speed v. Figure 15.20*a* shows the position of this compression at successive times. As the airplane moves at speed v_p from A to B to C to D, the compression it makes moves outward at the *slower* speed v to the corresponding positions A', B', C' and D'. The compression wavefront thus makes an angle θ with the direction of the airplane's velocity, where

$$\sin\theta = \frac{v}{v_p}$$

Superman does not seem to be affected by any laws of physics. In flying "faster than a speeding bullet," he never seems to produce a shock wave or a sonic boom.

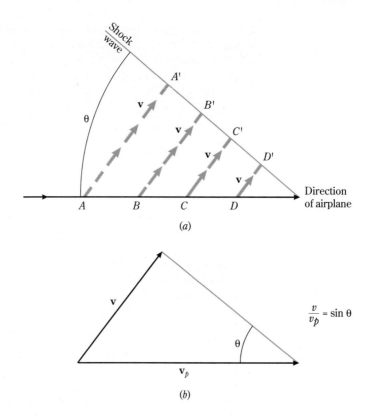

FIGURE 15.20

(a) The formation of a shock wave.
(b) The relation between sound velocity
v and airplane velocity $\mathbf{v}_p$.

$$\frac{v}{v_p} = \sin \theta$$

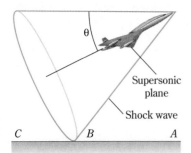

FIGURE 15.21

The sonic boom has already hit point
C and is moving through point B
toward point A.

This angle is shown in Fig. 15.20b.

Actually, the compression wave moves out in three dimensions, creating a *conical* wave, as shown in Fig. 15.21. This very concentrated region of sound energy is called a **shock wave** and causes a loud boom as it passes any given point such as B in Fig. 15.21. This *sonic boom* moves along the ground at the speed of the

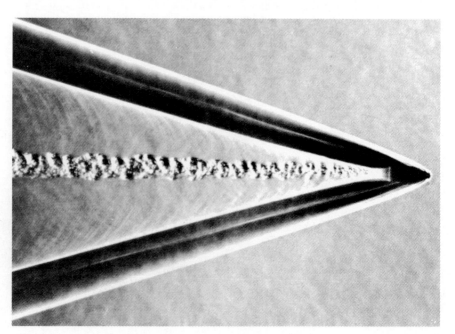

FIGURE 15.22

Shock waves generated by a bullet
shooting through the air.

airplane. There is a considerable air-pressure difference across the shock wave. Depending on the intensity of the shock wave, its effects can be damaging. This is particularly true of low-flying aircraft, where the energy of the shock wave has not had an opportunity to spread out before striking the ground.

Notice that the angle θ decreases as the speed of the airplane increases. The ratio of the airplane's speed to that of sound, v_p/v, is called the **Mach number:**

$$\text{Mach number} = \frac{v_p}{v} = \frac{1}{\sin \theta}$$

A plane traveling twice the speed of sound is said to be traveling at Mach 2. Figure 15.22 shows the shock wave created by a high-speed bullet. By measuring the shock angle in the photograph, can you show that this bullet is moving at approximately Mach 3?

LEARNING GOALS

Now that you have finished this chapter, you should be able to

1 Define (a) sound wave, (b) compression and rarefaction, (c) wavefront, (d) ray, (e) plane wave, (f) spherical wave, (g) sound intensity, (h) intensity level, (i) decibel, (j) inverse-square law, (k) infrasound and ultrasound, (l) constructive and destructive interference, (m) quality of sound, (n) beats and beat frequency, (o) harmonics and overtones, (p) Doppler effect, (q) shock wave, (r) Mach number.
2 Explain what a sound wave is and why sound cannot travel through a vacuum.
3 Recall the approximate velocity of sound in air at 0°C and 20°C. Calculate the speed of sound through various gases at given temperatures.
4 Calculate the decrease in sound intensity as a function of distance from a point source.
5 Convert from sound intensity in watts per square meter to sound level (intensity level) in decibels. Convert from decibels of sound level to intensity.

6 Sketch the approximate response curve for the normal ear as a function of frequency. Give the approximate decibel levels for very weak and very loud sounds. Identify the ultrasonic region.
7 Explain sound quality and point out why it is different from frequency.
8 Combine two waves of the same frequency and amplitude but of different phase so as to obtain destructive and/or constructive interference.
9 Use beats to find the difference in frequency between two sound sources.
10 Find the resonance frequencies of sound in a pipe.
11 Explain the Doppler effect and compute the frequency shift for an approaching or receding sound source.
12 Explain how a shock wave originates and why it gives rise to a sonic boom.

SUMMARY

DEFINITIONS AND BASIC PRINCIPLES

The Speed of Sound
Sound waves are longitudinal (compressional) waves. The speed of sound in any medium is

$v = \sqrt{Y/\rho}$ (one-dimensional medium)

$v = \sqrt{B/\rho}$ (two- and three-dimensional medium)

where Y = Young's modulus of medium
B = bulk modulus of medium
ρ = density of medium

In ideal gases, the speed of sound is

$$v = \sqrt{\frac{\gamma P}{\rho}} = \sqrt{\frac{\gamma RT}{M}}$$

INSIGHTS
1. The ratio of specific heats, $\gamma = C_p/C_v$, depends on the type of gas and on the temperature.
2. In SI units, $R = 8314\,\text{J/kmol} \cdot \text{K}$. Thus M should be expressed in kilograms per kilomole and T must be in kelvins.
3. For air, $M = 28.8\,\text{kg/kmol}$. The speed of sound $v = 331$ m/s at 0°C and increases approximately 0.61 m/s per degree above that for common ambient temperatures.

Sound Intensity (I)
Intensity = power/area   (W/m²)

The sound intensity from a point source diminishes as the inverse square of the distance from the source:

$$I(r) = \frac{P}{4\pi r^2}$$

where P = total power output of source
r = distance at which I is measured

Intensity Level or Sound Level (Decibel Scale)
Intensity (sound) level in decibels (dB) = $10 \log (I/I_o)$

INSIGHTS
1. The terms "sound level" and "intensity level" refer to the same phenomenon.
2. The threshold of hearing is at $I_o = 10^{-12}\,\text{W/m}^2$. This is usually taken to be the zero-dB level of intensity.
3. The decibel is a dimensionless number.

Interference between Two Sound Sources: Beats
Beats are periodic variations in the combined amplitude of waves from two sound sources of different frequencies f and f'. The frequency of the beats is equal to the difference in the two frequencies.

Beat frequency = $f' - f$

Resonance Conditions for Sound in Air Columns
Standing waves (resonances) occur in air columns of length L at the following wavelengths and frequencies:

1. One end open, one closed:

$$\lambda_n = \frac{4L}{n} \qquad f_n = n\frac{v}{4L} \qquad n = \text{odd integers only}$$

2. Both ends open:

$$\lambda_n = \frac{2L}{n} \qquad f_n = n\frac{v}{2L} \qquad n = \text{all integers}$$

INSIGHTS
1. In both cases, $n = 1$ is called the *fundamental*. It has the longest wavelength and the lowest frequency.
2. Each successively higher frequency is called an *overtone*. The number n is the *harmonic* number of the resonance. Pipes open at one end and closed at the other have only odd harmonics. Pipes open at both ends have both odd and even harmonics.

The Doppler Effect
The Doppler effect occurs whenever a sound source and a listener are moving relative to each other.

The measured (or heard) frequency of sound is increased when the source and the listener are approaching each other and decreased when they are receding from each other. The observed frequency (f') is related to the frequency of the source (f) by:

$$f' = f\frac{v}{v \pm v_s} \qquad \text{(moving source)}$$

$$f' = f\frac{v \pm v_l}{v} \qquad \text{(moving listener)}$$

where v = speed of sound
v_s = speed of source
v_l = speed of observer

INSIGHTS
1. Which sign to use can be determined from the qualitative description above.
2. In both cases, v_s and v_l are assumed to be less than v.

Supersonic Speed: Mach Number
When an object's speed v_p exceeds the speed of sound v, a shock wave is created. The shock wave makes a conical angle θ with the direction of the object, where

$$\sin \theta = v/v_p$$

$$\frac{v_p}{v} = \text{Mach number} = \frac{1}{\sin \theta}$$

QUESTIONS AND GUESSTIMATES

1 Explain why a bell ringing inside a vacuum chamber cannot be heard on the outside.

2 Would you expect a sound heard under water to have the same frequency as a sound heard in air if the sources vibrate identically? Explain.

3 When a person inhales helium and then speaks, the person's voice sounds high-pitched. Why?

4 Suppose a few pipes of a pipe organ were mounted close to a hot heater. Would this affect the performance of the organ? Explain.

5 A siren can be made by drilling equally spaced holes on a circle concentric to the axis of a solid metal disk. When the disk is rotated while a jet of air is blowing against it near the holes, a sirenlike tone is given off. Explain how this gives the sensation of sound to the ear, and state what factors influence the pitch and quality of the tone.

6 A singer claims to be able to shatter a wine glass by singing a particular note. Could this be true? Explain.

7 Suppose that on some distant planet there exist humanoids whose hearing mechanism is designed as follows. From the outside, their heads look like our own. However, a 1-cm-diameter hard-surfaced cylindrical hole passes through the head from ear to ear. At the midpoint of the channel, a thin, circular membrane acts like a drumhead, separating the two halves of the channel. These beings experience the sensation of sound when this drumhead vibrates. What can you infer about their hearing abilities and about the way they communicate orally with one another?

8 In an adult human, the ear canal, which is the hollow tube leading from the outer ear to the eardrum, has a length of about 2.5 cm. How does the resonance frequency of such a tube correspond with the sensitivity curve of the ear?

9 Estimate the frequencies at which a test tube 15 cm long resonates when you blow across its lip.

10 A sound source emits a 1000-Hz sound in all directions while a strong wind blows to the east past the source. How do the frequency, speed, and wavelength of the observed sound depend on the position of the observer?

11 Two speakers are to be connected to a stereo. The directions say, "Set the speakers side by side and connect the two red wires from the amplifier to the two terminals at the back of one speaker. Connect the two gray wires from the amplifier to the two terminals at the back of the other speaker. Listen to the sound. Reverse the red wires at the speaker so that the wire that was in the left speaker terminal is now in the right terminal and vice versa. Listen to the sound again. The final connections should be in the way that gives maximum sound." Explain the physical reasons for these instructions.

PROBLEMS

(Unless otherwise stated, use 343 m/s for the speed of sound in air.)

Section 15.3

1 You hear the thunder from a lightning flash 5.5 s after you see the flash. How far away was the lightning? Assume you can ignore the time it takes light to reach you, since the speed of light, 3×10^8 m/s, is so much greater than the speed of sound.

2 The sound from a pile driver is heard considerably after the hammer is seen to impact the piling. What is the delay time if you are 630 m from the pile driver? Make the same assumption about light speed as you did in Prob. 1.

3 A guitar string vibrates with a frequency of 530 Hz. What is the wavelength of the sound it emits? Repeat for frequencies of 180 and 1550 Hz.

4 In order to locate objects in the dark, a bat emits sound at an ultrasonic frequency of 57 kHz. What is the corresponding wavelength? What frequency must the bat emit in order to make the wavelength 1.33 mm?

5 Use the equation in the footnote of Table 15.1 to calculate the speed of sound in air at 20°C.

6 Use Eq. 15.4 to calculate the speed of sound in air at 0°C and at −20°C. Take the molecular mass of air to be 29. Repeat for helium gas.

7 Use the value of bulk modulus to find how fast sound travels in mercury.

8 Using the density of aluminum and the data in Table 15.1, calculate the bulk modulus of aluminum.

9 The velocity of sound in bone at 1 MHz is 3455 m/s, and the density of bone is about 1.85 g/cm³. Calculate the bulk modulus of bone at this frequency.

10 By what percentage does the speed of sound in air change as the temperature changes by 1 C° from 20°C to 21°C?

11 The depth finder on a fishing boat sends sound waves downward through the water and detects reflected waves. The instrument shows a school of fish at a depth of 3.85 m directly below the boat in water at 20°C. (*a*) How long after a pulse is sent out is the reflected wave from these fish received? (*b*) In order to interpret the distance, the instrument must receive a wave reflection from each pulse before the next pulse is sent. What is the highest frequency the instrument can send out in order to detect this school of fish? In order to detect fish at shallower depths, does the frequency requirement increase or decrease?

■12 You and a friend are walking along railroad tracks. Up ahead a railroad crew is working on the tracks. Your friend places her ear to the steel tracks and signals that she hears a hammer blow on the tracks. Standing next to her, you hear the hammer blow 2.56 s after her signal. How far from you is the railroad crew?

Section 15.4 and 15.5

13 A certain stereo set consumes power at a rate of 75 W. It has two speakers, each of which has a 50-cm² area from which the sound comes. If 0.085 W of sound power comes from each speaker, what is the sound intensity at the speaker? With what efficiency does the set convert electric energy to sound energy?

14 A certain loudspeaker has a circular opening with an area of 52 cm². Assume that the sound it emits is uniform and outward through this entire opening. If the sound intensity at the opening is 4.35×10^{-4} W/m², how much power is being radiated as sound?

▪**15** A beam of sound has an intensity of $4.25 \times 10^{-6} \, \text{W/m}^2$. What is the intensity level in decibels?

▪**16** What is the intensity level in decibels for a sound that has an intensity of $0.55 \, \text{W/m}^2$?

▪**17** (a) What is the intensity of a 33-dB sound? (b) If the sound level is 33 dB close to a speaker that has an area of $90 \, \text{cm}^2$, how much sound energy comes from the speaker each second?

▪**18** Eight typists in a room give rise to an average sound intensity level of 56 dB. What is the intensity level in the room when an additional three typists, each generating the same amount of noise, begin to type?

▪**19** The sound level in a room in which 35 people are conversing is 63 dB. How many people must leave to lower the sound level to 57 dB? Assume that the people speak with the same intensity.

▪▪**20** A tiny sound source sends sound equally in all directions. If the intensity is $3.5 \times 10^{-3} \, \text{W/m}^2$ 5.2 m from the source, (a) how much sound energy does the source emit each second? (b) What is the intensity 2.0 m from the source?

▪**21** What is the sound intensity at a location where the sound level is 25 dB?

▪**22** Find the sound intensity in a room where the intensity level is 88 dB.

▪**23** A sound beam has a cross-sectional area of $2.75 \, \text{cm}^2$ and a sound level of 105 dB. The beam strikes a sheet of sound absorber that completely absorbs it. How many joules are transferred in the absorber in 1 min?

▪▪**24** When the decibel level of a certain sound is increased by a factor of 6, its intensity increases fivefold. What is the original intensity of the sound?

25 An audio amplifier in a certain stereo set is guaranteed to have a gain of 35 dB. By what factor can this amplifier increase the intensity of a sound it receives?

▪**26** The intensity level 45 m from a small isotropic sound source is measured to be 85 dB. What is the total power output of the source?

Sections 15.8 and 15.9

27 The tiny loudspeaker in Fig. P15.1 is in an air-filled tube that is bent into a circle. If the radius of the circle is 1.35 m, what are the three lowest frequencies for which intense sound waves will be produced? (The drawing is not to scale.)

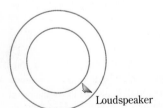

Loudspeaker

FIGURE P15.1

▪**28** The loudspeaker in Fig. P15.1 sends sound through the air-filled interior of a hollow tube that is bent into a circle. The tube resonates to loudspeaker frequencies of 66, 132, 198, and 264 Hz, as well as to higher frequencies. What is the circumference of the circle? Assume the loudspeaker to be much tinier than shown.

▪**29** The two identical loudspeakers in Fig. P15.2 vibrate in phase with a frequency of 3400 Hz. At what values of x is a (a) loud sound heard at P? (b) weak sound heard at P?

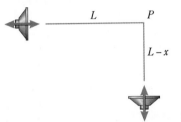

FIGURE P15.2

▪**30** In Fig. P15.2, the two identical loudspeakers vibrate in phase with the same frequency. The length L is 27.5 cm longer than the length $L - x$. The frequency of the waves sent out from the speakers is slowly increased from 15 Hz. At what frequency does an observer at P first hear (a) maximum loudness and (b) minimum loudness?

▪▪**31** Two tiny sound sources face each other and send out identical, in-phase waves. One source is at $x = 0$, and the other is at $x = 4.6 \, \text{m}$. If the wavelength of the waves sent out by the sources is 42 cm, at what points along the line from $x = 0$ to $x = 4.6 \, \text{m}$ does a detector register minimum sound?

▪▪**32** Assume the situation described in Prob. 31 but with variable-frequency sources. The source frequency starts low and then is increased slowly. At what frequencies is minimum sound heard at the point $x = 1.6 \, \text{m}$?

▪▪**33** In Fig. P15.3, two small speakers (small enough to be considered as point sources) are shown separated by a distance d. You are located at the position $y = 0$, a distance D from the midpoint between the speakers. The speakers are synchronized, both emitting the single frequency of 820 Hz. At $y = 0$, you are equidistant from the speakers, and hence hear a maximum sound intensity. You now move along the y axis to a position y_{min} where the sound intensity becomes minimum. (a) Find y_{min} if $D = 2.0 \, \text{m}$ and $d = 1.0 \, \text{m}$. (b) Find the angle θ shown in Fig. P15.3.

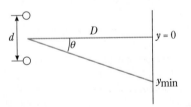

FIGURE P15.3

34 Two violins are slightly out of tune. When they are bowed so as to send out the supposedly same note, a beat frequency of 1.3 Hz is noticed. If the frequency emitted by one violin is 275 Hz, what are the possible frequencies emitted by the other?

35 Two pianos sound the same note, but the vibration from one is 320.4 Hz and that of the other is 321.1 Hz. What is the beat frequency between the two tones?

Section 15.10

36 A pipe open at one end and closed at the other is 76.4 cm long. What are the three lowest frequencies to which it will resonate? Draw the wave within the tube for each frequency. Repeat for a pipe open at both ends.

37 What are the three lowest frequencies to which a pipe open at both ends will resonate? The pipe is 90.5 cm long. Draw the wave within the tube for each frequency. Repeat for a pipe closed at one end.

38 In an experiment like that shown in Fig. 15.14, resonance occurs when the water is 31.55 cm high in the tube and again when it is 40.65 cm high. If no resonance occurs at any height between these two values, find the frequency of the tuning fork.

39 A man wishes to find out how far down the water level is in an iron pipe leading into an old well. Being blessed with perfect pitch, he merely hums musical sounds at the mouth of the pipe and notices that the lowest-frequency resonance is 81 Hz. About how far from the top of the pipe is the water level?

40 The Lincoln Tunnel under the Hudson River in New York City is about 2630 m long. To what sound frequencies does it resonate? What practical importance, if any, do you think this has?

41 A cylindrical tube resonates at the following consecutive frequencies: 415 Hz, 581 Hz, and 747 Hz. (*a*) What is its fundamental resonance frequency? (*b*) Is it open at just one end or at both ends?

42 The speed of sound in hydrogen is about 1270 m/s. If a pipe that resonates in its fundamental frequency to 550 Hz in air is filled with hydrogen, what will its fundamental resonance frequency be?

43 A certain organ pipe emits a fundamental frequency of 630 Hz when the temperature of the pipe is 18°C. A second identical pipe near a heater has a temperature of 27°C. What beat frequency is heard when these two pipes are sounded in unison?

44 You have two identical pipes open at one end and closed at the other, both 67 cm long. One pipe is placed in a room containing pure oxygen, the other is placed in a room with pure nitrogen. If you listened to a recording of them simultaneously sounding their fundamental tones, what beat frequency would you hear?

Section 15.11

45 How fast must a car be approaching you for its horn to be heard emitting a frequency 5 percent higher than when the car is standing still? How fast must the car be going away from you for the sound from its horn to be heard at a frequency 5 percent lower than when standing still?

46 A bird flying away from you at a speed of 21.3 m/s is emitting a pure note of 2040 Hz. The air temperature is 15°C. What is the frequency you hear?

47 A sound source at the origin of a coordinate system sends out waves of frequency f along the positive x axis. There is a 17.5 m/s wind blowing toward the positive x direction. (*a*) Find the frequency and wavelength of the sound wave observed by a stationary observer positioned on the x axis. Represent the speed of sound in still air by v. (*b*) Repeat for wind blowing at the same speed in the negative x direction.

48 A source of sound waves whose frequency is 440 Hz is approaching a wall with speed 12.5 m/s. The sound reflects back to an observer traveling with the source. What is the frequency of the reflected wave as heard by the observer?

49 An ambulance siren changes pitch from 850 to 770 Hz as it rushes past where you are standing by the curb. The air temperature is 10°C. How fast is the ambulance moving?

50 Two trains are moving in opposite directions on parallel tracks and are approaching a station. The horns on both trains are producing sound at a frequency of 550 Hz. One train is approaching at 32 m/s. What is the speed of the other train if an observer at the station hears a beat frequency of 4.4 Hz?

Section 15.12

51 An aircraft is flying horizontally at Mach 1.8 over a flat desert. A sonic boom is heard on the ground 8.1 s after the aircraft has passed directly overhead. Assume the speed of sound in the air is 350 m/s. At what altitude is the aircraft flying?

52 A plane is flying at supersonic speed at an altitude where the speed of sound is 320 m/s. The shock angle makes an angle of 33.5° with the direction of the plane. What is the plane's speed and its Mach number?

53 The temperature of earth's atmosphere varies with altitude, and hence so does the speed of sound. At a height of 20 km, atmospheric temperature is approximately 218 K, whereas at a height of 1 km, 280 K is a typical temperature. Suppose a spacecraft plunges into the atmosphere traveling at a speed of 8700 m/s as it passes through a height of 20 km and slows to 4800 m/s by the time it reaches an altitude of 1 km. Calculate the Mach number and the shock angle produced by the spacecraft at these two altitudes.

General Problems

54 A wire 4.0 m long having a linear density of 2.2 g/m is clamped on both ends and is placed under 340 N of tension.

This produces a standing wave pattern with five loops between the ends. A nearby narrow pipe has an adjustable piston closing one end. As the piston is moved, it is found that sound from the wire causes a resonance to occur when the piston is 1.07 m from the open end of the pipe. In what harmonic is the pipe resonating, and what is the frequency of the sound? Assume the air in the room is at 30°C.

■55 An organ pipe is tuned so that its third harmonic is at 1320 Hz at the start of a concert. The original temperature in the concert hall was 23°C, but the hall warms up as the concert proceeds. At intermission, the organist hears 5 beats per second when this same harmonic is compared with a 1320-Hz tuning standard. What has the temperature in the hall become by intermission time? (Assume the length of the pipe has not changed.)

■56 In Fig. P15.4, A and B are stationary sources of sound waves, both emitting the same frequency f. An observer is

A • $V = 100$ km/h B •

FIGURE P15.4

in a car, traveling toward A and away from B at 100 km/h. She measures a beat frequency of 20 Hz in the interference of the waves from the two sources. If the air temperature is 23°C, calculate the frequency being emitted.

■57 In order to determine the depth of a well, you drop a rock from the top of the well and hear the splash 3.34 s after releasing the rock. How deep is the well? (Ignore air resistance on the falling rock.)

■58 A stereo amplifier takes an input signal of 0.50 mW and amplifies the signal to an output of 90 W. What is the gain of the amplifier measured in decibels?

■■59 A very narrow tube open at one end and closed at the other is 45 cm long. A 205-Hz sound source is held just above the open end and accelerated away from the tube along the tube's axis. At what speed will a resonance first be detected? What harmonic will this resonance correspond to? Assume the air temperature is 0°C.

■■60 An aluminum rod is 10.6 m long. When struck along its axis by a vibrating mechanism, longitudinal sound waves travel along the rod and reflect back from its ends. What vibrator frequency would set up the lowest resonance of waves in the rod?

ELECTRICITY AND MAGNETISM

PART THREE

ELECTRICITY AND MAGNETISM

A problem is not solved in the laboratory. It is solved in some person's head, and what all the apparatus is for is to get their head turned around so they can see the thing right.

CHARLES KETTERING

So far our description of physical phenomena has required only four basic independently defined quantities—mass, length, time, and temperature. Yet the observation of other forces in nature—the natural magnetism of lodestones and the attraction of bits of matter by the mineral amber (*elektron* in Greek) that had been rubbed by cloth—have been recorded since ancient times.

It was during the late eighteenth century and early nineteenth century when researchers such as Coulomb in France and Franklin in the United States began to investigate the behavior of electrically charged matter, finding that two opposite kinds of charge exist and deriving the force law by which charges interact. Success followed success throughout the nineteenth century as scientists continued to develop an understanding of the field of electricity and magnetism. The deflection of a compass needle when placed near an electric current showed that a current produces a magnetic field. Changing magnetic fields were shown to produce electric fields. The existence of electromagnetic waves, consisting of oscillating electric and magnetic fields traveling at the speed of light, was predicted theoretically and subsequently demonstrated.

Of all the achievements of classical physics, perhaps none have had as far-reaching an effect as those mentioned above. They have enabled us to design and construct devices that have truly transformed our daily lives. A small sampling would include the electric light, electric generators and motors, and all the means of electronic communication: the telephone, radio, and TV. Other devices based on the principles of electricity and magnetism have made it possible to measure ever smaller and faster phenomena, expanding the frontiers of basic research. Great advances in medical diagnosis and treatment have also been made possible, as have new methods of processing materials and manufacturing goods.

The benefits which have accrued to us due to research in electricity and magnetism could not have been foreseen by those involved in that research, and were not the primary motivation behind their investigations. We would do well to look at these examples when people who base decisions only on short-term results question why it is necessary to continue investing in the quest for basic knowledge.

CHAPTER 16

ELECTRIC FORCES AND FIELDS

It is difficult for us to visualize the world of a century ago, when the use of electricity was just beginning. A few people had electric lights, but the electrical appliances and machines we are so accustomed to today were totally missing. Primitive electric motors and batteries were curiosities just beginning to show practical importance. What a contrast to today, when nearly everything we do uses electricity in some way. Because of its widespread use, electricity is a tool that all educated people must understand. We will spend the next several chapters learning about the ways in which electricity plays an influential role in the world about us.

16.1 THE CONCEPT OF ELECTRIC CHARGE

We have historical accounts that, as early as the sixth century B.C., the Greek Thales knew that sparks can be produced and lightweight objects attracted when the fossilized resin called amber is rubbed with fur. The Greek word for amber is *elektron,* from which we derive our word electricity. During the eighteenth century there was a great deal of experimentation in "electrifying" objects, including people, often with humorous results. One of the most influential and productive experimenters was the American Benjamin Franklin, whose demonstration of the equivalence of electricity and lightning by means of flying a kite into thunderclouds is legendary. Franklin also was first to propose the terms "positive" and "negative" for the two kinds of electrification an object can have.

The following two centuries witnessed the evolution of a complete theory of electrical and magnetic phenomena. It was near the end of the nineteenth century and into the early twentieth century that the fundamental discoveries of the electrical structure of the atom were made. In 1897 the Englishman J. J. Thomson measured the properties of the fundamental negative charge, the electron. Ernest Rutherford, also English, succeeded in 1911 in identifying the exceedingly small positive nucleus of the atom around which the electrons move. Finally, in a series of experiments performed between 1909 and 1917, the American Robert Millikan and his coworkers accurately measured the amount of charge on the electron.

Let us now proceed to discuss the nature of electrical charge in order to begin to explore the great variety of electrical phenomena.

16.2 ATOMS AS THE SOURCE OF CHARGE

The discoveries mentioned in the previous section showed that an atom is composed of a tiny positively charged nucleus around which are negatively charged particles called electrons. This is illustrated for an atom of carbon in Fig. 16.1. You may recall from courses in chemistry that atoms are electrically neutral. That is, the quantity of positive charge on the nucleus exactly equals the quantity of total negative charge on the electrons about the nucleus. In the case of the carbon atom, if $-e$ is the charge on each electron, the charge on the nucleus is exactly $+6e$. We shall postpone a detailed discussion of the atom to a later chapter and merely make use of its electrical constitution here.

It appears that the universe as a whole is nearly, if not completely, electrically neutral. The earth has very little, if any, excess of either positive or negative charge. For nearly all practical purposes, it can be considered to have zero net charge. The vast majority of the charges on and in the earth reside in atoms. When free negative or positive charges are found, they are usually assumed to have come from the tearing apart of an atom.

It is not at all difficult to remove an electron from an atom—under certain circumstances. For example, if a rod of ebonite (hard rubber) is rubbed with animal fur, some of the electrons from the atoms in the fur are rubbed off onto the ebonite rod. (The reason for this charge transfer is not simple to explain. It is covered in courses dealing with solid-state physics.) Hence, the rod acquires *a net excess of electrons,* making it negatively charged. When it is touched to a metal object, some of the excess electrons are transferred to the metal, as illustrated in Fig. 16.2.

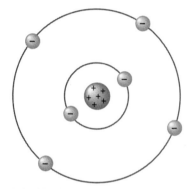

FIGURE 16.1

A schematic representation of a carbon atom. The negative charges on its six electrons are exactly balanced by the positive nuclear charge. (The nucleus and electrons are much smaller than shown.)

FIGURE 16.2

When the negatively charged ebonite rod touches the uncharged metal ball, electrons are conducted off the rod onto the ball.

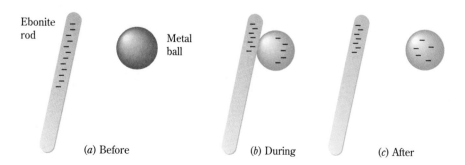

Ebonite rod

Metal ball

(a) Before   (b) During   (c) After

Similarly, if a glass rod is rubbed with silk, some of the electrons leave the atoms in the rod and leave *an excess of positive charge on it.* If the positively charged rod is touched to a neutral metal ball, electrons leave some of the atoms of the metal and replace those electrons lost by the atoms in the glass. As a result, the metal ball acquires a net positive charge.

Many other materials give rise to a separation of charge when rubbed together. The ones we have described were used to define positive and negative charge before the existence of the electron was even known.

16.3 FORCES BETWEEN CHARGES

Now that we know how to obtain positively and negatively charged bodies, we are ready to examine the forces between charges. One of the simplest ways of doing this is with very light metal-coated balls. Such balls can be charged by touching them with a charged glass or ebonite rod. If the balls are suspended by light threads, four interesting experiments can be performed. These are illustrated in Fig. 16.3, and from them we can conclude the following:

1 Like charges repel one another; that is, two positive charges repel each other, as do two negative charges.
2 Unlike charges attract one another; that is, positive charges attract negative charges, and vice versa.
3 The magnitude of the electric force between two charged objects often exceeds the gravitational attraction between them. (For example, the gravitational force between the two balls in (b), (c), and (d) is far too small to affect the way they hang.)

The strands of hair on this student's head all receive an electrical charge of the same sign from the static electricity generator. These charges repel each other, causing her hair to stand up in a wild pattern.

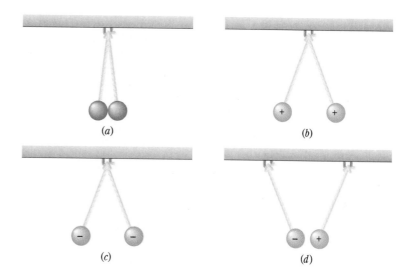

FIGURE 16.3

In (*a*) the balls are uncharged. The charged balls in (*b*), (*c*), and (*d*) show that like charges repel one another while unlike charges attract one another.

16.4 INSULATORS AND CONDUCTORS

Although all materials are made of atoms and all atoms are made of electrons and nuclei, we are well aware that the electrical properties of substances vary widely. There are two basic groups into which substances can be divided according to their electrical properties: *conductors* and *nonconductors* (or *insulators*).*

In insulators, the electrons of any given atom are bound tightly to that atom and are not free to move through the material. Hence, even if a charged rod is brought close to an insulator, the electrons and nuclei in the atoms of the insulator cannot move under the attraction or repulsion of the rod's charge.

Electrical conductors behave quite differently. These substances contain charges that are free to move throughout the material. Metals are familiar conductors; though each atom of the metal is normally neutral (that is, uncharged), the electrons farthest from the nucleus are easily freed from the atom. They then move through the metal, carrying their negative charge from place to place in the process. Therefore, when a negatively charged rod is brought close to a piece of metal (without touching it), the rod repels some of the free electrons in the metal to the most distant regions of the metal. Similarly, a positively charged rod attracts free electrons to the portion of the metal nearest the rod.

Metals are not the only electrical conductors. Many substances—ionic solutions, for example—contain ions (charged atoms) that can move relatively freely through the substance. All electrical conductors contain charges that can move over long distances when repelled or attracted by nearby charged objects.

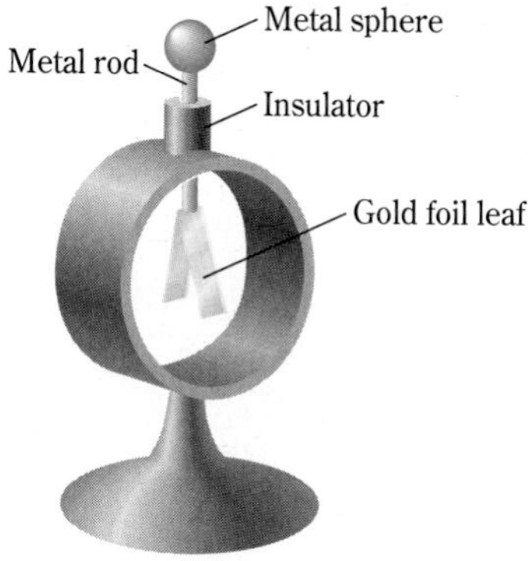

16.5 THE ELECTROSCOPE

The *electroscope* (Fig. 16.4) is a simple device used for detecting and measuring charges of small magnitude. A metal rod from which are suspended two very thin

Metal sphere
Metal rod
Insulator
Gold foil leaf

FIGURE 16.4

One type of gold-leaf electroscope. The portion consisting of the metal sphere, rod, and gold foil is insulated from the case.

*A third class of substances, called *semiconductors,* act as either insulators or conductors depending on temperature and other energy conditions imposed on them.

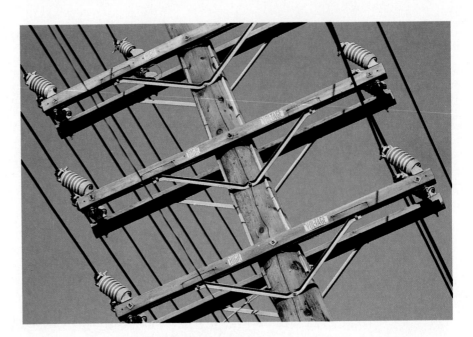

The transmission of electricity uses conductors to carry the electricity and insulators to protect the poles which hold up the conducting wires.

leaves of gold foil is held inside a metal case by an insulator that keeps the rod from touching the case. The two faces of the case are covered with glass, so that the positions of the leaves can be seen.

Suppose some negative charge is placed on the electroscope by touching a charged piece of ebonite to its metal sphere. The charge is confined entirely to the sphere, rod, and leaves since they are insulated. Because like charges repel one another, the negative charges originally on the rod distribute themselves more or less uniformly over the sphere, rod, and leaves. As a result, the leaves, being free to bend and being repelled by the like charges on one another, take the position shown in Fig. 16.5a.

If a negatively charged ball is now brought close to the metal sphere of the electroscope, as shown in Fig. 16.5b, many of the negative charges in the metal sphere are repelled down the rod, causing the leaves to separate farther. An opposite effect is observed if a positively charged ball is brought close to the electroscope (Fig. 16.5c). In addition, a ball with no net charge does not disturb the electroscope to any great extent. With this apparatus, the sign of a charge and its approximate magnitude can be determined. You should convince yourself that a similar procedure can be followed if the electroscope is charged positively.

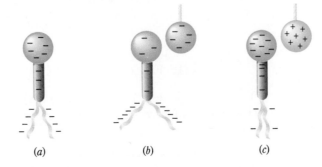

FIGURE 16.5

The charged electroscope is used to determine the sign and approximate magnitude of the charge on an object.

(a)   (b)   (c)

16.6 CHARGING BY CONDUCTION AND BY INDUCTION

There are two general ways of placing charge on a metal object using a second object that is already charged. As an example, consider the ways in which you might use a negatively charged ebonite rod to charge a metal ball. One way is to touch the ball to the rod, as in the previous section. On contact, some of the excess negative charge on the rod moves onto the ball. This process, shown in Fig. 16.2, is called **charging by conduction.**

The same rod can also be used in a different way to charge the ball. This is shown in Fig. 16.6. In this process, **charging by induction,** the rod is not touched to the ball at all. When the rod is brought close to the left side of the ball, some of the electrons in the metal are repelled to the right side of the ball, leaving a positive charge on the left side. Since no charge has been added to or subtracted from the ball, it is still electrically neutral, of course. Now, suppose you touch the ball with an object other than the charged ebonite rod, such as your finger. Because your body is a conductor (though not a good one), charge moves from the ball through your body to the earth. Thus the nearby negatively charged rod induces negative charge to leave the ball and move to the ground. (We say that the ball is **grounded,** and the symbol —||ı is used to show this. Usually, to ground an object, we attach it by a wire to a water pipe or some other object that goes into the earth.) Once negative charge has moved from the ball to the ground, the ball is no longer neutral. After the conducting path to ground is removed, the negative rod can be taken away and the ball will have a positive charge. (Why must the grounding object be removed *before* the rod is?)

If you compare Figs. 16.2 and 16.6, you can see that an ebonite rod can charge a metal object *negatively* by *conduction* but it charges the same object *positively* by *induction.* You may find it interesting to work out similar diagrams using a positively charged glass rod. The charges are reversed in that case.

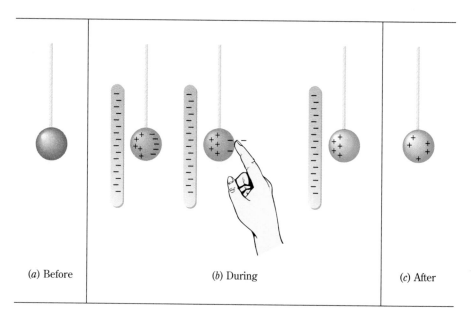

FIGURE 16.6

Charging a metal ball by induction. Note that the rod and the ball never touch, but the finger and ball do. As a result of this process, the rod and ball end up with unlike charges.

(a) Before   *(b)* During   *(c)* After

16.7   FARADAY'S ICE-PAIL EXPERIMENT

In 1843, Michael Faraday carried out a simple but highly instructive experiment. He attached a metal ice pail to an uncharged electroscope, as shown in Fig. 16.7*a*. When a positively charged metal ball suspended from a thread was lowered into the pail (without touching it), as in (*b*), the leaves of the electroscope diverged, indicating that some charge was induced on the outer surface of the pail.

Moreover, when the charge was moved around within the pail, the leaves did not change their amount of divergence. Only when the ball was removed from the pail did the leaves return to their original collapsed position, indicating that the pail had returned to electrical neutrality.

Faraday further noticed that if the charged metal ball was touched to the inside of the pail, as in (*c*), the electroscope leaves remained in their divergent position. In this case, however, when the ball was removed from the pail, the electroscope leaves remained divergent, as in (*d*), indicating that the pail remained charged. When the ball was brought close to a second electroscope, it had no effect on the leaves. Apparently, touching the inside of the pail had completely neutralized the original excess charge on the ball. Since the leaves of the electroscope attached to the outside of the pail did not move when the ball touched the inside of the pail, Faraday concluded that the inner surface of the pail had *precisely* enough charge on it to neutralize the ball, and that the pail now was left with a net charge on its exterior surface equal to that originally carried by the ball.

From these experiments, we can draw the following conclusions:

1 A charged metal object suspended inside a neutral metal container induces an equal and opposite charge on the inside of the container.

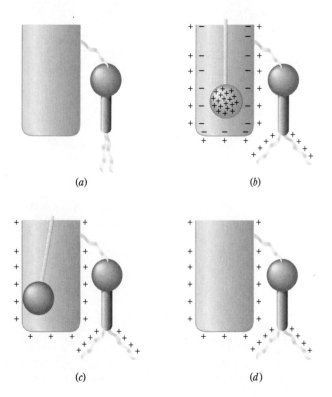

(*a*)   (*b*)

(*c*)   (*d*)

FIGURE 16.7
Faraday's ice-pail experiment.

2 When the charged metal object is touched to the inside of the container, the induced charge exactly neutralizes the excess charge on the object.

3 When a charged object is placed inside a neutral metal container, an equal charge of the same sign is forced to the outer surface of the container.

4 All the net charge on any metal object will reside on its outer surface if a conducting path is provided so that the charge can move there.

These are important facts concerning electric charges on conductors. We shall interpret them more fully when we examine Coulomb's law and the concept of electric fields in Secs. 16.9 to 16.13.

16.8 CONSERVATION OF CHARGE

In mechanics, we learned that nature conserves certain quantities. Among these are energy, linear momentum, and angular momentum. Each of these quantities obeys a conservation law, and, as we have seen, that fact is of great importance in our universe.

There are also conservation laws that apply to electrical quantities. One of these is the law of conservation of electric charge. By this we mean that the algebraic sum of all the charges in the universe remains constant. This fact has become clear to us only in this century. When physicists became able to create new particles by bombarding one high-energy particle with another in huge accelerators, they found that charges are always created (or annihilated) in pairs. Any reaction that creates an electron (charge $-e$) also creates a particle whose charge is $+e$. Similarly, when a particle with charge $+e$, such as a positron (a positive electron), combines with a particle of charge $-e$, both charges disappear; their algebraic sum at the outset was zero, and it remains zero after the reaction is completed. In every experiment, the algebraic sum of the charges before the reaction is the same as that afterwards. There appears to be no way in which net charge can be created or destroyed. We conclude that the net charge of the universe does not change. This is called the **law of conservation of charge.** A way of stating this law on a somewhat smaller scale is

Net positive or negative charge cannot be created or destroyed in any physical process.

Notice that the law does not state that the number of electrons or protons in the universe is constant. We know of many reactions in which pairs of oppositely charged particles are created or destroyed. Even though we do not yet know the net charge of the universe, or of our galaxy, or even of the earth and its atmosphere (it is possible that these net charges are close to zero), our knowledge of the conservation of charge is still of great use to us. We shall use this concept when we discuss electric circuits. Moreover, when particle physicists are trying to understand the particles that may be created in high-energy reactions, the conservation law guides them in deciding which reactions are possible.

16.9 COULOMB'S LAW

The mathematical law that describes how like charges repel and unlike charges attract each other was discovered in 1785 by Charles Augustin de Coulomb (1736–1806) and is called **Coulomb's law.** By means of a very sensitive balance, similar

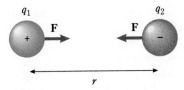

FIGURE 16.8

The two unlike charges attract each other with equal force, even if the charges are unequal in magnitude.

to that used by Cavendish in his study of gravitation, Coulomb was able to measure the force between two small charged objects (Fig. 16.8). Two balls, small enough to be considered points relative to the distance r between their centers, carry charges $+q_1$ and $-q_2$. After a number of experiments, Coulomb concluded that the force on ball 1 varied in proportion to the product of the two charges and inversely as the square of the distance between centers:

$$F \propto \frac{q_1 q_2}{r^2}$$

$$F = \text{constant} \, \frac{q_1 q_2}{r^2} \tag{16.1}$$

and that the force was in the direction shown in Fig. 16.8. If the two charges were either both positive or both negative, the magnitude of the force would be the same but the direction would be the reverse of what is shown in Fig. 16.8. According to Newton's law of action and reaction, the force on ball 2 must be identical in magnitude but oppositely directed.

Before we can evaluate the proportionality constant in Eq. 16.1, we must decide on a unit with which quantities of charge can be measured. Since charge and the electrical force it produces are fundamental physical properties new to us, the unit of charge cannot merely be derived from units we have already established. Like mass, length, time, and temperature, charge is a fundamental dimension whose unit must be defined. As we will see in Sec. 19.10, the SI unit of charge is defined in terms of electric current. For now, we simply state that the SI charge unit is the *coulomb* (C). When we use this unit for q_1 and q_2, Coulomb's law can be written

$$F = k \frac{q_1 q_2}{r^2} \tag{16.2}$$

where F is in newtons and r is in meters. The constant of proportionality k, determined by experiment, is $8.9874 \times 10^9 \, \text{N} \cdot \text{m}^2/\text{C}^2$ when the experiment is performed in vacuum (or air, to good approximation). We shall usually take k to be $9.0 \times 10^9 \, \text{N} \cdot \text{m}^2/\text{C}^2$.

The constant k is often written as $1/4\pi\epsilon_o$, where ϵ_o is called the *permittivity of vacuum*. It follows that ϵ_o has the value

$$\epsilon_o = 8.85 \times 10^{-12} \, \text{C}^2/\text{N} \cdot \text{m}^2$$

When we insert numerical values into Eq. 16.2, we can see that the coulomb is a very large quantity of charge. Two charges each of one coulomb separated by one meter would exert a force on each other of *nine billion newtons!* Amounts of static charge we encounter in everyday experience usually have magnitudes measured in microcoulombs or smaller.

The fundamental quantity of charge found in matter, as mentioned in Sec. 16.2, is that carried by the electron and proton, which we symbolize by e. The experimentally determined value of e is

$$e = 1.60218 \times 10^{-19} \, \text{C}$$

As implied in Sec. 16.2, the proton has a charge of $+e$, the electron $-e$. All the

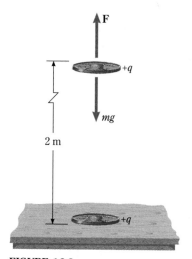

FIGURE 16.9

Only a tiny fraction of electrons needs to be removed from a penny to give rise to large electric forces.

fundamental charged particles thus far discovered in ordinary matter have the charge e or integral multiples of it. The charge e thus appears to be the smallest quantity, or **quantum,** of charge occurring in nature.*

Experiment shows another interesting feature of the electric force. When several charged particles exert forces on one another, the forces are additive. For example, suppose two charges are close to a third charge. Each of the two exerts a Coulomb's law force on the third, and the total force on the third is simply the vector sum of the two separate forces. We call this fact the *superposition principle* for Coulomb's law forces. Its use will become evident in some of the examples that follow.

Example 16.1

A copper penny has a mass of about 3 g and contains about 3×10^{22} copper atoms. Suppose two pennies have a fraction of their electrons removed, leaving them each with a net positive charge $+q$. When one penny is placed on a table, the other will be suspended against its weight by the electric force 2 m above the first, as shown in Fig. 16.9. (***a***) How large must q be to balance the weight of the penny? (***b***) How many electrons must each penny have lost to produce $+q$? (***c***) What fraction of the copper atoms must be missing an electron?

Reasoning   Part (*a*)

Question   What is the weight of a penny?
Answer   Weight = $mg = (3 \times 10^{-3}\,\text{kg})(9.8\,\text{m/s}^2) = 0.03\,\text{N}$

Question   What is the expression for the electric force on the upper penny?
Answer   Equation 16.2 tells us that

$$F = \frac{(9 \times 10^9\,\text{N} \cdot \text{m}^2/\text{C}^2)q^2}{(2\,\text{m})^2}$$

where q is the charge on each penny.

Question   What equation determines the charge q?
Answer   The magnitude of the force F must equal the weight, 0.03 N, so

$$\frac{(9 \times 10^9\,\text{N} \cdot \text{m}^2/\text{C}^2)\,q^2}{(2\,\text{m})^2} = 0.03\,\text{N}$$

Reasoning   Part (*b*)

Question   Once q is found, what determines the number of electrons removed?
Answer   Each electron removed leaves the penny with an excess charge $+e$. Thus the number of electrons removed is $n = q/e$.

---

*Recent theories on fundamental particles hold that some particles, such as the proton and neutron, consist of combinations of particles (called *quarks*) that have charges of $e/3$ or $2e/3$. Such particles have not yet been isolated experimentally. Even if they are found in the future, that discovery will not alter the fact that nature has a minimum quantity in which charge can exist.

Reasoning   Part (c)

Question   How is n related to the fraction of atoms losing an electron?
Answer   The penny has a total number $N = 3 \times 10^{22}$ atoms. The fraction that have lost a single electron is n/N.

Solution and Discussion   In part (a) we have

$$q^2 = \frac{(0.03\,\text{N})(2\,\text{m})^2}{9 \times 10^9\,\text{N} \cdot \text{m}^2/\text{C}^2}$$

This yields

$$q = 4 \times 10^{-6}\,\text{C} = 4\,\mu\text{C}$$

The number of electrons removed is

$$n = \frac{q}{e} = \frac{4 \times 10^{-6}\,\text{C}}{1.6 \times 10^{-19}\,\text{C}} = 2 \times 10^{13}$$

This represents a fraction

$$\frac{n}{N} = \frac{2 \times 10^{13}}{3 \times 10^{22}} = 7 \times 10^{-10}$$

of the atoms. Notice that charges as small as microcoulombs give rise to easily measurable forces between macroscopic objects.

Example 16.2

Find the force on the center charge q_2 in Fig. 16.10.

R   *Reasoning*

Question   What are the directions of the individual forces on q_2?
Answer   Both q_1 and q_3 exert an attractive force on q_2. These two forces thus oppose each other, as shown in Fig. 16.10. The force exerted by q_1 on q_2 we call $\mathbf{F}_1$. That exerted by q_3 we call $\mathbf{F}_3$.

Question   How do I calculate the individual forces?
Answer   By applying Coulomb's law separately to each interaction, as if the remaining charge were not present.

Question   How do I treat the signs of the charges?

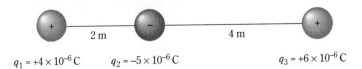

$q_1 = +4 \times 10^{-6}\,\text{C}$ $q_2 = -5 \times 10^{-6}\,\text{C}$ $q_3 = +6 \times 10^{-6}\,\text{C}$

FIGURE 16.10

The central charge is attracted to q_1 by force $\mathbf{F}_1$ and to q_3 by force $\mathbf{F}_3$.

Answer You have already used the signs to establish the directions of the forces. You can now calculate the magnitudes of the opposing forces, knowing that you will take the *difference* between these magnitudes.

Solution and Discussion Coulomb's law gives the following for the magnitudes of the individual forces:

$$F_1 = \frac{(9 \times 10^9 \ N \cdot m^2/C^2)(4 \times 10^{-6} \ C)(5 \times 10^{-6} \ C)}{(2 \ m)^2}$$

$$= 0.04 \ N$$

$$F_3 = \frac{(9 \times 10^9 \ N \cdot m^2/C^2)(5 \times 10^{-6} \ C)(6 \times 10^{-6} \ C)}{(4 \ m)^2}$$

$$= 0.02 \ N$$

The net force of these opposing forces is obtained by subtracting their magnitudes:

$$F_{net} \ (on \ q_2) = 0.04 \ N - 0.02 \ N = 0.02 \ N$$

This force is directed to the left in Fig. 16.10.

Exercise Find the force on the 4-μC charge. *Answer:* $4 \times 10^{-2} \ N.$

Example 16.3

Find the resultant force on the +10-μC charge of Fig. 16.11.

R   *Reasoning*

Question In what directions are the individual forces exerted on the +10-μC charge?

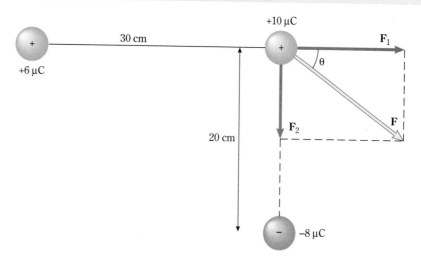

FIGURE 16.11

To find the resultant force **F** on the +10-μC charge, we must add the forces exerted on it by the other two charges.

Answer The force $\mathbf{F}_1$ exerted by the $+6$-μC charge is repulsive, to the right. The -8-μC charge exerts an attractive force $\mathbf{F}_2$ downward.

Question What expressions will give the magnitudes of these forces?
Answer Coulomb's law:

$$F_1 = \frac{(9 \times 10^9 \, \text{N} \cdot \text{m}^2/\text{C}^2)(+6 \, \mu\text{C})(+10 \, \mu\text{C})}{(0.3 \, \text{m})^2}$$

$$F_2 = \frac{(9 \times 10^9 \, \text{N} \cdot \text{m}^2/\text{C}^2)(-8 \, \mu\text{C})(+10 \, \mu\text{C})}{(0.2 \, \text{m})^2}$$

As in the previous example, once you determine the direction of the forces, all you need is the magnitude of each, without regard to algebraic sign.

Question How do I add these magnitudes?
Answer They are vectors at right angles, so the Pythagorean theorem applies. From Fig. 16.11 we see that

$$F_{\text{net}} = \sqrt{F_1^2 + F_2^2} \qquad \text{and} \qquad \theta = \tan^{-1}\frac{F_2}{F_1}$$

Solution and Discussion The magnitudes of the forces are

$$F_1 = 6 \, \text{N} \qquad \text{and} \qquad F_2 = 18 \, \text{N}$$

This gives

$$F_{\text{net}} = \sqrt{36 + 324} \, \text{N} = 19 \, \text{N}$$

$$\theta = \tan^{-1}\frac{18.0}{6.00} = 72°$$

Example 16.4

Find the force on the $+20$-μC charge in Fig. 16.12.

R *Reasoning*

Question What are the directions of the two forces on the $+20$-μC charge?
Answer Since all charges are positive, the forces are both repulsive. Thus one force ($\mathbf{F}_1$) is downward and the other ($\mathbf{F}_2$) is $37°$ below horizontal to the right.

Question How do I add these forces?
Answer Resolve $\mathbf{F}_2$ into x and y components. The y component will add to $\mathbf{F}_1$. You can then use the Pythagorean theorem to find the resultant force.

Solution and Discussion Coulomb's law gives the following magnitudes:

$$F_1 = 2.0 \, \text{N} \qquad F_2 = 1.8 \, \text{N}$$

F_2 has the components

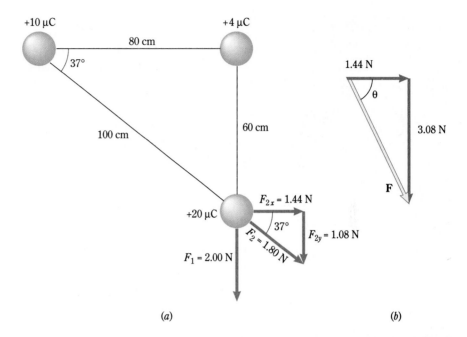

FIGURE 16.12

The vector forces acting on the 20-μC charge produce the resultant force **F** shown in (b).

$$F_{2x} = (1.8\,\text{N})\cos 37° = 1.4\,\text{N}$$

$$F_{2y} = (1.8\,\text{N})\sin 37° = 1.1\,\text{N}$$

Thus the components of the net force are

$$F_x = 1.4\,\text{N} \quad \text{and} \quad F_y = 2.0\,\text{N} + 1.1\,\text{N} = 3.1\,\text{N}$$

Then

$$F = \sqrt{1.4^2 + 3.1^2}\,\text{N} = 3.4\,\text{N}$$

and

$$\theta = \tan^{-1}\frac{3.1}{1.4} = 66°$$

Exercise Find the magnitude of the force on the 10-μC charge. *Answer:* 2.3 N.

16.10 THE ELECTRIC FIELD

We find it convenient to discuss electric forces in terms of a concept called the *electric field.* It serves much the same purpose in electricity that the concept of the gravitational field serves in mechanics. Before we discuss this new concept in detail, let us review the more familiar situation of the gravitational field.

We are familiar with the fact that the earth exerts a gravitational force directed toward its center on objects on and above its surface. The moon and other planets exert a similar force on objects near them. To describe this effect, we say that a

A lightning bolt is dramatic evidence that when the electric field between charges on the ground and in the clouds becomes great enough, charge will flow. Notice the small lightning stroke on the TV antenna at the left. Even that was enough to damage the TV set in the house. Imagine what could have happened if the main stroke had hit there instead of the tree.

gravitational field exists in these regions. At any point, the field is taken to be in the direction of the force an object would experience there. The strength of the field is proportional to the strength of that force.

It is convenient to sketch gravitational fields. That of the earth is shown in Fig. 16.13. We interpret this sketch in the following way. If an object is placed at point A, it will experience a force in the direction of the arrowhead, toward the earth's center. The lines, called *field lines,* show the direction of the earth's gravitational pull; this is taken to be the direction of the gravitational field. (To be truly representational, of course, Fig. 16.13 should be drawn in three dimensions, with lines of force directed from all sides toward earth's center.)

Field lines not only represent the direction of the force but also indicate its relative magnitude. You can see this in Fig. 16.13 by noticing that the lines are closer together near the earth, where the force is strong, than they are farther away from the earth, where the force is weaker. We shall return to this feature of field lines after we discuss the electric field, a field that describes the electrical forces charged objects exert on one another.

The electric field represents the electric force a stationary positive charge experiences. Consider how you might go about determining the electric field in a region. You could simply place a charged object (call it a test charge) in the region and determine the force on it due to all other charges. However, your test charge exerts forces on all other charges in the vicinity and, if these charges are in metals, could cause them to move. To eliminate this difficulty, we imagine that the test charge has a very special property: *the* **test charge** *is a fictitious charge that exerts no forces on nearby charges.* We represent it by q_t. In practice, we can approximate the test charge by using a very small charge that disturbs other charges in the vicinity by only a negligible amount.

We take the direction of the electric field at a point to be the same as the direction of the force on a *positive* test charge placed at that point. For example, suppose a positive test charge is placed at point A in Fig. 16.14a. It is attracted radially inward, as shown by the arrow at A. Indeed, the force on the positive test charge is directed radially inward no matter where in the neighborhood of the central negative charge it is placed. We therefore surmise that the electric field is directed as shown by the arrows: *the electric field near a negative charge is directed radially into the charge.*

We can determine the direction of the field near a positive charge in the same way, as shown in Fig. 16.14b. The positive test charge is repelled radially outward by the central positive charge. Hence, *the electric field near a positive charge is directed radially away from the charge.*

The directed lines we have drawn in Fig. 16.14 to show the direction of the electric field are called **electric field lines.** As we have seen, electric field lines

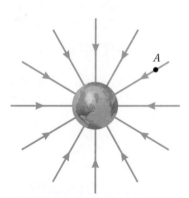

FIGURE 16.13

The gravitational field of the earth is directed radially inward and becomes stronger as one approaches the earth.

FIGURE 16.14

The electric field is directed radially inward toward a negative charge and radially outward from a positive charge.

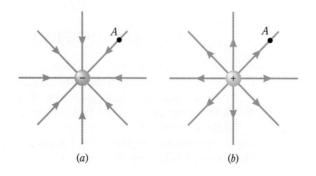

(a) (b)

originate on and are directed away from positive charges, and terminate on and are directed toward negative charges.

To make the electric field concept quantitative, we define a quantity called the **electric field strength E.** At any given point, the direction of **E**, a vector quantity, is taken to be the same as that of the electric field line through that point. The magnitude of **E** is equal to the force experienced by the test charge divided by the amount of charge q_t:

$$\mathbf{E} = \mathbf{F}/q_t \tag{16.3}$$

The units of **E** are thus defined to be N/C. Because **E** is a force per unit of charge, we frequently state that it is the force per unit positive test charge. However, we should realize that, in measuring the strength of an electric field, we would use a charge much smaller than 1 C so as not to disturb the other charges present.

As with the gravitational field, the relative strength of the electric field can be estimated by examining the field-line diagram. For example, the field lines in Fig. 16.14 are closest together near the charges. The force on a unit positive test charge (the electric field strength) is also largest close to the charges. Electric field strength is largest where the field lines are closest together. We often estimate the field strength in a region by noticing the density of the field lines in that region in a sketch of the electric field.

16.11 THE ELECTRIC FIELD OF A POINT CHARGE

We are often interested in the electric field generated by an ion or some other charged atomic-size particle. For most purposes, we can consider these to be point charges. Even a charged sphere acts like a point charge under certain circumstances, as we point out shortly. It is therefore important for us to know the electric field due to a point charge.

Suppose we wish to find the electric field strength at point P in Fig. 16.15, which is a distance r away from a positive point charge q. We know that the electric field due to q is radially outward, as we saw in Fig. 16.14b. Hence **E** at point P is in the direction shown. If we imagine a test charge q_t placed at P, the force on it is given by Coulomb's law:

$$F = k\frac{qq_t}{r^2}$$

Dividing through by q_t to obtain F/q_t, the electric field strength, we have

$$\frac{F}{q_t} = k\frac{q}{r^2}$$

from which

$$E = k\frac{q}{r^2} \qquad \text{for a point charge} \tag{16.4}$$

If q is positive, the electric field is directed radially outward; if q is negative, the field is directed radially inward.

FIGURE 16.15

To find the electric field **E** at point P, we must compute the force a positive test charge would experience if placed at that point.

We can extend this relation to another important situation, the field around a uniformly charged sphere. From a large distance away, a charged sphere (let it be positive) appears as a point charge, and so the field lines it generates extend radially outward from it into space. Because the charge on the sphere is uniform, the lines are uniformly spaced around the sphere. And, as we approach the sphere, the lines must remain uniformly spaced. Thus even close to the sphere, they remain radial and similar to those of a point charge. Hence, for a uniformly charged sphere, the field looks like that in Fig. 16.14 for a point charge. We therefore conclude that

Outside a uniformly charged sphere, the field is that of an equal point charge placed at its center.

Thus Eq. 16.4 applies to a uniformly charged sphere as well as to a point charge. Notice, however, that it applies only to the region outside the sphere.

Illustration 16.1

Find the electric field strength 50 cm from a positive point charge of 1×10^{-4} C.

Reasoning   We wish to find $\mathbf{E}$ at point P in Fig. 16.15 with $r = 0.50$ m and $q = 1 \times 10^{-4}$ C. Because q is positive, the test charge we mentally place at P is repelled outward by q. Hence the direction of $\mathbf{E}$ is as shown. To find the magnitude of $\mathbf{E}$, we use Eq. 16.4:

$$E = k\frac{q}{r^2} = (9 \times 10^9\,\text{N} \cdot \text{m}^2/\text{C}^2)\frac{1 \times 10^{-4}\,\text{C}}{(0.50\,\text{m})^2} = 3.6 \times 10^6\,\text{N/C}$$

Exercise   What would the field strength at P be if the point charge is a uniformly charged sphere of radius 3.0 cm?   *Answer:* 3.6×10^6 N/C.

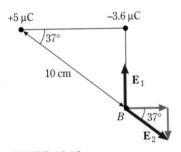

FIGURE 16.16

Justify the directions shown for $\mathbf{E}_1$ and $\mathbf{E}_2$. How do we find the total field at B resulting from the two charges?

Example 16.5

Find the magnitude of $\mathbf{E}$ at point B in Fig. 16.16 due to the two point charges.

Reasoning

Question   Does the principle of superposition apply to calculation of the total field at B?
Answer   Yes. The field at B due to each charge can be calculated from Eq. 16.4. The individual contributions then add vectorially.

Question   What determines the directions of the vector components of the field?
Answer   Remember that a field due to a positive charge is directed radially outward from the charge. A field due to a negative charge is directed radially toward that charge. Thus you have the contributions $\mathbf{E}_1$ and $\mathbf{E}_2$ in the directions shown in Fig. 16.16.

Question   What determines the magnitudes of $\mathbf{E}_1$ and $\mathbf{E}_2$?
Answer   Equation 16.4 gives the magnitude of the field from single point charges:

$$E_1 = \frac{(9 \times 10^9\,\text{N} \cdot \text{m}^2/\text{C}^2)(-3.6 \times 10^{-6}\,\text{C})}{(0.10\,\text{m}\,\sin 37°)^2}$$

$$E_2 = \frac{(9 \times 10^9\,\text{N} \cdot \text{m}^2/\text{C}^2)(5 \times 10^{-6}\,\text{C})}{(0.10\,\text{m})^2}$$

Solution and Discussion The magnitudes of the individual field strengths are

$$E_1 = 9.0 \times 10^6 \, \text{N/C} \qquad E_2 = 4.5 \times 10^6 \, \text{N/C}$$

The rectangular components of $\mathbf{E}_2$ are

$$\mathbf{E}_{2x} = E_2 \cos 37° = 3.6 \times 10^6 \, \text{N/C}$$

$$\mathbf{E}_{2y} = -E_2 \sin 37° = -2.7 \times 10^6 \, \text{N/C}$$

Then the components of $\mathbf{E}$ are

$$\mathbf{E}_x = 3.6 \times 10^6 \, \text{N/C}$$

$$\mathbf{E}_y = (9.0 - 2.7) \times 10^6 \, \text{N/C}$$

This gives

$$E = \sqrt{E_x^2 + E_y^2} = 7.3 \times 10^6 \, \text{N/C}$$

Exercise Show that the direction of $\mathbf{E}$ is 60.3° above the horizontal.

Illustration 16.2

If a charge $q = +4 \times 10^{-7} \, \text{C}$ were placed at point B in Example 16.5, what force would be exerted on it by the electric field?

Reasoning We could use Coulomb's law and calculate the force as in previous examples. However, once we have calculated the field $\mathbf{E}$ at point B, the force any charge q experiences when placed at that point is just $\mathbf{F} = q\mathbf{E}$. Thus the magnitude of the force is

$$F = (+4 \times 10^{-7} \, \text{C})(7.3 \times 10^6 \, \text{N/C}) = 2.9 \, \text{N}$$

The direction of the force is the direction of $q\mathbf{E}$. In the case of a positive charge, $\mathbf{F}$ is in the direction of $\mathbf{E}$. In the case of a negative charge, $\mathbf{F}$ is in the direction of $-\mathbf{E}$, or opposite $\mathbf{E}$.

16.12 THE ELECTRIC FIELD DUE TO VARIOUS DISTRIBUTIONS OF CHARGE: GAUSS' LAW

We can obtain a great deal of insight into a problem by examining the pertinent electric field mapping. Remember the following interpretations:

1. Electric field lines begin on positive charges and end on negative charges.
2. The electric field is strongest where the field lines are most dense.
3. The force on a positive charge placed at a point in the field is directed along the field at that point. The force on a negative charge is directed oppositely to the field.

We could in principle determine the direction of the electric field due to the charges in Fig. 16.16 at as many points in space as we desired to. In practice this would be a tedious task best handled by computer. If the points were sufficiently close, we could draw a map of the field direction that would graphically tell us much about the field in the space around the charges. Figure 16.17 is such a field

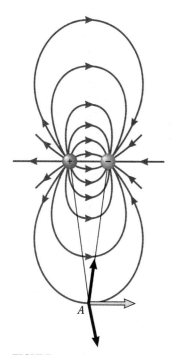

FIGURE 16.17

The electric field lines originate on the positive charge and end on the negative one. At any given point, such as A, the electric field is in a direction tangent to the field line passing through that point.

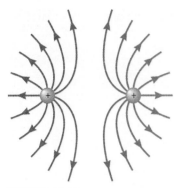

FIGURE 16.18

The lines of force about two like charges seem to repel one another. Why must this be so?

mapping for opposite but equal charges. Examine several points in the figure to convince yourself that a positive test charge placed there would experience a force in the direction indicated by the force lines. To see how this is done, consider point A. A positive test charge at A is repelled by the positive charge and attracted by the negative charge. The attractive force equals the repulsive force because the test charge is as close to the positive charge as it is to the negative charge. The resultant of these two forces is tangent to the line of force at A.

The field map in the neighborhood of two equal like charges is shown in Fig. 16.18. You should be able to show that the field is zero at the point midway between the charges.

In many cases of interest, charge is distributed in shapes having simple geometry, such as spheres, lines, or planes. There is a very powerful approach that simplifies the calculation of electric field strength in these cases, known as *Gauss' law*. To understand the concept behind this law, consider a closed surface in a region of electric field. This surface does not have to be the material surface of a real object. It can be any *hypothetical* surface (called a *gaussian surface*) you choose as long as it encloses some volume of space. Think of dividing this surface up into small area elements ΔA. Each ΔA has an orientation which can be described by the normal to ΔA, $\mathbf{n}$, which points outward from the region enclosed by the surface. The electric field lines which pass through ΔA have a component $E_\perp = E \cos \theta$ along $\mathbf{n}$, where θ is the angle between $\mathbf{E}$ and $\mathbf{n}$ (Fig. 16.19). We now multiply each ΔA by $E_\perp$, forming a quantity which we call the **electric flux** $(E_\perp \Delta A)$ through ΔA. Notice that $E_\perp$ (and hence the flux) can be positive or negative, depending on $\cos \theta$, as demonstrated in Fig. 16.20. Since the strength of the electric field is represented by the density of field lines, we can think of the flux as the number of field lines passing through the plane of ΔA.

Figure 16.21 shows an example of a gaussian surface segmented into small area elements. It is located in a region of uniform electric field indicated by the field lines. Notice that the flux through some area elements is negative; through others it is positive; and in some cases, where $\mathbf{n}$ and $\mathbf{E}$ are perpendicular, the flux is zero. What is the result when we add up all these flux contributions over the entire gaussian surface? **Gauss' law** gives the answer:

The sum of all contributions of electric flux over a closed surface is porportional to the total amount of charge enclosed by that surface.

In the case of Fig. 16.21, the surface encloses no charge, and so the total flux through the surface is zero; as many field lines leave the enclosed region as enter it.

FIGURE 16.19

If the area pictured is A, then the electric flux through the left area, where $\mathbf{n}$ is parallel to $\mathbf{E}$, is EA. Through the area on the right, where $\mathbf{n}$ is perpendicular to $\mathbf{E}$, the flux is zero.

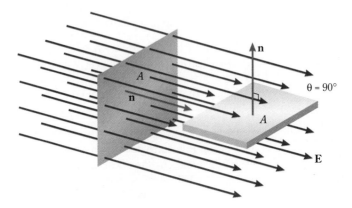

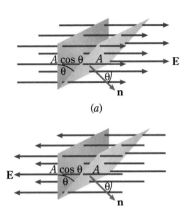

(a)

(b)

FIGURE 16.20

The sign of the electric flux depends on the angle between the normal to the area and the electric field. In (a) the flux is $+(E \cos \theta)A$, while in (b) it is $-(E \cos \theta)A$. Notice that this result is the same as if you considered the flux through the effective area $A \cos \theta$.

Since charges are the source (or termination) of electric field lines, the only way there can be a net flux of electric field through a closed surface is when there is net charge within the surface.

The precise mathematical form of Gauss' law is

$$\Sigma(E \, \Delta A) = 4\pi k \, \Sigma \, q \text{ (enclosed)} = \frac{Q_{\text{tot}} \text{ (enclosed)}}{\epsilon_o} \qquad (16.5)$$

At this stage, if you think Gauss' law is more confusing than helpful, you're probably right. Equation 16.5 can be solved algebraically only in cases where the charge distribution has simple geometry, allowing us to choose simple surfaces. Let us consider three such situations: spherical symmetry, cylindrical symmetry, and planar symmetry.

SPHERICAL SYMMETRY

Examples of spherical symmetry are point charges and charges distributed uniformly over spherical surfaces or volumes. Let us consider a total charge $+Q$ spread uniformly over an empty, hollow sphere of radius R as in Fig. 16.22a. At any point A outside the sphere, we can use symmetry arguments to show that all transverse components (components perpendicular to the radial direction) of the force on a positive test charge at A would cancel. The force on the test charge and hence the direction of **E** are radially outward from the center of the sphere. Symmetry also allows us to say that all points the same distance r from the center of the sphere are equivalent. If we choose our gaussian surface to be a sphere (colored green) of radius r (passing through A), we can make the following statements:

1 The field **E** has the same magnitude at all points on the gaussian surface, even though we don't know that value yet.
2 **E** is perpendicular to the gaussian surface at all points and hence directed radially outward from the center of the sphere.

This information makes it possible to calculate the left side of Gauss' law:

$$\Sigma(E_\perp \, \Delta A) = E \Sigma \, \Delta A = E(A_{\text{sphere}}) = E(4\pi r^2)$$

The sum on the right side of Gauss' law is just the total charge on the hollow sphere, Q. Gauss' law then yields the solution for E:

$$E = \frac{Q}{4\pi\epsilon_o r^2} \qquad r \geq R \qquad (16.6a)$$

FIGURE 16.21

The normal vector **n** is perpendicular to the plane of each small area element. For an element of a closed gaussian surface, **n** is directed out of the enclosed region. The flux through each area element is $(E \cos \theta) \, \Delta A$. (After Edward M. Purcell, "Electricity and Magnetism," *Berkeley Physics Course*, vol. 2, McGraw-Hill Book Co., New York, 1965, p. 22. Courtesy of Education Development Center, Inc., Newton, Mass.)

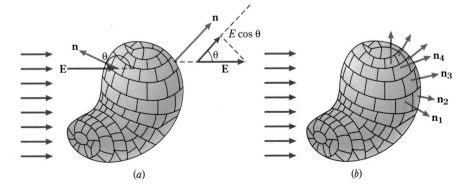

(a)

(b)

FIGURE 16.22

Gauss' law applied to a spherically symmetric distribution of charge, shown by the sphere of radius R. (a) The gaussian surface at $r > R$ (shown in green) encloses a total charge $+Q$. The electric field at point A would be the same as if Q was a point charge at the center of the sphere. (b) The gaussian surface at $r < R$ encloses zero charge, and therefore the electric field is zero at all points such as A. (c) The gaussian surface at $r > R$ encloses zero net charge, so the electric field at A would be zero. (d) At $r < R$, the gaussian surface encloses a charge $-Q$. The field at any point A inside the spherical charge would be the same as if the outer charge $+Q$ did not exist.

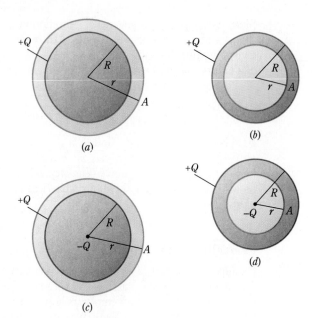

(a)

(b)

(c)

(d)

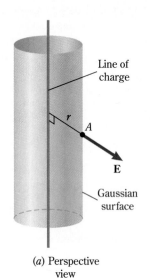

(a) Perspective view

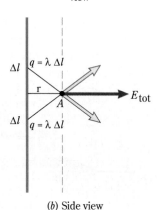

(b) Side view

This shows that, *for points either on or outside a spherical distribution of charge,* the electric field is the same as if the charge were all at the center of the sphere. If the charge on the sphere were $-Q$, we would obtain the same result, except that the direction of **E** would be radially inward.

Now let us choose a gaussian surface *inside* the hollow sphere ($r < R$) as in Fig. 16.22b. The same symmetry arguments hold, and the left side of Gauss' law is again $E(4\pi r^2)$. But since we consider the sphere to be empty, the surface *now encloses no charge,* and so Gauss' law becomes

$$E(4\pi r^2) = 0 \qquad r < R \tag{16.6b}$$

meaning that

$E = 0$ *at all points inside the charged hollow sphere*

In Fig. 16.22c and d we have placed a point charge $-Q$ at the center of the same charged sphere. This placement preserves all of the previous spherical symmetry. Considering the same gaussian surfaces as before, we see that now the outer surface encloses no net charge, whereas the inner one now encloses $-Q$. We can immediately conclude that the field produced by the charge distribution in Fig. 16.22c and d is

$$E = \begin{cases} 0 & r \geq R \\ \dfrac{-Q}{4\pi\epsilon_o r^2} & r < R \end{cases}$$

FIGURE 16.23

A very long line carrying uniform linear charge density λ. The appropriate gaussian surface is a cylinder centered on the line charge. Notice in (b) that the contributions to the electric field parallel to the line of charge from pairs of symmetrically chosen point elements of the line charge cancel out, and so the field is radially outward from the line charge.

CYLINDRICAL SYMMETRY

Now let us consider a straight line along which charge (either positive or negative) is uniformly distributed as shown in Fig. 16.23. We can characterize this charge by its *linear density,* or charge per meter of length. The symbol usually used for this linear charge density is λ, measured in coulombs per meter. We choose some point A at a perpendicular distance r from the line. If the line of charge extends "infinitely" in both directions, we can make some simplifying symmetry arguments. In practical terms, infinite length means that the length of the line of charge is very much greater than the distance r. The transverse force components exerted on a positive test charge at A by various sections of the linear charge would cancel, as shown in Fig. 16.23b. The force on q_t and hence **E** would then be purely in a radial direction outward from or toward the line, depending on whether the linear charge was positive or negative. Again, symmetry allows us to say also that *all* points at the same distance r are equivalent, and hence must have the same value of E. Such points are on the surface of a *cylinder* whose center is the line charge.

To apply Gauss' law to this charge distribution, we choose our gaussian surface to be a relatively short cylinder of length L and radius r, as shown in green in Fig. 16.23a. Using the symmetry arguments, we can conclude:

1 **E** has no perpendicular components at the surfaces of the ends of the cylinder, and so $\Sigma(E_\perp \, \Delta A) = 0$ for those parts of the surface.
2 $\Sigma(E_\perp \, \Delta A) = E(2\pi r L)$ on the side area of the cylinder.
3 The total charge enclosed by the cylinder is $Q = \lambda L$.

Gauss' law thus gives us the value of the electric field from an infinite uniform line of charge:

$$E = \frac{\lambda}{2\pi\epsilon_o r} \tag{16.7}$$

If the charge were distributed on a cylindrical shell of radius R, we could choose gaussian surfaces inside and outside of R to calculate E in a manner similar to the section on spherical charges.

PLANAR SYMMETRY

As our final example of the usefulness of Gauss' law, we now consider charge to be uniformly distributed on an infinite plane as shown in Fig. 16.24. Again, "infinite" means that we will stay close enough to the plane in our calculations that our distance x from the plane is very much smaller than the dimensions of the plane, and that we consider a region far from the edges of the plane. We can characterize the charge on the plane as having a uniform density per area. This area charge density is symbolized by σ (Greek "sigma"), measured in coulombs per square meter.

Once again, we can argue that the transverse components of force on a positive test charge at a distance x from the plane cancel. For each small area of charge above or to the right of q_t, there is an equal charge below or to the left that will cancel all but the component of force perpendicularly away from or toward the plane. Also, all points at the same distance from the infinite plane are equivalent. A

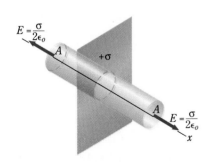

FIGURE 16.24

A plane containing uniform surface charge density σ.

convenient gaussian surface that takes advantage of this symmetry is shown in Fig. 16.24. It is a cylinder of cross-sectional area A whose axis is perpendicular to the charged plane. We make the following observations:

1 $\mathbf{E}$ has no component perpendicular to the cylindrical sides of this surface, and so $\Sigma\, E_\perp\, \Delta A = 0$ for that part of the surface.
2 $\mathbf{E}$ is entirely perpendicular to the end caps of the cylindrical surface and has a constant value across these areas. The end caps thus have $\Sigma\, (E_\perp\, \Delta A) = 2(EA)$.
3 The charge enclosed by the gaussian surface is σA.

Gauss' law gives us the following result for the electric field from a uniform plane of charge:

$$E = \frac{\sigma}{2\epsilon_o} \tag{16.8}$$

Notice that this result does not depend on the position x! The field from the plane has a constant value everywhere in the region close to the plane and far from its edges.

You should be aware of how difficult it would be to obtain these results by applying Coulomb's law directly. Yet the results of Gauss' law are simple and straightforward to use.

Example 16.6

Sparking occurs through air when the electric field strength exceeds about 3×10^6 N/C. (We call this the *electric strength* of air.) About how much charge can a 10.0-cm-diameter metal sphere hold before sparking occurs?

R

Reasoning

Question For a uniformly charged sphere, what is the expression for the electric field at the sphere's surface?
Answer Equation 16.6a shows that $E = Q/4\pi\epsilon_o r^2$ as long as $r \geq R$. Thus you can use this expression with $r = R$.

Question What condition will give me the maximum charge before sparking occurs?
Answer Use $r = R$ and put in the maximum field strength $E_{max} = 3 \times 10^6$ N/C. This will correspond to the maximum charge.

Solution and Discussion Using the numerical data given, we get

$$3 \times 10^6 \text{ N/C} = (9 \times 10^9 \text{ N} \cdot \text{m}^2/\text{C}^2)\frac{Q}{(0.050 \text{ m})^2}$$

$$Q = 8.3 \times 10^{-7} \text{ C}$$

Thus a sphere of this size can hold almost $1\ \mu$C of charge.

Exercise What is the field strength 75 cm from the center of the sphere if its charge is $0.50\ \mu$C? *Answer: 8000 N/C.*

Example 16.7

Figure 16.25 shows two large (infinite) sheets of charge facing each other. The sheets contain equal surface charge densities of opposite charges, $+\sigma$ and $-\sigma$. Find the expression for **E** due to these charges at 3 places: between the plates, to the right of the plate on the right, and to the left of the plate on the left.

R *Reasoning*

Question   Do I use Gauss' law for this calculation?
Answer   Since we've already used it for a single sheet of charge, you can use that result and the principle of superposition.

Question   What does the principle of superposition imply?
Answer   It implies that you can pick any point you want and add the contributions of each sheet to the field at that point as if the other sheet is not there.

Question   What are the individual contributions to **E**?
Answer   Equation 16.8 gives $E = \sigma/2\epsilon_o$ *at any point you pick*. The directions of the fields are toward negative charges and away from positive charges, as always.

Solution and Discussion   At any point, the two sheets contribute equally to the magnitude of E. As can be seen in Fig. 16.25, these contributions are oppositely directed and thus cancel at all points in the regions to the left and right of both sheets, such as points A and C. At any point B between the sheets the two contributions are in the same direction. Thus we have

$$E = \begin{cases} 0 & \text{at all points } \textit{not} \text{ between the sheets} \\ 2\left(\dfrac{\sigma}{2\epsilon_o}\right) = \dfrac{\sigma}{\epsilon_o} & \text{at all points between the sheets} \end{cases}$$

The direction of **E** between the sheets is from the positive sheet toward the negative one.

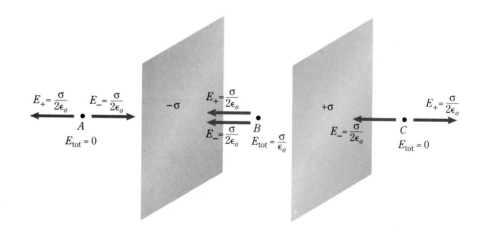

FIGURE 16.25

Oppositely charged plates. When their area is much greater than their separation distance, E is σ/ϵ_o between them and zero outside.

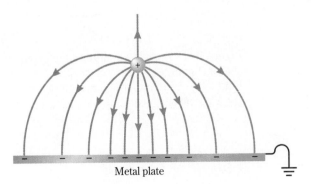

FIGURE 16.26

The positive charge attracts negative charges to the top of the metal plate. Why are the field lines perpendicular to the plate at its surface?

Metal plate

16.13 CONDUCTORS IN ELECTRIC FIELDS

As we discussed in Sec. 16.4, electrons are free to move throughout a conducting material in response to electric forces. Suppose a small, positively charged ball is held above a large metal plate, as in Fig. 16.26. Electrons in the metal plate are attracted by the positive charge. Although they cannot leave the plate, they tend to move toward the positive charge and congregate at the plate's surface nearest the ball. If the plate is connected to ground, negative charge will then flow from the ground into the plate and replace those electrons that were induced to move closest to the charged ball. The plate, originally neutral, thus acquires a net negative charge equal to the positive charge on the ball, resulting in the electric field pattern shown in Fig. 16.26.

This adjustment of charge on the plate takes place quickly and establishes a condition where no more movement of charge takes place within the metal. This is known as an **electrostatic condition,** and implies the following very important fact:

Under electrostatic conditions, no electric field can exist within a conductor.

An important corollary to the above statement is

Under electrostatic conditions, an external electric field is everywhere perpendicular to the surface of a conductor.

The proof of this statement lies in the fact that a component of **E** parallel to the conductor's surface would cause electrons to move along the surface, again until a static condition was established. The perpendicular component of **E** is not strong enough (except in extreme circumstances) to pull electrons out of the surface of the metal.

Notice that, in keeping with these observations, the electric field lines in Fig. 16.26 are perpendicular to the plate's surface and terminate at the surface. Remember that these rules presume the freedom of electrons to move and hence do not apply to insulators.

Example 16.8

Figure 16.27 shows a charge $+q$ suspended at the center of a hollow spherical metal shell. The outer radius of this shell is R_2 and its inner radius is R_1. Use

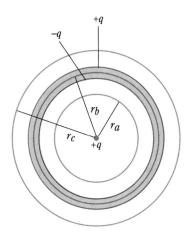

FIGURE 16.27

If a point charge $+q$ is placed at the center of a hollow sphere, a charge $-q$ is induced *on the inner surface* of the sphere. This can be shown by considering spherical gaussian surfaces inside the spherical cavity (r_a), between the inner and outer surfaces of the conductor (r_b), and outside the conductor (r_c). These gaussian surfaces are shown as green circles. Remember, the field must be everywhere zero within the conducting material.

Gauss' law to determine the electric field strength: (**a**) between the charge and the inner surface of the sphere (at r_a), (**b**) between the inner and outer surfaces of the sphere (r_b), and (**c**) outside the sphere (r_c). (**d**) Show that charges of $-q$ and $+q$ are induced on the inner and outer surfaces of the sphere, respectively.

R

Reasoning

Question   How do I decide what gaussian surfaces to use?
Answer   The problem has spherical symmetry, and so your gaussian surfaces should be spheres centered on $+q$. The radii of your gaussian surfaces should be taken in each of the regions for which you want to evaluate the field. These radii are labeled r_a, r_b, and r_c in Fig. 16.27.

Question   What does $\Sigma\, E_\perp\, \Delta A$ give me for these gaussian surfaces?
Answer   You can use the symmetry arguments discussed in Sec. 16.12. For all three regions, the result is

$$\Sigma\, E_\perp\, \Delta A = E(4\pi r^2)$$

where E is directed radially.

Question   What is the total charge enclosed by each gaussian surface?
Answer   This is the revealing question. For a gaussian surface of radius r_a, it is obvious that $Q_{encl} = +q$. For a gaussian surface of radius r_c outside the sphere, we get the same result, since the sphere itself contains no net charge. Inside the shell, at r_b, we can't tell simply from inspection.

Question   A gaussian surface of radius r_b is inside a conductor. What information can I infer from that?
Answer   The field must be zero in that region, since we are assuming an electrostatic situation.

Question   What can I infer about charge from that fact?
Answer   Since E must be zero everywhere in the conductor, Gauss' law requires that $Q_{encl} = 0$ for any gaussian surface between R_1 and R_2, right down to $r = R_1$. To have $Q_{encl} = 0$ means that a negative charge of $-q$ must reside somewhere within the gaussian surface in order to cancel the $+q$ at the sphere's center. By

shrinking the radius of the gaussian surface to $r = R_1$, we can eliminate the possibility of net negative charge residing anywhere in the interior of the conductor, and conclude that *the charge $-q$ is located on the conductor's inner surface.*

Question   What does this imply about the charge on the outer surface?
Answer   Since the sphere is neutral and carries no net charge, $+q$ must reside on the outer surface. This is consistent with the results of Gauss' law for the region outside the sphere.

Solution and Discussion   To summarize the electric field values:

$$E = \begin{cases} \dfrac{kq}{r^2} & r < R_1 \\ 0 & R_1 \leqslant r \leqslant R_2 \\ \dfrac{kq}{r^2} & r \geqslant R_2 \end{cases}$$

A charge $-q$ is induced to move to the inner surface of the metal sphere and a charge $+q$ is left on the outer surface. In this case of symmetry, the surface charges are uniformly distributed on the two surfaces of the sphere. Can you reason that the same charges would be induced regardless of the shape of the hollow conductor? (In an arbitrary shape, they would no longer be uniformly distributed.)

16.14   PARALLEL METAL PLATES

The electric field between two oppositely charged metal plates is of particular importance in electricity, as we shall see as our studies progress. We show a typical situation in Fig. 16.28a. The charges on the plates come from a battery (discussed in the next chapter). The battery gives one plate a positive charge and the other a negative charge as shown schematically in Fig. 16.28b. Because the charges attract one another, they reside for the most part on the inner surfaces of the plates. (Notice the symbol ⊣⊢, commonly used to signify a battery.)

The electric field due to this charge configuration was calculated in Example 16.7. Except for regions near the edges of the plates, the field is uniform and constant:

$$E = \frac{\sigma}{\epsilon_o}$$

where σ is the uniform charge per area on the plates.

Now recall from Eq. 16.3 that the force on a charge q placed in an electric field **E** is

$$\mathbf{F} = q\mathbf{E}$$

Since **E** is constant, the force on any charge between the plates is constant also. Parallel charged plates are thus a convenient way of producing constant forces on

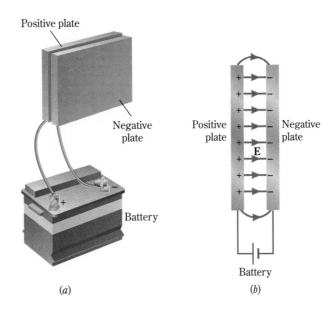

FIGURE 16.28

The battery places equal charges of opposite sign on the two metal plates.

(a)

Positive plate

Negative plate

Battery

Positive plate

Negative plate

E

Battery

(b)

charges. This is not true of any of the other charge distributions we have examined. As a result, free charges between parallel charged plates experience a constant acceleration according to Newton's second law, $\mathbf{a} = \mathbf{F}/m$. For positive charges, the force is along the direction of the field; for negative charges, the force is opposite the field direction.

Example 16.9

Parallel metal plates 3 mm apart carry equal and opposite charge densities of $\pm 2\ \mu\text{C/m}^2$. A proton ($q = e$ and $m = 1.67 \times 10^{-27}$ kg) is released from rest at the positive plate. What is the proton's speed just as it strikes the negative plate? Assume the space between the plates is a vacuum.

R *Reasoning*

Question   What principle will determine the speed acquired?
Answer   The equations of motion for constant acceleration derived from Newton's second law, specifically, the one relating change of speed to distance traveled: $v^2 = v_o^2 + 2ax$. You know $v_o = 0$. You want to find v when $x = 3$ mm.

Question   What gives the value of acceleration?
Answer   As always, $a = F_{\text{net}}/m$, and m is given.

Question   What determines the net force on the proton?
Answer   The only force in the problem is the electric force created by the electric field between the plates, $F = qE$, with $q = e$ in this case.

Question   What is the value of the field strength E?
Answer $E = \sigma/\epsilon_o$, and you know both σ and ϵ_o.

Solution and Discussion   First, calculate the field:

$$E = \frac{2 \times 10^{-6}\,\text{C/m}^2}{8.85 \times 10^{-12}\,\text{C}^2/\text{N} \cdot \text{m}^2}$$

$$= 2.26 \times 10^5\,\text{N/C}$$

(Be sure you can derive the units in the answer.) Next calculate the force on the proton:

$$F = eE = (1.6 \times 10^{-19}\,\text{C})(2.26 \times 10^5\,\text{N/C}) = 3.62 \times 10^{-14}\,\text{N}$$

Next, find the acceleration:

$$a = \frac{F}{m} = \frac{3.62 \times 10^{-14}\,\text{N}}{1.67 \times 10^{-27}\,\text{kg}}$$

$$= 2.17 \times 10^{13}\,\text{m/s}^2$$

Then determine the final speed:

$$v = (2ax)^{1/2} = [2(2.17 \times 10^{13}\,\text{m/s}^2)(0.003\,\text{m})]^{1/2}$$

$$= 3.61 \times 10^5\,\text{m/s}$$

Notice that even though the charge and resulting force are very small, the very small mass allows the proton to acquire a very large speed.

LEARNING GOALS

Now that you have finished this chapter, you should be able to

1 Define (*a*) conductor, (*b*) insulator, (*c*) free electron, (*d*) electrical ground, (*e*) induced charge, (*f*) Coulomb's law, (*g*) electric field lines, (*h*) electric field strength **E**.

2 Give the magnitude and sign of the charge on the electron and proton.

3 Describe qualitatively how charges within a metal object redistribute when a charged object is brought nearby. Explain how an object can be charged by conduction and by induction.

4 State the conclusions that can be drawn from the Faraday ice-pail experiment.

5 Use Coulomb's law to find the force on a charge due to nearby point charges.

6 Find the electric field strength at a point due to several specified point charges.

7 Sketch the electric field lines in the vicinity of simple charged objects.

8 State Gauss' law in words and in mathematical terms, and apply it to charge distributions possessing simple symmetry.

9 Find the electric field strength at any given point due to uniform spherical, linear, and planar charge distributions.

10 Specify the following under electrostatic conditions: (*a*) electric field in a metal, (*b*) origin of electric field lines, (*c*) termination points of field lines, (*d*) angle at which field lines strike metal surfaces.

11 Use the relation **F** = *q***E** in simple situations.

SUMMARY

DERIVED UNITS AND PHYSICAL CONSTANTS

Quantities of Electric Charge
SI unit of charge: coulomb (C)

Proton charge (*e*): $e = 1.6 \times 10^{-19}\,\text{C}$

Electron charge ($-e$): $-1.6 \times 10^{-19}\,\text{C}$

Coulomb Force Constant (k)
$k = 8.99 \times 10^9\,\text{N} \cdot \text{m}^2/\text{C}^2$

Permittivity of Free Space (ϵ_o)

$$\epsilon_o = \frac{1}{4\pi k} = 8.85 \times 10^{-12}\, C^2/N \cdot m^2$$

Units of Electric Field (E)

$$E = \frac{F}{q} \quad N/C$$

DEFINITIONS AND BASIC PRINCIPLES

Concepts of Electric Charge
1. There exist two kinds of electric charge, positive (+) charge and negative (−) charge.
2. Atoms contain charged fundamental particles. The proton carries a definite amount of positive charge and the electron carries an equal amount of negative charge.
3. Forces between charges of the same sign are repulsive; forces between charges of opposite signs are attractive.

Conservation of Charge
Net positive or negative charge cannot be created or destroyed in any physical process.

Coulomb's Law
The magnitude of the electrical force between two point charges q_1 and q_2 separated by a distance r is

$$F = \frac{kq_1q_2}{r^2}$$

where k is a universal physical constant known as the Coulomb force constant.

INSIGHTS
1. The electrical force is attractive if the charges have opposite sign and repulsive if the charges have the same sign.
2. If the charges have spherical symmetry, the distance r is the distance between their centers.

The Electric Field (E)
The electric field at a point in space is defined as the ratio of the electric force a small positive test charge q_t experiences at that point to the magnitude of the test charge:

$$\mathbf{E} = \frac{\mathbf{F}\,(\text{on } q_t)}{q_t}$$

INSIGHTS
1. The direction of $\mathbf{E}$ is the same as the direction of the force $\mathbf{F}$ on a positive charge.
2. The SI units of E are N/C.
3. A corollary of the definition of $\mathbf{E}$ is that the force on a

charge q placed at a point where the electric field has the value $\mathbf{E}$ is

$$\mathbf{F} = q\mathbf{E}$$

Electric Field of a Point Charge
The magnitude of the electric field of a point charge Q at a distance r from Q is

$$E = \frac{kQ}{r^2}$$

INSIGHTS
1. The direction of the electric field is radially outward from a positive charge, radially toward a negative charge.
2. The electric field due to a number of point charges can in principle be calculated at any point by applying the superposition principle: calculate the field due to each point charge separately and then add the individual contributions *vectorially*.
3. In a map of the electric field, the strength of the field is greatest where the field lines are densest and least where the lines are farthest apart.

Electric Flux
The electric flux through a small area element ΔA is

$$\text{Electric flux} = (E \cos \theta)\, \Delta A = E_\perp\, \Delta A$$

where

E = electric field strength passing through ΔA

θ = angle between $\mathbf{E}$ and the normal to ΔA, $\mathbf{n}$.

Gauss' Law
The sum of the flux contributions $E_\perp\, \Delta A$ over an entire closed surface—called a gaussian surface—is equal to the total charge enclosed by the surface divided by ϵ_o. Gauss' law is particularly useful in cases where the charge distribution has simple symmetry.

Electric Fields of Simple Geometries
UNIFORM SPHERICAL CHARGE Q (RADIUS R)

$$E = \frac{kQ}{r^2} \quad r > R$$

SPHERICAL SHELL OF CHARGE Q (RADIUS R)

$$E = \begin{cases} 0 & r < R \\ \dfrac{kQ}{r^2} & r \geq R \end{cases}$$

UNIFORM LINE CHARGE

$$E = \frac{2k\lambda}{r} \quad \text{where } \lambda = \text{charge per length (C/m)}$$

UNIFORM FLAT SHEET OF CHARGE

$$E = \frac{\sigma}{2\epsilon_o}$$ where σ = charge per area (C/m²)

REGION BETWEEN TWO FLAT PLATES

$$E = \frac{\sigma}{\epsilon_o}$$

Here the plates have equal and opposite charge densities σ.

Conductors in Electric Fields
Under electrostatic conditions,

1. No electric field can exist within a conducting material.
2. External electric field lines must be everywhere perpendicular to the surface of a conductor.

QUESTIONS AND GUESSTIMATES

1 A tiny charged ball hangs from a thread. How can you tell whether the charge on the ball is positive or negative?

2 You can place a static charge on nearly any dry piece of plastic by rubbing it with a piece of fabric, fur, or plastic wrap. How can you determine the sign of the charge placed on the plastic?

3 Static electricity produces sparks that can cause some volatile gases to explode. This used to be a real danger in hospital operating rooms because the anesthetic that was then used, ether, is combustible. What measures can be taken to minimize this danger?

4 The electric strength of air is about 3×10^6 N/C. That is, a spark will jump through the air if the electric field strength exceeds this value. Why do sparks jump preferentially from sharp metal points and edges? When your body becomes highly charged as you walk across a deep-pile carpet in dry weather, why does a spark jump from your fingernail to a metal object, such as a stove or doorknob?

5 Clothes often cling together when they are removed from the dryer. Why? What is often done to eliminate this effect?

6 Never try to wipe the dust off a phonograph record with an ordinary cotton or wool cloth. Why?

7 In dry climates, one frequently sees (or hears) sparks jump when hair is combed or when clothes are removed in darkness. Why?

8 Two equal-magnitude positive point charges are a distance D apart. Where can you place a third charge so that the resultant force on it is zero? Is it in stable equilibrium there?

9 A positive point charge and a much larger negative point charge are a distance D apart. Is there any place a third point charge can be placed where the resultant force on it is zero?

10 A tiny ball with charge q is suspended between two very large parallel metal plates that are grounded. Sketch the electric field between the plates. What can you infer about the induced charges on the plates?

11 Properly drawn electric field lines never cross each other. Why?

12 Sensitive apparatus is frequently shielded from unwanted electric fields by placing it inside a metal can or a fine-mesh wire box that is grounded. Explain why the field of a charge placed outside such a shield does not affect the interior region.

PROBLEMS

Sections 16.1–16.8

1 Calculate the net charge on the sample of a material consisting of (a) 8×10^{15} electrons and (b) a collection of 8×10^{15} electrons and 6×10^{14} protons.

2 Two point charges, $q_1 = -4.0 \, \mu\text{C}$ and $q_2 = +3.0 \, \mu\text{C}$, are 100 cm apart. Find the magnitude and direction of the electrostatic force on both.

3 Two protons are brought to a distance of 3.5×10^{-14} m from each other. (a) Find the magnitude and direction of electrostatic force on both. (b) What is the ratio of the magnitude of this force to the weight of a proton on earth? The proton can be considered a point charge with mass 1.67×10^{-27} kg.

4 What would be the mass of a proton if the magnitudes of

gravitational and electrostatic forces between a pair of protons were equal?

5 Two point charges are placed on the x axis: $+6.0 \, \mu\text{C}$ at $x = 0$ and $+8.0 \, \mu\text{C}$ at $x = +30$ cm. Find the magnitude and direction of the electrostatic force on the $+6.0$-μC charge.

6 A point charge of $+2.0 \, \mu\text{C}$ is placed on the x axis at $x = 25$ cm, and an unknown charge q is placed at $x = 65$ cm on the axis. The force on the 2.0-μC charge is 1.2 N in the positive x direction. What are the magnitude and sign of q?

■7 Two point charges q_1 and q_2 are 60 cm apart, and they repel each other with a force of 0.3 N. The algebraic sum of the two charges is $+7.2 \, \mu\text{C}$. Find q_1 and q_2.

■8 Repeat the previous problem if the two charges were attracting each other with a force of 0.3 N.

9 Two point charges of equal magnitude repel each other with a force of 2.4 N when separated by 6.0 cm. Find the magnitude of each charge.

■10 Two identical 240-g balls of diameter 2.0 cm are 6 cm apart (between centers). Each carries a uniform charge of 7.0 μC. One of the balls is released. Find the initial acceleration of the ball. Neglect gravity.

■11 Two identical pointlike balls, each with a mass 60 g, are 240 cm apart. They carry charges q equal in magnitude but of opposite sign. What would be the magnitude of the charge q if the electrostatic and gravitational attraction between the balls were equal?

12 The following three point charges are placed on the x axis: +5.0 μC at $x = 0$, +4.0 μC at $x = -40$ cm, and +6.0 μC at 80 cm. Find the force on (a) the 6.0-μC charge and (b) the 4.0-μC charge.

13 Three point charges -6μC, $+5\mu$C, and -5μC are placed at $x = 0$, $x = -60$, and $x = +60$ cm, respectively. Find the force on (a) the -6-μC charge and (b) the $+5$-μC charge.

14 A charge of 6 μC and another charge of -3 μC are separated by a distance of 60 cm. Find the position at which a third charge 12 μC should be placed so that the net electrostatic force on it is zero.

■15 Two point charges are placed on the x axis: a -5×10^{-9} C charge at $x = 0$ and a $+6 \times 10^{-9}$ C at $x = 100$ cm. At what position(s) in the vicinity of these charges can a $+4 \times 10^{-9}$ C charge be placed so that it experiences no net force?

■16 Two point charges $+5$ μC and $+7$ μC are placed on the x axis at $x = 0$ and $x = 120$ cm, respectively. At what position(s) in the vicinity of these two charges is the net electrostatic force on a -8-μC charge zero?

17 Three identical point charges of $+6$ μC are placed at three corners of a square whose sides are 8 cm long. What is the resultant electrostatic force experienced by a $+5$-μC charge placed at the fourth corner of the square?

18 Three point charges $+5.0$, $+6$, and $+4.0$ μC are placed at the three corners of an 8-cm square. Find the net electrostatic force on a -8.0-μC charge placed at the fourth corner, diagonally opposite the 6-μC charge.

19 Three identical point charges of $+5.0$ μC are placed at three corners of an equilateral triangle with sides 10.0 cm long. Find net electrostatic force on each charge.

■20 The four point charges in Fig. P16.1 are each $+4.0$ μC. Find the magnitude and direction of the electrostatic force on q_2 due to the other three charges ($a = 40$ cm and $b = 60$ cm).

■21 In Fig. P16.1, $q_1 = q_3 = +5.0$ μC and $q_2 = q_4 = -6.0$ μC. Find the magnitude and direction of the resultant electrostatic force on q_1. Take $a = 40$ cm and $b = 60$ cm.

■22 Three point charges in Fig. P16.2 are each $+6.0$ μC. Find the magnitude and direction of the electrostatic force on q_3 due to the other two charges. Use $a = 40$ cm.

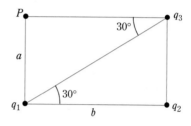

FIGURE P16.2

■23 In Fig. P16.2, $q_1 = q_3 = +5.0$ μC and $q_2 = -7.0$ μC. Find the magnitude and direction of the force on q_1. Take $a = 3.0$ m.

■■24 Two balls hang from a single support, as shown in Fig. P16.3. Each has a mass of 1.0 g and carries a charge q. The length of the string is 40 cm, and the balls come to equilibrium with $\theta = 30°$. Find the charge q on each ball.

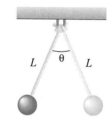

FIGURE P16.3

■■25 Repeat the previous problem if the balls carry unequal charges, with the ball on the left having one-half the charge of that on the right.

■26 Two small spherical charges experience an electrostatic force when separated by a distance R. If the charge on one is doubled and the charge on the other is tripled and at the same time their separation is halved, what is the ratio of the new electrostatic force to the original force between them?

Sections 16.9–16.11

27 Find the magnitude and direction of the electric field at a distance of 1.0 m from an electron. Repeat for a proton.

28 Find the electric field strength due to a point charge $q = -6.0$ μC at a distance of 90 cm from the charge. Is the field directed radially outward or inward?

29 Two charges are placed on the x axis: a $+5.0$-μC charge at $x = 90$ cm and a -4.0-μC charge at $x = 0$. Find the electric field $\mathbf{E}$ at (a) $x = 40$ cm and (b) $x = -60$ cm.

30 (a) Find the electric field at a point midway between two charges of 3.0 μC and 6.0 μC separated by 60 cm. (b) Repeat when the magnitude of the second charge is -5.0 μC.

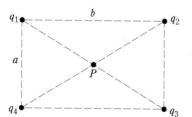

FIGURE P16.1

∎**31** Find the electric field **E** at the center of the rectangle in Fig. P16.1 if (*a*) $q_1 = q_2 = q_3 = q_4$ and (*b*) $q_1 = q_2 = -3.0\ \mu C$ and $q_3 = q_4 = 4.0\ \mu C$. Take $a = 40$ cm and $b = 60$ cm.

∎**32** Two charges of 6.0 μC and $-6.0\ \mu C$ are placed on the two vertices of an equilateral triangle of side 10.0 cm. What are the magnitude and direction of the electric field at the third vertex of the triangle?

33 In Fig. P16.2 $q_1 = q_3 = -5.0\ \mu C$ and $q_2 = 3.0\ \mu C$. Find the electric field **E** at the point *P*. Take $a = 40$ cm.

∎**34** Two charges 3.0 μC and $-5\ \mu C$ are placed on the *x* axis at $x = 0$ cm and $x = 40$ cm, respectively. Where on the *x* axis, if anywhere, is the electric field **E** zero?

∎**35** A tiny ball carrying a charge of -4.0×10^{-13} C experiences an eastward force of 1.0×10^{-9} N due to its charge when suspended at a certain point in space. What are the magnitude and direction of the electric field **E** at that point?

∎**36** The electric field in a certain region is directed eastward and has a strength of 3600 N/C. Find the magnitude and direction of the electrostatic force experienced by a -0.6-μC charge placed in this region.

∎**37** An electron is released in a region where the electric field is along the positive *x* direction and has a strength of 3600 N/C. Find the magnitude and direction of the electron's acceleration. (mass of the electron $m_e = 9.1 \times 10^{-31}$ kg)

∎**38** A tiny oil drop with a mass *m* has a charge $+q$ on it. When placed in a uniform electric field **E** directed vertically the drop "floats" in the free space. Express the magnitude of the electric field **E** in terms of *q* and the mass of the drop *m*.

∎**39** A tiny ball of mass 0.05 g is supported against gravity when placed in a uniform electric field of strength 600 N/C directed vertically downward. Find the charge on the ball.

∎**40** A 0.450-g sphere is suspended by a thread in an electric field of 6000 N/C that is directed vertically upward. The tension in the thread is 3.0×10^{-3} N. Find the charge on the sphere.

∎**41** In Fig. P16.4, a ball of mass *m* having a charge *q* is suspended by means of a thread in an electric field **E**. The ball hangs so that the thread makes an angle θ with the vertical. Find **E** in terms of *m*, *q*, and θ.

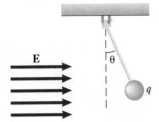

FIGURE P16.4

∎**42** In the previous problem if the mass of the ball is 0.500 g and the thread makes an angle of 15° when suspended in an electric field of strength 500 N/C, what is the charge *q* on the ball?

Additional Problems

∎**43** Two point charges are placed on the *x* axis: a +6-μC charge at $x = 0$ and a +8-μC charge at 100 cm. Where on the *x* axis between these charges should a third charge be placed so that the net electrostatic force on all three charges is zero? Evaluate the third charge.

∎**44** A -5.0-μC charge is placed at the origin on the *x* axis. Two other charges are placed on the *x* axis: q_1 at $x = 40$ cm and q_2 at $x = 50$ cm. Find the magnitudes and signs of q_1 and q_2 if the net electrostatic force on all three charges is to be zero.

∎∎**45** Two point charges q_1 and q_2 1.0 m apart exert a force of magnitude 0.090 N on each other. The algebraic sum of the two charges is $q_1 + q_2 = 7.0\ \mu C$. Find the magnitudes of q_1 and q_2. Is the force attractive or repulsive?

∎**46** A medium-size person contains about 3×10^{28} protons and a like number of electrons. Suppose two people are 40 m apart and 0.2 percent of the electrons on one person is transferred to the other. How large a force is required to hold the two persons apart at this distance?

47 Two tiny identical conducting spheres have charges $q_1 = 9.0\ \mu C$ and $-5.0\ \mu C$ on them, and their centers are separated by 0.5 m. (*a*) Find the electrostatic force between them. (*b*) The spheres are touched and then separated by the same distance. After equilibrium has occurred, what is the electrostatic force between the spheres?

∎**48** In the Bohr model of the hydrogen atom, an electron orbits a stationary proton at a radius of 0.53×10^{-10} m. (*a*) How large an electrostatic force does the proton exert on the orbiting electron? (*b*) If this force serves as the centripetal force to hold the electron in its circular orbit, how fast is the electron moving? (The mass of the electron $m_e = 9.1 \times 10^{-31}$ kg.)

∎**49** In the previous problem find the magnitude and direction of the electric field due to the proton at the position of the electron.

∎**50** Radium nuclei are radioactive and emit alpha particles $(m_\alpha = 4 \times 1.66 \times 10^{-27}$ kg, $q_\alpha = +2e)$. The nucleus left behind has a charge of $+86e$ and a very large mass. Find (*a*) the electrostatic force exerted on the alpha particle by the nucleus when they are 6×10^{-14} m apart and (*b*) the acceleration of the alpha particle at that instant.

∎**51** An insulating thin spherical shell of radius *R* has a charge *Q* uniformly distributed over its surface. What is the electric field **E** at the center of the shell?

∎∎**52** An isolated hollow metal sphere of radius 40 cm carries a charge of $-10.0\ \mu C$. What is the magnitude of the electric field **E** (*a*) in the empty space inside the sphere and (*b*) 60 cm from the center of the sphere?

∎∎**53** A proton traveling along the *x* axis is slowed down by a uniform electric field **E**. At $x = 0$ cm, the proton has a speed of 3.5×10^6 m/s, and at $x = 70$ cm it has completely stopped. Find the magnitude and direction of the field **E**. (The mass of the proton $m_p = 1.67 \times 10^{-27}$ kg.)

∎∎**54** At a certain instant, an electron is traveling out from the origin along the x axis with a speed of 6.0×10^6 m/s. A uniform electric field **E** parallel to the x axis causes the electron to slow, stop, reverse its direction of motion, and finally return to the origin in 40.0 μs. Determine the magnitude and direction of the electric field **E**. (The mass of the electron $m_e = 9.11 \times 10^{-31}$ kg.)

■■**55** An electron is shot from the coordinate origin out along the positive x axis with a speed v_{xo}. There is an electric field **E** directed along the y axis in this region. (*a*) Show that the y coordinate of the electron a time t later is $y = -eEt^2/2m_e$, where e is the charge on the electron and E is the magnitude of the electric field. (*b*) Show that the trajectory of the electron in the xy plane is given by $y = (eE/2m_ev_{xo}^2)(x^2)$.

■■**56** Two charges of equal magnitude 0.5 μC but of opposite sign are placed on the x axis, connected together by means of a massless (and electrically neutral) rod 5.4×10^{-8} m long. A uniform electric field of strength 400 N/C is applied along the y direction. (*a*) Find the net torque on the charges. (*b*) What would the net torque be if the electric field were directed at 60° from the x axis?

■**57** The electrons in a particle beam each have a kinetic energy of 1.2×10^{-16} J. How large an electric field will stop the electrons in the beam in a distance of 15 cm? What is the direction of the field? (The mass of the electron $m_e = 9.1 \times 10^{-31}$ kg.)

■■**58** Two charges equal in magnitude q but of opposite sign are placed along the x axis at $x = b$ and $x = -b$. Show that the electric field due to these charges at a point on the y axis is in a direction parallel to the x axis and its magnitude is given by $E = 2kqb/(y^2 + b^2)^{3/2}$.

■■**59** If the two charges in the previous problem were of same sign, what would be the direction and magnitude of the electric field?

■■**60** Two point charges q and $-q$ are located on the x axis very close to each other at a very small distance b on either side of the coordinate origin. Show that the magnitude of the electric field at a distant point along the x axis is given by $E = 4kqb/x^3$.

CHAPTER 17

ELECTRIC POTENTIAL

In our study of mechanics, we found the scalar concepts of work and energy to be of great utility because many situations were too involved to be solved in detail by using force vectors. The development of the concept of scalar energies often allowed us to obtain useful results quickly and simply. We see in this chapter that the concept of electrical potential energy is extremely useful in many electrical applications. It is indispensible for an understanding of such diverse topics as electrical circuits and elementary particle accelerators.

17.1 ELECTRICAL POTENTIAL ENERGY

When we discussed the movement of an object from one place to another within a gravitational field, we used the concept of gravitational potential energy (GPE). To lift an object of mass m, we must apply an upward force mg to it in order to balance the downward pull of gravity. The work done in lifting the object through a distance h is simply force times distance, which is mgh. We say that this work has accomplished an increase in the object's gravitational potential energy. When the object is allowed to fall freely through the height h, it acquires kinetic energy, and from the law of energy conservation, we can write

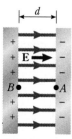

FIGURE 17.1

The electric field between the two oppositely charged parallel plates is uniform.

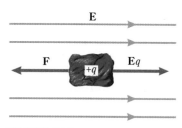

FIGURE 17.2

A force $\mathbf{F} = -\mathbf{E}q$ is required if the object with charge q is to remain suspended between the plates of Fig. 17.1.

GPE at height h = KE gained in falling through distance h

We made extensive use of gravitational potential energy and its interconversion with kinetic energy in our study of mechanics.

A similar situation exists in electricity. Charged objects often possess electrical potential energy that can be transformed to kinetic energy. To show this, consider a charged object between two charged parallel plates. (We ignore the gravitational force in this discussion because it is negligibly small compared with the electric forces we are concerned with.) The electric field in the central region between the plates is shown in Fig. 17.1; it has a constant value $\mathbf{E}$ and is directed as shown. Figure 17.2 shows the forces acting on a positively charged object between the plates. Because of the electric field, the object with charge q experiences a force $\mathbf{E}q$ directed to the right. If we are to hold the charged object in place, we must exert a force $\mathbf{F} = -\mathbf{E}q$ on it.

Suppose the charged object (which is much tinier than shown) is originally at point A in Fig. 17.1. If we are to move it to point B, we must pull it the entire way with the force $\mathbf{F}$. Hence we do work on the object as we pull it from A to B. Since $\mathbf{E}$ is constant in this situation, the work done by $\mathbf{F}$ in going from A to B is just

$$W_{AB} = Fd = qEd \qquad \text{(constant } \mathbf{E}\text{)}$$

This work is precisely analogous to the work done in lifting an object against a constant gravitational force. We say that the work done in pulling the charge against the electric force increases the charge's electrical potential energy. Remember that, in both gravitational and electrical cases, only *differences* in potential energy are physically important.

After we get the object to point B, we can release it and recover this potential energy in the form of kinetic energy. The charged object at B will be pulled toward A by the (now unbalanced) force $\mathbf{E}q$ that acts on it. Therefore, when it is released at B, the object accelerates toward A. Thus we define the electrical potential energy (EPE) of a charge at a point B relative to another point A:

The electrical potential energy of a charge at point B relative to point A is equal to the work done against electrical forces in moving the charge from A to B.

$$W_{AB} = \Delta \text{EPE} = \text{EPE}_B - \text{EPE}_A$$

One basic difference in comparing electrical potential energy with gravitational potential energy lies in the fact that there are two kinds of charge. Consider what would happen if the charge between the plates was *negative*. The direction of the electrical force on $-q$ is now opposite the direction of $\mathbf{E}$. Thus an applied force must do positive work on $-q$ in order to move it from B to A. Thus $-q$ would have higher electrical potential energy at A than at B. If allowed to move freely, it would "fall" from A toward B, in a direction opposite the direction of $\mathbf{E}$.

17.2 POTENTIAL DIFFERENCE

In electricity, we go one step further than we did in mechanics by defining another scalar quantity, called the **electric potential.** To illustrate this concept, we return to the positive charge moving between the charged plates of Fig. 17.1. The difference in potential *energy* of that charge between points A and B was

$$PE_B - PE_A = qEd$$

Now let us divide this expression by q, obtaining the difference in a new quantity that depends only on E and on the distance between A and B:

$$\frac{PE_B - PE_A}{q} = V_B - V_A \tag{17.1}$$

This new quantity, PE/q, is called the electric potential, symbolized by V. Electric potential has the units of joules per coulomb, which we call the **volt.** Notice that, unlike PE, electric potential does not depend on the specific charge q upon which the field acts. Also notice that the definition of the volt gives rise to an alternate interpretation of the units of E.

$$1\,N/C = 1\,(J/m)/C = 1\,V/m$$

Thus, in addition to measuring the force per charge (N/C), the electric field is also a measure of how fast the electric potential changes with position (per meter of distance). The potential decreases in the direction of the electric field. The potential difference between A and B, also referred to as the **voltage difference,** or sometimes just **voltage,** is thus

$$V_{AB} = V_B - V_A = Ed \qquad \text{(constant } \mathbf{E} \text{ field)} \tag{17.2}$$

Let us restate the definition of potential difference:

The **potential difference** (or **voltage**) between points A and B is the difference in the potential energy of a positive charge between those points, divided by the charge.

It is important at this point to observe (and remember) the following:

1 Equation 17.2 applies only to the case of a constant field such as that produced by parallel plates of charge.
2 Electric potential is defined in terms of the EPE of a *positive* charge. This means that in moving from high to low potential, a positive charge *loses* EPE. A negative charge, however, would have lower EPE at points of higher potential and would thus gain EPE as it was moved from high to low potential.

This latter point can be emphasized by borrowing some language from gravity. When talking about electric potential, free positive charge will "fall" *down* the potential hill to regions of lower potential, while negative charge will "fall" *uphill* in potential. In both cases, the charges are losing PE as they "fall."

If we know the voltage V_{AB} between A and B, we can calculate the work necessary to move a charge from A to B. Using Eq. (17.1),

$$\text{Work} = \Delta EPE = qV_{AB} \tag{17.3}$$

This applies equally to both $+$ and $-$ charges if we just remember to observe the signs of both q and V_{AB}. Negative work and ΔEPE have the same interpretation as they did in mechanics.

Illustration 17.1

Suppose the electric field between the two plates in Fig. 17.1 is 2400 N/C (or V/m). If the plates are separated by 0.50 cm, what is the potential difference between them?

Reasoning The field has a constant value, and so

$$V_{AB} = Ed = (2400 \text{ V/m})(0.50 \times 10^{-2} \text{ m}) = 12 \text{ V}$$

Plate B, the positive plate, is at a potential of 12 V above plate A. If we arbitrarily set $V = 0$ at plate A, then the potential at any point between the plates a distance x from plate A is given by

$$V(\text{at } x) = Ex$$

Example 17.1

A large, flat sheet carries a surface charge density of $-4.0 \ \mu\text{C/m}^2$. If we designate the electric potential at the sheet to be $V = 0$, what is the potential at a distance of 2.0 cm?

R

Reasoning

Question What determines how electric potential depends on distance?
Answer The electric field. Where the electric field is constant, $\Delta V = -E \Delta x$, if Δx is measured along the direction of E.

Question What is the expression for the electric field due to a single sheet of uniform charge?
Answer For points not very distant from the sheet and away from its edges, Eq. 16.8 tells us

$$E = \sigma/2\epsilon_o$$

This is constant, as we have already observed.

Question What is the direction of the field?
Answer Because the charge on the sheet is negative, the field is directed toward the sheet (a positive test charge would be pulled toward the sheet). Thus moving away from the sheet means moving to higher values of electric potential.

Solution and Discussion The magnitude of the field is

$$E = \frac{4.0 \times 10^{-6} \text{ C}}{2(8.85 \times 10^{-12} \text{ C}^2/\text{N} \cdot \text{m}^2)}$$

$$= 2.26 \times 10^5 \text{ N/C} = 2.26 \times 10^5 \text{ V/m}$$

The change in potential as we move 2.0 cm away from the sheet will be

$$\Delta V = V - 0 = -(2.26 \times 10^5 \text{ V/m})(-0.020 \text{ m}) = +4520 \text{ V}$$

Be sure you understand the use of signs here. It helps to remember that moving either away from a negative charge or toward a positive charge means an *increase* in potential. Moving either toward a negative charge or away from a positive charge means a *decrease* in potential.

Example 17.2

Suppose a proton is released from rest at point B in Fig. 17.1 while at the same time an electron is released from rest at point A. Find the speed with which each charge strikes the opposite plate. Assume $V_{AB} = 45$ V, and take $m_p = 1.67 \times 10^{-27}$ kg, $m_e = 9.1 \times 10^{-31}$ kg, and $e = 1.60 \times 10^{-19}$ C.

R

Reasoning

Question   What will determine the final speed of the particle?
Answer   Each charge starts with a certain amount of electrical potential energy relative to the opposite plate. That EPE will be all converted to KE as the particle reaches the opposite plate.

Question   What is the change in EPE that each particle undergoes?
Answer   Both particles have the same amount of charge, but opposite signs. The proton will "fall" through a *drop* of 45 V of potential. The electron will "fall" through a *gain* of 45 V. Thus they will both lose the same amount of PE. For the proton:

$$\Delta \text{PE} = (+e)(V_A - V_B) = (+1.6 \times 10^{-19} \text{ C})(-45 \text{ V})$$

For the electron:

$$\Delta \text{PE} = (-e)(V_B - V_A) = (-1.6 \times 10^{-19} \text{ C})(45 \text{ V})$$

Question   Does this imply that they strike the plates at the same speed?
Answer   No. They gain the same amount of KE, but the speed depends on mass, which is very different for the two particles. Remember, KE $= \frac{1}{2}mv^2$.

Solution and Discussion   The amount of PE lost in both cases is

$$\Delta \text{PE} = -7.2 \times 10^{-18} \text{ J}$$

This equals the gain in KE:

$$\Delta \text{KE} = \tfrac{1}{2}m_e v_e^2 = \tfrac{1}{2}m_p v_p^2 = -\Delta \text{PE}$$

Then

$$v_e = \sqrt{\frac{2(7.2 \times 10^{-18} \text{ J})}{9.1 \times 10^{-31} \text{ kg}}} = 4.0 \times 10^6 \text{ m/s}$$

$$v_p = \sqrt{\frac{2(7.2 \times 10^{-18}\,\text{J})}{1.67 \times 10^{-27}\,\text{kg}}} = 9.3 \times 10^4\,\text{m/s}$$

Notice how different the results are because of the great difference in mass.

17.3 EQUIPOTENTIALS

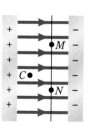

FIGURE 17.3

Points M and N are on an equipotential line.

Let us now look at points other than A and B in the region between two charged plates. For example, we might ask for the potential difference between points M and N in Fig. 17.3. Because potential difference is simply work per unit charge (Eq. 17.3), we must find the work required to move a unit positive test charge from M to N. Note that to hold the test charge in place, we must exert a force to the left on it. This force is needed to counterbalance the effect of the electric field on the test charge. If we move the charge from M to N, our balancing force does no work since the direction of motion is perpendicular to the force. Indeed, we see that no work is ever needed to move the test charge in a direction perpendicular to the electric field. Therefore, there is no potential difference between points M and N in Fig. 17.3. In fact, it should be clear that all points on the line passing through M and N are at the same potential; there is no potential difference between them. We call this line of constant potential an **equipotential line.** Moreover, the plane that lies through this line and is parallel to the plates is a constant-potential plane, which we call an equipotential plane. No work is done in moving a charge along an equipotential line or equipotential plane since such motion is always perpendicular to the lines of force, that is, the electric field. Conversely, *lines of force are always perpendicular to equipotential lines.*

As we did in the case of gravity, we can show that the work done in moving a charge between two points in the presence of an electric field is independent of the path taken between the points. Any path between points M and C in Fig. 17.3 can be reduced to a series of small steps either along or perpendicular to equipotential lines. Since no work is done on the segments along the equipotentials, the work is entirely proportional to the difference in the coordinates of M and C measured perpendicular to the plates. We therefore conclude that

We are all familiar with topographic maps, which show contours of equal elevation, as in the case of the map of this mountain. Points at equal elevation have the same gravitational potential, and so the contours are equipotential lines for the gravitational field.

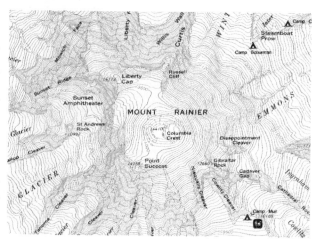

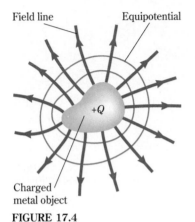

Field line   Equipotential

+Q

Charged
metal object

FIGURE 17.4

The equipotentials are perpendicular to
the field lines.

The electrical potential difference between two points is independent of the choice
of path taken between the points. This property shows us that the static electric
field is a conservative field.

Indeed, this observation is necessary for us to be able to define electrical potential
energy and to apply the conservation of energy to problems as we just did.

Before leaving our discussion of equipotentials, we should recall some of our
previous conclusions from Sec. 16.13 regarding conductors in electric fields. Since
no electric field can exist anywhere within a conductor under static conditions, we
conclude that

The volumes and surfaces of conductors are equipotential volumes and surfaces
under electrostatic conditions.

Illustration 17.2

Sketch the equipotentials and electric field lines near a charged solid metal object.

Reasoning   Consider the charged metal object shown in cross section in Fig.
17.4. The object is an equipotential volume, and so its surface is an equipotential
surface. Because lines of force must be perpendicular to equipotential lines and
surfaces, the electric field lines must be perpendicular to the object's surface. Also,
the equipotential surfaces near the conductor follow the contour of the surface fairly
closely.

Exercise   Suppose the object in Fig. 17.4 is viewed from a great distance, so that
it appears pointlike. Draw the equipotentials and field lines as they are now observed.   *Answer: The field lines are radial, and the equipotentials are circles.*

There are many types and sizes of
batteries, depending on the voltage and
power they are required to deliver. The
ones pictured here range from 1.5 V to
12 V.

17.4 BATTERIES AS SOURCES OF ELECTRICAL ENERGY

One of the easiest ways to supply a potential difference between two points is by using a battery. There are many kinds of batteries, most of them essentially chemical devices. The lead-cell battery in an automobile, for instance, uses a chemical reaction to supply energy. This is likewise true of the "dry cell," which is not dry inside despite its name. In addition to chemical batteries, other types are now becoming common. Perhaps you have heard of solar cells, which are used to supply energy to solar-powered watches and hand calculators as well as for more exotic purposes. Solar cells, which operate on quite different principles from chemical batteries, transform light directly into electric energy. Other types of nonchemical batteries are currently being developed. Despite this diversity, the purpose of any battery is to supply electrical energy.

Every simple battery has two terminals (metal posts) that provide a means for connecting wires to the battery. The quantity we commonly call the voltage of a battery is the potential difference between its two terminals, typically 1.5 V for a flashlight battery and 12 V for an automobile battery. When the battery terminals are connected by wires to two metal plates, as in Fig. 17.5, electrons flow out of the negative terminal to one plate (*B* in Fig. 17.5), charging it negatively. The source of these electrons is the other plate, *A*, which is thus left with a deficiency of electrons and hence acquires a net positive charge of equal magnitude. In this way a battery can be characterized as a "charge pump," utilizing various internal physical processes to produce the energy necessary to accomplish this transfer of charges.

As noted in Sec. 16.14, the symbol used for a battery is ——+|—— . Usually the plus and minus signs are left off the symbol, and you are expected to know that the longer line represents the positive terminal. Often the positive terminal of a battery is stamped with a plus sign or painted red.

The potential difference between the terminals of a battery depends somewhat on whether or not charge is flowing from the battery. Its potential difference when no charge is flowing is called the **electromotive force (emf)** of the battery. This term, which is a holdover from the last century, is really a misnomer, for emf is not a force at all, but represents a *voltage*. For many purposes, the emf of a battery and the potential difference between its terminals, even when charge is flowing from it, can be considered to be the same. We denote emf by the symbol $\mathcal{E}$. Do not confuse it with the symbol E used for electric field strength.

Let us examine the situation in Fig. 17.5 in more detail. When the originally uncharged metal plates are attached to the battery by metal wires, charge flows for a tiny instant as the battery establishes the charges on the plates. Thereafter, no charge flows and the situation is electrostatic. You will recall that metals are equipotential volumes under electrostatic conditions. Hence the wire from terminal *C* to plate *A* and the plate are at the same potential. Similarly, terminal *D*, which is at a potential 1.5 V lower than *C*, is at the same potential as plate *B*. Therefore, the potential difference between plates *A* and *B* is 1.5 V, with plate *A* being at the higher potential because it is positive. We thus conclude that, under electrostatic conditions, *the potential difference between a metal object connected to one battery terminal and another metal object connected to the other terminal is equal to the terminal potential difference of the battery.*

We saw in Sec. 17.2 that the charges on charged plates have electrical potential energy. Because the plates in Fig. 17.5 acquire their charge from the battery, the battery is the source of the energy that the charges on the plates possess. This is but one of many ways in which a battery acts as an energy source. When a

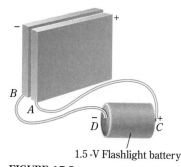

1.5 -V Flashlight battery

FIGURE 17.5

The potential difference from *B* to *A* is 1.5 V, the emf of the battery. Terminal *C* is positive and 1.5 V higher in potential than *D*.

flashlight battery lights a bulb, the heat and light energy that the bulb gives off is furnished by the battery. When a battery causes a motor to run, the mechanical energy output of the motor is furnished by the battery. As our study of electricity progresses, we shall learn of still other sources of electric energy.

Example 17.3

How much work is done by a 12.0-V battery in moving 1 C of charge from its negative terminal to its positive terminal?

R **Reasoning**

Question   How is work related to voltage?
Answer   From Eq. 17.3, $W = q \, \Delta V$.

Question   What is the voltage difference in going from the negative to the positive terminal?
Answer   +12.0 V.

Solution and Discussion   The work is positive:

$$W = (1 \text{ C})(+12.0 \text{ V}) = +12.0 \text{ J}$$

This is consistent with previous statements that positive charge increases its EPE as it is moved from lower to higher potential.

Exercise   How much work is done in moving 1 million electrons from the positive to the negative terminal?   *Answer: 1.92×10^{-12} J.*

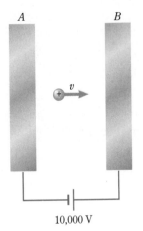

A   B

10,000 V

FIGURE 17.6

Will the proton speed up or slow down as it moves toward plate B?

Example 17.4

The proton in Fig. 17.6 is shot from plate A toward plate B. It leaves plate A at a speed of 8×10^6 m/s. A battery of 10,000 V is connected between the plates as shown. What speed does the proton have just as it strikes plate B? Repeat using the same numbers for an electron.

R **Reasoning**

Question   What principle connects a change of speed to voltage?
Answer   Moving through a voltage difference represents a change in EPE. This change in EPE in turn will change the KE and hence the speed because energy must be conserved:

$$\Delta \text{EPE} = q \, \Delta V = -\Delta \text{KE}$$

Question   Is the proton moving toward higher or lower potential?
Answer   The battery symbol shows plate B to be at 10,000 V higher potential than plate A.

Question   What is the specific equation for determining the proton's speed?
Answer $\frac{1}{2}m_p(v_B^2 - v_A^2) = -(+e)(V_B - V_A)$

Question   In what ways will the situation be different for the electron?
Answer m_e will replace m_p, and $-e$ will replace $+e$.

A high voltage between two electrodes in a vacuum can cause a beam of electrons to flow between the electrodes. The electrode emitting the electrons is referred to as the cathode, so the electrons in this beam are often called *cathode rays*.

Solution and Discussion   For the proton, the numbers give

$$v_B^2 = v_A^2 - \frac{2e(V_B - V_A)}{m_p} = (8 \times 10^6 \, \text{m/s})^2 - \frac{2(1.6 \times 10^{-19} \, \text{C})(10^4 \, \text{V})}{1.67 \times 10^{-27} \, \text{kg}}$$

Thus

$$v_B = 7.9 \times 10^6 \, \text{m/s}$$

The proton slows up, as it should when moving toward higher potential. For the electron, moving toward higher potential means speeding up:

$$v_B^2 = v_A^2 - \frac{2(-e)(V_B - V_A)}{m_e}$$

Notice how the second term on the right-hand side *adds* in this case. You should be able to show that

$$v_B = 6.0 \times 10^7 \, \text{m/s}$$

The electron's speed will actually be slightly less than this, since the equations of the relativity theory must be used when speeds approach the speed of light. (See Sec. 3.11).

Exercise   What must the potential difference between the plates be if the proton is to stop just before it reaches *B*?   *Answer: 3.34 × 10⁵ V.*

17.5   THE ELECTRONVOLT

As you well know by now, the SI unit for energy is the joule. In atomic and nuclear physics, however, there is another unit for energy that is so widely used that we must become familiar with it. This unit is defined in terms of the energy a charge of magnitude e gains as it moves through a potential difference of one volt:

One **electronvolt (eV)** is the energy acquired by a charge of magnitude $+e$ as it moves through a potential difference of one volt.

To see how the electronvolt is related to the joule, recall the kinetic energy that a charge of q coulombs acquires as it moves freely through ΔV volts of potential difference:

$$\Delta\text{KE (J)} = -\Delta\text{PE} = q \, (\text{C}) \, \Delta V \, (\text{V})$$

From the definition of eV given above,

$$\Delta\text{KE (eV)} = q \, (\text{in units of } e) \, \Delta V \, (\text{V}) \tag{17.4}$$

Comparing the two expressions for ΔKE, we have

$$\Delta KE\,(\text{eV}) = \Delta KE\,(\text{J})\frac{q\,(\text{in units of }e)}{q(\text{C})}$$

Since $1e = 1.602 \times 10^{-19}\,\text{C}$, it follows that

$$\Delta KE\,(\text{eV}) = \frac{\Delta KE\,(\text{J})}{1.602 \times 10^{-19}}$$

Thus the conversion factor between electronvolts and joules is

$$1\,\text{eV} = 1.602 \times 10^{-19}\,\text{J} \tag{17.5}$$

In atomic and nuclear physics, the particles carry charges that are integral multiples of $1.602 \times 10^{-19}\,\text{C}$, and so their charge, measured in units of e, is unity or some other small integer.

When a proton moves freely through a potential difference of 1000 V, say, its energy is, from Eq. 17.4,

$$\Delta KE = (1\,e)(1000\,\text{V}) = 1000\,\text{eV}$$

Similarly, if a particle that has a charge $3e$ moves through 1000 V, the energy it acquires is $3 \times 1000 = 3000\,\text{eV}$. Even though the electronvolt cannot be used in our SI-based equations, its convenience in dealing with the elementary particles we encounter in atomic and nuclear physics has established it firmly in science.

Illustration 17.3

To tear the single electron loose from a hydrogen atom requires an energy of 13.6 eV. Suppose we wish to knock an electron loose by bombarding hydrogen atoms with protons that have been accelerated through a potential difference V_{AB}. What is the minimum value for V_{AB} needed?

Reasoning   Each proton must have an energy of at least 13.6 eV. Since each has a charge of $1.602 \times 10^{-19}\,\text{C}$, its energy in electronvolts is numerically equal to the potential difference through which it moves. Hence the required potential difference is 13.6 V.

Exercise   Repeat if the bombarding particles are ions that have a charge $3e$. *Answer:*   4.53 V. ∎

17.6   ABSOLUTE POTENTIALS

So far, we have been concerned only with differences in potential because, as in gravitational potential, the choice of a position for zero potential energy is merely a matter of convenience. Gravitational potential energy can be measured with respect to any point we choose: a tabletop, the ground, the top of a building, or wherever. Similarly, in electrical potential energy problems, the zero potential energy location is a matter of choice. In electric circuit theory, a particular wire in the circuit may be attached to the ground (perhaps connected to a water pipe). This point is usually taken to have zero potential energy. However, a different zero for electrical potential is frequently taken, as we shall now see.

When dealing with point charges, such as atoms and molecules, it is often convenient to specify the zero of potential as being at a distance of infinity from the charge. In such cases the potential at any finite distance r is said to be the **absolute potential** at that point. In effect, what we are doing is as follows. So far, we have discussed situations in terms of potential differences V_{AB}. Now, however, we specify that point A is to be taken at infinity. Further, we specify that the potential at infinity is to be taken as zero, so that the potential at point B becomes what we refer to as the absolute potential at B. Note carefully that *when we speak of the absolute potential at a point, we are really speaking about the potential difference between that point and infinity.*

Let us find the expression for the absolute potential due to an isolated point charge $+q$ such as the one shown in Fig. 17.7. To do this, we need to calculate the work necessary to bring a positive test charge q_t from $r = \infty$ to some finite distance r from q. This is not as simple as finding potential differences between charged plates, because we no longer have a constant value of E. Instead, we must now calculate the work done by a force that varies as $1/r^2$. To do this correctly requires the methods of calculus, and so we shall simply quote the results of that calculation:

$$W(\infty \text{ to } r) = q_t \frac{kq}{r}$$

where k is the constant in the Coulomb force law. Upon dividing by q_t, we obtain the expression for absolute potential due to an isolated point charge q (or spherically symmetric charge):

$$V_{\text{abs}} = \frac{W(\infty \text{ to } r)}{q_t} = \frac{kq}{r} \tag{17.6}$$

This expression is also valid for a *negative* point charge. Equation 17.6 tells the following important information:

V_{abs} due to a positive charge q has a positive value for all distances r away from q. For a negative charge $-q$, V_{abs} is negative for all distances r.

We can use these results to calculate the absolute potential at a given point due to an assemblage of point charges. Since potential is a scalar, we need only to calculate the values of V_{abs} for each individual charge and add their contributions algebraically.

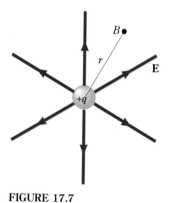

FIGURE 17.7

We define the absolute potential at B to be the work done in carrying a unit positive test charge from infinity up to B.

Example 17.5

Suppose in Fig. 17.7 that $r = 50$ cm and $q = 5.0 \times 10^{-6}$ C. If a proton is released at point B, how fast will it be moving when it gets far away?

R *Reasoning*

Question What principle connects speed to distance in this case?
Answer As before, the proton will gain KE as it loses PE by moving to lower potential.

Question What does the term "far away" mean?
Answer In practical terms, it means being far enough away that you can take the final value of potential to be essentially zero.

Question What gives me the initial value of potential?
Answer Evaluate V_{abs} at 50 cm away from a $+5\text{-}\mu C$ charge:

$$V_{abs} = \frac{kQ}{r} = \frac{(9 \times 10^9 \text{ N} \cdot \text{m}^2/\text{C}^2)(5 \ \mu C)}{0.50 \text{ m}}$$

Question What expression will give me the speed acquired?
Answer $e \, \Delta V = e V_{abs} = \frac{1}{2} m_p v^2$

Solution and Discussion First, the initial potential is

$$V_{abs} = \frac{kQ}{r} = \frac{(9 \times 10^9 \text{ N} \cdot \text{m}^2/\text{C}^2)(5 \ \mu C)}{0.50 \text{ m}} = +90 \text{ kV}$$

Hence the proton loses an amount of PE equal to

$$\Delta PE = (1.6 \times 10^{-19} \text{ C})(-90 \times 10^3 \text{ V}) = -1.44 \times 10^{-14} \text{ J}$$

The speed acquired is obtained from $\Delta KE = -\Delta PE$:

$$\tfrac{1}{2}(1.67 \times 10^{-27} \text{ kg})v^2 = 1.44 \times 10^{-14} \text{ J}$$
$$v = 4.15 \times 10^6 \text{ m/s}$$

Exercise To get a feeling for how far "far away" is in this example, calculate the distance at which the potential has dropped to 900 V (1 percent of the voltage at the proton's initial position). *Answer: 50 m.*

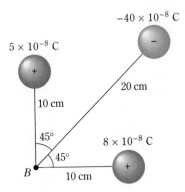

FIGURE 17.8

Find the absolute potential at point B due to the three charges.

Example 17.6

Compute the absolute potential at point B in the vicinity of the three point charges in Fig. 17.8.

R *Reasoning*

Question How do I calculate potential when there is more than one point charge present?
Answer You can calculate the potential at B due to each charge individually, as if the others weren't there. The total potential is the algebraic sum of the separate contributions. This is the principle of superposition again, using *scalar* quantities.

Question What is the expression for each contribution?
Answer $V = kQ/r$, where r is the distance from each charge to B.

Question What is the significance of the signs of the charges?
Answer Remember, positive charges produce only positive absolute potentials, negative charges only negative values. You need to keep the correct signs in the terms when you add them.

Solution and Discussion   The distances are given in Fig. 17.8. We have the following contributions to the potential at *B*:

$$V_1 = \frac{(9 \times 10^9\,\text{N} \cdot \text{m}^2/\text{C}^2)(+5.8 \times 10^{-8}\,\text{C})}{0.10\,\text{m}} = +4500\,\text{V}$$

$$V_2 = \frac{(9 \times 10^9\,\text{N} \cdot \text{m}^2/\text{C}^2)(-40 \times 10^{-8}\,\text{C})}{0.20\,\text{m}} = -18000\,\text{V}$$

$$V_3 = \frac{(9 \times 10^9\,\text{N} \cdot \text{m}^2/\text{C}^2)(+8 \times 10^{-8}\,\text{C})}{0.10\,\text{m}} = +7200\,\text{V}$$

The total potential at *B* is then

$$V_{\text{tot}} = V_1 + V_2 + V_3$$
$$= 4500\,\text{V} + (-18000\,\text{V}) + 7200\,\text{V} = -6300\,\text{V}$$

Notice how much simpler this calculation is than calculating the electric field. With potentials you do not have vector components, just positive and negative numbers to add.

Exercise   How much energy would be required to bring an electron up to point *B* from very far away?   *Answer: +6300 eV, or +1.01 × 10⁻¹⁵ J.*

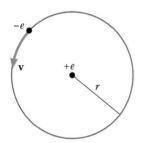

FIGURE 17.9

The Bohr model of the hydrogen atom. The electron moves in a circular orbit of 0.053 nm radius about the center of the atom.

Example 17.7

In the Bohr model of the hydrogen atom, sketched in Fig. 17.9, the pointlike electron ($q = -e$) moves in a circular orbit of radius $r = 0.053$ nm with a proton ($q = +e$) at the center. (*a*) Calculate the electrical potential energy and the kinetic energy of the electron in this orbit. (*b*) Show that, as mentioned in Illustration 17.3, 13.6 eV of energy is necessary from external sources to pull the electron loose from the atom, i.e., to *ionize* the atom.

Reasoning

Question   With the electron moving, can I use the static expression for EPE between point charges?
Answer   Yes. Even though the electron is moving, the distance *r* is constant. Other than charge, that is the only quantity EPE depends on.

Question   What is the expression for the electron's EPE?
Answer   We choose the EPE to be zero at $r = \infty$, where the electron and proton exert zero force on each other. Then

$$\text{EPE} = (-e)V_{\text{abs}}$$

where V_{abs} is the absolute potential due to the proton at the radius of the electron's orbit.

Question   What is the absolute potential at a distance *r* from the proton?
Answer $V_{\text{abs}} = ke/r$. Then $\text{EPE} = (-e)V_{\text{abs}} = -ke^2/r$. You should notice that the electron will have a negative value of EPE for all values of *r*.

Question   What type of motion is the electron undergoing?
Answer   It is circular motion at constant speed.

Question   What is the expression that describes this type of motion?
Answer   Newton's second law requires that the net force on the electron equal its mass times its centripetal acceleration:

$$F_{\text{net}} = m_e \frac{v^2}{r}$$

Question   What is the net force on the electron?
Answer   It is the electrical force, whose magnitude is given by Coulomb's law:

$$F = \frac{k(e)(e)}{r^2}$$

Question   How does the centripetal force equation relate to the electron's kinetic energy?
Answer   Notice that, since $KE = \frac{1}{2}mv^2$, the centripetal force equation can be written as

$$F_{\text{net}} = \frac{2(\text{KE})}{r}$$

Thus you can find the KE from

$$\text{KE} = \left(\frac{r}{2}\right)F_{\text{net}} = \frac{r}{2}\frac{ke^2}{r^2} = \frac{ke^2}{2r}$$

KE is a positive quantity, as it always must be. *Notice that its magnitude is just half the magnitude of the PE.*

Question   What has to happen for the electrons to be pulled free from the atom?
Answer   If the electron were being held fixed a distance r from the proton, work equal to the change in its PE would have to be done on the electron as it was pulled out to $r = \infty$:

$$W = \text{PE}(\infty) - \text{PE}(r) = 0 - \left(\frac{-ke^2}{r}\right) = \frac{ke^2}{r}$$

One way to do this would be to give the electron this much KE at its initial position so that it would be able to reach $r = \infty$ before stopping. However, the electron is *not* fixed. It already has $KE = \frac{1}{2}ke^2/r$. Thus the *additional* KE it would have to receive (perhaps by a collision with another atom) to free it would be only another $\frac{1}{2}ke^2/r$.

Solution and Discussion   We have, for the potential due to the proton,

$$V_{\text{abs}} = \frac{ke}{r}$$

$$= \frac{(9 \times 10^9 \, \text{N} \cdot \text{m}^2/\text{C}^2)(1.6 \times 10^{-19} \, \text{C})}{5.3 \times 10^{-11} \, \text{m}}$$

$$= 27.2 \, \text{V}$$

We can then say that the electron's PE is

$$PE = (-e)V_{abs} = -(1e)(27.2\,V) = -27.2\,eV$$

The electron has a KE equal in magnitude to half of this:

$$KE = +13.6\,eV$$

The *total* energy of the electron, KE + PE, is

$$KE + PE = -13.6\,eV$$

Thus the additional amount of KE needed to free the electron is $+13.6\,eV$. This is what we call the *ionization energy* (or binding energy) of hydrogen. Even the overly simple Bohr model gives accurate agreement with experimental values of this energy.

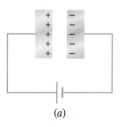

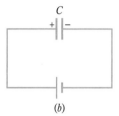

FIGURE 17.10

Equal and opposite charges reside on the inner faces of the capacitor plates. Notice the symbol used for a capacitor in *b*.

17.7 CAPACITORS

We have made frequent reference to two oppositely charged metal plates. This is one form of a device that is of considerable practical importance for storing electric charge and energy, as we shall see in later chapters. It is called a **capacitor.** Such a device is shown connected to a battery in Fig. 17.10. In Sec. 17.4 we discussed how the battery delivers positive and negative charges to the plates as shown in Fig. 17.10*a*. The plates are shown on edge; their flat surfaces are facing each other. An electrostatic condition is quickly established in which the potential difference between the plates is equal to the emf of the battery. If the battery is then disconnected, the plates remain charged to this potential. A capacitor is thus a device

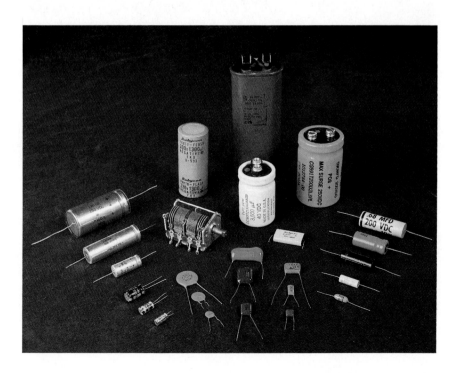

Capacitors are made in all sizes to fulfill a large variety of functions in electric circuits.

capable of *storing charge.* As shown in Fig. 17.10*b*, the symbol used for a capacitor is ┤├.

Let us represent the charges on the plates by $+q$ and $-q$. We assume these charges are spread uniformly over the area A of the plates. This means the plates have charge densities $\sigma = q/A$ and $-q/A$. In Chap. 16 we saw that the electric field between the charged plates is given by

$$E = \frac{\sigma}{\epsilon_o} = \frac{q}{A\epsilon_o}$$

The potential V between the plates is related to the field by

$$V = Ed = \frac{d}{A\epsilon_o}q \qquad (17.7)$$

where d is the distance between the plates. We thus see that V *is proportional to* q, which is a general result applicable to other forms of capacitors as well.

The **capacitance C** of the plates is the ratio of charge stored on the plates to the potential between them:

$$C = \frac{q}{V} \qquad (17.8)$$

The SI units of capacitance are thus coulombs per volt. We give this derived quantity the name farad (after the English physicist Michael Faraday).

One **farad** (F) = one coulomb per volt (C/V)

From Eq. 17.7 we can identify the capacitance of the parallel-plate arrangement:

$$C = \frac{q}{V} = \frac{\epsilon_o A}{d} \qquad \text{(parallel plates)} \qquad (17.9)$$

You should be able to show that this expression does give the units of farads.

An important point to observe is that the capacitance is a property of a particular device. *Once the dimensions and shape of a capacitor are given, its capacitance is determined,* independent of the amount of charge stored on it. For parallel plates, for example, C is determined entirely by the size (area) and separation of the plates.

A farad is an enormous amount of capacitance, and values of C for practical devices usually are of the order of μF or smaller. For example, plates of area $100\ \text{cm}^2$, separated by a distance of $1\ \text{mm}$, have a capacitance of

$$C = \frac{(8.85 \times 10^{-12}\ \text{C}^2/\text{N} \cdot \text{m}^2)(100 \times 10^{-4}\ \text{m}^2)}{10^{-3}\ \text{m}}$$

$$= 8.85 \times 10^{-11}\ \text{F} = 88.5\ \text{picofarads (pF)}$$

(a)

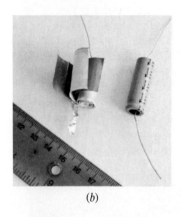

(b)

FIGURE 17.11

Two sheets of metal foil separated by an insulator act as the plates of a commercial capacitor. If the sheets are rolled or folded into a compact package, a parallel-plate capacitor can be reduced to a convenient size. We show two types of capacitor in both original and partly disassembled form. (*a*) A 100-pF capacitor that uses a thin plastic sheet as insulator. (*b*) A 470-μF electrolytic capacitor that uses a thin oxide coating on the metal foil as insulator. A paper spacer impregnated with moist electrolyte separates the metal sheets. Although they provide large capacitance, electrolytic capacitors usually cannot withstand high voltages.

---

*In Section 17.9 we shall see that the material surrounding the charged surfaces also affects the capacitance. Strictly speaking, Eq. (17.9) represents parallel plates in a *vacuum.*

In practice, most parallel-plate capacitors contain a sheet of nonconducting material between the plates. This sheet allows the plates to be placed very close together with no fear that they will touch and permit the charges to join together. Many commercial capacitors are formed by taking two thin sheets of metal foil and laying one on top of the other with a thin plastic film between them to keep them from touching. The layered sheets are then rolled up into a tight cylinder and packaged for convenience. The device is essentially a parallel-plate capacitor, but it looks very different from the sketch in Fig. 17.10. Capacitors with a capacitance of 0.1 μF, a common size, occupy a volume of about 1 cm^3 when made this way. Figure 17.11 shows two common capacitors.

Example 17.8

What is the capacitance of an isolated metal sphere whose radius is $R = 10$ cm?

Reasoning

Question How can an isolated conductor have a capacitance?
Answer "Isolated" means that other charges are effectively an infinite distance away. This is the same concept that allows you to define absolute potentials for point or spherical charges. If charge q is placed on a sphere, it produces an absolute potential V at every point outside the sphere. The general definition of capacitance, $C = q/V$, is applicable in all cases.

Question If the sphere carries charge q, what value of V applies to this problem?
Answer You want to find the voltage between the conducting sphere and infinity. Thus in this case it is the value of V_{abs} at the sphere's surface that applies.

Question What is q/V_{abs} for a sphere?
Answer For point and spherical charges, Eq. 17.6 applies. $V = kq/r = q/4\pi\epsilon_o r$ as long as r is at or outside of the sphere's surface.

Question What does this give me for the capacitance C?
Answer $C = q/V = q/(kq/R) = R/k = 4\pi\epsilon_o R$, where use has been made of the fact that $k = \dfrac{1}{4\pi\epsilon_o}$. R is the radius of the sphere.

Solution and Discussion Notice that $C = 4\pi\epsilon_o R$ has a constant value for a given sphere. This is another example showing that capacitance depends only on the size and geometry of the objects storing charge. Putting in values,

$$C = \frac{R}{k} = \frac{0.10 \text{ m}}{9 \times 10^9 \text{ N} \cdot \text{m}^2/\text{C}^2} = 11.1 \text{ pF}$$

17.8 DIELECTRICS

Despite the fact that nonconductors contain no free charges, they have a marked effect on electric fields in which they are placed. These materials, called **dielectrics** in this context, tend to partially cancel the electric fields established by charged objects. We now see how they do this.

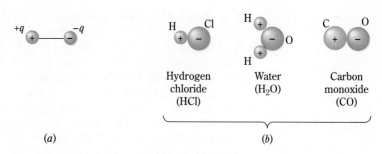

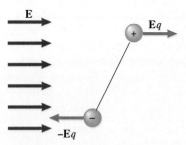

FIGURE 17.12

The dipolar molecules in (b) act like the dipole in (a).

FIGURE 17.13

An electric field causes a dipole to experience a torque that tends to align it in the field.

We can divide dielectrics into two groups, those that contain molecular dipoles and those that do not. A **dipole** consists of two equal-magnitude charges of opposite sign separated by a small distance, as shown in Fig. 17.12a. Many molecules, although electrically neutral (that is, uncharged), are in effect tiny dipoles. Examples are shown in Fig. 17.12b. Molecules such as these are called *dipolar molecules*. When a dipolar molecule is placed in an electric field, as in Fig. 17.13, its oppositely charged ends experience equal, oppositely directed forces ($\mathbf{E}q$ and $-\mathbf{E}q$). The resultant torque on the molecule tends to align it in the electric field. As a result, dipolar molecules between charged plates tend to align as shown in Fig. 17.14. In practice, thermal motion prevents them from becoming fully aligned except in extremely strong fields.

Atoms and many molecules are not ordinarily dipoles. Although they have negatively charged electrons and positively charged nuclei, the effective centers of the two types of charge coincide, as shown at the top of Fig. 17.15. As a result, these atoms and molecules behave as though the negative and positive charges were not separated, and so they possess no permanent dipole. However, when such an atom or molecule is placed in an electric field, as shown in the lower part of Fig. 17.15, the negatively charged electrons are attracted slightly to the left and the positively charged nucleus is repelled slightly to the right. This slight shift in the charges causes the atom (or molecule) to become a dipole; we say that it has become *polarized* and now possesses an *induced dipole*.

We see then that all materials, when placed in an electric field, become dipoles aligned with the field, as in Fig. 17.14. Notice how the positive plate induces the negative ends of the dipoles to come close to it and the negative plate attracts the positive ends. Notice further in Fig. 17.14 that the dipole alignment causes a layer

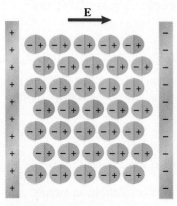

FIGURE 17.14

Dipoles align along the field lines.

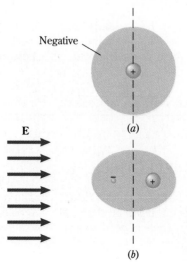

FIGURE 17.15

(a) Normally, in a nonpolar atom or molecule, the negative electrons form a symmetric charge distribution around the positive nucleus. (b) When placed in an external electric field the electron charge distribution is shifted away from the nucleus and in the direction opposite $\mathbf{E}$. (Why?) This causes the atom or molecule to become an induced dipole.

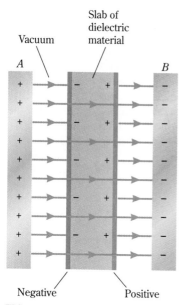

FIGURE 17.16

The electric field induces bound charges on the surface of the dielectric. They cause the field to be less inside the dielectric than outside.

of positive charges (the positive ends of the dipoles) to exist near the plate on the right. Similarly, there is a layer of negative charges near the plate on the left. When a slab of dielectric material is placed between the plates, as in Fig. 17.16, the dipole alignment causes charges to appear on the two faces of the slab. These charges are simply the charged ends of the dipoles at the surfaces of the dielectric. We refer to this type of charge as *induced polarization charge* or *bound charge*. The latter name reflects the fact that this charge is bound to atoms and molecules within the dielectric; it is not free to move from its parent atom or molecule.

The amount of bound charge that can be induced at the surface of an object differs from material to material. We know, for instance, that the volume of a conductor must be a field-free region. If a *metal* slab (a conductor) is inserted between the plates, the induced surface charge must *equal* the charge on the plates. This *completely* cancels the field inside the conductor, as shown in Fig. 17.17. Notice that all field lines end at the negative surface of the conductor and begin again at the positive surface. No field lines exist within the conductor.

For dielectrics, the induced charge is less than the charge on the plates. Thus not all of the field lines terminate on charges at the dielectric surface; some of them penetrate through the dielectric material, as shown in Fig. 17.16. *The general result is that the electric field inside a dielectric is less than the external field imposed on it.* The more easily a substance is polarized, the greater the difference between the internal and external fields.

The ability of a dielectric to decrease the electric field strength is characterized by its **dielectric constant** K, defined by reference to Fig. 17.16:

$$\text{Dielectric constant } K = \frac{\text{electric field in vacuum}}{\text{electric field in dielectric}}$$

The electric field is only $1/K$ as large inside the dielectric as outside it. Typical dielectric constants are given in Table 17.1. Notice that vacuum does not alter the field at all, and so its dielectric constant is unity. Because air has so few molecules per unit volume, its dielectric constant differs only slightly from that of vacuum. For most solids, K is in the range from 2 to 10. Although we do not consider metals to be dielectrics, you should be able to show that the dielectric constant for a metal is infinite.

TABLE 17.1
Dielectric constants (20°C)

Material	K
Vacuum	1.00000
Air	1.006
Paraffin	2.1
Petroleum oil	2.2
Benzene	2.29
Polystyrene	2.6
Ice (−5°C)	2.9
Mica	6
Acetone	27
Methyl alcohol	38
Water	81
Metal	∞

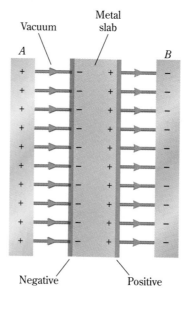

FIGURE 17.17

When the dielectric slab of Fig. 17.16 is replaced by a metal plate, enough charge is induced on the surfaces of the metal to reduce the field within the metal to zero.

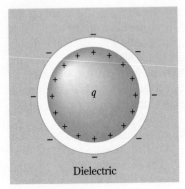

FIGURE 17.18

A charged sphere in an infinite dielectric. Why is the electric field decreased by the dielectric?

17.9 THE EFFECTS OF DIELECTRICS

Coulomb's law is altered when charges are immersed in a dielectric. To see why this occurs, refer to Fig. 17.18. We see there a sphere with charge q immersed in a dielectric that extends a great distance in all directions—in other words, an essentially infinite dielectric. Notice how the sphere induces a charge on the dielectric surface next to it. This induced charge in effect cancels some of the charge on the sphere. Thus the electric field in the dielectric is reduced from the value $E = kq/r^2$ that applies in a vacuum. The dielectric reduces the field by a factor of $1/K$, so that the field in the dielectric is

$$E = k\frac{q}{Kr^2} \qquad \text{(point charge)} \tag{17.10}$$

This is the electric field of a point charge immersed in a dielectric.

Suppose two point charges q_1 and q_2 that are a distance r apart are immersed in an infinite dielectric. The field due to q_1 at the position of q_2 is given by Eq. 17.10 after q has been replaced by q_1. This field causes a force Eq_2 on q_2, and so we find that the force on q_2 due to q_1 is

$$F = k\frac{q_1 q_2}{Kr^2} \qquad \text{(Coulomb's law)} \tag{17.11}$$

This is Coulomb's law for point charges in an infinite dielectric.

Because a dielectric greatly influences the forces between charges, chemical and biological reactions are markedly solvent-dependent. For example, two ions in solution exert forces given by Eq. 17.11 on each other. Water has $K = 81$, and so the force between two ions is much less in water than it is in a liquid such as benzene, which has $K = 2.3$. As a result, the Na^+ and Cl^- ions of sodium chloride can escape from each other in water, whereas they cannot do so in benzene. Hence water dissolves NaCl and benzene does not. There are many other similar situations in chemical and biological systems where the dielectric constant of the solvent is the controlling factor in chemical reactions.

Most capacitors are made with some dielectric material between the plates, as mentioned in Sec. 17.7. This not only provides increased structural strength but also *increases the capacitance,* as we now shall see.

We begin by charging a parallel plate capacitor with charges q and $-q$ on the plates. The plates have only vacuum between them. Let C_{vac} be the capacitance under these conditions. The voltage between the plates is $V_{vac} = q/C_{vac}$. Let us now insert a dielectric slab that completely fills the space between the plates. Even though the surfaces of the plates and the dielectric touch, charges cannot move across the boundary between the materials. The field between the plates is now decreased by $1/K$: $E = E_{vac}/K$. This in turn decreases the voltage between the plates:

$$V = Ed = \frac{E_{vac}}{K}d = \frac{V_{vac}}{K}$$

But the charge on the plates is not changed by the insertion of the dielectric. Thus the ratio of charge to voltage, the capacitance, is now

$$C = \frac{q}{V} = \frac{q}{V_{vac}/K} = KC_{vac} \tag{17.12}$$

Thus the same plates, with the same separation, can store more charge per volt with a dielectric between them.

A simple way to measure the dielectric constant of materials is to measure the voltage across charged plates in vacuum and again when filled with the dielectric. The ratio of these voltages yields the dielectric constant K:

$$\frac{V_{vac}}{V_{diel}} = K$$

Example 17.9

Two parallel plates having an area of 20 cm² and a separation of 0.4 mm are connected to a 120-V battery. What charge flows to the plates?

R

Reasoning

Question How is the plate charge related to voltage, plate size, and plate separation?

Answer The size and separation of the plates determine the capacitance of the plates (Eq. 17.9). If you know capacitance, the plate charge can be found from the definition: $C = q/V$.

Question What voltage will the plates acquire?
Answer Charge will flow from the battery until the voltage between plates equals that of the battery.

Solution and Discussion The capacitance has the value

$$C = \frac{(8.85 \times 10^{-12}\,C^2/N \cdot m^2)(20 \times 10^{-4}\,m^2)}{(0.4 \times 10^{-3}\,m)}$$

$$= 44.3\,pF$$

The charge that will flow to the plates is

$$q = VC = (120\,V)(44.3 \times 10^{-12}\,F) = 5.32 \times 10^{-9}\,C$$

Make sure you understand the units involved.

Example 17.10

If the charged plates of the previous example were disconnected from the battery and then immersed in water, which of the quantities C, V, and q would change, and by how much?

R

Reasoning

Question What effect does disconnecting the battery have?
Answer Without a battery, there is no longer a source of charge. The charge placed there originally is now trapped on the plates. It cannot leave, and no more charge can be supplied. Thus q must remain the same.

Question What happens to the value of C when the plates are immersed?
Answer $C_{\text{diel}} = KC_{\text{vac}}$

Question What must happen to the voltage between the plates?
Answer With the battery gone, the voltage can (and must) change:

$$V_{\text{diel}} = \frac{V_{\text{vac}}}{K}$$

Solution and Discussion The dielectric constant of water is $K = 81$. Thus we have the simple results that

$$C_{\text{diel}} = 81 C_{\text{vac}} = (81)(44.3 \times 10^{-12}\,\text{F}) = 3.59\,\text{nF}$$

$$V_{\text{diel}} = \frac{120\,\text{V}}{81} = 1.48\,\text{V}$$

Once you establish that charge must remain fixed, this means that the *product* $VC\,(= q)$ must remain constant. The polarization of the water cancels all but $1.48/120 = 1.2$ percent of the field between the plates.

Example 17.11

If the plates in Example 17.10 were immersed in the water with the battery *still connected*, how would these results differ?

R

Reasoning

Question Can charge change with the battery still connected?
Answer Yes. The battery can supply charge as long as it is connected.

Question What must the voltage be when the battery is still connected?
Answer The battery will supply charge until the voltage across the plates is the same as the battery voltage: $V_{\text{diel}} = V_{\text{vac}} = V_{\text{battery}}$.

Question What will happen to the capacitance?
Answer Capacitance is independent of the charge or voltage. It is a property of the materials and dimensions of the hardware. Thus we again have

$$C_{\text{diel}} = KC_{\text{vac}}$$

Solution and Discussion Here the quantity that is *forced* to stay the same is the voltage of 120 V across the plates. C is again increased to 3.59 nF. Charge must adjust so that

$$q_{\text{diel}} = C_{\text{diel}}V = (3.59 \times 10^{-9}\,\text{F})(120\,\text{V})$$
$$= 4.31 \times 10^{-7}\,\text{C}$$

Remember that the original charge was 5.32 nC. Thus an *additional* amount of 4.26×10^{-7} C would have to be supplied by the still-connected battery.

17.10 CAPACITORS CONNECTED IN SERIES AND PARALLEL

In many applications we encounter later, capacitors will be connected in various combinations. We will want to know what the effective total capacitances of such combinations are.

First, let us connect three capacitors to a battery of voltage V as shown in Fig. 17.19a. This is called a **parallel** connection. How do the individual capacitances add? In other words, what single capacitance C is equivalent to the parallel combination?

Notice that the three left plates are connected together by conducting wire attached to the positive terminal of the battery. Thus all three left plates must be at the same potential. Similarly, the capacitor plates on the right side must all have the same potential as the negative terminal of the battery. We can draw the following conclusion:

The potential across all capacitors connected in parallel must be the same.

In the case shown in Fig. 17.19a, the potential across each capacitor is V, the battery voltage.

The charge on each capacitor is given by the definition of capacitance:

$$q_1 = C_1 V \qquad q_2 = C_2 V \qquad q_3 = C_3 V$$

The *total* charge on the left plates is $q_{\text{tot}} = q_1 + q_2 + q_3$. The total charge on the right plates is just the negative of this. A capacitor equivalent to the three in Fig. 17.19 would store a charge q_{tot} at a voltage V:

$$C_{\text{eq}} = \frac{q_{\text{tot}}}{V} = \frac{q_1 + q_2 + q_3}{V} = \frac{q_1}{V} + \frac{q_2}{V} + \frac{q_3}{V} = C_1 + C_2 + C_3$$

Here we have used the fact that $\dfrac{q_1}{V} = C_1$, and similarly for C_2 and C_3.

FIGURE 17.19

(a) Capacitors connected in parallel to a voltage V acquire different charges, but all acquire the same voltage. The equivalent (total) capacitance of this combination is the sum of their individual capacitances. (b) Capacitors connected in series to a voltage V all acquire the same charge. Their individual capacitances add reciprocally to give the reciprocal of the equivalent (or total) capacitance of the combination.

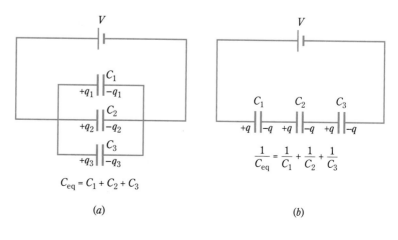

We can generalize this result for *n capacitors in parallel:*

$$C_{eq} = C_1 + C_2 + C_3 + \cdots + C_n \tag{17.13}$$

Figure 17.19*b* shows three capacitors hooked up end to end. This type of connection is called a **series connection.** We would like to find the equivalent total capacitance of this combination.

When the combination is hooked up to a battery of voltage V as shown, the left plate of C_1 is at the potential of the positive terminal of the battery and the right plate of C_3 is at the potential of the negative terminal. Thus the voltage V is the potential across the entire series arrangement. The two plates just mentioned acquire charges of $+q$ and $-q$, respectively. This causes charges of $+q$ and $-q$ to be induced on the remaining plates of the capacitors as shown in Fig. 17.19*b*. To see that this must be so, notice that, without an external connection, no *net* charge can be brought to the inner plates. The right plate of C_1 and the left plate of C_2 form a single neutral conductor when connected. The same can be said of the other two inner plates. All the external battery can do is to induce charge *separation* between these plates. We can thus draw the following conclusion about capacitors in series:

Each capacitor hooked to a battery in a series combination carries the same amount of charge.

Since the charges are equal, the individual capacitors must have different voltages across them:

$$V_1 = \frac{q}{C_1} \qquad V_2 = \frac{q}{C_2} \qquad V_3 = \frac{q}{C_3}$$

Furthermore the three voltages must add up to the total voltage V:

$$V = V_1 + V_2 + V_3 = \frac{q}{C_1} + \frac{q}{C_2} + \frac{q}{C_3}$$

An equivalent single capacitor would acquire the charge q from the battery voltage V, hence $V = q/C_{eq}$. Equating these two expressions for V gives

$$V = \frac{q}{C_{eq}} = \frac{q}{C_1} + \frac{q}{C_2} + \frac{q}{C_3}$$

We can cancel q in this equation and generalize the result to *n capacitors in series:*

$$\frac{1}{C_{eq}} = \frac{1}{C_1} + \frac{1}{C_2} + \frac{1}{C_3} + \cdots + \frac{1}{C_n} \tag{17.14}$$

When adding reciprocals, remember to invert the sum to get C_{eq}. This is the single most common error in this kind of calculation. A helpful check on your answer is the following:

C_{eq} must be smaller than any of the individual capacitors in the series combination.

Illustration 17.4

Suppose you have three capacitors: $C_1 = 3\,\text{nF}$, $C_2 = 4\,\text{nF}$, and $C_3 = 6\,\text{nF}$. Calculate the equivalent capacitance when these are connected (*a*) in parallel and (*b*) in series.

Reasoning   The parallel combination is very easy:

$$C_{\text{par}} = C_1 + C_2 + C_3 = 3\,\text{nF} + 4\,\text{nF} + 6\,\text{nF} = 13\,\text{nF}$$

In series you add the reciprocals:

$$\frac{1}{C_{\text{ser}}} = \frac{1}{3\,\text{nF}} + \frac{1}{4\,\text{nF}} + \frac{1}{6\,\text{nF}}$$

Find a suitable common denominator, for example, 12 nF.

$$\frac{1}{C_{\text{ser}}} = \frac{4 + 3 + 2}{12\,\text{nF}} = \frac{9}{12\,\text{nF}}$$

Notice that this has to be inverted:

$$C_{\text{ser}} = \frac{12\,\text{nF}}{9} = 1.33\,\text{nF}$$

This result is indeed a smaller capacitance than the smallest individual value (1.33 < 2).   ∎

17.11 THE ENERGY STORED IN A CAPACITOR

A charged capacitor has electrical potential energy stored in it. We know this to be true because one of its charges, when released from one plate, gains kinetic energy as it moves to the other plate. We can find out how much energy is stored in a charged capacitor by calculating the work a battery must do to deliver charge to the plates.

Let us view the charging process as one in which the final charge q is the result of small increments of charge Δq being delivered to the plates. To begin with, there is no voltage across the uncharged plates, and so the first Δq can be delivered free (no work). But the next Δq has to have work done on it, since there is now a voltage $\Delta q/C$ across the plates. Each successive increment of charge requires more work as the voltage builds up proportionally to the charge already in place. The final charge Δq requires work of $\Delta q\,V$, where V is the final voltage across the fully charged plates. The total work done is equivalent to delivering the entire charge in the presence of the average value of the voltage during the charging process. This average value is just $\tfrac{1}{2}V$. Thus the energy stored in a charged capacitor is

The energy that can be stored in a large charged capacitor becomes dramatically evident when the terminals of the capacitor are short-circuited.

$$\text{Energy} = \tfrac{1}{2}qV = \frac{\tfrac{1}{2}q^2}{C} = \tfrac{1}{2}CV^2 \tag{17.15}$$

where we have used the definition of capacitance $C = q/V$.

17.12 THE ENERGY STORED IN AN ELECTRIC FIELD

In the preceding section, we saw that the energy stored in a charged capacitor is $\frac{1}{2}CV^2$, where V is the voltage across a capacitor having a capacitance C. Although it is not necessary to specify exactly how and where this energy is stored, it is sometimes convenient to think of it as being stored in the electric field between the capacitor plates. With this in mind, it would be well to express the equation for the stored energy in terms of the electric field E between the plates. We can do this by recalling that, for a parallel-plate capacitor, $V = Ed$, where d is the separation of the plates.

We therefore have for the energy stored in a parallel-plate capacitor

$$\text{Energy} = \tfrac{1}{2}CV^2 = \tfrac{1}{2}CE^2d^2$$

From Eq. 17.7, however, the capacitance of a parallel-plate capacitor with plate area A is $C = \epsilon_o A/d$, provided the capacitor has vacuum between its plates. If it is filled with a dielectric with a constant K, the equation becomes $C = K\epsilon_o A/d$.

Substituting this value for C in the energy equation yields

$$\text{Energy} = (\tfrac{1}{2}K\epsilon_o E^2)(Ad)$$

The term Ad is the volume of the space between the capacitor plates—in other words, the volume in which the constant electric field E exists. Dividing both sides of the equation by the volume gives us an expression for the energy per unit volume, that is, the energy we picture as being stored in a unit volume of the region of space where the electric field is E:

$$\text{Energy density} = \text{energy per unit volume} = \tfrac{1}{2}K\epsilon_o E^2 \qquad (17.16)$$

Notice that the energy stored in a unit volume of space is proportional to the square of the electric field strength. It is often convenient to use Eq. 17.16 for assigning energy to an electric field. Although this expression was derived for a very special case, it is shown in more advanced texts that it has general validity.

LEARNING GOALS

Now that you have finished this chapter, you should be able to

1 Define (*a*) potential difference, (*b*) volt, (*c*) equipotential lines, surfaces, and volumes, (*d*) emf, (*e*) electronvolt, (*f*) absolute potential, (*g*) capacitor, (*h*) capacitance, (*i*) farad, (*j*) dielectric, (*k*) dipole, (*l*) dielectric constant, (*m*) parallel and series connections.

2 Find the potential difference between two points when the work required to carry a charge q from one point to the other is given (or vice versa).

3 Find the potential difference between any two points in a region in which a known uniform electric field exists.

4 Sketch the equipotentials and field lines in simple situations.

5 Use the relation $W = qV_{AB}$ in simple specified situations.

6 Find the energy change in electronvolts of a particle of known charge due to its movement through a given potential difference. Convert energies between electronvolts and joules.

7 Find the absolute potential at a point due to several specified point charges in the vicinity of the point.

8 Find the change in kinetic energy of a charged particle due to its motion through a given potential difference. If either the initial or final speed is known, find the other speed.

9 Calculate the capacitance of parallel plates and an isolated sphere using their dimensions. Give the relation connecting q, V, and C.

10 Explain the reason why some liquids or solids have large dielectric constants and some have small dielectric constants.

11 Compute the energy stored in a given capacitor charged to a known potential difference.
12 Calculate the effect of dielectrics on capacitance, voltage, and electric field.

13 Calculate the equivalent capacitance of capacitors connected in parallel and capacitors in series.
14 Calculate the energy per volume in an electric field.

SUMMARY

DERIVED UNITS AND PHYSICAL CONSTANTS

Unit of Electric Potential (V)
1 volt (V) = 1 J/C

The Electronvolt Unit of Energy (eV)
$1\,\text{eV} = 1.6 \times 10^{-19}\,\text{J}$

Unit of Capacitance (F)
1 farad (F) = 1 C/V

DEFINITIONS AND BASIC PRINCIPLES

Electric Potential (V) and Potential Energy
The difference in electric potential (voltage) between two points A and B is defined as the difference in the electric potential energy of a *positive* charge divided by that charge:

$$\Delta V = V_{AB} = V_B - V_A = \frac{\text{PE}_B - \text{PE}_A}{q}$$

INSIGHTS
1. In a region of constant electric field, the voltage difference between two points is simply

$$V_{AB} = Ed$$

where d = distance from A and B, measured *along E.*
2. Electric potential *decreases* in the direction of E.
3. An alternate SI unit for E is volt/meter:

$$1\,\text{N/C} = 1\,\text{V/m}$$

4. A free positive charge will "fall" from a region of higher voltage to one of lower voltage. Free negative charges will "fall" from regions of low to higher potential. In both cases, the free charges *lower* their potential *energy.*
5. The work required to move a charge q through a voltage difference V_{AB} is

$$W = \Delta\text{PE} = qV_{AB}$$

The correct sign of W follows from the correct observance of the signs of q and $V_{AB}.$
6. Lines or surfaces of constant values of potential are

equipotentials. Equipotentials are everywhere perpendicular to electric field lines.
7. All points of a conductor are equipotential under electrostatic conditions.

Absolute Potential of Point or Spherical Charges
The choice of where $V = 0$ is arbitrary. For charges of spherical symmetry (which includes point charges), the choice of $V = 0$ at $r = \infty$ is convenient. Then the absolute potential of such a charge Q is given by

$$V(r) = \frac{kQ}{r} = \frac{Q}{4\pi\epsilon_o r}$$

for points *at or outside* the charge distribution.

Capacitors
Two conducting surfaces can be charged with equal charges of opposite sign. Such a device is called a *capacitor.* Capacitance is defined as the ratio of charge stored on the surfaces to the resulting voltage between the surfaces:

$$C = \frac{Q}{V}$$

INSIGHTS
1. Once the dimensions and shape of a capacitor are given, its capacitance is determined.
2. A large value of C indicates the device can store large amounts of charge without building up a large voltage. Small values of C indicate a large potential difference for relatively small amounts of stored charge.
3. The most common capacitor is one with parallel plates of area A and separation d. The capacitance of a parallel-plate capacitor is

$$C = \frac{\epsilon_o A}{d}$$

4. The capacitance of an isolated sphere of radius R is

$$C = 4\pi\epsilon_o R$$

Dielectrics
Nonconductors, called dielectrics, between capacitor plates can alter the field between the plates because of the polarization

of their molecules. This has the effect of partially reducing the strength of the field from the value it had in vacuum. The degree to which a dielectric can decrease the field is characterized by its dielectric constant K, defined by

$$K = \frac{\text{field in vacuum}}{\text{field in dielectric}}$$

INSIGHTS
1. Values of K are equal to or greater than 1.
2. Materials which are more easily polarized have larger values of K.
3. In all equations where k or ϵ_o appear, the presence of dielectric materials filling the space can be accounted for by substituting k/K or $K\epsilon_o$, respectively.
4. The results of the above are that V and E are decreased in the presence of a dielectric by a factor of $1/K$, and C is increased by a factor of K.

Capacitors Connected in Series and Parallel
The equivalent total capacitance of n capacitors connected in parallel is

$$C_{\text{par}} = C_1 + C_2 + \cdots + C_n$$

The equivalent total capacitance of n capacitors connected in series is

$$\frac{1}{C_{\text{ser}}} = \frac{1}{C_1} + \frac{1}{C_2} + \cdots + \frac{1}{C_n}$$

INSIGHTS
1. Each capacitor in a parallel combination has the same voltage across it. Each carries a different charge (unless they have the same C).
2. Each capacitor in a series combination carries the same charge. Each has a different voltage across it (unless they have the same C).
3. In a series combination, remember to take the final reciprocal to find C_{ser}. As a check, the answer must be smaller than the least value of the individual capacitors.

Energy Stored in a Capacitor
The energy stored in a capacitor C carrying charge q is

$$\text{Energy} = \frac{\frac{1}{2}q^2}{C} = \tfrac{1}{2}qV = \tfrac{1}{2}CV^2$$

Energy Density in an Electric Field
The energy density (energy per unit volume) associated with a region of electric field strength E is

$$\text{Energy density} = \tfrac{1}{2}K\epsilon_o E^2$$

where K is the dielectric constant of the material filling the volume.

QUESTIONS AND GUESSTIMATES

1 Two points A and B are at the same potential. Does this necessarily mean that no work is done in carrying a positive test charge from one point to the other? Does it mean that no force has to be exerted to carry the test charge from one point to the other? Explain.

2 Can two equipotential surfaces intersect? Explain.

3 The absolute potential midway between two equal but oppositely charged point charges is zero. Can you find an obvious path along which no work would be done in carrying a positive test charge from infinity up to this point? Explain.

4 Starting from the fact that a piece of metal is an equipotential body under electrostatic conditions, prove that the electric field inside a hollow piece of metal is zero.

5 If the absolute potential is zero at a point, must the electric field be zero there as well?

6 What can be said about the electric field in a region in which the absolute potential is constant?

7 Prove that all points of a metal object are at the same potential under electrostatic conditions. Does this also apply in a hole inside the object? Does it matter if there is a charge suspended in the hole?

8 A parallel-plate capacitor has a fixed charge q on its plates. The plates are now pulled farther apart. The puller must do work. Why? Does the potential difference change during the process? What happens to the work done by the puller?

9 A hollow uniformly charged spherical metal ball has a charge $+q$. Where is the charge located? Is the absolute potential zero inside the sphere? Is it constant? What is it? Repeat for a charge $-q$.

10 Electrostatic methods are frequently used in industry to spray-paint metal objects. The sprayer is attached to one terminal of a high-voltage source, and the metal object to be painted is attached to the other. Explain the principle of operation for this method. Why does it generate less air pollution and use less paint than conventional methods?

11 Two identical metal spheres carry charges $+q$ and $-2q$. They are touched together and again separated. What are their final charges? If the two spheres have different radii, which has the larger final charge?

12 The electric strength of air is about 30,000 V/cm. By this we mean that when the electric field intensity exceeds this value, a spark will jump through the air. We say that "electric breakdown" has occurred. Using this value, estimate the potential difference between two objects where a spark jumps. A typical situation might be the spark that jumps between your body and a metal door handle after you have walked on a deep carpet or slid across a plastic car seat in very dry weather.

13 Refer to the data given in the previous question. About how much charge could you place on a metal sphere that has a diameter of 50 cm?

14 A simple electrostatic precipitator for removing smoke from air can be constructed as shown in Fig. P17.1. A very thin wire is placed along the axis of a much larger metal tube, and a high voltage is applied to these two elements, with the wire being made the negative terminal. If the wire is very thin and the voltage high, the electric field near the wire will be very high. Why? Tiny sparks (called corona) are formed

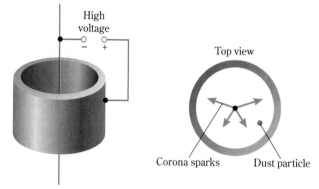

FIGURE P17.1

near the wire due to electric breakdown (see Question 12), and electrons shoot away from the wire. Why? They charge the smoke particles negatively. How? These particles then move to the tube and precipitate there. Why? As a result, the smoke is removed from the air.

PROBLEMS

Sections 17.1–17.4

1 How much work is required to carry a $+6.0$-μC charge from the negative to the positive terminal of a 9.0-V battery? From the positive to negative terminal?

2 How much work is needed to carry an electron from the positive terminal to the negative terminal of a 3.0-V battery? Repeat for a proton.

3 Two parallel metal plates 0.6 mm apart are connected to the terminals of a 1.5-V battery. (a) What is the electric field strength between the plates? (b) How large a force would an electron experience if it were between the plates?

4 Two parallel metal plates are 0.3 mm apart, and the electric field strength between them is 3000 V/m. (a) What is the potential difference between the plates? (b) How large a force would a proton experience if it were placed between them?

5 How much work is needed to move Avogadro's number (6.02×10^{23}) of electrons between two points where the potential difference is 24 V?

■ 6 Two points A and B on the x axis are separated by 40 cm in a region of constant **E** and have a potential difference of 60 V, A being at higher potential. (a) Find E_x, the constant electric field in the x direction, in this region. (b) Repeat if B is at higher potential.

■ 7 In a certain region of space, the electric field is directed in the positive z direction and has a magnitude of 4000 V/m. Find the potential difference between the origin of the coordinate system and the points having the following (x, y, z) coordinates expressed in meters: (a) $=(0, 0, 8)$; (b) $=(16, 0, 0)$; (c) $=(0, 0, -10)$; (d) $=(-12, 10, 12)$.

8 How much work is done when a proton moves a distance of 4 cm along a uniform electric field of strength 250 N/m?

■ 9 An electron is released at the coordinate origin in a region where there is an electric field of strength 2800 V/m in the positive y direction. (a) Find the time the electron takes to reach a speed of 7.2×10^6 m/s. (b) How far has the electron traveled during this time?

■10 A proton is traveling along the positive x axis with a speed of 6.0×10^5 m/s. An electric field is now switched on such that $E_x = -500$ V/m, $E_y = E_z = 0$. (a) What is the speed of the proton after it has traveled 3 m? (b) How much time does it take to reach this point?

11 A proton is released from rest and accelerates through a 60-V potential difference. What is the final speed of the proton? Repeat for an electron.

12 Through how large a potential difference must an alpha particle move if it is to be accelerated from rest to 6.0×10^5 m/s? (Mass of the alpha particle $m_\alpha = 4 \times 1.66 \times 10^{-27}$ kg and charge $q_\alpha = 2e$.)

13 The potential difference between the accelerating plates of a television set is about 25,000 V. If the plates are separated by 1.5 cm, estimate the magnitude of the uniform electric field between the plates.

■14 An electron is shot from one large metal plate toward a parallel plate. If the electron's initial speed is 6×10^6 m/s and its speed just before it hits the second plate is 4×10^6 m/s, what is the potential difference between the plates? Is the second plate at a higher or lower potential?

■15 A proton is shot with a speed v_o from one metal plate toward a second plate parallel to the first. If there is a potential difference V between the plates, find the speed of the proton just before it strikes the second plate. Is the answer unique? If not, give other possible answers.

Section 17.5

16 (a) What is the speed of a 2.4-keV proton? (b) What is the speed of a 0.2-keV electron?

17 What potential difference is needed to stop an electron moving with an initial speed of 5.0×10^5 m/s?

18 An alpha particle (mass $m_\alpha = 4 \times 1.66 \times 10^{-27}$ kg and charge $q_\alpha = 2e$) has a kinetic energy of 7.2 MeV. (a) What is its energy in joules? (b) How fast is it moving? (c) Through how large a potential difference must it move to attain this energy?

19 A triply ionized lithium ion (mass $m = 6.94 \times 1.66 \times 10^{-27}$ kg and charge $q = 3e$) is accelerated through a potential difference of 7200 V. What is its kinetic energy in electronvolts? How fast is it moving?

20 An ion after being accelerated through a potential difference of 417 V has a kinetic energy of 2.0×10^{-16} J. What is the charge on the ion?

21 In the Van de Graaff accelerator in a certain research laboratory protons are accelerated from rest through a potential difference of 250,000 V. (a) What is the kinetic energy of the protons in electronvolts? (b) What is the kinetic energy of the protons in joules? (c) What is the speed of the protons?

■22 Two parallel plates have a potential difference of 80 V. (a) A proton is shot from the negative plate toward the positive plate with an initial kinetic energy of 100 eV. What is the kinetic energy of the proton just before it strikes the positive plate? (b) Repeat if the proton is shot from the positive plate toward the negative plate.

23 How much energy is gained by a charged particle with a charge of 60 μC when accelerated through a potential difference of 100 V?

24 An electron moving with a speed of 4.0×10^6 m/s is accelerated through a potential difference of 30 V. What is the new speed of the electron?

■25 A proton is to be slowed from an initial speed of 6.0×10^7 m/s to a final speed of 4.0×10^7 m/s. How large a potential difference must it move through in order to change its speed by this amount?

26 A proton is shot with a kinetic energy of 4800 eV from a negative plate toward a positive plate. The potential difference between the plates is 2000 V. (a) How much kinetic energy (in electronvolts) does the proton lose as it shoots to the positive plate? (b) What is its kinetic energy (in electronvolts) just before it hits the plate? (c) Repeat for an alpha particle with the same initial kinetic energy. (The charge on the alpha particle, a helium nucleus, is 2e.)

Section 17.6

27 What is the absolute potential 3.2×10^{-14} m from an atomic nucleus if the nuclear charge is 76e? Ignore the electrons in the atom. If a proton is released at this radius, what will its kinetic energy be (in millions of electronvolts) when it is far from the nucleus?

28 At what distance from a −8-μC point charge would the electrical potential be -2.8×10^4 V?

29 In the Bohr model of the hydrogen atom an electron circles a proton in an orbit of radius 0.51×10^{-10} m. What is the potential due to the proton at the position of the electron? What is the potential energy of the electron?

30 Two point charges are placed on the x axis: a + 6.0-μC charge at the coordinate origin and a −5.0-μC charge at $x = 16.0$ cm. Find the absolute potential due to the two charges at (a) $x = 12$ cm and (b) $x = -6$ cm.

31 Four identical −5.0-μC charges are located at the four corners of a square that is 40 cm on each side. What is the absolute potential at the center of the square?

32 Repeat Prob. 31 if one of the four charges is positive.

■33 A -4.0×10^{-9} C charge is located at the origin of the coordinate system and another charge of 6.0×10^{-9} C is at $x = 2.4$ m. At what two locations on the x axis is the electrical potential due to the two charges equal to zero?

34 A 6.0-μC charge is located at the point (0, 1.0) where the coordinates are in meters. A second charge of −4.0 μC is located at (−3.0, 0). Find the absolute potential due to the two charges at (a) (0, −3.0) and (b) (1.0, 0).

■35 Two point charges $q_1 = -5$ nC and $q_2 = 4$ nC are separated by 40 cm. What is the absolute potential (a) at a point midway between the charges and (b) at a point 40 cm from both charges?

■36 A metal sphere of radius 30 cm carries a uniform charge of 8.0×10^{-9} C. Assuming that it is far from all other objects, what is the absolute potential at its surface?

Section 17.7

37 When the plates of a radio capacitor are charged with 1.8 μC, the potential difference between them is 9.0 V. What is the capacitance of the capacitor?

38 What is the charge on a 36-nF capacitor subject to a potential difference of 840 V?

■39 The charge on the plates of a capacitor increases by 24.0 μC when the potential difference between them is increased from 18.0 to 34.0 V. What is the capacitance?

40 The plates of a parallel-plate capacitor are separated by 0.05 mm and have a capacitance of 0.4 μF. What is the area of each plate if the space between the plates is vacuum?

■41 A parallel-plate capacitor has plates with an area of 280 cm² separated by 0.5 mm. What is the electric field between the plates when the capacitor is charged with a 1.0-μC charge?

42 If the gap between the plates of a parallel-plate capacitor is halved while their area is tripled, what is the ratio of the new capacitance to the original?

■43 Two identical metal plates are placed parallel to each other with a spacing of 0.05 mm. They have an area of 360 cm² each. (*a*) Find the capacitance if vacuum exists between the plates. (*b*) How much charge exists on the capacitor when it is connected to a 9.0-V battery?

Sections 17.8 and 17.9

44 Repeat parts (*a*) and (*b*) of Prob. 43 if the space between the plates is filled with a plastic material of dielectric constant $K = 4.0$.

45 What must be the surface area of a 12-μF capacitor if a 20-nm film of aluminum oxide fills the gap between the parallel plates? Take $K = 8$ for aluminum oxide.

■46 Sparking will occur through air if the electric field exceeds about 3.0×10^6 V/m. How large a charge can be placed on a 30-pF parallel-plate capacitor having air between the plates before the sparking takes place? Assume the area of each plate to be 30 cm².

■47 A parallel-plate air capacitor holds of 28 nC of charge when subjected to a potential difference V_o. When a fluid is placed between the plates, the charge increases to 84 nC if the potential difference is maintained at V_o. What is the dielectric constant of the fluid?

■48 A parallel-plate air capacitor is charged to 120 V and then disconnected from the battery. When the space between the plates is filled completely with a piece of glass, the voltage across the capacitor drops to 30 V. What is the dielectric constant of the glass?

Section 17.10

49 Two capacitors, $C_1 = 6\ \mu$F and $C_2 = 12\ \mu$F, are connected in parallel, and the combination is connected to a 9.0-V battery. (*a*) What is the equivalent capacitance of the combination? (*b*) What is the potential difference across each capacitor? (*c*) What is the charge stored on each of the capacitors?

50 The two capacitors of the previous problem are now connected in series and to a 9.0-V battery. Find (*a*) the equivalent capacitance of the combination, (*b*) the potential difference across each capacitor, and (*c*) the charge on each capacitor.

51 Three capacitors, $C_1 = 40$ pF, $C_2 = 60$ pF, and $C_3 = 120$ pF, are connected. (*a*) Find the equivalent capacitance of the group when they are connected in parallel. (*b*) What will be the equivalent capacitance if they are connected in series?

52 The combination in Prob. 51 is connected to a 9.0-V battery. Find the charge on each and the potential difference across each of the capacitors when they are connected in (*a*) series, (*b*) parallel.

■53 A series circuit contains a 0.5-μF capacitor, a 40-pF capacitor, and a 120-V battery. Find the charge on each of the capacitors. What would be the charge on each capacitor if they were connected in parallel across the battery?

■54 How many capacitance values can you obtain by combining the following three capacitors: 4, 8, and 16 μF? What are these values?

■55 Four capacitors are connected as shown in Fig. P17.2. Find (*a*) the equivalent capacitance of the combination and (*b*) the charge on and potential drop across each of the capacitors.

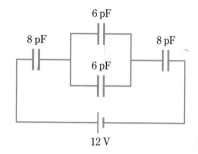

FIGURE P17.2

Sections 17.11 and 17.12

56. A capacitor is connected across a 120-V battery and stores a charge of 45 μC. (*a*) What is the capacitance of the capacitor? (*b*) How much energy does the capacitor store?

57 A parallel-plate capacitor is charged and then disconnected from the battery. How does the energy stored by the capacitor change when the plate separation is doubled?

■58 Determine the energy stored in each capacitor in Fig. P17.2.

59 A parallel-plate capacitor has plates of area 4 cm² each separated by 0.5 mm. The space between the plates is filled with a material of dielectric constant $K = 8$. If a 12-V battery is connected to the capacitor, how much energy does it store? By what factor will the stored energy be changed if the dielectric is removed and the gap is filled with air while the battery remains attached?

Additional Problems

■60 A tiny ball carrying a charge of +30 nC is held by a thread between two horizontal parallel plates that are 4.0 cm apart. (*a*) When the potential difference between the plates is 6000 V, the tension in the thread is zero. What is the mass of the ball? (*b*) What is the tension in the thread when the polarity of the plates is reversed?

■■61 Two vertical parallel plates are 5.0 cm apart and maintain a potential difference of 8000 V. A tiny ball (mass $m = 2.0 \times 10^{-4}$ g) is suspended as a pendulum between the plates. The thin, massless thread holding the ball reaches equilibrium at an angle of 15° to the vertical. Determine the charge on the ball.

■■62 A proton is shot from the lower plate in Fig. P17.3 with a

FIGURE P17.3

velocity $v_o = 4 \times 10^4$ m/s at the angle shown. What must the potential difference between the plates be if the proton is to just miss the upper plate?

▪▪**63** An electron is shot from the lower plate in Fig. P17.3 at the angle shown. The potential difference between the plates is 3000 V. What must the initial speed of the electron be if it is to just miss the upper plate? Should the top plate be positive or negative?

▪**64** An electron is shot from plate A, as shown in Fig. P17.4, toward a parallel plate B with an initial velocity $v_o = 4.0 \times 10^6$ m/s. Plates A, B, and C are at -120-V, 0-V, and -100-V potentials, respectively. Assuming that the electron travels perpendicularly to the plates, what will be its speed just before it strikes plate C? Take $a = 8.0$ cm and $b = 10.0$ cm.

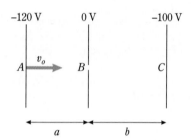

FIGURE P17.4

▪**65** A point charge of 10.0 μC is located at the coordinate origin. How much work is required to bring a positive charge of 3.0 μC from infinity to the location $x = 20.0$ cm?

▪▪**66** A test charge $q_1 = 0.2\,\mu$C is positioned on the y axis $+4.0$ cm away from a fixed charge $q_2 = 20.0\,\mu$C located at the coordinate origin. The test charge q_1 is moved 8.0 cm along the y axis and then moved 9.0 cm parallel to the x axis, both away from the fixed charge. What is the change in the electrical potential energy of the test charge q_1?

▪▪**67** A metal sphere of radius 3.0 cm hangs on a thin string in the center of a very large room. It carries a charge of -6×10^{-8} C. What is the approximate potential difference between the sphere and the walls of the room?

▪**68** How should four 3-μF capacitors be connected if the combinations is to have a total capacitance of (a) 12 μF, (b) 3 μF, (c) 1.2 μF, and (d) 1.5 μF?

▪**69** A 1.0-μF capacitor is charged by connecting it to a 12-V battery. The capacitor is then disconnected from the battery and connected across an uncharged 3.0-μF capacitor. What is the charge on each of the capacitors? What is the potential difference across each?

▪**70** The gap in a certain parallel-plate capacitor can be changed without otherwise disturbing the electrical system. With gap A the capacitance is 40 pF, and with gap B it is 36 pF. The capacitor is charged by a 9.0-V battery when it has gap A. The battery is then removed, and the capacitor is changed to gap B without changing the charge on it. (a) How much charge is on the capacitor when it has gap A? (b) What is the potential difference across it when it has gap B? (c) By how much does the stored energy change as it goes from gap A to gap B? (d) What is the minimum amount of work that a person holding the plates must have done to change the capacitor from gap A to gap B?

▪▪**71** Repeat Prob. 70 if the battery is left connected to the plates as the capacitor is changed from gap A to gap B.

▪▪**72** A pendulum of length L hangs from the ceiling of a room in which a downward electric field **E** exists. The pendulum ball has a mass m and carries a positive charge q. Find the frequency of the pendulum for small angle oscillations.

▪▪**73** Find the equivalent capacitance of the system shown in Fig. P17.5 when switch S is open.

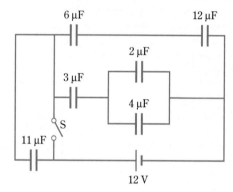

FIGURE P17.5

▪▪**74** Find the equivalent capacitance of the system shown in Fig. P17.5 when switch S is closed.

▪▪**75** Two capacitors, one 4 μF and the other 6 μF, are individually charged to 100 V by connecting them, one at a time, across a battery. After they are removed from the battery, the positive plate of one is connected to the positive plate of the other and the negative plate of one is connected to the negative plate of the other. Find (a) the potential across each capacitor and (b) the resultant charge on each capacitor. *Hint:* After they are connected, the potential drop is the same across the two capacitors.

▪▪**76** Repeat Prob. 75, but with the positive plate of one capacitor connected to the negative plate of the other.

CHAPTER 18

DIRECT-CURRENT CIRCUITS

In the previous two chapters we discussed the properties of electric charges at rest. Most practical applications of electricity, however, involve charges in motion, or electrical currents. For example, charge flowing through the coils of an electric motor causes its shaft to rotate. A light bulb emits light because of charges flowing through its filament. When we turn on a radio or television set, it operates because charge flows through its circuits. Although most common devices in industry and in our homes operate with alternating current (ac) circuits, where charges flow back and forth through conductors, we shall begin our study of charges in motion by discussing the simpler case of direct current (dc) circuits, in which charges flow through a conductor without reversing their motion. The electrically powered car (pictured above) is an example of a dc circuit.

18.1   ELECTRIC CURRENT

We begin our discussion of charges in motion by defining a quantity called electric current. Suppose we have a device, called a *charge gun,* that can shoot out a stream of charged particles, such as ions or electrons. (A television set uses such a gun to shoot an electron beam at the screen.) For our discussion, consider a gun that shoots a beam of positive particles through a hole in a plate, as in Fig. 18.1.

The beam passing through the hole constitutes a flow of charge, and we now

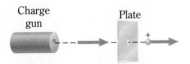

FIGURE 18.1

The beam of moving charges passes through the hole in the plate. If a charge Δq passes through the hole in time Δt, the current is $\Delta q / \Delta t$.

wish to characterize the magnitude of this flow. We do so by defining a quantity called **electric current,** which we designate by the symbol I:

If in a time Δt a beam carries a charge Δq past a given point (the hole in the plate in this case), then the current carried by the beam is

$$I = \frac{\Delta q}{\Delta t} \qquad (18.1)$$

The SI unit of current, the coulomb per second, is called the ampere.

One **ampere** (A) = one coulomb per second (C/s)

If the charges in the beam are positive, then both Δq and I are positive. If the beam is composed of negative charges, however, then both Δq and I are negative. For this reason, a flow of negative charge in one direction is equivalent to a positive current in the opposite direction. You might object that since it has been determined that the actual moving charges in conductors are electrons, current should be defined in terms of negative charge flow. Historically, however, before sign of the charge carriers was known, current was defined in the sense of the movement of positive charges. Once the nature of charge carriers became known, there was no great compulsion to change the definition, since the equivalence between positive and negative charge flow is so simple.

To see what our definition means for currents in wires, refer to Fig. 18.2. If an amount of charge Δq flows through the cross section at A in a time Δt, then the current in the wire is defined by Eq. 18.1 to be

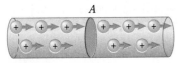

FIGURE 18.2

The current in amperes in the wire is defined to be the quantity of positive charge in coulombs flowing through a cross section such as A in 1 s.

$$I = \frac{\Delta q}{\Delta t}$$

just as in Fig. 18.1. The current is again taken to be in the direction of positive charge flow, consistent with our definition.

Illustration 18.1

The current through a flashlight bulb is 0.150 A. How many electrons flow through the bulb each second?

Reasoning   Since current is the charge per second that flows past a point, we know that 0.150 C of charge flows through the bulb each second. Each electron carries a charge of magnitude 1.60×10^{-19} C. Thus the number of electrons needed to make up a charge of 0.150 C is

$$\text{Number of electrons} = \frac{0.150 \, \text{C}}{1.60 \times 10^{-19} \, \text{C/electron}} = 9.36 \times 10^{17} \text{ electrons}$$

As we shall soon see, this tremendously large number of flowing charges is what causes electric currents in wires to be analogous to water flow in pipes. ∎

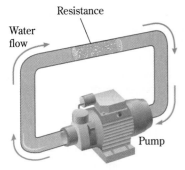

FIGURE 18.3

The water pump furnishes energy to the water and causes the water to flow through the tightly packed glass wool in the resistance section.

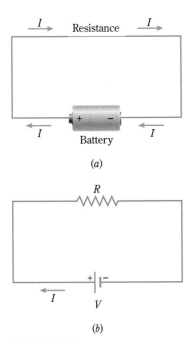

FIGURE 18.4

The battery causes charges to flow in the circuit. Energy given to the charges by the battery is released as heat in the resistance. Part (b) is a schematic diagram for the circuit in (a).

18.2 A SIMPLE ELECTRIC CIRCUIT

Before we discuss the way an electric circuit behaves, let us look at a more easily visualized situation, the flow of water molecules through a pipe. Figure 18.3 shows a pipe system completely filled with water. A pump provides energy to the water molecules and causes them to flow through the pipe. Because water fills the entire pipe system and is incompressible, all parts of the pipe carry the same water current. The pipe is large enough so that little viscous loss occurs, but the pipe section labeled "resistance" is packed with glass wool so that the water has great difficulty flowing through it. Obviously, the resistance section is the major obstacle to flow; nearly all the energy furnished to the water by the pump appears as viscous energy losses (that is, heat) in the resistance section. In effect, the water simply carries energy from the pump to the resistance section, where the energy is converted to thermal energy.

An analogous electrical system is shown in Fig. 18.4a. A battery is connected to two metal wires to form what is called an *electric circuit*. Because the red wire is much thinner than the blue wires, it offers a very large resistance to charge flow through it.* These metal wires contain a multitude of free electrons, which we can liken to water molecules in the pipe of Fig. 18.3. Just as the pump furnishes energy to the water molecules, the battery furnishes energy to the free charges in the metal and causes them to flow.

Notice that positive current flows *from* the positive terminal of the battery and *into* the negative terminal. In the previous chapter, we saw a battery perform this same function in charging a capacitor. There, however, the delivery of charge lasted only a very brief time as the capacitor quickly built up a potential equal to that of the battery. In the present circumstance, the flow of charge through the battery and the circuit is *continuous*.

Most of the energy furnished by the battery is lost as heat when the charges flow through the high-resistance wire. Thus the flowing charges simply carry energy from the battery to the resistance; there, their energy is converted to thermal energy as they collide with the atoms of the resistive material. In fact, if the amount of heat liberated in the resistance is large enough, the wire will become white-hot. An example of this is shown in Fig. 18.5, where a battery causes charge to flow through a flashlight bulb. The filament of the bulb, a hair-thin wire, glows white-hot as it releases the energy furnished by the battery.

---

*Alternatively, the red wire could be made of a metal that offers much more resistance to charge flow than does the metal used to construct the remainder of the circuit. An example of such a system is one using iron for the red wire and copper for the blue wire.

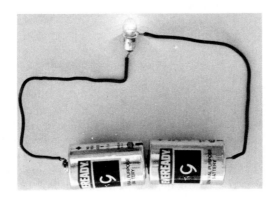

FIGURE 18.5

A simple circuit. Where does the heat and light energy emitted by the bulb come from?

Figure 18.4*b* shows the diagram used to represent the circuit in *a*. Notice the symbol -ww- used for the resistance wire. We call it a **resistor.** All other wires in a circuit are assumed to have negligible resistance to flow, and so no appreciable heat is generated in them. The energy furnished by the battery of voltage V is delivered to the resistor R and is there converted to thermal energy.

Before leaving this section, we should point out another similarity between water flow in a pipe and charge flow in an electric circuit. In the water circuit, it is obvious that when a certain amount of water enters one end of the pump, an equal amount flows out the other end. Because the pipe is filled, water cannot flow in one section unless it flows in all sections. Like the water molecules, the free charges in an electric circuit fill their confining "pipes," the wires. When any amount of charge flows into one end of the battery, a like amount must flow out the other end. Thus the current (the flow of charge per second) is the same everywhere in the circuit of Fig. 18.4.

18.3 RESISTANCE AND OHM'S LAW

Let us examine the circuit shown in Fig. 18.6. Because we assume that only negligible energy losses occur in the wire from p to a, the energy of the charges does not change as they move through this section of the wire. Hence wire pa is an equipotential, and so point a is at the same electrical potential as point p. Similarly, point b is at the same potential as point n. Hence we arrive at the fact that the potential difference across the resistor is the same as the potential difference across the battery, namely V.

Because end a of the resistor is connected to the positive terminal of the battery, point a is at a higher potential than point b. Any positive charge free to move through the resistor moves from a to b—in other words, from high to low potential. Hence the direction of the current through the resistor is from a to b. In fact,

The direction of the current through a resistor is always from the resistor's high-potential end to its low-potential end.

We characterize a resistor by its resistance R. If a potential difference V across the resistor causes a current I through it, then the resistance R is defined to be

$$R = \frac{V}{I} \quad \text{or} \quad V = IR \tag{18.2}$$

The unit for resistance is the volt per ampere, which is called the ohm (Ω). The definition of resistance expressed in Eq. 18.2 was first proposed by Georg Simon Ohm (1787–1854), whose experiments showed that I is proportional to V. Consequently, Eq. 18.2 is often called **Ohm's law.** Strictly speaking, however, Ohm's law pertains only to resistors for which I is proportional to V over a range of values of I and V. Such resistors are called *ohmic* resistors. They are characterized by the straight-line I versus V graph shown in Fig. 18.7. In many materials, however, the resistance as defined by Eq. 18.2 is not constant, but depends on the values of V and I. Such resistors are referred to as *nonohmic*. The graphs of I versus V for such materials are nonlinear, as shown in Fig. 18.7.

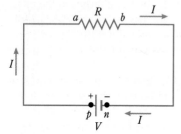

FIGURE 18.6

The direction of positive current is always from high to low potential through a resistance.

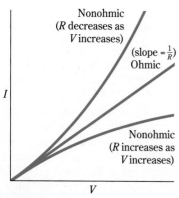

FIGURE 18.7

A graph of the dependence of current on applied voltage for ohmic and nonohmic resistors.

Illustration 18.2

A certain flashlight bulb draws a current of 0.160 A when the potential difference across it is 3.10 V. What is the resistance of the bulb?

Reasoning This situation is shown in Fig. 18.5. We are told that $V = 3.10$ V across the resistor (the filament of the bulb) and that I through it is 0.160 A. Using Ohm's law, $V = IR$, we have

$$R = \frac{V}{I} = \frac{3.10\text{ V}}{0.160\text{ A}} = 19.4\ \Omega$$

We shall see in the next section that the bulb's resistance is much lower if its filament is not white-hot. ∎

18.4 RESISTIVITY AND ITS TEMPERATURE DEPENDENCE

Wires that are identical in size but made from different metals have different resistances. For example, a copper wire has less resistance than an iron one of the same size. We therefore need a way to characterize the inherent resistance properties of a material.

To do this, let us consider the wire of length L and cross-sectional area A shown in Fig. 18.8. As you might guess, the resistance of the wire increases as L is made larger and decreases as A is made larger. Indeed, experiment shows that

$$R \propto \frac{L}{A}$$

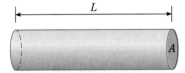

FIGURE 18.8

The resistance of a uniform wire varies directly with L and inversely with A.

We can change this proportionality to an equation by introducing a constant of proportionality ρ (Greek rho):

$$R = \rho \frac{L}{A} \qquad \text{or} \qquad \rho = R\frac{A}{L} \tag{18.3}$$

where ρ has the units ohm-meters and depends on the material from which the wire is made. We call ρ the **resistivity** of the material. For very good electrical conductors, such as copper, ρ is small. Typical values for resistivity are given in Table 18.1. Notice that values are given for insulators (nonconductors) as well as metals. Insulators, such as wood and glass, contain a few ions (usually impurities) that give rise to charge motion when a voltage is impressed across them. Therefore, the resistivity of these materials is very large, but not infinite.

The resistivity of any given substance changes with temperature. For example, the resistance of a metal filament in an incandescent light bulb increases more than tenfold as the filament changes from room temperature to white-hot. Over a limited temperature range, the fractional change in resistivity is proportional to the change in temperature:

$$\frac{\Delta\rho}{\rho_o} = \alpha\,\Delta T \tag{18.4}$$

TABLE 18.1
Resistivity at 20°C

Material	$\rho\,(\Omega \cdot m)$
Silver	1.6×10^{-8}
Copper	1.7×10^{-8}
Aluminum	2.8×10^{-8}
Tungsten	5.6×10^{-8}
Iron	10×10^{-8}
Graphite	3.5×10^{-5}
Blood	1.5
Fat	25
Wood	$10^8 - 10^{12}$
Glass	10^{12}
Polystyrene	$10^{15} - 10^{19}$

TABLE 18.2

Temperature coefficients of resistivity at 20°C

Material	α (per C°)
Silver	0.0038
Copper	0.0039
Aluminum	0.0040
Tungsten	0.0045
Iron	0.0050
Graphite	−0.0005
Germanium	−0.05
Silicon	−0.07

In this expression ρ_o is the resistivity at some reference temperature, often 20°C. The constant α, called the *temperature coefficient of resistivity,* depends on the material. The typical values given in Table 18.2 are correct only for modest temperature changes near the reference temperature. Although the resistivity of most metals increases with temperature, the opposite occurs for graphite and most semiconductors (notice the minus signs in Table 18.2).

As Eq. 18.3 shows, the resistance of a wire depends on its dimensions as well as on the material out of which it is made. These dimensions are also temperature-dependent, as we saw in Chap. 11. However, the coefficients of thermal expansion are usually orders of magnitude smaller than the coefficients of resistivity given in Table 18.2. Thermal changes in the dimensions of a resistor can therefore usually be ignored compared to changes in resistivity. Thus we can write the same expression for the temperature dependence of the resistance R of a specific resistor as we used for ρ:

$$\Delta R = R_o \alpha \, \Delta T \tag{18.5}$$

Because resistance varies with temperature, it can be used to measure temperature. The small electronic probes now widely used as fever thermometers make use of this fact. These devices use semiconductor resistors, materials that have exceptionally large temperature coefficients of resistivity.

Illustration 18.3

Number 12 copper wire has a cross-sectional area of 0.0331 cm². What is the resistance of a 40.0-m length?

Reasoning We make use of $R = \rho(L/A)$ with $L = 40.0$ m, $A = 0.0331 \times 10^{-4}$ m², and $\rho = 1.7 \times 10^{-8} \, \Omega \cdot$ m:

$$R = \rho \frac{L}{A} = \frac{(1.7 \times 10^{-8} \, \Omega \cdot \text{m})(40.0 \, \text{m})}{0.0331 \times 10^{-4} \, \text{m}^2} = 0.20 \, \Omega$$

This wire has a diameter typical of electrical connecting wires, and so you see why we can usually ignore the resistance of such wires.

Example 18.1

A light bulb whose filament is made of tungsten has a resistance of 240 Ω when white-hot (about 1800°C). Find the approximate resistance of the bulb at room temperature (20°C).

R

Reasoning

Question What expression relates a change in resistance to a change in temperature?
Answer Equation 18.5.

Question Since the values for α in Table 18.2 are constant only for a limited temperature range, and are referred to 20°C, how can I find a value of α applicable to this temperature range?

Answer This is a good question. In the absence of values of α which pertain to high temperatures, all you can do is to calculate an approximate result by assuming that α doesn't change significantly enough to make your result meaningless.

Question What should I use for the reference resistance R_o in Eq. 18.5?

Answer Since you have to assume that the value of α is that at 20°C, the unknown value of R at 20°C is the reference resistance, R_o.

Solution and Discussion Inserting the given values into Eq. 18.5, we have

$$\Delta R = 240\ \Omega - R_o = R_o(0.0045/\text{C}°)(1800°\text{C} - 20°\text{C}) = 8.0R_o$$

Thus

$$240\ \Omega = 9.0R_o \qquad \text{or} \qquad R_o = 27\ \Omega$$

18.5 POWER AND ELECTRICAL HEATING

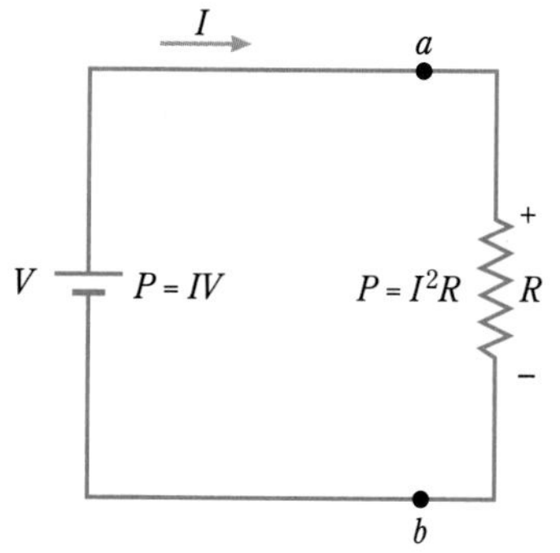

FIGURE 18.9

The power delivered by the battery appears as heat in the resistor.

When a battery sends a current through a resistor, as in Fig. 18.9, the battery is furnishing energy to the resistor.

In effect, the internal chemical process in the battery moves charge from the low electrical potential at the negative terminal to the high electrical potential at the positive terminal. In order to do this, the battery must do work on an amount of charge Δq equal to the increase in the charge's EPE:

$$W = \Delta \text{EPE} = \Delta q\, V$$

where V is the voltage of the battery. As the charge passes through the resistor R from point a to point b it loses the energy given to it by the battery, generating an equal amount of thermal energy in the resistor.

If charge Δq moves through the battery (and the resistor) in a time Δt, the *power* delivered by the battery is, from Eq. 5.2,

$$\text{Power} = \frac{\text{work done}}{\text{time taken}} = \frac{\Delta q\, V}{\Delta t}$$

But $\Delta q / \Delta t$ is simply the current I in the circuit. Hence the power delivered by a voltage source V as it furnishes a current I is

$$\text{Power} = IV \tag{18.6}$$

As the charge passes through the resistor, it falls through a potential difference V. Consequently, Eq. 18.6 also gives the electric power lost in the resistor. Thus we have the following relation for the electric power loss for a current I through a resistor R:

$$\text{Power loss in resistor} = IV = I^2R = \frac{V^2}{R} \qquad (18.7)$$

where the latter forms are obtained by use of $V = IR$.

We learned in Sec. 5.2 that the unit of power is the joule per second, which is given the name watt (W). We are all familiar with the use of this unit in electricity because we see it on light bulbs and electrical appliances. For example, if you examine a 60-W bulb, you will see stamped on it "60 W, 120 V." This means that the bulb consumes 60 W of power when 120 V is impressed across it. Another example is a 1500-W electric space heater designed for use on 120 V. Because power is work per unit time, the space heater furnishes 1500 J of heat each second when operated on 120 V.

Electrical energy to operate appliances is furnished by the power company, which charges us according to how many kilowatthours (kWh) of energy we use. You should recall from Sec. 5.2 that the energy unit the company uses, the kilowatt-hour, is equivalent to 3.60×10^6 J.

Equation 18.6 shows us that, in electrical terms, watts are the product of amperes times volts. Thus the current that an appliance will draw can be quickly calculated from its operating voltage and power consumption:

$$I = \frac{P}{V}$$

For example, a 100-W 120-V light bulb draws a current of

$$I = \frac{100 \text{ W}}{120 \text{ V}} = 0.83 \text{ A}$$

A 1-hp (746-W) electrical motor operating on 120 V draws

$$I = \frac{746 \text{ W}}{120 \text{ V}} = 6.2 \text{ A}$$

We will discuss further the importance of the current drawn by appliances in Sec. 18.11.

Illustration 18.4

How much heat does a 40-W light bulb generate in 20 min?

Reasoning Because power is work per unit time, a 40-W bulb generates 40 J of heat per second. Therefore, in 20 min it develops

$$\text{Heat} = (40 \text{ J/s})(20 \text{ min})(60 \text{ s/min}) = 48,000 \text{ J}$$

Exercise How many calories of heat is this? *Answer: 11,500 cal.*

Example 18.2

To brew 8 cups (approximately 1.6 kg) of coffee, you have to raise the temperature of the water from 20°C to about 90°C. Suppose you use a 700-W coffeemaker to do this. How long does the brewing take? At $0.10/kWh, how much does the electricity cost you?

R

Reasoning

Question How does time enter the problem?
Answer The 700-W rating means that the coffeemaker can supply energy at the rate of 700 J/s.

Question How much energy is required for this task?
Answer Recall the relation between mass, temperature change, and heat given in Eq. 11.1:

Heat Q = mass m × specific heat c × ΔT

The specific heat of water is 1 kcal/kg · C°.

Question How do I convert from kilocalories to joules?
Answer Use the mechanical equivalent of heat (Chap. 11):

1 kcal = 4184 J

Question What equation determines the time the brewing takes?
Answer Power = energy/time, so

$$\text{Time} = \frac{\text{energy required}}{\text{power}}$$

Question What is the relation between time and cost?
Answer You pay by the kilowatthour. Multiply the power of the device (0.700 kW) by the time it operates (in hours) to find the number of kilowatthours.

Solution and Discussion   The energy needed is

$(1.6 \text{ kg})(1 \text{ kcal/kg} \cdot \text{C}°)(70 \text{ C}°) = 112 \text{ kcal}$

This is equal to

$(112 \text{ kcal})(4184 \text{ J/kcal}) = 4.7 \times 10^5 \text{ J}$

At a rate of 700 J/s, the brewing time is

$$t = \frac{4.7 \times 10^5 \text{ J}}{700 \text{ J/s}} = 671 \text{ s} = 11.2 \text{ min}$$

This is equal to 0.187 h.
   To find the cost, first find the number of kWh of energy consumed:

No. of kWh = (0.700 kW)(0.187 h) = 0.13 kWh

At $0.10/kWh, this amounts to about 1.3 cents.

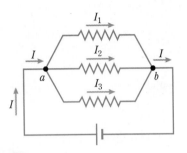

FIGURE 18.10

Kirchhoff's junction rule tells us that $I = I_1 + I_2 + I_3$.

18.6 KIRCHHOFF'S JUNCTION RULE

Until now, we have been discussing current in a single wire, in which case all charge must flow through the same path. In Sec. 18.2 we referred to a path which allows current to flow as an *electric circuit*. Circuits can have more than one path for current to follow, however. Figure 18.10 shows a circuit in which charges can take any one of three paths between point *a* and point *b*. Such circuits are more complicated than single-loop circuits. Their analysis requires the use of two basic rules, called *Kirchhoff's rules*. These rules are quite obvious and easy to understand.

To see what the first rule is, consider point *a* in Fig. 18.10. The current into this point is *I*, and the currents out of it are labeled I_1, I_2, and I_3. There is a simple relation between these currents. The law of conservation of charge requires that charge be neither created nor destroyed at any such junction. The current simply branches into the various possible paths. If you add up all the currents leaving the junction, you must obtain the same amount as the total current entering the junction. In the case of Fig. 18.10, this means

$$I = I_1 + I_2 + I_3$$

No matter how complicated the junction or where it occurs, this principle must hold. This simple observation is called **Kirchhoff's junction rule:**

The sum of all currents entering a junction must equal the sum of all currents leaving the junction.

The junction rule is of great importance in **circuit analysis,** where the goal is to find the currents flowing in each of the possible paths of a circuit. The first step in analyzing circuits having more than one current path, or *branch,* is to assign a different symbol on the circuit drawing for the current in each separate branch. You also need to assign a direction to each of these currents in order to apply the junction rule to each point where the current branches. For example, Fig. 18.11 shows a junction involving four branches. For the currents labeled as shown, the junction rule gives $I_4 + I_7 = I_5 + I_6$. An important point to realize is that *only one current can exist in a given branch, and charge must flow in either one direction or the opposite direction between junctions.* If you happen to label a current in the direction opposite to its actual direction, the worst that can happen is that your solution for that current will be a negative number.

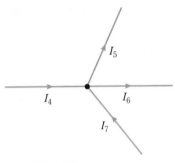

FIGURE 18.11

According to the junction rule, $I_4 + I_7 = I_5 + I_6$.

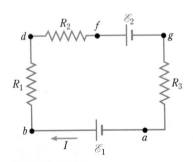

FIGURE 18.12

What does Kirchhoff's loop rule tell us about this circuit?

18.7 KIRCHHOFF'S LOOP RULE

To understand the second of Kirchhoff's rules, consider a single circuit loop, as shown in Fig. 18.12. A steady current *I* is flowing in the direction shown. Under these steady conditions we can analyze the variations in the potential as charge travels around the circuit.

Let us start at point *a* and follow a positive charge Δq through points *b, d, f,* and *g,* and back to *a*. The charge receives a boost of $\Delta q\, \mathscr{E}_1$ in energy as it goes through the first battery from the negative to positive terminal. It then loses energy in creating heat as it passes through R_1, R_2, and R_3. It also loses potential energy when it passes backward (from the positive terminal to the negative one) through

Circuits in modern devices use many resistors in complex combinations.

$\mathcal{E}_2$. The law of conservation of energy requires that the gains and losses in energy balance out by the time the charge gets back to its starting point, a. The dc current, which is the rate of charge flow, is constant. The charge cannot gain or lose a net amount of energy merely by going around the loop repeatedly, because where would that extra energy come from or go to?

Thus at each point in the circuit the charge Δq has a certain value of electrical potential energy. In turn, that means that each point has a fixed value of electric potential relative to some starting point. If you start and end at the same point in the circuit, you come back to the same value of the potential. This fact is summarized by **Kirchhoff's loop rule:**

The algebraic sum of the voltage changes around any closed loop in a circuit must equal zero.

As we see, the loop rule is intimately connected with potential rises and drops. For that reason, let us review what happens to the potential as we move across a resistor, a battery, and a capacitor.

Suppose we move from a to b through the resistor in Fig. 18.13. We know that the current direction is always from high to low potential through a resistor. Hence we know that the change from a to b is a drop in potential; therefore its sign is negative. Ohm's law tells us that its magnitude is IR. The change in potential in going from a to b is $-IR$.

The battery symbol tells us that the left side of the battery in Fig. 18.13 is positive. Therefore point a is $\mathcal{E}$ volts higher than point b. Going from a to b, the potential change is $-\mathcal{E}$.

For the capacitor, we must be told which plate is positively charged. According to the diagram, plate a is positive. It therefore is at the higher potential. Since the potential difference across a capacitor is given by Eq. 17.6 as q/C, the potential change in going from a to b is $-q/C$. Of course, one more thing should be obvious about a branch of a circuit that contains a capacitor. A capacitor does not allow dc current to pass through it. Therefore, any circuit branch containing a capacitor has

Resistor: a to $b \rightarrow \Delta V = -IR$

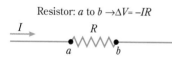

Battery: a to $b \rightarrow \Delta V = -\mathcal{E}$

Capacitor: a to $b \rightarrow \Delta V = -q/C$

FIGURE 18.13

In each case, going from a to b is a voltage drop, that is, a negative voltage change. Going from b to a, the voltage change would be positive.

TABLE 18.3

Voltage changes across circuit elements in dc circuits

---

1. *Resistors*

a. In the direction of the current — $V = -IR$
b. In the direction opposite the current — $V = +IR$

---

2. *Sources of emf $\mathscr{E}$*

a. Across the emf from − to + terminal — $V = +\mathscr{E}$
b. Across the emf from + to − terminal — $V = -\mathscr{E}$

---

3. *Capacitors carrying charge q*

a. Across C from + plate to − plate — $V = -q/C$
b. Across C from − plate to + plate — $V = +q/c$
c. In any branch containing a capacitor — $I = 0$

---

zero dc current. The potential change in each of these three cases is negative when we go from *a* to *b*. If we were going from *b* to *a*, the change would be positive. Table 18.3 summarizes our results.

Let us now use the loop rule in a few simple circuits before moving on to its more serious applications.

Illustration 18.5

Find the current in the circuit of Fig. 18.14.

Reasoning Let us guess that the current is in the direction shown. (Recalling from Sec. 18.2 that current flows out of a positive terminal, you might protest that this guess is wrong since the 12-V battery will certainly have more effect on current direction than the 3-V battery, but one of the nice things about Kirchhoff's rules is that even a poor guesser can use them, as we will see.) We pick a point such as *a* as a starting point and move around the circuit. The voltage changes are

$$
\begin{aligned}
a \to b & \quad +3\,\text{V} \\
b \to c & \quad -I(5\,\Omega) \\
c \to d & \quad -12\,\text{V} \\
d \to e & \quad 0\,\text{V} \\
e \to a & \quad -I(6\,\Omega)
\end{aligned}
$$

(It is very important that you understand the choice of the sign used in each term.) According to the loop rule, the algebraic sum of these voltage changes must be zero:

$$3 - I(5\,\Omega) - 12 - I(6\,\Omega) = 0$$

Solving for *I*, we find $I = -\frac{9}{11}\,\text{A}$. The negative sign tells us we guessed wrongly the direction of the current. No harm is done. The current is $\frac{9}{11}\,\text{A}$ in the direction opposite to our guess.

Suppose we choose the current *I* to be in the opposite direction. If we then move around the circuit in the same direction as before, the *signs of the voltage changes across the resistors would be reversed* (Rule 1b in Table 18.3). Current passes through a resistor in the direction of decreasing *V*. Our equation in this case would be

$$+3 + I(5\,\Omega) - 12 + I(6\,\Omega) = 0$$

Now the solution is $I = +\frac{9}{11}\,\text{A}$. This indicates a correct choice of current direction. *Notice that it makes no difference which way you choose to go around the loop in counting voltage changes.* Once you have chosen a current direction, you will get the same answer. If you choose the opposite current direction, the sign on the answer will be opposite. Of course, all this presumes that you assign the correct sign for each voltage change.

Exercise Find *I* if the 3-V battery is reversed. *Answer:* −1.36 A.

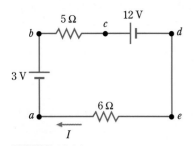

FIGURE 18.14

When we determine the magnitude of the current in this circuit, how will our answer tell us that we have chosen *I* in the wrong direction?

Example 18.3

Find the currents in all the branches of the circuit shown in Fig. 18.15.

R

Reasoning

Question How many independent equations do I need?
Answer You always need as many independent equations as there are un-knowns in the problem. In this case, everything in the circuit is given except the currents in the three branches. You will need three equations in order to be able to determine these currents.

Question What equation do I get from the junction rule?
Answer Points *a* and *c* are junctions. They will give you the same equation:

$$I_3 = I_1 + I_2$$

Notice that I_1 runs from *a* through *d* into *c*, I_3 runs from *c* through *b* into *a*, and I_2 runs from *a* into *c*.

Question What loop should I pick first and in what direction should I go around it?
Answer Pick any closed loop. Go around it in *either* direction. Just be very careful to apply the signs consistently each time you encounter a voltage change.

Question What equation does the loop rule give in going around the loop *acda*?
Answer You lose voltage $I_2(18\,\Omega)$ in going from *a* to *c*. You also lose 9 V across the emf going from *c* to *d*. Therefore,

$$-I_2(18\,\Omega) - 9\,\text{V} = 0$$

Question How do I deal with the current I_1 in the path *cda*?
Answer It doesn't pass through a resistor, so there is no *IR* voltage change. In applying the loop rule when you go across an emf, you just count the emf, regardless of the current passing through it. Thus I_1 will not appear in the loop rule equations, but is involved in the junction equation. (In a later section, we will modify this to take the resistance of a battery into account.)

Question How do I get a third equation?
Answer You still have not taken into account the branch *abc*, so you need another loop equation that includes it.

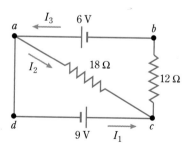

FIGURE 18.15

Determine the value of the current in each of the three branches of the circuit.

Question   If I go around the loop *abca*, what equation does the loop rule give?
Answer   From *a* to *b* you lose 6 V. From *b* to *c* you gain a voltage of $I_3(12\,\Omega)$ (it's a gain because you are going in the direction opposite I_3). From *c* to *a* you gain $I_2(18\,\Omega)$ for the same reason.

$$-6\,\text{V} + I_3(12\,\Omega) + I_2(18\,\Omega) = 0$$

Solution and Discussion   Generally, Kirchhoff's rules will give you a number of equations, each containing more than one unknown. Because two or more unknowns cannot be determined from a single equation, these *simultaneous equations* have to be manipulated until you arrive at a single equation where all but one unknown have been eliminated. This can be a tedious process that takes careful attention to the rules of algebra.

In this case, the first loop equation contains only one unknown, and so it can be solved directly.

$$-18I_2 = 9 \qquad I_2 = -0.50\,\text{A}$$

This minus sign tells you that I_2 is actually in the opposite direction than the incorrect guess shown in Fig. 18.15.

This value for I_2 can now be substituted in the second loop equation:

$$-6 + 12I_3 + (-0.50)(18) = 0 \qquad I_3 = +1.25\,\text{A}$$

(Be sure you notice the consistency in the use of the sign of I_2.) The plus sign of this result tells you that direction of I_3 indicated in Fig. 18.15 is correct.

The junction rule now yields I_1:

$$I_1 = I_3 - I_2 = 1.25\,\text{A} - (-0.50\,\text{A}) = +1.75\,\text{A}$$

The need to observe signs correctly cannot be overemphasized.

Exercise   Find I_2 and I_3 if the 9-V battery is reversed.   *Answer: 0.500 A, −0.250 A.*

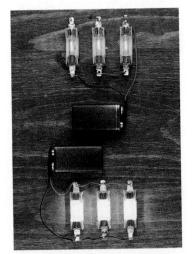

Three light bulbs are connected to the same voltage source. The bulbs connected in parallel glow brighter than those connected in series. Why?

18.8   RESISTORS IN SERIES AND IN PARALLEL

In Chap. 17 we recognized two ways of connecting capacitors together, in parallel and in series. We now examine the equivalent total resistance that results when a number of resistors are hooked together in these configurations.

Figure 18.16a shows three resistors connected in series. When they are connected to a battery as in Fig. 18.16b, some current *I* will flow. From what we have learned, we make the following observations:

RESISTORS IN SERIES

1 The same current *I* flows through all resistors connected in series.
2 The voltage drops across the resistors are IR_1, IR_2 and IR_3.

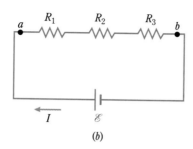

(b)

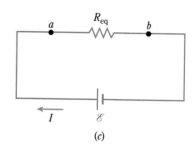

(c)

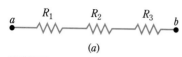

(a)

FIGURE 18.16

The three resistors are in series. Their equivalent resistance is $R_{eq} = R_1 + R_2 + R_3$.

Kirchhoff's loop rule applied to this simple circuit gives

$$\mathscr{E} - IR_1 - IR_2 - IR_3 = 0$$

Thus

$$\mathscr{E} = I(R_1 + R_2 + R_3)$$

We are looking for the resistance R_{eq} that would give the same current I if connected to the same emf, shown in Fig. 18.16c. Ohm's law for this circuit gives

$$\mathscr{E} = IR_{eq}$$

Upon comparing these two expressions for $\mathscr{E}$, we have the equation

$$R_{eq} = R_1 + R_2 + R_3$$

This result can be generalized to any number n of resistors in series:

$$R_{eq} = R_1 + R_2 + R_3 + \cdots + R_n \qquad \text{(in series)} \tag{18.8}$$

Figure 18.17a shows three resistors connected in parallel. When they are connected to a source of emf as in Fig. 18.17b, the same voltage is applied across each of them. Again we can apply what we have already learned to conclude the following.

RESISTORS IN PARALLEL

1 The voltage drop $\mathscr{E}$ across each resistor connected in parallel is the same.
2 The current in each resistor connected in parallel is determined by $I_1 = \mathscr{E}/R_1$, $I_2 = \mathscr{E}/R_2$, etc.

Kirchhoff's junction rule requires that

$$I = I_1 + I_2 + I_3 = \frac{\mathscr{E}}{R_1} + \frac{\mathscr{E}}{R_2} + \frac{\mathscr{E}}{R_3}$$

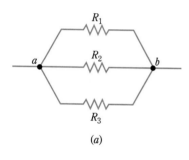

(a)

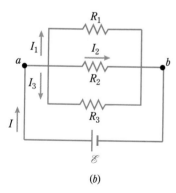

(b)

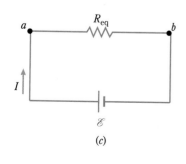

(c)

FIGURE 18.17

The three resistors are in parallel. Their equivalent resistance is given by $1/R_{eq} = 1/R_1 + 1/R_2 + 1/R_3$.

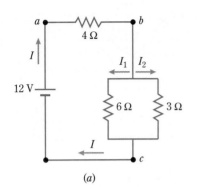

(a)

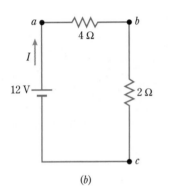

(b)

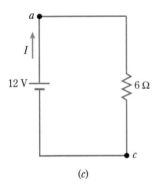

(c)

FIGURE 18.18

The parallel resistors between b and c are equivalent to $2\,\Omega$, as shown in (b). The two series resistors in (b) can be combined, as in (c).

Again we want the R_{eq} which draws the same current I from the emf $\mathcal{E}$ that the parallel arrangement does. Thus

$$I = \frac{\mathcal{E}}{R_{eq}} = \frac{\mathcal{E}}{R_1} + \frac{\mathcal{E}}{R_2} + \frac{\mathcal{E}}{R_3}$$

This immediately gives

$$\frac{1}{R_{eq}} = \frac{1}{R_1} + \frac{1}{R_2} + \frac{1}{R_3}$$

The generalization of this to n resistors in parallel is

$$\frac{1}{R_{eq}} = \frac{1}{R_1} + \frac{1}{R_2} + \frac{1}{R_3} + \cdots + \frac{1}{R_n} \qquad \text{(in parallel)} \qquad (18.9)$$

It is interesting to observe that resistors in series combine in the same way that capacitors in parallel do, and vice versa. You should again remember to take special care in adding reciprocals.

Illustration 18.6

Find the current I through the battery in Fig. 18.18a.

Reasoning We could solve this by applying Kirchhoff's rules to the circuit as drawn in part a. However, it is usually simpler to combine obvious series and parallel resistors before writing the loop equations. First we combine the two parallel resistors between points b and c:

$$\frac{1}{R_{bc}} = \frac{1}{6} + \frac{1}{3} = \frac{1}{6} + \frac{2}{6} = \frac{3}{6} \qquad R_{bc} = 2\,\Omega$$

An equivalent circuit is now drawn in Fig. 18.18b, with the parallel combination replaced by its equivalent resistance. We then see that the 4- and 2-Ω resistors are connected in series between points a and c. Their equivalent is

$$R_{ac} = 4\,\Omega + 2\,\Omega = 6\,\Omega$$

A new equivalent circuit, shown in part c, is now drawn. This is a situation to which Ohm's law can be applied. The voltage difference across the 6-Ω resistor is 12 V. Hence we have

$$I = \frac{V}{R} = \frac{12\,V}{6\,\Omega} = 2\,A$$

Example 18.4

Find the current through the battery in Fig. 18.19*a*.

R   *Reasoning*

Question   What does the current through the battery depend on?
Answer   Since $V = IR$, the current must depend on the battery voltage (6 V) and the total resistance between points *a* and *d*.

Question   What is the equivalent resistance between points *c* and *d*?
Answer   The two resistances are in parallel.

$$\frac{1}{R_{cd}} = \frac{1}{3} + \frac{1}{6} = \frac{3}{6} \qquad R_{cd} = 2\,\Omega$$

Question   What is the equivalent resistance between *b* and *d*?
Answer   The 4-Ω and 2-Ω resistors on the top branch are in series, totaling 6 Ω. This in turn is in parallel with a 6-Ω resistor. Thus

$$\frac{1}{R_{bd}} = \frac{1}{6} + \frac{1}{6} = \frac{2}{6} \qquad R_{bd} = 3\,\Omega$$

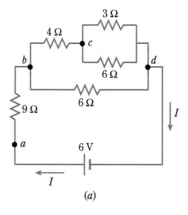

(a)

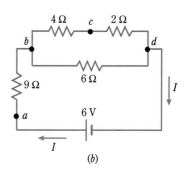

(b)

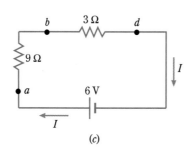

(c)

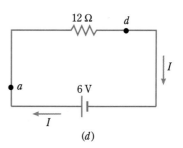

(d)

FIGURE 18.19

The complex circuit of (*a*) can ultimately be reduced to the simple equivalent circuit in (*d*).

Question What is the total resistance between *a* and *d*?
Answer As the simplified circuit of part *c* shows, a 9-Ω resistor is in series with
the 3-Ω resistance R_{bd}. Thus

$$R_{\text{tot}} = 9\,\Omega + 3\,\Omega = 12\,\Omega$$

Solution and Discussion Ohm's law gives

$$I = \frac{V}{R_{\text{tot}}} = \frac{6\,\text{V}}{12\,\Omega} = 0.50\,\text{A}$$

18.9 SOLVING CIRCUIT PROBLEMS

We now have at our disposal the tools needed to solve most dc circuit problems.
Before we use these tools in several examples, let us state a few facts we should
remember. Although each problem has its own peculiar features, the following
general approach is most often useful.

1 Draw the circuit.
2 Assign a current (both symbol and direction) to each branch of the circuit. Be
careful to use only one current designation for a given branch even though it
may contain several elements. At each junction the currents in each branch must
be labeled differently from each other.
3 Reduce series and parallel resistance systems whenever possible and convenient.
4 Write the loop equations for the simplified circuit. Each equation must contain
information from at least one new branch.
5 Write the junction equations for each junction that contains at least one new
current.

Steps 4 and 5 ought to give you as many equations as there are unknowns in the
circuit. Solve these equations simultaneously for the unknowns.

Example 18.5

Find the three currents in the circuit shown in Fig. 18.20.

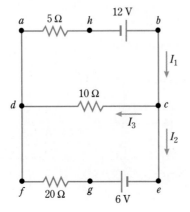

FIGURE 18.20

A circuit that is easily solved using
Kirchhoff's rules.

R

Reasoning

Question Can I simplify any series or parallel combinations?
Answer No. The resistors in branches *ab* and *ef* are not in simple parallel connection with *cd*. These branches also contain emfs.

Question What loop equations can I write?
Answer One loop is *abcda*. Starting at *a*:

$$-I_1(5\,\Omega) + 12\,\text{V} - I_3(10\,\Omega) = 0 \tag{1}$$

Another loop is *abefa*. Starting at *a*:

$$-I_1(5\,\Omega) + 12\,\text{V} + 6\,\text{V} - I_2(20\,\Omega) = 0 \tag{2}$$

which can be simplified to

$$-I_1(5\,\Omega) + 18\,\text{V} - I_2(20\,\Omega) = 0$$

Notice that some of the terms in (2) are the same as in (1), but there are some new terms as well.

Question What about loop *dcef*?
Answer The equation for this loop would contain no new terms. Equations (1) and (2) include all the circuit elements.

Question What is the junction equation at point *c*?
Answer $I_1 = I_2 + I_3$ \hfill (3)

Question How do I begin to solve these three equations?
Answer There are no equations containing only one unknown, and so you have to begin to eliminate unknowns by substitution. For example, use (3) to substitute for I_1 in (1) and (2):

$$-(I_2 + I_3)(5\,\Omega) + 12\,\text{V} - I_3(10\,\Omega) = 0 \tag{4}$$
$$-(I_2 + I_3)(5\,\Omega) + 18\,\text{V} - I_2(20\,\Omega) = 0 \tag{5}$$

Combining terms gives

$$-I_2(5\,\Omega) - I_3(15\,\Omega) + 12\,\text{V} = 0 \tag{6}$$
$$-I_2(25\,\Omega) - I_3(5\,\Omega) + 18\,\text{V} = 0 \tag{7}$$

Equation (6) can be solved for I_2 in terms of I_3:

$$I_2 = 2.4\,\text{V}/\Omega - 3I_3 \tag{8}$$

Substitute this into (7) and solve for I_3. Then use this value in (8) to get I_2. Finally, I_1 is obtained from (3).

Solution and Discussion If you keep your work neat and methodical, you lessen the risk of algebraic errors in solving simultaneous equations. From (7) we get

$-60\,\text{V} + I_3(75\,\Omega) - I_3(5\,\Omega) + 18\,\text{V} = 0 \qquad I_3 = 0.600\,\text{A}$

Then (8) gives

$I_2 = 2.4\,\text{V}/\Omega - (0.600\,\text{A})(3) = 0.600\,\text{A}$

Finally, (3) yields

$I_1 = 0.600\,\text{A} + 0.600\,\text{A} = 1.200\,\text{A}$

Notice that all three currents came out positive, indicating that their directions were chosen correctly.

Exercise Calculate the voltage difference between points e and f in Fig. 18.20.
Answer: $-6\,V$.

Example 18.6

The circuit in Fig. 18.21a contains two new symbols, $\text{\textcircled{V}}$ and $\text{\textcircled{A}}$, representing a voltmeter and an ammeter, respectively. The operation of these meters is discussed in more detail in Sec. 18.10. For the moment, assume the voltmeter is reading the potential drop across the 8.0-Ω resistor and the ammeter is reading the current I_{ab} flowing in the branch ab. These readings are 16 V and 0.50 A. The polarity of the 8-Ω resistor and the direction of the 0.50-A current are shown. We assume that the presence of the meters does not alter the circuit in any significant way. Find the values of $\mathcal{E}$, R, I_{acb}, and I_{bda} as indicated in Fig. 18.21a. Figure 18.21b shows the junction at point a.

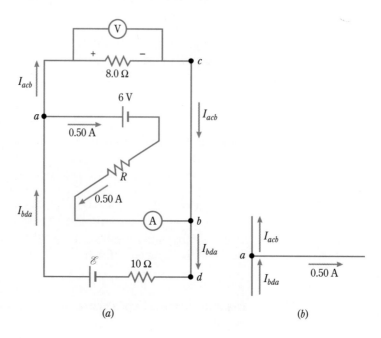

FIGURE 18.21
The ammeter and voltmeter readings are known. We wish to find I, R, and $\mathcal{E}$.

(a)   (b)

R

Reasoning

Question   There are four unknowns, so I will need four equations in order to solve the problem. Where will all these equations come from?
Answer   You have enough data about branch *acb* to find I_{acb} immediately. Then you must apply Kirchhoff's rules to each junction and loop until you have developed three independent equations.

Question   How is I_{acb} related to the measured voltage across the 8-Ω resistor?
Answer   The voltage drop across a resistor is $V = IR$. Here $V = 16$ V and $R = 8.0\ \Omega$, and so $I_{acb} = V/R = 2.0$ A.

Question   What is the result of applying the junction rule to point *a*?
Answer   Referring to Fig. 18.21*b*, I_{bda} is shown entering point *a* and the currents I_{acb} and 0.50 A leave *a*. Thus

$$I_{bda} = 2.0\ \text{A} + 0.50\ \text{A} = 2.5\ \text{A}$$

Question   Do I derive a second equation by using the junction rule at point *b*?
Answer   Point *b* will give you the same equation as point *a*, since no new currents are involved.

Question   What results from applying the loop rule to loop *acba*? Also, how do I choose the direction to go around the loop?
Answer   You can choose either of the two possible ways to go around the loop. As long as you observe the correct signs of the voltage changes, you will get the same result. Choosing the clockwise direction, you get

$$-16\ \text{V} + (0.50\ \text{A})R + 6.0\ \text{V} = 0 \tag{1}$$

Notice that this involves only one unknown, *R*. Be sure you understand the signs.

Question   What loop should I choose next?
Answer   Either of the two remaining loops, *acbda* or *abda*, will complete the problem, since all branches of the circuit will then have been used in the equations.

Question   What does going around loop *acbda* give?
Answer   Starting at *a*, you get

$$-16\ \text{V} - (I_{bda})(10\ \Omega) + \mathscr{E} = 0 \tag{2}$$

Since the junction rule already gave you the value of I_{bda}, this equation can be solved for $\mathscr{E}$.

Solution and Discussion   In equation (1), the voltage drop across *R* is entered with a plus sign because, in going around the loop clockwise, you go through *R* in the direction opposite the current, hence in the direction of voltage increase. Equation (1) gives $R = (10\ \text{V})/(0.50\ \text{A}) = 20\ \Omega$. Equation (2) gives $\mathscr{E} = 16\ \text{V} + (2.5\ \text{A})(10\ \Omega) = 41$ V.

Example 18.7

For the circuit in Fig. 18.22, find I_1, I_2, I_3, and the charge on the capacitor.

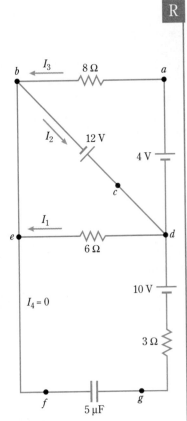

FIGURE 18.22

When the capacitor is fully charged, the current through the bottom wire is zero, and that portion of the circuit can be neglected.

Reasoning

Question   How does the capacitor figure into Kirchoff's rules?
Answer   Once the capacitor is charged, no current can flow through the branch containing it. (Notice in Fig. 18.22 that $I_4 = 0$.) From Eq. 17.6, the voltage drop across the capacitor is related to its charge by $V = q/C$. Once the rest of the circuit is analyzed, you can find this V and hence q.

Question   What does the junction rule give for point d?
Answer $I_2 = I_1 + I_3$

Question   What does the loop rule give for loop *abcda*?
Answer   Moving counterclockwise, you get $-I_3(8\,\Omega) + 12\,\text{V} + 4\,\text{V} = 0$. Thus, $I_3 = 2.00\,\text{A}$.

Question   What does the loop *badeb* give?
Answer   Moving clockwise, $+(2.00\,\text{A})(8\,\Omega) - 4\,\text{V} - I_1(6\,\Omega) = 0$. Thus $I_1 = 2.00\,\text{A}$.

Question   What expression can give me I_2?
Answer   The junction rule. For example, junction b shows that $I_2 = I_1 + I_3 = 4.00\,\text{A}$.

Question   What equation will yield the charge on the capacitor?
Answer   Apply the loop rule to a loop which contains the capacitor, even though the current in the branch containing the capacitor (I_4) must be zero. For example, move counterclockwise around loop *defgd*.

Question   If I_4 is zero, does this mean there is zero voltage change across the 3-Ω resistor?
Answer   Yes.

Question   How can I tell the direction of the voltage change across the capacitor?
Answer   You don't need to know in order to write down the equation. Just use the symbol V_{fg} to denote the voltage change from f to g. When you find the solution, the sign of V_{fg} will tell you the direction of the voltage change.

Solution and Discussion   The loop rule for *defgd* gives

$$-(2.00\,\text{A})(6.0\,\Omega) + V_{fg} + 10\,\text{V} = 0$$

Thus $V_{fg} = +2.0\,\text{V}$. The charge on the capacitor must then be

$$q = CV_{fg} = (5.0 \times 10^{-6}\,\text{F})(2.0\,\text{V}) = 1.0 \times 10^{-5}\,\text{C}$$

Notice that the plate connected to point g is the positive plate of the capacitor.

18.10   AMMETERS AND VOLTMETERS

In Example 18.6, we saw a typical situation in which an ammeter and a voltmeter are used in a circuit. Although we learn how these meters are constructed in Chap. 19, let us not postpone a discussion of how they are used, since you will be using them in the laboratory.

We use both analog and digital meters to measure current and voltage.

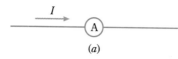

(a)

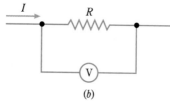

(b)

FIGURE 18.23

Why should an ammeter have little resistance and a voltmeter have nearly infinite resistance?

An ammeter is used to measure the current through a wire. We connect it directly in line with the wire, as shown in Fig. 18.23a. Notice that the current to be measured passes through the meter. If the meter had much resistance, it would alter the current in the circuit. Therefore, an ideal ammeter has zero resistance. The ammeters you will be using in the laboratory will usually have a resistance of only a fraction of an ohm.

Voltmeters are used to measure potential difference. To measure the potential difference $V = IR$ across the resistor in Fig. 18.23b, the voltmeter terminals must be connected to the two ends of the resistor as shown. Ideally, we want the voltmeter to leave the circuit undisturbed. This is possible only if the voltmeter resistance is very large. An ideal voltmeter has infinite resistance and therefore does not cause the current to branch at the points where the meter is connected.

Students who confuse the ammeter and the voltmeter in situations such as those shown in Fig. 18.23 face serious danger to life and happiness because of the severe displeasure of their laboratory teacher. An ideal voltmeter has an infinite resistance. As such, no current passes through it when its terminals are connected to two points that differ appreciably in potential. An ideal ammeter has zero resistance, however. If its terminals were inadvertently connected to two points of different potential, the current through the ammeter would be given by

$$I = \frac{V}{R} = \frac{\text{something}}{\text{zero}} \to \infty$$

This student error is accompanied by smoke issuing from the meter case, irreparable harm to the meter, and an antagonistic attitude on the part of the instructor. So beware.

18.11 HOUSE CIRCUITS

We are all familiar with the ordinary electric circuits that extend throughout our houses. The power company runs at least two wires to each house to provide a

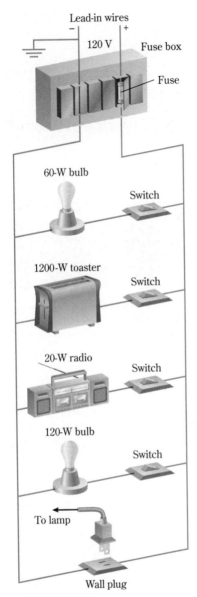

FIGURE 18.24

When a switch is closed, current passes through the device that the switch controls.

potential difference of about 120 V.* These lead-in wires usually have a large diameter so that they can carry considerable current without heating up. (The larger the cross-sectional area of the wire, the less its resistance. Since heat generated is proportional to I^2R, the low resistance ensures low heat dissipation.)

In most newer houses, the wires inside the house are capable of carrying about 20 A without undue heating. However, to protect against too large a current, a fuse or a circuit breaker is placed in series with the wire. Its purpose is to automatically disconnect the wire from the voltage source if more than the allowed current is drawn from the source.

A typical house circuit consists of two parallel wires strung through the house from the 120-V source provided by the lead-in wires (Fig. 18.24). One terminal of each light bulb, appliance, and so on, is connected to the high-potential wire, and the other terminal is connected to the low-potential wire. When the switch to that appliance is closed, charge can flow through the device. The low-potential wire is usually grounded.

Many 120-V appliances have a third prong on the plug. This furnishes a connection between a ground wire and the metal frame of the appliance. If, by accident, the high-voltage wire touches the metal frame of the appliance, a direct connection to ground is made. The effect is the same as connecting the high- and low-voltage wires directly. There is a large current through the high-voltage wire to ground, and the fuse in the high-voltage wire will blow. If the ground wire is absent, such a malfunction will leave the whole appliance "floating" at high potential. Anyone touching the metal frame will then suffer a shock.

Let us compute how much current is drawn by the 60-W bulb of Fig. 18.24 when it is turned on. Since power = VI and since $P = 60$ W and $V = 120$ V in this case, the current through the bulb is $I = 0.500$ A. Similarly, when turned on, the toaster draws 10.0 A, the radio draws 0.167 A, and the 120-W bulb draws 1.00 A. If they are all turned on at once, a total of 11.667 A passes through the fuse. Usually a house circuit is fused for no less than 15 A, and so there is no danger in this case.

A house with many electrical appliances requires more than one circuit. Most houses have several separate circuits, each with its own fuse, similar to the one shown in Fig. 18.24.

It is interesting to compute the resistance of a light bulb. When the bulb is cool, its resistance is not very large. However, when it is connected across the rated voltage, usually 120 V, its resistance element becomes white-hot. As discussed previously, its resistance increases considerably when the bulb heats up. When it is hot, it is operating at the wattage stamped on it. Suppose we have a 60-W 120-V bulb. We know that

$$P = VI = \frac{V^2}{R}$$

$$60 \text{ W} = \frac{(120 \text{ V})^2}{R}$$

$$R = 240 \text{ } \Omega$$

We saw in Example 18.1 that the resistance of this bulb at room temperature is about 27 Ω.

---

*The potential difference supplied by the power company is constantly reversing in a sinusoidal way. We discuss this type of voltage in detail in Chap. 21. For the purposes of the present discussion, the alternating voltage has the same effect as a dc voltage.

The metal lightning rods along the tops of these farm buildings are connected to ground by means of wires such as the one seen on the right-hand edge of the building on the right. The sharp tips of these rods allow charge induced in them by clouds to escape, inhibiting charge buildup and lessening the likelihood of a sudden damaging stroke of lightning occurring. In the event that lightning does strike the building, it will strike the rod and the charge will be routed to ground, protecting the structure from severe damage.

18.12 ELECTRICAL SAFETY

Since we use electrical appliances daily, we should understand the elements of electrical safety. There are two ways in which electricity can kill a person. It can cause the muscles of the heart and lungs (or other vital organs) to malfunction, or it can cause fatal burns.

Even a small electric current can seriously disrupt cell functions in that portion of the body through which it passes. When the current is 0.001 A or higher, a person can feel the sensation of shock. At 0.01 A, a person is unable to release an electric wire held in the hand because the current causes the hand muscles to contract violently. Currents larger than 0.02 A through the torso paralyze the respiratory muscles and stop breathing. Unless artificial respiration is started at once, the victim will suffocate. Of course, the victim must be freed from the voltage source before he or she can be touched safely; otherwise the rescuer, too, will be in great danger. A current of about 0.1 A passing through the region of the heart will shock the heart muscles into rapid, erratic contractions (ventricular fibrillation) so that the heart can no longer function. Finally, currents of 1 A and higher through body tissue cause serious burns.

To prevent injury, *the important quantity to control is current. Voltage is important only because it can cause charge to flow.* Even though your body can be charged to a potential thousands of volts higher than the metal of an automobile when you slide across the car seat, you feel only a harmless shock as you touch the door handle. Your body cannot hold much charge on itself, and so the current through your hand to the door handle is short-lived and the effect on your body cells is negligible.

In some circumstances, a 120-V house circuit is almost sure to cause death. One of the two wires of the circuit is usually attached to the ground, and so it is always at the same potential as the water pipes in the house. Suppose you are soaking in a bathtub; your body is effectively connected to the ground through the water and piping. If your hand accidentally touches the high-potential wire of the house circuit (by touching an exposed wire on a radio or heater, for example), charge will flow through your body to the ground. Because of the large, efficient contact your body makes with the ground, the resistance of the body circuit is low. Consequently, the current through your body is so large that you are in danger of being electrocuted.

Similar situations exist elsewhere. For example, if you accidentally touch an exposed wire while you are standing on the ground with wet feet, you are in far greater danger than if you are on a dry, insulating surface. The electric circuit through your body to the ground has a much higher resistance if your feet are dry. Similarly, if you sustain an electric shock by touching a bare wire or a faulty appliance, the shock is greater if your other hand is touching a faucet or is in water.

As you can see from these examples, the danger from electric shock can be eliminated by avoiding a current path through the body. When the voltage is greater than about 50 V, avoid touching any exposed metal portion of the circuit. If a high-voltage wire must be touched, for example, in a power-line accident when help is not immediately available, use a dry stick or some other substantial piece of insulating material to move it. When in doubt about safety, avoid all contact or close approach to metal or to the wet earth. *Above all, do not let your body become the connecting link between two points that are at widely different potentials.*

18.13   THE EMF AND TERMINAL POTENTIAL OF A BATTERY

Probably everyone has noticed at one time or another that the lights on an automobile dim when the motor is started. This happens because the electric starter draws considerable current from the battery. In so doing, it lowers the potential between the battery terminals, and the car lights dim. We now investigate this nonconstancy of the terminal potential difference of a battery.

As pointed out in Chap. 17, the emf of a battery is generated by the chemical action in the battery. However, a battery is a very complex chemical device, and charge cannot move through it without encountering internal resistance. As a result, a battery behaves in a circuit like a pure ($R = 0$) emf source in series with a resistor. This internal resistance r and the equivalent circuit element for a battery are shown in Fig. 18.25.

Notice that when no current is being drawn from the battery, there is no potential drop across the internal resistance r. Hence the potential difference between the terminals is equal to the emf. However, if the battery is connected across an external resistor, as in Fig. 18.26, the current is I and the potential difference across the terminals is

Terminal potential = $V_T = \mathcal{E} - Ir$ (discharging)

If a battery is being charged, that is, if current is flowing through the battery from the positive to the negative terminal, we have

Terminal potential = $V_T = \mathcal{E} + Ir$ (charging)

FIGURE 18.25

A battery acts as though it consists of a pure ($R = 0$) emf and a resistor in series.

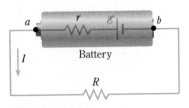

FIGURE 18.26

The voltage across the battery terminals is $\mathcal{E} - Ir$.

For a good 12-V battery, the internal resistance is only of the order of 0.01 Ω. If this battery is connected across a 3-Ω resistor, we have

$$I = \frac{12\,V}{3\,\Omega + 0.01\,\Omega} \approx 4\,A$$

The terminal potential V_T is the voltage difference between points a and b:

$$V_T = 12\,V - (4\,A)(0.01\,\Omega) = 11.96\,V$$

In this case, the terminal potential is nearly equal to the emf.

As a battery becomes older, however, its internal resistance increases. If the internal resistance of the 12-V battery increases to 1.0 Ω, the current when connected to the same 3-Ω resistor is

$$I = \frac{12\,V}{4\,\Omega} = 3.0\,A$$

and the terminal potential is

$$V_T = 12\,V - 3.0\,V = 9.0\,V$$

It should be clear that when a starter on a car draws 100 A from the battery, the terminal potential of even a new battery will decrease noticeably.

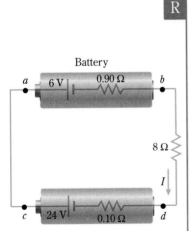

Battery

FIGURE 18.27

The 24-V battery is charging the 6-V battery. We find that the terminal potential difference of a discharging battery is less than its emf, while the reverse is true for a battery that is being charged.

Example 18.8

What is the terminal potential of each battery in Fig. 18.27?

Reasoning

Question What do I have to know in order to determine the terminal potentials?
Answer Since you know the internal resistances, you can calculate terminal potentials if you can find the current passing through the batteries.

Question What does Kirchhoff's loop rule say about this circuit?
Answer Going around clockwise, starting at a:

$$-6\,V - I(0.90\,\Omega) - I(8\,\Omega) - I(0.10\,\Omega) + 24\,V = 0$$

Thus

$$I = \frac{18\,V}{9.00\,\Omega} = 2.00\,A$$

Question What is the expression for terminal potential?
Answer $V_T = \mathscr{E} \pm Ir$. You use the plus sign if the battery is being charged, and the negative sign when the battery is discharging (delivering current to the circuit).

Solution and Discussion For the 24-V battery:

$$V_T = 24\,V - (2\,A)(0.10\,\Omega) = 23.8\,V$$

For the 6-V battery, which is being charged:

$$V_T = 6\,\text{V} + (2\,\text{A})(0.90\,\Omega) = 7.8\,\text{V}$$

A MODERN PERSPECTIVE
SUPERCONDUCTIVITY

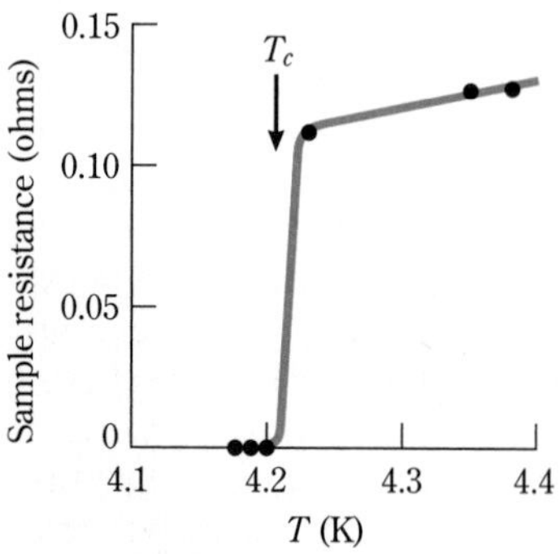

At the beginning of the twentieth century, considerable speculation existed about the behavior of materials, including resistivity, at temperatures approaching 0 K. Let us briefly review some of what we have learned about resistivity. In Sec. 18.2 we saw that resistivity is caused by collisions between the electrons comprising the electric current and the atoms of the conducting material. As the electrons move through the conductor, the energy given to them by the applied emf in the circuit is converted into thermal energy by these collisions. In Sec. 18.4 we noted that the resistivity of most conducting materials decreases with decreasing temperature. Thus we ask, "Would resistivity fall to zero if a conductor could be cooled to 0 K?"

Until 1908, the lowest temperatures available in the laboratory were provided by liquid hydrogen, which is liquid at atmospheric pressure from about 20 K down to approximately 14 K, where hydrogen solidifies. Evidence at temperatures below 25 K showed that the resistivity of many metals continues to decrease with decreasing temperature, but at a less rapid rate than at higher temperatures. A breakthrough in achieving lower temperatures occurred during 1908 when the Dutch physicist Kamerlingh Onnes succeeded in liquefying helium at 4.2 K.

In 1911 Onnes made the startling discovery that, instead of continuing to decrease gradually at liquid helium temperatures, the resistivity of mercury dropped abruptly (by a factor of over 10^6 within a temperature decrease $\Delta T \approx 0.01$ K) to zero at a temperature of 4.15 K (Fig. 18.28). Onnes described this transition as a new state of mercury—the *superconducting state*. In the years that followed, it became evident that most metals and a large number of alloys exhibit this type of sudden transition to zero resistivity at various temperatures (called *critical* temperatures, T_c). A brief list of critical temperatures is given in Table 18.4.

Onnes went on to devise a very sensitive test to determine whether the resistivity of a superconductor was truly zero or just extremely small. He established a current in a ring of lead by means of magnetic induction, a subject we discuss in Chap. 20. If the resistivity of the lead was not truly zero, the current would be expected to decrease eventually to zero as the kinetic energy of the electrons was gradually converted into the thermal energy of the lead. However, Onnes was unable to discern any decrease in the ring current over a period of many hours. This experiment has been tried many times by various researchers since, and currents in superconducting rings have been observed to persist for many years without measurably decreasing. We thus conclude that the sudden decrease of resistivity that occurs at T_c does result in a truly zero value.

It was not until 1957 that a complete theoretical explanation of superconductivity based on the dynamics of the electrons was put forth. J. Bardeen, L. Cooper, and J. R. Schrieffer, then at the University of Illinois, later shared the 1972 Nobel prize in physics for their explanation, which has become known as the BCS theory. As has been the case for most of the successful solutions to physics problems in the twentieth century, the BCS theory is based on the principles of quantum theory. Because the mathematics of quantum theory lies beyond the scope of this course, the following discussion of the BCS theory has to be qualitative.

FIGURE 18.28

The superconducting transition of mercury, as reported by Onnes in 1911.

TABLE 18.4

Examples of critical temperatures for superconductivity

Material	T_c (K)
Elements	
Tungsten	0.01
Titanium	0.40
Aluminum	1.19
Mercury	4.15
Lead	7.18
Niobium	9.46
Alloys	
50% nickel–50% bismuth	4.25
75% neodymium–25% aluminum	17.5
75% neodymium–25% tin	18.0

An electron moving through a conductor interacts with nearby atoms, altering their positions slightly and causing a local vibration of the lattice structure (the regular spacing of the atoms) of the conductor. The force exerted on the electron by the atom during this interaction disrupts the direction of the electron's motion, thereby temporarily removing its contribution to the current in the conductor. At "ordinary" temperatures, these local vibrations are quickly shared and randomized throughout the metal, representing an increase in the thermal energy of the metal, a process which we have referred to as *joule heating*.

According to the BCS theory, when the temperature of the conductor is below T_c, the energy of the lattice vibration produced by one electron is quickly (typically within 10^{-12} s) returned to a second electron instead of being shared throughout the conductor. Thus the total energy possessed by the electrons in the current remains constant, and there is no retention of energy by the atoms of the metal, hence no joule heating. It also means that the current carried collectively by the electrons is not diminished, and hence the conductor displays zero resistivity.

This electron-lattice-electron energy exchange is a process which cannot be explained by classical theory. The result of the energy exchange is to create an *attractive* interaction between the two electrons involved. The interacting electrons, referred to as a *correlated pair*, are typically $1\ \mu m$ apart (a very great distance in terms of the average separation between electrons in the metal) and have opposite linear momenta and opposite rotational (spin) angular momenta. Since two free electrons interact repulsively via the Coulomb force, a correlated pair has a lower energy than two uncorrelated electrons. As the temperature of the conductor approaches zero K, the thermal vibrations of the lattice become too weak to break these correlations, and *all* of the conduction electrons throughout the entire conductor become correlated pairs. In such a state, there can be no exchange of energy between the lattice atoms and the electrons, and so the resistance of the conductor becomes truly zero.

There are a very large number of practical applications of superconductivity, which we can better understand after learning about magnetism. We shall therefore defer discussion of applications until Sec. 20.12, where another Modern Perspective deals with the magnetic properties of superconductors. Meanwhile, we should observe that the extremely low critical temperatures for superconductivity require liquid helium as a coolant and are very expensive to achieve and maintain. Ever since Onnes first discovered the phenomenon of superconductivity, a continuous search for materials having higher and higher values of T_c has been conducted, so that the potential applications could become more affordable. An obvious target was to find materials with T_c above the boiling point of nitrogen, since liquid nitrogen, obtained inexpensively from the atmosphere, could then be used as the coolant. At 1 atm pressure, nitrogen boils at approximately 77 K, a temperature much higher than any T_c found until the mid-1980s.

Starting with a discovery in 1986 by K. A. Muller and J. G. Bednorz, new types of ceramic oxides that possess values of T_c above 77 K have been extensively explored. Some of these materials exhibit superconductivity at temperatures as high as 120 K and slightly above. Many researchers feel that even higher values of T_c can be found and that many applications of superconductivity will become feasible in the near future. Others, however, feel that many obstacles remain, pointing to the fact that these ceramic oxides are brittle and cannot be drawn into thin wires or easily formed into useful shapes. In any case, the search for higher-temperature superconductors is likely to remain vigorous well into the next century.

LEARNING GOALS

Now that you have finished this chapter, you should be able to

1 Define (a) dc circuit, (b) current, (c) ampere, (d) Ohm's law, (e) resistance, (f) resistivity, (g) ohm, (h) temperature coefficient of resistivity, (i) electric power, (j) Kirchhoff's rules, (k) series and parallel circuits, (l) equivalent resistance, (m) terminal potential difference and emf, (n) internal resistance.

2 Use the relation $I = \Delta q / \Delta t$ in simple situations.

3 Interpret a simple circuit diagram. State the potential difference between various points of the circuit.

4 State which end of a resistor is at the higher potential when the direction of current through the resistor is given.

5 Use Ohm's law in simple situations.

6 Compute the resistance of a given piece of wire if the resistivity of the wire material is known.

7 Find the resistance of a wire at a given temperature when its resistance and temperature coefficient at some reference temperature are known.

8 Use the power equation $P = IV$ to find the power loss or gain in a resistor, battery, and capacitor under dc conditions.

9 Apply Kirchhoff's junction rule.

10 Write Kirchhoff's loop equation for a series circuit that contains batteries, resistors, and capacitors.

11 Reduce a given set of series and parallel resistors to a single equivalent resistor.

12 Use Kirchhoff's rules to solve dc circuits that contain batteries, resistances, and capacitances.

13 Sketch a typical house circuit and point out the various elements in it. Compute the current drawn by various portions of a house circuit when the appliances it is running are given.

14 Analyze a given electrical situation from the viewpoint of safety.

15 Explain why the terminal potential of a battery is not always equal to the emf. Find the terminal potential if $\mathscr{E}$, I, and r are known.

SUMMARY

DERIVED UNITS AND PHYSICAL CONSTANTS

Unit of Electric Current (A)
1 ampere (A) = 1 C/s

Unit of Resistance (Ω)
1 ohm (Ω) = 1 V/A

DEFINITIONS AND BASIC PRINCIPLES

Electric Current (I)
Electric current (amperes) is defined as the rate of flow of electric charge.

$$I = \frac{\Delta q}{\Delta t}$$

The direction of the current is taken to be in the direction of positive charge flow.

Resistance (R) and Ohm's Law
The resistance (ohms) of a circuit element is defined as the ratio of the voltage across the element to the steady current passing through it:

$$R = \frac{V}{I}$$

Resistors that have a constant value of R over a range of values of V and I obey Ohm's law and are said to be **ohmic** resistors.

INSIGHTS

1. It follows from the definition of R that a resistor carrying a current I has a potential drop across its ends equal to $V = IR$. This drop in potential is in the direction of positive current flow.

2. The graph of I versus V for a resistor obeying Ohm's law is a straight line with slope $= 1/R$. Resistors whose I versus V graph is not a straight line are said to be nonohmic. Nevertheless, R is defined as the ratio V/I for any given pair of values.

Resistivity (ρ)
Different materials have different inherent resistance to current. The value of resistance R for a given sample of material depends on its length L, cross-sectional area A, and this inherent property, which is called the resistivity of the material. The resistivity ρ is defined by

$$R = \frac{\rho L}{A} \quad \text{or} \quad \rho = \frac{RA}{L}$$

The units of ρ are ohm-meters ($\Omega \cdot$ m).

Temperature Dependence of R and ρ

As the temperature of a resistor changes, the fractional change in its resistivity is proportional to ΔT over a limited range of temperatures:

$$\frac{\Delta\rho}{\rho_o} = \alpha\,\Delta T$$

α is called the temperature coefficient of resistivity. It has units of $(C°)^{-1}$.

Since values of α are usually many times larger than thermal expansion coefficients, the temperature variation of R of a given resistor also follows the same expression:

$$\frac{\Delta R}{R_o} = \alpha\,\Delta T$$

Power in DC Circuits

A battery or other source of emf voltage supplies power to a circuit given by

$$P = IV$$

where I is the current being delivered by the battery. When I is in amperes and V in volts, P is in watts.

The power converted to thermal form in a resistor R is

$$P = I^2R$$

where I is the current passing through R.

Since energy = power × time, a useful energy unit used in the electrical industry is the kilowatt-hour (kWh).

Kirchhoff's Rules

THE JUNCTION RULE

The sum of all currents entering a junction must equal the sum of all currents leaving the junction.

THE LOOP RULE

The algebraic sum of the voltage changes around any closed loop of a circuit must equal zero.

INSIGHTS

1. In using Kirchhoff's rules to analyze a circuit, you should first label a different current in each individual branch of the circuit. You can arbitrarily choose a direction for each current.
2. You can apply the junction rule to each junction that involves at least one new current. Application of the junction rule must follow your choice of current direction.
3. You can apply the loop rule to each different loop that involves at least one new circuit element.

4. The sign of individual voltage changes are to be taken as follows:
   a. Batteries or emfs: $\Delta V = +\mathscr{E}$ when going from the negative to the positive terminal.
   b. Resistors: $\Delta V = -IR$ when going across the resistor in the direction of your chosen current.
   c. Capacitors: $\Delta V = +q/C$ when going from the negatively charged plate to the positively charged plate.
5. If you choose a wrong direction for one of the currents, the solution for that current will have a negative sign.
6. The dc current in a branch containing a capacitor must be zero.

Resistors Connected in Series and Parallel

RESISTORS CONNECTED IN SERIES

The equivalent total resistance of n resistors connected in series is

$$R_{eq} = R_1 + R_2 + \cdots + R_n \qquad \text{(series)}$$

RESISTORS CONNECTED IN PARALLEL

$$\frac{1}{R_{eq}} = \frac{1}{R_1} + \frac{1}{R_2} + \cdots + \frac{1}{R_n} \qquad \text{(parallel)}$$

INSIGHTS

1. These are the same rules as for the addition of capacitors, except that the series and parallel rules are reversed.
2. All resistors in series in a circuit branch carry the same current and have different individual voltage drops.
3. All resistors in parallel in a circuit branch have the same voltage drop and carry different individual currents.

The EMF and Terminal Voltage of a Battery

A battery in a circuit acts like a source of emf $\mathscr{E}$ in series with an internal resistance r. When the battery delivers a current I, there is an internal voltage drop Ir which subtracts from $\mathscr{E}$ to give the effective terminal voltage, V_T:

$$V_T = \mathscr{E} - Ir$$

If the battery is receiving a current I (i.e., being charged), the terminal voltage becomes

$$V_T = \mathscr{E} + Ir$$

INSIGHTS

1. The voltage the battery can supply to a circuit diminishes as the current being supplied increases.
2. A fresh battery will have a very small internal resistance, a condition that allows it to furnish large currents at close to its rated emf. As a battery ages, its internal resistance increases.

QUESTIONS AND GUESSTIMATES

1 Sometimes a student insists that current is used up in a resistor. Arguing from the water analogy, how would you convince such a student that current is not lost in a resistor?

2 How do we know which end of a battery is at the higher potential, that is, positive, in a schematic diagram of a circuit? How do we know which end of a resistor is at the higher potential?

3 Fluorescent light bulbs are usually more efficient light emitters than incandescent bulbs. That is, for the same input energy, the fluorescent bulb gives off more light than the incandescent bulb. Carefully touch a fluorescent bulb and an incandescent one after each has been lit for a few minutes. Explain why the incandescent bulb is a less efficient light emitter.

4 In Fig. P18.1a, the pump lifts water to the upper reservoir at a rate such that the water level remains constant. The water slowly trickles out of the narrow tube into the lower reservoir. Point out the similarities between this water circuit and the electric circuit shown in part b.

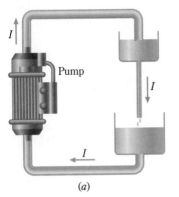

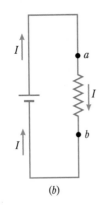

(a)   (b)

FIGURE P18.1

5 A resistor is connected from point a to point b. How does one tell whether there is a potential drop or a potential rise from a to b? Repeat for a battery and for a capacitor.

6 Explain the following statement: For series resistors, the equivalent resistance is always larger than the largest resistance in the series; for parallel resistors, the equivalent resistance is always smaller than the smallest resistance in the combination.

7 Using an ohmmeter (basically a battery in series with a very sensitive ammeter), measure your resistance from one hand to the other. A current of about 0.02 A through the midsection of the body is sufficient to paralyze the breathing mechanism. About how large a voltage difference between your hands is needed to electrocute you?

8 If you grasp the two wires leading from the two plates of a charged capacitor, you may feel a shock. The effect is much greater for a 2-μF capacitor than for a 0.02-μF capacitor, even though both are charged to the same potential difference. Why?

9 Birds perch on high-tension wires all the time. Why are they not electrocuted, even when they perch on a part of the wire where the insulation has worn off?

10 If there is a current of only a small fraction of an ampere into one of a person's hands and out the other, the person will probably be electrocuted. If the current path is into one hand and out the elbow on the same arm, the person can survive even if the current is large enough to burn the flesh. Explain.

11 Parents frequently worry about children playing near electrical outlets. Discuss the various factors that determine how badly shocked a child could be. What would happen if a child were to cut a lamp cord in two with a pair of noninsulated, wire-cutting pliers when the cord is plugged in? Is the child in any danger?

12 Explain why touching an exposed circuit wire when you are in a damp basement is much more dangerous than touching the same wire when you are on the second floor.

13 It is extremely dangerous to use a plug-in radio near a bathtub when you are taking a bath. Why? Does the same reasoning apply to a battery-operated radio?

PROBLEMS

Sections 18.1 and 18.2

1 There is a current of 0.5 A through a light bulb. (a) How much charge passes through the bulb in 4 h? (b) How many electrons flow through the bulb during this time?

2 A television picture tube has an electron beam current of 56 μA. How many electrons strike the screen every minute?

3 How long does it take for 64 C to pass through a cross-sectional area of a wire that carries a current of 72 A?

4 A battery charger sends a current of 3.6 A through a battery for 8.0 h. How much charge is furnished to the battery by the charger in this time?

5 A certain car battery maintains a current of 2.2 A for 12 h. How much charge flows from the battery during this time?

6 A single +1.8-μC charge travels around a circular orbit of 2.0-m radius with a speed of 1.0×10^5 m/s. What is the average current in the orbit?

7 In a certain cathode-ray tube 3.2×10^{12} electrons strike the screen per second. What is the current corresponding to the electron beam in the tube?

Sections 18.3 and 18.4

8 When a 3.0-V flashlight battery is connected across a bulb, the current is 40 mA. What is the resistance of the bulb?

9 The filament of a light bulb has a resistance of 300 Ω. What will be the current through it when operating at 120 V?

10 Suppose the resistance through your body from one hand to the other is 30 kΩ. How large is the current through your arms when you grasp the terminals of a 9-V battery?

11 A color television set draws a current of 2.4 A when operating at 120 V. What is the effective resistance of the set?

12 How large is the potential difference across a 240-Ω resistor when a current of 0.25 A flows through it?

▪13 A resistor carries a current of 0.4 A when connected to a 120-V source. What will the current carried by the resistor be if (a) the operating voltage is reduced to 96 V, (b) the voltage is increased to 144 V?

▪14 A flashlight uses three batteries of 1.5 V each, connected in series. What is the resistance of the light bulb if it draws a current of 0.6 A?

15 A charge of 6.0×10^4 C flows in 1 h through a resistor when the potential drop across it is 9 V. Determine the resistance of the resistor.

16 Find the resistance at 20°C of a 24-m length of aluminum wire that has a diameter of 1.6 mm.

17 Determine the resistance at 20°C of a 40-cm length of silver wire that has a diameter of 0.160 mm.

▪18 The resistance of a spool of insulated copper wire is measured by noticing that a 9-V battery causes a current of 0.3 A in the entire length of the wire. The diameter of the metal part of the wire is 0.80 mm. What length of wire is on the spool?

▪19 Number 18 copper wire has a diameter of 1.024 mm. The maximum safe current through this size of wire is 12 A. (At higher currents, the wire becomes too hot.) (a) Find the resistance of a 20-m length of a number 18 copper wire at 20°C. (b) How large a potential difference exists between the ends of the wire in (a) when it carries a current of 12 A?

20 A wire which has a resistance of 0.25 Ω per meter of its length is to be used for wiring a house. What is the maximum length of this wire which can be used if the total resistance of the wiring is not to exceed 750 Ω?

21 A coil of tungsten wire that has a resistance of 30 Ω at 20°C is to be used to measure temperature. How much does its resistance change for a 4-C° temperature change near 20°C?

22 The heating element of a room heater is made of nichrome wire 1.0 mm in diameter. Assume the resistivity of nichrome to be 10^{-6} Ω · m. If the heater has a resistance of 25 Ω, how long is the wire in the heating element?

23 The resistance of a coil of wire is 156.8 Ω at 20°C and 166.6 Ω at 50°C. What is the temperature coefficient of the resistance for the material of the wire?

24 What is the percentage change in the resistance of a tungsten wire as its temperature changes from 15 to 36°C?

25 At what temperature does aluminum have the same resistivity as tungsten does at 20°C?

▪26 A 3-m length of iron wire carries a current of 0.2 A when connected to a 6-V battery. What length of silver wire will carry this same current when connected to the 6-V battery?

Section 18.5

27 A light bulb is marked 100 W/120 V. (a) What current does it draw? (b) What is its resistance when operating at 120 V?

28 A 15-W fluorescent lamp is designed to operate on 120 V. (a) How much current does it draw? (b) What is its resistance?

▪29 The heating element of a room heater is made up of a 4-m length of tungsten wire. When connected to a 120-V source, the filament attains a temperature of 450°C and consumes 1500 W of power. What is the cross-sectional area of the wire?

30 How long will it take a 500-W submersible heater to heat 300 g of water from 23°C to 88°C? Assume that there are no heat losses to the surroundings.

31 A portable CD player draws a current of 280 mA when operated by a 9-V battery. How much power does it dissipate?

32 A 0.5-hp motor is connected to a 120-V line. How large is the current carried by the motor?

33 (a) How much energy (mostly heat) does a lighted 75-W bulb emit in 5 min? (b) How many kilowatthours does it consume in this time?

34 A voltage surge in a power line produces 132 V momentarily. By what percentage will the output of a 60 W/120 V light bulb increase, assuming that its resistance does not change?

35 A 200-W street lamp operates for 12 h every day. How many kilowatthours of energy will it consume in 30 days? How much will it cost if the electricity costs $0.068/kWh?

Sections 18.6–18.10

36 Find the equivalent resistance of four resistors, 2 Ω, 4 Ω, 6 Ω, and 10 Ω, when connected (a) in series and (b) in parallel.

37 Three resistors, 2 Ω, 3 Ω, and 6 Ω, are connected in parallel and their combination is then connected in series with a 4-Ω resistor. What is the equivalent resistance of the combination?

38 Three resistors, 6 Ω, 2 Ω, and 10 Ω, are connected in series with a 9-V battery. (a) Find the equivalent resistance of the combination. (b) What is the current in each resistor?

39 The resistors in the previous problem are connected in parallel across a 9-V battery. Find (a) the equivalent resistance of the combination and (b) the current in each resistor.

▪40 A 4-Ω resistor and a 6-Ω resistor are connected in parallel across a battery. The current through the equivalent resistor is measured to be 2.5 A. Find the voltage of the battery.

41 In Fig. P18.2, each resistor is 6 Ω and $\mathscr{E} = 3.0$ V. Find (a) the equivalent resistance of the combination and (b) the current drawn from the battery.

▪42 In Fig. P18.2 the vertical resistances are 4 Ω each and the horizontal resistances are 6 Ω each. Find (a) the equivalent resistance of the combination and (b) the current drawn from the battery if $\mathscr{E}$ = 3.0 V.

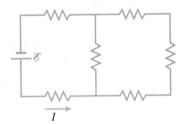

FIGURE P18.2

■43 For Prob. 42 find the current through the two vertical resistors.

■44 Two resistors, one 8 Ω and one 10 Ω, are connected in series with a power supply. The measured voltage drop across the 10-Ω resistor is found to be 25 V. What is the voltage supplied by the power supply?

■45 The two resistors in Prob. 44 are connected in parallel across a power supply. The measured current through the 8-Ω resistor is found to be 1.2 A. Find the voltage of the power supply.

■46 In Fig. P18.3 find the equivalent resistance seen by the battery (a) when the switch S is open and (b) when the switch is closed. (c) What is the current through the 4-Ω resistor when the switch is closed?

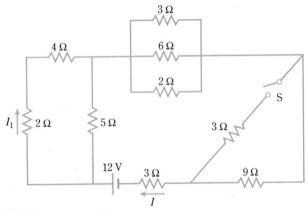

FIGURE P18.3

■47 In Prob. 46 find the current through the 9-Ω resistor (a) when the switch S is open and (b) when the switch is closed.

■48 In Fig. P18.3 find the voltage drop across the 3-Ω resistor just to the right of the battery (a) when the switch S is open and (b) when the switch is closed.

■49 (a) Find the equivalent resistance for the circuit shown in Fig. P18.4 if each resistor is 5 Ω. The circuit is shown in its entirety. (b) Find I. (c) Find I_1. (d) What is I_2?

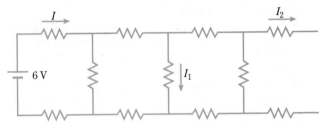

FIGURE P18.4

■50 Repeat the previous problem if all the horizontal resistances are 4 Ω each and all the vertical resistors are 6 Ω each.

■51 Find the currents I_1, I_2, and I_3 in Fig. P18.5.

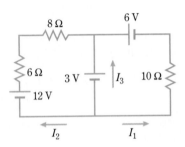

FIGURE P18.5

■52 Suppose, in Fig. P18.5, the polarity of the 6-V battery is reversed. What are the new values of the currents I_1, I_2, and I_3?

■53 In Fig. P18.6 the current I_3 is measured and found to be 3 A. Find (a) the currents I_1 and I_2, (b) the emf $\mathscr{E}$ of the battery, and (c) the voltage drop across the 4-Ω resistor.

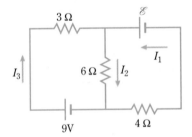

FIGURE P18.6

■54 In Fig. P18.6 if the emf $\mathscr{E}$ is 8 V find (a) the currents I_1, I_2, and I_3, and (b) the voltage drop across the 3-Ω resistor.

■■55 In Fig. P18.7 find (a) the currents in each part of the circuit and (b) the voltage drop across each resistor.

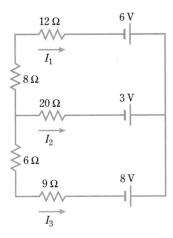

FIGURE P18.7

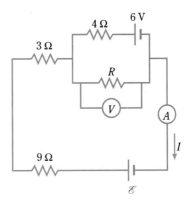

FIGURE P18.9

◾56 A Wheatstone bridge, as shown in Fig. P18.8, is sometimes used to measure resistance. When the circuit is balanced, there is no current through the meter G, and the voltage difference between the points B and D is zero. Show that for the balanced circuit (no current through G) the resistances should satisfy the relation $R_1/R_2 = R_3/R_4$.

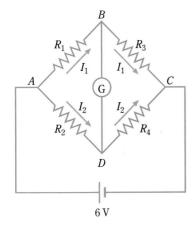

6 V   FIGURE P18.8

◾57 In Fig. P18.8 what is the value of the resistance R_4 if the bridge is balanced for $R_1 = 60\,\Omega$, $R_2 = 20\,\Omega$, and $R_3 = 19.6\,\Omega$? (See Prob. 56.)
◾58 The voltmeter in Fig. P18.9 reads 3.6 V, and the ammeter reads 2.2 A with the current direction as indicated. Find (*a*) R and (*b*) $\mathscr{E}$.
◾59 In Fig. P18.9, how large must $\mathscr{E}$ be if the current through the 6-V battery is zero when R is $14\,\Omega$?
◾60 In Fig. P18.9, if $\mathscr{E}$ were 28 V and R were $8\,\Omega$, what would (*a*) the ammeter and (*b*) the voltmeter read?

◾61 A voltmeter with an internal resistance $r = 4.0 \times 10^4\,\Omega$ is used to measure the voltage across the resistor $R_1 = 12\,\text{k}\Omega$ as in Fig. P18.10. Take $R_2 = 24\,\text{k}\Omega$. (*a*) What is the voltage drop across R_1 when the switch S is open? (*b*) What is the equivalent resistance of the circuit when the switch S is closed? (*c*) What is the voltage across R_1 with the switch S closed?

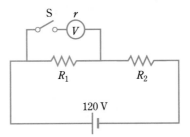

FIGURE P18.10

Section 18.11
62 A 192-Ω lamp, 16-Ω toaster, and a 60-Ω fan are connected in parallel in a 120-V household circuit. Find (*a*) the total current drawn in the circuit, (*b*) the voltage drop across the toaster, (*c*) the current in the fan, and (*d*) the power dissipated by the toaster.
63 A particular 120-V circuit has a 1200-W toaster, a 60-W lamp, and a 600-W soldering iron operating on it simultaneously. The fuse in the circuit blows when an additional 40-W bulb is turned on. What is the maximum rating of the fuse?
64 A household operates a 1500-W dryer, a 540-W washer, five 40-W lamps, and a 25-W TV set from the same 120-V line. For what minimum current must this line be fused?

65 How many 75-W light bulbs can you use in a 120-V household circuit without blowing a fuse rated at 15 A?

*66 An electric device is designed to consume 2000 W of power when operating at 240 V. (*a*) Assuming the resistance of the device remains constant, how much current will it draw when connected to a 120-V source? (*b*) How much power will it consume at this voltage?

*67 A household circuit operating at 120 V has a 30-A circuit breaker. A 1500-W iron, a 2000-W electric grill, and a lamp are all operated simultaneously. What is the maximum possible wattage of the bulb which can be used in the lamp without tripping the circuit breaker?

Section 18.13

68 When a current of 3.2 A is drawn from a certain battery, its terminal voltage drops from its zero current value of 1.57 V to 1.28 V. What is the internal resistance of the battery?

69 What maximum current can be drawn from a battery that has an emf of 1.57 V and an internal resistance of 0.16 Ω?

*70 A 7.0-Ω resistor draws a current of 0.2 A when connected to a battery. When the same battery is connected to a 4.5-Ω resistor, the current in the circuit is 0.3 A. Find (*a*) the emf and (*b*) the internal resistance of the battery.

*71 A flashlight operates with three AA batteries in series, each with an emf of 1.5 V. When the flashlight is turned on, it draws a current of 0.5 A and the terminal voltage of the three batteries drops to 3.3 V. What is the internal resistance of each battery?

*72 The terminal voltage of a certain battery is 11.52 V when connected across a 24-Ω resistor and 11.76 V when connected across a 50-Ω resistor. Find the emf and the internal resistance of the battery.

*73 A 58-Ω resistor draws a current of 150 mA when connected across a 9.0-V battery. (*a*) What is the internal resistance of the battery? (*b*) What would be the terminal voltage of the battery when connected to the resistor?

Additional Problems

*74 The current through a resistor increases by 2 A when the voltage across the resistor increases from 8 V to 12 V. What is the resistance of the resistor?

*75 You are given three resistors 3 Ω, 5 Ω, and 8 Ω. (*a*) How many different resistance values can you obtain using these resistors? (*b*) What are these values and how are the resistors connected in each case?

**76 The measured resistance of a certain metal wire with an initial length L_o and diameter d_o is found to be 4 Ω. The wire is drawn under a tensile stress to a new uniform diameter 0.4d_o. Find the value of the new resistance of the wire. *Hint:* Note that the total volume and mass of the metal in the wire do not change under tensile stress.

**77 A 40-Ω resistor that is temperature-independent is to be made by using a graphite resistor in series with a tungsten resistor. What should the resistance of each be at 20°C?

**78 A thin metallic wire of radius $r = 3.8 \times 10^{-2}$ mm has a uniform electric field of 84 V/m established along its length when connected to a battery and draws a current of 3.6 A. Find the resistivity of the material of the wire.

**79 A graphite resistor is placed in series with a 9-Ω (at 20°C) iron resistor. How large should the resistance of the graphite resistor be in order for the combination to be temperature-independent? What is the resistance of the combination?

**80 In Fig. P18.11 determine all the currents I_1, I_2, I_3, I_4, I_5, and I_6.

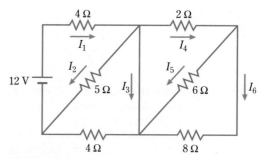

FIGURE P18.11

**81 In Fig. P18.12 calculate (*a*) the potential difference between points *A* and *B*, (*b*) the potential difference between points *A* and *C*, and (*c*) the power delivered to the 16-Ω resistor.

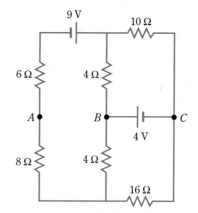

FIGURE P18.12

**82 In Fig. P18.13 the terminal potential of the battery is measured to be 5.8 V when the switch S is open and to be 5.76 V when the switch is closed. Find the emf $\mathscr{E}$ and the internal resistance r of the battery.

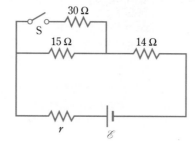

FIGURE P18.13

CHAPTER 19

MAGNETISM

As children in grade school, we performed simple experiments that involved magnetism. We learned that a bar magnet has two poles, a north pole and a south pole. Further, we found that unlike poles attract each other and like poles repel each other. We also learned that the earth acts like a huge magnet and that a magnetic compass needle always aligns itself along the earth's magnetic field.

By sprinkling iron filings on a glass plate placed above a magnet, we discovered that the filings formed a picture of the magnetic field that surrounds the magnet. Most of these facts were known thousands of years ago. Not until 1820, however, did scientists learn that magnetism is closely related to electric currents and fields. Even today, scientists are still making discoveries concerning magnetism and the materials from which magnets are made. As we shall see in the following chapters, magnets and their effects are only a small facet of magnetism.

19.1 MAGNETIC FIELD MAPPING

Much of the terminology of magnetism was developed centuries ago by those who first investigated the behavior of magnets. The first magnets were simply pieces of iron-bearing rock called lodestone. We now know that iron is one of a few materials

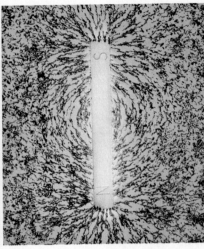

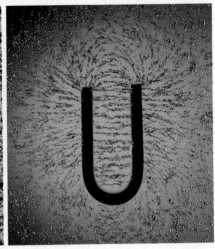

A piece of magnetite, called a lodestone, possesses a permanent magnetic field which attracts a compass needle.

Iron filings are oriented by the magnetic fields of a bar and horseshoe magnet, thereby displaying the patterns of the fields.

that have the property of being able to be permanently magnetized. These materials, which include nickel and cobalt, are called ferromagnetic (Latin *ferrum,* "iron").

It has long been known that elongated pieces of lodestone could be suspended by a thread and used as a crude compass to determine a direction which roughly corresponded to geographic north. Just as with today's magnetic compass needles, the lodestone would orient its length with the earth's magnetic field. The ends of the lodestone magnet were called magnetic *poles*; that pole which pointed approximately toward the geographic North Pole of the earth was called the north pole of the magnet. The opposite end was called the magnet's south pole. To this day we retain this terminology when we refer to the properties of bar magnets and compass needles. (See Figure 19.1.)

Further studies with magnets reveal that like poles (two north or two south poles) repel each other, whereas opposite poles attract. This behavior is reminiscent of the behavior of the two kinds of electric charge, and has led scientists to try to find magnetic "charges," or monopoles. However, if one tries to separate the poles of a magnet by breaking the magnet in two, the effort proves unsuccessful. The broken magnet becomes two new bar magnets, each with a south and a north pole.

Interesting things happen in the vicinity of magnets. Unmagnetized pieces of iron, such as nails or iron filings, are attracted to either pole of the magnet. Compass needles deflect when a magnet is brought near. A wire through which an electric current is passing is attracted or repelled by magnets, and streams of charged particles can be deflected by magnets. It is convenient to explain all these phenomena in terms of what we call the *magnetic field* of the magnet.

As always, we start by defining the field, however arbitrarily, in terms of some measurable property. In this case, we define the *direction* of the magnetic field at any given point to be the direction a compass needle points when placed at that point. For example, suppose we wish to sketch the direction of the magnetic field in the vicinity of the bar magnet shown in Fig. 19.2. We can do that by placing many tiny compass needles at various points around the magnet and observing their orientation. We assume the effects of the needles on each other are negligible compared to the effect of the magnet on each one.

Because the arrowhead end of a compass needle is a north pole, it is repelled by the north pole of the magnet; hence a compass needle near the north pole of the magnet points away from the magnet. Similarly, a needle near the south pole points

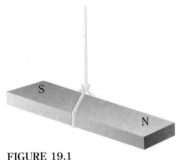

FIGURE 19.1

The north pole of a magnet is defined to be the pole that points northward on the earth when the magnet is freely suspended.

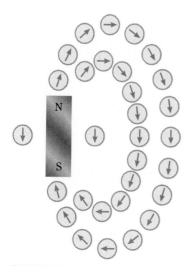

FIGURE 19.2

The direction of the magnetic field in the vicinity of a magnet can be mapped by using many tiny compass needles.

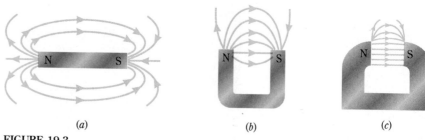

(a)   (b)   (c)

FIGURE 19.3

By definition, the magnetic field points away from the north pole of a magnet and toward the south pole.

toward the magnet because unlike poles attract. To sketch a magnetic field, we draw a series of lines about the magnet in such a way that the arrows on the lines show the direction in which a compass needle would point. These lines, called *magnetic field lines,* are shown for three shapes of magnet in Fig. 19.3. Like the compass needles that define them,

Magnetic field lines point out from the north pole of a magnet and enter the south pole.

Sketches such as those in Fig. 19.3 show not only the direction of the field but its strength as well. As with electric fields, the magnetic field lines are closest together where the field is strongest.

19.2 THE EARTH'S MAGNETIC FIELD

Figure 19.4 shows a sketch of the earth's magnetic field. The field pattern is very similar to that of a bar magnet. Notice that the magnetic poles do not coincide with the geographic poles, which are defined by the axis of earth's rotation.

Let us pause a moment to eliminate a potential source of confusion. Our custom of saying that the north pole of a compass needle points toward (and is therefore attracted to) the north magnetic pole of the earth appears to be in conflict with the

FIGURE 19.4

(a) The earth's magnetic field. (b) The dip angle is the angle between the magnetic field **B** and the horizontal.

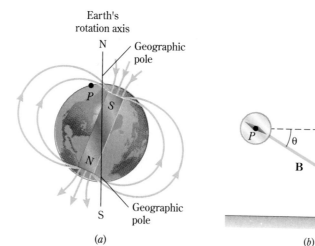

(a)   (b)

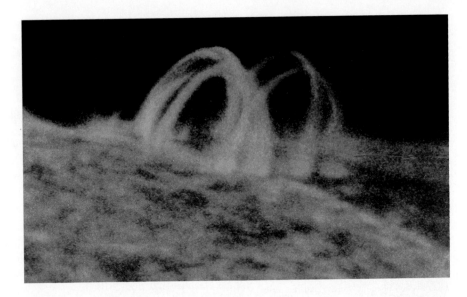

Magnetic fields trap charged particles, such as those in hot gases in the sun's atmosphere. As these hot gases emit light, they reveal the structure of the sun's magnetic field as shown in these loop prominences. The prominences bridge between sunspots, which are regions of intense magnetic fields of opposing magnetic polarity.

observation that like poles repel. The confusion arises because we refer to that magnetic pole which is near the geographic North Pole as the north magnetic pole of the earth's field. If we remain consistent with our definition of the north pole of the compass as the pole that points northward, we should call this the earth's south magnetic pole. However, changing the historic designation would probably cause more confusion than recognizing this misnomer and continuing to live with it.

The location of the magnetic poles varies over long periods of time. At present, the north magnetic pole is located approximately 1600 km south of the geographic North Pole along longitude 100° west. If you are at a longitude other than 100° west, your compass reading has to be corrected for east or west deviation in order to indicate true north. The amount of correction is noted on maps intended for navigation.

As can be seen in Fig. 19.4, the earth's magnetic field is nearly parallel to the earth's surface in equatorial regions and nearly perpendicular to the surface near the poles. In general, at a point P in the northern hemisphere a compass needle suspended on a horizontal axis will point at an angle θ below the horizontal. This is called the angle of dip of the magnetic field.

19.3   THE MAGNETIC FIELD CREATED BY AN ELECTRIC CURRENT

Magnets are not the only source of magnetic fields. In 1820, Hans Christian Oersted discovered that an electric current in a wire causes a nearby compass needle to deflect. This indicates *that an electric current in a wire is capable of generating a magnetic field.* Oersted's experiment was the first demonstration that electric and magnetic phenomena were intimately connected. We now know from many other types of experiments that electric currents do indeed produce magnetic fields. Further, the magnetic field of a magnet is also the result of the motion of charges, as we shall show later.

Oersted investigated the magnetic field surrounding a long, straight wire carrying current in the direction indicated in Fig. 19.5. When a compass is placed near the wire, the needle lies with its length tangent to a circle concentric with the wire,

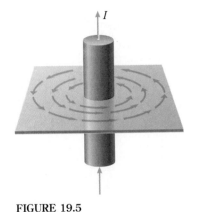

FIGURE 19.5

The magnetic field forms circles concentric to the current-carrying wire.

Physicists at Work DANIEL N. BAKER *Colorado University Laboratory for Atmospheric & Space Physics*

I work in an area of science called space physics. This type of research is dedicated to the study of the charged particles (electrons, protons, heavier nuclei) that populate our solar system and the magnetic and electric fields that control their motions. My particular area of concentration is the uppermost part of the earth's atmosphere, the *magnetosphere*. This region is populated by a very tenuous gas (10 to 1000 particles per cubic centimeter) made up mostly of electrons, protons, and atomic nuclei (such as charged oxygen that has moved up from the lower atmosphere), all held together by the magnetic field emanating from earth's molten iron-nickel core. The magnetosphere was discovered 35 years ago by the first artificial earth satellites and has continued to be studied by more and more sophisticated spacecraft instruments ever since.

I first became interested in space research as a 9-year old boy in 1957 when I read about the Russian Sputnik mission and about James Van Allen's discovery of the earth's radiation belts. I decided then that I would like to be a space physicist and perhaps even work with Prof. Van Allen some day. I was very fortunate to be able to study with Prof. Van Allen as I began graduate school in 1970, and I participated with him in the design and testing of instruments that later flew on the first missions to the outer solar system. These Pioneer 10 and 11 spacecraft proved that Jupiter and Saturn also have magnetospheres. We now believe that virtually all the planets have a magnetosphere-like region, and, in fact, we know that our sun, neutron stars, and even galaxies have regions around them that can properly be called magnetospheres.

One of the great benefits of studying earth's magnetosphere is that it is relatively close to us. To send spacecraft to other planets (as with the Pioneer or Voyager missions) can take years or even decades because the planets are so very far away. Just imagine trying to go to other stars: Even traveling at the speed of light, it would take tens or hundreds of years to get to nearby stellar systems. By studying the processes going on in our own magnetosphere, we can develop models of particle acceleration, energy conversion, and complex charged-particle motion. Most important, we can then send instruments into the magnetosphere and make measurements that test our theoretical ideas. The charged-particle gas and magnetic field in our magnetosphere (called a *plasma*) is characteristic of 99 percent of the universe. Our results can therefore be applied to other cosmic systems: In this sense we

say that earth's magnetosphere is a giant cosmic laboratory.

Since the beginning of the space age, we humans have begun to use our space environment more and more. We now have satellites in space that permit worldwide television broadcasts and nearly instantaneous intercontinental communication. We also use space for surveillance to help defend ourselves, and very sophisticated spacecraft warn us about hurricanes and other major, potentially disastrous, weather phenomena. Even the long-term changes in earth's atmosphere, oceans, and plant life are regularly monitored from space. What we find is that all of the spacecraft that serve these sophisticated functions are very susceptible to the hostile radiation of space. The Van Allen belt particles and intense bursts of radiation from solar flares can completely destroy satellite electronic components. Thus, a very practical aspect of my work is to understand and to predict the effects of the space environment on operational satellites.

I have been very lucky to be able to follow a dream that began with the dawn of the space age. I have had the chance to study Jupiter, Saturn, Mercury, and the Sun, as well as our own Earth. By comparing and contrasting our neighbors in space, we have come to a good understanding of our little corner of the universe. We now look outward farther and farther with ever more powerful telescopes, but we always come back to our own experience in earth's environment to understand what we see. Thus, perhaps the most interesting thing of all is that whatever window we open on the universe we will always look through that window from our place here on earth.

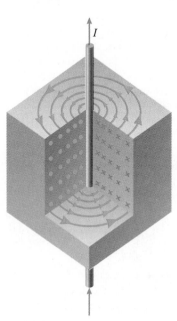

FIGURE 19.6

The magnetic field circles about a long, straight wire. The magnitude of the field decreases inversely with distance from the wire.

FIGURE 19.7

When you grasp a current-carrying wire in your right hand with the thumb pointing in the direction of the current, your fingers circle the wire in the same direction as the magnetic field.

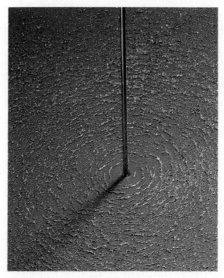

Iron filings are aligned by the magnetic field produced by a current in the straight wire.

Intense magnetic fields can be produced by large currents, as in this industrial electromagnet used to pick up scrap iron.

and the inference is that a magnetic field exists in a circular form about the wire. As is to be expected, the strength of the field is greatest close to the wire. A three-dimensional representation of the magnetic field is shown in Fig. 19.6. (In this and later diagrams, the symbol · indicates an arrow coming toward the reader and the × represents an arrow going away from the reader. The symbols are meant to suggest the point and tail of the arrow that shows the direction of the magnetic field.)

There is a simple **right-hand rule** for remembering the direction of the magnetic field about a wire. If you grasp the wire in your right hand with your thumb pointing in the direction of the current, your fingers will circle the wire in the direction of the field (Fig. 19.7).

19.4   THE FORCE ON A CURRENT IN AN EXTERNAL MAGNETIC FIELD; THE RIGHT-HAND RULE

Thus far we have discussed only the qualitative features of the magnetic field and how to determine its direction. To complete the description, we must seek some method of defining and measuring its magnitude. The key to this lies in the observation that when a wire carrying a current is placed in a region where an external magnetic field exists,* the wire experiences a force.

A wire carrying a current through a region in which an external magnetic field exists experiences a magnetic force due to that field.

As an example of this phenomenon, consider the situation shown in Fig. 19.8a. A

*An external magnetic field is one that is created by currents or magnets external to the current-carrying wire. This external field does not include the field created by the current in the wire itself.

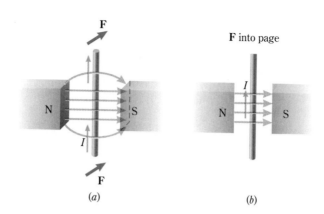

F into page

(a) (b)

FIGURE 19.8

The external magnetic field (orange lines) furnished by the poles of the bar magnets causes the current-carrying wire to experience a force. (a) Three-dimensional perspective. (b) Side view, showing that **B**, **I**, and **F** are mutually perpendicular.

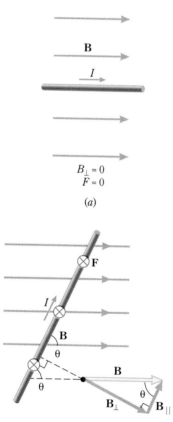

$B_\perp = 0$
$F = 0$

(a)

$F = B_\perp IL = (B \sin \theta)IL$

(b)

FIGURE 19.9

When a current-carrying wire is immersed in an external magnetic field, the force on the wire is proportional to the component of **B** that is perpendicular to the wire. In (b), verify the direction of **F**.

wire carrying a current I vertically upward is placed in the external magnetic field produced between the poles of a magnet. Experiment shows that the wire experiences a force that is perpendicular to both the external magnetic field and the current. If the direction of the current is reversed, the direction of the force on the wire also reverses; it is then directed out of the page. We can see this more clearly if we redraw the situation in two dimensions, as in Fig. 19.8b. Notice there that the line of the wire and a magnetic field line intersecting it determine a plane, the plane of the page. The force the wire experiences is always perpendicular to this plane; in this case, the force is into the page. We will deal more with the direction of this force in the next section. For now, let us concentrate on the question of its magnitude and the definition of the magnitude of the magnetic field.

For the sake of simplicity, we assume the strength of the external magnetic field is uniform over the length of the wire, L. If the current and the magnetic field are perpendicular to each other, as in Fig. 19.8, it is found that the force on the wire is proportional to the current and to the length of wire in the magnetic field. We use the letter **B** to symbolize the magnetic field and define the **magnitude** (or strength) of the field to be

$$B = \frac{F}{IL} \qquad (\textbf{B perpendicular to I})$$

As this equation shows, the units of B are force per meter per ampere, which is called the tesla (T) in the SI:

$$1\,\text{T} = 1\,\text{N/m} \cdot \text{A}$$

A non-SI unit of magnetic field sometimes encountered is the gauss (G), where $1\,\text{G} = 10^{-4}\,\text{T}$. For purposes of comparison, the earth's magnetic field is typically of the order of $5 \times 10^{-5}\,\text{T}$, whereas B near the end of a strong bar magnet might be 0.1 T.

The direction of **B** has already been established as the direction in which a compass needle points. Thus our description of the magnetic field vector **B** is now complete.

In Fig. 19.8, the field lines (and therefore **B**) are perpendicular to the current direction (that is, to the wire). Let us now see what happens when the two are not perpendicular. Suppose the field lines and the wire are parallel to each other, as in Fig. 19.9a. Then the wire experiences no force. *A current parallel (or antiparallel) to* *an external magnetic field line experiences no force due to the field.* Clearly, the relative orientation of the field lines and the current direction is very influential.

If we denote the angle between **I** and **B** as θ, the general expression for the force exerted on the wire by the field is

$$F = BIL \sin \theta$$

As shown in Fig. 19.9b, this relationship is equivalent to

$$F = B_\perp IL \qquad (19.1)$$

You should notice that this expression agrees with the two limiting cases; that is, $\theta = 0$ $(F = 0)$ and $\theta = 90°$ $(F = BIL)$.

Illustration 19.1

In Fig. 19.9b, suppose that $B = 2.0\,\text{G}$, $\theta = 53°$, and $I = 20\,\text{A}$. Find the magnetic force on a 30-cm length of the wire.

Reasoning We know that $B_\perp = B \sin \theta = B(0.799)$. Converting B to SI units, we have $B = 2.0\,\text{G} = 2.0 \times 10^{-4}\,\text{T}$. Then

$$F = B_\perp IL = (2.0 \times 10^{-4}\,\text{T})(0.799)(20\,\text{A})(0.30\,\text{m}) = 9.58 \times 10^{-4}\,\text{N}$$

Exercise Find F if the wire is perpendicular to the field lines. *Answer: $12.0 \times 10^{-4}\,N$*

19.5   AN EXTENSION OF THE RIGHT-HAND RULE

In the preceding section, it was pointed out that the direction of the force experienced by a wire carrying a current in an external magnetic field is perpendicular to the plane defined by the wire and the field. We now consider a simple intuitive extension of the right-hand rule (Sec. 19.4) that helps us state the direction of the force experienced by the wire. It is purely an intuitive aid for remembering the direction of the force. No real physical significance should be attached to the rule; it is simply a memory device.

Point the fingers of your right hand along the lines of the magnetic field and your thumb in the direction of the current. The force on the wire is in the direction in which your palm would push.

This rule is pictured in Fig. 19.10. Let there be no confusion on this point. The line of the magnetic field vector **B** and the line of the wire together define a plane (the plane of the page in Figs. 19.9 and 19.10). The force on the wire is always perpendicular to this plane. Once you know this, a pure guess allows you a 50 percent chance of obtaining the proper direction for the force. It must be either into or out of the plane. To find which is the proper alternative, use the rule illustrated in Fig. 19.10. The direction of the force in Fig. 19.10 is toward you, out of the page. Using the same rule, you see that the direction of the force in Figs. 19.8 and 19.9 is into the page.

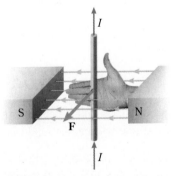

FIGURE 19.10

The right-hand rule: The fingers point in the direction of **B**, the thumb points in the general direction of the current, and the palm pushes in the direction of **F**.

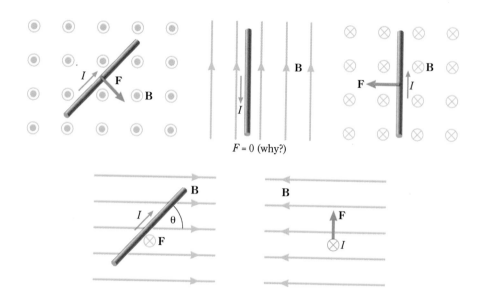

FIGURE 19.11

Verify the direction of the magnetic forces in each case.

Illustration 19.2

Use the right-hand rule to verify the direction of the magnetic forces in Figure 19.11. As mentioned earlier, an $\otimes$ denotes a vector into the page and a $\odot$ denotes a vector out of the page.

19.6 MAGNETIC FORCES ON MOVING CHARGES

Current as we have defined it is the result of the motion of positive charges. An obvious question to ask at this point is what effect an external magnetic field might have on freely moving charges if they were not constrained to move along a wire. To develop an answer to this question, let us begin by using what we have already developed concerning the magnetic force on currents to find the magnetic force on a single charge carrier within a wire.

To do this, we divide the total force on a length L of wire by the number of charge carriers in this length. If the wire has a cross-sectional area A, as in Fig. 19.12, the volume of a length L is AL. If there are n_u charge carriers per unit volume, the number of charge carriers in length L is $n_u AL$. Therefore

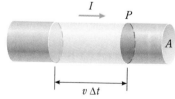

FIGURE 19.12

In a time Δt, all the charge in the length $v \Delta t$ will pass through the cross-sectional area at P.

$$\text{Force per charge} = \frac{\text{force on wire}}{\text{number of carriers}} = \frac{B_\perp IL}{n_u AL} = \frac{B_\perp I}{n_u A}$$

We still need to express the current in terms of the individual charges that cause it. A charge carrier moves a certain distance in the direction of the current in a time Δt. If the average speed of the carrier is v, then the distance moved in time Δt is $v \Delta t$. Hence, in a time Δt, all the charge carriers in a length $v \Delta t$ to the left of point P in Fig. 19.12 move through the cross section at P. Because the volume of this length is $Av \Delta t$ and because there are n_u charge carriers per unit volume, the number of charge carriers that pass P in time Δt is $n_u Av \Delta t$. Each carries a charge q, and so

$$I = \frac{\text{charge past } P \text{ in time } \Delta t}{\Delta t} = \frac{q n_u A v \, \Delta t}{\Delta t} = q n_u A v$$

We can now use this value for I in the expression for the force per unit charge:

$$F = \frac{B_\perp I}{n_u A} = q v B_\perp$$

We therefore conclude the following:

A charge q moving with speed v perpendicular to a magnetic field of magnitude $B_\perp$ experiences a magnetic force of magnitude

$$F = q v B_\perp \tag{19.2}$$

We can use the right-hand rule to determine the direction of this force. The key is to remember that the direction of a current is defined as the direction of the velocity of moving positive charges. Thus if we point the fingers of our right hand in the direction of **B** and our right thumb in the direction of the velocity **v**, our palm will push in the direction of the force on the charge.

As an example of such a situation, refer to Fig. 19.13. We see there a charge q moving with velocity **v** through a magnetic field **B** directed out of the page. The intersecting vectors **B** and **v** define a plane (horizontal), and the force **F** on q is perpendicular to this plane. Using the right-hand rule, we find that **F** has the direction shown in Fig. 19.13. Notice that Eq. 19.2 tells us that the direction of **F** reverses when the particle is charged negatively. Hence if the particle in Fig. 19.13 had been negative, **F** would be directed upward instead of downward.

An important observation concerns the fact that the force is always perpendicular to the velocity. Since the velocity vector is always instantaneously along the direction of motion, it follows that the force has no component along the direction of motion. This means that the force does no work on the charge and hence does not change its kinetic energy. The only effect of the force is to change the direction of the charge's motion.

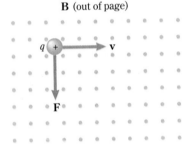

B (out of page)

FIGURE 19.13

Use the right-hand rule to find the direction of **F** on the charge.

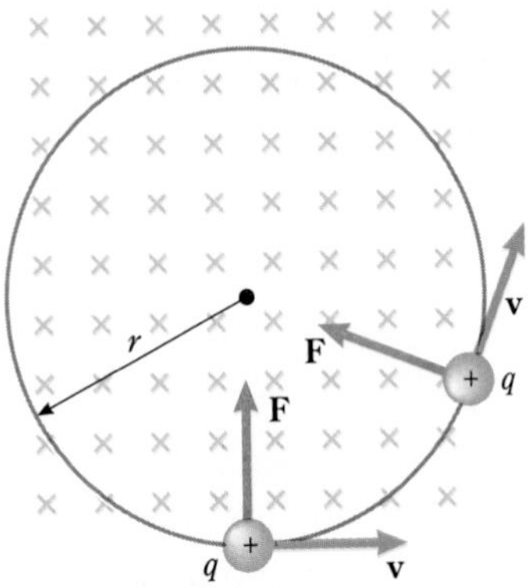

B (into page)

FIGURE 19.14

The charged particle follows a circular path in the uniform magnetic field.

19.7 PARTICLE MOTION IN A MAGNETIC FIELD

Let us now follow the motion of a charged particle in a magnetic field as shown in Fig. 19.14. We have already seen that the speed v will not be changed by the force (only the direction of the velocity changes). If we now assume that the magnetic field is uniform (has the same strength and direction everywhere), the magnitude of the magnetic force $F = q v B$ remains constant. You should verify that the force direction shown in Fig. 19.14 is correct.

We have encountered a dynamic situation like this a number of times before. Two cases where an object is subjected to a constant force continuously perpendicular to the direction of motion are (1) a ball being swung in a circle at the end of a string and (2) circular gravitational orbits. In each case the force causes the object to move in a circular path at constant speed. This motion is described by a centripetal acceleration v^2/r (Eq. 7.9), where r is the radius of the circular motion. In the

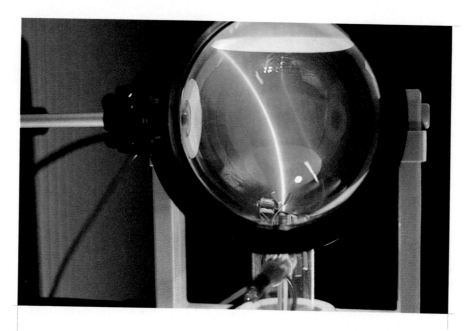

A beam of electrons is bent in a circle when the beam travels through a region in which there is an external magnetic field. Can you determine the direction of the magnetic field in this picture?

present case, the force responsible for this acceleration is qvB, the magnetic force on the charge q. Newton's second law allows us to write

$$qvB = \frac{mv^2}{r}$$

where m is the mass of the charged particle. Thus a charge q having mass m and moving in a uniform magnetic field **B** directed perpendicular to the charge velocity **v** will travel in a circle of radius

$$r = \frac{mv}{qB} \tag{19.3}$$

If the charge in Fig. 19.14 were negative, the direction of the force would be reversed, and the negative charge would move in a clockwise circle.

A very important difference to remember between electric and magnetic forces on charges is the following:

The electric force qE is in the direction of **E** (or opposite to **E** for negative charges). The magnetic force qvB is perpendicular to **B**. Thus E fields can do work on charges, whereas B fields cannot.

19.8 APPLICATIONS OF MAGNETIC FORCES ON CHARGES

The properties of the particles making up atoms and molecules can be studied by observing their behavior in the presence of E fields and B fields. These extremely small bits of matter carry charges which are usually from one to a few times the electronic charge e. Let us briefly discuss three such applications.

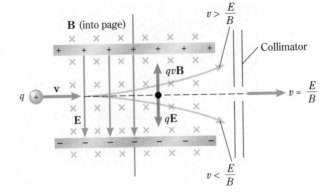

FIGURE 19.15

The velocity selector passes undeflected those particles for which the electric force qE equals the magnetic force qvB.

THE VELOCITY SELECTOR

Figure 19.15 shows a pair of charged parallel plates immersed in a uniform magnetic field directed into the page. As we have seen many times before, the charged plates create a uniform electric field between them, directed from the positive to the negative plate. Because of the directions of the magnetic and electric fields, this device is sometimes referred to as a *crossed-field selector*. We assume the apparatus is in a vacuum chamber, so that air resistance is negligible.

Suppose a charged particle $+q$ enters the region of crossed fields with a velocity **v** parallel to the plates, as shown in Fig. 19.15. You should be able to show that the electric and magnetic forces on q are oppositely directed, as shown. Thus in general, the particle will be deflected upward or downward as it traverses the region, depending on which force is greater.

Only if the opposing forces are equal will the charge pass through the region undeflected. This condition requires that

$$qE = qvB \qquad \text{or} \qquad v = \frac{E}{B}$$

Particles with precisely this speed will pass through small apertures aligned with the central axis of the device. Particles moving at any other speeds will be blocked. Thus the device allows us, by adjusting E and B, to select particles all having the same speed out of a beam of particles having different velocities. You should be able to convince yourself that the same result applies to negative charges. You should also take the time to show that the SI units obtained from E/B are indeed meters per second.

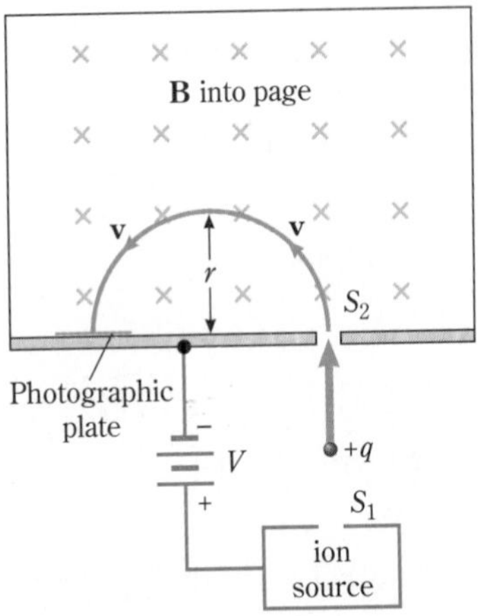

MASS SPECTROMETER

As discussed in Chap. 2, macroscopic masses are defined in relation to the international standard kilogram. The most precise mass measurements, however, are those of the atoms of the various elements. A device known as a *mass spectrometer* utilizes the magnetic force on charged atoms (ions) to measure masses to a precision of 7 or 8 significant digits. Figure 19.16 shows a sketch of this device. Enclosed in a vacuum chamber is an ion source, a region where a uniform magnetic field exists, and a voltage between the source and the region of magnetic field. Atoms of a gas are ionized by electron bombardment and then emerge from the ion source slit S_1.

FIGURE 19.16

A mass spectrometer. The mass of an ion can be determined from the position at which it strikes the photographic plate.

They are then accelerated toward the entrance slit S_2 by a known voltage V. From Eq. 17.3, the ions thus enter the magnetic field with kinetic energy

$$\text{KE} = \tfrac{1}{2}mv^2 = qV \tag{19.4}$$

Depending on the degree of ionization, q can be $+e$, $+2e$, etc. Very often, singly ionized atoms are used.

Once they enter the B-field region, the ions travel at constant speed, and are turned by the magnetic force in a circle the radius of which is given by Eq. 19.3: $r = mv/qB$. The ions travel through a semicircle, striking a detector such as a photographic plate at a distance $2r$ from S_2. Solving Eq. 19.4 for v and then substituting into Eq. 19.3, we can obtain an expression for the mass of the ions. First, we get $v^2 = 2qV/m$. Then

$$r^2 = \frac{m^2 v^2}{q^2 B^2} = \frac{m^2 \left(\dfrac{2qV}{m} \right)}{q^2 B^2}$$

This yields the expression

$$m = \frac{qB^2 r^2}{2V} \tag{19.5}$$

Since q, V, and B are known, careful measurement of the distance $2r$ will allow the determination of the mass of the ions. A particularly important use of the mass spectrometer has been to measure the difference in the masses of various isotopes of the same element.

Example 19.1

In the mass spectrometer shown in Fig. 19.16, singly ionized atoms of some element are accelerated through a potential difference of 1.000 kV and enter a 1.950-T magnetic field. The ions are observed to hit a screen at a distance of 2.088 cm from S_2. What is the mass of these ions, and what isotope do they represent? Use the information on isotopic masses in Appendix 2.

R *Reasoning*

Question How does the given information translate into the quantities involved in the equation for mass, Eq. 19.5?
Answer You have $V = 1.000$ kV and $B = 1.950$ T. The distance from S_2 is twice the radius r, and so $r = 1.044$ cm. Singly ionized atoms have a charge $q = e = 1.602 \times 10^{-19}$ C.

Question How do I obtain an isotopic mass from this?
Answer Isotopic masses are listed in Appendix 2 in terms of the atomic mass unit (u), which is defined to be one-twelfth of the mass of the carbon 12 isotope:

$$1\,\text{u} = 1.6606 \times 10^{-27}\,\text{kg}$$

Equation 19.5 will give you a mass in kilograms, which you'll have to convert.

Solution and Discussion The mass of a single ion is

$$m = \frac{(1.602 \times 10^{-19}\,\text{C})(1.950\,\text{T})^2(1.044 \times 10^{-2}\,\text{m})^2}{2(1.000 \times 10^3\,\text{V})} = 3.320 \times 10^{-26}\,\text{kg}$$

In atomic mass units this is

$$m = \frac{3.320 \times 10^{-26}\,\text{kg}}{1.6606 \times 10^{-27}\,\text{kg/u}} = 19.99\,\text{u}$$

Appendix 2 shows that the mass of ^{20}Ne is 19.992440 u.

Exercise Calculate how far apart ions of ^{22}Ne and ^{20}Ne would be as they struck the detector. *Answer: For ^{22}Ne, r = 1.095 cm. Therefore the separation distance is 2(1.095 − 1.044) cm = 0.102 cm. Mass spectrometers have sufficient precision to measure this separation easily.*

CYCLOTRON

Much of our knowledge about the structure of the atomic nucleus has come from bombarding them with ions, electrons, and protons of very large energies. When we "split" a nucleus by such a bombardment, some details of its inner structure are revealed to us. One of the early devices for attaining very high particle energies and using magnetic fields to control their paths is the **cyclotron,** developed by E. O. Lawrence at the University of California at Berkeley in 1930. The cyclotron was of such importance as a research tool that Lawrence received the 1939 Nobel Prize in Physics.

The basic elements of a cyclotron are shown in Fig. 19.17. As in the case of the mass spectrometer, a magnetic field is produced perpendicular to the region in which the charged particles move. The particles move in circular paths in a vacuum chamber in two D-shaped electrodes called *dees* that are separated by a small gap. Typically, protons are released from a source S near the center of the gap between the dees. As in the mass spectrometer, a voltage between the electrodes accelerates the protons toward one dee. Once inside the dee, the proton coasts in a circle, emerging from the dee just as the voltage is reversed. The proton is thus accelerated again, entering the opposite dee at higher speed and moving in a larger circle. This continues again and again, accelerating the protons to higher and higher speeds. They progressively move in ever larger circles within the dees, finally being diverted from the perimeter of the cyclotron as a high-energy beam aimed toward some desired target.

The key to this device lies in the fact that the time it takes a charged particle to move once around its circular path does not depend on either particle speed or path radius. This is easy to show. The period of revolution T is just

$$T = \frac{\text{distance}}{\text{speed}} = \frac{2\pi r}{v} = \frac{2\pi mv}{qBv} = \frac{2\pi m}{qB}$$

The frequency of revolution, f, is just $1/T$:

$$f = \frac{1}{T} = \frac{qB}{2\pi m}$$

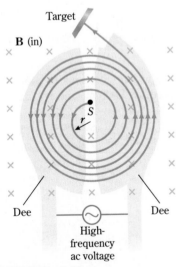

FIGURE 19.17

Schematic diagram of a cyclotron. Protons are accelerated by electric fields between the dees and kept moving in a circle by the strong magnetic field. The protons spiral outward, moving more rapidly as the radius of their orbit increases, and finally crash into a target outside the cyclotron.

E. O. Lawrence constructed this
original cyclotron in 1932.

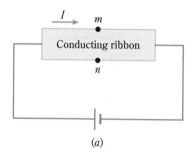

(a)

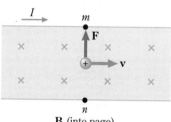

B (into page)

(b)

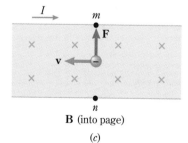

B (into page)

(c)

FIGURE 19.18

The Hall effect. Can you show that the
voltage between m and n reverses
signs if the charge carriers are
negative instead of positive?
(Remember that the right-hand rule
gives the magnetic force exerted on a
positive charge. The force on a
negative charge is in the opposite
direction.)

If the voltage on the dees is reversed at a frequency equal to half of this value, it
will be synchronized with the proton's arrival at the gap, regardless of how fast the
proton is moving or how large a radius its path has. Thus the proton is accelerated
many times before it leaves the cyclotron, attaining very large energies.

19.9 THE HALL EFFECT

There are very few electrical phenomena that show clearly the sign of the charge
carriers. Most experiments can be explained equally well by positive charges flowing in one direction or by negative charges flowing in the reverse direction. The
experiment we describe in this section is one of the few that distinguish between
positive and negative charge carriers.

Consider the circuit in Fig. 19.18a. A battery is connected to the ends of a thin,
uniform conducting ribbon made perhaps from a metal. The two symmetric points
m and n are at the same potential, and so there is no voltage difference between
them. When a magnetic field is impressed perpendicular to the broad face of the
ribbon, however, as in (b), points m and n differ in potential. Let us see how this
potential difference arises.

Suppose that the charges flowing through the ribbon are positive. One such
charge is shown in Fig. 19.18b. The right-hand rule tells us that the charge is forced
upward, toward m. Hence point m becomes positively charged and a potential
difference appears between m and n. We repeat, point m is positive when the
charge carriers are positive.

Suppose, alternatively, that the current consists of negative charges moving to
the left, as in Fig. 19.18c. The right-hand rule would show a downward force on
positive charges traveling to the left. However, since we now are considering negative charges, we find that these negative charges experience an upward force, toward m. Thus in this case point m becomes *negatively* charged.

Here, then, we have a clear-cut way of determining the sign of the charge carriers
in a material. This effect was discovered by the American physicist Edwin Hall in
1879, and it is called the **Hall effect.** Using it, present-day scientists are able to
tell the sign of the charge carriers in newly developed electronic materials for use in
solid-state electronics. The Hall effect also forms the basis for a commercially
produced device for measuring magnetic fields.

The Hall effect can also be used to determine the average velocity of the charge
carriers in a conductor. This average velocity in response to an applied voltage is
called the *drift velocity* of the charges. To see how this determination is made,
consider what happens as the charge carriers (+ or −) accumulate on side m. The
opposite side, n, is left with a deficiency of those charges, creating an electric field
and hence a voltage between sides m and n. This transverse voltage is called the
Hall voltage, V_H. Its value is determined by the balance achieved between the
magnetic and electric forces on the charge carriers:

$$qE_H = q\left(\frac{V_H}{d}\right) = qvB$$

which gives

$$V_H = vBd$$

Here v is the drift velocity of the charges and d is the width of the ribbon. Once B,
d, and V_H have been measured, v can be determined.

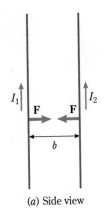

(a) Side view

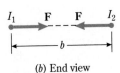

(b) End view

FIGURE 19.19

Two parallel currents attract each other. In (b), the green dots represent current coming toward you out of the page. What would happen if they were antiparallel?

19.10   FORCES BETWEEN PARALLEL CURRENTS; THE AMPERE

Let us briefly review the basic principles about magnetism we have learned so far. We have defined the direction of the magnetic field **B** in terms of the behavior of a compass. We know that a current-carrying wire placed in an external B field experiences a magnetic force. Furthermore, a current is a source of a magnetic field, since a compass is affected when brought near a current.

It logically follows that, when two currents are placed near each other, each should create a magnetic field that causes a force on the other. Experiments by Oersted and the French physicist and mathematician Andre Marie Ampere in the early nineteenth century proved this to be true. As a matter of fact, we use this fundamental interaction between currents to define the ampere.

Suppose two long, straight wires are carrying currents I_1 and I_2 parallel to each other, as shown in Fig. 19.19. The wires are separated by a distance b. It is found that the currents I_1 and I_2 exert on each other an attractive force per unit length that is proportional to both currents and inversely proportional to the distance b:

$$\frac{F}{L} = \frac{kI_1I_2}{b} \tag{19.6}$$

where k is some constant of proportionality.

In Fig. 19.20, we apply the right-hand rule to show that the force exerted on I_2 is directed toward I_1 (that is, the force is *attractive*). The right thumb extends out of the page in the direction of I_1. At the position of I_2 the right fingers extend toward the top of the page and the right palm pushes toward I_1. By Newton's third law, I_2 must exert an equal and opposite force on I_1, which can be similarly demonstrated by drawing the magnetic field due to I_2 at the position of I_1 and applying the right-hand rule there. We thus understand that Eq. 19.6 gives the magnitude of the magnetic force per length on each current.

We use the special case of equal currents ($I_1 = I_2 = I$) to define the unit of current and hence determine the constant of proportionality k:

When two equal parallel currents of one ampere (A) are placed one meter apart, they exert on each other a force equal to 2×10^{-7} N per meter of their length.

If this definition seems arbitrary, that's because it is! As we have discussed in Chap. 1, some of the quantities we measure in physics are basic to all others and their units must be arbitrarily defined. The ampere is one such unit. (Others we have encountered are mass, length, time, and temperature.)

Even though we introduced the coulomb of charge before we began our study of electric current, we now use the above definition of the ampere to define the coulomb:

One coulomb is equal to one ampere-second ($1\,\text{C} = 1\,\text{A} \cdot \text{s}$)

Let us now see what this definition says about the value of the constant of proportionality in Eq. 19.6. Solving Eq. 19.6 for k, we get

$$k = \frac{Fb}{I_1I_2L}$$

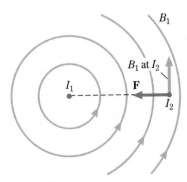

FIGURE 19.20

Current I_1 creates a magnetic field B_1, which exerts a force **F** on I_2. Can you show that the field B_2 produced by I_2 exerts an equal but oppositely directed force on I_1? (This is another example of Newton's third law.)

We see that the SI units of k are

$$\frac{\text{N} \cdot \text{m}}{\text{A}^2 \cdot \text{m}} = \frac{\text{N}}{\text{A}^2}$$

Using these units and the above definition of the ampere, we must have

$$k = \frac{(2 \times 10^{-7}\,\text{N})(1\,\text{m})}{(1\,\text{A})(1\,\text{A})(1\,\text{m})} = 2 \times 10^{-7}\,\text{N/A}^2 \tag{19.7}$$

Although it may seem awkward (and strange), the constant k is usually written in the form $\mu_o/2\pi$, where μ_o is another universal physical constant, called the **permeability of vacuum.** From this definition, μ_o has the value

$$\mu_o = 4\pi \times 10^{-7}\,\text{N/A}^2$$

Using this new symbol, Eq. 19.6 becomes

$$\frac{F(1\text{ on }2)}{L} = \frac{\mu_o I_1 I_2}{2\pi b} \tag{19.8}$$

Finally, what if we reverse the direction of one of the currents, so that they are antiparallel? It should come as no surprise that the forces on both currents also reverse, becoming *repulsive*.

Example 19.2

In Fig. 19.21 a 20-V battery is shown hooked up to an electrical device by means of two parallel wires that are 80 cm long and 2 cm apart. The device has a 4-Ω resistance, and all other wires have negligible resistance compared to this. Calculate the magnetic force the wires exert on each other. Is the force repulsive or attractive?

R

Reasoning

Question What does the magnetic force depend on?
Answer From Eq. 19.6, you know that the force per unit length is proportional to the currents in the wires and inversely proportional to their distance of separation.

Question What determines whether the force is attractive or repulsive?
Answer It is attractive if the currents are in the same direction, repulsive if they are oppositely directed.

Question Which of these is the case in this problem?

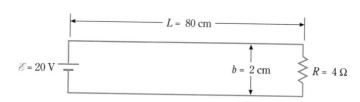

FIGURE 19.21

Calculate the force between the long wires of the circuit.

Answer   Charge is flowing around a closed circuit. Therefore the current in the two wires has the same magnitude and is oppositely directed.

Question   What determines the amount of current?
Answer   Ohm's law: $I = V/R$, where $V = 20$ V and $R = 4\,\Omega$.

Question   What is the precise mathematical expression for the force?
Answer   Eq. 19.8 gives $F = \mu_o I^2 L/2\pi b$. Both L and b are given.

Solution and Discussion   First, note that the forces on the wires are repulsive. The current is

$$I = V/R = 20\ V/4 = 5\ A$$

The force on each wire is then

$$F = \frac{(2 \times 10^{-7}\,\text{N/A}^2)(5\ \text{A})^2(0.80\ \text{m})}{0.02\ \text{m}}$$

$$= 2.00 \times 10^{-4}\,\text{N}$$

The constant μ_o is a very small number, and so the magnetic force between currents is very small unless the currents are extremely large or the distances between them very small.

19.11   MAGNETIC FIELDS PRODUCED BY CURRENTS

So far, we have merely stated without proof that electric currents produce magnetic fields and have examined one instance in which two currents exert a force on each other. We have yet to determine precisely the magnetic fields produced by various current configurations. To begin, let us use the results of the previous section to find the magnetic field created by a long, straight current.

We know from the behavior of a compass needle that the magnetic field created by a straight current forms concentric circles around the current (Figs. 19.5 and 19.6). The field B_1 due to the current I_1 of Fig. 19.19 is thus sketched in Fig. 19.20.

We can use Eq. 19.1 to write the force exerted by B_1 on I_2 in the situation shown in Fig. 19.20:

$$F \text{ (on } I_2) = (B_1)_\perp I_2 L$$

where $(B_1)_\perp$ is the component of the magnetic field perpendicular to I_2 *at the location of* I_2. Notice in Fig. 19.20 that B_1 is perpendicular to I_2 so that $(B_1)_\perp = B_1$.

Now let us use the experimental results of the last section:

$$F \text{ (on } I_2) = \frac{\mu_o I_1 I_2 L}{2\pi b}$$

By comparing these two equations, we can see that we have the expression for the magnetic field strength due to a long, straight current I:

$$B = \frac{\mu_o I}{2\pi r} \quad \text{(long, straight current)} \qquad (19.9)$$

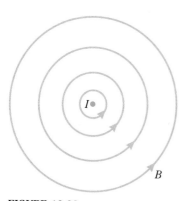

FIGURE 19.22

The magnetic field produced by a long, straight current.

Here we have dropped the subscripts on B_1 and I_1 and have generalized the distance of separation b to any distance r from the current I. The circular magnetic field of a long, straight current is shown in Fig. 19.22. Notice that the field lines become more separated as distance from I increases, indicating that B decreases as r increases (Eq. 19.9).

The calculation of the magnetic field produced by other current arrangements is more complicated than the above and requires the methods of calculus. Ampere devised a mathematical approach to the general problem of relating any currents to the fields they produce. This is known as **Ampere's law.** Since this law goes well beyond the mathematical level of this course, we simply state the results in a few simple and useful cases.

CIRCULAR LOOP OF WIRE

Suppose a circular loop of wire carries a current I, as shown in Fig. 19.23a. Apply the right-hand rule to the wire by grasping it with your right hand with your thumb pointing in the direction of the current. Notice that your fingers circle through the loop in the same sense as do the magnetic field lines in Fig. 19.23a. We show the field lines in more detail in (b). If the radius of the loop is a, then the magnitude of the field at the loop's center is

$$B = \frac{\mu_o I}{2a} \qquad (19.10)$$

You should remember that this expression applies only to the single point at the loop's center. A coil consisting of N loops tightly packed together in a plane produces at its center a field N times larger.

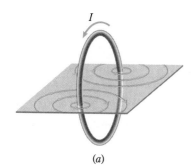

(a)

SOLENOIDS

If we wind a wire in a helix, like a spring, we create a *solenoid.* The one shown in Fig. 19.24 is more loosely wound than most. Usually the solenoid is wound so that the adjacent turns of the coil touch each other. Compare Fig. 19.24 with the field of a single coil in Fig. 19.23b. You can see how the fields of all the turns of the solenoid add together to form the total field. Part (b) shows in cross section a portion of a more closely wound solenoid. As indicated, the magnetic field within it is nearly uniform. Inside a long, hollow solenoid that carries a current I and has n loops of wire per meter of length, the magnetic field magnitude is

$$B = \mu_o n I \qquad (19.11)$$

This relation applies throughout the interior of the hollow solenoid, except near its ends. A solenoid is often used to produce a field that is approximately uniform. Notice that B does not depend on the diameter or the length of the solenoid. Remember that n is the number of turns *per meter.* If N is the *total* number of turns and L is the length of the solenoid, then $n = N/L$ and Eq. 19.11 can be written in an alternate form:

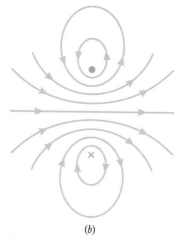

(b)

FIGURE 19.23

Two views of the magnetic field about a current-carrying loop. (a) Perspective view. (b) A cross section as seen from directly above in (a).

$$B = \frac{\mu_o N I}{L}$$

Solenoids wound on iron cores are used as electromagnets in doorbells and many other devices.

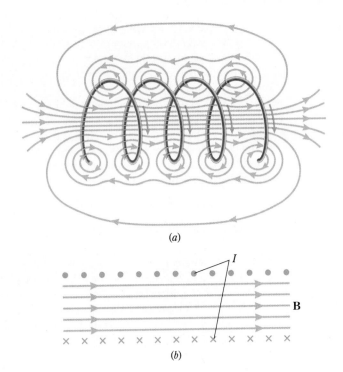

(a)

(b)

When a current flows through the windings of this solenoid, the resulting magnetic field pulls the steel core into the solenoid. Such solenoids are widely used as switching devices.

Illustration 19.3

Compare the magnetic field strengths of the following three situations:

1 At the center of a planar coil of radius $r = 2$ cm having 100 turns. The current is 5 A.
2 At the center of a 100-turn solenoid having radius $r = 2$ cm and length $L = 5$ cm. The current is 5 A.
3 At a distance of 2 cm from a long, straight wire. The current is 500 A.

Reasoning The expressions for the three cases are:

1 $B = \dfrac{N\mu_o I}{2r}$ where $N = 100$

2 $B = \dfrac{\mu_o N I}{L}$ (approximately)

3 $B = \dfrac{\mu_o I}{2\pi r}$

Evaluating the three expressions gives

1 $B = \dfrac{(100)(4\pi \times 10^{-7}\,\text{T} \cdot \text{m/A})(5\,\text{A})}{2(0.02\,\text{m})} = 1.57 \times 10^{-2}\,\text{T}$

2 $B = \dfrac{(100)(4\pi \times 10^{-7}\,\text{T} \cdot \text{m/A})(5\,\text{A})}{0.05\,\text{m}} = 1.26 \times 10^{-2}\,\text{T}$

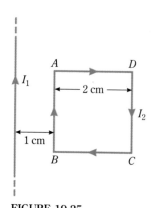

I_1

A D

2 cm

I_2

1 cm

B C

FIGURE 19.25

Find the net force on the square loop.

3 $$B = \frac{(4\pi \times 10^{-7}\,\text{T} \cdot \text{m/A})(500\,\text{A})}{2\pi(0.02)} = 5.0 \times 10^{-3}\,\text{T}$$

The use of multiple-turn coils is a way of essentially multiplying the field-producing effect of a modest current. This illustration also points out the differences that current geometry makes on the value of B.

Example 19.3

In Fig. 19.25 a very long, straight wire is carrying a current $I_1 = 50$ A upward. A square coil 2 cm on a side is placed so that the sides AB and CD are parallel to the wire, with side AB 1 cm away. This coil is carrying a current $I_2 = 30$ A clockwise as shown. Determine the directions of the magnetic forces on each side of the coil and the net force the straight wire exerts on the coil.

R

Reasoning

Question What determines the direction of the force on sides AB and CD?
Answer In side AB the current I_2 is parallel to I_1. The force between parallel currents is attractive. Thus AB is pulled toward the long wire. Because the current in CD is antiparallel to I_1, the force on CD will be away from the wire.

Question How can I determine the direction of the forces on sides AD and CB? In these cases the current I_1 is perpendicular to I_2.
Answer The forces on the square are due to the interaction of I_2 with the B field produced by I_1. With your right thumb pointing in the direction of I_1, your fingers should show that $\mathbf{B}_1$ points into the page in the region of the coil.

Question What results from applying the right-hand rule to sides AD and BC when I use this direction of $\mathbf{B}_1$?
Answer With the right thumb in the direction of I_2 along AD and the fingers in the direction of $\mathbf{B}_1$, you should conclude that the force (the direction in which the palm pushes) points upward. By the same rule, you can show that the force on CB is downward.

Question Will there be a net upward or downward component of force?
Answer The directions of the forces on AD and BC are opposite. If you select a small segment of each of these sides at the same distance from the wire, the forces on those segments cancel. In this way you can show that for every point on side AD experiencing an upward force, there is a corresponding point on BC experiencing an equal downward force. Thus the total forces on these sides cancel.

Question Why is there a net force on the coil?
Answer The magnetic field is stronger closer to the wire. Thus the force on AB will be larger than that on CD, even though the current I_2 through AB and CD is the same.

Question What is the specific expression for the forces on AB and CD?
Answer The magnitude of the force between parallel or antiparallel currents is given by Eq. 19.8:

$$\frac{F}{L} = \frac{\mu_o I_1 I_2}{2\pi d}$$

Here $d = 1$ cm for side AB and $d = 3$ cm for side CD.

Solution and Discussion   The forces per length on AB and CD are

$$\frac{F_{AB}}{L} = \frac{(4\pi \times 10^{-7}\,\text{T} \cdot \text{m/A})(20\,\text{A})(30\,\text{A})}{2\pi(0.01\,\text{m})} = 1.20 \times 10^{-2}\,\text{N/m} \qquad \text{(to the left)}$$

$$\frac{F_{CD}}{L} = \frac{(4\pi \times 10^{-7}\,\text{T} \cdot \text{m/A})(20\,\text{A})(30\,\text{A})}{2\pi(0.03\,\text{m})} = 4.00 \times 10^{-3}\,\text{N/m} \qquad \text{(to the right)}$$

The net force on the coil is then

$$F_{\text{net}} = (0.0120\,\text{N/m} - 0.0040\,\text{N/m})(0.02\,\text{m}) = 1.60 \times 10^{-4}\,\text{N}$$

toward the wire. Notice also that the forces on all sides of the coil tend to expand the coil.

19.12   THE TORQUE ON A CURRENT LOOP

Many practical devices, including motors and many meters, make use of the torque that a current loop experiences when placed in a magnetic field. To see how such a torque originates, refer to Fig. 19.26a, where we show a current-carrying coil in a magnetic field. The coil is mounted on an axle and can rotate. Using the right-hand rule, we find the forces on the various sides to be those illustrated. Notice that only the two forces $\mathbf{F}_h$ cause a torque about the axis of rotation. Even these two forces cause no net torque when the plane of the coil is perpendicular to the field of the magnet, since then their lever arms relative to the axis of rotation become zero. Maximum torque occurs when the magnetic field lines skim past the surface of the coil, that is, when they lie in the plane of the coil, because then the lever arm for $\mathbf{F}_h$ is maximum.

To obtain a quantitative expression for the torque on the coil, we note that each of the two forces $\mathbf{F}_h$ gives a torque

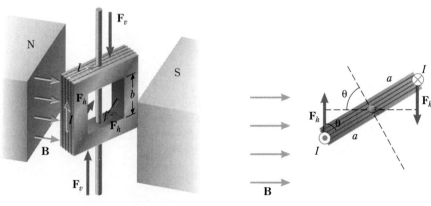

FIGURE 19.26

The external magnetic field causes the current-carrying coil to experience a torque.

(a) Perspective view

(b) Top view

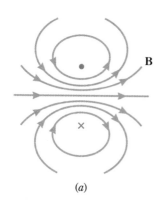

(a)

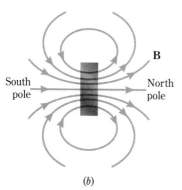

(b)

FIGURE 19.27

Notice how the current loop acts like a short bar magnet. The current direction is represented by the · and × in (a).

FIGURE 19.28

The coil in (a) acts like the bar magnet in (b). Notice how the direction of the magnetic moment μ is assigned in (c).

(F_h)(lever arm)

From Fig. 19.26b we see that the lever arm is $a \sin \theta$. Therefore the torque on the coil is

$$\text{Torque} = 2F_h a \sin \theta$$

where *θ is the angle between* **B** *and the perpendicular to the surface area of the coil.* But F_h is simply the force on the vertical side of the coil. If the vertical side has a length b and if the current is I, each vertical wire contributes a force BIb to F_h. There are N loops on the coil, however, and so $F_h = NBIb$ and the torque becomes

$$\text{Torque} = (2ab)(NI)(B \sin \theta)$$

Notice that $2ab$ is simply the area A of the coil. We can therefore write

$$\text{Torque} = (A)(NI)(B \sin \theta) \tag{19.12}$$

Although we have derived Eq. 19.12 for a very specially shaped coil, it turns out to be true for all flat coils. Because NI is the current around the coil, the important features of the coil (aside from its orientation) are its area and the current in it. In view of this, it is customary to define a quantity called the **magnetic moment** of a current loop:

$$\mu = \text{magnetic moment} = AI_{\text{tot}} = A(NI) \tag{19.13}$$

Notice that the units of magnetic moment are $A \cdot m^2$. It is important to avoid confusing the symbol for permittivity μ_o with that used for magnetic moments, μ. Even though we use the same Greek letter, these two symbols represent completely different quantities.

There is a definite advantage to thinking of a current loop as a bar magnet characterized by its magnetic moment, as we now shall see.

If we compare the field pattern produced by either a current loop (Fig. 19.27a) or a solenoid (Fig. 19.28a) with that of a bar magnet, we see that the fields are very similar. Notice that both the loop and the coil act like a short bar magnet. Moreover, when placed in a magnetic field, the coil and loop experience a torque in the same direction as the torque on a bar magnet. For example, if **B** is directed from left

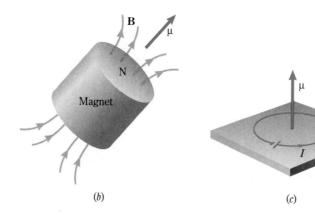

(a)   (b)   (c)

to right as shown in Figs. 19.27 and 19.28, all three devices experience a clockwise torque.

We can obtain maximum usefulness from the concept of magnetic moment if we assign it direction. The direction assigned to μ is shown in Fig. 19.28. Notice that μ is directed along the axis of the magnet, loop, or coil in such a way that it follows the central field line through the coil. A useful way of describing the direction of μ involves another right-hand rule: If you curl the fingers of your right hand in the direction of the current in a coil, your right thumb will point in the direction of μ. As a result, the magnetic-moment vector μ points out the north pole of the equivalent magnet. This has the following important consequence:

A current loop placed in a magnetic field rotates so as to align its magnetic-moment vector with the magnetic-field vector. The torque on the loop is

Torque $= \mu B \sin \theta$

where θ is the angle between μ and **B**.

You can appreciate why this is true by recalling that a compass needle is simply a bar magnet and that the field direction is defined to be the direction along which the needle aligns. We shall find it convenient from time to time to think of a current loop as a magnet with magnetic moment μ.

Example 19.4

A small 10-turn coil of radius 5 cm is inserted into a solenoid so that the plane of the coil is at 45° relative to the axis of the solenoid (Fig. 19.29). The solenoid is wound with 1000 turns per meter and carries a current of 25 A in the direction shown. If a 0.060-A current flows through the small coil in the direction shown, what torque does the small coil experience?

Reasoning

Question On what quantities does the torque on the coil depend?
Answer The magnetic moment of the coil, the magnetic field in which it is placed, and the orientation between μ and **B**. Specifically,

$$\tau = \mu B \sin \theta$$

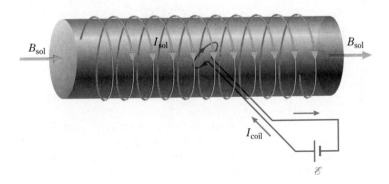

FIGURE 19.29

Find the torque on the small coil.

Question How is the magnetic moment defined?
Answer $\mu = AI_{tot} = A(NI)$. The area $A = \pi r^2$, where r is the radius of the coil.

Question What is the direction of $\boldsymbol{\mu}$?
Answer With the fingers of your right hand curling in the direction of the coil's current, your right thumb points in the direction of $\boldsymbol{\mu}$. This is perpendicular to the plane of the coil.

Question What is the expression for the magnetic field produced by a solenoid?
Answer $B = \mu_o NI$, where n is the number of turns per unit length of the solenoid.

Question What is the direction of the solenoid's field?
Answer It is determined in the same way as the magnetic moment of the small coil.

Question In what direction will the torque be exerted?
Answer The torque will tend to rotate the coil so that $\boldsymbol{\mu}$ lines up with **B**.

Solution and Discussion Your right hand should tell you that the direction of $\boldsymbol{\mu}$ is 45° below horizontal to the right in Fig. 19.29. Similarly, the direction of **B** in the solenoid is horizontally to the right.
The magnetic moment of the small coil is

$$\mu = (10 \text{ turns})(0.060 \text{ A})\pi(0.05 \text{ m})^2 = 4.7 \times 10^{-3} \text{ A} \cdot \text{m}^2$$

The magnetic field in the interior of the solenoid is

$$B = (4\pi \times 10^{-7} \text{ T} \cdot \text{m/A})(1000 \text{ turns/m})(25 \text{ A}) = 3.14 \times 10^{-2} \text{ T}$$

Thus the torque on the coil is

$$\tau = (4.7 \times 10^{-3} \text{ A} \cdot \text{m}^2)(3.14 \times 10^{-2} \text{ T}) \sin 45°$$

$$= 1.04 \times 10^{-4} \text{ m} \cdot \text{N}$$

It is not immediately obvious that these units result from the calculation. You should make sure you can verify that they do. You can think of the coil as representing a compass needle with $\boldsymbol{\mu}$ in the direction of the north pole. The torque would tend to rotate the coil counterclockwise around an axis perpendicular to the solenoid axis in the same way that a compass needle would tend to line up with **B**.

19.13 MOVING-COIL GALVANOMETERS, AMMETERS, AND VOLTMETERS

As we just saw, a current-carrying coil in a magnetic field experiences a torque. Because the torque is proportional to the current in the coil, this effect can be used to measure current.
To see how this effect is utilized, let us refer to Fig. 19.30, where we see a current-carrying coil between the poles of a magnet. When the current is in the direction shown, the coil acts as a magnet with its north pole on the back side. (Check this and see.) We indicate this fact by the magnetic-moment vector $\boldsymbol{\mu}$.

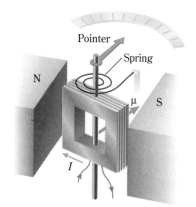

FIGURE 19.30

The current in the galvanometer coil causes it to rotate in the magnetic field provided by the permanent magnet.

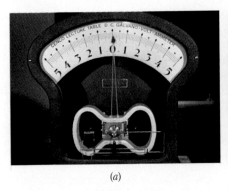

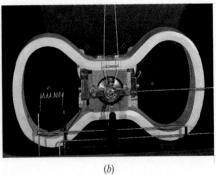

(a)   (b)

(a) A moving-coil galvanometer, showing the permanent magnet and the rotating coil and attached pointer. (b) A close-up of the axis of the coil inside the surrounding magnet. A very small current through the coil will produce an easily read deflection of the pointer.

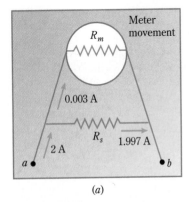

(a)

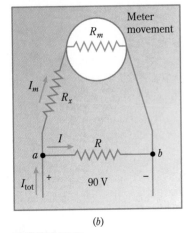

(b)

FIGURE 19.31

(a) Only a small portion of the current goes through the movement of an ammeter. Most of it goes through the parallel shunt resistor R_s. For the current values shown, $R_s = R_m/666$. This is an example of how small typical values of R_s are. (b) In order that the current passing through the voltmeter be very small, a very large resistor R_x is placed in series with the meter coil. For the application shown, R_x would typically be 90,000 Ω or more.

Because the magnetic-moment vector tries to align with the field, the coil rotates so as to point this vector toward the south pole of the permanent magnet. However, a spring attached to the coil provides a torque to stop this rotation. As a result, the coil rotates an amount proportional to the current in it. Therefore, the rotation of the coil, indicated by the pointer attached to it, can be used as a measure of the current in the coil.

GALVANOMETERS

We call the device sketched in Fig. 19.30 a *meter movement*. In practice, the coil often has an iron core to intensify the field and the torque. Many very sensitive meters, called **galvanometers,** are simply a movement such as this placed in an appropriate case. For that reason, it is common to use the terms *meter movement* and *galvanometer* interchangeably.

The sensitivity of a meter movement, that is, how large a deflection results from a given amount of current, depends on several factors. Of course, the stiffness of the restoring spring is of primary importance. The spring must be responsive enough to measure reasonably small currents, but on the other hand, it must not be too delicate if the instrument is to be rugged and portable. The sensitivity also depends on the number of turns of wire on the coil. If the number of turns is doubled, the torque is doubled as well.

A very sensitive meter gives full-scale deflection for a current of only a fraction of a microampere. Such a highly sensitive meter must have a large number of turns of wire on its coil, and so its resistance might easily be 100 Ω. Even so, a voltage of 10^{-4} V across its terminals would cause a current of 10^{-6} A. Table-model galvanometers commonly give a full-scale deflection for a current of about 1 mA (10^{-3} A) and have a resistance of about 20 Ω.

AMMETERS

A moving-coil meter can be inserted into a branch of a circuit to measure the current in that branch. In this case the meter is called an **ammeter.** In order to perform properly, the meter must satisfy two criteria. First, it is important that the presence of the meter in the circuit not cause a significant change in the current it is supposed to measure. Thus the resistance of the ammeter must be considerably less than the resistance in the branch when the meter is absent. Furthermore, the basic galvanometer movement swings full scale when $\approx$ 1 mA passes through it. If the meter is to measure currents larger than this, say 1 A, most of this current must

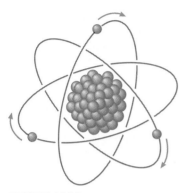

FIGURE 19.32

The orbiting electrons, acting like current loops, generate magnetic fields.

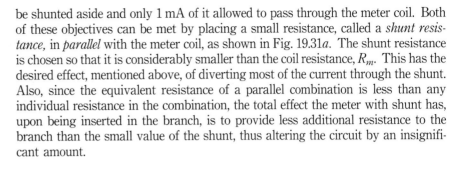

be shunted aside and only 1 mA of it allowed to pass through the meter coil. Both of these objectives can be met by placing a small resistance, called a *shunt resistance,* in *parallel* with the meter coil, as shown in Fig. 19.31*a*. The shunt resistance is chosen so that it is considerably smaller than the coil resistance, R_m. This has the desired effect, mentioned above, of diverting most of the current through the shunt. Also, since the equivalent resistance of a parallel combination is less than any individual resistance in the combination, the total effect the meter with shunt has, upon being inserted in the branch, is to provide less additional resistance to the branch than the small value of the shunt, thus altering the circuit by an insignificant amount.

VOLTMETERS

In order to measure the voltage drop across a circuit element, a moving-coil meter can be placed in parallel with the circuit element R as shown in Fig. 19.31*b*. In this case the meter is called a **voltmeter.** The voltage drop across the meter is then the same as that across the circuit element. Once again, in order that the meter coil draw no more than ≈ 1 mA, most of the current must be blocked from passing through the coil. This can be done by adding a large resistance R_x in *series* with the coil. This also ensures that the presence of the meter does not appreciably change the current in the branch containing R from that which would flow with no meter present.

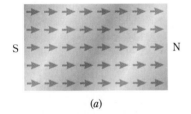

(a)

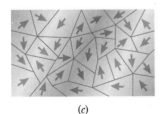

(b)

(c)

FIGURE 19.33

(*a*) A magnetized piece of iron; (*b*) the disordered, unmagnetized iron shown schematically; (*c*) a more realistic sketch of the domains.

19.14 MAGNETIC MATERIALS

We learned in grade school that magnets attract iron but not most other materials. It turns out that only a few *ferromagnetic* materials (iron, nickel, cobalt, gadolinium, dysprosium, and their alloys) are greatly influenced by a steady magnetic field.

Some atoms act like tiny bar magnets. To see why, refer to the often-used atom model shown in Fig. 19.32, which pictures the electrons as orbiting the nucleus. Since an electron in its orbit constitutes a circular current loop, each orbiting electron in Fig. 19.32 generates a magnetic field similar to that of the loop in Fig. 19.23.

There is a second phenomenon that causes atoms to act like magnets. Particles such as the electron and proton act as though they are spinning on an axis through their centers; we say that such particles have *spin*. Any spinning charge acts like a current loop and generates a magnetic field.

In many atoms, the magnetic effects of the various electrons cancel each other. In other atoms, the cancellation is nearly, but not quite, complete. Only in the transition-element atoms, which are the ferromagnetic elements mentioned earlier, do the contributions of enough electrons add to give each atom a significant total magnetic moment. These atoms therefore act like tiny compass needles. If a majority of these atoms become aligned in a bulk sample of ferromagnetic material, the sample becomes *magnetized*. Let us examine this condition more closely.

We know that if we place a group of tiny magnets close to one another, insofar as possible they arrange themselves in such a way that each south pole is close to a north pole. This is the result of the attraction of unlike poles and the repulsion of like poles. The lowest potential energy of the system is reached when the magnets are arranged somewhat as shown in Fig. 19.33*a*. Notice that magnets arranged in this way are equivalent to a single large magnet.

However, if the magnets are strongly agitated (perhaps by someone shaking the board on which they rest), they will break loose from this alignment as shown in Fig. 19.33*b*. Notice now that the individual magnets no longer align so as to produce a strong bar magnet.

An analogous situation exists with atoms in a solid. Thermal vibrations agitate the system and prevent the atoms from ordering themselves as in Fig. 19.33*a*. Only certain atomic magnets—iron and the other ferromagnetic materials—can preserve the alignment at ordinary temperatures. Even these atoms, when heated enough, acquire enough thermal energy to break loose and disorient as in (*b*). The temperature at which this happens is quite definite for each atomic species and is called the **Curie temperature.** There are, in addition to the magnetic forces between the ferromagnetic atoms, other forces that are much more complex. These forces can be understood only in terms of quantum mechanics, and so we are unable to discuss them further here. They play a major role in the aligning of atomic magnets.

Most materials have their atomic magnets, if there are any, randomly oriented, as in Fig. 19.33*b*. The ferromagnetic materials, however, consist of little regions within which the atoms are all aligned. Each of these oriented regions is called a **domain** (Fig. 19.33). In an ordinary piece of iron, each domain may contain as many as 10^{16} atoms and consist of a region a small fraction of a millimeter in linear dimension. However, the domains in an unmagnetized piece of iron are randomly oriented, as in Fig. 19.34*a*. For a bar of iron to be magnetized, the domains within it must be lined up. This can be done in the following way.

Suppose you start with the unmagnetized bar of iron shown in Fig. 19.34*a*. A solenoid with a current in it possesses a weak magnetic field within its windings. If now you place the iron bar in the solenoid, the magnetic field of the solenoid will exert forces on the domains. Those domains that are oriented along the field grow, and those oriented in other directions decrease in size. The net effect is to align the domains with the field, as shown in (*b*). The iron is now a bar magnet, with north and south poles. In what is referred to as *soft iron,* the domains are easily oriented, but in what is called *hard iron* the external field must be made quite strong or the domains must be agitated by heat or mechanical means to allow them to grow in the direction of the field. (The designations *hard* and *soft* refer only to the magnetic properties, not to the physical hardness.) It is possible, however, to align the domains nearly perfectly and form a strong bar magnet.

Once the domains have been aligned, the magnetic field consists of two parts: the original small field of the solenoid plus the field produced by the bar magnet, which is usually hundreds of times larger than the field of the solenoid. *The combination of a solenoid and a piece of soft iron is called an* **electromagnet.**

If the current in the solenoid is turned off, the domains in a bar of soft iron return nearly to their original random state. Thermal motion causes them to disarrange. This is a desirable situation in an electromagnet because it makes it possible to turn it on or off at will. A piece of hard iron used in the solenoid, on the other hand, would retain most of its alignment when it was removed from the solenoid and would be a permanent bar magnet.

The degree to which a material responds to an external magnetic field can be characterized by a quantity called the **relative magnetic permeability,** K_m. For example, suppose we have a very long solenoid carrying a current that produces a field B_o. We now fill the interior of this solenoid with some material. The total field B is now the sum of B_o and whatever field is produced by alignment of atomic magnets.

Relative magnetic permeability K_m of a material is defined as the ratio of the total field B to that of the magnetizing field B_o:

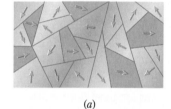

(*a*)

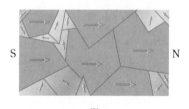

(*b*)

FIGURE 19.34

(*a*) Randomly oriented domains in an unmagnetized sample. (*b*) When placed in an external magnetic field, the domains aligned with the field grow at the expense of the nonaligned domains, which gives the sample a net magnetic field.

$$K_m = \frac{B}{B_o}$$

Ferromagnetic materials, some of which are listed in Table 19.1, have values of K_m that range typically from 100 to 100,000. Other materials fall into two categories, also listed in Table 19.1. Some, called diamagnetic, reduce the field and hence have values of K_m less than 1. A number of other materials, called *paramagnetic* materials, increase the total field slightly. They have values of K_m slightly greater than 1.

Let us summarize what we have found about the magnetic properties of materials. Most materials, when placed in a magnetic field, change the field scarcely at all. A very few substances, however, chiefly iron and its alloys, increase the magnitude of the magnetic field in which they are placed; often the field is strengthened by factors of hundreds. The ability of iron to greatly augment a magnetic field is of prime importance in many applications of magnetism.

TABLE 19.1   *Values of relative magnetic permeability at room temperature for selected materials*

Material	Relative permeability K_m
Ferromagnetic	
Cobalt	250
Nickel	600
Iron	5,000
Permalloy (55% Fe, 45% Ni)	25,000
Mumetal (77% Ni, 16% Fe, 5% Cu, 2% Cr)	100,000
Paramagnetic	
Air	1.0000004
Aluminum	1.000023
Magnesium	1.000012
Uranium	1.00040
Diamagnetic	
Bismuth	0.99983
Mercury	0.99997
Silver	0.99998
Copper	0.99999
Water	0.99999

LEARNING GOALS

Now that you have finished this chapter, you should be able to

1 Define (*a*) right-hand rule for a magnetic field, (*b*) right-hand rule for a magnetic force, (*c*) magnetic field strength, (*d*) tesla and gauss, (*e*) velocity selector, (*f*) solenoid, (*g*) Hall effect, (*h*) ferromagnetic material, (*i*) domain, (*j*) electromagnet, (*k*) magnetic moment, (*l*) meter movement, (*m*) shunt resistor, (*n*) relative magnetic permeability.

2 Sketch the magnetic field in the vicinity of (*a*) magnets of various shapes, (*b*) a straight current-carrying wire, (*c*) a loop of current-carrying wire, (*d*) a solenoid.

3 Use a compass to determine the direction of the magnetic field lines in a given region.

4 Find the magnitude and direction of the force on a given straight-wire current in a specified magnetic field.

5 Use $F = B_\perp IL$ to find one of the quantities when the others are given.

6 Use $F = qvB_\perp$ to find one of the quantities when the others

are given. Calculate the radius of the path followed by a particle of known charge and mass moving perpendicular to a known magnetic field.

7 Calculate the magnetic field (a) at a given distance from a straight wire carrying a given current, (b) at the center of a coil of N turns carrying a known current, (c) in the interior of a given solenoid carrying a known current. Calculate (c) when the solenoid is empty and when it is filled with a material whose K_m is given.

8 Given a list of common materials, select those that greatly alter the magnetic field into which they are placed.

9 Describe in terms of domains what happens when a bar of ferromagnetic material is magnetized or demagnetized.

10 Explain how the Hall effect allows us to determine the sign of charge carriers.

11 State which way a current-carrying coil will turn when placed in a given position in a magnetic field. Compute the torque on the coil when sufficient data are given.

12 Point out where the effective north and south poles are for a current-carrying loop. Explain what is meant by the magnetic-moment vector for a current loop.

13 Explain the major features of a meter movement. Tell how it is used to make an ammeter or a voltmeter.

QUESTIONS AND GUESSTIMATES

1 The north pole of a bar magnet is brought close to an unmagnetized iron nail. What does the field of the magnet do to the nail? Why is the nail attracted by the magnet?

2 Two concentric circular loops lie on a table. The larger loop carries a current of 10 A counterclockwise, and the smaller carries a current of 5 A clockwise. Describe the forces on each loop.

3 As shown in Fig. P19.1, two high-voltage leads cause a beam of charged particles to shoot to the right through a partially evacuated tube. Their path is shown by a fluorescent screen placed along the length of the tube. When a magnet is brought close, the beam deflects. How could you determine the sign of the charge on the particles?

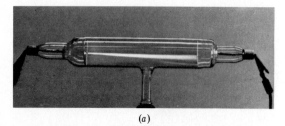

(a)

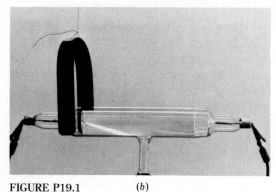

FIGURE P19.1   (b)

4 Describe the motion of an electron shot into a long solenoid at a small angle to the solenoid axis.

5 It is sometimes said that the earth's North Pole is a south pole, and vice versa. What does this mean?

6 When a beam of electrons is shot into a certain region of space, the electrons travel through the region in a straight line. Can we conclude there is no electric field in the region? No magnetic field?

7 In a certain experiment, a beam of electrons shot out along the positive x axis deflects toward positive y values in the xy plane. If this deflection is the result of a magnetic field, in what direction is the field? Repeat for an electric field.

8 A beam of charged particles is deflected as it passes through a certain region of space. By taking measurements on the beam, how could you determine which causes the deflection, a magnetic or an electric field?

9 A proton is shot from the coordinate origin out along the positive x axis. There is a uniform magnetic field in the positive y direction. (a) Describe the motion of the proton, paying particular attention to the quadrants in which it travels. (b) Repeat for an electron. (c) Repeat if the proton velocity is such that $v_x = v_y$ and $v_z = 0$. (d) Repeat if $v_x = v_y = v_z \neq 0$.

10 We know that in a television tube electrons are shot from one end of the tube to the other, where they strike the fluorescent screen. Suppose your little brother insists that his general science teacher says that protons are used, not electrons. How could you prove to him that he is wrong without dismantling the set?

11 Suppose you are given a material that is a poor conductor but conducts enough to obtain a measurable current through it. How could you decide whether the current is made up of positive or negative charges, or both? Give as many ways as you can.

12 It is proposed to furnish the propulsion force to a spaceship in the following way. Electricity is furnished by a nuclear reactor or some other means. Large currents are sent through copper bars in the ship, and the forces exerted on these bars by the earth's magnetic field propel the ship. What objections do you see to such a plan?

13 Cosmic rays (charged particles coming to the earth from

outer space) are unable to reach the surface of the earth unless they have very high energy. One reason is that they have to penetrate the earth's atmosphere. However, for particles coming toward the equator along a radius of the earth, magnetic effects are also important. Explain why, being careful to point out why particles can reach the poles of the earth without encountering this difficulty.

14 Give an order-of-magnitude estimate of the displacement of the electron beam on a television screen as a result of the earth's magnetic field.

SUMMARY

DERIVED UNITS AND PHYSICAL CONSTANTS

Units of Magnetic Field (B)
1 tesla (T) = $1\,\text{N/m}\cdot\text{A}$
1 gauss (G) = $10^{-4}\,\text{T}$

Permeability of Vacuum (μ_o)
$\mu_o = 4\pi \times 10^{-7}\,\text{T}\cdot\text{m/A}$

Magnetic Moment of a Current Loop (μ)
μ = area × total current in the loop
  = *ANI*

The units of μ are $\text{A}\cdot\text{m}^2$.

The direction of $\boldsymbol{\mu}$ is the direction of the right thumb when the fingers of the right hand curl in the direction of the current around the loop.

DEFINITIONS AND BASIC PRINCIPLES

Magnetic Field (**B**)
The direction of a magnetic field **B** at any point is the direction a free compass needle points when placed at that point.

The magnitude of a magnetic field **B** is defined by the force it exerts on a straight wire carrying a current perpendicular to the direction of **B**:

$$B = \frac{F/L}{I}$$

where *L* is the length of the wire. When *F*, *L*, and *I* are in SI units, *B* is measured in teslas (T).

Magnetic Force on a Current
The magnetic force per unit length exerted on a current-carrying wire in a magnetic field **B** is

$$\frac{F}{L} = BI \sin \theta$$

where θ is the angle between **B** and **I**. The direction of the force is perpendicular to both **B** and **I** in the sense given by the right-hand rule.

INSIGHT
1. The magnetic force on a current is maximum when **I** is perpendicular to **B** and zero when **I** is parallel (or antiparallel) to **B**.

Right-Hand Rule for Magnetic Force on a Current
Point the fingers of your right hand along the direction of **B** and your right thumb in the direction of the current. The magnetic force on the current is in the direction your right palm is facing.

Magnetic Force on a Moving Charge
A charge *q* moving with a velocity **v** in a magnetic field **B** experiences a force of magnitude

$$F = qvB_\perp$$

where $B_\perp$ is the component of **B** perpendicular to the velocity **v**. The direction of the force for positive charges is determined by using the right-hand rule for currents. The direction of the force on negative charges is opposite this.

INSIGHTS
1. Since the magnetic force is always perpendicular to the direction of motion, **v**, a steady magnetic field can do no work on a moving charge.
2. A uniform magnetic field will cause a moving charged particle to move in a circle of radius

$$r = \frac{mv}{qB}$$

where *m* is the mass of the particle.

The Hall Effect
When a conductor of rectangular cross-section is placed in a magnetic field perpendicular to its current, a voltage is created transverse to **I** and **B**. This is the Hall voltage, given by

$$V_\text{H} = vBd$$

where *v* = average "drift" velocity of charges carrying current
  d = dimension of the conductor perpendicular to **I** and **B**

The polarity of this voltage depends on the sign of the charge carriers.

Magnetic Force Between Parallel Currents

The magnetic force per unit length two parallel currents exert on each other is

$$\frac{F}{L} = \frac{\mu_o I_1 I_2}{2\pi b}$$

where b is the distance between the currents. If the currents are in the same direction, this force is attractive. If they are in opposite directions, the force is repulsive.

INSIGHTS

1. This phenomenon is the basis for the definition of the ampere. The coulomb of charge is then derived as $1\,C = 1\,A \cdot s$.

The Magnetic Field Produced by Certain Current Configurations

LONG, STRAIGHT CURRENT

$$B = \frac{\mu_o I}{2\pi r}$$

CENTER OF A CIRCULAR COIL (RADIUS a)

$$B = \frac{\mu_o I}{2a}$$

If the coil has N turns, multiply the right side of this expression by N.

SOLENOID

$$B = \frac{\mu_o NI}{L} = \mu_o nI$$

where L = length of solenoid
N = total turns
n = turns per length

The Torque on a Current Loop

A current loop placed in a magnetic field tends to rotate so as to align its magnetic moment μ with the field **B**. The torque on the loop is given by

$$\tau = \mu B \sin\theta$$

where θ is the angle between μ and **B**.

Magnetic Materials

The response of materials placed in magnetic fields is measured by their relative magnetic permeability, K_m. K_m is the ratio of the total magnetic field B resulting from their being placed in the external field B_o to the external field:

$$K_m = \frac{B}{B_o}$$

There are three categories of materials based on this factor:

1. Ferromagnetic: $K_m \gg 1$
2. Paramagnetic: $K_m > 1$
3. Diamagnetic: $K_m < 1$

PROBLEMS

Sections 19.1–19.5

1 A power line carries a current of 32 A straight west in a region where the magnetic field is parallel to the earth's surface and directed straight north, with $B = 8.2 \times 10^{-4}$ T. Find the magnitude and direction of the force due to the field on a 2.0-m length of the power line.

2 A conductor carries a current of 24 A straight up from the surface of earth. The earth's magnetic field in that region is horizontal and directed straight north, with a strength of 8.0×10^{-4} T. What are the magnitude and direction of the magnetic force on a 50-cm length of the conductor?

3 Calculate the force on a 1-m length of a wire carrying a current of 6 A in a region where a uniform magnetic field of magnitude 0.75 T is perpendicular to the wire.

4 A conductor carries a current of 12 A in a direction 45° with respect to the direction of a uniform magnetic field of magnitude 0.5 T. Calculate the magnetic force on a 2.5-m length of the conductor.

5 A circular loop of wire, radius $r = 10.0$ cm, lies on a table and carries a current of 1.8 A. The loop is in a uniform vertical magnetic field of strength 0.1 T and directed upward. (a) Find the total force on the loop because of the magnetic field. (b) Find the approximate force on a 0.2-mm length of the loop.

6 Calculate the direction and magnitude of the force the earth's magnetic field exerts on a 120-m length of wire stretched horizontally between two poles and carrying an 80-A current. The magnitude of earth's average magnetic field in the region is 4.0×10^{-5} T and its direction makes an angle of 50° with the current.

7 A certain horizontal wire lying in the east-west direction has a mass of 0.18 g per meter of its length and carries a current I. The wire is in a northward-directed horizontal magnetic field of strength 0.5 T. Find the minimum amount of current I if the magnetic force is to support the weight of the wire.

8 A current of 15 A in a certain wire is directed along the positive x axis and is perpendicular to the direction of a magnetic field. The wire experiences a magnetic force of

0.18 N/m per unit length in the positive *y* direction. Find the direction and magnitude of the magnetic field in the region.

■ **9** A thin conducting rod 1 m long has a mass of 24 g and carries a current of 0.3 A. What is the minimum strength of a magnetic field, applied perpendicular to the rod, that will cause the rod to float freely?

10 A wire in the *xy* plane makes an angle of 30° with the positive *x* axis and carries a current of 3 A toward positive *x* and *y* values. Impressed on it is a 0.04-T magnetic field. Find the magnitude and direction of the magnetic force on a 0.5-m length of the wire if the field is directed along (*a*) the positive *x* axis, (*b*) the negative *y* axis, and (*c*) the positive *z* axis.

11 A thin conducting wire lies in the *xy* plane making an angle of 24° with the positive *y* axis and carries a current of 6.0 A toward negative *x* and *y* values. A 0.04-T magnetic field exists in the region of the conductor. Find the magnitude and direction of the magnetic force on a 1-m length of the conductor if the field is directed along (*a*) the negative *x* axis, (*b*) the negative *y* axis, and (*c*) the positive *z* axis.

Section 19.6

12 A proton is shot along the positive *x* axis with a speed of 5.0×10^4 m/s. Find the direction and magnitude of the force exerted on it by a 0.04-T magnetic field if the field is in (*a*) the negative *y* direction, (*b*) the negative *z* direction, and (*c*) the negative *x* direction.

13 An electron is shot along the positive *y* axis with a speed of 6.0×10^5 m/s. Find the direction and magnitude of the force exerted on it by a magnetic field of strength 0.005 T if the field is in (*a*) the negative *y* direction, (*b*) the positive *x* direction, and (*c*) the negative *z* direction.

14 A proton is moving perpendicular to a magnetic field of strength 0.08 T. What speed does the proton have if the magnetic force on it has a magnitude of 5.0×10^{-14} N?

15 An electron moving with a speed of 4.8×10^5 m/s at right angles to a magnetic field experiences a force of magnitude 7.2×10^{-11} N. What is the magnitude of the magnetic field?

16 An electron is moving in the *xy* plane with a velocity of 4.0×10^5 m/s directed at an angle of 30° above the positive *x* axis. A magnetic field of strength 0.065 T is applied on the electron. Find the direction and magnitude of the magnetic force experienced by the electron if the direction of the field is along (*a*) the negative *x* axis, (*b*) the positive *y* axis, (*c*) the positive *z* axis.

17 A proton is moving in the *xy* plane with a velocity of 3.6×10^4 m/s in a direction making an angle of 60° above the positive *x* axis. Calculate the magnitude and direction of the magnetic force on the proton if a 0.004-T magnetic field is along (*a*) the negative *y* axis, (*b*) the negative *x* axis, (*c*) the positive *z* axis.

18 A proton is moving with a speed of 6.0×10^7 m/s through a magnetic field of strength 1.6 T. What is the angle between the proton's velocity and the direction of the magnetic field if the proton experiences a force of 1.3×10^{-11} N?

■ **19** A proton moves horizontally with a speed of 4×10^6 m/s at

right angles to a magnetic field. (*a*) What is the strength of the magnetic field necessary to just balance the weight of the proton and keep it moving horizontally? (*b*) What should be the direction of this field?

■ **20** A proton moving perpendicular to a uniform magnetic field with a speed of 2×10^6 m/s experiences an acceleration of 3×10^{12} m/s² in the negative *x* direction at the instant that its velocity is along the positive *z* direction. Find the magnitude and direction of the magnetic field.

■ **21** Repeat the previous problem for an electron.

■ **22** An electron accelerated through a potential difference of 3000 V enters a region where there is a uniform magnetic field of magnitude 1.5 T. What are the (*a*) maximum and (*b*) minimum values of the force the electron can experience in the magnetic field? At what angles between **B** and the electron's velocity do these maximum and minimum forces occur?

Sections 19.7–19.9

23 A proton with speed 4×10^5 m/s is traveling perpendicular to a uniform 24-mT magnetic field. Describe quantitatively the path it follows.

24 Describe quantitatively the path followed by an electron with a speed of 6×10^6 m/s traveling at right angles to a uniform 2-mT magnetic field.

25 An electron moves in a circular orbit of radius 1.2 m in a region of uniform magnetic field. What would be the radius of the orbit if the strength of the magnetic field were reduced to one-half the original value?

26 A singly charged positive ion of mass 4.56×10^{-27} kg moves counterclockwise in the *xy* plane in a circular path of radius 4 cm with a speed of 2×10^4 m/s. Determine the magnitude and direction of the magnetic field.

27 Alpha particles from a certain radioactive source are emitted with a speed of 1.66×10^7 m/s. What strength of magnetic field, applied perpendicular to the motion of alpha particles, will make them move in a circular path of radius 0.8 m? (The mass of the alpha particle is 6.64×10^{-27} kg and its charge is twice that of a proton.)

■ **28** A doubly charged positive ion ($q = +2e$) of mass 6.2×10^{-26} kg is accelerated through a potential difference of 300 V. It then enters a region perpendicular to a uniform 0.5-T magnetic field. Determine the radius of the circular path of the ion in the field.

■ **29** A proton is accelerated through a potential difference of 300 kV and then enters a uniform 0.4-T magnetic field with its velocity perpendicular to the field lines. What is the radius of the circle in which the proton travels?

■ **30** A proton accelerated through an unknown potential difference enters a region where there is a uniform 0.06-T magnetic field perpendicular to its velocity, and describes a circular path of radius 35 cm. Calculate the energy of the proton in electronvolts.

■ **31** A charged particle with charge *q* moving with a velocity **v** enters a region perpendicular to a uniform magnetic field **B** and follows a circular path of radius r. Show that the kinetic energy of the particle can be expressed as $KE = q^2r^2B^2/2m$, where m is the mass of the particle.

■32 Calculate the radius of the circular path of an electron having a kinetic energy 1 eV moving perpendicular to a magnetic field of strength 0.4 T.

33 A beam of protons traveling with a velocity of 2×10^5 m/s follows a straight-line path through the crossed magnetic and electric fields in a velocity selector. What is the strength of the magnetic field if the electric field is 8×10^5 N/C?

■34 A particular beam of electrons travels in a straight line through a region of crossed magnetic and electric fields in a velocity selector. The strength of the magnetic field in the region is 0.04 T; the plates are separated by 6 cm and have a potential difference of 120 V. Find (a) the speed of the electrons and (b) the radius of the circle in which the electron travels when the potential difference across the plates is zero.

■35 When moving perpendicular to a 0.04-T magnetic field, a proton beam follows a circular orbit of radius 1 m. How large an electric field perpendicular to both proton velocity $\mathbf{v}$ and the magnetic field $\mathbf{B}$ will cause the proton to follow a straight line path?

■36 A velocity selector uses a magnet to produce a uniform magnetic field of strength 0.050 T and a pair of parallel metal plates separated by 20 mm to produce a perpendicular electric field. How large should be the potential difference applied to the plates to permit a singly charged positive ion of speed 6.0×10^6 m/s to pass through the selector?

■37 When a fast-moving particle such as an electron shoots through superheated liquid hydrogen, bubbles form along the path of the particle. Figure P19.2 shows the paths of several particles in such a "bubble chamber." The paths are curved because a magnetic field is perpendicular to and into the page. Assume this field to have a magnitude of 4.0 mT. Is the particle that is leaving point a and moving to the right positive or negative? If you assume it to be an electron,

what is its approximate speed? (The tracks are actual size. Assume them to be in the plane of the page.)

■38 The particle that starts at point b in Fig. P19.2 slows as it moves through the liquid hydrogen. As a result, it spirals inward. Assume the same data as in the previous problem and assume the particle to be an electron. Find its speed at point c.

■39 A singly charged positive ion traveling with a speed of 5×10^5 m/s leaves a spiral track of radius 8 mm in a photograph in a plane perpendicular to the magnetic field of a bubble chamber. The magnetic field used in the bubble chamber has a magnitude of 2 T. Calculate the mass of the ion.

Sections 19.10 and 19.11

40 Determine the magnetic field strength at a distance of 20 cm from a long, straight wire carrying a current of 4 A.

41 A long, straight wire carries a current of 5 A. At what distance from the wire is the magnetic field due to the current in the wire equal in magnitude to the strength of the earth's magnetic field of 5×10^{-5} T?

42 Two long, straight, parallel wires are 20 cm apart, and each carries a current of 10 A. Find the magnitude and direction of the magnetic field at a point midway between the wires if the currents are (a) in the same direction and (b) in opposite directions.

■43 Two straight parallel wires are 30 cm apart, and each carries a current of 20 A. Find the magnitude and direction of the magnetic field at a point in the plane of the wires that is 10 cm from one wire and 20 cm from the other if the currents are (a) in the same direction and (b) in opposite directions.

■44 A long, straight wire carries a 6-A current along the positive x axis, and a second wire carries an 8-A current along the negative y axis. Find the magnitude and direction of their combined magnetic field at the point $x = 6$ cm, $y = 8$ cm.

45 How large is the current in a 15-cm-radius circular wire loop if the magnetic field at the center of the loop is equal in magnitude to the strength of the earth's magnetic field of 5×10^{-5} T?

46 One hundred turns of wire are wound into a coil 40 cm in diameter and are connected to a 9-V battery. The total resistance of the coil is 1.8 Ω. Find the strength of the magnetic field at the center of the coil.

47 A long solenoid has 2000 loops of wire on its 30-cm length. The solenoid diameter is 2.4 cm. Find the magnetic field inside the solenoid when a current of 250 mA flows through it.

48 A 50-cm-long solenoid with 1500 turns is used to generate a magnetic field of 0.2 T. How large is the current?

■49 A long, straight wire carrying a current of 50 A lies along the axis of a long solenoid in which the magnetic field is 4.0 mT. (a) How large is the force on a 1.0-cm length of the wire? (b) What is the value of the total magnetic field inside the solenoid at 0.5 cm from its axis?

50 Two parallel conductors are 8 cm apart and each carries a current of 5 A. Find the force per unit length exerted on one

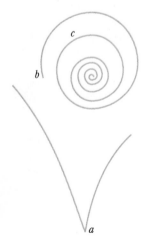

FIGURE P19.2

of the conductors by the other when the currents are (*a*) in the same direction and (*b*) in the opposite direction.

51 Two parallel wires attract each other with a force per unit length of 2.0×10^{-3} N when separated by 2 cm. If the current in one wire is 100 A, what is the current in the other wire?

Section 19.12

52 A flat coil of wire with 40 turns lies on a horizontal table. Its area is 120 cm², and it carries a current of 30 A. Find the torque on it due to an 80-mT magnetic field if the field lines are (*a*) parallel to the tabletop, (*b*) perpendicular to the tabletop, and (*c*) at an angle of 30° to the horizontal. (*d*) What is the magnetic moment of the coil?

53 A flat coil of wire with 40 loops lies flat against the north wall of a room. The coil has an area of 240 cm² and carries a current of 25 A. Find the torque on the coil due to an 80-mT field if the field lines are directed (*a*) westward, (*b*) southward, (*c*) vertically, and (*d*) in the plane of the wall at an angle of 60° to the vertical. (*e*) What is the magnetic moment of the coil?

▪**54** A rectangular 600-turn coil of dimensions 5 cm by 6 cm is suspended in a magnetic field of 0.8 T. What is the current in the coil if the maximum torque exerted on it by the magnetic field is 0.24 N · m?

Section 19.13

55 A certain meter movement has a resistance of 50 Ω and deflects full scale for a voltage of 250 mV across its terminals. How can it be made a 3-A ammeter?

56 How can the meter movement described in Prob. 55 be made into a 10-V voltmeter?

57 A meter movement deflects full scale to a current of 0.010 A and has a resistance of 100 Ω. How can it be made into a 5-A ammeter?

Additional Problems

▪**58** A square wire loop lies on a horizontal table and carries a current *I*. Impressed on the loop is a uniform magnetic field of strength *B* in a direction that makes an angle θ with the horizontal. Prove that the net force on the loop due to the field is zero.

▪**59** A charged particle is observed to follow a circular path of 8.3-cm radius in a plane perpendicular to a uniform 8.0-mT magnetic field. From a separate measurement the momentum of the particle is found to be 2.0×10^{-22} kg · m/s. Determine the electric charge of the particle.

▪▪**60** An electron is shot from the origin of coordinates with a velocity of 3×10^6 m/s at an angle of 60° above the positive *x* axis. In this region there is a 0.006-T magnetic field in the

positive *x* direction. Describe quantitatively the path followed by the electron. *Hint:* Separate the electron velocity into components along and perpendicular to the *x* axis.

▪▪**61** Two ions, one singly charged and the other doubly charged, each with mass 6.6×10^{-27} kg, move out of the slit of a mass spectrometer and enter a region where the magnetic field is perpendicular to their velocities and has a magnitude of 0.4 T. They both have a speed of 2×10^6 m/s. Find (*a*) the radius of the circular path followed by each in the field and (*b*) the distance of separation between them when they have moved through a semicircle and strike a photographic plate.

▪▪**62** A long solenoid has 50 turns per centimeter of length and carries a current of 10 A. A proton is shot from a point on the solenoid axis with a velocity of 4×10^5 m/s at an angle of 20° to the axis. Describe quantitatively the path followed by the proton. *Hint:* Separate the initial velocity of the proton into components parallel and perpendicular to the axis.

▪▪**63** A straight 1.6-m-long wire weighing 0.1 N per meter is suspended directly above and parallel to a fixed second wire. The top wire carries a current of 32 A, and the bottom wire carries a current of 65 A. If the top wire is held in place by the magnetic repulsion due to the bottom wire, how large must the separation between the wires be?

▪▪**64** A square coil of wire 15 cm on an edge has 50 loops and a mass of 100 g. It lies flat on a table and is subject to a horizontal field of 0.048 T parallel to one edge. How large must the current in the coil be if one edge of the coil is to lift from the table?

▪▪**65** A circular coil of wire of diameter 20 cm has 40 loops and a mass of 50 g. It lies flat on a table and is subject to a 60-mT magnetic field that makes an angle of 30° to the vertical. How large must the current in the coil be if one part of the coil is to lift from the table?

▪▪**66** The uniform magnetic field within a long solenoid of inner radius *R* has a magnitude *B* and is directed along the solenoid's axis. What is the largest speed an electron shot radially from the axis can have if it is to escape hitting the inner surface of the solenoid?

▪▪**67** A hollow, straight plastic pipe has charge uniformly distributed over its cylindrical face. The charge per unit length is *Q*, and the pipe is very long. If it is rotating on its axis with a frequency *f*, what is the magnetic field inside the pipe due to the charge on its surface?

▪▪**68** An electron moving with a speed of 2×10^4 m/s follows a circular path of radius 4 cm in a solenoid. The magnetic field of the solenoid is perpendicular to the plane of the electron's path. Find (*a*) the strength of the magnetic field inside the solenoid and (*b*) the current in the solenoid if it has 30 turns per centimeter of length.

CHAPTER 20

ELECTROMAGNETIC INDUCTION

The industrial revolution that transformed the world more than a century ago was based on three major scientific advances: the invention of the steam engine and the further development of heat engines through the use of thermodynamics, the discovery that forces to turn motors can be produced by the interaction of electric currents with magnetic fields, and the discovery that currents can be produced by changing magnetic fields. We have already discussed the first two advances. In this chapter, we investigate the third.

20.1 INDUCED EMF

The discovery that electric currents generate magnetic fields was made by the Danish physicist Hans Christian Oersted in 1820. As so often happens in science, this newly found facet of nature led to intense investigations of related phenomena. One avenue of experimentation was followed by those who attempted to answer the question "If currents produce magnetic fields, is it not possible that magnetic fields produce currents?" About 10 years later the answer to this question was shown to be affirmative by Michael Faraday (1791–1867) in England and independently by

A dramatic demonstration of induced current. (*a*) A coil, with a photoflash bulb attached, is held in a strong magnetic field. (*b*) When the coil is withdrawn quickly, the abrupt change in magnetic flux through the coil induces an emf sufficient to flash the bulb.

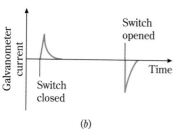

(*a*)

(*b*)

FIGURE 20.1

An induced current exists in the secondary coil only when the current in the primary coil is changing. The current pulses are actually much narrower than in (*b*).

Joseph Henry (1797–1878) in the United States.* Let us now discuss an experiment that shows this effect vividly.

The experiment makes use of the simple equipment shown in Fig. 20.1*a*. We see there two simple series circuits. One consists of a battery and switch connected in series by a long wire coiled around a soft iron rod. We call this coil the **primary coil** because it is attached to the battery. A second, independent wire is also coiled around the rod. This coil is in series with a galvanometer (symbolized by ⊙) but has no battery in its circuit. It is called the **secondary** coil.

Since there is no battery in the secondary-coil circuit, one might guess that the current through it is always zero. But a startling fact emerges if the switch in the primary circuit is suddenly either closed or opened. At that exact instant, the galvanometer needle suddenly deflects and then returns to zero. In other words, a current is induced in the secondary-coil circuit for an instant. It is as though the secondary circuit possesses a battery (a source of emf) only for the short time it takes the switch to open or to close. We say that an **induced emf** exists in the secondary coil during that instant.

Figure 20.1*b* shows another feature of this induced current and emf: the short-lived current is in one direction when the switch is pushed closed and in the opposite direction when the switch is pulled open. This tells us that the direction of the induced emf depends on whether the current in the primary coil is increasing or decreasing.

A second, somewhat similar experiment is shown in Fig. 20.2. It involves a bar magnet and a coil in series with a galvanometer. When the magnet is stationary beside the coil, as in *a* and *c*, there is no current in the coil. However, when the magnet is moving relative to the coil, a current exists in the coil, as indicated in *b*, *d*, and *e*. As we see, an induced emf exists in the coil only when the magnet and coil are in relative motion. *No induced emf exists when conditions are not changing.*

There are two ways to analyze this effect. One uses the fact that a charge

*Henry's work, carried out in relative obscurity in Albany, New York, was published only in the United States, and few people knew of it. As a result, his experiments had very little influence on scientific progress at that time.

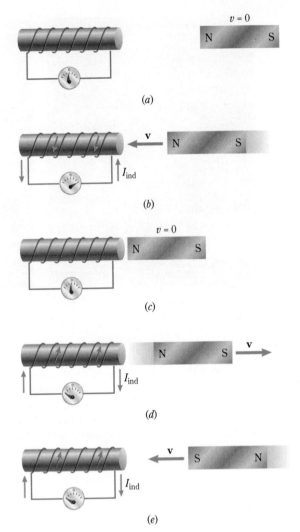

FIGURE 20.2

Current is induced in the coil only when the magnet is moving relative to the coil. The direction of the current depends on the direction of the magnet's motion and on the direction of the magnet's field.

moving through a magnetic field experiences a force. Although Fig. 20.2 shows the magnet in motion, exactly the same effects occur if the magnet is held fixed and the coil is moved.* Consider what happens when the coil moves toward the magnet. The free charges in the wire, since they are moving in the magnetic field of the magnet, experience a force $qvB_\perp$, as given by Eq. 19.2. They flow under the action of this force and thereby give rise to the induced current.

This approach shows us how induced emf is related to phenomena we have already studied, and we shall use this way of looking at the situation from time to time. In most practical situations, however, another approach is more useful. It involves the concept of magnetic flux, as we shall see in the next sections.

---

*This is an example of the fact that motion is a relative quantity. When relative motion occurs between two objects, the effect of one on the other is a function only of the relative motion. It makes no difference which object is considered to be at rest. More is said about this in Chap. 26 when we discuss the theory of relativity.

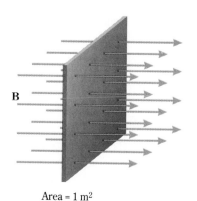

Area = 1 m²

FIGURE 20.3

We agree to draw a number of magnetic field lines proportional to the magnitude of **B** through a unit area erected perpendicular to the field lines.

20.2 MAGNETIC FLUX

Faraday explained the induced emf in a coil in terms of a quantity called *magnetic flux*. To do so, he made a rule about how to draw a map of magnetic field lines. If the magnetic field in a certain region of space has a magnitude B, we represent this magnitude graphically by agreeing to draw field lines a certain distance apart. In this way, stronger fields are represented by lines closer together and weaker fields are represented by lines farther apart. Another way of saying this is that the *density* of field lines in the drawing is proportional to the value of B.

The way we measure this line density is to construct an imaginary surface perpendicular to the lines and count the lines per unit area passing through the surface, as in Fig. 20.3. There we see 16 field lines passing through the 1-m² area. We could choose to let this line density represent any field strength we wished—for example, 1 T. Then on the same drawing a region showing 8 lines per square meter would represent a field half as strong (0.5 T), and a region having 32 lines per square meter would represent a field of 2 T, and so on. Thus the graphical interpretation of B is that B is proportional to the density of field lines, by which we mean the number of field lines passing through an area divided by that area.

It follows from this interpretation that the number of lines through the area A represents $B_\perp A$. This quantity is called the **magnetic flux** through A, which is given the symbol Φ:

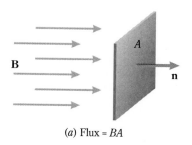

(a) Flux = BA

$$\text{Magnetic flux through } A = \Phi = B_\perp A \tag{20.1}$$

The SI units of magnetic flux are obviously T · m². This combination of units is given a special name, the *weber* (Wb). Thus

$$1\,\text{Wb} = 1\,\text{T} \cdot \text{m}^2$$

or, conversely,

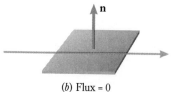

(b) Flux = 0

$$1\,\text{T} = 1\,\text{Wb/m}^2$$

Because of this latter expression, the magnetic field **B** is sometimes referred to as *flux density*.

It is very important to remember that we have considered **B** to be perpendicular to the plane of the area A, as shown in Fig. 20.4*a*. If we rotate A as shown in Fig. 20.4*b*, no field lines pass through it, and so $\Phi = 0$. Flux can also be negative, as is shown in Figure 20.4*c*, where **n** and **B** are oppositely directed. The simple way to write this dependence of Φ on orientation is to describe the direction of the *normal,*

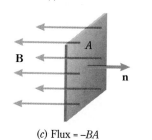

(c) Flux = −BA

FIGURE 20.4

The magnetic flux through an area depends on the relative orientation of the area and the field lines.

FIGURE 20.5

The magnetic flux Φ through an area A is the product of A times the component of the magnetic field **B** along the normal to the area, **n**. Thus $\Phi = (B\cos\theta)\,A$, where θ is the angle between **B** and **n**.

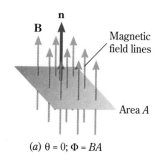

(a) θ = 0; Φ = BA

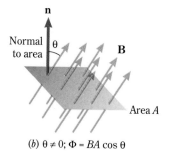

(b) θ ≠ 0; Φ = BA cos θ

n, to the area A. As shown in Fig. 20.5, the component of **B** along the normal is $B_\perp = B \cos \theta$. Thus the general expression for magnetic flux is

$$\Phi = (B \cos \theta)A = BA \cos \theta \qquad (20.2)$$

where θ is the angle between **B** and **n**.

Example 20.1

In a certain room, the earth's magnetic field has a magnitude of 4.0×10^{-5} T and is directed at an angle of 70° below the horizontal. Find the flux through a 400 cm × 80 cm tabletop in the room.

R

Reasoning

Question   What determines the flux?
Answer   The field strength B, the area A, and the angle between **B** and the normal to the area: $\Phi = BA \cos \theta$.

Question   What is the correct angle to use?
Answer   The normal to the tabletop is vertical. Since **B** is directed 70° below the horizontal, it is 20° from the vertical (and pointing downward).

Solution and Discussion   The area A is

$$A = (4.00 \,\text{m})(0.80 \,\text{m}) = 3.2 \,\text{m}^2$$

Then the flux is

$$\Phi = (4.0 \times 10^{-5} \,\text{T})(3.2 \,\text{m}^2) \cos 20° = 1.2 \times 10^{-4} \,\text{Wb}$$

Exercise   How much flux goes through the 15-m² north wall of the room if the field has no component in the east-west direction?   *Answer:* 2.1×10^{-4} Wb.

20.3   FARADAY'S LAW AND LENZ'S LAW

After performing many experiments similar to those in Figs. 20.1 and 20.2, Faraday concluded that an induced emf exists in a coil only if the magnetic flux through the coil is changing. As another example, look at the experiment shown in Fig. 20.6.

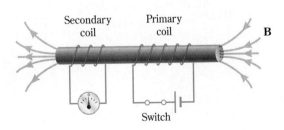

FIGURE 20.6

What happens in the secondary coil when the current in the primary coil is steady? What happens when the switch is pulled open? When it is again pushed closed?

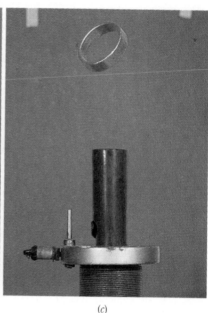

(a)   (b)   (c)

An effect of magnetic induction. (a) A small aluminum ring is sitting freely on the larger aluminum collar above a solenoid coil. The black pillar is a ferromagnetic rod. As current is delivered to the coil, its changing magnetic field creates a changing flux in the aluminum ring. The result, in order to oppose this increase in flux, is to induce a current in the ring which opposes the direction of the current in the solenoid. In (b) and (c), the coil shows the effect of the repulsive force between these opposite currents.

When the switch is closed, the current in the primary coil produces the magnetic field shown. Since the field lines follow the iron rod, considerable flux passes through the secondary coil. If the switch is pulled open, this flux decreases to zero because the current that generates it stops. Only when this *change* in flux is occurring does an induced emf exist in the secondary coil; no induced emf exists when the flux is not changing. The presence of an induced emf is evidenced by a pulse of current through the galvanometer connected to the secondary coil (Fig. 20.6).

Similarly, if we start with the switch open, the flux through the secondary is zero. Upon closing the switch, there is a brief period of time during which the flux builds up to its steady-state value. While this is happening, a current is again observed in the secondary. This time the current is in the direction opposite that which occurs when the switch is pulled open. Once the current in the primary reaches its steady-state value, the emf in the secondary disappears because the magnetic flux through the secondary is again constant.

The experiment shown in Fig. 20.7 also confirms Faraday's conclusion. Notice that because the field lines are denser near the magnet, the flux through the coil increases as the magnet is brought closer to it. A current is observed in the coil as long as the magnet approaches. When the magnet is stationary, there is no flux change, and so no current is induced in the coil. As the magnet is removed, as in Fig. 20.7c, the flux *decreases* and the galvanometer shows a current in the opposite direction. The two current directions indicate that the polarity of the emf induced by the approaching magnet is opposite that induced by the receding magnet. This is shown in Fig. 20.7b and c.

We can now give a quantitative statement of Faraday's results. Suppose that the magnetic flux through an N-loop coil is made to change from Φ_1 to Φ_2 in a time Δt. Faraday found that the average emf induced in the coil during this change is

$$\overline{\text{emf}} = N\frac{(\Phi_2 - \Phi_1)}{\Delta t} = N\frac{\Delta \Phi}{\Delta t} \tag{20.3}$$

This is called **Faraday's law of magnetic induction.** It is one of the most

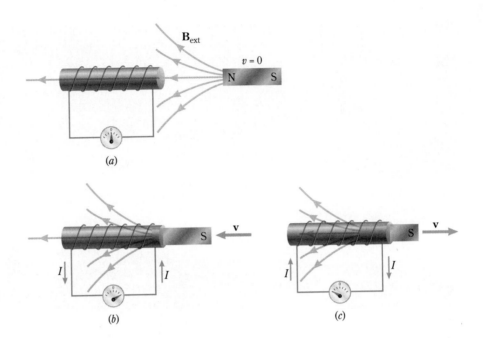

FIGURE 20.7

When the magnet is moved as shown in (b) and (c), the induced current is directed as shown. Why?

FIGURE 20.8

The induced current creates a magnetic flux through the coil that tends to oppose the change in flux produced by a changing external field (not shown here). (a) The north pole is approaching the coil, as in Fig. 20.7b. (b) The north pole is receding, as in Fig. 20.7c.

important principles of electricity and magnetism, and forms the basis for electrical generators and motors as well as a number of other important devices.

As does any other current, the induced current produces a magnetic field of its own. Figure 20.8 shows the directions of this induced field ($\mathbf{B}_{ind}$) resulting from the two magnet motions depicted in Figs. 20.7b and c. You should verify that the directions shown for $\mathbf{B}_{ind}$ in Fig. 20.8 are in agreement with the right-hand rule.

It is important at this point to realize that magnetic flux Φ can be either positive or negative, depending on whether the angle θ between $\mathbf{B}$ and $\mathbf{n}$ is between 0° and 90° (Fig. 20.4a shows an angle of 0°), or between 90° and 180° (Fig. 20.4c shows an angle of 180°). In other words, if the direction of the magnetic field through an area is reversed, the sign of Φ is also reversed. In the discussion that follows, let us say that the normal to the plane of the coil and the external magnet are both lying along the x axis. Then we can call the flux positive when the magnetic field has a component along the $+x$ direction and negative when it has a component along the $-x$ direction.

In the situations covered in Figs. 20.7 and 20.8, there are *two* sources of magnetic field and hence two sources of magnetic flux through the coil. The source of the flux Φ_{ext} is the field of the magnet ($\mathbf{B}_{ext}$), and the source of flux Φ_{ind} is the magnetic field ($\mathbf{B}_{ind}$) produced by the induced current. In Fig. 20.7b, notice that the field $\mathbf{B}_{ext}$ is directed to the left. Therefore Φ_{ext} is negative. As the magnet approaches, more lines of $\mathbf{B}_{ext}$ pass through the plane of the coil, and so this negative flux increases.

In Fig. 20.7c, the external magnetic field creating Φ_{ext} is also directed left; therefore Φ_{ext} is again negative. Here, however, this negative flux through the coil is *decreasing* because the magnet is moving away from the coil. Fig. 20.8b shows that

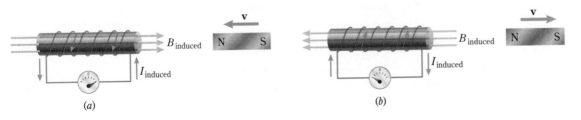

the induced magnetic field $\mathbf{B}_{ind}$ produced by the induced current is now directed to the *left*. Therefore the Φ_{ind} it creates is negative. This negative Φ_{ind} replaces some of the negative Φ_{ext} being removed by the receding magnet. Thus once again *the induced flux opposes the change in the external flux.*

What is common to these two cases is that a current is induced in whichever direction creates an induced flux that opposes the change in external flux that is being imposed by $\mathbf{B}_{ext}$. That is, *the induced flux tends to maintain the original flux condition.* This observation turns out to be a general principle and is called **Lenz's law:**

A change in the external magnetic flux Φ_{ext} through a coil induces an emf in the coil. The direction of the current produced by this emf is such that the magnetic field it creates, $\mathbf{B}_{ind}$, produces a flux Φ_{ind} that opposes the change taking place in Φ_{ext}.

As an additional example, suppose you approach a coil with the *south* pole of a magnet, as in Fig. 20.9*a*. In this case, the field $\mathbf{B}_{ext}$ is directed to the *right*. Therefore Φ_{ext} through the coil is positive, and is increasing as the magnet approaches. By applying Lenz's law, you should be able to show that now the direction of I_{ind} is as shown in Fig. 20.8*b*. This current induces a magnetic field $\mathbf{B}_{ind}$ directed to the left. Therefore the Φ_{ind} it creates is negative. This negative flux cancels some of the increase in positive Φ_{ext} taking place due to the motion of the magnet. Again, as it must, *the induced flux opposes the change taking place in the external flux.*

A parallel analysis of Fig. 20.9*b* should convince you that the induced magnetic field in this case is directed as shown in Fig. 20.8*a*.

Using Lenz's law to find the direction of an induced current is simplified if you remember the following procedure:

1 Determine the direction of the external magnetic field that is passing through the loop. Once you know the direction of $\mathbf{B}_{ext}$, you know the sign of Φ_{ext}. For $\mathbf{B}_{ext}$ having a component along the positive x axis, call Φ_{ext} positive; for $\mathbf{B}_{ext}$ having a component along the negative x axis, call Φ_{ext} negative.
2 Determine whether $\mathbf{B}_{ext}$ is decreasing or increasing.
3 Determine which sign Φ_{ind} must have in order to oppose the change in Φ_{ext} due to the change taking place in $\mathbf{B}_{ext}$. (Remember, the induced flux does not necessarily oppose the external flux, but it does always oppose *changes* in that flux.)
4 Determine the direction $\mathbf{B}_{ind}$ must have in order to produce a Φ_{ind} having the sign determined in step 3.
5 Determine (from the right-hand rule) which direction the induced current must have in order to create the direction of $\mathbf{B}_{ind}$ found in step 4.

Up to this point we have considered flux changes produced by changes in the magnetic field passing through a coil. Remember, however, that flux also depends on the area of the coil and on its orientation relative to the field. Thus the flux through a coil can change in any of three ways:

1 Changes in B
2 Changes in area A
3 Changes in θ

Faraday's and Lenz's laws have proven to hold *no matter how the flux is made to change.* In later sections we shall apply these laws to cases where area and orientation change.

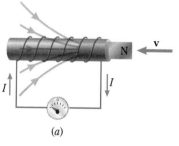

(a)

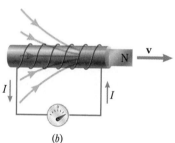

(b)

FIGURE 20.9

A south magnetic pole approaching and receding from a coil induces currents opposite those that occur for a north pole. Compare the drawings above with Figs. 20.7*b* and *c*.

Example 20.2

A solenoid with 100 loops has a cross-sectional area of 4.0 cm². It is suddenly transferred from a region where there is no magnetic field to one where the field is 0.50 T directed down the length of the solenoid. If the transfer requires 0.020 s, how large is the average emf induced in the solenoid?

R

Reasoning

Question What principle determines the induced emf?
Answer Faraday's law: $\overline{\text{emf}} = N(\Delta\Phi/\Delta t)$.

Question What is causing the flux to change?
Answer The general expression for flux is $\Phi = BA \cos \theta$. Here $\theta = 0$. Since A is the constant area of the solenoid, it is the change in B causing the change in flux.

Question What is $\Delta\Phi$?
Answer $\Delta\Phi = (B_2 - B_1)A$, where $B_1 = 0$ in this case.

Solution and Discussion

$$\Delta\Phi = (0.50 \text{ T})(4.0 \times 10^{-4} \text{ m}^2) = 2.0 \times 10^{-4} \text{ Wb}$$

The average induced emf is

$$\overline{\text{emf}} = \frac{(100 \text{ turns})(2.0 \times 10^{-4} \text{ Wb})}{0.020 \text{ s}} = 1.0 \text{ V}$$

If you go back to the definition of the tesla, you should be able to show that webers per second is equivalent to volts.

Example 20.3

In Fig. 20.10, a small 10-turn coil of radius $r = 5.00$ cm is inserted into a solenoid so that the axis of the coil is parallel to that of the solenoid. The coil is in a circuit containing a galvanometer and a resistance $R = 20.0 \, \Omega$. The solenoid has 2000 turns per meter of length and is carrying a current of 15.0 A in the direction shown. When the switch on the power supply to the solenoid is opened, the

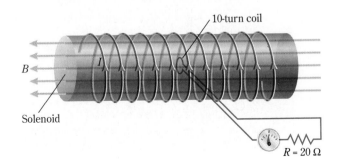

FIGURE 20.10

When the current in the solenoid changes, a current flows through the galvanometer. Why?

solenoid current becomes zero in 30.0 ms. (*a*) What is the average current through the galvanometer? (*b*) What is the direction of this current?

Reasoning Part (a)

Question Why will there be a current through the galvanometer?
Answer Because the initial magnetic field of the solenoid goes to zero as the current is interrupted. Some of this field contributes a magnetic flux through the 10-turn coil. As the solenoid's field decreases, this flux changes in time, inducing a current in the coil.

Question What determines the magnitude of the average induced current?
Answer The rate of change of flux induces an average emf in the coil: $\overline{\text{emf}} = N_{\text{coil}} (\Delta\Phi_{\text{coil}}/\Delta t)$. The average current that flows will be determined by Ohm's law: $\bar{I} = \overline{\text{emf}}/R$.

Question What is the initial flux through the small coil?
Answer The solenoid field has an initial value B_1. Thus $\Phi_1 = B_1 A_{\text{coil}} = B_1 \pi r^2$.

Question What is the expression for B_1?
Answer From Eq. 19.11, $B_1 = \mu_o n I_1$, where $n = 2000/\text{m}$ and $I_1 = 15.0$ A.

Solution and Discussion The change in flux is

$$\Delta\Phi = 0 - \Phi_1 = -(\mu_o n I_1)(\pi r^2)$$
$$= -(4\pi \times 10^{-7}\,\text{T} \cdot \text{m/A})(2000/\text{m})(15.0\,\text{A})\pi(0.0500\,\text{m})^2$$
$$= -2.96 \times 10^{-4}\,\text{Wb}$$

We can ignore the minus sign for now; it merely tells us that the flux is decreasing. We will examine the direction of the change in part (*b*). The magnitude of the average induced emf is

$$\overline{\text{emf}} = (10\,\text{turns})(2.96 \times 10^{-4}\,\text{Wb})/(30.0 \times 10^{-3}\,\text{s}) = 9.87 \times 10^{-2}\,\text{V}$$

The average induced current is

$$\bar{I} = \frac{\overline{\text{emf}}}{R} = \frac{9.87 \times 10^{-2}\,\text{V}}{20.0\,\Omega} = 4.93\,\text{mA}$$

Reasoning Part (b)

Question In what direction is the initial field passing through the coil?
Answer By using the right-hand rule, you can show that the magnetic field caused by the solenoid current is to the left in Fig. 20.10.

Question As the switch is opened, does the field in this direction increase or decrease?
Answer It decreases.

Question In what direction would an induced field from the small coil oppose the change of flux taking place?
Answer If the coil produces a magnetic field to the left, the flux created by this field will partially offset the decrease taking place in the solenoid's flux.

Question What current direction in the small coil would produce a magnetic field to the left?
Answer A current having the same direction as the original current in the solenoid. This induced current is in the left to right direction through the galvanometer and the resistor in Fig. 20.10.

20.4 MUTUAL INDUCTION

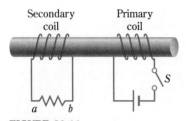

Secondary   Primary
coil   coil

a *b*

FIGURE 20.11

Why is the current direction in the secondary coil, at the instant the switch *S* is opened, from *a* to *b*?

Faraday's law for the induced emf in a coil applies to any method for changing the flux through the coil. Suppose we have two coils side by side, as in Fig. 20.11. When the switch is open, both coils have zero magnetic flux through them. When the switch is suddenly closed, the primary coil will act as an electromagnet and will generate flux in the region near it. Some of this flux will go through the secondary coil. Hence, the flux through the secondary will change when the switch is suddenly closed. According to Faraday's law, an induced emf will be generated in the secondary during the time the current in the primary rises from zero to its final value. You should be able to show that the direction of the induced current through the resistor in Fig. 20.11 will be from *b* to *a* just as the switch is closed. The current will be in the opposite direction just as the switch is opened.

The magnitude of the emf induced in the secondary will depend on many geometric factors. Among these are the number of turns of wire on each coil, how close together the coils are to each other, their orientation relative to each other, and their cross-sectional area. (Why?) In addition, since the flux through the secondary will be proportional to the current in the primary, the induced emf in the secondary will be proportional to the rate of change of current in the primary, $\Delta I_p/\Delta t$. We therefore write the following equation for the induced emf in the secondary:

$$\text{emf}_{\text{sec}} = M\frac{\Delta I_p}{\Delta t} \tag{20.4}$$

where the proportionality constant M contains the effects of the geometry of the two coils. M is called the **mutual inductance** of the two coils. If the emf is in volts, I in amperes, and t in seconds, the unit of inductance M is defined to be the **henry** (H), or $V \cdot s/A$.

Finally, a very important way to increase mutual inductance is to link the coils by means of a core of ferromagnetic material such as iron. Because of the large value of relative magnetic permeability K_m (Sec. 19.14), the field produced by a given current in the primary is greatly increased over what it would be without the iron core. This in turn greatly increases the magnetic flux linking the two coils for any given primary current. As the primary current is made to change, the flux change and hence the emf induced in the secondary is proportionately greater than would be the case without the core. This results in a large value of mutual inductance, as can be seen from the definition of M in Eq. 20.4.

Illustration 20.1

Two coils of wire wound on an iron core have a mutual inductance of 0.50 H. How large an average emf is generated in the secondary as the current in the primary is increased from 2.0 to 3.0 A in 0.010 s?

Reasoning Equation 20.4 gives

$$\overline{\text{emf}} = \frac{(0.50\,\text{H})(3.0\,\text{A} - 2.0\,\text{A})}{0.010\,\text{s}} = 50\,\text{V}$$

Remember, emf is being induced only during the 0.010 s that the primary current is changing. Once the current in the primary becomes steady, the flux linking the secondary is no longer changing, and emf is no longer induced.

20.5 SELF-INDUCTANCE

Faraday's law tells us that any change in the magnetic flux through a coil induces an emf in the coil. An isolated coil carrying a current creates a magnetic field whose flux passes through the plane of the coil. It follows that when the current in the coil changes, the flux passing through it does also. Thus any time the current in a coil is made to change, an emf is *self-induced* in the coil while this change is happening.

Suppose the current in the coil shown in Fig. 20.12 changes from zero to some finite value when the switch is first closed. An increasing magnetic field is generated by the rising current and is directed to the left through the coil. By Faraday's law, an emf is induced in the coil, and it tries to produce an opposing field to the right through the coil. Hence, the induced emf must be opposed to the emf of the battery. However, if the switch is suddenly opened, the induced emf will aid, rather than oppose, the battery. (You should be able to show this.)

The rate of change of magnetic flux through the coil will be proportional to the rate of change of current in the coil. If $\Delta I/\Delta t$ is the rate of change of current through the coil, we can write for the average induced emf

$$\overline{\text{emf}} = L\frac{\Delta I}{\Delta t} \tag{20.5}$$

The constant of proportionality L is called the **self-inductance** of the coil. It depends on the geometry of the coil as well as the core material around which the coil may be wound. L has the same units as mutual inductance: henrys.

If the coil is wound on an iron core, the flux through it is much greater than if no ferromagnetic material is present. Hence, if a large self-inductance is desired, the inductor should be wound on an iron core. We return to mutual and self-inductance in later sections. They are of particular importance in alternating-current (ac) circuits, where the current, and thus the flux, are changing continually.

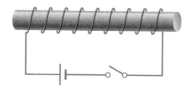

FIGURE 20.12

When the switch is first closed, the coil will induce an emf in itself. Will this emf aid or oppose the battery?

Example 20.4

A certain solenoid has cross-sectional area A, length l, and n loops per unit length. What is its self-inductance?

Reasoning

Question How is self-inductance defined?

Answer From Eq. 20.5, L is the proportionality constant between self-induced emf and the rate of change of current:

$$L = \frac{\text{emf}}{\Delta I / \Delta t}$$

Question What does the induced emf depend on?
Answer Faraday's law always applies:

$$\text{emf} = N \frac{\Delta \Phi}{\Delta t}$$

In this case, Φ is the flux through the solenoid created by its own field.

Question What is the self-flux for a solenoid?
Answer The field of an air-filled solenoid is, from Eq. 19.11,

$$B = \mu_o n I = \frac{\mu_o N I}{l}$$

Since the field is uniform throughout the inside of the solenoid, its flux is simply

$$\Phi = BA = \frac{\mu_o N I}{l} A$$

Question When the current changes, what change in flux results?
Answer Except for current, all the other factors in the expression for Φ are constant. Thus $\Delta \Phi / \Delta t$ must be proportional to $\Delta I / \Delta t$:

$$\frac{\Delta \Phi}{\Delta t} = \frac{\mu_o N A}{l} \frac{\Delta I}{\Delta t}$$

Question What expression for emf do I get by using this result in Faraday's law?
Answer $\text{emf} = N \dfrac{\Delta \Phi}{\Delta t} = N \dfrac{\mu_o N A}{l} \dfrac{\Delta I}{\Delta t} = L \dfrac{\Delta I}{\Delta t}$

Solution and Discussion From the last equation, you can see that for a solenoid

$$L = N \frac{\mu_o N A}{l}$$

Using $n = N/l$, we can also write this as

$$L = \mu_o n^2 l A$$

If the core of the solenoid was filled with a substance whose relative magnetic permeability was K_m, L would be multiplied by the factor K_m.

Exercise Evaluate L for a 500-loop air-filled coil that is 80.0 cm long and has a 1.20-cm diameter. *Answer:* $4.44 \times 10^{-5}\,H$

20.6 CIRCUITS CONTAINING INDUCTANCE AND RESISTANCE

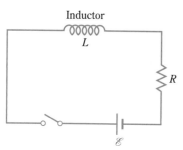

Inductor

L

R

$\mathscr{E}$

FIGURE 20.13

Why doesn't the current rise to a value $\mathscr{E}/R$ immediately after the switch is closed?

Some very interesting and useful properties of a self-inductance coil are explored in detail in Chap. 21. At present, we are concerned with only one facet of the inductor's behavior—its ability to store energy.

Consider the circuit in Fig. 20.13. It consists of an inductance coil (represented by the symbol ⌒⌒⌒), a resistor, a battery, and a switch. If the inductor were not in the circuit, the current would rise quickly just as the switch was closed and the final current would be $\mathscr{E}/R$. However, as the current rises, flux is generated in the inductance coil. This induces an emf in the coil in such a direction as to oppose the increasing current. In other words, the inductor acts as a battery whose polarity is the reverse of that of the real battery in the circuit.

The upshot of this is that the inductance of the coil diminishes the rate of increase of the current in the circuit. The larger the value of L, the greater the effect the coil has in delaying the increase in current. Despite this diminishing effect, the current rises until it ultimately reaches its steady Ohm's law value, $\mathscr{E}/R$. Calculus can be used to derive the exact time dependence of the current when the switch is closed at $t = 0$. The result is

$$I(t) = I_f(1 - e^{-t/(L/R)}) \tag{20.6}$$

where $I_f = \mathscr{E}/R$ and $e = 2.718 \ldots$ is the base of natural logarithms. Take some time to examine the behavior of this expression. Your pocket calculator has a key marked "e^x" that makes it simple to calculate e raised to any power.

A graph of the behavior of Eq. 20.6 is shown in Fig. 20.14. You should be able to show that at $t = 0$, Eq. 20.6 gives $I = 0$. (Remember that any number to the zero power equals 1.) The quantity L/R in the exponent has the units of time. You should verify this. It is called the **inductive time constant** τ_L of the circuit. Using your calculator, you can find that, at $t = L/R = \tau_L$,

$$I(t = \tau_L) = I_f(1 - e^{-1}) = I_f\left(1 - \frac{1}{2.718}\right) = 0.63I_f$$

Figure 20.14 shows this point on the graph of I versus t. At $t = 2\tau_L$,

$$I = I_f(1 - e^{-2}) = 0.865\ I_f$$

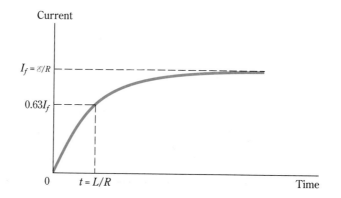

Current

$I_f = \mathscr{E}/R$

$0.63I_f$

0 $t = L/R$

Time

FIGURE 20.14

After the switch in Fig. 20.13 is closed, the current rises as shown here.

The larger the time constant L/R, the more slowly the current rises to its final value. Let us now compute how much work is done against the opposing emf of the coil.

The emf induced in the coil is, from Eq. 20.5, $L(\Delta I/\Delta t)$. Therefore, when there is a current through the coil, the charges flow through a potential difference $L(\Delta I/\Delta t)$. The work the current does as it carries a charge Δq through the inductor and through the potential difference $L(\Delta I/\Delta t)$ is, from Eq. 17.2,

$$\Delta W = (\Delta q)(V) = (\Delta q)\left(L\frac{\Delta I}{\Delta t}\right)$$

This expression can be simplified if we notice that $\Delta q/\Delta t$ is simply I. Then,

$$\Delta W = LI\,\Delta I$$

To summarize, this is the work required to increase the current from the value I to an amount $I + \Delta I$.

We wish to sum all these small quantities of work as the current in the circuit changes from zero to its final maximum value I_f. The result for the work expended while the current in the coil is changed from $I = 0$ to $I = I_f$ is found to be

$$W = \tfrac{1}{2}LI_f^2$$

This supplied work can be thought of as energy stored in the coil. Vivid evidence for this stored energy is shown when the switch in a circuit such as that of Fig. 20.13 is pulled open. If the inductance is large, a large spark jumps across the gap of the switch. Moreover, a very high voltage is induced in the coil as it tries unsuccessfully to oppose the loss of flux through itself. We therefore have found that

An inductor L through which there is a current I has stored in it an energy $\tfrac{1}{2}LI^2$.

20.7 THE ENERGY IN A MAGNETIC FIELD

You will recall that we computed the energy stored in an electric field by considering the energy stored in a capacitor (Sec. 17.12). Let us now find the energy stored in a magnetic field by considering the energy stored in an inductor. We assume the inductor to be a long solenoid. As was shown in Chap. 19, the magnetic field is confined mainly to the core of the solenoid and has a uniform value $B = \mu_o nI$.

We computed the inductance of a solenoid in Example 20.4.

$$L = \mu_o n^2 lA$$

where l is the length of the solenoid and A is its cross-sectional area. Notice, however, that lA is simply the volume of the core region of the solenoid.

The energy stored inside the solenoid is

$$\text{Energy} = \tfrac{1}{2}LI^2 = \tfrac{1}{2}\mu_o n^2 I^2 lA$$

from which the energy per unit volume is

$$\frac{\text{Energy}}{\text{Volume}} = \frac{\frac{1}{2}\mu_o n^2 I^2 lA}{lA} = \frac{1}{2}\mu_o n^2 I^2$$

The magnetic field in the solenoid is $B = \mu_o n I$, however, from which $I = B/\mu_o n$. Substituting this value in the above equation gives

$$\frac{\text{Energy}}{\text{Volume}} = \frac{1}{2}\mu_o n^2 \frac{B^2}{\mu_o^2 n^2}$$

$$\text{Energy per unit volume} = \frac{B^2}{2\mu_o} \tag{20.7}$$

for the energy density in a magnetic field of magnitude B. This should be compared with the value $\frac{1}{2}\epsilon_o E^2$ (Eq. 17.14) we found for the energy density in an electric field in vacuum.

If the solenoid is filled with a material whose relative magnetic permeability is K_m, Eq. 20.7 is still valid if μ_o is multiplied by K_m. Although Eq. 20.7 was derived for the case of a solenoid, it is a perfectly general result. The concept of energy stored in a magnetic field will be important when we study the way in which light and other electromagnetic waves carry energy.

Example 20.5

A certain coil has an inductance of 0.500 H and a resistance of 2.0 Ω. It is connected in series with a switch, a 12.0-V battery, and a 4.0-Ω resistor. Find (*a*) the time constant of the circuit, (*b*) the final value of the current, (*c*) the value of the current at the instant 0.050 s after the switch is closed, (*d*) the final energy stored in the inductor.

R

Reasoning

Question What is the expression for the time constant?
Answer $\tau_L = L/R$, where R is the *total* resistance in the circuit.

Question What determines the final current?
Answer Ohm's law: $I_f = \mathscr{E}/R$.

Question How does the current increase with time?
Answer Equation 20.6 shows that $I(t) = I_f(1 - e^{-t/\tau_L})$.

Question How do I use this expression to find the current at a particular time?
Answer To find the current at any instant of time, put that value of t into Eq. 20.6 and evaluate the expression with the help of the e^x (or inv ln) key on your calculator.

Question On what does the energy stored in an inductor depend?
Answer Energy = $\frac{1}{2}LI^2$

Solution and Discussion From the data given, we obtain

$$\tau_L = \frac{L}{R} = \frac{0.50\,\text{H}}{2.0\,\Omega + 4.0\,\Omega} = 0.083\,\text{s}$$

As always, you should convince yourself of the correctness of derived units, in this case that henrys per ohm is equivalent to seconds. The final current is

$$I_f = \frac{12.0\,\text{V}}{6.0\,\Omega} = 2.0\,\text{A}$$

The current at $t = 0.050\,\text{s}$ is

$$
\begin{aligned}
I(t = 0.05\,\text{s}) &= (2\,\text{A})[1 - e^{-(0.050\,\text{s})/(0.083\,\text{s})}] \\
&= (2\,\text{A})[1 - e^{-(0.60)}] = (2\,\text{A})(1 - 0.55) \\
&= (2\,\text{A})(0.45) = 0.91\,\text{A}
\end{aligned}
$$

The final energy stored is

$$\text{Energy} = \tfrac{1}{2}(0.50\,\text{H})(2.0\,\text{A})^2 = 1.0\,\text{J}$$

Again, you should prove these are the correct units.

20.8   MOTIONAL EMF

An emf can be induced in many ways. Until now, we have been dealing mainly with flux changes through stationary coils and the attendant induced emf. Sometimes, however, the induced emf is the result of the motion of a wire through the magnetic field. In such cases, it is often more convenient to derive a result that is not based directly on the concept of flux change through a loop.

Let us begin our discussion by referring to the simple experiment shown in Fig. 20.15. A rod of approximate length l rolls with velocity $\mathbf{v}$ along parallel wires that form a U-shaped loop from m through r and s back to n. Notice that the rod and wires form a loop *(pqrsp)* to the left of the rod. As the rod moves to the right, the area of this loop increases.

Suppose a magnetic field $\mathbf{B}$ directed out of the page exists in this region. As the rod moves along, the flux through the area of the loop increases because the area increases. Hence an emf is induced in the loop. To find this emf, we note that in time Δt the rod rolls a distance $v\,\Delta t$. As a result, the loop area increases by an amount $\Delta A = l(v\,\Delta t)$, the shaded area in the figure. The accompanying flux change is

$$\Delta \Phi = B_\perp \Delta A = B_\perp lv\,\Delta t$$

FIGURE 20.15

As the rod moves to the right, the area enclosed by the circuit *pqrsp* increases, thereby increasing the magnetic flux through this circuit. According to Lenz's law, this results in an emf being induced in the circuit.

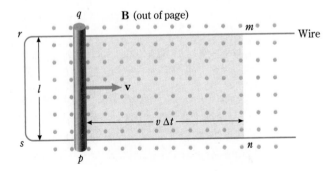

Then, according to Faraday's law, the magnitude of the induced emf in the loop is

$$\text{Induced emf} = \frac{\Delta\Phi}{\Delta t} = B_\perp vl$$

You should convince yourself that this induced emf will cause a clockwise current around the loop.

There is another way to analyze this situation. Consider a positive charge q in the moving rod as shown in Fig. 20.16. This charge, because of its motion with velocity $\mathbf{v}$ through the field $\mathbf{B}$, experiences a force $qvB_\perp$. In this case, the total field is perpendicular to the velocity of the charge and so $B = B_\perp$. We conclude that

$$F = \text{force on } q = qvB_\perp$$

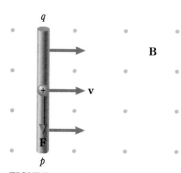

FIGURE 20.16

The force on a positive charge in a conducting rod moving transverse to a magnetic field.

Using the right-hand rule of Fig. 19.10, you can see that the force on q is directed from q to p along the rod.

From the definition of electric field as force per unit charge, however, we conclude that the charges moving with the rod experience an electric field directed from q to p along the rod. Thus*

$$E = \frac{F}{q} = vB$$

If we recall that the electric potential difference between two points equals the work done in carrying a unit test charge from one point to the other (Eq. 17.2), then we know that the potential difference from p to q due to the electric field E is

$$V = El = B\perp vl$$

Notice that this is exactly equal to the induced emf in the loop we found using Faraday's law. Moreover, the electric field induced by the charge motion causes a clockwise current in the loop, the same direction we found from Faraday's law.

We can summarize these latter results in the following way:

A wire (or rod) of length l moving with velocity $\mathbf{v}$ perpendicular to both a magnetic field $\mathbf{B}$ and its own length has an induced emf along its length of

$$\text{Induced emf} = B_\perp vl \tag{20.8}$$

This is called a *motional emf*. Notice that a loop or complete circuit is not necessary for the emf to be induced between the ends of the rod. In the more general case in which $\mathbf{B}$, $\mathbf{v}$, and the wire are not mutually perpendicular, the components of $\mathbf{B}$ and $\mathbf{v}$ that are perpendicular to each other and to the wire must be used.

The statement immediately preceding Eq. 20.8 is often paraphrased in terms of cutting across lines of magnetic field. When the rod in Fig. 20.15 moves so as to change the flux through the loop by an amount $\Delta\Phi$, the rod cuts magnetic field lines. But the induced emf in the rod is simply $\Delta\Phi/\Delta t$, proportional to the rate at which the rod is cutting the field lines. We can therefore state:

A moving wire has induced within itself an emf proportional to the rate at which the wire is cutting lines of magnetic field.

*Strictly speaking, this value for E applies only in a reference frame moving with the charge.

As we will see, the concept of a motional emf is convenient in many situations.

Before we leave this section, we should say a few words about the conservation of energy when current is generated by a motional emf. In the arrangement shown in Fig. 20.15, the magnitude of the current in the circuit *(pqrsp)* is determined by the circuit resistance. Thermal energy will be dissipated in R at a rate of I^2R, or $(\text{emf})^2/R$. Where does this energy come from? The answer lies in the fact that, once a current is generated in the moving rod, there is a magnetic force exerted on the rod. You should be able to show that, with the current directed from q to p in Fig. 20.15, this force is directed to the left, opposite the rod's velocity. This means that *an equal force must be applied in the direction of motion in order to maintain a constant speed.* Let us put this together in mathematical form:

Magnetic force on rod: $F = BIl$ (to the left)

The power delivered by an equal applied force to the right:

$F_{app}v = BIlv = Blv(\text{emf})/R = Blv(Bvl)/R = (Bvl)^2/R$

Thermal power dissipated in R: $P = \dfrac{(\text{emf})^2}{R} = (Bvl)^2/R$

It is obvious that the power supplied by the applied force is equal to that dissipated as heat in the resistor. Thus energy is conserved, as always.

Example 20.6

A rod of length 5.0 m is held horizontal with its axis in an east-west direction. It is allowed to fall straight down. What emf is induced in it when its speed is 3.0 m/s if the earth's magnetic field is 0.60 G at an angle of 53° below the horizontal?

R

Reasoning

Question What quantities does the induced emf depend on?
Answer The speed of the rod perpendicular to its length, its length, and the magnetic field strength perpendicular to its velocity. Equation 20.8 gives

$\text{emf} = B_\perp vl$

Question How do I calculate $B_\perp$?
Answer Since the velocity is vertical, $B_\perp$ is the horizontal component of **B**. A vector diagram of **B** would show that $B_\perp = B\cos 53°$.

Question The field is given as 0.60 G. What is this in SI units?
Answer The conversion is $1\,T = 10^4\,G$.

Solution and Discussion The induced emf is

$\text{emf} = (0.60 \times 10^{-4}\,T)(\cos 53°)(3.0\,\text{m/s})(5.0\,\text{m}) = 5.4 \times 10^{-4}\,V$

This hand-turned demonstration generator clearly shows the rotating coil, the surrounding magnet (the lower rectangular coil energizes this electromagnet), and the slip-ring brushes on the axis of the coil (connected to the red terminals).

20.9 AC GENERATORS

A generator is a device that converts mechanical energy to electrical energy. It does this by changing the flux through a coil, thereby inducing an emf between the two terminals of the coil. In theory, the flux could be changed either by moving a magnet with respect to the coil or by moving the coil with respect to a magnet. The latter procedure is more easily realized in practice and is the one ordinarily used.

A schematic representation of a simple generator is shown in Fig. 20.17. An external energy source causes the loop of wire to rotate in the magnetic field of the magnet. (In practice, the loop is replaced by a coil wound on an iron core so as to intensify the effects we discuss.) As the loop rotates, the flux through it changes continually. This changing flux induces an emf in the loop, and the emf causes a current through the loop in the direction indicated. The current can be used to do useful work, perhaps to light a bulb as shown.

In a well-designed generator, the external energy source that rotates the coil does minimal work against friction. Work it must do, however, because the generator produces current that can do work. To see how this interchange of input work with output work occurs, recall what happens to the current-carrying wire of the coil. Since the wire passes through a magnetic field, the current experiences a force due to the magnetic field. As in the example of the previous section, this force turns out to be in a direction that opposes the rotation of the coil; the larger the current, the larger the opposing force. Thus we see that the external energy source must do work to rotate the coil, and the more current drawn from the generator to do useful work, the more work the external energy source must do to rotate the coil. In this way, the energy source that drives the generator furnishes the energy that the generator current uses to do useful work. Examples of the external source might be a diesel motor or a waterfall. Let us now look at the operation of the generator in detail so that we can ascertain the form of the emf it produces.

In your mind's eye, replace the single loop of wire in Fig. 20.17 by a coil containing N loops. The coil rotates on axis aa' in a uniform magnetic field. Notice that one end of the coil is attached to ring R and the other end to ring R'. These rings, called *slip rings,* are fastened rigidly to the coil and rotate as a unit with it. Contact between the rotating rings and the stationary outside terminals is made by means of the brushes B and B' that slide along the rings. In a very simple motor, the brushes might be short ribbons of spring steel.

To see how the induced emf between the terminals of the coil is generated, let us

FIGURE 20.17

An alternating emf is produced between terminals B and B' as the loop rotates in an external magnetic field.

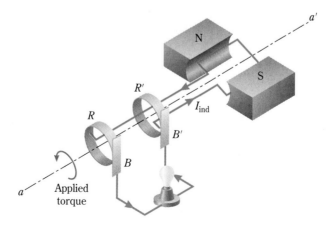

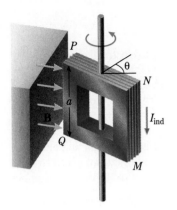

(a) Perspective view

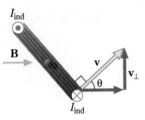

(b) Top view

FIGURE 20.18

An induced emf is produced in the rotating coil, resulting in an induced current in the coil.

refer to Fig. 20.18a. The coil is assumed to be rotating in the direction indicated. As you see, it is moving from a position in which the field lines are perpendicular to its plane to a position in which the lines just skim past it. In other words, the flux due to field lines directed to the right through the coil is decreasing. Because of this change in flux, an emf is induced in the coil.

Using Lenz's law, we see that the emf induced in the coil because of this changing flux in turn induces a current in the direction indicated, so as to try to maintain the flux through the coil, counteracting the change.

Notice, however, what happens once the coil has turned through 180° from the position shown. Everything in Fig. 20.18a is unchanged except that points M and N are interchanged with points Q and P. As a result, the induced current is now directed opposite what it was before. It is clear that the induced current in the coil keeps reversing as the coil continues to turn.

To analyze this situation quantitatively, we could find $\Delta\Phi/\Delta t$ for the coil and use Faraday's law to compute the induced emf. When calculus is used, this approach is very convenient. However, we can opt to analyze the system in terms of motional emfs.

We will compute the induced emf in side MN, which is moving with velocity $\mathbf{v}$ through the field, by first computing $v_\perp$, the magnitude of the component of $\mathbf{v}$ that is perpendicular to $\mathbf{B}$. From Fig. 20.18b, we see that $v_\perp = v \sin \theta$. Because the wire MN, the field $\mathbf{B}$, and $\mathbf{v}_\perp$ are all mutually perpendicular, the induced emf in MN is

$$\text{emf}_{MN} = B(v \sin \theta)(a)$$

Using the right-hand rule for the deflection of moving positive charges, you should satisfy yourself that the direction of the induced current is from N to M in side MN and from Q to P in side PQ. Therefore, an identical induced emf in PQ will augment the induced emf in MN. Notice that sides PN and MQ do not cut field lines as the coil rotates. Thus no motional emf is induced in these sides.

Induced emf in loop $= 2Bva \sin \theta$

We can put this in a more convenient form if we note that v is the tangential speed of point M as it describes a circle about the axis of rotation. Calling the radius of this circle r (where $r = \frac{1}{2}\overline{MQ}$), we have

$$v = \omega r = 2\pi f r$$

where ω is the constant angular velocity of the coil and f is its frequency of rotation. As you should remember from Chaps. 7 and 13, ω is measured in radians per second and f is measured in hertz. Moreover, θ is simply the rotation angle of the loop, and it increases continuously by the relation

$$\theta = \omega t = 2\pi f t$$

When these substitutions are made, the induced emf becomes

$$\text{emf} = 2\pi f B(2ra) \sin 2\pi f t$$

but $2ra$ is simply the area A of the loop, and so we have as our final result

$$\text{emf} = 2\pi f A B \sin 2\pi f t \qquad (20.9)$$

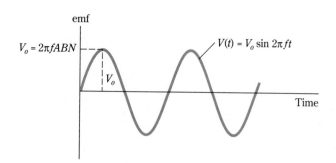

FIGURE 20.19

An alternating emf is induced in a coil rotating in a uniform magnetic field. The emf varies sinusoidally with time.

If instead of a single loop we have a coil consisting of N loops, the emf would be N times as large.

As we see, *the induced emf in a rotating coil varies sinusoidally with time.* A graph of this behavior is given in Fig. 20.19. The induced emf (or voltage) has its maximum value when $\sin 2\pi ft = 1$. Therefore the maximum emf is $2\pi fABN$. It is reasonable that the induced voltage should be large for large f (flux is changing rapidly), for large A and B (the flux is large), and for a large number of loops on the coil.

Equation 20.9 is frequently written in the alternative form

$$V = V_o \sin 2\pi ft$$

where V is the voltage at any instant and V_o is the maximum voltage. Clearly, the voltage in the rotating coil varies sinusoidally and reverses its direction twice during each rotation cycle.

From the above considerations, it becomes evident that a coil of wire rotating at constant angular speed in a magnetic field has an alternating emf generated between its terminals. If such a generator is used as the power source in the simple circuit shown in Fig. 20.20, the current through the resistor will reverse its direction $2f$ times per second. (Notice that the symbol for an alternating-voltage generator is $\ominus$.)

The ac generators used by power companies are usually more complex than the one discussed here, but their basic operation is the same. Mechanical energy to rotate the coil is usually furnished by steam turbines or by water power. Let us briefly consider the conversion of energy in a system such as that shown in Fig. 20.20.

If the circuit is open so that there is no current in the generator coil, very little force need be exerted to rotate the coil. As soon as current is drawn from the generator (the coil), however, the magnetic field exerts a force on the current-carrying wires of the generator, and these forces are in such a direction as to stop the coil from rotating. Hence, the mechanical energy fed into the generator is dependent on the current drawn from the generator—more current requires more mechanical energy.

At an instant when the voltage of the generator is V, the power being delivered to the resistor in Fig. 20.20 is VI (Eq. 18.7). Clearly, if I is very small, the power consumed by the resistor is small and the mechanical energy needed to operate the generator is small. We therefore see that *the energy needed to operate the generator depends directly on the energy being drawn from it. The mechanical energy is transformed to electrical energy by means of the interactions between magnetic field and charge motion within the coil of the generator.*

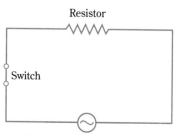

FIGURE 20.20

A simple ac circuit.

Electric motors are used in a wide variety of applications, which is reflected in a wide variety of sizes.

20.10   ELECTRIC MOTORS

An electric motor is a device that converts electrical energy to mechanical energy. A schematic diagram of a simple motor is shown in Fig. 20.21. A source of emf (a battery in this case) sends current through the loop of wire, part of which sits in the field created by a permanent magnet. This external magnetic field causes the loop

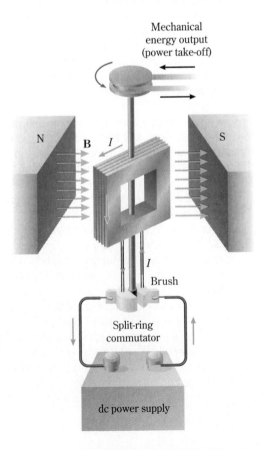

FIGURE 20.21

A simple dc motor. With the slip ring as shown, which way should the motor rotate?

to experience a torque that rotates the loop on its axis. (Use the right-hand rule of Fig. 19.10 to convince yourself that the loop rotates in the direction indicated.) Thus the energy furnished to the loop by the battery causes the loop to turn and thereby causes it to do external work by means of a pulley on its axle. The more work the motor performs, the harder it is to turn and so the more energy the battery must furnish to it.

In your mind's eye, replace the single loop in Fig. 20.21 by a coil wound on an iron core so as to make the motor more realistic. As we discussed previously, a current-carrying coil wound on an iron core acts as an electromagnet. Refer to Sec. 19.11 and Fig. 19.23 to convince yourself that the front side of the coil (as shown in the figure) is its north pole and the rear side is its south pole. Because of the nearby permanent-magnet poles, forces on the poles of the coil cause it to turn in the direction indicated. However, when the plane of the coil is perpendicular to the page, its south pole is as close as possible to the permanent north pole; the coil will then stop rotating if nothing further is done.

To keep the coil rotating, we reverse the direction of the current in it, thereby reversing its north and south poles. This reversal is accomplished by means of what is called a *split-ring commutator*. Electrical contact is made with the separate halves of the ring through stationary brushes that slide on the ring as it and the coil turn. (The brushes are typically slippery conducting blocks of graphite pushed against the ring by springs.) Notice that, as the coil turns, current enters it first through one half of the ring and then through the other half. In this way, the direction of the current through the coil is reversed at just the proper instant to keep the coil turning.

There are many kinds of electric motors. Many use electromagnets instead of permanent ones. Most have more than one coil so as to produce a more constant torque. Some run on both ac and dc voltage, whereas others run on only one or the other. In all motors, however, the source of emf furnishes energy to the coil by means of a current. It is this energy that the motor uses to do work.

Before leaving the subject of motors, we should point out that a motor is much like a generator running in reverse. The rotating coil of the motor acts like the coil of the generator and has an emf generated in it. This emf is in such a direction as to oppose the emf running the motor. For this reason, it is called a *back* or *counter emf*. Since the resistance of a motor is usually quite small, the chief limitation on the current through it is the back emf. When a motor is overloaded, it slows down. This in turn decreases the back emf (why?) and allows the motor to draw more current. This increased current through the overloaded motor may, on occasion, become large enough to burn it out. To protect against this, many motors have a thermal switch that inactivates them when they become too hot.

Example 20.7

The winding of a particular permanent-magnet dc motor has a resistance of 2.0 Ω. The motor is designed to deliver 500 W of mechanical power output when operating on a 120-V power line capable of supplying up to 20 A. (*a*) What current does the motor draw? (*b*) What is the back emf it develops?

R **Reasoning**

Question What is the power output of the motor related to?
Answer Some of the power supplied by the line is converted to waste heat in the motor's resistance. The rest can be converted to mechanical power.

Question   What is the expression for the power supplied by the battery?
Answer   Eq. 18.7: P (supplied) $= IV$.

Question   What is the expression for the power dissipated in a resistor?
Answer   Again from Eq. 18.7: P (heat) $= I^2R$.

Question   What is the expression for the power delivered by the motor?
Answer P (output) $= P$ (supplied) $- P$ (heat) $= 500$ W

This gives a quadratic equation that can be solved for the current:

$$IV - I^2R = 500 \text{ W}$$

Question   What determines the back emf of the motor?
Answer   As an element of the circuit, the motor can be treated as a resistance in series with its back emf (Fig. 20.22). Kirchhoff's loop rule yields

$$120 \text{ V} - \mathcal{E} - IR = 0$$

With I known, the back emf can be solved for.

Solution and Discussion   First, get the quadratic equation for I in standard form: $ax^2 + bx + c = 0$.

$$-RI^2 + VI - 500 \text{ W} = 0$$

You can see that the quadratic coefficients are $a = -R$, $b = V$, and $c = -500$. The solution of the quadratic equation is

$$I = \frac{-V \pm \sqrt{V^2 - 4(-R)(-500)}}{2(-R)}$$

After putting in the values of V and R, you get the two possible solutions

$$I = 4.5 \text{ A}, 55 \text{ A}$$

You always have to find the one solution that makes *physical* sense. The 55-A current is more than the line can supply. Taking $I = 4.5$ A, therefore, you find that the back emf is

$$\mathcal{E} = 120 \text{ V} - IR = 111 \text{ V}$$

You can see that the back emf of a motor can be quite large.

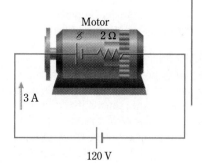

FIGURE 20.22

A motor acts as though it were a resistance in series with a counter emf.

Example 20.8

An ac generator is designed to produce a 60-Hz alternating voltage. Its coil has 500 turns and rotates in a magnetic field of 0.50 T. (*a*) How large an area must the coil have in order for the maximum emf to be 120 V? (*b*) If the emf has the time dependence shown in Fig. 20.19, what is the instantaneous voltage at $t = 10^{-3}$ s? How much later will the voltage have this same phase?

R

Reasoning

Question How does the maximum emf depend on the area of the coil?
Answer The maximum emf is directly proportional to A.

Question What else does the maximum emf depend on?
Answer Frequency, number of turns, and magnetic field:

$$\text{emf}_{\text{max}} = V_o = 2\pi fNAB$$

where V_o is 120 V.

Question What expression describes the time behavior in Fig. 20.19?
Answer You should remember from Chap. 14 that the general expression for sinusoidal time dependence is $\sin 2\pi ft$. Thus $V(t) = V_o \sin(2\pi ft)$.

Question What is the value of this at 10^{-3} s?
Answer You have to find the value of the sine, remembering that $2\pi ft$ is measured in radians. With $f = 60$ Hz, you have

$$2\pi ft = 2\pi(60/\text{s})(10^{-3}\,\text{s}) = 0.377\,\text{rad}$$

Question How often does the voltage go through the same value of its phase?
Answer Once per cycle, or the *period* of oscillation. The period is just $1/f$.

Solution and Discussion

$$A = \frac{V_o}{2\pi fNB} = \frac{120\ V}{2\pi(60/\text{s})(500)(0.50\ \text{T})}$$

$$= 1.27 \times 10^{-3}\,\text{m}^2$$

This is equivalent to a square about 3.6 cm on a side.
At $t = 10^{-3}$ s we have $\sin(0.377\,\text{rad}) = 0.368$. Thus at this instant the voltage has the value

$$V = (120\ \text{V})(0.368) = 44.2\ \text{V}$$

The period of the oscillating voltage is

$$T = \frac{1}{f} = 0.0167\,\text{s}$$

20.11 TRANSFORMERS

One of the most important applications of electromagnetic induction takes place in the transformer, a device that changes (transforms) one ac voltage to another ac voltage. For example, in the typical television set, a transformer changes the 120-V ac input voltage to the 15,000 V needed to operate the picture tube. As another example, the common doorbell requires a voltage of about 9 V, and a transformer is used to obtain this voltage from the 120-V house-line voltage. Transformers cannot be used to change dc voltages because a continually changing flux is basic to their operation.

Transformers (foreground) in this electric power substation are used to change the ac voltage from high voltage cross-country transmission lines to the lower voltages used in local distribution lines.

A typical transformer is shown in Fig. 20.23. The transformer consists of an iron core onto which are wound two coils, the primary (with N_p loops) and the secondary (with N_s loops). The primary coil is connected to the ac power source, and the alternating current in this coil sets up a changing magnetic field in the iron core. Because flux lines tend to follow iron, the lines circle through the secondary coil as indicated. Therefore the flux Φ through the primary and secondary coils is the same.

The changing flux through the secondary coil gives rise to an induced emf in it:

$$\text{Secondary emf} = -N_s \frac{\Delta\Phi}{\Delta t}$$

In most transformers, the resistance of the coils is negligible, and so the current-limiting factor in the primary is the back emf the primary coil induces in itself. In other words, the induced emf in the primary is equal to the voltage of the power source. We can therefore write that

$$\text{Primary emf} = -N_p \frac{\Delta\Phi}{\Delta t}$$

where Φ is the same flux that flows through the secondary coil.

Taking the ratio of these two induced emfs, we find

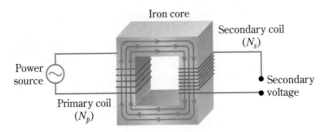

FIGURE 20.23

An iron core step-up transformer.

$$\frac{\text{Secondary emf}}{\text{Primary emf}} = \frac{N_s}{N_p} \tag{20.10}$$

This is the **transformer equation,** and it tells us how the secondary emf is related to the primary emf. The two are in the same ratio as the number of loops on the coils. A transformer that raises the input emf ($N_s > N_p$) is called a *step-up transformer* and one that lowers it ($N_p > N_s$) is called a *step-down transformer.* Notice carefully that transformers make use of ac, not dc, voltages.

If the secondary circuit is not closed, the current in it must be zero. Hence, there is no power loss in the secondary coil when it is not in use. Moreover, we show in the next chapter that there is also no power loss in an inductor that has no resistance. This fact makes it possible for the power company to keep transformers on line throughout a city even when no one is using the electricity they are providing. The transformers themselves consume very little energy.

However, if current is drawn from the secondary—to run an electric heater, for example—energy is consumed by the heater. This energy must be fed into the primary of the transformer so that it can be delivered to the secondary. Under these conditions, the loss in power at the secondary causes the primary to act as though it had resistance.

One of the most important uses of transformers has to do with power transmission. Many power companies provide power to cities that are perhaps 100 km from the generators. This proves to be quite a problem. Suppose that each person in a city of 100,000 people is using 150 W of power. This is the equivalent of one or two lighted light bulbs for each person. The power consumed is (150)(100,000) W, and at a voltage of 120 V (the usual house voltage) we have

$$P_{\text{tot}} = VI$$
$$(150 \text{ W})(100,000) = (120 \text{ V})(I)$$
$$I = 125,000 \text{ A}$$

Since an ordinary house wire can safely carry only about 20 A without overheating, the power company would need the equivalent of about 6500 of these wires to carry power to the city at this level of current. Although this is not impossible, the cost of the copper alone would be tremendous. The power companies get around this difficulty quite nicely by noticing that the important quantity in determining power is VI and not I alone. In the above example, if V is 100,000 V, we now have

$$(150 \text{ W})(100,000) = (100,000 \text{ V})(I)$$
$$I = 150 \text{ A}$$

As you see, the required current is much less in this case. High-voltage, low-current transmission also has the very important consequence that heating losses in the transmission wires are greatly reduced. You will recall that these power losses depend on the *square* of the current (I^2R), so that reducing the current by a factor of 1000 reduces the power lost in transmission by a factor of about 1 million! Thus power companies use high-voltage lines (sometimes referred to as high-tension lines) to transmit power over long distances. In some cases, the transmission voltage is greater than 500,000 V.

Of course, they would not dare to have such high voltages wired directly to a house because the danger from electrocution and fire would be tremendous. Instead, step-down transformers are used at distribution substations and again on local power poles to convert these high voltages to about 120 V.

Many houses also have 240-V lines because large appliances (air conditioners, dryers, stoves) are usually run on 240 V rather than 120 V. This is for essentially the same reason that the power companies use high voltages. You should be able to explain why these large power-consuming devices are more profitably run on 240 V than on 120 V.

A MODERN PERSPECTIVE
THE MAGNETIC PROPERTIES OF SUPERCONDUCTORS

In Sec. 18.13 we discussed the phenomenon of superconductivity. In Chaps. 19 and 20 we have studied the relationship between currents and magnetic fields and how currents can be induced by changing magnetic flux. We can now put together what we have learned and examine the magnetic consequences of superconductivity.

The ring-current test Onnes devised to measure whether or not his superconducting lead had zero resistance was the following: The lead ring was placed in an external magnetic field and then its temperature was lowered to 4.2 K, the temperature of liquid helium (T_c for lead is 7.2 K). The superconducting ring was then removed from the magnetic field, inducing a current which tended to maintain the original magnetic flux through the ring. Once out of the external field, no more emf would be induced in the ring, and so the current should decay exponentially with an inductive time constant $\tau_L = L/R$. By observing the magnetic field produced by the induced ring current, one could observe the decay of the current even if τ_L was very large (very small R) since the period of observation could extend over a very long time. As mentioned in Chap. 18, ring currents have persisted undiminished for years, demonstrating that R is truly zero.

In 1933 the German physicists Meissner and Ochsenfeld found a surprising new magnetic property of superconductors, now known as the **Meissner effect.** They placed a lead sphere in an external magnetic field and then lowered its temperature below T_c, keeping the sphere in the external field. Since this procedure produces no change of external magnetic flux, Faraday's law predicts no induced currents. Amazingly, however, as the material becomes superconducting, currents on the surface of the sphere spontaneously occur to completely cancel the external field in the interior of the material. This is evidence that a superconductor is *perfectly* diamagnetic, with a relative magnetic permeability $K_m = 0$ (Sec. 19.14). It is important to repeat that the currents that expel the external field in the Meissner effect *are not* a consequence of the induction we have just studied. A magnetic field can penetrate the interior of a conductor, even a "classically perfect" one. This spontaneous exclusion of an external magnetic field is a specific and unexpected phenomenon of the superconducting state.

It was further found that the Meissner effect is observed only for external fields below a certain critical value. In other words, strong external magnetic fields can destroy the superconducting state. External fields up to this critical value reduce T_c for a given material. For pure metals, these critical fields are small, ranging typically from 5 to 200 mT. Onnes tried to use large currents in superconductors to produce large magnetic fields, but quickly realized that the internal magnetic fields produced by currents could render superconductivity impossible. However, researchers later discovered alloys of niobium that could maintain the superconducting state in fields exceeding 15 T. These and other high-critical-field alloys are known as type II superconductors, and have been used to generate and maintain larger magnetic fields than are possible by other means.

There are many other applications of superconductivity still being developed. They include superconducting electrical generators and transmission lines to reduce power losses due to resistance. Devices known as *SQUIDS* (*s*uperconducting *q*uantum *i*nterference *d*evices) can measure magnetic field changes with extreme precision and are being used to measure magnetic patterns associated with brain, heart, and other organ functions. *Josephson junctions* allow extremely precise (seven-digit) measurements of voltages and can be used as very-high-speed switches in logic elements in computers. The cost of magnetically suspended transportation systems (*magnetic levitation,* or *maglev*) would be decreased greatly if high T_c materials could be used to generate the strong magnetic fields required. It seems certain that superconductivity will continue to play an increasingly important practical role in our lives. This has proven to be true of almost all phenomena which we have come to understand as a result of basic physics research.

LEARNING GOALS

Now that you have finished this chapter, you should be able to

1 Define (*a*) induced emf, (*b*) magnetic flux, (*c*) Faraday's law, (*d*) Lenz's law, (*e*) mutual and self-inductance, (*f*) inductive time constant, (*g*) motional emf, (*h*) ac voltage, (*i*) back emf, (*j*) transformer.

2 When presented with a simple situation that involves a change of magnetic flux through a coil, explain qualitatively how the induced emf behaves and identify the direction of induced current.

3 Apply Faraday's and Lenz's laws to simple situations.

4 Explain how the induced emf behaves in mutual and self-inductances. Describe what geometric factors influence mutual inductance.

5 Sketch a graph of current versus time for a circuit consisting of an inductor, a resistor, and a battery connected in series from the time the circuit switch is closed. Show the inductive time constant on the graph.

6 For the circuit in 5 above, calculate the inductive time constant from the values of R and L. Determine the value of the current in the circuit for any time t after the switch is closed.

7 Explain qualitatively why a conductor cutting through magnetic field lines has an emf induced between its ends. Compute this emf for a wire of length l moving perpendicularly to the field with a speed v.

8 Sketch a simple ac generator. Explain how it gives rise to a sinusoidal ac voltage and on what factors the amplitude of the voltage depends. Sketch a graph of the voltage versus time.

9 Explain why the back emf of a motor depends on the angular speed of the shaft of the motor.

10 Explain how a transformer changes ac voltage. Apply the transformer equation to simple situations.

QUESTIONS AND GUESSTIMATES

1 Two circular current loops lie on a table. Loop 1 has a battery and switch in it, and loop 2 is just a closed wire loop. Describe what happens in loop 2 when the switch in loop 1 is suddenly closed and suddenly opened (*a*) when the loops overlap and (*b*) when they do not overlap. Sketch a graph of current versus time in each case.

2 A long, straight wire carries a current along the top of a table. A rectangular loop of wire lies on the table. If the current in the straight wire is suddenly shut off, what is the direction of the current induced in the loop? Draw a diagram for several positions of the loop relative to the wire, showing in each case the direction of the induced current in the loop.

3 A copper ring lies on a table. There is a hole through the table at the center of the ring. If a bar magnet is held verti-

cally by its south pole high above the table and is then released so that it drops through the hole, describe the induced emf in the ring and the forces that act on the magnet.

4 What happens in the secondary coil in Fig. P20.1 when the switch in the primary-coil circuit is (*a*) pushed closed and (*b*) pulled open? Repeat for Fig. 20.1.

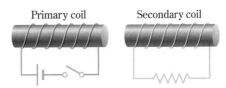

FIGURE P20.1

5 Suppose you are given two identical flat coils. How should the coils be placed so that their mutual inductance is (*a*) larg-

est and (b) smallest? When the coils are connected in series by a flexible wire, how should they be positioned to make the self-inductance (c) largest and (d) smallest?

6 A small coil is placed inside a long solenoid. How does the mutual inductance of the two change with the orientation of the coil?

7 A very long copper pipe is oriented vertically. Describe the motion of a bar magnet that is dropped lengthwise down the pipe. Why does the magnet reach a terminal speed?

8 Discuss the possibility of using induced emfs in earth satellites to power the various pieces of electronic equipment. Satellites travel with very high speed through the earth's magnetic field.

9 A closed wire loop experiences a large stopping force as it falls into a magnetic field. Justify this assertion by reference to Fig. P20.2. Does the same effect occur when a solid piece of metal supported by a string swings in a magnetic field?

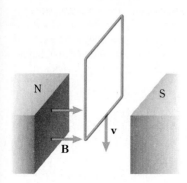

FIGURE P20.2

This general effect is referred to as *magnetic damping* of motion.

10 In Fig. P20.3, the metal ring sits on the end of a solenoid and is held in place there. An alternating current (produced by an alternating emf) is sent through the solenoid. The ring becomes hot. Why? A metal plate also becomes hot when it is held above the solenoid. Explain how eddy currents are induced in this plate and cause it to heat up.

Metal ring

Solenoid

FIGURE P20.3

11 The metal ring in Fig. P20.3 is made of copper, and the solenoid has an iron core to increase its field. When the current is turned on in the solenoid, the ring flies upward. Explain. Be particularly careful about directions.

12 Motors do work on external objects. Explain clearly how energy is transferred from the electric current to the rotating portion of the motor.

13 Explain how electric generators transform mechanical work into electric energy.

SUMMARY

DERIVED UNITS AND PHYSICAL CONSTANTS

Units of Magnetic Flux (Φ)
1 Weber (Wb) = $1\,\text{T} \cdot \text{m}^2$

Units of Inductance (M or L)
1 henry (H) = $1\,\text{V} \cdot \text{s/A}$

DEFINITIONS AND BASIC PRINCIPLES

Magnetic Flux (Φ)
The magnetic flux through area A is

$$\Phi = BA \cos \theta = B_\perp A$$

where θ is the angle between $\mathbf{B}$ and the normal to the area, $\mathbf{n}$.

Faraday's Law of Magnetic Induction
The average emf (volts) induced in a coil of N turns by a changing magnetic flux is

$$\overline{\text{emf}} = N\frac{\Delta\Phi}{\Delta t}$$

Lenz's Law
The direction of induced emf is such that a current in the direction of the emf creates a magnetic flux that tends to oppose the change of external flux taking place.

INSIGHTS
1. Notice that Wb/s = V.
2. Flux through a coil can change in three ways:
   a. Changes in the field $\mathbf{B}$ through the coil.
   b. Changes in the area of the coil.
   c. Changes in the orientation θ between the coil and $\mathbf{B}$.

Mutual Inductance (M)

The mutual inductance M between two coils is the constant of proportionality between the emf induced in the secondary coil and the rate of change of the current in the primary coil.

$$\overline{emf}_{sec} = M\left(\frac{\Delta I}{\Delta t}\right)_{prim}$$

Self-Inductance (L)

The self-inductance of a coil is the constant of proportionality between the emf induced in the coil and the rate of change of the current in the coil.

$$\overline{emf} = L\left(\frac{\Delta I}{\Delta t}\right)$$

INSIGHTS

1. Mutual inductance depends on the design of the two coils and on their relative orientation.
2. The self-inductance of a solenoid is

$$L_{sol} = \mu_o N^2 A / l = \mu_o n^2 A l$$

where N = total number of turns
l = length of solenoid
A = cross-sectional area of solenoid
n = number of turns per meter of length

3. If the solenoid is filled with material having relative magnetic permeability K_m, the above equation should be multiplied by K_m.

Series Circuit Containing a Resistor, Inductor, and EMF

Upon closing the switch, the current buildup in the circuit follows the mathematical relation

$$I(t) = I_f(1 - e^{-t/(L/R)})$$

where $I_f = \mathcal{E}/R$ is the final current in the circuit.

INSIGHTS

1. The quantity L/R has the units of seconds and is called the inductive time constant τ_L of the circuit.
2. The larger the time constant, the slower the circuit is in responding to an applied emf. Large L and/or small R cause a large time constant.

Energy Stored in an Inductor

The energy stored in an inductor carrying current I is

$$Energy = \tfrac{1}{2}LI^2$$

When L is in henrys and current in amperes, energy (E) is in joules.

Energy Density in a Magnetic Field

The energy per volume in a magnetic field $\mathbf{B}$ in vacuum is

$$\frac{Energy}{volume} = \frac{B^2}{2\mu_o}$$

INSIGHT

1. If the region of space is filled with a material whose relative magnetic permeability is K_m, μ_o should be replaced with $K_m\mu_o$. The same is true for an inductor filled with such a material.

Motional EMF

A conductor of length l moving through a region of uniform magnetic field $\mathbf{B}$ with velocity $\mathbf{v}$ perpendicular to its length and to the field has an emf induced between its ends equal to

$$emf = Bvl$$

INSIGHTS

1. The uniform magnetic field must extend at least over the length of the rod l.
2. If $\mathbf{v}$, $\mathbf{l}$ and $\mathbf{B}$ are not mutually perpendicular, you must use the components of $\mathbf{v}$ and $\mathbf{B}$ that are perpendicular to $\mathbf{l}$.

AC Generators

A coil of cross-sectional area A and N turns rotating at constant angular speed in a transverse magnetic field $\mathbf{B}$ generates an induced voltage of the following time dependence:

$$V(t) = V_o \sin 2\pi ft$$

where $V_o = 2\pi fNAB$
f = frequency of rotation (Hz)

INSIGHT

1. The voltage output of the generator alternates from $+V_o$ to $-V_o$, going through one phase cycle in the period of the rotation of the coil, $1/f$.

Transformers

A transformer consists of a primary and secondary coil wound on (usually) an iron core. As an alternating emf is applied to the primary, the emf induced in the secondary is given by the transformer equation

$$\frac{Secondary\ emf}{Primary\ emf} = \frac{N_s}{N_p}$$

where N_s = number of turns in secondary coil
N_p = number of turns in primary coil

PROBLEMS

Sections 20.1 and 20.2

1 A circular loop of wire of radius 20 cm is placed perpendicular to a 0.2-T magnetic field. Find the magnetic flux through the area of the loop.

2 A flat, rectangular piece of cardboard with an area of 240 cm^2 is placed in a magnetic field of 25 mT. The perpendicular to the surface of the cardboard makes an angle θ to the field lines. Find the magnetic flux through the area if θ is (a) 0°, (b) 90°, and (c) 30°.

3 A flat circular loop that has an area of 1800 cm^2 is placed in a 30-mT magnetic field. The perpendicular to the area of the loop makes an angle θ to the field lines. Find the magnetic flux through the area if θ is (a) 0°, (b) 90°, and (c) 30°.

4 Calculate the flux of the earth's magnetic field of strength $5 \times 10^{-5} \text{ T}$ through a flat square loop of area 25 cm^2 (a) when the field is perpendicular to the area of the loop, (b) when the field makes an angle of 30° with the normal to the plane of the loop, and (c) when the field makes an angle of 90° with the normal to the plane.

5 A flat, rectangular 5 cm by 6 cm loop of wire lies in the xy plane in a region that has a uniform magnetic field of strength 0.6 T. The magnetic field lines make an angle of 40° with the z axis. Find the magnetic flux through the area of the loop.

*6 A circular loop of wire of radius R is placed in a uniform magnetic field of strength B and is then rotated at a constant frequency f about an axis along its diameter. Find the magnetic flux through the area of the loop as a function of time if the axis of rotation is (a) perpendicular to the magnetic field and (b) parallel to the field.

*7 A hollow solenoid that has 600 turns on its 60-cm length carries a current of 4 A. Suspended within the central region of the solenoid is a loop of wire that has a cross-sectional area A. How much flux goes through the loop if the angle between the loop axis and the solenoid axis is (a) 90°, (b) 60°, and (c) 0°?

*8 In Fig. P20.4, the cross-sectional area of the solenoid is A_s and that of coil 1 is A_c. A current through the solenoid causes a magnetic field of 30 mT inside it and essentially zero field outside. There is no current in coil 1. Calculate the magnetic flux through each loop of (a) the solenoid and (b) coil 1.

Sections 20.3 and 20.4

9 The magnetic field strength in a single-loop coil of cross-sectional area 40 cm^2 changes from 6 to 9.6 mT in 0.5 s. Find the average induced emf in the coil.

10 A circular loop of two turns has the magnetic flux through its area changing at 6 Wb/s. Find the induced emf in the loop.

11 A five-turn circular coil of diameter 50 cm is placed in an external magnetic field of strength 0.4 T such that the magnetic field is perpendicular to the plane of the coil. The coil is pulled out of the field in 0.2 s. Find the average induced emf during this time.

12 A 20-loop square coil has sides 4.0 cm long. It rests flat against the north pole of a large electromagnet. The current in the electromagnet is slowly increased, so that the magnetic field increases from zero to 0.5 T in 4.0 s. (a) Find the average induced emf in the coil while the current is being changed. (b) If you look toward the north pole, is the induced emf in the coil clockwise or counterclockwise?

13 A flat, square 200-turn coil 6.0 cm on each side is designed such that it can rotate by 90° in 0.2 s. The coil is placed in a magnetic field so that the magnetic flux is zero. When the coil is rotated through 90° so that the flux is maximum, the average voltage on the coil due to the induced emf is 0.4 mV. What is the magnetic field strength?

14 A flat, 400-turn circular-loop coil 20 cm in diameter is initially aligned so that its axis is parallel to the earth's magnetic field. In 0.2 s the coil is flipped so that its axis is perpendicular to the earth's magnetic field. If an average voltage of 1.44 mV is induced in the coil, what is the value of the earth's magnetic field?

15 As a laboratory experiment, a student attempts to measure earth's magnetic field by connecting the two ends of a flat horizontal square loop 0.8 m on a side to the terminals of a sensitive voltmeter. She then moves the loop parallel to the earth at 4 m/s. (a) If the vertical component of the earth's field is $5 \times 10^{-5} \text{ T}$, what does the voltmeter read? (b) Suppose instead that she flips the coil completely over in 1.0 s. What is the average emf induced in the loop?

*16 The flexible wire loop in Fig. P20.5 has an area of 100 cm^2 and is in a magnetic field of 0.80 mT. By grasping the loop

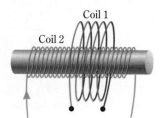

FIGURE P20.4

FIGURE P20.5

at m and n and pulling radially outward, a student collapses it to a straight line in 0.12 s. Find the average induced emf in the coil as it is collapsed.

◼17 When the wire loop in Fig. P20.5 is suddenly rotated through 90° about an axis through m and n in 0.16 s, an induced emf of 64 μV is induced in it. The area of the loop is 40 cm². (*a*) What is the magnitude of the magnetic field? (*b*) What is the average emf if the coil is rotated through 180° in the same amount of time?

◼18 A 400-turn solenoid 12 cm in diameter has a magnetic field of 0.3 T within its interior. How fast must the field within the solenoid be reduced to zero if the magnitude of the average induced emf within the coil during this time interval is 6 kV?

◼19 A flat circular coil of 60 turns of wire is placed in a magnetic field such that the normal to the plane of the coil makes an angle of 40° with the magnetic field lines. When the magnitude of the magnetic field is increased steadily from 0.2 to 0.7 mT in 0.5 s, an emf of 90 mV is induced in the coil. Find the total length of the wire used to wind the coil.

◼20 The wire loop in Fig. P20.6 is being pulled out of the magnetic field at a constant rate. If the field is uniform with a strength of 2.0 mT in the region shown and zero elsewhere, (*a*) what is the induced emf in the loop and (*b*) what is its direction?

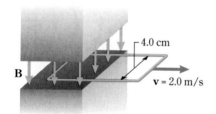

— 4.0 cm

B

v = 2.0 m/s

FIGURE P20.6

◼21 Magnetic transducers are often used to monitor small vibrations. For example, the end of a vibrating bar is attached to a coil that vibrates in and out of a uniform magnetic field **B** as shown in Fig. P20.7. Show that the induced emf in the coil is given in terms of the speed of the end of vibrating bar v as emf = $NBbv$.

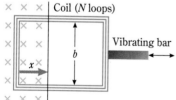

B (into page)

Coil (N loops)

Vibrating bar

b

x

FIGURE P20.7

◼22 The average field produced within a certain solenoid is 0.8 T. A secondary coil of 120 turns is wound on the outside of the solenoid. If the cross-sectional area of the solenoid is 0.60 cm², how large an average emf is induced in the secondary if the field in the solenoid is reduced to zero in 0.020 s?

◼23 A bar magnet can be moved into or out of a 240-turn coil into which it fits snugly. An average emf of 0.40 V is induced in the coil when the magnet is suddenly brought up and inserted into the coil in 0.36 s. If the cross-sectional area of the magnet is 3.0 cm², find the value of its magnetic field, B.

◼24 In Fig. P20.4, the radius of coil 1 is b and that of the solenoid is a. The solenoid is actually much longer than shown. (*a*) If the magnetic field in the solenoid is changing at a rate of 0.040 T/s, find the induced emf in coil 1. This coil has N turns on it, and the solenoid has n loops per meter of its length. (*b*) If a bar of iron with relative magnetic permeability K_m = 200 is placed inside the solenoid so as to fill it, what is the emf induced in coil 1?

◼25 The current in an air-core solenoid is increasing at a rate of 1.5 A/s. There are 10^6 turns of wire for each meter length of the solenoid, and its cross-sectional area is 2.0 cm². A secondary coil of 10^4 turns is wound over the solenoid. How large an emf is induced in the secondary?

◼26 Two coils are wound tightly and side by side on the same iron core. The cross-sectional area of both coils is 5.0 cm². When there is a current of 1.8 A in the primary, B = 0.40 T throughout the core. There are 120 turns on the secondary coil. (*a*) How large an emf is induced in the secondary if the current in the primary drops uniformly to zero in 0.050 s? (*b*) What is the mutual inductance of the coils?

Section 20.5

27 An air-core solenoid 25 cm in length and 2.0 cm in diameter has an inductance of 0.2 mH. How many loops of wire does it have?

28 A coil has an inductance of 4 mH and a current through it increases from 0.4 A to 2.0 A in 0.4 s. What is the average induced emf in the coil during this time?

29 How large is the magnetic flux through each turn of a 400-turn 8-mH coil when the current in the coil is 12 mA?

30 A solenoid with 500 turns has a radius of 2.0 cm and a length of 25 cm. (*a*) What is the inductance of the solenoid? (*b*) Find the rate at which the current through the solenoid must change in order to produce an emf of 72 mV.

31 In Prob. 30, find the rate of change of magnetic flux through the cross-sectional area of the solenoid at the instant the emf is 72 mV.

32 What is the rate of change of current in a 40-mH inductor when the induced emf across its coils is 0.020 V?

Section 20.6

33 Show that the unit of the inductive time constant $\tau = L/R$ is seconds.

34 A 40-mH coil with a resistance of 16 Ω is connected to a 24-V battery of negligible internal resistance. What is the time constant of the circuit?

35 A 6-V battery is connected in series with a resistor and an inductor. The circuit has a time constant of 500 μs, and the maximum current is 360 mA. Find the value of the inductance.

36 A potential difference of 60 V is suddenly applied to a 15-mH, 10-Ω coil by closing a switch. Find (*a*) the initial current and the initial rate of change of current, (*b*) the current when the rate of change of current is 2000 A/s, and (*c*) the final current and rate of change of current.

37 A 30-mH inductor, a 10-Ω resistor, and a 9-V battery are connected in series. The switch is closed at $t = 0$. Find the potential drop across the resistor (*a*) at $t = 0$, (*b*) after one-half time constant has elapsed, and (*c*) after one time constant has elapsed.

38 In Prob. 37, find the voltage drop across the inductor (*a*) at $t = 0$, (*b*) after one-third of a time constant has elapsed, and (*c*) after one time constant has elapsed.

39 A 300-turn inductor of radius of 4 cm and length of 25 cm is connected in series with a 1-kΩ resistor and a 12-V battery. Find the current through the circuit (*a*) after one time constant and (*b*) after two time constants.

Section 20.7

40 How much energy is stored in a 50-mH inductor at the instant when the current is 3 A?

41 Calculate the energy stored in the magnetic field of a 400-turn solenoid in which a current of 2 A produces a magnetic flux of 0.4 mWb in each turn.

42 A 2-mH inductor carries a current of 0.5 A. How much energy is stored in the field of the inductor? What is the rate of loss of energy if the inductor has a resistance of 2 Ω?

43 A 24-mH self-inductance carries a current of 3 A. (*a*) How much energy is stored in the inductance? (*b*) If this energy is stored in 3.0 cm^3 of air, what is the average value of the magnetic field B in the air?

44 The coil of a certain solenoid is wound uniformly on a wooden rod in such a way that the interior of the solenoid has a volume of 24 cm^3. When there is a current of 0.50 A in the solenoid, the field within it is 0.080 T. (*a*) What is the energy per unit volume of the magnetic field? (*b*) How much energy is stored in the solenoid? (*c*) What is the self-inductance of the solenoid?

45 A 10-Ω resistor and a 20-mH inductor are connected in series with a 12-V battery. How much energy is stored in the inductor when (*a*) the current reaches its maximum value, (*b*) one time constant has elapsed after the switch is closed, and (*c*) two time constants have elapsed after the switch is closed?

Section 20.8

46 A car is traveling at 25 m/s in a region where the vertical component of the earth's magnetic field is 3.0×10^{-5} T. How large is the potential difference induced between the ends of one of its 2.0-m-long axles?

47 A truck is traveling south at 30 m/s in a location where the vertical component of the earth's magnetic field is 5.0 μT. Find the emf induced in the truck's vertical antenna, which is 1.2 m long.

48 A 1.0-m metal bar is held in the east-west direction and is dropped from a height of 15 m at a place where the horizontal component of the earth's magnetic field has a value of 5.0×10^{-5} T. How large an emf is induced between the ends of the bar just before it strikes the ground?

49 A metal airplane with a wingspan of 30 m flies horizontally in a westerly direction at 250 m/s at a location where the downward vertical component of the earth's magnetic field is 0.8×10^{-4} T. (*a*) What is the potential difference between the tips of the wings? (*b*) Which wing tip is negative, the south or the north? (*c*) Can this voltage be measured? If so, how?

50 An engineer plans to light a train station by utilizing the emf induced in the axles of the train running on the tracks. (*a*) If the downward vertical component of the earth's magnetic field is 0.6 G and the tracks are 1.5 m apart, how large an emf is produced between the tracks by a train traveling at 40 m/s? (*b*) Could this voltage be utilized on the moving train? (*c*) Could it be utilized in the train station at the distant end of the tracks? Explain your answers to *b* and *c*.

51 A 1-m-long metal rod lies in a plane where there is a uniform magnetic field of strength 6.0×10^{-3} T. The axis of the rod makes an angle of 60° with the direction of the field. The rod moves perpendicularly to the plane at a speed of 1.6 m/s. What is the induced emf between the two ends of the rod?

52 Suppose that the speed with which the loop in Fig. P20.6 is being pulled is $v = 3$ m/s. The width of the loop is $d = 5$ cm, and the magnetic field has a constant value $B = 10$ G between the poles and zero elsewhere. (*a*) Find the induced emf in the loop. (*b*) If the resistance of the loop is $R = 8\,\Omega$, what is the current in it, at the instant shown? (*c*) With how large a force must the loop be pulled to keep its speed constant?

Sections 20.9 and 20.10

53 A coil consisting of 300 loops, each with a 5.0-cm^2 area, rotates with a frequency of 120 rev/s in a 0.040-T magnetic field. Write the induced voltage in the coil in the form $V = V_o \sin \omega t$.

54 A single coil rotating in a magnetic field gives a voltage $V = 40 \cos (1000t)$. Find the frequency of rotation of the coil and the maximum voltage output.

55 A 150-loop coil with an area of 100 cm^2 rotates at a frequency of 45 rev/s in a magnetic field. If the maximum induced emf in the coil is 5.0 V, what is the magnitude of the magnetic field?

56 A 300-loop generator coil has an area of 400 cm^2 and rotates in a field of magnitude $B = 40$ mT. How fast, in revolutions per second, must the coil be rotating in order to generate a maximum voltage of 2.0 V?

57 A flat, square 100-turn coil that is 20 cm on each edge rotates about a vertical axis at 2400 rev/m, at a location where the horizontal component of the earth's magnetic field is 2 G. Find the maximum emf induced in the coil by the earth's magnetic field.

∎58 The 500-loop rectangular coil of dimensions 10 cm by 20 cm of a generator rotates at 120 rev/s in a uniform magnetic field of magnitude 0.8 T. (*a*) What is the maximum emf induced in the coil? (*b*) What is the instantaneous value of the emf in the coil at $t = (\pi/30)$ s? Assume that the emf is zero at $t = 0$. (*c*) What is the smallest value of t when the emf will reach its maximum value?

59 The coil of a motor has a resistance of 5 Ω. When the motor is turning at its rated speed, it draws a current of 4.0 A from 120 V. (*a*) How large is the counter emf of the motor? (*b*) How much current would the coil draw if it stopped rotating?

60 A motor with a coil of resistance 20 Ω operates from a 240-V source. When the motor is operating at its maximum speed, the counter emf is 105 V. Find the current in the coil (*a*) when the motor is first turned on and (*b*) when the motor has reached maximum speed.

61 In the previous problem if the current in the motor is 6 A, what is the counter emf at this time?

∎62 The current in the coils of a motor is 12 A when the motor is first turned on and 4 A when it is rotating at maximum speed. If the motor is operating at 120 V, find the counter emf in the coil and the resistance of the coil.

∎63 Very large motors take nearly a minute to get up to speed after they are turned on. One such motor has a resistance of 1.2 Ω and normally draws 10 A on 120 V. (*a*) What resistance (called the starting resistance) must be placed in series with the motor if it is not to draw more than 20 A when it is first turned on? (This resistance is later removed, of course.) (*b*) What is the back emf of this motor when operating at normal speed?

Section 20.11

64 A certain transformer in a radio changes the 120-V ac line voltage to 6 V ac. (*a*) What is the turns ratio N_p/N_s for the transformer? (*b*) By mistake, the transformer is connected into the circuit backward. About what output voltage does it deliver before everything burns out?

65 A transformer in a neon sign is designed to change ac voltage from 120 V to 16,800 V. (*a*) What is the turns ratio N_p/N_s of the transformer? (*b*) If the transformer is connected backward (120 V to the secondary), what voltage appears across the primary?

66 A transformer at a distribution station reduces ac voltage from 36,000 V to 2400 V. The primary has 15,000 turns. (*a*) What is the number of turns in the secondary? (*b*) If the current in the secondary is 500 A, find the current in the primary. Assume that there is no loss of power.

67 The primary current in an ideal transformer (no power loss) is 5.0 A when the primary voltage is 90 V. Calculate the voltage across the secondary when a current of 0.9 A is drawn from it.

68 A welding machine uses a current of 360 A. The device uses a transformer whose primary has 720 turns that draws a 4.0-A current from a 240-V power line. (*a*) How many turns are there in the secondary? (*b*) What is the output voltage across the secondary? Assume no power loss in the transformer.

∎69 A transformer has 200 turns in its primary and 80 turns in the secondary and its primary is connected to a 120-V power line. The current in the primary is 0.25 A when a light bulb is connected to the secondary. Find the resistance of the bulb. Assume no power loss in the transformer.

Additional Problems

∎70 A 500-turn solenoid with a length of 25 cm and a radius of 2.0 cm carries a current of 4 A. A second coil of 5 turns is tightly wound around the solenoid such that it has the same radius as that of the solenoid. (*a*) Find the change in the magnetic flux through the coil and (*b*) the magnitude of the average induced emf in the coil when the current in the solenoid increases to 10 A in a period of 1.2 s.

∎71 An emf of 20 mV is induced in a 400-turn solenoid at the instant when the current through its coils is 4 A and is changing at a rate of 12 A/s. Calculate the magnetic flux through each turn of the solenoid.

∎∎72 A long iron-core solenoid with 2400 turns has a cross-sectional area of 4.0 cm². When a current of 4.0 A passes through the solenoid coil, the magnetic field inside the solenoid is $B = 0.60$ T. (*a*) Calculate the average induced emf in the solenoid if the current is reduced to zero in 0.2 s. (*b*) How large is the self-inductance of the solenoid coil?

∎∎73 Two inductances L_1 and L_2 are connected in series. Show that the equivalent inductance of the combination L_e is given by $L_e = L_1 + L_2$. *Hint:* The total potential difference across the two inductors is the sum of the induced emfs in each inductor, and the rate of change of current is the same in both the inductors.

∎∎74 Show that the equivalent inductance L_e of two inductors L_1 and L_2 connected in parallel is given by $1/L_e = 1/L_1 + 1/L_2$. *Hint:* Kirchhoff's law requires that the two emfs be equal and the total current be the sum of the individual currents.

∎∎75 The metal rod in Fig. P20.8 slides down the incline while a

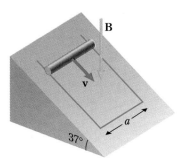

FIGURE P20.8

magnetic field $B = 2.5\,\text{T}$ is present. (*a*) Find the induced emf when its speed is $v = 20\,\text{m/s}$. (*b*) If the resistance of the loop is $R = 25\,\Omega$, what is the current in the loop? (*c*) Is the current clockwise or counterclockwise? (*d*) How large is the force acting up the incline on the rod because of the current in the magnetic field? (*e*) Does this force tend to slow down or speed up the rod?

▪▪**76** For the previous problem, shown in Fig. P20.8, find the terminal speed of the rod as it slides without friction down the incline. The mass of the rod is $m = 36\,\text{g}$, and the resistance of the loop is nearly zero except for the resistance $R = 25\,\Omega$ of the rod.

▪▪**77** The square loop of resistance $20\,\Omega$ in Fig. P20.2 has an edge length of 4.0 cm and a mass of 15 g. Assuming the magnitude of the magnetic field to be 2.4 T between the poles and zero elsewhere, find the terminal speed of the loop as it enters the region between the poles. Assume that the loop is in about the position shown when its terminal speed is reached.

▪▪**78** The resistance of a number 10 copper wire is $5.2 \times 10^{-3}\,\Omega/\text{m}$. It can carry a current of only about 30 A without overheating. A power company wants to use the wires of this size to deliver 36 MW of power to a city 50 km from the generating station. What fraction of power sent from the station is lost along the transmission lines if the transmission voltage is (*a*) 240 V and (*b*) 12,000 V? Assume that the 30-A restriction is not exceeded.

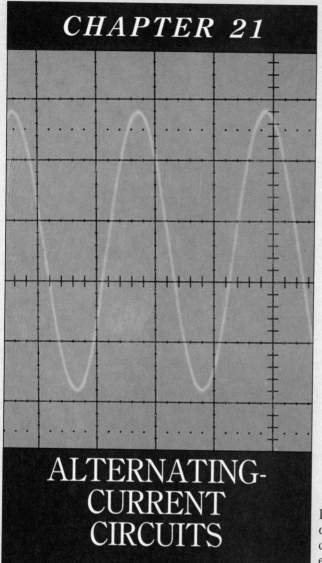

CHAPTER 21

ALTERNATING-CURRENT CIRCUITS

In the past few chapters, we were concerned mainly with direct currents, that is, those in which the charges flow continuously in one direction. We saw in Chap. 20, however, that a voltage source of alternating polarity is obtained by rotating a coil in a magnetic field. An alternating-voltage source such as this gives rise to alternating currents, and these, too, are of great importance. We see in this chapter how such currents behave when they are sent through resistances, capacitances, and inductances.

21.1 CHARGING AND DISCHARGING A CAPACITOR

To understand the behavior of alternating current circuits, we have to know how the current through elements of the circuit responds to continual changes in a source of emf. We know there is no time delay between the application of a voltage V across a pure resistor R and the establishment of a current $I = V/R$ through the resistor. In other words, *the current and the voltage in a resistor obey Ohm's law instantaneously:*

$$I(t) = \frac{V(t)}{R} \qquad \text{(pure resistor)} \tag{21.1}$$

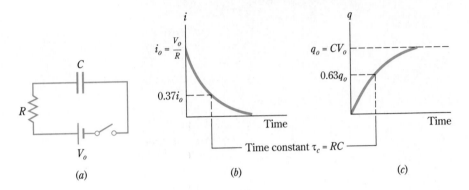

FIGURE 21.1

The time constant τ_c is a convenient measure of the time it takes a capacitor to charge or discharge.

In the previous chapter, we examined the way a current builds up over time when a switch connects a voltage source to an inductor. We saw that there is a time delay before the current attains its final Ohm's-law value. Thus we might expect that if the voltage applied to the circuit is always changing, the current is always "chasing" the voltage and consequently Ohm's law is *not* obeyed at every instant. This is indeed the case, and we will pursue the matter further in later sections.

The one circuit element whose time-dependent behavior we still have to examine is the capacitor. We know that capacitors do not permit a direct current to pass. However, charging and discharging a capacitor require the movement of charge to and from the plates. This transport of charge through the circuit containing the capacitor constitutes a transient current. Let us examine this behavior by considering the simple circuit in Fig. 21.1*a*. Suppose the switch is open initially and no charge exists on the capacitor, whose capacitance we represent by C. We wish to know what happens when the switch is suddenly closed.

When the switch is closed, the battery tries to send charge clockwise around the circuit. Since there is initially no charge on the capacitor, the current i is limited only by the resistor R. Therefore, just after the switch is closed (at time $t = 0$), the current is $i_o = V_o/R$, as shown in (*b*). As time goes on, however, the current decreases as charge builds up on the capacitor plates, since this charge produces a voltage across C that opposes that of the battery. The current must drop to zero when the capacitor's charge finally becomes $q_f = CV_o$.

The exact way the current behaves in this circuit is shown in Fig. 21.1*b*. The curve followed by the current is called an **exponential decay curve.** The mathematical form of this behavior is

$$i(t) = i_o e^{-t/RC} \tag{21.2}$$

where $i_o = V_o/R$. The product RC has units of time and is called the **capacitive time constant** τ_C. (Be sure you can show that $\Omega \cdot F = s$.) At $t = \tau_C$, the current has fallen to $1/e = 0.37$ of its initial value. By the time $t = 2\tau_C$, the current has fallen to $1/e^2 = 0.135$ of i_o, etc.

As long as there is a current, the capacitor is receiving charge. When the current finally stops, it is because the voltage on the capacitor has become equal to that of the battery, namely $V_C = q_f/C = V_o$, where q_f is the final charge on the capacitor. The time dependence of the charge on the capacitor is shown graphically in Fig. 21.1*c*. The mathematical equation of this graph is

$$q(t) = q_f(1 - e^{-t/\tau_C}) = CV_o(1 - e^{-t/\tau_C}) \tag{21.3}$$

Notice that this is precisely the same behavior as the *current* buildup in an inductor

(Eq. 20.6). The capacitive time constant RC plays the same role for the charge as the inductive time constant L/R does for the inductor current. The larger RC is, the longer it takes a capacitor to "fill up" with charge. This makes sense in a qualitative way. A larger value of C requires more accumulated charge to create each volt of potential across the plates. Also, a larger value of R restricts the current more and hence restricts the rate at which charge can be brought to the plates.

If a resistor R is connected directly across a charged capacitor C, the capacitor will discharge through the resistor. If we assume the initial potential difference between the capacitor plates to be V_o, the current from the capacitor as the capacitor discharges will vary as in Fig. 21.1*b*. The capacitor discharge current behaves in the same way as the charging current. The charge on the capacitor also decreases exponentially after the connection is made. Can you show that the voltage across the capacitor must decrease in the same way as the charge? The expressions for $q(t)$ and $V(t)$ are

$$q(t) = q_o e^{-t/RC} \tag{21.4a}$$

$$V(t) = V_o e^{-t/RC} = \frac{q_o}{C} e^{-t/RC} \tag{21.4b}$$

It turns out that the capacitor is about five-eighths discharged in one time constant.

Illustration 21.1

In most television sets, a capacitor is charged to a potential difference of about 20,000 V. As a safety measure, a resistor called a bleeder is connected across the capacitor's plates so that the capacitor will discharge after the set has been turned off. Suppose a bleeder resistor is $10^6 \, \Omega$ and $C = 10 \, \mu\mathrm{F}$. About how long must you wait after turning off the set before it is safe to touch the capacitor?

Reasoning The time constant for this circuit is

$$\tau_C = RC = (10^6 \, \Omega)(10^{-5} \, \mathrm{F}) = 10 \, \mathrm{s}$$

As a guess, we might say that it would be safe to touch the capacitor after 10 time constants have passed. After 10 time constants, Eq. 21.4a shows that $q = q_o e^{-10}$, which is $4.5 \times 10^{-5} \, q_o$. Thus at the time $t = 10 \, \tau_C$ the charge and voltage have been reduced to 4.5×10^{-5} times their original values, and so V is then 0.90 V, a perfectly safe value.

21.2 AC QUANTITIES; RMS VALUES

Electric power companies furnish what is known as ac (alternating-current) voltages. The companies generate these voltages with rotating-coil generators, and the voltage v thus provided is similar to the ac voltage shown in Fig. 20.18. You will recall that it is a sinusoidal voltage given by $v = v_o \sin 2\pi ft$, where f is the rotation frequency of the generator coil (60 Hz in the United States). This type of voltage, when impressed across a resistor, gives rise to a current such as the one shown in Fig. 21.2, a sinusoidal current. Its equation is $i = i_o \sin 2\pi ft$. As you see, we have changed notation from that used in previous chapters. In the remainder of this text, we use small letters, v and i, for voltages and currents that vary with time. We will

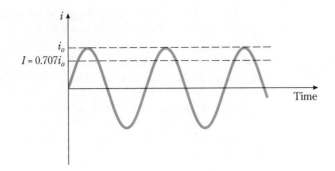

FIGURE 21.2

For an alternating current, the effective, or root-mean-square, current, is $I = i_o/\sqrt{2} = 0.707i_o$.

soon see that capital letters, V and I, are reserved for other quantities when discussing alternating voltages and currents.

It is interesting to notice that, over one cycle, an ac voltage or current has an average value of zero. As you can see from Fig. 21.2, a sine function (as well as a cosine function) is negative as much as it is positive. Hence its average value is zero. Therefore, for an ac voltage and current,

$$v_{av} = i_{av} = 0$$

Because of this, alternating currents cannot be used to charge batteries or for similar applications. If the battery is being charged when the current is positive, it will undergo an equal discharge when the current is negative.

There was considerable controversy in the late 1800s concerning which was more practical, ac or dc electricity. Both can be used for lighting and to run motors. In the end, alternating current triumphed because its voltage can easily be stepped up and down by transformers, as we saw in the last chapter.

The electric power furnished to our homes is often used to operate an electric stove or an incandescent lamp. Uses such as these involve the heat generated by the current in a resistor. Since the electric power provided in such situations is i^2R, it does not matter whether the current is negative or positive because i^2 is always positive. Alternating current is as useful in these applications as direct current.

The meter on the right is reading 70 V rms ac. The meter on the left is 70 V dc. The two voltages have the same effect on the two identical light bulbs.

Because i_{av} and v_{av} are zero for ac conditions, however, we need a special way to describe currents and voltages in ac circuits.

Consider an alternating current $i = i_o \sin 2\pi ft$ that is delivering power to a resistor R. The power delivered at any instant is

$$\text{Power} = i^2 R = R i_o^2 \sin^2 2\pi ft$$

In most applications, we are interested only in the average power:

$$\text{Average power} = R i_o^2 (\sin^2 2\pi ft)_{av}$$

It turns out* that the average value of $\sin^2 \theta$ is 0.50. Therefore

$$\text{Average power} = R\left(\frac{i_o}{\sqrt{2}}\right)^2$$

We can compare this expression with the expression for dc power, $P = I^2 R$. It appears that the alternating current that produces an equivalent average power is $i_o/\sqrt{2}$, or $0.707 i_o$. We call this value the *root-mean-square* (*rms* or *effective*) current and represent it by the symbol I. Similarly, we define the rms voltage as $V = v_o/\sqrt{2} = 0.707 v_o$. In summary:

The rms current I and rms voltage V are

$$I = \frac{i_o}{\sqrt{2}} \qquad \text{and} \qquad V = \frac{v_o}{\sqrt{2}}$$

where i_o and v_o are the amplitudes of the sinusoidally varying current and voltage.

Figure 21.2 shows the value of I. The average power loss in a resistor in an ac circuit is thus

$$P = \tfrac{1}{2} i_o^2 R = I^2 R$$

You should notice that, in a dc system, the rms, average, and instantaneous current are all the same.

21.3 RESISTANCE CIRCUITS

We can study ac circuits by considering in turn three different circuit elements connected in series with an ac voltage source. First let us consider the simple resistance circuit shown in Fig. 21.3a. At any particular instant, Ohm's law applied to the resistor tells us that $v = iR$. Because $v = v_o \sin 2\pi ft$, we have

*Recall that $\sin^2 \theta + \cos^2 \theta = 1$. Because the graphs of $\sin \theta$ and $\cos \theta$ have the same shape, $(\sin^2 \theta)_{av} = (\cos^2 \theta)_{av}$. Therefore $(\sin^2 \theta)_{av} + (\cos^2 \theta)_{av} = 1$ becomes $2 (\sin^2 \theta)_{av} = 1$, from which $(\sin^2 \theta)_{av} = 0.50$.

FIGURE 21.3

The current in a resistor is in phase with the voltage across the resistor.

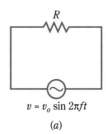

$v = v_o \sin 2\pi ft$

(a)

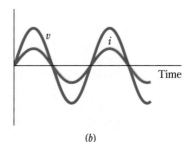

(b)

$$i = \frac{v}{R} = \frac{v_o}{R} \sin 2\pi ft$$

Hence both the voltage and the current vary with time in the same way. They go through zero and reach their maximum and minimum values at the same times. At every instant the ratio of v to i is the same, namely R. We can see that current and voltage in a purely resistive circuit are *in phase*.

As outlined in the previous section, the power loss in the resistor is I^2R. In this particular case, where only a resistance is present, $I = V/R$, and so the power loss could also be written IV, where I and V are the rms meter readings. We see in the next sections that there is no average power loss in pure capacitors or pure inductors. *All power losses in simple ac circuits occur in resistors.*

21.4   CAPACITANCE CIRCUITS; CAPACITIVE REACTANCE

Let us now consider the capacitance circuit in Fig. 21.4a. We saw in Sec. 21.1 how a current has to deliver charge to the capacitor in order to create a voltage across it. In other words, the current is a necessary precursor for the voltage and precedes it. In the case of a sinusoidal voltage signal, the current is continually reversing, bringing positive charge first to one plate and then to the other. Qualitatively, this might lead us to expect that the alternating current continually leads the ac voltage across the capacitor by some amount of time.

We can be much more quantitative than this, however. In Fig. 21.4 the potential difference between points a and b is given by the voltage source:

$$v(t) = v_o \sin 2\pi ft$$

This is the same as the voltage across C, given by $q(t)/C$. Since C is constant, the charge on the capacitor must oscillate in phase with the source voltage:

$$q(t) = Cv_o \sin 2\pi ft$$

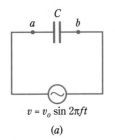

$v = v_o \sin 2\pi ft$

(a)

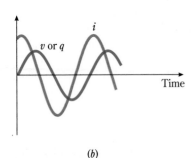

(b)

FIGURE 21.4

The voltage across a capacitor reaches its maximum $\frac{1}{4}$ cycle later than the current. The charge on the capacitor is in phase with the voltage across the capacitor.

The graph of $q(t)$ is shown in Fig. 21.4b.

To see how the current in the circuit varies with time, recall that current is defined as the rate of flow of charge. Hence the rate of change of the charge on the capacitor at any instant, $\Delta q/\Delta t$, is the current in the circuit at that instant. But $\Delta q/\Delta t$ is just the slope of the curve obtained when q is plotted against t, as in Fig. 21.4b. We need only examine the slope of this curve and plot the results to have a graph of the time dependence of the current.

Notice that at $t = 0$, the graph of q has the steepest positive slope. This positive slope diminishes until it reaches zero when q reaches its peak value, signifying $i = 0$. Then the slope of the q versus t curve turns *negative,* reaching a maximum negative value as q reaches zero. This is halfway through the voltage cycle. The slope continues to be negative, but less and less so, until the minimum in the charge curve is reached. At this point the current is again zero and becomes positive as the charge curve starts to increase once more. By filling in the details, we can plot an accurate curve of the current versus t. This is shown in Fig. 21.4b.

We see that q and i both go through a sinusoidal variation with the same frequency. But as we discussed qualitatively, the current and charge are *not in phase.*

As a matter of fact, i reaches its maxima and minima precisely $\frac{1}{4}$ cycle ahead of the corresponding values of q (and v). This leads us to the following important conclusion:

In a circuit containing only a capacitor, the alternating current leads the ac voltage by $\frac{1}{4}$ cycle.

The curve in Fig. 21.4b is the graph of a *cosine* function. The mathematical descriptions for $v(t)$ and $i(t)$ are

$$v(t) = v_o \sin 2\pi ft \qquad \text{and} \qquad i(t) = i_o \cos 2\pi ft$$

Now let us investigate the power consumption in this type of circuit. The instantaneous power furnished to the capacitor is given by the usual relation:

$$\text{Power} = vi = v_o i_o \sin 2\pi ft \cos 2\pi ft$$

This can be put in a more convenient form if you recall that

$$\sin 2\theta = 2 \sin \theta \cos \theta$$

Dividing this equation by 2 and substituting $2\pi ft$ for θ, we obtain

$$\text{Power} = \tfrac{1}{2} v_o i_o \sin 4\pi ft$$

This means that the instantaneous power furnished to the capacitor varies sinusoidally at twice the frequency of the ac voltage. Hence the average power furnished to the capacitor is zero because the sinusoidal function is negative as much as it is positive. During half of the cycle, the capacitor is being charged and so energy is being stored in it. During the other half of the cycle, however, the capacitor is discharging and returning its stored energy to the power source. The net effect is the following:

In an ac circuit, the average power consumed by a perfect capacitor is zero.

To complete our analysis of how a capacitor affects the current in an ac circuit, we need to obtain a relationship between i and v similar to Ohm's law for resistors. To do so, let us consider how the capacitor reacts to the frequency of the applied voltage. If the frequency is very low, such as one cycle per hour, the capacitor will become fully charged in a small fraction of the cycle. Most of the rest of the time the capacitor will simply block any further charge from flowing. At very high frequencies, however, the voltage alternates so rapidly that the capacitor spends most of the time being charged and discharged. This means that current is almost continually passing back and forth through the circuit. Hence we can say that

The ability of a capacitor to impede current is large at low frequencies and small at high frequencies.

We can also see that the value of C plays a part in limiting current. A large C requires more charge to build up a voltage equal to v_o. This means that more charge must flow to a large capacitance. Relatively little current is required to fully charge a capacitor with small C. Thus we can say

The ability of a capacitor to impede current is large if C is small and is small if C is large.

We designate the ability of a capacitor to impede the flow of charge by the term **capacitive reactance,** which we designate by X_C. It is related to the rms current and voltage in the circuit of Fig. 21.4 through a relation similar to Ohm's law:

$$V = IX_C \tag{21.5}$$

where X_C replaces R in Ohm's law. Using calculus, it is easy to show that

$$X_C = \frac{1}{2\pi fC} \tag{21.6}$$

It should be obvious from Eq. 21.5 that the units of X_C must be ohms. However, be sure you can derive that result from the definitions of the hertz and farad. Notice that the impeding effect of the capacitor expressed in X_C depends on f and C in a manner consistent with our qualitative discussion.

It is very important to observe the following difference between Eq. 21.5 and Ohm's law:

The capacitive reactance X_C is defined only in terms of the rms values of current and voltage. It does not apply to instantaneous values.

The reason for this is simply that at any instant, v and i are in different parts of their respective cycles.

Example 21.1

Suppose an ac voltmeter across the voltage source in Fig. 21.4 reads 80 V and $C = 0.40\ \mu\text{F}$. Find the rms current in the circuit if the frequency of the ac voltage is (**a**) 20 Hz and (**b**) 2×10^6 Hz.

R

Reasoning

Question How is the rms current related to the rms voltage for the circuit in Fig. 21.4?
Answer The ratio V/I is the capacitive reactance X_C:

$$\frac{V}{I} = X_C$$

Question What determines X_C?
Answer The frequency of the voltage and the value of the capacitance C:

$$X_C = \frac{1}{2\pi fC}$$

Solution and Discussion First, calculate X_C in terms of frequency f:

$$X_C = \frac{1}{2\pi f(4.0 \times 10^{-7}\,\text{F})} = \frac{4.0 \times 10^5}{f}\ \Omega/\text{s}$$

Thus

For $f = 20\,\text{Hz}$: $X_C = 2.0 \times 10^4\,\Omega$

For $f = 2 \times 10^6\,\text{Hz}$: $X_C = 0.20\,\Omega$

Then the rms currents are as follows:

For $f = 20\,\text{Hz}$: $I = \dfrac{V}{X_C} = \dfrac{80\,\text{V}}{2.0 \times 10^4\,\Omega} = 4.0\,\text{mA}$

For $f = 2\,\text{MHz}$: $I = \dfrac{80\,\text{V}}{0.20\,\Omega} = 400\,\text{A}$

Notice the dramatic effect that frequency has in determining the rms current.

Exercise   What are the amplitudes of the voltage and currents in this example?   *Answer: $v_o = 113\,V;\ i_o = 5.7\,mA;\ i_o = 570\,A.$*

21.5   INDUCTANCE CIRCUITS; INDUCTIVE REACTANCE

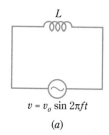

$v = v_o \sin 2\pi ft$

(a)

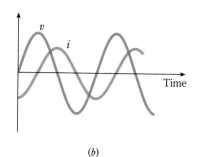

(b)

FIGURE 21.5

The voltage across the inductor leads the current through it by 90°, or $\frac{1}{4}$ cycle. Notice the symbol used for the inductor.

The behavior of the simple self-inductance circuit in Fig. 21.5a can be analyzed in a manner similar to that used for the capacitance circuit. Let us begin by assuming that the voltage has a sinusoidal dependence given by

$$v(t) = v_o \sin 2\pi ft$$

This behavior is plotted in Fig. 21.5b. We want to find out how the voltage across the inductor, $v(t)$, is related to $i(t)$. From Eq. 20.5, we know the voltage across the inductor to be $L(\Delta i / \Delta t)$. Thus the instantaneous values of v and i are related by

$$v(t) = L\frac{\Delta i}{\Delta t}$$

Therefore $v(t)$ is proportional to the slope of the i versus t curve. Notice that the curve shown for $i(t)$ in Fig. 21.5b has this property. We see there that the voltage leads the current by $\frac{1}{4}$ cycle.

In a circuit containing only an inductor, the ac voltage leads the alternating current by $\frac{1}{4}$ cycle.

This result is consistent with our qualitative observation at the beginning of this chapter that the current in an inductor always "chases" the voltage applied to it.

We can use the same reasoning as in Sec. 21.4 to show that the inductor consumes no energy on the average. Although the source stores energy in the inductor during part of the cycle, the inductor gives it back to the source in a later portion of the cycle. We showed in Chap. 20 that the energy stored in an inductor is $\frac{1}{2}Li^2$. You would do well to examine Fig. 21.5b and ascertain the part of the cycle during which the source is losing energy and the part during which energy is being returned to the source.

Inductors are made in many sizes to perform a wide variety of functions in business and industry.

In an ac circuit, the average power consumed by a pure inductor is zero.

As before, we seek to find a relationship between the voltage and current in an inductance circuit. The induced emf, which impedes the rise of the current, is equal to $L(\Delta i/\Delta t)$. The larger L is, the larger this effect is. Thus we can say

The ability of an inductor to impede current in an ac circuit is proportional to the inductance.

The factor $\Delta i/\Delta t$ is simply proportional to the frequency with which the current direction changes. Thus we conclude:

The ability of an inductor to impede current in an ac circuit is proportional to the frequency.

We represent the impeding effect of an inductor by the **inductive reactance** X_L, where X_L is the constant of proportionality between the rms voltage and the rms current in the circuit:

$$V = IX_L \tag{21.7}$$

It can be shown that

$$X_L = 2\pi fL \tag{21.8}$$

a result that is consistent with our qualitative discussion above. Equation 21.7 is the equivalent of Ohm's law for inductors. The units of X_L must be ohms, a fact you should be able to verify from the definitions of the hertz and the henry.

As with capacitive reactance, X_L relates only the rms values of I and V. It does not apply to instantaneous values.

The ac effects of the three types of circuit elements are summarized in Table 21.1.

TABLE 21.1   *Effects of L, R, and C in ac circuits*

	Resistor	Capacitor	Inductor
Phase relationship between v and i	In phase	i leads v by $\frac{1}{4}$ cycle	v leads i by $\frac{1}{4}$ cycle
Relationship between V and I	$V = IR$ R is independent of f	$V = IX_C$ $X_C = \dfrac{1}{2\pi fC}$	$V = IX_L$ $X_L = 2\pi fL$
Average power loss	$P = I^2R$	$P = 0$	$P = 0$

Example 21.2

Suppose the inductor in Fig. 21.5a has a value of 15 mH. The source voltage, as read by an ac meter, is 40 V, and its frequency is 60 Hz. Find the current through the inductor. Repeat for a frequency of 6.0×10^5 Hz.

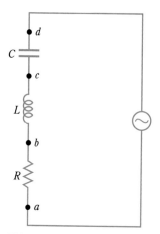

FIGURE 21.6

A series *LRC* circuit.

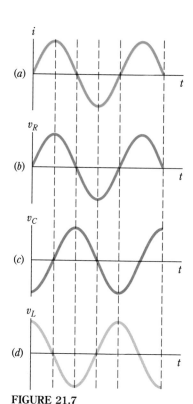

FIGURE 21.7

Only v_R is in phase with i. Both v_L and v_C are $\frac{1}{4}$ cycle (90°) out of phase with i.

[R]

Reasoning

Question What quantity relates the rms current to the rms voltage in an inductor?

Answer The inductive reactance X_L: $V = IX_L$, where $X_L = 2\pi f L$.

Solution and Discussion For the 60-Hz source, the reactance is

$$X_L = 2\pi(60 \text{ Hz})(15 \times 10^{-3} \text{ H}) = 5.6 \, \Omega$$

For 0.60 MHz, it is

$$X_L = 2\pi(0.60 \times 10^6 \text{ Hz})(15 \times 10^{-3} \text{ H}) = 5.7 \times 10^4 \, \Omega$$

The respective rms currents are

$$I = \frac{V}{X_L} = \frac{40 \text{ V}}{5.6 \, \Omega} = 7.1 \text{ A}$$

and

$$I = \frac{40 \text{ V}}{5.7 \times 10^4 \, \Omega} = 7.1 \times 10^{-4} \text{ A}$$

Notice how the inductor greatly impedes the current at higher frequencies.

21.6 COMBINED *LRC* CIRCUITS; CURRENT-VOLTAGE PHASE RELATION

Now let us consider a circuit in which all three elements are connected in series. This is known as an *LRC* series circuit, and one is shown in Fig. 21.6. As before, we would like to determine the relationship between rms current and rms voltage. We would also like to find out the phase relationship between the instantaneous values v and i and the power loss in the circuit.

To begin, every element in a series circuit must have the same instantaneous current. Let us assume this current to be of the form $i(t) = i_o \sin 2\pi f t$. A graph of i is shown in Fig. 21.7a. From our previous discussion we can immediately graph the voltage v across each of the elements. The voltage across R, v_R, is in phase with i (Fig. 21.7b); the voltage across C, V_C, lags the current by $\frac{1}{4}$ cycle (Fig. 21.7c); the voltage across the inductor, v_L, leads i by $\frac{1}{4}$ cycle (Fig. 21.7d).

Notice from these graphs that v_L and v_C are always of opposite sign. Hence they always subtract from each other. Suppose an ac voltmeter reads V_L across the inductor and V_C across the capacitor. If $V_C = V_L$, then the amplitudes of v_C and v_L are the same and v_C exactly cancels v_L. In this case, a voltmeter connected between points b and d in Fig. 21.6 would read zero, not $V_C + V_L$! Thus we see that ac voltmeter readings do not add to give proper voltage differences. Even though the instantaneous voltages add directly, the rms voltages read by ac meters are always positive and cannot show the cancellation effects that may be present.

Oscillating quantities that are out of phase with each other can be added with a

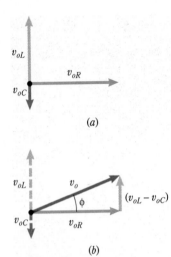

(a)

(b)

FIGURE 21.8

The root-mean-square voltages in an LRC circuit add vectorially.

simple geometrical diagram. The key to this diagram is in recognizing that $\frac{1}{4}$ cycle amounts to a difference of 90° in the phase of a sinusoidally varying quantity. Let us represent the amplitude of the voltage across R, v_{oR}, by a vector directed to the right in Fig. 21.8a. *We know that this voltage is in phase with the current in the circuit, and so this direction also represents the current.* To represent the voltage amplitude v_{oL} across L, we must draw a vector directed 90° away from the V_{oR} direction, as shown in Fig. 21.8a. This angle takes care of the $\frac{1}{4}$ cycle phase difference between v_{oL} and i_o. The voltage amplitude v_{oC} across the capacitor must then be drawn in the direction *opposite* the direction of v_{oL}. The *magnitudes* of these vectors are given by Ohm's law and its equivalents:

$$v_{oR} = i_oR \qquad v_{oL} = i_oX_L \qquad v_{oC} = i_oX_C$$

The total voltage amplitude v_o applied to the circuit can then be obtained by the usual vector addition. As shown in Fig. 21.8b, we first have to subtract the opposing v_{oL} and v_{oC}. Then we add this vector to v_{oR} using the Pythagorean theorem:

$$v_o^2 = v_{oR}^2 + (v_{oL} - v_{oC})^2 = i_o^2[R^2 + (X_L - X_C)^2]$$

Taking the square root of this gives the Ohm's law equivalent for the LRC circuit:

$$v_o = i_oZ \tag{21.9}$$

where Z is called the **impedance** of the circuit and is given by

$$Z = \sqrt{R^2 + (X_L - X_C)^2} \tag{21.10}$$

The units of Z are ohms, as you can readily verify. Fig. 21.9 shows the quadratic relationship between R, $(X_L - X_C)$, and Z in the sense of Eq. 21.10.

Of course Eq. 21.9 also holds for the rms values of I and V, since they are just the constant factor 0.707 times the amplitudes. Notice that it makes no difference whether you subtract X_L from X_C or vice versa. The difference always gets squared in calculating Z.

The angle ϕ in Fig. 21.9 is the phase difference between i and v in the circuit. To see this, notice that it is the angle between the total voltage and the voltage across R, which is in phase with i. We can see that ϕ is easily obtained from

$$\cos \phi = \frac{R}{Z} \tag{21.11}$$

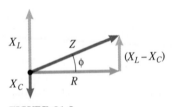

FIGURE 21.9

The impedance is the hypotenuse of a right triangle with sides R and $|X_L - X_C|$. The angle ϕ is the phase angle between i_o (along R) and v_o (along Z).

If $X_L > X_C$, the voltage *leads* the current by the phase angle ϕ. If $X_L < X_C$, the voltage *lags* the current by ϕ.

Although we already know that the power loss in the circuit occurs entirely in R and is equal to I^2R, there is another useful way to calculate power loss:

$$\text{Power loss} = I^2R = \frac{V}{Z}IR = VI\left(\frac{R}{Z}\right) = VI \cos \phi \tag{21.12}$$

where use has been made of Eq. 21.11. Here V and I are the rms values as usual. The factor $\cos \phi$ is called the *power factor* of the circuit.

There are two limiting cases of interest. If the circuit contains only R and C (an RC circuit), we can take $L = 0$, making $X_L = 0$. Then the circuit impedance is just

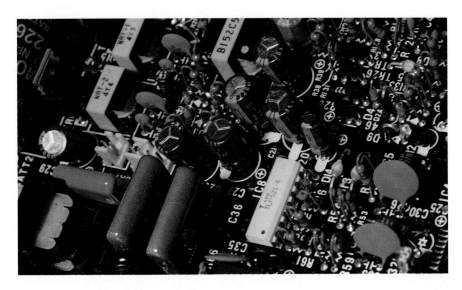

An electronic circuit board contains many resistors, capacitors, and inductors.

$$Z = \sqrt{R^2 + X_C^2} \qquad (RC \text{ circuit})$$

If the circuit contains no capacitor, what value of X_C should we use? It is not correct to say that $C = 0$ in this case, because doing so would give an infinite value for X_C. Having no capacitor is equivalent to $C = \infty$, since such a capacitor would be insatiable and therefore would never impede the current. Thus $X_C = 0$ is the correct choice for an *RL* circuit. The impedance is then just

$$Z = \sqrt{R^2 + X_L^2} \qquad (RL \text{ circuit})$$

Example 21.3

A power source ($V = 80.0$ V, $f = 2000$ Hz) is connected in series across a 300-Ω resistor and a 0.600-μF capacitor. Find (**a**) the current in the circuit, (**b**) the voltmeter readings across the resistor, (**c**) the reading across the capacitor, and (**d**) the power loss in the circuit.

R   *Reasoning*   *Part (a)*

Question   What is the relationship between V and I in this series *RC* circuit?
Answer $V = IZ$, with $Z = \sqrt{R^2 + X_C^2}$.

Question   What is the expression for X_C?

Answer   From Eq. 21.6: $X_C = \dfrac{1}{2\pi f C}$, with $f = 2000$ Hz and $C = 0.60 \times 10^{-6}$ F.

Solution and Discussion   Solve for X_C, Z, and I:

$$X_C = \frac{1}{2\pi(2000 \text{ Hz})(0.600 \times 10^{-6} \text{ F})} = 133 \ \Omega$$

$$Z = \sqrt{(300 \ \Omega)^2 + (133 \ \Omega)^2} = 328 \ \Omega$$

$$I = \frac{V}{Z} = \frac{80.0 \text{ V}}{328 \ \Omega} = 0.244 \text{ A}$$

R

Reasoning   Parts (b) and (c)

Question   What voltage will a meter read?
Answer   The rms voltage.

Question   What determines the rms voltage across R and C?
Answer $V_R = IR$ and $V_C = IX_C$

Solution and Discussion   Using the previously determined I and X_C,

$V_R = (0.244 \text{ A})(300 \text{ }\Omega) = 73.2 \text{ V}$
$V_C = (0.244 \text{ A})(133 \text{ }\Omega) = 32.6 \text{ V}$

Notice that $V_R + V_C$ does not equal the source voltage of 80.0 V. This is because the two voltages are out of phase. Their instantaneous values would always equal 80.0 V, but this is not what the meter reads.

R

Reasoning   Part (d)

Question   What does the average power loss depend on?
Answer   The rms current and the resistance: $P = I^2R$.

Solution and Discussion   The average power loss is

$P = (0.244 \text{ A})^2(300 \text{ }\Omega) = 17.9 \text{ W}$

An alternate method is to calculate the power factor first:

$$\cos \phi = \frac{R}{Z} = \frac{300 \text{ }\Omega}{328 \text{ }\Omega} = 0.915$$

The alternate expression for power loss is

$P = IV \cos \phi = (0.244 \text{ A})(80.0 \text{ V})(0.915) = 17.9 \text{ W}$

Example 21.4

Suppose the voltage source in Fig. 21.6 has an rms value of 50.0 V and a frequency of 600 Hz. Suppose further that $R = 20.0 \text{ }\Omega$, $C = 10.0 \text{ }\mu\text{F}$, and $L = 4.00 \text{ mH}$. Find (*a*) the current in the circuit and (*b*) the voltmeter reading across R, C, and L individually.

R

Reasoning

Question   What is the expression for current?
Answer $I = \dfrac{V}{Z}$

Question   What is Z for this circuit?
Answer   For an LRC circuit: $Z = \sqrt{R^2 + (X_L - X_C)^2}$, where $X_C = \dfrac{1}{2\pi fC}$ and $X_L = 2\pi fL$.

Question What are the expressions for the individual rms voltage readings?
Answer $V_R = IR$, $V_C = IX_C$, and $V_L = IX_L$.

Solution and Discussion The reactances are:

$$X_C = \frac{1}{2\pi(600\,\text{Hz})(10^{-5}\,\text{F})} = 26.5\,\Omega$$

$$X_L = 2\pi(600\,\text{Hz})(4.0 \times 10^{-3}\,\text{H}) = 15.1\,\Omega$$

The difference of these reactances is $X_C - X_L = 11.4\,\Omega$. The impedance is then

$$Z = \sqrt{(20\,\Omega)^2 + (11.4\,\Omega)^2} = 23.0\,\Omega$$

This gives the value of the current:

$$I = \frac{50\,\text{V}}{23.0\,\Omega} = 2.17\,\text{A}$$

The individual voltage drops are

$$V_R = (2.17\,\text{A})(20\,\Omega) = 43.4\,\text{V}$$
$$V_C = (2.17\,\text{A})(26.5\,\Omega) = 57.5\,\text{V}$$
$$V_L = (2.17\,\text{A})(15.1\,\Omega) = 32.8\,\text{V}$$

Notice that the potential difference across the capacitor is larger than the source voltage. Once again, rms voltages do not add as instantaneous ones do. They do add vectorially, however, when their phase difference is taken into account.

Exercise Find the phase difference between i and v. Which leads?
Answer: $\phi = 29.6°$; *since* $X_C > X_L$, i *leads* v *by this phase angle.*

21.7 ELECTRICAL RESONANCE IN SERIES *LRC* CIRCUITS

Let us consider a circuit containing only a capacitor C and an inductance L, shown in Fig. 21.10. This is not a very realistic situation, since any inductance coil generally has some resistance. Nevertheless, such an idealized circuit has much to teach us. With $R = 0$, Eq. 21.10 for impedance becomes

$$Z = |X_L - X_C|$$

We use the absolute-value sign because negative impedance has no physical significance. The current in such a circuit is

$$I = \frac{V}{Z} = \frac{V}{|X_L - X_C|}$$

Notice that when $X_L = X_C$, the current in the circuit becomes infinite.

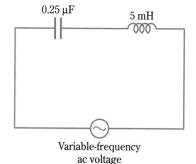

0.25 μF   5 mH

Variable-frequency
ac voltage

FIGURE 21.10

As the voltage frequency is changed, X_L and X_C change as shown in Fig. 21.11. The current in the circuit varies as shown in Fig. 21.12.

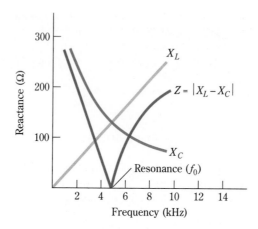

FIGURE 21.11

Both X_L and X_C, as well as Z, for the circuit in Fig. 21.10 vary with source frequency.

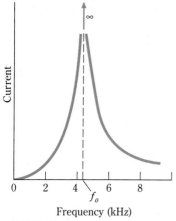

FIGURE 21.12

As the frequency of the source in Fig. 21.10 is varied, the current in the circuit behaves as shown here.

It is easy to obtain the condition $X_L - X_C = 0$ because X_L increases with frequency while X_C decreases with frequency. Figure 21.11 shows how these quantities vary for the C and L values given in Fig. 21.10. The impedance becomes zero at $f = 4500$ Hz in this case. This frequency, the frequency at which $X_L = X_C$, is called the **resonance frequency** of the circuit, and we denote it by f_o. Since $X_L = 2\pi f L$ and $X_C = 1/2\pi f C$, at resonance we have that

$$2\pi f_o L = \frac{1}{2\pi f_o C}$$

from which the resonance frequency is

$$f_o = \frac{1}{2\pi}\sqrt{\frac{1}{LC}} \tag{21.13}$$

Figure 21.12 shows how the current in the circuit of Fig. 21.10 varies as the frequency of the ac voltage is changed. (Of course, the voltage amplitude must be

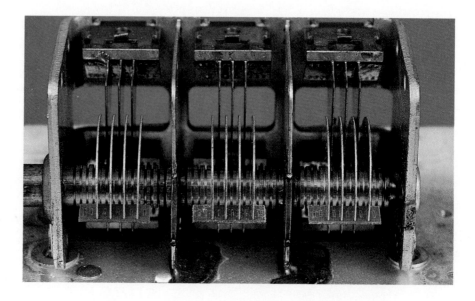

This shows a variable capacitor used in the antenna circuit of a radio receiver. The tuning knob is attached to the end of the brass shaft. As the knob is turned, the silver-edged metal plates move in or out of the space between the fixed red-edged plates. This varies the effective area of the capacitor and hence its capacitance. In turn, this changes the resonant frequency of the RLC antenna circuit, enabling the radio to "tune in" stations of different broadcast frequencies.

kept the same for all frequencies.) As we see, the current peaks sharply at the resonance frequency. In practical circuits, the peak is finite rather than infinite because all wires have some resistance.

Let us now apply this result to the *LRC* circuit whose impedance is given by Eq. 21.10. We see that at resonance X_C and X_L cancel each other, leaving $Z = R$. This also means that $\cos \phi = 1$ and power loss $= IV$. We thus see that

At the resonance frequency, an *LRC* series circuit behaves as if it was a purely resistive circuit.

We can understand electrical resonance better if we recognize that it is much like mechanical resonance. You know that mechanical systems often have a natural frequency at which they vibrate. If pushed with this frequency, they will vibrate with maximum amplitude; in other words, they resonate. A simple *LC* circuit also has a natural frequency of vibration. Let us explore this analogy between resonance in electrical and mechanical systems. Consider the *LC* circuit and the child on a swing shown in Fig. 21.13. Suppose that at the starting instant, the current in the circuit is zero and the swing is at its highest position. The charge on the capacitor is q_o, and an energy $\frac{1}{2}(q_o^2/C)$ is stored in it. By analogy, the swing possesses gravitational potential energy.

We know that in the electrical system, the capacitor will begin to discharge

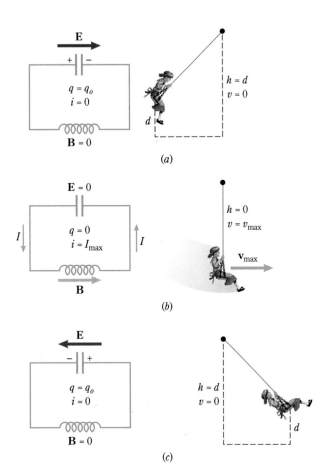

FIGURE 21.13

Just as the energy of the swing continually oscillates between potential and kinetic, the energy of the circuit is alternately stored in the capacitor and the inductor.

through the inductor. The current will rise rather slowly because the inductor opposes any change in current. Similarly, the swing will begin to pick up speed as its inertia is overcome by the accelerating forces acting on it. Both the swing and the capacitor lose their potential energy. Once the swing has reached the bottom of its path, all its potential energy has been changed to kinetic energy. Similarly with the circuit: once the capacitor has lost all its charge, the current in the circuit has its largest value and the original energy is now stored in the inductor. Its value is $Li^2/2$. This situation is shown in Fig. 21.13b.

Of course, the swing does not stop at the bottom of the path. Its inertia keeps it moving until it comes to rest in the position shown in Fig. 21.13c. Now its energy is all potential energy once again. Much the same thing happens in the electric circuit. The inductance, having inertia of a sort, opposes any change in current, and so the current does not stop at once. By the time the current finally stops, the capacitor is fully charged again, as in c. These processes repeat over and over.

The electric circuit undergoes an energy interchange much like that of the child and swing. The swing's energy alternates between potential and kinetic, and the energy in the circuit is alternately stored in the capacitor and the inductor. Both systems would oscillate back and forth forever if there were no energy losses. In the case of the swing, friction losses eventually damp out the oscillation. In the electrical case, resistive effects cause some of the energy to be lost and so the oscillation slowly damps down in amplitude.

The analogy can be extended further. Both the swing and the circuit possess natural resonance frequencies for their motion. The swing system constitutes a pendulum; we computed its natural frequency of vibration in Sec. 14.6. The natural resonance frequency of the circuit is the resonant frequency computed in Eq. 21.13.

If we wish to cause the child to swing very high, we must push on the swing at just the proper time and with the same frequency as the resonant frequency of the swing. We have seen that a very large current could be built up in the LC circuit if the oscillator "pushed" on the circuit at its resonant frequency. Hence, even the resonance behavior of the two systems is quite similar. It is shown in the next chapter that the LC resonant circuit discussed here forms an integral part of any radio or television receiver.

Example 21.5

Consider an LRC series circuit with $R = 10.0\,\Omega$, $L = 50.0\,\text{mH}$, and $C = 5.00\,\text{pF}$. Suppose a 20.0-V rms voltage can be applied to the circuit at various frequencies. (*a*) Find the resonance frequency of this circuit. (*b*) Find the rms current at the resonance frequency. (*c*) Calculate the impedance of the circuit and the current at a frequency 1 percent above the resonance frequency.

R *Reasoning   Parts (a) and (b)*

Question   What condition determines the resonant frequency?
Answer   Resonance occurs at the frequency where $X_C = X_L$.

Question   What is the expression for f_o?
Answer $f_o = \dfrac{1}{2\pi\sqrt{LC}}$

Question   What is the relationship between V and I at resonance?
Answer   At resonance, $Z = R$, and so $I = V/R$.

Solution and Discussion The resonance frequency is

$$f_o = \frac{1}{2\pi\sqrt{(50.0 \times 10^{-3}\,\text{H})(5.00 \times 10^{-12}\,\text{F})}} = 3.18 \times 10^5\,\text{Hz}$$

The rms current at this frequency

$$I = \frac{20.0\,\text{V}}{10.0\,\Omega} = 2.00\,\text{A}$$

R

Reasoning Part (c)

Question What frequency is 1 percent above f_o?
Answer $f = 1.01f_o = 1.01(3.18 \times 10^5\,\text{Hz}) = 3.21 \times 10^5\,\text{Hz}$

Question What are the values of X_C and X_L at this frequency?
Answer

$$X_C = \frac{1}{2\pi(3.21 \times 10^5\,\text{Hz})(5.00 \times 10^{-12}\,\text{F})} = 9.9 \times 10^4\,\Omega$$

$$X_L = 2\pi(3.21 \times 10^5\,\text{Hz})(50.0 \times 10^{-3}\,\text{H}) = 1.01 \times 10^5\,\Omega$$

Question What is the difference between these reactances?
Answer $X_L - X_C = 2000\,\Omega$

Question What is the impedance at the frequency $1.01f_o$?
Answer $Z = \sqrt{(10\,\Omega)^2 + (2000\,\Omega)^2} \approx 2000\,\Omega$

Solution and Discussion Notice that X_L and X_C are very large compared to R, *even at resonance.* Unless their cancellation is almost exact (at resonance), they impede the current much more than the small resistance does. The current at frequency $1.01f_o$ is only

$$I = \frac{20\,\text{V}}{2000\,\Omega} = 0.01\,\text{A}$$

This is 0.5 percent of the resonance current. The power consumption, which depends on I^2, is only $(0.005)^2 = 2.5 \times 10^{-5}$ as much as at resonance. Circuits with very sharp resonances like this are used in sensitive radio receivers, as we will see in the next chapter.

LEARNING GOALS

Now that you have finished this chapter, you should be able to

1 Define (a) RC time constant, (b) ac versus dc voltage or current, (c) effective and rms values, (d) capacitive reactance, (e) inductive reactance, (f) impedance, (g) power factor, (h) resonance in LC circuit.

2 Sketch the current and charge curves for an RC circuit during charging and discharging. Define the time constant for

the circuit and relate it to the curves.

3 Sketch a typical ac voltage or current curve, showing the peak, average, and rms values. Relate the rms value to the peak value in a quantitative way.

4 State the Ohm's law form that applies to an ac voltage impressed on a resistor. Sketch the current and voltage curves on the same graph. Compute the average power loss in the resistor when sufficient data are given.

5 Explain why the impeding effect of a capacitor should be higher at low frequencies than at high frequencies. Use $V = IX_C$ in simple situations.

6 Sketch the current and voltage curves for a capacitor connected across an ac power source. State the average power loss in the capacitor.

7 Explain why the impeding effect of an inductor should be larger at high frequencies than at low frequencies. Use $V = IX_L$ in simple situations.

8 Sketch the current and voltage curves for an inductor connected across an ac source. State the average power loss in the inductor.

9 Use the relation $V = IZ$ for simple problems involving series LRC circuits.

10 Use $V = IZ$ to explain why a resonance frequency exists for an LC circuit. Show how to find the resonance frequency.

QUESTIONS AND GUESSTIMATES

1 You are given a 2-μF capacitor, a dry cell, and an extremely sensitive and versatile current-measuring device. How could you use these to measure a resistance that is thought to be about $10^8 \, \Omega$? Could you make the measurement using an ordinary voltmeter in place of the current meter?

2 In some places, low-frequency ac voltage (considerably less than 60 Hz) is used. Electric lights operated on this voltage flicker rapidly. Explain the cause of this flickering.

3 For which of the following uses would dc and ac voltage be equally acceptable: incandescent light, electric stove, electrolysis, television set, fluorescent light, neon-sign transformer, battery charger, toaster, electric clock?

4 Draw an analogy between the vibration of a mass on a spring and the oscillation of an LC circuit. What quantities in the mechanical system correspond to L and C in the electrical system? Explain.

5 Compare the equation for the resonance frequency of a mass vibrating at the end of a spring with the resonance equation for an LC circuit. What analogy can you draw between them?

6 A dc voltmeter is connected across the terminals of a variable-frequency oscillator. How does the meter behave as the frequency of the oscillating voltage is slowly increased from 0.01 to 100 Hz? Explain.

7 In an LRC series circuit, when, if ever, is the current through the circuit in phase with the source voltage?

8 The following statement was published in a daily newspaper. "A warning that home electrical appliances can cause fatal injuries has been sounded by City Health Director J. R. Smith. His comments followed the death of an 18-year-old boy who accidentally electrocuted himself by inserting a fork in a toaster. Dr. Smith pointed out that even adults can be killed by such electrical shocks. Ordinary house current is 110 V, but the voltage is increased if the current is grounded, he said." What is wrong with the last sentence, and how should it have been worded?

9 The devices shown in Fig. P21.1 are called filters. When an ac voltage is put into the device, the output ac voltage depends on the frequency of the oscillating voltage. One of these devices lets the input voltage pass through undisturbed if the oscillation frequency is high. The other passes only low-frequency voltages. Explain which is which.

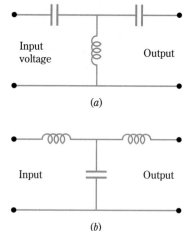

(a)

(b)

FIGURE P21.1

SUMMARY

DEFINITIONS AND BASIC PRINCIPLES

Root-Mean-Square (rms) Values
The relationship between the amplitude of a sinusoidally varying current or voltage (i_o and v_o) and the rms values (I and V) is

$$I = \frac{i_o}{\sqrt{2}} \quad \text{and} \quad V = \frac{v_o}{\sqrt{2}}$$

INSIGHTS

1. It is the rms values that give the correct expressions for the power delivered by the source voltage or converted to heat in a resistor:

$$P = IV_R = I^2 R$$

2. It follows from the above that the amplitudes are derived from the rms values by

$$i_o = (\sqrt{2})I = 1.414I \quad \text{and} \quad v_o = (\sqrt{2})V = 1.414 \, V$$

Time Dependence of Charge in an RC Series Circuit
Upon closing the switch in an *RC* circuit, the charge (and hence voltage) on *C* builds up according to

$$q(t) = q_f(1 - e^{-t/\tau_C})$$

where $q_f = CV_C$ and $\tau_C = RC$ is the capacitive time constant.

Phase Relationships Between Current and Voltage in ac Circuits
PURELY RESISTIVE CIRCUIT   The instantaneous current and voltage are in phase.

PURELY CAPACITIVE CIRCUIT   The instantaneous current leads the voltage by $\frac{1}{4}$ cycle.

PURELY INDUCTIVE CIRCUIT   The instantaneous voltage leads the current by $\frac{1}{4}$ cycle.

Relationship Between I and V in AC Circuits: Reactances
PURELY RESISTIVE CIRCUIT

$V = IR$ (Ohm's law)

PURELY CAPACITIVE CIRCUIT

$V = IX_C$

where $X_C = \dfrac{1}{2\pi fC}$ is the capacitive reactance.

PURELY INDUCTIVE CIRCUIT

$V = IX_L$

where $X_L = 2\pi fL$ is the inductive reactance.

INSIGHTS
1. The units of R, X_C, and X_L are ohms.
2. Resistance is independent of frequency, but the reactances X_C and X_L both depend on frequency as well as on the values of C and L, respectively.
3. Ohm's law for resistors applies to rms *and* instantaneous values of v and i, since they are in phase. The relationship between current and voltage equivalent to Ohm's law for capacitors and inductors applies *only* to rms and amplitude values but not to v and i, since they are out of phase across C and L.

Relationship Between V and I in AC LRC Series Circuits: Impedance
In an *LRC* series circuit, V and I are related through the impedance of the circuit, Z:

$V = IZ$

where $Z = \sqrt{R^2 + (X_L - X_C)^2}$.

INSIGHTS
1. The opposing phase relationships between v and i for capacitors and inductors cause their effects to subtract. Their difference then adds to the resistance vectorially.
2. In an *RL* circuit $X_C = 0$, making $Z = \sqrt{R^2 + X_L^2}$.
3. In an *RC* circuit $X_L = 0$, making $Z = \sqrt{R^2 + X_C^2}$.

Phase Angle in Series LRC Circuits
The phase angle ϕ between v and i in an *LRC* circuit is given by

$$\cos \phi = \frac{R}{Z} \qquad \text{or} \qquad \phi = \cos^{-1}\frac{R}{Z}$$

INSIGHTS
1. If $X_C > X_L$, the instantaneous voltage lags the current by this phase angle.
2. If $X_L > X_C$, the instantaneous voltage leads the current by this phase angle.

Resonance in LRC Circuits
At the frequency f_o where $X_C = X_L$, the reactances cancel, leaving $Z = R$. This is the least possible value of Z, which gives the greatest possible value of I. This is called resonance, and f_o is the resonance frequency:

$$f_o = \frac{1}{2\pi\sqrt{LC}}$$

INSIGHT
1. At resonance the *LRC* circuit acts as if it was purely resistive. $I = V/R$, and the current and voltage are in phase ($\phi = 0$).

Power in AC Circuits
POWER FACTOR $\cos \phi$ is called the power factor of an ac circuit. The average power delivered to an ac circuit is

$$P = IV \cos \phi = IV\left(\frac{R}{Z}\right)$$

POWER CONSUMPTION   The average power consumed (converted into heat) in an ac circuit occurs entirely in the resistance R:

$$P \text{ (in } R) = I^2R$$

There is no average power consumed in a capacitor or pure inductor.

PROBLEMS

Section 21.1

1 How large a resistor must be placed in series with a 0.50-μF capacitor if the time constant is to be 4.0 s?

2 About how long does it take for the charging current to drop to one-third its initial value when a 3.0-μF capacitor is being charged through a 10-MΩ resistor by a 9.0-V battery?

3 A series circuit consists of an uncharged 4.0-μF capacitor, a 6.0-MΩ resistor, a 12-V battery, and a switch. What is the current in the circuit and the charge on the capacitor (*a*) immediately after the switch is closed and (*b*) one time constant later?

4 A series circuit consists of a 6.0-μF capacitor charged to 9.0 V, a switch, and a 50-MΩ resistor. What is the current in the circuit and voltage across the capacitor (*a*) when the switch is first closed and (*b*) one time constant after the switch is closed?

5 A series circuit consists of a 9.0-V battery, a 4-MΩ resistor, a 5.0-μF capacitor, and an open switch. The capacitor is initially uncharged. The switch is now closed. (*a*) What is the time constant of the circuit? (*b*) About how long does it take for the capacitor to become two-thirds charged? (*c*) How much charge will flow into the capacitor in the time calculated in part *b*? (*d*) About what is the average current into the capacitor during this interval?

6 Suppose you measure the resistance of your body between your two hands with an ohmmeter and find it to be 62 kΩ. A 20.0-μF capacitor has been charged to 12.0 V and disconnected. You now grasp the two terminals of the capacitor with your two hands. (*a*) What is the time constant of the circuit? (*b*) What is the approximate potential difference across the capacitor after 0.8 s? (*c*) What is the charge on the capacitor when the potential across it is 9.0 V? (*d*) About what is the average current through your body in the 0.8 s?

7 A capacitor is being charged by a battery in series with a resistor. In two time constants after the switch is closed, what percentage of the final charge is on the capacitor?

8 Two capacitors, 3.0-μF and 6.0-μF respectively, are connected in series with a 5.0-MΩ resistor, a 9.0-V battery, and an open switch. (*a*) What is the time constant of the circuit? (*b*) What is the voltage across the 6.0-μF capacitor after one time constant? (*c*) How much charge has been delivered to the 3.0-μF capacitor during this time?

Sections 21.2, 21.3

9 An ac ammeter in series with an incandescent lamp reads 0.4 A, and an ac voltmeter across the lamp reads 110 V. What are the maximum (peak) current through the lamp and the maximum voltage across it?

10 An rms voltage of 110 V is applied to an electrical device that has a resistance of 15 Ω. Find the maximum current through the device and the rms current supplied.

11 The current through an incandescent bulb is read to be 0.72 A by an ac ammeter, and the voltage across it is measured to be 120 V by an ac voltmeter. (*a*) What are the peak current and voltage furnished to the bulb? (*b*) How much power is used by the bulb? (*c*) What is the resistance of the bulb?

12 What is the resistance of a light bulb that uses an average power of 60 W when connected to a 60-Hz ac power source with an rms voltage of 110 V?

13 How much current does a 900-W, 110-V toaster draw from a 110-V (rms) ac power line? What is the resistance of the toaster during normal operation? How many calories of heat does it generate in 5 min?

■14 Two 120-W light bulbs and one 90-W bulb are connected in parallel to the 110-V ac (rms) household supply. Find the rms value of the current and the resistance of each of the bulbs.

■15 The current through a 40-Ω resistor is given by the relation $i = 4 \sin 240t$ A. How much power does the current dissipate in the resistor?

■16 The output from an ac source is given by $v = 120 \sin 377t$ V. Find (*a*) the frequency of the source, (*b*) the rms voltage of the output, and (*c*) the voltage at $t = \frac{1}{15}$ s.

■17 What are the maximum and rms currents when the source in Prob. 16 is connected to a 60-Ω resistor? How much power does the resistor dissipate?

■18 A voltage $v = 60 \cos 300t$ V is impressed across a 25-Ω resistor. How much power is dissipated in the resistance?

■19 For a certain ac generator, the output voltage is $v = 0.3v_o$ and increasing when $t = 0.004$ s. What is the operating frequency of the generator? (Assume $v = 0$ at $t = 0$.)

■20 A 110-V 60-Hz ac source is connected across a 30-Ω resistor. (*a*) Find the current drawn from the voltage source. (*b*) Repeat for 5000 Hz. (*c*) How much power is dissipated in each case?

■21 The current in a resistance circuit at $t = 0.004$ s is increasing and equal to 72 percent of the maximum current. What is the frequency of the source? (Assume $i = 0$ at $t = 0$.)

Section 21.4

22 How much rms current does a 4.0-μF capacitor draw from a 110-V 60-Hz source connected directly across it? Repeat for a 110-V 60,000-Hz source.

23 A 3.0-μF capacitor is connected directly across a 60-V 240-Hz source. How much rms current does it draw from the source? Repeat for a source frequency of 0.4 MHz.

24 A capacitor in a circuit has a capacitive reactance of 40 Ω when the frequency of the source is 120 Hz. What capacitive reactance would the capacitor have if the source frequency were changed to 10,000 Hz?

25 How large an rms current will be delivered by an ac source whose rms voltage output is 42 V at a frequency of 90 Hz when directly connected across a 2.8-μF capacitor?

26 A 60-Hz ac source with a maximum output voltage of 170 V is connected directly across an unknown capacitor. What value of the capacitance will draw an rms current of 0.72 A?

27 The rms current in a circuit containing a 5.0-μF capacitor is 0.4 A when connected to an ac source with an rms voltage of 40 V. Find the frequency of the source.

28 An 8.0-μF capacitor is connected directly to a 220-V 50-Hz power source. (a) What is the average value of the power consumed by the capacitor? (b) What is the rms current to the capacitor? (c) What is the maximum charge on the capacitor?

29 By what factor does the current to a capacitor change as the frequency of the source of voltage across it is increased by a factor of (a) 10, (b) 1000, and (c) 10,000? Assume that the circuit has no resistance and that the magnitude of the source voltage remains constant.

30 Two capacitors, 2.0-μF and 6.0-μF, are connected in series across an ac source of rms voltage 240 V and frequency 50 Hz. Find the maximum charge on each capacitor.

Section 21.5

31 Find the inductive reactance of a 4.0-mH inductance coil at a frequency of (a) 60 Hz and (b) 600 kHz.

32 If an inductance coil is to have a reactance of 32 Ω for a frequency of 1200 Hz, what must its inductance be? What is its reactance at 6.0 Hz?

33 Calculate the inductance of a coil which has an inductive reactance of 60 Ω when the source has an angular frequency of 1508 rad/s.

34 An inductor is connected to a 30-Hz power supply with an rms voltage of 50 V. What inductance is needed in order that the maximum current in the circuit be below 90 mA?

35 In a purely inductive circuit, the rms voltage is 110 V. (a) Calculate the value of the inductance if the rms current is 8 A at a frequency of 60 Hz. (b) At what frequency will the rms current be reduced to one-half its original value?

36 An ac voltage source is connected directly across an ideal 30-mH inductance and causes an rms current of 0.8 A when the maximum voltage is 9 V. (a) What is the frequency of the source? (b) If the frequency is tripled and the maximum voltage is kept at 9.0 V, what is the rms current in the coil?

37 An inductance has a reactance of 78 Ω at 60 Hz. What will be the maximum current if this inductor is connected to a 50-Hz source with an rms voltage of 220 V?

38 An ac voltage source is connected directly across a resistanceless 1.2-mH inductance coil. How large must the voltage be to give a current of 1.80 A if the frequency is (a) 50 kHz and (b) 500 kHz?

Section 21.6

39 A 40-Ω resistor is connected in series with a 30-μF capacitor and an ac generator with rms voltage of 80 V at 60 Hz. Find (a) the rms current in the circuit, (b) the voltage drop across

the capacitor, and (c) the phase angle between instantaneous current and voltage.

40 A 4.0-μF capacitor and a 400-Ω resistor are connected in series across a 30-V 120-Hz power source. Find the current in the circuit and the power drawn from the source.

41 A 50-Ω resistor is connected in series with a 6.0-μF capacitor across a voltage source. At what frequency will the rms voltage across the resistor be the same as that across the capacitor?

42 A series circuit consists of a 60-V 1200-Hz power source, a 1-kΩ resistor, and an unknown capacitor. The rms voltage drop across the resistor is 42 V. What are the current in the circuit and the value of the capacitor?

43 A 4.0-mH inductor that has a resistance of 200 Ω is connected directly across a 30-V 6000-Hz power source. Find the current in the circuit and the power drawn from the source.

44 A 5-mH ideal inductor is connected in series with a 60-Ω resistor across a variable-frequency ac voltage source. At what frequency is the rms voltage across the resistor the same as that across the inductor?

45 An unknown inductor L is connected in series with an 800-Ω resistor and a 90-V 2000-Hz power source. The voltage drop measured across the resistor is found to be 40 V. What is the current in the circuit and the value of the inductance?

46 At what frequency will the capacitive reactance of a 70-μF capacitor equal the inductive reactance of a 70-mH inductance?

47 A 60-Hz source is connected across a 40-μF capacitor. What inductor connected across this same source would draw the same current?

48 A 3.0-μF capacitor and an inductor are connected in series across a 110-V 60-Hz source. The inductor has an inductance of 0.6 mH, and its resistance is 720 Ω. (a) Find the current in the circuit. (b) Repeat for a frequency of 6000 Hz.

49 How large an inductance must be connected in series with a 6-μF capacitor, a 40-Ω resistor, and a 240-V 50-Hz power source if the rms current in the circuit is to be 3.2 A?

50 A 50-Ω resistor, 80-mH inductor, and 40-μF capacitor are connected in series to a 90-V 60-Hz ac source. Find the voltage drop (a) across the RC combination and (b) across the LC combination.

51 An LRC circuit consists of a 50-Ω resistor, a 12-μF capacitor, and a 240-mH inductor, connected in series with a 110-V 60-Hz power supply. (a) What is the phase angle between the current and the applied voltage? (b) Does the current lead or lag the voltage?

52 A 100-Ω resistor, 20-μF capacitor, and 180-mH inductor are connected in series across a 110-V 60-Hz power source. Find (a) the current in the circuit, (b) the voltage drop across the LC combination, (c) the power loss in the circuit, and (d) the power factor.

53 An inductance coil draws 0.8 A when connected across a 12-V battery and 3.6 A when connected to a 110-V ac source with a frequency of 60 Hz. What is the inductance of the coil and how much power does it draw from the ac source?

54 A coil has an inductance of 300 mH and a resistance of $120\,\Omega$, which can be considered to be in series. At what frequency is the impedance of the coil $144\,\Omega$?

55 A series LRC circuit has a resistance of $100\,\Omega$ and an impedance of $210\,\Omega$. How much average power will be dissipated in the circuit if it is connected to an ac source with an rms voltage of 110 V?

■56 An inductor, a capacitor, and a resistor are connected in series across an ac power source. The rms voltages are 120 V across the inductor, 60 V across the capacitor, and 60 V across the resistor. Find (a) the peak voltage of the source and (b) the phase angle between i and v.

Section 21.7

57 (a) How large a capacitor must be connected in series with a 0.40-mH inductor if they are to resonate with a frequency of 60 Hz? (b) How large an inductor is needed to resonate at the same frequency with a 6-μF capacitor?

58 When connected in series with a 5.0-μF capacitor, an inductor resonates sharply to a frequency of 720 Hz. What is the value of the inductance?

■59 A series circuit with $R = 1600\,\Omega$, $L = 400$ mH, and $C = 10\,\mu$F is driven with an oscillating source of variable frequency. (a) What is the resonance frequency of the circuit? (b) What is the impedance of the circuit at the resonance frequency?

■60 In a radio an LRC circuit is used to tune an FM station broadcasting at 96.5 MHz. The inductance in the circuit is 1.44 μH and the resistance is $14\,\Omega$. What capacitance should be used to tune this station?

■61 A variable capacitor is used in a tuning circuit for the AM broadcast frequencies in the range of 500 to 1600 kHz. If an inductance of 4-μH is used in series with the capacitor, what

should the extreme values of the variable capacitor be to cover the entire frequency range?

■■62 A 30-Ω resistor, a 3-μF capacitor, and a 4-mH inductor are connected in series to an ac source with an rms output voltage of 60 V. Find (a) the resonance frequency of the circuit, (b) the current at resonance frequency, and (c) the power delivered to the circuit at one-half the resonance frequency.

Additional Problems

■63 The rms current through a 3.0-μF capacitor when connected across an ac source with an rms voltage of 9.0 V is 20.0 mA. (a) What is the operating frequency of the source? (b) If the capacitor is replaced by an ideal inductance coil with an inductance of 0.2 H, what is the rms current through coil?

■■64 A series LRC circuit in which $R = 60\,\Omega$ is connected across an ac source of frequency 300 Hz and rms voltage 180 V. The rms voltage is the same across each of the elements. (a) Find the voltage reading across the pure inductor. (b) Find the values of L and C.

■■65 A 0.8-H inductance is connected in series with a fluorescent lamp to limit the current through the lamp. The combination is connected to a 110-V 60-Hz power line. If the voltage across the lamp is to be 48 V, what is the current in the circuit? Assume that the lamp is a purely resistive load.

■■66 An LRC series circuit has a resonance frequency of $2400/\pi$ Hz. When operating at a certain frequency higher than the resonance frequency the circuit has an inductive reactance of $14\,\Omega$ and a capacitive reactance of $9\,\Omega$. What are the values of the inductance and capacitance in the circuit?

■■67 A resistor and an inductor are connected in series to a 120-V 60-Hz source. The voltage drop across the resistor is found to be 54 V and the power dissipated in the circuit is 16 W. What are the values of the resistance and the inductance in the circuit?

LIGHT AND OPTICS

PART FOUR

LIGHT AND OPTICS

Research is to see what everyone else has seen,
and to think what nobody else has thought.

ALBERT SZENT-CYÖRYI

Light and the process of seeing have proved to be among the most intriguing subjects science has studied. We depend more on our sense of vision and perception of color than on any other sense to gain detailed information about the world around us. Throughout the seventeenth century we discovered how light is bent (refracted) as it passes from one medium to another, how it is reflected, and how a spectrum of colors is contained in white light. As a result of this knowledge, lenses and mirrors were fabricated that enabled astronomy to become truly established as an observational science and to flourish throughout the eighteenth century.

As it did for the rest of classical physics, the nineteenth century produced an explosive increase in our understanding of the properties of light. Wave interference and polarization were discovered, and the speed of light was measured with precision in air and in water. Instruments using prisms and diffraction gratings gave rise to analysis of light spectra from various sources and the field of spectroscopy was born. These spectra were to become the key to understanding the structure of the atom during the early twentieth century. The culminating event in light theory came with Maxwell's equations, which unified the study of optics and electricity and magnetism, predicting the existence of electromagnetic waves over a very wide spectrum of wavelengths.

As we gained in our understanding of light, we have been able to create imaging systems that allow us to see clearer, farther, and in more detail than is possible with the unaided eye. We have been able to measure smaller distance and time intervals, which in turn makes possible more precision in manufacturing processes, faster control switches and more sensitive sensors, and faster, more reliable, and more efficient means of information-storage processing. The importance of light in our lives, obvious as it has always been through our sense of sight, has scarcely begun to be developed. Applications of light in communications, computing, and manufacturing, among many other areas, will continue to increase at an astonishing rate. Whereas the control of the electron—electronics—has characterized the present century, the control of the photon—photonics—will characterize the twenty-first century.

CHAPTER 22

ELECTROMAGNETIC WAVES

In our daily lives we encounter many types of wave phenomena. In some waves, like those on the surface of a lake and vibrating guitar strings, the wave structure is visible. In other types—sound waves, for example—the structure cannot be seen, but we know from our previous discussions that a sound wave consists of oscillations in the pressure of air molecules. There is another type of wave whose structure is not so apparent, examples of which are radio waves, light waves, infrared waves, and microwaves. These waves can travel and carry energy through empty space, which brings up the question of just what is waving in this vacuum. These latter waves are called *electromagnetic waves,* and their nature is the subject of this chapter.

22.1 OSCILLATING ELECTRIC AND MAGNETIC FIELDS; MAXWELL'S EQUATIONS

The explanation of electromagnetic waves by the Scottish physicist James Clerk Maxwell (1831–1879) is recognized as one of the greatest achievements in the history of science. Maxwell developed his theory in the 1860s. Before we proceed to discuss his work, let us review what was known about electricity and magnetism up to this time.

The following basic principles, all of which we have studied in previous chapters, had been established by the middle of the nineteenth century:

1 The existence of positive and negative charge and Coulomb's law of force between charges. Charges were recognized as a source of electric fields, with fields emanating from positive charges and terminating on negative charges.
2 Moving charges, or currents, were recognized as a source of magnetic fields. Ampere's law describes the connection between electric current and magnetic field.
3 Magnetic field lines are closed loops, having no beginning or end. This is a way of saying that magnetic monopoles do not exist, that magnetic poles always occur in opposite north-south pairs.
4 An electric field can be produced by a time-varying magnetic field. This is Faraday's law of induction.

It is important to remember that the mathematical form of these basic principles contains the two physical constants μ_o and ϵ_o, which we encountered in Chaps. 16 and 19. The measured values of these constants were known to Maxwell.

Let us examine the properties of one particular charge distribution, the electric dipole. As we discussed in Chap. 17, a dipole consists of two equal charges of opposite sign separated by a distance which we will call d. Figure 22.1 shows a simple way of creating a dipole, using a battery to charge two small conducting spheres connected to opposite terminals. A portion of the static electric field created by the dipole is shown in Fig. 22.1. At distances much greater than d, the strength of this field decreases in inverse proportion to the cube of the distance from the dipole.

Now suppose we reverse the polarity of the battery suddenly. We know that this action will cause the direction of the field shown in Fig. 22.1 to reverse. Let us ask a basic question: "Will this change in the field be felt immediately everywhere?" In other words, will a test charge at point A immediately experience a reversal of the electric force? Nothing we have studied so far enables us to answer this question, and so let us go on to another observation.

When we reverse the polarity of the battery, charge must flow along the dipole in the process of reversing the electric field. During this time a magnetic field must therefore be created by the current produced by this flow of charge. The question that arises is: "Is this magnetic field felt at point A immediately?"

Yet another question arises out of this observation. By reversing the battery voltage, we cause a change in the electric field. This change in turn creates a magnetic field, due to the current produced by charge flowing between the spheres. Is it possible to generalize this effect to a situation where no charge flows in the region of changing electric field? In other words: "Does a changing electric field induce a magnetic field even where no charges flow?"

To answer this question, let us consider the example of the capacitor plates in Fig. 22.2. When we change the polarity of the plates, no charge can flow between them. Current flows only in the external circuit, which we can arrange so that the wires connecting the battery to the plates and the charges the wires carry are far away from the space between the plates. If we hook the plates up to an oscillating voltage, does the changing electric field between the plates *induce* a magnetic field between the plates, even though no charge flows between them? Such a principle—the idea that a magnetic field could be induced by a changing electric field—was unknown at the time when Maxwell was considering this question.

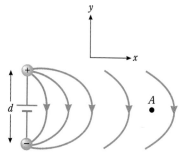

FIGURE 22.1

A portion of the instantaneous electric field close to two charged balls. If the charges oscillate back and forth between the balls, the electric field at point A will alternately point up and down.

FIGURE 22.2

Does the changing E field between the plates create a B field?

Maxwell noticed that the known laws of electricity and magnetism lacked symmetry between E and B fields: it was known that changing B fields induced E fields, but no counterpart to this law was known that would predict the induction of B fields by changing E fields. Maxwell took the bold step of assuming the latter. He postulated a fictitious current, which he called a *displacement current, I_D,* that is proportional to the time rate of change of an electric field in a given region. More specifically, Maxwell defined the **electric flux** Φ_E through an area A in the same way that we defined the magnetic flux in Eq. 20.1: for an E field that is uniform over the area A,

$$\Phi_E = E_\perp A$$

where $E_\perp$ is the component of **E** perpendicular to the area A. Maxwell then wrote his displacement current as

$$I_D = \epsilon_o \frac{\Delta \Phi_E}{\Delta t} = \epsilon_o A \frac{\Delta E_\perp}{\Delta t}$$

You should verify that the units of this expression come out to be amperes. The crucial point of Maxwell's new idea was that magnetic fields can be produced by *both* I_D and a real current I. He used the sum of the two terms $I_{\text{tot}} = I + I_D$ in Ampere's law in place of I alone.

Maxwell wrote the known laws plus his new postulate in a mathematical form known as differential equations. Even though we cannot discuss the mathematical details in this course, there are a number of interesting and important observations we can make qualitatively.

Because his equations incorporated what was already known about electricity and magnetism, they included the known physical constants ϵ_o and μ_o. By combining his differential equations, Maxwell derived time-dependent equations relating **E** and **B**. The solutions of these equations represent sinusoidal oscillations (waves) in the field strengths. Furthermore, the equations predicted that these oscillations— now called **electromagnetic waves**—travel through empty space with a wave speed v determined solely by the fundamental constants contained in the equations:

$$v = \frac{1}{\sqrt{\mu_o \epsilon_o}}$$

Since the values of these constants were already known, Maxwell could calculate this speed (and so can you!):

$$v = \frac{1}{\sqrt{(4\pi \times 10^{-7}\,\text{T} \cdot \text{m/A})(8.85 \times 10^{-12}\,\text{C}^2/\text{N} \cdot \text{m}^2)}} = 2.998 \times 10^8\,\text{m/s}$$

Astonishingly, this is just the speed of light, c! For the first time in history a connection was made between visible light (the field of study known as optics) and electricity and magnetism. Notice that since the speed of light is determined by two fundamental constants, *it also must be a universal physical constant.* Maxwell's postulate of induced magnetic fields not only resulted in the explanation of the nature of light waves. It also predicted that electromagnetic waves could exist at *any* frequencies, both above and below the frequencies of visible light ($\approx 10^{15}$ Hz).

In 1887, almost 10 years after Maxwell's death, the German physicist Heinrich Hertz (1857–1894) produced electromagnetic waves at frequencies near 10^8 Hz, waves which we now call *radio waves*. Hertz measured the wavelength of his waves and calculated their speed to be 3.2×10^8 m/s, sufficiently accurate to show experimentally that light and radio waves were examples of the same type of wave phenomena.

Now let us return to the questions we posed at the beginning of this section.

1 Are changes in electric and magnetic fields transmitted to all points instantaneously? The answer is *no*. A change in either field moves out from the source at a speed c, so that at a distance r from the source it takes a time $t = r/c$ for the change to be felt.
2 Does a changing electric field induce a magnetic field even in empty space where no charges can flow? *Yes*. Without this principle, the other laws of electricity and magnetism were incomplete, and could not account for electromagnetic waves. The fact that electromagnetic waves exist and the fact that their measured properties agree with the predictions of Maxwell are verifications of the correctness of his postulate.

All other forms of waves we have previously considered, such as sound, water waves, and waves on a string, involve oscillations in the material supporting the wave. Without some material to oscillate, therefore, such waves cannot exist, as can be demonstrated by causing a bell to ring inside a vacuum chamber. With no air to carry the vibrations of the bell, no sound wave is produced, and we cannot hear the bell. In the case of electromagnetic waves traveling through empty space, however, no material is needed to support the wave. A sinusoidally varying E field induces a sinusoidally varying B field, which in turn induces a sinusoidally varying E field, and on and on. The two oscillating fields keep renewing each other as the energy in the fields propagates through space at the speed c. As with all types of waves, the frequency f of an electromagnetic wave is determined by the frequency of the source. In the case of the electric dipole, this is the frequency of the applied oscillating voltage. The wavelength of the resulting wave is then

$$\lambda = \frac{v}{f} = \frac{c}{f} \tag{22.1}$$

Today we recognize that Maxwell's equations are as fundamental and important to electromagnetism as Newton's laws are to mechanics. Maxwell's equations form the basis for all theoretical work in electromagnetism.

22.2 ELECTROMAGNETIC WAVES FROM A DIPOLE ANTENNA

Now that we have discussed the general results of Maxwell's theory, let us look more closely at the electromagnetic waves generated by an oscillating voltage applied to an electric dipole. Let us attach an ac voltage source to two conducting rods as shown on the left in Fig. 22.3. The ac source alternates the applied voltage sinusoidally at a frequency f:

$$V_{\text{source}} = V_o \sin 2\pi ft$$

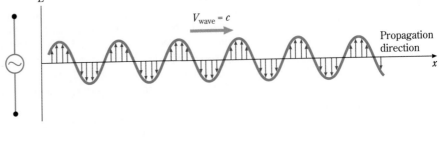

FIGURE 22.3

The alternating charges on the dipole antenna send an electric field disturbance out into space.

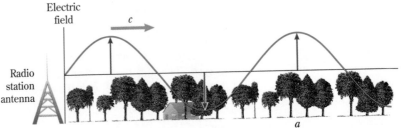

FIGURE 22.4

The electric field wave from the antenna blankets an area even quite distant from the station.

We can think of the electric field as being the disturbance sent out by the dipole source much the same way a wave on a string is the disturbance sent down the string by an oscillating source. At a certain instant, the field sent out along the x axis is as shown in Fig. 22.3. The field shows the history of the charge on the dipole. The downward-directed fields were sent out when the top of the dipole was positive; the upward-directed fields were sent out one-half cycle later, when the top of the dipole was negative. This waveform travels outward from the dipole at the speed of light, c.

For a radio station, the dipole (antenna) is often simply a long wire. If you visit a radio transmitting site, you will see the antenna as a long wire stretched between two towers or as a vertical wire held by a single tower. Charges are placed on the antenna by an ac voltage from a transformer system. The electric field wave sent out by the antenna blankets the earth around it, as in Fig. 22.4. At a point such as a in the path of the wave, the electric field reverses periodically as the wave passes. The frequency of the oscillating electric field at a is the same as the frequency of the source. Further, we notice that the quantity that vibrates, the electric field vector, is always perpendicular to the direction of propagation. Hence the electric field wave is a transverse wave (Sec. 14.11).

It is easy to see that a radio station's antenna necessarily generates a magnetic field wave as it generates an electric field wave. To see this, refer to Fig. 22.5. At the radio station, charges are sent up and down the antenna in Fig. 22.5a to produce the alternating charges we have been discussing. This charge movement constitutes an alternating current in the antenna, and because a magnetic field circles a current, an oscillating magnetic field is produced, as shown in Fig. 22.5b. As with the oscillating electric field, the magnetic field travels out along the x axis as a transverse wave. Because the direction of the current oscillates, so too does that of the magnetic field.

Notice, however, that the magnetic field is in the z direction, while the electric field is in the y direction. As shown in Fig. 22.6, the magnetic field is perpendicular to both the electric field and the direction of propagation. The two waves are drawn

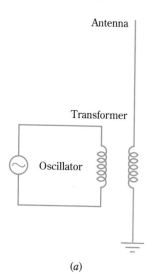

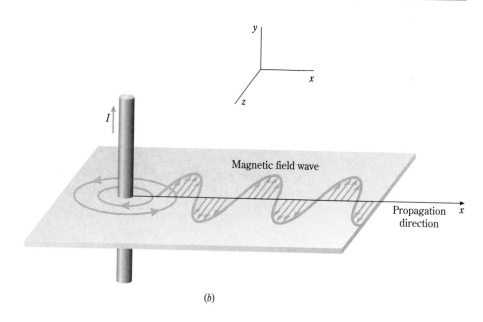

(a)

FIGURE 22.5

(a) As charge rushes up and down the antenna, (b) a magnetic field wave is sent out as shown.

in phase (that is, they reach their maxima together). That this is true for distances many wavelengths from the antenna is not obvious; it is the result of detailed computations.

There is one other feature of electromagnetic wave generation that we should point out. The charges that oscillate up and down the antenna are accelerating. It turns out that whenever a charge undergoes acceleration, it emits electromagnetic radiation; the larger the acceleration (or deceleration) of the charge, the more radiation it emits. Thus, if a fast-moving charged particle undergoes an impact, it will emit a burst of electromagnetic radiation as it suddenly stops.

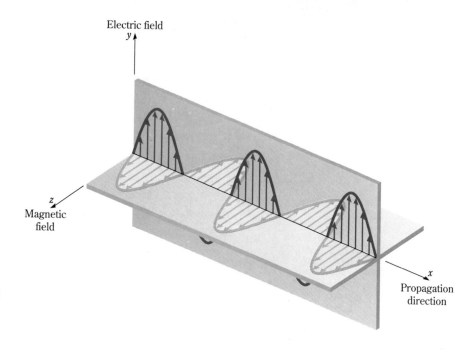

FIGURE 22.6

In an electromagnetic wave, the magnetic field wave is perpendicular to both the electric field wave and the direction of propagation.

Illustration 22.1

The oldest radio station in the United States is KDKA in Pittsburgh, which went on the air in 1920. It operates at a frequency of 1.02×10^6 Hz. What is the wavelength of its radio wave? Assume the speed of electromagnetic waves to be 3×10^8 m/s.

Reasoning   We know that $\lambda = v/f$ for any wave. In our case, $v = 3 \times 10^8$ m/s and $f = 1.02 \times 10^6$ Hz. Substitution gives $\lambda = 294$ m.

Exercise   Radar waves (microwaves) have wavelengths of several centimeters. What is the frequency of an electromagnetic wave that has a wavelength of 20 cm?
Answer: 1.50×10^9 Hz

22.3   TYPES OF ELECTROMAGNETIC WAVES

In addition to visible light and radio waves, there is a wide range of wavelengths (frequencies) of electromagnetic waves with which we have become familiar. This range is called the **electromagnetic wave spectrum.** Various wavelengths are produced by a variety of processes, both natural and engineered. Also, there are a number of different devices and techniques which we use to detect electromagnetic waves in different parts of the spectrum.

The electromagnetic spectrum is shown in Fig. 22.7. It has been convenient to divide the spectrum into the indicated categories of waves, even though the division is arbitrary and the categories sometimes overlap. Notice that wavelength increases to the right and frequency increases to the left. For all the waves, $f\lambda = c$. Notice also that the spectrum covers an enormous range of values, about 20 powers of 10. Let us discuss each category briefly.

RADIO WAVES

The radio wave region of the spectrum consists of all wavelengths longer than about 1 m. Notice that the corresponding frequency range goes up to about 10^9 Hz. Within that range, FM radio uses the range from $f = 88$ MHz to 108 MHz, as you can easily see on your FM radio dial. AM radio covers the range from $f = 550$ kHz

Dish-shaped antennas like this one are used to receive electromagnetic waves emitted at radio wavelengths from objects in space.

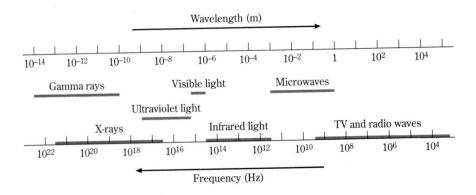

FIGURE 22.7

Types of electromagnetic radiation. The bars indicate the approximate wavelength range of each type of radiation.

(a)   *(b)*

Even at night, when the lack of visible light makes it look dark, warm objects still radiate infrared waves. These two photos are of the same scene taken at the same time. Picture (a) was taken on film sensitive to visible light, while picture (b) is a magnified image of the central portion of (a), taken by an infrared-sensitive night vision device. Both photos were 1/4 second exposures.

to 1600 kHz. Television occupies frequency bands on either side of the FM region. To avoid hopeless confusion, the allocation of frequency ranges for various purposes is controlled by federal regulation.

MICROWAVES

Microwaves, as the name implies, are very short radio waves. This category includes radar, microwave ovens, and communications links used for long-distance telephone message transmission. Microwave wavelengths are in the range of 1 millimeter to a few centimeters.

INFRARED WAVES

Infrared waves range all the way from the short-wavelength end of microwaves *(far infrared)* to the red limit of visible light *(near infrared)*. The wavelengths are often expressed in microns (μ) ($1 \mu = 1 \mu m = 10^{-6}$ m). Infrared radiation is radiated by all warm and hot objects. It is also absorbed strongly by many molecules, including carbon dioxide and water. As it is absorbed, the wave energy is converted to thermal energy, warming the absorbing object. For this reason, infrared is often erroneously referred to as "heat radiation."

VISIBLE LIGHT

That part of the electromagnetic spectrum the human eye can detect—what is commonly referred to as *light*—comprises an extremely small range of wavelengths between 400 and 700 nm. Within that range, we perceive what we call "colors," ranging from violet through blue, green, yellow, and orange to red. Figure 22.8

FIGURE 22.8
Sensitivity curve for the eye. The human eye is most sensitive to greenish-yellow light.

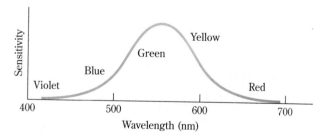

shows how the sensitivity of the human eye depends on wavelength. The human eye has a peak sensitivity to light of about 550 nm. Electrons undergoing energy changes within atoms provide the "antenna" that produces light.

ULTRAVIOLET WAVES

Beyond the short-wavelength (violet) limit of human eye sensitivity lies what is called the ultraviolet region of the spectrum. A type of ultraviolet light source (known as "black light") can be used to illuminate displays containing fluorescent paint: The paint absorbs the invisible ultraviolet waves and reradiates part of the energy in the form of waves from the visible spectrum. *Near ultraviolet* is strongly absorbed by ozone in the earth's atmosphere. *Far ultraviolet,* approaching $\lambda = 10$ nm, overlaps with the x-ray spectrum. Common forms of glass are opaque to most of the ultraviolet spectrum.

X-RAYS

One way x-rays can be created is when an energetic stream of electrons bombards a metal plate within a vacuum tube. These waves typically have wavelengths of the same size or less than the diameter of a single atom, about 0.1 nm. X-rays can penetrate through soft material such as flesh. Much of the x-ray spectrum overlaps with gamma rays, and the two types differ mainly in the manner of their production. We discuss x-rays in more detail in Chap. 27.

GAMMA RAYS

Gamma rays are the shortest of all electromagnetic waves. They include waves whose wavelengths are as small as the radius of the *nucleus* of an atom, 10^{-15} m.

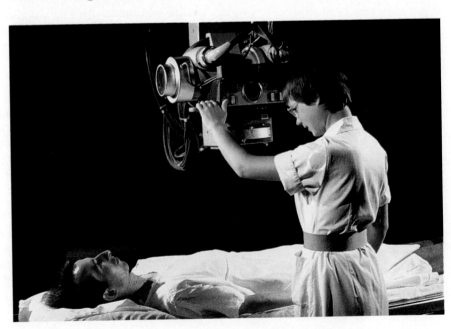

The most familiar application of x-rays, such as those produced by this device, is that of medical diagnostics.

Spontaneous changes in the structure of certain nuclei (radioactivity) and cosmic rays from outer space are the two main sources of gamma rays. A fuller discussion of gamma rays is presented in Chap. 28.

Notice that the electromagnetic spectrum encompasses waves with wavelengths extending from longer than 10^6 m to shorter than 10^{-15} m. Even though these waves are all electromagnetic waves, they differ considerably in their mode of interaction with matter. Much of the remainder of this book is concerned with various aspects of this subject.

22.4 RECEPTION OF RADIO WAVES

Television sets and radios are very sensitive electronic devices designed to detect electromagnetic waves in the radio range. Although we will not discuss the construction of such devices in detail, let us see how they detect and tune to radio waves.

An electromagnetic wave can be detected by either the electric or the magnetic portion. To detect the electric field part of the wave, we need only place a long wire (called the *receiving antenna*) in the path of the wave. Referring to Fig. 22.9a, we see that the electric field causes charges to oscillate in the antenna. When $\mathbf{E}_y$ is positive, the top of the antenna is positive. An instant later, the antenna's polarity reverses as the electric field vector in the wave reverses. This repeated action causes charge to flow up and down the antenna with a sinusoidal time dependence. As it does so, this varying current induces an oscillating voltage in the *RLC* circuit that is coupled to the antenna by means of a mutual inductance. If the *RLC* circuit is properly tuned, the circuit will resonate to the frequency of the incoming radio wave. Let us clarify this point.

Each radio or television station is assigned its own broadcast frequency and sends out waves of that frequency only. Since waves from many stations are incident simultaneously on an antenna, some means must be employed to select only the wave from the desired station. The capacitor in the *RLC* circuit in Fig. 22.9 has an arrow drawn through it, indicating that the value of C can be varied. Remember that a series *RLC* circuit has a resonance frequency f_r that depends on L and C: $f_r = 1/2\pi\sqrt{LC}$. As C is varied, the frequency at which the circuit will resonate changes. The process of varying C is called *tuning* the circuit. If f_r for the circuit matches the frequency of the incoming wave, a maximum ac current i will occur in the circuit, causing a large ac voltage iR across the resistor R. This ac voltage becomes the input signal to the radio receiver and is amplified by other stages of the receiver. If the resonance of the circuit is "sharp" (which means the resonant current has a very narrow peak in its frequency dependence), selecting one radio station by tuning the circuit to its frequency enables the receiver to ignore all nonresonance frequencies which strike the antenna.

We can also detect the electromagnetic wave by means of its oscillating magnetic field. Since this field is varying rapidly, the wave induces an emf in a loop such as the one in Fig. 22.9b.* Notice that the loop must be properly oriented so that the flux of the magnetic field passes through it. (It is for this reason that small radios that use this type of antenna show markedly different reception in different orientations.) The induced voltage in the loop antenna is impressed on an *RLC* circuit. Tuning is accomplished as described above.

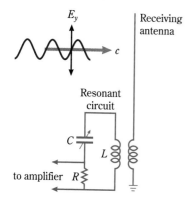

(a) Electric wave detection

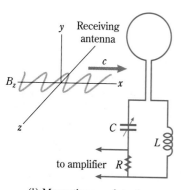

(b) Magnetic wave detection

FIGURE 22.9

Two methods for detecting radiowaves: (a) electric wave detection; (b) magnetic wave detection.

---

*In practice, the loop is a coil wound on a ferromagnetic rod.

 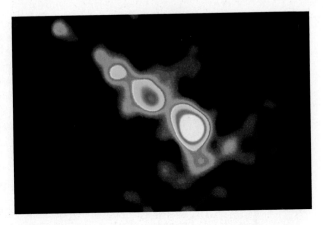

The galaxy NGC 5128 as seen in (a) visible-light emissions and (b) x-ray emissions. Notice that the jet of matter seen by the x-ray telescope is entirely invisible in the visible-light picture. This demonstrates that astronomers have to explore all parts of the electromagnetic wave spectrum in order to obtain the maximum information about our universe.

One might well ask why all waves, light and x-rays included, can't be detected by radio-type devices. The reason is quite simple. Very-high-frequency waves require *RLC* resonant circuits that are thus far impossible to build. Again, the resonance frequency of a circuit is given by $1/2\pi\sqrt{LC}$. To make this frequency very high, both L and C must be very small. In the cases of infrared waves, light waves, and x-rays, two tiny wires lying side by side already have L and C values that are too large. We shall see in later chapters that a circuit of atomic size is needed to detect these waves. Indeed, we shall learn that individual atoms and molecules, in effect, become the resonant circuit for detecting very-high-frequency electromagnetic waves.

22.5 THE SPEED OF ELECTROMAGNETIC WAVES

Now that we understand many of the qualitative features of electromagnetic waves, let us obtain an expression for their speed. We use a method that depends on a fact first pointed out clearly by Einstein in his theory of relativity, a fact that we discuss at greater length in Chap. 26: *only relative velocities can be determined.* An object can be said to be at rest relative to another object but not at rest in any absolute sense.

For example, as you read this, you are probably at rest relative to the earth, but the earth is in motion relative to the sun, and so too are you. Moreover, the sun is in motion in our galaxy, the Milky Way, and our galaxy is in motion relative to other galaxies in the universe. It makes sense to say that something is at rest relative to something else, but we cannot say which of two objects is at rest in any noncomparative way.

With this fact in mind, let us reconsider the force experienced by a charge q moving with speed v perpendicular to a magnetic field of magnitude $B_\perp$. We found in Chap. 19 that the force experienced by the charge has a magnitude

$$F = qvB_\perp$$

However, who can say that the charge is not at rest and the field moving instead? After all, only relative motion is observable. Therefore our experiment can be interpreted in the following alternative way: a field **B** moving with speed v perpendicular to the field lines past a charge q exerts a force $F = qvB_\perp$ on the charge.

Physicists at Work PAUL HOROWITZ *Harvard University*

For the last decade I've been consumed by the search for extraterrestrial intelligence (SETI), listening with a gigantic radio telescope dish festooned with ever more sophisticated listening equipment for radio signals from an advanced civilization. Although this sort of activity might once have been considered "unscientific," recent findings in astronomy and related fields support the idea of many life-bearing habitats. In particular, infrared and visible observations of planetary disks and indirect evidence of large planetary bodies called brown dwarfs suggest that planetary systems are commonplace in the universe. At the same time, organic constituents of life have been found in meteorites, in interstellar clouds of cold gas, and in the residue produced in tabletop experiments that subject the atmospheric ingredients of primitive earth to heat, sunlight, and electric discharges. Thus the raw materials and habitat for life appear to be in plentiful supply. We humans are but a bit of life on a probably average planet, circling an average star, one of 400 billion in our galaxy, itself one of 100 billion in the universe. It now seems preposterous to think that life could be unique to earth.

The nearest star is 4 lightyears away, and our galaxy spans 100,000 lightyears. Is contact with another civilization even *possible* over such mind-boggling distances? The surprising answer is *yes*. Using microwave radio astronomy technology available today, the earth could *communicate* with a sister planet anywhere in our galaxy. This is in stark contrast to the effort required to *travel* to another star system—where one would have to consume hundreds of years of the earth's total energy supply just to make a single round-trip to the nearest star.

So, if advanced life exists around other stars and the means of communication is at hand, why haven't we heard from them? Probably because we haven't really looked. There have been a few small searches here and there, done on a shoestring budget: We've ruled out the possibility that the sky is littered with transmitters directing powerful beams our way. But a proper job of scouring the heavens for that intentional radio "beacon" transmission, so very feeble by the time it reaches earth, requires complex signal-processing equipment, of the sort that is only blossoming now, thanks to the silicon revolution. In our lab at Harvard, and elsewhere around the world, receiving systems with millions of channels are beginning searches that many of us expect to succeed within a century. The detection of signals from another civilization will be the end of earth's cultural isolation, in a deep sense; it will be the greatest event in human history. The mere detection of such a signal will answer the question, "Are we alone?" And the information that accompanies the interstellar beacon may be a sort of *encyclopedia galactica,* telling of science and art, history and literature, beyond our wildest imaginings. Ours is the first generation that could realistically make contact. I cannot imagine a more exciting exploration, and that is why I put all my energy into these experiments.

From my earliest memories I always loved playing with gadgetry, following in the footsteps of my big brother. We played mostly with electronics, but we became junior-grade rocketeers in the Sputnik era. My brother studied electrical engineering at MIT (he now owns a high-tech communications company), but told me to major in physics instead, because it is more universal. My parents added that I should do it at Harvard. That's where I went, and that's where I have stayed. Doing experimental science in academia is absolutely wonderful, with the freedom to follow your nose, even to the point of doing slightly crackpot experiments. I probably qualify as a crackpot, having worked on a variety of experiments whose common thread is that they are off the beaten path: searches for starquakes on pulsars, new kinds of microscopes using x-rays and protons, the structure of the bacterial rotary engine in *Escherichia coli,* searches for ultraheavy atoms, and of course the ultimate scientific long shot—SETI.

I enjoy doing experiments that require a heavy dose of custom electronic instrumentation because I like to build things. But I find the community of physicists to be terrifically stimulating, and I'm glad I followed my brother's advice. When I'm doing engineering, I find myself "thinking like a physicist." And it's a perhaps not-so-curious fact that the best electronic circuit designers are "fallen" physicists! I would not hesitate to offer the following advice: If you aren't sure what you want to do for a career, but think it may border on physics, then major in physics. It's great preparation for doing something other than physics.

Because the force per unit charge F/q is defined to be the magnitude of the electric field E, however, we can restate this as follows:

A magnetic field **B** moving with speed v perpendicular to the field lines generates an electric field of magnitude

$$E = vB \tag{22.2}$$

in the region through which it passes.

To illustrate this, consider the situation in Fig. 22.10. The magnet poles are moving with speed v as shown. They carry the magnetic field lines with them, and so we have a moving magnetic field **B** in this region. Therefore at a point such as A an electric field **E** exists whose value is $E = vB$. You should be able to show that the direction of **E** is into the page.*

It seems from this line of reasoning that the magnetic field moving out from a radio antenna should generate an electric field in the region through which the magnetic field moves. At a given point, the electric field should be related to the speed v of the magnetic field wave and to its magnitude B through $E = vB$. The question now arises as to whether a moving *electric field* can generate a *magnetic field*. Answering this question leads us to a very important result.

Consider the very long, uniformly charged wire shown in Fig. 22.11. The wire is moving to the right in a direction along its length with speed v, and so the electric field lines from it are flying past point P with speed v. We know that the moving charged wire constitutes a current along the wire. The magnitude of this current can be found by noticing how much charge passes P each second. Suppose the wire has a charge ρ per unit length. (To avoid confusion with the use of λ for wavelength, we now use the symbol ρ for linear charge density instead of the λ we used in Chap. 16.) Since a length vt of the wire flies past P in time t, we have

$$\text{Current} = \frac{\text{charge passing } P}{\text{time taken}} = \frac{\rho vt}{t} = \rho v$$

The moving charged wire constitutes a current of magnitude ρv.

A current produces a magnetic field, however, and so the moving wire is surrounded by a magnetic field. (You should be able to show that this field circles the wire and is out of the page above the wire.) We learned in Chap. 19 that the magnetic field due to a current I in a long, straight wire is $B = \mu_o I / 2\pi r$ (Eq. 19.9). If we apply this result to the present case, we find the magnetic field at point P to be

$$B = \frac{\mu_o \rho v}{2\pi r} \tag{22.3}$$

We wish to relate this to the electric field outside the wire at P.

From Eq. 16.7 we know that the electric field outside a long, uniformly charged straight wire is

$$E = \frac{\rho}{2\pi \epsilon_o r} \tag{22.4}$$

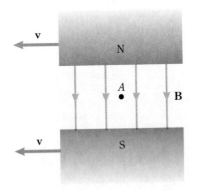

FIGURE 22.10

The magnetic field **B** (shown by the colored vertical lines) moves with the magnet poles past point A with speed v. This motion generates an electric field $E = Bv$ directed into the page.

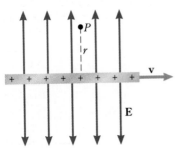

FIGURE 22.11

The moving charged wire carries the electric field lines past point P.

Hint: Place a positive charge at A and remember that the motion is relative.

where ϵ_o is the permittivity of free space, $8.85 \times 10^{-12}\,C^2/N \cdot m^2$. Let us eliminate ρ between Eqs. 22.3 and 22.4. The result is

$$B = \epsilon_o \mu_o v E \tag{22.5}$$

This is to be compared with the relation

$$B = \frac{E}{v} \tag{22.6}$$

which we obtained for a moving magnetic field.

Although this is a very special situation in which a moving charged wire generates a magnetic field, it is typical. Moving charges generate a magnetic field, but the moving charges have associated with them an electric field that travels along with them. The magnetic field generated by the motion of the charges can equally well be attributed to the motion of the electric field. We therefore are led to conclude that

An electric field **E** moving with speed v perpendicular to the field lines generates a magnetic field of magnitude $B = \epsilon_o \mu_o v E$ in the region through which it passes.

Let us now return to Fig. 22.6, in which we see the electric and magnetic fields generated by an antenna. The fields are rushing out along the line of propagation with speed v. Consider the magnetic field as it flies past a point in space. It generates an electric field at that point. Similarly, the electric field from the antenna also flies past the same point and generates a magnetic field there.

If you spend a little time considering the situation in Fig. 22.6, you can see that the electric field shown there is in the same direction as the electric field generated by the moving magnetic field. In addition, the magnetic field shown is in the same direction as the magnetic field generated by the moving electric field. We are therefore tempted to say that the electric and magnetic fields in an electromagnetic wave regenerate themselves as the wave travels out through space. Let us make this supposition and see where it leads us.

Suppose the electric and magnetic fields of an electromagnetic wave generate each other as the wave moves with speed v through space. Then both Eq. 22.5 and Eq. 22.6 must apply to the wave. If this is so, B and E must be related in the same way in the two equations, and so the proportionality constants between E and B must be the same. Hence

$$\epsilon_o \mu_o v = \frac{1}{v}$$

Solving this relation for v, the speed of the electromagnetic wave in vacuum, we find

$$v = \frac{1}{\sqrt{\mu_o \epsilon_o}} = 2.998 \times 10^8\,m/s \tag{22.7}$$

which is the value Maxwell obtained, as described in Sec. 22.1. Hence we conclude, as Maxwell did:

All electromagnetic waves travel through vacuum with the speed $c = 2.998 \times 10^8$ m/s, and light is one form of electromagnetic wave.

Further, Eq. 22.6 gives us the relation between B and E in an electromagnetic wave traveling through vacuum:

$$E = cB \qquad\qquad (22.8)$$

Example 22.1

As a certain electromagnetic wave passes a point in space, its electric field varies as

$$E = E_o \sin 2\pi ft$$

with $E_o = 0.0042$ V/m. Find the amplitude of the magnetic field in this wave.

Reasoning

Question   What is the expression for the magnetic field in the wave?
Answer   A detailed solution of Maxwell's equation shows that, far away from the source, the fields are in phase with each other. Thus

$$B = B_o \sin 2\pi ft$$

Question   Is there a constant relationship between the values of E and B in an electromagnetic wave?
Answer   Yes. $E/c = B$.

Solution and Discussion   We use the expression $E/c = B$ for the amplitudes:

$$B_o = \frac{0.0042 \text{ V/m}}{3 \times 10^8 \text{ m/s}} = 1.4 \times 10^{-11} \text{ T}$$

Notice how very small the magnetic field is in an electromagnetic wave. That the magnitude of the B field in electromagnetic waves is so small is the fundamental reason why induced magnetic fields had not been observed by the time Maxwell put forth his theory.

You should convince yourself of the correctness of the units in the solution.

Example 22.2

Assume the wave in Example 22.1 has a frequency of 5×10^8 Hz. As this wave passes a loop antenna such as that shown in Fig. 22.9b, the magnetic field induces an emf in the loop. The loop has a single turn and an area of 25 cm^2 oriented perpendicular to the magnetic field of the wave. Estimate the average emf induced in the loop.

R

Reasoning

Question   What principle deals with induced emf?
Answer   Faraday's law of induction.

Question   What information does this principle require?
Answer   Faraday's law (Eq. 20.3) states that $\overline{\text{emf}} = \Delta\Phi_B/\Delta t$. In this case the area of the loop is perpendicular to **B**, so the flux at any instant is just $\Phi_B = BA$. Since A is constant, you have $\Phi_B = (\Delta B)A$, and

$$\overline{\text{emf}} = \frac{\Delta B}{\Delta t}A$$

Question   How can I estimate the rate of change of the flux?
Answer   From Example 22.1 you have the amplitude of B. You know the magnetic field goes from B_o to zero within $\frac{1}{4}$ cycle. Even though $\Delta B/\Delta t$ is not constant over this time, we can get an average emf by assuming it is.

Solution and Discussion   From $f = 5 \times 10^8\,\text{Hz}$, we can find the time equal to $\frac{1}{4}$ cycle:

$$\frac{T}{4} = \frac{1}{4f} = \frac{1}{4(5 \times 10^8\,\text{s}^{-1})} = 5 \times 10^{-10}\,\text{s}$$

The average rate of change of B over this time is

$$\frac{\Delta B}{\Delta t} = \frac{B_o - 0}{T/4} = \frac{1.4 \times 10^{-11}\,\text{T}}{5 \times 10^{-10}\,\text{s}}$$

$$= 2.8 \times 10^{-2}\,\text{T/s}$$

The average induced emf is then

$$\overline{\text{emf}} = \frac{\Delta B}{\Delta t}A = (2.8 \times 10^{-2}\,\text{T/s})(25 \times 10^{-4}\,\text{m}^2) = 7.0 \times 10^{-5}\,\text{V}$$

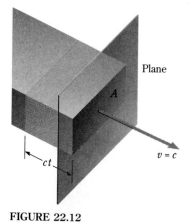

FIGURE 22.12

A volume Act of the beam goes through the plane in time t.

22.6   THE ENERGY CARRIED BY ELECTROMAGNETIC WAVES

We have seen that electromagnetic waves consist of moving electric and magnetic fields. Because these fields contain energy, the waves carry energy through space. For example, electromagnetic waves from the sun warm the earth and provide the energy for plant growth. Waves from a distant television station carry the energy that brings the picture and sound to your television. Let us now compute how much energy is carried to a surface by an electromagnetic wave incident on the surface.

Recall from Sec. 17.12 that the energy stored in unit volume of an electric field of magnitude E in vacuum is $\frac{1}{2}\epsilon_o E^2$. Similarly, we showed in Sec. 20.7 that the energy stored in unit volume of a magnetic field of magnitude B is $B^2/2\mu_o$. With these facts in mind, let us look at the beam of electromagnetic radiation in Fig. 22.12; it carries energy through the plane indicated. The beam has an end area A and travels to the right with the speed of light c. Because the beam travels a distance ct

Controversies in Physics: The Nature of Light

Of all the physical phenomena our senses perceive, light is the most important and perhaps the most perplexing. Our sense of sight gives us the shape, size, and color of the world around us with great precision. Throughout history, humans have observed that light is given off by the sun, fire, hot objects, and lightning. The creation of light appears in the genesis stories of the major religions. Yet while light enables us to see, we cannot see light itself. That is, we cannot directly sense the physical nature of light. Does light consist of some form of matter? Does it consist of a stream of particles, or is it some kind of vibration or wave? How fast does light travel? How do we perceive the image of an object with which we have no physical contact? Both the process by which we see and the nature of light are phenomena for which explanations have been sought since long before the beginnings of modern science.

By the end of the seventeenth century, the process of image formation by lenses was understood. Sight was accepted as a process by which the lens of the eye images incident light on the retina. The Danish astronomer Roemer, a contemporary of Newton's, experimentally established the fact that the speed of light, while very great, was finite. The present value of the speed of light is approximately 50 percent higher than Roemer's original result. The most difficult question to answer has proved to be *what light is*— whether it consists of a stream of particles or waves of some kind. Over the years there have been many arguments advanced to support each of these competing views.

Let us first examine just what we mean by the terms *particle* and *wave*. In a general sense, these terms refer to fundamentally opposite concepts. Particles are localized in position at any given time, in that they are either ide-

alized as points or have definite boundaries, and their momenta and energies are consequently also localized. Waves, on the other hand, represent coordinated movement which extends over great distances. Wave energy depends on wave amplitude; it is not localized but is a property of the entire wave.

Newton rejected the wave model of light, since he considered space to be an empty vacuum, with no material that could support and propagate vibrations. Particles, on the other hand, can move unimpeded through empty space in straight lines. The fact that light particles seemed not to be affected by gravity was attributed by Newton to their extremely high speed. The particle model also accounted for the law of reflection, since the direction at which an incident ray of light reflects from a mirror is the same as that of a ball when it rebounds elastically from a surface. Newton explained refraction (the change of direction as light goes from one medium to another) as an attraction that molecules of a transparent substance exerted on light particles as they entered the substance. This force would alter a particle's direction by increasing the component of its velocity normal to the surface of the substance, causing the particles to be deflected toward the normal. The fact that different colors are refracted by different amounts was ascribed to particles of various colors having slightly different masses.

A Dutch contemporary of Newton, Christian Huygens (1629–1695), developed the wave theory of light. Huygens found the speed required of light particles hard to accept, and observed that intersecting beams of light don't show any evidence of scattering as they collide, as one would expect from crossing streams of particles. He developed a geometric explanation (Huygens' principle) for the shape of waves as they spread through apertures and

around the edges of barriers, correctly describing the phenomenon of diffraction. Huygens' theory explained refraction on the basis of the *slowing* of light upon entering a medium, in opposition to Newton's model. Unfortunately, there was no available method of actually measuring the speed of light in a transparent material, which would have provided a definitive way to choose between the two contending theories. Newton's particle model, because of his great reputation and influence, prevailed through the eighteenth century.

In 1804, the Englishman Thomas Young provided the first definitive test of the competing models of light. He performed an experiment (Sec. 24.3) in which two point sources of light were shown to produce a pattern of intensity distributed precisely as the combined intensity of two superimposed waves is distributed due to interference between the waves. Since particles do not possess the property of amplitude interference, Young's results showed that light indeed possesses wave properties. However, this conclusion was not generally accepted until about 15 years later, after the mathematical theory of Young's experiment had been worked out by the French physicist Augustin Fresnel. Fresnel's theory proposed that light is a *transverse* wave, which was supported by observations (1808–1815) that light could be polarized. This observation also argues against the particle model, since a beam of particles, according to classical theory, does not have the property of being polarizable. Finally, the French physicist Fizeau was able to make a direct measurement on the speed of light in water, finding this speed to be *less* than that in air. This observation, which is in direct contradiction to Newton's particle model, affirmed Huygens' wave theory of refraction.

With all this accumulating evidence, one might think that all doubts as to the wave nature of light would vanish. Such was not the case, however. One major question remained: "How does light travel through vacuum, where there seems to be no material that can carry the wave?" The extremely high value of the speed of light would require the vibrating medium to be extremely rigid, yet this same medium would have to offer no resistance to the passage of planets through it. Even those who accepted the wave model could not answer the question of just *what* was waving.

As discussed in this chapter, Maxwell supplied the answer to this last question with his theory of oscillating electric and magnetic fields. He also predicted the existence of an entire spectrum of electromagnetic waves, of which light was only a small part. It was still widely held that some medium (referred to as the *aether*) must exist, and that its properties would determine the absolute speed of light. In the 1880s Michelson attempted to determine the speed of the earth through the surrounding aether, using his interferometer (Sec. 26.1) to measure the difference in the speed of light predicted by the aether theory as the earth moved in opposite directions in its orbit every 6 months. The fact that he and collaborator Edward Morley were unable to measure any difference in light speed, even though their interfer-

ometer was sufficiently sensitive to record the predicted difference of 36 mi/s, led most physicists to conclude that the aether did not exist. Thus at the close of the nineteenth century, it seemed that the age-old question of the nature of light had been conclusively settled. Light is a nonmaterial wave composed of oscillating electric and magnetic fields and, being nonmaterial, the wave can travel through vacuum without the aid of a material substance.

However, nature always seems to have a few humbling surprises just when we think we've got it right. The closing years of the nineteenth century and the first decade of the twentieth posed new challenges to our understanding of the nature of light. The spectrum of light radiated by heated objects (Sec. 26.7) could not be explained by the wave model. Nor could the wave model explain the photoelectric effect (Sec. 26.8), by which electrons are ejected from metal surfaces when the surfaces are illuminated by light. Both observations were explained rather neatly and precisely (by Planck and then Einstein) only by assuming that light consists of a stream of particles, called *photons,* that travel at the speed of light and carry an amount of energy proportional to the light frequency. In the 1920s, Compton observed that when x-rays strike elec-

trons, they exchange energy and momentum precisely as if the x-rays are particles which collide elastically with the electrons (Sec. 26.9).

As if the above developments weren't confusing enough, in 1924 the French physicist de Broglie theorized that material particles should possess a "matter wave" whose wavelength is inversely proportional to the particle's momentum (Sec. 26.10). If true, particles passing through narrow apertures should display wave effects such as diffraction and interference. In 1927, electron diffraction was observed in agreement with de Broglie's prediction (Sec. 26.10), and wave effects involving beams of protons and neutrons have since been observed also.

Thus we arrive at the present state of affairs, in which light has a *dual* nature: It displays a wave nature in some experiments and particle-like behavior in others. The same is true for those bits of matter we call particles. Importantly, only one of the two opposite types of behavior is shown in any given experiment. The answer to our original question on the nature of light has a more complicated answer (and to many, a more disturbing one) than anyone expected: Whether the nature of light is a wave or a stream of particles depends on the question a given experiment is designed to ask.

during time t, a length ct of the wave moves through the plane in this time. Hence the volume of the beam that travels through the plane in time t is Act. This volume is indicated by shading in the figure.

Let us take t short enough for ct to be much smaller than the wavelength of the beam's radiation. Then both **E** and **B** are substantially constant throughout the shaded volume. Therefore we can write that the energy carried through the plane by the beam volume Act is

$$\text{Energy in volume } Act = \left(\begin{array}{c} \text{magnetic field} \\ \text{energy density} \end{array} \right)(\text{volume}) + \left(\begin{array}{c} \text{electric field} \\ \text{energy density} \end{array} \right)(\text{volume})$$

$$\text{Energy in } Act = \frac{B^2}{2\mu_o}Act + \tfrac{1}{2}\epsilon_o E^2 Act$$

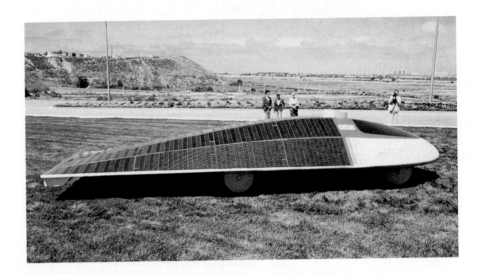

Photocells convert solar radiation into electrical current to power this experimental car.

To find the energy carried through a unit area of the plane in a unit time, we need only divide by t and the area A of the beam. We then have

$$\text{Energy per unit} \atop \text{area per second} = \frac{c}{2}\left(\frac{B^2}{\mu_o} + \epsilon_o E^2\right)$$

We call this the *intensity* I of the wave. Because $B^2 = E^2/c^2 = E^2\epsilon_o\mu_o$, this equation can be written in the form

$$I = \frac{\text{energy per unit}}{\text{area per second}} = \tfrac{1}{2}c\epsilon_o(E^2 + E^2) = c\epsilon_o E^2$$

We can see in this last equation that the magnetic and electric field terms are equal in magnitude. Hence we conclude that

The electric field and the magnetic field of an electromagnetic wave transport equal energies.

The intensity calculated above is an *instantaneous* value, since we took t to be a very small fraction of the wave's period. Of much greater interest is the average of the intensity over each cycle. We therefore need the average value of E^2 in one cycle. In our study of alternating currents we found that the per-cycle average of the square of any sinusoidally varying quantity is half the amplitude squared. Thus $\overline{E^2} = \tfrac{1}{2}E_o^2$, and the average intensity of the wave is

$$\overline{I} = \frac{\text{power per}}{\text{unit area}} = \frac{\text{average energy transported}}{\text{per unit time per unit area}} = \tfrac{1}{2}c\epsilon_o E_o^2 \qquad (22.9a)$$

Or, if we choose, we can write this in terms of B_o, the amplitude of the magnetic field wave. We recall that $E = cB$, and so we have

$$\overline{I} = \frac{\text{power per}}{\text{unit area}} = \tfrac{1}{2}c\epsilon_o c^2 B_o^2 = \frac{cB_o^2}{2\mu_o} \qquad (22.9b)$$

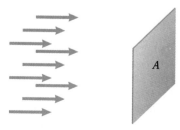

FIGURE 22.13

The intensity of a beam is the energy passing through unit area per second, provided the beam is perpendicular to the area.

where we have used $c = 1/\sqrt{\epsilon_o \mu_o}$. We therefore conclude that (see Fig. 22.13)

The average power transported through a unit area by an electromagnetic wave incident perpendicular to the area is $\frac{1}{2}c\epsilon_o E_o^2 = cB_o^2/2\mu_o$. This is called the intensity of the wave.

The SI unit for intensity is watts per square meter (W/m²). You should verify that the quantities in Eqs. 22.9a and 22.9b do indeed have this unit.

Example 22.3

A typical laboratory laser might emit a 1-mm-diameter beam of light with 1 mW of power. What is the intensity of this beam and what are the magnitudes of its electric and magnetic fields?

R

Reasoning

Question How is intensity defined?
Answer Intensity is power per area. Here the power in the beam is given, and the beam's area is $A = \pi r^2$.

Question How do the field magnitudes relate to the intensity?
Answer You have $I = \epsilon_o c E_o^2/2 = cB_o^2/2\mu_o$. You can choose either one.

Question How are E_o and B_o related?
Answer Simply by $E_o = cB_o$.

Solution and Discussion   The intensity is

$$I = \frac{10^{-3}\,\text{W}}{\pi(0.5 \times 10^{-3}\,\text{m})^2} = 1.27 \times 10^3\,\text{W/m}^2$$

Choosing $I = cB_o^2/2\mu_o$, we get

$$B_o^2 = \frac{2\mu_o I}{c}$$

$$= \frac{2(4\pi \times 10^{-7}\,\text{N/A}^2)(1.27 \times 10^3\,\text{W/m}^2)}{3 \times 10^8\,\text{m/s}}$$

$$= 1.07 \times 10^{-11}\,\text{T}^2$$

Thus $B_o = 3.27 \times 10^{-6}\,\text{T}$. Finally,

$$E_o = cB_o = (3 \times 10^8\,\text{m/s})(3.27 \times 10^{-6}\,\text{T}) = 9.8 \times 10^2\,\text{V/m}$$

It is interesting to note that the intensity of this laser beam is comparable to the intensity of sunlight at the top of earth's atmosphere, 1.4×10^3 W/m². Also note that the magnetic field in the beam is about one-tenth of a typical value of the earth's magnetic field.

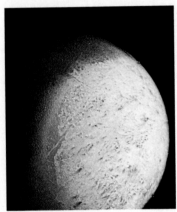

Images like this one of Neptune's moon Triton, which was sent by the Voyager 2 spacecraft, depend on our ability to receive and process electromagnetic signals of very low intensity. Voyager was 330,000 miles from Triton when this photo was taken, and the distance the signal containing the image had to travel to reach earth was over 3 billion miles!

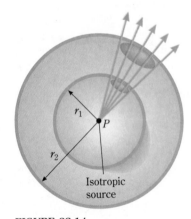

FIGURE 22.14

If the power output of the source is P, what are the values of radiation intensity at distances r_1 and r_2?

22.7   THE INVERSE-SQUARE LAW FOR RADIATION

As was pointed out in the preceding section, the intensity of a radiation beam is defined in the following way. We imagine an area A placed perpendicular to the beam, as in Fig. 22.13. Since the beam (let us say it is a beam of light) carries energy in the direction of propagation (in this case, to the right), a certain quantity of energy passes through the area in unit time. We define the intensity of the light I by the relation

$$I = \frac{\text{energy}}{\text{area} \times \text{time}} = \frac{\text{power}}{\text{area}}$$

Let us now discuss the energy emitted by a small source of light, such as the one shown in Fig. 22.14. We assume the source to be small enough that it acts as a *point source;* further, we assume it to be an *isotropic source,* that is, one that emits light equally in all directions. To describe the energy that flows out from the source, let us imagine a spherical surface of radius r_1 concentric to it. We will use the symbol I_1 to denote the intensity of the light at this surface. The intensity must be the same at all points of the sphere, since we have assumed the source to be isotropic. In other words, I_1 is the intensity of the light at a distance r_1 from the source.

Since our imaginary sphere completely encloses the source, all of the energy emitted by the source must flow through the sphere's surface. The area of this surface is $4\pi r_1^2$. The total rate at which the source is emitting energy is its power P, and so we conclude that the intensity at a distance r_1 from the source is

$$I_1 = \frac{\text{power}}{\text{area}} = \frac{P}{4\pi r_1^2}$$

If we now consider a second sphere of larger radius r_2 centered on the source, we can use the same reasoning to conclude that the intensity I_2 at the distance r_2 is

$$I_2 = \frac{P}{4\pi r_2^2}$$

(This of course assumes that no energy is being absorbed as it travels outward from the source.) If we now take the ratio of the intensities, we get

$$\frac{I_1}{I_2} = \frac{r_2^2}{r_1^2} \tag{22.10}$$

This is called the *inverse-square law for the radiation of energy from a point source.* It states that the light intensity from such a source decreases as the inverse square of the distance from the source. For example, if we triple the distance from the source, the light intensity decreases by a factor of 9.

Example 22.4

As mentioned in Example 22.3, the intensity of sunlight at the top of the earth's atmosphere is 1.4 kW/m^2. This is known as the **solar constant.** Assuming the sun to radiate isotropically, what is its total power output (also called the sun's **luminosity**)?

Reasoning

Question What relates the intensity we measure to the power of the source?
Answer Equation 22.10: $I = P/4\pi r^2$.

Question What is r?
Answer The earth-sun distance, given in the table of physical constants and data in the front endpapers: $r = 1.5 \times 10^{11}$ m.

Solution and Discussion

$$P = I(4\pi r^2) = (1.4 \times 10^3 \text{ W/m}^2)(4\pi)(1.5 \times 10^{11} \text{ m})^2$$
$$= 3.96 \times 10^{26} \text{ W}$$

Exercise The planet Neptune is about 30 times farther away from the sun than earth is. What would the intensity of sunlight be at Neptune's position? *Answer: 1.6 W/m².*

LEARNING GOALS

Now that you have finished this chapter, you should be able to

1 Define (a) electromagnetic wave, (b) electromagnetic spectrum, (c) radio wave, (d) radar or microwave, (e) infrared radiation, (f) visible light, (g) ultraviolet radiation, (h) x-rays, (i) gamma rays, (j) electromagnetic wave intensity.
2 Qualitatively describe Maxwell's postulate of displacement current.
3 Express the speed of light in terms of the universal constants μ_o and ϵ_o.
4 Calculate the wavelength of an electromagnetic wave knowing its frequency and vice versa.
5 Sketch the electric and magnetic fields in an electromagnetic wave.
6 Describe the relationship between the electric and magnetic field strengths in an electromagnetic wave.

7 Explain qualitatively how electromagnetic waves are sent out from a dipole antenna.
8 Describe two ways in which radio waves can be detected by a radio receiver. Explain the function of the *RLC* circuit in a radio and how it is used to select the signal from different radio stations.
9 Arrange a list of electromagnetic wave types in order of decreasing wavelength. State the type of wave to which a given wavelength belongs.
10 When the value of E_o or B_o for an electromagnetic wave is given, find the intensity of the wave.
11 Calculate the amplitudes of the electric and magnetic fields in an electromagnetic wave when the intensity of the wave is given.
12 Apply the inverse-square law for radiation to simple situations.

SUMMARY

DERIVED UNITS AND PHYSICAL CONSTANTS

Speed of Light (c)

$$c = \frac{1}{\sqrt{\mu_o\epsilon_o}} = 2.998 \times 10^8 \text{ m/s}$$

DEFINITIONS AND BASIC PRINCIPLES

Maxwell's Displacement Current (I_D)
Magnetic fields can be produced by time-varying electric

fields as well as by a current I. The effect of a time-varying E field can be treated as producing a fictitious current I_D, called the displacement current, where

$$I_D = \epsilon_o A \frac{\Delta E_\perp}{\Delta t}$$

Here $E_\perp$ is the component of **E** perpendicular to the plane of area A. I_D produces a magnetic field in the same way that a real current I does. If both I_D and I are present, the magnetic field is produced by an effective total current $I_{\text{tot}} = I + I_D$.

Relation Between Magnetic and Electric Field Amplitudes in Electromagnetic Waves

$$B = \frac{E}{c}$$

Energy Density in an Electromagnetic Wave

$$\frac{Energy}{Volume} = \frac{B^2}{2\mu_o} + \frac{\epsilon_o E^2}{2} = \epsilon_o E^2 = \frac{B^2}{\mu_o}$$

Thus the electric and magnetic fields represent equal energy densities.

Intensity of Electromagnetic Waves (I)
The intensity of a wave is the average power transported per unit area:

$$I = \tfrac{1}{2}c\epsilon_o E_o^2 = \frac{\tfrac{1}{2}cB_o^2}{\mu_o}$$

The electric and magnetic fields transport equal amounts of energy.

Inverse-Square Law of Radiation
The intensity of electromagnetic waves from a point source varies inversely as the square of the distance from the source. Thus if r_1 and r_2 represent two distances from a source, the ratio of intensities at those distances is

$$\frac{I_1}{I_2} = \frac{r_2^2}{r_1^2}$$

QUESTIONS AND GUESSTIMATES

1 Some radio stations have their transmitting antenna vertical, while others have theirs horizontal. Describe and compare the electromagnetic waves generated by these two types of antennas. In particular, how are **B** and **E** directed relative to the earth's surface?

2 If you open up a transistor radio, you can see how its coil antenna is mounted. How could you use the radio to tell whether a distant station's antenna is vertical or horizontal?

3 Electromagnetic waves from most of the radio stations in the world are passing through the region around you. How does a radio or television set select the particular station you want to listen to? When you turn the dial on a radio, what is happening inside to select the various stations?

4 There are two types of radio and television receiving antennas in use. One picks up the electric part of the electromagnetic wave, and the other picks up the magnetic part. Examine a pocket transistor radio or a table radio and see which method is used. Is it possible to use both?

5 From time to time, in the movies or on television, one sees the good guys trying to locate a clandestine radio transmitter by driving through the neighborhood with a device that has a slowly rotating coil on top. Explain how the device works.

6 It is claimed that, in the vicinity of a very powerful radio transmitting antenna, one can at times see sparks jumping along a wire fence. What do you think of this claim?

7 In microwave ovens, foods and utensils are subjected to very-high-frequency radar (electromagnetic) waves. If a spoon is left in such an oven, it becomes very hot. What heats it? Can you explain the heating action in terms of the electric part of the wave? The magnetic? How are nonmetallic substances heated in the oven? Will a glass dish heat up in such an oven?

8 There is some doubt about the safety of human exposure to intense radio waves and microwaves. Why would one expect the danger to depend on the frequency of the waves? Which would you expect to present more danger (if any), radio waves or microwaves?

9 Refer to Fig. 22.10. Find the direction of the electric field at *A* induced by the moving magnetic field.

10 Refer to Fig. 22.11. Find the direction of the magnetic field at *P* induced by the moving electric field.

11 Are the direction and phase of the magnetic part of the wave in Fig. 22.6 drawn properly if the magnetic field is generated by the moving electric field? Repeat for the electric field generated by the moving magnetic field.

12 Estimate the wavelength of the electromagnetic wave generated by the vibration of a positively charged ball suspended as a 1-meter-long pendulum. Compare this wavelength with the diameter of the earth, 12,700 km.

PROBLEMS

Sections 22.1–22.4

1 What is the wavelength of the electromagnetic waves radiated by a 50-Hz power source?

2 What is the frequency of electromagnetic waves with wavelengths of (*a*) 1.2 m, (*b*) 12 m, and (*c*) 120 m?

3 What is the wavelength range covered by the AM radio band with frequencies in the range of 540 to 1600 kHz?

4 What is the wavelength range for the electromagnetic waves emitted by the FM radio wave range of 88 to 108 MHz?

5 The human eye is most sensitive to the green-yellow part of

the electromagnetic spectrum having a wavelength of about 5.5×10^{-7} m. What is the frequency of this light?

■6 Your radio is tuned in to a radio station 144 km away. (*a*) How long does it take for an electromagnetic signal to travel from the radio station to your receiver? (*b*) If the station operates at 980 kHz, how many wavelengths away is the radio station from you?

■7 A radar pulse transmitted by a police car returns to the receiver, being reflected by a distant truck, after a total time of 5×10^{-4} s. How far away from the police car is the truck that reflected the wave?

8 An explosion occurs 4.0 km away from an observer. How long after the observer sees the explosion will the sound be heard? (Take the speed of sound to be 340 m/s.)

9 The tuning circuit in a radio is tuned to a station such that the value of the inductance in the circuit is 6.4 μH and the capacitance value is 1.9 pF. (*a*) What is the frequency of the waves being detected? (*b*) What is the wavelength of these waves?

10 A radio is used to tune a radio station which operates at 840 kHz. If the radio tuner has an inductance of 0.04 mH, what must the value of the capacitor be to tune in this station?

11 A certain television channel has a frequency of about 96 MHz. The tuning circuit of a television set uses an inductance of 6.0 μH. Find the value of the capacitor required for reception of this television channel.

■12 The coil in the tuning circuit of a radio set has an inductance of 3 μH. Find the range of values the tuning capacitor must have if it is to be able to tune in the complete range of FM frequencies (88 to 108 MHz).

Section 22.5

13 A certain bar magnet has $B = 0.85$ T at its end. The magnet is given a velocity of 10.0 m/s, perpendicular to its length. (*a*) As its end passes close to a certain point, how large an electric field is induced at the point? (*b*) Is this an easily observed electric field?

■14 Assume that, in Fig. 22.10, the magnetic poles are moved with a speed $v = 8.0$ m/s and that the magnetic field strength B between the pole pieces is 0.6 T. (*a*) How large is the electric field at point A at the instant shown? (*b*) Would this be easily observed? (*c*) What is the direction of the electric field at the point A?

■15 In Fig. 22.11, suppose the electric field at the point P due to the charge on the wire is 8×10^3 V/m and the speed of the wire is $v = 6.0$ m/s. (*a*) How large is the magnetic field induced at the point P? (*b*) Is this large enough to be easily measured? (*c*) What is the direction of the magnetic field at P?

■16 The electric field between the plates of an air-filled parallel-plate capacitor is 5×10^4 V/m. Suppose that the capacitor is moved parallel to the plates at 7.2 m/s. (*a*) What is the

magnetic field B at a point through which the electric field is moving? (*b*) What is the direction of this magnetic field?

17 If the amplitude of the magnetic field wave in an electromagnetic wave is to be 1.0 T, what must the amplitude of the electric field wave be?

18 The electric field of a radio wave has an amplitude of 0.90 mV/m at a certain point. What maximum voltage does the wave induce between the ends of a wire 20 cm long at that point? How must the wire be oriented to achieve this voltage? What is the amplitude of the accompanying magnetic field wave?

19 An electromagnetic wave has the electric field given by $E = 8.0 \times 10^{-4} \cos{(6 \times 10^{10} t)}$ V/m. Write the equation for the magnetic field wave. What is the frequency of the wave? The wavelength?

■20 A certain electromagnetic wave has the magnetic field given by $B = 4 \times 10^{-11} \sin{(8 \times 10^9 t)}$ T. (*a*) What is the frequency of the wave? (*b*) What is the period of the wave? (*c*) By how much does the value of B change as t goes from zero to $\tau/4$, where τ is the period of the wave calculated in part b?

■21 In Prob. 20 find the average emf induced during the interval $t = 0$ to $t = \tau/4$ in a loop of wire (area $A = 10.0$ cm^2) placed perpendicular to the magnetic field lines.

■22 A certain electromagnetic wave has the electric field wave given by $E = 1.0 \times 10^{-2} \sin{(3 \times 10^9)}$ V/m. (*a*) Find the period of the wave. (*b*) Write the expression representing the magnetic field in the wave. (*c*) Find the maximum emf induced in a metal rod 40 cm long held parallel to the electric field lines.

Sections 22.6 and 22.7

23 A 0.60-mW laser used in a laboratory experiment has a cylindrical beam with a cross-sectional area of 0.85 mm^2. Assuming the beam to consist of a single sinusoidal wave, find the peak values of the electric and magnetic fields, E_o and B_o, in the beam.

24 A searchlight sends out 4000 W of electromagnetic radiation by means of a cylindrical beam that has a diameter of 0.8 m. Assuming that the beam consists of a single sinusoidal wave, calculate E_o and B_o in the beam.

25 The average intensity of solar radiation received at the top of the earth's atmosphere is 1340 W/m^2. Calculate the magnitudes of the electric and magnetic field of an equivalent electromagnetic wave.

26 A 25-W light bulb emits radiation uniformly in all directions. Calculate the maximum values of the electric and magnetic fields of an equivalent electromagnetic wave (*a*) 2 m and (*b*) 5 m from the bulb.

27 The intensity of the wave from a distant 1.4-MHz radio station is 4.0×10^{-10} W/m^2. Write the equations for the electric and magnetic field waves in this region.

■28 A certain laser beam has a cross-sectional area of 3.6 mm^2 and a power of 1.2 mW. Assuming that the beam consists of a single sinusoidal wave, find the intensity of the beam and the maximum values of the electric and magnetic fields, E_o and B_o, in the beam.

29 A certain 96-MHz radio transmitter sends out 65 W of radiation. Assume that the radiation is uniform on a sphere with the transmitter at its center. (*a*) What is the intensity of the waves at 12 km from the transmitter? (*b*) What are the amplitudes of the electric and magnetic field waves at this distance?

30 A small light bulb hangs in the middle of a room. By what percentage does the light intensity from the bulb decrease as one moves from a point that is 4.0 m from the bulb to a point that is 9.0 m from it?

■31 Calculate the approximate light intensity on the surface of a dining table that is 1.8 m from a 150-W light bulb that is 10 percent efficient in generating light (that is, only 10 percent of the power consumed is converted to light). State any approximations you make and discuss their validity.

■32 The intensity of light measured at 2.0 m from a tiny intense light source is found to be 2.2 W/m². What will be the intensity from the same source if measured at 5.0 m?

General Problems

■33 Calculate the average power radiated uniformly in all directions by a source if the amplitude of the magnetic field is 6×10^{-8} T at a distance of 3 m from the source.

■34 A radio station broadcasts uniformly in all directions with an average power of 18 kW. Calculate the maximum value of the electric field at (*a*) 1 km, (*b*) 5 km, and (*c*) 25 km from the transmitter.

■35 A transmitter radiates electromagnetic waves uniformly in all directions with a power of 80 W. At a distant point the maximum value of the electric field due to this source is found to be 16 mV/m. How far is the transmitter from this point?

■■36 A household uses a dish antenna with a diameter of 22 m to receive the TV signals from a distant TV station. Assume that the TV signal is a single continuous sinusoidal wave with electric field amplitude $E_o = 0.1$ mV/m and that the antenna absorbs all the radiation that is incident on the circular dish. (*a*) What is the amplitude of the magnetic field in the wave? Calculate (*b*) the intensity of the radiation and (*c*) the power received by the antenna.

CHAPTER 23

GEOMETRIC OPTICS: THE REFLECTION AND REFRACTION OF LIGHT

In this and the next two chapters, we are concerned primarily with a very small but very important portion of the electromagnetic spectrum: the wavelengths to which the human eye is sensitive. This region of the spectrum is referred to as **visible light,** or often, just light. Even though our primary concern is with visible light, much of what we will learn is applicable to all electromagnetic radiation.

23.1 THE CONCEPT OF LIGHT

Of all the senses, vision supplies us with more information, both in quantity and in detail, than all the others combined. What we see depends fundamentally on the properties of light as well as the physical and psychological processes of interpretation. It is no wonder that the nature of light has always been a subject of great speculation and interest. In spite of this wide interest and many attempts at explanation, just what light is remained in dispute until the first decade of the twentieth century. Some of the outstanding details of the historic quest to understand light were given in the Controversies in Physics essay in Chap. 22. We will mention here only a few milestones as we examine what we now know about the properties of light.

The San Gabriel mountains (Mt. Wilson in the foreground) are where Michelson made his most precise measurements of the speed of light in the 1920s.

λ (nm)

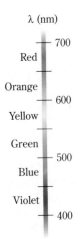

FIGURE 23.1

The correspondence between wavelengths and colors shown here is only approximate. Colors such as blue-green and orange occupy the intermediate regions. (See also Fig. 22.8.)

FIGURE 23.2

The electric field of an electromagnetic wave vibrates perpendicular to the direction of propagation. Hence the wave is transverse.

At the time of Newton, debate about light centered on the question of whether light consists of a stream of particles, or "corpuscles," or whether it is some sort of wave phenomenon. Newton favored the corpuscular idea, and because of his prestige, many others were inclined to accept this view. In 1803 Thomas Young presented the result of an experiment in which light from two sources displayed interference patterns similar to those one would expect from two overlapping waves. We will discuss Young's experiment in detail in Chap. 24. At about this same time, the speed of light passing through water was measured, and found to be slower than the speed of light in air. Since Newton's corpuscular theory held that light should move faster in water, this was a second piece of evidence contradicting that theory. As a result, the wave theory became the dominant concept of light, to be given a rigorous mathematical foundation in the 1860s by the extraordinary work of Maxwell (Chap. 22).

One would think, then, that by 1900 the wave nature of light would have been reasonably well understood and widely accepted. However, the interaction of light with matter, both in how light is emitted and in how it is absorbed, remained puzzling. The spectrum of light emitted by heated solids (blackbody radiation) and by simple atoms such as those of hydrogen, could not be explained adequately by a wave theory of light. A phenomenon known as the *photoelectric effect,* in which electrons are ejected from metal surfaces illuminated by light, was successfully explained in 1905 by Einstein, using the idea that light interacts with the electrons as if it consists of a stream of particles. Through the development of quantum theory during the twentieth century, we have come to an uneasy truce with the idea that under certain circumstances, light behaves as a wave, while under different circumstances, it behaves as a stream of massless particles called *photons.* This dual nature of light will be discussed more fully in Chap. 26. For the next few chapters, we will concentrate on those aspects of light that are evident from its electromagnetic wave characteristics.

Light waves are electromagnetic and thus consist of an oscillating electric field perpendicular to and in phase with an oscillating magnetic field, as discussed in the preceding chapter. The wavelengths of visible light lie in the range of 400 to 700 nm (Fig. 23.1). By using Eq. 22.1, we can see that this wavelength range corresponds to a frequency range of 4.3×10^{14} to 7.5×10^{14} Hz. The electric field for a wave propagating in the x direction is shown in Fig. 23.2. Notice that the oscillating E field is perpendicular to the x axis. Hence, *light waves are transverse waves,* since the wave oscillation is perpendicular to the direction of propagation. As such, they have many properties in common with other transverse waves such as waves on a string or waves on a water surface. One of the most direct pieces of evidence that

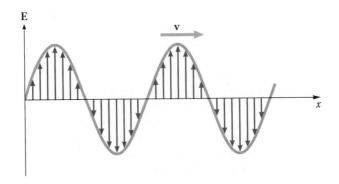

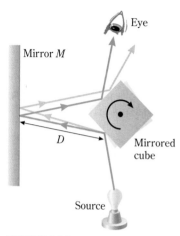

FIGURE 23.3

A simplified drawing of Michelson's method for measuring the speed of light. If the mirrored cube is rotating at just the proper speed, the beam will be reflected into the eye of the observer. In practice, the distance D is much larger than shown.

TABLE 23.1
Speed of light at 589 nm

Material	Speed (10^8 m/s)
Vacuum	2.99792
Air	2.9970
Water	2.25
Ethanol	2.20
Benzene	2.00
Crown glass	1.97
Polystyrene	1.89
Flint glass	1.81
Diamond	1.24

light is a transverse wave is that light can be *polarized*. Only transverse waves have this property. We will discuss polarization of light in Chap. 24.

23.2 THE SPEED OF LIGHT

You will recall from Sec. 2.1 that, in the SI, the speed of light in vacuum is defined to be exactly $c = 299,792,458$ m/s, which we usually round off to 3.0×10^8 m/s. This definition was chosen so as to agree with the measured speed of light in terms of the meter, defined in Sec. 2.1. Prior to the adoption of this standard, many attempts were made to measure c. One of the first to try was Galileo; he was unsuccessful, concluding only that light transmission "if not instantaneous, was extremely rapid." The first quantitative result came in 1675 when the Danish astronomer Roemer used the relative motion of earth and one of Jupiter's moons to conclude that light traveled at about 2.1×10^8 m/s. Most of Roemer's error can be attributed to an incorrect value for the radius of earth's orbit. In 1849 the French physicist Fizeau measured the time it took light to make a round-trip from one mountain to another 8.6 km away. Fizeau's results gave a value of $c = 3.1 \times 10^8$ m/s.

The first highly precise measurements were carried out by the American A. A. Michelson in the 1920s. Michelson measured the time of flight of a light beam over the round-trip distance of 70 km between Mt. San Antonio (now Mt. Baldy) and Mt. Wilson in California. He used an apparatus shown in simplified form in Fig. 23.3. A beam of light from the source is reflected from one side of a cube that has mirrored surfaces on four sides. The beam is then reflected from mirror M back to the cube, where it is reflected again as shown. If the cube is at just the right position, the beam will enter an observer's eye in the position indicated.

Suppose, however, that the cube is rotating about an axis through its center, perpendicular to the page. When the cube is in the position indicated by the heavy lines in Fig. 23.3, the beam is reflected to the mirror as shown. By the time the beam returns to the cube from the mirror, however, the cube will have rotated, perhaps to the position represented by the lighter lines, and so the beam will not be reflected into the observer's eye. If the beam is to be reflected to the eye, the cube must rotate through 0.25 rev during the time the beam takes to travel to the mirror M and back, for only then will the cube again be in a position given by the heavy lines in Fig. 23.3 and reflect the beam to the eye.

The measurement technique is to vary the speed of rotation of the cube until the reflected beam enters the eye. At that speed of rotation, we know that the time taken for one-fourth of a cube rotation is equal to the time the light takes to travel a distance $2D$. It is necessary to know only the speed of rotation of the cube and D in order to compute the speed of the light. Michelson's experiments resulted in a value of 2.99796×10^8 m/s.

The experiments that established the present value of c were carried out in the early 1970s, using wavelength and frequency measurements on light emitted by lasers. These remain among the most precise measurements made of any physical constant.

Light travels fastest through vacuum. Its speed in other materials is always less than c. Moreover, its speed in materials other than vacuum depends on the wavelength of the light as well as on the material. We list in Table 23.1 the speed of light in various materials.

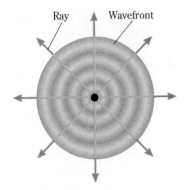

FIGURE 23.4

The rays are perpendicular to the wavefronts and show the direction of propagation of the wave.

23.3   THE REFLECTION OF LIGHT

When a stone is dropped into a pond, a set of circular waves, or *wavefronts,* moves out from the point where the stone hits the water. The wavefronts, shown in Fig. 23.4, travel radially outward from the center. The arrows in the direction of wavefront travel are called *rays.* Notice that the rays are always perpendicular to the wavefronts, just as we learned in Sec. 15.1. Hence we can specify the motion of a wave by drawing either rays or wavefronts. Both methods are of value.

In Fig. 23.5, we see how the pattern of wavefronts and rays looks far from the source. The wavefronts are segments of circles having a 1-m radius, indicating the source is 1 m away. Notice that the wavefronts are nearly straight lines and the rays are almost parallel to each other. In three dimensions, these wavefronts are nearly planar. Hence for a far distant source, we refer to such waves as *plane waves.* The term *parallel light,* which refers to the ray pattern, is synonymous with the term *plane waves,* which refers to the wavefront pattern.

Suppose that plane water waves are incident on a flat wall, as in Fig. 23.6*a.* The velocity of an incoming wave can be resolved into two components, $v_\perp$ perpendicular to the wall and $v_\parallel$ parallel to the wall. Upon striking the wall, $v_\perp$ is reversed in direction and $v_\parallel$ is unchanged. As a result, the wave is reflected from the surface. A reflected ray and its velocity components are shown in Fig. 23.6*b.* Let us now find how the angle of incidence θ_i, shown in part (*a*), is related to the angle of reflection θ_r, shown in (*b*).

As you can see, $\cos \theta_i = v_\perp/v$ (Fig. 23.6*a*) and $\cos \theta_r = v_\perp/v$ (Fig. 23.6*b*). Therefore, because the cosines are equal, the angle of incidence equals the angle of reflection.

The fact that a water wave is reflected in such a way that the angle of incidence

FIGURE 23.5

Rays from a distant source are nearly parallel. Notice that the wavefronts are nearly flat. For an infinitely far away source, these are plane waves and the rays are parallel.

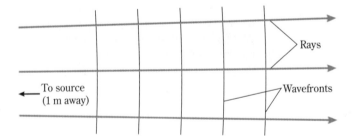

FIGURE 23.6

The incident wave is reflected in such a way that the angle of incidence *i* equals the angle of reflection *r.*

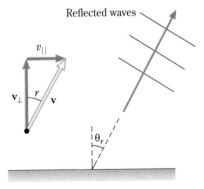

(*a*) Before reflection   (*b*) After reflection

equals the angle of reflection is of general validity. We could use the same reasoning to show that light waves are also reflected in this way. Notice that the only basic assumption made was that, upon reflection, the velocity component perpendicular to the surface was reversed and the parallel component was unchanged. Our result is true for any type of wave for which this assumption is true. Measurements on light and other forms of electromagnetic radiation confirm our deduction. We may therefore formulate the following rule, known as the **law of reflection:**

The angle of incidence equals the angle of reflection.

The type of reflection shown in Fig. 23.7a, where the reflecting surface is perfectly smooth as in a mirror, is called *specular reflection.* Rougher surfaces, such as paper or painted walls, give rise to *diffuse reflection,* shown in Fig. 23.7b. For such surfaces, even though the law of reflection applies to individual rays in a light beam, the nonsmooth surface causes the rays to be reflected at various angles from the average plane of the surface.

23.4 PLANE MIRRORS

Let us now apply what we have learned about reflection to the important topic of image formation by mirrors. First we consider how a plane mirror (one that is flat) forms an image.

Every day you look at yourself in the mirror and see an image of your face in front of you. If you stop to examine exactly what you are seeing, you perceive the image of your face as being behind the surface of the mirror. In fact, the image appears to be about as far behind the mirror as your face is in front of the mirror. Let us now examine such a reflection in order to understand clearly why the image is seen as it is.

Suppose you place an object in front of a mirror and wish to find where your eye will perceive the location of the image of the object. Each point of the object acts as a point source of light, either emitting or reflecting light in a diverging pattern of rays. When you look directly at the tip of the object, as shown in Fig. 23.8a, what

Multiple images are formed when objects are placed between two plane mirrors facing each other.

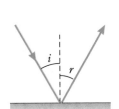

(a) Specular reflection

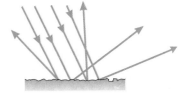

(b) Diffuse reflection
FIGURE 23.7
The angle of incidence equals the angle of reflection for each ray in a beam of light. A flat surface reflects all rays parallel to each other, giving rise to specular reflection. A rough surface diffuses the rays upon reflection, resulting in diffuse reflection.

FIGURE 23.8
(a) Rays from an object point O diverge in all directions. Those within the yellow area enter the eye and are "seen." (b) The reflected rays which are seen by the eye seem to come from point I on the image of the object O.

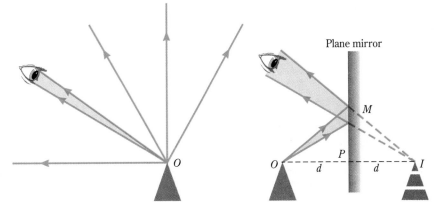

you are seeing is the small fraction of the light diverging from that point which enters the pupil of your eye. When you look at the same bundle of light rays reflected by the mirror, as in Fig. 23.8*b*, your mind interprets these rays as having come in straight lines from a point behind the mirror. This point, shown in Fig. 23.8*b*, is called the image of the object tip. You could choose any other point on the object and draw a similar ray diagram. *Each object point has a corresponding image point behind the mirror,* which is the point from which the rays leaving that object point appear to come after being reflected by the mirror. For the sake of clarity, we have not shown the rays from other object points, but you should realize that together, reflected rays from all points on the object form its complete image.

By applying the law of reflection to the rays in Fig. 23.8*b*, it is easy to show that the triangles *OMP* and *IMP* are congruent, so that corresponding object and image points are equal distances in front of and behind the mirror.

This type of image, one through which the observed rays do not actually pass, is called a **virtual image** or an **imaginary image.** In other words, the rays reaching the eye do not really come from the point at which we see the image. There is no possibility whatsoever that an image of the object would appear on a sheet of paper placed at *I* behind the mirror. The mind merely interprets that the light comes from *I*. It is always true, of course, that the image of an object seen by reflection in a plane mirror is a virtual image. The image is always exactly as far behind the mirror as the object is in front of it.

23.5 THE FOCAL LENGTH OF A SPHERICAL MIRROR

Plane mirrors are used by all of us, but spherical mirrors are not quite so common. However, makeup and shaving mirrors are portions of spheres, as are surveillance mirrors in stores. A spherical mirror is a portion of the surface of a hollow sphere, as shown in Fig. 23.9. The line *PA*, running through the center of the sphere and perpendicular to its surface, is the *principal axis* of the mirror. If light is reflected from the inside surface of the sphere, as in Fig. 23.10*a*, the mirror is called a *concave* mirror. If light is reflected from the outside surface of the sphere, as in Fig. 23.10*b*, the mirror is called a *convex* mirror.

In drawing Fig. 23.10, we assumed that the light comes from a distant source so that the incoming rays are parallel and the incoming wavefronts are planes. As indicated in *a*, parallel rays that travel in the direction of the principal axis of a

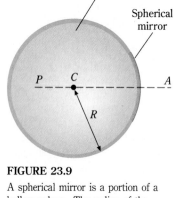

FIGURE 23.9

A spherical mirror is a portion of a hollow sphere. The radius of the mirror is *R*, its center of curvature is point *C*, and its principal axis is the line *PA*. The principal axis goes through the center of curvature and the center point of the mirror's surface.

FIGURE 23.10

(*a*) The reflected light from parallel rays incident on a concave mirror is converged to the focus *F* in front of the mirror; (*b*) parallel incident light reflected from a convex mirror appears to diverge from the focus *F* behind the mirror.

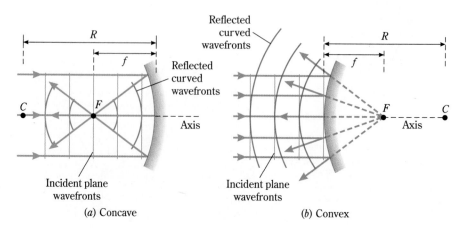

(*a*) Concave   (*b*) Convex

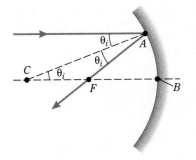

FIGURE 23.11

A ray parallel to and near the principal axis of a concave mirror is reflected through the focal point.

concave mirror are all reflected to a point F. (This is only approximately correct, as we will soon show.) This point to which the light from a distant object is reflected by a concave mirror is called the **focus** (or the **focal point**) of the mirror. Figure 23.10*b* shows what happens to parallel rays reflected from a convex mirror. The reflected rays appear to come from a point F **behind** the mirror. This point is the focus (or focal point) for a convex mirror. In both types of spherical mirrors, the distance from the central point of the mirror surface to the focal point F is called the **focal length,** f, of the mirror.

Let us now examine what determines the focal length of a concave mirror. Consider an incident (incoming) ray parallel to the principal axis CB and striking the mirror at point A in Fig. 23.11. The line CA is a radius of the mirror and hence is perpendicular to the mirror at A. Recall from Sec. 23.3 that the law of reflection was defined in terms of the angle between the incident ray and the perpendicular (or normal) to the reflecting surface. Thus the reflected ray leaving point A in Fig. 23.11 at an angle θ_i with line CA from the incident ray crosses the principal axis at the focal point F. Since the incident ray was assumed to be parallel to the principal axis CB, angle ACB must also equal θ_i. This means that triangle CFA is isosceles, so that distances CF and FA are equal. If the incident ray is not very far from the principal axis, so that point A is close to point B, FA (and hence CF) are approximately equal to FB. Since $CF + FB = R$, the radius of the sphere, we have the following result:

The focal length of a concave spherical mirror is half the mirror radius:

$$FB = f = \frac{R}{2} \tag{23.1}$$

It is not strictly true that all rays parallel to the principal axis are reflected back through the same point F. As an exercise you should draw the case of such a ray reflected from a point far away from the principal axis and show that it is not reflected through F. However, if we restrict the incident rays to that portion of the mirror where the arc length AB is small compared to the radius of the sphere, the approximation used in deriving Eq. 23.1 is quite good. We can make this restriction either by using a small aperture in a mask in front of the mirror or by making the mirror itself small in comparison with its radius of curvature. This defect in focusing is called *spherical aberration*. Mirrors having a *parabolic* cross-section do not have this disadvantage. They are more expensive to make than spherical mirrors, but are widely used in astronomical telescopes, where a wide aperture is desirable.

23.6 RAY DIAGRAMS; IMAGE FORMATION BY CONCAVE SPHERICAL MIRRORS

Of all the possible light rays we could draw from an object point to a mirror, there are three that are particularly useful in locating the corresponding image point. We already have discussed one of these: the incident ray that is parallel to and near the principal axis. We know that it is reflected back through the focal point F, which is midway between the mirror's center of curvature C and the point where the principal axis intersects the mirror. This is shown in Fig. 23.12*a*.

A second important ray is the one passing through the focal point on its way to the mirror. This ray is reflected back parallel to the principal axis, as seen in Fig.

Concave mirrors reflect light rays from the sun to a focus in this solar furnace in southern France. Temperatures exceeding 4000°C can be achieved at the focus of this mirror.

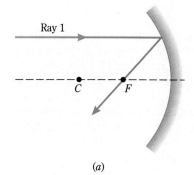

(a)

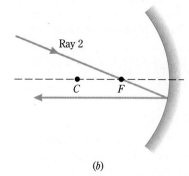

(b)

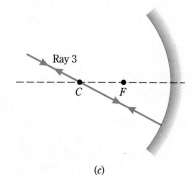

(c)

FIGURE 23.12

The three special rays used to locate an image by a concave spherical mirror.

23.12*b*. This is so because the law of reflection applies equally well when we reverse the direction of any light ray.

The third special ray is the one that passes from the object point through the center of curvature *C*. As shown in Fig. 23.12*c*, it strikes the mirror perpendicularly, reflecting back on itself. We can summarize these three special rays for concave mirrors as follows:

1 A ray parallel to the principal axis is reflected through the focus.
2 A ray passing through the focus is reflected parallel to the principal axis.
3 A ray passing through the center of curvature is reflected back through the center of curvature.

We will now apply these rules as we use ray diagrams to locate the position of images.

Suppose we wish to find the image of the object *O* formed by the mirror shown in Fig. 23.13. Let us say that the object is a light bulb. The bulb emits light in all directions, but we need to draw only three of the rays coming from it. These three rays are exactly those described by our rules, and you should trace each to see that it is properly drawn in Fig. 23.13. Once the positions of *C* and *F* are known, only a straightedge is needed to draw these rays.

If you place your eye as shown in Fig. 23.13, the three rays appear to come from point *I*. In other words, you see an image of the light bulb at *I*. Moreover, since the rays actually do converge on point *I* and pass through it, a sheet of paper placed at *I* will show a lighted picture of the original bulb. This is therefore a **real image:** *at a real image, the light actually passes through the image and reproduces the object.* Notice how this differs from the imaginary, or virtual, image found for the plane mirror.

We have used the three special rays to locate the image point *I* corresponding to the object point *O*. All other points on the object are also sources of either emitted or reflected light. To find the image points for other points on the object, we could proceed in the same way, ultimately reproducing the image of the entire object. If the object is vertical as in Fig. 23.13, we expect that all image points lie in a vertical

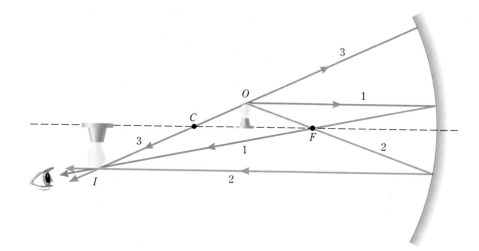

FIGURE 23.13

A real image *I* is formed from the object *O*. Trace the three rays from the object.

line also. Thus by locating the image point corresponding to the tip of the object, we can immediately fill in the rest of the image.

We can use these ray diagrams to tell us more about the image than just its location. As mentioned, when the reflected rays physically converge, the image is real. A screen or photographic film placed at the image location would record this real image. Also, notice that in Fig. 23.13 the reflected rays all cross the principal axis before they form the image. This causes the image to be *inverted* relative to the object. Finally, the ray diagram in Fig. 23.13 shows that the image is larger than the object. We say the image is *enlarged*. By examining Fig. 23.13, you can see that, for other placements of the object between *C* and *F*, you obtain these same image characteristics.

Consider what the situation would be like if the object were placed outside of *C*, for example at position *I* in Fig. 23.13. Again using the fact that the direction of the rays can be reversed, you should be able to convince yourself that the image would be formed at *O*. This image would again be real and inverted, but *reduced* in size. You should convince yourself that these image characteristics would result from **any** object placement outside *C*. Let us summarize these image characteristics of a concave mirror so far:

1 For an object placed between *C* and *F*, the image is real, inverted, and enlarged.
2 For an object placed farther away than *C*, the image is real, inverted, and reduced.

Consider now the situation in Fig. 23.14, where the object is placed very near the mirror, inside the point *F*. Once again we draw our three rays from the tip of the object. Now, however, ray 2 does not go through *F* on its way to the mirror since *O*

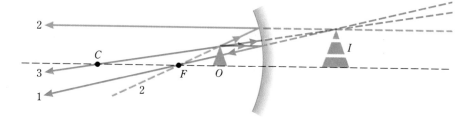

FIGURE 23.14

The three rays appear to come from the virtual image *I*. Notice especially rays 2 and 3 so that you can draw them in other situations.

Images formed by concave and convex mirrors. Notice that the image shown in (a) is inverted, while that in (b) is upright. Which of these images is virtual and which is real? Can the concave mirror form an upright image of the puppet? Can the convex mirror form an inverted image?

is inside the focal point. It still travels along a line in the direction from F, however, and is reflected back parallel to the principal axis as before. The result of reflection of the three rays is now quite different from before. As seen in Fig. 23.14, the reflected rays all *diverge* from each other. They will never come to a convergent point to form a real image as they did in Fig. 23.13. However, their pattern of divergence is the same *as if* they came directly from a point I behind the mirror. As we saw in the case of a plane mirror, this pattern of rays represents what we call a virtual image. Notice that the ray diagram also shows the image to be erect (upright) and enlarged. We can add this result to the previous two image characteristics of a concave mirror:

3 For an object closer than F, the image is virtual, upright, and enlarged.

23.7 THE MIRROR EQUATION

To derive a mathematical equation that describes the location of an image, let us refer to Fig. 23.15. The distance p from the object to the mirror is called the **object distance.** The height of the object is called O. The height of the image is called I, and its distance from the mirror, called the **image distance,** is i. Notice that the distance BF from the mirror to the focal point is the focal length f of the mirror. The ray ABE in part (a) is not one of our special three. However, it is reflected in such a way that angle ABH equals angle DBE. For this reason, the shaded triangles ABH and DBE in (a) are similar. Taking the ratios of corresponding sides gives

$$\frac{O}{I} = \frac{p}{i}$$

In Fig. 23.15b, the shaded triangles are also similar. The lengths AH and DE are the object and image heights, respectively. Also, notice that $DE = GJ$. Hence

$$\frac{O}{I} = \frac{AH}{DE} = \frac{AH}{GJ} = \frac{HF}{FG}$$

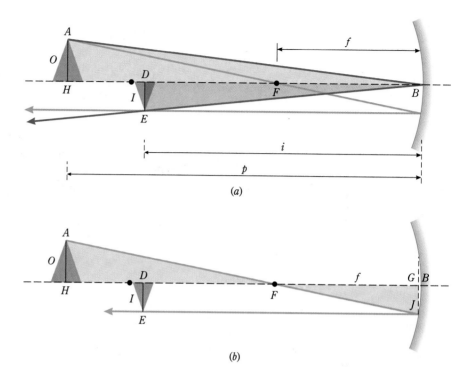

FIGURE 23.15

(a) Triangle *ABH* is similar to triangle *DBE*. (b) Triangle *AFH* is similar to triangle *JFG*. In the text, we assume the curvature of the mirror to be so small that distance *GB* is negligible.

But *HF* is just $p - f$, and *FG* is nearly f. (They differ by the negligible distance *GB*.) To this approximation, we have

$$\frac{O}{I} = \frac{p - f}{f}$$

Equating this to the expression found in part (a) gives

$$\frac{p}{i} = \frac{p - f}{f}$$

After dividing this equation by p and rearranging, we have

$$\frac{1}{p} + \frac{1}{i} = \frac{1}{f} = \frac{2}{R} \tag{23.2}$$

where we have used $f = R/2$ from Eq. 23.1.

Equation 23.2 is called the **mirror equation.** It allows us to calculate the distance i of the image from the mirror, provided that the object distance p and the focal length f are given. Conversely, it also allows us to determine where an object must be placed to form an image at a specified location. Notice that this equation involves the addition of *reciprocal* quantities. As we will soon see, f, p, and i can have positive and negative signs in various situations, so care must be taken to apply algebraic rules correctly.

To calculate the relative heights of object and image, notice that $O/I = p/i$, as shown previously. The ratio of image height to object height is called the **magnification** produced by the mirror:

$$\text{Magnification} = M = \frac{I}{O} = \frac{i}{p} \qquad\qquad (23.3)$$

As before, if I/O is less than unity, we say that the image is reduced. If I/O is greater than unity, the image is *enlarged*.

Example 23.1

An object 2.0 cm high is placed 30 cm from a concave mirror with a radius of curvature of 10 cm. Find the location and size of the image. Is the image upright or inverted? Real or virtual?

R *Reasoning*

Question How does image location depend on the known quantities p, O, and R?
Answer R determines the focal length f (Eq. 23.1). Then you can solve the mirror equation (Eq. 23.2) for i.

Question How does image location determine image size?
Answer Eq. 23.3 shows that the ratio of distances, i/p, is the same as the ratio of sizes, I/O.

Question How can I tell whether the image is (1) upright or inverted and (2) real or virtual?
Answer The object is placed outside of C. This is the second of the three object placements listed in Sec. 23.6, which summarizes the image characteristics derived from ray diagrams.

Solution and Discussion The focal length of the mirror is $f = R/2 = 5.0$ cm. Then the image distance is

$$\frac{1}{i} = \frac{1}{5.0 \text{ cm}} - \frac{1}{30 \text{ cm}} = \frac{5}{30 \text{ cm}}$$

Taking the reciprocal of this gives

$$i = \frac{30 \text{ cm}}{5} = 6.0 \text{ cm}$$

Notice that we do not have to change to meters, so long as all distances are expressed in the same units. The magnification is

$$M = \frac{i}{p} = \frac{6.0 \text{ cm}}{30 \text{ cm}} = \frac{1}{5}$$

The image is reduced to one-fifth the size of the object. Thus $I = O/5 = \frac{2}{5}$ cm = 0.40 cm. You can quickly check this solution by means of a ray diagram.

Section 23.6 shows that objects placed outside of C have images that are real, inverted, and reduced.

Example 23.2

An object is placed 5.0 cm in front of a concave mirror with a 10-cm focal length. Find the location of the image and its characteristics.

R **Reasoning**

Question   Here the object is placed inside the focal point of the mirror. I know from Fig. 23.14 that the ray diagram for this situation shows a virtual image. Does the mirror equation apply to this case?

Answer   Yes. Make sure you do the algebra correctly. Then you will see how a virtual image shows up in the answer. The mirror equation for this case is $1/i = 1/(10 \text{ cm}) - 1/(5.0 \text{ cm})$.

Solution and Discussion   The mirror equation gives a *negative* result for *i*:

$$\frac{1}{i} = \frac{1-2}{10 \text{ cm}} = \frac{-1}{10 \text{ cm}}; \qquad i = -10 \text{ cm}$$

Thus we can interpret the type of image by the algebraic sign of *i*. *If i is positive, the image is real and located in front of the mirror. If i is negative, the image is virtual and located behind the mirror.*

The image characteristics for an object placed inside *F* (Sec. 23.6) are: virtual, upright, and enlarged. The magnification is $I/O = i/p = -10 \text{ cm}/5.0 \text{ cm} = -2.0$. The meaning of the minus sign will be discussed in the next section. For now, this result simply tells us that the image is twice the height of the object.

23.8   IMAGE FORMATION BY CONVEX MIRRORS

A convex spherical mirror is a portion of a sphere that reflects rays from outside the sphere, as shown in Fig. 23.10*b*. We saw there how parallel rays are reflected from such a mirror. They appear to diverge from a point behind the mirror. *Rays incident on a convex mirror parallel to the principal axis are reflected as though they came from the focal point.* To prove this, we proceed in much the same way as for the concave mirror.

Referring to Fig. 23.16, we see from the law of reflection and the geometry

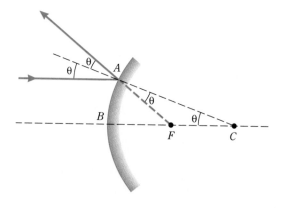

FIGURE 23.16

An incident ray parallel to the axis is reflected as though it came from the focal point of the convex mirror.

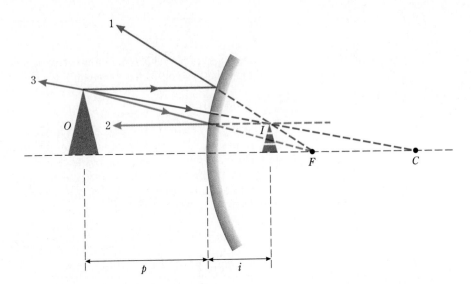

FIGURE 23.17

You should be able to draw the three
rays for any situation involving a
convex mirror.

involved that several angles are equal. The triangle *AFC* is isosceles, so that *AF* =
FC. If the length *AB* is small compared with the radius of curvature of the mirror,
then *AF* is nearly equal to *BF*. Consequently, *BF* nearly equals *FC*, and so here too
the focal point can be considered to be midway between the mirror and its center of
curvature.

We are therefore able to write rules for drawing our three special rays for a
convex mirror:

1 A ray parallel to the axis is reflected as though it came from the focal point.
2 A ray heading toward the focal point is reflected parallel to the axis.
3 A ray heading toward the center of curvature is reflected back on itself.

These three rays are illustrated in Fig. 23.17. You should trace them to see that they
conform to these rules. Notice that all three reflected rays appear to come from the
image *I* behind the mirror. As you see, the image is virtual, upright, and reduced in
size.

The algebraic relation used in locating the image for a convex mirror can be
obtained by reference to Fig. 23.18. You should be able to show that triangle *ABH*
is similar to *EBD* in part (*a*) and that triangle *JFG* is similar to triangle *EFD* in (*b*).
This being true, the following equations are found, as in the case of the concave
mirror:

$$\frac{O}{I} = \frac{p}{i} \quad \text{and} \quad \frac{O}{I} = \frac{f}{f-i}$$

In writing these equations, the distance *BG* was considered negligibly small.

Equating the two expressions, inverting, dividing by *i*, and rearranging yields

$$\frac{1}{p} - \frac{1}{i} = -\frac{1}{f}$$

Notice that, except for signs, this equation is the same as Eq. 23.2 for the concave
mirror. The difference in signs alerts us to the fact that the image in this case is

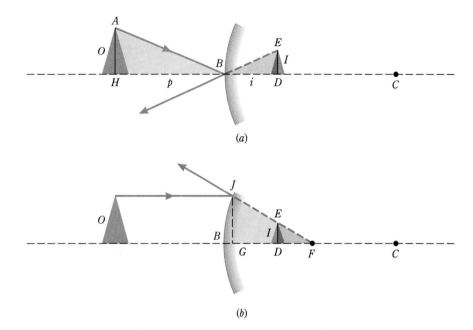

(a)

(b)

FIGURE 23.18

Triangles *ABH* and *EBD* are similar, as are triangles *EFD* and *JGF*. We make the assumption that the length *FG* is essentially the same as *FB*.

behind the mirror rather than in front of it. Also, the negative focal-length term is the result of the mirror's being convex rather than concave.

Rather than remembering two mirror equations, we can set up rules that allow us to use Eq. 23.2 even for convex mirrors. If we agree to always call image distances behind a mirror, that is, virtual-image distances, negative, we can omit the negative sign from the i term in the convex-mirror equation. Moreover, if we always say that the focal length of a convex mirror is negative, we can omit the other negative sign as well. Hence we can write for *all* mirrors

$$\frac{1}{p} + \frac{1}{i} = \frac{1}{f} \qquad \text{mirrors} \tag{23.2}$$

where we agree that

1 Object distances are positive if the object is in front of the mirror and negative otherwise.
2 Image distances are positive if the image is in front of the mirror (real images) and negative otherwise (virtual images).
3 The focal length is positive for a concave mirror and negative for a convex mirror.

We can extend this use of signs to interpret whether the image is erect or inverted relative to the object. We will rewrite the equation for magnification with a minus sign:

$$M = -\frac{i}{p} \tag{23.3a}$$

The arbitrary sign we gave to magnification in this equation has nothing to do with the relative sizes of object and image, but we can use it to tell us whether the image is upright or inverted. From our previous examples, we can see that when the image is real, it is also inverted. Furthermore, the image distance i is positive. Since p and i are both positive, the ratio $M = -i/p$ is negative. If the image is virtual, the image is upright, and its distance i is negative. This makes $M = -i/p$ a positive number. Let us summarize this information as follows:

If magnification M is positive, the image is upright relative to the object; if M is negative, the image is inverted.

You can see that in Eqs. 23.2 and 23.3a it is very important to use the correct signs and equally important to interpret the signs of the results of your calculations.

Example 23.3

A convex mirror with a 100-cm radius of curvature is used to reflect the light from an object placed 75 cm in front of the mirror. Find the location of the image and its magnification. Is the image upright or inverted?

R   *Reasoning*

Question   I am given $p = 75$ cm. If I can determine the focal length of the mirror, then I can use Eq. 23.2 to find i. How can I find f?
Answer   The focal length f is half the radius of curvature, but since the mirror is convex, you must use $f = -50$ cm in the mirror equation.

Question   What is the equation for image position?
Answer $1/i = 1/(-50$ cm$) - 1/(75$ cm$)$. Notice that both terms are negative, and so i must be negative.

Question   With i known, how do I determine magnification, and how can I tell whether the image is upright or inverted?
Answer   From Eq. 23.3a, $M = -i/p$. Once you calculate M, the sign rule for M tells you whether the image is erect or inverted.

Solution and Discussion   Solving Eq. 23.2 for i, we notice that the common denominator is 150 cm:

$$\frac{1}{i} = \frac{-3 - 2}{150 \text{ cm}} = \frac{-5}{150 \text{ cm}}$$

This gives $i = -30$ cm. The negative sign tells you that the image is virtual and located behind the mirror. Remember, this image position is the location from which the reflected rays appear to be diverging.
   The magnification is

$$M = -\frac{-30 \text{ cm}}{75 \text{ cm}} = +0.40$$

The image size is 40 percent of the object size, and the positive sign tells you that the image is upright.

Example 23.4

Suppose you have a concave mirror with a focal length of 40 cm. Where would you have to place an object in order to form an image 100 cm in front of the mirror?

R | *Reasoning*

Question   What relationship involving the known quantities determines the object location?
Answer   The mirror equation:

$$\frac{1}{p} = \frac{1}{f} - \frac{1}{i}$$

Question   What are the correct signs to use?
Answer   A concave mirror always has f positive, and an image in front of the mirror is real, with i positive.

Solution and Discussion

$$\frac{1}{p} = \frac{1}{40 \text{ cm}} - \frac{1}{100 \text{ cm}} = \frac{10 - 4}{400 \text{ cm}}$$

$$= \frac{6}{400 \text{ cm}}$$

This gives $p = +66.7$ cm. You should verify this result by drawing a ray diagram.

Exercise   If the image is to be 2.5 cm tall, how tall must the object be?
Answer: *1.67 cm.*

23.9   THE REFRACTION OF LIGHT: SNELL'S LAW

When a beam of light enters water from air, its path bends as shown in Fig. 23.19. This change in the direction of a ray as it passes from one medium to another is called **refraction.** The angle θ_1 is, of course, the angle of incidence, and angle θ_2 is called the **angle of refraction.** (Part of the beam is also reflected by the water surface, as shown by the dashed ray in Fig. 23.19, but we ignore this reflection in this section.)

The basic reason a light ray changes direction as it goes from one transparent medium to another is that, as mentioned in Sec. 23.2, *light travels at different speeds in different media.* From Table 23.1, we see that in a vacuum light has its greatest speed, c; in other materials light travels more slowly. We characterize the extent to which a medium reduces the speed of light by what is called the medium's **index of refraction** n:

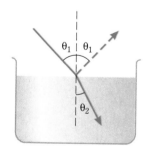

FIGURE 23.19

When a ray of light in air is incident on a water surface, part of the ray is refracted toward the normal to the surface. The other part of the incident ray obeys the law of reflection.

$$n = \frac{\text{speed of light in vacuum}}{\text{speed of light in the medium}} = \frac{c}{v} \qquad (23.4)$$

A common perception resulting from refraction is that a stick seems to bend when it enters the water.

TABLE 23.2   *Refractive indices at 589 nm wavelength*

Material	$c/v = n$	Material	$c/v = n$
Air*	1.0003	Crown glass	1.52
Water	1.33	Sodium chloride	1.53
Ethanol	1.36	Polystyrene	1.59
Acetone	1.36	Carbon disulfide	1.63
Fused quartz	1.46	Flint glass	1.66
Benzene	1.50	Methylene iodide	1.74
Lucite or Plexiglas	1.51	Diamond	2.42

*Standard temperature and pressure

Because light has its highest speed in vacuum, the index of refraction is always unity or larger. Typical values for n are listed in Table 23.2. As you see there, the refractive index for air is very close to unity, while the index for diamond is large, 2.42. Of course, the index of refraction for vacuum is exactly 1. The index of refraction varies slightly with wavelength, as we will see later; it is larger for blue light than for red light.

In order to find a relationship between the angle of incidence θ_1 and the angle of refraction θ_2, it is convenient to consider the motion of the wavefronts in a plane wave, as shown in Fig. 23.20. We assume that the speed of the wave is v_1 in medium 1 and v_2 in medium 2, with v_1 greater than v_2. The wavefronts have a bend in them at the interface between the two media because the wave travels more slowly in medium 2 than in medium 1.

Suppose it takes a time t for wavefront ABC to travel to the position $A'B'C'$. The distance the wavefront moves in medium 2 in time t is $b = v_2 t$, and the distance the wavefront moves in medium 1 is $d = v_1 t$. If we divide d by b, we find that

$$\frac{d}{b} = \frac{v_1}{v_2}$$

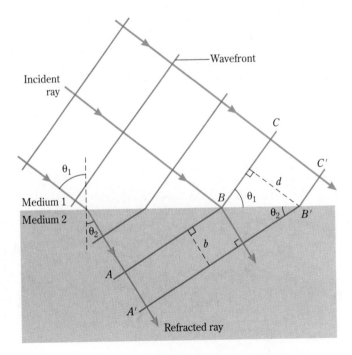

FIGURE 23.20

Because the wave travels more slowly in medium 2 than in medium 1, distance AA' is smaller than distance CC'.

Moreover, we see in the figure that

$$\frac{d}{BB'} = \sin \theta_1 \quad \text{and} \quad \frac{b}{BB'} = \sin \theta_2$$

which, after division of one by the other, gives

$$\frac{d}{b} = \frac{\sin \theta_1}{\sin \theta_2}$$

Since $d/b = v_1/v_2$, we have

$$\frac{\sin \theta_1}{\sin \theta_2} = \frac{v_1}{v_2} \tag{23.5}$$

Because $v = c/n$ from the definition of refractive index, we can rewrite Eq. 23.5 as

$$\frac{\sin \theta_1}{\sin \theta_2} = \frac{c/n_1}{c/n_2} = \frac{n_2}{n_1}$$

which can be rewritten as

$$n_1 \sin \theta_1 = n_2 \sin \theta_2 \tag{23.6}$$

which we refer to as **Snell's law.** An easy way to remember Snell's law is:

As light crosses a boundary from one medium to another, the product $n \sin \theta$ remains constant.

Remember that the angles of incidence and refraction are always measured relative to the *normal* to the boundary.

We can see from Eq. 23.6 that if n_2 is greater than n_1, then $\sin \theta_1$ is larger than $\sin \theta_2$, which means that θ_1 is larger than θ_2. This is the instance illustrated in Fig. 23.21*a*. Sometimes, however, we are interested in the reverse case, where n_2 is smaller than n_1. This is applicable to a beam of light going from glass to air, for example. Under these circumstances, Eq. 23.6 predicts that θ_2 is larger than θ_1, as shown in Fig. 23.21*b*.

FIGURE 23.21

(*a*) If $n_2 > n_1$, the beam bends toward the normal. (*b*) If $n_2 < n_1$, the reverse is true.

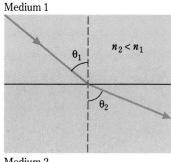

If $n_2 > n_1$, the beam bends toward the normal; if $n_2 < n_1$, the beam bends away from the normal.

Notice one important special case, that of *normal incidence* ($\theta_1 = 0$). In this case Eq. 23.6 has the solution $\theta_2 = 0$ regardless of what values n_1 and n_2 have. In general,

Light that is incident perpendicularly to a surface between one medium and another does not change direction as it enters the second medium.

Illustration 23.1

A diver beneath the surface of the ocean shines a laser beam up at an angle of 37° to the vertical. At what angle does the light emerge into the air?

Reasoning   The situation is shown in Fig. 23.22. Notice that medium 1 is water and medium 2 is air. Applying Snell's law and using $n_1 = 1.33$ and $n_2 = 1.00$ (from Table 23.2), we have

$$1.33 \sin 37° = 1.00 \sin \theta$$

$$\sin \theta = 0.80$$

$$\theta = 53°$$

Exercise   Find the refraction angle in water for light entering the water from air at an angle of incidence of 53°.   *Answer:*   37°

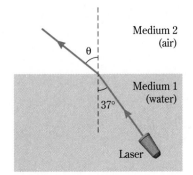

FIGURE 23.22

The underwater laser sends out a beam that bends away from the normal as it passes into the air.

Example 23.5

Light in air is incident at an angle of 30° relative to the normal on a slab of crown glass that has parallel sides as shown in Fig. 23.23a. At what angle does the light emerge from the lower glass surface into the air?

FIGURE 23.23

Light passing through a glass plate with parallel sides.

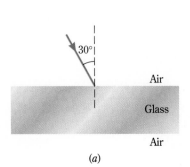
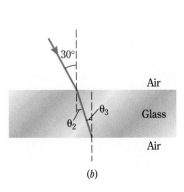
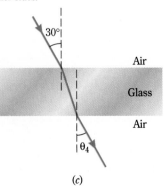

(a)   (b)   (c)

Reasoning

Question   What principle determines the direction of the emergent ray?
Answer   Snell's law applies to the point where the light ray emerges from the lower surface. Figure 23.23b is a sketch of the ray as it travels through the glass. You have to find the angle θ_3 with which this ray strikes the bottom surface.

Question   What is the relation between θ_3 and θ_2?
Answer   Because the sides of the glass are parallel, $\theta_3 = \theta_2$.

Question   How is θ_2 related to the original angle of incidence?
Answer   Snell's law gives $\sin \theta_2 = (n_1/n_2) \sin 30°$.

Solution and Discussion   Putting all the above relations together, we have

$$\sin \theta_4 = \frac{n_2}{n_1} \sin \theta_3 = \frac{n_2}{n_1} \sin \theta_2$$

$$= \frac{n_2}{n_1} \times \frac{n_1}{n_2} \sin 30° = \sin 30°$$

Thus $\theta_4 = 30°$, which shows that the light emerges from the glass in the same direction as it entered. This result can be generalized to any number of layers bounded by parallel sides. The beam moves sideways, but does not change direction, when it returns to the same medium that it started in.

Example 23.6

The beam of light in Fig. 23.24 is incident on an equilateral prism in a direction parallel to the prism's base. The prism is made of fused quartz. Find the angle θ_4 that the emerging beam makes with the normal to the right-hand face of the prism.

Reasoning

Question   What angle is θ_4 related to by Snell's law?
Answer   Snell's law relates θ_4 to θ_3:

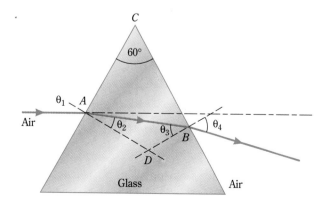

FIGURE 23.24

Light is deviated from its original direction by both faces of a prism.

$$\sin \theta_4 = \frac{n_{\text{quartz}}}{n_{\text{air}}} \sin \theta_3$$

Both angles are measured relative to the normal drawn through point B in Fig. 23.24.

Question   How can I find θ_3?
Answer   You have to recall a little geometry. First, the sum of angles in a triangle is 180°, so that

$$\theta_2 + \theta_3 + \text{angle } D = 180°$$

Also, the angles in any quadrilateral (four-sided figure) add up to 360°. Looking at the quadrilateral $ACBD$, we see that the angles at A and B are each 90°. Angle C is given as 60°. This leaves angle $D = 120°$. Combining this result with the previous equation gives the relation between θ_2 and θ_3:

$$\theta_2 + \theta_3 = 180° - 120° = 60°$$

Question   What determines θ_2?
Answer   Snell's law applied to point A will give you θ_2 if you know θ_1. Since each angle of the prism is 60° and the incident beam is parallel to the base, you should be able to conclude that $\theta_1 = 30°$.

Solution and Discussion   First, look up n_{quartz} in Table 23.2. Starting with θ_1 we find θ_2:

$$\sin \theta_2 = \frac{1.00}{1.46} \sin 30° = 0.342$$

$$\theta_2 = 20.0°$$

Then $\theta_3 = 60° - 20.0° = 40.0°$, and

$$\sin \theta_4 = \frac{1.46}{1.00} \sin 40.0° = 0.938$$

$$\theta_4 = 69.8°$$

Be sure you understand why the rays at A and B are bent as shown in Fig. 23.24.

Exercise   Suppose the same fused quartz prism is surrounded by flint glass instead of air. Sketch the passage of the same incident beam through the prism.
Answer:

23.10   TOTAL INTERNAL REFLECTION

Diamonds owe a great deal of their beauty to the optical phenomenon called **total internal reflection.** This phenomenon is also responsible for the ability of glass fibers to carry light and direct it around corners. Optical fibers are used in many

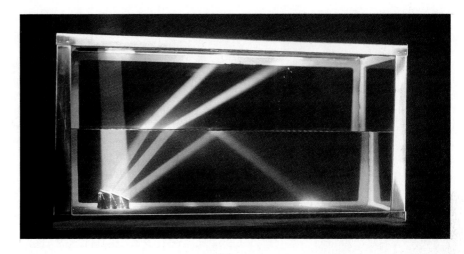

Light rays from a medium of greater index of refraction on the bottom refract into a medium of less index of refraction, bending away from the normal to the boundary as they do so. If the angle of incidence is sufficiently large, there is no refracted ray, and the incident ray is totally reflected.

important practical applications, including fiber-optic diagnostic tools in medicine and fiber-optic cables that are a communications revolution still in progress.

To understand total internal reflection, we begin by considering the passage of light from one medium into a second medium having a *smaller* index of refraction. For example, Fig. 23.25 shows a light source O below the surface of a pond. As ray B passes from water into air, it is refracted away from the normal to the water surface. Of course, some reflection also occurs at the surface, and B' is the reflected ray. The energy carried by the incident ray is divided between the reflected and refracted rays. We now consider a ray C that is incident at a greater angle to the normal. It turns out that the reflected ray C' carries a greater portion of the incident energy than the ray B', which was reflected closer to the normal.

There is a limiting ray, shown in Fig. 23.25b, for which the refracted ray is parallel to the surface ($\theta_2 = 90°$). This occurs at a *critical angle* of incidence, θ_c. If the incident angle is greater than θ_c, no refracted ray is possible. *All* the incident light is reflected back into the water, for total internal reflection. Snell's law gives us the value of the critical angle for any pair of media:

$$n_1 \sin \theta_c = n_2 \sin 90° = n_2 \, 1.00$$

Thus

$$\sin \theta_c = \frac{n_2}{n_1} \qquad \text{or} \qquad \theta_c = \sin^{-1}\frac{n_2}{n_1} \tag{23.7}$$

FIGURE 23.25

When θ_1 is greater than the critical angle θ_c, the beam undergoes total internal reflection.

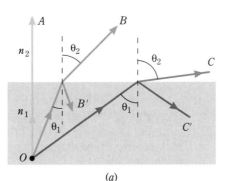

(a)

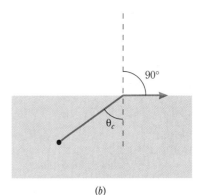

(b)

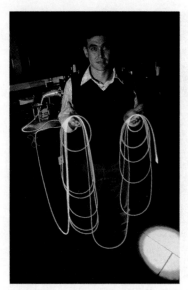

The ability of optical fibers to contain and conduct light by internal reflection is vividly demonstrated here. A tiny fraction of the laser light entering this fiber is scattered from microscopic irregularities in the fiber and its surface, which makes the fiber itself visible. Notice the great intensity of transmitted light exiting the end of the fiber and shining on the floor at the lower right.

It is very important to remember that total internal reflection can occur only when $n_2 < n_1$. No angle has a sine greater than 1, and so Eq. 23.7 has no solution unless $n_2 < n_1$.

When medium 2 is air, you can easily verify that $\theta_c = 49°$ for water, 41° for crown glass, and 24.4° for diamond. Light from any direction can enter a diamond (there is no critical angle for light to be refracted into a material of greater index of refraction), but the light can emerge only at angles close to the normal to a facet. Thus light is reflected internally many times before it escapes. Diamond cutters cut a great number of facets on each stone. Because light encounters many internal reflections, each facet eventually receives a portion of the incident light at an angle smaller than 24.4°. As you turn a diamond in your hand, it seems to sparkle because the light you see emerges nearly perpendicularly to each of the many facets.

Illustration 23.2

Figure 23.26 shows light incident upon a right-angle isosceles prism made of glass. Show that the light is totally reflected internally and emerges at an angle of 90° to its original direction. Assume the prism is surrounded by air.

Reasoning   At point A the light enters the prism without being bent, since it strikes the surface along a normal ($\theta_1 = 0°$). Since the prism has angles of 45°, the angle of incidence at B is 45°. The critical angle for a glass-air boundary, however, is $\sin^{-1}(1/1.52) = 41.1°$. Thus no part of the incident ray can be refracted into the air at B. The angle of reflection at B is equal to the angle of incidence, and so the ray is 100 percent reflected at 90° to its original direction. This ray strikes the glass-air boundary at C along the normal direction, and so emerges straight through as shown in Fig. 23.26.

Internal reflection is indeed total, more perfect than any silvered coating can produce. This kind of prism is therefore used in binoculars to make a precise right-angle change in the path of the light.

Exercise   If the glass prism is immersed in water, will the light still be bent in a right angle?   *Answer: The critical angle for a glass-water boundary is 61° (show this), so there would be a refracted ray into the water at point B. Some light would still be reflected as before, but much reduced in intensity.*

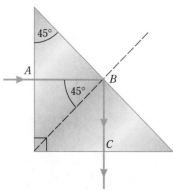

FIGURE 23.26

Light incident perpendicular to a face of a glass right angle prism undergoes total internal reflection and emerges at 90° to its original direction.

Total internal reflection makes it possible to "pipe" light around corners. Light that enters one end of a gently curved glass rod undergoes total internal reflection around the curve, as shown in Fig. 23.27a. When a bundle of such curved rods (which are called *optical fibers*) is used, the composite picture of an object can be piped from one place to another. Such a device is called a light pipe (Fig. 23.27b).

In recent years, optical fibers have been used in telecommunications. In this application, modulated laser beams are used to carry the signals in place of the electric currents and radio waves of the method formerly used by telephone companies. Such an application has been made feasible by the development of fibers with extremely low energy losses. Because the frequency of light waves is much higher than that of conventional electric currents and radio waves, much more information can be transmitted per unit time on an optical beam in a fiber than through a conventional wire or on a comparable radio wave beam.

FIGURE 23.27

(*a*) Light is caused to follow a glass fiber by total internal reflection. (*b*) A glass-fiber gastroscope attached to a camera. A light source (outside the picture at the left) supplies light to the fiber bundle at the bottom. This light pipe is inserted through the throat into the stomach. Light reflected from the stomach wall is reflected up through the central fibers of the bundle and forms an image on the film of the camera. Often the camera is not used and the light is observed directly by eye. (*American Optical Corp., Fiber Optics Div.*)

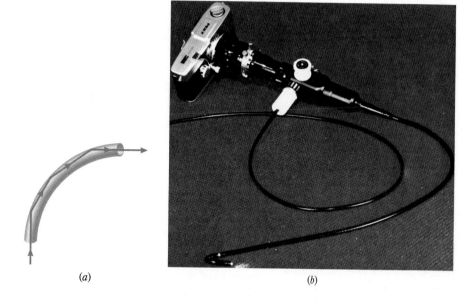

(*a*)

(*b*)

23.11 SPHERICAL LENSES

The phenomenon of refraction finds its most useful application in lenses and their ability to form images. A properly constructed lens is capable of focusing a beam of parallel light rays into a small region at a focal point. To see how this is done, let us consider how Snell's law applies to the refraction of light incident upon a spherical surface.

Figure 23.28 shows cross sections of glass spheres and some incident rays paral-

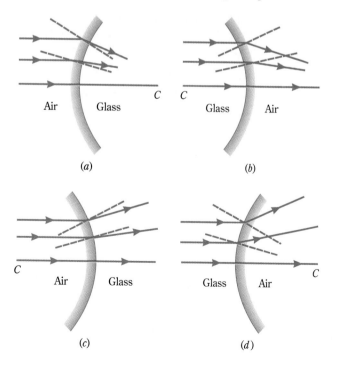

(*a*)

(*b*)

(*c*)

(*d*)

FIGURE 23.28

Refraction at different points of a spherical air-glass boundary.

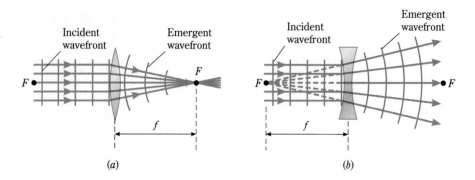

FIGURE 23.29

(a) Parallel rays are converged to the focal point by the converging lens. (b) They are diverged and appear to come from the focal point in a diverging lens.

lel to the principal axis of the spheres. The center of curvature of the spheres in each case is shown as the point C. We can see qualitatively that the effect of refraction at various points on the surfaces is either to converge the rays toward the axis (Fig. 23.28a and b) or to diverge them away from the axis (Fig. 23.28c and d). Although rays are shown striking only the upper half of the surfaces, the situation is symmetrical and there are undrawn rays striking the lower half as well.

Now let us form two lenses by combining the surfaces of Fig. 23.28a and b and combining those of Fig. 23.28c and d. The result is shown in Fig. 23.29. Let us call the space on the side of the lens containing the incident rays the "front" of the lens and the space containing the refracted rays "behind" the lens. Although we do not prove it here, if we use only a small part of the spherical surface, the refracted rays in Fig. 23.29a converge to a point F behind the lens and those in Fig. 23.29b diverge in a pattern that appears to come from the point F in front of the lens. These lenses are called **converging and diverging lenses,** respectively. The points F are called the focal points of the lenses, and the distance f from the center of the lens to F, measured along the principal axis, is called the **focal length** of the lens.*

The focal point of a converging lens is the point at which rays incident in a direction parallel to the principal axis converge after passing through the lens. The focal point of a diverging lens is the point from which rays incident in a direction parallel to the principal axis appear to diverge after passing through the lens.

Since light can pass through a lens in either direction, the lens has focal points on both sides. If the lens is a *thin* lens, i.e., if its thickness is much less than its focal length, these focal points will lie at equal distances on the two sides of the lens.

There is an alternate way of describing the refractive property of a lens. We learned earlier that refraction at a boundary is a consequence of the different speeds that light had in the two media. In a converging lens the central portion of the

---

*The focal length of a lens is determined by more factors than is that of a mirror, since a lens has two curved surfaces, and the amount of refraction at these surfaces depends also on the index of refraction of the lens relative to the surrounding medium. The expression which relates all these factors is known as the **lensmaker's equation:**

$$\frac{1}{f} = (n - 1)\left(\frac{1}{R_1} - \frac{1}{R_2}\right)$$

Here $n = \dfrac{n_{\text{lens}}}{n_{\text{surr medium}}}$, and R_1 and R_2 are the radii of curvature of the front and back surfaces, respectively. These radii are positive when the surface they represent is convex toward the light and negative when concave toward the light. This equation then gives the correct sign of f for all the lens shapes shown in Fig. 23.30.

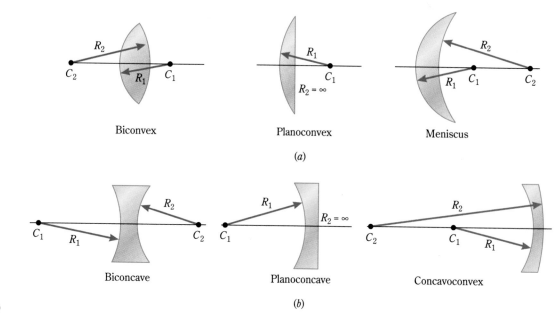

FIGURE 23.30

(a) Various types of converging lenses.
(b) Various types of diverging lenses.

incident plane wave falls behind the outer portions because the central portion travels a greater distance through the glass. As a result, the emerging wavefront is curved as shown in Fig. 23.29a. The rays, always perpendicular to the wavefronts, thus converge toward the focal point F. For a diverging lens, the outer portions of the wave are delayed more than the central portion, giving the emerging wavefronts the opposite curvature, as shown in Fig. 23.29b. This perspective allows us to generalize the distinction between converging and diverging rays beyond the two types shown in Fig. 23.29.

1 Spherical lenses consist of portions of the surfaces of two spheres.
2 If the lens is thicker at the principal axis than at the edges, it is converging. If it is thinner at the principal axis than at the edges, it is diverging.

Examples of both types of lenses are shown in Fig. 23.30, along with the centers and radii of curvature of their surfaces.

23.12 RAY DIAGRAMS FOR THIN LENSES; THE THIN-LENS FORMULA

Just as we did with mirrors, we can use three special rays to locate the image formed by a thin lens. Ray 1 we have already seen, in Fig. 23.29; it is the ray traveling parallel to the principal axis. It is refracted to the focal point behind the lens by a converging lens and refracted in a direction away from the focal point in front of the lens by a diverging lens. Ray 1 is shown in red in Fig. 23.31. Ray 2 is the ray that goes through the front focal point before striking a converging lens or is directed toward the focal point behind the lens before striking a diverging lens. In both cases, ray 2 emerges from the lens parallel to the principal axis, as shown by the green ray in Fig. 23.31. Ray 3 (the blue ray in Fig. 23.31) goes straight through the center of the lens without deflection. It is easy to see why it behaves

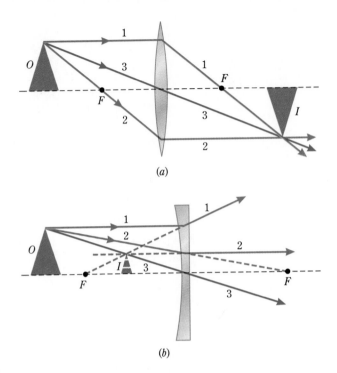

(a)

(b)

FIGURE 23.31

As with mirrors, three special rays are used to locate the image formed by the lens.

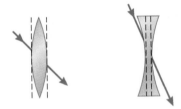

FIGURE 23.32

The ray passing through the center of the lens essentially passes through a flat plate (defined by the broken lines) and is therefore not deviated. A small displacement of the ray occurs, but this is not shown in the figure. Why is this displacement negligible for a thin lens?

this way by referring to Fig. 23.32. Notice that the ray enters and leaves the lens at surfaces that are parallel to each other. Hence the ray behaves as if it has gone through a flat plate of glass. Recall from Example 23.5 that a ray of light is not deviated in direction by a plate that has parallel faces. It is displaced slightly, but this effect is negligible when we ignore the thickness of the lens.

Any two of these three rays allow us to locate the position of the image of a given object. Notice in Fig. 23.31a that the image is real since the three refracted rays converge. A screen placed there would catch the light and show the image. In Fig. 23.31b, however, the image is virtual, since the refracted rays diverge in a pattern appearing to come from a point in front of the lens. This point is the position of the virtual image, but a screen placed there would not actually show an image.

Illustration 23.3

A converging lens of focal length 10.0 cm is used to form an image of an object placed 5.0 cm in front of the lens. Draw a ray diagram to locate the image.

Reasoning The appropriate ray diagram is shown in Fig. 23.33. Notice that an eye looking at the refracted rays from behind the lens will assume that the rays come from the position indicated. The image is virtual, upright, and enlarged.

Illustration 23.4

A diverging lens of focal length −10.0 cm is used to form an image of an object placed 5.0 cm in front of the lens. Find the image by means of a ray diagram.

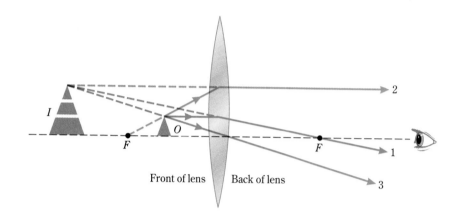

FIGURE 23.33

Virtual images are formed by convex lenses when the object is inside the focal point. The eye sees rays that appear to come from image I.

Front of lens   Back of lens

Reasoning   The appropriate ray diagram is shown in Fig. 23.34. Here too the image is virtual. It is upright and diminished in size.

THE THIN-LENS FORMULA

Ray diagrams are a useful way of sketching the object-image relationship, but we would like to develop an analytical way of treating this relationship. To start, consider the image formed by the converging lens in Fig. 23.35. In part (*a*), triangles *ABH* and *EBD* are similar, and so we can write

$$\frac{I}{O} = \frac{i}{p}$$

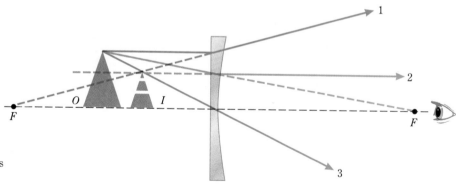

FIGURE 23.34

The image formed by a diverging lens is always virtual if the object is real.

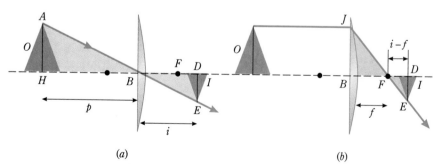

FIGURE 23.35

Triangles *ABH* and *EBD* are similar, as are triangles *JFB* and *EDF*.

(*a*)   (*b*)

Here we have used the same notation for object and image distances as we did for mirrors. From the two similar triangles *JFB* and *EDF* in (*b*), we get

$$\frac{I}{O} = \frac{i-f}{f}$$

Equating these two expressions and simplifying yields

$$\frac{1}{p} + \frac{1}{i} = \frac{1}{f} \tag{23.2}$$

This relation is exactly the same as the mirror equation, hence we have assigned it the same number. We have assumed p to be positive for an object in front of the lens and i to be positive for a real image formed behind the lens.

We can derive the relationship applicable to diverging lenses by referring to sets of similar triangles in Fig. 23.36. We find

$$\frac{I}{O} = \frac{i}{p} \qquad \text{and} \qquad \frac{I}{O} = \frac{f-i}{f}$$

Equating these expressions and simplifying gives

$$\frac{1}{p} - \frac{1}{i} = \frac{-1}{f}$$

We can make this equation identical with Eq. 23.2 if we make the following agreement on the signs to be used for f, p, and i:

To use Eq. 23.2 for all lens situations:

1 The object distance p is positive if the object is in front of the lens and negative if the object is behind the lens. (Negative object distances are treated in the following section.)
2 The image distance i is positive if the image is formed behind the lens (a real image) and negative if the image is formed in front of the lens (a virtual image).
3 The focal length is positive for a converging lens and negative for a diverging lens.

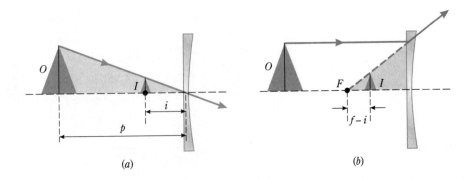

FIGURE 23.36

Consideration of the similar triangles leads to the thin-lens equation for diverging lenses.

(*a*)   (*b*)

With the above sign agreement, we can define magnification as we did with mirrors:

$$M = -\frac{i}{p}$$ (23.3a)

Once again, the minus sign allows us to identify inverted images as those having negative values of M and upright images as those having positive values of M.

The following general observations are helpful when dealing with lens problems:

1 A diverging lens *always* forms a virtual, upright, reduced image of a real object, regardless of where the object is placed in front of the lens.
2 A converging lens forms a real, inverted image of a real object if the object is placed outside the focal point of the lens. If the object is inside the focal point, a virtual, upright image is formed.

Example 23.7

A diverging lens with a 20-cm focal length forms an image of a 3.0-cm-high object placed 40 cm in front of the lens. Find the image position and magnification. Is the image upright or inverted?

R

Reasoning

Question What are the correct signs for the given quantities f and p?
Answer The object is in front of the lens, and so $p = +40$ cm. Because the lens is diverging, $f = -20$ cm.

Question What must I know to find the magnification of the image and to determine whether it is upright or inverted?
Answer The magnification I/O is equal to $-i/p$. The sign of this tells you whether the image is upright or inverted.

Question   Is there any way I can anticipate whether the image is upright or inverted?
Answer Since you have a real object and a diverging lens, you should expect to find a virtual image, upright and reduced in size.

Solution and Discussion The lens equation gives

$$\frac{1}{i} = \frac{-2 - 1}{40 \text{ cm}} = \frac{-3}{40 \text{ cm}}$$

$$i = \frac{-40}{3 \text{ cm}} = -13.3 \text{ cm}$$

The minus sign tells us the image is virtual, in front of the lens. The magnification is

$$M = -\frac{-13.3 \text{ cm}}{+40 \text{ cm}} = \frac{+1}{3}$$

The plus sign tells you that the image is erect. The size of the image is then

$$I = \tfrac{1}{3}O = \tfrac{1}{3}(3.0 \text{ cm}) = 1.0 \text{ cm}$$

23.13 COMBINATIONS OF LENSES

Most optical instruments contain more than one lens. These *lens systems* are easy to deal with if we proceed in a systematic fashion. Let us determine the final image formed by the two lenses in Fig. 23.37a. The object is 20 cm from the first lens, which is 30 cm from the second lens. Both lenses are converging. As a first step, let us ignore the second lens completely and find the image formed by the first lens. The ray diagram locates this image as I_1 in Fig. 23.37b. Using the lens equation we have, using centimeters,

$$\frac{1}{20} + \frac{1}{i_1} = \frac{1}{10}$$

$$i_1 = 20 \text{ cm}$$

We now use this image formed by the first lens as the object for the second lens. To see that this approach is valid, notice (Fig. 23.37c) that the rays incident on the second lens are the same rays an object placed at I_1 would produce. Thus ignoring the first lens now and using I_1 as the object for the second lens, we draw the ray diagram in Fig. 23.37c. The final image is located at I_2. In this example, the final image formed by the two lenses is real and erect.

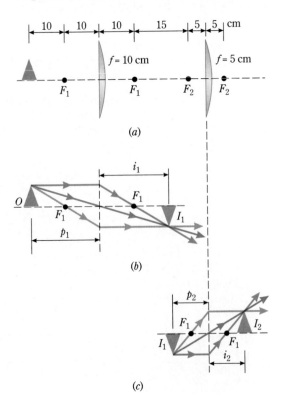

FIGURE 23.37

In order to find the image formed by a lens combination, we consider each lens in turn.

To use the lens equation for the second lens, note that the object distance p_2 is $(30\text{ cm} - 20\text{ cm}) = +10\text{ cm}$. The positive sign is consistent with this point being in front of the second lens. Then

$$\frac{1}{i_2} = \frac{1}{f_2} - \frac{1}{p_2} = \frac{1}{5\text{ cm}} - \frac{1}{10\text{ cm}} = \frac{1}{10\text{ cm}}$$

$$i_2 = +10\text{ cm}$$

To find the size of the final image, we apply our definition of magnification to each lens and multiply the individual magnifications:

$$M_{\text{tot}} = M_1 M_2 = \frac{-i_1}{p_1} \times \frac{-i_2}{p_2} = \frac{-20}{20} \times \frac{-10}{10} = 1$$

We have the unusual result that the final image is the same size as the original object. The plus sign tells us that the final image is upright relative to the original object. Each lens has formed an inverted image of its object.

With two lenses, we could have a situation where the first image is formed *behind* the second lens. For example, suppose the two lenses we just used were placed 15 cm apart instead of 30 cm. Then the rays from the first lens would still be *converging* when they encountered the second lens. This is clearly not the behavior of rays from a real object, which always *diverge*. We do not have to seek a new equation, however. As suggested in the previous section, this case can be handled by treating the object for the second lens as a **virtual object**, assigning a negative number to its distance p_2 from the second lens. *All possible lens situations can be handled with Eq. 23.2 if we carefully observe the signs agreed to in Sec. 23.12.*

Example 23.8

Find the location, size, and orientation (upright or inverted) of the image formed by the two lenses in the above discussion if they are 15 cm apart.

R | *Reasoning*

Question   Is anything about the first image changed from the previous discussion?
Answer   No. We completely ignored the second lens while treating the first lens, and so the first image is unaffected by the position of the second lens.

Question   What is the object distance for the second lens?
Answer   Since I_1 is now formed 5 cm *behind* lens 2, you must use $p_2 = -5\text{ cm}$ for the distance of this virtual object. The lens equation for the second lens is therefore $1/i_2 = 1/(5\text{ cm}) - 1/(-5\text{ cm})$. Be sure you notice how careful you have to be with the signs involved.

Solution and Discussion   The second image distance is now

$$\frac{1}{i_2} = \frac{2}{5\text{ cm}} \quad \text{or} \quad i_2 = 2.5\text{ cm}$$

The magnification is now

$$M_{tot} = \frac{-20\,cm}{20\,cm} \times \frac{-2.5\,cm}{-5.0\,cm} = -0.50$$

The image is real, inverted, and reduced.

Exercise Repeat the above example with the second lens a diverging lens with a 10-cm focal length. *Answer: $i_2 = +10\,cm$; $M_{tot} = -2$; the image is real, inverted, and enlarged.*

LENSES IN CLOSE COMBINATION

Perhaps you have had your eyes tested and noticed that the examiner often places several lenses in front of your eye at once. In order to make use of the observed best combination, he or she needs to know how to add the effects of thin lenses in close combination. The necessary simple formula can be derived readily, as we now show. We consider only the case in which the focal lengths of the lenses are much longer than the distances from one lens to the next.

The image location formed by the first lens is given by

$$\frac{1}{p_1} + \frac{1}{i_1} = \frac{1}{f_1}$$

Let us consider the case where lens 1 is converging and forms a real image. Since this image is located behind lens 1, it must also be behind lens 2 because we are considering the two lenses to be essentially at the same location. This is what we mean when we say the distance between the lenses is negligible compared to their focal lengths. Thus the first image provides a virtual object for lens 2, and we take $p_2 = -i_1$ according to our sign agreement. The lens equation for lens 2 then gives

$$\frac{1}{-i_1} + \frac{1}{i_2} = \frac{1}{f_2}$$

When we add the two lens equations together, i_1 drops out:

$$\frac{1}{p_1} + \frac{1}{i_2} = \frac{1}{f_1} + \frac{1}{f_2}$$

As can be seen in Fig. 23.38, p_1 is the location of the original object and i_2 is the location of the final image. So this equation is just the same as the lens equation for a *single lens* whose focal length f is given by

FIGURE 23.38

If the two lenses are very close together, their combined effect is to act as a single lens with a focal length $1/f = 1/f_1 + 1/f_2$.

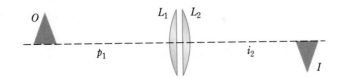

$$\frac{1}{f} = \frac{1}{f_1} + \frac{1}{f_2} \qquad (23.8)$$

Equation 23.8 can be extended to more than two lenses in contact as long as the thickness of the combination is negligible compared to each of their individual focal lengths. This equation also applies to any combination of converging and diverging lenses, as long as the correct signs of the focal lengths are used.

Illustration 23.5

Three lenses of focal lengths 20, −30, and 60 cm are placed in contact with each other. Find the focal length of the combination. Is the combination equivalent to a converging or a diverging lens?

Reasoning The effective focal length of the combination is given by Eq. 23.8:

$$\frac{1}{f} = \frac{1}{f_1} + \frac{1}{f_2} + \frac{1}{f_3} = \frac{1}{20 \text{ cm}} + \frac{1}{-30 \text{ cm}} + \frac{1}{60 \text{ cm}} = \frac{3 - 2 + 1}{60 \text{ cm}} = \frac{2}{60 \text{ cm}}$$

$$f = 30 \text{ cm}$$

Since f is positive, the combination is converging.

Exercise Show that if the third lens in the above illustration had $f = -60$ cm instead of $+60$ cm, the three lenses would effectively act like a flat plate of glass.

LEARNING GOALS

Now that you have finished this chapter, you should be able to

1 Define (a) laws of reflection and refraction, (b) wavefront, (c) ray, (d) plane wave, (e) real and virtual image, (f) real and virtual object, (g) focal point and focal length, (h) index of refraction, (i) total internal reflection, (j) critical angle, (k) object distance, (l) image distance, (m) radius of curvature, (n) converging and diverging mirrors and lenses, (o) magnification.

2 Give the approximate wavelength limits of the visible spectrum and state the approximate color associated with a given wavelength.

3 Compute the index of refraction for a medium when the speed of light in it is given, and vice versa.

4 Draw the appropriate rays for a given set of wavefronts, and vice versa. Explain why a distant source gives rise to parallel rays.

5 Draw the reflected ray when the incident ray on a smooth surface is given.

6 Use Snell's law in simple situations.

7 Show why total internal reflection occurs only when $n_2 > n_1$. Calculate the critical angle for total internal reflection for a boundary between two media whose refractive indexes are given.

8 Use ray diagrams to locate images for single spherical mirrors and thin lenses. Give the characteristics of the image in any given case.

9 Use the lensmaker's equation to calculate the focal length of a thin lens given the radii of curvature of the lens surfaces and the material the lens is made of.

10 Use the lens or mirror equation to obtain p, i, or f when any two of the three are given. Relate f to the radius of curvature of a spherical mirror. Interpret the signs of p, i, and f in any given case.

11 Determine the magnification and orientation of an image when you know p and i.

12 Tell by its shape whether a lens is diverging or converging when used in air.

13 Explain how the focal length of a concave mirror and a converging lens can be obtained by experiment.

14 Calculate the effective focal length of a number of thin lenses in contact with each other when their individual focal lengths are known.

SUMMARY

DEFINITIONS AND BASIC PRINCIPLES

Law of Reflection
Angle of incidence = angle of reflection $\theta_i = \theta_r$

Types of Objects and Images
REAL OBJECT
An object positioned in front of a lens or mirror. The incident rays from a real object form a diverging pattern.

REAL IMAGE
An image positioned behind a lens or in front of a mirror. The rays for a real image actually converge through a point.

VIRTUAL IMAGE
An image positioned in front of a lens or behind a mirror. The rays for a virtual image diverge from the image point.

VIRTUAL OBJECT
An object positioned in back of a lens or mirror. The rays for a virtual object form a converging incident pattern on the lens or mirror. This requires that these rays came from some previous lens or mirror.

Principal Rays for Concave Mirrors
A concave mirror has a focal point in front of the mirror at a distance $f = R/2$ from the mirror. Principal rays for locating the image are:

1. An incident ray parallel to the principal axis is reflected through the focal point.
2. An incident ray along a line passing through the focal point is reflected parallel to the principal axis.
3. An incident ray along a line passing through the center of curvature is reflected back upon itself.

Principal Rays for Convex Mirrors
A convex mirror has a focal point in back of the mirror at a distance $f = R/2$ from the vertex of the mirror. Principal rays for locating the image are:

1. A ray incident parallel to the principal axis is reflected along a line directed away from the focal point.
2. A ray incident along a line directed toward the focal point is reflected parallel to the principal axis.
3. A ray incident along a line directed toward the center of curvature is reflected back along the same line.

The Mirror Equation
The focal length f, object distance p, and image distance i are related by the mirror equation:

$$\frac{1}{f} = \frac{1}{p} + \frac{1}{i}$$

The following sign convention is used:

f: + for concave, − for convex

p: + for real object, − for virtual object

i: + for real image, − for virtual image

Magnification (M)

$$M = \frac{I}{O}$$

where I and O are the linear dimensions of the image and object, respectively. In terms of object and image position,

$$M = \frac{-i}{p}$$

This gives a positive value of M for an erect image, a negative value for an inverted image.

Index of Refraction (n)

$$n = \frac{\text{speed of light in vacuum}}{\text{speed of light in material}} = \frac{c}{v}$$

Light is slowed down when traveling through transparent materials, so $n > 1$ for all materials.

Law of Refraction (Snell's Law)
At any boundary between materials of refractive indexes n_1 and n_2, the directions of the incident and refracted rays are related by

$$n_1 \sin \theta_1 = n_2 \sin \theta_2$$

These angles are measured relative to the normal to the boundary at the point of incidence.

Total Internal Reflection
When light passes from a material of higher index of refraction n_1 into a medium of lower index n_2, no refraction is possible if the angle of incidence exceeds a certain critical value θ_c. The incident ray is then 100 percent reflected back in the incident material. This is called total internal reflection, and θ_c is given by

$$\theta_c = \sin^{-1} \frac{n_2}{n_1}$$

Thin Spherical Lenses and the Thin-Lens Equation
A thin lens is one whose focal length is much greater than the thickness of the lens. The two surfaces of the lens are spherical. There are two focal points, located symmetrically on both sides of the lens.

A *converging lens* is a lens thicker in the middle than at its edges. A *diverging lens* is a lens thicker at its edges than at its center.

The thin-lens equation is

$$\frac{1}{f} = \frac{1}{p} + \frac{1}{i}$$

The same sign convention is used for p and i as for mirrors. The focal-length signs are: f is $+$ for converging lenses, f is $-$ for diverging lenses.

The Lensmaker's Equation
$$\frac{1}{f} = (n-1)\left(\frac{1}{R_1} - \frac{1}{R_2}\right)$$

where n is the index of refraction of the lens material relative to the material surrounding the lens. R_1 is the radius of curvature of the front surface. R_2 is the radius of curvature of the back surface. R_1 and R_2 are $+$ if the surfaces are convex toward the incident light, $-$ if concave.

Principal Rays for Thin Lenses
CONVERGING LENSES
1. A ray incident parallel to the principal axis is refracted through the far focal point.
2. A ray incident through the near focal point is refracted parallel to the principal axis.
3. A ray incident at the middle of the lens passes straight through the lens.

DIVERGING LENSES
1. A ray incident parallel to the principal axis is refracted along a line coming from the near focal point.
2. A ray incident along a line through the far focal point is refracted parallel to the principal axis.
3. A ray incident at the middle of the lens passes straight through the lens.

Multiple Lens Systems
The lens equation is applied to each lens in the system. The image formed by the first lens acts as the object for the second, and so on. The total magnification of a multiple lens is the product of the individual magnifications of each lens.

Lenses in Close Combination
If the distance separating two lenses can be ignored compared to their focal lengths f_1 and f_2, the two lenses act as one lens, with an effective focal length given by

$$\frac{1}{f} = \frac{1}{f_1} + \frac{1}{f_2}$$

QUESTIONS AND GUESSTIMATES

1 Consider a concave mirror and an object at infinity. Where is the image formed? Is it erect or inverted? Is it larger or smaller than the object? Answer these questions as the object is slowly moved in toward the mirror. In particular, note the positions of the object at which any of the answers change.

2 Repeat Question 1 for a convex mirror.

3 Repeat Question 1 for a converging lens.

4 Repeat Question 1 for a diverging lens.

5 When you look down into a clear lake or vat of water, why does the water always appear shallower than it actually is?

6 Using a wavefront diagram, explain why a lens can be either converging or diverging depending on the material in which it is embedded.

7 Can an empty water glass focus a beam of light? A full water glass? Is it possible to start a fire by accident if a bowl of water is set in a sunlit window?

8 How can one use the mirror equation to find the position of the image of an object in a plane mirror?

9 As light goes from air to glass, which of the following change: f, λ, v?

10 Why can a smart fish in a calm lake see you on the bank by looking up at an angle of about 50° to the vertical?

11 How do so-called one-way mirrors (or windows) operate?

12 A "solar furnace" can be constructed by using a concave mirror to focus the sun's rays on a small region, the furnace region. How would you expect the temperature of the furnace to vary with the mirror area and focal length?

13 A spherical air bubble in a piece of glass acts like a small lens. Explain. Is it converging or diverging?

14 How can one determine the focal length of a converging lens? Of a diverging lens? Of a concave mirror? Of a convex mirror?

15 Two plane mirrors are placed together so that they form a right angle. An object is then placed between them. How many images are formed? Repeat for an angle of 30° between the mirrors.

16 About how much longer does it take for a pulse of light from

the moon to reach the earth because of the presence of air rather than a vacuum above the earth?

17 Newton believed that light consisted of a stream of particles and that these "light corpuscles" were strongly attracted by the water surface as light went from air to water. How would this lead to the observed effect of refraction?

18 In various science museums (as well as in some unexpected places), a room is so designed that a person can whisper at one particular point in the room and be heard clearly at a certain distant point. How must the room be constructed to achieve this effect?

PROBLEMS

Sections 23.1–23.4

1 A laser beam from earth is reflected back to the earth by a mirror on a space shuttle which is 4.2×10^6 m away. How long does it take for the beam to make the round-trip?

2 A radar beam is reflected by rain clouds that are 30 km away from the transmitting station. How long does it take the radar waves to make the round-trip?

3 Many cameras have a focusing mark that must be set to read the distance of the object from the camera. Suppose you wish to photograph yourself in a plane mirror. If you and the camera are 50 cm from the mirror, at what value should you set the distance indicator on the camera?

4 An interior decorator wishes to mount a plane wall mirror in such a way that a person 1.8 m tall will be able to see his or her full length in the mirror. What is the shortest length of mirror for this case, and how high above the floor should the bottom of the mirror be mounted?

5 If you walk toward a plane mirror at a speed of 1.2 m/s, how fast do you approach your image in the mirror?

6 A beam of light is reflected straight back on itself by a plane mirror that is perpendicular to the beam. The mirror is then rotated so that its normal makes an angle of 24° to the beam. What is the new angle between the incident and reflected beams?

7 A ray of light is reflected from a plane mirror so that the angle between the incident and the reflected ray is 64°. (a) If the mirror is rotated to increase the angle of incidence by 3°, what will be the new angle between the incident and the reflected rays? (b) If the mirror is moved to reduce the angle of incidence by 2°, what will be the new angle between the incident and the reflected rays?

8 If an object is placed between two parallel flat mirrors, an infinite number of images result. If the mirrors are 50 cm apart and the object is placed midway between them, find the distance of the first five images from the object.

∎9 A person stands inside a room with two plane parallel mirrors on opposite walls. If the person is 6 ft from one mirror and 12 ft from the other mirror, find the distance from the person to the first three images seen in the first mirror.

10 Two plane mirrors are standing perpendicular to a tabletop. The reflecting surfaces are oriented 90° relative to one another. A light ray parallel to the tabletop is reflected by one mirror and then by the second mirror. Show that the direc-

tion of the final reflected ray is exactly opposite to the original incident ray.

∎11 The two plane mirrors in Prob. 10 are reoriented so that they make an angle θ with each other. A ray of light parallel to the tabletop is incident on one of the mirrors at some angle and is successively reflected from the mirrors. Show that the angle between the incident ray and the outgoing ray is 2θ.

Sections 23.5–23.7

12 A concave mirror that has a 30-cm radius of curvature forms an image of a 2.0-cm-high object that is placed 45 cm in front of the mirror. (a) Find the position and size of the image. Is the image real or virtual? Upright or inverted? Repeat for object distances of (b) 30 cm, (c) 20 cm, and (d) 10 cm. Check with a ray diagram.

13 An object 1.0 cm high is placed 36 cm in front of a concave mirror that has a radius of curvature of 20 cm. Find the position and size of the image, and state whether the image is real or imaginary and upright or inverted. Check your calculations with a ray diagram.

14 Repeat Prob. 13 for object distances of (a) 20 cm, (b) 16 cm, and (c) 6 cm.

15 A coin 2.0 cm in diameter is held 10 cm in front of a concave mirror that has a radius of curvature of 30 cm. Find the location and size of the image of the coin. Is the image real or virtual?

16 A virtual image is formed 15 cm from a concave mirror whose radius of curvature is 30 cm. Find the position of the object.

17 A dentist uses a concave mirror of focal length 25 mm. What magnification does it produce when held 16 mm from a tooth?

18 A concave mirror that has a focal length of 120 cm is used to form a real image of an object. (a) Where must the object be placed if the image distance is equal to the object distance? (b) Are the object and image superimposed? (c) What is the magnification?

19 Where must an object be placed in front of a concave mirror whose radius of curvature is 1.00 m if the image is to be real and twice the size of the object?

20 In the previous problem where should the object be placed if the image is to be virtual and twice the size of the object?

21 Where must an object be placed if the image formed by a concave mirror is to be one-third as far from the mirror as the object? Is the image real or virtual?

22 A 2-cm-high object has a 5-cm-high virtual image when placed 3 cm from a concave mirror. What is the focal length of the mirror?

23 A concave mirror has a radius of curvature of 60 cm. Determine the position of an object for which the image will be inverted and three times the size of the object.

24 In Prob. 23 find the distance of the object if the image is to be upright and three times the size of the object.

Section 23.8

25 (a) Find the position, size, and nature (real or virtual, upright or inverted) of the image formed when a 3-cm-high object is placed 50 cm in front of a convex mirror that has a radius of curvature of 25 cm. Repeat for object distances of (b) 20 and (c) 10 cm. Check with a ray diagram.

26 What is the position of the image of an object placed 48 cm in front of a convex mirror that has a focal length of 24 cm? What is the magnification? Is the image real or virtual? Upright or inverted?

27 A virtual image is formed by a convex mirror that has 40-cm focal length. (a) Where must the object be placed if the image is to be half the size of the object? (b) Is it possible to obtain a virtual image that is larger than the object using this type of mirror?

28 A virtual image one-third of the size of the object is formed by a convex mirror of focal length 40 cm. Find the location of the object and the image.

29 You want a convex mirror of focal length 20 cm to form an image 12 cm from the mirror. What should be the location of the object? What is the magnification?

30 Where must an object be placed if the image formed by a convex mirror is to be one-half as far from the mirror as the object? How large is the magnification?

31 A wide-angle convex mirror with a 0.50-m radius of curvature is used in a grocery store to monitor the aisles. Find the position and describe the nature of the image of a customer standing in an aisle 8.0 m from the mirror. What is the magnification?

32 A shiny, spherical Christmas tree ornament is 8.0 cm in diameter. (a) What is the position of the image of a child standing 80 cm away from the ornament? (b) What is the magnification of the image?

Section 23.9

33 Yellow light from a sodium arc lamp has a wavelength of 589 nm. When a beam of this light travels through ethanol, what are its (a) speed, (b) wavelength, and (c) frequency?

34 Blue light from mercury has a wavelength of 436 nm. When a beam of this light travels through water, what are its (a) speed, (b) wavelength, and (c) frequency?

35 A beam of red light from a helium-neon laser ($\lambda = 633$ nm) enters a glass plate ($n = 1.56$) at an angle of 30° to the nor-

mal. (a) What is the speed of the beam in the glass? (b) What is its wavelength? (c) What angle does the beam make with the normal inside the glass?

36 Green light with $\lambda = 546$ nm enters water in a cup at an angle of 60° to the normal to the surface of water. (a) What is the wavelength of this light in the water? (b) At what angle to the normal does the beam travel inside the water?

37 Light enters a flat glass plate ($n = 1.56$) at an angle of 48° to the normal to the top surface. (a) What is the angle of refraction inside the glass plate? (b) After the beam leaves the bottom surface of the plate, what is the angle between it and the original beam incident on the plate?

38 How far does a beam of light travel through water ($n = 1.33$) in the same time it takes to travel 1 m in glass ($n = 1.56$)?

39 The index of refraction of glass varies slightly for different wavelengths. Flint glass has an index of refraction of 1.650 for blue light ($\lambda = 430$ nm) and 1.615 for red light ($\lambda = 680$ nm). A beam of light consisting of these two colors is incident on a flint glass plate at an incidence angle of 45°. Find the angle between the two colored beams inside the glass.

40 At what angle of incidence should a beam of light strike the flat surface of a glass plate ($n = 1.56$) if the refracted ray is to be perpendicular to the reflected ray?

41 A swimmer in a swimming pool notices that a beam of sunlight in water makes an angle of 27° with the vertical. At what angle does the incident beam in air strike the surface of water? Assume that the water surface is level and horizontal.

42 A beam of light is incident at an angle of 24° on the surface of a liquid. If the light travels with a speed of 2.3×10^8 m/s through the liquid, what is the angle of refraction of the beam inside the liquid?

43 A ray of light initially in water enters a transparent substance at an angle of 36° to the normal. The refracted ray inside the transparent material makes an angle of 24°. Find the speed of light inside the transparent material. Assume that the surface of contact between the water and transparent material is level and flat.

44 A scuba diver is 280 m horizontally away from the shore and 90 m beneath the surface of water. The diver shines a laser beam toward the water surface such that the beam strikes the surface of water at a point 190 m from the shore and after refraction just strikes the top of a building standing at the water's edge on the shore. What is the height of the building?

45 Viewed from directly above, a coin at the bottom of a swimming pool appears to be 2.4 m below the surface. What is the actual depth of the pool?

Section 23.10

46 A light beam from a source 4 m below the surface of a calm pool of water shines upward toward the surface. What is the maximum angle that the beam in the water can make with the vertical if part of the beam is to be able to escape into the air?

47 What is the critical angle for light inside a piece of diamond that has an index of refraction of 2.42 when (a) the diamond is surrounded by air and (b) the diamond is immersed in water?

48 The critical angle for a certain mineral surrounded by air is found to be 41°. What is the speed of light in the mineral?

49 A beam of light is incident from air on the surface of a liquid. The angle of incidence is 36° and the angle of refraction inside the liquid is 25°. Find the critical angle of the liquid with respect to air.

▪50 A rectangular fish tank filled with water is made of flat, vertical glass walls of a transparent plastic material with refractive index of 1.6. What is the maximum angle of incidence for a light ray within the water to strike the plastic wall and still emerge to the outside air?

▪51 A plastic fiber used for fiber-optic signal transmission has an index of refraction of 1.54. What is the minimum angle of incidence with respect to the normal to the cylindrical wall of the fiber which produces total internal reflection if the fiber is in (a) air and (b) water?

▪52 A lamp is placed in the center at the bottom of a 2.0-m-deep large swimming pool. The lamp emits light in all directions. Starting from a point directly above the lamp, a man in a canoe paddles until he no longer can see the lamp. How far did he paddle the canoe? Assume that the sides of the pool do not reflect light.

Sections 23.11 and 23.12

53 A converging lens with a focal length of 40 cm forms an image of a 2.0-cm-high object. Find the position, size, and nature of the image for the object distances (a) 100 cm, (b) 60 cm, and (c) 30 cm. Check your answers with ray diagrams.

54 Find the position, size, and nature of the image formed by a diverging lens with a focal length of −30 cm if the 2.0-cm-high object is placed at a distance of (a) 80 cm, (b) 50 cm, and (c) 15 cm. Check with ray diagrams.

55 An image of a 4.0-cm-high statuette is formed by a diverging lens of focal length −40 cm. Find the position and nature of the image and its magnification for the following object distances: (a) 90 cm, (b) 40 cm, and (c) 15 cm. Check with ray diagrams.

56 A converging lens of focal length 45 cm is used to form an image of a 3-cm-high object. Find the position, size, and nature of the image if the object is located at (a) 120 cm, (b) 90 cm, and (c) 20 cm. Check with ray diagrams.

57 A virtual image of an object placed 40 cm from a lens is formed 20 cm from the lens. (a) What is the focal length of the lens? (b) Is the lens converging or diverging?

58 An object is placed 60 cm to the left of a diverging lens. The image is formed at 30 cm to the left of the lens. Determine the focal length of the lens. What is the magnification?

59 A converging lens of focal length 2.4 cm is used to examine a biological sample on a microscope slide. The lens forms an image of the sample 12.6 cm from the slide. How far is the lens from the slide if the image is (a) real and (b) virtual?

▪60 A diverging lens of focal length −30 cm forms a virtual image that is one-half as far from the lens as the object is. (a) What must the object distance be? (b) How tall is the image relative to the height of the object?

▪61 (a) Where must an object be placed with respect to a converging lens if the image is to be one-third as large as the object? (Express your answer in terms of focal length f of the lens.) (b) Is the image real or virtual?

▪62 A diverging lens is used to form an image that is half the size of the object. Find the position of the object in terms of the focal length of the lens.

▪63 It is desired to use a single lens to form a virtual image of an object three times as tall as the object. (a) What kind of lens should be used? (b) Where should the object be placed? (Express in terms of the lens focal length f.)

▪64 An object is placed at a distance of 8f from a divergent lens of focal length f. (a) Find the position of the image. (b) What is the magnification?

▪65 Repeat Prob. 64 for a convergent lens.

Section 23.13

66 Two identical converging lenses with f = 30 cm are 80 cm apart. (a) Find the final image position for an object placed 100 cm in front of the first lens. (b) How large is the total magnification of the system? (c) Is the image real or virtual? Upright or inverted?

67 An object is placed 15 cm in front of a converging lens of focal length 10 cm. At a distance of 50 cm beyond the first lens is a diverging lens of focal length −8 cm. (a) Find the final position and magnification of the image formed by the combination. (b) Is the image real or virtual? Upright or inverted?

68 A converging lens of focal length 24 cm is placed 36 cm in front of a diverging lens of focal length −36 cm. A small object is placed 6 cm in front of the converging lens. Find the (a) position and (b) magnification of the final image.

69 A 1-cm-high object is placed 6 cm to the right of a converging lens with a focal length of 12 cm. A diverging lens with a focal length of −24 cm is placed 10 cm to the left of the converging lens. Find the position and size of the final image. Is the image real or virtual? Upright or inverted?

70 Two converging lenses of focal lengths 10 cm and 20 cm respectively are placed 50 cm apart. The image of an object is to be formed between the two lenses 30 cm beyond the first lens. (a) Where should the object be placed to form this image? (b) What is the final magnification? (c) Is the final image upright or inverted?

General Problems
(Problems 71 and 72 refer to the lensmaker's equation. See the footnote in Sec. 23.11.)

⁼71 Suppose you made a flint glass biconvex lens having spherical surfaces with radii of $R_1 = 100$ cm and $R_2 = 200$ cm. (a) What would the focal length of the lens be in air? (b) When surrounded by water?

⁼72 A diverging lens has a focal length of -55 cm when immersed in benzene. Its focal length when surrounded by air is -15 cm. If the front surface is flat ($R_1 = \infty$), what is the radius of curvature of the back surface?

⁼73 A rectangular tank filled with benzene has a colored stone at its bottom. The apparent depth of the stone when viewed directly from above is 40 cm. Find the real depth of the tank.

⁼74 An object is located 60 cm from a screen. A converging lens is placed between the object and the lens to form an image of the object on the screen. If the focal length of the lens is 12 cm find (a) the position of the lens and (b) the final magnification. Is there more than one answer to (a) and (b)?

⁼75 Show that the lens and mirror equations can be expressed in an alternative form as $s_i s_o = f^2$, where s_o and s_i are the distances of the object and the image from the focal point.

⁼⁼76 An object is placed a distance D from a screen. A converging lens is used to form an image of the object on the screen. Show that there are two possible positions of the lens x distance from the object, given by

$$x = \frac{1}{2}D\left(1 \pm \sqrt{1 - \frac{4f}{D}}\right)$$

Under what conditions will no image be formed?

⁼⁼77 An object is placed 40 cm in front of a converging lens of focal length 30 cm, which in turn is 80 cm in front of a plane mirror. Find all the images formed by this system and state whether each is real or virtual.

⁼⁼78 A diverging lens with a focal length of -10 cm is placed 20 cm to the left of a concave spherical mirror that has a radius of curvature 25 cm. If an object is placed 10 cm to the left of the lens, find all the images formed by the system and state whether each is real or virtual.

CHAPTER 24

WAVE OPTICS: INTERFERENCE AND DIFFRACTION

In the preceding chapter, we discussed the behavior of lenses and mirrors, using the concept of light rays. We did not need to know whether the light consisted of particles or waves for these discussions. This is not true of the topics treated in this chapter. We will see that the wave nature of light gives rise to interference phenomena much like those we encountered in our study of mechanical wave motion such as sound and waves on a stretched string. The mere existence of these phenomena, as well as other effects discussed in this chapter, led to the acceptance of the wave nature of light.

24.1 HUYGENS' PRINCIPLE AND DIFFRACTION

Have you ever watched carefully as gentle water waves lap against a post or some other obstacle in their path? If you have, you have noticed that the waves seem to bend around the post instead of casting a clear shadow of it. A related situation is shown in Fig. 24.1, where we see plane water waves generated in a ripple tank. They strike a barrier that has a small hole in it. Notice how the waves pass through the hole and spread out to fill the whole region beyond the barrier.

This general type of behavior can be observed not only with water waves but also with sound waves and electromagnetic waves. It is a characteristic behavior of waves, and we give it a special name, diffraction:

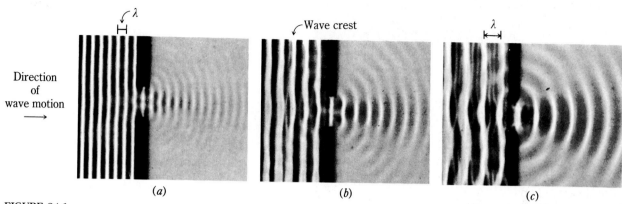

FIGURE 24.1

Plane water waves incident on a hole in a barrier. Diffraction causes the waves to spread into the entire region beyond the barrier. Notice that when the wavelength λ becomes approximately equal to the diameter of the hole, diffraction becomes more pronounced.

Waves are capable of bending around behind obstacles, a phenomenon called **diffraction.** In other words, obstacles do not cast perfectly sharp shadows of incident waves.

To explain diffraction, Christian Huygens, a contemporary of Newton's, postulated what is now known as **Huygens' principle:**

Each point on a wavefront acts as a source of new wavelets which spread out in all directions from that point at the speed of the wave in the medium.

For example, in Fig. 24.1, the portion of the wave crest that strikes the tiny hole in the barrier acts as a new wave source. As a result, waves spread outward from the hole and fill the whole region beyond the barrier.

Diffraction seems to disagree with what we know about light waves, because objects in the path of light cast shadows that are easily observed. A clue to the resolution of this discrepancy can be found in Fig. 24.1. Notice that in (a) the wavelength λ is only about one-third as large as the width of the opening and the waves spread only slightly into the region of shadow. When the wavelength is larger, however, as in (c), the waves spread into the shadow region much more. We will return to this phenomenon later.

24.2 INTERFERENCE

An interesting experiment involving water waves is shown in Fig. 24.2. There two oscillators act as point sources, sending two sets of identical water waves out along the water surface. Notice what happens as the waves from the two sources interact with one another. Along certain lines that radiate from the midpoint between the sources (labeled B), the interaction creates very large wave crests; along other lines (labeled D), no wave crests are seen. Apparently the waves from the two sources reinforce one another at certain points and cancel one another at other points. Let us now investigate such phenomena.

Recall from Chaps. 14 and 15 that identical waves can either reinforce or cancel each other. To review this fact, consider the two waves A and B in Fig. 24.3. In part (a), the two waves are in phase, crest on crest and trough on trough. When they are added, the resultant wave is twice as large as the original waves. The waves in (a) undergo **constructive interference.**

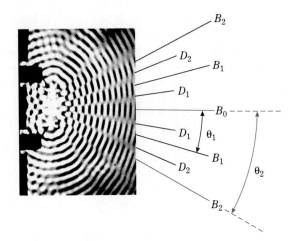

FIGURE 24.2

Two synchronized sources send out identical water waves that interfere constructively along the lines labeled *B* and destructively along those labeled *D*.

The situation in (*b*) is quite different, however. There wave *B* has been held back through one-half wavelength, $\lambda/2$, so that crest falls on trough for the two waves. The waves are $\lambda/2$, or 180°, out of phase. Now when they are added, they exactly cancel each other, and so their sum wave is zero. The waves undergo **destructive interference.**

In (*c*), wave *B* lags behind wave *A* by a distance equal to λ. The two waves are now λ, or 360°, out of phase. Now crest falls on crest, and so the two waves add to give a resultant wave that is twice as large. The waves again interfere constructively.

In general, we conclude (as we did in the case of mechanical waves) that *two identical waves interfere constructively if they are in phase with each other*. If one of the waves is retarded by a distance λ, 2λ, 3λ, and so on, relative to the other, they will still reinforce when combined because crest will still fall on crest. If the relative

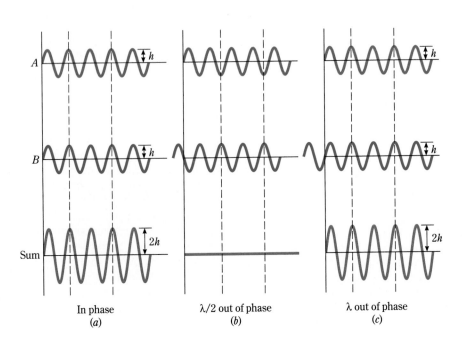

FIGURE 24.3

Identical waves can reinforce or cancel each other, depending on their relative phases.

retardation is $\lambda/2$, $3(\lambda/2)$, $5(\lambda/2)$, and so on, then crest will fall on trough and the two waves will interfere destructively; they will cancel each other.

Let us now return to our discussion of the combined effect of two wave sources. We wish to find out why the waves from these sources reinforce in certain regions and cancel in others. This question is easily settled if we refer to Fig. 24.4. The two sources are located at A and B. They send out identical waves in all directions. Consider first the waves they send in the horizontal direction, as shown in (a). These waves are in phase, crest on crest and trough on trough, and hence reinforce each other. This is why reinforcement occurs along line B_0 in Fig. 24.2.

Consider next the waves sent out in the direction shown in Fig. 24.4b. In this direction, the wave from B is retarded by $\lambda/2$ relative to the wave from A. Now the crests of one wave fall on the troughs of the other. As a result, the waves leaving the sources in this direction cancel, as along the two lines labeled D_1 in Fig. 24.2. (You should be able to explain why there are two lines labeled D_1.)

If the angle θ in Fig. 24.4 is increased, wave B will be held back still more relative to wave A. But if θ and thus the relative retardation are increased until the relative retardation is λ, the two waves reinforce each other again, as along lines B_1 in Fig. 24.2.

Reasoning in this way, you can convince yourself that the relative retardation is $3(\lambda/2)$ along cancellation lines D_2 and 2λ along reinforcement lines B_2. As we see, B_0, B_1, B_2, and all similar lines represent lines along which the waves from the two sources reinforce each other. Along them, the relative retardation between the waves from the two sources is 0, λ, 2λ, and so on.

Let us now obtain a mathematical relationship for the angles at which these reinforcing lines occur. To do this, examine the small shaded triangle in Fig. 24.4.

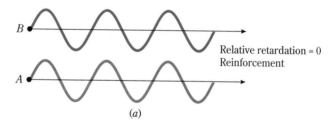

(a)

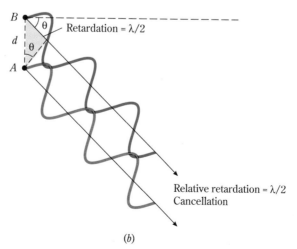

(b)

FIGURE 24.4

Reinforcement occurs at those angles for which the relative retardation is λ, 2λ, 3λ, and so on. Cancellation occurs at those angles for which the relative retardation is $\lambda/2$, $3\lambda/2$, $5\lambda/2$, and so on.

Notice that the angle θ in it is equal to the angle θ that the rays make with the horizontal. We see at once that in the shaded triangle,

Relative retardation $= d \sin \theta$

where d is the distance between the sources.

To find the angles at which reinforcement occurs, we recall that, for reinforcement, the relative retardation must be 0, λ, 2λ, 3λ, or, in general, $n\lambda$, where n is an integer. Therefore, if θ_n is the angle for which the relative retardation is $n\lambda$, we have

$$n\lambda = d \sin \theta_n \qquad (24.1)$$

For example, along line B_0 in Fig. 24.2, we have $n = 0$ (because the waves are not retarded relative to each other), and so $0 = d \sin \theta_0$ and $\theta_0 = 0$. Similarly, along line B_2 we have $n = 2$, and so $2\lambda = d \sin \theta_2$.

Illustration 24.1

Suppose the sources in Fig. 24.2 are 2.0 cm apart and the wavelength is 0.70 cm. What is the angle at which the reinforcement line B_2 would occur?

Reasoning   We are told that $d = 2.0$ cm and $\lambda = 0.70$ cm, and we are interested in the situation for which $n = 2$. Substituting in Eq. 24.1 gives

$$\sin \theta_2 = \frac{(2)(0.70 \text{ cm})}{2.0 \text{ cm}} = 0.70$$

from which $\theta_2 = 44°$. The lines B_2 in Fig. 24.2 would make $44°$ angles with the horizontal.

Exercise   At what angle would line B_1 be found?   *Answer:*   20.5°

24.3   YOUNG'S DOUBLE-SLIT EXPERIMENT

The experiment described in Sec. 24.2 on the interference of waves from two sources is not peculiar to water waves. You will recall from Sec. 15.8 that two identical sound sources can give rise to interference in sound waves. The explanation for this phenomenon is similar to the description of the interfering water waves, except that the soundwaves are longitudinal rather than transverse. Any identical waves, transverse or longitudinal, are capable of exhibiting interference phenomena.

As we mentioned in Chap. 23, Newton believed light to be corpuscular. He pictured light as a stream of particles emitted by light sources, traveling in straight lines. Although the Italian scientist Grimaldi had shown as early as 1660 that light can be diffracted, Newton was able to explain this observation in terms of his light corpuscles. His explanation was not very satisfying, but most people respected him highly and therefore accepted it. It was not until after 1803 that the wavelike nature of light became widely accepted.

In 1803 and 1807, the Englishman Thomas Young (1773–1829) published the results of his experiments demonstrating the interference of light waves. He al-

FIGURE 24.5

The two slits S_1 and S_2 act as sources for two waves synchronized in phase. In the case of light waves, the interference fringes are usually only a few millimeters apart. (Compare this experiment with Fig. 24.2 for water waves.)

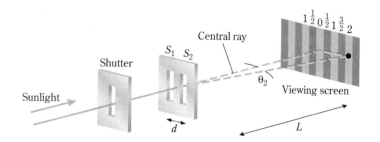

lowed a narrow beam of sunlight passing through a hole in a window shutter to fall on two narrow, parallel slits in a piece of cardboard, as illustrated in Fig. 24.5. On a viewing screen placed beyond the slits, he observed a pattern of alternating bright and dark regions called **fringes.** His observations of these fringes, and his inference that light is a wave phenomenon, allowed him to compute the wavelength of light for the first time. Let us see how he did it.

A vertical wall placed at the right-hand edge of Figure 24.2 would show a water wave pattern. At the points labeled B, the water wave crests are high, but where the lines labeled D intersect the wall, the water is calm. This interference pattern on the wall corresponds to the fringes in Fig. 24.5. The bright fringes in Fig. 24.5 correspond to the positions labeled B in the (imagined) water wave interference pattern of Fig. 24.2. As you might suspect, the D positions correspond to the dark fringes in Young's double-slit pattern.

Using this correspondence with the water interference experiment, we can interpret Young's pattern as follows. The two slits act as two light sources that send out identical waves. The fringe labeled 0 is bright because the waves traveling to this position reinforce each other; their relative retardation is zero.

At fringes 1 and 2, the two waves again reinforce. These fringes correspond to regions B_1 and B_2 in Fig. 24.2, and so the relative phase shift between the waves reaching there is λ for B_1, and 2λ for B_2. Because the situations in Figs. 24.2 and 24.5 are completely analogous, we can apply Eq. 24.1, $n\lambda = d \sin \theta_n$, to Young's double-slit fringes, where n is the fringe number as labeled in Fig. 24.5, d is the slit separation, and λ is the wavelength of the light. Just as θ_2 is the angle between the ray to the central fringe (0) and the ray to fringe 2, θ_n is the angle between the central ray and the ray to fringe n. Frequently we call n the *order number* of the fringe. Using this terminology, we call the θ_2 fringe the *second-order fringe.*

Young could thus use Eq. 24.1 to compute the wavelength of light. The light Young used was sunlight, which contains all visible wavelengths. Since Eq. 24.1 implies that each wavelength creates a bright fringe at a different angle, Young's fringes were bands consisting of all the colors of visible light, with the blue edge of the band closest to the middle and the outer edge red. Monochromatic (single-wavelength) light, such as that provided by a laser, produces more clearly defined single-color fringes, as shown in Fig. 24.6.

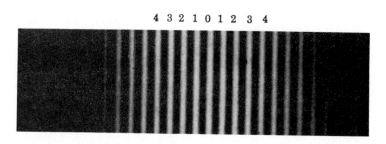

FIGURE 24.6

Interference fringes produced by a double-slit system using monochromatic (single-wavelength) light.

In a typical experiment, consider the distance L in Fig. 24.5 to be 120 cm. The slit separation $d = 0.025$ cm, and the distance from the center of the pattern to the approximate center of the fringe labeled 2 is 0.50 cm. To find θ_2, we see from Fig. 24.5 that

$$\tan \theta_2 = \frac{\text{distance } 0 \rightarrow 2}{L} = \frac{0.50 \text{ cm}}{120 \text{ cm}} = 0.00417$$

from which $\theta_2 = 0.24°$.

Young made use of data such as these to compute the wavelength of the light near the center of a typical fringe. Substituting in Eq. 24.1, he obtained

$$\lambda = \frac{d}{n} \sin \theta_n = \frac{0.025 \times 10^{-2} \text{ m}}{2} \sin 0.24 = 5.2 \times 10^{-7} \text{ m}$$

He was therefore able to conclude that visible light has a wavelength of about 500 nm, with the wavelength of blue light being somewhat shorter than this and that of red light somewhat longer.

It is difficult to overstate the importance of interference, particularly in the case of light. Waves of a single frequency have, in their wavelength, a built-in "apparatus" for measuring length. *We cannot directly detect the waveform when we see light, but interference patterns reveal the wavelength.* The wavelengths of visible light are very small compared with the precision of ordinary length-measuring devices, and so being able to use light as a "yardstick" has great advantages. Devices that use interference patterns to measure lengths are called **interferometers.** It is with interferometers that the most precise measurements of length have been attained.

In describing interference effects, we have used waves that are identical in shape and wavelength. Moreover, we have always assumed that the waves maintain a fixed phase relation to each other. Two such waves are said to be *coherent:*

Coherent waves have the same form and wavelength and have a fixed phase relation to each other.

Sources of coherent waves are called *coherent sources.*

Because two sources of light are almost always noncoherent, it is usually necessary to divide a single light beam into two parts to obtain an interference pattern. For example, in the double-slit experiment, the two slits are illuminated by the same beam, the same light wave, and this wave is divided into two distinct parts by the two slits. Because the two waves thus generated are part of the same wave, they are coherent and give rise to the interference effects discussed.

24.4 EQUIVALENT OPTICAL PATH

As we saw in Sec. 23.9, light travels fastest in vacuum, with a speed c. As light enters a transparent medium whose index of refraction is n, the light's speed is reduced to $v = c/n$. The frequency of the light is not changed,* however. Since the

*According to Huygens' principle, each wave crest striking the boundary between the vacuum and the medium acts as a new source of waves; the frequency of waves passing the boundary therefore remains the same as the frequency of the incident waves.

wavelength in the medium is given by $\lambda_m = v/f = c/nf$, and that in vacuum is $\lambda_{\text{vac}} = c/f$, we have

$$\lambda_m = \frac{\lambda_{\text{vac}}}{n} < \lambda_{\text{vac}} \tag{24.2}$$

The wavelength of light when traveling in a medium is shorter than when the same light is traveling in vacuum.

In slowing light down, the medium has the effect of bunching the waves closer together, as shown in Fig. 24.7. Remember, each wavelength represents one complete phase cycle of the light. This bunching of waves means that if light undergoes a certain number of cycles in a length L in vacuum, *the light will undergo the same number of cycles in a shorter length when in a transparent medium*.

This leads us to an important concept, which we call the **equivalent optical path length** of a medium. For a given light wavelength let us calculate how many wavelengths are contained in a thickness L of a medium whose refractive index is n:

$$\text{No. of wavelengths in thickness } L \text{ of medium} = \frac{L}{\lambda_m} = \frac{L}{\lambda_{\text{vac}}/n} = \frac{nL}{\lambda_{\text{vac}}}$$

where we have made use of Eq. 24.2. The number of wavelengths contained in a distance L of vacuum is just L/λ_{vac}. Thus, in terms of the number of wavelengths contained, and therefore of the amount of phase change which occurs, we conclude the following:

A path length L in a medium whose index of refraction is n creates the same phase change of light as a path length nL in vacuum would.

The optical equivalent to a path length L in a medium whose index of refraction is n is thus

$$L_{\text{opt}} = nL \tag{24.3}$$

Knowing Eq. 24.3 gives us an easy way to find the phase change in a light wave of vacuum wavelength λ_{vac} that is produced when the light passes through a given

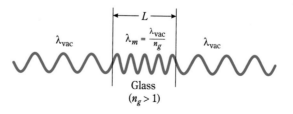

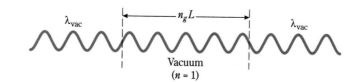

FIGURE 24.7

A thickness L of glass whose index of refraction is n_g contains the same number of wavelengths as a thickness n_gL of vacuum. The glass thus has an equivalent optical thickness n_gL.

thickness L of the medium: it is just the number of vacuum wavelengths contained in the equivalent optical thickness.

$$\text{Phase change (in cycles)} = \frac{L_{opt}}{\lambda_{vac}} = \frac{nL}{\lambda_{vac}}$$

This need not come out as an integer, of course; it may be some fraction of a cycle as well.

The concept of equivalent optical path is important in discussing many aspects of interference, as we will see in the next section.

Example 24.1

What thicknesses of crown glass and diamond are optically equivalent to 1.00 cm of vacuum? What wavelength would light of $\lambda = 600$ nm have in these two materials?

R

Reasoning

Question How does the first question translate into the quantities that define the equivalent optical path of a medium?
Answer The equivalent optical thickness of the material L_{opt} is related to the actual thickness L by $L_{opt} = nL$ (Eq. 24.3). You are asked to find the actual thickness of the material that acts optically like 1.00 cm of vacuum. In other words, find L for $L_{opt} = 1.00$ cm.

Question Should I expect the optical thicknesses to be greater or less than 1.00 cm?
Answer A material with $n > 1$ shortens the wavelength of light passing through it. Thus the same number of cycles fits into a *shorter* thickness than in vacuum.

Question How is the wavelength in a medium related to the index of refraction of the medium?
Answer In a medium, $\lambda_m = \lambda_{vac}/n$, where λ_{vac} is the vacuum wavelength.

Solution and Discussion For an optical thickness L_{opt} equal to 1.00 cm, the actual thicknesses L are:

$$L\text{ (glass)} = \frac{L_{opt}}{n} = \frac{1.00 \text{ cm}}{1.52} = 6.58 \text{ mm}$$

$$L\text{ (diamond)} = \frac{1.00 \text{ cm}}{2.42} = 4.13 \text{ mm}$$

In 1.00 cm of vacuum there are

$$\frac{1.00 \times 10^{-2} \text{ m}}{600 \times 10^{-9} \text{ m}} = 1.67 \times 10^4$$

wavelengths, or cycles. To be optically equivalent, the actual thicknesses contain the same number of wavelengths. In the present case, 6.58 mm of glass is optically equivalent to 4.13 mm of diamond. Both thicknesses contain 1.67×10^4 wavelengths.

We determine the wavelengths in the media from Eq. 24.2:

$$\lambda_m \text{ (glass)} = \frac{600 \text{ nm}}{1.52} = 395 \text{ nm}$$

$$\lambda_m \text{ (diamond)} = \frac{600 \text{ nm}}{2.42} = 248 \text{ nm}$$

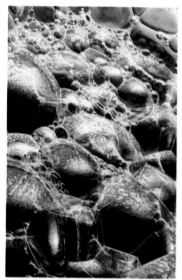

Soap bubbles show the effect of interference in thin films. The wavelength of light waves we see interfering constructively from the top and bottom surfaces of the bubble depends on the angle at which we view the bubble. Thus when we look at different areas of the bubble, we see different colors due to this interference effect.

24.5 **INTERFERENCE IN THIN FILMS**

One of the most common interference effects is the colored fringes we often see in soap and oil films. Let us now analyze this important type of interference.

Figure 24.8 shows a thin film of water, of thickness L, on a glass plate. The light we see reflected from the film is reflected partially from the top surface of the water and partially from the water-glass boundary. These two reflections are shown as rays a and b. To simplify the discussion, these rays are shown almost perpendicular to the film, since then we do not have to deal with refraction.

Because rays a and b are part of the same incident beam, they are coherent. Therefore when they strike the top of the water film, they are in phase. As it passes through the film, ray b is slowed down relative to ray a. Ray b has to travel through the thickness of the film twice (a round trip of $2L$) before it emerges from the water and recombines with ray a. A phase shift is thus introduced between the two rays, a shift that depends on the *equivalent optical path length* traveled by ray b. From our discussion in the preceding section, this phase shift is

$$\text{Phase shift between rays } a \text{ and } b = \frac{2L_{\text{opt}}}{\lambda_{\text{air}}} = \frac{2nL}{\lambda_{\text{air}}}$$

If this phase shift is equal to an integral number, ray b will recombine *in phase* with ray a as it (ray b) passes back through the top surface of the film and the light reflected by the two surfaces of the film will be *bright*. If $2L_{\text{opt}}$ is an *odd* number of *half* wavelengths ($\lambda/2, 3\lambda/2, 5\lambda/2$, etc.), the recombination will be half a cycle out of phase, resulting in destructive interference.

FIGURE 24.8

Light rays reflected from the top and bottom of a thin film travel different distances before they recombine. The eye sees the resulting interference condition.

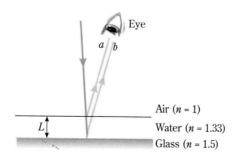

Films very often have thicknesses comparable to or even less than the wavelengths of visible light. Thus if a film is illuminated by white light, constructive interference may occur for only one of the wavelengths in the source. The film, seen by reflected light, will appear colored as a result.

There is one additional source of phase shifts in dealing with reflections. You may recall in our discussions of waves on strings that we observed the wave form to be inverted (a 180°, or half cycle, phase shift) when the wave reflected from a fixed end of the string. A wave reflected from a free end of the string does not experience any change of phase. A similar phenomenon occurs when light is reflected at a boundary between materials of different refractive indexes:

If light traveling in a medium of refractive index n_1 is reflected by a medium having a greater refractive index ($n_2 > n_1$), the reflected wave is phase-shifted by half a cycle from the incident wave. If $n_2 < n_1$, the reflected wave does not experience any phase shift.

This phase shift is in addition to the phase shift due to unequal optical paths.

How rays interfere when they recombine is determined by the total *difference* in phase that has occurred. If the rays both experience a phase shift of either zero or 180° upon reflection, the total phase shift is determined solely by the round-trip optical path length difference as discussed. However, if one or the other of the rays experiences a half-cycle reflection phase shift and the other does not, this difference must be added to that due to path length difference.

In summary, to calculate the interference condition for light reflected from a thin film:

1 Identify the indexes of refraction of the incident material, the film, and the material on which the film rests. Use this information to find out whether any reflective phase shifts occur.
2 If neither ray or both rays have a phase shift on reflection, a bright reflection results when the round-trip optical path through the film is equal to an integral number of wavelengths.
3 If only one of the two rays (either one) undergoes a reflective phase shift, a bright reflection results when the round-trip optical path through the film is equal to an odd number of half wavelengths.

An example of case 3 would be water film surrounded top and bottom by air. Then ray *a* would have a half-cycle phase shift upon reflection, but ray *b* would not.

Notice that the condition for constructive interference changes to destructive interference when the film thickness L changes by $\lambda/4$. This is a very small change in thickness, yet it is very easily observed. Thin-film interference thus has many applications based on its ability to detect very small changes in distance.

Flatness of surfaces is one such application. Standards of flatness, known as *optical flats,* can be manufactured out of glass plates so that the top and bottom of any one plate are parallel to each other to within a fraction of a wavelength of visible light. An optical flat is placed on top of a sample material that is to be made flat and illuminated with monochromatic light. If the sample material has an uneven surface, a thin film of air occurs between the optical flat and the sample. Variations in the thickness of this film, due to unevenness of the sample surface, show up as bright and dark interference fringes. An example is shown in Fig. 24.9.

FIGURE 24.9

Observed fringes for a wedge-shaped air film between two glass plates that are not flat. Each dark fringe corresponds to a region of equal thickness in the film; between two adjacent fringes the change in thickness is $\lambda/2n$. As always, n refers to the index of refraction of the film. In this case the film is air, so $n \approx 1$. The fringes indicate that the plates are approximately flat only near their left edge.

When no fringes show, the sample has been polished to the same flatness as the standard, to within approximately $\lambda/4$ of the illuminating light.

Another application of thin-film interference is the use of optical flats to measure the thickness of very thin objects. Suppose we sandwich a strand of hair between the ends of two optically flat glass plates as shown in Fig. 24.10a. This creates a wedge of air between the flat surfaces. When this wedge is illuminated from the top with monochromatic light, we observe a series of alternating bright and dark interference fringes across the plates and parallel to the strand of hair as seen in Fig. 24.10b. The fringe at the edge where the plates touch (D_1 in Fig. 24.10b) is dark because here the only phase shift is due to the reflection of ray b by the lower plate. The separation between the centers of two adjacent dark fringes represents an increase of $\lambda/2$ in the air wedge thickness. (You should be able to explain why this is true.) Thus if there are three dark fringes from the end where the plates are in contact to the end where the hair separates them, the thickness of the plate separation produced by the hair is $3\lambda/2$. If this interference pattern was produced by light of 600-nm wavelengths, for example, we would conclude that the thickness of the hair strand was 900 nm, or 0.900 μm.

The experiment shown in Fig. 24.11 was conducted by Newton and also illustrates thin-film interference. A planoconvex lens (much less curved than the one shown) is placed on a flat glass plate and illuminated from above with monochromatic light. Rays reflected to the eye from the surfaces of the air wedge formed between the lens and the plate give rise to the fringe pattern shown in Fig. 24.11, a pattern called *Newton's rings*. This pattern is produced for the same reason as in the case of Fig. 24.10, except that the fringes are circular because of the circular geometry of the air wedge formed by the lens.

FIGURE 24.10

(a) Two optical flats with a strand of hair separating the right edges. The hair forms a wedge-shaped air gap between the plates, which causes an interference pattern when the plates are illuminated by monochromatic light from above. (For the sake of clarity, only one incident and reflected ray are shown, but you should realize that light is incident and reflected all across the plates.) (b) A top view of the plates, showing the interference pattern seen by the eye.

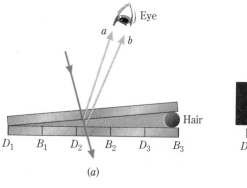

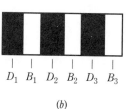

(a)

(b)

FIGURE 24.11

(a) Ray *a*, reflected from the lower surface of the lens, interferes with ray *b* reflected from the glass plate. (b) The interference pattern caused by this interference is called Newton's rings. Why is the center of the pattern dark? (The angles in part (a) are distorted so as to show the two reflected rays clearly.)

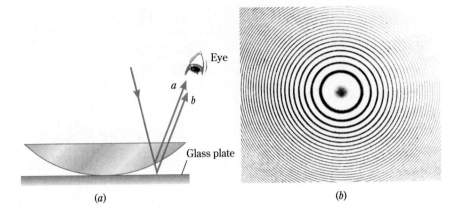

(a)   *(b)*

Eye

a

b

Glass plate

Example 24.2

In order to reduce reflection and thereby enhance the intensity of the transmitted light, lenses are often coated with a thin layer of magnesium fluoride ($n = 1.38$). What is the thinnest coating that will produce minimum reflection for 550-nm light?

Reasoning

R

Question   What does "minimum reflection" mean in the terms we used in discussing thin films?
Answer   The film thickness that causes destructive interference between the rays reflected from the two surfaces of the film.

Question   Which rays experience phase shifts upon reflection?
Answer   From Table 23.2, you can see that $n_{glass} > n_{MgF_2}$. Thus the reflection at both MgF_2 surfaces produces a half-cycle phase shift.

Question   What condition on path length will cause destructive interference?
Answer   The net phase difference is due entirely to the optical path length difference.

Question   What condition will yield the *thinnest* coating?
Answer   Various thicknesses will cause destructive interference. The *minimum* thickness corresponds to $2nL = \lambda/2$.

Solution and Discussion   Putting in numbers,

$$2(1.38)L = \frac{(550\ \text{nm})}{2}$$

$$L = 99.6\ \text{nm}$$

Antireflection coatings are often referred to as *quarter-wave coatings*. Notice that the condition for minimum thickness is the same as $nL = \lambda/4$. Thus the *optical* thickness of the film is equal to $\lambda/4$.

24.6   THE DIFFRACTION GRATING

Although Young used his double-slit experiment to measure the wavelength of light, the double-slit pattern he obtained is far too diffuse to give accurate results. It turns out that a large number of equally spaced slits give a much sharper fringe system. For example, Fig. 24.12 shows the interference pattern for 20 parallel slits illuminated by monochromatic light. Notice how very sharp the fringes are. To measure wavelengths with high precision, a large number of parallel, equally spaced slits is used. We call such a device a **diffraction grating.** A typical grating might consist of 10,000 parallel slits, each separated from the next by a distance $d = 10^{-4}$ cm. Let us now discuss the behavior of such a grating.

A diffraction grating is typically used in the manner shown in Fig. 24.13a. Let us assume for the moment that we use a monochromatic light source to illuminate the entrance slit. Because the entrance slit is at the focal point of the **collimating lens,*** a beam of parallel light exits from this lens and is incident perpendicularly on the grating. In Fig. 24.13a the grating slits are perpendicular to the page. The light passing through the grating can be observed with a small telescope. No matter what wavelength we use for illumination, we see a sharp image of the slit when we view the beam head-on.

Now suppose we swing the telescope through an angle θ from the head-on direction, as in Fig. 24.13b. At most values of θ we see no light at all. At certain values, however, we see a very sharp image of the entrance slit. These images are equivalent to the bright fringes in Fig. 24.12, but more sharply defined.

If we change the illuminating wavelength, we also change the values of θ at which the images occur. Thus if the illumination contains more than a single

***A *collimating lens* is a converging lens used to produce a parallel, or *collimated,* beam of light. This is accomplished by placing the lens a focal length away from a small light source. As we learned in Chap. 23, the incident rays that diverge from the source will all emerge from the lens parallel to the principal axis.

FIGURE 24.12

The interference pattern of monochromatic light produced by 20 parallel, equally spaced slits. Notice how narrow these fringes are compared to those in Fig. 24.6, which were produced by only two slits.

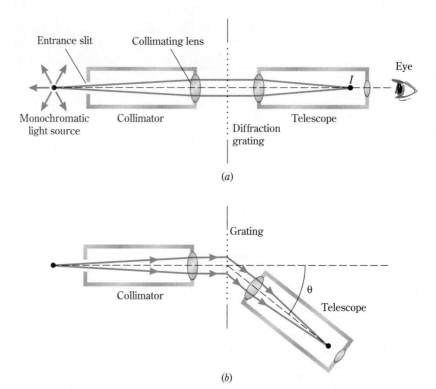

(a)

(b)

FIGURE 24.13

(a) A schematic diagram of a grating spectrometer, one of the most common applications of the diffraction grating. (b) When the telescope is rotated on the arc of a circle centered on the diffraction grating, an image of the slit is formed by constructive interference at an angle θ to the undeviated beam. The angle at which this occurs depends on the wavelength of the illuminating light.

wavelength, each wavelength will produce an image of the entrance slit at an angle separate from those produced by the other wavelengths. The light from the source will be separated into a number of sharp single-wavelength images, one for each wavelength contained in the illuminating light. These images are called **spectral lines,** and reveal the spectrum being emitted by the source. Because of this capability, a device such as that diagramed in Fig. 24.13 is known as a **grating spectrometer.** Let us now examine the connection between the illumination wavelength and the angles at which an image of the entrance slit is observed.

The first thing to realize is that each narrow slit in the grating acts like a source of light waves (Huygens' principle). When $\theta = 0$ in Fig. 24.13b, we see the undeviated beam shown in Fig. 24.14a. The light rays from all the slits travel the same distance to the telescope, and so they reinforce each other. This is true for any wavelength. Thus when the telescope is pointed along $\theta = 0$, we see an image of the entrance slit containing all the wavelengths of the source. This image is called by various names: the *central maximum,* the *zeroth-order maximum,* and the *central image*.

Suppose at a certain angle θ, shown in Fig. 24.14b, we see a bright entrance-slit image. The light rays from all the grating slits are again parallel to each other as they enter the telescope, but now they are no longer along the undeviated direction. As shown, each ray lags or leads the one adjacent to it by a distance Δ. We know that if $\Delta = \lambda, 2\lambda, 3\lambda$, etc., all the rays will reinforce each other. The condition for an image to be observed is thus $\Delta = m\lambda$, where m is an integer.

From each red triangle in Fig. 24.14b, we have that $\sin\theta = \Delta/d$, where d is the distance between grating slits, called the *grating spacing*. For an image to form, we must have $\Delta = m\lambda$, and so we find bright images of the entrance slit when θ is equal to values θ_m given by

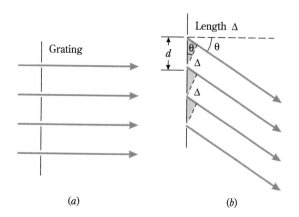

FIGURE 24.14

(a) The relative retardation of the straight-through rays is zero. (b) When the retardation Δ is a whole number of wavelengths long, the rays all reinforce one another. It is at these angles that the grating yields intensity maxima.

$$m\lambda = d \sin \theta_m \qquad m = 1, 2, 3 \ldots \qquad (24.4)$$

This is called the **grating equation**.

To better understand the grating equation, suppose the light source contains only two wavelengths: 500 nm and 600 nm. Suppose further that $d = 2 \times 10^{-6}$ m. Substituting these values in Eq. 24.4, we find the image positions listed in Table 24.1. These images are also shown in Fig. 24.15 along with the names we use to describe them. (Because the fourth-order spectral lines occur at $\theta = 90°$ and larger, they cannot be observed.) Notice that the lines appear on both sides of the central maximum. Also notice that the separation between lines increases as θ increases. One way to make the first-order image positions occur at larger angles is to make d as small as possible, as can be seen from Eq. 24.4. Doing so helps to separate closely spaced lines.

Since we can measure with high precision the angle θ_m at which the mth-order maximum occurs, it is necessary to know only the grating spacing d in order to determine λ accurately. For example, if the yellow light from a sodium arc lamp is used in even a simple spectrometer, it is not difficult to see that the sodium light gives *two* slit images (or lines) at each order position. These lines are very close together and have wavelengths of 589.0 and 589.6 nm. The mere fact that we are able to see these two lines as distinct images provides some measure of the potential accuracy of such a device.

TABLE 24.1

*Positions of Spectral Lines**

λ (nm)	m	θ_m (degrees)
500 and 600	0	0
500	1	14.5
600	1	17.5
500	2	30.0
600	2	36.9
500	3	48.6
600	3	64.2
500	4	90
600	4	missing

*Assuming $d = 2\ \mu$m

FIGURE 24.15

The first, second, and third spectral orders each contain two lines, one for the 500-nm light and one for the 600-nm light.

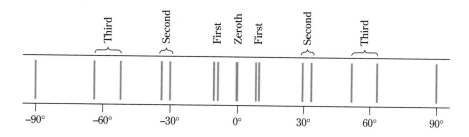

Example 24.3

A particular diffraction grating has 1.0000×10^4 lines per centimeter. At what angles does the 589.0-nm line of sodium light occur? To what precision do you have to be able to measure angles in order to observe the separation between this line and the 589.6-nm sodium line?

R | *Reasoning*

Question What is the equation for the angle of the first-order image?

Answer $\sin \theta_1 = \dfrac{\lambda}{d} = \dfrac{0.5890 \times 10^{-6}\,\text{m}}{d}$

Question What is the grating spacing d?

Answer $d = \dfrac{1}{1.0000 \times 10^4\,\text{lines/cm}} = 1.0000 \times 10^{-6}\,\text{m}$

Question What determines whether or not a second-order line occurs?
Answer The quantity $\sin \theta_m$ must always be less than unity. Thus the condition for a line of order m to appear is that $m\lambda/d < 1$.

Question At what angle will the 589.6-nm line appear?

Answer At an angle given by $\sin \theta_1 = \dfrac{(0.5896 \times 10^{-6}\,\text{m})}{d}$

Solution and Discussion The first-order lines lie at angles of $\sin^{-1} 0.5890$ and $\sin^{-1} 0.5896$. These are

$$\theta_1 = 36.09°, \ 36.13°$$

To separate these lines, you would need to measure angles to the nearest 0.01°.
The second-order lines require $\sin \theta_2 = 2 \sin \theta_1$. (Notice that this is not the same as twice the angle!) In both cases this is a number greater than 1, and so the second-order lines cannot appear.

24.7 DIFFRACTION BY A SINGLE SLIT

Until now, we have assumed that the width of the entrance slit was negligible relative to the wavelength of light being used. If you look at Fig. 24.1, you will see that the diffraction for longer wavelengths is greater than the diffraction for shorter wavelengths. Thus diffraction appears to depend on the size of the wavelength relative to the width of the slit. We wish now to investigate the reasons for this effect and to describe more fully how a single slit diffracts light. The result we obtain has fundamental importance and places limits on our ability to make measurements.

To see the effect of the diffraction of light waves, we can send light through a single slit and record the transmitted light on a photographic film, as shown in Fig. 24.16. The central bright fringe is considerably wider than the slit. Moreover, bright fringes separated by dark fringes occur on each side of the central image,

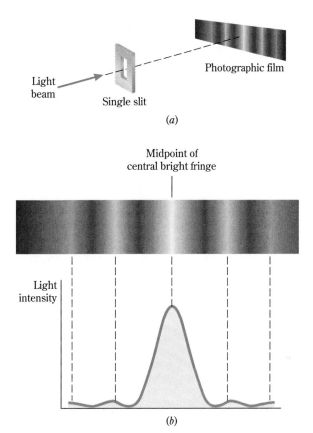

FIGURE 24.16

(a) Single-slit diffraction pattern. (Not to scale.) (b) The central bright region is much more intense than the higher-order fringes, as the graph indicates.

and these bright fringes must result from interference. Let us now see what is involved in this situation.

Consider a wave crest as it strikes the slit. According to Huygens' principle, each tiny point on the crest acts as a source of new waves. Thus light rays emanate from all points along the crest. Some rays travel straight forward, while others make an angle θ to the forward direction. As shown in Fig. 24.17a, the light rays that go straight through the slit are all in phase with one another. For this reason, the straight-through position is bright and gives rise to the central bright fringe in Fig. 24.16. However, at an angle θ to the straight-through beam, rays from various parts of the slit travel different distances to the film. The most important situations are shown in Fig. 24.17b, c, and d.*

In (b), ray B from the middle of the slit is half a wavelength behind ray A. As a result, these two rays cancel each other. That is not all, however, because we see that the rays leaving the slit from positions just above A and B also cancel since they, too, have a path difference of $\lambda/2$. In fact, for each ray leaving the lower half of the slit there is leaving the upper half a corresponding ray that cancels it. Hence, at this angle θ, no light reaches the film from the slit and we observe a dark fringe. As you can see from the figure, this situation occurs when $\sin \theta = \lambda/b$, where b is

---

*If the rays were exactly parallel, they could not meet; as a result, they would not interfere with each other. We consider here either of the following situations: (1) a lens focuses the parallel rays to a point or (2) slight nonparallelism causes the rays to meet at a point.

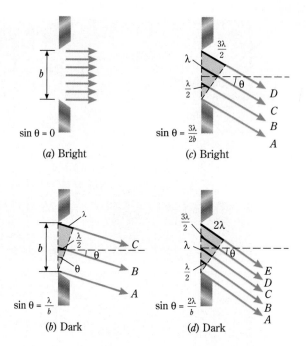

FIGURE 24.17

In analyzing the single-slit pattern qualitatively, we section the slit into portions whose rays differ by $\lambda/2$ in path length. Why?

the slit width. Notice that if b is equal to the wavelength of the light, the dark fringe occurs at $\theta = 90°$. In other words, *if the slit width is decreased until it is as small as λ, the image of the slit spreads to become infinitely wide.*

If b is considerably larger than λ, as in Fig. 24.17, a side bright fringe occurs for the angle θ shown in part (c). In this case, the rays from the bottom third of the slit cancel those from the center third while the top third is left uncanceled. Darkness is again achieved at the larger angle-shown in part (d). Here the slit can be thought of as being divided into fourths. The bottom one-fourth of the slit is canceled by the portion just above it. Similarly, the two upper sections also cancel. Hence darkness is observed at this angle.

The most important feature of the single-slit pattern, for our purposes, is the position of the first minimum next to the central maximum. Calling the angle between the central maximum and the first minimum θ_c, we have found that

$$\sin \theta_c = \frac{\lambda}{b} \qquad\qquad (24.5)$$

This relation is used in the next section.

24.8 DIFFRACTION AND THE LIMITS OF RESOLUTION

One of the most important consequences of diffraction is that it limits our ability to observe very fine details. We can appreciate this difficulty by referring to Fig. 24.18: two light sources sending light through a slit to a viewing screen. If the slit is quite small, the images cast on the screen are accompanied by noticeable diffraction fringes, as shown. These fringes are the result of the light's having passed through the slit, whose width we specify to be b.

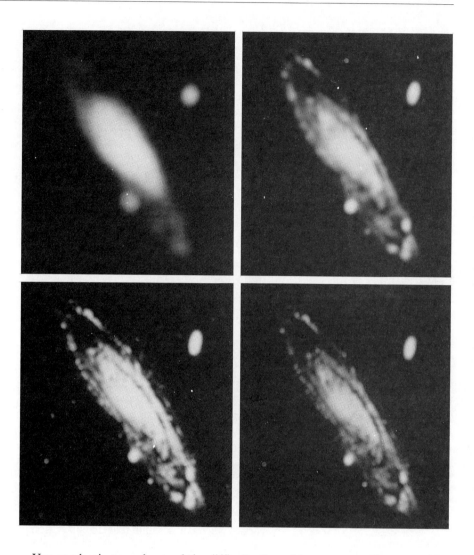

These photos of a galaxy were taken by a successively larger aperture telescope objectives, showing how resolution of detail improves with larger aperture.

You can begin to understand the difficulty these fringes present if you consider the following rough analogy, which we pursue further later. The pupil of the eye corresponds roughly to the slit, and two lines on an object being viewed by the eye correspond to the two sources in Fig. 24.18. The retina acts as a screen. Because the images on the retina are made fuzzy by the diffraction effects of the slit (the pupil), the eye is hindered in seeing fine detail in the object being viewed.

Returning to the situation in Fig. 24.18, we can see the images of the two sources on the screen as separate entities as long as the angle θ is not too small. Difficulty

FIGURE 24.18

The two sources are well resolved on the screen because their diffraction patterns do not overlap seriously.

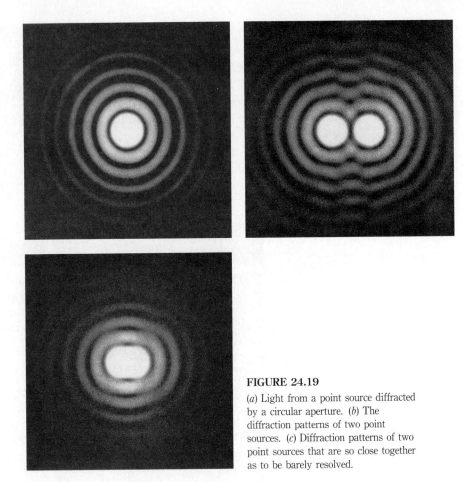

FIGURE 24.19

(*a*) Light from a point source diffracted by a circular aperture. (*b*) The diffraction patterns of two point sources. (*c*) Diffraction patterns of two point sources that are so close together as to be barely resolved.

arises when θ is so small that the diffraction patterns overlap appreciably. The two sources can no longer be seen as separate (that is, they can no longer be *resolved*) when they are close enough together for the central maximum of one pattern to fall on the first minimum of the other pattern. In that situation, the case of *minimum resolution*, angle $\theta = \theta_c$, where θ_c is as defined in Eq. 24.5. *Thus we can resolve the sources only if their angular separation θ is larger than θ_c.* As we expect, the smaller the slit width b, the farther apart the objects must be if they are to be resolved, for the interference pattern broadens with decreasing slit width.

Although our discussion has been in terms of diffraction of light sources produced by slits, a similar phenomenon occurs when the slit is replaced by a small circular hole, or aperture. Examples of such apertures include the pupil of the eye and the iris of a camera lens.

For a point source of light, the diffraction pattern produced by a circular aperture is shown in Fig. 24.19*a*. The angular diameter* of the central maximum of this pattern is given by

$$\sin \theta = 1.22\frac{\lambda}{D} \tag{24.6}$$

*The term *angular diameter* refers to the angle that the central maximum of the diffraction pattern subtends at the center of the aperture. In other words, it is the angle formed by lines drawn from the center of the aperture to points on opposite ends of the diameter of the central maximum.

where D is the aperture diameter. Notice the similarity between this equation and Eq. 24.5 for a slit of width b.

As two point sources come closer together, the diffraction patterns produced by their light passing through the aperture begin to overlap, ultimately blending into one. This can be seen in Fig. 24.19*b* and *c*. The limit of resolution of the sources is that the angular separation between their central maxima must be *at least* as great as the angular width of those maxima. Thus we have the following condition:

The limiting angle θ_c for the resolution of two point sources viewed through a circular aperture of diameter D is

$$\sin \theta_c = 1.22\frac{\lambda}{D} \tag{24.7}$$

Now let us see what sort of limit this diffraction effect places on our ability to view objects with a microscope.

We show in Fig. 24.20 the lens of a microscope and two details S_1 and S_2 of an object being viewed. The details are a distance s apart, where s is much smaller than shown, the lens has a diameter D, and the details are a distance d from the lens. How close together can the details be and still be resolved?

According to Eq. 24.7, the details can just be resolved if the angle θ_c they subtend is $\sin^{-1}(1.22\,\lambda/D)$. From Fig. 24.20, we see that

$$\sin \frac{\theta_c}{2} = \frac{s/2}{\sqrt{d^2 + (s/2)^2}} \approx \frac{s}{2d}$$

because s is actually very much smaller than d.

For small angles, the angle in radians is equal to its sine. Since θ_c is usually very small, we can therefore replace $\sin \theta_c$ by θ_c in radians to give

$$\theta_c = \frac{s}{d}$$

Making this same approximation in Eq. 24.7, we have

$$\theta_c = 1.22\frac{\lambda}{D}$$

Equating these two expressions for θ_c yields, after a little arithmetic,

$$s = 1.22\left(\frac{d}{D}\right)\lambda$$

If we look at Fig. 24.20, we see that d/D is the ratio of the object distance to the lens diameter. In all normal uses of microscopes, this ratio is approximately unity. As a result, to a rough approximation, $s \approx \lambda$. In other words, *the smallest detail that can be seen in a microscope is about the same size as the wavelength of light being used.* This is a fundamental restriction imposed by diffraction; it cannot be circumvented by use of perfect lenses or ingenious microscope design.

As we see, diffraction effects cause images to be fuzzy. Another example of this fact is shown in Fig. 24.21. The shadow of the washer shown there is surrounded by diffraction fringes, and the situation becomes even worse for a smaller object. In

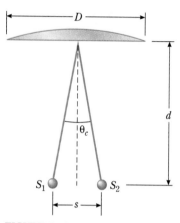

FIGURE 24.20

The two details S_1 and S_2 can just be resolved if $\theta = \theta_c$.

FIGURE 24.21

Shadow of a star-shaped washer. Diffraction bands are seen inside the hole and around the outer edge. A smaller object would show greater fuzziness due to more pronounced diffraction effects.

the case of objects whose size is comparable to the wavelength of light being used, details of the object are completely obscured by diffraction. We must therefore conclude that *it is impossible to obtain images of objects with detail comparable in size to the wavelength of radiation being used.*

Example 24.4

The Hale telescope at Mt. Palomar, California, has an aperture diameter of 5.0 m. What is the smallest angle between two stars that can be resolved by this telescope?

R

Reasoning

Question   What determines the minimum angle that can be resolved?
Answer   The wavelength of the light that is being imaged and the diameter of the aperture through which the light passes (Eq. 24.7).

Question   What wavelength should I use?
Answer   We have been discussing visible light, which has a narrow range of wavelengths. You should use a wavelength near the middle of the visible spectrum, such as 550 nm.

Solution and Discussion   Equation 24.7 gives

$$\sin \theta_c = \frac{1.22(550 \times 10^{-9}\,\text{m})}{5.0\,\text{m}} = 0.134 \times 10^{-6}$$

As just mentioned, for small values of θ (measured in radians), $\sin \theta \approx \theta$. The above result certainly qualifies as being a small value. Thus

$$\theta_c = 0.134 \times 10^{-6}\,\text{rad} = 7.68 \times 10^{-6}\,\text{deg}$$

To emphasize what a small angle this is, the Hale telescope could theoretically resolve a 1-in object which was over 100 mi away!

Stresses in transparent material are revealed by the use of polarized light. The stress is greatest in those regions where the color changes most rapidly.

24.9 POLARIZED LIGHT

Both transverse and longitudinal waves show diffraction and interference effects. There is one wave property, however, that only transverse waves display: **polarization.** We can visualize polarization by imagining transverse waves on a rope. There can be many waves vibrating on the rope simultaneously, in various orientations. That is, some waves may be in the horizontal plane, others may be in the vertical plane, and still others may have components of wave motion in both planes. A mixed wave such as this is said to be *unpolarized.* Now suppose the rope passes through a vertical slot, which we call a *polarizer,* as shown in Fig. 24.22. This slot will stop all horizontal components of the waves, and let only vertical wave motion pass through. Thus the wave beyond the slot has vibrations in only one plane and is said to be **plane-polarized.** This polarization can be checked by passing the polarized wave through a second slot oriented at 90° relative to the first. As shown in Fig. 24.22, the second slot, called an *analyzer,* will block the wave, and no wave energy will be detected beyond it.

Longitudinal waves such as sound, on the other hand, consist of molecules vibrating along the direction of the wave, and so the slot has no effect on the longitudinal motions. Thus a longitudinal wave is not polarizable. To prove that a wave is transverse, therefore, all we need to do is demonstrate that it can be polarized.

There are a number of ways that light can be polarized. Two of the most common are by reflection, which we discuss later in this section, and by transmission of light through polarizing material. This latter process is very similar to the way a slot polarizes a wave on a rope. A transparent film is made in which needle-shaped crystals of quinine iodosulfate are oriented in one certain direction.* These crystals have the property of allowing electric fields to pass through them only in the direction transverse to the crystal length. Thus a nonpolarized light wave will be plane-polarized after passing through such a material. This can be demonstrated by passing this light through a second polarizing sheet whose crystals are oriented 90° with respect to those of the first sheet. This should (and does) block all the remaining light.

Figure 24.23 shows this effect. In (*a*), unpolarized light is incident on the first polarizer, which permits only vertically polarized light to pass. The purple arrows show the *transmission axis* of the polarizer, perpendicular to the alignment of the iodosulfate crystals. When the transmission axis of the second polarizer, the analyzer, is oriented 90° to the first polarizer (Fig. 24.23*a*), all the light is blocked out. In

*Such films are known by the trade name Polaroid and were invented in 1934 by Edwin H. Land.

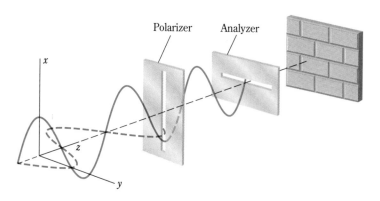

FIGURE 24.22

Polarization of waves in a stretched rope is analogous to the polarization of light.

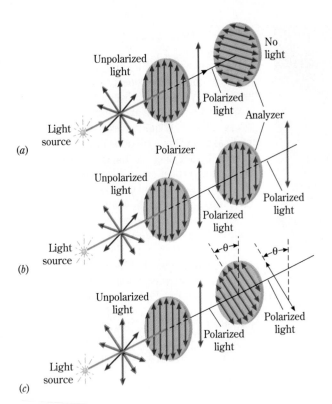

FIGURE 24.23

Light is plane-polarized by passing it through the polarizer. All, part, or none of this light passes through the analyzer, depending on the relative orientation of the two transmission axes. The arrows on the polarizing plates indicate the directions of the components of the transverse electric field vectors that each plate allows to pass.

(b) the analyzer is aligned with the first polarizer, allowing all the vertically polarized light to pass. If the analyzer has its transmission axis at an angle θ to that of the polarizer, as in (c), only light polarized in the plane of the analyzer's axis passes through.

In unpolarized light the electric field is oriented equally in all directions transverse to the direction of the light. A single polarizer, in allowing only one plane of vibration to pass, reduces the intensity of the transmitted light to half that of the unpolarized incident light. When the transmission axis of the analyzer is oriented at an angle θ relative to the electric field of the light incident upon the analyzer, only the field component $E \cos \theta$ is allowed to pass. Since light intensity is proportional to the square of the field amplitude, it follows that the intensity transmitted by an analyzer such as that shown in (c) is

$$I_{\text{transmitted}} = I_{\text{incident}} \cos^2\theta \tag{24.8}$$

A very common application of the principles used in making Polaroid film is found in some sunglasses. In addition to being tinted to reduce light transmission, they are made so that the transmission axes of the films are vertical when the glasses are worn. Such glasses reduce "glare" because, in reflecting off flat surfaces, light becomes partially polarized parallel to the reflecting surface. Water and road surfaces provide such reflecting surfaces, and so polarizing glasses are especially popular with people who fish and spend considerable time driving.

The degree to which reflected light is polarized depends on the angle of incidence on the reflecting surface and the index of refraction of the reflecting material. There is one particular angle of incidence, called **Brewster's angle** (θ_B), at which the polarization of the reflected light becomes 100 percent. This happens when the

FIGURE 24.24

(a) Partial polarization of originally unpolarized light by reflection from the surface of a glass plate. (b) Complete polarization of light by reflection from the plate at Brewster's angle θ_B, where $\tan \theta_B = n_2/n_1$. Here the reflected and refracted waves are at right angles to each other, and all electric field vectors in the reflected light are parallel to the plate surface.

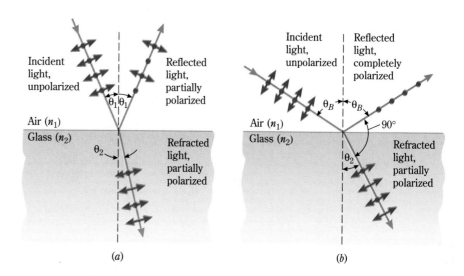

(a)

(b)

direction of the reflected light is perpendicular to that of the light refracted into the surface. This is shown in Fig. 24.24 for an air-glass boundary.

We can apply Snell's law to find how θ_B depends on the materials involved. Referring to Fig. 24.24b, we have

$$n_1 \sin \theta_B = n_2 \sin \theta_2$$

Using some trigonometric identities (you should verify these if they seem unfamiliar) we get

$$n_1 \sin \theta_B = n_2 \sin (90° - \theta_B) = n_2 \cos \theta_B$$

By dividing one side of the equation by the other and remembering that $\tan \theta = (\sin \theta)/(\cos \theta)$, we arrive at a simple expression for Brewster's angle:

$$\tan \theta_B = \frac{n_2}{n_1} = n \qquad \text{or} \qquad \theta_B = \tan^{-1} n \qquad (24.9)$$

As with all other applications of Snell's law, θ_B is measured from the *normal* to the reflecting surface. In Eq. 24.9, n is the index of refraction of the refracting medium relative to the incident medium. Notice in Fig. 24.24 that the reflected light is polarized with its electric field parallel to the surface. Also notice that the refracted ray is partially polarized.

Example 24.5

At what angle of incidence will light reflected off the surface of a lake be completely polarized? If you have polarizing sunglasses on and tilt your head 20° away from vertical, what fraction of the reflected light intensity will enter your eyes? Assume the polarizing lenses are not tinted.

R

Reasoning

Question   What is the condition for complete polarization by reflection?
Answer   That the direction of reflected light be 90° to the refracted light. This condition is satisfied if the light is incident at Brewster's angle.

Question   What does Brewster's angle depend on?
Answer   Equation 24.9 shows that $\theta_B = \tan^{-1} n$ where $n = n_2/n_1$. Table 23.2 gives the values $n_2 = 1.33$ (water) and $n_1 = 1.00$ (air).

Question   What determines the fraction of intensity the polarizing sunglasses will let through?
Answer   If the transmission axis of the analyzer is at an angle of θ relative to the plane of polarization, the fraction of light transmitted is $\cos^2 \theta$ (Eq. 24.8). Since the glasses are untinted, you can assume that no other factor inhibits light transmission.

Question   What is θ when the head is tilted 20° from the vertical?
Answer   The sunglasses are made so that the transmission axis is vertical when your head is vertical. The plane of polarization is horizontal. Thus $\theta = 70°$.

Solution and Discussion   Brewster's angle for a water-air boundary is

$$\theta_B = \tan^{-1} 1.33 = 53.1°$$

Remember, this is with respect to the vertical. The fraction of polarized light that passes through the lenses is

$$\frac{I_{\text{transmitted}}}{I_{\text{incident}}} = \cos^2 70° = 0.117 = 11.7\%$$

It should also be noted that the intensity of the totally polarized light coming off the water is 50 percent of the intensity incident on the water. If you have worn polarizing sunglasses, you have probably noticed the variation of light intensity as you tilted your head, even with partially polarized light.

The polarization of light is used in many technical and scientific applications. For example, under a microscope, details can often be seen more clearly if they are examined between crossed polarizing sheets. Portions of the object that appear the same in ordinary light may differ considerably in their ability to change the polarization of the transmitted light. Hence, these otherwise unobservable details can easily be seen. When a transparent object is under high stress, the stress often

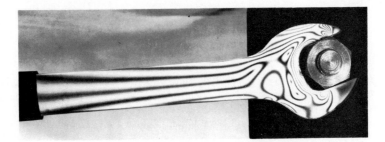

FIGURE 24.25

A strained transparent object viewed through crossed Polaroids shows alternate dark and bright bands. The stress variation is greatest where the bands are closest together.

rotates the plane of polarization of the transmitted light. As a result, a nonuniformly stressed object observed between crossed polarizers shows alternate dark and bright bands, as in Fig. 24.25. Where the bands are closest together, the stress is most uneven. By examining plastic models of strained objects like that in Fig. 24.25, it is possible to tell exactly how the stress is distributed. This is of great importance in the design of various parts for machines.

LEARNING GOALS

Now that you have finished this chapter, you should be able to

1 Define (*a*) diffraction, (*b*) Huygens' principle, (*c*) order number of fringe or spectral line, (*d*) coherent waves, (*e*) Newton's rings, (*f*) equivalent optical path length, (*g*) diffraction grating, (*h*) limiting angle of resolution, (*i*) Brewster's angle.

2 Describe a water-wave experiment that illustrates the phenomenon of diffraction.

3 Show the phase relation of two identical waves if they are to interfere (*a*) constructively and (*b*) destructively.

4 Describe Young's experiment and how two coherent beams are obtained in it. Using a diagram, show why these two beams can interfere destructively and constructively at various points. Consider the diagram and justify the relation $n\lambda = d \sin \theta_n$ for the bright-fringe positions.

5 Use a double-slit interference pattern to determine λ if sufficient data are given.

6 Calculate the equivalent optical path of a thickness L of material whose index of refraction is n.

7 Explain how interference is produced by a thin film or wedge. Tell why the fringes formed in white light are colored. Compute the wedge-thickness difference between adjacent dark or bright fringes.

8 Explain how a diffraction grating is used to measure the wavelength of a spectral line.

9 Describe what happens to a beam of light transmitted through a slit as the slit is made very narrow. Pay particular attention to what happens when the slit width approaches λ. Explain the importance of this effect to our ability to observe details.

10 Calculate the angle of incidence that will produce a completely polarized reflected ray, given the index of refraction of the incident and refracting material.

11 Calculate the fraction of light intensity allowed to pass through polarizing plates whose transmission axes are oriented at an angle θ relative to each other.

SUMMARY

DEFINITIONS AND BASIC PRINCIPLES

Diffraction
Diffraction is the phenomenon whereby waves are able to bend around into the region behind obstacles. Diffraction becomes pronounced when the diameter of the obstacle is comparable to the wavelength of the waves.

Huygens' Principle
Each point on a wavefront acts as a point source of new waves.

Interference
Interference describes the superposition of the amplitudes of two or more waves at any given point and time. Where two identical waves are a half cycle out of phase, their amplitudes cancel. Where the two waves are in phase, their amplitudes add constructively.

Two-Source Interference (Young's Experiment)
When two wave sources are separated by a distance d and are emitting identical waves in phase, constructive interference between the waves occurs in directions given by

$$\sin \theta_m = \frac{m\lambda}{d}$$

where θ is measured from a line equidistant from the two sources, using a point midway between the sources as the origin and m is any integer. The value of m is called the *order* of the constructive interference.

Destructive interference occurs in directions given by

$$\sin \theta_m = (m + \tfrac{1}{2})\lambda$$

Equivalent Optical Path Length
A length L of material having index of refraction n has an equivalent optical path length L_{opt} given by

$$L_{opt} = nL$$

This means that as many waves are contained in a thickness L of the material as are contained in a thickness L_{opt} of vacuum.

Phase Shift of Reflected Waves
When a wave traveling in a medium with refractive index n_1 is reflected by a medium having index $n_2 > n_1$, the reflected wave undergoes a half-cycle phase shift relative to the incident wave. If $n_2 < n_1$, the reflection causes no phase shift.

Thin-Film Interference
For normal incidence, interference occurs between light reflected at the top and bottom of a thin film (thickness L and index of refraction n) according to the following:

If neither or both rays undergo a phase shift upon reflection, a bright reflection results when the round-trip optical path through the film is equal to an integral number of wavelengths.

If only one of the rays (either one) undergoes a phase shift upon reflection, a bright reflection results when the round-trip optical path through the film is equal to an odd number of half wavelengths.

The Diffraction Grating
A diffraction grating consists of many closely spaced, narrow slits. Light passing through the grating interferes constructively only for certain very sharply defined angles, given by the grating equation

$$\sin \theta_m = \frac{m\lambda}{d}$$

where d is the spacing between adjacent slits.

Single-Slit Diffraction
The angle θ_c between the central maximum and the center of the first minimum in a single-slit diffraction pattern is

$$\sin \theta_c = \frac{\lambda}{b}$$

where b is the width of the slit.

Diffraction Limit of Angular Resolution
The angular width of the central circle of brightness in a diffraction pattern produced by a circular aperture is the ultimate limit of resolution of the images of two point sources. This limit is given by

$$\sin \theta_c = 1.22\frac{\lambda}{D}$$

where D is the aperture diameter.

Polarization by Reflection (Brewster's Angle, θ_B)
Light is completely polarized by reflection from a boundary when the angle between the reflected ray and the refracted ray is 90°. The incident angle for this to occur is called Brewster's angle, θ_B, given by

$$\theta_B = \tan^{-1} n$$

where n is the index of refraction of the reflecting material relative to the incident material.

QUESTIONS AND GUESSTIMATES

1 The two loudspeakers in Fig. P24.1 are connected to the same oscillator and therefore send out identical sound waves. Under what conditions would you be able to notice an interference effect as you walked along line AB? What if the loudspeakers were replaced by light bulbs?

2 Two cars sit side by side in a large, vacant parking lot with their horns blowing. Would you expect to be able to notice interference effects from the two sound sources? What if the horns were replaced by two violins playing the same note?

3 A telephone pole casts a clear shadow in the light from a distant source. Why is no such effect noticed for the sound from a distant car horn?

4 Why is it impossible to obtain interference fringes in a double-slit experiment if the slit separation is less than the wavelength of the light being used?

5 Devise a Young's double-slit experiment for sound, using a single loudspeaker as a wave source.

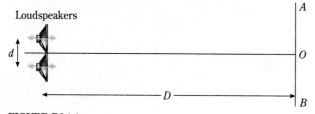

FIGURE P24.1

6 Mercury light consists of several distinct wavelengths. Suppose that in a double-slit experiment, filters are placed over the slits so that $\lambda = 436$ nm (blue) light goes through one slit and $\lambda = 546$ nm (green) light goes through the other. Is it possible to see an interference pattern on the screen?

7 What change occurs in a Young's double-slit experiment when the whole apparatus is immersed in water rather than air? What change is observed in an arrangement of Newton's rings if the space between plate and lens is filled with water?

8 Very thin films are sometimes deposited on glass plates. The thickness of the film can be controlled by observing the change in color of white light reflected from the surface as the film's thickness is increased. Explain.

9 Why does a glass or metal surface that has a thin oil film on it often reflect a rainbow of color when white light is reflected from it?

10 Figure P24.2 shows the interference fringes that are observed when glass plates are placed on top of optically flat surfaces (called *optical flats*). Tell as much as you can about the surface of the two plates used here.

11 Suppose two additional slits are added to the two slits in a Young's double-slit experiment, one on each side of the original two, so that there are four equally spaced slits. For a certain slit-to-screen distance, it is noticed that the center point of the fringe pattern is dark. Explain how this could occur.

12 Explain the following statement: The difference in thickness between the position of two adjacent bright fringes in a thin-film interference pattern is zero or $\lambda/2n$, where λ is the wavelength of light used and n is the index of refraction of the film.

13 Should a microscope have better resolving power when blue rather than red light is used? Explain.

14 Suppose you are given a diffraction grating whose characteristics are unknown. How can it be used to determine the wavelength of an unknown spectral line?

15 Press two pieces of flat glass (microscope slides are ideal) together in various ways and estimate from the interfering reflected light how close together the surfaces are. (You can see the interference pattern easily in any lighted room *provided* you get the plates close enough together.)

16 Assuming that diffraction caused by the pupil of your eye is the limiting factor, about how far away from you is an oncoming car if its headlights are just resolved?

17 What happens to the light energy that a polarizing sheet does not transmit when unpolarized light is incident on it? Can you think of any drawback this might pose to using a polarizing sheet?

18 How can one determine whether a beam of light is polarized? Whether it is composed of two beams, one polarized and the other not?

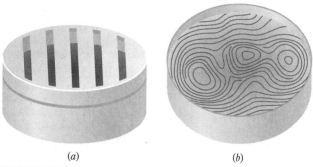

(a) (b)

FIGURE P24.2

PROBLEMS

Sections 24.1–24.2

1 Two identical wave sources at the coordinate origin send in-phase waves of wavelength 60 cm toward an observer on the x axis at $x = 6.0$ m. One of the sources is now moved slowly along the axis away from the observer. What are its first three x-coordinate positions at which the observer detects (a) constructive and (b) destructive interference?

2 Suppose the sources in the previous problem are at the origin and send out in-phase waves of known wavelength. As one source is slowly moved to negative x values, the observer notices constructive interference at several points on the x axis and that the distance between adjacent points is 20 cm. What is the wavelength of the waves?

3 A radio station sends out waves of wavelength 320 m. A home receiver 16 km away from the station receives these waves arriving by two paths. One is the direct path from the station, and the second is after reflection of the waves from a mountain directly behind the home receiver. Find the minimum distance from the mountain to the receiver such that destructive interference occurs at the receiver. Assume that no phase change occurs on reflection from the mountain.

4 Figure P24.1 shows two identical sound sources that vibrate in phase and send out waves of wavelength 20 cm. Maxima and minima of sound are heard as a receiver is moved along line AB. What is the path difference from the two sources at (a) first maximum to one side O and (b) the second minimum to one side of O?

5 The identical sound sources in Fig. P24.1 send out in-phase waves. An observer at A notices that loud sound occurs when a line from midway between the sources to A makes an angle of 30° with a line from midway between the sources

to O. If $d = 30$ cm, what possible wavelengths does the sound wave have? Assume $d \ll D$.

■ **6** Two identical sound sources vibrate in phase and send out waves of wavelength 60 cm toward each other along the x axis. The sources are located at $x = 0$ and $x = 6.0$ m. At what points along the x axis between them is the combined sound (a) maximum and (b) minimum?

■ **7** Two identical variable-frequency radio wave sources vibrate in phase and send out waves toward each other along the x axis. The two sources are separated, along the x axis, by 4.0 km. A home receiver is located between them at 2.5 km from one source. The equal frequencies of the sources are increased from zero at the same time. The combined intensity of the radio waves at the detector decreases with increasing frequency until it reaches a minimum and then begins to increase again. What is the wavelength of the radio waves at the minimum?

Section 24.3

8 Monochromatic 436-nm light is used in a Young's double-slit experiment, and the first-order maximum is found at 3.2°. (a) What is the slit separation? (b) At what angle does the second-order maximum occur?

9 The slits in a Young's double-slit experiment are 0.10 mm apart, and light of wavelength 600 nm is used. (a) At what angle does the third-order maximum occur? (b) The fifth-order maximum?

10 Green light of wavelength 550 nm is incident on a pair of two narrow slits separated by 0.5 mm. At what angle will the second-order maximum be observed?

11 The sound sources in Fig. P24.1 send out identical in-phase waves of wavelength 60 cm. If $d = 6.0$ m and $D = 30$ m, how far along AB from O are (a) the first-order maximum and (b) the second-order minimum?

12 In a double-slit experiment, the slits are separated by 0.2 cm and the slit-to-screen distance is 1.2 m. Light of wavelength 480 nm is used to illuminate the slits. With respect to the position of the central bright fringe locate the position of the first three (a) maxima and (b) minima on both sides of the central maxima.

13 A light of 460-nm wavelength is incident on two slits spaced 0.4 mm apart. What is the distance of the screen from the slits if the spacing between the first and second dark fringes is 3.6 mm?

14 What slit separation in a double-slit experiment gives a second-order maximum at 6.5 mm from the central bright fringe? The screen-to-slit distance is 2.0 m, and the wavelength of the light used is 550 nm.

■ **15** The two slits in a double-slit experiment are illuminated with blue light of wavelength 434 nm. The successive interference maxima are separated by 1.00 mm on a screen placed 1.0 m from the slits. What will be the separation of successive maxima if red light of wavelength 656 nm is used to illuminate the slits?

■ **16** When mercury light ($\lambda = 436$ nm) is used in a double-slit experiment, the first-order maximum occurs at an angle of 4.0×10^{-4} rad. When this light is replaced by a source of unknown wavelength, the second-order maximum occurs at 6.0×10^{-4} rad. (a) What is the wavelength of the second source? (b) In what region of the spectrum is it found?

■ **17** White light, which spans the wavelength range between about 400 nm to 700 nm, is incident on a pair of slits 0.3 mm apart. The interference pattern is observed on a screen 1.8 m from the slits. Find the distance between the first-order maxima of violet ($\lambda = 400$ nm) and red ($\lambda = 700$ nm) colors.

■ **18** A Young's double-slit experiment uses slits 0.30 mm apart. The entire apparatus is placed under water. At what angles do the first two interference maxima occur if light of 550-nm vacuum wavelength is used?

Sections 24.4 and 24.5

19 A flat glass plate is coated with a thin layer of a material with index of refraction 1.3. How thick should the coating be if normally incident light of wavelength 450 nm is to be transmitted with minimum reflection?

20 In Prob. 19 how thick should the coating be if a light of wavelength 560 nm is to undergo maximum reflection?

■ **21** A crown glass plate is coated with a thin film of thickness 140 nm. When light of wavelength 520 nm is incident normal to the film it is totally transmitted with minimum reflection. Find the index of refraction of the film. (*Hint:* Consider what reflective phase shifts are necessary to make $n > 1$.)

22 White light is incident on a thin glass sheet 400 nm thick, surrounded by air. What wavelengths in the visible spectrum of white light will be most strongly reflected at near normal incidence? Take the index of refraction of glass to be 1.5.

■ **23** A soap bubble strongly reflects both red ($\lambda = 700$ nm) and green ($\lambda = 500$ nm) when illuminated by white light. If the index of refraction of the soap bubble is 1.40, what thickness of the bubble allows this reflection to occur?

■ **24** A transparent oil of index of refraction 1.26 spills on the surface of ocean. Orange light of wavelength 600 nm is found to have maximum reflection when incident normal to the oil film. Find the minimum thickness of the oil film. Take the index of refraction of the ocean water to be that of pure water, $n = 1.33$.

■ **25** A metal mirror has a thin layer of plastic (index of refraction $n = 1.6$) coating its surface. It is found that the reflected intensity is minimum for a light of 550-nm wavelength. Find the two smallest possible thicknesses for the coating. (*Hint:* In effect, $n \to \infty$ for a metal.)

26 Two flat glass plates form a very thin air wedge between them. When the combination is viewed with light of wavelength 500 nm, a dark fringe exists at the line of contact. What is the thickness of the air wedge at (a) the first bright fringe and (b) the third bright fringe?

27 When blue light (wavelength 589 nm) is reflected from an air wedge formed by two flat glass plates, the bright fringes are 0.6 cm apart. (a) How thick is the air wedge 5.0 cm from the line of contact of the plates? Assume the wedge is viewed at normal incidence. (b) Repeat for the wedge filled with an oil of index of refraction 1.4 rather than air.

28 A wedge-shaped sliver of glass has an index of refraction of 1.56. When it is viewed from directly above with 460-nm light, the sharp edge of the wedge is dark. What is the thickness of the wedge at the fourth bright fringe?

29 An oil slick on a water puddle shows interference fringes. What is the difference in thickness of the oil slick at adjacent green fringes? Take the index of refraction of the oil to be 1.40 and the wavelength of green light to be 500 nm.

30 Sodium light of wavelength 590 nm is used to produce Newton's rings, and the radius of the tenth dark ring is found to be 1.64 cm. (a) How large is the air gap at this position? (b) If the gap is now filled with water, how big is the gap at the new position of this tenth dark ring? The center point of the pattern is dark.

31 The convex side of a planoconvex lens (flat on one side and convex on the other) of radius of curvature 4.0 m is in contact with a flat plate of glass. The front face of the lens is illuminated, at normal incidence, by a light of unknown wavelength. The radius of the 30th dark ring is found to be 7.6 mm. The center point of the pattern is dark. What is the wavelength of the light producing the pattern?

Section 24.6

32 Light of wavelength 680 nm is directed on a grating having 4000 lines per centimeter. What is the angular deviation of this light in (a) first order and (b) third order?

33 To calibrate a diffraction grating, a student sends red light from helium-neon laser (632.8 nm) through it. The first-order maximum occurs at an angle of 19°. (a) What is the grating spacing? (b) At what angle does the third-order maximum occur?

34 A sodium-arc yellow light is a doublet composed of two wavelengths, 588.995 and 589.592 nm. Compute the angular separation between these two lines in the first-order spectrum produced by a grating with 5000 lines per centimeter. Repeat for the second-order spectrum.

35 A diffraction grating has 6000 lines per centimeter. Calculate the angular separation between the blue (435.8 nm) and green (546.1 nm) lines of mercury in (a) the first-order spectrum and (b) the second-order spectrum.

36 Calculate the angular position of the second-order spectrum of the sodium yellow line (589 nm) produced by a grating with 5600 lines per centimeter.

37 For a certain grating, a second-order green line (546 nm) is found at 41.0°. At what angle will a first-order yellow line (589 nm) be found?

38 Light of wavelength 579 nm is normally incident on a grating with 5000 lines per centimeter. How many different orders of diffraction can be observed in transmission?

39 A diffraction grating with 6000 lines per centimeter is used in a large water tank. What are the three smallest angles (in water) at which a green mercury line (546.1 nm) will be seen?

40 White light that covers wavelengths from 400 to 700 nm is incident on a grating with 4000 lines per centimeter. How wide is the first-order spectrum on a screen 1.6 m from the grating?

Sections 24.7 and 24.8

41 Find the angular width of the central maximum (the angle between the two first-order minima) for a single slit that is 0.030 cm wide and is illuminated with a 590-nm light.

42 A single slit is illuminated with a 436-nm light, and its first-order diffraction minimum occurs at an angle of 1.8° from the center of the diffraction pattern. What is the width of the slit?

43 The diffraction pattern produced by a light of wavelength 589 nm passing through a narrow slit of width 0.2 mm is observed on a screen located 1.0 m away from the slit. Find the width of the central maximum as observed on the screen.

44 A single-slit diffraction pattern is formed by passing light through a narrow slit 0.060 mm wide. The width of the central maximum observed on a screen 2.0 m away from the slit is found to be 4.25 cm. What is the wavelength of the light?

45 Infrared radiation of wavelength 12.4 μm is allowed to pass through a narrow slit. The diffraction pattern observed on a screen 1.2 m away from the slit shows that the separation of the two first-order minima on both sides of the central maximum is 0.6 mm. What will be the new separation between the first-order minima if the width of the slit is reduced by half?

46 A man is looking at the headlights of a distant truck. If the pupil of his eye has a diameter of 0.24 cm, how far away is the truck when the two headlights are just resolved? Assume that the limiting factor is diffraction caused by the pupil. Take the wavelength of the light to be 490 nm and the separation of the lights to be 1.6 m. What can you conclude from your result?

47 A 3.0-cm-diameter lens is used to project the image of a photographic slide onto a screen 2.8 m away. The lens is 10 cm from the slide. Assume the lens is perfect, so that diffraction is the only limiting factor on its imaging ability. A light of wavelength 490 nm is used. How close together can two tiny spots on the slide be if they are to be resolved on the screen? What is their separation on the screen?

48 The Hale telescope at Mount Palomar Observatory in California uses a concave mirror 5.0 m in diameter. What must be the minimum distance between two points on the surface of the moon in order to be resolved by this telescope? The distance of the moon from the earth is 3.8×10^8 m. Assume the wavelength of light being imaged is 500 nm.

Section 24.9

49 Two polarizers with their polarization directions aligned transmit light of intensity I_o. What percentage of this intensity will transmitted if they are oriented at an angle of 50°?

50 Two identical polarizers with their polarization axis aligned transmit light of intensity I_o. At what angle should they be oriented if the transmitted intensity is to be $\frac{1}{2}I_0$?

51 Two polarizers oriented at 40° pass light of intensity I_1. What will the intensity of the transmitted light be if the polarizers are aligned with their polarization axes parallel?

52 An ideal polarizer passes 50 percent of the incident light intensity when the incident light is unpolarized. Unpolarized light of intensity I_o is incident on an ideal polarizer with its polarizing axis vertical. The transmitted light passes through a second polarizer whose axis makes an angle of 30° with the vertical. Finally, the light passes through a third polarizer whose polarization direction is horizontal. Find the intensity of the light emerging out of the second and third polarizers.

53 Unpolarized light is incident, from air, on a glass surface whose refractive index is 1.54. What is the angle of incidence for maximum polarization for the reflected light?

54 What is the Brewster angle for maximum polarization for light reflected at a water-air interface? Assume the incident light is within the water.

55 Show that, for a transparent medium surrounded by air, the Brewster's angle for maximum polarization by reflection θ_B and the critical angle for total internal reflection θ_c satisfy the relation $\cot \theta_B = \sin \theta_c$.

56 Calculate the angle of incidence for maximum polarization for light reflected at a water-glass interface, assuming the light is incident from within the water. Take the refractive index for glass as 1.52.

57 A beam of light is incident at Brewster's angle on a piece of transparent plastic material of refractive index 1.62. What is the angle of refraction for the transmitted beam?

General Problems

58 Radio waves of wavelength 200 m are received by a home receiver 20 km away from the station by two different paths. One is a direct path from the station and the second is reflected by a truck approaching the receiver, from the side opposite to the transmitter, along the straight line joining the transmitter and the receiver. Two successive destructive interferences of waves are observed at the receiver at a time interval of 18 s. How fast is the truck moving?

59 Light with wavelength 560 nm, together with light of unknown wavelength, falls on two slits of unknown separation. The fourth-order maximum of the 560-nm light falls exactly at the same position as the fifth-order maximum of the light of unknown wavelength. (*a*) What is the wavelength of the unknown light in air? (*b*) Repeat for the entire system in water.

60 A double-slit system immersed in water is illuminated by a 620-nm light. An interference pattern is formed on a screen 2.0 m away in the same tank of water. What is the distance on the screen from the central maximum to the second-order maximum if the slits are separated by 0.5 mm?

61 Two parallel glass plates are originally in contact and viewed from directly above with 590-nm light (yellow) reflected nearly perpendicular by the surfaces. As the distance between plates is slowly increased, darkness is observed at certain separation distances. (*a*) What are the values of the first three separation distances? *Hint:* Darkness is observed when the plate separation is zero. (*b*) Repeat for the gap between the plates filled with water.

62 (*a*) Is it possible to design a grating in such a way that the first-order 600-nm line will overlap with the second-order 400-nm violet line? (*b*) If so, how? (*c*) If not, could it be done for any other combination of orders? (*d*) If so, how?

63 Steel sheds often have a corrugated metal surface with corrugation repeating every 10 cm or so. Under appropriate conditions, this type of wall can act as a reflection grating for sound waves. What value of λ for sound waves at normal incidence will give rise to a first-order maximum at an angle of 41° to the normal?

64 A thin, opaque plastic sheet floating on the surface of a 4.0-m-deep swimming pool has a narrow slit of width 0.15 mm. Laser light of wavelength 633 nm is normally incident on the sheet. What is the width of the central maximum of the diffraction pattern at the bottom of the pool?

CHAPTER 25

OPTICAL DEVICES

Now that we understand the principles of reflection, refraction, and dispersion, we can discuss how these principles are applied in some common optical devices. We discuss imaging devices, such as the eye and the microscope, and devices used to measure light spectra. Not only will we obtain more practice in solving problems, but we will also become better able to use these devices in a wide variety of applications.

25.1 THE EYE

A simplified diagram of the eye is shown in Fig. 25.1. As you probably already know, the cornea is a protective cover, the iris diaphragm controls the amount of light entering, and the retina is a sensitive surface that changes the image formed on it to electrical energy, which is then transmitted to the brain. A ray of light entering the eye is refracted at the cornea. Lesser refraction effects occur in the pupil and lens since the refractive indices of the cornea, pupil, lens, and fluid portions of the eye are quite similar.

For the normal relaxed eye, these combined refraction effects form an image of distant objects on the retina. Hence, the focal length of the eye is about the distance from the retina to the lens along the lens's principal axis. From ray diagrams and

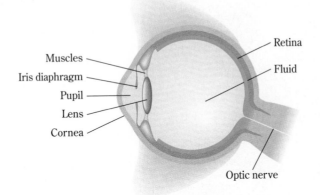

FIGURE 25.1

Diagram of the human eye.

also from the lens formula (Eq. 23.2), we know that for a fixed focal length, image distance must increase as the object is brought closer. In the eye, however, the image must always be formed on the retina, and so the image distance must remain fixed. This requires that the focal length of the eye be variable. This is the primary function of the eye's lens. Even though it contributes only about 20 to 25 percent of the total refraction, it is the ability to alter the shape of the lens which produces the required change in focal length. As a person focuses on a near object, the *ciliary muscles* attached to the lens cause it to thicken. Thickening produces a more converging lens, thus a shorter focal length. In the normal eye, this adjustment is limited to objects placed at a minimum distance of approximately 25 cm in front of the eye. The normal eye is thus capable of focusing on objects from a **far point** of infinity (relaxed eye muscles) up to a **near point*** of 25 cm.

In many people the eye lens cannot relax sufficiently to focus a very distant object on the retina. This is called **myopia**, or **nearsightedness.** The lens remains too converging, forming the image of the distant object well in front of the retina as shown in Fig. 25.2a. The myopic eye is able to focus only on objects

---

*You can find the near-point distance for your eyes by seeing how close you can hold a page and read it easily.

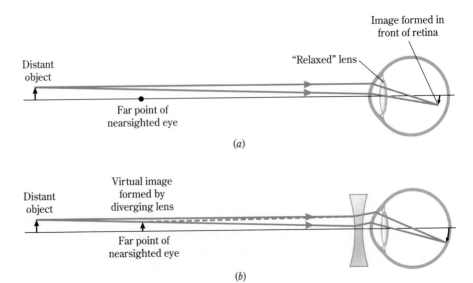

FIGURE 25.2

(a) The lens of a nearsighted eye (myopia) cannot focus on objects beyond a certain far point. (b) To correct for myopia, a diverging corrective lens is used to produce a virtual image of a distant object at the far point of the eye.

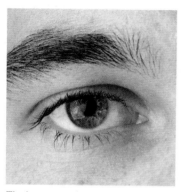

The human eye is a marvelous example of a simple camera. Its lens focuses light on the retina, and the iris adjusts the entrance aperture to varying conditions of light intensity.

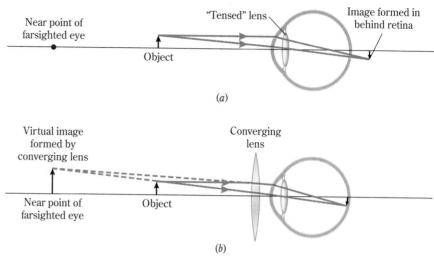

(a)

(b)

FIGURE 25.3

(a) The lens of a farsighted eye (hyperopia) cannot focus on objects placed closer than 25 cm, the normal near point. (b) To correct for hyperopia, a converging lens is used to produce a virtual image at the eye's near point when an object is placed at a distance of 25 cm. The focal length of the corrective lens must be greater than 25 cm. Why?

nearer than a certain *finite* far point. The correction for myopia is to add a *diverging* lens in front of the eye, which delays the formation of the image until the light reaches the retina.

Another way of understanding the function of a corrective lens is to remember that the image it forms is the object for the eye lens. The corrective lens should thus form a virtual image of an infinitely distant object at the far point of the myopic eye. This is shown in Fig. 25.2*b*.

A second vision defect is **farsightedness, or hyperopia** (Fig. 25.3). In this case the eye lens cannot be made sufficiently converging to focus objects located at the normal near point. People with hyperopia have the normal far point, but need a *converging* corrective lens in order to focus objects as close as 25 cm. The corrective lens should be chosen so that when an object is placed 25 cm from the eye, it forms a virtual image at the more distant near point of the hyperopic eye.

With advancing age, the eye lens in most people becomes more rigid and the ciliary muscles are no longer able to adjust the lens to focus objects at either the normal far point or the near point. We say that the eye has lost its *ability to accommodate*. The use of bifocal spectacles allows one to look through diverging lenses when looking straight ahead and converging lenses when looking down. Some people have three types of lenses built into a single spectacle lens, a trifocal lens. It furnishes good visibility for objects that are at far, intermediate, and near distances.

Example 25.1

A farsighted man is able to read the newspaper only when it is held at least 75 cm from his eyes. What focal length must the lenses of his reading glasses have? Assume that the distance between the glasses and his eyes is negligible.

R

Reasoning

Question   What does the 75 cm represent?
Answer   The near point of his eyes. He cannot focus on objects held closer than this.

Question   What must the corrective lenses do?
Answer   For an object held 25 cm in front of his eyes, the lenses must produce a virtual image at his near point, 75 cm. His eyes then can accommodate to focus on that image.

Question   What relationship exists between these data and the focal length of his reading glasses?
Answer   The thin-lens formula.

Question   What are the object and image distances?
Answer   As long as you neglect the distance between corrective lens and eye, you have the object and image at positions 25 cm and 75 cm, respectively. Both positions are in front of the lens.

Question   What signs must I use for p and i?
Answer   The object is real, so $p = +25$ cm. The image is virtual, so $i = -75$ cm.

Solution and Discussion   The thin-lens equation yields

$$\frac{1}{f} = \frac{1}{25\text{ cm}} + \frac{1}{-75\text{ cm}} = \frac{+2}{75\text{ cm}}$$

$$f = +37.5\text{ cm}$$

This positive focal length means a converging lens. You should be able to show that if the corrective lenses are actually 2 cm in front of the eyes, the required focal length is $f = +33.6$ cm. *Hint:* in this case, $p = +23$ cm and $i = -73$ cm.

Exercise   If your eyeglasses have $f = 60$ cm, what is your near point?
Answer: 43 cm

Example 25.2

What must the focal length of a corrective lens be for a woman whose far point is 75 cm?

R

Reasoning

Question   What sort of defect does this example describe and what must the corrective lenses do?
Answer   The far point for a normal eye is at infinity. The woman cannot see objects farther away than 75 cm. She has myopia. When she looks at a very distant object, the lenses must produce a virtual image of that object at her far point.

Question   What values must I use for p and i in the thin-lens equation?
Answer $p = +\infty$ and $i = -75$ cm.

Solution and Discussion

$$\frac{1}{f} = \frac{1}{\infty} + \frac{1}{-75\,\text{cm}} = 0 - \frac{1}{75\,\text{cm}}$$

$$f = -75\,\text{cm} \qquad \text{(a diverging lens)}$$

25.2   THE SIMPLE CAMERA

A camera (Fig. 25.4) operates very much like the human eye. It uses a lens to produce an image of an object on a film. The film serves the purpose of the retina in the eye—that is, the camera lens produces a real image on the film in much the same way that the eye lens produces a real image on the retina. The image is inverted on the film, and its size I is related to the object size O by the usual relation: $I/O = i/p$.

Unlike that of the eye, the lens of a simple camera is not made with a variable focal length. Hence, to achieve good focus on the film, the lens must be moved back and forth as the distance to the object changes.

Expensive cameras possess very complex lens systems instead of a single lens. The complexity is necessary if a camera is to give sharp images with fast shutter speeds. It is clear why sharp images are advantageous, and fast shutter speeds allow one to take sharp pictures of swiftly moving objects. Any movement will blur the image somewhat, but the shorter the time the camera shutter is open, the less blurred the image will be. Since the shutter must be open long enough to allow sufficient light to hit the film, fast shutter speeds mean that the lens must be large so that a large amount of light can enter the camera during a very short time.

As we saw in Sec. 23.11, only the central portion of a spherical lens can be used if a clear image is desired. This restriction becomes even more important if a camera is to be used to take closeup pictures, since then the lens must be very convex. It is only by making a complicated combination lens that the focusing errors inherent in a single lens can be eliminated. We say that such a lens has been corrected for **spherical aberration.**

Another lens defect causes images to have colored edges. This is called **chromatic aberration.** It results from the fact that the speed of light in glass varies with wavelength. As a result, the index of refraction of the glass is not the same for all colors. Blue light is refracted more strongly by the lens than red light is. This

The flexible bellows on this studio camera allows a large range of lens-to-film distances. This enables the photographer to position the lens close to an object so as to obtain greatly enlarged images. In this particular camera, the image fills a 20 × 24-in sheet of film.

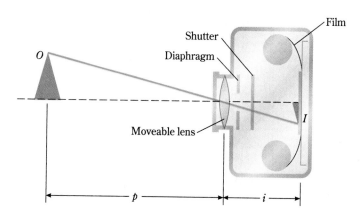

FIGURE 25.4

A simple camera. How is the image brought to focus on the film?

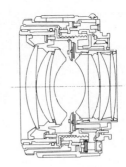

A modern high-performance camera lens is a combination of many lenses, designed with the help of complex computer calculations.

causes the colors in a beam of ordinary light to separate, and the image is therefore colored. To correct for this defect, two or more types of glass are layered together to form the lens. A lens that has been partly corrected for chromatic aberration is called an **achromatic lens.** However, it is impossible to free a lens of this defect completely.

Example 25.3

You have a camera whose lens has a focal length of +55 mm. As an object moves from very far away up to a position 25 cm in front of the lens, how far must the lens move in order to keep the image focused on the film? Must the lens move away from the film or toward the film? (Assume the lens can be considered to be a thin lens.)

R

Reasoning

Question   What tells me the distance the lens must travel?
Answer   The distance the lens has to move is the difference between the lens-to-film distances (image distances) required to produce images for these two object positions.

Question   What lens-to-film distance i is required to form these two images?
Answer   With $p = +\infty$ for the distant object, $1/f = 0 + 1/i$, or $i = f = 55$ mm. For the object at 25 cm

$$\frac{1}{55\,\text{mm}} = \frac{1}{250\,\text{mm}} + \frac{1}{i}$$

Question   Is there any way I can anticipate whether the lens moves toward or away from the film?
Answer   Since f is constant, the thin-lens equation requires that if you decrease p, you must increase i, and vice versa.

Solution and Discussion   For the object at 25 cm, the lens-to-film distance is

$$\frac{1}{i} = \frac{1}{55\,\text{mm}} - \frac{1}{250\,\text{mm}}$$

$$i = +70.5\,\text{mm}$$

Magnifying glasses are used for many purposes. Examples shown are (clockwise from top right): a general reading glass, a glass used for examining the thread count in textiles, a geologist's glass, and a magnifier for examining stereo photographs.

FIGURE 25.5

Why is only the magnified text in focus for the camera that took this photo?

This is $70.5 - 55 = 15.5$ mm farther from the lens than for the distant object. The lens must travel 15.5 mm away from the film to accommodate to the near object.

25.3 THE MAGNIFYING GLASS

One of the simplest optical instruments is the magnifying glass (Fig. 25.5). It is simply a converging lens, and is one of the basic parts of many optical devices. The function of the magnifier is to form an enlarged image of a small object placed close to the eye.

We can understand how a magnifying glass works by referring to Fig. 25.6. The size of the image formed on the retina increases as the object is brought closer and closer to the eye. However, the human eye is unable to focus well on objects that are closer than the near point. If we use a converging lens in front of the eye, as in Fig. 25.7, we can view the virtual image it forms. Even though the object is inside the near point (and therefore too close to be seen clearly), the image is formed at the near point. The eye uses this enlarged image as its object. The image then formed by the eye lens on the retina is the same as if an enlarged version of the object were placed at the near point. This retinal image is much larger than it would be if the actual small object were being viewed with the naked eye, and so much more detail can be seen.

Two methods are used to measure the magnifying effect in this case. The magnification we defined in Eq. 23.3, $M = I/O$, we call the **linear magnification.** This was shown to be equivalent to the ratio $-i/p$ (Eq. 23.3a). To use the magnifying glass we place the eye right behind the magnifying lens. Let us call the distance between the magnifier and the eye's near point p_n. As we see in Fig. 25.7, $i = -p_n$ when the image formed by the magnifier is at the near point. We then have

$$M = \frac{-i}{p} = -i\left(\frac{1}{f} - \frac{1}{i}\right) = p_n\left(\frac{1}{f} - \frac{1}{-p_n}\right) = \frac{p_n}{f} + 1 \tag{25.1}$$

where we have used the lens equation to substitute $i/f - 1/i$ for $1/p$.

The second method for describing magnification is to use a quantity called the **angular magnification.** We define it in reference to Fig. 25.8. Notice that when the object is placed at the near point of the eye, as in Fig. 25.8a, it subtends an angle ϕ at the eye. However, when it is placed inside the near point and viewed through

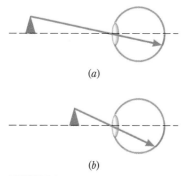

(a)

(b)

FIGURE 25.6

When an object is brought closer to the eye, the image on the retina becomes larger.

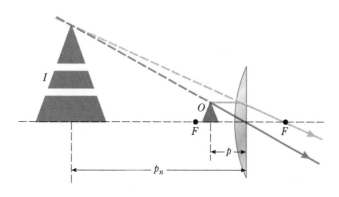

FIGURE 25.7

A magnifying glass allows one to place the object being examined far inside the eye's near point. Doing so enlarges the image on the retina.

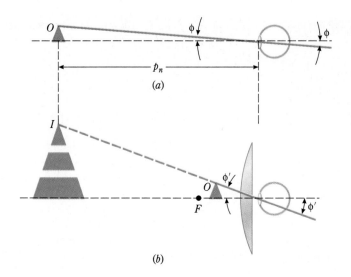

FIGURE 25.8
In both cases the eye is focused at the near point. (a) When the object is at the near point, the angle it subtends at the eye (and on the retina) is ϕ. (b) When the object is far inside the near point, the much larger angle it subtends is ϕ'. Because the image formed by the magnifying glass is at the near point, it can be seen clearly by the eye.

the magnifier, the object subtends an angle ϕ' at the eye. We define

$$\text{Angular magnification} = \frac{\phi'}{\phi} \qquad (25.2)$$

To obtain an expression for angular magnification in the present case, we note from Fig. 25.8 that

$$\tan \phi = \frac{O}{p_n} \qquad \text{and} \qquad \tan \phi' = \frac{I}{p_n}$$

Then, since the angles usually encountered in such situations are small, the tangents can be replaced by the angles themselves, giving us

$$\text{Angular magnification} = \frac{\phi'}{\phi} = \frac{I/p_n}{O/p_n} = \frac{I}{O} = \frac{p_n}{p}$$

This expression is identical to Eq. 25.1 for the linear magnification.

As we see, the two definitions give the same results under the present conditions. In practice, the image is often viewed at infinity with the relaxed eye rather than at the near point p_n. Then $p = f$, and the magnification is simply

$$M = \frac{p_n}{f} \qquad \text{(image viewed at infinity)} \qquad (25.1a)$$

As you see, M depends upon how the magnifier is used.

A typical simple magnifying glass might have a focal length of 5 or 10 cm. Since $p_n \approx 25$ cm, such a magnifier would provide a magnification of between 2.5 and 5. In other words, if all other factors remained constant, such a lens would allow you to observe details with dimensions as small as one-fifth the size that would be possible with the naked eye. Usually, however, other factors must also be considered. Among these are the blurring of the image due to spherical and chromatic aberrations of the lens. Also, as we saw in the preceding chapter, even with perfect lenses diffraction ultimately limits the detail that can be resolved.

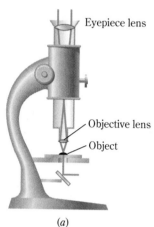

Eyepiece lens

Objective lens

Object

(a)

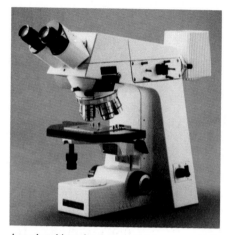

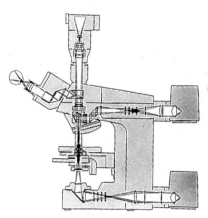

A modern binocular compound microscope. Notice the turret at the bottom, which affords a selection of objective lenses.

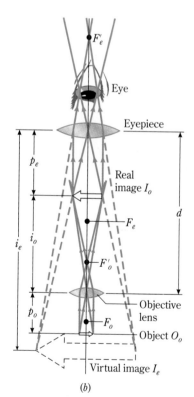

(b)

FIGURE 25.9

In the compound microscope, the eyepiece is used as a magnifying glass to observe the real image cast by the objective lens.

25.4 THE COMPOUND MICROSCOPE

The compound microscope achieves greater magnification than the simple magnifier by using a system of two lenses, each of which enlarges the object (Fig. 25.9). A lens called the **objective** gives an enlarged, real image I_o of an object placed very close to it on the stage of the microscope. In order to do this, the objective must be strongly converging and have a very short focal length f_o, often only a few millimeters. The second lens, called the *eyepiece,* or *ocular,* functions as a simple magnifier. The image I_o formed by the objective lies inside f_e, the focal point of the eyepiece, and becomes the object for the eyepiece. A final enlarged virtual image I_e is formed at the near point of the eye.

Let us now find an expression for the linear magnification of the microscope. We start with the linear magnification of the objective, which we label M_o. Combining the definition of linear magnification with the lens formula, we get

$$M_o = i_o/p_o = i_o\left(\frac{1}{f_o} - \frac{1}{i_o}\right) = \frac{i_o}{f_o} - 1$$

For the magnification of the eyepiece, M_e, we can use Eq. 25.1:

$$M_e = 1 + \frac{p_n}{f_e} = 1 + \frac{i_e}{f_e}$$

where, as before, p_n is the near point of the eye.

The *total* magnification M is the product of the magnifications produced by each lens. Thus

$$M = M_o M_e = \left(\frac{i_o}{f_o} - 1\right)\left(1 + \frac{p_n}{f_e}\right) \approx \frac{i_o p_n}{f_o f_e} \tag{25.3}$$

The latter approximation is justified when both focal lengths are very short, which is usually the case. In practice, i_o is about equal to the length of the microscope body (≈ 18 cm) and $p_n = i_e$ is approximately 25 cm.

As we see, f_o and f_e should be small for highest magnification. In order to accomplish this without serious distortion from various lens aberrations, carefully designed complex systems of lenses must be used for the eyepiece and objective instead of the simple lenses depicted in Fig. 25.9. The focal lengths used in Eq. 25.3 are then the equivalent focal lengths of these complex lenses.

Exercise 25.4

Suppose the objective of a compound microscope has a focal length of 5 mm, and its eyepiece has a focal length of 30 mm. The distance between objective and eyepiece is 230 mm. If the final image is to be at the near point of a normal eye, where must the object be placed? What is the linear magnification of the object?

Reasoning

Question   How are the positions of the final image and the original object related?
Answer   The objective forms an image I_o of the object that then acts as the object for the eyepiece. The lens equation applies to each lens.

Question   What lens-equation quantities are known?
Answer   For the objective, $f_o = +5$ mm with p_o and i_o unknown. For the eyepiece $f_e = +30$ mm and $i_e = -250$ mm. (You should be able to state why i_e is negative.) You have enough information to solve for p_e.

Question   Once p_e is found, how is it related to the position of the objective?
Answer   From Fig. 25.9b you can see that the distance d between the lenses is $d = p_e + i_o$. This gives you i_o, which then allows you to find p_o from the objective-lens equation.

Question   Do I then have enough information to calculate the linear magnification?
Answer   Yes. All the quantities in Eq. 25.3 are then known, if you realize that $p_n = i_e = 250$ mm.

Solution and Discussion   The eyepiece-lens equation is

$$\frac{1}{p_e} = \frac{1}{30 \text{ mm}} - \frac{1}{-250 \text{ mm}} = \frac{25 + 3}{750 \text{ mm}}$$

Thus $p_e = 26.8$ mm. With $d = 230$ mm, we get $i_o = 230$ mm $- 26.8$ mm $= 203$ mm. The objective-lens equation is then

$$\frac{1}{p_o} = \frac{1}{5 \text{ mm}} - \frac{1}{203 \text{ mm}} = \frac{203 - 5}{1015 \text{ mm}}$$

$$p_o = 5.13 \text{ mm}$$

This location is just outside the focal point of the objective. The magnification is

$$M = \frac{i_o p_n}{f_o f_e} = \frac{(203 \text{ mm})(250 \text{ mm})}{(5 \text{ mm})(30 \text{ mm})} = 340$$

25.5 THE ASTRONOMICAL TELESCOPE

In contrast to the microscope, the purpose of a telescope is to magnify objects that are far away. This is particularly true of astronomical telescopes, where the objects we study literally span the universe. Astronomers need the telescope to have abilities other than just forming an enlarged image. A good telescope must also (1) gather enough light from faint sources to make a bright image and (2) resolve as much detail as possible in the image.

The most important element of a telescope is the primary lens or mirror, the objective, which collects the light from a distant object and forms an image of the object. Since the distance to the object is essentially infinity, the image is formed at a distance f_o from the objective.

Telescopes which use an objective lens are called **refractors;** those which use a curved mirror for the objective are called **reflectors.** It is much easier and cheaper to build large mirrors than it is to build large lenses. Mirrors can be made lighter in weight, and they require only one precisely machined surface. For this reason, all large modern telescopes are reflectors. Among the largest single-mirror reflectors are the Hale telescope at Mt. Palomar, California, and one in Ukraine. These have objective mirrors of 5- and 6-m diameters, respectively. The largest refractor in the world is the 1-m-diameter lens at the Yerkes observatory in Williams Bay, Wisconsin, which was built approximately a century ago.

Telescopes can be used for direct viewing, in which case an eyepiece is used to magnify and view the image made by the objective, just as in a microscope. Usually viewing is direct only with small telescopes for casual use, however. Major telescopes involved in research are almost always used without an eyepiece. They function exactly as large cameras, with the objective lens or mirror forming an image on a photographic plate or an electronic sensor.

Let us discuss the performance criteria of astronomical telescopes in more detail. Although we will use lens diagrams, all the results we obtain apply to reflecting telescopes as well.

First, the size, or **scale,** of the image formed by the objective is proportional to its focal length f_o. We can see from Fig. 25.4 that, in a camera, both object and image subtend the same angle ϕ at the lens. Thus the size of the image on the film is $I = i \tan \phi$. For astronomical sources, $p_o = \infty$, $i_o = f_o$, and $\tan \phi \approx \phi$, where ϕ is

(*a*) A typical small personal telescope for casual viewing. (*b*) The 375-ton Mayall telescope at Kitt Peak National Observatory. The 4-meter diameter objective mirror is under a protective cover at the bottom of the picture.

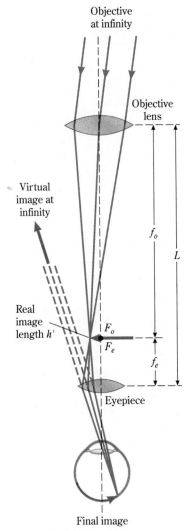

FIGURE 25.10

The astronomical telescope used with an eyepiece. Notice the differences between this telescope and the microscope in Fig. 25.9.

measured in radians. From this we can derive an expression for the image size:

$$I_o = 0.0175 f_o \phi \qquad (25.4)$$

where we have supplied the conversion factor enabling us to express the value of ϕ in degrees instead of radians.

The **brightness** B of the image is proportional to the area of the objective aperture, which is proportional to the square of the objective's diameter d. Brightness is also inversely proportional to the square of the focal length, f_o. Thus $B \sim (d/f_o)^2$.

A third performance criterion is the telescope's ability to resolve fine detail. Ultimately, this is the diffraction limit given by Eq. 24.7:

$$\sin \theta \approx \theta = \frac{1.22\lambda}{d}$$

If λ and d are expressed in the same units, θ is obtained in radians.

We can summarize these three criteria as follows:

1 A long objective focal length gives a large image having relatively little brightness. If brightness is no problem, as in a solar telescope dedicated to imaging the sun, one can afford to make a large image without worrying about being able to see it.
2 Both brightness and resolution of detail benefit from a large objective diameter, or aperture. If excellent resolution is achieved, image size becomes a secondary issue. Thus large objective diameter is the most important determinant of telescope performance.

Figure 25.10 shows how a telescope is used with an eyepiece. An objective forms a real image of an infinitely distant object at a distance f_o behind the objective. The objective focal length f_o is much longer than in a microscope. An eyepiece, acting as a simple magnifier, is placed so that its focal point F_e essentially coincides with F_o. The eyepiece focal length f_e is much shorter than f_o. The eyepiece thus forms a final virtual image of the object at infinity. The relaxed eye then views this enlarged image.

We can derive the expression for the angular magnification of a telescope fitted with an eyepiece with the help of Fig. 25.11. The angle ϕ subtended by the distant object at the objective is the same as the angle subtended by the image I_o at the objective. This relationship gives

$$\tan \phi \approx \phi = \frac{I_o}{f_o}$$

The magnified angle ϕ' seen by the eye is given by

$$\tan \phi' \approx \phi' = \frac{I_o}{f_e}$$

Taking the ratio of these two expressions gives

$$M_\phi = \frac{\phi'}{\phi} = \frac{I_o/f_e}{I_o/f_o} = \frac{f_o}{f_e} \qquad (25.5)$$

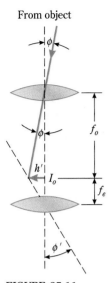

FIGURE 25.11

The telescope magnifies the angle subtended by very distant objects.

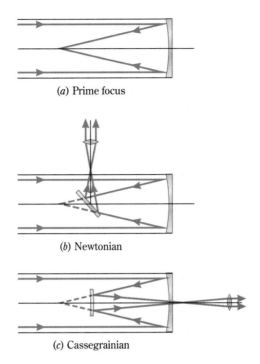

(*a*) Prime focus

(*b*) Newtonian

(*c*) Cassegrainian

FIGURE 25.12

Variations of mirror arrangements in reflecting telescopes.

Since mirrors reflect the light back up the axis of the telescope, astronomers have devised various ways of redirecting the light in reflectors to a convenient location. Some of these variations are shown in Fig. 25.12. The largest telescopes can accommodate instruments and even the astronomer right at the focus of the objective (called the **prime focus**) inside the telescope, as in Fig. 25.12*a*.

A second approach is the *newtonian* arrangement, employed for the first time by Isaac Newton, which is particularly useful in smaller telescopes. This design (Fig. 25.12*b*) employs a small, flat mirror mounted diagonally on the telescope axis slightly closer to the objective than the prime focus. This mirror intercepts the rays from the objective before they reach the prime focus, diverting them perpendicularly to the axis of the telescope. The rays pass through a small hole, coming to a focus as shown at the side of the telescope. Because most of the area of the objective mirror and hence most of the light it collects involve the outer parts of the mirror, the small, centrally located secondary mirror doesn't interrupt much light.

Figure 25.12*c* shows another mirror arrangement, the *cassegrainian*, which has a convex secondary mirror that redirects the light back down the axis of the telescope through a central hole in the objective mirror. The image is formed just beyond this exit hole. You can see that the cassegrainian arrangement extends the focal length of the objective by "folding" the path of the light. This makes it possible to reduce the physical length of the telescope while still maintaining the advantage of a longer-focal-length objective.

As we have discussed, both resolution and light-gathering ability are enhanced by making the diameter of the objective very large. When this is done, however, spherical aberration becomes serious because much of the light is reflected from parts of the mirror far from the axis. In order to eliminate this problem, most of the large objectives are made with a *parabolic*, rather than a spherical, cross section. Parabolic surfaces can accurately focus parallel rays even when they impinge on the mirror far away from the central axis.

While not used for astronomical observing except in a very casual way, **binocu-**

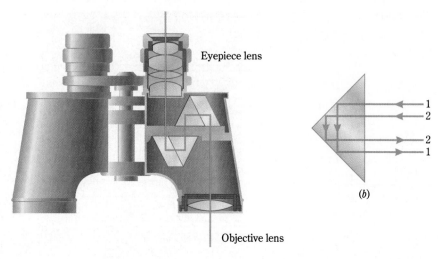

Eyepiece lens

Objective lens

(b)

(a)

FIGURE 25.13
The prism binocular.

lars are essentially a system of two telescopes side by side (Fig. 25.13). This allows the observer to see enlarged images with the depth perception that using both eyes provides. Moreover, prisms between the objectives and eyepieces are used to invert the image by total internal reflection as shown in Fig. 25.13*b*. This inversion counteracts the up-down and right-left inversions already caused by the objective. As a result the viewer sees an enlarged image that preserves the orientation of the original object.

Illustration 25.1

The full moon subtends an angle of 0.5° to an observer on earth. The Hale telescope at Mt. Palomar has an objective focal length of 16.8 m. What is the diameter of the image of the full moon at the prime focus of this telescope? Compare this with the size of the moon's image you would get using a camera with a typical lens of 50-mm focal length.

Reasoning Equation 25.4 gives the image size for a given focal length and subtended angle. For the Hale telescope,

$$I = 0.0175 f_o \phi = 0.0175(16.8 \text{ m})(0.5°) = 0.147 \text{ m}$$
$$= 14.7 \text{ cm}$$

For the camera,

$$I = 0.0175(50 \text{ mm})(0.5°) = 0.44 \text{ mm}$$

The moon would show up as a dot less than half a millimeter wide on your film!

A prism spectrometer. The prism is seen on the center pedestal. Light enters through a slit in the fixed arm at upper left, is dispersed through the prism, and images of the slit at the various wavelengths contained in the illuminating light source are viewed through the small telescope in the arm at the right. This arm can be moved and its angle relative to the fixed arm is read off through the small magnifier (black circle) above the base.

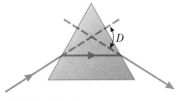

FIGURE 25.14

A prism deviates the light beam through the angle D.

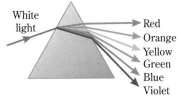

FIGURE 25.15

The angle of deviation by a prism is not the same for all wavelengths of light. Hence the prism disperses white light into its constituent colors.

25.6 THE PRISM SPECTROMETER

A prism, usually made of glass, is frequently used to separate light into its various colors. A beam of light is usually bent twice in a prism, once when it enters and once when it leaves. We call the total angle through which the ray is bent the **angle of deviation.** It is shown as angle D in Fig. 25.14.

With the incident angle, the angles of the prism, and the refractive index of the glass known, it is possible to compute D by using Snell's law. The higher the index of refraction of the glass, the larger the deviation of the beam. This has important consequences, as we now see.

We mentioned in Sec. 23.9 that the speed of light in most materials varies with wavelength. This is equivalent to saying that the index of refraction of the material depends on the color of the light. For most materials, the index of refraction for violet light is larger than that for red light. Hence, violet light is bent more by a glass prism than red light is. Consequently, if a beam of white light enters a prism, as in Fig. 25.15, the light is dispersed into its colors.

The ability of a medium to disperse light, called the **dispersion** of the medium, depends on the extent to which the refractive index varies with wavelength. Dispersion varies from material to material, as shown in Table 25.1. Flint glass, which is an example of a high-dispersion medium, shows a variation in its refractive index of slightly more than 3 percent over the visible spectrum.

TABLE 25.1 *Variation of index of refraction with wavelength (dispersion) for glass and quartz*

λ (nm)	Color	Crown glass	Flint glass	Fused quartz
360	u.v.	1.539	1.705	
434	Violet	1.528	1.675	1.467
486	Blue-green	1.523	1.664	1.463
589	Yellow	1.517	1.650	1.458
656	Red	1.514	1.644	1.456

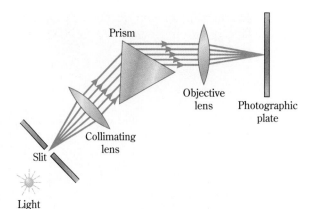

The dispersive property possessed by prisms is of great importance in scientific research and industrial applications. Because each atom and molecule can be induced to emit its own characteristic wavelengths of electromagnetic radiation, the wavelengths emitted by a substance help us to determine what the substance is. A device that uses a prism to separate a beam of light into the wavelengths that compose it is called a *spectroscope* or *spectrometer*. The simple prism spectrometer sketched in Fig. 25.16 is being used to analyze the wavelengths emitted by the light source. Let us suppose for this discussion that the source emits a single wavelength. (Sodium vapor lamps, the yellow lights often used on highways, emit essentially one visible wavelength, 589 nm.)

Light from the source enters the spectrometer through a narrow slit that is placed at the focal point of the collimating lens. Because the slit acts as an object placed at this focal point, the lens produces parallel light. Since they are of a single wavelength, the light rays are all deviated through the same angle by the prism and therefore emerge from it as parallel rays. As they pass through the objective lens, the parallel rays are brought to a focus at its focal point. There they produce an image of the object that produced them, namely, the slit. If a photographic plate or film is placed at the focal point of the objective, the image of the slit appears as a spectral line on the plate or film.

Each type of light source emits its own characteristic wavelengths, and we learn about the inner workings of atoms and molecules from them (Chap. 27). If a mercury vapor lamp (the bluish lamps often used as yard lamps) is used as the light source for the spectrometer, several spectral lines appear on the photographic plate, as shown in Fig. 25.17. Each line represents a wavelength in the spectrum of light emitted by mercury atoms. Atoms of each chemical element produce a spectrum unique to that element. These individual spectra can be thought of as a kind of identifying "fingerprint" of each element. Thus examining which wavelengths are

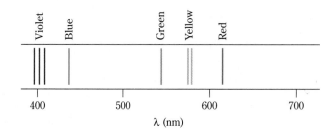

present in the spectrum produced by a source of unknown composition allows us to identify the elements present in the source.

Illustration 25.2

Suppose a beam of light in air is incident at an angle of 30° relative to the normal on a plate of flint glass. What would be the angle between the refracted rays of 434-nm light and 656-nm light? Use data from Table 25.1.

Reasoning
Snell's law gives the direction of the refracted rays:

$$n \sin \theta_r = \sin \theta_i \qquad \text{or} \qquad \theta_r = \sin^{-1}\frac{\sin \theta_i}{n}$$

In both cases, $\theta_i = 30°$, so that $\sin \theta_i = 0.500$. For $\lambda = 434$ nm, $n = 1.675$, giving

$$\theta_r = \sin^{-1}\frac{0.500}{1.675} = 17.37°$$

For $\lambda = 656$ nm, $n = 1.644$, giving

$$\theta_r = \sin^{-1}\frac{0.500}{1.644} = 17.71°$$

Thus, as they pass into the glass, these two colors would be separated by

$$17.71 - 17.37 = 0.34° \quad \blacksquare$$

LEARNING GOALS

Now that you have finished this chapter, you should be able to

1 Define (*a*) myopia and hyperopia, (*b*) near point and far point, (*c*) spherical aberration, (*d*) chromatic aberration, (*e*) linear and angular magnification, (*f*) magnifying power, (*g*) resolution, (*h*) brightness, (*i*) image scale, (*j*) dispersion, (*k*) spectral line.

2 Sketch the important features of the eye and explain the function of each.

3 Explain how corrective lenses can remedy myopia and hyperopia. Calculate the focal length of the corrective lens required when given the actual near or far points of the defective eye.

4 Explain the operation of the simple magnifier and compute its magnification.

5 Show how the compound microscope operates by sketching the placement of its objective and eyepiece and the location of the object. Draw a ray diagram to locate the image.

6 Sketch the optical arrangement for an astronomical telescope and locate the images it produces.

7 Explain how a binocular produces an image with the same orientation as the object.

8 Calculate the magnifying power of a compound microscope and of an astronomical telescope, given the focal lengths of objective and eyepiece.

9 Calculate the image scale and resolution limit for an astronomical telescope, given the focal length and diameter of its objective.

10 Explain how a prism spectrometer gives rise to line spectra. Describe how it separates wavelengths and can be used to analyze a beam of light.

SUMMARY

DEFINITIONS AND BASIC PRINCIPLES

Near and Far Points of the Eye
The near point is the closest point at which an object can be placed and be imaged on the retina when the eye is at maximum accommodation. This is normally 25 cm.

The far point is the farthest away an object can be placed and be imaged on the retina by the fully relaxed eye. This is normally at infinity.

Myopia and Hyperopia
Myopia, or nearsightedness, is a condition where the eye's far point is less than infinity.

Hyperopia, or farsightedness, is a condition where the eye's near point is greater than the normal 25 cm.

The Simple Camera
The simple camera is a single lens system. The lens is able to move toward or away from the focal plane (the film) in order to accommodate various object distances.

The Simple Magnifier
The simple magnifier is a converging lens used to create an enlarged virtual image of an object held close to the eye. The image is usually placed at the near point or the far point of the eye.

The Compound Microscope
The compound microscope is a two-lens system used to magnify objects placed very close to the objective lens. The objective is a short-focal-length, converging lens that forms a real image close to the eyepiece. The eyepiece is a simple magnifier.

The Astronomical Telescope
The astronomical telescope consists of a long-focal-length, converging lens or mirror (the objective) that forms a real image of a very distant object at its focal point.

When a telescope is used for direct viewing, a short-focal-length eyepiece is used as a simple magnifier to view the image formed by the objective.

Angular Magnification (M_ϕ)
The ratio of the angle ϕ' subtended at the eye by the image formed by an optical device to the angle ϕ the object itself would subtend to the naked eye.

Simple Camera: $M_\phi = 1$

Simple Magnifier: $M_\phi = 1 + \dfrac{p_n}{f}$ (image at near point)

$$M_\phi = \dfrac{p_n}{f}$$ (image at infinity)

Astronomical Telescope (only when eyepiece is used):
$$M_\phi = \dfrac{f_o}{f_e}$$

Linear Magnification (M)
The ratio of the height of the final image produced by an optical device to the height of the object.

Simple Magnifier: $M = M_\phi$

Compound Microscope: $M = \dfrac{i_o p_n}{f_o f_e}$

Image Scale or Size (I)
The image scale in a camera or telescope is the linear dimension I of the image from an object which subtends an angle ϕ at the objective.

$$I = 0.0175 f_o \phi$$

where f_o is the objective focal length and ϕ is the subtended angle measured in degrees.

Image Brightness (B)
The brightness of the image formed by an objective lens or mirror is proportional to the square of the ratio of objective diameter D to objective focal length.

$$B \propto \left(\dfrac{D}{f_o}\right)^2$$

Angular Resolution
The minimum angle that can be resolved by a perfect objective is given by the diffraction limit, discussed in Chap. 24. It is repeated here for the sake of completeness.

$$\theta_m = 1.22\dfrac{\lambda}{D}$$

where D is the diameter of the objective.

The Prism Spectrometer
The prism spectrometer uses the phenomenon of dispersion to separate light of various wavelengths. It consists of a prism whose index of refraction varies with wavelength (dispersion) and of lenses or mirrors that form an image of the slit for each wavelength emitted by the light source.

QUESTIONS AND GUESSTIMATES

1 In Chap. 23, we learned that the image in a plane mirror is the same size as the object. Why then do we place our face very close to the mirror when we examine our bloodshot eyes?

2 Show that a real image of a woman formed by a converging lens is inverted but that she and her image still have the same right hand. Show that exactly the reverse is true for an image formed by a plane mirror.

3 Clearer images are obtained in optical instruments when only a small portion of the lens is used. In the case of the pinhole camera, no lens is needed. To see how this is possible, draw a small, bright object about 1 mm high 10 cm from a 1-cm opening in a large, opaque screen. Show how the bright spot cast by the object on a screen 5 cm behind the opening decreases in size as the opening is made smaller. Show that in the limit of a pinhole opening, two objects 1 cm apart that are both 10 cm from the opening give rise to well-defined images on the screen.

4 Show why a pinhole placed in front of a lens leads to a good image even when the image is not quite in focus. (See Question 3.)

5 A glass prism deviates a beam of blue light somewhat more than a beam of red light. Show by means of wavefronts how this leads us to conclude that red light travels faster in glass.

6 Which of the following, as normally used, form real images: (*a*) eye, (*b*) camera, (*c*) microscope, (*d*) telescope, (*e*) binocular, (*f*) slide projector, (*g*) plane mirror, (*h*) concave shaving mirror, (*i*) searchlight mirror?

7 Explain clearly why spectral lines are called lines.

8 One can buy a cheap microscope for use by children. Invariably, the images seen in such a microscope have colored edges. Why is this so?

9 Suppose that the inside of a box camera is filled with water and that the lens is made stronger so that the image still falls on the film surface. Will the pictures that the camera takes be changed in any way? Repeat for a box with only a pinhole and no lens.

10 Why are the shutter speed and lens speed of a camera important? What design factors influence these speeds?

11 For a commercial camera in which the diameter of the lens opening, that is, the aperture diameter, is 5 mm, the proper exposure time for a scene is $\frac{1}{60}$ s. What is the proper exposure time for a pinhole camera using the same type of film if the diameter of the pinhole is 0.50 mm?

12 You have available a long, cylindrical, cardboard mailing tube and two lenses with focal lengths of 60 and 10 cm that can be fitted into the tube. Use these to design a toy telescope. What is the magnification of this telescope when used to view distant objects? How would you arrange these lenses for use as a microscope? Evaluate the performance of this microscope.

PROBLEMS

Section 25.1   (In the following, neglect the distance between the eye and the corrective lens.)

1 A tree 4.0 m high is 16 m from a person. What is the height of the image on the person's retina? Assume that the eye lens is 1.5 cm from the retina.

2 If the height of an image of an object on the retina of a person is 0.54 mm when the object is at the eye's near point (25 cm), what is it when the object is 4.0 m away?

3 The eye of a farsighted person has a near point of 90 cm. Objects nearer than 90 cm cannot be seen clearly. A converging lens is used to correct the vision of a book placed 25 cm from the eye. Find the focal length of this lens.

4 A myopic student is able to see the letters on the blackboard in the classroom clearly only when sitting less than 1.6 m away from the board. What should the focal length of the student's eyeglasses be in order to see distant objects clearly?

5 A person wears eyeglasses of focal length −80 cm. Where is the far point of the person?

6 The eyeglass prescription of a person is $f = +60$ cm. What type of eye defect does the person have?

7 During a year the prescription of a student's eyeglasses has changed from $f = -120$ cm to $f = -90$ cm. By how much has the student's near point shifted?

■8 A little boy wears thick, magnifying-glass-type eyeglasses. His older brother holds the eyeglasses in sunlight and obtains images of the sun. Each lens gives an image about 42 cm from the lens. What are the boy's probable far point and near point without glasses?

■9 A farsighted person whose near point is 60 cm without glasses wears eyeglasses with $f = +35$ cm. What is the corrected near point of the person?

Section 25.2

10 A simple camera uses a single lens of focal length 10 cm, and the size of the image formed on the film is to be 35 mm. What should the distance of a 3-m-tall object from the camera be if the image is to fit on the film?

11 A single-lens camera has a lens-to-film distance of 6 cm and takes pictures of size 6 cm × 4 cm. How far from a painting of size 80 cm × 80 cm must the camera be placed if the image of the painting is to fit on the film?

12 When the camera of Prob. 11 is used to photograph a tower from a distance of 20 m, the image on the film is 1.8 cm high. How high is the tower?

■13 A single-lens camera forms a clear image of a distant object when the lens is 7 cm from the film. (a) What is the focal length of the lens? (b) How far should the lens be moved to best focus an object 3.0 m away?

■14 A fixed-lens box camera uses a lens of focal length 25 cm, and the photographic plate is 25 cm from the lens. An object 4.0 m away from the camera is photographed. How far from the photographic plate is the image formed?

■15 A camera having a 50-mm-diameter aperture (lens opening) photographs an object adequately when the exposure time is 1/50 s. If the aperture is decreased to 35 mm, what exposure time should be used to produce a photograph with the same image quality?

Section 25.3

16 A lens having a focal length of 6 cm is used as a magnifying glass. (a) What should the position of the object be to obtain maximum magnification? (b) What is this value of magnification?

17 A magnifying glass enlarges the image of an object by an angular magnification of 5. Find the approximate focal length of the lens.

18 A person with near point of 20 cm uses a magnifying glass of focal length 6 cm. What magnification does he obtain if the image is at (a) his near point and (b) infinity?

■19 A student whose near point is 25 cm can view a 0.3-mm insect with her naked eye. She then uses a magnifying glass of focal length 8 cm to view the same insect. What is the approximate ratio of the two image sizes on the retina?

20 A magnifying lens of focal length 7.0 cm is used by a near-sighted student in such a way that the final image is formed at her near point, 15 cm. How much magnification does she obtain?

■21 A stamp collector uses a magnifying glass whose angular magnification is 8. He places a stamp 5 cm away from the magnifying glass. (a) Where is the image of the stamp formed? (b) Is it virtual or real?

Section 25.4

22 What is the approximate magnification of a microscope which uses an objective lens of focal length 3 cm and an eyepiece of focal length 9 cm? Assume the lenses are 18 cm apart.

23 The objective of a compound microscope alone produces a magnification of 20. Determine what focal length of the eyepiece is required to achieve an overall magnification of 2000. Assume that the final image is formed 25 cm from the eye and that the lenses are 18 cm apart.

24 A microscope is required to have a total magnification of 900. It has a tube of length 18 cm and uses an objective of focal length 0.90 cm. Find the focal length of the required eyepiece.

25 A microscope has a tube length of 18 cm and uses an eyepiece lens of focal length 4.0 cm and an objective lens of focal length 1.0 cm. What is the approximate magnification of the microscope?

■26 The objective lens of a compound microscope with a 20-cm-long tube has a magnification of 40. It uses an eyepiece that has a magnification of 16. What is the focal length of (a) the eyepiece and (b) the objective? (c) What is the total magnification of the microscope?

■27 A student makes a microscope by cementing a lens of focal length 6.0 cm to one end of an 18-cm-long tube and a lens of focal length 3.0 cm as an eyepiece on the other end. (a) About how far in front of the objective must she place the specimen she is examining? (b) What is the approximate magnification of her microscope?

■28 The first image of an insect in a laboratory microscope is formed inside the microscope 16 cm from the objective lens. The insect is 4.0 mm from the objective when the image is in focus. Find the focal length of the objective lens.

Section 25.5

29 An astronomical telescope used to view the moon has an objective of 60-cm focal length and uses an eyepiece of 3.0-cm focal length. What is the angular magnification of the moon obtained by the telescope?

30 An astronomical telescope has a 15-cm-diameter objective lens with a focal length of 75 cm. What is the magnification of the telescope if it is used with an eyepiece of focal length 2.5 cm?

31 A telescope uses an eyepiece that has a magnification of 5. The distance between the eyepiece and the objective is 55 cm. Find the overall magnification of the telescope.

32 A telescope at an observatory has an objective lens that has a focal length of 16 m. When this telescope is used to observe the moon, what distance on the surface of the moon will correspond to 1.0 cm on the image formed by the objective lens? (The distance of the moon from earth is 3.8×10^8 m.)

■33 What is the magnifying power of a telescope that uses an objective lens of focal length 100 cm and an eyepiece that has a magnifying power of 6?

■34 In a certain telescope, the distance between the objective lens and the eyepiece is 100 cm. The angular magnification of the telescope is 70. Find the focal lengths of the two lenses.

■35 You are looking at an 18-m-tall building 600 m away through a telescope with an overall magnifying power of 12. What angle, in radians, does the building subtend at your eye?

36 A reflecting telescope uses as its objective a mirror having a focal length of 80 cm. (a) How large an image of the moon does this mirror produce? (b) If the telescope uses an eyepiece that has a focal length of 5.0 cm, what is the magnifying power of the telescope? (Take the distance to the moon to be 3.8×10^8 m and the moon's diameter to be 3.5×10^6 m.)

37 A telescope with an objective lens that is 20 cm in diameter requires a 2.5-min exposure to properly photograph a distant star. What would the proper exposure time be if the objective of the telescope had a diameter of 25 cm?

38 A refracting telescope uses an objective lens of focal length 1.8 m and an eyepiece that has a focal length of +10 cm.

How much larger does a distant tower appear to be when viewed through this telescope?

Section 25.6

39 A beam of light consisting of only two wavelengths, $\lambda_1 = 434$ nm (violet) and $\lambda_2 = 589$ nm (yellow), is incident at an angle of 40° on a flat plate of flint glass. Find the angle between the two beams inside the glass plate. The index of refraction of flint glass is 1.528 for the violet and 1.517 for the yellow wavelengths.

40 A beam of light from a source which emits three wavelengths—434 nm, 656 nm, and 768 nm—is incident at an angle of 60° on a flat surface of a crown glass plate. This glass has indices of refraction of 1.546, 1.520, and 1.517 respectively for the three wavelengths. Calculate the angular separation between each pair of adjacent beams inside the glass plate.

■41 For this problem refer to Sec. 25.6 and Fig. 25.14. As the angle of incidence of light on the front face of the prism is varied, the angle of deviation D also varies. It can be shown that the angle D is a minimum when the ray of light inside the prism is parallel to the base of the prism. A measurement of the minimum angle of deviation D_{min} enables one to find the index of refraction of the material of the prism. Show that the index of refraction of the prism is given by

$$n = \frac{\sin\left[\frac{1}{2}(A + D_{min})\right]}{\sin(A/2)}$$

where A is the apex angle of the prism.

■42 The index of refraction of a certain glass is 1.4650 for $\lambda = 440$ nm and 1.4570 for $\lambda = 580$ nm. Calculate the minimum angle of deviation for each of these two wavelengths when they are incident on a prism of this material with an apex angle of 60°. *Hint:* Use the result of Prob. 41.

■43 Show that for a very thin prism whose apex angle A is very small, the angle of deviation D can be expressed $D = A(n - 1)$ for small angles of incidence.

■44 A light beam is incident at an angle of incidence of 48° on a face of a prism with an apex angle of 60°. The index of refraction of the material of the prism for this light is 1.590. Find (*a*) the angle at which the beam leaves the prism and (*b*) the angle of deviation D for this beam.

■■45 Yellow light of wavelength 589 nm is incident on the face of a fused quartz prism at an angle of incidence of 72°. The prism has an apex angle of 60°. The index of refraction of the prism material for the yellow light is 1.458. Find (*a*) the angle of refraction at the first face, (*b*) the angle of incidence at the second face, (*c*) the angle of refraction at the second face, and (*d*) the angle of deviation between the incident and emerging rays.

General Problems

■46 Show that the length of the image of an object on the retina is inversely proportional to the distance of the object from the eye.

■47 A teacher notices that a child in his class holds pages 15 cm from her eyes when reading. (*a*) Is the child myopic or hyperopic? (*b*) What kind of lens should be used to correct the child's vision, and what should its focal length be?

■■48 A private detective whose near point is 16 cm tries to use a diverging lens as a magnifier. (*a*) What must the focal length of the lens be if the detective is to see a distinct image? (*b*) If the lens has the focal length $f = -50$ cm, what maximum magnification can be achieved?

■■49 Two marks which are 0.0300 mm apart are viewed through a microscope. What angle (in degrees) do they subtend at the eye when viewed through a microscope having a magnifying power of 360?

■■50 A standard microscope (tube length 18 cm) normally uses an objective that produces a magnification of 20 and an eyepiece that has a magnification of 5. Suppose the 20× objective and 5× eyepiece are placed in a microscope with a tube of length 18.75 cm. Determine the ratio of the overall magnification of the latter arrangement to that of the standard microscope.

■■51 The diameter of the objective of a telescope is changed from 0.8 cm to 4.0 cm. (*a*) By what factor is the light intensity in the telescope increased if all the other dimensions are kept constant? (*b*) By what factor is the light intensity changed if the focal length of the objective lens is also doubled along with the increase in diameter?

■■52 A student has available two eyeglass lenses that have focal lengths of +100 cm and +36 cm. She wishes to place them in a cylindrical cardboard tube in order to make a telescope that is as short as possible in length and yet has the largest possible angular magnification. (*a*) About how far apart should the lenses be? (*b*) What will be the approximate magnification of the telescope?

■■53 It was shown in Sec. 25.5 that an astronomical telescope gives an inverted image. This is objectionable if one wishes to view an opera from a distant seat in an opera house. Instead, one can use a type of opera glass called a Galilean telescope. An example of a Galilean telescope uses an objective lens of focal length +40 cm and an eyepiece lens of focal length −20 cm placed 10 cm away from the objective lens. Locate the position of the final image of a distant object formed by the combination. Is the image real or virtual? Erect or inverted? What is the overall magnifying power of this telescope?

■■54 A certain type of glass has an index of refraction of 1.650 for blue light of wavelength 430 nm and an index of refraction of 1.615 for red light of wavelength 680 nm. A beam of light containing these two wavelengths is incident at an angle of incidence of 70° on one face of a prism made of this glass material. The apex angle of the prism is 60°. Find the angular separation $D_b - D_r$ (also called the dispersion) of the two wavelengths as they emerge out of the prism from the opposite face.

PART FIVE

MODERN PHYSICS

I think and think for months and years. Ninety-nine times, the conclusion is false. The hundredth time I am right.

ALBERT EINSTEIN

As the nineteenth century came to a close, many observers felt that, with the great successes made in understanding chemistry, electromagnetism, and thermodynamics, the subject of physics was nearly complete. Light had been shown to be a wave, and the electron had been discovered as a component of matter, indicating that atoms were electromagnetic in structure. Newtonian mechanics and his law of gravity were virtually unchallenged in their ability to predict the outcome of experiment. The classical universe was viewed as fully deterministic, operating according to a few simple principles with clockwork precision.

As the twentieth century began, however, many new experiments yielded results that were unexplainable by the tried-and-true classical laws. These results included the discovery of the nuclear atom, the way in which light interacts with electrons in metals, and the observation that the speed of light does not vary with the speed of the observer.

To explain all these confusing new observations, a fundamental revolution in our concepts of physical laws was necessary. The resulting framework of explanation, which we call *modern physics,* has two major components: relativity and quantum mechanics. The theory of relativity is necessary in order to explain observations on objects moving very fast (nearing the speed of light). Quantum mechanics has been able to explain the structure and behavior of atoms and nuclei by showing that particles on a very small scale are dominated by wave properties. This has had the effect of replacing the certainty of classical physics with the uncertainty of a probabilistic description of the interaction of matter and light on the atomic scale.

Classical physics still is valid in our everyday, "ordinary" experience—that is why it is valuable to study it. What has happened is that when we leave the realm of the ordinary to examine very small or very fast phenomena, we must leave our common sense prejudices behind and explain nature on its own terms. That we have been able to do so to such an extent in such a short period of history as the twentieth century is nothing less than a triumph for the human spirit and intellect. The quest is far from over. We realize this far better now than we did one century ago.

CHAPTER 26

THREE REVOLUTIONARY CONCEPTS

By 1900, many scientists felt that most of the great discoveries in physics had been made. To be sure, a few vexing problems remained, but it appeared that nearly all the fundamental physical laws had been found. As we will see in this chapter, such a view was completely incorrect. Vast areas of nature's physical behavior were still unknown at that time.

As we look through the history of science, we see that each truly great scientific advance is associated with the name of a single person. Galileo is recognized as the leader in our understanding of how objects undergo translational motion. Newton's name is enshrined in his three laws of motion and in the law of gravitation. Faraday pioneered the way to an understanding of magnetism, and Maxwell unified all electricity with his four fundamental equations. These and many other similar examples attest to the fact that the intellect of a single individual has the power to illuminate large areas of science for us all.

This is not to say that these individuals made their discoveries in isolation. Quite the contrary. Historians of science show clearly that each of these discoveries was the culmination of years of work by many others. Indeed, Newton once wrote, "If I have seen further than other men, it is because I stood on the shoulders of giants." Even so, other people stood on the shoulders of these same giants and saw nothing. While we must pay due respect to their predecessors, the insight and genius of these great scientists should not be underestimated. We should not stand in such awe of our scientific ancestors that we underestimate our own capabilities, however.

The discoveries we discuss in this chapter and in those that follow often came from unexpected sources.

PART I: RELATIVITY

26.1 **THE POSTULATES OF RELATIVITY**

Over the centuries, multitudes of experiments have been carried out to learn the laws of nature. In 1905, Albert Einstein became convinced that the experimental data force us to accept two seemingly innocuous facts of nature:

1 The speed of light in vacuum is always measured to be the same ($c = 2.998 \times 10^8$ m/s), no matter how fast the light source or observer may be moving.
2 Absolute speeds cannot be measured. Only speeds relative to some other object can be determined.

Assuming the truth of these statements, Einstein was able to show that many unexpected facets of the world about us were yet to be discovered. His line of reasoning is known as the *theory of relativity,** and the two statements of apparent fact above are its basic postulates.

It is not possible to prove these postulates directly. They are the consensus of all the experimental facts known. We consider it possible, though unlikely, that some experiment will someday disprove one of them, but they are as of now supported by many unsuccessful attempts to disprove them. Moreover, as we will see, Einstein's postulates lead to astounding conclusions that have been well verified by experiment.

The first postulate was the result of a series of experiments begun in 1887 by A. A. Michelson and his colleague E. W. Morley in the United States. At this time, most scientists believed that light waves were vibrations in a substance that filled all of space. This substance, described as early as the fourth century B.C. by Aristotle, was called the *aether*. On the one hand, the aether had to be very thin so as to allow planets and stars to move freely through it. Yet in order to carry the transverse vibrations of light at such a high speed, the aether must have the properties of a very rigid substance. These contradictions were hard to accept, but scientists held to the aether concept in part because it provided a rest frame of reference in which it should be possible to measure absolute motion.

Using an interferometer of his own design, Michelson surmised that he should be able to detect the motion of the earth through the aether by comparing the speed of light along the direction of earth's motion around the sun with the speed of light transverse to this motion. Light entering the interferometer would be split into two directions. Part of the light would travel along the direction of earth's motion. The other part would travel perpendicular to the earth's motion. Supposedly, the aether was like a river flowing through the device, carrying the light with it. Just as it takes a different length of time for a boat to travel a round-trip along a river's direction than it does to travel an equal distance across the river and back, the aether theory predicted that these two light beams would take different times to

---

*We discuss here Einstein's *special* theory of relativity. It applies only to measurements on objects that are not accelerating. In 1916, Einstein extended the theory to objects that are accelerating in his *general* theory.

return to the point where they had split. This difference in time would introduce a phase difference between the beams, a phase difference that should be observed as interference fringes when the beams are recombined.

The speed of the earth as it moves in its orbit around the sun is about $10^{-4}c$, well within the detection limit of Michelson's interferometer. Yet repeated attempts to measure the predicted effect showed *no interference effects whatsoever*. Michelson concluded that there was no aether flowing through the device and that the speed of light was the same in both paths. Increasingly precise measurements throughout the twentieth century have repeatedly confirmed this conclusion, which Einstein took as his first postulate.

The second postulate needs some explanation, perhaps. It is easy to measure the relative speeds of objects. A car's speedometer tells us at once how fast the car is moving relative to the roadway, but this is not an absolute speed. The earth is moving because of both its rotation on its axis and its motion around the sun. Since we know both these speeds, we could, if required, find the car's motion relative to the sun.

The sun itself is moving in our galaxy, the Milky Way, however, and the center of the galaxy is in motion relative to more distant stars. There seems to be no way to define a definite, absolute speed of an object since everything appears to be moving. We can state only how fast one object is moving relative to another.

There is another way to state the second postulate, a way that gives us an inkling of its fundamental importance. This alternate statement is usually made in terms of reference frames. *A* **reference frame** *is any coordinate system relative to which measurements are taken.* For example, the position of a sofa, table, and chairs can be described relative to the walls of a room. The room is then the reference frame used. Or, perhaps a fly is sitting on a window in a moving car. We can describe the fly's position in the car, using the car as a reference frame. Alternatively, we can describe the position of a spaceship relative to the positions of the distant stars. A coordinate system based on these stars is then the reference frame.

The second postulate can be stated in terms of reference frames in the following way:

2′ The basic laws of nature are the same in all reference frames moving with constant velocity relative to each other.

Often this statement is shortened by using the term inertial reference frame. *An* **inertial reference frame** *is a coordinate system in which the law of inertia applies:* a body at rest remains at rest unless an unbalanced force on it causes it to be accelerated. The other laws of nature also apply in such a system. To a very good approximation, all reference systems moving with constant velocity relative to the distant stars are inertial frames. Thus we have a third version of the second postulate:

2″ The basic laws of nature are the same in all inertial reference frames.

You can understand the relationship between these two alternative ways of stating the second postulate by considering the following. When we say that only relative speeds can be measured, we are assuming a lack of bias in reference frames. For example, a spaceship may be heading for the moon at a speed of 10^5 km/day relative to the moon, but it is also true that the moon is heading toward the ship at 10^5 km/day relative to the ship. The fact that one is moving relative to the other is easily ascertained, but the statements are equivalent to each other, and neither object can be said to be at rest in an absolute sense.

Suppose, though, that some law of nature depended on the speed of the reference frame. The people in the spaceship could use such a law to determine their speed. People on the moon could do likewise. The two measured speeds would be different. As a result, people would be capable of measuring more than just their relative speeds. In fact, the law could be used to set up an absolute ranking of speeds. This would contradict the second postulate, however, which we, along with Einstein, assume to be correct. We therefore conclude that all nature's laws must be the same in all inertial reference frames.

26.2 THE SPEED OF LIGHT AS A LIMITING SPEED

Using Einstein's two postulates, we can prove by logic alone that

No material object can be accelerated to speeds in excess of the speed of light in vacuum.

The validity of this statement is easily demonstrated in the following simple way. We prove it by the technique called *reductio ad absurdum,* in which we disprove a proposition (in this case, that an object can travel faster than c) by showing that it leads to a known false result (in this case, that an observer will measure a value different from c for the speed of light).

Suppose we have two nonaccelerating stations in space, shown as A and B in Fig. 26.1. They act as inertial reference frames. Observers at A and B have instructed

Spaceship   Light pulse

Station A Station B

FIGURE 26.1

What is the maximum speed with which the ship can pass between the two space stations?

Modern particle accelerators like this one at Fermilab at Batavia, Illinois, can accelerate protons to speeds nearly equal to the speed of light. Strong magnetic fields are used to bend the path of the protons in circles. The red and blue magnets (upper ring) are made of conventional copper coils. The yellow and red magnets are superconducting. In this accelerator the circular paths are four miles in circumference.

the spaceship operator to follow a straight-line path between A and B, traveling at the spaceship's top constant speed. Just as it passes A, it is to send a light pulse from the front of the ship toward B. Of course, A and B, working in partnership, can determine the speed of the spaceship by timing its flight from A to B. Let us make the false assumption that they find the speed to be $2c$.

The spaceship sent out a pulse of light as it passed A, and since the laws of nature must apply to all three inertial observers (A, B, and the person in the ship), the light pulse must behave in a normal way for each of them. Remember, the observer in the ship cannot tell whether or not the ship is moving, except in a relative sense. Therefore, the observer in the ship must see the light pulse precede the ship at the speed c and reach B before the ship does. Therefore A and B, working together, would see the light pulse moving faster than the ship. But they measure the ship as moving with speed $2c$, and so they find that the speed of the light pulse is greater than $2c$. But this is an impossible result, since it contradicts the known fact that all observers will always obtain c for the speed of light. We therefore conclude that our original assumption was false; the spaceship could not have been moving between A and B with a speed of $2c$.

This experiment will always lead to this contradiction as long as we insist that the speed of the ship exceed c. We therefore conclude that the spaceship cannot exceed the measured speed of light c. Indeed, we can enlarge this line of reasoning to include all material objects and signals that carry energy. As a result we can state:

Nothing that carries energy can be accelerated to the speed of light c.

As we proceed, we will see that this result of Einstein's theory also has repeatedly been tested carefully and has been found correct in every test.

26.3   SIMULTANEITY

Ordinarily we expect that two observers will agree as to whether or not two events occur at the same time. Einstein showed, however, that under certain circumstances the expected result does not correspond to reality. The basic postulates of relativity force us to conclude that events that are simultaneous in one inertial reference frame may not be simultaneous in another. To show this simply, we again resort to a thought experiment. The progress of a light pulse, as noted by two inertial observers, forms the basis for our experiment.

Suppose that a boxcar is traveling to the right at a very high constant velocity, as in Fig. 26.2a. At the exact center of the car is a high-speed flashbulb that will send out light pulses to the right and left when it flashes. The boxcar is fitted with photocells at each end, so that a man in the boxcar can detect when the light pulses strike the ends of the car. By some ingenious device, a woman at rest on the earth is also able to measure the progress of the two pulses. Because both observers are in inertial reference frames (the boxcar and the earth), each must see the light pulses behave according to the same laws of physics. Both the man and the woman must observe that both pulses travel from the flashbulb at the speed c. In addition, the man must observe that the pulses strike the detectors at the opposite ends of the boxcar simultaneously, since they travel the same distances in his reference frame.

Consider the man first. To him, the experiment is very simple. The flashbulb is

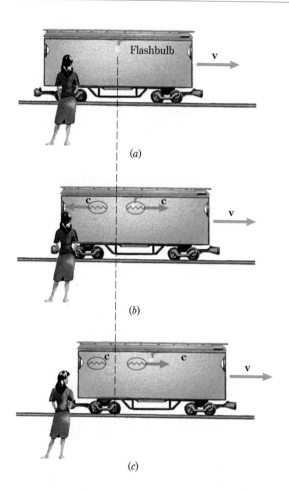

(a)

(b)

(c)

FIGURE 26.2

Unlike the inertial observer in the moving frame, the stationary observer on earth does not see the light pulses strike the ends of the car simultaneously.

at rest relative to him in the center of the car. When the bulb flashes, two pulses travel the equal distances to the two ends of the car in equal times. (Remember, for him the experiment must be the same whether or not the car is moving, because postulate 2 implies identical results for any inertial reference frame.) Hence the light pulses hit the two ends of the car simultaneously.

Now consider how the woman sees the experiment. Her measurements show that the experiment proceeds according to the laws of physics, and so the situation progresses as in Figs. 26.2*b* and *c*. Notice that the pulses travel equal distances to the right and left in equal time. But since the boxcar is moving to the right, the distance the light has to travel to reach the left photocell is shortened. As a result, the woman measures the pulse on the left as striking the left end of the boxcar before the other pulse strikes the right end. According to her, the light pulses do *not* hit the two ends of the car simultaneously.

We therefore conclude that time is not a simple quantity, because

Events that are observed to be simultaneous in one inertial system are not observed to be simultaneous in another inertial system moving relative to the first.

Further considerations show that this situation exists only if the two events occur at different locations. In our example, the events took place at the opposite ends of the car.

A corollary to the lack of simultaneity in two reference frames is that, to two observers in different inertial reference frames, events at two locations can appear *reversed in sequence*. That is, if one observer sees event A followed by event B, it is possible that another observer moving relative to the first observer will see event B followed by event A. This can happen only if one event did not physically cause the other. If A caused B, this cause-effect relationship (A preceding B) will be seen by all observers, although with different time lags.

26.4 MOVING CLOCKS RUN TOO SLOWLY

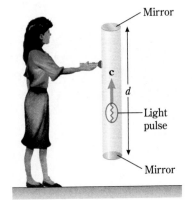

FIGURE 26.3

The light clock registers one click each time the light pulse is reflected from the lower mirror.

As we observed from the results of the previous section, time is not a simple quantity. Einstein pointed this out when he showed that the rate at which a hand-held clock ticks out time for the person holding the clock is different from the rate observed by a person moving past the clock. We demonstrate this effect in a thought experiment using a very special clock, but it was proved to be true in general by Einstein.

Consider the clock held by the woman in Fig. 26.3. It consists of a pulse of light reflecting between two mirrors in a cylindrical vacuum tube. Each time the light pulse strikes the lower mirror, it clicks out a unit of time that we will call a "click." If the tube is $d = 1.5\,\text{m}$ long, the woman computes that

$$1\,\text{click} = \frac{2d}{c} = \frac{3.0\,\text{m}}{3.0 \times 10^8\,\text{m/s}} = 10^{-8}\,\text{s}$$

Suppose that a man in a spaceship is using an identical clock. The woman looks out the window of her laboratory (which is in another spaceship) and sees the man shoot past her with speed v. She is pleased to see that he is using a clock like hers and contacts him by radio. He tells her that the clock is functioning well and is ticking out time as usual, one click each $2d/c$ seconds.

After thinking about it a bit, the woman discovers that there is something very peculiar about this. She concludes that the man's clock must be ticking out time more slowly than hers. We can understand her reasoning as follows.

Since the man's clock is operating properly for him, she knows it must be operating as in Fig. 26.4. We see there the clock in its position at two consecutive clicks. The woman observes that the light pulse moves along the path indicated. Although

FIGURE 26.4

The light pulse in the moving clock must travel a distance larger than $2d$ during a one-click interval. The light-pulse path length is $2\sqrt{d^2 + (\frac{1}{2}vt_w)^2}$.

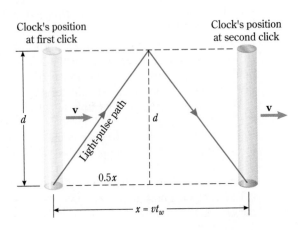

Clock's position at first click

Clock's position at second click

Light-pulse path

d

d

$0.5x$

$x = vt_w$

the man sees the pulse move straight up and down in the clock, the woman asserts that the pulse moves to the right as well, because of the movement of the clock to the right.* The woman computes the time between clicks on the man's clock as follows.

According to her, the pulse moves a distance given by the blue line in the figure. From the pythagorean theorem and the dimensions given in the figure, we see that

$$\text{Pulse path length} = 2\sqrt{d^2 + (\tfrac{1}{2}x)^2}$$

The woman knows that the man's clock is traveling past her with speed v. Further, according to her clock, it will take the man's clock a time t_w to move from one position to the other. Therefore, she knows that $x = vt_w$. As a result, according to the woman,

$$\text{Pulse path length} = 2\sqrt{d^2 + (\tfrac{1}{2}vt_w)^2}$$

Further, she knows that a light pulse always travels through vacuum with speed c. According to her, then, the time taken for the change in position shown in Fig. 26.4 should be

$$t_w = \frac{\text{pulse path length}}{c} = \frac{2\sqrt{d^2 + (\tfrac{1}{2}vt_w)^2}}{c}$$

We can solve for t_w in this equation and find

$$t_w = \frac{2d/c}{\sqrt{1 - (v/c)^2}}$$

But we recognize $2d/c$ as the time that the man insists it takes for his clock to make one click. We therefore have the following result:

$$\begin{array}{l}\text{Time interval on} \\ \text{stationary clock}\end{array} = \left[\frac{1}{\sqrt{1 - (v/c)^2}}\right] \left(\begin{array}{l}\text{time interval on} \\ \text{moving clock}\end{array}\right)$$

The quantity $\sqrt{1 - (v/c)^2}$ is called the **relativistic factor.** Figure 26.5 is a graph of the relativistic factor as a function of v/c. Notice that this factor is essentially equal to 1 until speed becomes greater than about 10 percent of the speed of light. Even for $v = 0.10c$, the factor is 0.995. In almost all everyday observations, we are not aware of the effects of relativity because we do not encounter speeds this great. However, when experimenting with atomic particles, relativistic effects are very common, and experimental results cannot be explained without taking Einstein's equations into account.

As an example of the effect of the relativistic factor, suppose that the man is moving past the woman at a speed of 0.75c. Then $\sqrt{1 - (v/c)^2}$ has a value 0.66, and the inverse of this is 1.51. Under these conditions the woman's clock will tick out 1.51 clicks during the time she knows the man's clock takes to tick out 1 click. As we see, the moving clock ticks out time more slowly than the stationary clock.

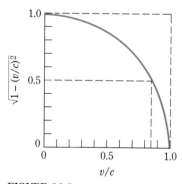

FIGURE 26.5

The relativistic factor differs appreciably from unity only at speeds that approach the speed of light.

*You might ask: "Who is correct?" They both are, as we will soon see. Each person is describing the behavior correctly as measured in her or his reference frames.

A clock that is moving with speed v ticks out a time of $\sqrt{1 - (v/c)^2}$ seconds during 1 s on a stationary clock.

After arriving at this unexpected result, the woman contacts the man by radio and informs him that she has discovered that moving clocks tick out time too slowly. Before she can give him the details, he states that he has been thinking along the same lines. He has discovered that her clock, which was moving past him with speed v, was ticking out time too slowly. Then they both recall that only relative motion has meaning. Neither clock is special.

Any clock that is moving relative to an observer will appear to tick out time more slowly than a clock that is stationary with respect to the observer.

We call this effect **time dilation,** since time is stretched out, so to speak, for moving clocks.

This astonishing result applies to all timing mechanisms, no matter how complex. If the man had been using the growth rate of a fungus as a clock, the woman would have found the fungus growth rate to be slowed by its motion. Even aging of the human body will be slowed by motion at high speed, as we see in one of the following examples.

There is one point we should always remember, however. A good clock always behaves normally to a person at rest relative to it. Observers moving past the clock may claim that it ticks out time too slowly. In spite of this, the clock still ticks out time properly as viewed by an observer stationary relative to it. The time ticked out by a clock that is stationary relative to the observer is called the **proper time.**

Illustration 26.1

One striking example of time dilation is obtained by measuring how long unstable particles "live." For example, a particle called the pion lives on the average only about 2.6×10^{-8} s when at rest in the laboratory. It then changes to another form. How long would such a particle live when shooting through the laboratory at $0.95c$?

Reasoning   The pion is moving with speed $0.95c$ relative to observers in the laboratory. Experiments should show that the internal clock of the pion, which controls how long it lives, should be slowed because of its motion. A time of 2.6×10^{-8} s read by the moving clock should be as follows when timed by the laboratory clock:

$$\text{Life according to lab clock} = \frac{2.6 \times 10^{-8}\,\text{s}}{\sqrt{1 - (0.95)^2}} = 8.3 \times 10^{-8}\,\text{s}$$

As we see, the moving pion should live about 3 times as long as a stationary one. This experiment and variations of it have been carried out. The results found by experiment agree with the computed results.

Exercise   How fast must the pion be moving if it is to "live" 10^{-7} s? *Answer:* $v/c = 0.966$.

Example 26.1

The star closest to our solar system is Alpha Centauri, which is 4.1×10^{16} m from earth. Since light moves with a speed of 3×10^8 m/s, it would take a pulse of light 1.37×10^8 s, or 4.3 yr, to reach there from the earth. (We say that the distance to the star is 4.3 light-years.) How long would it take, according to earth clocks, for a spaceship to make the round-trip if its speed is $0.9990c$? According to clocks on the spaceship, how long would it take?

Reasoning

Question What is the ship's speed measured relative to?
Answer Consider the earth and Alpha Centauri to be a fixed distance apart. The ship's speed is $0.9990c$ measured relative to earth. Measurements by people aboard the ship and by people on earth will agree on this value.

Question Whose clocks will measure "proper" time?
Answer Earth clocks, since they are at rest in the earth–Alpha Centauri reference system.

Question How fast will the spaceship clocks appear to run?

Answer They will appear to run slower by the relativistic factor, $\sqrt{1 - \left(\dfrac{v}{c}\right)^2}$.

Solution and Discussion Earth clocks would measure a round-trip time of

$$t = \frac{d}{v} = \frac{2(4.3 \text{ light-years})}{1.0 \text{ light-year/yr}} = 8.6 \text{ yr}$$

The relativistic factor is

$$\sqrt{1 - (0.999)^2} = 0.045$$

Thus the spaceship clocks will have ticked off only

$$(8.6 \text{ yr})(0.045) = 0.39 \text{ yr}$$

This is a little less than 5 months!
   Incidentally, the twin of one of the crew who was left behind on the earth would age 8.6 yr during the voyage. The twin in the spaceship, however, would age only 0.39 yr. This phenomenon, the **twin paradox,** has been discussed at length by scientists. They generally agree that this result is valid and that the two twins actually will age differently.*

*To test this effect, an extremely precise clock transported around the earth by plane has been compared with a stationary "twin" clock. The expected result was found. For a discussion of the experiment, see J. Hafele, *Physics Teacher,* **9**:416 (1971).

26.5 RELATIVISTIC LENGTH CONTRACTION

The time-dilation effect implies a peculiar effect involving measured lengths. To see what this effect is, consider once again the man and woman of the previous section. Let us say that the woman is on the earth while the man is traveling with speed v along a straight line from earth to Alpha Centauri. Astronomers on earth tell us that this star is $d = 4.1 \times 10^{16}$ m away from the earth. Because relative speeds can be measured easily, the man and woman agree that their speed relative to each other is v as the man in the spaceship shoots from the earth to the star. The woman is at rest in a reference frame in which the earth and the star are also at rest. She sees the man shooting past her at speed v.

The man is at rest relative to his spaceship, and he takes the ship itself to be his reference frame. Relative to the ship, both the earth and the star are moving with speed v. Let us now examine the man's flight from the earth to the star from the woman's vantage point.

The woman knows that the distance from earth to star, both at rest relative to her reference frame, is $d_e = 4.1 \times 10^{16}$ m, where the subscript e stands for "earth." Using $x = vt$, she computes that the time registered by her earth clock for the man's trip to the star will be

$$t_e = \text{earth time} = \frac{d_e}{v}$$

Indeed, when the ship turns around at the star and returns to earth, the total time the ship has been in flight is $2t_e = 2d_e/v$.

However, the man's computation will be different. Using spaceship clocks, he times his flight from earth to star as taking a time t_s. He can then compute the distance to the star by use of $x = vt$ and obtain

$$d_s = vt_s$$

where the subscript s refers to measurements in a reference frame at rest relative to the spaceship. A similar computation for the return trip tells him that his total flight covered a distance $2d_s$ in $2t_s$.

We therefore have the following equations that are undeniably correct for the two observers who formulated them:

$$2d_e = v(2t_e) \quad \text{and} \quad 2d_s = v(2t_s)$$

But we know that time dilation influences the spaceship clock in such a way that, when it is compared with the earth clock after the spaceship returns to earth, we have

$$t_s = \sqrt{1 - \left(\frac{v}{c}\right)^2}\, t_e$$

The spaceship clock ticked out time more slowly than the clock on earth. Substituting this value for t_s in the expression for d_s yields

$$d_s = v\sqrt{1 - \left(\frac{v}{c}\right)^2}\, t_e$$

However, $d_e = vt_e$, and so $t_e = d_e/v$. Using this value for t_e gives

$$d_s = \sqrt{1 - \left(\frac{v}{c}\right)^2}\, d_e$$

In other words, the distance from the earth to the star measured by the man in the spaceship is smaller than the distance measured by astronomers on earth. Apparently, if you are in motion relative to two points that are a fixed distance apart, the distance between the two points appears shorter than if you were at rest relative to them. The ratio of the two distances is the relativistic factor, $\sqrt{1 - (v/c)^2}$.

Einstein found this to be a general result. We can summarize it as follows:

If an object and an observer are in relative motion with speed v, then the observer will measure the length of the moving object as having contracted along the line of motion by a factor of $\sqrt{1 - (v/c)^2}$.

Notice that the contraction occurs only along the line of motion. No such contraction is observed perpendicular to the direction of motion. The length of an object measured by an observer at rest relative to it is called the **proper length.**

We can now reconcile the measurements of the earth-based observers in Example 26.1 with those made by the spaceship occupants. The length-contraction factor is the same as the time-dilation factor, 0.045. The distance from earth to Alpha Centauri can be thought of as a very long roadway moving past the spaceship. Measured on earth, the length of this roadway is a proper length, but measured from the spaceship this length is contracted to a value of

$$d_{\text{ship}} = 0.045 d_{\text{earth}} = (0.045)(4.3 \text{ light-years})$$

$$= 0.19 \text{ light-years}$$

The occupants of the ship see themselves as traveling this road at a speed $v = 0.999c$. Thus they conclude without any surprise that the round-trip would take them only

$$t_{\text{ship}} = \frac{2(0.194 \text{ light-year})}{0.999 \text{ light-year/year}} = 0.39 \text{ year}$$

Illustration 26.2

An astronaut traveling at high speed in a spaceship holds a meterstick in her hand. What does she notice about the length of the stick as she rotates it from a position that is parallel to the line of motion to a position that is perpendicular?

Reasoning She notices no change in the stick's length. The length-contraction effect concerns objects moving at high speed relative to the observer. The meterstick is at rest relative to the woman. ∎

26.6 THE RELATIVISTIC MASS-ENERGY RELATION

In Sec. 3.11 we stated that Einstein's theory of relativity predicts that an object's mass depends on its speed, an effect that becomes very noticeable when that speed approaches c. At the time we did not introduce the postulates of relativity to explain this effect, so let us do so now.

As we saw in Sec. 26.2, these postulates tell us that no object can be accelerated to speeds in excess of the speed of light. This limit on speed conflicts with Newton's laws of motion, as pointed out in Chap. 3. Newton's laws would predict that the speed of an object would continue to increase without limit as long as a net force continued to act on the object:

$$v = v_o + at = v_o + \frac{F}{m}t$$

where the mass m is considered to be constant. This relationship violates the speed limit postulate of Einstein, because given enough time, $v_o + (F/m)t$ can become larger than c. Einstein determined that, to be consistent with the relativity postulates and the law of conservation of momentum, an object's mass must increase with its speed. In this way, the term F/m decreases as t increases, so that v approaches the limiting value of c as t becomes very large. Einstein's postulates led him to conclude that the mass-speed relationship must be

$$m = \frac{m_o}{\sqrt{1 - v^2/c^2}} \qquad (26.1)$$

where m_o is called the **rest mass,** equal to the mass we have used in Newton's laws. The speed-dependent mass m is called the **apparent mass** of the object. A graph showing this variation of mass with speed is shown in Fig. 3.25. As we saw there, the apparent mass m is close to the rest mass m_o as long as v/c is less than a few tenths. As v approaches c, $v/c \rightarrow 1$ and $\sqrt{1 - v^2/c^2} \rightarrow \sqrt{1 - 1} = 0$. This makes the apparent mass approach infinity:

$$m \rightarrow \frac{m_o}{\sqrt{1 - 1}} = \infty$$

The charge and energy of nuclear particles can be studied by the tracks they leave in going through a bubble chamber like this one. For charged particles, the tracks are curved because of a magnetic field transverse to the direction of motion. Often the oppositely curved tracks of an electron-positiron pair, oppositely charged particles of equal masses, are observed to be created from a single gamma ray. The total energy of the pair is calculated to be equal to the gamma ray energy, in accordance with Einstein's prediction that $E = mc^2$.

The variation of mass with speed can be used to justify the fact that no object can be accelerated to a speed in excess of the speed of light. An infinite mass would require an infinite force to accelerate it. Because infinite forces are not available, it is apparent that an object with a speed $v \to c$ cannot be accelerated to the speed of light, a speed at which its mass would be infinite.

The force that acts to accelerate an object gives energy to the object. At low speeds, we know that the work done by a net applied force equals the increase in the kinetic energy of the object, provided changes in potential energy and friction work are negligible. This is still true at speeds close to c, but the kinetic energy of an object is no longer given by $\frac{1}{2}m_o v^2$. Nor is it, *as one might guess,* $\frac{1}{2}mv^2$. Instead it is found that the kinetic energy of an object is

$$\text{KE} = (m - m_o)(c^2) \tag{26.2}$$

When $v \ll c$, Eq. 26.2 reduces* to the classical expression for kinetic energy, $\text{KE} = \frac{1}{2}m_o v^2$.

When you don't know an object's speed, but know the amount of energy it has been given, there is a very useful way of determining whether you must use Eq. 26.2 or $\frac{1}{2}m_o v^2$ for the KE of the object. Calculate the rest mass energy of the object $m_o c^2$. Compare this with the amount of energy given to the object. If the energy given is more than one- or two-tenths of the rest mass energy, we say that the object is "relativistic," and you must use Eq. 26.2. If the energy that is given is less than this, the object is "classical," and $\text{KE} = \frac{1}{2}m_o v^2$ will usually suffice. (As always, this depends on the precision you need in your calculations.)

Equation 26.2 states that kinetic energy is the difference between the terms mc^2 and $m_o c^2$. Furthermore, it implies that an object at rest (KE = 0) still contains some fundamental energy, $m_o c^2$, which we call the **rest mass energy.** Einstein was able to show that a relation similar to Eq. 26.2 applies to all types of energy. He showed that *for any change in the energy of an object, there is a corresponding change in the object's mass,* given by

$$\Delta E = \Delta m\, c^2 \tag{26.3}$$

(This is often written as $E = mc^2$, the most-often cited equation of Einstein.) Notice that Eq. 26.3 can also be written as $\Delta m = \Delta E/c^2$. Because c^2 is a very large number, this implies that significant changes in mass require enormous changes in energy. In our everyday "classical" world, energy changes from chemical reactions or small changes in KE or potential energy are too small to provide measurable mass changes. It is only when we observe energy changes involving nuclear reactions that mass changes become dramatically evident, as we will see.

---

*To show this, make use of the mathematical fact that, for $x \ll 1$, the quantity $1/\sqrt{1-x} \cong 1 + \frac{1}{2}x$. Then, if we call $(v/c)^2$ the quantity x, in the case when $(v/c)^2 \ll 1$,

$$m = \frac{m_o}{\sqrt{1 - (v/c)^2}} \cong m_o\left(1 + \frac{1}{2}\frac{v^2}{c^2}\right)$$

Therefore Eq. 26.2 becomes

$$\text{KE} = mc^2 - m_o c^2 \cong m_o c^2 + \frac{m_o}{2}\frac{v^2}{c^2}c^2 - m_o c^2 = \frac{1}{2}m_o v^2$$

R

Example 26.2

Electrons are routinely accelerated in the laboratory through 1 million volts of electric potential. They thus acquire a kinetic energy of 1 MeV. How fast are these electrons going, and what is their mass as measured in our frame of reference?

Reasoning

Question   How can I tell the correct relationship between KE and speed to use here?
Answer   Calculate the electron's rest mass energy. With $m_o = 9.1 \times 10^{-31}$ kg, $m_o c^2 = 8.2 \times 10^{-14}$ J $= 0.511$ MeV. Since KE $= 1$ MeV, these electrons are certainly relativistic, and you must use Eq. 26.2 for their KE.

Question   How is speed involved in Eq. 26.2?
Answer   Mass depends on speed according to Eq. 26.1. Putting the value for m from Eq. 26.1 into Eq. 26.2 will give you an equation for v/c:

$$\mathrm{KE} = m_o c^2 \left(\frac{1}{\sqrt{1 - v^2/c^2}} - 1 \right)$$

Solution and Discussion   The above equation gives us

$$\frac{\mathrm{KE}}{m_o c^2} + 1 = \frac{1}{\sqrt{1 - v^2/c^2}}$$

Squaring and inverting, we get

$$1 - \frac{v^2}{c^2} = \frac{1}{(\mathrm{KE}/m_o c^2 + 1)^2} = \frac{1}{8.74} = 0.114$$

This gives $(v/c)^2 = 0.886$, and thus $v/c = 0.941$. The electrons are traveling at 94 percent of the speed of light. Their mass is almost three times the rest value:

$$m = \frac{m_o}{\sqrt{.114}} = 2.96 \, m_o$$

Exercise   Determine the speed that would be predicted classically for these electrons.   *Answer: 5.93×10^8 m/s $\approx 2c$!*

Illustration 26.3

The available chemical energy in a 100-g apple is about 100 kcal (the nutritionists leave off the prefix *kilo-* and call them Calories). We learned in our study of heat that 1 cal is 4.184 J of energy, and so an apple contains about 420 kJ of available energy. Compare this with the energy one could obtain by changing all the mass to energy.

Reasoning According to the mass-energy relation,

$$\text{Energy} = \Delta m \, c^2$$

In this case $\Delta m = 0.10\,\text{kg}$ and $c = 3 \times 10^8\,\text{m/s}$, giving

$$\text{Energy} = 9 \times 10^{15}\,\text{J}$$

We see from this that when we eat an apple, we obtain only a very small fraction (5×10^{-11}) of its total energy. ∎

PART II: PHOTONS

26.7 PLANCK'S DISCOVERY

In 1900, five years before Einstein proposed his theory of special relativity, Max Planck (1858–1947) made a discovery which seemed less than earth-shaking at the time but which we now recognize as the first of a Pandora's box of surprises. Planck, along with others, had been trying to interpret the radiation given off by hot, nonreflecting objects, so-called *blackbodies* (Sec. 11.11). Careful measurements of the intensity of the visible, infrared, and ultraviolet radiation given off by hot objects indicated that the intensity varies with wavelength as shown in Fig. 26.6. As we see, only a small fraction of the emitted radiation has wavelengths in the visible range. Most is in the infrared range. Furthermore, the radiation maximum shifts from the infrared to the visible as the temperature is increased. This agrees with our experience that a white-hot body is hotter than a red-hot one.

In order to interpret these curves, let us ask what sort of transmitting antenna could be sending out electromagnetic radiation from the hot object. Since the wavelengths involved are very short, the frequency of the vibrating charges must be very large. For example, at a wavelength of 1000 nm we have

$$\text{Frequency} = f = \frac{c}{\lambda} = \frac{3 \times 10^8\,\text{m/s}}{10^{-6}\,\text{m}} = 3 \times 10^{14}\,\text{Hz}$$

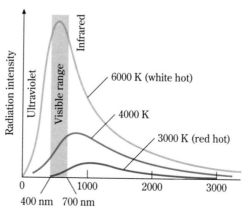

FIGURE 26.6

Blackbody radiation. For comparative purposes, the temperatures correspond as follows: 6000 K (sun's surface), 4000 K (carbon arc), 3000 K (very hot tungsten lamp).

Molten steel radiates energy at a high rate, providing a vivid demonstration of the T^4 dependence of the Stefan-Boltzmann radiation law.

Notice how very high this frequency is. Only in atomic-size antennas can charges be oscillated this fast. As a result, we expect that the electromagnetic radiation is being emitted by the vibrating charges within the atoms and molecules that compose the hot object.

There are many models we could postulate for these atomic or molecular oscillators. For example, if the object were composed of diatomic polar molecules, the vibrating molecule would be represented as shown in Fig. 26.7. The two atoms are held together by a springlike force, and since the molecule is polar, its two atoms carry equal and opposite charges. As the atoms vibrate back and forth, they act as vibrating charges on an antenna and therefore emit electromagnetic radiation of frequency f_o, the natural frequency of vibration of the molecular spring system. At least, this is the way Planck and his contemporaries reasoned.

It turns out, however, that all theories of radiation based on this model failed to describe the radiation from hot objects accurately. The theories were capable of duplicating the curves of Fig. 26.6 at long wavelengths but gave completely incorrect predictions at short wavelengths. It was Planck who discovered how the theory could be modified to agree with experiment. His modification is easily understood but difficult to justify. In fact, his only justification for it was that it gives the correct answer. Let us now see what he had to assume to get agreement between theory and experiment.

As we know, the amplitude of vibration of an oscillating system depends on the energy of the system. Although the frequency of vibration is always f_o, the amplitude increases as the energy increases. According to the accepted concepts of Planck's time, an oscillator could have *any energy* in a continuous range of values.

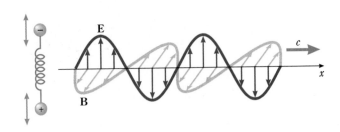

FIGURE 26.7

Before 1900, it was thought that the dipolar molecule acts like a radio antenna and sends out electromagnetic waves as it vibrates.

FIGURE 26.8

The radiation spectrum of a blackbody at $T = 1600\,\mathrm{K}$. Experimental data points are shown as circles. Classical radiation theory (dashed line) approaches the data at long wavelengths, but completely fails to explain the decrease at short wavelengths. Planck's theory (solid curve), assuming quantized energies of oscillating molecules, conforms remarkably to the observed behavior.

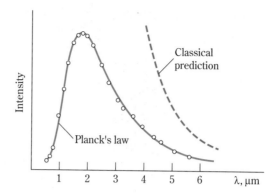

Since this assumption led to disagreement with experiment, Planck asked a "what if" question. Without any justification, he decided to assume that oscillators could have only *discrete* values of energy:

An oscillator of frequency f_o can vibrate only with energies hf_o, $2hf_o$, $3hf_o$, . . . , nhf_o. No other values of energy are possible.

The quantity h is a constant of proportionality now called **Planck's constant.** Planck found that with this assumption he could obtain excellent agreement with the observed spectrum of radiation by hot objects (Fig. 26.8) if h had the value

$$h = 6.626 \times 10^{-34}\,\mathrm{J \cdot s}$$

Planck's assumption was truly astonishing. It had the effect of **quantizing** the energies an oscillator is allowed to have. This concept that energy comes in indivisible "lumps," or **quanta,** instead of being able to occur in any conceivable amount, had never been encountered before. No experience with mechanical systems had given anyone reason to suspect this.

To see why energy quantization is not readily perceived in the laboratory, let us consider the oscillation of a pendulum. Its energy is mgH, where H is its highest vertical position. Planck's idea says that the energies of the pendulum can occur only in integral multiples of the basic quantum hf_o. To see what this means, let us consider a pendulum that has a natural frequency $f_o = 1\,\mathrm{Hz}$ and a bob of mass $100\,\mathrm{g}$. The heights to which this pendulum could swing would be

$$H_1 = \frac{hf_o}{mg} = \frac{(6.6 \times 10^{-34}\,\mathrm{J \cdot s})(1\,\mathrm{s^{-1}})}{(0.10\,\mathrm{kg})(9.8\,\mathrm{m/s^2})} = 6.7 \times 10^{-34}\,\mathrm{m}$$

or $H_2 = 2H_1 = 13 \times 10^{-34}\,\mathrm{m}$, or $H_3 = 3H_1 = 20 \times 10^{-34}\,\mathrm{m}$, and so on. No intermediate maximum vibration heights are possible.

Notice that the difference between successive allowed heights of vibration as predicted by Planck is only about $10^{-33}\,\mathrm{m}$. In comparison, the diameter of an atom is on the order of $10^{-10}\,\mathrm{m}$, and that of the atomic nucleus is approximately $10^{-14}\,\mathrm{m}$. The gaps between allowed heights are far too small to be measured. This turns out to be the case for all commonly encountered examples of oscillation. Thus we cannot observe the effects of the quantized energies using large-scale (laboratory-size) vibrating systems.

Planck was therefore faced with a disturbing situation. He could obtain a suitable theory for the radiation from a hot object provided that he was willing to make the assumption outlined above. The experimental test of it for other vibrating systems appeared to be impossible. Therefore, at the time, it was viewed by both Planck and his contemporaries as a rather curious result but one of doubtful validity. We shall see, however, that it appears to be correct and of extreme importance.

26.8 EINSTEIN'S USE OF PLANCK'S CONCEPT

Five years after Planck's discovery, Einstein showed that another natural phenomenon involved Planck's constant h. In explaining the results of an experiment first performed by Heinrich Hertz, Einstein postulated that light had particle as well as wave properties. Einstein's postulate, later verified, has become an integral part of modern physics.

In 1887, Hertz (who also produced and detected the first radio waves) discovered that light could dislodge electrons from a metal plate. We now know this to be a general phenomenon: *short-wavelength electromagnetic energy incident on a solid can cause the solid to emit electrons.* This is called the **photoelectric effect,** and the emitted electrons are called **photoelectrons.**

An experiment for observing the photoelectric effect is shown in Fig. 26.9. A metal plate is sealed in a vacuum tube, together with a small wire called the collector. (Such an arrangement is called a photocell.) These elements are connected in a circuit with a battery and galvanometer, as shown. When the tube is covered so that no light enters it, the current through the galvanometer is zero, since the portion of the circuit from plate to collector inside the bulb lacks a connection. The vacuum space has essentially infinite resistance.

If short-wavelength light is incident on the plate, the galvanometer needle deflects. The direction of the current shows that electrons are leaving the plate and traveling to the collector. One might guess that the light heats up the plate and that when it becomes hot, electrons with high thermal energy escape from it. This is not the case, however. Careful experiments have shown that no matter how feeble the

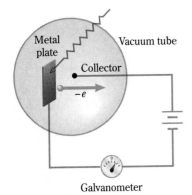

FIGURE 26.9

When light strikes the plate, electrons are emitted from it.

The exposure meter and pocket calculator are examples of devices whose operation is based on the photoelectric effect.

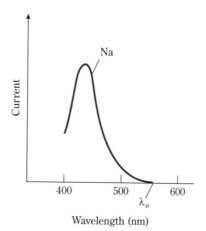

FIGURE 26.10

The current in the circuit of Fig. 26.9 varies with wavelength as shown for the metal sodium. What is the meaning of the λ_o value indicated?

light and no matter how massive the metal plate, a stream of electrons is emitted from the plate the instant the light reaches it. No heating is required.

It is further observed that, for a given light source, the number of electrons emitted from the plate is proportional to the intensity of the light (its energy per unit area per second). If the battery voltage is large enough to attract all the emitted electrons to the collector, then the current in the galvanometer is directly proportional to the light intensity. (It is for this reason that a photoelectric cell is used to measure light intensity.)

A more startling feature is shown in Fig. 26.10. Suppose that the wavelength of a light beam can be varied while its intensity is kept constant. Then the current in the circuit of Fig. 26.9 can be monitored as the variable-wavelength beam is incident on the plate of the photocell. It is found that the current varies with wavelength in the way shown in Fig. 26.10. Other plate materials yield similar curves but with different values for λ_o, the wavelength at which the current in the circuit becomes zero.

The most startling feature of these curves is that no electrons are emitted if the wavelength of the light is larger than λ_o, called the **photoelectric-threshold wavelength.** No matter how intense the light, no electrons are emitted if its wavelength is even slightly longer than λ_o. No matter how weak the light, if the light has a wavelength shorter than λ_o, electrons are emitted essentially as soon as the light is turned on. The particular value of λ_o, the critical wavelength for electron emission, depends on the material from which the plate is made.

Another experiment involving the circuit of Fig. 26.9 yields further important data. In this experiment, a beam of known wavelength and intensity is directed at the plate. The energy of the fastest electron emitted by the plate is then measured. To carry out that measurement, the battery is replaced by a variable-voltage source *of reversed polarity*. Because the collector is now negative instead of positive, it repels the photoelectrons. The current in the circuit drops to zero when this reverse voltage is made large enough. At the voltage V_o (the *stopping potential*) that results in zero current, the work done by the fastest photoelectron as it travels from the plate to the collector is $V_o e$ because it moves through a voltage difference V_o. This work must equal the kinetic energy of the most energetic photoelectron. We can therefore determine the maximum kinetic energy of the photoelectrons by measuring the stopping potential V_o: $(KE)_{max} = V_o e$.

When V_o is measured for various incident wavelengths, an interesting result is found. When $(KE)_{max}$ is plotted against $1/\lambda$, the result is a straight-line relation, as

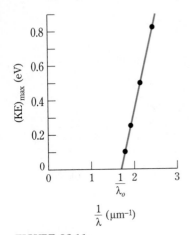

FIGURE 26.11

The photoelectron energy varies inversely as the wavelength. This particular graph is for sodium.

shown in Fig. 26.11. Moreover, the value of λ at which $(KE)_{max}$ becomes zero is the threshold wavelength, λ_o. The equation of a straight line, $y = mx + b$, becomes in this case

$$(KE)_{max} = \frac{A}{\lambda} - B \qquad (26.4)$$

where $1/\lambda$ replaces x, A replaces m, and the intercept b is $-B$. The constant B varies from substance to substance, but A, the slope of the line, is the same for all materials and has a value of 2.0×10^{-25} J · m.

Many attempts have been made to explain all these observations in terms of the wave nature of light. None has been successful. Two basic difficulties are encountered by any wave interpretation:

1 How can we conceive of waves giving rise to a threshold wavelength? Light with λ just slightly less than λ_o does not differ appreciably from light with λ just slightly greater than λ_o. Yet wavelengths slightly shorter than λ_o cause electrons to be emitted, whereas those that are just slightly longer do not.
2 How can even the weakest possible beam of light cause electrons to be emitted as soon as the light is turned on? The light energy seems to localize on one electron instantaneously and causes it to break free from the solid.

Thus it appeared that a new approach was needed to explain the photoelectric effect. This bold, imaginative step was taken by Einstein, who seized on Planck's ideas of special oscillator energies. Einstein reasoned that if the atomic oscillators in a hot object were to emit radiation in the way Planck visualized, the energy must be emitted in little bursts or packets. For example, since electromagnetic radiation carries energy, an oscillator that is emitting light must be sending out energy. However, since an oscillator can have only certain discrete energies, it cannot throw out energy continuously. It must throw out the energy in bursts of magnitude hf_o because this is the spacing between the allowed energies of the oscillator.

To be specific, suppose an oscillator has energy $37hf_o$. If it loses energy by sending out radiation, its energy can change to $36hf_o$ but not to anything in between these two values, since the oscillator's energies are quantized. But in so doing, the oscillator must have thrown out a pulse of light or other radiation with energy hf_o. We call such a pulse of electromagnetic energy a **light quantum** or **photon.** Hence we see that there is some justification for thinking that a beam of light consists of a series of energy packets called photons. These photons would act like *particles* of light, traveling at speed c and carrying energy hf.

Einstein therefore postulated the following character for light:

A beam of light with wavelength λ (and frequency $f = c/\lambda$) consists of a stream of photons. Each photon carries energy hf.

We see later how the photon energy is related to the structure of atoms and molecules. Let us now apply Einstein's model for a light beam to the photoelectric effect.

If light does consist of photons, they will collide with individual electrons as the light beam strikes a substance. When the photon energy is greater than the energy needed to tear an electron loose from the substance, electrons are emitted the instant the light is turned on. When the photon energy is less than that value, no electrons are emitted no matter how intense the light. (The chance of two photons hitting the

TABLE 26.1 *Work function and photoelectric threshold wavelength for selected materials*

| Material | Work function (ϕ) | | Threshold wavelength | |
	(eV)	10^{-19} J	nm	Spectral region
Rubidium	2.10	3.36	592	Visible
Cesium	2.14	3.42	581	Visible
Potassium	2.30	3.68	541	Visible
Aluminum	4.28	6.85	290	Ultraviolet
Tungsten	4.55	7.28	273	Ultraviolet
Copper	4.65	7.44	267	Ultraviolet
Gold	5.10	8.18	244	Ultraviolet
Platinum	5.65	9.04	220	Ultraviolet

same electron simultaneously is practically zero.) We see at once that the energy needed to tear an electron out of the plate is exactly equal to the energy of a photon with the threshold wavelength. Hence the minimum work needed to tear an electron loose from a solid is

$$\text{Minimum work} \equiv \phi = \frac{hc}{\lambda_o} = hf_o$$

where this minimum work is represented by ϕ and is called the **work function** of the material. Values of the work function for a few metals are given in Table 26.1. Notice that in many cases, ultraviolet light is needed to eject electrons from the metal.

In the event that a photon has energy greater than ϕ, that is, if λ is smaller than λ_o, not only can an electron be knocked out of the plate, but also it can have kinetic energy to spare. That is, part of the energy hc/λ of the photon is lost in doing work ϕ, or tearing the electron loose, and the remainder appears as the kinetic energy of the electron. We may therefore write, for nonrelativistic energies,

$$(\tfrac{1}{2}mv^2)_{\text{max}} = \frac{hc}{\lambda} - \phi \qquad (26.5)$$

This is called the **photoelectric equation.**

Most of the emitted photoelectrons have less kinetic energy than the $(\tfrac{1}{2}mv^2)_{\text{max}}$ given by Eq. 26.5 because they undergo collisions before they escape from the material. Thus $\tfrac{1}{2}mv^2$ in Eq. 26.5 is the same as $(\text{KE})_{\text{max}}$ in Eq. 26.4. Comparison of Eq. 26.5 with Eq. 26.4 tells us that A in Eq. 26.4 should be hc. Experiment shows that the numerical value of A is indeed equal to hc. As a final confirmation of Eq. 26.5, the work function ϕ determined by equating it to the experimental value of B in Eq. 26.4 is the same as the work function determined by entirely different experiments.

Thus we can conclude that photoelectrons are ejected from a material if an incident photon has enough energy to eject one. The photon energy is hf, which is the same as hc/λ. A photon with the threshold wavelength λ_o has an energy hc/λ_o, and this energy is equal to ϕ, the work function. Such a photon is just barely capable of ejecting photoelectrons. Photons of wavelength shorter than λ_o have more than enough energy to eject photoelectrons, and the excess energy appears as the kinetic energy of the photoelectron.

Illustration 26.4

What is the energy of a photon in a beam of infrared radiation whose wavelength is 1240 nm?

Reasoning

$$\text{Photon energy} = \frac{hc}{\lambda} = \frac{(6.626 \times 10^{-34}\,\text{J}\cdot\text{s})(3.00 \times 10^{8}\,\text{m/s})}{1240 \times 10^{-9}\,\text{m}}$$
$$= 1.602 \times 10^{-19}\,\text{J} = 1.00\,\text{eV}$$

where we have used the conversion factor $1\,\text{eV} = 1.602 \times 10^{-19}\,\text{J}$.

It is convenient to remember this result: *The photons in 1240-nm radiation have an energy of 1 eV.* For example, light of wavelength 1240/4 nm has photon energies of 4×1 eV. ∎

Illustration 26.5

Find the energy of a photon in each of the following: (*a*) radio waves for which $\lambda = 100$ m; (*b*) green light with $\lambda = 550$ nm; (*c*) x-rays with $\lambda = 0.200$ nm.

Reasoning Using the result of Illustration 26.4, we have

(*a*) $\dfrac{1240 \times 10^{-9}\,\text{m}}{100\,\text{m}} \times 1\,\text{eV} = 1.24 \times 10^{-8}\,\text{eV}$

(*b*) $\dfrac{1240\,\text{nm}}{550\,\text{nm}} \times 1\,\text{eV} = 2.25\,\text{eV}$

(*c*) $\dfrac{1240\,\text{nm}}{0.200\,\text{nm}} \times 1\,\text{eV} = 6200\,\text{eV}$

Notice what high energies x-ray photons have.

Exercise A laser beam ($\lambda = 633$ nm) has a power of 2.0 mW; that is, it carries an energy of 2.0 mJ past a given point each second. How many photons pass a point in the path of the beam each second? *Answer:* 6.4×10^{15}.

Example 26.3

When 500-nm light is incident on a particular surface, the stopping potential for photoelectrons is found to be 0.44 V. Find the work function for this material and the longest wavelength that will eject electrons from its surface.

R | *Reasoning*

Question What does the stopping potential represent?
Answer The photoelectrons are ejected from the surface with energy in excess

of the minimum required energy. The stopping potential V_o is the retarding voltage necessary to stop the most energetic electrons from reaching the collector. eV_o is thus equal to $(KE)_{max}$ of the electrons.

Question How is V_o related to the work function?
Answer The work function ϕ is the minimum energy required to eject electrons. Photon energy in excess of this goes into electron KE. This is shown by Eq. 26.5:

$$(\tfrac{1}{2}mv^2)_{max} = eV_o = \frac{hc}{\lambda} - \phi$$

Question What condition determines the longest wavelength for electron emission?
Answer That the photon energy is able to eject an electron with no excess KE. This is the case when $hc/\lambda_o = \phi$.

Solution and Discussion

$$\phi = \frac{hc}{\lambda} - eV_o$$

$$= \frac{(6.63 \times 10^{-34}\,\text{J} \cdot \text{s})(3 \times 10^8\,\text{m/s})}{5 \times 10^{-7}\,\text{m}} - (1.6 \times 10^{-19}\,\text{C})(0.44\,\text{V})$$

$$= 3.27 \times 10^{-19}\,\text{J} = 2.05\,\text{eV}$$

Then

$$\lambda_o = \frac{hc}{\phi}$$

$$= \frac{(6.63 \times 10^{-34}\,\text{J} \cdot \text{s})(3 \times 10^8\,\text{m/s})}{3.27 \times 10^{-19}\,\text{J}}$$

$$= 608\,\text{nm}$$

From Table 26.1, we might conclude that the material was rubidium.

26.9 THE COMPTON EFFECT: THE MOMENTUM OF THE PHOTON

Since both light and x-rays are electromagnetic waves, the photon concept should apply to x-rays as well. Direct evidence for the x-ray photon was first provided by A. H. Compton in 1923. He noticed that when a monochromatic beam of x-rays was shone on a block made of graphite, two kinds of x-rays were scattered from the block (Fig. 26.12). One kind had the same wavelength as the incoming radiation, and the other kind had a longer wavelength than the incident rays. The unchanged-wavelength portion can be pictured as arising in the following way: the oscillating electric field in the incident beam causes the charges in the atoms to oscillate with the same frequency as the wave. These oscillating charges act as antennas, radiat-

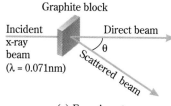

Graphite block

(a) Experiment

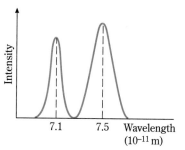

(b) Spectrum of scattered beam
($\theta = 135°$)

FIGURE 26.12

The Compton effect. When x-rays ($\lambda = 0.071$ nm in this case) are scattered, the scattered beam has two components. One has the same wavelength as the original beam; the other has a longer wavelength.

ing waves of the same frequency and wavelength. Hence the scattered x-rays are reradiated waves from the oscillating atomic charges.

As we stated, in addition to this beam of scattered x-rays, there is another type of scattered x-ray, one that has a slightly longer wavelength. The exact wavelength of these x-rays depends on the angle θ at which they are scattered in a precise and relatively simple way.

No explanation for their existence appeared possible using the wave picture of x-rays. However, a simple explanation was simultaneously and independently presented by Compton and Peter Debye. They postulated that the basic scattering interaction was elastic collisions between x-ray photons and electrons of the graphite atoms in which the kinetic energy and momentum of the electron-photon system are conserved. Because the electron's binding energy in the graphite is negligible relative to the energy of an x-ray photon, the electron behaves essentially like a free particle when struck by the photon.

In order to analyze the photon-electron collision, we have to figure out how to express the momentum of the photon. We already have two pieces of information about photons: (1) since they represent light, their speed must be c, and (2) their energies depend on their wavelengths, $E = hc/\lambda$. It is tempting to remember the classical definition of momentum, mv, and write $p = mc$ for the photon, but we don't have a value for the photon's mass. As a matter of fact, we can show that *the rest mass of a photon must be zero!* Since it travels with speed c in vacuum, we have

$$m = \frac{m_o}{\sqrt{1 - (v/c)^2}} = \frac{m_o}{\sqrt{1 - 1}} = \frac{m_o}{0}$$

If m_o were anything but zero, the photon would have infinite mass. Since $E = mc^2$, however, infinite mass implies infinite photon energy, and we know this to be untrue. Therefore, we must conclude that $m_o = 0$. If this seems strange, remember that *a photon is never at rest.* It is both emitted and absorbed at the speed of light. *A photon moving through vacuum never travels at any speed other than c.* The only mass such a particle has is due to its kinetic energy. Thus

$$E_{\text{photon}} = (m - m_o)c^2 = mc^2 = \frac{hc}{\lambda}$$

From this we can identify an expression for the photon momentum equivalent to mc:

$$\text{Photon momentum} = p = mc = \frac{mc^2}{c} = \frac{E}{c} = \frac{h}{\lambda} \tag{26.6}$$

In Compton scattering, the x-ray photon gives up some of its energy and momentum to the electron it hits. Since both of these properties involve wavelength, the scattered x-ray photon must have a wavelength different from that of the incident x-ray photon. By applying the principles of kinetic energy and momentum conservation, using $E = hc/\lambda$ and $p = h/\lambda$ for the photon, Compton and Debye obtained the wavelength change:

$$\Delta\lambda = \frac{h}{m_e c}(1 - \cos\theta) \tag{26.7}$$

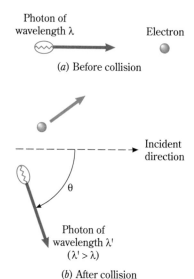

Photon of
wavelength λ   Electron

(a) Before collision

- - - - - - - - - → Incident
                      direction

θ

Photon of
wavelength λ'
(λ' > λ)

(b) After collision

FIGURE 26.13

In the Compton effect, a photon
collides with an electron. Both energy
and momentum are conserved in the
collision.

where m_e is the rest mass of the electron and θ is the angle of the observed scattered
x-ray relative to the incident beam (Fig. 26.13). Notice that the change of wavelength depends *only* on the angle at which the x-ray is scattered. The quantity
$h/m_e c$ is a constant with the dimensions of length, known as the **Compton wave-
length** of the electron. Its value is 2.43×10^{-12} m. $\Delta\lambda$ ranges from 0 at $\theta = 0°$ to
$2h/m_e c$ at $\theta = 180°$.

Equation 26.7 was found to be in complete agreement with Compton's experimental results. Here again was a striking confirmation of the particle-like interaction properties of electromagnetic waves with matter.

Exercise   Show that the data in Fig. 26.12 obey Eq. 26.7.

PART III: QUANTUM MECHANICS

26.10   THE DE BROGLIE WAVELENGTH

As we have seen, electromagnetic radiation has a dual nature. It has a wavelike
character that causes it to show interference and diffraction effects. It also has
particlelike behavior, as shown by its photon properties. Given this duality, it is
natural to speculate that the electron, and perhaps other particles, may have wave
properties.

Louis de Broglie was the first to seriously propose the dual nature of the electron.
He was led to his proposal by a suggestive theory for the hydrogen atom presented
earlier by Niels Bohr. De Broglie found (in 1923) that he could rationalize one of
Bohr's major assumptions by assuming the electron to have wave properties.
Rather than follow through the historical chain of events we will proceed directly to
de Broglie's result.

The photon has a momentum h/λ (Eq. 26.6), and therefore its wavelength is

Assuming they have approximately the
same density, which of these two
"particles" would you expect to exhibit
greater wave effects if they were given
the same velocity? (Actually, both
objects are completely "classical" in
their behavior.)

$\lambda = h/p_{\text{photon}}$. By analogy, if a particle has wave properties, perhaps its associated wavelength and momentum can also be related by a similar equation. De Broglie postulated that particles have wave properties and that their wavelength is

$$\text{de Broglie wavelength} = \lambda = \frac{h}{p} \tag{26.8}$$

where h is Planck's constant and p is the momentum of the particle in question.

Experimental confirmation of de Broglie's supposition was obtained somewhat by accident in 1927 by C. J. Davisson and L. H. Germer. They were investigating the scattering of a beam of electrons by a metal crystal (nickel). Their apparatus, which was enclosed in a vacuum chamber, is sketched schematically in Fig. 26.14. A beam of electrons was given a known energy by accelerating the electrons through the potential difference V, and then measurements were made of the number of electrons scattered by a nickel crystal upon which the beam was incident. The unexpected result was that the electrons scattered very strongly only at certain special angles. These results were reported as unexplained by Davisson and Germer.

It was suggested to the two investigators that perhaps this was evidence of de Broglie's radical ideas. They therefore undertook further measurements with properly oriented crystals to see whether the sharply defined angles of scattered electrons could be explained in terms of interference effects produced by the regular spacing of the atoms in the crystal acting as a sort of diffraction grating. Physicists W. H. Bragg and his son W. L. Bragg had developed the theory of x-ray diffraction by crystals in 1913. This phenomenon is the basis for x-ray crystallography, by which the structures of crystals and other complex molecules such as DNA have been determined. Bragg's law for x-ray diffraction is identical in form to the grating equation we used in Chap. 24.

If the spacing between planes in the crystal is d and if the waves have wavelength λ, strong reflection (constructive interference) should occur at angles given by

$$m\lambda = 2d \sin \theta_m \qquad m = 1, 2, 3, \dots$$

where in this case θ is the angle between the scattered beam and the scattering plane. The spacing d in most crystals is on the order of 0.1 nm. You may recall that interference effects become pronounced only when the wavelength of incident light is approximately the same as the grating spacing. Crystal diffraction thus requires wavelengths of approximately 0.1 nm, which is in the x-ray region of the electromagnetic spectrum.

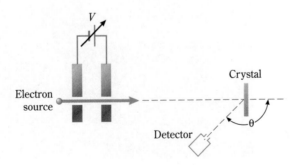

FIGURE 26.14

Davisson and Germer measured the numbers of electrons reflected from the crystal at various angles.

Since Davisson and Germer knew the value of d and the positions of strong reflection θ of the electrons, they could compute λ. Alternatively, since $\frac{1}{2}mv^2 = Ve$, they could calculate the momentum of the electrons:

$$p = mv = \sqrt{2Vme}$$

where V is the potential difference through which the beam is accelerated. The de Broglie wavelength $\lambda = h/p$ could be obtained. Davisson and Germer found that these two wavelengths were identical. In other words, the electrons are reflected in the same way as their de Broglie waves should be reflected. This is direct evidence for de Broglie's idea that electrons have wave properties.

Over the years, it has been found that neutrons, protons, atoms, and molecules, as well as other particles, show the same wave effects that can be obtained with electrons. We are therefore compelled to believe that particles, when moving through space, behave like waves of wavelength h/p, where h is Planck's constant and p is the momentum of the particle in question. Why this behavior had not previously been noted for macroscopic particles is discussed in Illustration 26.7.

Illustration 26.6

An electron in a television tube may have a speed of 5×10^7 m/s. If we neglect relativistic effects, what is the de Broglie wavelength associated with this electron?

Reasoning Substituting in Eq. 26.8, we find $\lambda = 0.145 \times 10^{-10}$ m. Apparently, the wavelength associated with an electron is in the x-ray range of lengths. (We do not mean to imply that de Broglie waves are related to electromagnetic waves. They most certainly are not electromagnetic in nature. More will be said about this in the following section.) ∎

Illustration 26.7

Describe the diffraction pattern that would be obtained by shooting a bullet ($m = 0.10$ g, $v = 200$ m/s) through a slit 0.20 cm wide.

Reasoning The wavelength of the de Broglie wave associated with the bullet is

$$\lambda = \frac{h}{p} = \frac{6.63 \times 10^{-34}}{(10^{-4})(2 \times 10^2)} = 3.30 \times 10^{-32} \text{ m}$$

Knowing that interference and diffraction effects become large only if λ is comparable with the slit width or separation (Sec. 24.8), we can conclude that interference effects are negligible. To show this clearly, however, let us find the angle θ between the straight-through beam and the first diffraction minimum. This minimum occurs when (Eq. 24.5)

$$\sin \theta = \frac{\lambda}{\text{slit width}} = 1.6 \times 10^{-29}$$

In other words, the diffraction angles will be so small that all the particles will travel essentially straight through the slit. Straight-line motion results, and the wave effects are unobservable. This situation always occurs for macroscopic experiments, and it is for this reason that the de Broglie wave effects are unobservable in the motion of macroscopic particles. ∎

26.11   WAVE MECHANICS VERSUS CLASSICAL MECHANICS

The discovery of the wave properties of particles has serious implications both for the interpretation of particle motion and for mechanics in general. We must investigate under what circumstances the wave nature of particles is important enough to modify the classical description of particle behavior. In this investigation, we can rely on our previous knowledge about wave behavior, such as diffraction and interference.

The interpretation of the diffraction pattern of light using the photon concept is that the pattern tells us the distribution of the paths of the photons which pass through the slit. The regions of maximum intensity in the diffraction pattern are where most of the photons go. Fig. 26.15a shows a diffraction pattern for a beam of x-rays passing through aluminum foil. Fig. 25.15b shows the pattern formed when electrons are shot through the same foil. The similarity between x-ray and electron diffraction patterns indicates that a similar condition exists for de Broglie waves. Using the de Broglie wavelength for the electrons, we can predict where the electrons have a high probability of striking the screen beyond a narrow slit.

Consider the two situations in Fig. 26.16. If a wave passes through a barrier that has two slits much wider than the wavelength, the situation will be that of Fig. 26.16a, in which two distinct shadows of the edges of the slits are cast. We saw in Illustration 26.7 that this would be the case for macroscopic particles. However, particles of very small mass (such as electrons), have very small values of momentum even when traveling very fast. This means that their de Broglie wavelength can easily be comparable to the dimensions of a macroscopic experiment, and so their wave properties may become discernible. Electrons passing through the same

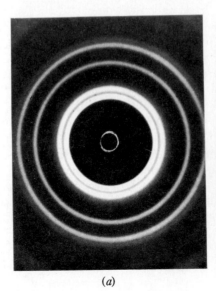

FIGURE 26.15

Diffraction pattern produced by a beam of (a) x-rays and (b) electrons incident on an aluminum foil target.

(a)

(b)

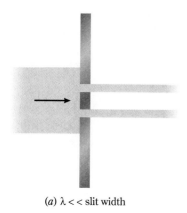

(a) λ << slit width

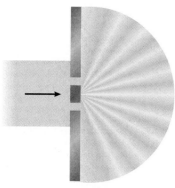

(b) λ comparable
to slit dimension

FIGURE 26.16

(a) When the wavelength associated with a particle is much smaller than the slit width, clear images of the slits are formed by the particles passing through. *(b)* When λ is comparable to the slit width, however, typical wave interference phenomena are observed in the distribution of emerging particles.

two slits could display a distribution such as that in Fig. 26.16*b*, in which their trajectories are governed by their wave nature rather than classical particle mechanics.

Returning to our original question of when classical mechanics fails, we can state the following:

Classical mechanics becomes invalid when the particle's de Broglie wavelength is comparable to or smaller than the smallest dimension of the experiment.

This is likely to happen only when we are dealing with atomic and subatomic particles. In particular, wave effects dominate the behavior of electrons within atoms. Under these conditions we must replace classical mechanics with *wave mechanics*. For reasons we will soon encounter, wave mechanics is often referred to as *quantum mechanics*.

Soon after de Broglie suggested the wave nature of particles, the German physicist Erwin Schrödinger developed an equation to describe a particle's wavelike properties. **Schrödinger's equation,** analogous to the equation used to describe the behavior of electromagnetic waves, forms the basis of quantum mechanics. Newtonian (classical) principles still can be used to solve most macroscopic problems. Relativistic effects become important only when particle speeds approach the speed of light or when very accurate results are required. Quantum mechanics replaces newtonian mechanics only when we deal with dimensions comparable to particle wavelengths. We see in the next chapter that quantum mechanics must be used to explain the internal workings of atoms.

26.12 RESONANCE IN DE BROGLIE WAVES: STATIONARY STATES

When we dealt with mechanical waves, such as waves on a string and sound waves in tubes, we found that wave resonance is of great importance. This is also true for de Broglie waves. Let us examine a simple situation in which de Broglie wave resonance should occur.

Case 1: A Particle in a Tube

Consider a particle of mass m confined to a narrow tube of length L, and closed at both ends, as shown in Fig. 26.17*a*. If the particle acts like a wave, then its de Broglie wave should resonate in the tube, as shown in the lower parts of the figure. We call such a resonance a **stationary state.** Because the particle cannot leave the tube, the ends must be nodes. (Remember, the amplitudes of the de Broglie waves tell us where the particle is likely to be found.) Hence the particle will resonate within the tube when its de Broglie wave has the following wavelengths (remember, the distance between nodes is $\frac{1}{2}\lambda$):

$$L = \tfrac{1}{2}\lambda_1 \qquad L = 2(\tfrac{1}{2}\lambda_2) \qquad L = 3(\tfrac{1}{2}\lambda_3) \qquad \dots$$

Or, in general, a stationary state for the particle occurs when

$$\lambda_n = \frac{2L}{n} \qquad n = 1, 2, 3, \dots$$

Only if a particle has one of these resonance wavelengths will it resonate in the tube.

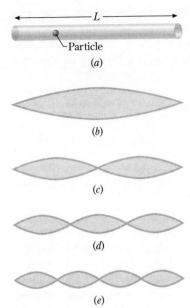

FIGURE 26.17

Stationary states for a particle in a tube. The amplitude of the wave pattern at a given position indicates the relative probability of finding the particle at that position.

In analogy with other forms of resonance we know about, we can infer the following. Only at wave resonance will a very large wave build up in the tube. Otherwise, the amplitude of the wave will be so small that it is essentially negligible. Because the amplitude of the de Broglie wave of a particle is a measure of where the particle is likely to be found, we expect the particle to be found in the tube only under resonance conditions. Moreover, the particle is most likely to be found where the resonance waves in Fig. 26.17 have the largest amplitude, that is, at the antinodes. And, more startling, the particle will never be found at the nodes. Before we examine this curious result further, let us examine the energy of the particle in the tube.

The particle has only kinetic energy, $\frac{1}{2}mv^2$. (We assume nonrelativistic conditions.) If we call the particle's energy E_n when it is in the nth resonance state, we have

$$E_n = \tfrac{1}{2}mv_n^2$$

Momentum p is mv, however, and so this expression can be written as

$$E_n = \frac{p_n^2}{2m}$$

However, the particle's de Broglie wavelength is $\lambda_n = h/p_n$, and so

$$E_n = \frac{h^2}{2m\lambda_n^2}$$

And finally, we have shown that the resonance wavelengths $\lambda_n = 2L/n$, and so

$$E_n = \frac{n^2h^2}{8mL^2} \qquad n = 1, 2, 3, \ldots \tag{26.9}$$

Thus we get the astounding result that, if the particle is to be found in the tube, it must have one of the energies given by Eq. 26.9. We say that *the particle's energy is* **quantized.** It is for this reason that wave mechanics is often referred to as quantum mechanics. The particle in the tube can have only these energies and no others. This startling finding contradicts classical mechanics, which predicts that a particle in a tube can have any and all kinetic energies, including zero. Doesn't this discrepancy between wave mechanical results and our known experience cause us to disbelieve wave mechanics? The answer is "no," for a reason we now explain.

Let us compute the resonance energies for a tiny dust particle ($m = 1 \times 10^{-15}$ kg) in a tube 50 cm long:

$$E_n = \frac{n^2h^2}{8mL^2} = (2 \times 10^{-52}\,\text{J})(n^2)$$

Thus the energies the particle can have are 2×10^{-52} J, $4(2 \times 10^{-52})$ J, $9(2 \times 10^{-52})$ J, and so on. Notice how very small these energies are and how tiny the energy gap between them is. The gap is only about 2×10^{-52} J, which is so small compared with the thermal energy of a gas particle (10^{-21} J) that we would never be able to tell that an energy gap existed. This is even more obvious for a particle of larger mass. Hence we conclude that for all ordinary particles in visible-size tubes, the particle energy is essentially continuous; experiment does not allow us to see the quantum character of energy that is predicted by wave mechanics.

The situation becomes different for atomic-size tubes, however. Suppose we have an electron ($m = 9 \times 10^{-31}$ kg) in a tube that is only 2×10^{-10} m long. Then

$$E_n = n^2(1.5 \times 10^{-18}\,\text{J}) = 9n^2\ \text{eV}$$

This energy is large enough for the energy gaps to be measurable. We therefore conclude that the wave nature of particles and the quantized character of their energies will be discernible in atomic-size systems.

Case 2: A Harmonic Oscillator

A mass m vibrating under a Hooke's law spring force is called a *harmonic oscillator*. To a first approximation, the vibrating atoms in molecules are harmonic oscillators. In many ways, the harmonic oscillator is like the particle in a tube that we have just discussed, but the problem is complicated by the fact that the system has a variable potential energy as the spring is distorted. Even so, the resonance motion of the system can be found by solving Schrödinger's equation. The end result is not too different from that for the particle in a tube. In particular, the energy is quantized, with values

$$E_n = (n + \tfrac{1}{2})\left(\frac{h}{2\pi}\right)\sqrt{\frac{k}{m}} \qquad n = 0,\ 1,\ 2,\ \ldots$$

where k is the spring constant.

We can place this result in an interesting form if we recall that the resonance frequency f_o for a mass at the end of a spring is

$$f_o = \frac{1}{2\pi}\sqrt{\frac{k}{m}}$$

Substitution of this value in the expression for E_n gives

$$E_n = (n + \tfrac{1}{2})(hf_o) \qquad n = 0,\ 1,\ 2,\ \ldots \tag{26.10}$$

Therefore the energies of a Hooke's law oscillator are quantized; the gaps between allowed energies are hf_o.

This astounding result is simply the property that Planck had to attribute to oscillators in order to explain blackbody radiation. Some 25 years after Planck guessed that this should be the case, the use of de Broglie wave concepts showed why it must be true. We saw in Sec. 26.7 that Planck's supposition could not be tested for laboratory-size oscillators. Now we see that his unsubstantiated guess is buttressed by the many other successes of quantum theory. Further substantiation for wave mechanics will be found in the next chapter.

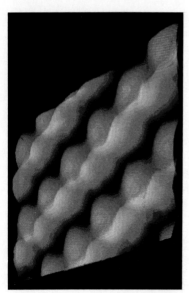

This image of the surface of a gallium-arsenide crystal was made by a device known as a scanning-tunneling microscope. Color coding has been applied to show the individual gallium atoms as blue and the arsenic atoms as red. Even though the lattice structure of the atoms is discernable, the individual atoms still appear fuzzy rather than point-like.

26.13   THE UNCERTAINTY PRINCIPLE

Since the discovery of the wavelike nature of the electron, many experiments have been carried out to see whether other particles also exhibit this behavior. Particles of atomic or subatomic size can be investigated relatively easily for wavelike effects. No exception to de Broglie's wavelength equation has ever been found. In fact, the use of electrons and neutrons as well as x-rays in diffraction experiments designed to investigate crystal structure is now commonplace.

The wave nature of all particles leads to a great philosophical principle. Prior to this discovery, philosophers had often argued about whether the fate of the universe was completely determined. Could we, in principle at least, determine the position, speed, and energy of all the particles in the universe and then predict the course of all future events? It appears that the wave nature of all particles requires us to give a negative answer to this question. This fact is embodied in the Heisenberg uncertainty principle, which we now examine.

Let us see how we would locate the position of a particle with the maximum possible precision. In order to locate it, we must scatter at least one second particle (we shall call it a probe particle) off the target particle and detect the angle at which the probe particle is scattered. So as to minimize the effect the probe particle has on the location of the target, we shall use a single photon of wavelength λ as our probe. This photon carries momentum $p = h/\lambda$ and energy $E = hc/\lambda$. Our detector (which could be a lens, for instance) subtends an angle α at the target, as shown in Fig. 26.18, and the photon is assumed to be directed along the y axis. As the photon scatters off the target, it transfers some of its momentum to the target particle. In the process, the photon will acquire some x component of momentum, but, in order for it to enter the lens and be detected, this momentum component has a maximum possible value of $\Delta p_x = p \sin \alpha$. Since momentum must be conserved, the target must acquire an x component of momentum equal and opposite to that acquired by the photon. However, since we have no way of knowing where the photon enters the lens, we do not know precisely how much momentum the target particle has acquired. All we can say is that, in order for the scattering photon to be detected, the momentum of the target is *uncertain* by an amount

$$\Delta p_x = p \sin \alpha = \frac{h}{\lambda} \sin \alpha$$

In Chapter 24 we saw that diffraction effects limit the accuracy with which we can locate a point source. We can write this limit as approximately $\Delta x \approx \lambda/\sin \alpha$. Therefore the detection of the photon can locate the target particle only to within this uncertainty in position. If we now multiply the uncertainties in position and momentum of the target particle, we obtain

$$\Delta p_x \, \Delta x = \left(\frac{h}{\lambda} \sin \alpha \right) \left(\frac{\lambda}{\sin \alpha} \right) = h$$

In other words, when we use the most precise experiment imaginable to locate the position of a particle and measure its momentum simultaneously, the product of the intrinsic uncertainties in these two measurements must be at least as large as Planck's constant. This appears to be a perfectly general relation, and is one form of Heisenberg's uncertainty principle.

A second form of the uncertainty principle can be obtained through similar reasoning. If the uncertainty in the target particle's position is $\Delta x \approx \lambda$, the time it

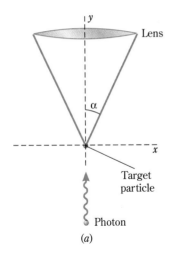

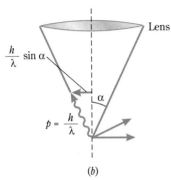

FIGURE 26.18

(a) A photon is incident on a target particle. (b) In order to detect the presence of the target particle, the scattered photon must pass through the lens, which subtends an angle α with respect to the target particle. As a result the target particle can acquire an x component of momentum as large as $(h/\lambda)\sin\alpha$.

takes the photon to pass through this distance is $\Delta t = \lambda/c$. The amount of energy the target particle can receive from the photon ranges from zero up to a maximum equal to the photon's entire energy hc/λ. Thus the energy acquired by the target particle has an uncertainty $\Delta E = hc/\lambda$. If we multiply the uncertainties in energy and time, we obtain

$$\Delta E\,\Delta t = \frac{hc}{\lambda}\frac{\lambda}{c} = h$$

We have thus arrived at two uncertainty relations, one involving momentum and the other involving energy. These relations were first proposed by Werner Heisenberg in 1927. Let us now restate them in a more exact form. According to the Heisenberg uncertainty principle

In a simultaneous measurement of coordinate x and momentum p_x of a particle,

$$\Delta x\,\Delta p_x \geqslant \frac{h}{4\pi} \tag{26.11}$$

where Δx and Δp are the uncertainties in x and p. Similarly, if the energy E of a particle at time t is measured, then the uncertainties ΔE and Δt are such that

$$\Delta E\,\Delta t \geqslant \frac{h}{4\pi} \tag{26.12}$$

The reason for the $\geqslant$ sign is that any realistic measurement will disturb the target particle to a greater extent than our idealized single-photon measurement.

It is impossible, then, even in principle, to know everything about an object. There will always be uncertainty about its exact energy at a given time and its exact momentum at a given place. This is one of the fundamental results inherent in the concepts of light quanta and particle waves. Clearly, a new formalism is needed to describe atomic particles and light quanta in situations in which these effects are important. The methods of quantum or wave mechanics must be used to handle these phenomena.

Example 26.4

Suppose that an electron is trapped inside a cube 10^{-10} m on a side. This is a volume roughly the same as that occupied by an atom. Estimate the minimum kinetic energy this electron must have if it is localized to this extent. Assume the KE can be treated classically. As a point of comparison, the KE of the electron of a hydrogen atom is 13.6 eV. Is your answer consistent with this?

Reasoning

Question What principle requires that the electron have some minimum KE?
Answer In classical physics, there is no requirement that KE have any particular value. It can be zero. But the uncertainty principle requires that momentum, which is related to KE, becomes more uncertain as the location of the electron

becomes more precisely known. Thus you cannot say that p (and hence KE) = 0 exactly.

Question What is the expression for the momentum uncertainty?
Answer Δp_x must be larger than $h/4\pi\,\Delta x$, where Δx is the dimension in which the electron is localized. A similar expression holds for the y and z directions.

Question How does this expression say that there is a *minimum* value of momentum?
Answer It says that there is no way of knowing or measuring momentum in any direction that is less than this uncertainty. Thus we can say that the minimum value of p_x is

$$p_{x,\,\min} = \frac{h}{4\pi\,\Delta x}$$

and similarly for p_y and p_z.

Question How is momentum related to KE?
Answer Classically, KE = $p^2/2m$. In three dimensions, $p^2 = p_x^2 + p_y^2 + p_z^2$

Question What expression do I get for minimum KE when I use the expressions for minimum momenta?
Answer

$$(KE)_{\min} = \frac{3p_{x,\,\min}^2}{2m} = \frac{3(h/4\pi\,\Delta x)^2}{2m}$$

where m is the rest mass of the electron.

Solution and Discussion Using the value 10^{-10} m for Δx, we arrive at

$$(KE)_{\min} = \frac{3(6.63 \times 10^{-34}\,\text{J}\cdot\text{s})^2}{32\pi^2(10^{-10}\,\text{m})^2(9.1 \times 10^{-31}\,\text{kg})}$$

$$= 4.6 \times 10^{-19}\,\text{J} = 2.9\,\text{eV}$$

This is the same order of magnitude as the KE of the electron in the hydrogen atom, which is a slightly different example of an electron localized in about the same volume. Put another way, the hydrogen electron has more than the minimum KE resulting from the above analysis.

LEARNING GOALS

Now that you have finished this chapter, you should be able to

1 Define (*a*) reference frame, (*b*) inertial reference frame, (*c*) relativistic factor, (*d*) proper length and proper time, (*e*) time dilation, (*f*) length contraction, (*g*) rest mass and apparent mass, (*h*) mass-energy relation, (*i*) rest mass energy, (*j*) Planck's constant, (*k*) photoelectric effect, (*l*) threshold wavelength, (*m*) work function, (*n*) photon, (*o*) Compton wavelength (*p*) de Broglie wavelength, (*q*) stationary state, (*r*) quantized energy, (*s*) uncertainty principle

2 State the two basic postulates of relativity.

3 State the conclusions that relativity gives rise to with respect to the following: maximum speed of objects, simultaneous events, time dilation, length contraction, mass variation with speed, kinetic energy, and mass-energy conversion. Compute the answers to simple problems involving these conclusions.

4 State under what conditions the relativistic expressions must be used to describe a particle's mass and KE.
5 Compute the allowed energies (according to Planck) for an oscillator with a known natural frequency provided Planck's constant is given. Explain why the energy of a pendulum appears to be continuous.
6 Sketch a graph of radiation intensity versus λ for a hot object. Show how the graph changes with temperature.
7 Describe the photoelectric effect and point out what is meant by the photoelectric threshold. State the energy of a photon in terms of its wavelength. Explain how the photon concept applies to the photoelectric effect. Compute the threshold wavelength from a knowledge of the work function. Use the photoelectric equation in simple situations.
8 Describe the Compton effect and explain how it can be interpreted in terms of photon scattering.

9 State the relation between the momentum of a photon and (*a*) its energy, (*b*) its wavelength, and (*c*) its frequency.
10 Give the de Broglie wavelength of a particle of known mass moving with a known speed. Explain why the wave properties of electrons are easily noticed, while those of a baseball are not noticeable.
11 Describe the Davisson-Germer experiment and explain how it verified the existence of de Broglie waves.
12 Describe the stationary states of a particle in a tube. Detail the novel predictions of the wave theory with regard to position and energy. Explain why these predictions do not violate common experience.
13 Explain under what conditions classical newtonian mechanics must be replaced by quantum mechanics. Reasoning from the interference effects observed for light, show why newtonian mechanics breaks down under these conditions.

SUMMARY

DERIVED UNITS AND PHYSICAL CONSTANTS

Planck's Constant (h)
$h = 6.626 \times 10^{-34} \, \text{J} \cdot \text{s}$

Compton Wavelength (λ_c)
$\lambda_c = \dfrac{h}{m_e c} = 2.43 \times 10^{-12} \, \text{m}$

DEFINITIONS AND BASIC PRINCIPLES

Inertial Reference Frame
An inertial reference frame is one in which Newton's law of inertia is valid. This essentially means a reference frame that is not accelerating.

Postulates of the Theory of Relativity
1. The speed of light is the same to all observers, regardless of their motion relative to the source of light.
2. Absolute speeds cannot be measured. Only speeds relative to some chosen reference frame can be measured.

Consequences of the Postulates of Relativity
1. The laws of nature are the same in all inertial reference frames.
2. Events observed as simultaneous in one inertial frame may not be observed as simultaneous in any other inertial frame moving relative to the first one.
3. No object can be accelerated to the speed of light in a vacuum, *c*.
4. No energy can be transmitted at a speed faster than *c*.

Proper Measurements of Length and Time
A proper measurement of length or time is one in which the measuring devices are at rest relative to the objects or events being measured.

Relation Between Proper and Improper Measurements
TIME
If an observer measures the time *t* between two events in an inertial frame which is moving at a speed *v* relative to him or her, this time interval is longer than the proper time interval, t_o, measured by someone at rest relative to the events. The two measured times are related by

$$t = \frac{t_o}{\sqrt{1 - v^2/c^2}}$$

LENGTH
If an observer measures a distance *d* between points that are moving at a speed *v* relative to him or her, this distance will be shorter than the proper distance, d_o, measured by someone at rest relative to the points. The two measured lengths are related by (provided *v* and the points are in line)

$$d = \sqrt{1 - \frac{v^2}{c^2}} \, d_o$$

INSIGHTS
1. Use of the term "proper" does not imply a more *correct* measurement than the "improper" measurement. It is assumed that both measurements are done "correctly."
2. The dimensionless factor $\sqrt{(1 - v^2/c^2)}$ is called the relativity factor. It has a value of nearly 1 except when *v* approaches *c*.
3. The measurements made by observers moving relative to each other will agree on the values of their relative speed *v* and the speed of light *c*.

Relativistic Mass

An object whose mass when measured at rest is m_o will have an increased mass m when observed to be moving at speed v. The relation between m and m_o is

$$m = \frac{m_o}{\sqrt{1 - v^2/c^2}}$$

INSIGHT

1. This increase in mass simply indicates the increased inertia the object has at high speeds. As v approaches c, it takes a larger and larger force to change the object's velocity.

Relativistic Energy

The energy of an object is related to its mass by the expression $E = mc^2$, where m depends on the object's speed as shown above. At rest, an object has an energy $E_o = m_o c^2$. The object's kinetic energy is

$$KE = (m - m_o)c^2$$

INSIGHTS

1. When v is much less than c, the expression for kinetic energy reduces to the classical expression $KE = \frac{1}{2}mv^2$.
2. Any process that changes the energy of an object by ΔE must be accompanied by a change in mass Δm, given by

$$\Delta m = \frac{\Delta E}{c^2}$$

Photon Energy

A photon of light of wavelength λ (frequency f) has an energy

$$E = \frac{hc}{\lambda} = hf$$

Photoelectric Effect

Electrons can be emitted from a surface if the surface is illuminated with light of wavelength below a certain threshold wavelength λ_o, which depends on the surface material.

WORK FUNCTION (ϕ)

The work function of a surface is the energy with which an electron is bound to the surface. It is equal to the energy of a photon of light having the threshold wavelength λ_o:

$$\phi = \frac{hc}{\lambda_o}$$

STOPPING POTENTIAL (V_o)

V_o is the retarding voltage necessary to stop the most energetic photoelectrons resulting from illumination by light of wavelength greater than λ_o.

THE PHOTOELECTRIC EQUATION

eV_o is equal to the maximum KE of the emitted photoelectrons.

$$eV_o = (\tfrac{1}{2}mv^2)_{\max} = \frac{hc}{\lambda} - \phi$$

Photon Momentum

The momentum of a photon is

$$p = \frac{h}{\lambda}$$

The relation between the energy and momentum of photons is

$$p = \frac{E}{c}$$

INSIGHT

1. Photons always have the speed c. Thus their properties are inherently nonclassical. For a photon, there is no meaning to the concept of rest mass.

Compton Scattering

When x-rays impinge on a surface, those scattered at an angle θ from the incident direction have their wavelength increased by

$$\Delta\lambda = \frac{h}{m_e c}(1 - \cos\theta)$$

The quantity $h/m_e c$ is known as the Compton wavelength of the electron. The change in wavelength is due to elastic scattering of the x-ray photon by an electron, in which the photon loses momentum.

de Broglie Wavelength

A particle having momentum p has an associated de Broglie wavelength given by

$$\lambda = \frac{h}{p}$$

INSIGHTS

1. Since h has such an extremely small value, the wave nature of material particles cannot be observed unless the particle mass is extremely small.

2. Classical mechanics becomes invalid when the de Broglie wavelength is comparable to or larger than the smallest dimension of an experiment.

The Uncertainty Principle

There are inherent limits to the precision with which we know both the position and momentum of a particle. The product of the uncertainties must obey the following inequality:

$$\Delta p_x \, \Delta x \geq \frac{h}{4\pi}$$

A corollary to this is a similar relation between uncertainties in measurement of energy and the time interval in which the energy is measured:

$$\Delta E \, \Delta t \geq \frac{h}{4\pi}$$

INSIGHTS

1. These inequalities show that the more precisely we measure one of the quantities, the less we know about the other.
2. These uncertainties are not due to defects or limitations of precision in measuring instruments. They are fundamental limits on what we can observe even in perfect experiments.

QUESTIONS AND GUESSTIMATES

1 Suppose you are in a spaceship traveling away from earth at a speed of 0.90c. A laser beam is directed at the ship from the earth. If you measure the speed of the laser light relative to your ship, what will be the light's speed?

2 Suppose an astronaut has perfect pitch and so can recognize at once that a particular tuning fork gives off a frequency of middle C when struck. What frequency would she hear if she listened to the tuning fork inside her spaceship while traveling through space at a speed of 0.9c?

3 Most human beings live less than 100 yr. Since the maximum velocity one can acquire relative to the earth is c, the speed of light, a person from earth can travel no farther than 100 light-years into space in 100 yr. Does this necessarily mean that no person from earth will ever be able to travel farther from earth than 100 light-years? (One light-year is the distance light travels in one year, or 9.46×10^{15} m.)

4 Suppose that the speed of light were only 20 m/s, and all the relativistic results applied when this speed was used for c. Discuss how our lives would be changed.

5 It should be clear from this chapter that the statement "matter can be neither created nor destroyed" is false. What can one say instead?

6 Discuss how our world would be affected if nature were suddenly to change in such a way that Planck's constant became 10^{32} times larger than it is. Consider the situation from two different aspects: (a) quantization of energy of oscillators, and (b) the uncertainty principle.

7 How does the photon picture of light explain the following features of the photoelectric effect: (a) the critical wavelength; (b) the stopping potential being inversely proportional to wavelength?

8 How can the work function of a metal be measured? Planck's constant?

9 Make a list of experiments in which light behaves as a wave and a list of experiments in which its quantum character is important. Is there any experiment in your list that can be explained from both standpoints?

10 When light shines on a reflecting surface in vacuum, a pressure is exerted on the surface by the light. Explain. Would the pressure be different if the surface were black, so that it absorbed the light?

11 If all the mass energy of a fuel could be utilized, about how many kilograms of fuel would be needed to furnish the energy required by a city of 300,000 people for 1 yr?

12 Estimate the power change for a local radio-station antenna system as it changes from one quantized oscillation energy state to an adjacent state. What wavelength and frequency photons does it emit in the change?

13 Ultraviolet light causes sunburn, whereas visible light does not. Explain why. Some people insist that they sunburn most easily when their skin is wet. Do you see any reason for this?

PROBLEMS

Sections 26.1–26.3

1 An airplane is flying at a speed of 360 m/s parallel to the earth's surface when a screw drops from the ceiling of the plane. Relative to the point on the floor directly below its original position, where does the screw land? The ceiling is 3.2 m above the floor.

2 Suppose you are in an elevator that is rising at a constant speed of 2.8 m/s. You drop a penny from your hand, which

is 1.4 m above the floor. How long will it take for the penny to reach the floor? Repeat when the elevator is standing still.

3 Two trains are running side by side on two parallel tracks. One train (call it train *A*) is slowly overtaking the other train (call it train *B*) at 1.2 m/s. A train employee in train *A* is walking at a speed of 0.5 m/s toward the front of the train, and a passenger is walking at 0.5 m/s toward the rear of the train. What are their velocities as observed by a passenger in train *B*?

4 A train is moving forward slowly at 3 m/s. A passenger inside one of the cars runs at 3 m/s toward the rear of the train. (*a*) What is the passenger's velocity as observed by a person standing on the station? (*b*) What would be the observed velocity if the passenger reversed velocity direction?

5 A boy inside a train going eastward at 16 m/s throws a ball westward with a speed of 4 m/s. (*a*) What is the velocity of the ball for a stationary observer standing nearby the train tracks? (*b*) For a passenger inside the train?

■ **6** Suppose you are on moon and want to synchronize your watch to a time signal on the earth. You receive a radio message saying that the time at the tone is exactly 5:00. To what time should you set your clock at the instant of the tone? Take the distance of the moon from the earth to be 3.8×10^8 m.

■ **7** At low speeds, if a person moving with a speed v relative to the earth shoots a projectile out along his line of motion with a speed u relative to himself, then the speed of the projectile relative to the earth will be simply $u + v$. This cannot be correct at high relativistic speeds near c, since speeds in excess of c would be predicted. (For example, if $v = 0.7c$ and $u = 0.6c$, the predicted speed relative to the earth would be $1.3c$, which is impossible according to the special theory of relativity.) Einstein showed the relative speed to be given by the expression

$$\frac{u + v}{1 + uv/c^2}$$

If a spaceship moving at a speed $v = 0.7c$ past the earth shoots a projectile out along the line of its motion with a speed of $u = 0.8c$, what is the relative speed of the projectile relative to the earth?

■ **8** Assuming the conditions stated in Prob. 7, suppose an astronaut in the spaceship sends a light pulse. Find the speed of this light pulse relative to the earth. (Before you work the problem, can you give the answer from the considerations of the postulates of special relativity?)

Section 26.4

9. You are traveling through space in a spacecraft at a speed of 0.88*c*. Using a very accurate stopwatch, you find your pulse to be 68 beats per minute. What will your pulse rate

be as measured (*a*) by a fellow passenger in the spacecraft and (*b*) by an observer on the earth?

10. An apprentice astronaut has been given a special dispensation to take her physics test, which is normally 2.0 h long, while she is on a spacecraft moving at a speed of 0.92*c* relative to the earth. How long should she be allowed for the test by a monitor (*a*) in the ship and (*b*) on the earth?

11. A simple pendulum is found to have a period of 2 s when measured in its own inertial reference frame. An observer passes by the pendulum at a very high speed and measures the period of the same pendulum to be 6 s. How fast was the observer moving?

■ **12** Suppose the spaceship *Enterprise* has a rotating antenna attached to it which takes 5.0 s, as measured on the *Enterprise*, to make one complete revolution. If the *Enterprise* is moving away from the earth at 0.84*c*, how much time is required for one complete revolution of the antenna according to an observer on earth?

13. A certain unstable substance disintegrates such that half of the material is lost in 960 days. When this substance is in a spaceship traveling at a speed of 0.90*c*, how long will it take for half of the material to disintegrate according to (*a*) an observer in the spaceship and (*b*) an observer on Earth?

■ **14** The pion, a subnuclear particle, has a lifetime of 2.6×10^{-8} s. How fast must a beam of pions be moving if they are to travel a distance of 20 m in the laboratory before disintegrating?

■ **15** Scientists at a research laboratory discover a new type of particle beam which travels 5.6 m before the particles disintegrate. Their speed in the laboratory is found to be 0.9880*c*. What is the lifetime of these new particles when they are observed at rest in the laboratory?

■ **16** 40-year-old Captain Picard visits his 30-year-old younger brother before leaving for his mission on the spaceship *Enterprise*. After 3 years, according to clocks on the *Enterprise*, Capt. Picard returns to earth as his brother is celebrating his 45th birthday. How long was he gone according to earth clocks, and what was his average speed during the mission?

Section 26.5

17. The length of a spaceship at rest on earth is measured to be 40 m. What will the length of the ship be measured by an observer on earth when the ship flies by the earth at a speed of (*a*) 0.3*c* and (*b*) 0.9885*c*?

18. The length of a meterstick is measured by a stationary observer as the sticks flies past her at high speed parallel to its length. The measurement gives the value 0.6 m. How fast was the meterstick moving?

19. In a research laboratory a subnuclear particle travels a straight 25-m section of a particle accelerator at a speed of 0.9980*c*. If you could ride along with the particle, how long would that straight section of the accelerator appear to you?

20. A cube has an edge length of 4 cm when at rest. It is set in motion with a high speed at $0.82c$ parallel to one of its edges. (*a*) What shape would it appear to have to a stationary observer? (*b*) What will its observed volume be as it hurtles through the laboratory?

21 The nearest star to earth is approximately 4.1×10^{16} m away. If you travel at a speed of $0.84c$ in a spaceship, how long will it take to get there according (*a*) to an observer at rest on the earth and (*b*) to an observer on the spaceship?

22 A spaceship is moving with a speed of $0.92c$ relative to a space platform that has a 6000-m-long landing strip. What is the length of the landing strip measured by an observer in the spaceship as it flies parallel to and past the strip?

23 A 5-m-long semitrailer truck is traveling at 100 km/h. How long will the truck appear to a stationary observer standing by the side of the highway? *Hint:* For $v/c \ll 1$, use the approximation $\sqrt{(1 - v^2/c^2)} = 1 - v^2/2c^2$.

24 You measure the length of two spaceships, one stationary and the other moving at a speed of $0.920c$, and find that they have the same length. Your friend is traveling in the moving ship. Find the ratio of the lengths of the two ships as observed by your friend. Assume that you are stationary on the earth.

Section 26.6

25. At what speed is the mass of a particle equal to 100 times its rest mass?

26. The rest mass of an electron is $m_o = 9.11 \times 10^{-31}$ kg. Find the ratio m/m_o for an electron when its speed is (*a*) $0.1c$, (*b*) $0.001c$, (*c*) $0.6c$, and (*d*) $0.99c$.

27. Find the mass and speed of an electron that has been accelerated through a potential difference of (*a*) 300 V and (*b*) 30,000 V.

28. Find the kinetic energy of an electron when it is moving with the velocities in parts (*a*) to (*d*) of Prob. 26.

29. What is the velocity of a particle whose kinetic energy is 8 times its rest mass energy?

30 In modern nuclear accelerators, particles are sometimes accelerated to extremely high energies. (*a*) Calculate the mass of a proton whose kinetic energy is 6×10^9 eV. (*b*) How fast is it moving? Take the rest mass of a proton, m_o, to be 1.67×10^{-27} kg.

31 Suppose that 100 g of matter were to be completely converted to energy. (*a*) How much energy would be produced? (*b*) If this energy were used to operate a 75-W bulb, how long would it keep operating?

32 Approximately 334 kJ of energy is needed to melt 1.0 kg of ice. By what percentage will the mass of the ice increase because of the added energy necessary for its melting?

33 When 2.0 g of hydrogen is burned with 16 g of oxygen, it

results in the formation of 18 g of water. This chemical reaction releases approximately 572 kJ of energy. How much mass is lost in this chemical process? Can the change in mass be detected?

Section 26.7

34 Calculate the energy, in electronvolts and in joules, of a photon corresponding to (*a*) a radio wave frequency, 95 MHz and (*b*) ultraviolet light, 10^{16} Hz.

35 Calculate the energy of a photon, in electronvolts and in joules, that has a wavelength (*a*) 5 cm, (*b*) 955 nm, (*c*) 489 nm, and (*d*) 10 nm.

36 Find the wavelength of a photon having an energy of (*a*) 3 eV, (*b*) 3 keV, and (*c*) 1.2 MeV.

37 The average thermal translational kinetic energy of a particle is $\frac{3}{2}kT$. (*a*) What photon wavelength is equivalent to this thermal energy at 30°C? (*b*) What type of radiation is it?

38 A solid ball of mass 1 kg falls from a 5-m height. If it was possible to convert all the translational kinetic energy of this ball to photons of visible light of wavelength 589 nm, how many photons would be produced?

39 Through what height would the ball in Prob. 38 have to fall in order for it to have the energy of a single photon of wavelength 434 nm?

40 A 0.500-mW helium-neon laser emits radiation with a wavelength of 633 nm. (*a*) What is the energy of a photon in such a beam? (*b*) How many photons per second pass a given point along the beam?

Section 26.8

41 The critical wavelength for photoelectric emission from a certain material is 432 nm. Find the work function (in electronvolts) for this material.

42 What is the work function (in electronvolts) of a material that has threshold wavelength of 465 nm?

43 The work function for a certain material (silver) is 4.74 eV. (*a*) Find the threshold wavelength for silver. (*b*) What region of the spectrum is this in?

44 A certain metal has a work function of 1.25 eV. Yellow light of wavelength 589 nm is incident on the surface of this metal. Find (*a*) the maximum kinetic energy of the photoelectrons emitted from the surface and (*b*) the threshold wavelength for the metal.

45 Light of wavelength 434 nm is shone on the surface of a material that has a work function of 1.4 eV. What is the speed of the most energetic photoelectrons emitted from the surface?

46 Light of unknown wavelength is incident on the surface of sodium, whose work function is 2.3 eV. The photoelectrons emitted by the surface have a maximum speed of 1.2×10^6 m/s. What is the wavelength of the light?

47 When the surface of a certain material is illuminated with the light of frequency 1.3×10^{15} Hz, the stopping potential for photoelectrons is measured to be 2.4 V. (*a*) What is the work function of the material? (*b*) What is the frequency corresponding to the threshold wavelength?

■48 Radiation of wavelength 340 nm is shone on the surface of potassium (work function 2.3 eV). Calculate the value of the photoelectric stopping potential in this case.

■49 The dissociation energy (the energy necessary to break apart the constituent atoms) for the CN (cyanogen) molecule is approximately 1.22×10^{-18} J. (*a*) What is the maximum wavelength of the radiation that would be capable of breaking apart the CN molecule? (*b*) What is the frequency of this radiation? (*c*) In what region of the spectrum is this?

Section 26.9

50 (*a*) What is the momentum of a 16-eV photon? (*b*) How does it compare with the momentum of a 16-eV electron?

51 Find the impulse exerted by a 486-nm photon on a surface when it is (*a*) absorbed and (*b*) reflected back at the surface.

52 Calculate the fractional Compton wavelength shift $(\lambda' - \lambda)/\lambda$ for a photon striking a free electron head-on and being scattered straight backward if (*a*) $\lambda = 489$ nm and (*b*) $\lambda = 0.45$ nm.

■53 A photon of wavelength 0.45 nm strikes a free electron at rest and is scattered straight backward. What is the speed of the electron after the collision? Is the electron relativistic?

■54 A 0.50-mW helium-neon laser emits a wavelength of 633 nm in a beam of 3.6-mm^2 cross-sectional area. (*a*) Find the number of photons striking a surface perpendicular to the beam each second. What is the force exerted by the beam on the surface (*b*) if it is totally absorbed and (*c*) totally reflected?

■55 X-ray photons of wavelength 0.800 nm strike free electrons in a carbon target. (*a*) Find the wavelength of the scattered photons which come out at 90° relative to the direction of the incident radiation. (*b*) How much momentum is transferred to the free electrons?

56 When a 0.680-nm x-ray photon is scattered from a free electron at rest, the electron recoils with a speed of 1.2×10^6 m/s. (*a*) What was the Compton shift $\Delta\lambda$ in the photon's wavelength? (*b*) At what angle is the scattered photon observed?

Sections 26.10 and 26.11

57 Find the de Broglie wavelength of an electron that has been accelerated from rest through a potential difference of 1200 V.

58 What is the de Broglie wavelength of a proton moving with a speed of (*a*) 10^4 m/s and (*b*) 10^6 m/s?

59 What is the de Broglie wavelength of a 1600-kg automobile moving at 120 km/h?

60 What is the speed of a particle that has a de Broglie wavelength of 0.4 nm if it is (*a*) an electron and (*b*) a proton?

61 What is the potential difference necessary to accelerate an electron from rest in order to give a de Broglie wavelength of 6×10^{-9} m?

62 An alpha particle (helium nucleus with a mass $m = 4 \times 1.67 \times 10^{-27}$ kg and a charge $q = +2e$) is accelerated from rest through a potential difference of 1500 V. Find the de Broglie wavelength of this alpha particle.

■63 The average kinetic energy of a free electron in a metal is given by $3kT/2$ at high temperatures. (*a*) What is the de Broglie wavelength of a free electron in a metal at 27°C? (*b*) At what temperature would the electron's de Broglie wavelength be 0.8 nm?

■64 An electron is accelerated from rest through a potential difference V (in volts). Show that if relativistic effects are neglected, the de Broglie wavelength of the electron can be expressed as λ (in nanometers) $= 1.228/\sqrt{V}$.

Section 26.12

65 Consider an electron confined to a one-dimensional potential box of size $L = 0.53$ nm. (*a*) Calculate the first three resonance wavelengths for the electron. (*b*) Compute the energy of the first three energy levels of the electron.

66 A proton is confined to a one-dimensional box of width 1.0×10^{-5} nm (roughly the size of an atomic nucleus). Find the energy of the first three levels of the proton in the box.

67 The lowest energy level of an electron confined to a one-dimensional box is 4 eV. The next energy level ($n = 2$) is 15 eV. Find the approximate length of the box.

68 A 100-g mass hangs at the end of a spring that has a spring constant of 0.040 N/m. (*a*) What is the natural vibration frequency of this system? (*b*) What is the energy gap between allowed energies for this oscillator? Express your answer in both joules and electronvolts.

69 The hydrogen bromide molecule acts in many ways like an oscillator (two balls joined by a spring and performing back-and-forth vibrations) with a natural frequency of 8.66×10^{13} Hz. Find, in electronvolts and in joules, the energy gap between allowed energy levels for this oscillator.

■70 The lowest energy level (also called zero-point energy) of a certain quantized harmonic oscillator is 6 eV. (*a*) What is the frequency of this oscillator? (*b*) What is the energy gap between allowed energy levels for this oscillator?

■71 Find the zero-point energy (energy of the lowest level) of the NO molecule if it can be considered as a harmonic oscillator with a natural frequency of 5.63×10^{13} Hz.

Section 26.13

72 A 15-g baseball is moving with a speed of 24 m/s. If its speed can be measured to an accuracy of 0.5 percent, what is the minimum uncertainty in its position?

73 An electron is localized in a region within 0.53 nm. What is the uncertainty in the measurement of its momentum?

74 The energy of an electron in a certain atom is approximately 2.3 eV. What minimum time would it take to measure this energy to a precision of 0.5 percent?

75 A proton is confined within a nucleus that has a typical radius of about 5×10^{-15} m. If this can be considered as the uncertainty in position of the proton, what will the smallest uncertainty in the proton's momentum be? In its energy in electronvolts? Assume the proton is nonrelativistic.

76 A certain proton has a kinetic energy of 5 MeV. Assuming that the momentum of the proton can be measured with an uncertainty of 1 percent, calculate the uncertainty in its position. *Hint:* A 5-MeV proton can be considered nonrelativistic.

77 The time taken by an atom in radiating a photon of wavelength 510 nm is approximately 1×10^{-9} s. What is the uncertainty in the energy of the photon?

78 If the value of Planck's constant was 66 J · s instead of 6.63×10^{-34} J · s, how large would be the de Broglie wavelength of an 80-kg baseball player running at 6 m/s? Approximately what would be the uncertainty of the player's location according to an umpire trying to make the right call at home plate?

General Problems

79 Suppose superior beings on a planet near Alpha Centauri, which is about 4.1×10^{16} m away from the earth, send a spaceship toward us at a speed of 0.9970c. It is contaminated by a pair of microbes that reproduce such that their population is doubled every 8.4×10^5 s. How many microbes will be on board the ship as it passes by the earth? Answer this as observed by beings on the ship and on earth.

80 The roads in rural Iowa are directed mostly north-south or east-west and are 1.6 km apart. (*a*) An airplane is flying westward above some rural area and to the passengers on it the north-south roads appear to be only 1.0 km apart. How fast is the plane moving? (*b*) An Iowan looking up at the plane as it flies overhead measures its length to be 20 m. What is the length of the airplane when it is at rest at the airport? (*c*) A passenger on the plane uses an elaborate watch to measure the time the plane takes to move from one road to the next. What value does she get from her watch? (*d*) The Iowan measures the time taken by the plane to move from one road to the next. What value does he obtain?

81 The star Alpha Centauri is about 4.1×10^{16} m away from earth. Suppose a spaceship can be sent to this star with a speed of 2.1×10^8 m/s. (*a*) How long will the trip take according to earth's clocks? (*b*) How long a time will the clocks on the spaceship record for this journey? (*c*) How much distance will the occupants of the ship measure from the earth to the star? (*d*) What will be the apparent speed of the ship as computed by the occupants of the ship as based on the results of parts (*b*) and (*c*)?

82 A solid cube 1 m on each side has a mass of 10 kg. Suppose the cube is moving with a speed of 0.88c parallel to one of its edges. (*a*) What is its density (mass per unit volume) for a stationary observer when the cube flies by the observer? (*b*) What is the density of the cube for an observer moving with the cube?

83 Aliens in a spaceship approaching the earth at a speed of 0.4c send a probe toward the earth. The observers on earth record the speed of the approaching probe at 0.5c. What is the speed of the probe as measured from the spaceship? *Hint:* See Prob. 7.

84 The rate of solar energy reaching the earth's upper atmosphere is approximately 1.8×10^{17} W. Suppose all of this energy were absorbed by the earth and converted to mass. By how much would the mass of the earth increase over a period of 100 years?

85 What is the maximum voltage for which the expression for wavelength derived in Prob. 64 is valid to within a precision of 5 percent?

CHAPTER 27

ENERGY LEVELS AND ATOMIC SPECTRA

The five years from 1923 to 1928 were of exceptional importance in physics. The 1923 discovery of the wave properties of particles cleared the way for an understanding of how electrons behave in atoms. By 1928, thanks to the Schrödinger representation of wave mechanics, atomic energy-level structures and the way atoms emit and absorb light were no longer a mystery. In this chapter we learn how the wave picture explains the internal workings of atoms.

27.1 THE MODERN HISTORY OF ATOMS

Although there had been much prior speculation about atoms, it was not until 1911 that the nuclear model of the atom was validated. In that year, Ernest Rutherford and his associates carried out the experiment sketched in Fig. 27.1. He used particles emitted from the radioactive element radium as projectiles. These particles, called alpha (α) particles, are now known to be the nuclei of helium atoms. A beam of such particles was shot at a thin film of gold known to be only a few hundred atoms thick.

Rutherford expected the result shown in part (a): like bullets passing through cardboard, the particles would be slowed by the atoms and perhaps deflected

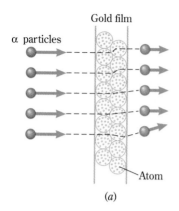

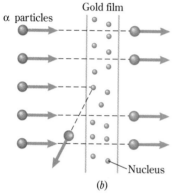

FIGURE 27.1

Rutherford shot α particles through a thin film of gold. (*a*) The original prediction of what would happen. (*b*) The concept required to explain the experimental results.

slightly. Instead, the result was as shown in part (*b*): although most of the particles were not deflected by the film, a very few were strongly deflected, as though they had struck a very tiny but massive object. From these observations Rutherford developed our modern concept of the atom, the so-called **nuclear atom.**

At the center of the atom is a tiny *nucleus;* it is about 10^{-15} m in radius and contains about 99.9 percent of the atom's mass. The nucleus carries a positive charge Ze, where e is the absolute value of the charge on the electron and Z is the atomic number of the element in question, equal to the number of protons in the nucleus ($Z = 1$ for H, 2 for He, 3 for Li, and so on.) The radius of the atom is of the order of 40,000 times that of the nucleus. Hence the nucleus is truly a tiny speck at the center of the atom. Orbiting in the (relatively) vast reaches of the atom outside the nucleus are Z electrons. They carry a combined charge $-Ze$, and so the atom is electrically neutral. Today we know that the wave nature of the electron predominates over its particle nature in determining the physical properties of the atom. As we see, the atom's volume is mostly empty space.

The simplest atom is hydrogen, consisting of a single proton as nucleus and a single electron. The model shown in Fig. 27.2 is in agreement with Rutherford's results. The electron orbits the nucleus, with the coulomb attraction exerted on it by the nucleus furnishing the required centripetal force. Such a model, however, should act like an electromagnetic wave antenna because it is much like an oscillating dipole. If it did behave like an antenna, an atom would "run down" as it lost energy in radiation, and its electron would spiral into the nucleus. However, hydrogen atoms don't behave this way. Usually hydrogen atoms do not radiate, and they never seem to collapse. Therefore this model must be in error in some way.

Hydrogen atoms can be induced to emit radiation under certain circumstances, however. Long before 1900, it had been shown that gases and even vaporized solids can be made to emit light (that is, their atoms can be *excited*) by sending a spark or high-voltage discharge through them. (For example, neon gas in the familiar neon sign emits light when a gas discharge is set up in it by high-voltage electrodes at the tube ends.) The wavelengths of light given off by these hot gases, their spec-

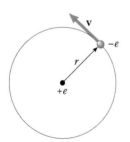

FIGURE 27.2

Classical model of the hydrogen atom. The electron is pictured as moving in a circular orbit around the one-proton nucleus.

During an eclipse, the red chromosphere of the sun becomes visible, as seen on the right edge in this photo. The red color is due to the strong red emission line of hydrogen gas.

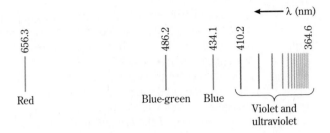

FIGURE 27.3

The Balmer series of spectral lines for hydrogen.

trum, can be investigated by use of a spectrometer, as discussed in Sec. 25.6 and shown in Fig. 25.17.

The spectral lines emitted by many atoms had been measured in detail even before 1900. Not knowing the structure of atoms, however, scientists were unable to give a meaningful interpretation of these spectra. Hydrogen atoms (but not H_2 molecules) give the simplest spectra; the visible part of the hydrogen emission spectrum consists of the series of spectral lines shown in Fig. 27.3. (Recall from Sec. 25.6 that a spectral line is actually an image of the spectrometer slit. Each wavelength gives a separate image.) The lines in the near ultraviolet were visible only in photographs, of course, because the eye cannot see ultraviolet waves.

Notice that the lines get closer and closer together as λ decreases. However, there are no lines at wavelengths shorter than $\lambda = 364.6$ nm, and this shortest wavelength of the series is called the **series limit.** According to the theory we will present shortly, there should be an infinite number of lines in this series. About 40 have been resolved; the remainder are too close together to be seen distinctly.

Since these spectral lines seem to have a definite pattern, it is natural to try to fit their wavelengths to an empirical formula. This was first done by Balmer in about 1885, and this series is now known as the **Balmer series.** He found that the wavelengths of the lines could be expressed by the following remarkably simple formula:

$$\text{Balmer:} \qquad \frac{1}{\lambda} = R\left(\frac{1}{2^2} - \frac{1}{n^2}\right) \qquad n = 3, 4, 5, \ldots \qquad (27.1)$$

where $R = 1.0974 \times 10^7 \text{ m}^{-1}$ and is called the **Rydberg constant** in honor of the man who determined its value. The integers from 3 to infinity, when placed in Eq. 27.1, yield the lines of the Balmer series shown in Fig. 27.3. When n is set equal to infinity, the formula yields the series limit, 364.6 nm.

Later, it was found that hydrogen atoms emit series of wavelengths other than those found in the Balmer series. The **Lyman series** occurs in the far ultraviolet, and the **Paschen series** is in the infrared (Fig. 27.4). These series follow formulas very much like Balmer's:

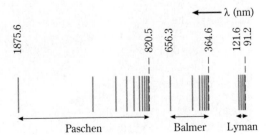

FIGURE 27.4

The three shortest-wavelength spectral series given off by hydrogen atoms.

$$\text{Lyman:} \quad \frac{1}{\lambda} = R\left(\frac{1}{1^2} - \frac{1}{n^2}\right) \quad n = 2, 3, \ldots$$

$$\text{Balmer:} \quad \frac{1}{\lambda} = R\left(\frac{1}{2^2} - \frac{1}{n^2}\right) \quad n = 3, 4, \ldots$$

$$\text{Paschen:} \quad \frac{1}{\lambda} = R\left(\frac{1}{3^2} - \frac{1}{n^2}\right) \quad n = 4, 5, \ldots$$

and so on, with $R = 1.0974 \times 10^7 \, \text{m}^{-1}$, *the same constant for every series.*

It is apparently more than mere coincidence that such simple formulas apply to a phenomenon as complicated as light emission. Some great simplicity in atomic behavior must be responsible for this remarkable set of relationships.

In 1912 Niels Bohr, a student from Denmark who was spending a postdoctoral year in Rutherford's laboratory in England, devised the first reasonable interpretation of the hydrogen spectrum. He started with the classical model of Fig. 27.2. To circumvent the difficulty associated with the fact that this model predicts antenna-like radiation, he simply accepted as fact that, in certain stable orbits, the atom does not radiate. Why this should be was not clear, but with this assumption Bohr was able to show how the observed hydrogen spectral lines originate.

Bohr's theory, although important in its time as an inspiration and guide for later workers, has now been supplanted. Its greatest drawback was that his bold assumption that stable orbits exist was not buttressed by an explanation of *why* they exist. Such an explanation became possible in 1923, when de Broglie found out that the electron has wave properties. We shall therefore jump forward in history and describe an early model of the hydrogen atom that makes use of the wave nature of the electron. We call it the *semiclassical theory* of the atom. Although a proper treatment of the atom using quantum mechanics has supplanted it, we will look at the semiclassical theory because it prepares us to understand the presently accepted model.

27.2 THE SEMICLASSICAL HYDROGEN ATOM

Let us assume that the hydrogen atom consists of an electron of mass m orbiting the nucleus, as in Fig. 27.2. (So that we may apply our calculations to other atoms with $Z > 1$, we take the nuclear charge to be Ze. For hydrogen, $Z = 1$.) We know that the electron has wave properties and that its de Broglie wavelength is $\lambda = h/mv$. However, the electron will not be in a stable, or stationary, state unless its de Broglie wave forms a standing wave within the orbit. For such a resonance, the orbit length, $2\pi r$, must be a whole number of wavelengths long.

For an example of the resonance of an electron's de Broglie wave on a circular orbit, refer to Fig. 27.5, which shows an orbit exactly four wavelengths long. As the wave winds itself around the orbit again and again, crest falls on crest and trough on trough; this is the condition for a stationary state and resonance. Thus the resonance condition for an orbit n de Broglie wavelengths long is

$$n\lambda_{\text{electron}} = 2\pi r_n \quad n = 1, 2, 3, \ldots \tag{27.2}$$

Detailed analysis using wave mechanics shows that an electron orbit satisfying this resonance condition is stable. An electron in such an orbit does not continually

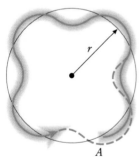

FIGURE 27.5

Resonance of electron waves determines stable orbits in the semiclassical model. If the orbit length $2\pi r$ is an integral number of wavelengths, the wave reinforces itself when it returns to the starting point A. In the case shown, $2\pi r = 4\lambda$.

radiate energy the way an orbiting point charge would in the classical model.

Because $\lambda_{\text{electron}} = h/mv$, we can rewrite Eq. 27.2 and solve for the angular momentum $r_n mv_n$ of the electron in the nth orbit:

$$r_n mv_n = n\left(\frac{h}{2\pi}\right) \tag{27.3}$$

This equation for the angular momentum is the same criterion Bohr had to use to select his stable orbits. He, however, could give no physical justification for it. Now we see why it must be true: It is the condition for electron wave resonance within the atom. Unfortunately, neither v_n nor r_n in Eq. 27.3 is known; we need yet another equation to evaluate these two quantities for the electronic orbits. Bohr showed how this can be done.

We can find a second equation by noticing that classically, the coulomb force between the electron and the positively charged nucleus supplies the centripetal force that holds the electron in orbit. If we assume that the massive nucleus remains at rest, we can write for the orbiting electron

Centripetal force = coulomb force

$$\frac{mv_n^2}{r_n} = k_e \frac{(Ze)(e)}{r_n^2} \tag{27.4}$$

where k_e is the coulomb force constant ($k_e = 8.99 \times 10^9 \, \text{N} \cdot \text{m}^2/\text{C}^2$).

We can solve Eqs. 27.3 and 27.4 simultaneously for the electron's speed v_n and the orbital radius r_n:

$$r_n = n^2 r_1 \quad \text{and} \quad v_n = \frac{h}{2\pi nmr_1} \qquad n = 1, 2, 3, \ldots \tag{27.5}$$

where r_1 is the radius of the smallest possible orbit ($n = 1$), given by

$$r_1 = \frac{h^2}{4\pi^2 Ze^2 mk_e} \tag{27.6}$$

For hydrogen, $Z = 1$ and $r_1 = 0.53 \times 10^{-10} \, \text{m}$. This is called the **Bohr radius,** since it is the radius that Bohr predicted for the unexcited hydrogen atom. Further, Bohr predicted the stable orbits whose radii are given by Eq. 27.5, and these orbits are called the **Bohr orbits.** Experiments show that unexcited hydrogen atoms do indeed have the 0.053-nm radius predicted by this theory. In the next two sections, we will see how the theory explains the observed emission spectrum of hydrogen.

27.3   HYDROGEN ENERGY LEVELS

We have found that the hydrogen atom should have certain stationary states in which it is stable. The theory we have outlined finds that the stable states consist of circular orbits whose radii, for hydrogen, are

$$r_n = n^2(0.53 \times 10^{-10} \, \text{m}) \qquad n = 1, 2, \ldots$$

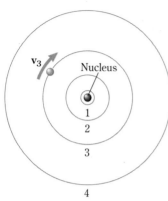

FIGURE 27.6

The electron can orbit the nucleus in a series of stable orbits that satisfy the resonance condition. No other orbits are stable. The size of the nucleus is much exaggerated.

Figure 27.6 shows the first few stable orbits. Let us see what energy the atom has in each of these states.

Each stationary state we have found for the atom must have its own characteristic energy. The energy of the atom consists of two parts. One part is the kinetic energy of the electron as it moves in its orbit. This energy is, for the nth stationary state,

$$(KE)_n = \tfrac{1}{2}mv_n^2$$

where relativistic effects can be neglected. Using Eq. 27.4, this becomes

$$(KE)_n = \frac{Ze^2k_e}{2r_n} \tag{27.7}$$

In addition to its kinetic energy, the electron has negative electrical potential energy. It is negative because we define the potential energy of two charges to be zero when the charges are infinitely far apart. As the electron comes closer to the nucleus, however, it is going "downhill" in potential energy because the nucleus attracts it. It therefore moves to potential energies that are less than zero, or negative. The potential energy of an electron that is a distance r_n from a positive charge Ze is

$$(PE)_n = \frac{-Ze^2k_e}{r_n} \tag{27.8}$$

Adding this to the kinetic energy of the electron in the nth orbit (Eq. 27.7) gives the total energy of the atom in the nth stationary state:

$$E_n = \frac{-Ze^2k_e}{2r_n} \tag{27.9}$$

Notice that the energy of the atom is negative and becomes more negative as r_n becomes smaller (in other words, as the electron gets closer to the nucleus).

We can put Eq. 27.9 in a more convenient form by using Eqs. 27.5 and 27.6 to replace r_n:

$$E_n = -\left(\frac{1}{n^2}\right)\left(\frac{2\pi^2Z^2e^4k_e^2m}{h^2}\right) \tag{27.10}$$

Evaluation of the constants in this expression yields, for $Z = 1$,

$$E_n = -\frac{2.18 \times 10^{-18}\,\text{J}}{n^2} = -\frac{13.6}{n^2}\,\text{eV} \tag{27.11}$$

The *negative total energy* means that the electron is *bound* to the nucleus. If it acquires enough energy from some external source (a collision, for example) to make its total energy positive, the electron is no longer bound: it is *free*.

Remember that each value of n corresponds to one stationary state of the atom. In terms of our semiclassical model, $n = 1$ corresponds to the electron orbiting in its smallest possible orbit, r_1. The atom's energy in this state, called the **ground**

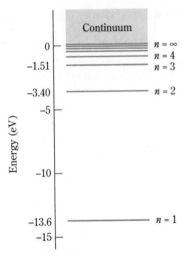

FIGURE 27.7

The energy-level diagram for hydrogen. There are an infinite number of levels between the $n = 4$ and $n = \infty$ levels.

state, is $E_1 = -13.6\,\text{eV}$. Because systems left to themselves fall to their lowest possible energy, hydrogen atoms are usually found in the $n = 1$ state. For $n = 2$, the next higher energy state, the orbit radius is (from Eq. 27.5) $4r_1$. The energy of the atom is then

$$E_2 = -\frac{13.6}{2^2}\,\text{eV} = -3.4\,\text{eV}$$

Notice that E_2 is greater than E_1. The atom has a higher energy in state 2 than in state 1. To summarize,

$$r_n = n^2 r_1 \quad \text{and} \quad E_n = \frac{E_1}{n^2} = -\frac{13.6}{n^2}\,\text{eV} \qquad (27.12)$$

As we see, like the energies of a particle in a tube, the energies of an electron in an atom are quantized.

It is convenient to represent the energies of quantized systems (such as atoms) on what is called an **energy-level diagram.** That for hydrogen is shown in Fig. 27.7. It is a vertical energy scale with horizontal lines drawn at the energies of the stationary states of the atom. We show only the first several levels because at higher values of n the levels are too close together to draw. This is obvious from the fact that all the levels from $n = 3$ to $n = \infty$ must fall within the small gap between $-1.51\,\text{eV}$ and zero. Since the orbit radius increases rapidly with n, the electron is free from the nucleus when $n = \infty$; under that condition, the atom is ionized.

Notice the region labeled *continuum* at energies above zero. At the $n = \infty$ level, the electron is free from the atom and is at rest. Higher energies represent the translational kinetic energy of the free electron. This energy is not quantized, and so all energies above $E = 0$ are allowed.

Example 27.1

How much energy is needed to ionize a hydrogen atom in its ground state?

R   *Reasoning*

Question What does the process of ionization consist of?
Answer Freeing the electron from the atom.

Question What does this mean in terms of electron energy?
Answer It means giving the electron enough energy to make $E_{\text{tot}} \geq 0$. The ionization energy is that energy required to make $E_{\text{tot}} = 0$.

Solution and Discussion Figure 27.7 shows the ground state ($n = 1$) to have $E_{\text{tot}} = -13.6\,\text{eV}$. Thus an energy of $+13.6\,\text{eV}$ is the minimum energy required to ionize the atom.

27.4   LIGHT EMISSION FROM HYDROGEN

Hydrogen atoms are usually in their lowest energy state, where $n = 1$. In this state, they are said to be unexcited. However, if you bombard the atoms with particles such as electrons or protons, collisions can excite them. In other words, a collision may give an atom enough energy to change it from its ground state to a higher stationary state.

From Fig. 27.7, we see that the difference in energy between the $n = 1$ and $n = 2$ states of hydrogen is

$$E = E_2 - E_1 = 13.6 - 3.4 = 10.2 \text{ eV}$$

Therefore the bombarding particle must provide an energy of 10.2 eV to excite the atom from the $n = 1$ state to the $n = 2$ state. Similarly, to excite it from the $n = 1$ state to the $n = 3$ state requires an energy

$$E = E_3 - E_1 = 13.6 - 1.51 = 12.1 \text{ eV}$$

In one of the most common ways of exciting gas atoms (Fig. 27.8), a high voltage is applied to a low-pressure gas. The few free electrons and ions that are always present (because of natural radioactivity and cosmic rays—see Chap. 28) are accelerated by the voltage, collide with the gas atoms, and generate an avalanche of charged particles. As a result, the gas in the tube, called a *discharge tube*, contains a large number of ionized and highly excited atoms. Neon signs and fluorescent lights are typical examples of discharge tubes. As you may know, they give off characteristic colors of light. Let us now see why a hydrogen gas discharge tube should emit light.

Like all other physical systems, atoms tend to fall to the lowest energy state possible. Excited electrons in hydrogen atoms spontaneously lose energy and fall to lower-energy states. For example, an excited electron in the $n = 3$ state may fall to the $n = 2$ state. In doing so, it must somehow lose the difference in energy between these two states, namely, $3.4 - 1.5 = 1.9 \text{ eV}$. It is possible for the atom to lose this energy through collisions with other atoms. Much of the energy lost in this way eventually appears as thermal energy. However, there is another important means by which the atom can rid itself of excess energy: it can emit a photon.

Suppose a hydrogen atom emits a photon as its electron falls from the $n = j$ level to the $n = i$ level. The difference in energy between the two levels is $E_j - E_i$, and the emitted photon must have this energy. The energy of a photon is hc/λ, however, and so we have

$$\text{Photon energy} = E_j - E_i = \frac{hc}{\lambda}$$

If we use Eq. 27.10 to replace E_i and E_j we find

$$\frac{1}{\lambda} = \frac{2\pi^2 Z^2 e^4 k_e^2 m}{h^3 c} \left(\frac{1}{i^2} - \frac{1}{j^2} \right) \tag{27.13}$$

Equation 27.13 has the same form as the empirical formulas of the Lyman, Balmer, and other series. Comparing Eqs. 27.1 and 27.13, we see that the experimentally measured Rydberg constant R should be equal to the coefficient in Eq. 27.13 for $Z = 1$ (hydrogen):

$$R = \frac{2\pi^2 e^4 k_e^2 m}{h^3 c}$$

This expression involves no less than *five* fundamental physical constants. It is

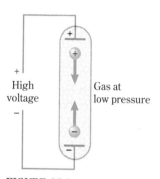

FIGURE 27.8

The high voltage across the discharge tube causes free electrons and ions within it to accelerate. If the voltage is high enough, these moving charges will ionize other atoms by collision.

High voltage

Gas at low pressure

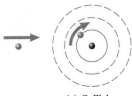

(a) Collision

(b) Excited atom

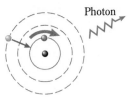

Photon

(c) Return to ground state,
emitting energy

FIGURE 27.9

A hydrogen atom in the ground ($n =$ 1) state is excited to the $n = 3$ state. It emits a photon as it falls back to the ground state. (Orbits not to scale.)

well worth your while to do the calculation that yields $R = 1.0974 \times 10^7 \, \text{m}^{-1}$. This was an astonishing accomplishment of Bohr's theory, which at the time rested on a very shaky physical basis.

With Eq. 27.13, the hydrogen spectrum can be explained in terms of energy changes of the electron as it jumps between the allowed stationary states. We can write the general form of allowed wavelengths as

$$\frac{1}{\lambda} = R\left(\frac{1}{i^2} - \frac{1}{j^2}\right) \tag{27.14}$$

Suppose, for example, that a collision has knocked the electron into the $n = 3$ orbit, as shown in Fig. 27.9. If the electron falls back to the $n = 1$ orbit, a photon will be emitted to carry away the lost energy. From Eq. 27.14, we would have

$$\frac{1}{\lambda} = R\left(\frac{1}{1^2} - \frac{1}{3^2}\right)$$

which turns out to be the second line in the Lyman series. Indeed, we can obtain the whole Lyman series if in Eq. 27.14 we let $i = 1$ and $j = 2, 3, 4, \ldots$; *the Lyman series of lines is emitted as electrons fall from outer orbits to the $n = 1$ orbit.*

Similarly, if electrons fall from outer orbits to the $n = 2$ orbit, we obtain a series of wavelengths given by

$$\frac{1}{\lambda} = R\left(\frac{1}{2^2} - \frac{1}{j^2}\right) \qquad j = 3, 4, \ldots$$

which is the Balmer series. Thus *the Balmer series of lines are emitted as electrons fall to the $n = 2$ orbit.* As you might expect, the Paschen series arises from transitions to the $n = 3$ orbit. These facts are summarized in Fig. 27.10, where only a few possible transitions are shown.

The energy difference between levels decreases rapidly as we go to higher and higher orbits. Hence, nearly as much energy is emitted when the electron falls from orbit 10 to 2 as when it falls from orbit 100 to 2. This means that the lines in the Balmer series become very closely spaced as we go to the wavelengths emitted by transitions from the outermost orbits to orbit 2. Of course, the most energy is emitted if the electron falls from outside the atom ($n = \infty$) to orbit 2. This results in the emission of the series-limit wavelength.

We can further clarify the origin of these spectral series by referring to Fig. 27.7 again. The diagram is redrawn in Fig. 27.11, with vertical arrows showing possible electronic transitions. There is a way to see at a glance how the wavelengths of the emitted lines vary. The energy of a transition is proportional to the length of its transition arrow. Hence the Lyman-series arrows (not all of which are shown) are longer than those of the Balmer series, telling us at once that the Lyman-series wavelengths will be shorter. We can also easily see from this diagram that the spectral lines in a series that correspond to transitions from the higher n values will lie very close together, since these energy levels have nearly the same values.

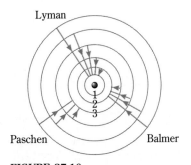

Lyman

1
2
3

Paschen   Balmer

FIGURE 27.10

Origin of the various spectral series for hydrogen. (Orbits not to scale.)

Exercise Using the known values of m, k_e, h, c, and e, calculate the value of R. Use values to four significant digits.

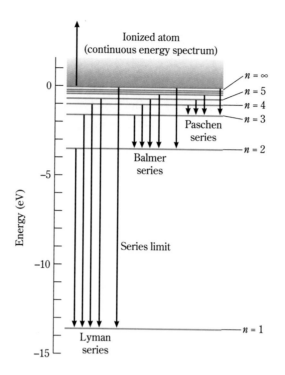

FIGURE 27.11

An energy-level diagram of the various hydrogen spectral series.

Illustration 27.1

Find the wavelength of the fourth line in the Paschen series.

Reasoning We know that the Paschen series results from transitions to the $n = 3$ state (Fig. 27.11). The fourth line occurs when the atom falls from the $n = 7$ state. Therefore, from Eq. 27.14,

$$\frac{1}{\lambda} = R\left(\frac{1}{3^2} - \frac{1}{7^2}\right)$$

Using $R = 1.0974 \times 10^7 \, \text{m}^{-1}$ gives $\lambda = 1005$ nm, a wavelength in the near-infrared.

Exercise What is the wavelength of the second line of the Paschen series?
Answer: 1281 nm.

Example 27.2

Singly ionized helium is a helium atom that has lost one of its two electrons. We might expect its remaining electron to act like a hydrogen electron. (*a*) Draw an energy-level diagram for this ion similar to Fig. 27.11. (*b*) Find the wavelength of the first line of its Balmer series.

R

Reasoning

Question   What is the difference between this ion and a hydrogen atom?
Answer   Helium has two protons in its nucleus, and so $Z = 2$. Equation 27.10 shows that the electron energies depend on Z^2. Since $Z = 2$, each of the hydrogen energies is multiplied by 4.

Question   What is the expression for the energy levels in ionized helium?

Answer $E_n = 4 \dfrac{-13.6\,\text{eV}}{n^2} = \dfrac{-54.4\,\text{eV}}{n^2}$

Question   What defines the Balmer series of wavelengths?
Answer   The energy transitions that end on the state $n = 2$.

Solution and Discussion   The above value for E_n shows that the ionization energy for the one remaining electron is 54.4 eV. Furthermore, the first excited state ($n = 2$) is bound with the same energy as the hydrogen electron, 13.6 eV. The energy levels are summarized in Fig. 27.12.

The first (longest-wavelength) line in the Balmer series is that for the $n = 3$ to $n = 2$ transition. We see in Fig. 27.12 that the energy lost by the electron in this transition is $-6.04\,\text{eV} - (-13.6\,\text{eV}) = 7.6\,\text{eV}$. The emitted photon therefore has a wavelength

$$\lambda = \frac{hc}{\Delta E}$$

$$= \frac{(6.63 \times 10^{-34}\,\text{J} \cdot \text{s})(3 \times 10^8\,\text{m/s})}{(7.6\,\text{eV})(1.6 \times 10^{-19}\,\text{J/eV})}$$

$$= 163\,\text{nm}$$

This is in the far ultraviolet part of the spectrum.

Another way of finding this wavelength would be to use Eq. 27.14 with $R = 4(1.0974 \times 10^7\,\text{m}^{-1})$. Why is the factor of 4 necessary?

Exercise   Find the wavelength limit for the Paschen series of singly ionized helium.   *Answer: 205 nm.*

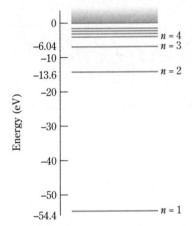

FIGURE 27.12

The energy-level diagram for singly ionized helium atoms.

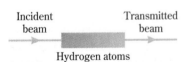

(a) Absorption experiment

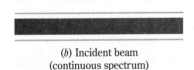

(b) Incident beam
(continuous spectrum)

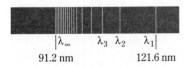

(c) Transmitted beam
(absorption spectrum)

FIGURE 27.13

Hydrogen atoms absorb only certain wavelengths out of a continuous incident spectrum. What are they?

27.5   THE ABSORPTION SPECTRUM OF HYDROGEN

Not only do atoms emit light, they also absorb it. To learn about light absorption, let us examine what happens in the experiment sketched in Fig. 27.13a. A beam of ultraviolet light passes through a tube filled with hydrogen atoms. The incident beam contains a continuous spectrum (a continuous range of wavelengths), as shown in Fig. 27.13b. However, certain discrete wavelengths are observed to be missing from the transmitted beam. Consequently, when the transmitted beam is examined with a spectrograph, the spectrum in Fig. 27.13c is obtained. We wish to find out which wavelengths the hydrogen atoms absorb from the beam.

To do so, we examine what happens as photons in the incident beam collide with the hydrogen atoms. Normally the atoms are in the ground state. When a photon strikes an atom, the photon loses either all its energy or none of it.* In other words,

*We ignore the Compton effect (Sec. 26.9) in this discussion because it is negligible in comparison with the effect we are discussing.

The spectrum of light from the sun shows many dark lines, indicating wavelengths which atoms in the sun's atmosphere have absorbed from the continuous spectrum emitted by the sun's photosphere.

the photon cannot be partially absorbed. The primary factor determining which of these two possibilities will prevail is the following: when the energy of the colliding photon is exactly equal to the energy difference between the $n = 1$ level and some other level, the photon can be absorbed. Otherwise, it must continue with its original energy.

The reason for this is quite simple. Since the electron in a hydrogen atom can exist in only one of the discrete energy levels, it can take on only increments of energy that will transfer it from one level to another. As we see in Fig. 27.11, these transitions correspond to energies that (in emission) give the Lyman series of lines. Therefore, photons with wavelength equal to that of the first line of the Lyman series (121.6 nm) will have enough energy to excite the atom from the $n = 1$ level to the $n = 2$ level. They can therefore be absorbed by the atom.

Similarly, photons with wavelength equivalent to any of the other lines in the Lyman series can be absorbed by hydrogen atoms in the ground state. No intermediate-wavelength photons can be absorbed, since their energies will not correspond to an allowed transition for the electron. However, photons with wavelengths less than the Lyman-series limit, 91.2 nm, can be absorbed. These photons have enough energy to excite the electron into the continuum, the region of continuous energy levels, where $E_{tot} \geq 0$. Photons with this much energy tear the electron completely loose from the atom (that is, ionize the atom) and give the freed electron additional kinetic energy. This type of photon absorption process is similar to photoelectric emission of electrons from a solid, and is referred to as the *atomic photoelectric effect*.

From what has been said, we can predict what will happen when a continuous spectrum of radiation is passed through a gas of atomic hydrogen. Most of the wavelengths will not be absorbed, since their photons do not have the proper energies to excite the atom to an allowed energy state. However, wavelengths that correspond to lines in the Lyman series will be absorbed, since the corresponding photons do have the proper energy to excite the atom to an allowed energy state. We call such an absorption spectrum a **line absorption spectrum.** All wavelengths shorter than the Lyman-series limit will be absorbed, since these photons will ionize the atom and carry the electron into the energy continuum. The absorption in this wavelength region, not shown in Fig. 27.13, is called a **continuous absorption spectrum** because a continuous range of wavelengths is absorbed.

Finally, we should note that absorption lines that correspond to the Balmer-series lines do not exist, except perhaps extremely weakly. The reason for this is as follows. We know the Balmer series corresponds to transitions between the $n = 2$ and higher levels. Since very few electrons are normally in the state $n = 2$, only a very few atoms are capable of having an electron knocked from the state at which $n = 2$ to higher states. Therefore photons that correspond to these energies will not be strongly absorbed. Of course, in highly excited hydrogen gas, the situation becomes more favorable for detecting absorption at the Balmer-line wavelengths. Why?

Example 27.3

When a hydrogen atom is excited by absorbing an ultraviolet photon, the atom can subsequently emit light of various wavelengths which depend on the way the electron returns to the ground state. For example, consider hydrogen atoms that have absorbed a photon of wavelength $\lambda = 97.23\,\text{nm}$. Which wavelengths (other than 97.23 nm) can these atoms subsequently emit?

Reasoning

Question   What principle determines the wavelengths an atom emits?
Answer   Conservation of energy. The energies of emitted photons are determined by the energies the electron can lose in jumping from an excited state to more tightly bound states.

Question   How do I find n for the excited state resulting from the absorption of the photon?
Answer   For the Lyman series, you have $\dfrac{1}{\lambda_n} = R\left(\dfrac{1}{1} - \dfrac{1}{n^2}\right)$. From this expression you can find the value of n that corresponds to $\lambda_n = 97.23\,\text{nm}$.

Question   What determines which electron transitions can emit photons?
Answer   The transitions must be to *lower* values of n, eventually ending up at $n = 1$. However, you don't want to count a *direct* transition to $n = 1$, which just gives back the 97.23-nm photon.

Solution and Discussion   The excited state corresponding to absorption of a 97.23-nm photon is found from a rearrangement of the Lyman-series formula:

$$\frac{1}{n^2} = 1 - \frac{1}{R\lambda_n}$$

$$= 1 - \frac{1}{(1.097 \times 10^7\,\text{m}^{-1})(0.9723 \times 10^{-7}\,\text{m})}$$

$$= 0.0625$$

$n^2 = 16.0$, giving $n = 4$.

Photons can be emitted from this state when the electron undergoes the following transitions:

1 $n = 4$ to $n = 3$
2 $n = 4$ to $n = 2$

3 $n = 3$ to $n = 2$
4 $n = 3$ to $n = 1$
5 $n = 2$ to $n = 1$

Transition (1) is the first line of the infrared Paschen series; (2) and (3) are lines in the visible Balmer series; and (4) and (5) belong to the ultraviolet Lyman series.

The energy levels are $E_1 = -13.6$ eV; $E_2 = E_1/4 = -3.4$ eV; $E_3 = E_1/9 = -1.51$ eV, and $E_4 = E_1/16 = -0.85$ eV. The energy changes involved in the transitions and the corresponding photon wavelengths are

1 $\Delta E = 0.66$ eV $\qquad \lambda = \dfrac{hc}{\Delta E} = 1879$ nm

2 $\Delta E = 2.55$ eV $\qquad \lambda = 486$ nm
3 $\Delta E = 1.89$ eV $\qquad \lambda = 656$ nm
4 $\Delta E = 12.1$ eV $\qquad \lambda = 103$ nm
5 $\Delta E = 10.2$ eV $\qquad \lambda = 122$ nm

27.6 THE WAVE THEORY OF THE ATOM

As we have seen, the Bohr theory predicts the correct energy levels for the hydrogen atom. It also explains the spectrum that hydrogen atoms emit and absorb. Using the wave properties of the electron, we have been able to justify Bohr's assumption that the electron exists only in certain stable states. Bohr assumed that these stable states consist of circular orbits around the nucleus. A better approach would be to start with Schrödinger's equation (Sec. 26.12) for the behavior of de Broglie waves and determine its resonance solutions for an electron in the coulomb potential of the nucleus.

Recall from Sec. 26.13 that the wave resonances of a particle in a tube told us where the particle is likely (and unlikely) to be found. Each resonance form was characterized by a quantum number, an integer between 1 and ∞. It turns out that resonances in three dimensions require three quantum numbers to specify their form. Thus we should expect the hydrogen atom's resonance forms to be characterized by three quantum numbers, not the single one we used in connection with the Bohr theory. Even so, the resonance forms should tell us where the electron is likely to be found when the atom is in a given resonance state. Let us now discuss the results one obtains for the hydrogen atom when the Schrödinger equation is used to find its resonance states.

The wave theory of the hydrogen atom gives the same energy levels found previously:

$$E_n = -\frac{13.6}{n^2} \text{ eV}$$

This result ensures that the wave-theory result will predict the observed hydrogen spectrum. Each energy state is characterized by the quantum number n, which we call the **principal quantum number.**

The resonance form found for the $n = 1$ state differs substantially from the circular orbit Bohr postulated for the $n = 1$ state. It turns out that the electron has a certain probability of being found somewhere in a fuzzy circular shell centered on

FIGURE 27.14

In the ground state, the hydrogen electron is most likely to be found somewhere in a fuzzy spherical shell centered on the nucleus. Shown is a cross section through the shell. The probability of finding the electron is greatest where the shading is most dense.

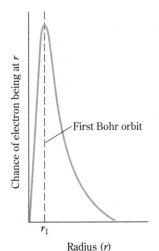

FIGURE 27.15

Wave theory predicts the relative probabilities shown that in the ground state the electron will be found at various radii from the hydrogen atom's center.

the nucleus. Figure 27.14 shows a cross section of this shell, and the electron is most likely to be found where the shading is heaviest. Although it is most likely to be found at a radius r_1 from the nucleus, the electron has some probability of existing anywhere in the shaded region. Be certain that you understand how this result differs from Bohr's concept of a single circular orbit. The wave theory replaces the circle with a spherical shell and, in addition, does not restrict the electron to a definite radius; the shell is very fuzzy. This is shown graphically in Fig. 27.15.

The resonance form predicted by wave theory for the $n = 2$ state is much more complicated than that for the $n = 1$ state. It turns out that there are three resonances that have the $n = 2$ energy. We can picture these resonances, and those for still larger values of n, by means of diagrams that show the chance of finding the electron at various positions in the atom. These diagrams, called **orbitals,** no longer portray the electron as moving in an orbit as in Bohr's semiclassical model. Figure 27.16 shows some of the orbitals for the $n = 1$, 2, and 5 states in two dimensions. To get the three-dimensional picture, these patterns should be rotated about a vertical axis through the center; the intensity of the pattern at any given point gives the relative probability of finding the electron at that point. At larger values of n, the orbitals become quite complicated, as illustrated by Fig. 27.16e.

We see from all this that the Bohr theory is indeed a gross oversimplification of the electron behavior in the hydrogen atom. In particular, Bohr's concept of fixed orbits is untenable. The energy levels of the atom are predicted correctly by the Bohr theory, however, and the principal quantum number n that Bohr introduced has great importance. Although we should always keep the limitations of the Bohr model in mind, it offers a framework for a systematic description of atoms, and we make frequent reference to it.

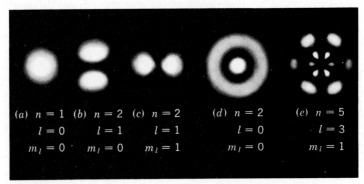

FIGURE 27.16

To obtain the electron distribution in three-dimensional space, the above patterns must be rotated about a vertical axis through the center of each.

27.7 QUANTUM NUMBERS AND THE PAULI EXCLUSION PRINCIPLE

As we have seen, the hydrogen atom and its electron can exist in certain discrete energy levels characterized by an integer n, given by the expression

$$E_n = \frac{-13.6Z^2}{n^2} \text{ eV}$$

where $Z = 1$ for hydrogen. The integer n ranges from 1 to infinity as the atom assumes its various allowed energies. Although we arrived at this result by use of Bohr's model, the wave picture, based on the solution of the Schrödinger equation, leads to the same result. Hence it is seen that n is a fundamental parameter needed to describe the state of a hydrogen atom. As mentioned earlier, it is called the *principal quantum number*. It characterizes the energy level in which the electron is to be found. Bohr pictured each value of n as being associated with a particular orbit for the electron, but this proves untenable, as pointed out in the preceding section. Nevertheless, it is common to say that each value of n corresponds to a particular **energy shell** (rather than orbit) about the nucleus. For example, when the atom is in the $n = 3$ energy level, it is customary to say that the electron is in the $n = 3$ shell.

We also saw in the preceding section that more than one wave resonance form is possible for the same value of the principal quantum number n. The wave theory shows that two other quantum numbers must be specified in order to designate a particular wave resonance within the atom. One of these, the **orbital quantum number,** is related to the angular momentum of the Bohr electron in its resonance orbit. It is represented by the letter l and can assume integer values from 0 to $n - 1$. For example, when $n = 1$, the possible values for l are limited to a single value, namely $l = 0$. When $n = 2$, it is apparent that l can take on the values 0 and 1, since $n - 1 = 1$ in this case. Notice that l is always less than n.

The third quantum number is called the **magnetic quantum number,** m_l, which can assume the values $0, \pm1, \pm2, \ldots, \pm l$. It describes the possible orientations of the orbital angular momentum of the electron in an external magnetic field. When $n = 4$, for example, the largest possible value for l is 3, and m_l can take on the values of $0, \pm1, \pm2,$ and ±3. In other words, when the electron is in the $n = 4$ energy level, seven different $l = 3$ orbitals are possible. In addition, there are five $l = 2$ orbitals possible, three $l = 1$ orbitals, and one with $l = 0$. Hence the atom can exist in 16 different resonant electron configurations when in the $n = 4$ energy level.

Finally, a quantum condition exists for the electron itself. The electron possesses a small magnetic moment because it is a spinning charged particle. Its magnetic moment can assume only two orientations relative to an external magnetic field: parallel or antiparallel. We characterize this by assigning to the electron a **spin quantum number,** m_s, with the two possible values $\pm\frac{1}{2}$; the two signs represent the parallel and antiparallel orientations. Table 27.1 summarizes the four quantum numbers needed to describe the state of an electron in an atom. We call each combination of the four quantum numbers an electronic **state** of the atom. We will now see that an extremely important principle applies to the behavior of electrons in the available states.

The importance of designating these electronic states was first appreciated fully by Wolfgang Pauli in 1925. He wished to extend these concepts to atoms other than hydrogen. In order to properly assign states to the various electrons in multielectron atoms, he arrived at the following conclusion, known as the **Pauli exclusion principle:**

No two electrons in an atom can have the same set of four quantum numbers; that is, no two electrons can exist in the same state.

This principle is basic to an understanding of the electronic structure of atoms, as we see in the next section.

TABLE 27.1

The four quantum numbers for the electron

Principal	$n = 1, 2, 3 \ldots$
Orbital	$l = 0, 1, 2, \ldots, n - 1$
Magnetic	$m_l = 0, \pm1, \pm2, \ldots, \pm l$
Spin	$m_s = \pm\frac{1}{2}$

27.8 THE PERIODIC TABLE

Until now we have been primarily concerned with an atom that has only one electron. This might be hydrogen, singly ionized helium, doubly ionized lithium, and so on. We are now in a position to discuss how the additional electrons are arranged in the multielectron atoms found in nature and listed in the periodic table. To do this, we once again use the concept of electron shells about the nucleus; each value of n has a shell associated with it. Moreover, we assume that the same resonances found for the single-electron atom can be carried over qualitatively to more complex atoms. That is, we use electronic states specified by the four quantum numbers described in the previous section.

The question we must now answer is: "How do the electrons arrange themselves in the various states when there is more than one electron in an atom?" For example, there are six electrons in a carbon atom. In which energy levels and electron states are they to be found? We can answer this question by using the following three rules, which we have already discussed:

1 The number of electrons in any neutral atom is equal to the atomic number Z of the atom.
2 In an unexcited atom, all electrons are in the lowest possible energy states. The atom is then said to be in its ground state.
3 No two electrons in an atom can have the same four quantum numbers (the Pauli exclusion principle).

Let us now use these rules to determine the electronic structure of the unexcited atoms in the periodic table.

HYDROGEN ($Z = 1$)

Its single electron will be in the $n = 1$ level. This is the lowest possible energy level, and no violation of the exclusion principle occurs.

HELIUM ($Z = 2$)

TABLE 27.2

Electron	n	l	m_l	m_s
1	1	0	0	$\frac{1}{2}$
2	1	0	0	$-\frac{1}{2}$

Its two electrons can both exist in the $n = 1$ level since they can have the nonidentical quantum numbers shown in Table 27.2. The table lists the only combinations of quantum numbers possible for the $n = 1$ level. A third electron cannot exist in this level. We call each value of n an **energy shell,** and say that the $n = 1$ shell is filled when occupied by two electrons.

LITHIUM ($Z = 3$)

TABLE 27.3

Electron	n	l	m_l	m_s
1	1	0	0	$\frac{1}{2}$
2	1	0	0	$-\frac{1}{2}$
3	2	0	0	$\frac{1}{2}$

This atom has three electrons, and so the third electron must go into the next highest energy shell, the one with $n = 2$ (Table 27.3). Since this electron is in the second energy level, it is less tightly bound to the atom than the $n = 1$ electrons are. Hence lithium easily shares one electron in chemical reactions, which in chemical terminology makes lithium a univalent (valence-one) element.

TABLE 27.4

n	l	m_l	m_s
2	0	0	$\pm\frac{1}{2}$
2	1	0	$\pm\frac{1}{2}$
2	1	+1	$\pm\frac{1}{2}$
2	1	−1	$\pm\frac{1}{2}$

LARGER Z ATOMS

There are quite a few possible combinations for the quantum numbers when $n = 2$. If you count them, you will find there are eight (Table 27.4).* Therefore eight electrons can exist in the $n = 2$ shell. This means that the shell will not become filled until element $Z = 10$, neon, is reached. Neon is chemically unreactive because it has filled shells. The next element, $Z = 11$, is sodium. This atom is univalent, since its eleventh electron is alone in the $n = 3$ shell and is rather easily removed.

As one proceeds to the very-high-Z elements in the table, the concept of shells becomes less useful. The trouble arises primarily because the separation between energy levels is relatively small at high n values. In these cases the repulsions between the various electrons in the atom sometimes contribute energies large enough to cancel the influence of energy differences between shells. Despite this complication, the shell approach still proves useful for qualitative considerations.

Illustration 27.2

Use the Pauli exclusion principle to determine the ground-state electron configuration of argon ($Z = 18$) and rubidium ($Z = 37$).

Reasoning The $n = 1$ and $n = 2$ shells can hold 2 and 8 electrons, respectively, so 10 electrons are in these shells in both argon and rubidium. For the $n = 3$ shell, we have 18 separate sets of quantum numbers, as shown in Table 27.5. The eight remaining electrons in argon fill the $l = 0$ and $l = 1$ subshells of the $n = 3$ level. When the ground-state electrons fill either a shell or a subshell, the electrons are tightly bound, rendering the atom chemically inert. Argon is another of the chemically inactive noble gases.

The first 18 electrons in rubidium occupy states with the same quantum numbers as the 18 argon electrons. The next 10 electrons fill the $n = 3$, $l = 2$ subshell. That leaves nine electrons which must go into the $n = 4$ level. Two occupy the $n = 4$,

*In multielectron atoms, electrons with the same value of n (same shell) are said to be in the same **subshell** if they have the same value of l. Thus the six electrons in the second, third, and fourth rows in Table 27.4 occupy the same subshell. The two electrons in the first row of the table occupy a different subshell.

TABLE 27.5 *Quantum numbers for the $n = 3$ shell*

n	l	m_l	m_s	Number of subshell states
3	0	0	$\pm\frac{1}{2}$	2
3	1	−1	$\pm\frac{1}{2}$	
		0	$\pm\frac{1}{2}$	6
		+1	$\pm\frac{1}{2}$	
3	2	−2	$\pm\frac{1}{2}$	
		−1	$\pm\frac{1}{2}$	
		0	$\pm\frac{1}{2}$	10
		+1	$\pm\frac{1}{2}$	
		+2	$\pm\frac{1}{2}$	

$l = 0$ subshell, and six more fill the $n = 4$, $l = 1$ subshell. The one remaining electron must occupy one of the $n = 4, l = 2$ states. Because this outermost electron (called a *valence* electron) is relatively loosely bound, rubidium readily forms chemical bonds with other elements.

27.9 X-RAYS AND THE SPECTRA OF MULTIELECTRON ATOMS

As we have seen, the Pauli exclusion principle tells us how the electrons pack into an atom in the ground state. To a first approximation, Eq. 27.10 gives the energy of any electron in the nth state. Therefore the energy of an electron in a multielectron atom is the same as Z^2 times the energy of an electron in the same state in the hydrogen atom. This approximation breaks down for the outer electrons of the atom, however, because the interaction energies between these electrons are comparable to the energy differences between the Bohr energy levels. Therefore the Bohr energies cannot apply to these outer electrons.

However, the interaction energy between electrons is small relative to the energy differences between the $n = 1$ and $n = 2$ states. In zinc ($Z = 30$), for example, the Bohr energies are

$$E_n = -\frac{13.6Z^2}{n^2}\,\text{eV} = -\frac{12{,}240}{n^2}\,\text{eV}$$

The situation is even more startling for gold ($Z = 79$), where

$$E_n = -\frac{84{,}900}{n^2}\,\text{eV}$$

As we see, the energy differences between states E_1 and E_2 in these atoms are tens of thousands of electronvolts. Compared with these inner-shell energies, the coulomb interaction energies between electrons are small. Hence the Bohr energies are nearly correct for electrons in the $n = 1$ and $n = 2$ shells of atoms that have high atomic numbers.

For an outer-shell electron the situation is quite different. First, the inner-shell electrons, being closer to the nucleus, appear to cancel part of the nuclear charge. Hence the $n = 2$ electrons "see" a nuclear charge of about $(Z - 2)e$ rather than Ze; similarly, the $n = 3$ electrons see a nuclear charge of about $(Z - 10)e$ because of the two $n = 1$ electrons and the eight $n = 2$ electrons. We say that the inner electrons *screen* the nuclear charge from the outer electrons.

In addition to this effect, the outer-shell electrons are subject to the energies from the repulsive electron-electron interaction involving all the other electrons of the atom. As mentioned, these energies are comparable to the small energy differences between outer shells, and the Bohr energy formula does not apply to them.

To cause an atom to emit radiation, some of its electrons must be excited to higher energies. Because the outer-shell electrons require only small amounts of energy to excite them to empty states, it is not difficult to obtain visible light from high-Z atoms. The material is simply vaporized and used in a discharge tube much like the one we saw in Fig. 27.8. However, the spectral lines emitted by transitions between these outer-shell levels are extremely numerous and complex.

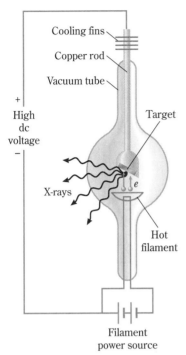

FIGURE 27.17

Electrons emitted by the hot filament bombard the target, which then emits x-rays.

The situation is quite different for transitions involving inner-shell electrons. From Illustration 27.2, we can see that the $n = 1$, $n = 2$, and $n = 3$ shells are filled in the unexcited zinc atom. Therefore we cannot excite an inner ($n = 1$) electron to the occupied $n = 2$ or $n = 3$ shells because of the exclusion principle. In order to excite an $n = 1$ electron, the energy supplied to the atom must be at least enough to allow the electron to jump to the $n = 4$ shell. In the case of zinc, nearly 12,000 eV is needed. Once this jump happens, a vacancy exists in the $n = 1$ shell, and so an electron from the $n = 2$ or $n = 3$ shells can readily jump into that vacancy, releasing a photon whose energy equals the energy difference between the electron's final and initial states. If an $n = 2$ electron falls back to the $n = 1$ state, it will release a photon whose energy is about 9000 eV. Recall from Illustration 26.4 that 1240-nm photons have an energy of 1 eV. Therefore, 9000 eV corresponds to a wavelength of

$$\lambda = \frac{1\,\text{eV}}{9000\,\text{eV}} \, 1240\,\text{nm} = 0.14\,\text{nm}$$

This wavelength is in the x-ray region. Thus we see that transitions between inner shells of a high-Z atom give rise to x-rays. To generate x-rays, we need to excite the inner-shell electrons to unoccupied outer shells, and, as we have seen, this requires large amounts of energy.

A typical x-ray tube circuit is shown in Fig. 27.17. Electrons emitted from the hot filament are accelerated through potential differences of the order of 10^5 V. When these high-energy electrons strike the high-Z atoms in the target, electrons are knocked out of the inner shells of the atoms. As other electrons fall into the vacancies, x-ray photons are emitted. The x-rays so generated have wavelengths that are characteristic of the energy differences between the various shells in the atom. That is, the emitted photons carry an energy equal to the difference between the energies of the two shells that act as the starting and end points for the electron that falls into the vacancy. X-rays emitted by this process are referred to as **characteristic x-rays.**

Another type of x-ray emitted from a target bombarded by electrons is referred to as **bremsstrahlung** ("braking radiation"). As the name implies, these x-rays are emitted by the bombarding electrons as they are suddenly slowed on impact with the target. We know that any accelerating charge emits electromagnetic radiation. Hence these impacting electrons emit radiation as they are strongly decelerated by the target. Since the rate of deceleration is so large, the emitted radiation is correspondingly of short wavelength, and so the bremsstrahlung is in the x-ray region. However, unlike characteristic x-rays, the bremsstrahlung has a continuous range of wavelengths. This reflects the fact that deceleration can occur in a nearly infinite number of different ways, so that the energy released varies widely from one impact to another.

Figure 27.18 is a graph of the radiation emitted from a molybdenum target bombarded by 35,000-eV electrons. The two sharp peaks are the characteristic x-rays emitted as electrons fall to the $n = 1$ shell from the $n = 2$ and $n = 3$ shells. The shorter wavelength, of course, corresponds to the higher-energy transition, the $n = 3$ to $n = 1$ transition. Bremsstrahlung is the cause of the lower-intensity radiation spread over all wavelengths longer than λ_m. Since the energy of the electrons in the impacting beam was 35,000 eV, the emitted photons cannot have energies larger than this value. Using our conversion based on 1240 nm being equivalent to 1 eV (Illustration 26.4), we find that 35,000 eV corresponds to $1240/35,000 \approx 0.035$ nm. As we see from Fig. 27.18, the highest-energy bremsstrahlung does indeed have this wavelength.

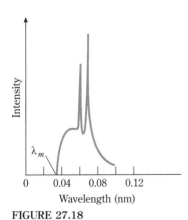

FIGURE 27.18

The spectrum of x-rays emitted from a molybdenum target bombarded by 35,000-eV electrons.

The narrow, intense beams of lasers provide spectacular visual effects.

Illustration 27.3

From the data in Fig. 27.18, find the energy difference between the $n = 1$ and $n = 2$ levels in molybdenum.

Reasoning   As we saw in our discussion of Fig. 27.18, the 0.070-nm peak results from the $n = 2$ to $n = 1$ transition. Therefore the photon of wavelength 0.070 nm carries away the energy lost by an electron as it falls from the $n = 2$ to the $n = 1$ shell. Since 1240 nm corresponds to 1 eV, 0.070 nm corresponds to an energy of 1240/0.070, or about 18,000 eV. Therefore the energy difference between these two shells in molybdenum atoms must be about 18,000 eV.

Exercise   A zinc target is bombarded by 13,000-eV electrons. What is the shortest-wavelength x-ray emitted by the target? What approximate wavelength corresponds to the $n = 3$ to $n = 1$ transition?   *Answer: 0.095 nm; 0.114 nm.*

27.10   LASER LIGHT

A beam of ordinary light is a multitude of individual waves sent out by the individual atoms of the source. Although the waves in a monochromatic light beam all have the same wavelength, the waves sent out by the individual atoms are not in phase. Indeed, they do not maintain a fixed phase relation relative to each other. In other words, the waves are not *coherent*. Statistical analysis shows us that, if the amplitude of each wave is A, then the amplitude of the wave that results from the sum of N such waves is $A\sqrt{N}$.

Suppose, however, that we could synchronize the atoms in a light source so that they emit monochromatic light waves that are in phase with one another, so that all the waves would be coherent. The resultant wave due to N in-phase coherent waves, each of amplitude A, is simply the direct sum of the waves, namely AN. If we compare this amplitude with the amplitude of N incoherent waves, $A\sqrt{N}$, we see that the amplitudes are in the ratio of $AN/A\sqrt{N}$. Because the intensity of a wave is proportional to its amplitude squared, we find that

$$\frac{\text{Intensity (coherent)}}{\text{Intensity (incoherent)}} = \left(\frac{AN}{A\sqrt{N}}\right)^2 = N$$

A beam consisting of N waves is N times more intense if the waves are coherent than if they are incoherent. Because a typical beam might consist of perhaps a million individual waves at a given point, a coherent beam might be a million times more intense than a similar incoherent beam.

It was not until the 1950s that a light source for coherent waves was devised. This type of source, called a **laser** (an acronym for *l*ight *a*mplification by *s*timulated *e*mission of *r*adiation) makes use of a fact pointed out in 1917 by Einstein: *Atoms in an excited state can be stimulated to jump to a lower energy level when they are struck by a photon of incident light whose energy is the same as the energy-level difference involved in the jump. The electron thus emits a photon of the same wavelength as the incident photon. The incident and emitted photons travel away from the atom in phase.*

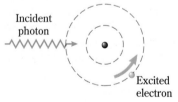

(a) Before

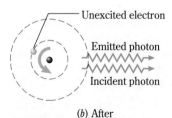

(b) After

FIGURE 27.19

Stimulated emission produces coherent waves.

This process, called **stimulated emission,** is shown in Fig. 27.19. Let us see how the laser makes use of this phenomenon.

To be able to emit energy, the electrons must be in an excited state when they are stimulated by the incident photons. Thus some method of excitation must be present. Moreover, in order to produce a large intensity of stimulated emission, there must be a greater number of electrons in the excited state than in the ground state. We call this situation a **population inversion**. To achieve a population inversion, an electron in an excited state must stay in that state for some time before spontaneously returning to the ground state. Such a state is said to be metastable: A **metastable** state is one in which an excited electron is unusually stable and from which the electron spontaneously falls to a lower state only after a relatively long time.

With these requirements in mind, we can summarize the fundamental operation of a laser in an energy-level diagram such as that in Fig. 27.20. By some means, electrons are excited from the ground state E_1 into an excited state E_2 (Fig. 27.20a and b). Most of these electrons jump to a metastable state E_2 where they cannot readily return spontaneously to E_1. This causes a buildup of the number of electrons in E_2, a population inversion relative to the ground state. If a photon of energy $E_2 - E_1$ passes through the atom, it can *stimulate* an electron to jump from E_2 to E_1 (Fig. 27.20c). This jump creates a photon, identical to and *in phase* with the incident photon (Fig. 27.20d). As this process repeats itself many times, the number of photons increases geometrically and the intensity of the light is amplified.

One common laser is the helium-neon laser, which consists of a very straight electrical discharge tube containing 15 percent helium gas and 85 percent neon gas. The atomic system of the helium and neon atoms possesses three energy levels of particular interest: E_1, E_2 and E_3 as shown in Fig. 27.21. The state E_3 is a metastable state of **helium** lying 20.61 eV above E_o. The E_2 state is a metastable state of **neon** lying 20.66 eV above E_o. The E_1 state is an energy level in **neon** 1.96 eV below E_2.

Before the electrical discharge is activated, the electrons are almost all in the ground state. Some of them are then excited to levels E_2 and E_3 by a high-voltage discharge. Collisions between helium and neon atoms transfer the energy of the excited helium electrons to E_2, creating a population inversion between E_2 and E_1.

Now suppose that a few excited neon atoms spontaneously make the transition from E_2 to E_1, thereby emitting photons. The wavelength of these photons will be 632.8 nm, corresponding to the energy jump of 1.96 eV. These photons can be absorbed by the few electrons in level E_1 and excite them to E_2. Also, in Fig. 27.21, they can cause electrons to fall from E_2 to E_1, giving rise to stimulated emission of

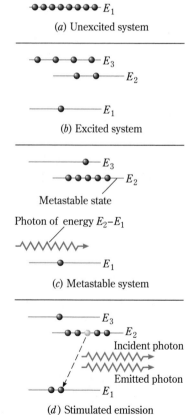

(a) Unexcited system

(b) Excited system

Metastable state

Photon of energy E_2-E_1

(c) Metastable system

Incident photon

Emitted photon

(d) Stimulated emission

FIGURE 27.20

In a laser, a population inversion, a metastable state, and stimulated emission are required.

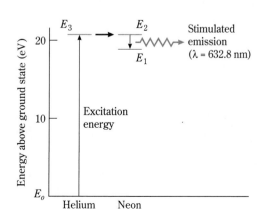

FIGURE 27.21

Energy level diagram for a helium-neon laser. Electrons are excited into energy levels E_2 and E_3 by an electrical discharge. Collisions between the He and Ne atoms cause the excited He electrons to excite more Ne electrons into the E_2 level, creating a population inversion in this metastable state. The E_2 electrons are then stimulated to jump to E_1, 1.96 eV below E_2.

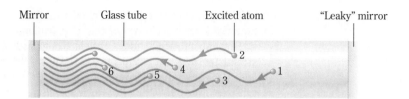

FIGURE 27.22

Schematic diagram showing how stimulated emission builds up a strong coherent wave in a laser tube.

waves that are identical to the incident wave. Because of the population inversion, the stimulated emission overwhelms further photon absorption, and so the intensity of the emitted waves grows as they pass through the gas. The end result is a coherent beam traveling through the discharge tube.

The ends of the discharge tube consist of accurately parallel plane mirrors (Fig. 27.22). However, the mirror at the right is only lightly metallized so that it reflects perhaps only 99 percent of the light. A multitude of excited neon atoms emit identical photons, and these photons are all in phase, as indicated in Fig. 27.22. In a short time, the tube is filled with coherent waves moving back and forth between the two mirrors at the ends of the tube, resulting in a very intense, monochromatic coherent beam in the tube. A small fraction of the coherent beam exits from the tube through the "leaky" mirror at one end.

Because all the waves issuing from the end of the laser tube are coherent, the beam is of high intensity. Its wavelength is sharply defined, 632.8 nm, because all the waves are identical. Not only is the beam intense and coherent, but also it is a very thin, straight beam that diverges very little. Any rays within the tube that diverged much from its axis were lost out the sides during their many trips back and forth. The fact that the beam does not diverge appreciably is of great practical importance. Unlike light from a bulb, the laser beam's energy does not fan out into space. Instead, it flows out into space through a thin cylinder and maintains its intensity over very long distances.

(a) A hologram of the Palace of Discovery at La Villette, Paris.

(b) (right) The coherent light of a laser can be used to recognize patterns, as in the bar code scanners used at many checkout stands in stores.

(a)

(b)

Although you are probably most accustomed to the helium-neon laser, which has an output of only a milliwatt or so, there are many types of lasers available today. All make use of a metastable state to achieve a population inversion so that stimulated emission can give rise to an in-phase coherent set of waves. They differ in wavelength, ranging from the far infrared to the long x-ray range. Their energies range from a small fraction of a milliwatt (in lasers used in laser disk audio systems) to millions of watts (in lasers being used in nuclear fusion research, which we discuss in the next chapter).

In the 40 years since its inception, the laser has become one of the most widely applied products of physics research. The coherence of its light allows phase information as well as intensity to be recorded photographically in the process known as *holography*. Holographic images capture the full three-dimensionality of the photographed object. Coherence also allows the laser beam to be focused into an exceptionally small area, providing both an extremely precise and extremely intense beam. Thus surgeons can destroy diseased tissue at a well-defined point or "weld" ruptured tissue, such as a detached retina. The beam can be made to burn through materials faster and more precisely than a drill. The straightness of the beam makes it useful in surveying and in controlling machining and robotics operations in manufacturing.

Lasers now interface with computers in many ways, as in the reading of bar codes on most of the goods we buy. Tiny solid-state lasers are used in compact-disk audio and video systems. The beam is reflected from the digitized patterns encoded on the disk, and converted to electronic signals that are analyzed by a computer and transformed into a pattern of voltage signals that drive loudspeakers and the output of a VCR.

New laser applications will be seen increasingly in the future, such as signal transmission by modulation of visible light and optical memory storage in computers.

LEARNING GOALS

Now that you have finished this chapter, you should be able to

1 Define (*a*) nuclear atom, (*b*) line and continuous spectrum, (*c*) series limit, (*d*) Rydberg constant, (*e*) Lyman, Balmer, and Paschen series, (*f*) Bohr orbits and Bohr radius, (*g*) energy-level diagram, (*h*) ground state, (*i*) ionization energy, (*j*) principal, orbital, magnetic, and spin quantum numbers, (*k*) energy shell and subshell, (*l*) Pauli exclusion principle, (*m*) characteristic x-rays and bremsstrahlung, (*n*) coherent waves, (*o*) stimulated emission, (*p*) metastable state, (*q*) population inversion, (*r*) laser.

2 Explain how Rutherford's experiment presented evidence for the concept of the nuclear atom.

3 Give the approximate diameter of an atom.

4 Sketch the lines of the Balmer series and write the Balmer formula. Compute the wavelength of a given line of the Balmer series when the Rydberg constant is given. Repeat for both the Lyman series and the Paschen series.

5 Explain how the wave properties of the electron lead to the Bohr orbits and the Bohr energy levels.

6 Give the general formula for the energy levels of the hydrogen atom in electronvolts. Sketch the energy-level diagram for hydrogen.

7 Compute the wavelength emitted by the hydrogen atom in any specified transition. Show on an energy-level diagram how the Lyman, Balmer, and Paschen series arise.

8 Explain why hydrogen atoms normally absorb the Lyman-series wavelengths but not those of the Balmer series.

9 Explain the meaning of an electron-distribution diagram, such as those in Figs. 27.15 and 27.16.

10 Use the Pauli exclusion principle to determine the ground-state electron configuration of simple elements. Explain how the exclusion principle predicts the chemical activity (valence) of these elements.

11 Describe how x-rays are produced in an x-ray tube. Compute the shortest-wavelength x-rays emitted by a target impacted by electrons of a given energy.

12 Explain the principle of a gas laser in terms of metastable states, population inversion, and stimulated emission. State the important features of a laser beam with regard to coherency, phase, and shape. Point out how these features lead to a wide variety of uses for lasers.

SUMMARY

DERIVED UNITS AND PHYSICAL CONSTANTS

Rydberg Constant (R)

$$R = \frac{2\pi^2 e^4 k_e^2 m}{h^3 c}$$

For the hydrogen atom, $Z = 1$ and $R = 1.0974 \times 10^7 \, \text{m}^{-1}$.

Bohr Radius (r_1)

$$r_1 = \frac{h^2}{4\pi^2 Z e^2 m k_e}$$

For hydrogen, $Z = 1$ and $r_1 = 0.53 \times 10^{-10} \, \text{m}$.

DEFINITIONS AND BASIC PRINCIPLES

Hydrogen Atom Spectral Series
The hydrogen atom emits and absorbs electromagnetic radiation in wavelength series given by the following general formula:

$$\frac{1}{\lambda} = R\left(\frac{1}{i^2} - \frac{1}{j^2}\right)$$

The symbols i and j are integers. Each wavelength series has a given value of i. Individual wavelengths within a series are obtained by j taking on all integer values $> i$.

INSIGHTS
1. The first three wavelength series are

   $i = 1$ Lyman series (ultraviolet)

   $i = 2$ Balmer series (visible)

   $i = 3$ Paschen series (infrared)

2. Each series has a shortest wavelength, called the series limit, corresponding to $j = \infty$. This limit is

$$\frac{1}{\lambda_\infty} = \frac{R}{i^2}$$

Stable Orbits and Energy Levels in the Bohr Hydrogen Atom
Stable orbits have radii given by

$$r_n = n^2 r_1$$

where n is any integer and r_1 is the first Bohr radius.

Electrons in the Bohr atom can have total energies given by

$$E_n = \frac{-Z e^2 k_e}{2r_n} = \frac{E_1}{n^2}$$

where $E_1 = -Z e^2 k_e / 2 r_1$ is the lowest energy, known as the ground state.

INSIGHTS
1. E_n is the sum of kinetic and electric potential energies. For each energy level, KE $= +Z e^2 k_e / 2 r_n$ and PE $= -Z e^2 k_e / r_n$.
2. When E_n is negative, the electron is in a bound state. To free the electron (ionizing the atom), a minimum positive energy equal to E_n must be given to the electron.
3. For hydrogen, $E_1 = -13.6 \, \text{eV}$.

Quantum Numbers and the Pauli Exclusion Principle
Four quantum numbers define a state of an electron in an atom:

Principal: $n = 1, 2, 3, \ldots$
Orbital: $l = 0, 1, 2, \ldots, n - 1$
Magnetic: $m_l = 0, \pm 1, \pm 2, \ldots, \pm l$
Spin: $m_s = \pm \frac{1}{2}$

The exclusion principle states that no two electrons in an atom can have the same four quantum numbers; i.e., no two electrons can occupy the same state in an atom.

INSIGHTS
1. The principal quantum number, n, determines the energy of the state. Since there are a number of possible values of l, m_l, and m_s for a given n, a number of electrons can possess the same energy without occupying the same quantum state.
2. The exclusion principle explains the ground-state electron configurations and valences of the periodic table of the elements.

QUESTIONS AND GUESSTIMATES

1 Why doesn't the hydrogen gas prepared by students in the laboratory glow and give off light?

2 Suppose you are given a glass tube containing two electrodes that is sealed at both ends. The gas inside is either hydrogen or helium. How can you tell which it is without breaking the tube? If the gas is at high pressure, what difficulty might you have?

3 When white light is passed through a vessel containing hydrogen gas, it is found that wavelengths of the Balmer series as well as those of the Lyman series are absorbed. We conclude from this that the gas is very hot. Why can we draw this conclusion? (This is actually the basis for one method of measuring the temperature of a hot gas.)

4 Explain clearly why x-ray emission lines in the range of 0.1 nm are not observed from an x-ray tube when a low-atomic-number metal is used as the target in the tube.

5 A steel company suspects that one of its competitors is adding a fraction of a percent of a rare-earth element to its (the competitor's) product. How can the element quickly be identified and its concentration determined?

6 In the helium atom, the two electrons are in the same shell but avoid each other well enough for their interaction to be of only secondary importance. Estimate the ionization energy (in electronvolts) for helium, that is, the energy required to tear one electron loose. Also, estimate the energy needed to tear the second electron loose. Which of these two values is more reliable?

7 The ionization energies for lithium, sodium, and potassium are 5.4, 5.1, and 4.3 eV, respectively, while those for helium, neon, and argon are 24.6, 21.6, and 15.8 eV, respectively. Explain qualitatively in terms of atomic structure why these values are to be expected.

8 Estimate how much energy a photon must have if it is to be capable of expelling an electron from the innermost shell of a gold atom.

PROBLEMS

Section 27.1

1 The radius of the gold nucleus is about 6×10^{-15} m and its atomic radius is about 0.150 nm. Suppose you wish to draw a scale diagram of the gold atom using a dot of diameter 0.1 mm for the nucleus. How far away from the center of the dot would be the outer edge of the atom?

2 A uniform beam of 10,000 tiny projectiles is directed at a 0.5-m^2 window whose pane is partly broken out. The beam has the same area as the window. (a) Only 800 of the projectiles go through the window. How large is the area of the hole in the pane? (b) The window pane is now completely removed, and 400 tiny spheres suspended by threads are placed in the window opening. Now 9200 of the 10,000 projectiles go straight through the window. About what is the cross-sectional area of each sphere? (c) To what in Rutherford's experiment do the spheres of part (b) correspond?

■ 3 Rutherford and his coworkers shot alpha particles (charge $q = 2e$) at gold atoms ($Z = 79$). Some of the particles had a kinetic energy of 4.8 MeV. (a) What is the potential energy (in terms of r) of an alpha particle at a distance r from the gold nucleus? (b) How close can Rutherford's particles come to the center of the gold nucleus? Assume that the gold nucleus remains essentially stationary, and neglect the effect of the distant atomic electrons.

4 The density of gold is 19.3 g/cm^3 and its atomic mass is 197 kg/mol. (a) What is the mass of a gold atom? (b) How many gold atoms are there in a 1-cm^2 area of gold film that is 0.040 mm thick? (c) The diameter of a gold nucleus is about 10^{-14} m. Assuming no overlap, how much of the 1-cm^2 total area do the gold nuclei cover? (d) If he had used a film of this thickness, about what fraction of the alpha particles would Rutherford have observed to be strongly deflected?

5 Suppose alpha particles of speed 2.0×10^7 m/s are shot at lead atoms ($Z = 82$). How close can the alpha particles come to the center of the lead nucleus?

6 What is the distance of closest approach of alpha particles of speed 1.8×10^7 m/s to a copper ($Z = 29$) nucleus?

Sections 27.2 and 27.3

7 Calculate the radius of the first, second, and third Bohr orbits for the hydrogen atom.

■ 8 Use the semiclassical model of the hydrogen atom to show that the speed of the electron v_n in the nth Bohr orbit can be expressed as $v_n = 2\pi k e^2/nh$.

■ 9 Calculate the classically predicted speed of an electron in the first and second Bohr orbits. Compare these with the speed of light c.

■10 Calculate the angular momentum of an electron in the first Bohr orbit.

■11 What is the kinetic energy of an electron in the first and second Bohr orbits of the hydrogen atom?

12 Calculate the potential energy of the electron in a hydrogen atom when it is in its ground state.

13 In the singly ionized helium atom, a single electron orbits the nucleus, which has a charge $+2e$. Calculate the radius of the first ($n = 1$) and second ($n = 2$) Bohr orbit for this ion.

14 Calculate the lowest three energy levels of the singly ionized helium atom of Prob. 13.

15 Suppose that the semiclassical theory of the atom can be applied to the innermost electron in a gold atom ($Z = 79$) if the presence of all the other electrons is neglected. (This is really not too bad an approximation.) (a) Show that the energy needed to remove this electron from the atom is 13.6×79^2 eV. (b) What is the radius of the first Bohr orbit for this atom?

■16 Suppose an electron orbits about the hydrogen nucleus in a circular path of radius 0.50×10^{-10} m. (a) What speed must the electron have if the coulomb force is to provide the centripetal force? (b) What is the frequency of the electron in the orbit? (c) On the basis of classical theory, what wavelength of radiation should this electron emit?

■17 Consider a doubly ionized lithium atom ($Z = 3$). (a) Calculate the lowest three energy levels for this ion. (b) How much energy is needed to remove the last electron from doubly ionized lithium?

■18 Suppose the nitrogen atom ($Z = 7$) is stripped of six electrons. Calculate the radius of the first Bohr orbit, the energy of the ground state, and the energy needed to remove the last electron for this atom.

■19 Repeat Prob. 18 for sodium ($Z = 11$) stripped of ten electrons.

Sections 27.4 and 27.5

20 Calculate the wavelength of the first four lines of the Balmer series.

21 Compare the wavelengths of the thirteenth and fourteenth lines of the Balmer series. What conclusion do you draw from these numbers?

22 Compare the wavelengths of the sixth line of the Balmer series and the first line of the Lyman series.

23 Calculate the wavelengths of the shortest- and longest-wavelength photons in the Paschen series.

24 Compare the wavelengths of the longest-wavelength photon in the Balmer series and the shortest-wavelength photon in the Paschen series.

25 Calculate the energy of the photon that, when absorbed by a hydrogen atom, causes an electronic transition from the initial state $n = 2$ to the final state $n = 5$.

■26 Electrons with energy 10.9 eV are shot into a gas of hydrogen atoms. What wavelength of radiation will be strongly emitted by the gas?

■27 Electrons with energy 12.9 eV are shot into a gas of hydrogen atoms. What wavelength of radiation will be strongly emitted by the gas?

■28 If a continuous spectrum is passed through unheated hydrogen gas, what are the five lowest-energy photons absorbed by the gas?

■29 What are the energies of the three lowest-energy photons absorbed by unexcited singly ionized helium atoms? What are their wavelengths?

■30 A beam of ultraviolet light of wavelength 72 nm passes through a gas of unexcited hydrogen atoms. When a photon strikes an atom and ejects an electron, what is the kinetic energy of the electron once it is free from the atom? (This is called the *atomic photoelectric effect*.)

■31 A beam of 5.0-nm-wavelength x-rays is incident on a gas of unexcited hydrogen atoms. It expels the atomic photoelectrons from the hydrogen atoms. (a) What is the energy of the expelled electrons? (b) What is their speed?

■32 The ionization energy of unexcited helium atoms is 24.6 eV. Suppose that 40-nm ultraviolet radiation is incident on such atoms. (a) What is the energy of the fastest electron ejected from the atoms by the ultraviolet radiation? (b) What is the speed of this electron?

■33 A room-temperature gas of hydrogen atoms is bombarded by a beam of electrons that has been accelerated through a potential difference of 13.3 eV. What wavelengths of light will the gas emit as a result of the bombardment?

Sections 27.6–27.8

34 What is the de Broglie wavelength of an electron in the fourth Bohr orbit?

35 Calculate the number of electrons that can exist in the shells (a) $n = 3$ and (b) $n = 5$ of a Bohr-type atom.

36 Calculate the de Broglie wavelength of the electrons in the Bohr orbits of Prob. 35.

37 How many orbital subshells are possible for the atomic level characterized by the principal quantum number $n = 3$?

■38 An atomic subshell is defined as the group of electrons in an atom that have the same principal quantum number n and orbital quantum number l, but different magnetic quantum numbers m_l and spin quantum numbers m_s. Use these facts to find the number of electrons that exist in the subshell $n = 3$, $l = 2$ in gold.

■39 How many magnetic substates are possible for the subshell that has the quantum numbers $n = 3$, $l = 1$? How many electrons are needed to fill this subshell?

■40 For the Bohr-like state with principal quantum number $n = 4$, how many different values are possible for (a) the orbital quantum number l and (b) the magnetic quantum number m_l?

■41 How many different sets of quantum numbers (l, m_l, and m_s) are possible for an electron with principal quantum numbers (a) $n = 3$, (b) $n = 4$, and (c) $n = 5$?

■42 Consider two electrons in the same system each having the quantum numbers $n = 3$, $l = 0$. (a) Suppose the electrons have spins but that the exclusion principle is not applicable. How many states will be possible for the two electrons? (b) How many states are permitted if the exclusion principle is applicable?

■43 Consider a system in which the electrons do not have spin and a spin quantum number does not exist. How many electrons can there be in the state with principal quantum number $n = 3$?

■44 Under the circumstances of Prob. 43, what would be the first four elements in the periodic table that would show a valence of +1?

■45 Make a table showing the quantum numbers for the various electrons in the sodium ($Z = 11$) atom.

■46 Write the values for the set of quantum numbers n, l, m_l, and m_s for the electrons in an oxygen atom ($Z = 8$).

■47 Write the sets of quantum numbers for electrons in (a) neon ($Z = 10$) and (b) potassium ($Z = 19$) atoms.

Section 27.9

48 Modern color television sets often use electron beams accelerated through more than 20,000 V. What are the shortest-wavelength x-rays generated by a 24,000-V beam as it hits the end of the television tube? (Some early television sets

were not properly shielded and leaked appreciable amounts of x-rays outside the set.)

49 To reach tumors deep within a person's body, so-called "hard" x-rays are used. These are generated using very high voltages. What is the shortest-wavelength x-ray generated by an x-ray tube operating at 184 kV?

50 What is the minimum possible voltage for an x-ray tube that produces x-rays of wavelength 0.045 nm?

■**51** An x-ray tube uses tungsten ($Z = 74$) as the target. (*a*) Estimate the minimum voltage needed if the $n = 1$ electron is to be excited. (*b*) Estimate the longest wavelength of the x-ray emitted as the atom undergoes an $n = 2$ to $n = 1$ transition.

■**52** The most intense line in the x-ray spectra of materials used as targets in x-ray tubes is called the K_α line. According to Bohr's theory, this line arises when the atom undergoes a transition from the state $n = 2$ to the state $n = 1$. Estimate the wavelength of the K_α line for a target made of chromium ($Z = 24$).

■**53** Estimate the wavelengths of the K_α x-ray lines from (*a*) lead ($Z = 82$) and (*b*) zirconium ($Z = 40$).

■**54** Estimate the minimum voltage needed across an x-ray tube to excite the $n = 1$ electron if the target is made of (*a*) nickel ($Z = 28$) and (*b*) aluminum ($Z = 13$).

■**55** Estimate the energy difference between the $n = 2$ and $n = 3$ levels in molybdenum ($Z = 42$). What is the wavelength of the x-ray emitted when the molybdenum atom undergoes the transition from the $n = 3$ to $n = 2$ level?

Section 27.10

■**56** A pulse of light from an argon laser ($\lambda = 456.5$ nm) is used to "weld" the detached retina in a person's eye. If the pulse lasts 1×10^{-8} s and contains an energy of 1.6×10^{-3} J, what is the instantaneous power delivered to the weld point?

■■**57** A laser beam diverges slightly because of diffraction effects at the end of the laser tube. Assume that a helium-neon laser beam ($\lambda = 633$ nm) has a diameter of 2.8 mm as it leaves the laser tube. About how large will the beam diameter be when the beam strikes a target 160 m away? Assume that beam spreading is due solely to diffraction.

■■**58** Two different laser beams of the same type will be coherent if the lasers emit exactly the same wavelengths. Even if the wavelengths are slightly different, the two beams will show interference effects. When joined, they will give a resultant beam that fluctuates over time from brightness to darkness. This is similar to the phenomena of beats of sound waves discussed in an earlier chapter. If one beam has a wavelength of exactly 632 nm, what must the wavelength of the other beam be to produce maximum brightness once each second? *Hint:* Use the fact that for $x \ll 1$, $1/(1 \pm x) \cong 1 \mp x$.

General Problems

■■**59** Suppose that the angular momentum of the earth's rotation about the sun obeyed the resonance condition for the de

Broglie waves $n\lambda_n = 2\pi r_n$. What would be the value of the quantum number n in this case? (This is an example of the Bohr correspondence principle which states the fact that macroscopic systems, like the earth, normally correspond to very large quantum numbers and so behave classically.)

■■**60** Classically a hydrogen atom with an orbital diameter of a few meters should act like a radio antenna and emit radiation with a frequency that is equal to the frequency of the electron in the orbit. Wave theory must also predict this result, since it applies to radio antennas as well as atoms. Show that the orbital frequency of the electron is

$$f_{\text{orb}} = \frac{me^4}{4\epsilon_o^2 h^3 n^3}$$

Compute the frequency emitted by the hydrogen atom as it falls from the state n to state $n - 1$. Show that when n is very large ($n \gg 1$), this frequency is the same as the orbital frequency f_{orb}.

■**61** Consider the following four possible electronic transitions for the hydrogen atom: (1) from $n = 2$ to $n = 5$, (2) from $n = 3$ to $n = 6$, (3) from $n = 7$ to $n = 4$, and (4) from $n = 4$ to $n = 1$. (*a*) Which transition will emit the longest-wavelength photon? (*b*) For which transition will the atom absorb maximum energy?

■■**62** Suppose that an atomic nucleus consists of noninteracting protons and neutrons traveling in circular paths within the nucleus. Since the radius of a typical large nucleus is about 5×10^{-15} m, assume that the particles in the ground state have the orbit radius of 5×10^{-15} m. What must be the de Broglie wavelength of a proton that resonates in such an orbit in its ground state? What is the kinetic energy (in electronvolts) of the proton? Neglect relativistic effects.

■■**63** The perimeter of the benzene molecule is a hexagon, with each side having a length of 0.140 nm. Since the molecule has three double bonds, it is not totally unreasonable to assume that one electron in the molecule can circulate freely around this perimeter much as though it were a free electron restricted to the hexagonal path. Using reasoning based on resonance and de Broglie wavelength, show that the energy levels for such an electron should be (to this approximation)

$$E_n = (7.1 \times 10^{17}) \frac{n^2 h^2}{m}$$

with all quantities in SI units. If the result of this computation is correct, at what wavelength would you expect benzene to absorb light? Does this contradict the fact that benzene is a crystal-clear liquid?

■■**64** (*a*) Calculate the recoil speed of a hydrogen atom due to its emission of a photon with wavelength 486 nm, the second line of the Balmer series. (*b*) Find the ratio of this recoil energy of the atom to the difference in energy between the two states that give rise to the emission line.

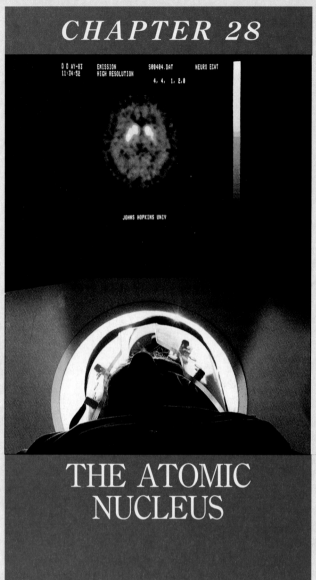

CHAPTER 28

JOHNS HOPKINS UNIV

THE ATOMIC NUCLEUS

As Rutherford demonstrated in 1911, at the very center of each atom is a positively charged nucleus. Although it constitutes only about 10^{-13} percent of the atom's volume, the nucleus contains 99.9 percent of its mass. In this chapter, we examine the salient features of the nucleus, how it is constructed, and what influences its stability. We also discuss a few of the many applications of nuclear physics in our present-day world.

28.1 ATOMIC NUMBER AND MASS NUMBER

Rutherford's investigations, discussed in Chap. 27, have been extended in many ways as the years have gone by. Today we know that the nucleus is composed of protons (p) and neutrons (n). Because they both reside in the nucleus, protons and neutrons are collectively called **nucleons.** You will recall that the charge on the proton is $+e$ and that the neutron has no charge. Furthermore, the masses of these particles are

$$m_p = 1.673 \times 10^{-27} \, \text{kg} = 1.007276 \, \text{u}$$

and

$$m_n = 1.675 \times 10^{-27}\,\text{kg} = 1.008665\,\text{u}$$

where the mass unit u is the **atomic mass unit** (sometimes written amu). We shall define this unit precisely in Sec. 28.2. For now we simply assert that

$$1\,\text{u} = 1.660566 \times 10^{-27}\,\text{kg}$$

Notice that the neutron and proton masses are nearly, but not exactly, the same. Like the electron, the proton and neutron have spin of $\frac{1}{2}$ and obey the Pauli exclusion principle. In comparison, the electron's mass is

$$m_e = 9.1094 \times 10^{-31}\,\text{kg} = 5.486 \times 10^{-4}\,\text{u}$$

As stated in Chap. 27, the atomic number Z designates the number of protons in the nucleus of an atom. Neutral (un-ionized) atoms have Z electrons in the space external to the nucleus. The chemical behavior of an atom is determined by these electrons, and so all atoms having the same Z value are atoms of the same *element*. Every carbon atom has 6 electrons, for instance, and every gold atom has 79 electrons. The atomic numbers of the elements are given in Appendix 1.

The mass of the nucleus (except in the case of hydrogen) is greater than the mass of the Z protons, due to the presence of neutrons. The number of neutrons in a nucleus is given the symbol N.

Since each nucleon has a mass very close to 1 u, we would expect the nuclear mass to be nearly an integer when measured in atomic mass units. This is indeed the case. For example, the mass of the helium nucleus, containing two protons and two neutrons, is 4.0026 u, ≈ 4 u, and the mass of the argon nucleus ($Z = 18$ and $N = 22$) is 39.96 u, ≈ 40 u. With this in mind, we assign to each nucleus a **mass number** A, which is equal to the number of nucleons it contains: $A = Z + N$. This mass number is very close to the nuclear mass measured in atomic mass units.

28.2 NUCLEAR MASSES; ISOTOPES

The masses of nuclei have been measured to high precision. These measurements are carried out with mass spectrometers, which we studied in Sec. 19.8. A schematic diagram of one type is given in Fig. 28.1. As indicated, ions of the element under consideration are allowed to escape from the ion source. After being accelerated through the potential difference V, the ion beam is collimated by means of slits such as S_2. When they leave S_2, the ions are moving with speed v and are deflected

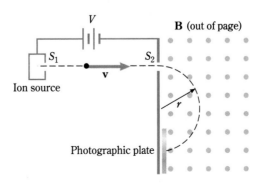

FIGURE 28.1

In the mass spectrometer, ions are deflected by a magnetic field.

This plant in Canada manufactures heavy water, $(^2H)_2O$, (the vertical towers), which is used in certain types of nuclear reactors. The water in the foreground is natural water, whose molecules contains about 1/100 of a percent of the ^2H (deuterium) isotope.

into a circular path by the magnetic field, as shown. The radius r of this path is measured by noting the positions at which the ions strike a photographic plate or some other detector.

The relation between the radius of curvature r and the mass of the ion is (Eq. 19.5)

$$m = \frac{r^2 B^2 q}{2V} \tag{19.5}$$

We can therefore compute the mass of the ion if r, B, q, and V are known. To find the mass of the nucleus, we subtract from m the mass of the electrons associated with the ion.

When the mass spectrometer is used to measure nuclear masses, an interesting effect is observed. Very frequently, an element will give rise to two or more different beams in the spectrometer. That is, particles will appear at the detector at two or more very well defined radii. Combining this finding with Eq. 19.5, we conclude that *nuclei of the same element may have different masses*.

As an illustration, when chemically pure chlorine is sent through the mass spectrometer, it appears to consist of two types of nuclei:

Species 1:   Mass = 34.97 u   Relative percentage = 75.4

Species 2:   Mass = 36.97 u   Relative percentage = 24.6

We say that the **natural abundance** of species 1 is 75.4 percent, and the natural abundance of species 2 is 24.6 percent. Both these species behave exactly the same chemically, and so their electronic structures must be the same. Therefore their nuclear charges must be the same, equal to the atomic number Z multiplied by the charge quantum e. We call nuclei such as this, which have the same charge but different masses, **isotopes** of the element in question.

Isotopic nuclei have the same number of protons but a different number of neutrons.

In order to classify nuclei in terms of their mass, charge, and nucleon number, it is customary to designate an element whose symbol is X as $^A_Z X$. For example, the chlorine isotopes we have just discussed are represented by $^{35}_{17}Cl$ and $^{37}_{17}Cl$; both isotopes have the same atomic number, $Z = 17$, but one has a mass number $A = 35$ and the other has $A = 37$. We refer to these two isotopes as chlorine 35 and chlorine 37. As another example, $^{238}_{92}U$ is called uranium 238. Its nucleus has a charge $+92e$ and contains 92 protons and $238 - 92 = 146$ neutrons. Uranium 235, $^{235}_{92}U$, has the same number of protons, 92, but only 143 neutrons in its nucleus.

In the periodic chart of the elements you are probably familiar with from your chemistry classes, one usually finds listed the *atomic* masses, defined as the average value of the *isotopic* masses found in nature. For example, the average mass of the two isotopes of chlorine is

$$m_{av} = 35(0.754) + 37(0.246) = 35.5 \, u$$

which is the value given in the periodic chart inside the back cover. The atomic masses* of many isotopes are given in Appendix 1. Remember that these are the masses of the nuclei *plus the atomic electrons,* given in atomic mass units, defined in terms of the mass of the $^{12}_{6}C$ atom:

One atomic mass unit (u) is exactly one-twelfth the mass of a single atom of carbon 12 ($^{12}_{6}C$).

All other masses are referred to this standard. Mass spectrometer data give us the value stated in Sec. 28.1.

Illustration 28.1

What fraction of the atomic mass of ^{235}U is due to its electrons?

Reasoning   From Appendix 1, the atomic mass of ^{235}U is 235.04 u. Since the atomic number of uranium is 92, this atom has 92 electrons. Using the fact that the mass of the electron is 9.11×10^{-31} kg, or 0.000549 u, we have

$$\frac{92(0.000549) \, u}{235 \, u} = 2.15 \times 10^{-4}$$

Therefore, for many purposes, the mass of the electrons can be ignored.   ■

28.3   NUCLEAR SIZE AND DENSITY

There are several methods we can use to estimate the size of a nucleus. One is to shoot particles of various types at the nucleus as Rutherford did and see how they are scattered. One must use very-high-energy particles to overcome the coulomb repulsion of the nucleus if the bombardment is to be done with protons or alpha

---

*Recall from Chap. 11 that it is customary to use the terms *atomic mass* and *atomic weight* interchangeably.

particles. The results of such measurements show that the nucleus cannot be pictured as a simple hard sphere of uniform constitution.

In spite of the fact that the nucleus has no sharp cutoff radius for its charge or its mass, its edges are well enough defined for a meaningful approximate radius to be given. As one would expect, bombardment with charged particles measures primarily the charge distribution in the nucleus, whereas bombardment with neutrons measures primarily the mass distribution. Other methods can also be used to measure the nuclear radius. They all agree approximately with each other, and from them it can be inferred that the nuclear radius R of the various elements is

$$R \approx (1.2 \times 10^{-15}\,\text{m})(A^{1/3}) \tag{28.1}$$

where A is the mass number of the atom concerned.

Notice in Eq. 28.1 that the radius of a typical nucleus is of the order of 10^{-15} m. For that reason, it is customary in nuclear work to measure lengths in femtometers (fm), where 1 fm $= 10^{-15}$ m. Originally this length was designated a *fermi* in honor of the distinguished nuclear physicist Enrico Fermi. It is customary to use the designations fermi and femtometer interchangeably.

The fact that the nuclear radius varies as $A^{1/3}$ gives important information as to how the A nucleons pack together in the nucleus. If we compute the volume of the nucleus, we have

$$V = \tfrac{4}{3}\pi R^3 = \tfrac{4}{3}\pi(1.2\,\text{fm})^3(A) = (7.2 \times 10^{-45}\,\text{m}^3)(A)$$

Notice what this says. If the factor 7.2×10^{-45} m^3 is taken as the volume of a single nucleon, then V is simply the sum of the individual volumes of the A nucleons. As a result, all large nuclei have about the same density, as we shall see in the following illustration.

Illustration 28.2

Find the density ρ of the gold nucleus.

Reasoning If we neglect the mass of the atomic electrons, the mass of a gold nucleus is equal to its atomic mass, given in Appendix 1 as 197 u. The volume of the nucleus is

$$V = \tfrac{4}{3}\pi R^3 = (7.2 \times 10^{-45}\,\text{m}^3)(A)$$

Since $A = 197$, and the mass of a gold atom is 197 u,

$$\rho = \frac{\text{mass}}{\text{volume}} = \frac{(197\,\text{u})(1.66 \times 10^{-27}\,\text{kg/u})}{(7.2 \times 10^{-45}\,\text{m}^3)(197)} \approx 2.3 \times 10^{17}\,\text{kg/m}^3$$

Notice that, because the mass number ($A = 197$) is nearly equal to the atomic mass (197 u), the 197s cancel, and so this value is the approximate density within all nuclei. Such extremely high densities are never encountered on a large scale on earth. Only in the interior of certain stars (*neutron* stars) are such high densities found. In these stars, the electron shells of the atoms have been collapsed by the huge gravitational forces at the star's center. ∎

28.4 NUCLEAR BINDING ENERGY

Because like charges repel one another, the electrostatic force between the protons in a nucleus tends to cause it to explode. The gravitational force between nucleons is many orders of magnitude too small to counterbalance this repulsive force. A third force must exist between nucleons to cause them to attract each other and hold the nucleus together. It is the **nuclear binding force,** often called simply the **nuclear force** or the **strong force.**

The nuclear force is unlike the electrostatic and gravitational forces in that it does not obey an inverse-square law. Instead, it has only a limited range. Experiment shows that it is essentially zero for separations larger than about 5×10^{-15} m—in other words, a distance equal to about twice the diameter of a nucleon. At separations only slightly smaller than this, the nuclear force overpowers the electrostatic repulsive force between any two protons and binds the protons together. To a first approximation, the nuclear force is the same between two neutrons, between two protons, and between a neutron and a proton. However, the nuclear force does not have any effect on electrons. This point will be important to remember later when we discuss nucleon changes that create electrons in the nucleus.

Let us consider what happens to the energy of a group of widely separated nucleons as they are assembled into a nuclear structure. When they are far apart, their energy of interaction can be taken to be zero, and their total energy is just the sum of their individual rest mass energies. As they are made to approach each other, the protons feel an increasing repulsion because of the coulomb force, but the neutrons experience no force. At a distance of about 5 fm, both protons and neutrons feel the strong nuclear binding force. This force overwhelms the coulomb repulsion, and as a result both neutrons and protons are drawn toward each other, forming a nucleus. For a given nucleus, each proton and neutron is bound within the nucleus with the same binding energy, $-E_o$. (Why is the binding energy negative?) Figure 28.2 summarizes the energies of a proton and a neutron at various distances from the nucleus (the separated rest mass energies are not included.) We conclude that

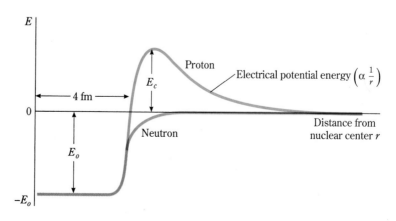

FIGURE 28.2

Potential-energy curves of a neutron and proton in a stable nucleus. Typical values might be $E_o = 50$ MeV, $E_c = 8$ MeV.

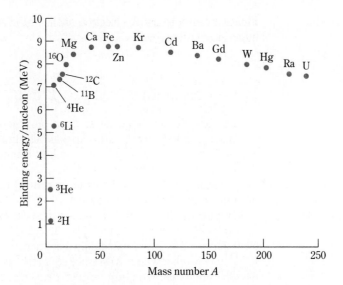

FIGURE 28.3

Binding energy per nucleon for representative elements.

The energy of a stable nucleus is less than the total rest mass energies of the separated nucleons which compose it.

The value of E_o varies from one nuclear structure to another, as shown in Fig. 28.3. In contrast to the few eV binding energy for atomic electrons, you see that nucleons are bound with energies roughly a million times greater. Notice that E_o is greatest for elements around iron ($Z = 26$), and is smaller for nuclei with lower and higher values of Z. Figure 28.3 can thus be interpreted as an indicator of nuclear stability.

Since, according to relativity theory, changes in energy are associated with changes in mass, we should expect that an assembled nucleus has a smaller mass than the sum of the rest masses of its separated nucleons. This mass difference, known as the **mass defect** of the nucleus, can be written as

$$\Delta m = Z m_p + N m_n - M_{\text{nuc}}$$

where m_p and m_n are the masses of single free protons and neutrons and M_{nuc} is the actual mass of the assembled nucleus. Relativity tells us that the mass defect is related to the total binding energy of the nucleus:

Total binding energy $= \Delta m\, c^2$

The fact that measured masses and binding energies of nuclei agree with this statement is one of the direct confirmations of relativity. We shall discuss this further when we consider methods of energy production from nuclei.

Illustration 28.3

How much energy is required to change the mass of a system by 1 u?

Reasoning We make use of Einstein's mass-energy relation $\Delta E = \Delta m\, c^2$. In the present case, $\Delta m = 1\,\text{u} = 1.6606 \times 10^{-27}\,\text{kg}$:

$$\Delta E = 1.492 \times 10^{-10}\,\text{J} = 931.5\,\text{MeV}$$

This is a convenient fact to remember: *one atomic mass unit of mass is equivalent to 931.5 MeV of energy.* ∎

Example 28.1

Appendix 1 gives the atomic mass of $^4_2$He as 4.002604 u. Determine the total binding energy of its nucleus and the average binding energy per nucleon.

[R]

Reasoning

Question   What information gives me the binding energy?
Answer   The difference between the total mass of the nucleons when they are separate and their mass when bound together. This mass defect times c^2 is equal to the total binding energy. Or, as we have just seen, each u of mass is equivalent to 931.5 MeV of energy.

Question   What is the total mass of the separated nucleons?
Answer   Each separate proton has a mass of 1.007276 u, and each separate neutron a mass of 1.008665 u. Thus the total mass of the four nucleons when separate is

$$m_{\text{tot}} = 2(1.008665 \text{ u}) + 2(1.007276 \text{ u}) = 4.031882 \text{ u}$$

Question   What is the mass of just the ^{4}He nucleus?
Answer   The given atomic mass minus the mass of two electrons. We can neglect the mass equivalent of the few electronvolts of electron binding energy.

Solution and Discussion   The nuclear mass of ^{4}He is

$$4.002604 \text{ u} - 2(0.000549 \text{ u}) = 4.001506 \text{ u}$$

The mass defect is then

$$\Delta m = 4.031882 \text{ u} - 4.001506 \text{ u} = 0.030376 \text{ u}$$

The total binding energy is

$$\text{Binding energy} = (0.030376 \text{ u})(931.5 \text{ MeV/u}) = 28.29 \text{ MeV}$$

Dividing by 4 gives

$$\text{Average binding energy per nucleon} = \frac{28.27 \text{ MeV}}{4} = 7.074 \text{ MeV}$$

You should compare this result with Fig. 28.3.

Exercise   The binding energy of the electron in the hydrogen atom is 13.6 eV. How much mass, in atomic mass units, is created when a hydrogen atom is ionized? *Answer: 1.46×10^{-8} u. This mass is too small to be measured. Thus chemical reactions do not make us aware of the conversion between mass and energy.*

28.5 RADIOACTIVITY

As we have seen, nucleons are subject to two competing forces: the attractive nuclear force between all nucleons and the repulsive coulomb force between the protons. A combination that contains too many protons relative to the number of neutrons will experience too large an explosive force as a result of the coulomb repulsions. It cannot exist as a stable entity. Other factors also influence the stability of a nucleus, as we will see later. Only those few combinations of protons and neutrons shown in Fig. 28.4 are relatively stable.

As you can see from Fig. 28.4, large nuclei are stable only if they contain more neutrons than protons. The extra neutrons are needed to "dilute" the positively charged protons and thereby decrease the repulsive effect of coulomb forces. Although most of the nuclei indicated in Fig. 28.4 are completely stable, those with Z larger than 83 are somewhat unstable.

Unstable nuclei can spontaneously undergo an internal change toward a more stable, lower energy state. They can get rid of excess energy by ejecting particles and electromagnetic radiation in the process we call **radioactivity.** Early investigators of radioactivity (in the 1890s) detected the energy emitted and were able to show with the help of magnetic fields that three distinct types of energy are present: positively charged, negatively charged, and electrically neutral. Beyond that, the initial observers were unable to identify the nature of the radiations, so they were called α-, β-, and γ-rays (the Greek equivalent of a, b, and c rays). We now know the

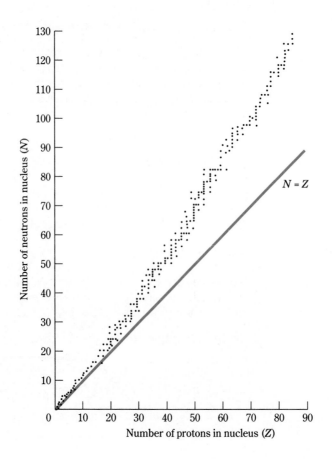

FIGURE 28.4

Each dot represents a nucleus that is either completely stable or nearly so. The solid line represents the positions of nuclei that have equal numbers of protons and neutrons.

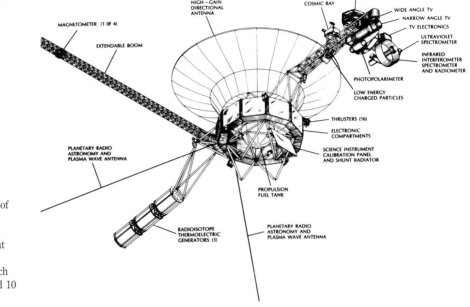

The energy released by radioactive decay is used to generate electrical power for spacecraft such as the Voyager pictured here. The isotope used is ^{238}Pu, which has a half-life of 89 years. The radioisotope thermal generators (RTG's) are designed to produce 160 W of electrical power at 30 V dc at the beginning of the mission. Can you estimate how much the power output will have lessened 10 years into the spacecraft's mission?

α particles to be ^{4}He nuclei, β particles to be electrons, and γ-rays to be extremely-short-wavelength electromagnetic waves (or photons).

Scientists believe that nucleons are in constant motion, engaged in a continuous attempt to escape from the nucleus. In stable nuclei, they never succeed in this attempt. But an unstable nucleus can reduce its energy and become more stable by ejecting a particle and/or energy. It does so on a purely random basis. We can think of a particle trying to escape from the nucleus, making many attempts each second. Once in a great while, the nucleus is in an internal configuration such that the particle can escape, and we say that the nucleus has undergone *radioactive decay*.

This continuous game of chance within all unstable nuclei means that each nucleus has a certain chance for decay during a time interval Δt. Let us say that the chance, or probability, that a given nucleus will decay in the time Δt is $\lambda\,\Delta t$, where λ is called the **disintegration, or decay, constant.** (Do not confuse this definition of λ with that for wavelength.) In a sample consisting of N such nuclei, the number that will decay in time Δt is $N\lambda\,\Delta t$. We can therefore write

$$\Delta N = -N\lambda\,\Delta t \tag{28.2}$$

where the negative sign arises because ΔN is a negative number since N is decreasing. We call the quantity $\Delta N/\Delta t$ the **activity** of the sample. It is the number of decays that occur in unit time, and we discuss it further in Sec. 28.12.

Suppose we have N_o radioactive atoms at a time $t = 0$. We can use Eq. 28.2 to show how the number N of undecayed atoms varies with time. The result is shown in Fig. 28.5. This type of curve is called an *exponential decay curve,* and we give its equation in the next section.

Exponential decay has the following simple alternate description:

The amount of a substance undergoing exponential decay is reduced by a factor of $\frac{1}{2}$ in successive equal time intervals, called the **half-life** of the substance.

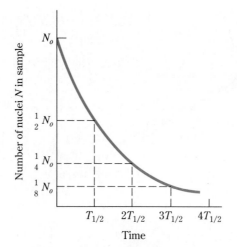

FIGURE 28.5

A radioactive element decays
exponentially.

The half-life $T_{1/2}$ is shown in Fig. 28.5. Notice that in each successive half-life the number of remaining nuclei is cut in half. Thus after n half-lives the number of undecayed nuclei remaining will be $(\frac{1}{2})^n N_o$.

Half-lives of radioactive substances differ widely. The half-life of uranium 238 is 4.47 billion years, while that of radium 226 is 1600 yr. Radon gas, the element to which radium decays, has a half-life of only 3.8 days. Many artificially produced radioactive substances have half-lives of only a fraction of a second. Even so, all these elements decay in conformity with the exponential decay law.

It is important to understand that the half-life is a *statistical* behavior of a large number of nuclei. There is no way to predict when any individual nucleus will decay. For example, a single radium nucleus may take a million years to change while another may take only 1 h, but in a statistically large sample (any measurable amount of an element contains trillions upon trillions of nuclei) half of the radium will undergo radioactive decay in 1600 years.

We now have two ways to characterize rate of decay: λ or $T_{1/2}$. Of course, these two quantities must be related. By using calculus, the relation can be shown to be

$$\lambda T_{1/2} = 0.693 = \ln 2 \qquad (28.3)$$

We will have occasion to use this relation often.

Illustration 28.4

Iodine 131 is a radioactive isotope made in nuclear reactors for use in medicine. When it is taken into the body, it becomes concentrated in the thyroid gland. There it acts as a radiation source in the treatment of hyperthyroidism. Its half-life is 8 days. Suppose a hospital orders 20 mg of ^{131}I and stores it for 48 days. How much of the original ^{131}I is present after the 48 days?

Reasoning   Each 8 days, the iodine decays by one-half. We can therefore make the following table:

Time (days)	0	8	16	24	32	40	48
Iodine (mg)	20	10	5	2.5	1.25	0.625	0.313

After 48 days, only 0.313 mg of the original 20 mg will remain. ∎

Example 28.2

A vial holds 1 g of ^{60}Co, whose half-life is 5.25 yr. What is the activity of the sample **(a)** initially, and **(b)** after the vial has been in storage for 21 yr?

Reasoning

Question   What is the expression for activity?
Answer   From Eq. 28.2, we have $\Delta N/\Delta t = -\lambda N$. Here N is the amount of sample present at the time the activity is being calculated and λ is the decay constant of the substance. The negative sign simply indicates that the number of ^{60}Co nuclei, N, is decreasing.

Question   How is the decay constant related to half-life?
Answer   From Eq. 28.3, $\lambda = 0.693/T_{1/2}$. Often, activity is expressed in decays per second, which requires you to express $T_{1/2}$ in seconds.

Question   How does the mass of the sample give me the initial value of N?
Answer   Recall that there is 1 mol (Avogadro's number N_A) of any substance in a mass of the substance (in grams) numerically equal to its atomic mass. To 3 significant digits, you can take the atomic mass of ^{60}Co to be equal to its mass number, $A = 60$. Thus 1 g of ^{60}Co has $(1/60)N_A$ nuclei.

Question   For part (*b*), what determines the number of remaining ^{60}Co nuclei after 21 yr?
Answer   Notice that 21 yr is 4 half-lives. Therefore N at 21 yr is $(\frac{1}{2})^4 = \frac{1}{16}$ of its initial value.

Solution and Discussion   Initially, $N = (\frac{1}{60})(6.022 \times 10^{23}) = 1.00 \times 10^{22}$ nuclei. The initial activity is then (using 5.25 yr = 1.66×10^8 s)

$$\left|\frac{\Delta N}{\Delta t}\right| = \frac{0.693}{1.66 \times 10^8 \text{ s}}(1.00 \times 10^{22})$$

$$= 4.19 \times 10^{13} \text{ decays/s}$$

After 21 yr there are $(\frac{1}{16})(1.00 \times 10^{22}) = 6.25 \times 10^{20}$ nuclei left. Thus the activity of the sample after 21 yr is simply $\frac{1}{16}$ of the original activity, or 2.62×10^{12} decays/s. Notice that the decay constant (and half-life) remains a constant characteristic of ^{60}Co decay, regardless of the amount N.

28.6   EXPONENTIAL DECAY

The exponential decay curve in Fig. 28.5 is well known in science. As we saw in the last section, its height decreases by one-half for each half-life along the horizontal axis. The curve can be stated in mathematical terms as

$$N = N_o e^{-\lambda t}$$

(28.4)

where λ is the decay constant. The function $e^{-\lambda t}$ is called an *exponential function,* and e is the base for natural logarithms, 2.7183.

Use of Eq. 28.4 is facilitated by the fact that most hand calculators have a key for this function. If you do not have a calculator with this capability, you can find a table of exponential functions in most handbooks.

Illustration 28.5

Uranium 238 has a half-life of 4.5×10^9 yr. It is believed that the earth solidified about 4.0×10^9 yr ago. What fraction of the uranium 238 then found on the earth remains undecayed today?

Reasoning   We use the decay law, Eq. 28.4, with

$$\lambda = \frac{0.693}{T_{1/2}} = \frac{0.693}{4.5 \times 10^9 \text{ yr}} = 1.54 \times 10^{-10} \text{ yr}^{-1}$$

This gives

$$\frac{N}{N_o} = e^{-\lambda t} = e^{-(1.54 \times 10^{-10} \text{ yr}^{-1})(4 \times 10^9 \text{ yr})}$$

$$= e^{-0.616} = 0.54$$

Today 54 percent of the uranium 238 is still in existence.   ∎

Example 28.3

Ninety percent of a sample of a certain radioactive substance remains after 12.0 h. What are the decay constant and the half-life for this substance?

Reasoning

Question   What is the significance of the 90 percent?
Answer   It is the ratio of the number of nuclei remaining to the initial number, N_o. In other words, it is the value of N/N_o at $t = 12$ h.

Question   What expression relates N/N_o to t?
Answer   Equation 28.4: $N/N_o = e^{-\lambda t}$.

Question   How do I find an unknown that is in the exponent?
Answer   You should recall the following general property of logarithms: $\log_a a^x = x$. Thus $\ln(N/N_o) = \ln e^{-\lambda t} = -\lambda t$, which can be solved algebraically for λ.

Solution and Discussion   Using a calculator, we find that

$$\ln\left(\frac{N}{N_o}\right) = \ln 0.90 = -0.105$$

Notice that the natural logarithm of a fraction is always negative. As we proceed, notice how this sign gives a positive result for λ. Now λ can be found from

$$-\lambda(12\,\text{h}) = -0.105 \qquad \lambda = 8.75 \times 10^{-3}/\text{h} = 2.43 \times 10^{-6}/\text{s}$$

The half-life is

$$T_{1/2} = \frac{0.693}{8.75 \times 10^{-3}/\text{h}} = 79.2\,\text{h}$$

Exercise What are the decay constant and the half-life if 20 percent decays in 40 s? *Answer:* $0.00558\,\text{s}^{-1}$, $124\,\text{s}$.

28.7 EMISSIONS FROM NATURALLY RADIOACTIVE NUCLEI

As stated previously, all nuclei with $Z > 83$ are radioactive. Early workers used the experiment sketched in Fig. 28.6 to examine the radiation from naturally radioactive substances. A small sample is placed in the center of a block of lead. The block has a thin hole drilled in it through which the radiation emitted by the sample can escape in a directional beam. When the beam of radiation is allowed to pass into a region of transverse magnetic field, it splits into three components, as shown. From the directions in which the rays are bent, we conclude that one component has no charge, one is positively charged, and the third is negatively charged. As mentioned before, since these radiations were originally unidentified, they were given the names α-, β-, and γ-rays. Let us now discuss each in turn.

GAMMA RADIATION

On occasion, a nucleus finds itself in an excited energy state. To reach its ground state, it may emit a high-energy γ-ray. If the nucleus makes a transition from a state with energy E_2 to a state with energy E_1 then it will emit a γ-ray photon of frequency

$$hf = E_2 - E_1$$

This is analogous to the emission of a photon by an atom as its electronic structure adjusts to a lower energy state. Gamma-ray photons are basically the same as x-ray photons, although many nuclear transitions are more energetic than electronic transitions and hence give rise to shorter-wavelength photons than the x-rays produced by inner-shell electron transitions. In any case, the term γ-ray usually is given to a photon emitted by the nucleus, while an identical photon emitted during an atomic electron transition would be termed an x-ray.

BETA-PARTICLE EMISSION

Many radioactive nuclei emit beta (β) particles, which are simply electrons. The process that occurs within a nucleus when β-particle emission occurs is quite com-

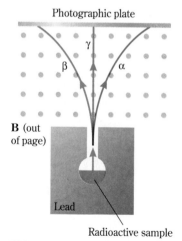

FIGURE 28.6

The radiation from a radioactive sample is separated into three components by a magnetic field.

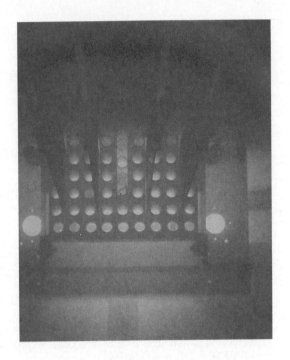

The blue light in this photograph of the underwater core of a fission reactor is called Cerenkov radiation. It is caused by very fast neutrons, produced by fission, entering water at speeds greater than that of light in water.

plex. There are no electrons in the nucleus, and so, in effect, the process transforms a neutron into a proton plus an electron. The new proton is retained by the nucleus while the electron is emitted.

We can represent the emission of a β particle from a nucleus whose symbol is ^A_ZX in the following way:

$$^A_ZX \rightarrow ^{\ \ A}_{Z+1}Y + ^{\ 0}_{-1}e + ^{0}_{0}\bar{\nu}$$

where $^{\ 0}_{-1}e$ represents the emitted β particle (electron), $^{\ \ A}_{Z+1}Y$ represents the transformed nucleus, and $^{0}_{0}\bar{\nu}$ represents a neutrino, a particle we will say more about in a moment. The transformed nucleus contains one more proton than the original nucleus did. Its atomic number is therefore $Z + 1$. Its mass number is still A because there are still the same number of nucleons in the nucleus. Because of its very small mass, the mass number of a β particle is considered to be zero.

Unlike the case of γ-ray emission, where only γ-rays with definite energies corresponding to differences in energy states of the nucleus are found, β particles of widely varying energies are emitted. A typical β-particle energy spectrum is shown in Fig. 28.7. This is not what one would expect, since, if a β particle is emitted, one would think it should carry away a reproducible energy corresponding to the difference in energy between the initial and final states of the nucleus. Another puzzling fact about β-particle emission is that the momentum of the ejected electron is not equal and opposite to the recoil momentum of the nucleus. In order to explain this, it was postulated that a second, undetected particle is emitted with the β particle. This particle should have zero rest mass* and zero charge; it was given the name *neutrino*. Direct experimental evidence for the existence of this particle was obtained in the mid-1950s.

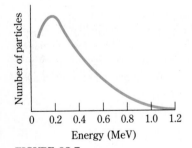

FIGURE 28.7

The energy distribution for β particles emitted from $^{210}_{83}$Bi.

*There is controversy as to whether or not the rest mass of the neutrino is exactly zero. However, its mass, if any, is many orders of magnitude smaller than the electron's mass.

ALPHA-PARTICLE EMISSION

Some radioactive nuclei emit alpha (α) particles. These are simply helium nuclei (two protons + two neutrons) and are represented by $_2^4\alpha$ or $_2^4$He. A typical α-particle decay is exhibited by radium nuclei:

$$_{88}^{226}\text{Ra} \rightarrow \, _{86}^{222}\text{Rn} + \, _2^4\text{He} + \gamma$$

This decay process has a half-life of 1600 yr. We call the original nucleus (radium in this case) the *parent* nucleus and the final nucleus (the unreactive gas radon) the *daughter* nucleus.

Alpha decay is often accompanied by γ-ray emission. In these cases, the daughter nucleus is formed in an excited state, which subsequently reaches the ground state by γ-ray production. These γ-rays give information on the energy levels of the daughter nucleus.

Illustration 28.6

Radon 222 decays to polonium 218 by α emission. Find the approximate energy of the emitted α particle. Pertinent atomic masses are ^{222}Rn = 222.01753 u, ^{218}Po = 218.00893 u, ^{4}He = 4.00263 u.

Reasoning   The mass loss in the reaction is

Mass loss = 222.01753 − (218.00893 + 4.00263) = 0.00597 u

Since 1 u is equivalent to 931.5 MeV, the energy released is

Energy = (931.5 MeV/u)(0.00597 u) = 5.56 MeV

Most of this energy is carried away by the α particle, the observed energy of which is 5.49 MeV. This value differs from the total energy lost because of the recoil energy of the daughter nucleus. ∎

28.8 NUCLEAR REACTIONS

The α- and β-particle decay schemes we have described in the previous section are simple nuclear reactions. Like chemical reaction equations, nuclear reaction equations must be balanced. To maintain balance, nuclear reactions must satisfy the conservation laws of physics. For the present, we are concerned only with the conservation of charge and the nucleon number.

In any nuclear reaction, the sum of all the nucleons (or A values) on one side of the reaction must equal that on the other side. Thus, in α decay,

$$_{88}^{226}\text{Ra} \rightarrow \, _{86}^{222}\text{Rn} + \, _2^4\text{He}$$

we see that A values are equal; 226 = 222 + 4. Moreover, because charge must be conserved, the sums of the Z values must also be equal. In the present reaction, these sums are $86 + 2$ and 88.

There are other conserved quantities besides nucleon number and charge, and nuclear reactions must also obey these conservation laws. As was pointed out previously, a neutrino is emitted in β decay. Without it, the β-decay reaction would not conserve linear and angular momentum and energy. Energy, including the energy equivalent of mass, must also be conserved in nuclear reactions.

The fact that the total energy before reaction (including the equivalent energy of the rest masses) must equal the total energy after reaction is a useful tool in the study of nuclear reactions. For example, when Rutherford performed one of the very first induced nuclear reactions in 1918, he shot α particles at nitrogen nuclei and observed the reaction

$$^{14}_{7}\text{N} + {}^{4}_{2}\text{He} \rightarrow {}^{17}_{8}\text{O} + {}^{1}_{1}\text{H}$$

In other words, the α particle entered the ^{14}N nucleus, which then disintegrated by ejecting a proton. The original nitrogen nucleus was *transmuted* into oxygen.

To learn more about this reaction, we can use the table in Appendix 1 to calculate the masses of the reacting nuclei before and after the reaction:

Masses before reaction

Mass of ^{14}N $= 14.0031\text{ u} - 7\,m_e$
Mass of ^4He $= \underline{4.0026\text{ u} - 2\,m_e}$
Total mass $= 18.0057\text{ u} - 9\,m_e$

Masses after reaction

Mass of ^{17}O $= 16.9991\text{ u} - 8\,m_e$
Mass of ^1H $= \underline{1.0078\text{ u} - 1\,m_e}$
Total mass $= 18.0069\text{ u} - 9\,m_e$

The products have more mass than the original reactants, the difference being 0.0012 u. This mass could be created only *if energy was added to the system*. Since 1.0 u is equivalent to 931.5 MeV, as shown in Illustration 28.3, we see that the increase in mass in this reaction required an external energy of $(931.5)(0.0012) = 1.1\text{ MeV}$. The incident α particle must have had at least this amount of kinetic energy to make the reaction occur. Actually, since momentum must also be conserved in such a reaction, the end products will not be standing still. As a result, the particle must have more than 1.1 MeV of kinetic energy if the reaction is to be feasible.

Spontaneous nuclear reactions, such as radioactivity, occur because the nucleus is more stable after the reaction (more tightly bound) than before. To determine whether or not a particular nucleus is stable, we can first identify what products it might decay to, based on conservation of A and Z. Then we can examine the masses of these products and compare the total with the mass of the original nucleus. If mass is decreased by the reaction, the reaction will occur spontaneously with a certain probability, releasing the energy represented by the mass defect.

Example 28.4

Consider a nucleus composed of 9 protons and 11 neutrons, having an atomic mass of 19.99999 u. (*a*) What element is this? (*b*) What daughter nucleus would result if this nucleus underwent α decay? β decay? (*c*) Are either of these decay processes possible, or is the original nucleus stable?

R

Reasoning

Question What element has $Z = 9$ and what is its mass number A?
Answer Fluorine, F. $A = 20$, so you have ^{20}F.

Question What do α and β decay do to the Z and A values of the parent?
Answer α decay lowers Z by 2 and A by 4. β decay increases Z by 1 and leaves A unchanged. Thus the daughters would be $^{16}_{7}\text{N}$ and $^{20}_{10}\text{Ne}$, respectively.

Question What principle determines the possibility of decay?
Answer Whether or not total mass before the decay is greater or less than afterward. If it is less before decay, spontaneous decay cannot take place.

Question What are the masses involved in α and β decay?
Answer In a number of handbooks or a chart of nuclides, you can find the following: for α decay, $M\,(^4\text{He}) = 4.00260\,\text{u}$ and $M\,(^{16}\text{N}) = 16.00610\,\text{u}$; for β decay $M\,(e^-) = 0.00055\,\text{u}$ and $M\,(^{20}\text{Ne}) = 19.99244\,\text{u}$.

Solution and Discussion For the α decay, the masses afterward would total

$$M_{\text{tot}} = 4.00260\,\text{u} + 16.00610\,\text{u} = 20.00870\,\text{u}$$

This is 0.00871 u *greater* than the original mass of ^{20}F, so α decay won't occur. It would take an energy *input* of $(0.00871)(931.5) = 8.11\,\text{MeV}$ to create the end products of α decay.
  On the other hand, the masses after β decay would total

$$M_{\text{tot}} = 0.00055\,\text{u} + 19.99244\,\text{u} = 19.99299\,\text{u}$$

This is 0.00700 u *less* than the original mass. ^{20}F can (and will) β-decay into ^{20}Ne, a stable nucleus. The energy given off during the process will be $(0.00700\,\text{u})(931.5\,\text{MeV/u}) = 6.52\,\text{MeV}$.

28.9 NATURAL RADIOACTIVE SERIES

You may have been puzzled by the fact that radium 226 is found on earth today. After all, it has a half-life of only 1600 years, while the earth is several billion years old. From the decay law, the ratio of the present-day number of radium nuclei to the number that existed 4×10^9 years ago should be

$$\frac{N}{N_o} = e^{-0.693t/T_{1/2}} = e^{-0.693(4 \times 10^9)/1600} \approx 10^{-740,000}$$

which is a truly negligible fraction. We must conclude that new radium nuclei are being furnished to the earth as the original nuclei are depleted. Similar calculations show that many other sources of natural radioactivity have half-lives far too short to explain their present-day existence. Let us now see how the existence of these nuclei is explained.
  Radium and similar radioactive nuclei are found on earth because they are decay products of extremely long-lived isotopes. For example, uranium 238 has a half-life of 4.47×10^9 years, comparable to the age of the earth. It is the parent nucleus for a whole series of radioactive nuclei. A uranium 238 nucleus decays according to the scheme

$$^{238}_{92}\text{U} \rightarrow {}^{4}_{2}\text{He} + {}^{234}_{90}\text{Th}$$

where the daughter nucleus is thorium. Thus ^{238}U acts as a continuous source of this isotope of thorium, which decays by β emission:

$$^{234}_{90}\text{Th} \rightarrow ^{\ 0}_{-1}\text{e} + ^{234}_{91}\text{Pa} + ^0_0\bar{\nu}$$

where the daughter nucleus is protactinium.

The protactinium in turn β-decays to ^{234}U:

$$^{234}_{91}\text{Pa} \rightarrow ^{\ 0}_{-1}\text{e} + ^{234}_{92}\text{U} + ^0_0\bar{\nu}$$

Several other steps occur in this radioactive series before the final stable element of the series is reached. In this case, it is an isotope of lead, ^{206}Pb. This series is shown in detail in Fig. 28.8. Notice that in the latter stages of the decay scheme, alternative possibilities for decay exist. Each possibility has a certain probability of occurring, as indicated in Fig. 28.8 in the case of ^{214}Bi. The percentages for various decay possibilities are called the **branching ratio** for the decay of the parent.

There are two other natural radioactive decay series found on earth. They, together with the one we have been discussing, are summarized in Table 28.1. Notice that they all start with an element that has a very long half-life and eventually decay to a stable isotope of lead. Presumably, other decay series existed on earth at earlier times, but they have decayed too rapidly to be detected at this late date.

TABLE 28.1
The natural radioactive series

Series	Starting element	Half-life (yr)	Stable end product
Uranium	$^{238}_{92}$U	4.47×10^9	$^{206}_{82}$Pb
Thorium	$^{232}_{90}$Th	1.41×10^{10}	$^{208}_{82}$Pb
Actinium	$^{235}_{92}$U	7.04×10^8	$^{207}_{82}$Pb

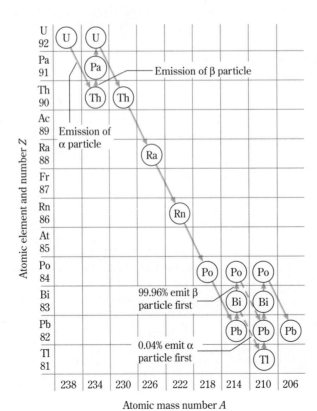

FIGURE 28.8
A typical radioactivity series. It is called the uranium series because the parent nucleus is uranium.

Illustration 28.7

If the age of the earth is 5.0×10^9 years, what fraction of the original amount of ^{232}Th is still in existence on the earth? (The earth is thought to have been molten prior to about 4×10^9 yr ago.)

Reasoning The half-life of ^{232}Th is 1.41×10^{10} yr. We know that $N/N_o = e^{-\lambda t}$. However, $\lambda T_{1/2} = 0.693$, and so $\lambda = 4.91 \times 10^{-11}$ per year. Therefore

$$\text{Fraction} = \frac{N}{N_o} = e^{-(4.91 \times 10^{-11}/\text{yr})(5.0 \times 10^9 \text{ yr})} = e^{-0.246} = 0.782$$

Thus about 78 percent of the ^{232}Th originally on the earth still exists today.

Exercise How many years will it take for the ^{232}Th on earth to decrease to one-fourth its present value? *Answer: 2.82×10^{10} yr.*

28.10 INTERACTIONS OF RADIATION WITH MATTER

As we use nuclear power and other sources of radiation, the effects of radiation on the human body and on materials become important. When a particle shoots through flesh or other material, it strikes* atoms along its path. It is in this way that the major effects of radiation occur.

The effects caused by a high-energy particle depend primarily on three factors: the mass of the particle, its energy, and its charge. An α particle, because it has a mass of 4 u, can cause more damage than an electron (0.00055 u) traveling at the same speed when it collides with an atom, much as a 10-ton truck can cause more damage than a child's wagon. Moreover, the α particle has a charge of $+2e$ compared with the electron's charge of $-e$; it therefore exerts a larger coulomb force on nearby charges than an electron does. For these reasons, an α particle ionizes atoms along its path much more frequently than an electron of the same energy does. However, because both the α particle and the electron continue moving until they have lost all their energy, the electron travels much farther before it stops than does an α particle with the same initial energy. In other words, the *range* of an electron is greater than that of an equal-energy α particle.

Typical approximate ranges for a 2-MeV particle in air are 1 cm for an α particle, 10 cm for a proton, and 1000 cm for an electron. The more dense the material through which the particle moves, the shorter its range will be. As a rough approximation, the range varies inversely with density. Therefore, an α particle that has a range of 10 cm in air ($\rho = 1.29$ kg/m^3) will have a range of only about 0.005 cm in aluminum ($\rho = 2700$ kg/m^3). It should be apparent to you why lead, a material of very high density, is used as a shield against high-energy particles.

Neutrons, which have no charge, are extremely penetrating particles. Coulomb forces do not act on them as they traverse a material. To be stopped or slowed, a

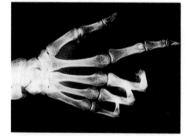

X-rays are easily detected by photographic film.

A geiger counter provides a very sensitive test for levels of radioactivity.

*We use the word *strike* in an imprecise fashion; the particle, if charged, need not actually hit an electron or nucleus to cause damage. The coulomb force exerted on the electrons and nuclei by the charged particle is often strong enough to cause damage even if the particle only passes close to the atom. Even in a near-collision with an atom or molecule, a particle can ionize an atom or cause a molecule to break apart.

neutron must undergo a direct collision with a nucleus or some other particle that has a mass comparable to that of the neutron. Materials such as water and plastic, which contain many low-mass nuclei per unit volume, are used to stop neutrons.

Gamma rays (and x-rays) are not easily stopped because they have neither charge nor rest mass. They lose energy as they penetrate material mainly through the photoelectric and Compton effects, processes that lead to ion formation. You have seen x-ray photographs of teeth and bones, so you know that x-rays can penetrate flesh and cast shadows of bones. The more electrons an atom of an absorbing material has, and the more dense the absorbing material is, the greater its ability to stop x-rays and γ-rays.

28.11   THE DETECTION OF RADIATION

Most detectors of high-energy particles and radiation make use of the fact that ions are formed along particle paths. The original radiation detector was the photographic emulsion, which Henri Becquerel used to detect radiation from uranium in 1896. Emulsions have the disadvantage of not being reusable, however, and lack the extreme sensitivity of newer methods.

One device that allows us to see the path of an ionizing particle is the *Wilson cloud chamber*. It makes use of the fact that droplets of a supersaturated vapor form preferentially on ions in the vapor. Therefore, if an ionizing particle passes through a region in which cloud droplets are about to form, the droplets will form first along the particle's path, showing the path as a trail of droplets. A somewhat similar device, called a *bubble chamber,* makes use of a superheated liquid, one that is ready to boil. Vapor bubbles form preferentially on ions, and so particle paths are shown as bubble tracks.

Electronic devices for detecting high-energy particles are convenient to use and are the most common type of particle detector. Typical of these is the *Geiger counter,* illustrated in Fig. 28.9. When no radiation is entering, no charge exists in the gas within the metal tube. No electric current is able to pass from the center wire to the metal tube, and therefore there is no current in the circuit. When an ionizing particle enters the tube, the ions and electrons that it liberates move across the tube under the influence of the electric field between the cylinder and the central wire. The field is made large enough so that the ions and electrons ionize other gas atoms as they move across the tube, causing an avalanche of charge. As a result, the current across the tube is much larger than the current that would result from the original ions alone. Soon after the particle has passed through, all the ions have been collected, and the current disappears. Therefore, each ionizing particle gives rise to a current pulse in the resistor. The resulting voltage pulses are applied to a recording electronic system, which then gives a record of the number of ionizing particles that entered the counter.

Scintillation counters use materials that emit light when struck by energetic particles. Examples of such materials include sodium iodide crystals containing a trace of thallium and certain organic plastics. The photons emitted by the particle impacts strike the cathode of a photomultiplier tube, emitting photoelectrons (Fig. 28.10). These electrons are accelerated through typically 100 V to a second electrode, where they each produce several additional electrons. This process is repeated through as many as 12 to 15 stages, creating an avalanche of electrons and hence an amplified pulse of current at the tube's output. This pulse signals the original particle impact on the detector.

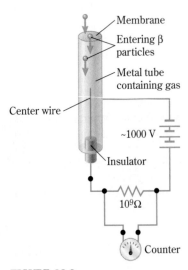

FIGURE 28.9

The Geiger counter.

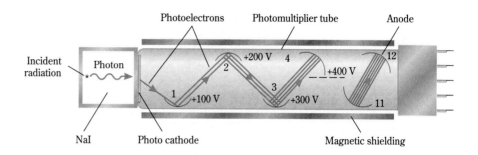

FIGURE 28.10

The photomultiplier tube converts a photon produced by incident radiation into an amplified pulse of electrons. This device is known as a scintillation counter.

Another type of detector is the *pn*-junction semiconductor, which utilizes pulses generated when a particle or γ-ray creates charges within the semiconductor. These detectors have fast response times, are efficient, and are relatively inexpensive.

Which of the many types of detectors to use depends on the type of particle (or radiation) being measured, the energy range involved, and the lack of convenience that can be tolerated.

28.12 RADIATION UNITS

In our modern world we are more and more concerned about the effects of radiation. Whether it is the result of a medical diagnostic test, of a nuclear accident, or of the radon that seeps into our dwellings from the earth below, radiation has become an important factor in our lives. Over the years, many radiation units have been used to describe its effects, and this has led to confusion. Now, however, SI units are becoming predominant, resulting in simplification. We shall list the most important measured quantities and their units.

SOURCE ACTIVITY

As stated earlier, the *activity* of a radiation source is the number of disintegrations that take place in the source in unit time:

$$\text{Source activity} = \frac{\Delta N}{\Delta t} \tag{28.5}$$

where ΔN is the number of nuclei that decay in time Δt.

The SI unit for activity is the *becquerel* (Bq); a source that has an activity of 1 Bq undergoes one disintegration per second. An older unit that is still in widespread use is the *curie* (Ci), where $1\,\text{Ci} = 3.7 \times 10^{10}\,\text{Bq}$ exactly. To give you an idea of the numbers involved, one gram of radium has an activity of $3.7 \times 10^{10}\,\text{Bq}$ (1 Ci), millions of times more radioactive than many medical sources of radiation.

We can use Eqs. 28.2 and 28.3 to obtain the following equation for activity in terms of the decay constant and the half-life:

$$\text{Activity} = \frac{\Delta N}{\Delta t} = \lambda N = \frac{0.693\,N}{T_{1/2}} \tag{28.6}$$

Use of this equation is illustrated in Illustration 28.8.

ABSORBED DOSE

The *absorbed dose* is the energy per unit mass absorbed by a material in the path of the radiation beam. Its SI unit is joules per kilogram, which, in this instance, is given the name *gray* (Gy). Suppose a radiation beam passes through a mass m and deposits therein an energy E. Then the absorbed dose given to the material constituting the mass is

$$\text{Absorbed dose (Gy)} = \frac{E}{m} \quad \text{J/kg}$$

In other words, 1 Gy is equivalent to an absorbed energy of 1 J/kg. Another unit frequently used for absorbed dose is the *rad,* where 1 rad = 0.01 Gy.

BIOLOGICALLY EQUIVALENT DOSE

The effect of radiation on the human body depends not only on the energy and type of radiation, but also on the region of the body it strikes. To describe the biological effects of radiation, we introduce another measure for radiation dose, the *biologically equivalent dose.* It is simply the absorbed dose multiplied by a factor that compares the effect of the radiation being used with the effect of 200-keV x-rays on flesh. Its unit is the *sievert* (Sv). For example, because an α-particle beam is about 15 times more damaging to flesh than are 200-keV x-rays, for a 1.0-Gy dose of α particles the biologically equivalent dose of x-rays would be 15 Sv. When discussing radiation damage to humans and animals, the biologically equivalent dose is the appropriate measure of radiation damage. An older unit that is still used frequently is the *rem,* where 1 rem = 0.010 Sv.

Illustration 28.8

Strontium 90 has a half-life of 28 yr and is a dangerous product of nuclear explosions. What is the activity of 1 g of ^{90}Sr?

Reasoning From Eq. 28.6,

$$\text{Activity} = \frac{0.693N}{T_{1/2}}$$

In this case, $T_{1/2} = 28 \text{ yr} = 8.8 \times 10^8 \text{ s}$. To find N, the number of atoms in 1 g of ^{90}Sr, we recall that 1 kmol of ^{90}Sr (which is 90 kg) contains 6.02×10^{26} atoms. Then

$$N = \frac{0.001 \text{ kg}}{90 \text{ kg}} (6.02 \times 10^{26}) = 6.7 \times 10^{21}$$

Using these values, the activity is found to be 5.3×10^{12} Bq.

Exercise How much ^{90}Sr would produce one disintegration per second?
Answer: 1.89×10^{-16} kg.

28.13 RADIATION DAMAGE

Since radiation can tear molecules apart, it is capable of damaging any material, including the materials that compose our bodies. Let us now examine the effects of different levels of radiation dose upon the body.

One of the most common types of radiation damage to humans is due to the ultraviolet rays in sunlight. These lead to sunburn and tanning of the skin. The high-energy photons disrupt skin molecules upon impact and cause these easily observed effects. In this case, the damage is usually of little importance. Most of the sun's ultraviolet rays are absorbed by ozone in the upper atmosphere. However, in recent years there has been growing evidence of a serious depletion in the ozone layer, due in part to the release of chlorofluorocarbons from aerosol sprays and refrigeration units. There is a danger that the increased ultraviolet radiation reaching the surface of earth could increase the incidence of skin cancer.

We are continuously exposed to other radiation in addition to sunlight. Nearly all materials contain a slight amount of radioactive substances. As a result, your body is unavoidably exposed to a low level of background radiation. Typically, each person experiences a background radiation dose of about 1 mSv each year.

High levels of radiation covering the whole body disrupt the blood cells so seriously that life cannot be maintained. For whole-body doses in excess of 5.0 Sv, death is likely. Even a whole-body dose of 1.0 Sv can cause radiation sickness of a very serious, although nonfatal, nature. Whole-body doses in the range of 0.30 Sv and above cause blood abnormalities. At still lower whole-body doses, the overall effects on the body are less apparent, but the consequences can nevertheless be serious.

Even very low radiation doses reaching the reproductive regions are potentially dangerous. The DNA molecules in our bodies that carry reproductive information can be disrupted by a single radiation impact. If enough of these molecules are damaged, defective reproductive information will be furnished to a fetus as it develops. As a result, birth abnormalities will occur. Even though there is some evidence that a low level of reproductive abnormalities may be beneficial to the human species, most birth defects are not desirable. For this reason, *no one of childbearing age should be exposed to unnecessary radiation of the reproductive organs.* Of course, a properly given x-ray of an arm, for example, presents no danger.

In addition to causing birth abnormalities, low levels of radiation present two other hazards. First, there appears to be a delayed cancer effect. Although cancer may not appear at once, low levels of radiation may cause it to develop many years later. Second, a child is particularly vulnerable to radiation. Because the child is growing rapidly, any cell mutations caused by radiation may have serious consequences. For this reason, most doctors are reluctant to prescribe x-ray scans for children unless absolutely necessary.

Since we are all subjected to a background radiation of about 1.0 mSv/yr, there is no reason to disrupt our lives to avoid radiation doses less than this. Maximum occupational doses are of value and have been specified. As a rough rule, the maximum yearly dose, except for the eyes and reproductive organs, is about 0.050 Sv.

28.14 MEDICAL USES OF RADIOACTIVITY

One of the earliest applications of radioactivity was the use of the radiation from radium and its decay products in the treatment of cancer. Since that time, methods of radiation therapy have greatly advanced because many new radioactive materi-

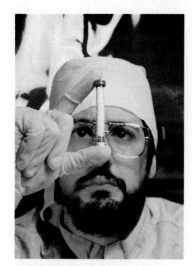

Manufactured radioisotopes, such as technetium-99, are widely used as radioactive tracers in nuclear medicine.

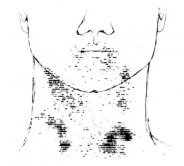

FIGURE 28.11
Radioactive gallium 67, traced here in a photoscan, settles preferentially in tumorous tissue.

als have been produced through the use of nuclear reactors and nuclear accelerating machines.

One of the most important isotopes available for research and technological use is cobalt 60. It has a half-life of 5.27 yr and is an intense source of γ-rays with energies near 1.2 MeV. Its radiation is extremely penetrating and is used to kill cancer cells deep within the body.

The radiation from iodine 131 is used in the treatment of thyroid cancer. This radioactive isotope has a half-life of 8 days. When foods containing iodine are eaten, much of the iodine localizes in the thyroid gland. Therefore, iodine 131 in foods is carried directly to the point in the body where its radiation is needed in the treatment of thyroid cancer. This is but one of many situations in which a radioactive isotope is carried to a specific point within the body for highly efficient localized radiation.

Sometimes radioactive isotopes are used as tracers, to follow the path that various chemicals take in the body. For example, if we did not already know that iodine localizes in the thyroid, we could ascertain that fact by observing the location of radioactivity in the body after iodine 131 had been ingested. Biologists use similar techniques to find out how plants utilize various chemicals.

Another medical use of radioactivity is shown in Fig. 28.11. The patient shown there has had gallium 67 injected into the bloodstream. This isotope lodges preferentially in certain types of tumorous tissue. As you see in the figure, the radioactivity (shown as the darkened regions) has localized in the lymph tissue of the throat and neck. This furnishes a strong clue to the location of cancer in this patient.

28.15   RADIOACTIVE DATING

One of the most interesting uses of radioactivity is in determining the age of ancient materials. For example, the age of uranium-bearing rocks can be found in the following way. Since uranium 238 decays to lead 206 (see Fig. 28.8), we surmise that the lead 206 that is intimately mixed with uranium 238 in a rock came from the uranium that has decayed over the years. Suppose that analysis of the rock shows that the numbers of uranium and lead atoms per unit volume are N_U and N_{Pb}, respectively. Then the ratio of the amount of uranium now present to that present at time t when the rock first solidified is

Radioactive dating using the rubidium/strontium method has established the age of this meteorite to be 4.5 billion years.

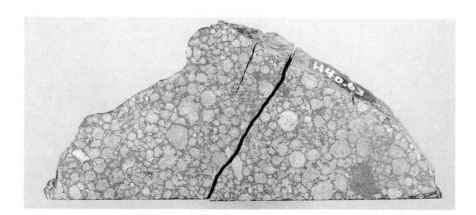

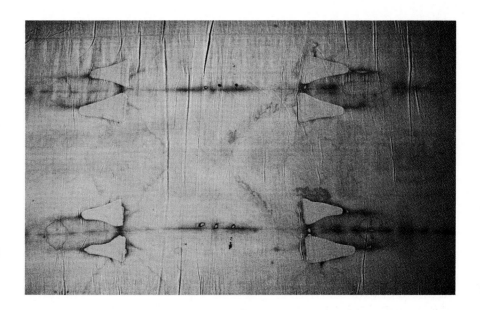

The origin of the shroud of Turin has long been an intriguing mystery. Studies in using carbon-14 dating have revealed that the material of the shroud dates from approximately the 11th century A.D.

$$\frac{N_{\text{U}}}{N_{\text{U}} + N_{\text{Pb}}} = e^{-\lambda t} = e^{-0.693t/T_{1/2}}$$

where $T_{1/2}$ is the half-life of uranium 238, which is 4.5×10^9 yr. The oldest rocks found on earth have $N_{\text{U}} \approx N_{\text{Pb}}$, and so we estimate that the earth solidified about one uranium 238 half-life ago.

Another radioactive decay scheme that has been used extensively in dating lunar and meteoritic rock samples is the β decay of ^{87}Rb ($T_{1/2} = 4.88 \times 10^9$ yr) to ^{87}Sr. Since the oldest lunar rocks and meteorites are assumed to have been formed in the very early stages of the solar system, astronomers look to the results of this method to obtain an estimate of the age of the sun and planets. Samples of both types of rocks yield maximum ages of approximately 4.6×10^9 yr, within an uncertainty of about $\pm 0.1 \times 10^9$ yr. There are other radioactive isotopes that independently yield the same ages.

To find the age of objects that were once alive, such as wood and bone, scientists employ a technique called *radiocarbon dating,* in which the radioactive isotope carbon 14 is used. This isotope, which has a half-life of 5730 yr, is continuously being produced on earth as cosmic rays from outer space bombard atmospheric nitrogen. Since the radioactive carbon is chemically identical to carbon 12, all living things have the two isotopes in intimate combination. Over the ages, the ratio of carbon 14 to carbon 12 has averaged about 1.30×10^{-12}. However, when a tree, for example, dies, the carbon 14 in its wood cannot be replenished; the carbon 14 content decays with a half-life of 5730 yr. Thus as time passes, the ratio ^{14}C/^{12}C decreases and so does the activity per gram of a sample. This fact can be used to determine the length of time since the tree died.

Illustration 28.9

How many counts per minute would you obtain from a 1-g sample of carbon taken from a new piece of wood or fiber?

Reasoning   The abundance of ^{14}C is 1.30×10^{-12}. One gram of carbon contains $(\frac{1}{12})N_A$ atoms, so there are

$$(1.30 \times 10^{-12})(\tfrac{1}{12})(6.02 \times 10^{23}) = 6.52 \times 10^{10}$$

atoms of ^{14}C in a new 1-g sample of carbon. The activity of this number of radioactive nuclei is

$$\frac{\Delta N}{\Delta t} = -\lambda N = \frac{0.693}{5730 \text{ yr}}(6.52 \times 10^{10})$$

$$= 7.89 \times 10^6 \text{ counts/yr} = 15.0 \text{ counts/min} \quad \blacksquare$$

Example 28.5

Suppose you obtained a sample of human bone from a cave. When reduced to a sample of pure carbon, 1 g of the sample had an activity of 4 counts per minute from ^{14}C. How long ago had this cave dweller lived?

R

Reasoning

Question   How is the activity related to the age of the sample?
Answer   The activity is proportional to the abundance of ^{14}C present at the time the counts are made:

$$\frac{\Delta N}{\Delta t} = -\left(\frac{0.693}{T_{1/2}}\right)N$$

The half-life is a constant for any given radioactive isotope.

Question   What is the relationship between the observed activity and the 15 count per minute activity of a contemporary carbon sample (Illustration 28.9)?
Answer   The ratio of the activities is equal to the ratio of the number of ^{14}C atoms in the two samples:

$$\frac{N \text{ (old sample)}}{N \text{ (new sample)}} = \frac{4}{15}$$

Question   What is the relationship between the activities of the old sample and a contemporary one?
Answer $\dfrac{4}{15} = \dfrac{N_o e^{-\lambda t}}{N_o} = e^{-\lambda t}$

Solution and Discussion   Using $\lambda = 0.693/T_{1/2}$, we have

$$\tfrac{4}{15} = 0.267 = e^{-(0.693/5730 \text{ yr})t}$$

$$\ln 0.267 = -1.32 = -\left(\frac{0.693}{5730 \text{ yr}}\right)t$$

Thus

$$t = \frac{(1.32)(5730 \text{ yr})}{0.693} = 10,900 \text{ yr}$$

The exceedingly small counting rates per gram for samples more than 4 to 5 half-lives old require larger samples and great care. At present the upper limit of radiocarbon dating is about 8 to 9 half-lives, or 40,000 to 50,000 yr.

28.16 **THE FISSION REACTION**

After the discovery of the neutron (in 1930), it became apparent that this neutral particle could be used to induce nuclear reactions. Because it has no charge, the neutron can easily enter a nucleus. Enrico Fermi was the leader in the use of this new projectile and, in the mid-1930s, produced many previously unknown isotopes. One of his major ambitions was to bombard massive nuclei to produce elements with Z larger than any then known. His efforts met with some success; others have carried on where he left off, and nuclei up to $Z = 107$ have now been produced.

When Fermi bombarded uranium with very-low-energy neutrons called *thermal neutrons*,* an energy-releasing reaction was indeed found to occur. Carrying on where Fermi left off, Otto Hahn and Fritz Strassman (in 1939) carried out a chemical analysis of the reaction products. To their surprise, they found many elements with atomic numbers near $Z = 50$ among the reaction products. Barium, in particular,

*Thermal neutrons (sometimes referred to as "slow" neutrons) have energies approximately equal to the average thermal energy represented by the temperature of their surroundings, kT. At room temperature, this is only about 1/40 eV, very much less than the energies they are given when formed as products of nuclear reactions. On the other hand, "fast" neutrons are those with energies of 1 MeV or greater. Fast neutrons become thermal neutrons when they undergo many energy-losing collisions with surrounding material.

(a) Nuclear reactors are used for many beneficial purposes, including commercial electrical power generation, the production of radioisotopes for medical diagnosis and treatment, and basic physics research. The reactor pictured is the advanced test reactor at Idaho Falls, which has contributed substantially to improved reactor design and technology.

(b) A fuel element is removed from the core of the high flux isotope reactor at Oak Ridge National Laboratories. Again, the blue Cerenkov radiation is due to faster-than-light (in water) neutrons emitted by fission reactions. This 100-MW reactor produces the world's highest flux of neutrons, and is a key facility in the production and research program on elements heavier than plutonium.

(a)

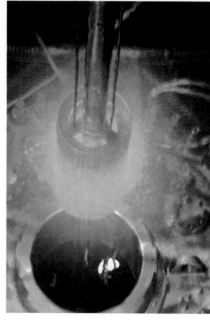

(b)

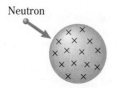

(*a*) Before reaction, n + ^{235}U

(*b*) Compound nucleus, ^{236}U*

(*c*) Vibrating droplet

(*d*) Coulomb forces stretch nucleus

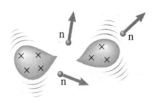

(*e*) Fission is complete

FIGURE 28.12

Vibration of the compound nucleus leads to its eventual fission.

was one of the reaction products. What could possibly be going on? They had added a neutron to a uranium nucleus ($Z = 92$) and ended up with an element (barium) that has $Z = 56$. Moreover, this nuclide appeared to be highly radioactive, even though ordinary barium is stable.

Seizing upon the work of Hahn and Strassman, Lise Meitner and her nephew, Otto Frisch, found the explanation for these puzzling results. They showed that a uranium nucleus captures a neutron, holds onto it for a fraction of a second, and then explodes into two roughly equal-size nuclei (see Fig. 28.12). The intermediate nucleus is called a *compound nucleus.* Energy and two or three neutrons are also released in the reaction. We call this splitting of a nucleus into two fragments of roughly equal size† *nuclear fission.* The discovery of nuclear fission, a simple scientific curiosity at the time, has greatly altered the course of history.

Further analysis of the reaction shows that only one uranium isotope found in quantity in nature undergoes fission in this way. It is uranium 235, which constitutes only about 0.7 percent of the natural mixture of uranium isotopes. The first step in the fission reaction is the capture of a neutron (n) by ^{235}U to form a compound nucleus:

$$n + {}^{235}_{92}U \rightarrow {}^{236}_{92}U*$$

where we represent the compound nucleus by U*. The compound nucleus then quickly decays by one of many possible reactions. The following is only one possibility:

$$ {}^{236}_{92}U* \rightarrow {}^{140}_{56}Ba + {}^{92}_{36}Kr + 4n + energy$$

The reaction products here are not the stable isotopes ^{84}Kr and ^{138}Ba found in nature. Hence they decay to other isotopes, and these to still others, until stability is reached. As a result, the products of the fission reaction are highly radioactive, and the reacting material is a strong source of radiation. Even more important, however, the reaction releases large amounts of energy.

We can obtain an understanding of this energy release by referring back to Fig. 28.3, which shows the binding energy per nucleon for various nuclei. Recall that nuclei with high binding energy have less mass per nucleon than do nuclei with lower binding energy. The graph tells us that the mass per nucleon in barium (Ba), for example, is less than that in uranium. Therefore, if a uranium nucleus is split into two nuclei with Z near 50, nucleons lose mass in the process. This lost mass is released as various forms of energy, including radiation as well as the kinetic energy of the neutrons and other reaction products. In the average fission of ^{235}U, the energy released is about 200 MeV, a tremendous energy indeed.

The mechanism of the fission process is best understood by noticing that a massive nucleus behaves in many ways like a drop of liquid. As shown in Fig. 28.12, addition of a neutron to the nucleus sets the nucleus into vibration. Because of the random nature of the vibrations, the situation in Fig. 28.12*d* may arise. In that case, the effect of the nuclear attractive force is decreased because of the much-increased surface area of the nucleus. Further, the coulomb repulsive force drives the two portions of the nucleus still farther apart, and the nucleus undergoes fission, as shown in Fig. 28.12*e*. Neutrons are released, and the two fission fragments are highly excited and unstable.

† The fission fragments produced by a large sample of fissions statistically fall into a low-mass group centered around 40 percent of the original mass and a high-mass group centered around 60 percent.

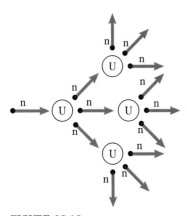

FIGURE 28.13

A chain reaction can be initiated by a single neutron.

Since the fission of one ^{235}U nucleus gives rise, on the average, to about 3 neutrons, and since neutrons induce ^{235}U nuclei to undergo fission, a self-sustaining reaction is possible. Consider a mass of ^{235}U so large that the number of neutrons that escape from its surface is negligible compared with the total number of neutrons. Then, when a single neutron enters a ^{235}U nucleus, it will give rise to, let us say, 3 neutrons as the nucleus undergoes fission. (The average number found by experiment is 2.47.) These 3 neutrons in turn cause three more nuclei to split, thereby liberating a total of $3^2 = 9$ neutrons. These 3^2 neutrons cause the fission of other nuclei to produce 3^3 neutrons, and so on. This process, illustrated in Fig. 28.13, is called a **chain reaction**. After q steps in the chain reaction have occurred, 3^q neutrons will be available. If each step of the reaction takes 0.01 s, at the end of 1 s the total number of neutrons will be $3^{100} \approx 10^{48}$. Since 235 kg of ^{235}U contains only 6×10^{26} atoms, it is clear that a reaction such as this could occur with explosive violence.

Another important fissionable nucleus in addition to ^{235}U is an isotope of plutonium, $^{239}_{94}$Pu, which readily fissions when struck by the fast neutrons emitted by the fission process. A fission chain reaction can thus be sustained in a large enough mass of Pu. Plutonium does not occur as a natural element, however, and must be manufactured in what is called a **breeding reaction**. When ^{238}U is struck by neutrons, the following series of reactions can occur:

$$^{238}_{92}\text{U} + ^{1}_{0}\text{n} \rightarrow ^{239}_{92}\text{U} \rightarrow ^{239}_{93}\text{Np} + ^{0}_{-1}\text{e}$$

$$^{239}_{93}\text{Np} \rightarrow ^{239}_{94}\text{Pu} + ^{0}_{-1}\text{e}$$

In words, what happens is that ^{239}U is formed by absorption of the neutron. Instead of fissioning, this nucleus undergoes β decay into Np, which in turn β-decays into Pu. These β decays happen very quickly, with half-lives of 23.5 minutes and 2.35 days, respectively. ^{239}Pu, however, is relatively stable, undergoing a decay with a 24,400-yr half-life. Thus a *fissionable* nucleus, ^{239}Pu, is *bred* from the nonfissionable ^{238}U. ^{239}Pu is the material used in practically all of the nuclear fission weapons in the world.

The fission chain reaction forms the basis for the operation of nuclear reactors. In practical applications, several complications arise. In order to maintain a steady, nonexplosive reaction in a reactor, each fission process should cause one additional fission process (not two, because then the reaction would explode, and not fewer than one, because then the reaction would die out). In order to retain enough neutrons in the reaction chamber, the size of the fissionable material must be large enough so that not too many neutrons stray through its surface and become lost to the reaction. There is a **critical mass** for the fissionable material. If too little material is available, not enough neutrons can be retained in it to produce a self-sustaining chain reaction.

In addition, the ability of neutrons to be captured by a ^{235}U nucleus depends upon the speed of the neutrons. Slow neutrons are much more likely to cause fission than are fast neutrons. For this reason, a large part of the total volume of a nuclear reactor consists of a moderator, a nonreactive material used to slow down the neutrons that are emitted in the fission process. Since neutrons have a mass of 1 u, and since, upon collision, a particle is slowed best by particles of nearly the same mass, the moderating material in reactors usually consists of low-atomic-mass substances. Common examples are carbon, water, and hydrocarbon plastics.

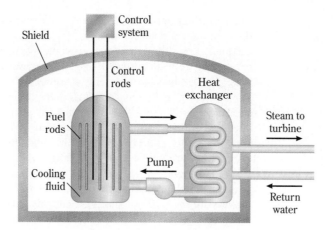

FIGURE 28.14

A schematic diagram of a nuclear fission reactor.

28.17   NUCLEAR REACTORS

The reactor in a nuclear power station serves the same purpose as the furnace in a steam generator. It acts as a source of intense heat, and that heat is used to generate steam, which is used to drive the turbines of the electric generator system. A schematic diagram of a typical reactor is shown in Fig. 28.14.

The reactor **core** contains the fissionable material sealed in long, narrow metal tubes called **fuel rods.** The fuel used in commercial reactors in the United States is UO_2. The percentage of ^{235}U has been increased from the naturally occurring 0.7 percent to about 3 percent by a process called **enrichment.** This step is necessary in order to provide a sufficient number of fissionable targets to achieve efficient operation of the reactor.* The rods are immersed in water, which acts both as a moderator and as a coolant. As moderator, the water slows down the neutrons produced by the fissions, thereby increasing the efficiency with which they produce subsequent fissions. The high specific heat capacity of water allows it to keep the

*Weapons-grade uranium, where the purpose is to achieve an explosive uncontrolled reaction, requires enrichment to at least 85 percent ^{235}U. The huge difference between this concentration of fissionable uranium and that in fuel rods is the reason why a commercial reactor cannot under any circumstances explode with the force of a nuclear bomb.

By means of the fusion of hydrogen, we can release energy on an unprecedented and almost unbelievable scale.

fuel rods at operating temperature, extracting the heat produced in the rods and delivering it to the heat exchanger, where steam is produced.

To keep the fission rate steady, a neutron-absorbing substance, such as boron or cadmium, is introduced in a series of **control rods.** These rods can be inserted or withdrawn from the core. The more they are inserted, the more neutrons they absorb, reducing the number of fissions that can be produced. If the control rods are fully inserted, the reaction stops completely.

As the fuel rods produce power, important changes take place within them. First, highly radioactive fission fragments build up. They emit energetic β particles at such a rate that 7 percent of the total thermal power output of the reactor is due to this radioactivity. In an emergency, such as the disruption of coolant flow, the other 93 percent of the power output due to fissions can be stopped immediately by insertion of the control rods. However, there is no way the fission-fragment radioactivity can be stopped. This source of energy is sufficient to melt the fuel rod assembly and cause an excessive temperature and pressure buildup, possibly rupturing the containment structure. To guard against this, a separate emergency core cooling system is built into the reactor. Reactors in the United States have an excellent record of safe operation over the past 30 years.

A second important change taking place in the fuel rods is the buildup of plutonium as some of the fast neutrons impact ^{238}U nuclei and cause the breeding reactions to occur. This plutonium buildup is an inescapable by-product of the reactor's operation. Typically, for every 100 ^{235}U fissions, 50 to 55 ^{239}Pu nuclei are formed.

These two changes in the fuel rods require that they be removed while there still is an appreciable amount of unused ^{235}U left. When reactors were first developed, the plan was to reprocess these spent rods. The uranium could be reclaimed, the plutonium separated chemically, and the highly radioactive fission fragments disposed of underground in sealed containers. Reprocessing has turned out to be highly dangerous and costly, however, and has been abandoned. Methods of disposal of the waste material have been developed, but unfortunately no politically feasible solution has been found for permanent disposal.

Unlike uranium, plutonium does not have to be enriched to be used in nuclear weapons. The fact that plutonium can be separated chemically from spent fuel rods allows the accumulation of many critical masses of Pu from the operation of ordinary uranium reactors. Thus proliferation of plutonium weapons is possible under the guise of peaceful production of electrical power from today's fission reactors.

Specialized reactors create the radioactive isotopes used in medical diagnosis and treatment and industrial processes. Many of the radiation sources presently used by hospitals, industry, and research laboratories are made by placing suitable materials within the core of the reactor. In addition, research reactors exist in many parts of the world. The intense radiation in their cores can be "piped" outside the reactor to act as powerful beams of radiation. As we see, the fission process possesses vast potential as well as hazards for humankind.

Example 28.6

A typical fission reactor converts one-third of the heat produced by fission to 1000 MW of electrical power. How many ^{235}U fissions per second are required for this? How much mass of ^{235}U will this reactor fission in one year of operation?

R

Reasoning

Question   How much heat must be released by fission to produce the 1000 MW?
Answer   Since the conversion efficiency is one-third, a 1000-MW output requires 3000 MW from fission.

Question   How much energy is released per fission?
Answer   Approximately 200 MeV, on average.

Question   What is the connection between the number of fissions and the mass of ^{235}U involved?
Answer   235 g of ^{235}U contains 6.02×10^{23} nuclei.

Solution and Discussion   First, convert 3000 MW to MeV/s:

$$3000 \text{ MW} = \frac{3000 \times 10^6 \text{ J/s}}{1.6 \times 10^{-13} \text{ J/MeV}}$$

$$= 1.88 \times 10^{22} \text{ MeV/s}$$

At 200 MeV per fission, the production of this much power requires

$$\frac{1.88 \times 10^{22} \text{ MeV/s}}{200 \text{ MeV/fission}} = 9.4 \times 10^{19} \text{ fissions/s}$$

The number of moles fissioned per second is

$$\frac{9.4 \times 10^{19}/\text{s}}{6.02 \times 10^{23}} = 1.56 \times 10^{-4} \text{ mol/s}$$

In one year this amounts to

$$(1.56 \times 10^{-4} \text{ mol/s})(3.16 \times 10^7 \text{ s/yr}) = 4930 \text{ mol/yr}$$

One year's operation of the reactor thus requires

$$(4930 \text{ mol/yr})(0.235 \text{ kg/mol}) = 1.16 \times 10^3 \text{ kg/yr}$$

This is a little over 1 metric ton (1000 kg). Since the amount of ^{235}U is only about 3 percent of the mass of the fuel, the reactor uses a total of about 35 metric tons of enriched UO_2 per year.

28.18 NUCLEAR FUSION

If we look back to Fig. 28.3, we see that the small-atomic-number nuclei, such as lithium, have even less binding energy per nucleon than uranium has. This means that the nucleons in low-atomic-number nuclei have more mass per nucleon than do those in nuclei with larger Z. We can therefore envision joining small nuclei together to form larger nuclei and, in the process, converting mass into energy. This type of reaction, in which small nuclei are joined together to form larger ones, is called **nuclear fusion**. To illustrate the tremendous energies released in fusion reactions, consider the following set of reactions, which furnishes a large fraction of the sun's energy.

$$^1_1H + \,^1_1H \rightarrow \,^2_1H + \,^0_{+1}e + \,^0_0\nu$$

where $^0_{+1}e$ is a positive electron (called a positron) and $^0_0\nu$ is a neutrino. The deuterium ^2_1H reacts further:

$$^2_1H + \,^1_1H \rightarrow \,^3_2He$$

and then

$$^3_2He + \,^3_2He \rightarrow \,^4_2He + 2^1_1H$$

As we see, in effect four protons are fused together to form a helium 4 nucleus.

To find the energy liberated in this process, we must find the mass loss. The starting mass is that of four protons, $4 \times 1.007276 = 4.029104$ u, while the final mass is that of the helium 4 nucleus, namely, $4.002604 - 2m_e = 4.001506$ u. Thus the mass loss is 0.0276 u. This mass has an energy equivalent of

$$(0.0276 \text{ u})(931 \text{ MeV/u}) = 25.7 \text{ MeV}$$

But in 1 kg of helium, there are $N_A/4$ atoms. So the energy lost in the formation of 1 kg of helium is

$$\text{Energy} = \tfrac{1}{4}(6 \times 10^{26})(25.7 \text{ MeV}) = 3.86 \times 10^{33} \text{ eV} = 6.2 \times 10^{14} \text{ J}$$

It is interesting to compare this amount of energy with the total mass energy contained in 1 kg of matter: $E = mc^2 = 9 \times 10^{16}$ J. The energy released by fusion is about 0.7 percent of this, and so we can say that approximately 0.7 percent of matter is converted to energy in hydrogen fusion. A similar calculation for the fission of 1 kg of ^{235}U gives an energy release of about 8×10^{13} J, which is a conversion of 0.1 percent of the mass into energy. By contrast, chemical combustion yields only about 3.3×10^7 J/kg of fuel and oxygen. Chemical reactions thus release only 10^{-7} as much energy per kilogram as do fission and fusion reactions.

Although the source of energy in the sun and stars is a fusion process, the fusion reaction has not yet been made a practical, steady energy source on earth. In principle, fusion is an extremely attractive energy source. Its by-product, ^4He, is not a radioactive waste, but rather a very useful and rare element. Fuel is plentiful, since hydrogen is a constituent of water. This availability, coupled with the large amount of energy available per kilogram, would provide an almost inexhaustible energy supply.

The difficulty in obtaining a steady fusion reaction is basically that fusion can't occur unless the protons are within the range of the nuclear strong force, about 5×10^{-15} m. At this range, their coulomb repulsion is enormous. Another way of stating this is that the electrical potential energy at this distance is very large, on the order of 1 MeV. This is approximately the kinetic energy protons must be given if they are to fuse before being repelled by the coulomb force. This energy is easily attainable by means of huge particle accelerators. However, the efficiency of these machines is far too low to make them practical. We must make use of *thermal*

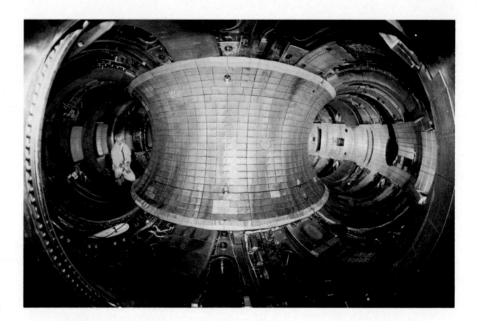

The inside of the tokamak fusion reactor. Physicists use this device to study the characteristics of a magnetically contained fusion reaction, with the goal of developing a practical commercial fusion reactor in the future.

collisions between protons in an extremely hot gas. Let us see what sort of temperature might be required to cause fusion in this way.

The kinetic theory of gases tells us that the average translational kinetic energy of a particle in a gas whose temperature is T is $\frac{3}{2}kT$. Setting this equal to 1 MeV, or 1.6×10^{-13} J, we have $\frac{3}{2}(1.38 \times 10^{-23}$ J/K$)T = 1.6 \times 10^{-13}$ J. Thus

$$T = 7.6 \times 10^9 \, \text{K}$$

The requirement of very high temperatures is why this reaction is called **thermonuclear fusion.** Particle energies are of course distributed over a large range of values around this average. At the densities that occur in the core of the sun, namely 150×10^3 kg/m^3, significant amounts of fusion energy are produced at 15 million K by particles in the high-energy tail of the thermal distribution. The

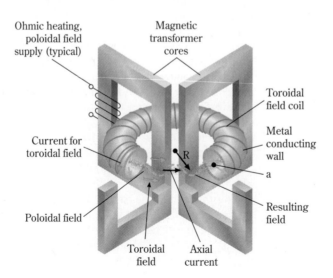

FIGURE 28.15

A tokamak magnetic confinement fusion system. A complex magnetic field confines the high-temperature gas (plasma) within a doughnut-shaped region.

observed power output of the sun, due to fusion, is 4×10^{26} watts! This requires that approximately 655 million tons of hydrogen (protons) be fused into 650 million tons of ^{4}He every second in the core of the sun. The sun is able to contain this high-temperature reaction because of the strength of its gravity. Gravity thus provides a stable containment of the fusion reaction in stars.

On earth, however, we have to seek other means of containing this extremely high-temperature reaction. We can produce fusion *explosively*, as in hydrogen bombs, but have not yet been successful in achieving a *controlled* thermonuclear reaction. Our attempts at containment involve the fact that material at these very high temperatures is highly ionized, consisting of separate ions and electrons in a state called a **plasma.** Charged particles can be confined by strong magnetic fields, but at very high temperatures and pressures, instabilities quickly occur and ruin the containment. What has evolved over many years of research in many countries is an attempt to heat a plasma very rapidly and to contain it long enough in magnetic fields so that more energy is produced than consumed before the containment is disrupted. One of the most promising approaches uses a device called a **tokamak,** illustrated schematically in Fig. 28.15. Containment times are approaching 1 s, and the break-even point in energy is anticipated as tokamak size increases.

Researchers are concentrating on two fusion reactions that can proceed at lower temperatures than the proton-proton reaction. Deuterium-deuterium (^{2}H–^{2}H) fusion occurs at 10^8 K, and deuterium-tritium (^{2}H–^{3}H) fusion needs "only" 4×10^7 K. A number of methods of heating are being used, and temperatures in this range have been achieved. Present results in the United States and Great Britain indicate that commercial utilization of the fusion reaction will be feasible, perhaps in 25 to 50 years.

LEARNING GOALS

Now that you have finished this chapter, you should be able to

1 Define (*a*) nucleon, (*b*) atomic mass unit, (*c*) atomic number and mass number, (*d*) isotope, (*e*) natural abundance, (*f*) nuclear binding energy, (*g*) radioactive decay, (*h*) activity, (*i*) decay constant, (*j*) half-life, (*k*) α decay and β decay, (*l*) branching ratio, (*m*) dose, (*n*) biologically equivalent dose, (*o*) becquerel, curie, gray, and sievert units, (*p*) nuclear fission, (*q*) nuclear fusion, (*r*) chain reaction, (*s*) fuel rods, (*t*) control rods, (*u*) moderator, (*v*) fission fragment, (*w*) breeding reaction, (*x*) breeding ratio.

2 Estimate the size of a nucleus when its mass number is given.

3 Sketch the graph of binding energy per nucleon versus mass number A.

4 Given the mass of a nucleus, calculate the binding energy of the nucleus.

5 Sketch a graph of N versus t for a radioactive sample. Given the half-life or the decay constant of the material in the sample, compute the fraction of the original sample left after a given length of time.

6 Write the nuclear reaction equation for a given nucleus for α and β, decay. Given the masses of the initial and final nuclei for these reactions, determine which, if any, will occur spontaneously.

7 Prepare a diagram such as Fig. 28.8 for a series in which the initial nucleus and the emitted particles are given.

8 Compare the range and ionization effects of α, β, and γ radiation passing through matter.

9 Explain, by reference to the nuclear binding-energy graph, why nuclear fission of uranium should release energy. State what is meant by a fission chain reaction, and relate this to why ^{235}U is usable in a bomb.

10 Sketch a diagram of a fission reactor showing fuel rods, control rods, moderator, heat exchanger, and turbine and explain the function of each. Explain the importance of fuel enrichment.

11 Explain the breeding reaction by which plutonium is made from uranium and explain how plutonium fission differs from that of ^{235}U.

12 Explain the source of thermal energy which remains in a fission reactor even after fission reactions have been shut down by the control rods. Explain the danger this heat poses.

13 Explain, by reference to the nuclear binding-energy graph, why the nuclear fusion of hydrogen should release energy. State why fusion is more difficult to achieve in the laboratory than fission is. State some of the potential advantages of fusion as an energy source compared to fission.

SUMMARY

DERIVED QUANTITIES AND PHYSICAL CONSTANTS

The Atomic Mass Unit (u)
$1\,u = \frac{1}{12}$ the mass of an atom of ^{12}C
    $= 1.660566 \times 10^{-27}\,kg$

Activity
1 becquerel (Bq) = 1 decay/s
1 curie (Ci) = $3.7 \times 10^{10}\,Bq$

Absorbed Dose
1 gray (Gy) = 1 J/kg
1 rad = 0.01 Gy

Biologically Equivalent Dose
1 sievert (Sv) = 1 Gy × RBE

where RBE = the relative biological effectiveness of the type of radiation being absorbed.

1 rem = 0.010 Sv

DEFINITIONS AND BASIC PRINCIPLES

Isotope Designation
For any given nucleus,

Z = number of protons (atomic number)
N = number of neutrons
$A = Z + N$ = number of nucleons (mass number)

INSIGHTS
1. All nuclei having the same Z are nuclei of the same chemical element.
2. Nuclei with the same Z but different N (hence different A) are isotopes of the chemical element.
3. A given isotope of element X is designated by $^{A}_{Z}X$.
4. As found in nature, elements generally occur in a mixture of various isotopes. The percentages of various isotopes making up the element is called the natural isotopic abundance.

Nuclear Size and Density
The radius of a nucleus of mass number A is approximately

$R \approx (1.2 \times 10^{-15}\,m)A^{1/3}$

INSIGHT
1. The dependence of R on A implies that the nuclear volume is proportional to A, and hence that all nuclei have about the same mass density.

Nuclear Binding Force and Energy
The nuclear binding force has the following characteristics:

1. Extremely short range. It becomes zero for separations between nucleons greater than approximately $5 \times 10^{-15}\,m$.
2. Extremely strong. Within its range, it is able to hold protons together, overcoming the very strong repulsion between the proton charges.
3. Applies equally to protons and neutrons, but does not have any effect on electrons. Thus electrons do not exist in the nucleus.

The binding energy of a nucleus is the energy necessary to separate the nucleus into its component protons and neutrons. If the difference between the total mass of the separate nucleons and the mass of the assembled nucleus is the mass defect Δm, the binding energy is

Binding energy = $\Delta m\,c^2$

Radioactivity
Radioactivity is a process whereby unstable nuclei get rid of excess energy by emitting particles and electromagnetic radiation. Common among these emissions are α particles (^{4}He nuclei), β particles (electrons), and γ-rays.

HALF-LIFE $(T_{1/2})$
A radioactive substance decays exponentially, which is characterized statistically by a time interval during which half of the amount present at the beginning of the time interval undergoes radioactive change. This time interval, which varies widely from one isotope to another, is called the half-life of the isotope.

DECAY CONSTANT (λ)
An alternate description of radioactive decay is the analytical expression

$N(t) = N_o e^{-\lambda t}$

where N_o is the original number of nuclei in a sample, $N(t)$ is the number left at time t, and λ is the decay constant of the isotope. λ is related to the half-life by

$$\lambda = \frac{0.693}{T_{1/2}}$$

ACTIVITY OF A SAMPLE
The activity of a radioactive sample is the rate of radioactive decays, i.e., the number of decays per second.

Activity = $\lambda N(t)$

Nuclear Reactions

When nuclei undergo reactions that change their structure, the changes must obey the conservation laws of physics:

1. The total charge on all particles before and after the reaction must remain constant.
2. The total number of nucleons before and after must remain constant.
3. Energy conservation requires that the difference in the total masses before and after a reaction be related to the energy absorbed or given off by

$$\Delta m \, c^2 = \text{energy absorbed or given off}$$

INSIGHTS

1. If $\Delta m < 0$, energy is given off. The nuclei resulting from the reaction thus have lower energy and are more stable. This reaction can occur spontaneously, as is the case with radioactivity.

2. If $\Delta m > 0$, energy must be supplied to make the reaction occur. This type of reaction cannot happen spontaneously.

Nuclear Fission

Fission is a process whereby a nucleus is split into two main fragments, of roughly equal size, with the release of energy. Approximately 1 percent of the original mass is converted into energy in a fission process. A few heavy isotopes have a significant probability of fissioning when struck by neutrons. The two most notable are $^{235}_{92}U$ and $^{239}_{94}Pu$. Two or more neutrons are also emitted when fissioning occurs. This provides a multiplying factor in the number of neutrons available to cause further fissions, and a chain reaction can occur, causing the rate of fissioning to grow exponentially.

Nuclear Fusion

Under certain conditions, light nuclei can be joined or fused into heavier nuclei with the release of energy, a process called nuclear fusion. Typically, approximately 8 percent of the original mass is converted into energy in this process.

QUESTIONS AND GUESSTIMATES

1 Cobalt 60 is widely used as a source of γ-rays for radiation therapy for cancer. How many protons, neutrons, and electrons does one $^{60}_{27}Co$ atom possess?

2 Why do chemists consider different isotopes to be the same element even though their nuclei are not the same?

3 Would the optical spectra of ^{235}U and ^{238}U atoms differ in any major way?

4 Estimate the atomic mass of $^{64}_{30}Zn$ from the fact that the binding energy per nucleon for it is about 8.7 MeV.

5 Tritium is the ^3H isotope of hydrogen. Its atomic mass is 3.016 u; the atomic mass of ^1H is 1.0078 u, and that for the neutron is 1.00867 u. What do you predict about the stability of tritium? Repeat for ^2H, deuterium, which has an atomic mass of 2.0141 u.

6 A certain metal decays to a stable element by emission of α particles whose energy is about 9 MeV. A tiny sphere of the pure metal is mounted on the end of a pin. Describe how you would find the half-life of the metal if its half-life is about (*a*) five days and (*b*) 2000 years.

7 A beam of α particles is absorbed in a block of lead. What happens to the α particles? Rutherford proved the nature of α particles by heating the irradiated lead.

8 Radon gas, being radioactive, is a dangerous contaminant of air. Because radon seeps into houses from the earth below the house, what factors lead to dangerous radon levels?

9 What is the source of the earth's helium gas?

10 A piece of uranium 235 that is smaller than the critical mass may explode if it is placed in a vat of water. Explain. Why may ^{235}U in the form of a wire not explode, even though the mass of wire exceeds the critical mass?

11 Most radiologists feel that women beyond childbearing age can safely be exposed to much more x-radiation than can young women. How can they justify such an opinion?

12 It is possible for a man working with x-rays to burn his hand so seriously that he must have it amputated, and yet suffer no other consequences. However, an x-ray overexposure so slight as to cause no observable damage to his body could cause one of his subsequent offspring to be seriously deformed. Explain why.

PROBLEMS

Sections 28.1–28.3

1 Evaluate the following quantities for the nucleus $^{14}_{7}N$: (*a*) nuclear charge, (*b*) number of neutrons, (*c*) approximate radius, and (*d*) nuclear density.

2 Find the following properties for the $^{202}_{80}Hg$ nucleus: (*a*) number of protons, (*b*) number of neutrons, (*c*) number of nucleons, (*d*) radius, and (*e*) nuclear density.

3 A certain nuclear isotope has the atomic mass number 43 and it has three more neutrons than protons. Identify the isotope.

4 An isotope has 10 neutrons and its atomic mass number is 18. What isotope is it?

5 Which stable nucleus has a radius approximately one-half of the radius of the nucleus of $^{216}_{84}Po$?

6 Compare the nuclear radii and the nuclear densities of the following nuclides: $^{7}_{3}\text{Li}$, $^{93}_{41}\text{Nb}$, and $^{220}_{86}\text{Rn}$.

▪7 The earth is approximately a sphere with a radius of 6.4×10^{6} m. It has an average density of about 5320 kg/m³. If the earth were to shrink into a sphere with the density equal to that of a nucleus ($\approx 2 \times 10^{17}$ kg/m³), what would the earth's radius be?

▪8 The total mass of the observable universe is estimated to be of the order of 10^{51} kg. If all this mass were compressed into a sphere of nuclear density ($\approx 2 \times 10^{17}$ kg/m³), what would its radius be? Compare this radius with that of the sun, 7×10^{8} m.

9 In a certain mass spectrometer, a velocity selector (Chap. 19) is used to obtain a beam of ions with velocity 2.9×10^{5} m/s. Find the radius of the path that a univalent ^{12}C ion will follow when the magnetic field B within the spectrometer is $0.080 T$.

10 For the mass spectrometer in Prob. 9, what will be the difference of the radii of the paths followed by a ^{12}C isotope and a ^{14}C isotope?

11 A certain mass spectrometer (Chap. 19) accelerates ions through 1700 V and deflects them in a magnetic field of $0.070 T$. A beam of certain univalent ions follows a path of radius 12.0 cm in the spectrometer. What is the mass of these ions in both kilograms and atomic mass units?

▪12 A beam of a mixture of univalent ions of two isotopes is examined in a mass spectrometer. The radii of the circular paths followed by the ions are found to be 12.0 and 14.0 cm, respectively. Find the ratio of the atomic masses of the two isotopes.

▪13 In a mass spectrometer it is found that the radius for the path followed by univalent ^{12}C ions is 10.0 cm. How large would the radius be for ^{16}O? (Assume identical charges and accelerating potentials.)

14 Chlorine found in nature essentially contains only two isotopes. One isotope, $^{35}_{17}\text{Cl}$, constitutes about 75.5 percent and the other, $^{37}_{17}\text{Cl}$, about 24.5 percent. Find the atomic mass of a natural sample of chlorine to three significant digits.

15 Natural potassium occurs as a mixture of two isotopes. One isotope, with atomic mass 38.964 u, occurs with a relative abundance of 93.3 percent; the other, with atomic mass 40.975, constitutes 6.7 percent. Compute the atomic mass of a natural sample of potassium.

16 Neon is found to exist in nature as three isotopes. The isotope ^{20}Ne with a relative abundance of 90.9 percent, the isotope ^{21}Ne with a relative abundance of 0.3 percent, and the isotope ^{22}Ne with an abundance of 8.8 percent. Estimate the atomic mass of Ne as listed in periodic tables.

▪17 Uranium found on the earth consists of two principal isotopes, ^{235}U and ^{238}U. The atomic masses of these two isotopes are 235.044 u and 238.051 u, respectively, while the mass of a natural sample is 238.030 u. Find the approximate percentage of each isotope in a natural sample of uranium.

▪18 Natural carbon has two principal isotopes; ^{12}C, with an atomic mass of 12.00000 u, has a relative abundance of 98.892 percent. What is the atomic mass of the other isotope if the mass of natural carbon is 12.01115 u?

Section 28.4

19 Use the data shown in Fig. 28.3 to find how much mass is lost as a zinc 64 nucleus is assembled from free protons and neutrons. What is the percentage mass loss?

20 From the data in Fig. 28.3 calculate how much energy is required to tear a mercury 202 nucleus apart into free protons and neutrons. What is the mass equivalence (in u) of this energy?

21 Calculate the total binding energy for the carbon 12 nucleus. What is the binding energy per nucleon? *Hint:* Recall that the mass for the carbon 12 atom is exactly 12 u.

22 Calculate the total binding energy and the binding energy per nucleon for the ^{40}Ca nucleus. The atomic mass for this nucleus is 39.96259 u.

23 The atomic mass of ^{14}N is 14.00307 u and that of ^{15}N is 15.00011 u. From this information estimate the binding energy for the extra neutron in the ^{15}N nucleus.

24 From the data of Fig. 28.3 and the masses of proton and neutron, find the mass of an atom of krypton 84.

25 Two isotopes having the same mass number but different atomic numbers are called *isobars*. Calculate the difference in binding energy per nucleon for the isobars $^{36}_{16}\text{S}$ and $^{36}_{18}\text{Ar}$. ($M_S = 36.012074$ u; $M_{Ar} = 35.972182$ u)

▪26 How much energy is required to remove a neutron from the nucleus of the isotope ^{13}C? What isotope results after this removal?

Sections 28.5 and 28.6

27 A Geiger counter placed above a radioactive sample registers 678 counts per minute. How many counts per minute will it register after four half-lives have passed?

28 A radioactive sample gives 840 counts per minute at one time; 48 h later, it has a count rate of 44 counts per minute. What is the half-life of the sample?

29 A sample of radioactive material contains 4.5×10^{12} nuclei and has a half-life of 0.84 yr. (*a*) What is its decay constant? (*b*) How many nuclei of the original sample undergo decay in 2 min?

30 Polonium has a half-life of 140 days. How long would it take for a sample of polonium to decay to one-eighth its original amount?

31 A tiny vial of radon gas contains 8.0×10^{12} atoms of radon. The half-life of radon is 3.8 days. How many disintegrations occur in the vial each minute?

32 Watches with numerals that are visible in the dark sometimes have radioactive material in the paint used for the numerals. A student estimates from the measurement of the Geiger counter that 750 disintegrations occur on the watch face each second. If the student's figures are correct, how many curies of radioactivity exist on the watch?

33 A Geiger counter placed above a tiny piece of radioactive rock registers 194 counts per minute. Assuming that the counter intercepts radiation from half of the decaying nuclei, what is the activity of the rock?

34 Tritium, a radioactive isotope of hydrogen, has a half-life of 12.33 yr. What percentage of the nuclei in a sample of tritium will disintegrate in 6 yr?

■35 It is observed that after 3 h only 0.25 mg of a pure radioactive material that was originally 2 mg is left. What is the half-life of the material?

■36 What fraction of a radioactive sample decays in 90 yr if the half-life of the material is 156 yr?

■37 Measurements show that only 14 percent of a radioactive material remains after 24.0 h. What is the half-life of the material?

■38 Strontium 90 is a radioactive fission product from nuclear reactors and bombs. Since its half-life is quite long (about 28 yr, or 8.8×10^8 s), it is a persistent contaminant and presents serious disposal problems. What fraction of the original strontium still remains 100 years after a nuclear bomb explodes?

■■39 Uranium 238 has a half-life of 4.5×10^9 yr. Calculate the activity of 0.1-g sample of pure uranium.

Sections 28.7–28.9

40 Identify the product nuclei denoted by X in the following radioactive decays:

$$^{226}_{88}\text{Ra} \rightarrow X + {}^4_2\text{He}$$

$$^{95}_{36}\text{Kr} \rightarrow X + {}^0_{-1}\text{e}$$

$$^{59}_{26}\text{Fe} \rightarrow X + \gamma$$

41 Complete the following radioactive decay formulas by identifying X:

$$^{233}_{91}\text{Pa} \rightarrow X + {}^0_{-1}\text{e}$$

$$^{234}_{90}\text{Th} \rightarrow {}^{230}_{88}\text{Ra} + X$$

$$X \rightarrow {}^{140}_{58}\text{Ce} + {}^4_2\text{He}$$

42 What is the resultant isotope when $^{210}_{84}\text{Po}$ decays by emitting a 5.3-MeV α particle and an 0.80-MeV γ-ray?

43 Identify the resulting isotope when $^{209}_{82}\text{Pb}$ decays by emitting a β particle. Repeat for $^{223}_{86}\text{Rn}$, which also emits a β particle.

44 What is the resultant isotope when $^{211}_{83}\text{Bi}$ decays by emission of a 6.62-MeV α particle?

45 $^{220}_{86}\text{Rn}$ emits a 0.54-MeV γ-ray. By what percentage does its nuclear mass change in this process?

46 What is the fractional change in the nuclear mass of $^{226}_{90}\text{Th}$ when it emits a 1.11-MeV γ-ray?

■47 Identify the resultant isotope in the decay of $^{234}_{92}\text{U}$ (M = 234.040947 u) by emission of an α particle. The energy released in this decay is 4.773 MeV. Calculate the mass of the daughter nucleus.

48 What is the particle emitted when $^{14}_6\text{C}$ decays to $^{14}_7\text{N}$?

■49 Uranium 238 decays by emission of an α particle, with a half-life of 4.5×10^9 yr, according to the reaction:

$$^{238}_{92}\text{U} \rightarrow {}^{234}_{90}\text{Th} + {}^4_2\text{He} + \text{energy}$$

If all the energy becomes kinetic energy of the α particle, what will its energy be in MeV? Its actual energy is 4.19 MeV. How can you account for the discrepancy? Take the masses of the isotopes involved in the reaction as $M({}^{238}_{92}\text{U}) = 238.05077$ u, $M({}^{234}_{90}\text{Th}) = 234.04358$ u, and $M({}^4_2\text{He}) = 4.00260$ u.

50 Which of the following decays occurs spontaneously (apply energy considerations)? ($M_{\text{Ce}} = 139.90528$ u)

$$^{11}_5\text{B} \rightarrow {}^7_3\text{Li} + {}^4_2\text{He}$$

$$^{144}_{60}\text{Nd} \rightarrow {}^{140}_{58}\text{Ce} + {}^4_2\text{He}$$

■51 Consider the reaction

$$^1_1\text{H} + {}^{13}_6\text{C} \rightarrow {}^{13}_7\text{N} + {}^1_0\text{n}$$

where n represents the neutron. Can this reaction be initiated by a proton (^1_1H) that has a kinetic energy of 2.2 MeV? Take the atomic masses of the nuclei in the reaction as $M({}^1_1\text{H}) = 1.007825$ u, $M({}^1_0\text{n}) = 1.008665$ u, $M({}^{13}_6\text{C}) = 13.00336$ u, and $M({}^{13}_7\text{N}) = 13.00574$ u.

■52 From energy considerations, is the following reaction possible if the incident proton has a kinetic energy of 1.6 MeV?

$$^1_1\text{H} + {}^7_3\text{Li} \rightarrow {}^7_4\text{Be} + {}^1_0\text{n}$$

Take the atomic masses of the nuclei in the reaction as $M({}^1_1\text{H}) = 1.007825$ u, $M({}^1_0\text{n}) = 1.008665$ u, $M({}^7_3\text{Li}) = 7.01600$ u, and $M({}^7_4\text{Be}) = 7.01693$ u.

■53 Consider the reaction

$$^4_2\text{He} + {}^{27}_{13}\text{Al} \rightarrow {}^{30}_{15}\text{P} + {}^1_0\text{n}$$

The atomic masses of the nuclei in the reaction are $M({}^4_2\text{He}) = 4.00260$ u, $M({}^1_0\text{n}) = 1.008665$ u, $M({}^{27}_{13}\text{Al}) = 26.98154$ u, and $M({}^{30}_{15}\text{P}) = 29.97831$ u. Is this reaction possible if the ^4_2He has 2.0 MeV of kinetic energy?

■54 Polonium 210 decays by emitting an 0.080-MeV γ-ray, along with an α particle with a kinetic energy of 5.3 MeV, through the reaction

$$^{210}_{84}\text{Po} \rightarrow {}^4_2\text{He} + {}^{206}_{82}\text{Pb} + \gamma$$

The atomic masses of the product nuclei are $M({}^4_2\text{He}) = 4.00260$ u and $M({}^{206}_{82}\text{Pb}) = 205.97447$ u. (*a*) Knowing the kinetic energy of the α particle to be 5.3 MeV, find the approximate recoil energy of the lead atom. (*b*) Calculate the expected atomic mass of the polonium 210. The measured mass is 209.9829 u.

55 Suppose 1 kg of deuterium (heavy hydrogen, ^2_1H) is combined to form 1 kg of helium according to the reaction

$$^2_1H + ^2_1H \rightarrow ^4_2He$$

The atomic masses are $M(^2_1H) = 2.0141$ u and $M(^4_2He) = 4.00260$ u. (a) How much energy (in joules) is liberated? (b) If the confined helium has a heat capacity of 0.75 cal/g · C°, by how much does its temperature increase as this energy is added to it?

56 The thorium series mentioned in Table 28.1 starts with $^{232}_{90}Th$ and emits in succession one α, two β, four α, one β, one α, and one β particle(s). Verify that the final resulting isotope in the series is the same as shown in the table.

57 Uranium 238 in the uranium series mentioned in Table 28.1, undergoes five successive α decays. Identify the resulting daughter nucleus at each decay step.

58 The actinium series mentioned in Table 28.1 starts with $^{235}_{92}U$ and emits in succession one α, one β, two α, one β, three γ, two β, and one α particle(s). Make a diagram of the series and identify daughter nuclei at each decay step.

Sections 28.10–28.15

59 Iodine 131 is used to treat thyroid disorders because, when ingested, it localizes in the thyroid gland. Its half-life is 8.1 days. (a) What is the activity of 0.80 μg of ^{131}I? (b) How much ^{131}I would have an activity of 0.2 μCi?

60 The half-life of phosphorus 32 is 14.3 days and it is used in medicine because it tends to localize in bone. What is the activity of 0.7 g of ^{32}P?

61 How many grams of iron 59 are there in a 1-mCi sample? Its half-life is 46.3 days.

62 A medical laboratory buys a sample of a radioactive isotope with 260-mCi activity. The sample has a half-life of 18 days. How long can the laboratory use the sample before its activity decreases to 26 mCi?

63 The isotope tritium (^3_1H) has a half-life of 4600 days. How many grams of tritium are there in a sample whose activity is 2.3 mCi?

64 By how much is the temperature of water raised when water is given a radiation dose of 10.0 mGy?

65 How large a radiation dose must be deposited in lead to raise its temperature 8°C? The specific heat of lead is 0.031 cal/g · C°.

66 A 70-kg worker in a nuclear lab is exposed to a radiation dose of 0.25 Gy. How many joules of energy are deposited in the worker's body?

67 In an effort to date a piece of bone, its count rate due to ^{14}C is computed and found to be only 0.048 that of a similar fresh piece of bone. Estimate the age of the bone. The half-life of ^{14}C is 5700 years.

68 Thorium 232 has a half-life of 1.39×10^{10} yr and decays through a number of steps to ^{208}Pb. The ratio of ^{208}Pb to ^{232}Th in a certain rock sample is found to be 0.17. Estimate the time since the rock solidified.

69 Tritium (^3_1H), an isotope of hydrogen with a half-life of 12.3 yr, is produced in the upper atmosphere by cosmic rays and is intimately mixed with the hydrogen in the air. In order to determine the age of a bottle of wine found in an ancient cave, the tritium in the wine is measured and found to be 6.9 percent that found in new wine. How old is the wine in the bottle?

Sections 28.16–28.18

70 Consider the nuclear fission reaction

$$^1_0n + ^{235}_{92}U \rightarrow ^{144}_{56}Ba + ^{92}_{36}Kr$$

Calculate the energy released in this reaction. The atomic masses for the nuclei involved in the reaction are $M(^{235}U) = 235.04392$ u, $M(^1n) = 1.008665$ u, $M(^{144}Ba) = 143.92285$ u, and $M(^{92}Kr) = 91.92627$ u.

71 (a) If the ^{235}U fission process yields an energy of 210 MeV, how much energy will be given off by the fission of 1 g of ^{235}U? (b) If the energy costs 8 cents per kilowatthour, what is the cost of the energy found in a?

72 How many grams of uranium 235 must be used to operate a 1500-MW power plant for a period of 1 h if the reactor has an overall efficiency of 30 percent? *Hint:* Assume that each fission event releases approximately 210 MeV.)

73 In a typical nuclear power plant, the reactor core produces 3600 MW of thermal power. If the ^{235}U core material is reduced by 28 percent in 6 yr, how much ^{235}U existed originally in the core? Assume that about 210 MeV is released per fission event.

74 A neutron with a speed of 4×10^6 m/s hits a stationary deuterium atom (^2_1H) head-on in a perfectly elastic collision. (a) Find the speed of the neutron after collision. (b) Repeat if the deuterium atom is replaced by an oxygen atom, $^{16}_8O$. Notice that small-mass nuclei are most effective in slowing neutrons.

75 Neutrons are most effectively slowed down by collisions with particles of equal mass. Suppose a neutron with speed 10^7 m/s strikes a free, stationary proton head-on. What will be the final speed of the neutron? Repeat if the neutron collides elastically with a free, stationary gold atom.

76 One possible reaction on which a fusion reactor can be based is

$$^2_1H + ^3_1H \rightarrow ^4_2He + ^1_0n$$

where ^3_1H is tritium. How many grams of deuterium and tritium must fuse each second to yield 1000 MW of power?

Pertinent atomic masses are $M(^2\text{H}) = 2.014102$ u, $M(^3\text{H}) = 3.016050$ u, and $M(^4\text{He}) = 4.002604$ u.

■■**77** Find the energy released in the fusion reactions

$$^2_1\text{H} + ^2_1\text{H} \rightarrow ^3_1\text{H} + ^1_1\text{H}$$

$$^2_1\text{H} + ^3_2\text{He} \rightarrow ^4_2\text{He} + ^1_1\text{H}$$

General Problems

■■**78** The half-life of cobalt 60 is 5.3 yr. (*a*) How many atoms of cobalt 60 are there in a 1-g sample of the material? (*b*) What is the decay constant of the material? (*c*) How many disintegrations occur each second in a 1-g sample of the material?

■■**79** A sample contains N_1 nuclei of a material with half-life $(T_{1/2})_1$ and N_2 nuclei of another material with half-life $(T_{1/2})_2$. What is the effective half-life of the sample in terms of $(T_{1/2})_1$, $(T_{1/2})_2$, N_1, and N_2? Assume both half-lives are much longer than the time of observation.

■■**80** In Sec. 28.16 we stated that one possible fission process for the compound nucleus ^{236}U is

$$^{236}\text{U} \rightarrow ^{140}_{56}\text{Ba} + ^{92}_{36}\text{Kr} + 4^1_0\text{n} + \text{energy}$$

The product isotopes then decay in several steps by β emission. The decays can be written as

$$^{140}_{56}\text{Ba} \rightarrow ^{140}_{58}\text{Ce} + 2\beta + \text{energy}$$

$$^{92}_{36}\text{Kr} \rightarrow ^{92}_{40}\text{Zr} + 4\beta + \text{energy}$$

Find the total energy released when the ^{235}U nucleus undergoes this form of fission. The atomic masses of the nuclei are $M(^{235}\text{U}) = 235.04392$ u, $M(^{140}\text{Ce}) = 139.90539$ u, and $M(^{92}\text{Zr}) = 91.90503$ u.

APPENDIX 1

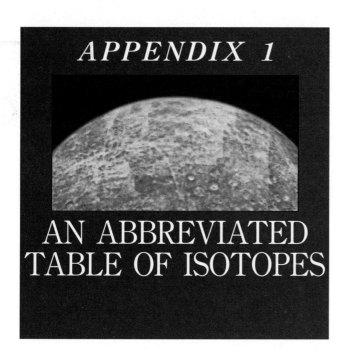

AN ABBREVIATED TABLE OF ISOTOPES

The values listed are based on $^{12}_{6}C = 12$ u exactly. Electron masses are included.

*The atomic masses of unstable elements are not listed unless the isotope given constitutes the major isotope.

Atomic number Z	Symbol	Average atomic mass	Element	Mass number A	Relative abundance %	Mass of isotope
0	n	1.008665	Neutron	1		
1	H	1.00797	Hydrogen	1	99.985	1.007825
				2	0.015	2.014102
2	He	4.0026	Helium	3	0.00015	3.016030
				4	100–	4.002604
3	Li	6.939	Lithium	6	7.52	6.015126
				7	92.48	7.016005
4	Be	9.0122	Beryllium	9	100–	9.012186
5	B	10.811	Boron	10	19.78	10.012939
				11	80.22	11.009305
6	C	12.01115	Carbon	12	98.892	12.0000000
				13	1.108	13.003354
7	N	14.0067	Nitrogen	14	99.635	14.003074
				15	0.365	15.000108
8	O	15.9994	Oxygen	16	99.759	15.994915
				17	0.037	16.999133
				18	0.204	17.999160
9	F	18.9984	Fluorine	19	100	18.998405
10	Ne	20.183	Neon	20	90.92	19.992440
				22	8.82	21.991384
11	Na	22.9898	Sodium	23	100–	22.989773
12	Mg	24.312	Magnesium	24	78.60	23.985045
13	Al	26.9815	Aluminum	27	100	26.981535
14	Si	28.086	Silicon	28	92.27	27.976927
				30	3.05	29.973761
15	P	30.9738	Phosphorus	31	100	30.973763
16	S	32.064	Sulfur	32	95.018	31.972074
17	Cl	35.453	Chlorine	35	75.4	34.968854
				37	24.6	36.965896

Atomic number Z	Symbol	Average atomic mass	Element	Mass number A	Relative abundance %	Mass of isotope
18	Ar	39.948	Argon	40	99.6	39.962384
19	K	39.102	Potassium	39	93.08	38.963714
20	Ca	40.08	Calcium	40	96.97	39.962589
21	Sc	44.956	Scandium	45	100	44.955919
22	Ti	47.90	Titanium	48	73.45	47.947948
23	V	50.942	Vanadium	51	99.76	50.943978
24	Cr	51.996	Chromium	52	83.76	51.940514
25	Mn	54.9380	Manganese	55	100	54.938054
26	Fe	55.847	Iron	56	91.68	55.934932
27	Co	58.9332	Cobalt	59	100	58.93319
28	Ni	58.71	Nickel	58	67.7	57.93534
				60	26.23	59.93032
29	Cu	63.54	Copper	63	69.1	62.92959
30	Zn	65.37	Zinc	64	48.89	63.92914
31	Ga	69.72	Gallium	69	60.2	68.92568
32	Ge	72.59	Germanium	74	36.74	73.92115
33	As	74.9216	Arsenic	75	100	74.92158
34	Se	78.96	Selenium	80	49.82	79.91651
35	Br	79.909	Bromine	79	50.52	78.91835
36	Kr	83.30	Krypton	84	56.90	83.91150
37	Rb	85.47	Rubidium	85	72.15	84.91171
38	Sr	87.62	Strontium	88	82.56	87.90561
39	Y	88.905	Yttrium	89	100	88.90543
40	Zr	91.22	Zirconium	90	51.46	89.90432
41	Nb	92.906	Niobium	93	100	92.90602
42	Mo	95.94	Molybdenum	98	23.75	97.90551
43	Tc	*	Technetium	98		97.90730
44	Ru	101.07	Ruthenium	102	31.3	101.90372
45	Rh	102.905	Rhodium	103	100	102.90480
46	Pd	106.4	Palladium	106	27.2	105.90320
47	Ag	107.870	Silver	107	51.35	106.90497
48	Cd	112.40	Cadmium	114	28.8	113.90357
49	In	114.82	Indium	115	95.7	114.90407
50	Sn	118.69	Tin	120	32.97	119.90213
51	Sb	121.75	Antimony	121	57.25	120.90375
52	Te	127.60	Tellurium	130	34.49	129.90670
53	I	126.9044	Iodine	127	100	126.90435
54	Xe	131.30	Xenon	132	26.89	131.90416
55	Cs	132.905	Cesium	133	100	132.90509
56	Ba	137.34	Barium	138	71.66	137.90501
57	La	138.91	Lanthanum	139	99.911	138.90606
58	Ce	140.12	Cerium	140	88.48	139.90528
59	Pr	140.907	Praseodymium	141	100	140.90739
60	Nd	144.24	Neodymium	144	23.85	143.90998
61	Pm	*	Promethium	145		144.91231
62	Sm	150.35	Samarium	152	26.63	151.91949
63	Eu	151.96	Europium	153	52.23	152.92086
64	Gd	157.25	Gadolinium	158	24.87	157.92410
65	Tb	158.924	Terbium	159	100	158.92495
66	Dy	162.50	Dysprosium	164	28.18	163.92883
67	Ho	164.930	Holmium	165	100	164.93030
68	Er	167.26	Erbium	166	33.41	165.93040
69	Tm	168.934	Thulium	169	100	168.93435
70	Yb	173.04	Ytterbium	174	31.84	173.93902
71	Lu	174.97	Lutetium	175	97.40	174.94089

Atomic number Z	Symbol	Average atomic mass	Element	Mass number A	Relative abundance %	Mass of isotope
72	Hf	178.49	Hafnium	180	35.44	179.94681
73	Ta	180.948	Tantalum	181	100	180.94798
74	W	183.85	Tungsten	184	30.6	183.95099
75	Re	186.2	Rhenium	187	62.93	186.95596
76	Os	190.2	Osmium	192	41.0	191.96141
77	Ir	192.2	Iridium	193	61.5	192.96328
78	Pt	195.09	Platinum	195	33.7	194.96482
79	Au	196.967	Gold	197	100	196.96655
80	Hg	200.59	Mercury	202	29.80	201.97063
81	Tl	204.37	Thallium	205	70.50	204.97446
82	Pb	207.19	Lead	208	52.3	207.97664
83	Bi	208.980	Bismuth	209	100	208.98042
84	Po	210	Polonium	210		209.98287
85	At	*	Astatine	211		210.98750
86	Rn	*	Radon	211		210.99060
87	Fr	*	Francium	221		221.01418
88	Ra	226.054	Radium	226		226.02536
89	Ac	*	Actinium	225		225.02314
90	Th	232.038	Thorium	232	100	232.03821
91	Pa	231.036	Protactinium	231		231.03594
92	U	238.03	Uranium	233		233.03950
				235	0.715	235.04393
				238	99.28	238.05076
93	Np	*	Neptunium	239		239.05294
94	Pu	*	Plutonium	239		239.05216
95	Am	*	Americium	243		243.06138
96	Cm	*	Curium	245		245.06534
97	Bk	*	Berkelium	248		248.070305
98	Cf	*	Californium	249		249.07470
99	Es	*	Einsteinium	254		254.08811
100	Fm	*	Fermium	252		252.08265
101	Md	*	Mendelevium	255		255.09057
102	No	*	Nobelium	254		254
103	Lw	*	Lawrencium	257		257

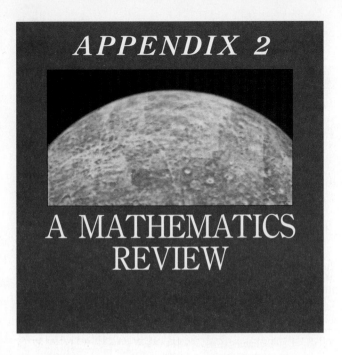

APPENDIX 2

A MATHEMATICS REVIEW

A. Addition

The order in which quantities are added is unimportant. For example, $8 + 7$ and $7 + 8$ are both 15. If we represent any two numbers by the symbols a and b, we have $a + b = b + a$.

B. Subtraction

If $a = 6$ and $b = 4$, then we know that $a - b = 6 - 4 = 2$ and $b - a = 4 - 6 = -2$.
 To subtract a negative number, we proceed as follows: if $a = 7$ and $b = -3$, then

$$a - b = 7 - (-3) = 7 + 3 = 10$$

Or, using two numbers c and $-e$, we have $c - (-e) = c + e$.

To subtract a negative number, we change its sign and add it.

C. Multiplication

The order in which common numbers are multiplied is unimportant. For example, $6 \times 3 = 3 \times 6 = 18$, and $a \times b$, which we write as $a \cdot b$ or ab, is the same as ba. In carrying out multiplications, the following sign rules must be used:

Rule	*Example*
$a \times b = ab$	$5 \times 6 = 30$
$(a) \times (-b) = -ab$	$(5) \times (-6) = -30$
$(-a) \times (-b) = ab$	$(-5) \times (-6) = 30$

D. Division

The sign rules for division are as follows:

Rule	*Example*
$a \div b = \dfrac{a}{b}$	$15 \div 3 = \dfrac{15}{3} = 5$
$a \div (-b) = \dfrac{a}{-b} = \dfrac{-a}{b} = -\dfrac{a}{b}$	$15 \div (-3) = -5$
$(-a) \div (-b) = \dfrac{-a}{-b} = \dfrac{a}{b}$	$(-15) \div (-3) = 5$

A-4

E. Parentheses

Parentheses may be manipulated as in the following examples:

$$(a + b) = (b + a) = a + b$$
$$d(a + b + c) = da + db + dc$$
$$(e + d)(a + b + c) = e(a + b + c) + d(a + b + c)$$
$$-(a - b) = -(a) - (-b) = -a + b$$

F. Fractions

In a fraction a/b, a is the numerator and b is the denominator. Notice the following identities:

$$\frac{a + b}{c} = \frac{a}{c} + \frac{b}{c} \quad \text{and} \quad \frac{a - b}{c} = \frac{a}{c} - \frac{b}{c}$$

$$\frac{a + b}{c + d} = \frac{a}{c + d} + \frac{b}{c + d} \quad \left(\text{not } \frac{a}{c} + \frac{b}{d}\right)$$

$$\frac{a}{b} \times \frac{c}{d} = \frac{ac}{bd} \quad \text{and} \quad \frac{a}{b} = \frac{a}{b} \times 1 = \frac{a}{b} \times \frac{c}{c} = \frac{ac}{bc}$$

The last example shows that we can multiply both the numerator and the denominator by the same quantity without changing the value of a fraction.

$$a \div \frac{c}{d} = a \times \frac{1}{c/d} = a \times \left(\frac{d}{c}\right) = \frac{ad}{c}$$

This latter identity is often stated in words as follows:

To divide a number by a fraction, invert the fraction and multiply by it.

As a more general example of this,

$$\frac{a}{b} \div \frac{c}{d} = \frac{a}{b} \times \frac{d}{c} = \frac{ad}{bc}$$

Multiplying the numerator and denominator by the same quantity does not change the value of the fraction. Thus we can cancel like factors from the numerator and the denominator. As examples,

$$\frac{ax}{ay} = \frac{a}{a} \times \frac{x}{y} = 1 \times \frac{x}{y} = \frac{x}{y} \quad \text{and} \quad \frac{ab + bc}{bd} = \frac{b(a + c)}{bd} = \frac{a + c}{d}$$

Be careful, however; $\dfrac{ab + c}{bd}$ is not $\dfrac{a + c}{d}$

Each term in the numerator and the denominator must possess the same factor, or the factor cannot be canceled.

G. Exponents

In the expression a^c, we call c the exponent of a. From the definition of exponent, it follows that

$$a^0 = 1 \qquad a^3 = a \cdot a \cdot a \qquad a^{-1} = \frac{1}{a} \qquad a^{-3} = \frac{1}{a^3} = \frac{1}{a \cdot a \cdot a}$$

The following rules apply:

$$a^n a^m = a^{n+m} \qquad (ab)^n = a^n b^n \qquad a^{-n} = \frac{1}{a^n} \quad \text{and} \quad \frac{1}{a^{-n}} = \frac{1}{1/a^n} = a^n$$

$$(a^n)^m = a^{nm} \qquad \text{from which} \qquad (a^n)^{1/2} = a^{n/2} = \sqrt{a^n}$$

H. Equations

Suppose we wish to solve the following equation for x:

$$5x + 7x^2 = 3(2 - x)$$

By use of the rule for parentheses, this becomes $5x + 7x^2 = 6 - 3x$. Using the rules given below, we can rewrite this as $7x^2 + 8x - 6 = 0$. This equation is of the form $ax^2 + bx + c = 0$ and can be solved for x using the quadratic formula:

$$x = \frac{-b \pm \sqrt{b^2 - 4ac}}{2a}$$

In the present case, this gives $x = -1.66$ and $x = 0.516$.

We can use the following rules to simplify equations.

Rule 1: Quantities equal to the same quantity are equal to each other. Therefore, if

$$a + b = c \quad \text{and} \quad x = c \quad \text{then} \quad a + b = x$$

Rule 2: Because equal quantities multiplied (or divided) by the same quantity remain equal, we can multiply (or divide) each side of an equation by the same quantity.

$$\frac{a}{b}x = c \quad \text{becomes} \quad \frac{\cancel{b}}{\cancel{a}} \cdot \frac{\cancel{a}}{\cancel{b}}x = \frac{b}{a} \cdot c \quad \text{and so} \quad x = \frac{bc}{a}$$

Rule 3: Cross multiplication can be used as follows:

$$\text{If} \quad \frac{a}{b} = \frac{c}{d} \quad \text{then} \quad da = bc$$

This follows directly from Rule 2 if we multiply both sides of the equation by bd and cancel like factors.

Rule 4: We can raise both sides of an equation to the same power. For example, if $9x^2 = 5$, then we can square each side to give $81x^4 = 25$. Or we can take the square root of each side (that is, raise each to the $\frac{1}{2}$ power) to give $3x = \sqrt{5}$. To prove these facts, use Rules 1 and 2.

Rule 5: A single term can be moved from one side of an equation to the other (transposed), provided we change its sign. Thus $x^2 - 5x = 8$ becomes $x^2 - 5x - 8 = 0$. This rule simply involves adding or subtracting the same quantity (8 in this case) from both sides of the equation.

ANSWERS TO ODD-NUMBERED PROBLEMS

Chapter 1

1 (a) 26.8 m/s; (b) 3.16×10^7 s; (c) 402 m; (d) 4920 ft; (e) 667 m/min
3 (a) 6.28×10^4 m; (b) 2.26×10^{-6} m; (c) 3.33×10^{-8} m; (d) 1.358×10^{-4} m; (e) 3.0002×10^1 m
5 1.31×10^{-1}
7 (a) 4; (b) 5; (c) 3; (d) 2; (e) 4, 3, or 2
9 5.83×10^9
11 152 in
13 (a) 4.0×10^{-18}; (b) 1.17×10^{-9}; (c) 6×10^{29}; (d) 2.96×10^{-2}
15 7.0 blocks; 27° south of east
17 128 paces; 61° south of west
19 1430 km; 32° north of east
21 78 cm at 337°; −78 cm at 203°
23 553 m at 134°; 553 m at −46°
25 128 steps; 61° south of west
27 1430 km; 32° north of east
29 $B_x = 48$ m and $B_y = 28$ m
31 $A_x = -0.925$ cm; $A_y = 0.563$ cm; $A_z = +1.15$ cm
33 135 m; 35° with vertical
35 571 mi at 225°; 697 mi at 260°

Chapter 2

1 480 mi/h; 215 m/s
3 1.34×10^{-9} s
5 (a) 2.6 cm/s at 69° (b) 3.4 cm/s
7 (a) 1.00 cm/s; (b) 0.86 cm/s; (c) −0.40 cm/s; (d) −1.0 cm/s; (e) 0 cm/s
9 11.2 s
11 0, 16.7, 0, and −32.5 m/min east

13 $v_{av} = 7.3$ cm/s; $v_A = 20$ cm/s; $v_D = -7.0$ cm/s
15 59 m
17 0.97 m/s²
19 $a = -2.19$ m/s²; $x = 128$ m
21 $a = 2.10 \times 10^{-2}$ mi/s²; $v_f = 0.103$ mi/s
23 $t = 1.79 \times 10^{-10}$ s; $a = 7.0 \times 10^{17}$ m/s²
25 $a = -5.85 \times 10^5$ m/s²; $t = 1.88 \times 10^{-4}$ s
27 (a) 113 s; (b) 5.6 m/s
29 (a) 106 s; (b) 20.3 s
31 $a = 3.72$ m/s²
33 $y = 8.54$ m; $v_f = 12.9$ m/s
35 $t = 1.08$ s; $v_f = 29.2$ m/s
37 $t = 3.2$ s; $v_f = 23$ m/s
39 $h_1 = 191$ m and $h_2 = 143$ m
41 0.43 s
43 115 m
45 4.23×10^3 m; 25.8 s
47 62.5 mi/h 77.0° north of west; 20.8 mi
49 83 km/h
51 4370 ft
53 $t = (v_o/g)[1 + \sqrt{1 + (2gh/v_o^2)}]$
55 9.1 m
57 13.9 s

Chapter 3

1 157 N; 29.5 m/s
3 (a) 3.03 m/s²; (b) 4100 N
5 325 N
7 (a) 706 N; (b) 770 N
9 2510 N
11 −2540 N
13 (a) zero; (b) 5 N down, 5 N up

15 −5650 N
17 (a) 9.8 N, 2.21 lb; (b) 590 N, 132 lb; (c) 13.2 kN, 2980 lb; (d) 8.9 kN, 2000 lb; (e) 4.45 N, 1.00 lb
19 58.2 N
21 (a) 160 N; (b) and (c) 98.0 kg
23 (a) 47.0 N; (b) 69.4 N; (c) 37.5 N
25 (a) 23, 13, and 20 N; (b) 2.9, 3.3, and zero m/s²
27 $\mu = 0.80$; it is static friction
29 35.3 m
31 71.9 m
33 -2.05×10^5 N
35 $a = 36.1$ m/s²
37 107 m
39 (a) $a = 5.27$ m/s² and $T' = 15.8$ N; (b) $a = 2.04$ m/s² and $T' = 15.8$ N
41 (a) $a = 3.62$ m/s²; $T = 11.8$ N; (b) $a = 1.63$ m/s²; $T = 15.5$ N
43 $T = 163$ N; $t = 0.82$ s
45 $T = 0.268$ N; $t = 0.98$ s
47 0.892 N; 2.82 N; 2.07 m/s²
49 $F = (M + m)g/\mu$
51 (a) 13.8 m/s²; (b) 3.49 m/s²
53 35 N; 0.36
55 (a) $T_t = 4.85$ N and $T_b = 2.70$ N; (b) $T_t = 13.0$ N and $T_b = 7.23$ N; (c) $T_t = 0.99$ N and $T_b = 0.55$ N; (d) $T_t = 0$ N and $T_b = 0$ N; (e) $T_t = 4.85$ N and $T_b = 2.70$ N
57 (a) 46 m; (b) 5.4 m
59 $a = (\mu \cos \theta + \sin \theta)g$
61 (a) $F_N = 14.2$ N; (b) $a = 8.7$ m/s²; (c) $F_N = 64.2$ N; (d) $a = 0$ m/s²
63 $m_1 = 0.107$ kg

Chapter 4

1 $T_u = 60$ N and $T_l = 25$ N
3 400 N at 263° with x axis
5 (a) $T_1 = 180$ N; $T_2 = T_3 = 90$ N;
    (b) $T_1 = 205$ N; $T_2 = T_3 = 90$ N
7 1072 N
9 $\theta = 28°$
11 50 N
13 $T_L = 650$ N; $T_R = 860$ N
15 $W_1 = 230$ N; $W_3 = 193$ N
17 $W_1 = 1128$ N; $W_2 = 489$ N; $W_3 = 639$ N
19 $T_1 = T_2 = T_4 = W_2 = 300$ N; $T_3 = 600$ N
21 $\tau_{90} = \tau_{50} = 0$; $\tau_{80} = +120$ N · m; $\tau_{70} = +303$ N · m; $\tau_{60} = -150$ N · m
23 (a) 2.00 m, zero, zero, 2.00 m, 2.83 m;
    (b) $-2.00F_1$, zero, zero, $+2.00F_4$, $+2.83F_5$
25 400 N
27 $\tau = 2.00 \times 10^{-3}$ N · m
29 120 N
31 $W_1 = 253$ N and $T_2 = 293$ N
33 4000 N
35 (a) $T = 1280$ N; (b) $H = 820$ N and $V = 140$ N
37 6.0 m
39 (a) $T = 1990$ N; (b) $H = -320$ N and $V = 2490$ N
41 $\theta = 21.8°$
43 $w = W[1 + (L/2b)]\cos\theta$
45 23.4°
47 1510 N
49 1.13 m

Chapter 5

1 70 J
3 8200 J
5 720 J
7 4.0 m/s
9 3.53 kJ
11 1.87 kJ
13 0.134 hp
15 12.0 W
17 4660 N
19 0.24 m/s
21 1.31 s
23 4.0×10^5 J
25 89.4 m
27 70,000 J; 90,000 J; 0.777
29 3.0 kW
31 1.62×10^5 J
33 (a) 900 N; (b) 0.60 s
35 17.6 kJ
37 9.8 J
39 (a) 1.88×10^8 J; (b) 93 hp
41 4.43 m/s
43 14.4 N
45 13.3 N

47 (a) 840 MJ; (b) 0.94; (c) 384 s;
    (d) 2900 hp
49 (a) 10.0 m; (b) 1.27 N; (c) 7.7 m/s
51 0.54 or 54%
53 $v_B = 10.0$ m/s; $v_C = 7.81$ m/s
55 5.9 m/s
57 19.2 km/gal
59 (a) 14.3; (b) 21.3; (c) 67 percent
61 AMA = 8.4; IMA = 9.4
63 0.59 hp
65 (a) 340 N; (b) 0.034 N
67 7.8°
69 (a) 70.1 J; (b) -21.2 J; (c) -9.25 J;
    (d) 39.7 J
71 (a) 480 kJ; (b) 470 kJ
73 9.0°

Chapter 6

1 (a) 35,600 kg · m/s north; (b) 9.3 kg · m/s up; (c) 6.5×10^8 kg · m/s west
3 $p = m\sqrt{2gh}$ downward
5 KE $= p^2/2m$
7 28,500 N
9 -7240 N
11 $\Delta t = 9.0 \times 10^{-11}$ s; $\overline{F} = 7.9 \times 10^{-10}$ N
13 (a) Impulse $= 0.0546t$ N · t; (b) $F = 0.0546$ N
15 $V = vM_1/(M_1 + M_2)$
17 12 m/s to the right
19 (a) $v = 0.141$ m/s; (b) $F = 0.064$ N
21 (a) 34.8 m/s; (b) 60.4 m/s
23 $m_2 = 1.45m_1$
25 $v_l = 5v_0/9$ and $v_r = 14v_0/9$
27 $V = 0.0351v_o$
29 $\Delta KE/KE_o = 4k/(k + 1)^2$ is largest when $k = 1$ or $m_1 = m_2$
31 $t = 65$ s
33 28.8 cm
35 $t = 1.89$ s
37 $v = 0.72v_o$ at 124°
39 $(-v_o/2, -v_o/2)$, it is perfectly elastic
41 $3.48v$, at 207°
43 20.2 m/s at 29.7°
45 $t = 850$ s
47 4.05 m
49 $V = \sqrt{gL/8}$
51 (a) 38.8 m/s; (b) 178 N

Chapter 7

1 (a) 0.089 rev, 0.56 rad; (b) 0.42 rev, 152°;
    (c) 241°, 4.2 rad
3 0.0329 rad; 1.89°; 5.24×10^{-3} rev
5 0.105 rad/s
7 (a) 3.5 rad/s; (b) 45°
9 4.53 rad/s², 2600 rev/min²
11 (a) 0.00284 rev/s²; (b) 0.69 rev
13 1.03 rev/s
15 400 cm/s

17 22.3 rev
19 1.15 rad/s²
21 (a) 463 m/s; (b) $v_T = 0$
23 15.8 m
25 48 rad/s = 7.7 rev/s = 2760 deg/s
27 (a) 125 m; (b) 63.9 rev
29 (a) 1.73 rev/s², 224 rev; (b) 221 rev, 87 m
31 1.85 rad/s²; 81.5 rev
33 17,000 N
35 1.07
37 0.34 rev/s
39 17.1 m/s
41 190 rev/s
43 1.86×10^{-40} N; 1.14×10^{-14}
45 0.314
47 $R_m/R_E = 0.27$
49 1.87×10^{27} kg
51 3.2°
53 30 rev/h
55 2.57 N
57 4070 N
59 inner breaks first; 2.57 rev/s

Chapter 8

1 0.339 J
3 49.3 J
5 (a) 1.67×10^{-2} J; (b) 2.83×10^{-4} N · m
7 0.50 rad/s²
9 12.8 kg · m²
11 107 N
13 12.6 s
15 22.6 rev
17 3.26 m
19 (a) $4ma^2$; (b) $4mb^2$; (c) $4m(a^2 + b^2)$
21 $Ma^2 + 2m(a^2 + b^2)$
23 (a) $2MR^2$; (b) $3MR^2/2$
25 $ML^2/3$; $ML^2/9$
27 70.7 J
29 89 J
31 0.030 N · m
33 324 J
35 $I = 0.0153$ kg · m²; $T = 0.58$ N
37 (a) 0.58 rev/s; (b) 0.650 J
39 (a) 8.05 rad/s; (b) 7.5 s
41 (a) 2.21 m/s; (b) 5.9 rev/s
43 1.45 cm
45 14.3 J
47 $v_s = 1.20\sqrt{gh}$, $v_d = 1.15\sqrt{gh}$, and $v_r = \sqrt{gh}$; sphere reaches the bottom first
49 11.3 kg · m²/s
51 (a) 0.60 kg · m²; (b) 27.8 rev/s
53 43 rev/min
55 (a) $4v_o/3$; (b) $2v_o$; (c) $3v_o$
57 5.8 rev/min
59 2.2×10^{-4} s; 1.00×10^{10}
61 $\Delta KE/KE = 2m/(M + 2m)$
63 $v = 5.71$ m/s; $\omega = 29$ rad/s
65 4.52 m

Chapter 9

1 $0.867 \, \text{g/cm}^3$
3 90 kg
5 58.9 g
7 31.6 percent
9 $1750 \, \text{kg/m}^3$
11 0.86 m
13 $3.51 \times 10^{11} \, \text{N/m}^2$
15 $6.54 \times 10^{-6} \, \text{m}$
17 $1.62 \times 10^5 \, \text{N}$
19 4630 Pa
21 $8.33 \times 10^6 \, \text{Pa}$
23 $\Delta V/V = 2.70 \times 10^{-6}$
25 500 N
27 $1.18 \times 10^5 \, \text{Pa}$
29 102 kPa
31 (a) $1.61 \times 10^7 \, \text{Pa}$; (b) 0.74 percent
33 6.00 m
35 11.5 cm
37 99.4 kPa
39 1.41 N
41 10.3 m
43 $P/W = 272$
45 0.067 N
47 $0.872 \, \text{g/cm}^3$
49 $964 \, \text{kg/m}^3$
51 $750 \, \text{kg/m}^3$
53 7060 N
55 23.9 g
57 0.152 kg
59 $Q/Q_o = 16.2$
61 $2.15 \, \text{cm}^3/\text{s}$
63 $0.278 \, \text{cm}^3/\text{s}$
65 0.361 cm
67 (a) 39.9 kPa; (b) 22.8 kPa
69 475 kN
71 $4.9 \times 10^4 \, \text{Pa}$
73 222 m/s = 800 km/h
75 4.1 cm/s
77 460 m/s; no
79 $7.1 \times 10^{-7} \, \text{m}$
81 $1:4:9$
83 $1.20 \times 10^{-5} \, \text{m}$
85 $3.18 \times 10^4 \, \text{N}$
87 77.65 cmHg
89 (a) $P_2 - P_1 = -2895 - 45.0(r_1/r_2)^4 \, \text{Pa}$; (b) same as in part (a)
91 $F = -500 \, \text{N}$, opposite the water flow
93 $W = 3.65 \times 10^5 \, \text{N}$; on a colder day, it can lift more weight

Chapter 10

1 (a) $23.3° \, \text{C}$, 296 K; (b) $-18.4° \, \text{F}$, 245 K; (c) $944° \, \text{F}$, $507° \, \text{C}$
3 $-423.17° \, \text{F}$; 20.28 K
5 (a) $T_m = 193.0° \, \text{C}$: (b) $T_b = 907.9° \, \text{F}$; $T_m = 379.5° \, \text{F}$
7 $58° \, \text{C}$, 331 K; $-89.2° \, \text{C}$, 184.0 K

9 574 K = $574° \, \text{F}$
11 3.27, 1.79, and $0.93 \times 10^{-25} \, \text{kg}$
13 3.86×10^{23} molecules
15 2.68×10^{24} molecules
17 (a) $7.7 \times 10^{-26} \, \text{kg}$; (b) $N = 1.03 \times 10^{25}$
19 30 g
21 (a) 630 kg; (b) 66 kg
23 34.9 g/mol
25 $V = 2.61 V_o$
27 9.40 liter
29 $T_2/T_1 = 1.50$
31 $810° \, \text{C}$
33 73.1 kPa
35 10.6 cm
37 $1.72 \, \text{m}^3$
39 $1.43 \, \text{kg/m}^3$
41 2100 K = $1827° \, \text{C}$
43 $6.21 \times 10^{-21} \, \text{J}$
45 901 K
47 $P = \rho \bar{v}^2/3$
49 $1.94 \times 10^{-24} \, \text{kg}$; no
51 (a) $8.14 \times 10^{-21} \, \text{J}$; (b) 2210 m/s, 1570 m/s; (c) $3.92 \times 10^9 \, \text{Pa}$
53 24.6 cm; 148 kPa
55 Air column is 20.0 cm
57 300 kPa
59 (a) 6.45 kg; (b) 730 kPa
61 Yes; 1.40 liter at 750 K
63 $64 \, \text{kg/m}^3$; 650 m/s

Chapter 11

1 49.7 kJ; 11.9 kcal
3 2.20 kcal = 9.2 kJ
5 6.0 kcal = 25.1 kJ
7 202 kcal = 845 kJ
9 $29.1° \, \text{C}$
11 1.84 g
13 $1.84 \times 10^{-6} \, \text{g}$
15 17.4 g
17 $6.74 \times 10^{-20} \, \text{J}$; $E/kT = 13.1$
19 $2.30 \times 10^4 \, \text{s}$
21 9.6 MJ; 1630 m
23 $x = 0.0208$
25 $\Delta L/L = 8.75 \times 10^{-4}$
27 1.13 cm
29 gap = $(1 \, \text{mm})\alpha \Delta t$; no
31 $\Delta V = 27.0 \times 10^{-3} \, \text{cm}^3$
33 $L_s/L_b = 1.58$
35 $3.21 \times 10^6 \, \text{J}$
37 $1.15 \times 10^6 \, \text{J/s}$
39 (a) 58.3 W; (b) 57.3 W
41 354 K = $81° \, \text{C}$
43 (a) $0.0175 \, \text{m}^2 \cdot \text{K/W}$; (b) $0.164 \, \text{m}^2 \cdot \text{K/W}$
45 $0.088 : 0.69 : 1.00$
47 (a) $170° \, \text{C}$; (b) $165° \, \text{C}$
49 $437° \, \text{C}$
51 $0.9923 \, \omega_o$; KE is unchanged
53 $0.025° \, \text{C}$

Chapter 12

1 737 kJ
3 $6.1° \, \text{C}$
5 $W_{CA} = 1.03 \, \text{MJ}$
7 (a) $\Delta U = 0$; (b) $\Delta Q = -167 \, \text{J}$
9 0.25 g; 0.063 moles
11 (a) 3.9 kJ, zero; (b) 6.4 kJ, 2.6 kJ
13 $720 \, \text{J/kg} \cdot \text{K}$
15 153 kJ
17 6.63 kJ; 6.63 kJ lost
19 $C_P = 4.57R$; $C_V = 3.57R$
21 $\Delta U = 5990 \, \text{J}$; $W = 2400 \, \text{J}$; $Q = 8390 \, \text{J}$
23 $W = 3150 \, \text{J}$; $\Delta Q = 6940 \, \text{J}$; $\Delta U = 3790 \, \text{J}$
25 106 kJ
27 $V_2/V_1 = 1.61$
29 $476° \, \text{C}$; 16 atm
31 (a) 169 kJ or 7.5%; (b) zero
33 $\Delta U/Q = 0.78$; $W/Q = 0.22$
35 1.6 J; $W/Q = 3.19 \times 10^{-6}$
37 (a) 22.6 kg/kmol; (b) 20.7 and $29.0 \, \text{kJ/kmol} \cdot \text{K}$; (c) 1.40

Chapter 13

1 (a) There are 8 possible ways. (b) 0.125; (c) 0.375
3 (a) 64 arrangements are possible; (b) there are 6 ways to get a sum of 5, the probability is 0.09375; there are 3 ways to get a sum of 11, the probability is 0.04688; most probable sums are 7 or 8, the probability is 0.1875
5 (a) 8; (b) 32; (c) 1.13×10^{15}
7 $-15.8 \, \text{J/K}$
9 $-0.34 \, \text{J/K}$
11 24.7 J/K
13 (a) $2.22 \times 10^{-23} \, \text{J/K}$; (b) $4.13 \times 10^{-23} \, \text{J/K}$; (c) $S = 0$
15 0.61
17 610 MJ
19 1.56
21 (a) $Q = 16.3 \, \text{kJ}$, 0, $-12.5 \, \text{kJ}$; $W = 0$, 8.9 kJ, $-5.0 \, \text{kJ}$; (b) 0.24; (c) 0.69
23 345 cal/s = 4900 Btu/h; 1.60
25 (a) 7.81; (b) 4.47
27 $T_f = 273$ and $T_b = 373$ which is same as Kelvin scale
29 (a) 0.25 and 0.34; (b) 0.51; (c) 0.87
31 (a) \$4.63; (b) \$8.72; (c) \$2.18

Chapter 14

1 (a) $f = 0.195 \, \text{Hz}$; (b) $\tau = 5.12 \, \text{s}$; (c) $A = 9.60 \, \text{cm}$
3 $3.74 \times 10^{-4} \, \text{J}$
5 (a) 1.10 m/s; (b) $24 \, \text{m/s}^2$
7 24.6 m/s
9 (a) 11.3 N/m; (b) $1.05 \, \text{m/s}^2$; (c) $v = 0.147 \, \text{m/s}$, $a = 0.75 \, \text{m/s}^2$

11 (a) 1.17×10^{-5} Hz; (b) 0.23 m/s^2; (c) 3100 m/s
13 (a) 2.17×10^4 N/m; (b) 2.0 cm
15 $f = (1/2\pi)\sqrt{AY/Lm}$
17 (a) $k = 7.6$ N/m; (b) $y_0 = 9.4$ cm; (c) $f = 1.08$ Hz; (d) $\tau = 0.92$ s
19 (a) $x(t) = 0.0295 \cos (6.74t)$ m; $v(t) = -0.199 \sin (6.74t)$ m/s; (b) $x(t) = -0.0287, +.0265,$ and 0.0180 m; $v(t) = 0.0451, -0.0878,$ and -0.158 m/s; (c) $t = 2.80$ s; (d) $t = 0.312$ s
21 $L_1/L_2 = 0.111$
23 $g_{pole} = 9.75$ m/s^2; $g_{eq} = 9.81$ m/s^2
25 13.5 N/m
27 0.60 cm; 7.6 Hz
29 5.77×10^{14} Hz
31 1.80 s
33 116 N
35 118 N
37 1300 N
39 $n \times 48.3$ Hz, $n = 1, 2, \ldots$
41 (a) 550 Hz; (b) $L_1 = 2L/5$
43 (a) $m_4 = 774$ g; (b) $m_5 = 495$ g; (c) $m_6 = 344$ g
45 (a) 49.7 J; (b) 2.95×10^4 N/m
47 0.35 Hz
49 The clock runs slow; error in 12 h is 2.9 s

Chapter 15

1 1870 m
3 0.65, 1.91, and 0.22 m
5 343.7 m/s
7 1430 m/s
9 22.1 GPa
11 (a) 5.20×10^{-3} s; (b) 192 Hz; for shallower depths f increases
13 17.0 W/m^2; 0.227 percent
15 66.3 dB
17 (a) 2.00×10^{-9} W/m^2; (b) 1.80×10^{-11} W
19 26 persons must be removed
21 3.16×10^{-10} W/m^2
23 5.22×10^{-4} J
5 3160
7 40.4, 80.9, and 121.3 Hz
29 (a) $\pm 10.1n$ cm, $n = 0, 1, 2, \ldots$ (b) $\pm 5.05n$ cm, $n = 1, 3, 5, \ldots$
31 $x = 2.30 \pm (0.1050 + 0.210n)$ m, with $n = 0, 1, 2, 3, \ldots$
33 (a) $y_{min} = 0.44$ m; (b) $\theta = 12.7°$
35 0.70 Hz
37 For pipe open at both ends: $f_1 = 190$, $f_2 = 379$, and $f_3 = 569$ Hz; for pipe closed at one end: $f_1 = 94.8$, $f_2 = 284$, and $f_3 = 474$ Hz
39 1.06 m
41 (a) 83.0 Hz; (b) open at one end
43 10 Hz

45 16.3 m/s; 18.1 m/s
47 (a) $f' = f$, $\lambda = (v + 17.5)/f$; (b) $f' = f$, $\lambda = (v - 17.5)/f$
49 16.7 m/s
51 3400 m
53 14.3; $\theta_s = 4.0°$
55 $t = 25.2°$ C
57 50.4 m
59 38 m/s

Chapter 16

1 (a) -1.28 mC; (b) -1.18 mC
3 (a) 0.188 N; (b) 1.15×10^{25}
5 4.80 N, in the $-x$ direction
7 $q_1 = 4.58 \,\mu$C and $q_2 = 2.62 \,\mu$C
9 $q = 0.98 \,\mu$C; both of same sign
11 5.2 pC
13 (a) -1.50 N (in $-x$ direction); (b) 0.91 N (in $+x$ direction)
15 -10.5 m
17 81 N along the diagonal passing through the $+5 \,\mu$C charge, away from origin
19 39 N along the bisector of the angle, directed outward
21 1.50 N at $-75.0°$ with x axis
23 7.0×10^{-3} N, 26.4° below x-axis
25 7.9×10^{-8} C, 15.8×10^{-8} C
27 1.44×10^{-9} N/C, toward electron; same magnitude, away from proton
29 (a) 410 kN/C to left; (b) 80 kN/C to right
31 (a) $E = 0$; (b) 5.4×10^5 N/C
33 2.66×10^5 N/C at $-77.8°$
35 2500 N/C westward
37 6.3×10^{14} m/s^2, along $-x$ axis
39 $q = -0.82 \,\mu$C
41 $E = (mg/q) \tan \theta$
43 $-1.72 \,\mu$C at $x = 0.46$ m
45 $5.0 \,\mu$C and $2.0 \,\mu$C
47 (a) 1.62 N; (b) 0.144 N
49 5.13×10^{11} N/C
51 $E = 0$
53 $E_x = 9.1 \times 10^4$ N/C, along $-x$ direction
55 (a) $y = (-eE/2m_e)t^2$; (b) $y = (eE/em_e v_{xo}^2)x^2$
57 5000 N/C, parallel to the beam
59 $E_y = 2kqy/(b^2 + y^2)^{3/2}$

Chapter 17

1 54 μJ; $-54 \,\mu$J
3 (a) 2.50×10^3 V/m; (b) -4.00×10^{-16} N
5 2.31×10^6 J
7 (a) -32 kV; (b) 0; (c) 40 kV; (d) -48 kV
9 (a) 1.46×10^{-8} s; (b) 5.3 cm
11 (a) 1.07×10^5 m/s; 4.6×10^6 m/s
13 1.67×10^6 V/m
15 $v = \sqrt{v_o^2 \pm 1.92 \times 10^8}$

17 0.71 V
19 21.6 keV; 7.75×10^5 m/s
21 (a) 250 keV; (b) 4.00×10^{-14} J; (c) 6.9×10^6 m/s
23 0.0060 J
25 10.4 MV
27 3.42 MV; 3.42 MeV
29 28.2 V; -28.2 eV
31 -6.36×10^5 V
33 -4.80 m and 0.96 m
35 (a) -45 V; (b) -22.5 V
37 0.200 μF
39 1.50 μF
41 4.0×10^6 V/m
43 (a) 6.37 nF; (b) 57.3 nC
45 3.4×10^{-3} m^2
47 3.00
49 (a) 18.0 μF; (b) 9.0 V; (c) 54 and 108 μC
51 (a) 220 pF; (b) 20 pF
53 4.4 μC; 60 μC and 4.8 μC
55 (a) 3.0 pF; (b) 36, 18, 18, 36 pC; 4.5, 3.0, 3.0, 4.5 V
57 The energy doubles
59 4.0 nJ; 1/8
61 3.28 pC
63 6.5×10^7 m/s; negative
65 1.35 J
67 $-18,000$ V
69 $q_1 = 3.0 \,\mu$C; $q_2 = 9.0 \,\mu$C; $V_1 = V_2 = 3$ V
71 (a) 360 pC; (b) 9.0 V; (c) -162 pJ; (d) $+162$ pJ
73 3.88 μF
75 (a) 100 V; (b) 600 μC and 400 μC

Chapter 18

1 (a) 7200 C; (b) 4.5×10^{22} electrons
3 0.89 s
5 9.5×10^4 C
7 0.51 μA
9 0.40 A
11 50 Ω
13 (a) 0.32 A; (b) 0.48 A
15 0.54 Ω
17 0.32 Ω
19 (a) 0.41 Ω; (b) 4.95 V
21 0.54 Ω
23 2.1×10^{-3}/C°
25 270° C
27 (a) 0.83 A; (b) 144 Ω
29 6.8×10^{-8} m^2
31 2.52 W
33 (a) 22.5 kJ; (b) 6.25×10^{-3} kWh
35 72 kWh; $4.90
37 5.0 Ω
39 (a) 1.30 Ω; (b) $I_6 = 1.50$, $I_2 = 4.50$, and $I_1 = 0.90$ A
41 (a) 16.5 Ω; (b) 0.182 A

43 0.158 and 0.039 A

45 9.6 V

47 (*a*) 0.76 A; (*b*) 0.33 A

49 (*a*) 13.7 Ω; (*b*) 0.44 A; (*c*) 0.088 A; (*d*) 0

51 $I_1 = 0.30$, $I_2 = 0.64$, and $I_3 = -0.94$ A

53 (*a*) $I_1 = -3.0$, $I_2 = 0$, and $I_3 = 3.0$ A; (*b*) -12.0 V; (*c*) 12.0 V

55 $I_1 = 0.0050$, $I_2 = -0.145$, and $I_3 = 0.140$ A

57 6.5 Ω

59 11.1 V

61 (*a*) 40 V; (*b*) 33.2 kΩ; (*c*) 33.3 V

63 Between 15.5 and 15.8 A

65 24

67 100 W

69 9.8 A

71 0.80 Ω

73 (*a*) 2.0 Ω; (*b*) 8.7 V

75 Seventeen different ways

77 4.0 Ω and 36 Ω

79 90 Ω; 99 Ω

81 (*a*) 6.0 V; (*b*) 2.0 V; (*c*) 0.183 W

Chapter 19

1 0.0525 N; down

3 4.5 N

5 (*a*) 0; (*b*) 3.60×10^{-5} N

7 3.5 mA

9 0.784 T

11 (*a*) 0.20 N in $-z$; (*b*) 0.098 N in $+z$; (*c*) 0.24 N at 156° in *x,y* plane

13 (*a*) 0; (*b*) 4.80×10^{-16} N along $+z$ axis; (*c*) 4.80×10^{-16} N along $+x$ axis

15 940 T

17 (*a*) 1.15×10^{-17} N along $-z$; (*b*) 2.00×10^{-17} N along $+z$; (*c*) 2.30×10^{-17} N in the *xy*; at $-30.0°$ with the $+x$ axis

19 (*a*) 2.56×10^{-14} T; (*b*) Horizontal, perpendicular to $\mathbf{v}_p$

21 8.5×10^{-6} T along the $-y$ axis

23 17.4 cm radius circle

25 2.40 m

27 0.43 T

29 19.8 cm

31 KE $= q^2r^2B^2/2m$

33 4.0 T

35 1.53×10^5 N/C

37 negative; 3.2×10^7 m/s

39 5.1×10^{-27} kg

41 2.0 cm

43 (*a*) 2.0×10^{-5} T; (*b*) 6.0×10^{-5} T both $\perp$ to plane of wires.

45 11.9 A

47 2.09 mT

49 (*a*) 0; (*b*) 4.5×10^{-3} T

51 2.0 A

53 (*a*) 1.92 N · m; (*b*) 0; (*c*) 1.92 N · m; (*d*) 1.92 N · m; (*e*) $\mu = 24$ A · m^2

55 0.0835 Ω shunt

57 0.200 Ω shunt

59 3.0×10^{-19} C

61 (*a*) 20.6 cm, 10.3 cm; (*b*) 20.6 cm

63 0.416 cm

65 1.30 A

67 $B = \mu_o Q f$

Chapter 20

1 0.025 Wb

3 (*a*) 5.4 mWb; (*b*) 0; (*c*) 4.7 mWb

5 1.38×10^{-3} Wb

7 (*a*) 0; (*b*) $2.51A$ mWb; (*c*) $5.03A$ mWb

9 2.9×10^{-5} V

11 1.96 V

13 0.111 mT

15 (*a*) Zero; (*b*) 64 μV

17 (*a*) 2.56 mT; (*b*) 128 μV

19 300 m

21 $NBbv$

23 2.0 T

25 3.8 V

27 360

29 2.4×10^{-7} Wb

31 1.44×10^{-4} Wb/s

35 8.3 mH

37 (*a*) 0; (*b*) 3.5 V; (*c*) 5.7 V

39 (*a*) 7.6 mA; (*b*) 10.4 mA

41 0.160 J

43 (*a*) 0.108 J; (*b*) 0.30 T

45 (*a*) 14.4 mJ; (*b*) 5.7 mJ; (*c*) 10.8 mJ

47 Zero

49 (*a*) 0.60 V; (*b*) north; (*c*) not easily

51 8.3 mV

53 $V = 4.52 \sin(754t)$ volts

55 11.8 mT

57 0.201 V

59 (*a*) 100 V; (*b*) 24 A

61 120 V

63 (*a*) 4.8 Ω; (*b*) 108 V

65 (*a*) 7.1 mV; (*b*) 0.86 V

67 500 V

69 77 Ω

71 1.67×10^{-5} Wb

73 $L_e = L_1 + L_2$

75 (*a*) 40*a* volts; (*b*) 1.60*a* A; (*c*) cw; (*d*) $3.2a^2$; (*e*) slow

77 320 m/s

Chapter 21

1 8.0 MΩ

3 (*a*) 2.0 μA, zero; (*b*) 0.74 μA, 46 μC

5 (*a*) 20.0 s; (*b*) 22 s; (*c*) 30 μA; (*d*) 1.36 A

7 86.5%

9 0.57 A; 156 V

11 (*a*) $I_o = 1.02$ A and $V_o = 170$ V; (*b*) 86 W; (*c*) 167 Ω

13 8.18 A; 13.4 Ω; 64.5 kcal

15 320 W

17 2.0 A, 1.42 A; 120 W

19 12.1 Hz

21 32 Hz

23 0.271 A; 452 A

25 66.5 mA

27 318 Hz

29 (*a*) 10; (*b*) 1000; (*c*) 10,000

31 (*a*) 1.51 Ω; (*b*) 15.1 kΩ

33 39.8 mH

35 (*a*) 36.5 mH; (*b*) 120 Hz

37 4.79 A

39 (*a*) 0.824 A; (*b*) 72.9 V; (*c*) $-66°$

41 531 Hz

43 0.120 A; 2.88 W

45 0.050 A; 128 mH

47 176 mH

49 1.89 or 1.49 H

51 (*a*) 69.0°; (*b*) current leads the voltage

53 70.7 mH; 194 W

55 27.4 W

57 (*a*) 17.6 mF; (*b*) 1.17 H

59 (*a*) 79.6 Hz; (*b*) 1600 Ω

61 2.47 to 25.3 nF

63 (*a*) 118 Hz; (*b*) 61 mA

65 0.33 A

67 182 Ω; 0.96 H

Chapter 22

1 6.0×10^6 m

3 188 to 556 m

5 5.45×10^{14} Hz

7 75.0 km

9 (*a*) 46 MHz; (*b*) 6.6 m

11 0.458 pF

13 (*a*) 8.5 V/m; (*b*) yes

15 (*a*) 0.53 pT; (*b*) no; (*c*) out of page

17 3.0×10^8 V/m

19 $B = 2.67 \times 10^{-12} \cos(6.00 \times 10^{10} t)$ T; $f = 9.55$ GHz; $\lambda = 0.0314$ m

21 200 μV

23 $E_o = 730$ V/m; $B_o = 2.43 \times 10^{-6}$ T

25 $E_o = 1000$ V/m; $B_o = 3.35 \times 10^{-6}$ T

27 $E = 5.49 \times 10^{-4} \sin(8.8 \times 10^6 t)$ V/m; $B = 1.83 \times 10^{-12} \sin(8.8 \times 10^6 t)$ T

29 (*a*) 3.6×10^{-8} W/m^2; (*b*) 5.2 mV/m, 1.73×10^{-11} T

31 0.368 W/m^2

33 48.6 W

35 4330 m

Chapter 23

1 0.028 s

3 1.00 m

5 2.40 m/s

7 (*a*) 70.0°; (*b*) 60.0°

9 12.0, 36.0, and 48.0 ft

11 $\phi = 2\theta$
13 13.8 cm, 3.9 mm; real, inverted
15 -30 cm, 6.0 cm; virtual, upright
17 2.8
19 75 cm
21 $p = 4f$; image is real
23 $p = 40$ cm
25 (*a*) -10.0 cm, up, vir, 0.60 cm;
(*b*) -7.7 cm, up, vir, 1.15 cm;
(*c*) -5.6 cm, up, vir, 1.67 cm
27 (*a*) $p = 40$ cm; (*b*) no
29 $p = 30.0$ cm and $M = 0.400$
31 $i = -24.2$ cm; vir, up; 0.030
33 (*a*) 2.21×10^8 m/s; (*b*) $\lambda = 433$ nm;
(*c*) $f = 5.08 \times 10^{14}$ Hz
35 (*a*) 1.92×10^8 m/s; 406 nm; (*c*) 18.7°
37 (*a*) $\theta_2 = 28.4°$; (*b*) zero
39 $\Delta\theta = 0.59°$
41 37°
43 1.56×10^8 m/s
45 3.20 m
47 (*a*) 24.4°; (*b*) 33.3°
49 46.0°
51 (*a*) 40.5°; (*b*) 59.7°
53 (*a*) 67 cm, re, inv, 1.33 cm; (*b*) 120 cm, re, inv, 4.0 cm; (*c*) -120 cm, vir, up, 8.0 cm
55 (*a*) -28, 1.23 cm; (*b*) -20, 2.0 cm; (*c*) -10.9, 2.9 cm. All vir and up.
57 (*a*) $f = -40$ cm; (*b*) diverging
59 (*a*) 2.96 cm; (*b*) 2.02 cm
61 (*a*) $p = 4f$; (*b*) real and inverted
63 (*a*) Converging lens; (*b*) $p = 2f/3$
65 (*a*) $i = -8f/9$; (*b*) $M = 1/9$
67 (*a*) 5.7 cm in front of diverg. lens, -0.57; (*b*) virtual, inverted
69 -11.5 cm from second lens, 1.05 cm; virtual, upright
71 (*a*) 101 cm; (*b*) 267 cm
73 53 cm
75 $s_i s_o = f^2$
77 ±40 cm and 200 cm from mirror; final is real, upright

Chapter 24

1 (*a*) $x = -60.0$, -120, and -180 cm; (*b*) $x = -30.0$, -90, and -150 cm
3 80.0 m
5 $\lambda = 15.0/n$ cm
7 3000 m
9 (*a*) $\theta_3 = 1.03°$; (*b*) $\theta_5 = 1.72°$
11 (*a*) 3.0 m; (*b*) 4.5 m
13 3.13 m
15 1.51 mm
17 $y_r - y_v = 1.80$ mm
19 86.5 nm
21 $n_f = 1.86$

23 1.25 μm
25 85.9 and 258 nm
27 (*a*) 2.45 μm; (*b*) 1.75 μm
29 179 nm
31 481 nm
33 (*a*) 1.94 μm; (*b*) 77.6°
35 (*a*) 3.97°; (*b*) 9.41°
37 20.7°
39 14.3°, 29.5°, and 47.6°
41 0.225°
43 5.89 mm
45 1.20 mm
47 2.0 μm; 5.60×10^{-5} m
49 $I = 0.413 I_o$
51 $I_o = 1.70 I$
53 $\theta_B = 57.0°$
55 $\cot\theta_B = \sin\theta_c$
57 $\theta_2 = 31.7°$
59 (*a*) 448 nm; (*b*) same as in part (*a*)
61 (*a*) 295, 590, and 885 nm; (*b*) 222, 444, and 666 nm
63 6.6 cm

Chapter 25

1 3.75 mm
3 34.6 cm
5 80 cm from eye
7 From 20.7 to 19.6 cm
9 22 cm
11 120 cm from the camera
13 (*a*) 7.00 cm; (*b*) 0.17 cm
15 1/25 s
17 $f = 6.25$ cm
19 4.1
21 (*a*) $i = 40$ cm; (*b*) the image is virtual
23 3.8 cm
25 $M = 113$
27 (*a*) $p_o = 10.0$ cm; (*b*) $M = 12.5$
29 $M_\phi = 20.0$
31 $M_\phi = 11$
33 $M_\phi = 24$
35 0.36 rad
37 1.6 minutes
39 $\Delta r = 0.193°$
45 (*a*) 40.7°; (*b*) 19.3°; (*c*) 28.9°; (*d*) 41°
47 (*a*) nearsighted; (*b*) $f = -37.5$ cm
49 $2\theta = 2.47°$
51 (*a*) 25; (*b*) 6.3
53 -60 cm; virtual; erect; 1.33

Chapter 26

1 Directly below its original position
3 1.70 m/s and 0.70 m/s
5 (*a*) 12.0 m/s; (*b*) 4.0 m/s
7 $0.962c$
9 (*a*) 68/min; (*b*) 32/min

11 $0.94c$
13 (*a*) 960 days; (*b*) 2.20×10^3 days
15 2.92×10^{-9} s
17 (*a*) 38.2 m; (*b*) 6.05 m
19 1.58 m
21 (*a*) 1.63×10^8 s; (*b*) 8.8×10^7 s
23 $5.00(1 - 4.29 \times 10^{-15})$ m
25 0.999, $950c$
27 (*a*) $1.00059\,m_o$, $0.0342c$; (*b*) $1.059\,m_o$, $0.328c$
29 $0.9938c$
31 (*a*) 9.0×10^{15} J; (*b*) 1.20×10^{14} s
33 6.36×10^{-12} kg; no
35 (*a*) 2.48×10^{-5} eV; (*b*) 1.30 eV; (*c*) 2.53 eV; (*d*) 124 eV
37 (*a*) 3.17×10^4 nm; (*b*) infrared
39 4.7×10^{-20} m
41 2.87 eV
43 (*a*) 262 nm; (*b*) ultraviolet
45 7.16×10^5 m/s
47 (*a*) $\phi = 2.97$ eV; (*b*) $f_o = 7.18 \times 10^{14}$ Hz
49 (*a*) $\lambda = 163$ nm; (*b*) 1.84×10^{15} Hz; (*c*) far ultraviolet
51 (*a*) 1.36×10^{-27} N $\cdot$ s; (*b*) 2.73×10^{-27} N $\cdot$ s
53 3.22×10^6 m/s; no
55 (*a*) 0.8024 nm; (*b*) 1.17×10^{-24} N $\cdot$ s
57 3.54×10^{-11} m
59 1.24×10^{-38} m
61 41.9 mV
63 (*a*) 6.23 nm; (*b*) 17900° C
65 (*a*) 1.06, 0.530, and 0.353 nm; (*b*) 1.34, 5.37, and 12.1 eV
67 3.07×10^{-10} m
69 5.74×10^{-20} J; 0.359 eV
71 0.117 eV
73 10×10^{-26} kg $\cdot$ m/s
75 1.06×10^{-20} kg $\cdot$ m/s; 420 keV
77 3.30×10^{-7} eV
79 9410
81 (*a*) 2.05×10^8 s; (*b*) 1.46×10^8 s; (*c*) 3.07×10^{16} m; (*d*) 2.10×10^8 m/s
83 $0.125c$
85 25.6 kV

Chapter 27

1 125 m
3 (*a*) $3.64 \times 10^{-26}/r$ J; (*b*) 4.7×10^{-14} m
5 2.82×10^{-14} m
7 0.053, 0.212, 0.477 nm
9 2.18×10^6 m/s $= 0.0073c$; 1.09×10^6 m/s $= 0.0036c$
11 13.6 and 3.40 eV
13 2.65×10^{-11} m; 1.06×10^{-10} m
15 (*a*) 85 keV; (*b*) 6.7×10^{-13} m
17 (*a*) -122, -30.6, and -13.6 eV; (*b*) 122 eV

19 4.8 pm; -1646 eV; 1646 eV
21 $\lambda_{13}/\lambda_{14} = 1.0022$
23 $\lambda_s = 820$ nm and $\lambda_l = 1875$ nm
25 2.86 eV
27 97, 102, 122, 486, 656, 1880 nm
29 40.8, 48.4, 51.0 eV; 30.4, 25.6, 24.3 nm
31 (*a*) 234 eV; (*b*) 9.1×10^6 m/s
33 Lyman series: 122, 102, and 96.9, 94.7, and 93.8 nm; Balmer series: 656, 486, 434, and 410 nm; Paschen series: 1880, 1280, and 1095 nm; and 4050, 2627, and 7470 nm
35 (*a*) 18; (*b*) 50
37 Three with $l = 0$, 1, and 2
39 3; 6
41 (*a*) 18; (*b*) 32; (*c*) 50
43 9
49 6.74 pm
51 (*a*) 71.9 keV; (*b*) 0.0226 nm
53 (*a*) 0.0184 nm; (*b*) 0.079 nm
55 $\Delta E = 3890$ eV; $\lambda = 0.32$ nm
57 8.8 cm
59 2.5×10^{74}
61 (*a*) $n = 7$ to $n = 4$; (*b*) $n = 2$ to $n = 5$

63 Three lowest energies absorbed correspond to the wavelengths 193, 72.5, and 38.6 nm which is in ultraviolet, thus benzene is a clear liquid

Chapter 28

1 (*a*) 7e; (*b*) $N = 7$; (*c*) $R = 2.89 \times 10^{-15}$ m; (*d*) $\rho = 2.29 \times 10^{17}$ kg/m^3
3 $^{43}_{20}$Ca
5 Aluminum 27
7 191 m
9 0.452 m
11 3.3×10^{-27} kg $= 1.99$ u
13 11.5 cm
15 39.1 kg/mol
17 ^{238}U: 99.3%; ^{235}U: 0.70%
19 0.945 percent
21 92.2 MeV; 7.7 MeV/nucleon
23 10.84 MeV
25 7.64 and 8.40 MeV/nucleon
27 42 counts per minute
29 (*a*) 0.825/yr; (*b*) 1.41×10^7
31 1.01×10^9 disintegrations per minute

33 6.47 Bq $= 175$ pCi
35 1.00 h
37 8.46 h
39 1.24×10^3 Bq
41 $^{233}_{92}$U; $^4_2$He; $^{144}_{60}$Nd
43 Bismuth 209; Francium 223
45 2.64×10^{-4} percent
47 $^{230}_{90}$Th; 230.03321 u
49 4.28 MeV; nuclear recoil
51 No, the proton need a KE of 3.00 MeV
53 No. Needs 2.64 MeV
55 (*a*) 5.75×10^{14} J; (*b*) 1.8×10^{11} C°
57 ^{234}Th, ^{230}Ra, ^{226}Rn, ^{222}Po, ^{218}Pb
59 (*a*) 98.5 mCi; (*b*) 1.63×10^{-12} g
61 2.09×10^{-8} g
63 0.243 μg
65 1040 Gy
67 2.50×10^4 yr
69 47.5 yr
71 (*a*) 8.60×10^{10} J; (*b*) \$1911
73 28,300 kg
75 $v = 0$; $v = 9.89 \times 10^6$ m/s
77 4.03 MeV; 18.4 MeV
79 $(N_1 + N_2)/[(N_1/\tau_1) + (N_2/\tau_2)]$

INDEX

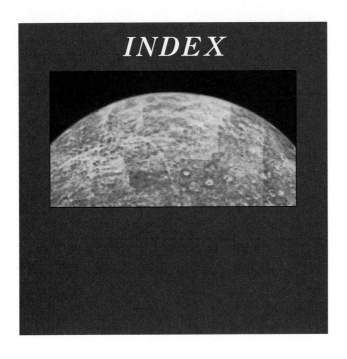

CREDITS FOR PHOTOGRAPHS
(referenced by page)

CHAPTER 1

1. NASA. **2.** Stock Montage/Historical Pictures Service. **3.** Courtesy of David Jerde. **5.** Courtesy of David Jerde. **12.** Frank Siteman/Stock, Boston. **15.** Bob Daemmrich/Stock, Boston. **24.** Mimi Forsyth/Monkmeyer Press.

CHAPTER 2

26. Jeff Albertson/The Picture Cube. **27.** *left,* US Dept. of Commerce; *right,* National Institute of Standards & Technology, Boulder Laboratories, US Dept. of Commerce. **29.** *top left,* Mike Douglas/The Image Works; *top right,* Joseph Schuyler/Stock, Boston; *bottom,* Loren M. Winters. **30.** Ulrike Welsch/Photo Researchers. **31.** Bob Daemmrich/Stock, Boston. **34.** Arthur Grace/Stock, Boston. **37.** Loren M. Winters.

CHAPTER 3

65. Joe Dimaggio/Joanna Kalish/Stock Market. **66.** James Sugar/Black Star. **68.** *bottom,* Margot Granitsas/The Image Works; *top,* Bob Daemmrich/The Image Works. **69.** National Institute of Standards & Technology, Boulder Laboratories, US Dept. of Commerce. **71.** Vandystadt/Photo Researchers. **74.** Vandystadt/Photo Researchers. **76.** Paul Silverman/Fundamental Photographs. **78.** Irven De Vore/Anthro-photo. **79.** Bill Bachmann/Stock, Boston. **90.** Alexander Lowry/Photo Researchers. **96.** Courtesy of Stanford Linear Accelerator Center, Stanford University, CA.

CHAPTER 4

106. Guy Gillette/Photo Researchers.   **108.** George Goodwin/The Picture Cube.   **112.** Robert J. Ashworth/Photo Researchers.   **115.** Bob Daemmrich/Stock, Boston.   **117.** James R. Fisher/Photo Researchers. **124.** Leonard Harris/Stock, Boston.

CHAPTER 5

132. Carl Purcell/Photo Researchers.   **136.** *left,* Mark Alcarez/The Picture Cube; *right,* Stephen Frisch/ Stock, Boston.   **139.** Charles Krebs/Stock Market.   **143.** Gerard Vandystadt/Photo Researchers. **144.** Bob Daemmrich/Stock, Boston.   **147.** *top,* Richard Megna/Fundamental Photographs; *bottom,* Jim Hamilton/The Picture Cube.   **149.** Focus on Sports.   **150.** Mark Antman/The Image Works.   **157.** Ellis Herwig/The Picture Cube.   **162.** NASA.

CHAPTER 6

170. Focus on Sports.   **172.** Joe Strunk/Loren Winters.   **175.** Focus on Sports.   **180.** *left,* Dr. Harold Edgerton/M.I.T., Cambridge, Mass.; *right,* Dan Overcash.   **184.** NASA.   **187.** Dan Overcash.   **190.** Richard Parsley/Stock, Boston.   **193.** AIP Emilio Segre Visual Archives.

CHAPTER 7

199. Frank Siteman/The Picture Cube.   **201.** Inga Spence/The Picture Cube.   **203.** Mark Antman/The Image Works.   **209.** Focus on Sports.   **211.** Focus on Sports.   **214.** Courtesy of Daytona International Speedway.   **216.** Miro Vintoniv/The Picture Cube.   **220.** NASA.

CHAPTER 8

232. Porterfield-Chickering/Photo Researchers.   **235.** Rodger Kingston/Stock, Boston.   **239.** Frank Pedrick/The Image Works.   **244.** Owen Franken/Stock, Boston.   **247.** Focus on Sports.   **248.** Frank Siteman/The Picture Cube.   **258.** Tom Martin/Stock Market.

CHAPTER 9

260. Randall Hyman/Stock, Boston.   **261.** Francie Manning/The Picture Cube.   **262.** Joseph P. Sinnot/ Fundamental Photographs.   **264.** ZAO-Grimberg/The Image Bank.   **270.** Cameraman/The Image Works. **271.** Richard Megna/Fundamental Photographs.   **275.** *top,* Fundamental Photographs; *bottom,* CENCO. **278.** Yoav Levy/Phototake.   **281.** Gary Stewart/Wide World Photo.   **286.** Takeshi Takahara/Photo Researchers.   **288.** Diane Schiumo/Fundamental Photographs.   **290.** Focus on Sports.

CHAPTER 10

299. D. Morrison/The Picture Cube.   **300.** Dan Overcash.   **302.** NOAO.   **304.** George A. Dillon/Stock, Boston.   **314.** Focus on Sports.

CHAPTER 11

320. Joe Carini/The Image Works.   **323.** *top,* Charles Kennard/Stock, Boston; *bottom,* Courtesy of David Jerde.   **329.** Paul Silverman/Fundamental Photographs.   **335.** Mark Burnett/Stock, Boston.   **336.** Wide World Photo.   **339.** Bob Daemmrich/The Image Works.   **342.** Bob Daemmrich/The Image Works. **343.** Peter Menzel/Stock, Boston.   **345.** Bob Daemmrich/The Image Works.

CHAPTER 12

352. Charles Krebs/Stock Market.   **355.** Alvis Upitis/The Image Bank.   **361.** NASA.

CHAPTER 13

379. Audrey Gibson/Stock Market.   **380.** Beitel Lottery Products.   **385.** Nikolai Ignatiev/Network Matrix.   **389.** Bob Daemmrich/The Image Works.

CHAPTER 14

398. L. Villota/Stock Market.   **401.** Capece/Monkmeyer Press.   **405.** Mark Burnett/Stock, Boston. **407.** Dr. Charles L. Nelson.   **411.** Richard Hutchings/Photo Researchers.   **414.** Bob Daemmrich/The Image Works.   **418.** *all,* Loren M. Winters.   **419.** Richard Megna/Fundamental Photographs.

CHAPTER 15

432. J. Zerschiling/Photo Researchers.   **435.** *left,* David Simson/Stock, Boston; *right,* John DeWaele/Stock, Boston.   **442.** Beth Bergman.   **448.** Joe DiStefano/Photo Researchers.   **449.** Gregg Mancuso/Stock, Boston.   **454.** J. Barry O'Rourke/Stock Market.   **459.** Kobal Collection.   **460.** Peter B. Wegener **468.** Gabriel Covian/The Image Bank.

CHAPTER 16

470. Hank Morgan/Rainbow. **472.** Mark C. Burnett/Photo Researchers. **474.** Bill Gallery/Stock, Boston. **484.** Johnny Autery.

CHAPTER 17

504. Murray & Assoc./Stock Market. **509.** Stan Osolinski/Stock Market. **513.** Richard Megna/Fundamental Photographs. **519.** Paul Silverman/Fundamental Photographs. **529.** Paul Silverman/Fundamental Photographs.

CHAPTER 18

537. General Motors. **547.** Ted Horowitz/Stock Market. **550.** Paul Silverman/Fundamental Photographs. **559.** Courtesy of David Jerde. **561.** Dick Poe/Visuals Unlimited.

CHAPTER 19

573. Fujifotos/The Image Works. **574.** *left,* Paul Silverman/Fundamental Photographs; *center and right,* Richard Megna/Fundamental Photographs. **576.** NASA. **578.** *left,* Richard Megna/Fundamental Photographs; *right,* Rogers/Monkmeyer Press. **583.** Central Scientific Company. **587.** Brenneis/Photo Researchers. **592.** The Image Works. **598.** Courtesy of David Jerde.

CHAPTER 20

608. Courtesy of Princeton University, Plasma Physics Laboratory. **609.** Courtesy of David Jerde. **613.** Courtesy of David Jerde. **627.** Courtesy of David Jerde. **630.** Grunnitus/Monkmeyer. **634.** Bill Gallery/Stock, Boston.

CHAPTER 21

645. Courtesy of David Jerde. **648.** Courtesy of David Jerde. **654.** Archives/The Image Works. **657.** Larry Mulvehill/Photo Researchers. **660.** Archives/The Image Works. **670.** Roger Ressmeyer/Starlight.

CHAPTER 22

672. Martin Bond/Photo Researchers. **678.** Joe Sohm/The Image Works. **679.** Moonlight Products. **680.** James Stevenson/Photo Researchers. **682.** *left,* Anglo-Australian Telescope Board 1980; *right,* Smithsonian Institution. **690.** General Motors. **692.** NASA.

CHAPTER 23

697. National Optical Astronomy Observatories. **699.** Roger Ressmeyer/Starlight. **701.** Richard Megna/Fundamental Photographs. **704.** Peter Menzel/Stock, Boston. **706.** *both,* Richard Megna/Fundamental Photographs. **714.** Richard Megna/Fundamental Photographs. **719.** Ken Kay/Fundamental Photographs. **720.** Hank Morgan/Photo Researchers.

CHAPTER 24

738. Dan McCoy/Rainbow. **739.** D.C. Health/Education Development Center. **740.** D.C. Health/Education Development Center. **743.** After Jenkins and White. **747.** Paul Silverman/Fundamental Photographs. **750.** Bausch & Lomb. **751.** After Jenkins and White. **757.** Photograph courtesy of Vincent Icke and Leiden Observatory, The Netherlands. **758.** *all,* "Atlas of Optical Phenomena" by Cagnet, Fragon, and Thrierr. Springer-Verlag/Prentice-Hall 1962, plate 16 a.b & c. **760.** Bausch & Lomb. **761.** Fundamental Photographs.

CHAPTER 25

771. NASA. **773.** George Hausman/Stock Market. **775.** Polaroid Corporation. **776.** *both,* Minolta Corporation. **777.** Courtesy of David Jerde. **779.** Carl Zeiss Inc., New York. **781.** *left,* Sotographs/Stock Market; *right,* National Optical Astronomy Observatories. **785.** *left,* Paul Silverman/Fundamental Photographs; *right,* Courtesy of David Jerde. **792.** Energy Technology Visuals Collection, Department of Energy, Washington, DC.

794. Richard Hutchings/Photo Researchers. **797.** Fermi National Accelerator Laboratory. **806.** Science Photo Library/Photo Researchers. **810.** Donald Deitz/Stock, Boston. **812.** Courtesy of David Jerde. **819.** Courtesy of David Jerde. **822.** Education Development Center. **826.** Randall M. Feenstra & Joseph A. Struscio, IBM Corporation.

CHAPTER 27

836. Berenholtz/Stock Market.   **837.** Dr. Fred Espenak/Science Photo Library/Photo Researchers.
847. National Optical Astronomy Observatories.   **856.** Ermakoff/Image Works.   **858.** *left,* Philippe
Plailly/Science Photo Library/Photo Researchers; *right,* John Eastcott/Yva Momatiuk/Image Works.

CHAPTER 28

864. Alexander Tsiaras/Stock, Boston.   **866.** Paolo Koch/Photo Researchers.   **873.** NASA.   **878.** Energy Technology Visuals Collection, Department of Energy, Washington, DC.   **883.** *top,* Biophoto Associates/Science Source/Photo Researchers; *bottom,* Roger Ressmeyer/Starlight.   **888.** *top,* David Parker/Science Photo Library/Photo Researchers; *center,* Oak Ridge Associated Universities; *bottom,* University of New Mexico.   **889.** G. Tortoli/H. Armstrong Roberts.   **891.** Energy Technology Visuals Collection, Department of Energy, Washington, DC.   **894.** Archive Photos.   **898.** *bottom,* Courtesy of Princeton University, Plasma Physics Laboratory

THE PERIODIC TABLE OF THE ELEMENTS

The values listed are based on $^{12}_{6}C = 12$ u exactly. For radioactive elements, the approximate atomic weight of the most stable isotope is given in brackets.

Period	I_A	II_A	III_B	IV_B	V_B	VI_B	VII_B	VIII			I_B	II_B	III_A	IV_A	V_A	VI_A	VII_A	0
1	1 H 1.00797	2 He 4.003																2 He 4.003
2	3 Li 6.939	4 Be 9.012											5 B 10.81	6 C 12.011	7 N 14.007	8 O 15.9994	9 F 19.00	10 Ne 20.183
3	11 Na 22.990	12 Mg 24.31											13 Al 26.98	14 Si 28.09	15 P 30.974	16 S 32.064	17 Cl 35.453	18 Ar 39.948
4	19 K 39.102	20 Ca 40.08	21 Sc 44.96	22 Ti 47.90	23 V 50.94	24 Cr 52.00	25 Mn 54.94	26 Fe 55.85	27 Co 58.93	28 Ni 58.71	29 Cu 63.54	30 Zn 65.37	31 Ga 69.72	32 Ge 72.59	33 As 74.92	34 Se 78.96	35 Br 79.909	36 Kr 83.80
5	37 Rb 85.47	38 Sr 87.62	39 Y 88.905	40 Zr 91.22	41 Nb 92.91	42 Mo 95.94	43 Tc [99]	44 Ru 101.1	45 Rh 102.905	46 Pd 106.4	47 Ag 107.870	48 Cd 112.40	49 In 114.82	50 Sn 118.69	51 Sb 121.75	52 Te 127.60	53 I 126.90	54 Xe 131.30
6	55 Cs 132.905	56 Ba 137.34	†	72 Hf 178.49	73 Ta 180.95	74 W 183.85	75 Re 186.2	76 Os 190.2	77 Ir 192.2	78 Pt 195.09	79 Au 196.97	80 Hg 200.59	81 Tl 204.37	82 Pb 207.19	83 Bi 208.98	84 Po [210]	85 At [210]	86 Rn [222]
7	87 Fr [223]	88 Ra [226]	‡															

†Lanthanide series														
57 La 138.91	58 Ce 140.12	59 Pr 140.91	60 Nd 144.24	61 Pm [147]	62 Sm 150.35	63 Eu 152.0	64 Gd 157.25	65 Tb 158.92	66 Dy 162.50	67 Ho 164.93	68 Er 167.26	69 Tm 168.93	70 Yb 173.04	71 Lu 174.97

‡Actinide series														
89 Ac [227]	90 Th 232.04	91 Pa [231]	92 U 238.03	93 Np [237]	94 Pu [242]	95 Am [243]	96 Cm [247]	97 Bk [247]	98 Cf [251]	99 Es [254]	100 Fm [253]	101 Md [256]	102 No [254]	103 Lw [257]